Work It Out
Like your instructors, authors Ratti and McWaters encourage you to [work?] this material frequently. Each chapter includes ample opportunities [to?] the material by solving problems and applying your understanding.

FINDING THE SOLUTION: A PROCEDURE

EXAMPLE 2   Solving Exponential Equations Using the Logarithms

OBJECTIVE
Solve exponential equations when both sides are not expressed with the same base.

EXAMPLE
Solve for x: $5 \cdot 2^{x-3} = 17$.

Step 1 Isolate the exponential expression on one side of the equation.

1. $2^{x-3} = \dfrac{17}{5} = 3.4$

Step 2 Take the common or natural logarithm of both sides.

2. $\ln 2^{x-3} = \ln (3.4)$

Step 3 Use the power rule, $\log_a M^r = r \log_a M$.

3. $(x - 3) \ln 2 = \ln (3.4)$

Step 4 Solve for the variable.

4. $x - 3 = \dfrac{\ln (3.4)}{\ln 2}$

$x = \dfrac{\ln (3.4)}{\ln 2} + 3 \approx 4.766$

Practice Problem 2 Solve for x: $7 \cdot 3^{x+1} = 11$.

Finding the Solution

Finding the Solution boxes present important multi-step procedures in a two-column format, illustrating the numbered steps of a procedure with a worked example.

Practice Problems

All examples conclude with a **Practice Problem**, so you can try a new problem and make sure you grasp the concept before moving ahead.

End-of-Section Exercises

Each section ends with three levels of exercises for you to practice the math and apply your understanding:

- **Basic Skills and Concepts**
- **Applying the Concepts**
- **Beyond the Basics**

A EXERCISES Basic Skills and Concepts

1. The domain of the function $y = \log_a x$ is _____ and its range is _____.

2. The logarithmic form $y = \log_a x$ is equivalent to the exponential form _____.

3. The logarithm with base 10 is called the _____ logarithm, and the logarithm with base e is called the _____ logarithm.

4. $a^{\log_a x} = $ _____, and $\log_a a^x = $ _____.

5. *True or False* The graph of $\log_a x, a > 0, a \neq 1$, is an increasing function.

6. *True or False* The graph of $y = \log_a x, a > 0$, and $a \neq 1$, has no horizontal asymptote.

In Exercises 31–40, evaluate each expression without using a calculator.

31. $\log_5 125$

32. $\log_9 81$

33. $\log 10,000$

34. $\log_3 \dfrac{1}{3}$

35. $\log_2 \dfrac{1}{8}$

36. $\log_4 \dfrac{1}{64}$

37. $\log_3 \sqrt{27}$

38. $\log_{27} 3$

39. $\log_{16} 2$

40. $\log_5 \sqrt{125}$

In Exercises 41–52, solve each equation.

41. $\log_5 x = 2$

42. $\log_e x = -2$

B EXERCISES Applying the Concepts

In Exercises 95–102, use the model $A = A_0 e^{kt}$.

95. *Doubling your money.* How long would it take to double your money if you invested P dollars at the rate of 8% compounded continuously?

96. *Investment goal.* How long would it take to grow your investment from $10,000 to $120,000 at the rate of 10% compounded continuously?

97. *Rate for doubling your money.* At what annual rate of return, compounded continuously, would your investment double in six years?

98. *Rate-of-investment goal.* At what annual rate of return, continuously, would your investment grow [to?] $50,000 in 25 years?

[...] Canada. The population of Canada was [...] 2000 and 33.4 million in 2007 Deter[...] from 2000 until Canada's population [...]nd (b) triples.

[...] Canada. In Exercise 99, find the time from [...]ada's population reaches 50 million.

C EXERCISES Beyond the Basics

107. Sketch the graph of $f(x) = \begin{cases} \ln x & \text{if } x > 0 \\ \ln(-x) & \text{if } x < 0 \end{cases}$

108. Sketch the graph of $f(x) = \begin{cases} \ln(1 - x) & \text{if } x < 1 \\ e^{-x} & \text{if } x \geq 1 \end{cases}$

109. Find the domain of $f(x) = \log_3 \left(\dfrac{x - 2}{x + 1} \right)$.

110. Find the domain of $g(x) = \log_2 \left(\dfrac{x + 3}{x - 2} \right)$.

111. *Finding the domain.* Find the domain of each function.
a. $f(x) = \log_2 (\log_3 x)$
b. $f(x) = \log (\ln (x - 1))$
c. $f(x) = \ln (\log (x - 1))$
d. $f(x) = \log (\log (\log (x - 1)))$

112. *Finding the inverse.* Find the inverse of each function in Exercise 111.

113. *Present value of an investment.* Recall that if P dollars is invested in an account at an interest rate r compounded continuously, then the amount A (called the *future value* of P) in the account t years from now will be $A = Pe^{rt}$.

PRACTICE TEST A

1. Solve the equation $5^{-x} = 125$.

2. Solve the equation $\log_2 x = 5$.

3. State the range of $y = -e^x + 1$ and find the asymptote of its graph.

4. Evaluate $\log_2 \dfrac{1}{8}$.

5. Solve the exponential equation $\left(\dfrac{1}{4}\right)^{2-x} = 4$.

6. Evaluate $\log 0.001$ without using a calculator.

7. Rewrite the expression $\ln 3 + 5 \ln x$ in condensed form.

8. Solve the equation $5x^3 e^x - x^4 e^x = 0$.

9. Solve the equation $e^{2x} + e^x - 6 = 0$.

10. Rewrite the expression $\ln \dfrac{2x^3}{(x + 1)^5}$ in expanded logarithmic form.

11. Evaluate $\ln e^{-5}$.

12. Give the equation for the graph obtained by shifting the graph of $y = \ln x$ three units up and one unit right.

13. Sketch the graph of $y = 3^{x-1} + 2$.

14. State the domain of the function $f(x) = \ln (-x) + 4$.

15. Write $3 \ln x + \ln (x^3 + 2) - \dfrac{1}{2} \ln (3x^2 + 2)$ in condensed form.

16. Solve the equation $\log x = \log 6 - \log (x - 1)$.

17. Find x if $\log_x 9 = 2$.

18. Rewrite the expression $\ln \sqrt{2x^3 y^2}$ in expanded logarithmic form.

19. Suppose $15,000 is invested in a savings account paying 7% interest per year. Write the formula for the amount in the account after t years if the interest is compounded quarterly.

20. Suppose the number of Hispanic people living in the United States is approximated by $H = 15,000 e^{0.02t}$, where $t = 0$ represents 1960. According to this model, about how many Hispanic people were living in the United States in 1980?

PRACTICE TEST B

1. Solve the equation $3^{-x} = 9$.
a. $\{2\}$ b. $\{-2\}$ c. $\left\{\dfrac{1}{2}\right\}$ d. $\left\{-\dfrac{1}{2}\right\}$ e. $\{\ln 2\}$

2. Solve the equation $\log_5 x = 2$.
a. $\{10\}$ b. $\{25\}$ c. $\left\{\dfrac{5}{2}\right\}$ d. $\left\{\dfrac{2}{5}\right\}$ e. $\{2\}$

3. State the range and asymptote of $y = e^{-x} + 1$.
a. $(0, \infty); y = -1$ b. $(-1, \infty); y = -1$
c. $(-\infty, \infty); y = -1$ d. $(-1, \infty); y = 1$
e. $(\infty, 1); y = 1$

4. Evaluate $\log_4 64$.
a. 16 b. 8 c. 2 d. 3 e. 4

5. Find the solution of the exponential equation $\left(\dfrac{1}{3}\right)^{1-x} = 3$.
a. $\left\{-\dfrac{1}{3}\right\}$ b. $\left\{\dfrac{1}{3}\right\}$ c. $\{1\}$ d. $\{2\}$

6. Evaluate $\log 0.01$.
a. -1.999999999 b. -2 c. 2 d. 100

7. Which of the following expressions is equivalent to $\ln 7 + 2 \ln x$?
a. $\ln (7 + 2x)$ b. $\ln (7x^2)$ c. $\ln (9x)$ d. $\ln (14x)$

8. Solve the equation $3x^2 e^x + x^3 e^x = 0$.
a. $\{-3\}$ b. $\{0, -3\}$ c. $\{0\}$ d. $\varnothing$

Preparation and Review

End-of-Chapter Material

Each chapter concludes with a **Sumary of Definitions, Concepts, and Formulas**; **Review Exercises**; two **Practice Tests**; and a **Cumulative Review** to prove your mastery of the concepts presented in the chapter.

College Algebra

Second Edition

J. S. Ratti
University of South Florida

Marcus McWaters
University of South Florida

Addison-Wesley

Boston San Francisco New York
London Toronto Sydney Tokyo Singapore Madrid
Mexico City Munich Paris Cape Town Hong Kong Montreal

Publisher:	Greg Tobin
Executive Editor:	Anne Kelly
Sr. Project Editor:	Joanne Dill
Associate Project Editor:	Leah Goldberg
Editorial Assistant:	Sarah Gibbons
Senior Managing Editor:	Karen Wernholm
Senior Production Project Manager:	Peggy McMahon
Text Designer:	Carolyn Deacy Design
Cover Designer:	Barbara T. Atkinson
Cover Photo:	Jack Hollingsworth/Getty Images
Photo Researcher:	Beth Anderson
Senior Media Producer:	Ceci Fleming
Math XL Project Supervisor:	Edward Chappell
QA Manager, Assessment Content:	Marty Wright
Senior Marketing Manager:	Katherine Greig
Market Development Manager:	Dona Kenly
Marketing Assistant:	Katherine Minton
Senior Media Buyer:	Ginny Michaud
Digital Assets Manager:	Marianne Groth
Senior Author Support/	
Technology Specialist:	Joe Vetere
Rights and Permissions Advisor:	Michael Joyce
Sr. Manufacturing Manager:	Carol Melville
Production Coordination, Technical	
Illustrations, and Composition:	Pre-Press PMG
Situational Art:	Scientific Illustrators

Library of Congress Cataloging-in-Publication Data
Ratti, J. S.
  College algebra / J.S. Ratti, Marcus McWaters. — 2nd ed.
    p. cm.
  Includes index.
  ISBN 0-321-64031-4 (student edition) — ISBN 0-321-65506-0
(annotated instructor's edition)
  1. Algebra—Textbooks.   I. McWaters, Marcus M.   II. Title.
  QA154.3.M36 2011
  512.9—dc22   2009043503

1 2 3 4 5 6 7 8 9 10—QWV—11 10 09

Addison-Wesley
is an imprint of

www.pearsonhighered.com

ISBN-13 978-0-321-64031-4
ISBN-10 0-321-64031-4

To Our Wives,

Lata and Debra

Foreword

Many challenges face today's college algebra students and instructors. Students arrive in this course with varying levels of comprehension from their previous courses. Instead of really learning the concepts presented, students often resort to

memorization to pass the course. As a result, a textbook needs to get students to a common starting point and engage them in becoming active learners, without sacrificing the solid mathematics that is necessary for conceptual understanding. Instructors in this course are faced with the task of providing students with an understanding of college algebra, preparing them for the next step, and ensuring that they find mathematics useful and interesting. Our goal of helping students and instructors achieve these goals has been aided considerably by the many suggestions we have received from users of the first edition of the text.

In this text, there is a strong emphasis on both concept development and real-life applications. The clearly explained and well developed in-depth coverage of topics such as functions, graphing, the difference quotient, and limiting processes provides thorough preparation for the study of calculus and will improve students' comprehension of algebra. Just-in-time review throughout the text assures that all students are beginning with the same foundation of algebra skills. Numerous applications are used to motivate students to apply the concepts and skills they learn in college algebra to other courses, including the physical and biological sciences, engineering, and economics, and to on-the-job and everyday problem solving. Students are given ample opportunities throughout this book to think about important mathematical ideas and to practice and apply algebraic skills.

Another goal was to create a book that clearly shows the relevance of what students are learning. Throughout the text, we emphasize why the material being covered is important and how it can be applied. By thoroughly developing mathematical concepts with clearly defined terminology, students see the "why" behind those concepts, paving the way for a deeper understanding, better retention, less reliance on rote memorization, and ultimately more success.

This focus on students does not mean that we neglected instructors. As instructors ourselves, we know how essential it is to use a book that you believe in and that helps you teach mathematics. To that end, the level of the book was carefully selected so that the material would be accessible to students and provide them with an opportunity to grow. Here again, insightful comments from first edition users has resulted in improved choices in both exposition and content. It is our hope that once you have looked through this textbook, you will see that we were able to fulfill our initial goals of writing for today's students and for you, the instructor.

Contents

CHAPTER P

Basic Concepts of Algebra 1

CHAPTER 1

Equations and Inequalities 82

CHAPTER 2

Graphs and Functions 169

CHAPTER 5

Systems of Equations and Inequalities 478

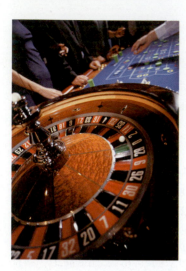

CHAPTER 8

Further Topics in Algebra 658

Preface

Students begin college algebra classes with widely varying backgrounds. Some haven't taken a math course in several years and may need to spend time reviewing prerequisite topics, while others are ready to jump right into new and challenging material. In Chapter P and in some of the early sections of other chapters, we have provided review material in such a way that it can be used or omitted as appropriate for your course. In addition, students may follow several different paths after completing a college algebra course. Many will continue their study of mathematics in courses such as trigonometry, finite mathematics, statistics, and calculus. For others, college algebra may be their last mathematics course.

Responding to the current and future needs of all of these students was essential in creating this text. Based on comments from users of the first edition, we now introduce each exercise set with several concept exercises. These exercises consist of fill-in-the-blank and true–false exercises. They are not computation-reliant, but rather test whether students have absorbed the basic concepts and vocabulary of the section. The second edition contains over 1000 new or revised exercises, some involving new data, some involving new applications, and some introducing new ways to test basic skills. Exercises asking students to extrapolate information from a given graph now appear in much greater number and depth throughout the text. We continue to present our content in a systematic way that illustrates how to study and what to review. We believe that if students use this textbook well, they will succeed in this course.

Features

CHAPTER OPENER Each chapter opener includes a description of applications relevant to the content of the chapter, a series of images related to these and other applications, and the list of topics that will be covered in the chapter. In one page, students see what they are going to learn and why they are learning it.

APPLICATION The discussion in each section begins with a motivating anecdote or piece of information that is tied to an application problem. This problem is solved in an example later in the section, using the mathematics covered in the section.

SECTION OPENERS These section openers lend continuity to the section and its content, utilizing material from a variety of fields: the physical and biological sciences (including health sciences), economics, art and architecture, the history of mathematics, and more. Of special interest are contemporary topics such as the greenhouse effect and global warming, CAT scans, and computer graphics.

REVIEW On the first page of each section is a list of topics that students should review prior to starting the chapter. Section references with page numbers accompany the suggested review material so that students can readily find the material. The **Objectives** of the section are then clearly stated and numbered. Each numbered objective is paired with a similarly numbered subsection so that students can quickly find the section material for an objective.

DEFINITIONS AND THEOREMS All are boxed for emphasis and titled for ease of reference, as are lists of rules and properties.

FIGURES All figures are titled to make it easy to identify what is being illustrated.

EXAMPLES The examples include a wide range of computational, conceptual, and modern applied problems that are carefully selected to build confidence, competency, and understanding. Every example has a title that indicates its purpose. For every example, clarifying side comments are provided for each step in the detailed solution.

PRACTICE PROBLEM All examples are followed by a Practice Problem for students to try so that they can check their understanding of the concept covered.

PROCEDURE BOXES These boxes, interspersed throughout the text, present important procedures in numbered steps. Special **Finding the Solution** boxes present important multistep procedures, such as the steps for doing synthetic division, in a two-column format. The steps of the procedure are given in the left column, and an example is worked, following these steps, in the right column. This approach provides students with a clear model with which they can compare when encountering difficulty in their work. These boxes are a part of the numbered examples.

MAIN FACTS These boxes summarize information related to equations and their graphs, such as those of the conic sections.

HISTORICAL NOTES When appropriate, historical notes appear in the margin, giving students information on key people or ideas in the history and development of mathematics. This information is included to add flavor to the subject matter.

TECHNOLOGY CONNECTIONS Although the use of graphing calculators is optional in this book, Technology Connections give students tips on using calculators to solve problems, check answers, and reinforce concepts.

WARNINGS These boxes appear as appropriate throughout the text to let students know of common errors and pitfalls that can trip them up in their thinking or calculations.

RECALL Located in the margins, Recall Notes periodically remind students of a key idea they learned earlier in the text that will help them with the new problems at hand.

STUDY TIPS Students are given hints for handling newly introduced concepts.

BY THE WAY These marginal notes provide students with additional interesting information on nonessential topics to keep them engaged in the mathematics presented.

EXERCISES The heart of any textbook is its exercises. Knowing this, we made sure that the quantity, quality, and variety of exercises meet the needs of all students. The problems in each exercise set are carefully graded to strengthen the skills developed in the associated section. Exercises are divided into three categories: **A: Basic Skills and Concepts** (developing fundamental skills), **B: Applying the Concepts** (using the section's material to solve real-world problems), and **C: Beyond the Basics** (providing more challenging exercises that give students an opportunity to reach beyond the material covered in the section). Exercises are paired so that the even-numbered A Exercises closely follow the preceding odd-numbered exercises. All application exercises are titled and relevant to the topics of the section. The A and B Exercises are intended for a typical student, while the C Exercises, generally more theoretical in nature, are suitable for honors students, special assignments, or extra credit. **Critical Thinking** exercises, appearing as appropriate, are designed to develop students' higher-level thinking skills, and calculator problems are included where needed. Finally, **Group Projects** are provided at the end of many exercise sets so that students can work together to reinforce and extend one another's comprehension of the material.

END-OF-CHAPTER The chapter-ending material includes a **Summary of Definitions, Concepts, and Formulas**; **Review Exercises**; and **two Practice Tests**. The chapter summary, a brief description of key topics indicating where the material occurs in the text, encourages students to reread sections rather than memorize definitions out of context. The Review Exercises provide students with an opportunity to practice what they have learned in the chapter. (Students should then be prepared for the Practice Test A in the usual open-ended format and Practice Test B, covering the same topics, in multiple-choice format.) All tests are designed to increase student comprehension and verify that students have mastered the skills and concepts in the chapter. Mastery of these materials should indicate a true comprehension of the chapter and the likelihood of success on the associated in-class examination.

CUMULATIVE REVIEW EXERCISES Starting with Chapter 2, these exercises appear at the end of every chapter to remind students that mathematics is not modular and that what is learned in the first part of the book will be useful in later parts of the book and on the final examination.

Key Chapter Content Changes

You'll see that we have improved the balance of exercises and added more challenging exercises. New exercises are included in every exercise set at the end of every section in every chapter, including new "concept exercises" at the beginning of each set of A Exercises: Basic Skills and Concepts. Exercises asking students to extract information from given graphs are weaved throughout the text where appropriate. Overall, 20% of the exercise sets have been updated and we added over 1000 additional exercises to the revision, but that is just a small part of the overall changes.

CHAPTER P In this first chapter, we introduce the concepts of universal set and complement of a set, with new material on arithmetic with fractions. There are more warnings about common errors that students should find very helpful.

CHAPTER 1 Section 1.5 contains new material on solving rational equations with variables in the denominator along with new coverage on equations with rational exponents. We increased the number of traditional applied exercises requiring linear equations and quadratic equations. Section 1.7 provides more examples and exercises on absolute value.

CHAPTER 2 This chapter has more emphasis on modeling, with a new subsection in Section 2.3 on linear regression. Section 2.4 features a new subsection explaining how information about a function can be obtained from its graph. Expanded material on the properties of functions is organized into one section, and in Section 2.6, the discussion of piecewise functions now follows a new subsection on graphing the square root and cube root functions. The problem set on arithmetic of functions is also expanded.

CHAPTER 3 Former Sections 3.3 and 3.4 are now divided into three new sections. New Section 3.3 contains the material on dividing polynomials, while Section 3.4 provides the material on the real zeros of a polynomial function, and Section 3.5 discusses the complex zeros of a polynomial function. Section 3.7 now covers polynomial and rational inequalities.

CHAPTER 4 This chapter provides new exercises asking for verbal descriptions of the transformations required to produce the graph of a given exponential or logarithmic function from a specified basic exponential or logarithmic function. Also, new applied exercises ask students to write an equation for a given graph and explain features of the graph in terms of the real-world problem it modeled. There is a new subsection on the half-life concept, and new material is provided on exponential and logarithmic inequalities, with exercises on logarithmic scales.

CHAPTER 5 Two of the sections were reorganized and expanded to develop the topics: Systems of Nonlinear Equations (that now immediately follows the section Systems of Linear Equations in Three Variables), Systems of Inequalities, and a separate section Linear Programming. Partial-Fraction Decomposition is now the final section in this chapter.

CHAPTER 6 This chapter has an improved balance of exercises with the various types of solutions.

CHAPTER 7 This chapter has additional challenge exercises.

CHAPTER 8 This last chapter has new exercises and updated data-dependent exercises.

Supplements

Student's Solutions Manual

- By Beverly Fusfield
- Provides detailed worked-out solutions to the odd-numbered end-of-section and Chapter Review exercises and solutions to all of the Practice Problems, Practice Tests, and Cumulative Review exercises
- ISBN-13: 978-0-321-66471-6; ISBN-10: 0-321-66471-X

Graphing Calculator Manual

- By Darryl Nester, *Bluffton University*
- Provides instructions and keystroke operations for the TI-83/83+, TI-84+, TI-86, and TI-89
- Keyed directly to text Examples and Technology Connections
- ISBN-13: 978-0-321-65431-1; ISBN-10: 0-321-65431-5

Video Resources on DVD-Rom with Optional Subtitles

- Videos feature Section Summaries and Example Solutions. Section Summaries cover key definitions and procedures for all sections. Example Solutions walk students through the detailed solution process for many examples in the textbook.
- There are over 25 hours of video instruction specifically filmed for this book, making it ideal for distance learning and for supplemental instruction on your home computer or in a campus computer lab.
- Videos include optional subtitles in English and Spanish.
- ISBN-13: 978-0-321-66529-4; ISBN-10: 0-321-66529-5

A Review of Algebra

- By Heidi Howard, *Florida Community College at Jacksonville*
- Provides additional support for those students needing further algebra review
- ISBN-13: 978-0-201-77347-7; ISBN-10: 0-201-77347-3

Annotated Instructor's Edition

- Answers included on the same page beside the text exercises where possible for quick reference
- ISBN-13: 978-0-321-65506-6; ISBN-10: 0-321-65506-0

Instructor's Solutions Manual

- By Beverly Fusfield
- Complete solutions provided for all end-of-section exercises, including the Critical Thinking and Group Projects, Practice Problems, Chapter Review exercises, Practice Tests, and Cumulative Review exercises
- ISBN-13: 978-0-321-66469-3; ISBN-10: 0-321-66469-8

Instructor's Testing Manual

- By James Lapp
- Includes diagnostic pretests, chapter tests, and additional test items, grouped by section, with answers provided
- Available online within MyMathLab or from the Instructor Resource Center at www.pearsonhighered.com/irc

TestGen®

- Enables instructors to build, edit, print, and administer tests
- Features a computerized bank of questions developed to cover all text objectives
- Available for download from Pearson Education's online catalog: www.pearsonhighered.com/testgen

PowerPoint Lecture Slides

- Features presentations written and designed specifically for this text, including figures and examples from the text
- Available online within MyMathLab or from the Instructor Resource Center at www.pearsonhighered.com/irc

Pearson Math Adjunct Support Center

The **Pearson Math Adjunct Support Center** (www. pearsontutorservices.com/math-adjunct.html) is staffed by qualified instructors with more than 50 years of combined experience at both the community college and university level. Assistance is provided for faculty in the following areas:

- Suggested syllabus consultation
- Tips on using materials packed with your book
- Book-specific content assistance
- Teaching suggestions, including advice on classroom strategies

Media Resources

MyMathLab® Online Course (access code required)

MyMathLab® is a text-specific, easily customizable online course that integrates interactive multimedia instruction with textbook content. MyMathLab gives you the tools you need to deliver all or a portion of your course online, whether your students are working in a lab setting or working from home.

- **Interactive homework exercises**, correlated to your textbook at the objective level, are algorithmically generated for unlimited practice and mastery. Most exercises are free-response and provide guided solutions, sample problems, and learning aids for extra help.
- **Personalized Study Plan**, generated when students complete a test or quiz, indicates which topics have been mastered and links to tutorial exercises for topics students have not mastered.
- **Multimedia learning aids** such as video lectures, animations, and a complete multimedia textbook help students independently improve their understanding and performance.
- **Assessment Manager** lets you create online homework, quizzes, and tests that are automatically graded. Select just the right mix of questions from the MyMathLab exercise bank, instructor-created custom exercises, and/or TestGen test items.
- **Gradebook**, designed specifically for mathematics and statistics, automatically tracks students' results and gives you control over how to calculate final grades. You can also add off-line (paper-and-pencil) grades to the gradebook.
- **MathXL Exercise Builder** allows you to create static and algorithmic exercises for your online assignments. You can use the library of sample exercises as an easy starting point.
- **Pearson Tutor Center** (www.pearsontutorservices.com) access is automatically included with MyMathLab. The Tutor Center is staffed by qualified math instructors who provide textbook-specific tutoring for students via toll-free phone, fax, e-mail, and interactive Web sessions.

MyMathLab is powered by CourseCompass™, Pearson Education's online teaching and learning environment, and by MathXL®, our online homework, tutorial, and assessment system. MyMathLab is available to qualified adopters. For more information, visit www.mymathlab.com or contact your Pearson sales representative.

MathXL® Online Course (access code required)

MathXL® is an online homework, tutorial, and assessment system that accompanies Pearson's textbooks in mathematics and statistics.

- **Interactive homework exercises**, correlated to your textbook at the objective level, are algorithmically generated for unlimited practice and mastery. Most exercises are free-response and provide guided solutions, sample problems, and learning aids for extra help.
- **Personalized Study Plan**, generated when students complete a test or quiz, indicates which topics have been mastered and links to tutorial exercises for topics students have not mastered.

- **Multimedia learning aids** such as video lectures and animations help students independently improve their understanding and performance.
- **Assessment Manager** lets you create online homework, quizzes, and tests that are automatically graded. Select just the right mix of questions from the MathXL exercise bank, instructor-created custom exercises, and/or TestGen test items.
- **Gradebook**, designed specifically for mathematics and statistics, automatically tracks students' results and gives you control over how to calculate final grades.
- **MathXL Exercise Builder** allows you to create static and algorithmic exercises for your online assignments. You can use the library of sample exercises as an easy starting point.

MathXL is available to qualified adopters. For more information, visit our website at www.mathxl.com or contact your Pearson sales representative.

MathXL® Tutorials on CD (ISBN-13: 978-0-321-67349-7; ISBN-10: 0-321-67349-2)

This interactive tutorial CD-ROM provides algorithmically generated practice exercises that are correlated at the objective level to the exercises in the textbook. Every practice exercise is accompanied by an example and a guided solution designed to involve students in the solution process. Selected exercises may also include a video clip to help students visualize concepts. The software provides helpful feedback for incorrect answers and can generate printed summaries of students' progress.

Video Resources on DVD

Video Resources on DVD-Rom with Optional Subtitles (ISBN-13: 978-0-321-66529-4; ISBN-10: 0-321-66529-5)

The video lectures for this text are available on DVD-Rom, making it easy and convenient for students to watch the videos from a computer at home or on campus. The videos feature an engaging team of mathematics instructors who present Section Summaries and Example Solutions. Section Summaries cover key definitions and procedures from all sections. Example Solutions walk students through the detailed solution process for many examples in the textbook. The format provides distance-learning students with comprehensive video instruction, but also allows students needing less review to watch instruction on a specific skill or procedure. The videos have optional text subtitles, which can be easily turned off or on for individual student needs. Subtitles are available in English and Spanish.

InterAct Math Tutorial Web site: www.interactmath.com

Get practice and tutorial help online! This interactive tutorial website provides algorithmically generated practice exercises that correlate directly to the exercises in the textbook. Students can retry an exercise as many times as they like with new values each time for unlimited practice and mastery. Every exercise is accompanied by an interactive guided solution that provides helpful feedback for incorrect answers, and students can view a worked-out sample problem that takes them through an exercise similar to the one they're working on.

Acknowledgments

We would like to express our gratitude to the reviewers of this edition, who provided such invaluable insights and comments. Their contributions helped shape the development of the text and carry out the vision stated in the preface.

Reviewers

Alison Ahlgren, *University of Illinois at Urbana–Champaign*
Mohammad Aslam, *Georgia Perimeter College*
Ratan Barua, *Miami-Dade College*
Sam Bazzi, *Henry Ford Community College*
Diane Burleson, *Central Piedmont Community College*
Maureen T. Carroll, *University of Scranton*
Melissa Cass, *State University of New York–New Paltz*
Jason Cates, *Lone Star College–Montgomery*
Charles Conrad, *Volunteer State Community College*
Baiqiao Deng, *Columbus State University*
Gene Garza, *Samford University*
Bobbie Jo Hill, *Coastal Bend College*
Fran Hopf, *University of South Florida*
Natalie Hutchinson, *Old Dominion University*
Yvette Janecek, *Coastal Bend College*
Mohammad Kazemi, *University of North Carolina at Charlotte*
Mile Krajcevski, *University of South Florida*
David Keller, *Kirkwood Community College*
Rebecca Leefers, *Michigan State University*
Paul Morgan, *College of Southern Idaho*
Kathy Nickell, *College of DuPage*
Catherine Pellish, *Front Range Community College*
Betty Peterson, *Mercer County Community College*
Marshall Ransom, *Georgia Southern University*
Dr. Traci Reed, *St. Johns River Community College*
Linda Reist, *Macomb Community College*
Jeri Rogers, *Seminole Community College–Oviedo Campus*
Jason Rose, *College of Southern Idaho*
Mehdi Sadatmousavi, *Pima Community College–West*
Delphy Shaulis, *University of Colorado–Boulder*
Cindy Shaw, *Moraine Valley Community College*
Cynthia Sikes, *Georgia Southern University*
James Smith, *Columbia State Community College*
Linda Snellings-Neal, *Wright State University*
Jacqueline Stone, *University of Maryland–College Park*
Kay Stroope, *Phillips County Community College*
Robert J. Strozak, *Old Dominion University*
Frances Tishkevich, *Massacusetts Maritime Academy*
Jo Tucker, *Tarrant County College–Southeast*
Roger Werbylo, *Pima Community College–West*
Tom Worthing, *Hutchinson Community College*
Vivian Zabrocki, *Montana State University–Billings*

Marketing Reviewers

Linda E. Barton, *Ball State University*
Antanas Gilvydis, *Richard J. Daley College*
Robert A. Johnson, *Salisbury University*
Jennifer R. McNeilly, *University of Illinois at Urbana-Champaign*
Robin Ruffato, *Ball State University*
Linda Tansil, *Southeast Missouri State University*

Focus Group Reviewers

J. Michael Albanese, *Central Piedmont Community College*
Gus Brar, *Delaware County Community College*
Irene Brown, *Mercyhurst College*
Alicia Collins, *Mesa Community College*
Said Fariabi, *San Antonio College*
Nina R. Girard, *University of Pittsburgh at Johnstown*
Gloria P. Hernandez, *Louisiana State University–Eunice*
Alicia K. Jefferson, *Jackson State University*
Christine S. Mirbaha, *The Community College of Baltimore County*
Denise M. Nunley, *Glendale Community College*
Paul O'Heron, *Broome Community College*
Ruth Seidel, *Blinn College*
Deoki N. Sharma, *DeVry University*
Leslie A. Soltis, *Mercyhurst College*
Katrina N. Staley, *North Carolina A&T State University*
Marilyn Toscano, *University of Wisconsin–Superior*

Class Testers

Shana Funderburk, *University of North Carolina at Charlotte*
Bobbie Jo Hill, *Coastal Bend College*
Anna Katsoulis, *East Carolina University*
Marcus McWaters, *University of South Florida*
Jeff Norris, *Paris Junior College*
Paul Nuñez, *Mesa Community College*
William Radulovich, *Florida Community College–Jacksonville*
Julie Turnbow, *Collin County Community College–Preston*

Our sincerest thanks go to the legion of dedicated individuals who worked tirelessly to make this book possible. We express special thanks to Elka Block and Frank Purcell of Twin Prime Editorial, Lenore Parens, Abby Tanenbaum, and Carrie Green for the excellent work they did as the development editors on the text and art development. We would also like to express our gratitude to our typist, Beverly DeVine-Hoffmeyer, for her amazing patience and skill. We must also thank Dr. Praveen Rohatgi, Dr. Nalini Rohatgi, and Dr. Bhupinder Bedi for the consulting they provided on all material relating to medicine. We particularly want to thank Professors Mile Krajcevski, Viktor Maymeskul, and Scott Rimbey, for many helpful discussions and suggestions, particularly for improving the exercise sets. Further gratitude is due to Irena Andreevska, Gokarna Aryal, Ferene Tookos, and Christine Fitch for their assistance on the answers to the exercises in the text. In addition, we would like to thank Charles Davis, Douglas Ewert, Beverly Fusfield, Patricia Nelson, Ellena Reda, John Samons, Lauri Semarne, Cindy Trimble, and Tom Wegleitner for their meticulous accuracy in checking the text. Thanks are due as well to Laura Hakala and Pre-Press PMG for their excellent production work. Finally, our thanks are extended to the professional and remarkable staff at Addison-Wesley. In particular, we would like to thank Greg Tobin, Publisher; Anne Kelly, Executive Editor; Joanne Dill, Senior Project Editor; Leah Goldberg, Associate Project Editor; Sarah Gibbons, Editorial Assistant; Peggy McMahon, Senior Production Project Manager; Katherine Greig, Senior Marketing Manager; Dona Kenly, Market Development Manager; Katherine Minton, Marketing Assistant; Barbara Atkinson, Cover Designer; Ceci Fleming, Senior Media Producer; Karen Wernholm, Senior Managing Editor; and Joe Vetere, Senior Author Support/Technology Specialist.

We invite all who use this book to send suggestions for improvements to Marcus McWaters at mmm@cas.usf.edu.

About the Authors

J. S. Ratti

EDUCATION

PhD Mathematics Wayne State University

TEACHING

Wayne State University (teaching assistant and instructor)

University of Nevada at Las Vegas (instructor)

Oakland University (assistant professor)

University of South Florida (professor and past chair)

Undergraduate courses taught: college algebra, trigonometry, finite mathematics, calculus (all levels), set theory, differential equations, linear algebra

Graduate courses taught: number theory, abstract algebra, real analysis, complex analysis, graph theory

AWARDS

USF Research Council Grant

USF Teaching Incentive Program (TIP) Award

USF Outstanding Undergraduate Teaching Award

Academy of Applied Sciences grants

RESEARCH

Complex analysis, real analysis, graph theory, probability

PERSONAL INTERESTS

Fan of Tampa Bay Buccaneers

Marcus McWaters

EDUCATION

BS Major: mathematics Minor: physics
Louisiana State University New Orleans

PhD Major: mathematics Minor: philosophy
University of Florida

TEACHING

Louisiana State University (teaching assistant)

University of Florida (teaching assistant and graduate fellow)

University of South Florida (associate professor and department chair)

Undergraduate courses taught: college algebra, trigonometry, finite mathematics, calculus, business calculus, life science calculus, advanced calculus, vector calculus, set theory, differential equations, linear algebra

Graduate courses taught: graph theory, number theory, discrete mathematics, geometry, advanced linear algebra, topology, mathematical logic, modern algebra, real analysis

AWARDS

Graduate Fellow, University of Florida (Center of Excellence Grant)

USF Research Council grant

USF Teaching Incentive Program (TIP) Award

Provost's Award

RESEARCH

Topology, algebraic topology, topological algebra

Founding member, USF Center for Digital and Computational Video

PERSONAL INTERESTS

Traveling with his wife and two daughters, theater, water-skiing, racquetball

Basic Concepts of Algebra

P

Many fascinating patterns in human and natural processes can be described in the language of algebra. We investigate events ranging from chirping crickets to the behavior of falling objects.

The Real Numbers and Their Properties

Before Starting this Section, Review from Your Previous Mathematics Texts

1. Arithmetic of signed numbers
2. Arithmetic of fractions
3. Long division involving integers
4. Decimals
5. Arithmetic of real numbers

Objectives

1 Classify sets of real numbers.
2 Use the ordering of the real numbers.
3 Specify sets of numbers in roster or set-builder notation.
4 Use interval notation.
5 Relate absolute value and distance on the real number line.
6 Identify the order of operations in arithmetic expressions.
7 Identify and use properties of real numbers.
8 Evaluate algebraic expressions.

CRICKET CHIRPS AND TEMPERATURE

Crickets are sensitive to changes in air temperature; their chirps speed up as the temperature gets warmer and slow down as it gets cooler. It is possible to use the chirps of the male snowy tree cricket (*Oecanthus fultoni*), common throughout the United States, to gauge temperature. (The insect is found in every U.S. state except Hawaii, Alaska, Montana, and Florida.) By counting the chirps of this cricket, which lives in bushes a few feet from the ground, you can gauge temperature. Snowy tree crickets are more accurate than most cricket species; their chirps are slow enough to count, and they synchronize their singing. To convert cricket chirps to degrees Fahrenheit, count the number of chirps in 14 seconds and then add 40 to get the temperature. To convert cricket chirps to degrees Celsius, count the number of chirps in 25 seconds, divide by 3, and then add 4 to get temperature. In Example 10, we evaluate algebraic expressions to learn the temperature from the number of cricket chirps. ■

---

1 Classify sets of real numbers.

Classifying Numbers

In algebra, we use letters such as a, b, x, and y to represent numbers. A letter that is used to represent one or more numbers is called a **variable**. A **constant** is a specific number such as 3 or $\frac{1}{2}$ or a letter that represents a fixed (but not necessarily specified) number. Physicists use the letter c as a constant to represent the speed of light ($c = 300,000,000$ meters per second).

We use two variables, a and b, to denote the results of the operations of addition ($a + b$), subtraction ($a - b$), multiplication ($a \cdot b$ or ab), and division $\left(a \div b \text{ or } \dfrac{a}{b} \right)$.

These operations are called **binary operations** because each is performed on two numbers. We frequently omit the multiplication sign when writing a product involving two

STUDY TIP

Here is one difficulty with attempting to divide by 0: If, for example, $\dfrac{5}{0} = a$, then $5 = a \cdot 0$. However, $a \cdot 0 = 0$ for all numbers a. So $5 = 0$ and so there is no appropriate choice for $\dfrac{5}{0}$.

variables (or a constant and a variable) so that $a \cdot b$ and ab indicate the same product. Both a and b are called **factors** in the product $a \cdot b$. This is a good time to recall that we never divide by zero. For $\dfrac{a}{b}$ to represent a real number, b cannot be zero.

Equality of Numbers

The **equal sign**, $=$, is used much like we use the word *is* in English. The equal sign means that the number or expression on the left side is equal or equivalent to the number or expression on the right side. We write $a \neq b$ to indicate that a is not equal to b.

Classifying Sets of Numbers

The idea of a set is familiar to us. We regularly refer to "a set of baseball cards," a "set of CDs," or "a set of dishes." In mathematics, as in everyday life, a **set** is a collection of objects. The objects in the set are called the **elements**, or **members**, of the set. In the study of algebra, we are interested primarily in sets of numbers.

In listing the elements of a set, it is customary to enclose the listed elements in braces, { }, and separate them with commas.

We distinguish among various sets of numbers.

The numbers we use to count with constitute the set of **natural numbers**: $\{1, 2, 3, 4, \ldots\}$.

The three dots … may be read as "and so on" and indicate that the pattern continues indefinitely.

The **whole numbers** are formed by adjoining the number 0 to the set of natural numbers to obtain $\{0, 1, 2, 3, 4, \ldots\}$.

The **integers** consist of the set of natural numbers together with their opposites and 0: $\{\ldots, -4, -3, -2, -1, 0, 1, 2, 3, 4, \ldots\}$.

The **rational numbers** consist of all numbers that *can* be expressed as the quotient, $\dfrac{a}{b}$, of two integers, where $b \neq 0$.

Examples of rational numbers are $\dfrac{1}{2}, \dfrac{5}{3}, -\dfrac{4}{17}$, and $\dfrac{7}{100}$. Any integer a can be expressed as the quotient of two integers by writing $a = \dfrac{a}{1}$. Consequently, every integer is also a rational number.

$\pi = \dfrac{\text{circumference}}{\text{diameter}}$

FIGURE P.1

Rational Numbers and Decimals

The rational number $\dfrac{a}{b}$ can be written as a decimal by using long division. When any integer a is divided by an integer $b, b \neq 0$, the result is always a **terminating decimal** such as $\left(\dfrac{1}{2} = 0.5\right)$ or a **nonterminating repeating decimal** such as $\left(\dfrac{2}{3} = 0.666\ldots\right)$.

We sometimes place a bar over the repeating digits in a nonterminating repeating decimal. Thus, $\dfrac{2}{3} = 0.666\ldots = 0.\overline{6}$ and $\dfrac{13}{11} = 1.181818\ldots = 1.\overline{18}$.

Some decimals neither terminate nor repeat. Decimals that neither terminate nor repeat represent **irrational numbers**. We can construct such a decimal using only the digits 0 and 1 as follows: $0.01001000100001.\ldots$ Because each group of zeros contains one more zero than the previous group, no group of digits repeats. Other numbers such as π (pi, see Figure P.1) and $\sqrt{2}$ (the square root of 2, see Figure P.2) can also be expressed as decimals that neither terminate nor repeat; so they are

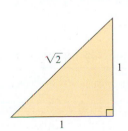

FIGURE P.2

irrational numbers as well. We can obtain an approximation of an irrational number by using an initial portion of its decimal representation. For example, we can write $\pi \approx 3.14159$ or $\sqrt{2} \approx 1.41421$, where the symbol $\approx$ is read "is approximately equal to."

No familiar process, such as long division, is available for obtaining the decimal representation of an irrational number. However, your calculator can provide a useful approximation for irrational numbers such as $\sqrt{2}$. (Try it!) Because a calculator displays a fixed number of decimal places, it gives a **rational approximation** of an irrational number.

It is usually not easy to determine whether a specific number is irrational. One helpful fact in this regard is that *the square root of any natural number that is not a perfect square is irrational.*

Because rational numbers have decimal representations that either terminate or repeat, whereas irrational numbers do not have such representations, *no number is both rational and irrational.*

The rational numbers together with the irrational numbers form the **real numbers**.

The diagram in Figure P.3 shows how various sets of numbers are related. For example, every natural number is also a whole number, an integer, a rational number, and a real number.

> **RECALL**
>
> An integer is a *perfect square* if it is a product $a \cdot a$, where a is an integer. For example, $9 = 3 \cdot 3$ is a perfect square.

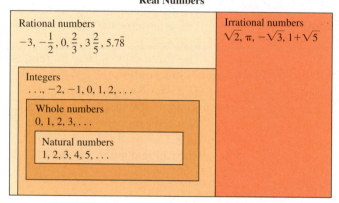

FIGURE P.3 **Relationships among sets of real numbers**

2 Use the ordering of the real numbers.

The Real Number Line

We associate the real numbers with points on a geometric line (imagined to be extended indefinitely in both directions) in such a way that each real number corresponds to exactly one point and each point corresponds to exactly one real number. The point is called the **graph** of the corresponding real number, and the real number is called the **coordinate** of the point. By agreement, *positive numbers* lie to the right of the point corresponding to 0 and *negative numbers* lie to the left of 0. See Figure P.4.

Notice that $\frac{1}{2}$ and $-\frac{1}{2}$, 2 and -2, and π and $-\pi$ correspond to pairs of points exactly the same distance from 0 but on opposite sides of 0.

FIGURE P.4 **The real number line**

When coordinates have been assigned to points on a line in the manner just described, the line is called a **real number line**, a **coordinate line**, a **real line**, or simply a **number line**. The point corresponding to 0 is called the **origin**.

Inequalities

The real numbers are **ordered** by their size. We say that *a* **is less than** *b* and write $a < b$, provided that $b = a + c$ for some *positive* number *c*. We also write $b > a$, meaning the same thing as $a < b$, and say that *b* **is greater than** *a*. On the real line, the numbers increase from left to right. Consequently, *a* **is to the left of** *b* **on the number line when** $a < b$. Similarly, *a* is to the right of *b* on the number line when $a > b$. We sometimes want to indicate that at least one of two conditions is correct: Either $a < b$ or $a = b$. In this case, we write $a \le b$ or $b \ge a$. The four symbols $<, >, \le,$ and $\ge$ are called **inequality symbols**.

EXAMPLE 1   **Identifying Inequalities**

Decide whether each of the following is true or false.

a. $5 > 0$ **b.** $-2 < -3$ **c.** $2 \le 3$ **d.** $4 \le 4$

SOLUTION

a. $5 > 0$ is true because 5 is to the right of 0 on the number line. See Figure P.5.

b. $-2 < -3$ is false because -2 is to the right of -3 on the number line.

c. $2 \le 3$ is true because 2 is to the left of 3 on the number line. (Recall that $2 \le 3$ is true if either $2 < 3$ or $2 = 3$.)

d. $4 \le 4$ is true because $4 \le 4$ is true if either $4 < 4$ or $4 = 4$.

-2 right of -3   2 left of 3

-3   -2   -1   0   1   2   3   4   5

5 right of 0

FIGURE P.5   ■ ■ ■

Practice Problem 1   Decide whether each of the following is true or false.

a. $-2 < 0$ **b.** $5 \le 7$ **c.** $-4 < -1$ ■

The following properties of inequalities for real numbers are used throughout this text.

Trichotomy Property: Exactly one of the following is true:

$$a < b, a = b, \text{ or } a > b$$

Transitive Property: If $a < b$ and $b < c$, then $a < c$.

The trichotomy property says that if two real numbers are not equal, then one is larger than the other. The transitive property says that "less than" works like "smaller than" or "lighter than." Frequently, we read $a > 0$ as "*a* is positive" instead of "*a* is greater than 0." We can also read $a < 0$ as "*a* is negative." If $a \ge 0$, then either $a > 0$ or $a = 0$, and we may say that "*a* is nonnegative."

3 Specify sets of numbers in roster or set-builder notation.

Sets

To specify a set, we do one of the following:

1. List the elements of the set (**roster method**)

2. Describe the elements of the set (often using **set-builder notation**)

Variables are helpful in describing sets when we use set-builder notation. The notation $\{x \mid x \text{ is a natural number less than } 6\}$ is in set-builder notation and describes the set $\{1, 2, 3, 4, 5\}$ using the roster method.

We read $\{x \mid x \text{ is a natural number less than } 6\}$ as "the set of all x such that x is a natural number less than six." Generally, $\{x \mid x \text{ has property } P\}$ designates the set of all x such that (the vertical bar is read "such that") x has property P.

It may happen that a description fails to describe any number. For example, consider $\{x \mid x < 2 \text{ and } x > 7\}$. Of course, no number can be simultaneously less than 2 and greater than 7, so this set has no members. We refer to a set with no elements as the **empty set**, or **null set**, and use the special symbol $\varnothing$ to denote it.

We say that two sets A and B are **equal** if they have exactly the same elements. If each element of a set A is also an element of a set B, we say that A is a **subset** of B. The symbol $\subseteq$ is used to denote the subset relation so that $A \subseteq B$ is read "A is a subset of B." For example, $\{0, 3\} \subseteq \{0, 1, 2, 3, 4\}$ because both 0 and 3 are elements of the set $\{0, 1, 2, 3, 4\}$.

The **union** of two sets A and B, denoted $A \cup B$, is the set consisting of all elements that are in A *or* B (or both). The **intersection** of A and B, denoted $A \cap B$, is the set consisting of all elements that are in both A *and* B. In other words, $A \cap B$ consists of the elements common to A and B.

EXAMPLE 2   Forming Set Unions and Intersections

Find $A \cap B$ and $A \cup B$ assuming that $A = \{-2, -1, 0, 1, 2\}$ and $B = \{-4, -2, 0, 2, 4\}$.

SOLUTION

$A \cap B = \{-2, 0, 2\}$, the set of elements common to both A and B.
$A \cup B = \{-4, -2, -1, 0, 1, 2, 4\}$, the set of elements that are in A or B (or both).

■ ■ ■

Practice Problem 2   Find $A \cap B$ and $A \cup B$ if $A = \{-3, -1, 0, 1, 3\}$ and $B = \{-4, -2, 0, 2, 4\}$. ■

When working with sets, we frequently designate a **universal set**, the set consisting of all elements under consideration. Then every set is a subset of the universal set, and the set of elements in the universal set that are not in a set A form the **complement of A**, denoted A'.

EXAMPLE 3   Finding the Complement of a Set

If $U = \{0, 1, 2, 3, 4, 5, 6, 7, 8, 9\}$ is the universal set and $A = \{0, 2, 4, 6, 8\}$, find A'.

SOLUTION

A' is the set of elements in U that are not in A; so $A' = \{1, 3, 5, 7, 9\}$.   ■ ■ ■

Practice Problem 3   If $U = \{0, 1, 2, 3, 4, 5, 6, 7, 8, 9\}$ is the universal set and $A = \{0, 3, 6, 9\}$, find A'. ■

4 Use interval notation.

Intervals

We now turn our attention to graphing certain sets of numbers. That is, we graph each number in a given set. We are particularly interested in sets of real numbers, called **intervals**, whose graphs correspond to special sections of the number line.

$(a, b) = \{x \mid a < x < b\}$

FIGURE P.6 An open interval

$[a, b] = \{x \mid a \le x \le b\}$

FIGURE P.7 A closed interval

Interval $(2, \infty)$

FIGURE P.8 An unbounded interval

Interval (a, ∞)

FIGURE P.9 An unbounded interval

If $a < b$, then the set of real numbers between a and b, but not including either a or b, is called the **open interval** from a to b and is denoted by (a, b). See Figure P.6. Using set-builder notation, we can write

$$(a, b) = \{x \mid a < x < b\}.$$

We indicate graphically that the endpoints a and b are excluded from the open interval by drawing a left parenthesis at a and a right parenthesis at b. These parentheses enclose the numbers between a and b.

The **closed interval** from a to b is the set

$$[a, b] = \{x \mid a \le x \le b\}.$$

The closed interval includes both endpoints a and b. We replace the parentheses with square brackets in the interval notation and on the graph. See Figure P.7. Sometimes we want to include only one endpoint of an interval and exclude the other. Table P.1 below shows how this is done.

We may also want to graph all numbers to the left or right of a given number, such as

$$\{x \mid x > 2\}.$$

The number 2 is not included in this set, and we again indicate this fact graphically by drawing a left parenthesis at 2. We then graph all points to the right of 2, as shown in Figure P.8.

For any number a, we call $\{x \mid x > a\}$ an **unbounded** (or infinite) **interval** and denote this interval by (a, ∞). The symbol ∞ ("infinity") does not represent a number. The notation (a, ∞) is used to indicate the set of all real numbers that are greater than a, and the symbol ∞ is used to indicate that the interval extends indefinitely to the right of a. See Figure P.9.

The symbol $-\infty$ is another symbol that does not represent a number. The notation $(-\infty, a)$ is used to indicate the set of all real numbers that are less than a. The notation $(-\infty, \infty)$ represents the set of all real numbers.

Table P.1 lists various types of intervals that we use in this book. In the table, when two points a and b are given, we assume that $a < b$. This is because if $a > b$, then (a, b) is the empty set.

TABLE P.1

Interval Notation	Set-Builder Notation	Graph
(a, b)	$\{x \mid a < x < b\}$	
$[a, b]$	$\{x \mid a \le x \le b\}$	
$(a, b]$	$\{x \mid a < x \le b\}$	
$[a, b)$	$\{x \mid a \le x < b\}$	
(a, ∞)	$\{x \mid x > a\}$	
$[a, \infty)$	$\{x \mid x \ge a\}$	
$(-\infty, b)$	$\{x \mid x < b\}$	
$(-\infty, b]$	$\{x \mid x \le b\}$	
$(-\infty, \infty)$	$\{x \mid x \text{ is a real number}\}$	

Open interval (a, b)

(a)

Closed interval $[a, b]$

(b)

FIGURE P.10 Endpoint inclusion and exclusion

An alternative notation for indicating whether endpoints are included uses closed circles to show inclusion and open circles to show exclusion. See Figure P.10.

5 Relate absolute value and distance on the real number line.

Absolute Value

The *absolute value* of a number a, denoted by $|a|$, is the distance between the origin and the point on the number line with coordinate a. The point with coordinate -3 is three units from the origin, so we write $|-3| = 3$ and say that the absolute value of -3 is 3. See Figure P.11.

FIGURE P.11 Absolute value

ABSOLUTE VALUE

For any real number a, the **absolute value** of a, denoted $|a|$, is defined by

$$|a| = a \quad \text{if } a \geq 0 \qquad \text{and} \qquad |a| = -a \quad \text{if } a < 0.$$

STUDY TIP

Finding the absolute value requires knowing whether the number or expression inside the absolute value bars is positive, zero, or negative. If it is positive or zero, you can simply remove the absolute value bars. If it is negative, you remove the bars and change the sign of the number or expression inside the absolute value bars.

EXAMPLE 4   Determining Absolute Value

Find the value of each of the following expressions.

a. $|4|$ **b.** $|-4|$ **c.** $|0|$ **d.** $|(-3) + 1|$

SOLUTION

a. $|4| = 4$ — Because the number inside the absolute value bars is 4 and $4 \geq 0$, just remove the absolute value bars.

b. $|-4| = -(-4) = 4$ — Because the number inside the absolute value bars is -4 and $-4 < 0$, remove the bars and change the sign.

c. $|0| = 0$ — Because the number inside the absolute value bars is 0 and $0 \geq 0$, just remove the absolute value bars.

d. $|(-3) + 1| = |-2|$ — Because the expression inside the absolute value
$= -(-2) = 2$ — bars is $(-3) + 1 = -2$ and $-2 < 0$, remove the bars and change the sign. ■ ■ ■

Practice Problem 4   Find the value of each of the following.

a. $|-10|$ **b.** $|3 - 4|$ **c.** $|2(-3) + 7|$ ■

◆ **WARNING**

The absolute value of a number represents a distance. Because distance can never be negative, the absolute value of a number is never negative; it is always positive or zero. However, if a is not 0, $-|a|$ is always negative. Thus, $-|5.3| = -5.3$, $-|-4| = -4$, and $-|1.\overline{18}| = -1.\overline{18}$.

Distance Between Two Points on a Real Number Line

The absolute value is used to define the distance between two points on a number line.

DISTANCE FORMULA ON A NUMBER LINE

If a and b are the coordinates of two points on a number line, then the distance between a and b, denoted by $d(a, b)$, is $|a - b|$. In symbols, $d(a, b) = |a - b|$.

The absolute value function on your graphing calculator will find the value of the expression entered and then compute its absolute value.

FIGURE P.13

EXAMPLE 5   **Finding the Distance Between Two Points**

Find the distance between -3 and 4 on the number line.

SOLUTION

Figure P.12 shows that the distance between -3 and 4 is seven units. The distance formula gives the same answer.

$$d(-3, 4) = |-3 - 4| = |-7| = -(-7) = 7.$$

$3 + 4 = 7$ units

FIGURE P.12 Distance on the number line

Notice that reversing the order of -3 and 4 in this computation gives the same answer. That is, the distance between 4 and -3 is $|4 - (-3)| = |4 + 3| = |7| = 7$. It is always true that $|a - b| = |b - a|$. See Figure P.13. ■ ■ ■

Practice Problem 5   Find the distance between -7 and 2 on the number line. ■

We summarize the properties of absolute value next.

PROPERTIES OF ABSOLUTE VALUE

If a and b are any real numbers, the following properties apply.

Property	Example
1. $\|a\| \geq 0$	$\|-5\| = 5$ and $5 \geq 0$
2. $\|a\| = \|-a\|$	$\|3\| = \|-3\|$
3. $\|ab\| = \|a\|\|b\|$	$\|3(-5)\| = \|3\|\|-5\|$
4. $\left\|\dfrac{a}{b}\right\| = \dfrac{\|a\|}{\|b\|}$	$\left\|\dfrac{-7}{3}\right\| = \dfrac{\|-7\|}{\|3\|}$
5. $\|a - b\| = \|b - a\|$	$\|2 - 7\| = \|7 - 2\|$
6. $a \leq \|a\|$	$-2 \leq \|-2\|, 2 \leq \|2\|$
7. $\|a + b\| \leq \|a\| + \|b\|$ (the triangle inequality)	$\|-2 + 5\| \leq \|-2\| + \|5\|$

---

6 Identify the order of operations in arithmetic expressions.

Arithmetic Expressions

When we write numbers in a meaningful combination of the basic operations of arithmetic, the result is called an **arithmetic expression**. The real number that results from performing all operations in the expression is called the **value** of the expression. In arithmetic and algebra, parentheses ( ) are **grouping symbols** used to indicate which operations are to be performed first. Other common grouping symbols are square brackets [ ], braces { }, fraction bars — or /, and absolute value bars | |.

Evaluating arithmetic expressions requires carefully applying the following conventions for the order in which the operations are performed.

TECHNOLOGY CONNECTION

The order of operations is built into your graphing calculator.

THE ORDER OF OPERATIONS

Whenever a fraction bar is encountered, work separately above and below it.
    When parentheses or other grouping symbols are present, start inside the innermost pair of grouping symbols and work outward.

Step 1 Do operations involving exponents.

Step 2 Do multiplications and divisions in the order in which they occur, working from left to right.

Step 3 Do additions and subtractions in the order in which they occur, working from left to right.

EXAMPLE 6 **Evaluating Arithmetic Expressions**

Evaluate each of the following.

a. $3 \cdot 2^5 + 4$ **b.** $5 - (3 - 1)^2$

SOLUTION

a. $3 \cdot 2^5 + 4 = 3 \cdot 32 + 4$ Do operations involving exponents first.

$= 96 + 4$ Multiplication is done next.

$= 100$ Addition is done next.

b. $5 - (3 - 1)^2 = 5 - (2)^2$ Work inside the parentheses first.

$= 5 - 4$ Next, do operations involving exponents.

$= 1$ Subtraction is done next. ■ ■ ■

Practice Problem 6 Evaluate.

a. $(-3) \cdot 5 + 20$ **b.** $7 - 2 \cdot 3$ **c.** $\dfrac{9 - 1}{4} - 5 \cdot 7$ ■

7 Identify and use properties of real numbers.

TECHNOLOGY CONNECTION

On your graphing calculator, the **opposite** of a is denoted with a shorter horizontal hyphen than that used to denote subtraction. It is also placed somewhat higher than the subtraction symbol.

Properties of the Real Numbers

When doing arithmetic, we intuitively use important properties of the real numbers. We know, for example, that if we add or multiply two real numbers, the result is a real number. This fact is known as the **closure** property of real numbers. *Throughout this section, unless otherwise stated, a, b, and c represent real numbers.*

The numbers 0 and 1 have special roles among the real numbers. Because 0 preserves the identity of any number under addition ($0 + 3 = 3, 3 + 0 = 3$) and 1 preserves the identity of any number under multiplication ($5 \cdot 1 = 5, 1 \cdot 5 = 5$), we call 0 the **additive identity** and 1 the **multiplicative identity**.

IDENTITY PROPERTIES

$$a + 0 = 0 + a = a \qquad a \cdot 1 = 1 \cdot a = a$$

Two other useful properties of 0 involve multiplication rather than addition.

> **ZERO-PRODUCT PROPERTIES**
>
> $$0 \cdot a = 0 \qquad a \cdot 0 = 0$$
>
> If $a \cdot b = 0$, then $a = 0$ or $b = 0$.

The Additive Inverse

Every real number a has an **opposite** number, denoted $-a$, with the property that the sum of a and $-a$ is 0. The *opposite* of a, $-a$, is also called the **additive inverse** of a. We state this property of real numbers as follows.

> **ADDITIVE INVERSE PROPERTY**
>
> $$a + (-a) = (-a) + a = 0$$

Reciprocals

Every nonzero real number a has an **inverse**, $\dfrac{1}{a}$, with the property that the product of a and $\dfrac{1}{a}$ is 1. The *inverse* $\dfrac{1}{a}$ is also called the **multiplicative inverse**, or **reciprocal**, of a.

> **MULTIPLICATIVE INVERSE PROPERTY**
>
> $$\text{If } a \neq 0, \qquad a \cdot \frac{1}{a} = \frac{1}{a} \cdot a = 1$$

Because $0 \cdot b = 0$ for any number b, the number 0 does not have a multiplicative inverse

EXAMPLE 7   **Finding the Reciprocal**

Find the reciprocal of the following numbers.

a. 3   **b.** $-\dfrac{3}{4}$

SOLUTION

a. The reciprocal of 3 is $\dfrac{1}{3}$ because $3 \cdot \dfrac{1}{3} = 1$.

b. The reciprocal of $-\dfrac{3}{4}$ is $-\dfrac{4}{3}$ because $-\dfrac{3}{4} \cdot -\dfrac{4}{3} = 1$.   ■ ■ ■

Practice Problem 7   Find the reciprocal of the following numbers.

a. $\dfrac{1}{2}$ **b.** -7 ■

PROPERTIES OF THE REAL NUMBERS

Name	Property	Example
Closure	$a + b$ is a real number. $a \cdot b$ is a real number.	$1 + \sqrt{2}$ is a real number. $2 \cdot \pi$ is a real number.
Commutative	$a + b = b + a$ $a \cdot b = b \cdot a$	$4 + 7 = 7 + 4$ $3 \cdot 8 = 8 \cdot 3$
Associative	$(a + b) + c = a + (b + c)$ $(a \cdot b) \cdot c = a \cdot (b \cdot c)$	$(2 + 1) + 7 = 2 + (1 + 7)$ $(5 \cdot 9) \cdot 13 = 5 \cdot (9 \cdot 13)$
Distributive	$a \cdot (b + c) = a \cdot b + a \cdot c$ $(a + b) \cdot c = a \cdot c + b \cdot c$	$3 \cdot (2 + 5) = 3 \cdot 2 + 3 \cdot 5$ $(2 + 5) \cdot 3 = 2 \cdot 3 + 5 \cdot 3$
Identity	There is a unique real number 0 such that $a + 0 = a$ and $0 + a = a$. There is a unique real number 1 such that $a \cdot 1 = a$ and $1 \cdot a = a$.	$3 + 0 = 3$ and $0 + 3 = 3$ $7 \cdot 1 = 7$ and $1 \cdot 7 = 7$
Inverse	There is a unique real number $-a$ such that $a + (-a) = 0$ and $-a + a = 0$. If $a \neq 0$, there is a unique real number $\dfrac{1}{a}$ such that $a \cdot \dfrac{1}{a} = 1$ and $\dfrac{1}{a} \cdot a = 1$.	$5 + (-5) = 0$ and $(-5) + 5 = 0$ $4 \cdot \dfrac{1}{4} = 1$ and $\dfrac{1}{4} \cdot 4 = 1$

Subtraction and Division of Real Numbers

You may have noticed that all of the properties discussed to this point apply to the operations of addition and multiplication. What about subtraction and division? We see next that subtraction and division can be defined in terms of addition and multiplication.

Subtraction of the number b from the number a is defined with the use of a and $-b$; to subtract b from a, add the opposite of b to a.

DEFINITION OF SUBTRACTION

$$a - b = a + (-b)$$

If a and b are real numbers and $b \neq 0$, **division** of a by b is defined with the use of a and $\dfrac{1}{b}$; to divide a by b, multiply a by the reciprocal of b.

DEFINITION OF DIVISION

$$\text{If } b \neq 0, \qquad a \div b = \frac{a}{b} = a \cdot \frac{1}{b}$$

For $b \neq 0$, $\dfrac{a}{b}$ is called the **quotient**, "the **ratio** of a to b," or the **fraction** with **numerator** a and **denominator** b. Here are some useful properties involving opposites, subtraction, and division.

PROPERTIES OF OPPOSITES

$$(-1)a = -a \qquad\qquad -(-a) = a \qquad\qquad (-a)b = a(-b) = -(ab)$$

$$(-a)(-b) = ab \qquad\qquad -(a + b) = -a - b \qquad\qquad -(a - b) = b - a$$

$$\frac{-a}{b} = \frac{a}{-b} = -\frac{a}{b} \qquad\qquad \frac{-a}{-b} = \frac{a}{b} \qquad\qquad a(b - c) = ab - ac$$

To combine real numbers by means of the division operation, we use the following properties. We write $x \pm y$ as shorthand for "$x + y$ or $x - y$".

PROPERTIES OF FRACTIONS

All denominators are assumed to be nonzero.

$$\frac{a}{c} \pm \frac{b}{c} = \frac{a \pm b}{c} \qquad\qquad \frac{a}{b} \pm \frac{c}{d} = \frac{ad \pm bc}{bd}$$

$$\frac{a}{b} \cdot \frac{c}{d} = \frac{ac}{bd} \qquad\qquad \frac{a}{b} \div \frac{c}{d} = \frac{a}{b} \cdot \frac{d}{c}$$

$$\frac{ac}{bc} = \frac{a}{b} \qquad\qquad \frac{a}{b} = \frac{c}{d} \text{ means } a \cdot d = b \cdot c$$

The property $\dfrac{ac}{bc} = \dfrac{a}{b}$ can be used to produce a common denominator when adding or subtracting fractions.

EXAMPLE 8   **Adding, Subtracting, Multiplying, and Dividing Fractions**

Perform the indicated operations.

a. $\dfrac{5}{3} + \dfrac{3}{2}$ 　　**b.** $\dfrac{5}{6} - \dfrac{2}{3}$ 　　**c.** $\dfrac{4}{3} \cdot \dfrac{15}{8}$ 　　**d.** $\dfrac{\frac{2}{5}}{\frac{9}{25}}$

SOLUTION

a. $\dfrac{5}{3} + \dfrac{3}{2} = \dfrac{5 \cdot 2}{3 \cdot 2} + \dfrac{3 \cdot 3}{2 \cdot 3}$ 　　　$\dfrac{ac}{bc} = \dfrac{a}{b}$

　　$= \dfrac{5 \cdot 2 + 3 \cdot 3}{3 \cdot 2} = \dfrac{19}{6}$ 　　　$\dfrac{a}{c} + \dfrac{b}{c} = \dfrac{a + b}{c}$

b. $\dfrac{5}{6} - \dfrac{2}{3} = \dfrac{5}{6} - \dfrac{2 \cdot 2}{3 \cdot 2}$ 　　　$\dfrac{ac}{bc} = \dfrac{a}{b}$

　　$= \dfrac{5 - 2 \cdot 2}{6} = \dfrac{1}{6}$ 　　　$\dfrac{a}{c} - \dfrac{b}{c} = \dfrac{a - b}{c}$

c. $\dfrac{4}{3} \cdot \dfrac{15}{8} = \dfrac{4 \cdot 15}{3 \cdot 8}$ 　　　$\dfrac{a}{b} \cdot \dfrac{c}{d} = \dfrac{a \cdot c}{b \cdot d}$

　　$= \dfrac{\cancel{4} \cdot \cancel{3} \cdot 5}{\cancel{3} \cdot \cancel{4} \cdot 2} = \dfrac{5}{2}$ 　　　$\dfrac{a \cdot c}{b \cdot c} = \dfrac{a}{b}$

d. $\dfrac{\frac{2}{5}}{\frac{9}{25}} = \dfrac{2}{5} \cdot \dfrac{25}{9}$ $\qquad$ $\dfrac{a}{b} \div \dfrac{c}{d} = \dfrac{a}{b} \cdot \dfrac{d}{c}$

$\qquad\qquad = \dfrac{2 \cdot 25}{5 \cdot 9}$ $\qquad\qquad$ $\dfrac{a}{b} \cdot \dfrac{c}{d} = \dfrac{a \cdot c}{b \cdot d}$

$\qquad\qquad = \dfrac{2 \cdot \cancel{5} \cdot 5}{\cancel{5} \cdot 9} = \dfrac{10}{9}$ $\qquad$ $\dfrac{a \cdot c}{b \cdot c} = \dfrac{a}{b}$ $\qquad\qquad$ ■ ■ ■

Practice Problem 8 Perform the indicated operations.

a. $\dfrac{7}{4} + \dfrac{3}{8}$ $\quad$ **b.** $\dfrac{8}{3} - \dfrac{2}{5}$ $\quad$ **c.** $\dfrac{9}{14} \cdot \dfrac{7}{3}$ $\quad$ **d.** $\dfrac{\frac{5}{8}}{\frac{15}{16}}$ $\qquad$ ■

8 Evaluate algebraic expressions.

Algebraic Expressions

In algebra, any number, constant, variable, or parenthetical group (or any product of them) is called a **term**. A combination of terms using the ordinary operations of addition, subtraction, multiplication, and division (as well as exponentiation and roots, which are discussed later in this chapter) is called an **algebraic expression**, or simply an **expression**. If we replace each variable in an expression with a specific number and get a real number, the resulting number is called the **value of the algebraic expression**. Of course, this value depends on the numbers we use to replace the variables in the expression. Here are some examples of algebraic expressions:

$$x - 2 \qquad \frac{1}{x} + \sqrt{7} \qquad \frac{10}{y + 3} \qquad \sqrt{x} + |y| \div 5$$

◆ **WARNING**

Incorrect	**Correct**
$5(x + 1) = 5x + 1$	$5(x + 1) = 5x + 5$
$\left(\dfrac{1}{3}x\right)\left(\dfrac{1}{3}y\right) = \dfrac{1}{3}xy$	$\left(\dfrac{1}{3}x\right)\left(\dfrac{1}{3}y\right) = \dfrac{1}{9}xy$
$x - (4y + 3) = x - 4y + 3$	$x - (4y + 3) = x - 4y - 3$

EXAMPLE 9 **Evaluating an Algebraic Expression**

Evaluate the following expressions for the given values of the variables.

a. $[(9 + x) \div 7] \cdot 3 - x$ $\qquad$ for $x = 5$

b. $|x| - \dfrac{2}{y}$ $\qquad\qquad\qquad$ for $x = 2$ and $y = -2$

SOLUTION

a. $[(9 + x)] \div 7] \cdot 3 - x = [(9 + 5) \div 7] \cdot 3 - 5$ $\qquad$ Replace x with 5.

$\qquad\qquad\qquad\qquad\quad = [14 \div 7] \cdot 3 - 5$ $\qquad$ Work inside the parentheses.

$\qquad\qquad\qquad\qquad\quad = 2 \cdot 3 - 5$ $\qquad\qquad\quad$ Work inside the brackets.

$\qquad\qquad\qquad\qquad\quad = 6 - 5$ $\qquad\qquad\qquad$ Multiply.

$\qquad\qquad\qquad\qquad\quad = 1$ $\qquad\qquad\qquad\qquad$ The value of the expression

b. $|x| - \dfrac{2}{y} = |2| - \dfrac{2}{-2}$ Replace x with 2 and y with -2.

$\qquad\qquad = 2 - \dfrac{2}{-2}$ Eliminate the absolute value, $|2| = 2$.

$\qquad\qquad = 2 - (-1)$ $\dfrac{2}{-2} = -1$; division occurs before subtraction.

$\qquad\qquad = 3$ The value of the expression   ■ ■ ■

Practice Problem 9   Evaluate.

a. $(x - 2) \div 3 + x$ for $x = 3$ **b.** $7 - \dfrac{x}{|y|}$ for $x = -1, y = 3$ ■

EXAMPLE 10   **Finding the Temperature from Cricket Chirps**

Write an algebraic expression for converting cricket chirps to degrees Celsius. Assuming that you count 48 chirps in 25 seconds, what is the Celsius temperature?

SOLUTION

Recall from the introduction to this section that to convert cricket chirps to degrees Celsius, you count the chirps in 25 seconds, divide by 3, and then add 4 to get the temperature.

$$\boxed{\begin{array}{c}\text{Temperature in}\\\text{Celsius degrees}\end{array}} = \left[\left(\dfrac{\text{Number of chirps in 25 seconds}}{3}\right) + 4\right]$$

If we let C = temperature in degrees Celsius and N = number of chirps in 25 seconds, we get the expression

$$C = \frac{N}{3} + 4.$$

Suppose we count 48 chirps in 25 seconds. The Celsius temperature is

$$C = \frac{48}{3} + 4$$

$$= 16 + 4 = 20 \text{ degrees.} \qquad \blacksquare\ \blacksquare\ \blacksquare$$

Practice Problem 10   Use the temperature expression in Example 10 to find the Celsius temperature assuming that 39 chirps are counted in 25 seconds.   ■

The **domain** of an algebraic expression having only one variable is the set of all possible numbers that may be used for the variable. For example, division by 0 is not defined, so no number can replace a variable if a denominator of 0 results. The number 1 is excluded from the domain of the expression $\dfrac{1}{x - 1}$ because replacing x with 1 would result in a denominator of 0. The domain of the expression $\dfrac{1}{x - 1}$ is thus the set of all real numbers x, $x \neq 1$.

EXAMPLE 11   **Finding the Domain of an Algebraic Expression**

Find the domain of the expression $\dfrac{5}{3 - x}$.

SOLUTION

Notice that $3 - x = 0$ if $x = 3$.

Because replacing x with 3 in $\dfrac{5}{3-x}$ results in a denominator of 0, the number 3 is not in its domain. If $x \neq 3$, the value of $\dfrac{5}{3-x}$ is a real number. The domain can be expressed in set notation as $\{x \mid x \neq 3\}$ or in interval notation as $(-\infty, 3) \cup (3, \infty)$.

■ ■ ■

Practice Problem 11   Find the domain of the variable x in the expression $\dfrac{2-x}{x+2}$.   ■

SECTION P.1 ■ Exercises

A EXERCISES Basic Skills and Concepts

1. Whole numbers are formed by adding the number _____ to the set of natural numbers.

2. The number -3 is an integer, but it is also a(n) _____ and a(n) _____ .

3. If $a < b$, then a is to the _____ of b on the number line.

4. If a real number is not a rational number, it is a(n) _____ number.

5. *True or False* If $-x$ is positive, then $-x > 0$.

6. *True or False* $\dfrac{-5}{2} < -2\dfrac{1}{2}$.

In Exercises 7–14, write each of the following rational numbers as a decimal and state whether the decimal is repeating or terminating.

7. $\dfrac{1}{3}$

8. $\dfrac{2}{3}$

9. $-\dfrac{4}{5}$

10. $-\dfrac{3}{12}$

11. $\dfrac{3}{11}$

12. $\dfrac{11}{33}$

13. $\dfrac{95}{30}$

14. $\dfrac{41}{15}$

In Exercises 15–24, classify each of the following numbers as rational or irrational.

15. -207

16. -114

17. $\sqrt{81}$

18. $-\sqrt{25}$

19. $\dfrac{7}{2}$

20. $-\dfrac{15}{12}$

21. $\sqrt{12}$

22. $\sqrt{3}$

23. 0.321

24. $5.8\overline{2}$

In Exercises 25–34, use inequality symbols to write the given statements symbolically.

25. 3 is greater than -2.

26. -3 is less than -2.

27. $\dfrac{1}{2}$ is greater than or equal to $\dfrac{1}{2}$.

28. x is less than $x + 1$.

29. 5 is less than or equal to $2x$.

30. $x - 1$ is greater than 2.

31. $-x$ is positive.

32. x is negative.

33. $2x + 7$ is less than or equal to 14.

34. $2x + 3$ is not greater than 5.

In Exercises 35–38, fill in the blank with one of the symbols =, <, or > to produce a true statement.

35. 4 _____ $\dfrac{24}{6}$

36. -5 _____ -2

37. -4 _____ 0

38. $-\dfrac{9}{2}$ _____ $-4\dfrac{1}{2}$

In Exercises 39–44, use the roster method to write the given sets of numbers. Interpret "between a and b" to include both a and b.

39. The first three natural numbers

40. The first five natural numbers

41. The natural numbers between 3 and 7

42. The natural numbers between 8 and 11

43. The negative integers between -3 and 5

44. The whole numbers between -2 and 4

In Exercises 45–54, find each set if the universal set $U = \{-4, -3, -2, -1, 0, 1, 2, 3, 4\}$, $A = \{-4, -2, 0, 2, 4\}$, $B = \{-3, 0, 1, 2, 3, 4\}$, and $C = \{-4, -3, -2, -1, 0, 2\}$.

45. $A \cup B$

46. $A \cap B$

47. A'

48. C'

49. $A' \cup (B \cap C)$

50. $A \cap B'$

51. $(A \cup B)'$

52. $(A \cap B)'$

53. $(A \cup B) \cap C$

54. $(A \cup B) \cup C$

In Exercises 55–68, rewrite each expression without absolute value bars.

55. $|20|$

56. $|12|$

57. $-|-4|$

58. $-|-17|$

59. $\left|\dfrac{5}{-7}\right|$

60. $\left|\dfrac{-3}{5}\right|$

61. $|5 - \sqrt{2}|$

62. $|\sqrt{2} - 5|$

63. $\dfrac{8}{|-8|}$

64. $\dfrac{-8}{|8|}$

65. $|5 + |-7||$

66. $|5 - |-7||$

67. $||7| - |4||$

68. $||4| - |7||$

In Exercises 69–76, use the absolute value to express the distance between the points with coordinates a and b on the number line. Then determine this distance by evaluating the absolute value expression.

69. $a = 3$ and $b = 8$

70. $a = 2$ and $b = 14$

71. $a = -6$ and $b = 9$

72. $a = -12$ and $b = 3$

73. $a = -20$ and $b = -6$

74. $a = -14$ and $b = -1$

75. $a = \dfrac{22}{7}$ and $b = -\dfrac{4}{7}$

76. $a = \dfrac{16}{5}$ and $b = -\dfrac{3}{5}$

In Exercises 77–88, graph each of the given intervals on a separate number line and write the inequality notation for each.

77. $[1, 4]$

78. $[-2, 2]$

79. $(14, 28)$

80. $\left(\dfrac{1}{2}, \dfrac{9}{2}\right)$

81. $(-3, 1]$

82. $[-6, -2]$

83. $[-3, \infty)$

84. $[0, \infty)$

85. $(-\infty, 5]$

86. $(-\infty, -1]$

87. $\left(-\dfrac{3}{4}, \dfrac{9}{4}\right)$

88. $\left(-3, -\dfrac{1}{2}\right)$

In Exercises 89–92, use the distributive property to write each expression without parentheses.

89. $4(x + 1)$

90. $(-3)(2 - x)$

91. $5(x - y + 1)$

92. $2(3x + 5 - y)$

In Exercises 93–96, find the additive inverse and reciprocal of each number.

	Number	Additive Inverse	Reciprocal
93.	5		
94.	$-\dfrac{2}{3}$		
95.	0		
96.	1.7		

In Exercises 97–108, name the property of real numbers that justifies the given equality. All variables represent real numbers.

97. $(-7) + 7 = 0$

98. $5 + (-5) = 0$

99. $(x + 2) = 1 \cdot (x + 2)$

100. $3a = 1 \cdot 3a$

101. $7(xy) = (7x)y$

102. $3 \cdot (6x) = (3 \cdot 6)x$

103. $\dfrac{3}{2}\left(\dfrac{2}{3}\right) = 1$

104. $2\left(\dfrac{1}{2}\right) = 1$

105. $(3 + x) + 0 = 3 + x$

106. $x(2 + y) + 0 = x(2 + y)$

107. $(x + 5) + 2y = x + (5 + 2y)$

108. $(3 + x) + 5 = x + (3 + 5)$

In Exercises 109–130, perform the indicated operations.

109. $\dfrac{3}{5} + \dfrac{4}{3}$

110. $\dfrac{7}{10} + \dfrac{3}{4}$

111. $\dfrac{6}{5} + \dfrac{5}{7}$

112. $\dfrac{9}{2} + \dfrac{5}{12}$

113. $\dfrac{5}{6} + \dfrac{3}{10}$

114. $\dfrac{8}{15} + \dfrac{2}{9}$

115. $\dfrac{5}{8} - \dfrac{9}{10}$

116. $\dfrac{7}{8} - \dfrac{1}{5}$

117. $\dfrac{5}{9} - \dfrac{7}{11}$

118. $\dfrac{5}{8} - \dfrac{7}{12}$

119. $\dfrac{2}{5} - \dfrac{1}{2}$

120. $\dfrac{1}{4} - \dfrac{1}{6}$

121. $\dfrac{3}{4} \cdot \dfrac{8}{27}$

122. $\dfrac{9}{7} \cdot \dfrac{14}{27}$

123. $\dfrac{\frac{8}{5}}{\frac{16}{15}}$

124. $\dfrac{\frac{5}{6}}{\frac{15}{6}}$

125. $\dfrac{\frac{7}{8}}{\frac{21}{16}}$

126. $\dfrac{\frac{3}{10}}{\frac{7}{15}}$

127. $5 \cdot \dfrac{3}{10} - \dfrac{1}{2}$

128. $2 \cdot \dfrac{7}{2} - \dfrac{3}{2}$

129. $3 \cdot \dfrac{2}{15} - \dfrac{1}{3}$

130. $2 \cdot \dfrac{5}{3} - \dfrac{3}{2}$

In Exercises 131–140, evaluate each expression for $x = 3$ and $y = -5$.

131. $2(x + y) - 3y$

132. $-2(x + y) + 5y$

133. $3|x| - 2|y|$

134. $7|x - y|$

135. $\dfrac{x - 3y}{2} + xy$

136. $\dfrac{y + 3}{x} - xy$

137. $\dfrac{2(1 - 2x)}{y} - (-x)y$

138. $\dfrac{3(2 - x)}{y} - (1 - xy)$

139. $\dfrac{\frac{14}{x} + \frac{1}{2}}{\frac{-y}{4}}$

140. $\dfrac{\frac{4}{-y} + \frac{8}{x}}{\frac{y}{2}}$

In Exercises 141–152, determine the domain of each expression. Write each domain in interval notation.

141. $\dfrac{3}{x-1}$

142. $\dfrac{-5}{1-x}$

143. $\dfrac{x+2}{x}$

144. $\dfrac{x}{x+7}$

145. $\dfrac{1}{x}+\dfrac{1}{x+1}$

146. $\dfrac{2}{x}-\dfrac{3}{x-5}$

147. $\dfrac{7}{(x-1)(x+2)}$

148. $\dfrac{x}{x(x+3)}$

149. $\dfrac{x}{2-x}+\dfrac{1}{x}$

150. $\dfrac{5}{x-3}+\dfrac{4}{3-x}$

151. $\dfrac{5}{x^2+1}$

152. $\dfrac{x}{x^2+2}$

B EXERCISES Applying the Concepts

153. Media players. Let $A =$ the set of people who own MP3 players and $B =$ the set of people who own DVD players.
  a. Describe the set $A \cup B$.
  b. Describe the set $A \cap B$.

154. Standard car features. The table below indicates whether certain features are "standard" for each of three types of cars.

	Navigation System	Automatic Transmission	Leather Seats
2009 Lexus SC 430	yes	yes	yes
2009 Lincoln Town Car	no	yes	yes
2009 Audi A3	no	no	yes

Source: http://autos.yahoo.com

  Use the roster method to describe each of the following sets.
  a. $A =$ cars in which a navigation system is standard
  b. $B =$ cars in which an automatic transmission is standard
  c. $C =$ cars in which leather seats are standard
  d. $A \cap B$
  e. $B \cap C$
  f. $A \cup B$
  g. $A \cup C$

155. Blood pressure. A group of college students had systolic blood pressure readings that ranged from 119.5 to 134.5 inclusive. Let x represent the value of the systolic blood pressure readings. Use inequalities to describe this range of values and graph the corresponding interval on a number line.

156. Population projections. Population projections suggest that by 2050, the number of people 60 years and older in the United States will be about 107 million. In 1950,

the number of people 60 years and older in the United States was 30 million. Let x represent the number (in millions) of people in the United States who are 60 years and older. Use inequalities to describe this population range from 1950 to 2050 and graph the corresponding interval on a number line. *Source:* U.S. Census Bureau

157. Heart rate. For exercise to be most beneficial, the optimum heart rate for a 20-year-old person is 120 beats per minute. Use absolute value notation to write an expression that describes the difference between the heart rate achieved by each of the following 20-year-old people and the ideal exercise heart rate. Then evaluate that expression.
  a. Latasha: 124 beats per minute
  b. Frances: 137 beats per minute
  c. Ignacio: 114 beats per minute

158. Downloading music. To download a 4 MB song with a 56 Kbs modem takes an average of 15 minutes. Use absolute value notation to write an expression that describes the difference between this average time and the actual time it took to download the following songs. Then evaluate that expression.
  a. *Irreplaceable* (Beyoncé): 14 minutes
  b. *Fallin'* (Alicia Keys): 17.5 minutes
  c. *Your Song* (Elton John): 9 minutes

In Exercises 159–160 use the physical interpretation to determine the domain of the variable.

159. Depth and pressure. The pressure p, in pounds per square inch, is related to the depth d, in feet below the surface of an ocean, by $p = \dfrac{5}{11}d + 15$. What is the domain of the variable d?

160. Volume. The volume V of a box that has a square base with a 2-foot side and a height of h feet is $V = 4h$. What is the domain of the variable h?

161. Weight vs. height. The average weight w, in pounds, of a male between 5 feet and 5 feet 10 inches tall is related to his height h by $w = \left(\dfrac{11}{2}\right)h - 220$. What is the domain (in inches) of the variable h?

162. Pay per hour. The weekly pay A a worker receives for working a 40-hour week or less is given by $A = 10h$, where h represents the number of hours worked in that week. What is the domain of the variable h?

Negative calories. In Exercises 163 and 164, use the fact that eating 100 grams of broccoli (a negative calorie food) actually results in a net *loss* of 55 calories.

163. If a cheeseburger has 522.5 calories, how many grams of broccoli would a person have to consume to have a net gain of zero calories?

164. Carmen ate 600 grams of broccoli and now wants to eat just enough French fries so that she has a net calorie intake of zero. If a small order of fries has 165 calories, how many orders does she have to eat?

Integer Exponents and Scientific Notation

Before Starting this Section, Review

1. Properties of opposites (Section P.1, page 13)
2. Variables (Section P.1, page 15)

Objectives

1. Use integer exponents.
2. Use the rules of exponents.
3. Simplify exponential expressions.
4. Use scientific notation.

COFFEE AND CANDY CONSUMPTION IN AMERICA

In 2002, Americans drank over 2,649,000,000 pounds of coffee and spent more than $13 billion on candy and other confectionery products. The American population at that time was about 280 million. Because our day-to-day activities bring us into contact with more manageable quantities, most people find large numbers like those just cited a bit difficult to understand. Using exponents, the method of scientific notation allows us to write large and small quantities in a manner that makes comparing such quantities fairly easy. In turn, these comparisons help us get a better perspective on these quantities. In Example 10 of this section, without relying on a calculator, we see that if we distribute the cost of the candy and other confectionery products used in 2002 evenly among all individuals in the United States, each person will spend over $46. Distributing the coffee used in 2002 evenly among all individuals in the United States gives each person about $9\frac{1}{2}$ pounds of coffee. ■

---

1. Use integer exponents.

Integer Exponents

The area of a square with side 5 feet is $5 \cdot 5 = 25$ square feet.
The volume of a cube that has sides of 5 feet each is $5 \cdot 5 \cdot 5 = 125$ cubic feet.

5 ft

5 ft

Area = 5·5 square feet

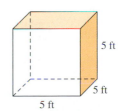

5 ft

5 ft

5 ft

Volume = 5·5·5 cubic feet

A shorter notation for $5 \cdot 5$ is 5^2 and for $5 \cdot 5 \cdot 5$ is 5^3. The number 5 is called the **base** for both 5^2 and 5^3. The number 2 is called the **exponent** in the expression 5^2 and indicates that the base 5 appears as a factor twice.

DEFINITION OF A POSITIVE INTEGER EXPONENT

If a is a real number and n is a positive integer, then

$$a^n = \underbrace{a \cdot a \cdot \ldots \cdot a}_{n \text{ factors}}$$

where a is the *base* and n is the *exponent*. We adopt the convention that $a^1 = a$.

The notation for 5^3 on a graphing calculator is 5^3. Any expression on a calculator enclosed in parentheses and followed by ^n is raised to the nth power. A common error is to forget parentheses when computing an expression such as $(-3)^2$, with the result being -3^2.

EXAMPLE 1   **Evaluating Expressions That Use Exponents**

Evaluate each expression.

a. 5^3 **b.** $(-3)^2$ **c.** -3^2 **d.** $(-2)^3$

SOLUTION

a. $5^3 = 5 \cdot 5 \cdot 5 = 125$

b. $(-3)^2 = (-3)(-3) = 9$ $(-3)^2$ is the opposite of 3, squared.

c. $-3^2 = -3 \cdot 3 = -9$ Multiplication of $3 \cdot 3$ occurs first. (-3^2 is the opposite of 3^2.)

d. $(-2)^3 = (-2)(-2)(-2) = -8$ ■ ■ ■

Practice Problem 1   Evaluate the following.

a. 2^3 **b.** $(3a)^2$ **c.** $\left(\dfrac{1}{2}\right)^4$ ■

In Example 1, pay careful attention to the fact (from **b** and **c**) that $(-3)^2 \neq -3^2$. In $(-3)^2$, the parentheses indicate that the exponent 2 applies to the base -3, whereas in -3^2, the absence of parentheses indicates that the exponent applies only to the base 3. When n is even, $(-a)^n \neq -a^n$ for $a \neq 0$.

DEFINITION OF ZERO AND NEGATIVE INTEGER EXPONENTS

For any nonzero number a and any positive integer n,

$$a^0 = 1 \text{ and } a^{-n} = \frac{1}{a^n}.$$

Negative exponents indicate the ***reciprocal*** of a number. Zero cannot be used as a base with a negative exponent because zero does not have a reciprocal. Furthermore, **0^0 is not defined**. We assume throughout this book that the base is not equal to zero if any of the exponents are negative or zero.

EXAMPLE 2   **Evaluating Expressions That Use Zero or Negative Exponents**

Evaluate.

a. $(-5)^{-2}$ **b.** -5^{-2} **c.** 8^0 **d.** $\left(\dfrac{2}{3}\right)^{-3}$

SOLUTION

a. $(-5)^{-2} = \dfrac{1}{(-5)^2} = \dfrac{1}{25}$ The exponent -2 applies to the base -5.

b. $-5^{-2} = -\dfrac{1}{5^2} = -\dfrac{1}{25}$ The exponent -2 applies to the base 5.

c. $8^0 = 1$ By definition

d. $\left(\dfrac{2}{3}\right)^{-3} = \dfrac{1}{\left(\dfrac{2}{3}\right)^3} = \dfrac{1}{\dfrac{8}{27}} = \dfrac{27}{8}$ $\left(\dfrac{2}{3}\right)^3 = \dfrac{2}{3} \cdot \dfrac{2}{3} \cdot \dfrac{2}{3} = \dfrac{8}{27}$ ■ ■ ■

Practice Problem 2   Evaluate.

a. 2^{-1} **b.** $\left(\dfrac{4}{5}\right)^0$ **c.** $\left(\dfrac{3}{2}\right)^{-2}$ ■

In Example 2, we see that parts **a** and **b** give different results. In part **a**, the base is −5 and the exponent is −2, whereas in part **b**, we evaluate the *opposite* of the expression with base 5 and exponent −2.

2 Use the rules of exponents.

Rules of Exponents

We now review the rules of exponents.

PRODUCT RULE OF EXPONENTS

If *a* is a real number and *m* and *n* are integers, then

(1) $$a^m \cdot a^n = a^{m+n}.$$

EXAMPLE 3   **Using the Product Rule of Exponents**

Simplify. Use the product rule and (if necessary) the definition of a negative exponent or reciprocal to write each answer without negative exponents.

a. $2x^3 \cdot x^5$ **b.** $x^7 \cdot x^{-7}$ **c.** $(-4y^2)(3y^7)$ **d.** $\dfrac{1}{3^{-2}}$

SOLUTION

a. $2x^3 \cdot x^5 = 2x^{3+5} = 2x^8$ Add exponents: $3 + 5 = 8$.
b. $x^7 \cdot x^{-7} = x^{7+(-7)} = x^0 = 1$ Add exponents: $7 + (-7) = 0$; simplify.
c. $(-4y^2)(3y^7) = (-4)3y^2y^7$ Group the factors with variable bases.
$\qquad\qquad\quad = -12y^{2+7}$ Add exponents:
$\qquad\qquad\quad = -12y^9$ $2 + 7 = 9$.

d. $\dfrac{1}{3^{-2}} = \dfrac{1}{\frac{1}{3^2}} = 3^2 = 9$ Negative exponents result in reciprocals.   ■ ■ ■

Practice Problem 3   Simplify. Write each answer without using negative exponents.

a. $x^2 \cdot 3x^7$ **b.** $\dfrac{1}{4^{-2}}$ ■

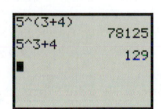

QUOTIENT RULE FOR EXPONENTS

If *a* is a nonzero real number and *m* and *n* are integers, then

(2) $$\dfrac{a^m}{a^n} = a^{m-n}.$$

EXAMPLE 4   **Using the Quotient Rule of Exponents**

Simplify. Use the quotient rule to write each answer without negative exponents.

a. $\dfrac{5^{10}}{5^{10}}$ **b.** $\dfrac{2^{-1}}{2^3}$ **c.** $\dfrac{x^{-3}}{x^5}$

SOLUTION

a. $\dfrac{5^{10}}{5^{10}} = 5^{10-10} = 5^0 = 1$

b. $\dfrac{2^{-1}}{2^3} = 2^{-1-3} = 2^{-4} = \dfrac{1}{2^4} = \dfrac{1}{16}$

c. $\dfrac{x^{-3}}{x^5} = x^{-3-5} = x^{-8} = \dfrac{1}{x^8}$ ■ ■ ■

Practice Problem 4 Simplify. Write each answer without negative exponents.

a. $\dfrac{3^4}{3^0}$ **b.** $\dfrac{5}{5^{-2}}$ **c.** $\dfrac{2x^3}{3x^{-4}}$ ■

POWER RULE FOR EXPONENTS

If a is a real number and m and n are integers, then

(3) $$(a^m)^n = a^{mn}.$$

EXAMPLE 5 **Using the Power Rule of Exponents**

Simplify. Use the power rule to write each answer without negative exponents.

a. $(5^2)^0$ **b.** $[(-3)^2]^3$ **c.** $(x^3)^{-1}$ **d.** $(x^{-2})^{-3}$

SOLUTION

a. $(5^2)^0 = 5^{2 \cdot 0} = 5^0 = 1$

b. $[(-3)^2]^3 = (-3)^{2 \cdot 3} = (-3)^6 = 729$ For $(-3)^6$, the base is -3.

c. $(x^3)^{-1} = x^{3(-1)} = x^{-3} = \dfrac{1}{x^3}$

d. $(x^{-2})^{-3} = x^{(-2)(-3)} = x^6$ ■ ■ ■

Practice Problem 5 Simplify. Write each answer without negative exponents.

a. $(7^{-5})^0$ **b.** $(7^0)^{-5}$ **c.** $(x^{-1})^8$ **d.** $(x^{-2})^{-5}$ ■

POWER-OF-A-PRODUCT RULE

If a and b are real numbers and n is an integer, then

(4) $$(a \cdot b)^n = a^n \cdot b^n.$$

EXAMPLE 6 **Using the Power-of-a-Product Rule**

Simplify. Use the power-of-a-product rule to write each expression without negative exponents.

a. $(3x)^2$ **b.** $(-3x)^{-2}$ **c.** $(-3^2)^3$ **d.** $(xy)^{-4}$ **e.** $(x^2y)^3$

SOLUTION

a. $(3x)^2 = 3^2x^2 = 9x^2$

b. $(-3x)^{-2} = \dfrac{1}{(-3x)^2} = \dfrac{1}{(-3)^2x^2} = \dfrac{1}{9x^2}$ Negative exponents result in reciprocals.

c. $(-3^2)^3 = (-1 \cdot 3^2)^3 = (-1)^3 (3^2)^3 = (-1)(3^6) = -729$

d. $(xy)^{-4} = \dfrac{1}{(xy)^4} = \dfrac{1}{x^4 y^4}$ Negative exponents result in reciprocals.

e. $(x^2 y)^3 = (x^2)^3 y^3 = x^{2 \cdot 3} y^3 = x^6 y^3$ Recall that $(x^2)^3 = x^{2 \cdot 3}$.   ■ ■ ■

Practice Problem 6   Simplify. Write each answer without negative exponents.

a. $\left(\dfrac{1}{2}x\right)^{-1}$ **b.** $(5x^{-1})^2$ **c.** $(xy^2)^3$ **d.** $(x^{-2}y)^{-3}$ ■

POWER-OF-A-QUOTIENT RULES

If a and b are real numbers and n is an integer, then

(5) $\left(\dfrac{a}{b}\right)^n = \dfrac{a^n}{b^n}$ (6) $\left(\dfrac{a}{b}\right)^{-n} = \left(\dfrac{b}{a}\right)^n = \dfrac{b^n}{a^n}.$

EXAMPLE 7   **Using the Power-of-a-Quotient Rules**

Simplify. Use the power-of-a-quotient rules to write each answer without negative exponents.

a. $\left(\dfrac{3}{5}\right)^3$ **b.** $\left(\dfrac{2}{3}\right)^{-2}$

SOLUTION

a. $\left(\dfrac{3}{5}\right)^3 = \dfrac{3^3}{5^5} = \dfrac{27}{125}$ Equation (5)

b. $\left(\dfrac{2}{3}\right)^{-2} = \left(\dfrac{3}{2}\right)^2 = \dfrac{3^2}{2^2} = \dfrac{9}{4}$ Equations (6) and (5)   ■ ■ ■

Practice Problem 7   Simplify. Write each answer without negative exponents.

a. $\left(\dfrac{1}{3}\right)^2$ **b.** $\left(\dfrac{10}{7}\right)^{-2}$ ■

◆ WARNING

Incorrect	**Correct**
$(3x)^2 = 3x^2$	$(3x)^2 = 3^2 x^2 = 9x^2$
$(x^2)^5 = x^7$	$(x^2)^5 = x^{2 \cdot 5} = x^{10}$
$x^3 x^5 = x^{15}$	$x^3 x^5 = x^{3+5} = x^8$

---

3 Simplify exponential expressions.

Simplifying Exponential Expressions

RULES FOR SIMPLIFYING EXPONENTIAL EXPRESSIONS

An exponential expression is considered **simplified** when

(i) each base appears only once,

(ii) no negative or zero exponents are used, and

(iii) no power is raised to a power.

Simplifying Exponential Expressions

Simplify the following.

a. $(-4x^2y^3)(7x^3y)$ **b.** $\left(\dfrac{x^5}{2y^{-3}}\right)^{-3}$

SOLUTION

a. $(-4x^2y^3)(7x^3y) = (-4)(7)x^2x^3y^3y$ Group factors with the same base.

$\qquad\qquad\qquad = -28x^{2+3}y^{3+1}$ Apply the product rule to add exponents; remember that $y = y^1$.

$\qquad\qquad\qquad = -28x^5y^4$

b. $\left(\dfrac{x^5}{2y^{-3}}\right)^{-3} = \dfrac{(x^5)^{-3}}{(2y^{-3})^{-3}}$ The exponent -3 is applied to both the numerator and the denominator.

$\qquad\qquad = \dfrac{x^{5(-3)}}{2^{-3}(y^{-3})^{-3}}$ Multiply exponents in the numerator; apply the exponent -3 to each factor in the denominator.

$\qquad\qquad = \dfrac{x^{-15}}{2^{-3}y^{(-3)(-3)}}$ Power rule for exponents

$\qquad\qquad = \dfrac{x^{-15}}{2^{-3}y^9}$ $5(-3) = -15; (-3)(-3) = 9$

$\qquad\qquad = \dfrac{x^{-15}x^{15}2^3}{x^{15}2^32^{-3}y^9}$ Multiply numerator and denominator by $x^{15}2^3$.

$\qquad\qquad = \dfrac{2^3}{x^{15}y^9}$ $x^{-15}x^{15} = x^0 = 1; 2^32^{-3} = 2^0 = 1$

$\qquad\qquad = \dfrac{8}{x^{15}y^9}$ $2^3 = 8$ ■ ■ ■

Practice Problem 8   Simplify each expression.

a. $(2x^4)^{-2}$ **b.** $\dfrac{x^2(-y)^3}{(xy^2)^3}$ ■

4 Use scientific notation.

Scientific Notation

Scientific measurements and calculations often involve very large or very small positive numbers. For example, 1 gram of oxygen contains approximately

$$37,600,000,000,000,000,000,000 \text{ atoms}$$

and the mass of one oxygen atom is approximately

$$0.0000000000000000000000266 \text{ gram.}$$

Such numbers contain so many zeros that they are awkward to work with in calculations. Fortunately, scientific notation provides a better way to write and work with such large and small numbers.

   Scientific notation consists of the product of a number smaller than 10, but no smaller than 1, and a power of 10. That is, scientific notation of a number has the form

$$c \times 10^n,$$

where c is a real number in decimal notation with $1 \le c < 10$ and n is an integer.

> **CONVERTING A DECIMAL NUMBER TO SCIENTIFIC NOTATION**
>
> 1. Count the number, n, of places the decimal point in the given number must be moved to obtain a number c with $1 \leq c < 10$.
> 2. If the decimal point is moved n places to the left, the scientific notation is $c \times 10^n$. If the decimal point is moved n places to the right, the scientific notation is $c \times 10^{-n}$.
> 3. If the decimal point does not need to be moved, the scientific notation is $c \times 10^0$.

EXAMPLE 9   Converting a Decimal Number to Scientific Notation

Write each decimal number in scientific notation.

a. 421,000   **b.** 10   **c.** 3.621   **d.** 0.000561

SOLUTION

a. Because $421{,}000 = 421000.0$, count five spaces to move the decimal point between the 4 and the 2 and produce a number between 1 and 10, namely, 4.21.

$$421{,}000 = 4\,2\,1\,0\,0\,0.$$

5 places

Because the decimal point is moved five places to the *left*, the exponent is positive and we write

$$421{,}000 = 4.21 \times 10^5.$$

b. The decimal point for 10 is to the right of the units digit. Count one place to move the decimal point between 1 and 0 and produce a number between 1 and 10 but less than 10, namely, 1.

$$10 = 1\,0.0$$

1 place

Because the decimal point is moved one place to the *left*, the exponent is positive and we write

$$10 = 1.0 \times 10^1.$$

c. The number 3.621 is already between 1 and 10, so the decimal does not need to be moved. We write

$$3.621 = 3.621 \times 10^0.$$

d. The decimal point in 0.000561 must be moved between the 5 and the 6 to produce a number between 1 and 10, namely, 5.61. We count four places as follows:

$$0.0\,0\,0\,5\,6\,1$$

4 places

Because the decimal point is moved four places to the *right*, the exponent is negative and we write

$$0.000561 = 5.61 \times 10^{-4}.$$

■ ■ ■

Practice Problem 9   Write 732,000 in scientific notation.

■

TECHNOLOGY CONNECTION

 To write numbers in scientific notation on a graphing calculator, you first must change the "mode" to "scientific." Then on most graphing calculators, you will see the number displayed as a decimal number, followed by the letter E, followed by the exponent for 10. So 421000 is shown as 4.21E5 and 0.0018 is shown as 1.8E-3. Some calculators omit the E.

```
421000
           4.21E5
0.0018
           1.8E-3
■
```

STUDY TIP

Note that in scientific notation, "small" numbers (positive numbers less than 1) have negative exponents and "large" numbers (numbers greater than 10) have positive exponents.

EXAMPLE 10   **Distributing Coffee and Candy in America**

At the beginning of this section, we mentioned that in 2002, Americans drank 2649 million pounds of coffee and spent more than $13 billion on candy and other confectionery products. To see how these products would be evenly distributed among the population, we first convert those numbers to scientific notation.

$$2649 \text{ million is } 2{,}649{,}000{,}000 = 2.649 \times 10^9.$$
$$13 \text{ billion is } 13{,}000{,}000{,}000 = 1.3 \times 10^{10}.$$

The U.S. population in 2002 was about 280 million, and 280 million is $280{,}000{,}000 = 2.8 \times 10^8$.

To distribute the coffee evenly among the population, we divide:

$$\frac{2.649 \times 10^9}{2.8 \times 10^8} = \frac{2.649}{2.8} \times 10^{9-8} \approx 0.95 \times 10^1 = 9.5, \text{ or } 9\tfrac{1}{2} \text{ pounds per person}$$

To distribute the cost of the candy evenly among the population, we divide:

$$\frac{1.3 \times 10^{10}}{2.8 \times 10^8} = \frac{1.3}{2.8} \times 10^{10-8} \approx 0.46 \times 10^2 = 46, \text{ or about } \$46 \text{ per person.}$$

■ ■ ■

Practice Problem 10 If the amount spent on candy consumption remains unchanged when the U.S. population reaches 300 million, what is the cost per person when cost is distributed evenly throughout the population? ■

SECTION P.2 ■ Exercises

A EXERCISES Basic Skills and Concepts

1. In the expression 7^2, the number 2 is called the _____.

2. In the expression -3^7, the base is _____.

3. The number $\dfrac{1}{4^{-2}}$ simplifies to the positive integer _____.

4. The power-of-a-product rule allows us to rewrite $(5a)^3$ as _____.

5. *True or False* $(-11)^{10} = -11^{10}$.

6. *True or False* The exponential expression $(x^2)^3$ is considered simplified.

In Exercises 7–16, name the exponent and the base.

7. 17^3
8. 10^2
9. 9^0
10. $(-2)^0$
11. $(-5)^5$
12. $(-99)^2$
13. -10^3
14. $-(-3)^7$
15. a^2
16. $(-b)^3$

In Exercises 17–42, evaluate each expression.

17. 6^1
18. 3^4
19. 7^0
20. $(-8)^0$
21. $(2^3)^2$
22. $(3^2)^3$
23. $(3^2)^{-2}$
24. $(7^2)^{-1}$
25. $(5^{-2})^3$
26. $(5^{-1})^3$
27. $(4^{-3}) \cdot (4^5)$
28. $(7^{-2}) \cdot (7^3)$
29. $3^0 + 10^0$
30. $5^0 - 9^0$
31. $3^{-2} + \left(\dfrac{1}{3}\right)^2$
32. $5^{-2} + \left(\dfrac{1}{5}\right)^2$
33. $\dfrac{2^{11}}{2^{10}}$
34. $\dfrac{3^6}{3^8}$
35. $\dfrac{(5^3)^4}{5^{12}}$
36. $\dfrac{(9^5)^2}{9^8}$
37. $\dfrac{2^5 \cdot 3^{-2}}{2^4 \cdot 3^{-3}}$
38. $\dfrac{4^{-2} \cdot 5^3}{4^{-3} \cdot 5}$
39. $\dfrac{-5^{-2}}{2^{-1}}$
40. $\dfrac{-7^{-2}}{3^{-1}}$
41. $\left(\dfrac{11}{7}\right)^{-2}$
42. $\left(\dfrac{13}{5}\right)^{-2}$

In Exercises 43–80, simplify each expression. Write your answers without negative exponents. Whenever an exponent is negative or zero, assume that the base is not zero.

43. $x^4 y^0$
44. $x^{-1} y^0$
45. $x^{-1} y$
46. $x^2 y^{-2}$

47. $-8x^{-1}$

48. $(-8x)^{-1}$

49. $x^{-1}(3y^0)$

50. $x^{-3}(3y^2)$

51. $x^{-1}y^{-2}$

52. $x^{-3}y^{-2}$

53. $(x^{-3})^4$

54. $(x^{-5})^2$

55. $(x^{-11})^{-3}$

56. $(x^{-4})^{-12}$

57. $-3(xy)^5$

58. $-8(xy)^6$

59. $4(xy^{-1})^2$

60. $6(x^{-1}y)^3$

61. $3(x^{-1}y)^{-5}$

62. $-5(xy^{-1})^{-6}$

63. $\dfrac{(x^3)^2}{(x^2)^5}$

64. $\dfrac{x^2}{(x^3)^4}$

65. $\left(\dfrac{2xy}{x^2}\right)^3$

66. $\left(\dfrac{5xy}{x^3}\right)^4$

67. $\left(\dfrac{-3x^2y}{x}\right)^5$

68. $\left(\dfrac{-2xy^2}{y}\right)^3$

69. $\left(\dfrac{-3x}{5}\right)^{-2}$

70. $\left(\dfrac{-5y}{3}\right)^{-4}$

71. $\left(\dfrac{4x^{-2}}{xy^5}\right)^3$

72. $\left(\dfrac{3x^2y}{y^3}\right)^5$

73. $\dfrac{x^3y^{-3}}{x^{-2}y}$

74. $\dfrac{x^2y^{-2}}{x^{-1}y^2}$

75. $\dfrac{27x^{-3}y^5}{9x^{-4}y^7}$

76. $\dfrac{15x^5y^{-2}}{3x^7y^{-3}}$

77. $\dfrac{5a^{-2}bc^2}{a^4b^{-3}c^2}$

78. $\dfrac{(-3)^2a^5(bc)^2}{(-2)^3a^2b^3c^4}$

79. $\left(\dfrac{xy^{-3}z^{-2}}{x^2y^{-4}z^3}\right)^{-3}$

80. $\left(\dfrac{xy^{-2}z^{-1}}{x^{-5}yz^{-8}}\right)^{-1}$

In Exercises 81–88, write each number in scientific notation.

81. 125

82. 247

83. 850,000

84. 205,000

85. 0.007

86. 0.0019

87. 0.00000275

88. 0.0000038

B EXERCISES Applying the Concepts

In Exercises 89 and 90, use the fact that when you uniformly stretch or shrink a three-dimensional object in every direction by a factor of a, the volume of the resulting figure is scaled by a factor of a^3. For example, when you uniformly scale (stretch or shrink) a three-dimensional object by a factor of 2, the volume of the resulting figure is scaled by a factor of 2^3, or 8.

89. A display in the shape of a baseball bat has a volume of 135 cubic feet. Find the volume of the display that results by uniformly scaling the figure by a factor of 2.

90. A display in the shape of a football has a volume of 675 cubic inches. Find the volume of the display that results by uniformly scaling the figure by a factor of 3.

91. The area A of a square with side of length x is given by $A = x^2$. Use this relationship to
  a. verify that doubling the length of the side of a square floor increases the area of the floor by a factor of 2^2.
  b. verify that tripling the length of the side of a square floor increases the area of the floor by a factor of 3^2.

92. The area A of a circle with diameter d is given by $A = \pi\left(\dfrac{d}{2}\right)^2$. Use this relationship to
  a. verify that doubling the length of the diameter of a circular skating rink increases the area of the rink by a factor of 2^2.
  b. verify that tripling the length of the diameter of a circular skating rink increases the area of the rink by a factor of 3^2.

93. The cross section of one type of weight-bearing rod (the surface you would get if you sliced the rod perpendicular to its axis) used in the Olympics is a square, as shown in the accompanying figure. The rod can handle a stress of 25,000 pounds per square inch (psi). The relationship between the stress S that the rod can handle, the load F the rod must carry, and the width w of the cross section is given by the equation $Sw^2 = F$. Assuming that $w = 0.25$ inch, find the load the rod can support.

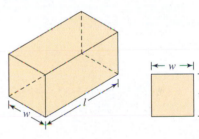

94. The cross section of one type of weight-bearing rod (the surface you would get if you sliced the rod perpendicular to its axis) used in the Olympics is a circle, as shown in the figure. The rod can handle a stress of 10,000 pounds per square inch (psi). The relationship between the stress S that the rod can handle, the load F the rod must carry, and the diameter d of the cross section is given by the equation $S\pi(d/2)^2 = F$. Assuming that $d = 1.5$ inches, find the load the rod can support. Use $\pi \approx 3.14$ in your calculation.

95. Complete the following table.

Celestial Body	Equatorial Diameter (km)	Scientific Notation
Earth	12,700	
Moon		3.48×10^3 (km)
Sun	1,390,000	
Jupiter		1.34×10^5 (km)
Mercury	4800	

In Exercises 96–101, express the number in each statement in scientific notation.

96. One gram of oxygen contains about

37,600,000,000,000,000,000,000 atoms.

97. One gram of hydrogen contains about

602,000,000,000,000,000,000,000 atoms.

98. One oxygen atom weighs about

0.0000000000000000000000266 kilogram.

99. One hydrogen atom weighs about

0.00000000000000000000000167 kilogram.

100. The distance from Earth to the moon is about

380,000,000 meters.

101. The mass of Earth is about

5,980,000,000,000,000,000,000,000 kilograms

Polynomials

Leaning Tower of Pisa

GALILEO AND FREE-FALLING OBJECTS

At one time, it was widely believed that heavy objects should fall faster than lighter objects, or, more exactly, that the speed of falling objects ought to be proportional to their weights. So if one object is four times as heavy as another, the heavier object should fall four times as fast as the lighter one. Legend has it that Galileo Galilei (1564–1642) disproved this theory by dropping two balls of equal size, one made of lead and the other of balsa wood, from the top of Italy's Leaning Tower of Pisa. Both balls are said to have reached the ground at the same instant after they were released simultaneously from the top of the Tower.

In fact, such experiments "work" properly only in a vacuum, where objects of different mass do fall at the same rate. In the absence of a vacuum, air resistance may greatly affect the rate at which different objects fall. Whether or not Galileo's experiment was actually performed, astronaut David R. Scott successfully performed a version of Galileo's experiment (using a feather and a hammer) on the surface of the moon on August 2, 1971. It turns out that on Earth (ignoring air resistance), regardless of the weight of the object we hold at rest and then drop, the expression $16t^2$ gives the distance in feet the object falls in t seconds. If we throw the object down with an initial velocity of v_0 feet per second, the distance the object travels in t seconds is given by the expression

$$16t^2 + v_0 t.$$

In Example 1, we evaluate such expressions. ∎

---

1 Learn polynomial vocabulary.

Polynomial Vocabulary

The height (in feet) of a golf ball above a driving range t seconds after being driven from the tee (at 70 feet per second) is given by the polynomial $70t - t^2$. After two seconds ($t = 2$), the ball is $70(2) - (2)^2 = 136$ feet above the ground. *Polynomials* such as $70t - t^2$ appear frequently in applications.

We begin by reviewing the basic vocabulary of polynomials. A **monomial** is the simplest polynomial in the variable x; it contains one term and has the form ax^k, where a is a constant and k is either a positive integer or zero. The constant a is called the **coefficient** of the monomial. For $a \neq 0$, the integer k is called the **degree** of the monomial. If $a = 0$, the monomial is 0 and has **no degree**. Any variable may be used in place of x to form a monomial in that variable.

Following are examples of monomials.

$2x^5$	The coefficient is 2, and the degree is 5.
$-3x^2$	The coefficient is -3, and the degree is 2.
-7	$-7 = -7(1) = -7x^0$. The coefficient is -7, and the degree is 0.
$8x$	The coefficient is 8, and the degree is $1(x = x^1)$.
x^{12}	The coefficient is $1(x^{12} = 1 \cdot x^{12})$, and the degree is 12.
$-x^4$	The coefficient is $-1(-x^4 = (-1) \cdot x^4)$, and the degree is 4.

The expression $5x^{-3}$ is not a monomial because the exponent of x is negative.

Two monomials in the same variable with the same degree can be combined into one monomial by the distributive property. For example, $-3x^5$ and $9x^5$ can be added: $-3x^5 + 9x^5 = (-3 + 9)x^5 = 6x^5$. Or $-3x^5$ and $9x^5$ can be subtracted: $-3x^5 - 9x^5 = (-3 - 9)x^5 = -12x^5$. Two monomials in the same variable with the same degree are called **like terms**.

POLYNOMIALS IN ONE VARIABLE

A **polynomial** in x is any sum of monomials in x. By combining like terms, we can write any polynomial in the form

$$a_n x^n + a_{n-1} x^{n-1} + \cdots + a_2 x^2 + a_1 x + a_0,$$

where n is either a positive integer or zero and $a_n, a_{n-1}, \ldots a_1, a_0$ are constants, called the **coefficients** of the polynomial. If $a_n \neq 0$, then n, the largest exponent on x, is called the **degree** and a_n is called the **leading coefficient** of the polynomial. The monomials $a_n x^n, a_{n-1} x^{n-1} \ldots, a_2 x^2, a_1 x$, and a_0 are the **terms** of the polynomial. The monomial $a_n x^n$ is the **leading term** of the polynomial, and a_0 is the **constant term**.

Polynomials can be classified according to the number of terms they have. Polynomials with one term are called **monomials**, polynomials with two *unlike* terms are called **binomials**, and polynomials with three *unlike* terms are called **trinomials**. Polynomials with more than three unlike terms do not have special names.

By agreement, the only polynomial that has *no* degree is the **zero polynomial**, which results when all of the coefficients are 0. It is easiest to find the degree of a polynomial when it is written in **descending order**—that is, when the exponents decrease from left to right. In this case, the degree is the exponent of the leftmost term. A polynomial written in descending order is said to be in **standard form**.

The symbols $a_0, a_1, a_2, \ldots a_n$ in the general notation for a polynomial

$$a_n x^n + a_{n-1} x^{n-1} + \cdots + a_2 x^2 + a_1 x + a_0$$

are just constants; the numbers to the lower right of a are called subscripts. The notation a_2 is read "a sub 2," a_1 is read "a sub 1," and a_0 is read "a sub 0." This type of notation is used when a large or indefinite number of constants are required. The terms of the polynomial $5x^2 + 3x + 1$ are $5x^2$, $3x$, and 1; the coefficients are 5, 3, and 1; and the degree is 2.

Note that the terms in the general form for a polynomial are connected by the addition symbol $+$. How do we handle $5x^2 - 3x + 1$? Because subtraction is defined in terms of addition, we rewrite

$$5x^2 - 3x + 1 = 5x^2 + (-3)x + 1$$

and see that the terms of $5x^2 - 3x + 1$ are $5x^2$, $-3x$, and 1 and that the coefficients are 5, -3, and 1. (The degree is 2.) Once we recognize this fact, we no longer rewrite the polynomial with the $+$ sign separating terms; we just remember that the "sign" is part of both the term and the coefficient.

EXAMPLE 1   **Examining Free-Falling Objects**

In our introductory discussion about Galileo and free-falling objects, we mentioned two well-known polynomials.

1. $16t^2$ gives the distance in feet a free-falling object falls in t seconds.

2. $16t^2 + 10t$ gives the distance an object falls in t seconds when it is thrown down with an initial velocity of 10 feet per second. Use these polynomials to

    a. Find how far a wallet dropped from the 86th-floor observatory of the Empire State Building will fall in five seconds.

    b. Find how far a quarter thrown down with an initial velocity of 10 feet per second from a hot air balloon will travel after five seconds.

SOLUTION

a. The value of $16t^2$ for $t = 5$ is $16(5)^2 = 400$. The wallet has fallen 400 feet.

b. The value of $16t^2 + 10t$ for $t = 5$ is $16(5)^2 + 10(5) = 450$. The quarter has traveled 450 feet on its downward path after five seconds.   ■ ■ ■

Practice Problem 1   If $16t^2 + 15t$ gives the distance an object falls in t seconds when it is thrown down at an initial velocity of 15 feet per second, find how far a quarter thrown down with an initial velocity of 15 feet per second from a hot air balloon has traveled after seven seconds.   ■

2 Add and subtract polynomials.

Adding and Subtracting Polynomials

Like monomials, polynomials are added and subtracted by combining like terms. By convention, we write polynomials in standard form.

EXAMPLE 2   **Adding Polynomials**

Find the sum of the polynomials

$$-4x^3 + 5x^2 + 7x - 2 \quad \text{and} \quad 6x^3 - 2x^2 - 8x - 5.$$

Horizontal Method: Group like terms and then combine them.

$(-4x^3 + 5x^2 + 7x - 2) + (6x^3 - 2x^2 - 8x - 5)$
$= (-4x^3 + 6x^3) + (5x^2 - 2x^2) + (7x - 8x) + (-2 - 5)$ Group like terms.
$= 2x^3 + 3x^2 - x - 7$ Combine like terms.

A second method for adding polynomials is to arrange them in columns so that like terms appear in the same column.

Column Method:

$$
\begin{array}{r}
-4x^3 + 5x^2 + 7x - 2 \\
(+)\ \ 6x^3 - 2x^2 - 8x - 5 \\
\hline
2x^3 + 3x^2 - \ \ x - 7 \ \text{(answer)}
\end{array}
$$
■ ■ ■

Practice Problem 2   Find the sum of

$$7x^3 + 2x^2 - 5 \quad \text{and} \quad -2x^3 + 3x^2 + 2x + 1.$$
■

EXAMPLE 3   **Subtracting Polynomials**

Find the difference of the polynomials

$$9x^4 - 6x^3 + 4x^2 - 1 \quad \text{and} \quad 2x^4 - 11x^3 + 5x + 3.$$

Horizontal Method: First, using the definition of subtraction, change the sign of each term in the second polynomial. Then add the resulting polynomials by grouping like terms and combining them.

$$(9x^4 - 6x^3 + 4x^2 - 1) - (2x^4 - 11x^3 + 5x + 3)$$
$$= (9x^4 - 6x^3 + 4x^2 - 1) + (-2x^4 + 11x^3 - 5x - 3)$$
$$= (9x^4 - 2x^4) + (-6x^3 + 11x^3) + 4x^2 - 5x + (-1 - 3) \qquad \text{Group like terms.}$$
$$= 7x^4 + 5x^3 + 4x^2 - 5x - 4 \qquad \text{Combine like terms.}$$

The second method for finding the difference of polynomials requires arranging them in columns so that like terms appear in the same column.

Column Method: As before, to find the difference, we change the sign of each term in the second polynomial and then add.

$$
\begin{array}{l}
9x^4 - 6x^3 + 4x^2 \qquad\;\; - 1 \\
\underline{(+)-2x^4 + 11x^3 \qquad\;\; - 5x - 3} \qquad \text{Change signs and then add.} \\
7x^4 + 5x^3 + 4x^2 - 5x - 4 \;\text{(answer)}
\end{array}
$$

■ ■ ■

Practice Problem 3   Find the difference of

$$3x^4 - 5x^3 + 2x^2 + 7 \qquad \text{and} \qquad -2x^4 + 3x^2 + x - 5.$$

■

3 Multiply polynomials.

Multiplying Polynomials

Previously, we multiplied two monomials, such as $-3x^2$ and $5x^3$, by using the product rule for exponents: $a^m \cdot a^n = a^{m+n}$. We multiply a monomial and a polynomial by combining the product rule and the distributive property.

EXAMPLE 4   **Multiplying a Monomial and a Polynomial**

Multiply $4x^3$ and $3x^2 - 5x + 7$.

$$4x^3(3x^2 - 5x + 7)$$
$$= (4x^3)(3x^2) + (4x^3)(-5x) + (4x^3)(7) \qquad \text{Distributive property}$$
$$= (4 \cdot 3)x^{3+2} + 4(-5)x^{3+1} + (4 \cdot 7)x^3 \qquad \text{Multiply monomials.}$$
$$= 12x^5 - 20x^4 + 28x^3 \qquad \text{Simplify.}$$

■ ■ ■

Practice Problem 4   Multiply $-2x^3$ and $4x^2 + 2x - 5$.

■

We can use the distributive property repeatedly to multiply any two polynomials. As with addition, there are two methods.

EXAMPLE 5   **Multiplying Polynomials**

Multiply $4x^2 + 3x$ and $x^2 + 2x - 3$.

Horizontal Method: We treat the second polynomial as a single term and use the distributive property, $(a + b)c = ac + bc$, to multiply it by each of the two terms in the first polynomial.

$$(4x^2 + 3x)(x^2 + 2x - 3)$$
$$= 4x^2(x^2 + 2x - 3) + 3x(x^2 + 2x - 3) \qquad \text{Distributive property}$$
$$= 4x^2(x^2) + 4x^2(2x) + 4x^2(-3) + 3x(x^2) + 3x(2x) + 3x(-3) \qquad \text{Distributive property}$$
$$= 4x^4 + 8x^3 - 12x^2 + 3x^3 + 6x^2 - 9x \qquad \text{Product rule}$$
$$= 4x^4 + (8x^3 + 3x^3) + (-12x^2 + 6x^2) - 9x \qquad \text{Group like terms.}$$
$$= 4x^4 + 11x^3 - 6x^2 - 9x \qquad \text{Combine like terms.}$$

The column method resembles the multiplication of two positive integers written in column form.

Column Method:

$$x^2 + 2x - 3$$

$$(*) \qquad\qquad 4x^2 + 3x$$

$$\overline{\qquad 3x^3 + \; 6x^2 - 9x \qquad}$$ Multiply by $3x$:
$(3x)(x^2 + 2x - 3)$.

$$\underline{4x^4 + \; 8x^3 - 12x^2 \qquad\qquad}$$ Multiply by $4x^2$:
$(4x^2)(x^2 + 2x - 3)$.

$$4x^4 + 11x^3 - \; 6x^2 - 9x \;\text{(answer)}$$ Combine like terms.   ■ ■ ■

Practice Problem 5 Multiply $5x^2 + 2x$ and $-2x^2 + x - 7$.   ■

The basic multiplication rule for polynomials can be stated as follows.

MULTIPLYING POLYNOMIALS

To multiply two polynomials, multiply each term of one polynomial by every term of the other polynomial and combine like terms.

---

4 Use special-product formulas.

Special Products

Particular polynomial products called *special products* occur frequently enough to deserve special attention. We introduce a very useful method, called FOIL, for multiplying two binomials. Notice that

$$(x + 2)(x - 7) = x(x - 7) + 2(x - 7)$$ Distributive property
$$= x^2 - 7x + 2x - 14$$ Distributive property
$$= x^2 - 5x - 14$$ Combine like terms.

Now consider the relationship between the terms of the factors in $(x + 2)(x - 7)$ and the second line, $x^2 - 7x + 2x - 14$, in the computation of the product.

Product of First terms

$$(x + 2)(x - 7) = x^2 - 7x + 2x - 14 \qquad \boxed{F}\text{irst terms: } x \cdot x = x^2$$

Product of Outside terms

$$(x + 2)(x - 7) = x^2 - 7x + 2x - 14 \qquad \boxed{O}\text{utside terms: } x \cdot (-7) = -7x$$

Product of Inside terms

$$(x + 2)(x - 7) = x^2 - 7x + 2x - 14 \qquad \boxed{I}\text{nside terms: } 2 \cdot x = 2x$$

Product of Last terms

$$(x + 2)(x - 7) = x^2 - 7x + 2x - 14 \qquad \boxed{L}\text{ast terms: } 2 \cdot (-7) = -14$$

FOIL METHOD FOR $(A + B)(C + D)$

$$(A + B)(C + D) = \overset{F}{A \cdot C} + \overset{O}{A \cdot D} + \overset{I}{B \cdot C} + \overset{L}{B \cdot D}$$

EXAMPLE 6 **Using the FOIL Method**

Use the FOIL method to find the following products.

a. $(2x + 3)(x - 1) = \overset{F}{(2x)(x)} + \overset{O}{(2x)(-1)} + \overset{I}{(3)(x)} + \overset{L}{(3)(-1)}$

$= 2x^2 - 2x + 3x - 3$

$= 2x^2 + x - 3$ Combine like terms.

b. $(3x - 5)(4x - 6) = \overset{F}{(3x)(4x)} + \overset{O}{(3x)(-6)} + \overset{I}{(-5)(4x)} + \overset{L}{(-5)(-6)}$

$= 12x^2 - 18x - 20x + 30$

$= 12x^2 - 38x + 30$ Combine like terms.

c. $(x + a)(x + b) = \overset{F}{x \cdot x} + \overset{O}{x \cdot b} + \overset{I}{a \cdot x} + \overset{L}{a \cdot b}$

$= x^2 + bx + ax + a \cdot b$ $xb = bx$ (commutative property)

$= x^2 + (b + a)x + ab$ Combine like terms. ■ ■ ■

Practice Problem 6 Use FOIL to find each product.

a. $(4x - 1)(x + 7)$ **b.** $(3x - 2)(2x - 5)$ ■

Squaring a Binomial Sum or Difference

We can find a general formula for the square of any binomial sum $(A + B)^2$ by using FOIL.

$(A + B)^2 = (A + B)(A + B) = \overset{F}{A \cdot A} + \overset{O}{A \cdot B} + \overset{I}{B \cdot A} + \overset{L}{B \cdot B}$

$= A^2 + AB + AB + B^2$ $AB = BA$

$= A^2 + 2AB + B^2$ $AB + AB = 2AB$

> **SQUARING A BINOMIAL SUM**
>
> $(A + B)^2 = A^2 + 2AB + B^2$

A formula such as $(A + B)^2 = A^2 + 2AB + B^2$ can be used two ways.

EXAMPLE 7 **Finding the Square of a Binomial Sum**

Find $(2x + 3)^2$.

SOLUTION

Pattern Method: In $(2x + 3)^2$, the first term is $2x$ and the second term is 3. Rewrite the formula as

Square of sum	=	Square of first term	+ 2	Product of first and second terms	+	Square of second term

$(2x + 3)^2 = (2x)^2 + 2 \cdot (2x) \cdot 3 + 3^2$

$= 2^2 x^2 + 2 \cdot 2 \cdot 3 \cdot x + 3^2$ Group numerical factors.

$= 4x^2 + 12x + 9$ Simplify.

Substitution Method: In the formula $(A + B)^2 = A^2 + 2AB + B^2$, substitute $2x = A$ and $3 = B$. Then

$$(2x + 3)^2 = (2x)^2 + 2 \cdot (2x) \cdot 3 + 3^2$$
$$= 2^2 x^2 + 2 \cdot 2 \cdot 3 \cdot x + 3^2 \qquad \text{Group numerical factors.}$$
$$= 4x^2 + 12x + 9 \qquad \text{Simplify.} \qquad \blacksquare\blacksquare\blacksquare$$

Practice Problem 7   Find $(3x + 2)^2$.   ■

The definition of subtraction allows us to write any difference $A - B$ as the sum $A + (-B)$. Because $A - B = A + (-B)$, we can use the formula for the square of a binomial sum to find a formula for the square of a binomial difference $(A - B)^2$.

$$(A - B)^2 = [A + (-B)]^2 \qquad \text{First term} = A; \text{second term} = -B$$
$$= A^2 + 2A(-B) + (-B)^2$$
$$= A^2 - 2AB + B^2 \qquad 2A(-B) = -2AB: (-B)^2 = B^2$$

◆ **WARNING**

Incorrect	Correct
$(x + y)^2 = x + y^2$	$(x + y)^2 = x^2 + 2xy + y^2$
$(x + y)^2 = x^2 + y^2$	$(x + y)^2 = x^2 + 2xy + y^2$
$(x + 4)(x + 5) = x + 4x + 5$	$(x + 4)(x + 5) = x^2 + 9x + 20$

The Product of the Sum and Difference of Terms

> **PRODUCTS GIVING A DIFFERENCE OF SQUARES**
>
> $$(A + B)(A - B) = A^2 - B^2$$

We leave it for you to use FOIL to verify the formula for finding the product of the sum and difference of terms.

EXAMPLE 8   **Finding the Product of the Sum and Difference of Terms**

Find the product $(3x + 4)(3x - 4)$.

SOLUTION

We use the substitution method; substitute $3x = A$ and $4 = B$ in the formula $(A + B)(A - B) = A^2 - B^2$. Then

$$(3x + 4)(3x - 4) = (3x)^2 - 4^2$$
$$= 9x^2 - 16 \qquad (3x)^2 = 3^2 x^2 = 9x^2 \qquad \blacksquare\blacksquare\blacksquare$$

Practice Problem 8   Find the product $(1 - 2x)(1 + 2x)$.   ■

The special-products formulas, such as the formulas for squaring a binomial sum or difference, are used often and should be memorized. However, you should be able to derive them from FOIL if you forget them. We list several of these formulas next.

SPECIAL-PRODUCT FORMULAS

A and *B* represent any algebraic expression.

Formula   **Example**

Products giving a difference of squares

$(A + B)(A - B) = A^2 - B^2$

$$(5x + 2)(5x - 2) = (5x)^2 - 2^2$$
$$= 25x^2 - 4$$

Squaring a binomial sum or difference

$(A + B)^2 = A^2 + 2AB + B^2$

$$(3x + 2)^2 = (3x)^2 + 2(3x)(2) + 2^2$$
$$= 9x^2 + 12x + 4$$

$(A - B)^2 = A^2 - 2AB + B^2$

$$(2x - 5)^2 = (2x)^2 - 2(2x)(5) + 5^2$$
$$= 4x^2 - 20x + 25$$

Cubing a binomial sum or difference

$(A + B)^3 = A^3 + 3A^2B + 3AB^2 + B^3$

$$(x + 5)^3 = x^3 + 3x^2(5) + 3x(5)^2 + 5^3$$
$$= x^3 + 15x^2 + 75x + 125$$

$(A - B)^3 = A^3 - 3A^2B + 3AB^2 - B^3$

$$(x - 4)^3 = x^3 - 3x^2(4) + 3x(4)^2 - 4^3$$
$$= x^3 - 12x^2 + 48x - 64$$

Products giving a sum or difference of cubes

$(A + B)(A^2 - AB + B^2) = A^3 + B^3$

$$(x + 2)(x^2 - 2x + 4) = x^3 + 2^3 = x^3 + 8$$

$(A - B)(A^2 + AB + B^2) = A^3 - B^3$

$$(x - 3)(x^2 + 3x + 9) = x^3 - 3^3 = x^3 - 27$$

To this point, we have investigated polynomials only in a single variable. We need to extend our previous vocabulary to discuss polynomials in more than one variable. Any product of a constant and two or more variables raised to whole-number powers is a **monomial** in those variables. The constant is called the **coefficient** of the monomial, and the **degree** of the monomial is the sum of all the exponents appearing on its variables. For example, the monomial $5x^3y^4$ is of degree $3 + 4 = 7$ with coefficient 5.

EXAMPLE 9   **Multiplying Polynomials in Two Variables**

Multiply. **a.** $(7a + 5b)(7a - 5b)$ **b.** $(2x + 3y)^2$

SOLUTION

a. $(7a + 5b)(7a - 5b) = (7a)^2 - (5b)^2$ $(A + B)(A - B) = A^2 - B^2$
$$= 49a^2 - 25b^2$$ Simplify.

b. $(2x + 3y)^2 = (2x)^2 + 2(2x)(3y) + (3y)^2$ $(A + B)^2 = A^2 + 2AB + B^2$
$$= 2^2x^2 + 2 \cdot 2 \cdot 3xy + 3^2y^2$$ Apply exponents to each factor.
$$= 4x^2 + 12xy + 9y^2$$ Simplify.   ■ ■ ■

Practice Problem 9   Multiply $x^2(y + x)$.   ■

A EXERCISES Basic Skills and Concepts

1. The polynomial $-3x^7 + 2x^2 - 9x + 4$ has leading coefficient _____ and degree_____.

2. When a polynomial is written so that the exponents in each term decrease from left to right, it is said to be in _____ form.

3. When a polynomial in x of degree 3 is added to a polynomial in x of degree 4, the resulting polynomial has degree _____.

4. When a polynomial in x of degree 3 is multiplied by a polynomial in x of degree 4, the resulting polynomial has degree _____.

5. *True or False* When FOIL is used, the terms 7 and x are called the inside terms of the product $(4x + 7)(x - 2)$.

6. *True or False* There are values of A and B for which $(A + B)^2 = A^2 + B^2$.

In Exercises 7–10, determine whether the given expression is a polynomial. If it is, write it in standard form.

7. $1 + x^2 + 2x$

8. $x - \dfrac{1}{x}$

9. $x^{-2} + 3x + 5$

10. $3x^4 + x^7 + 3x^5 - 2x + 1$

In Exercises 11–14, find the degree and list the terms of the polynomial.

11. $7x + 3$

12. $-3x^2 + 7$

13. $x^2 - x^4 + 2x - 9$

14. $x + 2x^3 + 9x^7 - 21$

In Exercises 15–24, perform the indicated operations. Write the resulting polynomial in standard form.

15. $(x^3 + 2x^2 - 5x + 3) + (-x^3 + 2x - 4)$

16. $(x^3 - 3x + 1) + (x^3 - x^2 + x - 3)$

17. $(2x^3 - x^2 + x - 5) - (x^3 - 4x + 3)$

18. $(-x^3 + 2x - 4) - (x^3 + 3x^2 - 7x + 2)$

19. $(-2x^4 + 3x^2 - 7x) - (8x^4 + 6x^3 - 9x^2 - 17)$

20. $3(x^2 - 2x + 2) + 2(5x^2 - x + 4)$

21. $-2(3x^2 + x + 1) + 6(-3x^2 - 2x - 2)$

22. $2(5x^2 - x + 3) - 4(3x^2 + 7x + 1)$

23. $(3y^3 - 4y + 2) + (2y + 1) - (y^3 - y^2 + 4)$

24. $(5y^2 + 3y - 1) - (y^2 - 2y + 3) + (2y^2 + y + 5)$

In Exercises 25–58, perform the indicated operations.

25. $6x(2x + 3)$

26. $7x(3x - 4)$

27. $(x + 1)(x^2 + 2x + 2)$

28. $(x - 5)(2x^2 - 3x + 1)$

29. $(3x - 2)(x^2 - x + 1)$

30. $(2x + 1)(x^2 - 3x + 4)$

31. $(x + 1)(x + 2)$

32. $(x + 2)(x + 3)$

33. $(3x + 2)(3x + 1)$

34. $(x + 3)(2x + 5)$

35. $(-4x + 5)(x + 3)$

36. $(-2x + 1)(x - 5)$

37. $(3x - 2)(2x - 1)$

38. $(x - 1)(5x - 3)$

39. $(2x - 3a)(2x + 5a)$

40. $(5x - 2a)(x + 5a)$

41. $(x + 2)^2 - x^2$

42. $(x - 3)^2 - x^2$

43. $(x + 3)^3 - x^3$

44. $(x - 2)^3 - x^3$

45. $(4x + 1)^2$

46. $(3x + 2)^2$

47. $(3x + 1)^3$

48. $(2x + 3)^3$

49. $(5 - 2x)(5 + 2x)$

50. $(3 - 4x)(3 + 4x)$

51. $\left(x + \dfrac{3}{4}\right)^2$

52. $\left(x + \dfrac{2}{5}\right)^2$

53. $(2x - 3)(x^2 - 3x + 5)$

54. $(x - 2)(x^2 - 4x - 3)$

55. $(1 + y)(1 - y + y^2)$

56. $(y + 4)(y^2 - 4y + 16)$

57. $(x - 6)(x^2 + 6x + 36)$

58. $(x - 1)(x^2 + x + 1)$

In Exercises 59–68, perform the indicated operations.

59. $(x + 2y)(3x + 5y)$

60. $(2x + y)(7x + 2y)$

61. $(2x - y)(3x + 7y)$

62. $(x - 3y)(2x + 5y)$

63. $(x - y)^2(x + y)^2$

64. $(2x + y)^2(2x - y)^2$

65. $(x + y)(x - 2y)^2$

66. $(x - y)(x + 2y)^2$

67. $(x - 2y)^3(x + 2y)$

68. $(2x + y)^3(2x - y)$

B EXERCISES Applying the Concepts

69. **Ticket prices.** Theater ticket prices (in dollars) x years after 1997 are described by the polynomial $-0.025x^2 + 0.44x + 4.28$. Find the value of this polynomial for $x = 6$ and state the result as the ticket price for a specific year.

70. Box-office grosses. Theater box-office grosses (in billions of dollars) x years after 1993 are described by the polynomial $0.035x^2 + 0.15x + 5.17$. Find the value of this polynomial for 1997 and state the result as box-office gross for a specific year.

71. Paper shredding. A business that shreds paper products finds that it costs $0.1x^2 + x + 50$ dollars to serve x customers. What does it cost to serve 40 customers?

72. Gift baskets. A company will produce x (for $x \geq 10$) gift baskets of wine, meat, cheese, and crackers at a cost of $(x - 10)^2 + 5x$ dollars per basket. What is the per basket cost for 15 baskets?

In Exercises 73 and 74, use the fact (from Example 1) that an object thrown down with an initial velocity of v_0 feet per second will travel $16t^2 + v_0 t$ feet in t seconds.

73. Free fall. A sandwich is thrown from a helicopter with an initial downward velocity of 20 feet per second. How far has the sandwich fallen after five seconds?

74. Free fall. A ring is thrown off the Empire State Building with an initial downward velocity of 10 feet per second. How far has the ring fallen after two seconds?

75. Cruise ship revenue. A cruise ship decides to reduce ticket prices to its theater by x dollars from the current price of $22.50.
   a. Write a polynomial that gives the new price after the x-dollar deduction.

b. The revenue is the number of tickets sold times the price of each ticket. If 30 tickets were sold when the price was $22.50 and 10 additional tickets are sold for each dollar the price is reduced, write a polynomial that gives the revenue in terms of x when the original price is reduced by x dollars.

76. Used-car rentals. A dealer who rents used cars wants to increase the monthly rent from the current $250 in n increases of $10.
   a. Write a polynomial that gives the new price after n increases of $10.
   b. The revenue is the number of cars rented times the monthly rent for each car. If 50 cars can be rented at $250 per month and 2 fewer cars can be rented for each $10 rent increase, write a polynomial that gives the monthly revenue in terms of n.

Factoring Polynomials

Before Starting this Section, Review

1. Exponents (Section P.2)
2. Polynomials (Section P.3)
3. Value of algebraic expressions (Section P.1, page 14)

Objectives

1. Identify and factor out the greatest common monomial factor.
2. Factor trinomials with a leading coefficient of 1.
3. Factor perfect-square trinomials.
4. Factor the difference of squares.
5. Factor the difference and sum of cubes.
6. Factor trinomials.
7. Factor by grouping.

PROFIT FROM RESEARCH AND DEVELOPMENT

The finance department of a major drug company projects that an investment of x million dollars in research and development for a specific type of vaccine will return a profit (or loss) of $(0.012x^3 - 32.928)$ million dollars.

The company would have to invest $12 million to get started and is prepared to invest up to $24 million at a rate of $1 million per year. You could probably stare at this polynomial for a long time without getting a clear picture of what kind of return to expect from different investments. In Example 7, we use the methods of this section to rewrite this profit–loss polynomial, making it easy to explain the result of investing various amounts. ■

The Greatest Common Monomial Factor

The process of factoring polynomials that we study in this section "reverses" the process of multiplying polynomials that we studied in the previous section. To factor *any sum of terms* means to write it as a *product of factors*. Consider the product

$$(x + 3)(x - 3) = x^2 - 9.$$

The polynomials $(x + 3)$ and $(x - 3)$ are called *factors* of the polynomial $x^2 - 9$. When factoring, we start with one polynomial and write it as a product of other polynomials. The polynomials in the product are the **factors** of the given polynomial.

The first thing to look for when factoring a polynomial is a factor that is common to every term. This **common factor** can be "factored out" by the distributive property, $a(b + c) = ab + ac$.

1 Identify and factor out the greatest common monomial factor.

Factoring Out a Monomial

In this section, we are concerned only with polynomials having integer coefficients. This restriction is called **factoring over the integers**.

Polynomial	Common Factor	Factored Form
$9 + 18y$	9	$9(1 + 2y)$
$2y^2 + 6y$	$2y$	$2y(y + 3)$
$7x^4 + 3x^3 + x^2$	x^2	$x^2(7x^2 + 3x + 1)$
$5x + 3$	No common factor	$5x + 3$

You can verify that any factorization is correct by multiplying the factors. We factored these polynomials by finding the *greatest common monomial factor* of their terms.

FINDING THE GREATEST COMMON MONOMIAL FACTOR

The term ax^n is the **greatest common monomial factor (GCF)** of a polynomial in x (with integer coefficients) if

1. a is the *greatest* integer that divides each of the polynomial coefficients.
2. n is the *smallest* exponent on x found in every term of the polynomial.

EXAMPLE 1   **Factoring by Using the GCF**

Factor. **a.** $16x^3 + 24x^2$ **b.** $5x^4 + 20x^2 + 25x$

SOLUTION

a. The greatest integer that divides both 16 and 24 is 8. The exponents on x are 3 and 2, so 2 is the smallest exponent. So, the **GCF** of $16x^3 + 24x^2$ is $8x^2$. We write

$$16x^3 + 24x^2 = 8x^2(2x) + 8x^2(3).$$ 　Write each term as the product of the GCF and another factor.

$$= 8x^2(2x + 3)$$ 　Use the distributive property to factor out the GCF.

We can check the results of factoring by multiplying.

Check: $8x^2(2x + 3) = 8x^2(2x) + 8x^2(3) = 16x^3 + 24x^2$.

b. The greatest integer that divides 5, 20, and 25 is 5. The exponents on x are 4, 2, and 1 $(x = x^1)$, so 1 is the smallest exponent. So, the **GCF** of $5x^4 + 20x^2 + 25x$ is $5x$. We write

$$5x^4 + 20x^2 + 25x = 5x(x^3) + 5x(4x) + 5x(5) = 5x(x^3 + 4x + 5).$$

You should check this result by multiplying. ■ ■ ■

Practice Problem 1　Factor. **a.** $6x^5 + 14x^3$　**b.** $7x^5 + 21x^4 + 35x^2$ ■

Polynomials that cannot be factored as a product of two polynomials (excluding the constant polynomials 1 and -1) are said to be **irreducible**. A polynomial is said to be **factored completely** when it is written as a product consisting of only irreducible factors. By agreement, the greatest integer common to all of the terms in the polynomial is not factored. We also allow repeated irreducible factors to be written as a power of a single factor.

2 Factor trinomials with a leading coefficent of 1.

Factoring Trinomials of the Form $x^2 + bx + c$

Recall that $(x + a)(x + b) = x^2 + (a + b)x + ab$.

We reverse the sides of this equation to get the factored form

$$x^2 + (a + b)x + ab = (x + a)(x + b).$$

STUDY TIP

To factor a trinomial in standard form with a leading coefficient of 1, we need two integers a and b whose product is the constant term and whose sum is the coefficient of the middle term.

EXAMPLE 2   **Factoring $x^2 + bx + c$ by Using the Factors of c**

Factor. **a.** $x^2 + 8x + 15$ **b.** $x^2 - 6x - 16$ **c.** $x^2 + x + 2$

SOLUTION

a. We want integers a and b such that $ab = 15$ and $a + b = 8$.

Factors of 15	1, 15	−1, −15	3, 5	−3, −5
Sum of factors	16	−16	8	−8

Because the factors 3 and 5 of 15 in the third column have a sum of 8,

$$x^2 + 8x + 15 = (x + 3)(x + 5).$$

You should check this answer by multiplying.

b. We must find two integers a and b with $ab = -16$ and $a + b = -6$.

Factors of −16	−1, 16	−2, 8	−4, 4	1, −16	2, −8	4, −4
Sum of factors	15	6	0	−15	−6	0

Because the factors 2 and −8 of −16 in the fifth column give a sum of −6,

$$x^2 - 6x - 16 = (x + 2)[x + (-8)] = (x + 2)(x - 8).$$

You should check this answer by multiplying.

c. We must find two integers a and b with $ab = 2$ and $a + b = 1$.

Factors of 2	1, 2	−1, −2
Sum of factors	3	−3

The coefficient of the middle term, 1, does not appear as the sum of any of the factors of the constant term, 2. Therefore, $x^2 + x + 2$ is irreducible.   ■ ■ ■

Practice Problem 2   Factor. **a.** $x^2 + 6x + 8$ **b.** $x^2 - 3x - 10$ ■

Factoring Formulas

We investigate how other types of polynomials can be factored by reversing the product formulas on page 36.

FACTORING FORMULAS

A and B represent any algebraic expression.

$A^2 - B^2 = (A + B)(A - B)$	Difference of squares
$A^2 + 2AB + B^2 = (A + B)^2$	Perfect square, positive middle term
$A^2 - 2AB + B^2 = (A - B)^2$	Perfect square, negative middle term
$A^3 - B^3 = (A - B)(A^2 + AB + B^2)$	Difference of cubes
$A^3 + B^3 = (A + B)(A^2 - AB + B^2)$	Sum of cubes

---

3 Factor perfect-square trinomials.

Perfect-Square Trinomials

> **EXAMPLE 3**   **Factoring a Perfect-Square Trinomial**

Factor. **a.** $x^2 + 10x + 25$ **b.** $16x^2 - 8x + 1$

SOLUTION

a. The first term, x^2, and the third term, $25 = 5^2$, are perfect squares. Further, the middle term is twice the product of the terms being squared, that is, $10x = 2(5x)$.

$$x^2 + 10x + 25 = x^2 + 2 \cdot 5 \cdot x + 5^2 = (x + 5)^2$$

b. The first term, $16x^2 = (4x)^2$, and the third term, $1 = 1^2$, are perfect squares. Further, the middle term is the *negative* of twice the product of terms being squared, that is, $-8x = -2(4x)(1)$.

$$16x^2 - 8x + 1 = (4x)^2 - 2(4x)(1) + 1^2 = (4x - 1)^2$$ ■ ■ ■

Practice Problem 3   Factor. **a.** $x^2 + 4x + 4$ **b.** $9x^2 - 6x + 1$ ■

4 Factor the difference of squares

Difference of Squares

To factor the difference of squares, we use the special product
$A^2 - B^2 = (A + B)(A - B)$.

> **EXAMPLE 4**   **Factoring the Difference of Squares**

Factor. **a.** $x^2 - 4$ **b.** $25x^2 - 49$

SOLUTION

a. We write $x^2 - 4$ as the difference of squares.

$$x^2 - 4 = x^2 - 2^2 = (x + 2)(x - 2)$$

b. We write $25x^2 - 49$ as the difference of squares.

$$25x^2 - 49 = (5x)^2 - 7^2 = (5x + 7)(5x - 7)$$ ■ ■ ■

Practice Problem 4   Factor. **a.** $x^2 - 16$ **b.** $4x^2 - 25$ ■

What about the *sum* of two squares? Let's try to factor $x^2 + 2^2 = x^2 + 4$. Notice that $x^2 + 4 = x^2 + 0 \cdot x + 4$ so that the middle term is $0 = 0 \cdot x$. We need two factors, a and b, of 4 whose sum is zero ($a + b = 0$).

Factors of 4	1, 4	−1, −4	2, 2	−2, −2
Sum of factors	5	−5	4	−4

Because no sum of factors is 0, $x^2 + 4$ is irreducible. A similar result holds regardless of what integer replaces 2 in the process. Consequently, $x^2 + a^2$ is irreducible for any integer a.

SOME IRREDUCIBLE POLYNOMIALS

If a and c are integers *having no common factors,* then

$$ax + c \quad \text{is irreducible} \quad \text{and} \quad x^2 + a^2 \text{ is irreducible.}$$

EXAMPLE 5   **Factoring the Difference of Squares**

Factor. $x^4 - 16$.

SOLUTION

We write $x^4 - 16$ as the difference of squares.

$$x^4 - 16 = (x^2)^2 - 4^2 = (x^2 + 4)(x^2 - 4)$$

From Example 4, part **a**, we know that $x^2 - 4 = (x + 2)(x - 2)$. Consequently,

$$x^4 - 16 = (x^2 + 4)(x^2 - 4)$$
$$= (x^2 + 4)(x + 2)(x - 2) \quad \text{Replace } x^2 - 4 \text{ with } (x + 2)(x - 2).$$

Because $x^2 + 4$, $x + 2$, and $x - 2$ are irreducible, $x^4 - 16$ is completely factored.

■ ■ ■

Practice Problem 5   Factor $x^4 - 81$.   ■

5 Factor the difference and sum of cubes.

Difference and Sum of Cubes

We can also use the special-product formulas to factor the difference and sum of cubes.

$$A^3 - B^3 = (A - B)(A^2 + AB + B^2) \quad \text{Difference of cubes}$$
$$A^3 + B^3 = (A + B)(A^2 - AB + B^2) \quad \text{Sum of cubes}$$

EXAMPLE 6   **Factoring the Difference and Sum of Cubes**

Factor. **a.** $x^3 - 64$ **b.** $8x^3 + 125$

SOLUTION

a. We write $x^3 - 64$ as the difference of cubes.

$$x^3 - 64 = x^3 - 4^3 = (x - 4)(x^2 + 4x + 16)$$

b. We write $8x^3 + 125$ as the sum of cubes.

$$8x^3 + 125 = (2x)^3 + 5^3 = (2x + 5)(4x^2 - 10x + 25)$$

■ ■ ■

Practice Problem 6   Factor. **a.** $x^3 - 125$ **b.** $27x^3 + 8$ ■

EXAMPLE 7   **Determining Profit from Research and Development**

In the beginning of this section, we introduced the polynomial $0.012x^3 - 32.928$ that gave the profit (or loss) a drug company would have after an investment of x million dollars. If the company had an initial investment of \$12 million and invests an additional \$1 million each year, determine when this investment returns a profit. What will the profit (or loss) be in six years?

SOLUTION

First, notice that 0.012 is a common factor of both terms of the polynomial because $32.928 = (0.012)(2744)$. We now factor $0.012x^3 - 32.928$ as follows:

$$0.012x^3 - 32.928 = 0.012x^3 - (0.012)(2744)$$
$$= 0.012(x^3 - 2744) \quad \text{Factor out } 0.012.$$
$$= 0.012(x^3 - 14^3) \quad 2744 = 14^3$$
$$= 0.012(x - 14)(x^2 + 14x + 196) \quad \text{Difference of cubes}$$

This product has three factors, and the factor $x^2 + 14x + 196$ is positive for positive values of x because each term is positive. The sign of the factor $x - 14$ then determines the sign of the entire polynomial. The factor $x - 14$ is negative if x is less than 14, zero if x equals 14, and positive if x is greater than 14. Consequently, the company receives a profit when x is greater than 14. Because the initial investment is $12 million, with an additional $1 million invested each year, it will be two years before the company breaks even on this venture. Every year thereafter it will profit from the investment. In six years, the company will have invested the initial $12 million plus an additional $6 million ($1 million per year). So, the total investment is $18 million. To find the profit (or loss) with $18 million invested, we let $x = 18$ in the profit–loss polynomial.

$$(0.012)(x - 14)(x^2 + 14x + 196)$$
$$= 0.012(18 - 14)(18^2 + (14)(18) + 196) \qquad \text{Replace } x \text{ with 18.}$$
$$= 0.012(4)(772)$$
$$= \$37.056 \text{ million}$$

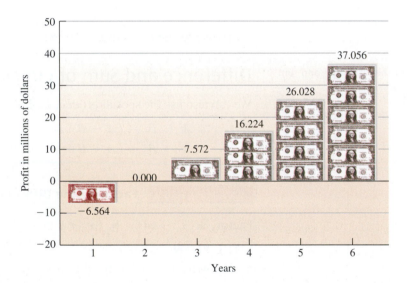

The company will have made a profit of over $37 million in six years.   ■ ■ ■

Practice Problem 7   In Example 7, what will the profit be in four years?   ■

6 Factor trinomials.

Factoring Trinomials of the Form $Ax^2 + Bx + C$

To factor the general trinomial $Ax^2 + Bx + C$, we begin by supposing that

$$Ax^2 + Bx + C = (ax + b)(cx + d) = acx^2 + (ad + bc)x + bd.$$

We can factor $Ax^2 + Bx + C$ if we find integers $a, b, c,$ and d such that

$$A = ac \qquad C = bd \qquad B = ad + bc.$$

EXAMPLE 8   **Factoring $Ax^2 + Bx + C$ When $A \neq 1$**

Factor. **a.** $6x^2 + 17x + 7$ **b.** $4x^2 - 8x - 5$

SOLUTION

a. When the leading coefficient is positive, we must consider only its positive factors. For example, $(-6x + 2)(-x + 3) = (6x - 2)(x - 3)$. The positive-factor pairs for 6 are 6, 1 and 2, 3. There are two possible forms for a factorization:

$$6x^2 + 17x + 7 = (6x + \square)(x + \square)$$

or

$$6x^2 + 17x + 7 = (2x + \square)(3x + \square)$$

The numbers to be filled in are factors of 7, the constant term. However, because the middle coefficient is positive, these numbers cannot be -7 and -1. We arrange the factors so that the first- and last-term products work out automatically and then test for the correct middle term, which is the sum of the inside and outside products.

The following table lists the possible factorizations of $6x^2 + 17x + 7$.

Possible Factorizations	Outside Terms	Inside Terms	Sum of Outside and Inside Products
$(6x + 1)(x + 7)$	$6x, 7$	$1, x$	$(6x)(7) + (1)(x) = 43x$
$(6x + 7)(x + 1)$	$6x, 1$	$7, x$	$(6x)(1) + (7)(x) = 13x$
$(2x + 1)(3x + 7)$	$2x, 7$	$1, 3x$	$(2x)(7) + (1)(3x) = 17x$
$(2x + 7)(3x + 1)$	$2x, 1$	$7, 3x$	$(2x)(1) + (7)(3x) = 23x$

The correct middle term, $17x$, occurs in the *third* row of the table. So,

$$6x^2 + 17x + 7 = (2x + 1)(3x + 7).$$

b. Because 4, the leading coefficient of $4x^2 - 8x - 5$, is positive, we must consider only its possible positive factors, namely 4, 1 and 2, 2. There are two possible forms for a factorization:

$$4x^2 - 8x - 5 = (4x + \square)(x + \square)$$

or

$$4x^2 - 8x - 5 = (2x + \square)(2x + \square)$$

The numbers to be filled in are factors of -5, the constant term. Because that term is negative, we use factors that are opposite in sign—that is, 5, -1 and -5, 1.

The following table lists the possible factorizations of $4x^2 - 8x - 5$.

Possible Factorizations	Outside Terms	Inside Terms	Sum of Outside and Inside Products
$(4x - 1)(x + 5)$	$4x, 5$	$-1, x$	$(4x)(5) + (-1)(x) = 19x$
$(4x + 5)(x - 1)$	$4x, -1$	$5, x$	$(4x)(-1) + (5)(x) = x$
$(4x + 1)(x - 5)$	$4x, -5$	$1, x$	$(4x)(-5) + (1)(x) = -19x$
$(4x - 5)(x + 1)$	$4x, 1$	$-5, x$	$(4x)(1) + (-5)(x) = -x$
$(2x - 1)(2x + 5)$	$2x, 5$	$-1, 2x$	$(2x)(5) + (-1)(2x) = 8x$
$(2x + 1)(2x - 5)$	$2x, -5$	$1, 2x$	$(2x)(-5) + (1)(2x) = -8x$

> **STUDY TIP**
>
> After using the techniques just explained to factor several trinomials, you may find that you can guess the factors without going through all of the steps. Guessing is fine as long as you check your guesses by using FOIL to multiply. If you guessed incorrectly, you can just guess again.

Notice that the correct middle term, $-8x$, occurs in the last row of the table. So,

$$4x^2 - 8x - 5 = (2x + 1)(2x - 5). \qquad ■ ■ ■$$

Practice Problem 8   Factor.   **a.** $5x^2 + 11x + 2$ **b.** $9x^2 - 9x + 2$ ■

7 Factoring by grouping.

Factoring by Grouping

For polynomials having four terms and a GCF of 1, when none of the preceding factoring techniques apply, we can sometimes group the terms in such a way that each group has a common factor. This technique is called **factoring by grouping**.

EXAMPLE 9   **Factoring by Grouping**

Factor.   **a.** $x^3 + 2x^2 + 3x + 6$ **b.** $6x^3 - 3x^2 - 4x + 2$ **c.** $x^2 + 4x + 4 - y^2$

SOLUTION

a. $x^3 + 2x^2 + 3x + 6 = (x^3 + 2x^2) + (3x + 6)$ Group terms so that each group can be factored.

$= x^2(x + 2) + 3(x + 2)$ Factor the grouped binomials.

$= (x + 2)(x^2 + 3)$ Factor out the common binomial factor $x + 2$.

b. $6x^3 - 3x^2 - 4x + 2 = (6x^3 - 3x^2) + (-4x + 2)$ Group terms so that each group can be factored.

$= 3x^2(2x - 1) + (-2)(2x - 1)$ Factor the grouped binomials.

$= (2x - 1)(3x^2 - 2)$ Factor out the common binomial factor $2x - 1$.

c. $x^2 + 4x + 4 - y^2 = (x^2 + 4x + 4) - y^2$ Group terms so that each group can be factored.

$= (x + 2)^2 - y^2$ Factor the trinomial.

$= (x + 2 - y)(x + 2 + y)$ Difference of squares. ▪▪▪

Practice Problem 9   Factor. **a.** $x^3 + 3x^2 + x + 3$ **b.** $28x^3 - 20x^2 - 7x + 5$

c. $x^2 - y^2 - 2y - 1$ ▪

SECTION P.4 ▪ Exercises

A EXERCISES Basic Skills and Concepts

1. The polynomials $x + 2$ and $x - 2$ are called _____ of the polynomial $x^2 - 4$.

2. The polynomial $3y$ is the _____ factor of the polynomial $3y^2 + 6y$.

3. The GCF of the polynomial $10x^3 + 30x^2$ is _____ .

4. A polynomial that cannot be factored as a product of two polynomials (excluding the constant polynomials ± 1) is said to be _____ .

5. *True or False* When factoring a polynomial whose leading coefficient is positive, we need to consider only its positive factors.

6. *True or False* The polynomial $x^2 + 7$ is irreducible.

In Exercises 7–18, factor each polynomial by removing any common monomial factor.

7. $8x - 24$
8. $5x + 25$
9. $-6x^2 + 12x$
10. $-3x^2 + 21$
11. $7x^2 + 14x^3$
12. $9x^3 - 18x^4$
13. $x^4 + 2x^3 + x^2$
14. $x^4 - 5x^3 + 7x^2$
15. $3x^3 - x^2$
16. $2x^3 + 2x^2$
17. $8ax^3 + 4ax^2$
18. $ax^4 - 2ax^2 + ax$

In Exercises 19–26, factor by grouping.

19. $x^3 + 3x^2 + x + 3$
20. $x^3 + 5x^2 + x + 5$
21. $x^3 - 5x^2 + x - 5$
22. $x^3 - 7x^2 + x - 7$
23. $6x^3 + 4x^2 + 3x + 2$
24. $3x^3 + 6x^2 + x + 2$
25. $12x^7 + 4x^5 + 3x^4 + x^2$
26. $3x^7 + 3x^5 + x^4 + x^2$

In Exercises 27–42, factor each trinomial or state that it is irreducible.

27. $x^2 + 7x + 12$
28. $x^2 + 8x + 15$
29. $x^2 - 6x + 8$
30. $x^2 - 9x + 14$
31. $x^2 - 3x - 4$
32. $x^2 - 5x - 6$
33. $x^2 - 4x + 13$
34. $x^2 - 2x + 5$
35. $2x^2 + x - 36$
36. $2x^2 + 3x - 27$
37. $6x^2 + 17x + 12$
38. $8x^2 - 10x - 3$
39. $3x^2 - 11x - 4$
40. $5x^2 + 7x + 2$
41. $6x^2 - 3x + 4$
42. $4x^2 - 4x + 3$

In Exercises 43–50, factor each polynomial as a perfect square.

43. $x^2 + 6x + 9$
44. $x^2 + 8x + 16$
45. $9x^2 + 6x + 1$
46. $36x^2 + 12x + 1$
47. $25x^2 - 20x + 4$
48. $64x^2 + 32x + 4$
49. $49x^2 + 42x + 9$
50. $9x^2 + 24x + 16$

In Exercises 51–60, factor each polynomial using the difference of squares pattern.

51. $x^2 - 64$
52. $x^2 - 121$
53. $4x^2 - 1$
54. $9x^2 - 1$
55. $16x^2 - 9$
56. $25x^2 - 49$
57. $x^4 - 1$
58. $x^4 - 81$
59. $20x^4 - 5$
60. $12x^4 - 75$

In Exercises 61–70, factor each polynomial using the sum or difference of cubes pattern.

61. $x^3 + 64$ **62.** $x^3 + 125$

63. $x^3 - 27$ **64.** $x^3 - 216$

65. $8 - x^3$ **66.** $27 - x^3$

67. $8x^3 - 27$ **68.** $8x^3 - 125$

69. $40x^3 + 5$ **70.** $7x^3 + 56$

In Exercises 71–102, factor each polynomial completely. If a polynomial cannot be factored, state that it is irreducible.

71. $1 - 16x^2$ **72.** $4 - 25x^2$

73. $x^2 - 6x + 9$ **74.** $x^2 - 8x + 16$

75. $4x^2 + 4x + 1$ **76.** $16x^2 + 8x + 1$

77. $2x^2 - 8x - 10$ **78.** $5x^2 - 10x - 40$

79. $2x^2 + 3x - 20$ **80.** $2x^2 - 7x - 30$

81. $x^2 - 24x + 36$ **82.** $x^2 - 20x + 25$

83. $3x^5 + 12x^4 + 12x^3$ **84.** $2x^5 + 16x^4 + 32x^3$

85. $9x^2 - 1$ **86.** $16x^2 - 25$

87. $16x^2 + 24x + 9$ **88.** $4x^2 + 20x + 25$

89. $x^2 + 15$ **90.** $x^2 + 24$

91. $45x^3 + 8x^2 - 4x$ **92.** $25x^3 + 40x^2 - 9x$

93. $ax^2 - 7a^2x - 8a^3$ **94.** $ax^2 - 10a^2x - 24a^3$

95. $x^2 - 16a^2 + 8x + 16$ **96.** $x^2 - 36a^2 + 6x + 9$

97. $3x^5 + 12x^4y + 12x^3y^2$ **98.** $4x^5 + 24x^4y + 36x^3y^2$

99. $x^2 - 25a^2 + 4x + 4$ **100.** $x^2 - 16a^2 + 4x + 4$

101. $18x^6 + 12x^5y + 2x^4y^2$ **102.** $12x^6 + 12x^5y + 3x^4y^2$

B EXERCISES Applying the Concepts

103. **Garden area.** A rectangular garden must have a perimeter of 16 feet. Write a completely factored polynomial whose values give the area of the garden, assuming that one side is x feet.

104. **Serving tray dimensions.** A rectangular serving tray must have a perimeter of 42 inches. Write a completely factored polynomial whose values give the area of the tray, assuming that one side is x inches.

105. **Box construction.** A box with an open top is to be constructed from a rectangular piece of cardboard with length 36 inches and width 16 inches by cutting out squares of equal side length x at each corner and then folding up the sides, as shown in the accompanying figure. Write a completely factored polynomial whose values give the volume of the box (volume = length × width × height).

106. **Surface area.** For the box described in Exercise 105, write a completely factored polynomial whose value gives its surface area (surface area = sum of the areas of the five sides).

107. **Area of a washer.** As shown in the figure, a washer is made by removing a circular disk from the center of another circular disk of radius 2 cm. Assuming that the disk that is removed has radius x, write a completely factored polynomial whose values give the area of the washer.

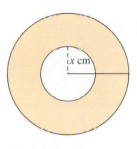

108. **Insulation.** Insulation must be put into the space between two concentric cylinders. Write a completely factored polynomial whose values give the volume between the inner and outer cylinder if the inner cylinder has radius 3 feet, the outer cylinder has radius x feet, and both cylinders are 8 feet high.

109. **Grazing pens.** A rancher has 2800 feet of fencing that will fence off a rectangular grazing pen for cattle. One edge of the pen borders a straight river and needs no fence. Assuming that the edges of the pen perpendicular to the river are x feet long, write a completely factored polynomial whose values give the area of the pen.

110. **Centerpiece perimeter.** A decorative glass centerpiece for a table has the shape of a rectangle with a semi-circular section attached at each end, as shown in the figure. The perimeter of the figure is 48 inches. Write a completely factored polynomial whose values give the area of the centerpiece.

Rational Expressions

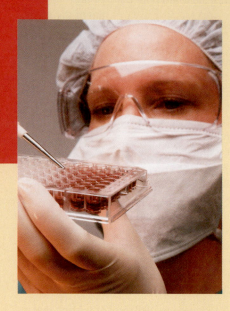

Before Starting this Section, Review

1. Domain of an expression (Section P.1, page 15)

2. Rational numbers (Section P.1, page 3)

3. Properties of fractions (Section P.1, page 13)

4. Factoring polynomials (Section P.4)

Objectives

1 Reduce a rational expression to lowest terms.

2 Multiply and divide rational expressions.

3 Add and subtract rational expressions.

4 Identify and simplify complex fractions.

FALSE POSITIVE IN DRUG TESTING

You are probably aware that the nonmedical use of steroids among adolescents and young adults is an ongoing national concern. In fact, many high schools have required students involved in extracurricular activities to submit to random drug testing. Many methods for drug testing are in use, and those who administer the tests are keenly aware of the possibility of what is termed a "false positive," which occurs when someone who is not a drug user tests positive for using drugs. How likely is it that a student who tests positive actually uses drugs? For a test that is 95% accurate, an expression for the likelihood that a student who tests positive is a nonuser of drugs is

$$\frac{(0.05)(1-x)}{0.95x + (0.05)(1-x)},$$

where x is the percent of the student population that uses drugs. This is an example of a rational expression, which we study in this section. In Example 1, we see that in a student population among which only 5% of the students use drugs, the likelihood that a student who tests positive is a nonuser is 50% when the test used is 95% accurate. ■

Rational Expressions

Recall that the quotient of two integers, $\frac{a}{b}$ $(b \neq 0)$, is a rational number. When we form the quotient of two polynomials, the result is called a **rational expression**. Following are examples of rational expressions.

$$\frac{10}{17}, \quad \frac{x+2}{3}, \quad \frac{x+6}{(x-3)(x+4)}, \quad \frac{7}{x^2+1}, \quad \text{and} \quad \frac{x^2+2x-3}{x^2+5x}$$

We use the same language to describe rational expressions that we use to describe rational numbers. For $\frac{x^2+2x-3}{x^2+5x+6}$, we call x^2+2x-3 the **numerator** and x^2+5x+6 the **denominator**. The domain of a rational expression is the set of all real numbers except those that result in a zero denominator. For example, the domain of $\frac{2}{x-1}$ is all real x, $x \neq 1$. We write $\frac{x+6}{(x-3)(x+4)}$, $x \neq 3$, $x \neq -4$ to indicate that 3 and −4 are not in the domain of this expression.

EXAMPLE 1   **Calculating False Positives in Drug Testing**

With a test that is 95% accurate, an expression for the likelihood that a student who tests positive for drugs is a nonuser is

$$\frac{(0.05)(1 - x)}{0.95x + (0.05)(1 - x)},$$

where x is the percent of the student population that uses drugs. In a student population among which only 5% of the students use drugs, find the likelihood that a student who tests positive is a nonuser.

SOLUTION

We convert 5% to a decimal to get $x = 0.05$, and we replace x with 0.05 in the given expression.

$$\frac{(0.05)(1 - x)}{0.95x + (0.05)(1 - x)} = \frac{(0.05)(1 - 0.05)}{0.95(0.05) + (0.05)(1 - 0.05)}$$

$$= 0.50 \qquad \text{Use a calculator.}$$

So, the likelihood that a student who tests positive is a nonuser when the test used is 95% accurate is 50%. Because of the high rate of false positives, a more accurate test is needed.   ■ ■ ■

Practice Problem 1   Rework Example 1 for a student population among which 10% of the students use drugs.   ■

1 Reduce a rational expression to lowest terms.

Lowest Terms for a Rational Expression

EXAMPLE 2   **Reducing a Rational Expression to Lowest Terms**

Simplify each expression.

a. $\dfrac{x^4 + 2x^3}{x + 2}, x \neq -2$ **b.** $\dfrac{x^6 - x - 6}{x^3 - 3x^2}, x \neq 0, x \neq 3$

SOLUTION

Factor each numerator and denominator and remove the common factors.

a. $\dfrac{x^4 + 2x^3}{x + 2} = \dfrac{x^3(x + 2)}{x + 2} = \dfrac{x^3\cancel{(x + 2)}}{\cancel{(x + 2)}} = x^3$

b. $\dfrac{x^2 - x - 6}{x^3 - 3x^2} = \dfrac{(x + 2)(x - 3)}{x^2(x - 3)} = \dfrac{(x + 2)\cancel{(x - 3)}}{x^2\cancel{(x - 3)}} = \dfrac{x + 2}{x^2}$ ■ ■ ■

Practice Problem 2   Simplify each expression.

a. $\dfrac{2x^3 + 8x^2}{3x^2 + 12x}$ **b.** $\dfrac{x^2 - 4}{x^2 + 4x + 4}$ ■

◆ **WARNING**   Because x^2 is a *term*, not a *factor*, in both the numerator and denominator of $\dfrac{x^2 - 1}{x^2 + 2x + 1}$, you cannot remove x^2. You can remove only *factors* common to both the numerator and denominator of a rational expression.

2 Multiply and divide rational expressions.

Multiplication and Division of Rational Expressions

We use the same rules for multiplying and dividing rational expressions as we do for rational numbers.

MULTIPLICATION AND DIVISION

For rational expressions $\dfrac{A}{B}$ and $\dfrac{C}{D}$,

$$\frac{A}{B} \cdot \frac{C}{D} = \frac{A \cdot C}{B \cdot D} \qquad \text{and} \qquad \frac{\dfrac{A}{B}}{\dfrac{C}{D}} = \frac{A}{B} \div \frac{C}{D} = \frac{A}{B} \cdot \frac{D}{C} = \frac{AD}{BC}$$

if $B \neq 0, D \neq 0$. $\qquad\qquad$ if $B \neq 0, C \neq 0, D \neq 0$.

EXAMPLE 3 **Multiplying and Dividing Rational Expressions**

Multiply or divide as indicated. Simplify your answer and leave it in factored form.

a. $\dfrac{x^2 + 3x + 2}{x^3 + 3x} \cdot \dfrac{2x^3 + 6x}{x^2 + x - 2}, \qquad x \neq 0, x \neq 1, x \neq -2$

b. $\dfrac{\dfrac{3x^2 + 11x - 4}{8x^3 - 40x^2}}{\dfrac{3x + 12}{4x^4 - 20x^3}}, \qquad x \neq 0, x \neq 5, x \neq -4$

SOLUTION

Factor each numerator and denominator and remove the common factors.

a. $\dfrac{x^2 + 3x + 2}{x^3 + 3x} \cdot \dfrac{2x^3 + 6x}{x^2 + x - 2} = \dfrac{(x+2)(x+1)}{x(x^2 + 3)} \cdot \dfrac{2x(x^2 + 3)}{(x - 1)(x + 2)}$

$\qquad\qquad = \dfrac{\cancel{(x+2)}(x+1)(2x)\cancel{(x^2+3)}}{x\cancel{(x^2+3)}(x-1)\cancel{(x+2)}} = \dfrac{2(x+1)}{x-1}$

b. $\dfrac{\dfrac{3x^2 + 11x - 4}{8x^3 - 40x^2}}{\dfrac{3x + 12}{4x^4 - 20x^3}} = \dfrac{3x^2 + 11x - 4}{8x^3 - 40x^2} \cdot \dfrac{4x^4 - 20x^3}{3x + 12} = \dfrac{(x + 4)(3x - 1)}{8x^2(x - 5)} \cdot \dfrac{4x^3(x - 5)}{3(x + 4)}$

$\qquad\qquad = \dfrac{\cancel{(x+4)}(3x-1)\overset{x}{\cancel{(4x^3)}}\cancel{(x-5)}}{\underset{2}{\cancel{8x^2}}\cancel{(x-5)}(3)\cancel{(x+4)}} = \dfrac{(3x-1)x}{6} = \dfrac{x(3x-1)}{6}$

$\blacksquare\ \blacksquare\ \blacksquare$

Practice Problem 3 Multiply or divide as indicated. Simplify your answer.

$$\dfrac{\dfrac{x^2 - 2x - 3}{7x^3 + 28x^2}}{\dfrac{4x + 4}{2x^4 + 8x^3}} \qquad x \neq 0, x \neq -4, x \neq -1 \qquad \blacksquare$$

3 Add and subtract rational expressions.

Addition and Subtraction of Rational Expressions

To add and subtract rational expressions, we use the same rules as for adding and subtracting rational numbers.

ADDITION AND SUBTRACTION OF RATIONAL EXPRESSIONS

For rational expressions $\dfrac{A}{D}$ and $\dfrac{C}{D}$ (same denominator $D \neq 0$),

$$\frac{A}{D} + \frac{C}{D} = \frac{A+C}{D} \qquad \text{and} \qquad \frac{A}{D} - \frac{C}{D} = \frac{A-C}{D}.$$

For rational expression $\dfrac{A}{B}$ and $\dfrac{C}{D}$ $(B \neq 0, D \neq 0)$,

$$\frac{A}{B} + \frac{C}{D} = \frac{AD+BC}{BD} \qquad \text{and} \qquad \frac{A}{B} - \frac{C}{D} = \frac{AD-BC}{BD}.$$

EXAMPLE 4   **Adding and Subtracting Rational Expressions with the Same Denominator**

Add or subtract as indicated. Simplify your answer and leave both the numerator and denominator in factored form.

a. $\dfrac{x-6}{(x+1)^2} + \dfrac{x+8}{(x+1)^2}, x \neq -1$

b. $\dfrac{3x-2}{x^2-5x+6} - \dfrac{2x+1}{x^2-5x+6}, x \neq 2, x \neq 3$

SOLUTION

a. $\dfrac{x-6}{(x+1)^2} + \dfrac{x+8}{(x+1)^2} = \dfrac{x-6+x+8}{(x+1)^2} = \dfrac{2x+2}{(x+1)^2}$

$$= \frac{2(x+1)}{(x+1)(x+1)} = \frac{2}{x+1}.$$

b. $\dfrac{3x-2}{x^2-5x+6} - \dfrac{2x+1}{x^2-5x+6} = \dfrac{(3x-2)-(2x+1)}{x^2-5x+6} = \dfrac{3x-2-2x-1}{x^2-5x+6}$

$$= \frac{x-3}{(x-2)(x-3)} = \frac{x-3}{(x-2)(x-3)} = \frac{1}{x-2}.$$

■ ■ ■

Practice Problem 4   Add or subtract as indicated. Simplify your answers.

a. $\dfrac{5x+22}{x^2-36} + \dfrac{2(x+10)}{x^2-36}$ $x \neq 6, x \neq -6$

b. $\dfrac{4x+1}{x^2+x-12} - \dfrac{3x+4}{x^2+x-12}$ $x \neq -4, x \neq 3$ ■

When adding or subtracting rational expressions with different denominators, we proceed (as with fractions) by finding a common denominator. The one most convenient to use is called the **least common denominator (LCD)**, and it is the polynomial of least degree that contains each denominator as a factor. In the simplest case, the LCD is the product of the denominators.

EXAMPLE 5 **Adding and Subtracting When Denominators Have No Common Factor**

Add or subtract as indicated. Simplify your answer and leave it in factored form.

a. $\dfrac{x}{x+1} + \dfrac{2x-1}{x+3}$, $x \neq -1, x \neq -3$ **b.** $\dfrac{2x}{x+1} - \dfrac{x}{x+2}$, $x \neq -1, x \neq -2$

SOLUTION

a.
$$\dfrac{x}{x+1} + \dfrac{2x-1}{x+3} = \dfrac{x(x+3)}{(x+1)(x+3)} + \dfrac{(2x-1)(x+1)}{(x+1)(x+3)} \qquad \text{LCD} = (x+1)(x+3)$$

$$= \dfrac{x(x+3) + (2x-1)(x+1)}{(x+1)(x+3)}$$

$$= \dfrac{x^2 + 3x + 2x^2 + 2x - x - 1}{(x+1)(x+3)}$$

$$= \dfrac{3x^2 + 4x - 1}{(x+1)(x+3)}$$

b.
$$\dfrac{2x}{x+1} - \dfrac{x}{x+2} = \dfrac{2x(x+2)}{(x+1)(x+2)} - \dfrac{x(x+1)}{(x+1)(x+2)} \qquad \text{LCD} = (x+1)(x+2)$$

$$= \dfrac{2x(x+2) - x(x+1)}{(x+1)(x+2)} = \dfrac{2x^2 + 4x - x^2 - x}{(x+1)(x+2)}$$

$$= \dfrac{x^2 + 3x}{(x+1)(x+2)} = \dfrac{x(x+3)}{(x+1)(x+2)} \qquad \blacksquare\blacksquare\blacksquare$$

Practice Problem 5 Add or subtract as indicated. Simplify your answers.

a. $\dfrac{2x}{x+2} + \dfrac{3x}{x-5}$, $x \neq -2, x \neq 5$ **b.** $\dfrac{5x}{x-4} - \dfrac{2x}{x+3}$, $x \neq 4, x \neq -3$ ■

TO FIND THE LCD FOR RATIONAL EXPRESSIONS

1. Factor each denominator polynomial completely.

2. Form a product of the different irreducible factors of each polynomial. (Each distinct factor is used once.)

3. Attach to each factor in this product the largest exponent that appears on this factor in *any* of the factored denominators.

EXAMPLE 6 **Finding the LCD for Rational Expressions**

Find the LCD for each pair of rational expressions.

a. $\dfrac{x+2}{x(x-1)^2(x+2)}, \dfrac{3x+7}{4x^2(x+2)^3}$ **b.** $\dfrac{x+1}{x^2-x-6}, \dfrac{2x-13}{x^2-9}$

SOLUTION

a. Step 1 The denominators are already completely factored.

    Step 2 $4x(x-1)(x+2)$ Product of the different factors

    Step 3 $\text{LCD} = 4x^2(x-1)^2(x+2)^3$ The largest exponents are 2, 2, and 3.

b. Step 1 $x^2 - x - 6 = (x + 2)(x - 3)$
$\qquad\qquad x^2 - 9 = (x + 3)(x - 3)$

Step 2 $(x + 2)(x - 3)(x + 3)$ Product of the different factors

Step 3 LCD $= (x + 2)(x - 3)(x + 3)$ The largest exponent on each is 1.

                                                                     ■ ■ ■

Practice Problem 6 Find the LCD for each pair of rational expressions.

a. $\dfrac{x^2 + 3x}{x^2(x + 2)^2(x - 2)}, \dfrac{4x^2 + 1}{3x(x - 2)^2}$ **b.** $\dfrac{2x - 1}{x^2 - 25}, \dfrac{3 - 7x^2}{(x^2 + 4x - 5)}$ ■

In adding or subtracting rational expressions with different denominators, the first step is to find the LCD. Here is the general procedure.

PROCEDURE FOR ADDING OR SUBTRACTING RATIONAL EXPRESSIONS WITH DIFFERENT DENOMINATORS

Step 1 Find the LCD.

Step 2 Using $\dfrac{A}{B} = \dfrac{AC}{BC}$, write each rational expression as a rational expression with the LCD as the denominator.

Step 3 Following the order of operations, add or subtract numerators, keeping the LCD as the denominator.

Step 4 Simplify.

EXAMPLE 7   **Using the LCD to Add and Subtract Rational Expressions**

Add or subtract as indicated. Simplify your answer and leave it in factored form.

a. $\dfrac{3}{x^2 - 1} + \dfrac{x}{x^2 + 2x + 1}, \quad x \neq 1, x \neq -1$

b. $\dfrac{x + 2}{x^2 - x} - \dfrac{3x}{4(x - 1)^2}, \quad x \neq 0, x \neq 1$

SOLUTION

a. Step 1 Note that $x^2 - 1 = (x - 1)(x + 1)$ and $x^2 + 2x + 1 = (x + 1)^2$; the LCD is $(x + 1)^2(x - 1)$.

Step 2 $\dfrac{3}{x^2 - 1} = \dfrac{3}{(x - 1)(x + 1)} = \dfrac{3(x + 1)}{(x - 1)(x + 1)^2}$ Multiply numerator and denominator by $x + 1$.

$\dfrac{x}{x^2 + 2x + 1} = \dfrac{x}{(x + 1)^2} = \dfrac{x(x - 1)}{(x + 1)^2(x - 1)}$ Multiply numerator and denominator by $x - 1$.

Steps 3–4 Add.

$$\dfrac{3}{x^2 - 1} + \dfrac{x}{x^2 + 2x + 1} = \dfrac{3(x + 1)}{(x - 1)(x + 1)^2} + \dfrac{x(x - 1)}{(x - 1)(x + 1)^2}$$

$$= \dfrac{3(x + 1) + x(x - 1)}{(x - 1)(x + 1)^2} = \dfrac{3x + 3 + x^2 - x}{(x - 1)(x + 1)^2}$$

$$= \dfrac{x^2 + 2x + 3}{(x - 1)(x + 1)^2}$$

b. Step 1 The LCD is $4x(x-1)^2$.

$x^2 - x = x(x - 1)$

Step 2

$$\frac{x+2}{x^2-x} = \frac{x+2}{x(x-1)} = \frac{(x+2)\cdot 4(x-1)}{x(x-1)\cdot 4(x-1)} = \frac{4(x+2)(x-1)}{4x(x-1)^2}$$

Multiply numerator and denominator · by $4(x-1)$.

$$\frac{3x}{4(x-1)^2} = \frac{(3x)x}{4(x-1)^2 x} = \frac{3x^2}{4x(x-1)^2}$$

Multiply numerator and denominator by x.

Steps 3–4 Subtract.

$$\frac{x+2}{x^2-x} - \frac{3x}{4(x-1)^2} = \frac{4(x+2)(x-1)}{4x(x-1)^2} - \frac{3x^2}{4x(x-1)^2} = \frac{4(x+2)(x-1) - 3x^2}{4x(x-1)^2}$$

$$= \frac{4x^2 + 4x - 8 - 3x^2}{4x(x-1)^2} = \frac{x^2 + 4x - 8}{4x(x-1)^2} \qquad \blacksquare\;\blacksquare\;\blacksquare$$

Practice Problem 7 Add or subtract as indicated. Simplify your answers.

a. $\dfrac{4}{x^2 - 4x + 4} + \dfrac{x}{x^2 - 4}$, $x \ne 2, x = -2$

b. $\dfrac{2x}{3(x-5)^2} - \dfrac{6x}{2(x^2 - 5x)}$, $x \ne 0, x \ne 5$ ■

---

4 Identify and simplify complex fractions.

Complex Fractions

A rational expression that contains another rational expression in its numerator or denominator (or both) is called a **complex rational expression** or **complex fraction**. To **simplify** a complex fraction, we write it as a rational expression in lowest terms.

There are two effective methods of simplifying complex fractions.

PROCEDURES FOR SIMPLIFYING COMPLEX FRACTIONS

Method 1 Perform the operations indicated in both the numerator and denominator of the complex fraction. Then multiply the resulting numerator by the reciprocal of the denominator.

Method 2 Multiply the numerator and the denominator of the complex fraction by the LCD of all rational expressions that appear in either the numerator or the denominator. Simplify the result.

EXAMPLE 8 **Simplifying a Complex Fraction**

Simplify $\dfrac{\dfrac{1}{2} + \dfrac{1}{x}}{\dfrac{x^2 - 4}{2x}}$, $x \ne 0, x \ne 2, x \ne -2$, using each of the two methods.

SOLUTION

Method 1
$$\frac{\dfrac{1}{2}+\dfrac{1}{x}}{\dfrac{x^2-4}{2x}}=\frac{\dfrac{x+2}{2x}}{\dfrac{x^2-4}{2x}}=\frac{x+2}{2x}\cdot\frac{2x}{x^2-4}=\frac{x+2}{2x}\cdot\frac{2x}{(x+2)(x-2)}$$

$$=\frac{\cancel{(x+2)}\cancel{(2x)}}{\cancel{(2x)}\cancel{(x+2)}(x-2)}=\frac{1}{x-2}$$

Method 2 The LCD of $\dfrac{1}{2}$, $\dfrac{1}{x}$, and $\dfrac{x^2-4}{2x}$ is $2x$.

$$\frac{\dfrac{1}{2}+\dfrac{1}{x}}{\dfrac{x^2-4}{2x}}=\frac{\left(\dfrac{1}{2}+\dfrac{1}{x}\right)(2x)}{\left(\dfrac{x^2-4}{2x}\right)(2x)}$$ Multiply numerator and denominator by the LCD.

$$=\frac{\dfrac{1}{2}(2x)+\dfrac{1}{x}(2x)}{\left[\dfrac{(x^2-4)}{2x}\right](2x)}=\frac{x+2}{x^2-4}=\frac{\cancel{x+2}}{(x-2)\cancel{(x+2)}}=\frac{1}{x-2}$$ ■■■

Practice Problem 8 Simplify: $\dfrac{\dfrac{5}{3x}+\dfrac{1}{3}}{\dfrac{x^2-25}{3x}}$, $x\neq0$, $x\neq5$, $x\neq-5$ ■

EXAMPLE 9 **Simplifying a Complex Fraction**

Simplify: $\dfrac{x}{2x-\dfrac{7}{3-\dfrac{1}{2}}}$, $x\neq\dfrac{7}{5}$

SOLUTION
We use Method 1.

Begin with $\dfrac{7}{3-\dfrac{1}{2}}=\dfrac{7}{\dfrac{5}{2}}=7\cdot\dfrac{2}{5}=\dfrac{14}{5}$.   Replace $3-\dfrac{1}{2}$ with $\dfrac{5}{2}$.

Then $\dfrac{x}{2x-\dfrac{7}{3-\dfrac{1}{2}}}=\dfrac{x}{2x-\dfrac{14}{5}}$ Replace $\dfrac{7}{3-\dfrac{1}{2}}$ with $\dfrac{14}{5}$.

$$=\frac{x}{\dfrac{10x-14}{5}}=x\cdot\frac{5}{10x-14}=\frac{5x}{10x-14}=\frac{5x}{2(5x-7)}.$$ ■■■

Practice Problem 9 Simplify: $\dfrac{5x}{3x-\dfrac{4}{2-\dfrac{1}{3}}}$, $x\neq\dfrac{4}{5}$ ■

A EXERCISES Basic Skills and Concepts

1. The least common denominator for two rational expressions is the polynomial of least degree that contains _____ as a factor.

2. The first step in finding the LCD for two rational expressions is to _____ the denominators completely.

3. If the denominators for two rational expressions are $x^2 - 2x$ and $x^2 - x - 2$, the LCD is _____.

4. A rational expression that contains another rational expression in its numerator or denominator is called a(n) _____.

5. *True or False* The expression $\dfrac{1}{2} + \dfrac{1}{x}$ is called a complex fraction.

6. *True or False* A polynomial is also a rational expression.

In Exercises 7–22, reduce each rational expression to lowest terms. Specify the domain of the rational expression by identifying all real numbers that must be excluded from the domain.

7. $\dfrac{2x + 2}{x^2 + 2x + 1}$

8. $\dfrac{3x - 6}{x^2 - 4x + 4}$

9. $\dfrac{3x + 3}{x^2 - 1}$

10. $\dfrac{10 - 5x}{4 - x^2}$

11. $\dfrac{2x - 6}{9 - x^2}$

12. $\dfrac{15 + 3x}{x^2 - 25}$

13. $\dfrac{2x - 1}{1 - 2x}$

14. $\dfrac{2 - 5x}{5x - 2}$

15. $\dfrac{x^2 - 6x + 9}{4x - 12}$

16. $\dfrac{x^2 - 10x + 25}{3x - 15}$

17. $\dfrac{7x^2 + 7x}{x^2 + 2x + 1}$

18. $\dfrac{4x^2 + 12x}{x^2 + 6x + 9}$

19. $\dfrac{x^2 - 11x + 10}{x^2 + 6x - 7}$

20. $\dfrac{x^2 + 2x - 15}{x^2 - 7x + 12}$

21. $\dfrac{6x^4 + 14x^3 + 4x^2}{6x^4 - 10x^3 - 4x^2}$

22. $\dfrac{3x^3 + x^2}{3x^4 - 11x^3 - 4x^2}$

In Exercises 23–40, multiply or divide as indicated. Simplify and leave the numerator and denominator in your answer in factored form.

23. $\dfrac{x - 3}{2x + 4} \cdot \dfrac{10x + 20}{5x - 15}$

24. $\dfrac{6x + 4}{2x - 8} \cdot \dfrac{x - 4}{9x + 6}$

25. $\dfrac{2x + 6}{4x - 8} \cdot \dfrac{x^2 + x - 6}{x^2 - 9}$

26. $\dfrac{25x^2 - 9}{4 - 2x} \cdot \dfrac{4 - x^2}{10x - 6}$

27. $\dfrac{x^2 - 7x}{x^2 - 6x - 7} \cdot \dfrac{x^2 - 1}{x^2}$

28. $\dfrac{x^2 - 9}{x^2 - 6x + 9} \cdot \dfrac{5x - 15}{x + 3}$

29. $\dfrac{x^2 - x - 6}{x^2 + 3x + 2} \cdot \dfrac{x^2 - 1}{x^2 - 9}$

30. $\dfrac{x^2 + 2x - 8}{x^2 + x - 20} \cdot \dfrac{x^2 - 16}{x^2 + 5x + 4}$

31. $\dfrac{2 - x}{x + 1} \cdot \dfrac{x^2 + 3x + 2}{x^2 - 4}$

32. $\dfrac{3 - x}{x + 5} \cdot \dfrac{x^2 + 8x + 15}{x^2 - 9}$

33. $\dfrac{x + 2}{6} \div \dfrac{4x + 8}{9}$

34. $\dfrac{x + 3}{20} \div \dfrac{4x + 12}{9}$

35. $\dfrac{x^2 - 9}{x} \div \dfrac{2x + 6}{5x^2}$

36. $\dfrac{x^2 - 1}{3x} \div \dfrac{7x - 7}{x^2 + x}$

37. $\dfrac{x^2 + 2x - 3}{x^2 + 8x + 16} \div \dfrac{x - 1}{3x + 12}$

38. $\dfrac{x^2 + 5x + 6}{x^2 + 6x + 9} \div \dfrac{x^2 + 3x + 2}{x^2 + 7x + 12}$

39. $\left(\dfrac{x^2 - 9}{x^3 + 8} \div \dfrac{x + 3}{x^3 + 2x^2 - x - 2} \right) \dfrac{1}{x^2 - 1}$

40. $\left(\dfrac{x^2 - 25}{x^2 - 3x - 4} \div \dfrac{x^2 + 3x - 10}{x^2 - 1} \right) \dfrac{x - 2}{x - 5}$

In Exercises 41–58, add and subtract as indicated. Simplify and leave the numerator and denominator in your answer in factored form.

41. $\dfrac{x}{5} + \dfrac{3}{5}$

42. $\dfrac{7}{4} - \dfrac{x}{4}$

43. $\dfrac{x}{2x + 1} + \dfrac{4}{2x + 1}$

44. $\dfrac{2x}{7x - 3} + \dfrac{x}{7x - 3}$

45. $\dfrac{x^2}{x + 1} - \dfrac{x^2 - 1}{x + 1}$

46. $\dfrac{2x + 7}{3x + 2} - \dfrac{x - 2}{3x + 2}$

47. $\dfrac{4}{3 - x} + \dfrac{2x}{x - 3}$

48. $\dfrac{-2}{1 - x} + \dfrac{2 - x}{x - 1}$

49. $\dfrac{5x}{x^2 + 1} + \dfrac{2x}{x^2 + 1}$

50. $\dfrac{x}{2(x - 1)^2} + \dfrac{3x}{2(x - 1)^2}$

51. $\dfrac{7x}{2(x - 3)} + \dfrac{x}{2(x - 3)}$

52. $\dfrac{4x}{4(x + 5)^2} + \dfrac{8x}{4(x + 5)^2}$

53. $\dfrac{x}{x^2 - 4} - \dfrac{2}{x^2 - 4}$

54. $\dfrac{5x}{x^2 - 1} - \dfrac{5}{x^2 - 1}$

55. $\dfrac{x-2}{2x+1} - \dfrac{x}{2x-1}$ **56.** $\dfrac{2x-1}{4x+1} - \dfrac{2x}{4x-1}$

57. $\dfrac{-x}{x+2} + \dfrac{x-2}{x} - \dfrac{x}{x-2}$

58. $\dfrac{3x}{x-1} + \dfrac{x+1}{x} - \dfrac{2x}{x+1}$

In Exercises 59–66, find the LCD for each pair of rational fractions.

59. $\dfrac{5}{3x-6}, \dfrac{2x}{4x-8}$ **60.** $\dfrac{5x+1}{7+21x}, \dfrac{1-x}{3+9x}$

61. $\dfrac{3-x}{4x^2-1}, \dfrac{7x}{(2x+1)^2}$ **62.** $\dfrac{14x}{(3x-1)^2}, \dfrac{2x+7}{9x^2-1}$

63. $\dfrac{1-x}{x^2+3x+2}, \dfrac{3x+12}{x^2-1}$

64. $\dfrac{5x+9}{x^2-x-6}, \dfrac{x+5}{x^2-9}$

65. $\dfrac{7-4x}{x^2-5x+4}, \dfrac{x^2-x}{x^2+x-2}$

66. $\dfrac{13x}{x^2-2x-3}, \dfrac{2x^2-4}{x^2+3x+2}$

In Exercises 67–82, perform the indicated operations and simplify the result. Leave the numerator and denominator in your answer in factored form.

67. $\dfrac{5}{x-3} + \dfrac{2x}{x^2-9}$ **68.** $\dfrac{3x}{x-1} + \dfrac{x}{x^2-1}$

69. $\dfrac{2x}{x^2-4} - \dfrac{x}{x+2}$ **70.** $\dfrac{3x-1}{x^2-16} - \dfrac{2x+1}{x-4}$

71. $\dfrac{x-2}{x^2+3x-10} + \dfrac{x+3}{x^2+x-6}$

72. $\dfrac{x+3}{x^2-x-2} + \dfrac{x-1}{x^2+2x+1}$

73. $\dfrac{2x-3}{9x^2-1} + \dfrac{4x-1}{(3x-1)^2}$

74. $\dfrac{3x+1}{(2x+1)^2} + \dfrac{x+3}{4x^2-1}$

75. $\dfrac{x-3}{x^2-25} - \dfrac{x-3}{x^2+9x+20}$

76. $\dfrac{2x}{x^2-16} - \dfrac{2x-7}{x^2-7x+12}$

77. $\dfrac{3}{x^2-4} + \dfrac{1}{2-x} - \dfrac{1}{2+x}$

78. $\dfrac{2}{5+x} + \dfrac{5}{x^2-25} + \dfrac{7}{5-x}$

79. $\dfrac{x+3a}{x-5a} - \dfrac{x+5a}{x-3a}$ **80.** $\dfrac{3x-a}{2x-a} - \dfrac{2x+a}{3x+a}$

81. $\dfrac{1}{x+h} - \dfrac{1}{x}$ **82.** $\dfrac{1}{(x+h)^2} - \dfrac{1}{x^2}$

In Exercises 83–96, perform the indicated operations and simplify the result. Leave the numerator and denominator in your answer in factored form.

83. $\dfrac{\frac{2}{x}}{\frac{3}{x^2}}$ **84.** $\dfrac{\frac{-6}{x}}{\frac{2}{x^3}}$

85. $\dfrac{\frac{1}{x}}{1-\frac{1}{x}}$ **86.** $\dfrac{\frac{1}{x^2}}{\frac{1}{x^2}-1}$

87. $\dfrac{\frac{1}{x}-1}{\frac{1}{x}+1}$ **88.** $\dfrac{2-\frac{2}{x}}{1+\frac{2}{x}}$

89. $\dfrac{\frac{1}{x}-x}{1-\frac{1}{x^2}}$ **90.** $\dfrac{x-\frac{1}{x}}{\frac{1}{x^2}-1}$

91. $x - \dfrac{x}{x+\frac{1}{2}}$ **92.** $x + \dfrac{x}{x+\frac{1}{2}}$

93. $\dfrac{\frac{1}{x+h}-\frac{1}{x}}{h}$ **94.** $\dfrac{\frac{1}{(x+h)^2}-\frac{1}{x^2}}{h}$

95. $\dfrac{\frac{1}{x-a}+\frac{1}{x+a}}{\frac{1}{x-a}-\frac{1}{x+a}}$ **96.** $\dfrac{\frac{1}{x-a}+\frac{1}{x+a}}{\frac{x}{x-a}-\frac{a}{x+a}}$

B EXERCISES Applying the Concepts

97. Bearing length. The length (in centimeters) of a bearing in the shape of a cylinder is given by $\dfrac{125.6}{\pi r^2}$, where r is the radius of the base of the cylinder. Find the length of a bearing with a diameter of 4 centimeters, using 3.14 as an approximation of π.

98. Toy box height. The height (in feet) of an open toy box that is twice as long as it is wide is given by $\dfrac{13.5-2x^2}{6x}$, where x is the width of the box. Find the height of a toy box that is 1.5 feet wide.

99. Diluting a mixture. A 100-gallon mixture of citrus extract and water is 3% citrus extract.
 a. Write a rational expression in x whose values give the percentage (in decimal form) of the mixture that is citrus extract when x gallons of water are added to the mixture.
 b. Find the percentage of citrus extract in the mixture assuming that 50 gallons of water are added to it.

100. Acidity in a reservoir. A half-full 400,000-gallon reservoir is found to be 0.75% acid.
   a. Write a rational expression in x whose values give the percentage (in decimal form) of acid in the reservoir when x gallons of water are added to it.
   b. Find the percentage of acid in the reservoir assuming that 100,000 gallons of water are added to it.

101. Diet lemonade. A company packages its powdered diet lemonade mix in containers in the shape of a cylinder. The top and bottom are made of a tin product that costs 5 cents per square inch. The side of the container is made of a cheaper material that costs 1 cent per square inch. The height of any cylinder can be found by dividing its volume by the area of its base.
   a. Assuming that x is the radius of the base (in inches), write a rational expression in x whose values give the cost of each container, assuming that the capacity of the container is 120 cubic inches.
   b. Find the cost of a container whose base is 4 inches in diameter, using 3.14 as an approximation of π.

Volume = 120 cubic inches

102. Storage containers. A company makes storage containers in the shape of an open box with a square base. All five sides of the box are made of a plastic that costs 40¢ square foot. The height of any box can be found by dividing its volume by the area of its base.
   a. Assuming that x is the length of the base (in feet), write a rational expression in x whose values give the cost of each container, assuming that the capacity of the container is 2.25 cubic feet.
   b. Find the cost of a container whose base is 1.5 feet long .

Rational Exponents and Radicals

Objectives

1. Define and evaluate square roots.

2. Simplify square roots.

3. Rationalize denominators and numerators.

4. Define and evaluate *n*th roots.

5. Combine like radicals.

6. Define and evaluate rational exponents.

FRICTION AND TAP-WATER FLOW

Suppose your water company had a large reservoir or water tower 80 meters or more above the level of your cold-water faucet. If there was no friction in the pipes and you turned on your faucet, the water would come pouring out of the faucet at nearly 90 miles per hour! This fact follows from a result known as Bernoulli's equation, which predicts the velocity of the water (ignoring friction and related turbulence) in this situation to be equal to $\sqrt{2gh}$, where g is the acceleration due to gravity and h is the height of the water level above a hole from which the water escapes.

We can use Bernoulli's equation to calculate the velocity of the water from the faucet by replacing h with 80 meters and g with 10 meters per second per second (a convenient approximation) in the expression $\sqrt{2gh}$ and verifying our statement about the water pouring out at "nearly 90 miles per hour."

Velocity of water:

$$= \sqrt{2gh}$$

$$= \sqrt{2(10 \text{ m/s}^2)(80 \text{ m})}$$

$$= \sqrt{1600 \text{ m}^2/\text{s}^2}$$

$$= 40 \text{ m/s (approximately 90 miles per hour)}$$

In Example 3, we use these same ideas to find the rate at which water is being lost from a punctured water tower. ■

---

1 Define and evaluate square roots.

Square Roots

When we raise the number 5 to the power 2, we write $5^2 = 25$. The reverse of this squaring process is called finding a **square root**. In the case of 25, we say that 5 is a square root of 25 and write $\sqrt{25} = 5$. In general, we say that b is a square root of a if $b^2 = a$. We can't call 5 *the* square root of 25 because it is also true that $(-5)^2 = 25$, so that -5 is *another* square root of 25.

The symbol $\sqrt{}$, called a **radical sign**, is used to distinguish between 5 and -5, the two square roots of 25. We write $5 = \sqrt{25}$ and $-5 = -\sqrt{25}$ and call $\sqrt{25}$ the *principal square root* of 25.

DEFINITION OF PRINCIPAL SQUARE ROOT

$\sqrt{a} = b$ means   (1) $b^2 = a$ and   (2) $b \geq 0$.

If $\sqrt{a} = b$, then $a = b^2 \geq 0$. Consequently, the symbol $\sqrt{a}$ denotes a real number only when $a \geq 0$. The domain of the expression $\sqrt{x}$ is then $x \geq 0$, or in interval notation, $[0, \infty)$. The number or expression under the radical sign is called the **radicand**, and the radicand together with the radical sign is called a **radical**.

According to the definition of the square root, there are two tests that the number b must pass for the statement $\sqrt{a} = b$ to be correct: (1) $b^2 = a$ and (2) $b \geq 0$. Let's see how these tests are used to decide whether the statements below are true or false.

Statement $\sqrt{a} = b$	Test (1) $b^2 = a$?	Test (2) $b \geq 0$?	Conclusion True when b passes both tests
$\sqrt{4} = 2$	$2^2 = 4$ ✓	$2 \geq 0$ ✓	**True**; 2 passed both tests.
$\sqrt{4} = -2$	$(-2)^2 = 4$ ✓	$-2 \geq 0$ ✗	**False**; -2 failed Test (2).
$\sqrt{\frac{1}{9}} = \frac{1}{3}$	$\left(\frac{1}{3}\right)^2 = \frac{1}{9}$ ✓	$\frac{1}{3} \geq 0$ ✓	**True**; $\frac{1}{3}$ passed both tests.
$\sqrt{7^2} = 7$	$7^2 = 7^2$ ✓	$7 \geq 0$ ✓	**True**; 7 passed both tests.
$\sqrt{(-7)^2} = -7$	$(-7)^2 = (-7)^2$ ✓	$-7 \geq 0$ ✗	**False**; -7 failed Test (2).

The last example (letting x represent -7) shows that $\sqrt{x^2}$ is not always equal to x. Let's apply Test (1) and Test (2) to the statement $\sqrt{x^2} = |x|$, where x represents any real number.

Statement	Test (1)	Test (2)	Conclusion
$\sqrt{x^2} = |x|$	$|x|^2 = x^2$ ✓	$|x| \geq 0$ ✓	True; $|x|$ passed both tests.

For any real number x,

$$\sqrt{x^2} = |x|.$$

The square root symbol on a graphing calculator does not have a horizontal bar that goes completely over the number or expression whose square root is to be evaluated. You must enclose a rational number in parentheses before taking its square root and then use the "Frac" feature to have the answer appear as a fraction.

```
√(121)
              11
√(25/81)
       .5555555556
Ans▶Frac
             5/9
```

If $\sqrt{x} = b$, then by Test (1), $b^2 = x$; replacing b with $\sqrt{x}$ in the equation $b^2 = x$ yields the following result.

For any nonnegative real number x,

$$(\sqrt{x})^2 = x.$$

EXAMPLE 1 Finding the Square Roots of Perfect Squares and Their Ratios

Find. **a.** $\sqrt{121}$ **b.** $\sqrt{\dfrac{25}{81}}$ **c.** $\sqrt{\dfrac{16}{169}}$

SOLUTION

Write each radicand as a perfect square.

a. $\sqrt{121} = \sqrt{11^2} = 11$ **b.** $\sqrt{\dfrac{25}{81}} = \sqrt{\dfrac{5^2}{9^2}} = \sqrt{\left(\dfrac{5}{9}\right)^2} = \dfrac{5}{9}$

c. $\sqrt{\dfrac{16}{169}} = \sqrt{\dfrac{4^2}{13^2}} = \sqrt{\left(\dfrac{4}{13}\right)^2} = \dfrac{4}{13}$

■■■

Practice Problem 1 Find.

a. $\sqrt{144}$ **b.** $\sqrt{\dfrac{1}{49}}$ **c.** $\sqrt{\dfrac{4}{64}}$ ■

2 Simplify square roots.

Simplifying Square Roots

When working with square roots, we can simplify a radical expression by removing perfect squares that are factors of the radicand from under the radical sign. For example, to simplify $\sqrt{8}$, we write $\sqrt{8} = \sqrt{4 \cdot 2} = \sqrt{4}\sqrt{2} = 2\sqrt{2}$. We rely on the following rules.

PRODUCT AND QUOTIENT PROPERTIES OF SQUARE ROOTS

$$\sqrt{a \cdot b} = \sqrt{a}\sqrt{b} \qquad a \geq 0, b \geq 0$$

$$\sqrt{\frac{a}{b}} = \frac{\sqrt{a}}{\sqrt{b}} \qquad a \geq 0, b > 0$$

EXAMPLE 2   **Simplifying Square Roots**

Simplify. **a.** $\sqrt{117}$ **b.** $\sqrt{6}\,\sqrt{12}$ **c.** $\sqrt{48x^2}$

d. $\dfrac{\sqrt{36}}{\sqrt{3}}$ **e.** $\sqrt{\dfrac{25y^3}{9x^2}}$, $x \neq 0, y \geq 0$

SOLUTION

a. $\sqrt{117} = \sqrt{9 \cdot 13} = \sqrt{9}\sqrt{13} = 3\sqrt{13}$

b. $\sqrt{6}\sqrt{12} = \sqrt{6 \cdot 12} = \sqrt{6 \cdot 6 \cdot 2} = \sqrt{6^2 \cdot 2} = \sqrt{6^2}\sqrt{2} = 6\sqrt{2}$

c. $\sqrt{48x^2} = \sqrt{48}\,\sqrt{x^2} = \sqrt{16 \cdot 3}\,\sqrt{x^2} = \sqrt{16}\sqrt{3}\sqrt{x^2} = 4\sqrt{3}\,|x|$

d. $\dfrac{\sqrt{36}}{\sqrt{3}} = \sqrt{\dfrac{36}{3}} = \sqrt{12} = \sqrt{4 \cdot 3} = \sqrt{4}\sqrt{3} = 2\sqrt{3}$

e. $\sqrt{\dfrac{25y^3}{9x^2}} = \dfrac{\sqrt{25y^3}}{\sqrt{9x^2}} = \dfrac{\sqrt{25 \cdot y^2 \cdot y}}{\sqrt{9 \cdot x^2}} = \dfrac{\sqrt{25}\,\sqrt{y^2}\sqrt{y}}{\sqrt{9}\sqrt{x^2}} = \dfrac{5|y|\sqrt{y}}{3|x|}$ $x \neq 0, y \geq 0$

$$= \dfrac{5y\sqrt{y}}{3|x|} \qquad |y| = y$$
                                                                                       ■ ■ ■

Practice Problem 2 Simplify.

a. $\sqrt{20}$ **b.** $\sqrt{6}\sqrt{8}$ **c.** $\sqrt{12x^2}$ **d.** $\sqrt{\dfrac{20y^3}{27x^2}}$ ■

◆ **WARNING**   It is important to note that $\sqrt{a + b} \neq \sqrt{a} + \sqrt{b}$. For example, with $a = 1$ and $b = 4$, $\sqrt{1 + 4} \neq \sqrt{1} + \sqrt{4}$ because $\sqrt{1 + 4} = \sqrt{5}$, whereas $\sqrt{1} + \sqrt{4} = 1 + 2 = 3$ and $\sqrt{5} \neq 3$. Similarly, $\sqrt{a - b} \neq \sqrt{a} - \sqrt{b}$.

EXAMPLE 3   **Calculating Water Flow from a Punctured Water Tower**

The son of the manager of a water company accidentally shoots a hole in the base of the company's large, full water tower with his new .22 Long Rifle. The diameter of the bullet used in the rifle is 223 thousandths of an inch, and the water tower is

20 meters high. What is the initial rate of water flow, in gallons per minute, from the punctured hole? (Use 1 in³ = 0.004 gallon.)

SOLUTION

We can use Bernoulli's equation to predict the velocity of the water from the hole. If we ignore friction, the velocity of water will be equal to $\sqrt{2gh}$, where g is the acceleration due to gravity and h is the height of the water level above the hole.

We replace h with 20 meters and roughly approximate g by 10 meters per second per second ($g \approx 10 \text{ m/s}^2$) in the expression $\sqrt{2gh}$.

Thus, the velocity of the water is

$$
\begin{aligned}
&= \sqrt{2gh} \\
&\approx \sqrt{2(10 \text{ m/s}^2)(20 \text{ m})} = 20 \text{ m/s} \\
&\approx (20 \text{ m/s})(39.37 \text{in./m})(60 \text{ s/min}) \qquad 1 \text{ m} \approx 39.37 \text{ in.} \\
&= 47{,}244 \text{ in./min.}
\end{aligned}
$$

Now we need to know two things:

1. That the hole made by the bullet is circular, with diameter 0.223 inch and the radius is one-half the diameter. Consequently, the area of the hole is $\pi\left(\dfrac{0.223}{2}\right)^2$ square inches.

2. That the rate of water leaving the water tower is given approximately by the area of the hole multiplied by the velocity of the water flowing through it.

The velocity of the water is 47,244 in./min; so the rate, R, of water leaving the water tower is given by

$$
\begin{aligned}
R &= \pi\left(\frac{0.223}{2}\right)^2 (47{,}244) \text{ in.}^3/\text{min} \qquad R = \text{area} \times \text{velocity} \\
&\approx 1845 \text{ in.}^3/\text{min} \approx 7.4 \text{ gallons/min} \qquad 1 \text{ in.}^3 = 0.004 \text{ gallon}
\end{aligned}
$$

The initial rate at which water flows from the punctured hole is approximately 7.4 gallons per minute.   ■ ■ ■

Practice Problem 3   Repeat Example 3 if the bullet's diameter is 0.240 inch and the water tower is 25 m high.   ■

3 Rationalize denominators and numerators.

Rationalizing Denominators

It is customary to rewrite a quotient involving square roots so that no radical appears in the denominator. The process for eliminating square roots in a denominator is called **rationalizing the denominator**. The special product $(A - B)(A + B) = A^2 - B^2$ is helpful when at least one term of a denominator binomial $A + B$ or $A - B$ is a square root. Here are some examples:

Denominator Factor	Multiply by	New Denominator Contains the Factor
$2 + \sqrt{3}$	$2 - \sqrt{3}$	$(2 + \sqrt{3})(2 - \sqrt{3}) = 2^2 - (\sqrt{3})^2 = 4 - 3 = 1$
$\sqrt{5} - \sqrt{2}$	$\sqrt{5} + \sqrt{2}$	$(\sqrt{5} - \sqrt{2})(\sqrt{5} + \sqrt{2}) = (\sqrt{5})^2 - (\sqrt{2})^2 = 5 - 2 = 3$
$1 + \sqrt{x}$	$1 - \sqrt{x}$	$(1 + \sqrt{x})(1 - \sqrt{x}) = 1^2 - (\sqrt{x})^2 = 1 - x, x \geq 0$

EXAMPLE 4   **Rationalizing Denominators**

Rationalize each denominator.

a. $\dfrac{1}{\sqrt{2}}$ **b.** $\dfrac{3}{7 - \sqrt{5}}$

SOLUTION

a. If we multiply the numerator and denominator by $\sqrt{2}$, the denominator becomes $\sqrt{2} \cdot \sqrt{2} = (\sqrt{2})^2 = 2$. Consequently,

$$\frac{1}{\sqrt{2}} = \frac{1 \cdot \sqrt{2}}{\sqrt{2} \cdot \sqrt{2}} = \frac{\sqrt{2}}{2}.$$

b. We multiply the numerator and denominator by $7 + \sqrt{5}$.

$$\frac{3}{7 - \sqrt{5}} = \frac{3 \cdot (7 + \sqrt{5})}{(7 - \sqrt{5})(7 + \sqrt{5})} = \frac{21 + 3\sqrt{5}}{7^2 - (\sqrt{5})^2} = \frac{21 + 3\sqrt{5}}{49 - 5} = \frac{21 + 3\sqrt{5}}{44}$$

■ ■ ■

Practice Problem 4   Rationalize each denominator.

a. $\dfrac{7}{\sqrt{8}}$ **b.** $\dfrac{3}{1 - \sqrt{7}}$ **c.** $\dfrac{x}{\sqrt{x} - 3}$ $(x \geq 0)$ ■

4 Define and evaluate *n*th roots.

Other Roots

If n is a positive integer, we say that b is an *n*th root of a if $b^n = a$. We next define the **principal *n*th root** of a real number a, denoted by the symbol $\sqrt[n]{a}$ next.

THE PRINCIPAL *n*th ROOT OF A REAL NUMBER

1. If a is positive ($a > 0$), then $\sqrt[n]{a} = b$ provided that $b^n = a$ and $b > 0$.
2. If a is negative ($a < 0$) and n is odd, then $\sqrt[n]{a} = b$ provided that $b^n = a$.
3. If a is negative ($a < 0$) and n is even, then $\sqrt[n]{a}$ is not a real number.

We use the same vocabulary for *n*th roots that we used for square roots. That is, $\sqrt[n]{a}$ is called a **radical expression**, $\sqrt{}$ is the **radical sign**, and a is the **radicand**. The positive integer n is called the **index**. The index 2 is not written; we write $\sqrt{a}$ rather than $\sqrt[2]{a}$ for the principal square root of a. It is also common practice to call $\sqrt[3]{a}$ the **cube root** of a.

EXAMPLE 5   **Finding Principal *n*th Roots**

Find each root.   **a.** $\sqrt[3]{27}$ **b.** $\sqrt[3]{-64}$ **c.** $\sqrt[4]{16}$ **d.** $\sqrt[4]{(-3)^4}$ **e.** $\sqrt[8]{-46}$

SOLUTION

a. The radicand, 27, is *positive*; so $\sqrt[3]{27} = 3$ because $3^3 = 27$ and $3 > 0$.

b. The radicand, -64, is *negative* and 3 is odd; so $\sqrt[3]{-64} = -4$ because $(-4)^3 = -64$.

c. The radicand, 16, is *positive*; so $\sqrt[4]{16} = 2$ because $2^4 = 16$ and $2 > 0$.

d. The radicand, $(-3)^4$, is *positive*; so $\sqrt[4]{(-3)^4} = 3$ because $3^4 = (-3)^4$ and $3 > 0$.

e. The radicand, -46, is *negative* and the index, 8, is even; so $\sqrt[8]{-46}$ is not a real number.

■ ■ ■

Practice Problem 5   Find each root.

a. $\sqrt[3]{-8}$ **b.** $\sqrt[5]{32}$ **c.** $\sqrt[4]{81}$ **d.** $\sqrt[6]{-4}$ ■

The fact that the definition of the principal *n*th root depends on whether n is even or odd results in two rules to replace the square root rule $\sqrt{x^2} = |x|$.

> **SIMPLIFYING** $\sqrt[n]{a^n}$
>
> If n is *odd*, then $\sqrt[n]{a^n} = a$.
>
> If n is *even*, then $\sqrt[n]{a^n} = |a|$.

The product and quotient rules for square roots can be extended to apply to *n*th roots as well.

> **PRODUCT AND QUOTIENT RULES FOR *n*th ROOTS**
>
> If a and b are real numbers and all indicated roots are defined, then
> $$\sqrt[n]{ab} = \sqrt[n]{a} \cdot \sqrt[n]{b}$$
> and
> $$\sqrt[n]{\frac{a}{b}} = \frac{\sqrt[n]{a}}{\sqrt[n]{b}}.$$

EXAMPLE 6   **Simplifying *n*th Roots**

Simplify.   **a.** $\sqrt[3]{135}$ **b.** $\sqrt[4]{162a^4}$

SOLUTION

a. We find a factor of 135 that is a perfect cube. Here $135 = 27 \cdot 5$ and 27 is a perfect cube ($3^3 = 27$). Then
$$\sqrt[3]{135} = \sqrt[3]{27 \cdot 5} = \sqrt[3]{27} = \sqrt[3]{5} = 3\sqrt[3]{5}.$$

b. We find a factor of 162 that is a perfect fourth power. Now $162 = 81 \cdot 2$ and 81 is a perfect fourth power: $3^4 = 81$. Thus,
$$\sqrt[4]{162a^4} = \sqrt[4]{162}\sqrt[4]{a^4} = \sqrt[4]{81 \cdot 2}|a| = \sqrt[4]{81}\sqrt[4]{2}|a| = 3\sqrt[4]{2}|a|. \quad ■ ■ ■$$

Practice Problem 6   Simplify.

a. $\sqrt[3]{72}$ **b.** $\sqrt[4]{48a^2}$ ■

5 Combine like radicals.

Like Radicals

Radicals that have the same index and the same radicand are called **like radicals**. Like radicals can be combined with the use of the distributive property.

EXAMPLE 7   **Adding and Subtracting Radical Expressions**

Simplify.   **a.** $\sqrt{45} + 7\sqrt{20}$ **b.** $5\sqrt[3]{80x} - 3\sqrt[3]{270x}$

SOLUTION

a. We find perfect-square factors for both 45 and 20.
$$\begin{aligned}
\sqrt{45} + 7\sqrt{20} &= \sqrt{9 \cdot 5} + 7\sqrt{4 \cdot 5} \\
&= \sqrt{9}\sqrt{5} + 7\sqrt{4}\sqrt{5} \quad \text{Product rule} \\
&= 3\sqrt{5} + 7 \cdot 2\sqrt{5} \\
&= 3\sqrt{5} + 14\sqrt{5} \quad \text{Like radicals} \\
&= (3 + 14)\sqrt{5} \quad \text{Distributive property} \\
&= 17\sqrt{5}
\end{aligned}$$

b. This time we find perfect-cube factors for both 80 and 270.

$$5\sqrt[3]{80x} - 3\sqrt[3]{270x} = 5\sqrt[3]{8 \cdot 10x} - 3\sqrt[3]{27 \cdot 10x}$$
$$= 5\sqrt[3]{8} \cdot \sqrt[3]{10x} - 3\sqrt[3]{27} \cdot \sqrt[3]{10x} \qquad \text{Product rule}$$
$$= 5 \cdot 2 \cdot \sqrt[3]{10x} - 3 \cdot 3\sqrt[3]{10x}$$
$$= 10\sqrt[3]{10x} - 9\sqrt[3]{10x} \qquad \text{Like radicals}$$
$$= (10 - 9)\sqrt[3]{10x} \qquad \text{Distributive property}$$
$$= \sqrt[3]{10x} \qquad\qquad ■■■$$

Practice Problem 7 Simplify.

a. $3\sqrt{12} + 7\sqrt{3}$ **b.** $2\sqrt[3]{135x} - 3\sqrt[3]{40x}$ ■

Radicals with Different Indexes

Suppose a is a positive real number and m, n, and r are positive integers. Then the property

$$\sqrt[n]{a^m} = \sqrt[nr]{a^{mr}}$$

is used to multiply radicals with different indexes.

EXAMPLE 8   **Multiplying Radicals with Different Indexes**

Write $\sqrt{2}\sqrt[3]{5}$ as a single radical.

SOLUTION
Because $\sqrt{2}$ can also be written as $\sqrt[2]{2}$, we have

$$\sqrt{2} = \sqrt[2]{2} = \sqrt[2\cdot3]{2^3} = \sqrt[6]{2^3},$$

and similarly,

$$\sqrt[3]{5} = \sqrt[2\cdot3]{5^2} = \sqrt[6]{5^2},$$

So, $$\sqrt{2}\sqrt[3]{5} = \sqrt[6]{2^3}\sqrt[6]{5^2} = \sqrt[6]{2^3 5^2} = \sqrt[6]{200}.$$ ■■■

Practice Problem 8 Write $\sqrt{3}\sqrt[5]{2}$ as a single radical.   ■

6 Define and evaluate rational exponents.

Rational Exponents

Previously, we defined such exponential expressions as 2^3, but what do you think $2^{1/3}$ means? If we want to preserve the power rule of exponents, $(a^n)^m = a^{nm}$, for rational exponents such as $\dfrac{1}{3}$, we must have

$$(2^{1/3})^3 = 2^{1/3 \cdot 3} = 2^1 = 2.$$

Because $(2^{1/3})^3 = 2$, it follows that $2^{1/3}$ is the cube root of 2; that is, $2^{1/3} = \sqrt[3]{2}$. Such considerations lead to the following definition.

THE EXPONENT $\dfrac{1}{n}$

For any real number a and any integer $n > 1$,

$$a^{1/n} = \sqrt[n]{a}.$$

When n is even and $a < 0$, $\sqrt[n]{a}$ and $a^{1/n}$ are not real numbers.

Notice that the denominator n of the rational exponent is the index of the radical.

EXAMPLE 9 **Evaluating Expressions with Rational Exponents**

Evaluate **a.** $16^{1/2}$ **b.** $(-27)^{1/3}$ **c.** $\left(\dfrac{1}{16}\right)^{1/4}$ **d.** $32^{1/5}$ **e.** $(-5)^{1/4}$

SOLUTION

a. $16^{1/2} = \sqrt{16} = 4$

b. $(-27)^{1/3} = \sqrt[3]{-27} = \sqrt[3]{(-3)^3} = -3$

c. $\left(\dfrac{1}{16}\right)^{1/4} = \sqrt[4]{\dfrac{1}{16}} = \sqrt[4]{\left(\dfrac{1}{16}\right)^4} = \dfrac{1}{2}$

d. $32^{1/5} = \sqrt[5]{32} = \sqrt[5]{2^5} = 2$

e. Because the base -5 is negative and $n = 4$ is even, $(-5)^{1/4}$ is not a real number.

■ ■ ■

Practice Problem 9 Evaluate.

a. $\left(\dfrac{1}{4}\right)^{1/2}$ **b.** $(125)^{1/3}$ **c.** $(-32)^{1/5}$

■

The box below shows a summary of our work so far.

RATIONAL EXPONENTS AND RADICALS

If a is any real number and n is any integer greater than 1, then

	n **even**	n **odd**
If $a > 0$,	$\sqrt[n]{a} = a^{1/n}$ is positive.	$\sqrt[n]{a} = a^{1/n}$ is positive.
If $a < 0$,	$\sqrt[n]{a} = a^{1/n}$ is not a real number.	$\sqrt[n]{a} = a^{1/n}$ is negative.
If $a = 0$,	$\sqrt[n]{0} = 0^{1/n} = 0.$	$\sqrt[n]{0} = 0^{1/n} = 0.$

We have defined the expression $a^{1/n}$. How do we define $a^{m/n}$, where m and n are integers, when $n > 1$ and $\sqrt[n]{a}$ is a real number? If we insist that the power-of-a-product rule of exponents holds for rational numbers, then we must have

$$a^{m/n} = a^{1/n \cdot m} = (a^{1/n})^m = (\sqrt[n]{a})^m$$

and

$$a^{m/n} = a^{m \cdot 1/n} = (a^m)^{1/n} = \sqrt[n]{a^m}.$$

From this, we arrive at the following definition.

RATIONAL EXPONENTS

$$a^{m/n} = (\sqrt[n]{a})^m = \sqrt[n]{a^m},$$

provided that m and n are integers with no common factors, $n > 1$, and $\sqrt[n]{a}$ is a real number.

Note that the numerator m of the exponent $\dfrac{m}{n}$ is the *exponent of the radical expression* and the denominator n is the *index of the radical*. When n is even and $a < 0$, the symbol $a^{m/n}$ is not a real number.

When rational exponents with negative denominators such as $\dfrac{3}{-5}$ and $\dfrac{-2}{-7}$ are encountered, we use equivalent expressions such as $\dfrac{-3}{5}$ and $\dfrac{2}{7}$, respectively, before applying the definition. Note that $(-8)^{2/6} \neq [(-8)^{1/6}]^2$ because $(-8)^{1/6} = \sqrt[6]{-8}$ is not a real number but $(-8)^{2/6}$ is a real number. Writing $\dfrac{2}{6} = \dfrac{1}{3}$ we have

$$(-8)^{2/6} = (-8)^{1/3} = \sqrt[3]{-8} = -2.$$

EXAMPLE 10 **Evaluating Expressions Having Rational Exponents**

Evaluate. **a.** $8^{2/3}$ **b.** $-16^{5/2}$ **c.** $100^{-3/2}$ **d.** $(-25)^{7/2}$

SOLUTION

a. $8^{2/3} = (\sqrt[3]{8})^2 = 2^2 = 4$

b. $-16^{5/2} = -(\sqrt{16})^5 = -4^5 = -1024$

c. $100^{-3/2} = (100^{1/2})^{-3} = (\sqrt{100})^{-3} = \dfrac{1}{(\sqrt{100})^3} = \dfrac{1}{10^3} = \dfrac{1}{1000}$

d. $(-25)^{7/2} = [(-25)^{1/2}]^7$ Not a real number

$\qquad\qquad = (\sqrt{-25})^7$ because $\sqrt{-25}$ is not a real number ■ ■ ■

Practice Problem 10 Evaluate

a. $(25)^{2/3}$ **b.** $-36^{3/2}$ **c.** $16^{-5/2}$ **d.** $(-36)^{-1/2}$ ■

All of the properties of exponents we learned for integer exponents hold true for rational exponents. For convenience, we list these properties next.

PROPERTIES OF RATIONAL EXPONENTS

If r and s are rational numbers and a and b are real numbers, then

$$a^r \cdot a^s = a^{r+s} \qquad (a^r)^s = a^{rs} \qquad (ab)^r = a^r b^r$$

$$\frac{a^r}{a^s} = a^{r-s} \qquad \left(\frac{a}{b}\right)^r = \frac{a^r}{b^r} \qquad a^{-r} = \frac{1}{a^r} \qquad \left(\frac{a}{b}\right)^{-r} = \left(\frac{b}{a}\right)^r$$

provided that all of the expressions used are defined.

EXAMPLE 11 **Simplifying Expressions Having Rational Exponents**

Express the answer with only positive exponents. Assume that x represents a positive real number. Simplify.

a. $2x^{1/3} \cdot 5x^{1/4}$ **b.** $\dfrac{21x^{-2/3}}{7x^{1/5}}$ **c.** $(x^{3/5})^{-1/6}$

SOLUTION

a. $2x^{1/3} \cdot 5x^{1/4} = 2 \cdot 5x^{1/3} \cdot x^{1/4}$ Group factors with the same base.

$\qquad\qquad\qquad = 10x^{1/3 + 1/4}$ Add exponents of factors with the same base.

$\qquad\qquad\qquad = 10x^{4/12 + 3/12} = 10x^{7/12}$

b. $\dfrac{21x^{-2/3}}{7x^{1/5}} = \left(\dfrac{21}{7}\right)\left(\dfrac{x^{-2/3}}{x^{1/5}}\right)$ Group factors with the same base.

$$= 3x^{-2/3-1/5} = 3x^{-10/15-3/15} = 3x^{-13/15} = 3 \cdot \left(\dfrac{1}{x^{13/15}}\right) = \dfrac{3}{x^{13/15}}$$

c. $(x^{3/5})^{-1/6} = x^{(3/5)(-1/6)} = x^{-1/10} = \dfrac{1}{x^{1/10}}$ ■ ■ ■

Practice Problem 11   Simplify. Express your answer with only positive exponents.

a. $4x^{1/2} \cdot 3x^{1/5}$ **b.** $\dfrac{25x^{-1/4}}{5x^{1/3}}, x > 0$ **c.** $(x^{2/3})^{-1/5}, x > 0$ ■

EXAMPLE 12   **Simplifying an Expression Involving Negative Rational Exponents**

Simplify $x(x + 1)^{-1/2} + 2(x + 1)^{1/2}$. Express the answer with only positive exponents and then rationalize the denominator. Assume that $(x + 1)^{-1/2}$ is defined.

SOLUTION

$$x(x + 1)^{-1/2} + 2(x + 1)^{1/2} = \dfrac{x}{(x + 1)^{1/2}} + \dfrac{2(x + 1)^{1/2}}{1} \qquad (x + 1)^{-1/2} = \dfrac{1}{(x + 1)^{1/2}}$$

$$= \dfrac{x}{(x + 1)^{1/2}} + \dfrac{2(x + 1)^{1/2}(x + 1)^{1/2}}{(x + 1)^{1/2}}$$

$$= \dfrac{x + 2(x + 1)}{(x + 1)^{1/2}} = \dfrac{3x + 2}{(x + 1)^{1/2}} \qquad \begin{array}{l}(x + 1)^{1/2}(x + 1)^{1/2} \\ = x + 1\end{array}$$

To rationalize the denominator $(x + 1)^{1/2} = \sqrt{x + 1}$, multiply both numerator and denominator by $(x + 1)^{1/2}$.

$$\dfrac{3x + 2}{(x + 1)^{1/2}} = \dfrac{(3x + 2)(x + 1)^{1/2}}{(x + 1)^{1/2}(x + 1)^{1/2}}$$

$$= \dfrac{(3x + 2)(x + 1)^{1/2}}{x + 1} \qquad (x + 1)^{1/2}(x + 1)^{1/2} = x + 1 \quad ■ ■ ■$$

Practice Problem 12   Simplify $x(x + 3)^{-1/2} + (x + 3)^{1/2}$.   ■

EXAMPLE 13   **Simplifying Radicals by Using Rational Exponents**

Assume that x represents a positive real number. Simplify.

a. $\sqrt[8]{x^2}$ **b.** $\sqrt[4]{9}\sqrt{3}$ **c.** $\sqrt{\sqrt[3]{x^8}}$

SOLUTION

a. $\sqrt[8]{x^2} = (x^2)^{1/8} = x^{2/8} = x^{1/4} = \sqrt[4]{x}$

b. $\sqrt[4]{9}\sqrt{3} = 9^{1/4} \cdot 3^{1/2} = (3^2)^{1/4} \cdot 3^{1/2} = 3^{2/4} \cdot 3^{1/2} = 3^{1/2} \cdot 3^{1/2} = 3^{1/2+1/2} = 3^1 = 3$

c. $\sqrt{\sqrt[3]{x^8}} = \sqrt{(x^8)^{1/3}} = \sqrt{x^{8/3}} = (x^{8/3})^{1/2} = x^{8/3 \cdot 1/2} = x^{4/3} = $
$x^{1+1/3} = x^1 \cdot x^{1/3} = x\sqrt[3]{x}$ ■ ■ ■

Practice Problem 13   Simplify. Assume that $x > 0$.

a. $\sqrt[6]{x^4}$ **b.** $\sqrt[4]{25}\sqrt{5}$ **c.** $\sqrt[3]{\sqrt{x^{12}}}$ ■

SECTION P.6 ■ Exercises

A EXERCISES Basic Skills and Concepts

1. Any positive number has _____ square roots.

2. To rationalize the denominator of an expression with denominator $2\sqrt{5}$, multiply the _____ and _____ by _____ .

3. Radicals that have the same index and the same radicand are called _____ .

4. The radical notation for $7^{1/3}$ is _____ .

5. *True or False* For all real x, $\sqrt{x^2} = x$.

6. *True or False* The radicals $-3\sqrt{5}$ and $12\sqrt{5}$ are "like radicals."

In Exercises 7–22, evaluate each root or state that the root is not a real number.

7. $\sqrt{64}$ 8. $\sqrt{100}$

9. $\sqrt[3]{64}$ 10. $\sqrt[3]{125}$

11. $\sqrt[3]{-27}$ 12. $\sqrt[3]{-216}$

13. $\sqrt[3]{-\dfrac{1}{8}}$ 14. $\sqrt[3]{\dfrac{27}{64}}$

15. $\sqrt{(-3)^2}$ 16. $-\sqrt{4(-3)^2}$

17. $\sqrt[4]{-16}$ 18. $\sqrt[6]{-64}$

19. $-\sqrt[5]{-1}$ 20. $\sqrt[7]{-1}$

21. $\sqrt[5]{(-7)^5}$ 22. $-\sqrt[5]{(-4)^5}$

In Exercises 23–46, simplify each expression using the product and quotient properties for square roots. Assume that x represents a positive real number.

23. $\sqrt{32}$ 24. $\sqrt{125}$

25. $\sqrt{18x^2}$ 26. $\sqrt{27x^2}$

27. $\sqrt{9x^3}$ 28. $\sqrt{8x^3}$

29. $\sqrt{6x}\sqrt{3x}$ 30. $\sqrt{5x}\sqrt{10x}$

31. $\sqrt{15x}\sqrt{3x^2}$ 32. $\sqrt{2x^2}\sqrt{18x}$

33. $\sqrt[3]{-8x^3}$ 34. $\sqrt[3]{64x^3}$

35. $\sqrt[3]{-x^6}$ 36. $\sqrt[3]{-27x^6}$

37. $\sqrt{\dfrac{5}{32}}$ 38. $\sqrt{\dfrac{3}{50}}$

39. $\sqrt[3]{\dfrac{8}{x^3}}$ 40. $\sqrt[3]{\dfrac{7}{8x^3}}$

41. $\sqrt[4]{\dfrac{2}{x^5}}$ 42. $-\sqrt[4]{\dfrac{5}{16x^5}}$

43. $\sqrt{\sqrt{x^8}}$ 44. $\sqrt{\dfrac{4x^4}{25y^6}}$

45. $\sqrt{\dfrac{9x^6}{y^4}}$ 46. $\sqrt[5]{x^{15}y^5}$

In Exercises 47–60, simplify each expression. Assume that all variables represent nonnegative real numbers.

47. $2\sqrt{3} + 5\sqrt{3}$

48. $7\sqrt{2} - \sqrt{2}$

49. $6\sqrt{5} - \sqrt{5} + 4\sqrt{5}$

50. $5\sqrt{7} + 3\sqrt{7} - 2\sqrt{7}$

51. $\sqrt{98x} - \sqrt{32x}$

52. $\sqrt{45x} + \sqrt{20x}$

53. $\sqrt[3]{24} - \sqrt[3]{81}$

54. $\sqrt[3]{54} + \sqrt[3]{16}$

55. $\sqrt[3]{3x} - 2\sqrt[3]{24x} + \sqrt[3]{375x}$

56. $\sqrt[3]{16x} + 3\sqrt[3]{54x} - \sqrt[3]{2x}$

57. $\sqrt{2x^5} - 5\sqrt{32x} + \sqrt{18x^3}$

58. $2\sqrt{50x^5} + 7\sqrt{2x^3} - 3\sqrt{72x}$

59. $\sqrt{48x^5y} - 4y\sqrt{3x^3y} + y\sqrt{3xy^3}$

60. $x\sqrt{8xy} + 4\sqrt{2xy^3} - \sqrt{18x^5y}$

In Exercises 61–74, rationalize the denominator of each expression.

61. $\dfrac{2}{\sqrt{3}}$ 62. $\dfrac{10}{\sqrt{5}}$

63. $\dfrac{7}{\sqrt{15}}$ 64. $\dfrac{-3}{\sqrt{8}}$

65. $\dfrac{1}{\sqrt{2} + x}$ 66. $\dfrac{1}{\sqrt{5} + 2x}$

67. $\dfrac{3}{2 - \sqrt{3}}$ 68. $\dfrac{-5}{1 - \sqrt{2}}$

69. $\dfrac{1}{\sqrt{3} + \sqrt{2}}$ 70. $\dfrac{6}{\sqrt{5} - \sqrt{3}}$

71. $\dfrac{\sqrt{5} - \sqrt{2}}{\sqrt{5} + \sqrt{2}}$ 72. $\dfrac{\sqrt{7} + \sqrt{3}}{\sqrt{7} - \sqrt{3}}$

73. $\dfrac{\sqrt{x + h} - \sqrt{x}}{\sqrt{x + h} + \sqrt{x}}$ 74. $\dfrac{a\sqrt{x} + 3}{a\sqrt{x} - 3}$

In Exercises 75–84, evaluate each expression without using a calculator.

75. $25^{1/2}$ 76. $144^{1/2}$

77. $(-8)^{1/3}$ 78. $(-27)^{1/3}$

79. $8^{2/3}$ 80. $16^{3/2}$

81. $-25^{-3/2}$ 82. $-9^{-3/2}$

83. $\left(\dfrac{9}{25}\right)^{-3/2}$ 84. $\left(\dfrac{1}{27}\right)^{-2/3}$

In Exercises 85–96, simplify each expression, leaving your answer with only positive exponents. Assume that all variables represent positive numbers.

85. $x^{1/2} \cdot x^{2/5}$

86. $x^{3/5} \cdot 5x^{2/3}$

87. $x^{3/5} \cdot x^{-1/2}$

88. $x^{5/3} \cdot x^{-3/4}$

89. $(8x^6)^{2/3}$

90. $(16x^3)^{1/2}$

91. $(27x^6y^3)^{-2/3}$

92. $(16x^4y^6)^{-3/2}$

93. $\dfrac{15x^{3/2}}{3x^{1/4}}$

94. $\dfrac{20x^{5/2}}{4x^{2/3}}$

95. $\left(\dfrac{x^{-1/4}}{y^{-2/3}}\right)^{-12}$

96. $\left(\dfrac{27x^{-5/2}}{y^{-3}}\right)^{-1/3}$

In Exercises 97–106, convert each radical expression to its rational exponent form and then simplify. Assume that all variables represent positive numbers.

97. $\sqrt[4]{3^2}$

98. $\sqrt[4]{5^2}$

99. $\sqrt[3]{x^9}$

100. $\sqrt[3]{x^{12}}$

101. $\sqrt[3]{x^6y^9}$

102. $\sqrt[3]{8x^3y^{12}}$

103. $\sqrt[4]{9}\sqrt{3}$

104. $\sqrt[4]{49}\sqrt{7}$

105. $\sqrt{\sqrt[3]{x^{10}}}$

106. $\sqrt{\sqrt[3]{64x^6}}$

In Exercises 107–114, convert the given product to a single radical. Assume that all variables represent positive numbers.

107. $\sqrt[3]{2} \cdot \sqrt[5]{3}$

108. $\sqrt[3]{3} \cdot \sqrt[4]{2}$

109. $\sqrt[3]{x^2} \cdot \sqrt[4]{x^3}$

110. $\sqrt[5]{x^2} \cdot \sqrt[3]{x^5}$

111. $\sqrt[4]{2m^2n} \cdot \sqrt[12]{5m^5n^2}$

112. $\sqrt{3xy} \cdot \sqrt[3]{9x^2y^2}$

113. $\sqrt[5]{x^4y^3} \cdot \sqrt[6]{x^3y^5}$

114. $\sqrt[5]{3a^4b^2} \cdot \sqrt[10]{7a^2b^6}$

B EXERCISES Applying the Concepts

115. Sign dimensions. A sign in the shape of an equilateral triangle of area A has sides of length $\sqrt{\dfrac{4A}{\sqrt{3}}}$ centimeters. Find the length of a side assuming that the sign has an area of 692 square centimeters. (Use the approximation $\sqrt{3} \approx 1.73$.)

116. Return on investment. For an initial investment of P dollars to mature to S dollars after two years when the interest is compounded annually, an annual interest rate of $r = \sqrt{\dfrac{S}{P}} - 1$ is required. Find r assuming that $P = \$1,210,000$ and $S = \$1,411,344$.

117. Terminal speed. The terminal speed V of a steel ball that weighs w grams and is falling in a cylinder of oil is given by the equation $V = \sqrt{\dfrac{w}{1.5}}$ centimeters per second. Find the terminal speed of a steel ball weighing 6 grams.

118. Model airplane speed. A spring balance scale attached to a wire that holds a model airplane on a circular path indicates the force F on the wire. The speed of the plane V, in m/sec, is given by $V = \sqrt{(rF)/m}$, where m is the mass of the plane (in kilograms) and r is the length of the wire (in meters). What is the top speed of a 2-kilogram plane on an 18-meter wire if the force on the wire at top speed is 49 newtons?

119. Electronic game current. The current I (in amperes) in a circuit for an electronic game using W watts and having a resistance of R ohms is given by $I = \sqrt{\dfrac{W}{R}}$. Find the current in a game circuit using 1058 watts and having a resistance of 2 ohms.

120. Radius of the moon. The radius of a sphere having volume V is given by $r = \left(\dfrac{3V}{4\pi}\right)^{1/3}$. Assuming that we treat the moon as a sphere and are told that its volume is 2.19×10^{19} cubic meters, find its radius. Use 3.14 as an approximation for π.

121. Atmospheric pressure. The atmospheric pressure P, measured in pounds per square inch, at an altitude of h miles ($h < 50$) is given by the equation $P = 14.7(0.5)^{h/3.25}$. Compute the atmospheric pressure at an altitude of 16.25 miles.

122. Force from constrained water. The force F that is exerted on the semicircular end of a trough full of water is approximately $F = 42r^{3/2}$ pounds, where r is the radius of the semicircular end. Find the force assuming that the radius is 3 feet.

Topics in Geometry

Before Starting this Section, Review

1. Exponents (Section P.2, page 19)

Objectives

1. Use the Pythagorean Theorem and its converse.

2. Use geometry formulas.

MATH IN THE MOVIES

Mistakes in mathematics are not unusual and even occur in movies. For example, in MGM's 1939 film version of *The Wizard of Oz*, the scarecrow went to see the wizard to ask him for some brains. Oz, the great and powerful wizard, bestowed upon the scarecrow a diploma with an honorary title of Th.D., "Doctor of Thinkology." The scarecrow, overjoyed at receiving his "brains," immediately decided to impress his friends by reciting the following mathematical statement: "The sum of the square roots of any two sides of an isosceles triangle is equal to the square root of the remaining side." Unfortunately, this statement is incorrect. Homer Simpson makes the same mistake in the introductory scene from the "Springfield" episode of *The Simpsons*. The correct statement that the scarecrow and Homer were attempting is probably the Pythagorean Theorem, a statement about right triangles, not isosceles triangles. (Moreover, it's the sum of the *squares*, not square roots, and the *square* of the remaining side, not its square root, that the theorem states are equal.) In Example 2, we use the Pythagorean Theorem to solve a practical problem. ■

Angles of a Triangle

Frequently, formulas we encounter in algebra come from geometry. We review some topics from geometry in this section, beginning with angles and triangles.

An **acute angle** is an angle of measure greater than $0°$ and less than $90°$.

An **obtuse angle** is an angle of measure greater than $90°$ and less than $180°$.

Any two angles whose measures sum to $180°$ are called **supplementary angles**.

Any two angles whose measures sum to $90°$ are called **complementary angles**.

Theorem. The sum of the measures of the angles of any triangle is $180°$.

Triangles

Triangles may be classified by the number of sides they have of equal length:

Scalene Triangle A triangle in which *no* two sides have equal length.

Isosceles Triangle A triangle with two sides of equal length.

Equilateral Triangle A triangle with all three sides having equal length.

The intuitive idea that two triangles have exactly the same shape (but not necessarily the same size) is captured in the concept of *similar triangles*:

Similar Triangles Two triangles are **similar** if and only if they have equal angles and their corresponding sides are proportional. See Figure P.14.

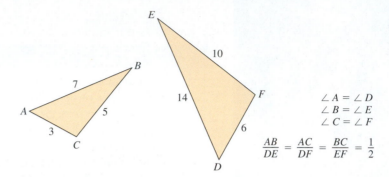

$$\angle A = \angle D$$
$$\angle B = \angle E$$
$$\angle C = \angle F$$

$$\frac{AB}{DE} = \frac{AC}{DF} = \frac{BC}{EF} = \frac{1}{2}$$

FIGURE P.14 **Similar triangles**

Congruent Triangles Two triangles are *congruent* if and only if they have equal corresponding angles and sides.

Included Side A side of a triangle that lies between two angles is called the **included side** of the angles.

Included Angle The angle formed by two sides is the **included angle** of the sides.

CONGRUENT-TRIANGLE THEOREMS

SAS (Side–Angle–Side). If two sides and the included angle of one triangle are equal to the corresponding sides and the included angle of a second triangle, then the two triangles are congruent.

ASA (Angle–Side–Angle). If two angles and the included side of one triangle are equal to the corresponding two angles and the included side of a second triangle, then the two triangles are congruent.

SSS (Side–Side–Side). If the three sides of one triangle are equal to the corresponding sides of a second triangle, then the two triangles are congruent.

---

1 Use the Pythagorean Theorem and its converse.

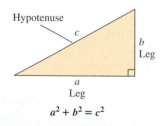

Hypotenuse
c
b
Leg
a
Leg

$$a^2 + b^2 = c^2$$

FIGURE P.15

Pythagorean Theorem

The *Pythagorean Theorem* states a useful fact about right triangles. A **right triangle** is any triangle that contains a right angle—that is, an angle of 90°. The side of the triangle opposite the 90° angle is called the **hypotenuse**; the other two sides are the **legs** of the triangle. Figure P.15 shows a right triangle with legs of length a and b and hypotenuse of length c. The symbol ⌐ is used to identify the right angle.

The Pythagorean Theorem says that the square of the length of the hypotenuse is equal to the sum of the squares of the lengths of the legs. See Figure P.15. We prove the theorem later in this section.

PYTHAGOREAN THEOREM

If a, b, and c represent the lengths of the two legs and the hypotenuse, respectively, of a right triangle, then

$$a^2 + b^2 = c^2.$$

EXAMPLE 1   **The Pythagorean Theorem**

a. A right triangle has legs of length 3 and 4. Find the length of the hypotenuse.

b. A right triangle has one leg of length 15 and a hypotenuse of length 39. Find the length of the other leg.

SOLUTION

a. Let c represent the length of the hypotenuse and a and b represent the lengths of the legs of the right triangle. Then

$$c^2 = a^2 + b^2 \qquad \text{Pythagorean Theorem}$$
$$c^2 = 3^2 + 4^2 \qquad \text{Replace } a \text{ with 3 and } b \text{ with 4.}$$
$$c^2 = 25 \qquad \text{Simplify.}$$
$$c = 5 \qquad c \text{ must be positive because it is a length.}$$

The length of the hypotenuse is 5.

b. Again, let c represent the length of the hypotenuse and a and b represent the legs of the right triangle. Then

$$c^2 = a^2 + b^2 \qquad \text{Pythagorean Theorem}$$
$$39^2 = 15^2 + b^2 \qquad \text{Replace } c \text{ with 39 and } a \text{ with 15.}$$
$$39^2 - 15^2 = b^2 \qquad \text{Subtract } 15^2 \text{ from both sides.}$$
$$1296 = b^2 \qquad \text{Simplify.}$$
$$36 = b \qquad b \text{ must be positive because it is a length.}$$

The length of the remaining leg is 36.   ■ ■ ■

Practice Problem 1 A right triangle has legs of lengths 8 and 6. Find the length of the hypotenuse.   ■

EXAMPLE 2   **The Pythagorean Theorem**

You want to change some prices advertised on a letter display fastened 15 feet high on the front wall of your business building. A garden along the front wall forces you to place a ladder 8 feet from the wall. What length ladder is required?

SOLUTION

If a ladder is placed 8 feet from the wall and rests 15 feet high on the wall, the ladder can be viewed as the hypotenuse of a right triangle with legs having lengths 8 feet and 15 feet. (See Figure P.16.)

Let c represent the length of the ladder. Then

$$c^2 = a^2 + b^2 \qquad \text{Pythagorean Theorem}$$
$$c^2 = 8^2 + 15^2 \qquad \text{Replace } a \text{ with 8 and } b \text{ with 15.}$$
$$c^2 = 289$$
$$c = 17$$

A 17-foot ladder will allow you to change the display.   ■ ■ ■

FIGURE P.16

Practice Problem 2 In Example 2, suppose the display is fastened 8 feet high and the ladder must be placed 6 feet from the wall. What length ladder is required?   ■

Converse of the Pythagorean Theorem

It is also true that if the square of the length of one side of a triangle equals the sum of the squares of the lengths of the other two sides, the triangle is a right triangle. The longest side of the triangle is the hypotenuse, and the angle opposite the longest side is the right angle.

CONVERSE OF THE PYTHAGOREAN THEOREM

Suppose a triangle has sides of length a, b, and c. If

$$c^2 = a^2 + b^2,$$

then the triangle is a right triangle with a right angle opposite the longest side.

EXAMPLE 3   **Converse of the Pythagorean Theorem**

A triangle has sides of length 20, 21, and 29. Is it a right triangle?

SOLUTION

To determine whether the triangle is a right triangle, we first square the length of each side.

$$20^2 = 400, \quad 21^2 = 441, \quad \text{and} \quad 29^2 = 841$$
$$841 = 400 + 441$$

Because the square of the length of the largest side equals the sum of the squares of the lengths of the other two sides, the triangle is a right triangle.   ■ ■ ■

Practice Problem 3   A triangle has sides of length 2, 5, and 6. Is it a right triangle? ■

2 Use geometry formulas.

Geometry Formulas

Although the area of a rectangle is just the product of its length and width, the general concept of area (and the related concepts of perimeter and volume) is developed with the use of calculus. Fortunately, the formulas for calculating these quantities require only algebra. Figure P.17 lists some useful formulas from geometry.

Formulas for area (A), circumference (C), perimeter (P), and volume (V)

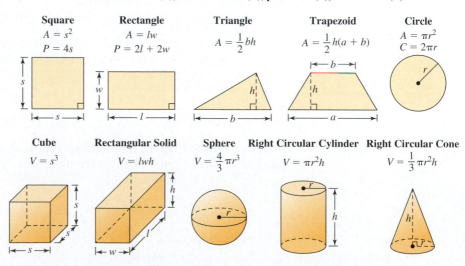

Square
$A = s^2$
$P = 4s$

Rectangle
$A = lw$
$P = 2l + 2w$

Triangle
$A = \frac{1}{2}bh$

Trapezoid
$A = \frac{1}{2}h(a + b)$

Circle
$A = \pi r^2$
$C = 2\pi r$

Cube
$V = s^3$

Rectangular Solid
$V = lwh$

Sphere
$V = \frac{4}{3}\pi r^3$

Right Circular Cylinder
$V = \pi r^2 h$

Right Circular Cone
$V = \frac{1}{3}\pi r^2 h$

FIGURE P.17 Area and volume formulas

A Proof of the Pythagorean Theorem

There are hundreds of proofs of the Pythagorean theorem. We offer one that is based on the formulas for the areas of a rectangle and a triangle. We write the area of the square of length $a + b$ in two ways. See Figures P.18 and P.19.

Area of the square in Figure P.18 = Area of the square in Figure P.19.

Areas of the two red squares	+	Areas of the two blue rectangles	=	Area of the red square	+	Areas of the four blue triangles
$a^2 + b^2$	+	$2(ab)$	=	c^2	+	$4\left(\dfrac{1}{2}ab\right)$
$a^2 + b^2$	+	$2ab$	=	c^2	+	$2ab$

$a^2 + b^2 = c^2$ Subtract $2ab$ from both sides.

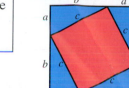

FIGURE P.18

FIGURE P.19

So the right triangle of the lower right corner of Figure P.19 with legs of lengths a and b and hypotenuse of length c satisfies the relation $a^2 + b^2 = c^2$. Since a and b can be the lengths of the legs of any right triangle, the relation $a^2 + b^2 = c^2$ holds for all right triangles.

SECTION P.7 ■ Exercises

A EXERCISES Basic Skills and Concepts

1. A triangle with two sides of equal length is a(n) _____ triangle.

2. A triangle with all three sides of equal length is a(n) _____ triangle.

3. Two triangles are _____ if and only if they have equal angles and their corresponding sides are proportional.

4. Two triangles are _____ if and only if they have equal corresponding angles and sides.

5. *True or False* Congruent triangles are also similar triangles.

6. *True or False* If $a, b,$ and c represent the lengths of the three legs of a triangle and $a^2 + b^2 = c^2$, the triangle is a right triangle.

In Exercises 7–12, the lengths of the legs of a right triangle are given. Find the length of the hypotenuse.

7. $a = 7, b = 24$ 8. $a = 5, b = 12$

9. $a = 9, b = 12$ 10. $a = 10, b = 24$

11. $a = 3, b = 3$ 12. $a = 2, b = 4$

In Exercises 13–16, the lengths of one leg, a, and the hypotenuse, c, of a right triangle are given. Find the length of the remaining leg.

13. $a = 15, c = 17$ 14. $a = 35, c = 37$

15. $a = 14, c = 50$ 16. $a = 10, c = 26$

In Exercises 17–20, find the area A and the perimeter P of a rectangle with length l and width w.

17. $l = 3, w = 5$ 18. $l = 4, w = 3$

19. $l = 10, w = \dfrac{1}{2}$ 20. $l = 16, w = \dfrac{1}{4}$

In Exercises 21–24, find the area A of a triangle with base b and altitude h.

21. $b = 3, h = 4$ 22. $b = 6, h = 1$

23. $b = \dfrac{1}{2}, h = 12$ 24. $b = \dfrac{1}{4}, h = \dfrac{2}{3}$

In Exercises 25–28, find the area A and the circumference C of a circle with radius r. (Use $\pi \approx 3.14$.)

25. $r = 1$ 26. $r = 3$

27. $r - \dfrac{1}{2}$ 28. $r = \dfrac{3}{4}$

In Exercises 29–32, find the volume V of a box with length l, width w, and height h.

29. $l = 1, w = 1, h = 5$

30. $l = 2, w = 7, h = 4$

31. $l = \dfrac{1}{2}, w = 10, h = 3$

32. $l = \dfrac{1}{3}, w = \dfrac{3}{7}, h = 14$

B EXERCISES Applying the Concepts

33. Garden area. Find the area of a rectangular garden assuming that the width is 12 feet and the diagonal measures 20 feet.

34. Perimeter of a mural. A mural is 6 feet wide, and its diagonal measures 10 feet. Find the perimeter of the mural.

35. Dimensions of a computer monitor. The monitor on a portable computer has a rectangular shape with a 15-inch diagonal. If the width of the monitor is 12 inches, what is its height?

12 inches

15 inches

36. Dimensions of a baseball diamond. A baseball diamond is actually a square with 90-foot sides. How far does the catcher have to throw the ball to reach second base from home plate?

37. Repair problem. A repairman has to work on an air conditioner's overflow vent located 25 feet above the ground on the side of a house. Shrubs along the house require that a ladder be placed 10 feet from the house. How long must the ladder be for the repairman to reach the overflow vent?

38. Building a ramp. You must build a ramp that allows you to roll a cart from a storeroom into the back of a truck. The rear of the truck must remain 8 feet from the storeroom because of a curb. If the back of the truck is 4 feet above the storeroom floor, how long (in feet) must the ramp be?

39. Football field. The playing surface of a football field (including the end zones) is 120 yards long and 53 yards wide. How far will a player jogging around the perimeter travel?

40. Reflecting pool. The reflecting pool of the Lincoln Memorial has an approximate perimeter of 4400 feet. The length is approximately 1860 feet more than the width. Find the area of the reflecting pool.

41. Crop circle. A crop circle with diameter of 787 feet occurred at Milk Hill in Wiltshire, UK, in August 2001. If you walk around the entire edge of the circle, how far will you have walked? (Use $\pi \approx 3.14$.)

42. Bicycle tires. The diameter of a bicycle tire is 26 inches. How far will the bicycle travel when the wheel makes one complete turn? (Use $\pi \approx 3.14$.)

43. Fishpond border. A rectangular border a foot and a half wide is built around a rectangular fishpond. If the pond is 6 feet long and 4 feet wide, what is the area of the border?

44. Garden fence. Casper wants to put a decorative fence around his rectangular garden. The garden is 5 feet wide, and its diagonal measures 13 feet. How many feet of fencing is required?

45. Oven dimensions. An oven in the shape of a rectangular box is 2 feet wide, 2.5 feet high, and 3 feet deep. What is the volume of the oven?

46. Speedway. A speedway track is constructed by adding semicircular ends to a rectangle, as shown in the figure. Find the area enclosed by the track (use $\pi \approx 3.14$).

740 ft

185 ft

SUMMARY ▪ Definitions, Concepts, and Formulas

P.1 The Real Numbers and Their Properties

i. Classifying numbers.

Natural numbers	$1, 2, 3, \ldots$
Whole numbers	$0, 1, 2, 3, \ldots$
Integers	$\ldots -2, -1, 0, 1, 2, \ldots$
Rational numbers	Numbers that can be expressed as the quotient $\dfrac{a}{b}$ of two integers, with $b \neq 0$; a terminating or a repeating decimal
Irrational numbers	Decimals that neither terminate nor repeat
Real numbers	All of the rational and irrational numbers

ii. Coordinate line. Real numbers can be associated with a geometric line called a number line or coordinate line. Numbers associated with points to the right (left) of zero are called positive (negative) numbers.

iii. Ordering numbers. $a < b$ means that $b = a + c$ for some positive number c.
Trichotomy property. For two numbers a and b, exactly one of the following is true: $a < b, a = b, b < a$.
Transitive property. If $a < b$ and $b < c$, then $a < c$.

iv. Absolute value.

$$|a| = a \text{ if } a \geq 0, \text{ and } |a| = -a \text{ if } a < 0.$$

The **distance** between points a and b on a number line is $|b - a|$.

v. Algebraic properties.

Commutative	$a + b = b + a,$	$ab = ba$
Associative	$(a + b) + c = a + (b + c)$	
	$(ab)c = a(bc)$	
Distributive	$a \cdot (b + c) = a \cdot b + a \cdot c$	
	$(a + b) \cdot c = a \cdot c + b \cdot c$	
Identity	$a + 0 = 0 + a = a,$	$a \cdot 1 = 1 \cdot a = a$
Inverse properties:	$a + (-a) = 0,$	$a \cdot \dfrac{1}{a} = 1,$
	where $a \neq 0$	

Zero-product properties: $0 \cdot a = 0 = a \cdot 0$.
If $a \cdot b = 0$, then $a = 0$ or $b = 0$.

P.2 Integer Exponents and Scientific Notation

i. If n is a positive integer, then

$$a^n = \underbrace{a \cdot a \cdots a}_{n \text{ factors}}$$

$$\left. \begin{array}{l} a^0 = 1 \\ a^{-n} = \dfrac{1}{a^n} \end{array} \right\} \quad a \neq 0$$

ii. Rules of exponents. Denominators and exponents attached to a base of zero are assumed not to be zero.

Product Rule	$a^m \cdot a^n = a^{m+n}$
Quotient Rule	$\dfrac{a^m}{a^n} = a^{m-n}$
Power Rule	$(a^m)^n = a^{mn}$
Power of a product	$(a \cdot b)^n = a^n \cdot b^n$
Power of a quotient	$\left(\dfrac{a}{b}\right)^n = \dfrac{a^n}{b^n}$
	$\left(\dfrac{a}{b}\right)^{-n} = \left(\dfrac{b}{a}\right)^n = \dfrac{b^n}{a^n}$

P.3 Polynomials

An algebraic expression of the form

$$a_n x^n + a_{n-1} x^{n-1} + \cdots + a_1 x + a_0,$$

where n is a nonnegative integer, is a polynomial in x. If $a_n \neq 0$, then n is called the degree of the polynomial. The term $a_n x^n$ is the **leading term** and a_0 is the **constant term**.

P.4 Factoring Polynomials

Factoring formulas:

$$A^2 - B^2 = (A - B)(A + B)$$
$$A^2 + 2AB + B^2 = (A + B)^2$$
$$A^2 - 2AB + B^2 = (A - B)^2$$
$$A^3 - B^3 = (A - B)(A^2 + AB + B^2)$$
$$A^3 + B^3 = (A + B)(A^2 - AB + B^2)$$

P.5 Rational Expressions

The quotient of two polynomials is called a rational expression.

i. Multiplication and division.

$$\frac{A}{B} \cdot \frac{C}{D} = \frac{A \cdot C}{B \cdot D}$$

$$\frac{\dfrac{A}{B}}{\dfrac{C}{D}} = \frac{A}{B} \div \frac{C}{D} = \frac{A}{B} \cdot \frac{D}{C} = \frac{A \cdot D}{B \cdot C}$$

All of the denominators are assumed not to be zero.

ii. Addition and subtraction.

$$\frac{A}{B} \pm \frac{C}{D} = \frac{A \cdot D \pm B \cdot C}{B \cdot D}$$

The denominators are assumed not to be zero.

P.6 Rational Exponents and Radicals

i. If $a > 0$, then $\sqrt[n]{a} = b$ means that $a = b^n$ and $b > 0$.

ii. If $a < 0$ and n is odd, then $\sqrt[n]{a} = b$ means that $a = b^n$.
   If $a < 0$ and n is even, then $\sqrt[n]{a}$ is not defined.

iii. When all expressions are defined,

$$a^{m/n} = \sqrt[n]{a^m} = \left(\sqrt[n]{a}\right)^m ; a^{-m/n} = \frac{1}{a^{m/n}}.$$

iv. $\sqrt{a^2} = |a|$

P.7 Topics in Geometry

i. An **acute angle** is an angle of measure greater than 0° and less than 90°.

ii. An **obtuse angle** is an angle of measure greater than 90° and less than 180°.

iii. Any two angles whose measures sum to 180° are called **supplementary angles.**

iv. Any two angles whose measures sum to 90° are called **complementary angles.**

v. Theorem. The sum of the measures of the angles of any triangle is 180°.

vi. Scalene triangle. A triangle in which *no* two sides have equal length.

vii. Isosceles triangle. A triangle with two sides of equal length.

viii. Equilateral triangle. A triangle with all three sides having equal length.

ix. Similar triangles. Two triangles are **similar** if and only if they have equal angles and their corresponding sides are proportional.

x. Congruent triangles. Two triangles are **congruent** if and only if they have equal corresponding angles and sides.

xi. Included side. A side of a triangle that lies between two angles is called the **included side** of the angles.

xii. Included angle. The angle formed by two sides is the **included angle** of the sides.

xiii. SAS (Side–Angle–Side). If two sides and the included angle of one triangle are equal to the corresponding sides and the included angle of a second triangle, then the two triangles are congruent.

xiv. ASA (Angle–Side–Angle). If two angles and the included side of one triangle are equal to the corresponding two angles and the included side of a second triangle, then the two triangles are congruent.

xv. SSS (Side–Side–Side). If the three sides of one triangle are equal to the corresponding sides of a second triangle, then the two triangles are congruent.

xvi. A right triangle is any triangle that contains a right angle—that is, an angle of 90°. The side of the triangle opposite the 90° angle is called the **hypotenuse**; the other two sides are the **legs** of the triangle.

xvii. Pythagorean Theorem. If a, b, and c represent the lengths of the two legs and the hypotenuse, respectively, of a right triangle, then $a^2 + b^2 = c^2$.

xviii. Converse of the Pythagorean Theorem. If a, b, and c represent the lengths of the legs of a triangle, and $c^2 = a^2 + b^2$, then the triangle is a right triangle, with a right angle opposite the longest leg.

REVIEW EXERCISES

Basis Skills and Concepts

1. List all of the numbers in the set
$$\left\{4, \sqrt{7}, -5, \frac{1}{2}, 0.\overline{31}, 0, 0.2, \sqrt{12}\right\} \text{ that are}$$

a. natural numbers.

b. whole numbers.

c. integers.

d. rational numbers.

e. irrational numbers.

f. real numbers.

In Exercises 2–5, state which property of the real numbers is being used.

2. $3x + 1 = 1 + 3x$

3. $\pi(x + y) = \pi x + \pi y$

4. $(x^2 + 1) \cdot 0 = 0 \cdot (x^2 + 1)$

5. $(x^2 + 1) \cdot 1 = x^2 + 1$

In Exercises 6 and 7, write each interval in inequality notation and then graph each interval.

6. $[-2, 3)$ **7.** $(-\infty, 1]$

In Exercises 8 and 9, write each inequality in interval notation and then graph the interval.

8. $1 \le x < 4$ **9.** $x > 0$

In Exercises 10–13, rewrite each expression without absolute value bars.

10. $\dfrac{-12}{|2|}$ **11.** $|2 - |-3||$

12. $|-5 - |-7||$ **13.** $|1 - \sqrt{15}|$

In Exercises 14–37, evaluate each expression.

14. -5^0 **15.** $(-4)^3$

16. $\dfrac{2^{18}}{2^{17}}$ **17.** $2^4 - 5 \cdot 3^2$

18. $(2^3)^2$ **19.** $-25^{1/2}$

20. $\left(\dfrac{1}{16}\right)^{1/4}$ **21.** $8^{4/3}$

22. $(-27)^{2/3}$ **23.** $\left(\dfrac{25}{36}\right)^{-3/2}$

24. $81^{-3/2}$ **25.** $2 \cdot 5^3$

26. $3^2 \cdot 2^5$ **27.** $\dfrac{21 \times 10^6}{3 \times 10^7}$

28. $\sqrt[4]{10,000}$ **29.** $\sqrt{5^2}$

30. $\sqrt{(-9)^2}$ **31.** $\sqrt[3]{-125}$

32. $3^{1/2} \cdot 27^{1/2}$ **33.** $64^{-1/3}$

34. $\sqrt{5}\sqrt{20}$ **35.** $(\sqrt{3} + 2)(\sqrt{3} - 2)$

36. $\left(\dfrac{1}{16}\right)^{-3/2}$ **37.** $\dfrac{3^{-2} \cdot 7^0}{18^{-1}}$

In Exercises 38–41, evaluate each expression for the given values of x and y.

38. $4(x + 7) + 2y; x = 3, y = 4$

39. $\dfrac{y - 6\sqrt{x}}{xy}; x = 4, y = 2$

40. $\dfrac{|x|}{x} + \dfrac{|y|}{y}; x = 6, y = -3$

41. $x^y; x = 16, y = -\dfrac{1}{2}$

In Exercises 42–49, simplify each expression. Write each answer with only positive exponents. Assume that all variables represent positive numbers.

42. $\dfrac{x^{-6}}{x^{-9}}$ **43.** $(x^{-2})^{-5}$

44. $\dfrac{x^{-3}}{y^{-2}}$ **45.** $\left(\dfrac{x^{-3}}{y^{-1}}\right)^{-2}$

46. $\left(\dfrac{xy^3}{x^5y}\right)^{-2}$ **47.** $(16x^{-2/3}y^{-4/3})^{3/2}$

48. $(49x^{-4/3}y^{2/3})^{-3/2}$ **49.** $\left(\dfrac{64y^{-9/2}}{x^{-3}}\right)^{-2/3}$

In Exercises 50–66, simplify each expression. Assume that all variables represent positive numbers.

50. $(2x^{2/3})(5x^{3/4})$ **51.** $(7x^{1/4})(3x^{3/2})$

52. $\dfrac{32x^{2/3}}{8x^{1/4}}$ **53.** $\dfrac{x^5(2x)^3}{4x^3}$

54. $\dfrac{(64x^4y^4)^{1/2}}{4y^2}$ **55.** $\left(\dfrac{x^2y^{4/3}}{x^{1/3}}\right)^6$

56. $7\sqrt{3} + 3\sqrt{75}$ **57.** $\dfrac{\sqrt{64}}{\sqrt{11}}$

58. $\sqrt{180x^2}$ **59.** $7\sqrt{6} - 3\sqrt{24}$

60. $7\sqrt[3]{54} + \sqrt[3]{128}$ **61.** $\sqrt{2x}\sqrt{6x}$

62. $\sqrt{75x^2}$ **63.** $\dfrac{\sqrt{100x^3}}{\sqrt{4x}}$

64. $5\sqrt{2x} - 2\sqrt{8x}$ **65.** $4\sqrt[3]{135} + \sqrt[3]{40}$

66. $y\sqrt[3]{56x} - \sqrt[3]{189xy^3}$

In Exercises 67–70, rationalize the denominator of each expression.

67. $\dfrac{7}{\sqrt{3}}$ **68.** $\dfrac{4}{9 - \sqrt{6}}$

69. $\dfrac{1 - \sqrt{3}}{1 + \sqrt{3}}$ **70.** $\dfrac{2\sqrt{5} + 3}{3 - 4\sqrt{5}}$

71. Write $3.7 \times (6.23 \times 10^{12})$ in scientific notation.

72. Write $\dfrac{3.19 \times 10^{-9}}{0.02 \times 10^{-3}}$ in decimal notation.

In Exercises 73–82, perform the indicated operations. Leave the resulting polynomial in standard form.

73. $(x^3 - 6x^2 + 4x - 2) + (3x^3 - 6x^2 + 5x - 4)$

74. $(10x^3 + 8x^2 - 7x - 3) - (5x^3 - x^2 + 4x - 9)$

75. $(4x^4 + 3x^3 - 5x^2 + 9) + (5x^4 + 8x^3 - 7x^2 + 5)$

76. $(8x^4 + 4x^3 + 3x^2 + 5) + (7x^4 + 3x^3 + 8x^2 - 3)$

77. $(x - 12)(x - 3)$

78. $(x - 7)^2$

79. $(x^5 - 2)(x^5 + 2)$

80. $(4x - 3)(4x + 3)$

81. $(2x + 5)(3x - 11)$

82. $(3x - 6)(x^2 + 2x + 4)$

In Exercises 83–104, factor each polynomial completely.

83. $x^2 - 3x - 10$ **84.** $x^2 + 10x + 9$

85. $24x^2 - 38x - 11$ **86.** $15x^2 + 33x + 18$

87. $x(x + 11) + 5(x + 11)$ **88.** $x^3 - x^2 + 2x - 2$

89. $x^4 - x^3 + 7x - 7$ **90.** $9x^2 + 24x + 16$

91. $10x^2 + 23x + 12$ **92.** $8x^2 + 18x + 9$

93. $12x^2 + 7x - 12$ **94.** $x^4 - 4x^2$

95. $4x^2 - 49$ **96.** $16x^2 - 81$

97. $x^2 + 12x + 36$ **98.** $x^2 - 10x + 100$

99. $64x^2 + 48x + 9$ **100.** $8x^3 - 1$

101. $8x^3 + 27$ **102.** $7x^3 - 7$

103. $x^3 + 5x^2 - 16x - 80$ **104.** $x^3 + 6x^2 - 9x - 54$

In Exercises 105–118, perform the indicated operation. Simplify your answer.

105. $\dfrac{4}{x-9} - \dfrac{10}{9-x}$

106. $\dfrac{2}{x^2 - 3x + 2} + \dfrac{6}{x^2 - 1}$

107. $\dfrac{3x+5}{x^2 + 14x + 48} - \dfrac{3x-2}{x^2 + 10x + 16}$

108. $\dfrac{x+7}{x^2 - 7x + 6} - \dfrac{5-3x}{x^2 - 2x - 24}$

109. $\dfrac{x-1}{x+1} - \dfrac{x+1}{x-1}$

110. $\dfrac{x^3 - 1}{3x^2 - 3} \cdot \dfrac{6x+6}{x^2 + x + 1}$

111. $\dfrac{x-1}{2x-3} \cdot \dfrac{4x^2 - 9}{2x^2 - x - 1}$

112. $\dfrac{x^2 + 2x - 8}{x^2 + 5x + 6} \cdot \dfrac{x+2}{x+4}$

113. $\dfrac{x^2 - 4}{4x^2 - 9} \cdot \dfrac{2x^2 - 3x}{2x + 4}$

114. $\dfrac{3x^2 - 17x + 10}{x^2 - 4x - 5} \cdot \dfrac{x^2 + 3x + 2}{x^2 + x - 2}$

115. $\dfrac{\dfrac{1}{x^2} - x}{\dfrac{1}{x^2} + x}$

116. $\dfrac{\dfrac{x}{x-2} + x}{\dfrac{1}{x^2 - 4}}$

117. $\dfrac{\dfrac{x}{x-3} + x}{\dfrac{x}{3-x} - x}$

118. $\dfrac{x + 2 - \dfrac{18}{x-5}}{x - 1 - \dfrac{12}{x-5}}$

Applying the Concepts

119. Assuming that a right triangle has legs a and b of lengths 20 and 21, respectively, find the length of the hypotenuse.

120. One leg of a right triangle measures 5 inches. The hypotenuse is 1 inch longer than the other leg. Find the length of the hypotenuse.

121. Find the area A and the perimeter P of a right triangle with hypotenuse of length 20 and a leg of length 12.

122. Find the volume V of a box of length $\dfrac{1}{3}$, width 7, and height 60.

123. Find the area of a rectangular rug if its width is 6 feet and its diagonal measures 10 feet.

124. The amount of alcohol in the body of a person who drinks x grams of alcohol every hour over a relatively long period is $\dfrac{4.2x}{10 - x}$ grams. How many grams of alcohol will be in the body of a person who drinks 2 grams of alcohol every hour (over a relatively long period)?

125. Because of the earth's curvature, the maximum distance a person can see from a height h above the ground is $\sqrt{7920h + h^2}$, where h is given in miles. What is the maximum distance an airplane pilot can see when the plane is 0.7 mile above the ground?

126. A contaminated reservoir contains 2% arsenic. The percentage of arsenic in the reservoir can be reduced by adding water. Assuming that the reservoir contains a million gallons of water, write a rational expression whose values give the percentage of arsenic in the reservoir when x gallons of water are added to it.

127. The height (in feet) of an open fruit crate that is three times as long as it is wide is given by $\dfrac{36 - 3x^2}{8x}$, where x is the width of the box. Find the height of a fruit crate that is 2 feet wide.

128. An object thrown down with an initial velocity of v_0 feet per second will travel $16t^2 + v_0 t$ feet in t seconds. How far has an orange that is dropped from a hot air balloon traveled after 9 seconds if it has an initial downward velocity of 25 feet per second?

PRACTICE TEST

Rewrite each expression without absolute value bars.

1. $|7 - |-3||$ **2.** $|\sqrt{2} - 100|$

3. Evaluate $\dfrac{x - 3y}{2} + xy$ for $x = 5$ and $y = -3$.

4. Determine the domain of the variable x in the expression $\dfrac{2x}{(x-3)(x+9)}$ and write it in interval notation.

Simplify each expression. Assume that all expressions containing variables are defined.

5. $\left(\dfrac{-3x^2 y}{x}\right)^3$ **6.** $\sqrt[3]{-8x^6}$

7. $\sqrt{75x} - \sqrt{27x}$

8. $-16^{-3/2}$

9. $\left(\dfrac{x^{-2}y^4}{25x^3y^3}\right)^{-1/2}$

10. Rationalize the denominator of the expression $\dfrac{5}{1 - \sqrt{3}}$.

Perform the indicated operations. Write the resulting polynomial in standard form.

11. $4(x^2 - 3x + 2) + 3(5x^2 - 2x + 1)$

12. $(x - 2)(5x - 1)$

13. $(x^2 + 3y^2)^2$

14. A circular rug with radius 5 feet has a border of uniform width. The distance from the center of the rug to the outer edge of the border is x feet. Write a completely factored polynomial that gives the area of the border.

Factor each polynomial completely.

15. $x^2 - 5x + 6$

16. $9x^2 + 12x + 4$

17. $8x^3 - 27$

Perform the indicated operations and simplify the result. Leave your answer in factored form.

18. $\dfrac{6 - 3x}{x^2 - 4} - \dfrac{12}{2x + 4}$

19. $\dfrac{\dfrac{1}{x^2}}{9 - \dfrac{1}{x^2}}$

20. Find the length of the hypotenuse of a right triangle with legs a and b of lengths 35 and 12, respectively.

Equations and Inequalities

Equations and inequalities are useful in analyzing designs found in the art and architecture of ancient civilizations, in understanding the dazzling geometric constructions that abound in nature, and in solving problems common to industry and daily life.

Linear Equations in One Variable

Before Starting this Section, Review

1. Algebraic expressions (Section P.1, page 14)
2. Like terms (Section P.3, page 30)
3. Least common denominator (Section P.5, page 51)

Objectives

1. Learn the vocabulary and concepts used in studying equations.
2. Solve linear equations in one variable.
3. Solve rational equations with variables in the denominators.
4. Solve formulas for a specific variable.
5. Solve applied problems by using linear equations.

BUNGEE-JUMPING TV CONTESTANTS

England's Oxford Dangerous Sports Club started the modern version of the sport of "bungee" (or bungy) jumping on April 1, 1978, from the Clifton Suspension Bridge in Bristol, England. Bungee cords consist of hundreds of continuous-length rubber strands encased in a nylon sheath. One end of the cord is tied to a solid structure such as a crane or bridge, and the other end is tied via padded ankle straps to the jumper's ankle. The jumper jumps, and if things go as planned, the cord extends like a rubber band and the jumper bounces up and down until coming to a halt. Suppose a television adventure program has its contestants bungee jump from a bridge 120 feet above the water. The bungee cord that is used has a 140%–150% elongation, meaning that its extended length will be the original length plus a maximum of an additional 150% of its original length. If the show's producer wants to make sure the jumper does not get closer than 10 feet to the water and if no contestant is more than 7 feet tall, how long can the bungee cord be? Using the information from this section, we learn in Example 10 that the cord should be about 41 feet long. ■

---

1 Learn the vocabulary and concepts used in studying equations.

Definitions

An **equation** is a statement that two mathematical expressions are equal. For example,

$$7 - 5 = 2$$

is an equation. In algebra, however, we are generally more interested in equations that contain variables.

An **equation in one variable** is a statement that two expressions, with at least one containing the variable, are equal. For example, $2x - 3 = 7$ is an equation in the variable x. The expressions $2x - 3$ and 7 are the *sides* of the equation. The **domain** of the variable in an equation is the set of all real numbers for which both sides of the equation are defined.

EXAMPLE 1 **Finding the Domain of the Variable**

Find the domain of the variable x in each of the following equations.

a. $2 = \dfrac{5}{x - 1}$ **b.** $x - 2 = \sqrt{x}$ **c.** $2x - 3 = 7$

STUDY TIP

When finding the domain of a variable, remember that

1. Division by 0 is undefined.
2. The square root (or any even root) of a negative number is not a real number.

SOLUTION

a. In the equation $2 = \dfrac{5}{x-1}$, the left side 2 does not contain x; so it is defined for

all values of x. Because division by 0 is undefined, $\dfrac{5}{x-1}$ is not defined if $x = 1$.

The domain of x is the set of all real numbers except the number 1. Frequently, when only a few numbers are excluded from the domain of the variable, we write the exceptions on the side of the equation. In this example, if we wanted

to specify the domain of the variable, we would simply write $2 = \dfrac{5}{x-1}, x \neq 1$.

b. The square root of a negative number is not a real number; so $\sqrt{x}$, the right side of the equation $x - 2 = \sqrt{x}$, is defined only when $x \geq 0$. The left side is defined for all real numbers. So, the domain of x in this equation is $\{x | x \geq 0\}$, or in interval notation, $[0, \infty)$.

c. Because both sides of the equation $2x - 3 = 7$ are defined for all real numbers, the domain is $\{x | x \text{ is a real number}\}$, or in interval notation, $(-\infty, \infty)$. ■ ■ ■

Practice Problem 1   Find the domain of the variable x in each of the following equations.

a. $\dfrac{x}{3} - 7 = 5$ **b.** $\dfrac{3}{2-x} = 4$ **c.** $\sqrt{x-1} = 0$ ■

When the variable in an equation is replaced with a specific value from its domain, the resulting statement may be true or false. For example, in the equation $2x - 3 = 7$, if we let $x = 1$, we obtain

$$2(1) - 3 = 7 \text{ or } -1 = 7,$$

which is obviously false. However, if we replace x with 5, we obtain $2(5) - 3 = 7$, a true statement. Those values (if any) of the variable that result in a true statement are called **solutions** or **roots** of the equation. Thus, 5 is a solution (or root) of the equation $2x - 3 = 7$. We also say that 5 **satisfies** the equation $2x - 3 = 7$. To **solve** an equation means to find all solutions of the equation; the set of all solutions of an equation is called its **solution set**.

An equation that is satisfied by every real number in the domain of the variable is called an **identity**. The equations

$$2(x + 3) = 2x + 6$$
$$\text{and} \quad x^2 - 9 = (x + 3)(x - 3)$$

are examples of identities. You should recognize that these equations are examples of the distributive property and the difference of squares, respectively.

An equation that is not an identity but is satisfied by at least one real number is called a **conditional equation**. To verify that an equation is a conditional equation, you must find at least one number that *is* a solution of the equation and at least one number that is *not* a solution of the equation. The equation $2x - 3 = 7$ is an example of a conditional equation: 5 satisfies the equation, but 0 does not.

An equation that *no* number satisfies is called an **inconsistent equation**. The solution set of an inconsistent equation is designated by the empty set symbol, $\varnothing$. The equation $x = x + 5$ is an example of an inconsistent equation. It is inconsistent because no number is 5 more than itself.

Equivalent Equations

Equations that have the same solution set are called **equivalent equations**. For example, equations $x = 4$ and $3x = 12$ are equivalent equations with solution set $\{4\}$.

In general, to solve an equation in one variable, we replace the given equation with a sequence of equivalent equations until we obtain an equation whose solution is obvious, such as $x = 4$. The following operations yield equivalent equations.

Generating Equivalent Equations

Operations	Given Equation	Equivalent Equation
Simplify expressions on either side by eliminating parentheses, combining like terms, etc.	$(2x - 1) - (x + 1) = 4$	$2x - 1 - x - 1 = 4$ or $x - 2 = 4$
Add (or subtract) the same expression on *both* sides of the equation.	$x - 2 = 4$	$x - 2 + 2 = 4 + 2$ or $x = 6$
Multiply (or divide) *both* sides of the equation by the same *nonzero* expression.	$2x = 8$	$\dfrac{1}{2} \cdot 2x = \dfrac{1}{2} \cdot 8$ or $x = 4$
Interchange the two sides of the equation.	$-3 = x$	$x = -3$

2 Solve linear equations in one variable.

Solving Linear Equations in One Variable

LINEAR EQUATIONS

A conditional linear equation in one variable, such as x, is an equation that can be written in the *standard form*

$$ax + b = 0,$$

where a and b are real numbers with $a \neq 0$.

Because the highest exponent on the variable in a linear equation is 1 $(x = x^1)$, a linear equation is also called a *first-degree equation*. We can see that the linear equation $ax + b = 0$ with $a \neq 0$ has exactly one solution by using the following steps:

$$ax + b = 0 \qquad \text{Original equation, } a \neq 0$$
$$ax = -b \qquad \text{Subtract } b \text{ from both sides.}$$
$$x = -\frac{b}{a} \qquad \text{Divide both sides by } a.$$

In the Finding the Solution box, we describe a step-by-step procedure for solving a linear equation in one variable.

FINDING THE SOLUTION: A PROCEDURE

EXAMPLE 2   **Solving Linear Equations in One Variable**

OBJECTIVE	**EXAMPLE**
Solve a linear equation in one variable.	**Solve** $\frac{1}{3}x - \frac{2}{3} = \frac{1}{2}(1 - 3x)$.

Step 1 Eliminate fractions. Multiply both sides of the equation by the least common denominator (LCD) of all of the fractions.

$$\frac{1}{3}x - \frac{2}{3} = \frac{1}{2}(1 - 3x)$$ Original equation

$$6\left(\frac{1}{3}x - \frac{2}{3}\right) = 6 \cdot \frac{1}{2}(1 - 3x)$$ Multiply both sides by 6 (LCD) to eliminate fractions.

Step 2 Simplify. Simplify both sides of the equation by removing parentheses and other grouping symbols (if any) and combining like terms.

$$6 \cdot \frac{1}{3}x - 6 \cdot \frac{2}{3} = 3(1 - 3x)$$ Distribute on the left side; simplify on the right.

$$2x - 4 = 3 - 9x$$ Simplify on the left side; distribute on the right.

Step 3 Isolate the variable term. Add appropriate expressions to both sides so that when both sides are simplified, the terms containing the variable are on one side and all constant terms are on the other side.

$$2x - 4 + 9x + 4 = 3 - 9x + 9x + 4$$ Add $9x + 4$ to both sides.

Step 4 Combine terms. Combine terms containing the variable to obtain one term that contains the variable as a factor.

$$11x = 7$$ Collect like terms and combine constants.

Step 5 Isolate the variable. Divide both sides by the coefficient of the variable to obtain the solution.

$$x = \frac{7}{11}$$ Divide both sides by 11.

Step 6 Check the solution. Substitute the solution into the original equation to check for computational errors.

Check:

Substitute $\frac{7}{11}$ for x in both sides of the original equation.

The result should be $-\frac{5}{11} = -\frac{5}{11}$.   ■ ■ ■

Practice Problem 2 Solve $\frac{2}{3} - \frac{3}{2}x = \frac{1}{6} - \frac{7}{3}x$.   ■

EXAMPLE 3   **Solving a Linear Equation**

Solve $6x - [3x - 2(x - 2)] = 11$.

SOLUTION

Step 2 $6x - [3x - 2(x - 2)] = 11$ Original equation

$6x - [3x - 2x + 4] = 11$ Remove innermost parentheses by distributing $-2(x - 2) = -2x + 4$.

$6x - 3x + 2x - 4 = 11$ Remove square brackets and change the sign of each enclosed term.

$5x - 4 = 11$ Combine like terms.

Step 3	$5x - 4 + 4 = 11 + 4$	Add 4 to both sides.
Step 4	$5x = 15$	Simplify both sides in Step 3.
Step 5	$\dfrac{5x}{5} = \dfrac{15}{5}$	Divide both sides by 5.
	$x = 3$	Simplify.

The apparent solution is 3.

Step 6 **Check:** Substitute $x = 3$ into the original equation. You should obtain $11 = 11$. Thus, 3 is the only solution of the original equation; so $\{3\}$ is its solution set. ■ ■ ■

Practice Problem 3   Solve $3x - [2x - 6(x + 1)] = -1$. ■

E X A M P L E   4   **Solving an Equation**

Solve $1 - 5y + 2(y + 7) = 2y + 5(3 - y)$.

SOLUTION

Step 2	$1 - 5y + 2(y + 7) = 2y + 5(3 - y)$	Original equation
	$1 - 5y + 2y + 14 = 2y + 15 - 5y$	Distributive property
	$15 - 3y = 15 - 3y$	Collect like terms and combine constants on each side.
Step 3	$15 - 3y + 3y = 15 - 3y + 3y$	Add $3y$ to both sides.
	$15 = 15$	Simplify.
	$0 = 0$	Subtract 15 from both sides.

We have shown that $0 = 0$ is equivalent to the original equation. The equation $0 = 0$ is always true; its solution set is the set of all real numbers. Therefore, the solution set of the original equation is also the set of all real numbers. Thus, the original equation is an identity. ■ ■ ■

Practice Problem 4   Solve the equation $2(3x - 6) + 5 = 12 - (x + 5)$. ■

TYPES OF EQUATIONS

There are three types of linear equations.

1. An equation that is satisfied by all values in the domain of the variable is an *identity*. For example, $2(x - 1) = 2x - 2$. When you try to solve an identity, you get a true statement such as $3 = 3$ or $0 = 0$.

2. An equation that is not an identity, but is satisfied by at least one number in the domain of the variable is a *conditional* equation. For example, $2x = 6$ is a linear conditional equation.

3. An equation that is not satisfied by any value of the variable is an *inconsistent* equation. For example, $x = x + 2$ is an inconsistent equation. Because no number is 2 more than itself, the solution set of the equation $x = x + 2$ is $\varnothing$. When you try to solve an inconsistent equation, you obtain a false statement such as $0 = 2$.

3 Solve rational equations with variables in the denominators.

Rational Equations

If at least one rational expression appears in an equation, then the equation is called a **rational equation**.

In Example 2, we solved a rational equation that contained constants in the denominators. We now consider rational equations with variables in the denominators. We follow the same procedure (page 86) for solving such equations. However, in Step 1, when we multiply both sides of the equation by the LCD containing the variable, we may introduce a solution that satisfies the new equation but does not satisfy the original equation. Such a solution is called an **extraneous solution** or **extraneous root**. So *whenever we multiply an equation by an expression containing the variable, we must check all solutions obtained to reject extraneous solutions (if any).*

EXAMPLE 5 Solving a Conditional Rational Equation

Solve $\dfrac{1}{x} + \dfrac{1}{3} = \dfrac{3}{4x} + \dfrac{1}{2}$.

SOLUTION

Step 1
$$\frac{1}{x} + \frac{1}{3} = \frac{3}{4x} + \frac{1}{2}$$
Original equation

$$12x\left(\frac{1}{x} + \frac{1}{3}\right) = 12x\left(\frac{3}{4x} + \frac{1}{2}\right)$$
Multiply both sides by $12x$, the LCD of x, 3, $4x$, and 2 to eliminate fractions.

Step 2 $\quad 12x \cdot \dfrac{1}{x} + 12x \cdot \dfrac{1}{3} = 12x \cdot \dfrac{3}{4x} + 12x \cdot \dfrac{1}{2}$ $\quad$ Distributive property

$$12 + 4x = 9 + 6x$$
Simplify by dividing out common factors in each term.

Step 3 $\quad 12 + 4x - 12 - 6x = 9 + 6x - 12 - 6x$ $\quad$ Add $-12 - 6x$ to both sides.

Step 4 $\qquad\qquad\qquad -2x = -3$ $\qquad\qquad\qquad$ Simplify.

Step 5 $\quad x = \dfrac{-3}{-2} = \dfrac{3}{2}$ $\qquad\qquad\qquad$ Divide both sides by -2 to solve for x.

Apparent solution is $x = \dfrac{3}{2}$.

Step 6 **Check:**

$$\frac{1}{\frac{3}{2}} + \frac{1}{3} \overset{?}{=} \frac{3}{4\left(\frac{3}{2}\right)} + \frac{1}{2}$$
Substitute $x = \dfrac{3}{2}$ in the original equation.

$$\frac{2}{3} + \frac{1}{3} \overset{?}{=} \frac{1}{2} + \frac{1}{2}$$
Simplify each term.

$$1 \overset{?}{=} 1 \qquad \text{Yes}$$

The solution set is $\left\{\dfrac{3}{2}\right\}$. ■ ■ ■

Practice Problem 5 Solve $\dfrac{3}{x} - \dfrac{1}{2x} = \dfrac{7}{2} + \dfrac{3}{4x}$. ■

EXAMPLE 6   **Solving a Rational Equation**

Solve for x: $\dfrac{1}{x-1} - \dfrac{1}{x+1} = \dfrac{2}{x^2-1}$

SOLUTION

Because $x^2 - 1 = (x-1)(x+1)$, the LCD of $x-1$, $x+1$, and x^2-1 is $(x-1)(x+1)$.

Step 1 $\dfrac{1}{x-1} - \dfrac{1}{x+1} = \dfrac{2}{x^2-1}$ Original equation

$(x-1)(x+1)\left[\dfrac{1}{x-1} - \dfrac{1}{x+1}\right] = (x-1)(x+1)\left[\dfrac{2}{x^2-1}\right]$ Multiply both sides by the LCD.

Step 2

$\cancel{(x-1)}(x+1)\dfrac{1}{\cancel{x-1}} - (x-1)\cancel{(x+1)}\dfrac{1}{\cancel{x+1}} = \cancel{(x^2-1)}\cdot\dfrac{2}{\cancel{x^2-1}}$

Distributive property; $(x-1)(x+1) = x^2 - 1$

$(x+1) - (x-1) = 2$ Simplify by dividing out common factors in each term.

Step 3 $x + 1 - x + 1 = 2$ $-(x-1) = -x + 1$

Step 4 $2 = 2$ Simplify.

The equation $2 = 2$ is equivalent to the original equation and is always true for all values of x in its domain.

The domain of x is all real numbers except -1 and 1. Therefore, the original equation is an identity. The solution set is $(-\infty, -1) \cup (-1, 1) \cup (1, \infty)$.   ■ ■ ■

Practice Problem 6   Solve for x: $\dfrac{1}{x-2} - \dfrac{1}{x+2} = \dfrac{4}{x^2-4}$ ■

EXAMPLE 7   **Solving an Inconsistent Rational Equation**

Solve for m: $\dfrac{5}{3} + \dfrac{6}{m-3} = \dfrac{2m}{m-3}$

SOLUTION

The least common denominator of 3 and $m-3$ is $3(m-3)$.

Step 1 $\dfrac{5}{3} + \dfrac{6}{m-3} = \dfrac{2m}{m-3}$ Original equation

$3(m-3)\left[\dfrac{5}{3} + \dfrac{6}{m-3}\right] = 3(m-3)\left[\dfrac{2m}{m-3}\right]$ Multiply both sides by the LCD: $3(m-3)$.

Step 2

$3(m-3)\dfrac{5}{\cancel{3}} + 3\cancel{(m-3)}\cdot\dfrac{6}{\cancel{m-3}} = 3\cancel{(m-3)}\cdot\dfrac{2m}{\cancel{m-3}}$ Distributive property; simplify.

$5(m-3) + 18 = 6m$

$5m - 15 + 18 = 6m$ Distributive property

$5m + 3 = 6m$ Simplify.

Steps 3–5 $\qquad 5m + 3 - 5m = 6m - 5m \qquad$ Subtract $5m$ from both sides.

$$3 = m \qquad\qquad\quad \text{Simplify.}$$

Step 6 **Check:** The apparent solution is $m = 3$.

Checking $m = 3$ in the original equation, we have

$$\frac{5}{3} + \frac{6}{3 - 3} \overset{?}{=} \frac{2(3)}{3 - 3}$$

$$\frac{5}{3} + \frac{6}{0} \overset{?}{=} \frac{6}{0}.$$

Because dividing by zero is undefined, we reject $m = 3$ as a solution. So there is no number m that satisfies the equation. The equation is inconsistent. The solution set is $\varnothing$. ■ ■ ■

Practice Problem 7 Solve for t: $\dfrac{t}{t - 4} - \dfrac{4}{t - 4} = 8$ ■

4 Solve formulas for a specific variable.

Formulas

Solving equations that arise from everyday experiences is one of the most important uses of algebra. An equation that expresses a relationship between two or more variables is called a **formula**.

EXAMPLE 8 **Converting Temperatures**
_____

The formula for converting the temperature in degrees Celsius (C) to degrees Fahrenheit (F) is

$$F = \frac{9}{5}C + 32.$$

If the temperature shows 86° Fahrenheit, what is the temperature in degrees Celsius?

SOLUTION

We first substitute 86 for F in the formula $\qquad F = \dfrac{9}{5}C + 32$

to obtain $\qquad\qquad\qquad\qquad\qquad\qquad\quad 86 = \dfrac{9}{5}C + 32.$

We now solve this equation for C.

$$5(86) = 5\left(\frac{9}{5}C + 32\right) \qquad \text{Multiply both sides by 5.}$$

$$430 = 9C + 160 \qquad\quad \text{Simplify.}$$

$$270 = 9C \qquad\qquad\quad \text{Subtract 160 from both sides.}$$

$$\frac{270}{9} = C \qquad\qquad\qquad \text{Divide both sides by 9.}$$

$$30 = C \qquad\qquad\qquad \text{Simplify.}$$

Thus, 86° F converts to 30° C. ■ ■ ■

Practice Problem 8 Use the formula in Example 8 to find the temperature in degrees Celsius assuming that the temperature is 50° Fahrenheit. ■

In Example 8, we specified $F = 86$ and solved the equation for C. A more general result can be obtained by assuming that F is a fixed number, or constant, and solving for C. This process is called **solving for a specified variable.**

EXAMPLE 9   Solving for a Specified Variable

Solve $F = \dfrac{9}{5}C + 32$ for C.

SOLUTION

$$5F = 5\left(\dfrac{9}{5}C + 32\right) \qquad \text{Multiply both sides by 5.}$$

$$5F = 9C + 5 \cdot 32 \qquad \text{Distributive property}$$

$$5F - 5 \cdot 32 = 9C \qquad \text{Subtract } 5 \cdot 32 \text{ from both sides.}$$

$$\dfrac{5F - 5 \cdot 32}{9} = C \qquad \text{Divide both sides by 9.}$$

$$\dfrac{5}{9}(F - 32) = C \quad \text{or} \quad C = \dfrac{5}{9}(F - 32) \qquad \text{Factor.} \qquad ■ ■ ■$$

Practice Problem 9   Solve $P = 2l + 2w$ for w.

5 Solve applied problems by using linear equations.

Applications

In the next example, we show how the skills for solving linear equations are used in solving real-world problems.

EXAMPLE 10   Bungee-Jumping TV Contestants

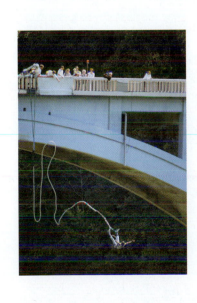

The introduction to this section described a television show on which a bungee cord with 140%–150% elongation is used for contestants' bungee jumping from a bridge 120 feet above the water. The show's producer wants to make sure the jumper does not get closer than 10 feet to the water. How long can the bungee cord be given that no contestant is more than 7 feet tall?

SOLUTION

We want the extended cord plus the length of the jumper's body to be a minimum of 10 feet above the water. To be safe, we determine the maximum extended length of the cord, add the height of the tallest possible contestant, and ensure that this total length is 10 feet above the water.

Let x = length, in feet, of the cord to be used.

Then $x + 1.5x = 2.5x$ is the maximum extended length (in feet) of the cord (with 150% elongation). We have

$$\boxed{\text{Extended cord length}} + \boxed{\text{Body length}} + \boxed{\text{10-foot buffer}} = \boxed{\text{Height of bridge above the water}}$$

$$2.5x + 7 + 10 = 120 \qquad \text{Replace descriptions with arithmetic expressions.}$$

$$2.5x + 17 = 120 \qquad \text{Combine constants.}$$

$$2.5x = 103 \qquad \text{Subtract 17 from both sides.}$$

$$x = 41.2 \text{ feet} \qquad \text{Divide both sides by 2.5.}$$

If the length of the cord is no more than 41.2 feet, all conditions are met. ■ ■ ■

Practice Problem 10   What length of cord should be used in Example 10 if the elongation is 200% instead of 140%–150%? ■

SECTION 1.1 ■ Exercises

A EXERCISES Basic Skills and Concepts

1. The domain of the variable in an equation is the set of all real numbers for which both sides of the equation are _____.

2. Standard form for a linear equation in x is
_____.

3. Two equations with the same solutions set are called
_____.

4. A conditional equation is one that is not true for
_____ values of the variables.

5. *True or False* The equation $ax + b = 0$ for $a \neq 0$ has exactly one solution.

6. *True or False* If you multiply both sides of an equation by an expression containing the variable, you may get an equation with more roots than the original equation.

In Exercises 7–11, determine whether the given value of the variable is a solution of the equation.

7. $x - 2 = 5x + 6$
   a. $x = 0$
   b. $x = -2$

8. $8x + 3 = 14x - 1$
   a. $x = -1$
   b. $x = \dfrac{2}{3}$

9. $\dfrac{2}{x} = \dfrac{1}{3} + \dfrac{1}{x + 2}$
   a. $x = 4$
   b. $x = 1$

10. $(x - 3)(2x + 1) = 0$
   a. $x = \dfrac{1}{2}$
   b. $x = 3$

11. $2x + 3x = 5x$
   a. $x = 157$
   b. $x = -2046$

In Exercises 12–16, find the domain of the variable in each equation. Write the answer in interval notation.

12. $(2 - x) - 4x = 7 - 3(x + 4)$

13. $\dfrac{y}{y - 1} = \dfrac{3}{y + 2}$

14. $\dfrac{1}{y} = 2 + \sqrt{y}$

15. $\dfrac{3x}{(x - 3)(x - 4)} = 2x + 9$

16. $\dfrac{1}{\sqrt{x}} = x^2 - 1$

In Exercises 17–20, determine whether the given equation is an identity. If the equation is not an identity, give a value that demonstrates that fact.

17. $2x + 3 = 5x + 1$

18. $3x + 4 = 6x + 2 - (3x - 2)$

19. $\dfrac{1}{x} + \dfrac{1}{2} = \dfrac{2 + x}{2x}$

20. $\dfrac{1}{x + 3} = \dfrac{1}{x} + \dfrac{1}{3}$

In Exercises 21–46, solve each equation.

21. $3x + 5 = 14$
22. $2x - 17 = 7$

23. $-10x + 12 = 32$
24. $-2x + 5 = 6$

25. $3 - y = -4$
26. $2 - 7y = 23$

27. $7x + 7 = 2(x + 1)$
28. $3(x + 2) = 4 - x$

29. $3(2 - y) + 5y = 3y$
30. $9y - 3(y - 1) = 6 + y$

31. $4y - 3y + 7 - y = 2 - (7 - y)$

32. $3(y - 1) = 6y - 4 + 2y - 4y$

33. $3(x - 2) + 2(3 - x) = 1$

34. $2x - 3 - (3x - 1) = 6$

35. $2x + 3(x - 4) = 7x + 10$

36. $3(2 - 3x) - 4x = 3x - 10$

37. $4[x + 2(3 - x)] = 2x + 1$

38. $3 - [x - 3(x + 2)] = 4$

39. $3(4y - 3) = 4[y - (4y - 3)]$

40. $5 - (6y + 9) + 2y = 2(y + 1)$

41. $2x - 3(2 - x) = (x - 3) + 2x + 1$

42. $5(x - 3) - 6(x - 4) = -5$

43. $\dfrac{2x + 1}{9} - \dfrac{x + 4}{6} = 1$

44. $\dfrac{2 - 3x}{7} + \dfrac{x - 1}{3} = \dfrac{3x}{7}$

45. $\dfrac{1 - x}{4} + \dfrac{5x + 1}{2} = 3 - \dfrac{2(x + 1)}{8}$

46. $\dfrac{x + 4}{3} + 2x - \dfrac{1}{2} = \dfrac{3x + 2}{6}$

In Exercises 47–64, solve each rational equation containing a variable in the denominator.

47. $\dfrac{2}{x} - 3 = \dfrac{5}{x}$
48. $\dfrac{4}{x} + 3 = \dfrac{10}{x}$

49. $\dfrac{1}{x} + \dfrac{1}{3} = 4$
50. $\dfrac{2}{x} - 3 = \dfrac{1}{2}$

51. $\dfrac{1 + x}{x} = \dfrac{1}{x} + 1$
52. $\dfrac{x - 1}{x - 2} - 1 = \dfrac{1}{x - 2}$

53. $\dfrac{1}{3x} + \dfrac{1}{2x} = \dfrac{1}{6} - \dfrac{1}{x}$
54. $\dfrac{5}{2x} - \dfrac{4}{7} = \dfrac{3}{14} - \dfrac{2}{x}$

55. $\dfrac{2}{x - 1} = \dfrac{3}{x + 1}$
56. $\dfrac{3}{y + 2} = \dfrac{4}{y - 1}$

57. $\dfrac{1}{3 - m} + \dfrac{7}{2m + 3} = 0$
58. $\dfrac{t}{t - 2} = -\dfrac{2}{3}$

59. $\dfrac{2}{x+1} + 3 = \dfrac{8}{x+1}$ **60.** $\dfrac{3}{x+2} = \dfrac{5}{x+2} - 4$

61. $\dfrac{5x}{x-1} = \dfrac{5}{x-1} + 3$

62. $\dfrac{3x}{x+1} - 3 + \dfrac{3}{2(x+1)} = 0$

63. $\dfrac{2}{x-2} + \dfrac{3}{2x-4} = \dfrac{1}{3} + \dfrac{5}{6x-12}$

64. $\dfrac{1}{2x+2} + \dfrac{1}{6} = \dfrac{5}{18} - \dfrac{1}{3x+3}$

In Exercises 65–76, solve each formula for the indicated variable.

65. $d = rt$ for r **66.** $F = ma$ for a

67. $C = 2\pi r$ for r **68.** $A = 2\pi rx + \pi r^2$ for x

69. $I = \dfrac{E}{R}$ for R **70.** $A = P(1 + rt)$ for t

71. $A = \dfrac{(a+b)h}{2}$ for h **72.** $T = a + (n-1)d$ for d

73. $\dfrac{1}{f} = \dfrac{1}{u} + \dfrac{1}{v}$ for u **74.** $\dfrac{1}{R} = \dfrac{1}{R_1} + \dfrac{1}{R_2}$ for R_2

75. $y = mx + b$ for m **76.** $ax + by = c$ for y

B EXERCISES Applying the Concepts

77. Swimming pool dimensions. The volume of a swimming pool is 2808 cubic feet. Assuming that the pool is 18 feet long and 12 feet deep, find its width.

78. Hole dimensions. If a box-shaped hole 7 feet long and 3 feet wide holds 168 cubic feet of dirt, how deep must the hole be?

79. Geometry. A circle has a circumference of 114π centimeters. Find the radius of the circle.

80. Geometry. A rectangle has a perimeter of 28 meters and a width of 5 meters. Find the length of the rectangle.

81. Storage barrel dimensions. A barrel has a surface area of 6π square meters and a radius of 1 meter. Find the height of the barrel.

82. Can dimensions. A can has a volume of 148π cubic centimeters and a radius of 2 centimeters. Find the height of the can.

83. Geometry. A trapezoid has an area of 66 square feet and a height of 6 feet. If one base is 3 feet, what is the length of the other base?

Trapezoid

|◄— 3 feet —►|

6 feet

84. Geometry. A trapezoid has an area of 35 square centimeters. Assuming that one base is 9 centimeters and the other base is 11 centimeters, find the height of the trapezoid.

85. Investment. If P dollars are invested at a simple interest rate r (in decimals), the amount A that will be available after t years is

$$A = P + Prt.$$

If \$500 is invested at a rate of 6%, how long will it be before the amount of money available is \$920? [*Hint:* Use $r = 0.06$.]

86. Investment. Using the formula from Exercise 85, determine the amount of money that was invested assuming that \$2482 resulted from a 10-year investment at 7%.

87. Estimating a mortgage note. A rule of thumb for estimating the maximum affordable monthly mortgage note M for prospective homeowners is

$$M = \dfrac{m - b}{4},$$

where m is the gross monthly income and b is the total of the monthly bills. What must the gross monthly income be to justify a monthly note of \$427 if monthly bills are \$302?

88. Engineering degrees. Among a group of engineers who had received master's degrees, the annual salary S in dollars was related to two numbers b and a, where

b = the number of years of work experience before receiving the degree and

a = the number of years of work experience after receiving the degree.

The relationship between S, a, and b is given by

$$S = 48{,}917 + 1080b + 1604a.$$

If an engineer had worked eight years before earning her master's degree and was earning $67,181, how many years did she work after receiving her degree?

89. Digital cameras. The price P (in dollars) for which a manufacturer will sell a digital camera is related to the number q of cameras ordered by the formula

$$P = 200 - 0.02q, \text{ for } 100 \le q \le 2000.$$

How many cameras must be ordered for a customer to pay $170 per camera?

90. Boyle's Law. If a volume V_1 of a dry gas under pressure P_1 is subjected to a new pressure P_2, the new volume V_2 of the gas is given by

$$V_2 = \frac{V_1 P_1}{P_2}.$$

Assuming that V_2 is 200 cubic centimeters, V_1 is 600 cubic centimeters, and P_1 is 400 millimeters of mercury, find P_2.

91. Transistor voltage. The voltage gain V of a transistor is related to the generator resistance R_1, the load resistance R_2, and the current gain α by

$$V = \alpha \frac{R_1}{R_2}.$$

If V is 950 and R_1 and R_2 are 100,000 and 100 ohms, respectively, find α.

92. Quick ratio. The quick ratio Q of a business is related to its current assets A, its current inventory I, and its current liabilities L, by

$$Q = \frac{A - I}{L}.$$

Assuming that $Q = 37{,}000$, $L = \$1500$, and $I = \$3200$, find the current assets.

93. Return on investment. Use the formula $A = P + Prt$ from Exercise 85 to find the amount resulting from a principal of $1247.65 invested at a rate of 13.91% for a period of 567 days. (Assume a 365-day year.)

94. Rate of return. Use the formula from Exercise 85 to find the rate of interest assuming that an investment of $2400 amounts to $3264 in four years.

95. Circular track. Joe can run a 1-mile circular track in 8 minutes, while Dick can run it in 12 minutes. If they start from the same place at the same time in the same direction, when will Joe lead Dick by exactly one lap?

96. In Exercise 95, assume that Joe and Dick run in opposite directions. How long must they run before they meet?

C EXERCISES Beyond the Basics

In Exercises 97–106, solve each rational equation.

97. $\dfrac{x}{x - 2} - \dfrac{1}{x + 2} = 1$

98. $4 - \dfrac{2 - x}{x + 1} = \dfrac{5x + 3}{x + 2}$

99. $\dfrac{1}{x - 3} - \dfrac{4}{x + 3} = \dfrac{6}{x^2 - 9}$

100. $\dfrac{1 + x}{1 - x} - \dfrac{1 - x}{1 + x} = \dfrac{1}{1 - x^2}$

101. $\dfrac{5}{x + 2} - \dfrac{8}{x^2 - x - 6} = \dfrac{4}{x - 3}$

102. $\dfrac{2x - 1}{x + 2} + \dfrac{x + 1}{2 - x} = \dfrac{x^2 + 3x + 1}{x^2 - 4}$

103. $\dfrac{x + 3}{x - 1} - \dfrac{5}{x + 1} = \dfrac{x^2 - 3x + 11}{x^2 - 1}$

104. $\dfrac{x}{x - 3} + \dfrac{x^2 - 2}{9 - x^2} = \dfrac{1}{x + 3}$

105. $\dfrac{1}{x - 4} - \dfrac{1}{x - 3} = \dfrac{1}{x - 2} - \dfrac{1}{x - 1}$

106. $\dfrac{2}{x + 1} + \dfrac{4}{x^2 - 1} = \dfrac{1}{x - 1}$

107. Explain whether the equations in each pair are equivalent.

    a. $x^2 = x$, $x = 1$

    b. $x^2 = 9$, $x = 3$

    c. $x^2 - 1 = x - 1$, $x = 0$

    d. $\dfrac{x}{x-2} = \dfrac{2}{x-2}$, $x = 2$

108. What is wrong with the following argument?

    Let $x = 1$.

    $x^2 = x$ Multiply both sides by x.

    $x^2 - 1 = x - 1$ Subtract 1 from both sides.

    $\dfrac{x^2-1}{x-1} = \dfrac{x-1}{x-1}$ Divide both sides by $x - 1$.

    $\dfrac{(x+1)(x-1)}{x-1} = 1$ Factor $x^2 - 1$.

    $x + 1 = 1$ Remove common factor.

    $1 + 1 = 1$ Replace x with 1 in the previous equation.

    $2 = 1$ Add.

109. Find a value of k so that $3x - 1 = k$ and $7x + 2 = 16$ are equivalent.

110. Find a value of k so that the equation $\dfrac{5}{y-4} = \dfrac{6}{y+k}$ has solution set $\{9\}$.

111. Find a value of k so that the equation $\dfrac{3}{y-2} = \dfrac{4}{y+k}$ is inconsistent.

112. Find a value of k so that the equation $\dfrac{1}{x^2-9} = \dfrac{1}{(x-3)(x+k)}$ is an identity.

In Exercises 113–118, solve each equation for x. Assume $a \neq 0$ and $b \neq 0$.

113. $a(a + x) = b^2 - bx$

114. $9 + a^2x - ax = 6x + a^2$

115. $\dfrac{ax}{b} - \dfrac{bx}{a} = \dfrac{(a+b)^2}{ab}$

116. $\dfrac{2}{3} - \dfrac{x}{3b} - \dfrac{2x+b}{2a} + \dfrac{6b^2+a^2}{6ab} = 0$

117. $\dfrac{b(bx-1)}{a} - \dfrac{a(1+ax)}{b} = 1$

118. $\dfrac{x-2a}{b} - \dfrac{3}{2} = \dfrac{b-x}{2a} + \dfrac{1}{2}$

Critical Thinking

119. A pawnshop owner sells two watches for $499 each. On one watch, he gained 10%, and on the other watch, he lost 10%. What is his percentage gain or loss?

    a. No loss, no gain   **b.** 10% loss

    c. 1% loss   **d.** 1% gain

120. The price of gasoline increased by 20% from July to August, while your consumption of gas decreased by 20% from July to August. What is the percentage change in your gas bill from July to August?

    a. No change   **b.** 5% decrease

    c. 4% increase   **d.** 4% decrease

Applications of Linear Equations

BOMB THREAT ON THE *QUEEN ELIZABETH 2*

While traveling from New York to Southampton, England, the captain of the *Queen Elizabeth 2* received a message that a bomb was on board and that it was timed to go off during the voyage. A search by crew members proved fruitless, so members of the Royal Marine Special Boat Squadrons bomb disposal unit were flown out on a Royal Air Force Hercules aircraft and parachuted into the Atlantic near the ship, which was 1000 miles from Britain, traveling toward Britain at 32 miles per hour. If the Hercules aircraft could average 300 miles per hour, how long would the passengers have to wait for help to arrive? We learn the answer in Example 5 using the methods of this section. ■

Solving Applied Problems

Practical problems are often introduced through verbal descriptions that do not contain explicit requests for algebraic activity, such as "Solve the equation...." Frequently, they describe a situation, perhaps in a physical or financial setting, and ask for a value for some specific quantity that will favorably resolve the situation. The process of translating a practical problem into a mathematical one is called **mathematical modeling**. Generally, we use variables to represent the quantities identified in the verbal description and represent relationships between these quantities with algebraic expressions or equations. Ideally, we reduce the problem to an equation that, when solved, also solves the practical problem. Then we solve the equation and check our result against the practical situation presented in the problem.

You might visualize the modeling process as illustrated in Figure 1.1.

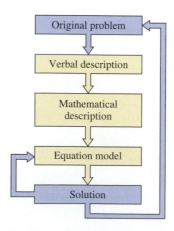

FIGURE 1.1 Modeling process

1 Learn procedures for solving applied problems.

Applied Problems

We provide a general strategy to follow when solving applied problems.

> **PROCEDURE FOR SOLVING APPLIED PROBLEMS**
>
> **Step 1** Read the problem as many times as needed to understand it thoroughly. Pay close attention to the question asked to help identify the quantity the variable should represent.
>
> **Step 2** Assign a variable to represent the quantity you are looking for and when necessary, express all other unknown quantities in terms of this variable. Frequently, it is helpful to draw a diagram to illustrate the problem or to set up a table to organize the information.
>
> **Step 3** Write an equation that describes the situation.
>
> **Step 4** Solve the equation.
>
> **Step 5** Answer the question asked in the problem.
>
> **Step 6** Check the answer against the description of the original problem (not just the equation solved in Step 4).

2 Use linear equations to solve applied problems.

Geometry

See page 74 for common geometric figures and formulas.

EXAMPLE 1   **Solving a Geometry Problem**

The length of a rectangle is 3 feet less than four times its width. Find the dimensions of the rectangle assuming that its perimeter is 64 feet.

SOLUTION

Step 2   Let $x =$ the width of the rectangle.
Then $4x - 3 =$ the length of the rectangle. (See the following figure.)

Step 3   Perimeter $=$ sum of the lengths of the four sides of the rectangle
$= (4x - 3) + x + (4x - 3) + x$
Perimeter $= 64$ ft   Given
$(4x - 3) + x + (4x - 3) + x = 64$ Replace the verbal description with the algebraic expression.

Step 4 $10x - 6 = 64$ Combine terms.
$10x = 70$ Add 6 to both sides and simplify.
$x = \dfrac{70}{10} = 7$ Divide each side by 10.

Step 5   The width of the rectangle $= 7$ ft, and
the length of the rectangle $= 4(7) - 3 = 28 - 3 = 25$ ft.

Step 6   **Check:** The perimeter is $25 + 7 + 25 + 7 = 64$ ft.   ■ ■ ■

Practice Problem 1   The length of a rectangle is 5 m more than twice its width. Find the dimensions of the rectangle assuming that its perimeter is 28 m.   ■

Finance

EXAMPLE 2   **Analyzing Investments**

Tyrick invests $15,000, some in stocks and the rest in bonds. If he invests twice as much in stocks as he does in bonds, how much does he invest in each?

SOLUTION

Step 2   Let x = the amount invested in stocks. The rest of the $15,000 investment ($15,000 − x$) is invested in bonds. We have one more important piece of information to use, as follows:

Amount invested in stocks, x	=	Twice the amount invested in bonds, $2(15,000 - x)$

Step 3 $x = 2(15,000 - x)$ Replace the verbal descriptions with algebraic expressions.

Step 4 $x = 30,000 - 2x$ Distributive property

             $3x = 30,000$ Add $2x$ to both sides.

               $x = 10,000$ Divide both sides by 3.

Step 5   Tyrick invests $10,000 in stocks and $15,000 − $10,000 = $5000 in bonds.

Step 6   Check in the original problem. If Tyrick invests $10,000 in stocks and $5000 in bonds, then because $2($5000$) = $10,000$, he invested twice as much in stocks as he did in bonds. Further, Tyrick's total investment is $10,000 + $5000 = $15,000.   ■ ■ ■

Practice Problem 2   Repeat Example 2, but Tyrick now invests three times as much in stocks as he does in bonds.

    Interest is money paid for the use of borrowed money. For example, while your money is on deposit at a bank, the bank pays you for the right to use the money in your account. The money you are paid by the bank is interest on your deposit. On any loan, the money borrowed for use (and on which interest is paid) is called the **principal**. The **interest rate** is the percent of the principal charged for its use for a specific time period. The most straightforward type of interest computation is described next.   ■

SIMPLE INTEREST

If a principal of P dollars is borrowed for a period of t years with interest rate r (expressed as a decimal) computed yearly, then the total interest paid at the end of t years is

$$I = Prt.$$

Interest computed with this formula is called **simple interest**.
    When interest is computed yearly, the rate r is called an *annual* interest rate (or *per annum* interest rate).

EXAMPLE 3   **Solving Problems Involving Simple Interest**

Ms. Sharpy invests a total of $10,000 in blue-chip and technology stocks. At the end of a year, the blue chips returned 12% and the technology stocks returned 8% on the original investments. How much was invested in each type of stock if the total interest earned was $1060?

Step 1   We are asked to find two amounts: that invested in blue-chip stocks and that invested in technology stocks. If we know how much was invested in blue-chip stocks, then we know that the rest of the $10,000 was invested in technology stocks.

Step 2   Let x = amount invested in blue-chip stocks. Then

$$10,000 - x = \text{amount invested in technology stocks.}$$

We organize our information in the following table.

Investment	Principal P	Rate r	Time t	Interest $I = Prt$
Blue Chip	x	0.12	1	$0.12x$
Technology	$10,000 - x$	0.08	1	$0.08(10,000 - x)$

Interest from blue chip	+	Interest from technology	=	Total interest

Step 3 $0.12x + 0.08(10,000 - x) = 1060$ Replace descriptions with algebraic expressions.

$12x + 8(10,000 - x) = 106,000$ Multiply by 100 to eliminate decimals.

Step 4 $12x + 80,000 - 8x = 106,000$ Distributive property

$4x = 26,000$ Combine like terms; subtract 80,000 from both sides.

$x = 6500$ Divide by 4 to get the amount in blue-chip stocks

$10,000 - x = 10,000 - 6500$ Replace x with 6500.

$= 3500$ Substitute to get the amount in technology stocks

Step 5   Ms. Sharpy invests $3500 in technology stocks and $6500 in blue-chip stocks.

Step 6   Check in the original problem. Now $3500 + $6500 = $10,000, and

$$12\% \text{ of } \$6500 \text{ is } 0.12(6500) = \$780$$
$$\text{and } 8\% \text{ of } \$3500 \text{ is } 0.08(3500) = \$280.$$

Thus, the total interest earned is $780 + $280 = $1060.   ■ ■ ■

Practice Problem 3   One-fifth of my capital is invested at 5%, one-sixth of my capital at 8%, and the rest at 10% per year. If the annual interest on my capital is $130, what is my capital?   ■

Uniform Motion

If you drive 60 miles in two hours, your average speed is $\dfrac{60}{2}$ (=30) miles per hour.

If you multiply your average speed for the trip (30 miles per hour) by the time elapsed (two hours), you get the distance driven, 60 miles ($60 = 2 \cdot 30$). When solving problems where "rate" refers to the average speed of an object, use the following formula.

UNIFORM MOTION

If an object moves at a rate (average speed) r per unit time, then the distance d traveled in time t units is

$$d = rt.$$

◆ **WARNING**   Care must be taken to make sure that units of measurement are consistent. If, for example, the rate is given in miles per hour, then the time should be given in hours. If the rate is given in miles per hour and the time is given as 15 minutes, change 15 minutes to $\frac{1}{4}$ hour before using the uniform-motion formula.

EXAMPLE 4   **Solving a Uniform-Motion Problem**

A motorcycle police officer is chasing a car that is speeding at 70 miles per hour. The police officer is 3 miles behind the car and is traveling 80 miles per hour. How long will it be before the officer overtakes the car?

SOLUTION

Step 2   We are asked to find the amount of *time* before the officer overtakes the car. Draw a sketch to help visualize the problem.

FIGURE 1.2

Let x = distance in miles the car travels before being overtaken. Then $x + 3$ = distance in miles the motorcycle travels before overtaking the car. We organize our information in a table.

Object	d (in miles)	r (miles per hour)	$t = \left(\dfrac{d}{r}\right)$ (hours)
Car	x	70	$\dfrac{x}{70}$
Motorcycle	$x + 3$	80	$\dfrac{x + 3}{80}$

Step 3   Because the time from the start of the chase to the interception point is the same for both the car and the motorcycle, we have

$$\frac{x}{70} = \frac{x + 3}{80} \qquad \text{The time is the same for both the car and the motorcycle.}$$

Step 4
$$8x = 7(x + 3) \qquad \text{Multiply both sides by the LCD, 560.}$$
$$8x = 7x + 21 \qquad \text{Distributive property}$$
$$x = 21 \qquad \text{Subtract } 7x \text{ from both sides.}$$

Step 5 The time required to overtake the car is

$$t = \frac{x}{70} = \frac{21}{70} = \frac{3}{10} \qquad \text{Replace } x \text{ with } 21.$$

In $\frac{3}{10}$ of an hour, or 18 minutes, the police officer overtakes the car.

Step 6 Check in the original problem. If the police officer travels for $\frac{3}{10}$ of an hour at 80 miles per hour, he travels $d = (80)\left(\frac{3}{10}\right) = 24$ miles. The car traveling for this same $\frac{3}{10}$ of an hour at 70 miles per hour travels $(70)\left(\frac{3}{10}\right) = 21$ miles. Because the officer travels 3 more miles during that time than the car does (because he started 3 miles behind the car), the officer does overtake the car. ■ ■ ■

Practice Problem 4 A train 130 meters long crosses a bridge in 21 seconds. The train is moving at a speed of 25 meters per second. What is the length of the bridge? ■

EXAMPLE 5 Dealing with a Bomb Threat on the *Queen Elizabeth II*

In the introduction to this section, the *Queen Elizabeth II* was 1000 miles from Britain, traveling toward Britain at 32 miles per hour. A Hercules aircraft was flying from Britain directly toward the ship, averaging 300 miles per hour. How long would the passengers have to wait for the aircraft to meet the ship?

SOLUTION

Step 1 The initial separation between the ship and the aircraft is 1000 miles.

Step 2 Let t = time elapsed when ship and aircraft meet.

Then $32t$ = distance the ship traveled and $300t$ = distance the aircraft traveled.

Step 3

| Distance the ship traveled | + | Distance the aircraft traveled | = | 1000 miles |

$$32t + 300t = 1000 \qquad \text{Replace descriptions with algebraic expressions.}$$

Step 4 $\qquad\qquad 332t = 1000 \qquad \text{Combine like terms.}$

$$t = \frac{1000}{332} \approx 3.01 \qquad \text{Divide both sides by 332.}$$

Step 5 The aircraft and ship meet after about three hours. ■ ■ ■

Practice Problem 5 How long would it take for the aircraft and ship in Example 5 to meet if the aircraft averaged 350 miles per hour and was 955 miles away? ■

Work Rate

Work-rate problems use an idea much like that of uniform motion. In problems involving rates of work, we assume that the work is done at a uniform rate.

WORK RATE

The portion of a job completed per unit of time is called the **rate of work**.

If a job can be completed in x units of time (seconds, hours, days, and so on), then the portion of the job completed in *one unit of time* (1 second, 1 hour, 1 day, and so on) is $\dfrac{1}{x}$. The portion of the job completed in t *units of time* (t seconds, t hours, t days, and so on) is $t \cdot \dfrac{1}{x}$. When the portion of the job completed is 1, the job is done.

EXAMPLE 6   **Solving a Work-Rate Problem**

One copy machine copies twice as fast as another. If both copiers work together, they can finish a particular job in two hours. How long would it take each copier, working alone, to do the job?

SOLUTION

Step 1   Because the speed of one copier is twice the speed of the other, if we find how long it takes the faster copier to do the job working alone, the slower copier will take twice as long.

Step 2   Let x = number of hours for the faster copier to complete the job alone.

$2x$ = number of hours for the slower copier to complete the job alone.

$\dfrac{1}{x}$ = portion of the job the faster copier does in one hour.

$\dfrac{1}{2x}$ = portion of the job the slower copier does in one hour.

The time it takes both copiers, working together, to do the job is two hours. Multiplying the portion of the job each copier does in one hour by 2 will give the portion of the completed job done by each copier. We organize our information in a table.

	Portion Done in One Hour	Time to Complete Job Working Together	Portion Done by Each Copier
Faster copier	$\dfrac{1}{x}$	2	$2\left(\dfrac{1}{x}\right) = \dfrac{2}{x}$
Slower copier	$\dfrac{1}{2x}$	2	$2\left(\dfrac{1}{2x}\right) = \dfrac{1}{x}$

Because the job is *completed* in two hours, the sum of the portion of the job completed by the faster copier in two hours and the portion of the job completed by the slower copier in two hours is 1. The model equation can now be written and solved.

Step 3 $\dfrac{2}{x} + \dfrac{1}{x} = 1$ Faster copier portion + slower copier portion = 1.

Step 4 $2 + 1 = x$ Multiply both sides by x.

$3 = x$

Step 5   The faster copier could complete the job, working alone, in three hours. The slower copier would require $2(3) = 6$ hours to complete the job.

Step 6   Check in the original problem. The faster copier completes $\dfrac{1}{3}$ of the job in one hour. Similarly, the slower copier completes $\dfrac{1}{6}$ of the job in one hour.

So, in two hours, the faster copier has completed $2\left(\dfrac{1}{3}\right) = \dfrac{2}{3}$ of the job and the slower copier has completed $2\left(\dfrac{1}{6}\right) = \dfrac{2}{6} = \dfrac{1}{3}$ of the job. Because $\dfrac{2}{3} + \dfrac{1}{3} = 1$, the copiers have completed the job in two hours.   ■ ■ ■

Practice Problem 6   A couple took turns washing their sports car on weekends. Jim could usually wash the car in 45 minutes, whereas Anita took 30 minutes to do the same job. One weekend they were in a hurry to go to a party, so they worked together. How long did it take them?   ■

Mixtures

Mixture problems require combining various ingredients, such as water and antifreeze, to produce a blend with specific characteristics.

> **EXAMPLE 7**   **Solving a Mixture Problem**

A full 6-quart radiator contains 75% water and 25% pure antifreeze. How much of this mixture should be drained and replaced by pure antifreeze so that the resulting 6-quart mixture is 50% pure antifreeze?

SOLUTION

Step 1   We are asked to find the quantity of the radiator mixture that should be drained and replaced by pure antifreeze. The mixture we drain is only 25% pure antifreeze, but it is replaced with 100% pure antifreeze.

Step 2   Because we want to find the quantity of mixture drained, let

$$x = \text{number of quarts of mixture drained.}$$

Then x = number of quarts of pure antifreeze added and
$0.25x$ = number of quarts of pure antifreeze drained.
Now we track the pure antifreeze.

$$\boxed{\begin{array}{c}\text{Pure antifreeze}\\\text{in final mixture}\end{array}} = \boxed{\begin{array}{c}\text{Pure antifreeze in}\\\text{original mixture}\end{array}} - \boxed{\begin{array}{c}\text{Pure antifreeze}\\\text{drained from}\\\text{original mixture}\end{array}} + \boxed{\begin{array}{c}\text{Pure}\\\text{antifreeze}\\\text{added}\end{array}}$$

$$(50\% \text{ of } 6) = (25\% \text{ of } 6) - (25\% \text{ of } x) + x$$

Step 3 $(0.5)(6) = (0.25)(6) - 0.25x + x$ Replace descriptions with algebraic expressions.

Step 4
$$3 = 1.5 + 0.75x \quad \text{Collect like terms; combine constants.}$$
$$1.5 = 0.75x \quad \text{Subtract 1.5 from both sides.}$$
$$2 = x \quad \text{Divide both sides by 0.75.}$$

Step 5   Drain 2 quarts of mixture from the radiator.

Step 6   Check in the original problem. Draining 2 quarts from the original 6 quarts leaves 4 quarts of mixture in the radiator. This mixture consists of 1 quart of pure antifreeze (25%) and 3 quarts of water (75%). Adding 2 quarts of antifreeze to the radiator produces 3 quarts of antifreeze mixed with 3 quarts of water. This is the desired 50% pure antifreeze solution.   ■ ■ ■

Practice Problem 7   How many gallons of 40% sulfuric acid solution should be mixed with 20% sulfuric acid solution to obtain 50 gallons of 25% sulfuric acid solution?   ■

SECTION 1.2 ■ Exercises

A EXERCISES Basic Skills and Concepts

1. If the sides of a rectangle are a and b units, the perimeter of the rectangle is _____ units.

2. The formula for simple interest is $I = Prt$, where $P =$ _____, $r =$ _____, and $t =$ _____.

3. The distance d traveled by an object moving at rate r for time t is given by $d =$ _____.

4. The portion of a job completed per unit of time is called the _____.

5. *True or False* If $100 is invested at 5% for three years, the interest $I = (100)(5)(3)$ dollars.

6. *True or False* If an object is traveling at a uniform rate of 60 ft/sec, the distance covered in 15 minutes is $d = (60)(15)$ ft.

In Exercises 7–16, write an algebraic expression for the specified quantity.

7. Leroy buys in-line skates for $327.62, including sales tax of $6\frac{1}{2}$%. Let $x =$ the price of in-line skates before tax. Write an algebraic expression in x for "the tax paid on the in-line skates."

8. A lamp manufacturing company produces 1600 reading lamps a day when it operates in two shifts. The first shift produces only 5/6 as many lamps as the second shift. Let $x =$ the number of lamps produced per day by the second shift. Write an algebraic expression in x for the number of lamps produced per day by the first shift.

9. Kim invests $22,000, some in stocks and the rest in bonds. Let $x =$ amount invested in stocks. Write an algebraic expression in x for "the amount invested in bonds."

10. An air conditioning repair bill for $229.50 showed a charge of $72.00 dollars for parts, with the remainder of the charge for labor. Let $x =$ number of hours of labor it took to repair the air conditioner. Write an algebraic expression in x for the labor charge per hour.

11. Natasha can pick an eight-tree section of orange trees in four hours working alone. It takes her brother five hours to pick the same section working alone.

Let $t =$ amount of time it takes Natasha and her brother to pick the eight-tree section working together. Write an algebraic expression in t for
   a. the portion of the job completed by Natasha in t hours.
   b. the portion of the job completed by Natasha's brother in t hours.

12. Jermaine can process a batch of tax forms in three hours working alone. Ralph can process a batch of tax forms in two hours working alone.

Let $t =$ amount of time it takes Jermaine and Ralph to process a batch of tax forms working together. Write an algebraic expression in t for
   a. the portion of the batch of tax forms completed by Jermaine in t hours.
   b. the portion of the batch of tax forms completed by Ralph in t hours.

13. Walnuts sell for $7.40 per pound, and raisins sell for $4.60 per pound. Let $x =$ number of pounds of raisins in a 10-pound mixture of walnuts and raisins. Write an algebraic expression in x for
   a. the number of pounds of walnuts in the 10-pound mixture.
   b. the value of the raisins in the 10-pound mixture.
   c. the value of the walnuts in the 10-pound mixture.

14. Dried pears sell for $5.50 per pound, and dried apricots sell for $6.25 per pound. Let $x =$ number of pounds of dried pears in a 15-pound mixture of dried pears and dried apricots. Write an algebraic expression in x for
   a. the number of pounds of dried apricots in the 15-pound mixture.
   b. the value of the dried apricots in the 15-pound mixture.
   c. the value of the dried pears in the 15-pound mixture.

15. A pharmacist has 8 liters of a mixture that is 10% alcohol. To strengthen the mixture, the pharmacist adds pure alcohol.

Let $x =$ number of liters of pure alcohol added to the 8-liter mixture. Write an algebraic expression in x for
   a. the number of liters of mixture the pharmacist has after the pure alcohol is added.
   b. the number of liters of alcohol in the mixture obtained by adding the pure alcohol to the original 8-liter mixture.

16. A gallon of a 25% saline solution is to be diluted by adding distilled water to it. Let $x =$ number of gallons of distilled water added to the gallon of 25% saline solution. Write an algebraic expression in x for
   a. the number of gallons of saline solution after the distilled water is added.
   b. the percent saline solution obtained by adding the distilled water to the original 25% solution.

B EXERCISES Applying the Concepts

In Exercises 17–60, use mathematical modeling techniques to solve the problems.

17. **Numbers.** The sum of two numbers is 28, and one number is three times the other. Find the numbers.

18. **Numbers.** The sum of three consecutive even integers is 42. Find the integers.

19. **Geometry.** The perimeter of a rectangle is 80 ft. Find its dimensions assuming that its length is 5 ft less than twice its width.

20. **Geometry.** The width of a rectangle is 3 ft more than one-half its length, and the perimeter of the rectangle is 36 ft. Find its dimensions.

21. **Chain store properties.** A fast-food chain bought two pieces of land for new stores. The more expensive piece of land costs $23,000 more than the less expensive piece of land, and the two pieces together cost $147,000. How much did each piece cost?

22. **Administrative salaries.** The manager at a wholesale outlet earns $450 more per month than the assistant manager. If their combined salary is $3700 per month, what is the monthly salary of each?

23. **Lottery ticket sales.** The lottery ticket sales at Quick Mart for August were 10% above the sales for July. If Quick Mart sold a total of 1113 lottery tickets during July and August, how many tickets were sold each month?

24. **Sales commission.** Jan's commission for February was $15 more than half her commission for March, and her total commission for the two months was $633. Find her commission for each month.

25. **Inheritance.** An estate valued at $225,000 is to be divided between two sons so that the older son receives four times as much as the younger son. Find each son's share of the estate.

26. **TV game show.** Kevin won $735,000 on a TV game show. He decided to keep a certain amount for himself, give one-half the amount he kept for himself to his daughter, and give one-fourth the amount he kept for himself to his dad. How much did each person get?

27. **Grades.** Your grades in the three tests in College Algebra are 87, 59, and 73.
   a. How many points do you need on the final to average 75?
   b. Find part (a) assuming that the final carries double weight.

28. **Real estate investment.** A real estate agent invested a total of $4200. Part was invested in a high-risk real-estate venture, and the rest was invested in a savings and loan. At the end of a year, the high-risk venture returned 15% on the investment and the savings and loan returned 8% on the investment. Find the amount invested in each assuming that the agent's total income from the investments was $448.

29. **Tax shelter.** Mr. Mostafa received an inheritance of $7000. He put part of it in a tax shelter paying 9% interest and part in a bank paying 6%. If his annual interest totals $540, how much did he invest at each rate?

30. **Interest.** Ms. Jordan invested $4900, part at 6% interest and the rest at 8%. If the yearly interest on each investment is the same, how much interest does she receive at the end of the year?

31. **Interest.** If $5000 is invested in a bank that pays 5% interest, how much more must be invested in bonds at 8% to earn 6% on the total investment?

32. **Retail profit.** A retailer's cost for a microwave oven is $480. If he wants to make a profit of 20% of the selling price, at what price should the oven be sold?

33. **Profit from sales.** The Beckly Company manufactures shaving sets for $3 each and sells them for $5 each. How many shaving sets must be sold for the company to recover an initial investment of $40,000 and earn an additional $30,000 as profit?

34. **Jogging.** Angelina jogs 100 meters in the same time that Harry bicycles 150 meters. If Harry bicycles 15 meters per minute faster than Angelina jogs, find the rate of each.

35. **Overtaking a lead.** A car leaves New Orleans traveling 50 kilometers per hour. An hour later a second car leaves New Orleans following the first car, traveling 70 kilometers per hour. How many hours after leaving New Orleans will it take the second car to overtake the first?

36. **Separating planes.** Two planes leave an airport traveling in opposite directions. One plane travels at 470 kilometers per hour and the other at 430 kilometers per hour. How long will it take them to be 2250 kilometers apart?

37. **Overtaking a bike.** Lucas is 1/3 of a mile from home, bicycling at 20 miles per hour, when his brother leaves home on his bike to catch up. How fast must Lucas's brother go to catch Lucas a mile from home?

38. **Overtaking a lead.** Karen notices that her husband left for the airport without his briefcase. Her husband drives 40 miles per hour and has a 15-minute head start. If Karen (with the briefcase) drives 60 miles per hour and the airport is 45 miles away, will she catch her husband before he arrives at the airport?

39. **Increasing distances.** Two cars start out together from the same place. They travel in opposite directions, with one of them traveling 7 miles per hour faster than the other. After three hours, they are 621 miles apart. How fast is each car traveling?

40. **Travel time.** Aya rode her bicycle from her house to a friend's house, averaging 16 kilometers per hour. It turned dark before she was ready to return home, so her friend drove her home at a rate of 80 kilometers per hour. If Aya's total traveling time was three hours, how far away did her friend live?

41. Draining a pool. An old pump can drain a pool in six hours working alone. A new pump can drain the pool in four hours working alone. How long will it take both pumps to empty the pool working together?

42. Draining a pool. A swimming pool has two drain pipes. One pipe can empty the pool in three hours, and the other pipe can empty the pool in seven hours. If both drains are open, how long will it take the pool to drain?

43. Filling a blimp. One type of air blower can fill a blimp (or dirigible) in six hours working alone, while a second blower fills the same blimp in nine hours working alone. Working together, how long will it take both blowers to fill the blimp?

44. Shredding hay. A shredder distributes hay across 5/9 of a field in ten hours working alone. With the addition of a second shredder, the entire field is finished in another three hours. How long would it take the second shredder to do the job working alone?

45. Sorting letters. A new letter sorting machine works twice as fast as the old one. Both sorters working together complete a job in eight hours. How long would it have taken the new letter sorter to do the job alone?

46. Plowing a field. A farmer can plow his field by himself in 15 days. If his son helps, they can do it in 6 days. How long would it take his son to plow the field by himself?

47. Job completion. Suppose an average professor can do a certain job in nine hours and an average student can do the same job in six hours. If three professors and two students work together, how long will the job take?

48. Butterfat in milk. Two gallons of milk contain 2% butterfat. How many quarts of milk should be drained and replaced by pure butterfat to produce a mixture of 2 gallons of milk containing 5% butterfat? (There are 4 quarts in a gallon.)

49. Gold alloy. A goldsmith has 120 grams of a gold alloy (a mixture of gold and one or more other metals) containing 75% pure gold. How much pure gold must be added to this alloy to obtain an alloy that is 80% pure gold?

50. Coffee blends. A coffee wholesaler wants to blend a coffee containing 35% chicory with a coffee containing 15% chicory to produce a 500-kilogram blend of coffee containing 18% chicory. How much of each type is required?

51. Solder. How many pounds of 60-40 solder (60% tin, 40% lead) must be mixed with 40-60 solder to produce 600 grams of 55-45 solder?

52. Whiskey proof. How many gallons of 90 proof whiskey (45% alcohol) must be mixed with 70 proof whiskey (35% alcohol) to produce 36 gallons of 85 proof whiskey?

53. Pharmacy. A pharmacist needs 20% boric acid solution. How much 60% boric acid solution must be added to 7.5 liters of an 8% boric acid solution to get the desired strength?

54. Mixture of nuts. A mixture of nuts contains almonds worth $0.50 a pound, cashews worth $1.00 per pound, and pecans worth $0.75 per pound. If there are three times as many pounds of almonds as cashews in a 100-pound mixture worth $0.70 a pound, how many pounds of each nut does the mixture contain?

55. Vending machine coins. A vending machine coin box contains nickels, dimes, and quarters. The coin box contains 3 times as many nickels and 4 times as many quarters as it does dimes. If the box contains a total of $96.25, how many coins of each kind does it contain?

56. Gumdrops. In a jar of red and green gumdrops, 12 more than half the gumdrops are red and the number of green gumdrops is 19 more than half the number of red gumdrops. How many of each color are in the jar?

57. Age computation. Eric's grandfather is 57 years older than Eric. Five years from now, Eric's grandfather will be 4 times as old as Eric is at that time. How old is Eric's grandfather?

58. Real estate investment. A real estate investor bought acreage for $7200. After reselling three-fourths of the acreage at a profit of $30 per acre, she recovered the $7200. How many acres did she sell?

59. Box construction. An open box is to be constructed from a rectangular sheet of tin 3 meters wide by cutting out a 1-meter square from each corner and folding up the sides. The volume of the box is to be 2 cubic meters. What is the length of the tin rectangle?

60. Interest rates. Mr. Kaplan invests one sum of money at a certain rate of interest and then half that sum at twice the first rate of interest. This turns out to yield 8% on the total investment. What are the two rates of interest?

C EXERCISES Beyond the Basics

You can solve many problems by modeling procedures introduced in this section other than those presented in the examples in the section. Here are some types for you to try.

61. Suppose you average 75 miles per hour over the first half of a drive from Denver to Las Vegas, but your average speed for the entire trip is 60 miles per hour. What was your average speed for the second half of the drive?

62. Davinder (D) and Mikhail (M) were 2 miles apart when they began walking toward each other. D walks at a constant rate of 3.7 miles per hour, and M walks at a constant rate of 4.3 miles per hour. When they started, D's dog, who runs at a constant rate of 6 miles per hour, ran to M and then turned back and ran to D. If the dog continued to run back and forth until D and M met and the dog lost no time turning around, how far did it run?

63. A mixture contains alcohol and water in the ratio $5:1$. After the addition of 5 liters of water, the ratio of alcohol to water becomes $5:2$. Find the quantity of alcohol in the original mixture.

64. An alloy contains zinc and copper in the ratio $5:8$, and another alloy contains zinc and copper in the ratio $5:3$. Equal amounts of both alloys are melted together. Find the ratio of zinc to copper in the new alloy.

65. Democritus has lived $\dfrac{1}{6}$ of his life as a boy, $\dfrac{1}{8}$ of his life as a youth, and $\dfrac{1}{2}$ of his life as a man and has spent 15 years as a mature adult. How old is Democritus?

66. A man and a woman have the same birthday. When he was as old as she is now, the man was twice as old as the woman. When she becomes as old as he is now, the sum of their ages will be 119. How old is the man now?

67. How many minutes is it before 6 P.M. if, 50 minutes ago, four times this number was the number of minutes past 3 P.M.?

68. Two pipes A and B can fill a tank in 24 minutes and 32 minutes, respectively. Initially, both pipes are opened simultaneously. After how many minutes should pipe B be turned off so that the tank is full in 18 minutes?

69. A small plane was scheduled to fly from Atlanta to Washington, D.C. The plane flies at a constant speed of 150 miles per hour. However, the flight was against a head wind of 10 miles per hour. The threat of mechanical failure forced the plane to turn back, and it returned to Atlanta with a tail wind of 10 miles per hour, landing 1.5 hours after it had taken off.
   a. How far had the plane traveled before turning back?
   b. What was the plane's average speed?

70. Three airports A, B, and C are located on an east-west line. B is 705 miles west of A, and C is 652.5 miles west of B. A pilot flew from A to B, had a stopover at B for three hours, and continued to C. The wind was blowing from the east at 15 miles per hour during the first part of the trip, but the wind changed to come from the west at 20 miles per hour during the stopover. The flight time between A and B and between B and C was the same. Find the airspeed of the plane.

71. Two trains are traveling toward each other on adjacent tracks. One of the trains is 230 feet long and is moving at a speed of 50 miles per hour. The other train is 210 feet long and is traveling at the rate of 60 miles per hour. Find the time interval between the moment the trains first meet until they completely pass each other. (Remember, 1 mile = 5280 feet.)

72. Suppose you are driving from point A to point B at x miles per hour and return from B to A at y miles per hour. What is your average speed for the round trip?
   Extend the forgoing problem: You travel along an equilateral triangular path ABC from A to B at x miles per hour, from B to C at y miles per hour, and from C to A at z miles per hour. What is your average speed for the round trip?

Critical Thinking

73. Suppose $1\frac{1}{2}$ men can do $1\frac{1}{2}$ jobs in $1\frac{1}{2}$ days. How long does it take one man to do the job?

74. Alex and Ben undertake a job for $800. This amount is to be distributed in proportion to the work done. Alex can do the job in six days, and Ben can do it in eight days. With assistance from Chris, they finish the job in three days. How much money did Chris get?

Complex Numbers

20c Charles Steinmetz

Before Starting this Section, Review

1. Properties of real numbers (Section P.1, page 12)
2. Special products (Section P.3, page 36)
3. Factoring (Section P.4, page 41)

Objectives

1. Define complex numbers.
2. Add and subtract complex numbers.
3. Multiply complex numbers.
4. Divide complex numbers.

ALTERNATING CURRENT CIRCUITS

In the early days of the study of alternate current (AC) circuits, scientists concluded that AC circuits were somehow different from the battery-powered direct current (DC) circuits. However, both types of circuits obey the same physical and mathematical laws. In 1893, Charles Steinmetz provided the breakthrough in understanding AC circuits. He explained that in AC circuits, the voltage, current, and resistance (called *impedance* in AC circuits) were not scalars, but alternate in direction, and possessed frequency and phase shift that must be taken into account. He advocated the use of the polar form of complex numbers to provide a convenient method of symbolically denoting the magnitude, frequency, and phase shift simultaneously for the AC circuit quantities voltage, current, and impedance. Steinmetz eventually became known as "the wizard who generated electricity from the square root of minus one." In Example 8, we use complex numbers to compute the total impendance in an AC circuit. ■

---

1 Define complex numbers.

Complex Numbers

Because the square of a real number is nonnegative (that is, $x^2 \geq 0$ for any real number x), the equation $x^2 = -1$ has no solution in the set of real numbers. To solve such equations, we extend the real numbers to a larger set called the set of complex numbers. We first introduce a new number i whose square is -1.

DEFINITION OF i

The square root of -1 is called i.

$$i = \sqrt{-1} \text{ so that } i^2 = -1.$$

The number i is called the **imaginary unit**.

With the introduction of the number i, the equation $x^2 = -1$ has two solutions, $x = \pm\sqrt{-1} = \pm i$.

COMPLEX NUMBERS

A complex number is a number of the form

$$z = a + bi,$$

where a and b are **real numbers** and $i^2 = -1$.

The number a is called the **real part** of z, and we write $\text{Re}(z) = a$. The number b is called the **imaginary part** of z, and we write $\text{Im}(z) = b$.

Most graphing calculators allow you to work with complex numbers by changing from "real" to $a + bi$ mode. Once the $a + bi$ mode is set, the $\sqrt{-1}$ is recognized as i. The i key is used to enter complex numbers such as $2 + 3i$.

The real and imaginary parts of a complex number can then be found.

BY THE WAY ...

In 1777, the great Swiss mathematician Leonhard Euler introduced the i symbol for $\sqrt{-1}$. However, electrical engineers often use the letter j for $\sqrt{-1}$ because in electrical engineering problems, the symbol i is traditionally reserved for the electric current.

A complex number z written in the form $a + bi$ is said to be in **standard form**. Where convenient, we use the form $a + ib$ as equivalent to $a + bi$. A complex number with $a = 0$ and $b \neq 0$, written as bi, is called a **pure imaginary number**. If $b = 0$, then the complex number $a + bi = a$ is a real number. So real numbers form a subset of complex numbers (with imaginary part 0). Figure 1.3 shows how various sets of numbers are contained within larger sets.

FIGURE 1.3 **Complex numbers**

We can express the square root of any negative number as a product of a real number and i.

SQUARE ROOT OF A NEGATIVE NUMBER

For any positive number b,

$$\sqrt{-b} = (\sqrt{b})i = i\sqrt{b}.$$

EXAMPLE 1 **Identifying the Real and Imaginary Parts of a Complex Number**

Identify the real and imaginary parts of each complex number.

a. $2 + 5i$ **b.** $7 - \dfrac{1}{2}i$ **c.** $3i$ **d.** -9 **e.** 0 **f.** $3 + \sqrt{-25}$

SOLUTION

To identify the real and imaginary parts, we express each number in the standard form $a + bi$.

a. $2 + 5i$ is already written as $a + bi$; real part 2, imaginary part 5

b. $7 - \dfrac{1}{2}i = 7 + \left(-\dfrac{1}{2}\right)i$; real part 7, imaginary part $-\dfrac{1}{2}$

c. $3i = 0 + 3i$; real part 0, imaginary part 3

d. $-9 = -9 + 0i$; real part -9, imaginary part 0

e. $0 = 0 + 0i$; real part 0, imaginary part 0

f. $3 + \sqrt{-25} = 3 + (\sqrt{25})i = 3 + 5i$; real part 3, imaginary part 5 ■ ■ ■

Practice Problem 1 Identify the real and imaginary parts of each complex number.

a. $-1 + 2i$ **b.** $-\dfrac{1}{3} - 6i$ **c.** 8 ■

EQUALITY OF COMPLEX NUMBERS

Two complex numbers $z = a + bi$ and $w = c + di$ are equal if and only if

$$a = c \quad \text{and} \quad b = d.$$

That is, $z = w$ if and only if $\text{Re}(z) = \text{Re}(w)$ and $\text{Im}(z) = \text{Im}(w)$.

EXAMPLE 2 **Equality of Complex Numbers**

Find x and y assuming that $(3x - 1) + 5i = 8 + (3 - 2y)i$.

SOLUTION

Let $\quad z = (3x - 1) + 5i \quad$ and $\quad w = 8 + (3 - 2y)i.$

Then, $\text{Re}(z) = \text{Re}(w) \quad$ and $\quad \text{Im}(z) = \text{Im}(w).$

So, $\quad 3x - 1 = 8 \quad$ and $\quad 5 = 3 - 2y.$

Solving for x and y we have

$$3x = 8 + 1 \quad \text{Isolate } x. \qquad 5 - 3 = -2y \quad \text{Isolate } y.$$
$$x = 3 \quad \text{Solve for } x. \qquad -1 = y \quad \text{Solve for } y.$$

So $x = 3$ and $y = -1$. ■■■

Practice Problem 2 Find a and b assuming that $(1 - 2a) + 3i = 5 - (2b - 5)i$. ■

2 Add and subtract complex numbers.

Addition and Subtraction

We find the sum or difference of two complex numbers by treating them as binomials in which i is the variable.

ADDITION AND SUBTRACTION OF COMPLEX NUMBERS

For real numbers $a, b, c,$ and d, let $z = a + bi$ and $w = c + di$.

Sum: $z + w = (a + bi) + (c + di) = (a + c) + (b + d)i$

To add complex numbers, add their real parts, add their imaginary parts, and express the sum in standard form.

Difference: $z - w = (a + bi) - (c + di) = (a - c) + (b - d)i$

To subtract complex numbers, subtract their real parts, subtract their imaginary parts, and express the difference in standard form.

TECHNOLOGY CONNECTION

In $a + bi$ mode, most graphing calculators perform addition and subtraction of complex numbers using the ordinary addition and subtraction keys.

```
(3+7i)+(2-4i)
           5+3i
(5+9i)-(6-8i)
          -1+17i
```

EXAMPLE 3 **Adding and Subtracting Complex Numbers**

Write the sum or difference of two complex numbers in standard form.

a. $(3 + 7i) + (2 - 4i)$ **b.** $(5 + 9i) - (6 - 8i)$
c. $(2 + \sqrt{-9}) - (-2 + \sqrt{-4})$

SOLUTION

a. $(3 + 7i) + (2 - 4i) = (3 + 2) + [7 + (-4)]i = 5 + 3i$

b. $(5 + 9i) - (6 - 8i) = (5 - 6) + [9 - (-8)]i = -1 + 17i$

c. $(2 + \sqrt{-9}) - (-2 + \sqrt{-4}) = (2 + 3i) - (-2 + 2i) \quad \sqrt{-9} = 3i, \sqrt{-4} = 2i$
$$= (2 - (-2)) + (3 - 2)i$$
$$= 4 + i$$

■■■

Practice Problem 3   Write the following complex numbers in standard form.

a. $(1 - 4i) + (3 + 2i)$ **b.** $(4 + 3i) - (5 - i)$

c. $(3 - \sqrt{-9}) - (5 - \sqrt{-64})$ ■

3 Multiply complex numbers.

Multiplying Complex Numbers

We multiply complex numbers by using FOIL (as we do with binomials, see page 33) and then replacing i^2 with -1. for example,

$$\overset{\text{F \qquad O \qquad I \qquad L}}{(2 + 5i)(4 + 3i) = 2\cdot4 + 2\cdot3i + 5i\cdot4 + 5i\cdot3i}$$

$$= 8 + 6i + 20i + 15i^2$$

$$= 8 + 26i + 15(-1) \qquad \qquad \text{Replace } i^2 \text{ with } -1.$$

$$= (8 - 15) + 26i$$

$$= -7 + 26i.$$

You should always use the FOIL method to multiply complex numbers in standard form, but for completeness, we state the product rule formally.

MULTIPLYING COMPLEX NUMBERS

For all real numbers $a, b, c,$ and $d,$

$$(a + bi)(c + di) = (ac - bd) + (ad + bc)i.$$

EXAMPLE 4   **Multiplying Complex Numbers**

Write the following products in standard form.

a. $(3 - 5i)(2 + 7i)$ **b.** $-2i(5 - 9i)$

SOLUTION

$$\overset{\text{F \quad O \quad I \quad L}}{\textbf{a.} \;\; (3 - 5i)(2 + 7i) = 6 + 21i - 10i - 35i^2}$$

$$= 6 + 11i + 35 \qquad \qquad \text{Because } i^2 = -1, -35i^2 = 35.$$

$$= 41 + 11i \qquad \qquad \text{Combine terms.}$$

$$\textbf{b.} \;\; -2i(5 - 9i) = -10i + 18i^2 \qquad \qquad \text{Distributive property}$$

$$= -10i - 18 \qquad \qquad \text{Because } i^2 = -1, 18i^2 = -18.$$

$$= -18 - 10i \qquad\qquad\qquad ■■■$$

Practice Problem 4   Write the following products in standard form.

a. $(2 - 6i)(1 + 4i)$ **b.** $-3i(7 - 5i)$ ■

◆ **WARNING**   Recall from algebra that if a and b are positive real numbers, then

$$\sqrt{a}\sqrt{b} = \sqrt{ab}.$$

However, this property is not true for all complex numbers. For example,

$$\sqrt{-9}\sqrt{-9} = (3i)(3i) = 9i^2 = 9(-1) = -9,$$

but
$$\sqrt{(-9)(-9)} = \sqrt{81} = 9.$$

Thus,
$$\sqrt{-9}\sqrt{-9} \neq \sqrt{(-9)(-9)}.$$

When performing multiplication (or division) involving square roots of negative numbers (say, $\sqrt{-b}$ with $b > 0$), always write $\sqrt{-b} = i\sqrt{b}$ before performing the operation.

EXAMPLE 5 **Multiplication Involving Square Roots of Negative Numbers**

Perform the indicated operation and write the result in standard form.

a. $\sqrt{-2}\sqrt{-8}$ **b.** $\sqrt{-3}(2 + \sqrt{-3})$ **c.** $(-2 + \sqrt{-3})^2$
d. $(3 + \sqrt{-2})(1 + \sqrt{-32})$

SOLUTION

a. $\sqrt{-2}\sqrt{-8} = i\sqrt{2} \cdot i\sqrt{8} = i^2\sqrt{2 \cdot 8} = i^2\sqrt{16} = (-1)(4) = -4$

b.
$$
\begin{aligned}
\sqrt{-3}(2 + \sqrt{-3}) &= i\sqrt{3}(2 + i\sqrt{3}) && \sqrt{-3} = i\sqrt{3} \\
&= 2i\sqrt{3} + i^2\sqrt{9} && \text{Distributive property} \\
&= 2i\sqrt{3} + (-1)(3) && i^2 = -1, \sqrt{9} = 3 \\
&= -3 + 2i\sqrt{3} && \text{Standard form}
\end{aligned}
$$

c.
$$
\begin{aligned}
(-2 + \sqrt{-3})^2 &= (-2 + i\sqrt{3})^2 && \sqrt{-3} = i\sqrt{3} \\
&= (-2)^2 + 2(-2)(i\sqrt{3}) + (i\sqrt{3})^2 && (a + b)^2 = a^2 + 2ab + b^2 \\
&= 4 - 4i\sqrt{3} + 3i^2 && (i\sqrt{3})^2 = (i\sqrt{3})(i\sqrt{3}) = 3i^2 \\
&= 4 - 4i\sqrt{3} + 3(-1) && i^2 = -1 \\
&= 1 - 4i\sqrt{3} && \text{Simplify.}
\end{aligned}
$$

d.
$$
\begin{aligned}
&(3 + \sqrt{-2})(1 + \sqrt{-32}) \\
&= (3 + i\sqrt{2})(1 + i\sqrt{32}) && \sqrt{-2} = i\sqrt{2}, \sqrt{-32} = i\sqrt{32} \\
&= 3 + 3i\sqrt{32} + i\sqrt{2} + (i\sqrt{2})(i\sqrt{32}) && \text{FOIL} \\
&= 3 + 3i(4\sqrt{2}) + i\sqrt{2} + i^2\sqrt{64} && \sqrt{32} = \sqrt{16 \cdot 2} = 4\sqrt{2} \\
&= 3 + 12i\sqrt{2} + i\sqrt{2} + (-1)(8) && i^2 = -1 \\
&= (-5) + 13i\sqrt{2} && \text{Add.}
\end{aligned}
$$

■ ■ ■

Practice Problem 5 Perform the indicated operation and write the result in standard form.

a. $(-3 + \sqrt{-4})^2$ **b.** $(5 + \sqrt{-2})(4 + \sqrt{-8})$ ■

---

4 Divide complex numbers.

Complex Conjugates and Division

To perform the division of complex numbers, it is helpful to learn about the *conjugate* of a complex number.

CONJUGATE OF A COMPLEX NUMBER

If $z = a + bi$, then the conjugate (or complex conjugate) of z is donated by $\bar{z}$ and defined by $\bar{z} = \overline{a + bi} = a - bi$.

Note that the conjugate $\bar{z}$ of a complex number has the same real part as z does, but the sign of the imaginary part is changed. For example, $\overline{2 + 7i} = 2 - 7i$ and $\overline{5 - 3i} = \overline{5 + (-3)i} = 5 - (-3)i = 5 + 3i$.

EXAMPLE 6   **Multiplying a Complex Number by Its Conjugate**

Find the product $z\bar{z}$ for each complex number z.

a. $z = 2 + 5i$ **b.** $z = 1 - 3i$

SOLUTION

a. If $z = 2 + 5i$, then $\bar{z} = 2 - 5i$.

$$\begin{aligned} z\bar{z} = (2 + 5i)(2 - 5i) &= 2^2 - (5i)^2 && \text{Difference of squares} \\ &= 4 - 25i^2 && (5i)^2 = 5^2 i^2 = 25i^2 \\ &= 4 - (-25) && \text{Because } i^2 = -1, 25i^2 = -25. \\ &= 29 && \text{Simplify.} \end{aligned}$$

b. If $z = 1 - 3i$, then $\bar{z} = 1 + 3i$.

$$\begin{aligned} z\bar{z} = (1 - 3i)(1 + 3i) &= 1^2 - (3i)^2 && \text{Difference of squares} \\ &= 1 - 9i^2 && (3i)^2 = 3^2 i^2 \\ &= 1 - (-9) && \text{Because } i^2 = -1, 9i^2 = -9. \\ &= 10 && \text{Simplify.} \end{aligned}$$

Practice Problem 6   Find the product $z\bar{z}$ for each complex number z.

a. $1 + 6i$ **b.** $-2i$

The results in Example 6 correctly suggest the following theorem.

COMPLEX CONJUGATE PRODUCT THEOREM

If $z = a + bi$, then

$$z\bar{z} = a^2 + b^2.$$

To write the reciprocal of a nonzero complex number or the quotient of two complex numbers in the form $a + bi$, multiply the numerator and denominator by the conjugate of the denominator. By the Complex Conjugate Product Theorem, the resulting denominator is a real number.

DIVIDING COMPLEX NUMBERS

To write the quotient of two complex numbers w and z ($z \neq 0$), write

$$\frac{w}{z} = \frac{w\bar{z}}{z\bar{z}} \qquad \text{Multiply numerator and denominator by } \bar{z}.$$

and then write the right side in standard form.

EXAMPLE 7   **Dividing Complex Numbers**

Write the following quotients in standard form.

a. $\dfrac{1}{2 + i}$ **b.** $\dfrac{4 + \sqrt{-25}}{2 - \sqrt{-9}}$

SOLUTION

a. The denominator is $2 + i$, so its conjugate is $2 - i$.

$$\frac{1}{2 + i} = \frac{1(2 - i)}{(2 + i)(2 - i)} \quad \text{Multiply numerator and denominator by } 2 - i.$$

$$= \frac{2 - i}{2^2 + 1^2} \quad (2 + i)(2 - i) = 2^2 + 1^2$$

$$= \frac{2 - i}{5} = \frac{2}{5} + \frac{-1}{5}i \quad \text{Standard form}$$

b. We write $\sqrt{-25} = 5i$ and $\sqrt{-9} = 3i$ so that $\dfrac{4 + \sqrt{-25}}{2 - \sqrt{-9}} = \dfrac{4 + 5i}{2 - 3i}$.

$$\frac{4 + 5i}{2 - 3i} = \frac{(4 + 5i)(2 + 3i)}{(2 - 3i)(2 + 3i)} \quad \text{Multiply numerator and denominator by } 2 + 3i.$$

$$= \frac{8 + 12i + 10i + 15i^2}{2^2 + 3^2} \quad \text{Use FOIL in the numerator;}$$
$$(2 - 3i)(2 + 3i) = 2^2 + 3^2.$$

$$= \frac{-7 + 22i}{13} \quad 15i^2 = -15, 8 - 15 = -7$$

$$= -\frac{7}{13} + \frac{22}{13}i \quad \text{Standard form} \qquad ■■■$$

Practice Problem 7   Write the following quotients in standard form.

a. $\dfrac{2}{1 - i}$ **b.** $\dfrac{-3i}{4 + \sqrt{-25}}$ ■

EXAMPLE 8   **Using Complex Numbers in AC Circuits**

In a parallel circuit, the total impendence Z_t is given by

$$Z_t = \frac{Z_1 Z_2}{Z_1 + Z_2}.$$

Find Z_t assuming that $Z_1 = 2 + 3i$ ohms and $Z_2 = 3 - 4i$ ohms.

SOLUTION

We first calculate $Z_1 + Z_2$ and $Z_1 Z_2$.

$$Z_1 + Z_2 = (2 + 3i) + (3 - 4i) = 5 - i$$

$$Z_1 Z_2 = (2 + 3i)(3 - 4i)$$

$$= 6 - 8i + 9i - 12i^2 \quad \text{FOIL}$$

$$= 18 + i \quad \text{Set } i^2 = -1 \text{ and simplify.}$$

$$Z_t = \frac{Z_1 Z_2}{Z_1 + Z_2} = \frac{18 + i}{5 - i}$$

$$= \frac{(18 + i)(5 + i)}{(5 - i)(5 + i)} \quad \text{Multiply numerator and denominator by conjugate of the denominator.}$$

$$= \frac{90 + 18i + 5i + i^2}{5^2 + 1^2} \quad \text{Use FOIL in the numerator;}$$
$$(5 - i)(5 + i) = 5^2 + 1^2.$$

$$= \frac{89 + 23i}{26} \quad \text{Set } i^2 = -1 \text{ and simplify.}$$

$$= \frac{89}{26} + \frac{23}{26}i \quad \text{Standard form} \qquad ■■■$$

Practice Problem 8   Repeat Example 8 assuming that $Z_1 = 1 + 2i$ and $Z_2 = 2 - 3i$.

■

SECTION 1.3 ■ Exercises

A EXERCISES Basic Skills and Concepts

1. We define $i =$ _____ so that $i^2 =$ _____.

2. A complex number in the form $a + bi$ is said to be in

   _____.

3. For $b > 0$, $\sqrt{-b} =$ _____.

4. The conjugate of $a + bi$ is _____, and the conjugate of $a - bi$ is _____.

5. *True or False* The product of a complex number and its conjugate is a real number.

6. *True or False* Division by a nonzero complex number z is accomplished by multiplying the numerator and denominator by $\bar{z}$.

In Exercises 7–10, use the definition of equality of complex numbers to find the real numbers x and y so that the equation is true.

7. $2 + xi = y + 3i$

8. $x - 2i = 7 + yi$

9. $x - \sqrt{-16} = 2 + yi$

10. $3 + yi = x - \sqrt{-25}$

In Exercises 11–34, perform each operation and write the result in the standard form $a + bi$.

11. $(5 + 2i) + (3 + i)$

12. $(6 + i) + (1 + 2i)$

13. $(4 - 3i) - (5 + 3i)$

14. $(3 - 5i) - (3 + 2i)$

15. $(-2 - 3i) + (-3 - 2i)$

16. $(-5 - 3i) + (2 - i)$

17. $3(5 + 2i)$

18. $4(3 + 5i)$

19. $-4(2 - 3i)$

20. $-7(3 - 4i)$

21. $3i(5 + i)$

22. $2i(4 + 3i)$

23. $4i(2 - 5i)$

24. $-3i(5 - 2i)$

25. $(3 + i)(2 + 3i)$

26. $(4 + 3i)(2 + 5i)$

27. $(2 - 3i)(2 + 3i)$

28. $(4 - 3i)(4 + 3i)$

29. $(3 + 4i)(4 - 3i)$

30. $(-2 + 3i)(-3 + 10i)$

31. $(\sqrt{3} - 12i)^2$

32. $(-\sqrt{5} - 13i)^2$

33. $(2 - \sqrt{-16})(3 + 5i)$

34. $(5 - 2i)(3 + \sqrt{-25})$

In Exercises 35–40, write the conjugate $\bar{z}$ of each complex number z. Then find $z\bar{z}$.

35. $z = 2 - 3i$

36. $z = 4 + 5i$

37. $z = \dfrac{1}{2} - 2i$

38. $z = \dfrac{2}{3} + \dfrac{1}{2}i$

39. $z = \sqrt{2} - 3i$

40. $z = \sqrt{5} + \sqrt{3}i$

In Exercises 41–54, write each quotient in the standard form $a + bi$.

41. $\dfrac{5}{-i}$

42. $\dfrac{2}{-3i}$

43. $\dfrac{-1}{1 + i}$

44. $\dfrac{1}{2 - i}$

45. $\dfrac{5i}{2 + i}$

46. $\dfrac{3i}{2 - i}$

47. $\dfrac{2 + 3i}{1 + i}$

48. $\dfrac{3 + 5i}{4 + i}$

49. $\dfrac{2 - 5i}{4 - 7i}$

50. $\dfrac{3 + 5i}{1 - 3i}$

51. $\dfrac{2 + \sqrt{-4}}{1 + i}$

52. $\dfrac{5 - \sqrt{-9}}{3 + 2i}$

53. $\dfrac{-2 + \sqrt{-25}}{2 - 3i}$

54. $\dfrac{-5 - \sqrt{-4}}{5 - \sqrt{-9}}$

B EXERCISES Applying the Concepts

55. **Series circuits.** Assuming that the impedance of a resistor in a circuit is $Z_1 = 4 + 3i$ ohms and the impedance of a second resistor is $Z_2 = 5 - 2i$ ohms, find the total impedance of the two resistors when placed in series (sum of the two impedances).

Series circuit

56. **Parallel circuits.** If the two resistors in Exercise 55 are connected in parallel, the total impedance is given by

$$\frac{Z_1 Z_2}{Z_1 + Z_2}.$$

Find the total impedance assuming that the resistors in Exercise 55 are connected in parallel.

Parallel circuit

As with impedance, the current I and voltage V in a circuit can be represented by complex numbers. The three quantities (voltage, V; impedance, Z; and current, I) are related by the equation $Z = \dfrac{V}{I}$. Thus, if two of these values are given, the value of the third can be found from the equation.

In Exercises 57–62, use the equation $V = ZI$ to find the value that is not specified.

57. **Finding impedance:** $I = 7 + 5i$ $V = 35 + 70i$

58. **Finding impedance:** $I = 7 + 4i$ $V = 45 + 88i$

59. **Finding voltage:** $Z = 5 - 7i$ $I = 2 + 5i$

60. **Finding voltage:** $Z = 7 - 8i$ $I = \dfrac{1}{3} + \dfrac{1}{6}i$

61. **Finding current:** $V = 12 + 10i$ $Z = 12 + 6i$

62. **Finding current:** $V = 29 + 18i$ $Z = 25 + 6i$

C EXERCISES Beyond the Basics

In Exercises 63–72, find each power of i and simplify the expression.

63. i^{17}

64. i^{125}

65. i^{-7}

66. i^{-24}

67. $i^{10} + 7$

68. $9 + i^3$

69. $3i^5 - 2i^3$

70. $5i^6 - 3i^4$

71. $2i^3(1 + i^4)$

72. $5i^5(i^3 - i)$

73. Prove that the reciprocal of $a + bi$, where both a and b are nonzero, is $\dfrac{a}{a^2 + b^2} - \dfrac{b}{a^2 + b^2}i$.

In Exercises 74–77, let $z = a + bi$ and $w = c + di$. Prove each statement.

74. $\text{Re}(z) = \dfrac{z + \bar{z}}{2}$

75. $\text{Im}(z) = \dfrac{z - \bar{z}}{2i}$

76. $\text{Re}\left(\dfrac{z}{z + w}\right) + \text{Re}\left(\dfrac{w}{z + w}\right) = 1$

77. The product $z\bar{z} = 0$ if and only if $z = 0$.

In Exercises 78–81, use the definition of equality of complex numbers to solve for x and y.

78. $\dfrac{x}{i} + y = 3 + i$

79. $x - \dfrac{y}{i} = 4i + 1$

80. $\dfrac{x + yi}{i} = 5 - 7i$

81. $\dfrac{5x + yi}{2 - i} = 2 + i$

82. Show that $\dfrac{1 - 2i}{5 - 5i} = \dfrac{2 - 3i}{9 - 7i}$.

83. Write in standard form:

$$\dfrac{1 + i}{1 - i} \div \dfrac{2 + i}{1 + 2i}$$

84. Solve for z:

$$(1 + 3i)z + (2 + 4i) = 7 - 3i$$

85. Let $z = 2 - 3i$ and $w = 1 + 2i$.
   a. Show that $\overline{(zw)} = (\bar{z})(\bar{w})$.
   b. Show that $\overline{\left(\dfrac{z}{w}\right)} = \dfrac{\bar{z}}{\bar{w}}$.

Critical Thinking

86. State whether each of the following is true or false. Explain your reasoning.
   a. Every real number is a complex number.
   b. Every complex number is a real number.
   c. Every complex number is an imaginary number.
   d. A real number is a complex number whose imaginary part is 0.
   e. The product of a complex number and its conjugate is a real number.
   f. The equality $z = \bar{z}$ holds if and only if z is a real number.

GROUP PROJECT

Show that the set of complex numbers does not have the ordering properties of the set of real numbers. [*Hint:* Assume that ordering properties hold. Then by the law of trichotomy, $i = 0$ or $i < 0$ or $i > 0$. Show that each leads to a contradiction.]

Quadratic Equations

Before Starting this Section, Review

1. Factoring (Section P.4, page 41)
2. Square roots (Section P.6, page 59)
3. Rationalizing a denominator (Section P.6, page 62)
4. Linear equations (Section 1.1, page 85)

Objectives

1 Solve a quadratic equation by factoring.

2 Solve a quadratic equation by the square root method.

3 Solve a quadratic equation by completing the square.

4 Solve a quadratic equation by using the quadratic formula.

5 Solve a quadratic equations with complex solutions.

6 Solve applied problems.

The Parthenon

The Pyramid of Giza

THE GOLDEN RECTANGLE

The ancient Greeks, in about the sixth century B.C., sought unifying principles of beauty and perfection, which they believed could be described mathematically. In their study of beauty, the Greeks used the term *golden ratio*. The golden ratio is frequently referred to as "phi," after the Greek sculptor Phidias, and is symbolized by the Greek letter Φ (phi). A geometric figure that is commonly associated with phi is the **golden rectangle**. (See Example 11.) This rectangle has sides of lengths a and b that are in proportion to the golden ratio. It has been said that the golden rectangle is the most pleasing rectangle to the eye. Numerous examples of art and architecture have employed the golden rectangle. Here are three of them:

The Mona Lisa

1. The exterior dimensions of the Parthenon in Athens, built about 440 B.C., form a perfect golden rectangle.
2. Leonardo Da Vinci called the golden ratio the "divine proportion" and featured it in many of his paintings, including the famous *Mona Lisa*. (Try drawing a rectangle around her face.)
3. The Great Pyramid of Giza is believed to be 4600 years old, a time long before the Greeks. Its dimensions are also based on the golden ratio.

In Example 11, we show that the numerical value of the golden ratio can be found by solving the quadratic equation $x^2 - x - 1 = 0$. ∎

BY THE WAY . . .

The word *quadratic* comes from the Latin *quadratus*, meaning square.

> **QUADRATIC EQUATION**
>
> A quadratic equation in the variable x is an equation equivalent to the equation
>
> $$ax^2 + bx + c = 0,$$
>
> where a, b, and c are real numbers and $a \neq 0$.

A quadratic equation written in the form $ax^2 + bx + c = 0$ is said to be in **standard form**. Here are some examples of quadratic equations that are not in standard form.

$$2x^2 - 5x + 7 = x^2 + 3x - 1$$
$$3y^2 + 4y + 1 = 6y - 5$$
$$x^2 - 2x = 3$$

We discuss four methods for solving quadratic equations: (1) by factoring, (2) by taking square roots, (3) by completing the square, and (4) by using the quadratic formula.

We begin by studying real number solutions of quadratic equations.

1 Solve a quadratic equation by factoring.

Factoring Method

Some quadratic equations in the standard form $ax^2 + bx + c = 0$ can be solved by factoring and using the **zero-product property**.

> **ZERO–PRODUCT PROPERTY**
>
> Let A and B be two algebraic expressions. Then $AB = 0$ if and only if $A = 0$ or $B = 0$.

EXAMPLE 1   **Solving a Quadratic Equation by Factoring**

Solve by factoring: $2x^2 + 5x = 3$

SOLUTION

$$
\begin{aligned}
2x^2 + 5x &= 3 & &\text{Original equation} \\
2x^2 + 5x - 3 &= 0 & &\text{Subtract 3 from both sides.} \\
(2x - 1)(x + 3) &= 0 & &\text{Factor the left side.}
\end{aligned}
$$

We now set each factor equal to 0 and solve the resulting linear equations. The vertical line separates the computations leading to the two solutions.

$$
\begin{array}{c|c l}
2x - 1 = 0 & x + 3 = 0 & \text{Zero-product property} \\
2x = 1 & x = -3 & \text{Isolate the } x \text{ term on one side.} \\
x = \dfrac{1}{2} & x = -3 & \text{Solve for } x.
\end{array}
$$

Check: You should check the solutions $x = \dfrac{1}{2}$ and $x = -3$ in the original equation.

The solution set is $\left\{ \dfrac{1}{2}, -3 \right\}$.   ■ ■ ■

Practice Problem 1   Solve by factoring: $x^2 + 25x = -84$ ■

EXAMPLE 2   Solving a Quadratic Equation by Factoring

Solve by factoring: $3t^2 = 2t$

SOLUTION

$$
\begin{array}{ll}
3t^2 = 2t & \text{Original equation} \\
3t^2 - 2t = 0 & \text{Subtract } 2t \text{ from both sides.} \\
t(3t - 2) = 0 & \text{Factor.} \\
t = 0 \quad | \quad 3t - 2 = 0 & \text{Zero-product property} \\
 \quad 3t = 2 & \text{Isolate the } t \text{ terms on one side.} \\
 \quad t = \dfrac{2}{3} & \text{Solve for } t.
\end{array}
$$

Check: You should check the solutions $t = 0$ and $t = \dfrac{2}{3}$ in the original equation.

The solution set is $\left\{0, \dfrac{2}{3}\right\}$.   ■ ■ ■

Practice Problem 2   Solve by factoring: $2m^2 = 5m$ ■

SOLVING A QUADRATIC EQUATION BY FACTORING

Step 1 Write the given equation in standard form so that one side is 0.

Step 2 Factor the nonzero side of the equation from Step 1.

Step 3 Set each factor obtained in Step 2 equal to 0.

Step 4 Solve the resulting equations in Step 3.

Step 5 Check the solutions obtained in Step 4 in the original equation.

◆ **WARNING**   The factoring method of solving an equation works *only when one of the sides of the equation is* 0. If neither side of an equation is 0, as in $(x + 1)(x - 3) = 10$, we cannot know the value of either factor. For example, we could have $x + 1 = 5$ and $x - 3 = 2$, or $x + 1 = 0.1$ and $x - 3 = 100$. There are countless possibilities for two factors whose product is 10.

EXAMPLE 3   Solving a Quadratic Equation by Factoring

Solve by factoring: $x^2 + 16 = 8x$

SOLUTION

Step 1   Write the equation in standard form.

$$
\begin{array}{ll}
x^2 + 16 = 8x & \text{Original equation} \\
x^2 + 16 - 8x = 0 & \text{Subtract } 8x \text{ from both sides.} \\
x^2 - 8x + 16 = 0 & \text{Standard form}
\end{array}
$$

Step 2   Factor the left side of the equation.

$$(x - 4)(x - 4) = 0 \qquad \text{Perfect-square trinomial}$$

Steps 3 and 4   Set each factor equal to 0 and solve each resulting equation.

$$
\begin{array}{ll}
x - 4 = 0 \quad | \quad x - 4 = 0 \\
 x = 4 \quad | \quad x = 4
\end{array}
$$

Step 5   Check the solution in the original equation.

Check: $x = 4$ because $x = 4$ is the only possible solution.

$$x^2 + 16 = 8x \qquad \text{Original equation}$$
$$(4)^2 + 16 \overset{?}{=} 8(4) \qquad \text{Replace } x \text{ with } 4.$$
$$32 = 32 \checkmark$$

Thus, 4 is the only solution of $x^2 + 16 = 8x$ and the solution set is $\{4\}$.

■ ■ ■

Practice Problem 3   Solve by factoring: $x^2 - 6x = -9$ ■

In Example 3, because the factor $(x - 4)$ appears twice in the solution, the number 4 is called a **double root**, or a **root of multiplicity 2**, of the given equation.

2 Solve a quadratic equation by the square root method.

The Square Root Method

The solutions of an equation of the type $x^2 = d$ are obtained by the square root method. Suppose we are interested in solving the equation $x^2 = 3$, which is equivalent to the equation $x^2 - 3 = 0$. Applying the factoring method but *removing the restriction that coefficients and constants represent only integers*, we have

$$x^2 - 3 = 0$$
$$x^2 - (\sqrt{3})^2 = 0 \qquad \text{Because } 3 = (\sqrt{3})^2$$
$$(x + \sqrt{3})(x - \sqrt{3}) = 0 \qquad \text{Factor (using real numbers).}$$
$$x + \sqrt{3} = 0 \quad \text{or} \quad x - \sqrt{3} = 0 \qquad \text{Set each factor equal to zero.}$$
$$x = -\sqrt{3} \quad \text{or} \qquad x = \sqrt{3} \qquad \text{Solve for } x.$$

Thus, the solutions of the equation $x^2 = 3$ are $-\sqrt{3}$ and $\sqrt{3}$. Because the solutions differ only in sign, we use the notation $\pm\sqrt{3}$ to write the two numbers $\sqrt{3}$ and $-\sqrt{3}$. In a similar way, we see that for any nonnegative real number d, the solutions of the quadratic equation $x^2 = d$ are $\pm\sqrt{d}$.

SQUARE ROOT PROPERTY

Suppose u is any algebraic expression and $d \geq 0$. If $u^2 = d$, then $u = \pm\sqrt{d}$.

EXAMPLE 4   **Solving an Equation by the Square Root Method**

Solve: $(x - 3)^2 = 5$

SOLUTION

$$(x - 3)^2 = 5 \qquad \text{Original equation}$$
$$u^2 = 5 \qquad \text{Let } u = x - 3.$$
$$u = \pm\sqrt{5} \qquad \text{Square root property}$$
$$x - 3 = \pm\sqrt{5} \qquad \text{Replace } u \text{ with } x - 3.$$
$$x = 3 \pm \sqrt{5} \qquad \text{Add 3 to both sides.}$$

The solution set is $\{3 + \sqrt{5}, 3 - \sqrt{5}\}$.

■ ■ ■

Practice Problem 4   Solve: $(x + 2)^2 = 5$ ■

3 Solve a quadratic equation by completing the square.

Completing the Square

A method called **completing the square** can be used to solve quadratic equations that cannot be solved by factoring and are not in the correct form to use the square root method. The method requires two steps. First, we write a given quadratic equation in the form $(x + k)^2 = d$. Then we solve the equation $(x + k)^2 = d$ by the square root method.

Recall that

$$(x + k)^2 = x^2 + 2kx + k^2.$$

Notice that the coefficient of x on the right side is $2k$, and half of this coefficient is k; that is, $\frac{1}{2}(2k) = k$. We see that the constant term, k^2, in the trinomial $x^2 + 2kx + k^2$ is the *square of one-half the coefficient of x*.

PERFECT-SQUARE TRINOMIAL

A quadratic trinomial in x with coefficient of x^2 equal to 1 is a **perfect-square trinomial** if the constant term is the square of one-half the coefficient of x.

When the constant term is *not* the square of half the coefficient of x, we subtract the constant term from both sides and add a new constant term that will result in a perfect-square trinomial.

Let's look at some examples to learn what number should be added to $x^2 + bx$ to create a perfect-square trinomial.

$x^2 + bx$	b	$\dfrac{b}{2}$	Add $\left(\dfrac{b}{2}\right)^2$	Result is $\left(x + \dfrac{b}{2}\right)^2$
$x^2 + 6x$	6	3	$3^2 = 9$	$x^2 + 6x + 9 = (x + 3)^2$
$x^2 - 4x$	-4	-2	$(-2)^2 = 4$	$x^2 - 4x + 4 = (x - 2)^2$
$x^2 + 3x$	3	$\dfrac{3}{2}$	$\left(\dfrac{3}{2}\right)^2 = \dfrac{9}{4}$	$x^2 + 3x + \dfrac{9}{4} = \left(x + \dfrac{3}{2}\right)^2$
$x^2 - x$	-1	$-\dfrac{1}{2}$	$\left(-\dfrac{1}{2}\right)^2 = \dfrac{1}{4}$	$x^2 - x + \dfrac{1}{4} = \left(x - \dfrac{1}{2}\right)^2$

Thus, to make $x^2 + bx$ into a perfect square, we need to add

$$\left[\frac{1}{2}(\text{coefficient of } x)\right]^2 = \left[\frac{1}{2}(b)\right]^2 = \frac{b^2}{4}$$

so that

$$x^2 + bx + \frac{b^2}{4} = \left(x + \frac{b}{2}\right)^2.$$

We say that $\dfrac{b^2}{4}$ was added to $x^2 + bx$ to *complete the square*. (See Figure 1.4.)

Area of shaded portion
$$= x \cdot x + x \cdot \frac{b}{2} + x \cdot \frac{b}{2}$$
$$= x^2 + bx$$

Area of unshaded portion
$$= \frac{b}{2} \cdot \frac{b}{2} = \frac{b^2}{4}$$

FIGURE 1.4 Area of a square $= \left(x + \dfrac{b}{2} \right)^2$

EXAMPLE 5   Solving a Quadratic Equation by Completing the Square

Solve by completing the square: $x^2 + 10x + 8 = 0$

SOLUTION

First, isolate the constant on the right side.

$$x^2 + 10x + 8 = 0 \qquad \text{Original equation}$$
$$x^2 + 10x = -8 \qquad \text{Subtract 8 from both sides.}$$

Next, add $\left[\dfrac{1}{2}(10) \right]^2 = 5^2 = 25$ to both sides to complete the square on the left side.

$$x^2 + 10x + 25 = -8 + 25 \qquad \text{Add 25 to each side.}$$
$$(x + 5)^2 = 17 \qquad x^2 + 10x + 25 = (x + 5)^2$$
$$u^2 = 17 \qquad \text{Let } u = x + 5.$$
$$u = \pm\sqrt{17} \qquad \text{Solve for } u.$$
$$x + 5 = \pm\sqrt{17} \qquad \text{Replace } u \text{ with } x + 5.$$
$$x = -5 \pm \sqrt{17} \qquad \text{Subtract 5 from each side.}$$

Thus, $x = -5 + \sqrt{17}$ or $x = -5 - \sqrt{17}$, and the solution set is $\{-5 + \sqrt{17}, -5 - \sqrt{17}\}$. ■ ■ ■

Practice Problem 5 Solve by completing the square: $x^2 - 6x + 7 = 0$ ■

We summarize the procedure for solving a quadratic equation by completing the square.

METHOD OF COMPLETING THE SQUARE

Step 1 Rearrange the quadratic equation so that the terms in x^2 and x are on the left side of the equation and the constant term is on the right side.

Step 2 Make the coefficient of x^2 equal to 1 by dividing both sides of the equation by the original coefficient. (Steps 1 and 2 are interchangeable.)

Step 3 Add the square of one-half the coefficient of x to both sides of the equation.

Step 4 Write the equation in the form $(x + k)^2 = d$ using the fact that the left side is a perfect square.

Step 5 Take the square root of each side, prefixing $\pm$ to the right side.

Step 6 Solve the two equations resulting from Step 5.

EXAMPLE 6   **Solving a Quadratic Equation by Completing the Square**

Solve by completing the square: $3x^2 - 4x - 1 = 0$

SOLUTION

Step 1 $3x^2 - 4x - 1 = 0$ Note that the coefficient of x^2 is 3.

$3x^2 - 4x = 1$ Add 1 to both sides.

Step 2 $x^2 - \dfrac{4}{3}x = \dfrac{1}{3}$ Divide both sides by 3.

Step 3 $x^2 - \dfrac{4}{3}x + \left(-\dfrac{2}{3}\right)^2 = \dfrac{1}{3} + \left(-\dfrac{2}{3}\right)^2$ Add $\left[\dfrac{1}{2}\left(-\dfrac{4}{3}\right)\right]^2 = \left(-\dfrac{2}{3}\right)^2$ to both sides.

Step 4 $\left(x - \dfrac{2}{3}\right)^2 = \dfrac{7}{9}$ $x^2 - \dfrac{4}{3}x + \left(-\dfrac{2}{3}\right)^2 = \left(x - \dfrac{2}{3}\right)^2$

Step 5 $x - \dfrac{2}{3} = \pm\sqrt{\dfrac{7}{9}}$ Take the square root of both sides.

$x - \dfrac{2}{3} = \pm\dfrac{\sqrt{7}}{3}$ $\sqrt{\dfrac{7}{9}} = \dfrac{\sqrt{7}}{\sqrt{9}} = \dfrac{\sqrt{7}}{3}$

Step 6 $x = \dfrac{2}{3} \pm \dfrac{\sqrt{7}}{3}$ Add $\dfrac{2}{3}$ to both sides.

$x = \dfrac{2 \pm \sqrt{7}}{3}$ Add fractions.

The solution set is $\left\{\dfrac{2 - \sqrt{7}}{3}, \dfrac{2 + \sqrt{7}}{3}\right\}$.   ■ ■ ■

Practice Problem 6   Solve by completing the square: $4x^2 - 24x + 25 = 0$ ■

4 Solve a quadratic equation by using the quadratic formula.

The Quadratic Formula

We can generalize the method of completing the square to derive a formula that gives a solution of *any* quadratic equation.

We solve the standard form of the quadratic equation by completing the square.

$$ax^2 + bx + c = 0, \ a \neq 0$$

Step 1 $ax^2 + bx = -c$ Isolate the constant on the right side.

Step 2 $x^2 + \dfrac{b}{a} = -\dfrac{c}{a}$ Divide both sides by a.

Step 3 $x^2 + \dfrac{b}{a}x + \left(\dfrac{b}{2a}\right)^2 = -\dfrac{c}{a} + \left(\dfrac{b}{2a}\right)^2$ Add the square of one-half the coefficient of x to both sides.

Step 4 $\left(x + \dfrac{b}{2a}\right)^2 = -\dfrac{c}{a} + \dfrac{b^2}{4a^2}$ The left side in Step 3 is a perfect square.

$\left(x + \dfrac{b}{2a}\right)^2 = \dfrac{b^2 - 4ac}{4a^2}$ Rearrange and combine fractions on the right side.

Step 5 $x + \dfrac{b}{2a} = \pm\sqrt{\dfrac{b^2 - 4ac}{4a^2}}$ Take the square roots of both sides.

Step 6   Solve the equations in Step 5.

$$x + \frac{b}{2a} = \pm\sqrt{\frac{b^2 - 4ac}{4a^2}}$$ Equations from Step 5

$$x = -\frac{b}{2a} \pm \frac{\sqrt{b^2 - 4ac}}{2|a|}$$ Add $-\dfrac{b}{2a}$ to both sides; $\sqrt{4a^2} = \sqrt{4}\sqrt{a^2} = 2|a|$.

$$x = -\frac{b}{2a} \pm \frac{\sqrt{b^2 - 4ac}}{2a}$$ $\pm 2|a| = \pm 2a$ because $|a| = a$ or $|a| = -a$.

$$x = \frac{-b \pm \sqrt{b^2 - 4ac}}{2a}$$ Combine fractions.

The last equation in Step 6 is called the **quadratic formula**.

QUADRATIC FORMULA

The solutions of the quadratic equation in the standard form
$ax^2 + bx + c = 0$ with $a \neq 0$ are given by the formula

$$x = \frac{-b \pm \sqrt{b^2 - 4ac}}{2a}.$$

◆ **WARNING**   To use the quadratic formula, write the given quadratic equation in standard form. Then determine the values of a (coefficient of x^2), b (coefficient of x), and c (constant term).

EXAMPLE 7   **Solving a Quadratic Equation by Using the Quadratic Formula**

Solve $3x^2 = 5x + 2$ by using the quadratic formula.

SOLUTION

The equation $3x^2 = 5x + 2$ can be rewritten in the standard form by subtracting $5x + 2$ from both sides.

$$3x^2 - 5x - 2 = 0$$ Rewrite in standard form.

$$\underset{\substack{\uparrow \\ a}}{3x^2} + \underset{\substack{\uparrow \\ b}}{(-5)x} + \underset{\substack{\uparrow \\ c}}{(-2)} = 0$$ Identify values of a, b, and c to be used in the quadratic formula.

Then $a = 3$, $b = -5$, and $c = -2$, and we have

$$x = \frac{-b \pm \sqrt{b^2 - 4ac}}{2a}$$ Quadratic formula

$$x = \frac{-(-5) \pm \sqrt{(-5)^2 - 4(3)(-2)}}{2(3)}$$ Substitute 3 for a, -5 for b, and -2 for c.

$$= \frac{5 \pm \sqrt{25 + 24}}{6}$$ Simplify.

$$= \frac{5 \pm \sqrt{49}}{6} = \frac{5 \pm 7}{6}$$ Simplify.

Then

$$x = \frac{5+7}{6} = \frac{12}{6} = 2 \text{ or } x = \frac{5-7}{6} = \frac{-2}{6} = -\frac{1}{3}, \text{ and the solution set is } \left\{-\frac{1}{3}, 2\right\}.$$

Now you should solve $3x^2 = 5x + 2$ by factoring to confirm that you get the same result. ■ ■ ■

Practice Problem 7   Solve $6x^2 - x - 2 = 0$ by using the quadratic formula. ■

5 Solve quadratic equations with complex solutions.

Quadratic Equations with Complex Solutions

The methods of solving quadratic equations are also applicable to solving quadratic equations with complex solutions.

> **EXAMPLE 8**   **Solving Quadratic Equations with Complex Solutions**

Solve each equation.

a. $x^2 + 4 = 0$ **b.** $x^2 + 2x + 2 = 0$

SOLUTION

a. Use the square root method.

$$x^2 + 4 = 0 \qquad \text{Original equation}$$
$$x^2 = -4 \qquad \text{Add } -4 \text{ to both sides.}$$
$$x = \pm\sqrt{-4} = \pm(\sqrt{4})i = \pm 2i \qquad \text{Solve for } x.$$

The solution set is $\{-2i, 2i\}$. You should check these solutions.

b. $x^2 + 2x + 2 = 1 \cdot x^2 + 2x + 2 = 0$

$$x = \frac{-2 \pm \sqrt{2^2 - 4(1)(2)}}{2(1)} \qquad \text{Use the quadratic formula with } a = 1, b = 2, \text{ and } c = 2.$$

$$= \frac{-2 \pm \sqrt{-4}}{2} \qquad \text{Simplify.}$$

$$= \frac{-2 \pm 2i}{2} = \frac{2(-1 \pm i)}{2}$$

$$= -1 \pm i$$

The solution set is $\{-1 - i, -1 + i\}$. You should check these solutions. ■ ■ ■

Practice Problem 8   Solve.

a. $4x^2 + 9 = 0$ **b.** $x^2 = 4x - 13$ ■

The Discriminant

If a, b, and c are real numbers and $a \neq 0$, we can predict the number and type of solutions that the equation $ax^2 + bx + c = 0$ will have without solving the equation. In the quadratic formula

$$x = \frac{-b \pm \sqrt{b^2 - 4ac}}{2a},$$

the quantity $b^2 - 4ac$ under the radical sign is called the **discriminant** of the equation. The discriminant reveals the type of solutions of the equation, as shown in the table on page 126.

Discriminant	Description of Solutions
$b^2 - 4ac > 0$	There are two unequal real solutions.
$b^2 - 4ac = 0$	There is one real solution (a root of multiplicity 2).
$b^2 - 4ac < 0$	There are two nonreal unequal complex solutions.

EXAMPLE 9   **Using the Discriminant**

Use the discriminant to determine the number and type of solutions of each quadratic equation.

a. $x^2 - 4x + 2 = 0$ **b.** $2t^2 + 2t + 19 = 0$ **c.** $4x^2 + 4x + 1 = 0$

SOLUTION

Equation	$b^2 - 4ac$	Conclusion
a. $x^2 - 4x + 2 = 0$	$(-4)^2 - 4(1)(2) = 8 > 0$	Two unequal real solutions
b. $2t^2 + 2t + 19 = 0$	$(2)^2 - 4(2)(19) = -148 < 0$	Two nonreal complex solutions
c. $4x^2 + 4x + 1 = 0$	$(4)^2 - 4(4)(1) = 0$	Exactly one real solution

▪ ▪ ▪

Practice Problem 9   Determine the number and type of solutions.

a. $9x^2 - 6x + 1 = 0$ **b.** $x^2 - 5x + 3 = 0$ **c.** $2x^2 - 3x + 4 = 0$.   ▪

6 Solve applied problems.

Applications

EXAMPLE 10   **Partitioning a Building**

A rectangular building whose depth (from the front of the building) is three times its frontage is divided into two parts by a partition that is 45 feet from and parallel to the front wall. Assuming that the rear portion of the building contains 2100 square feet, find the dimensions of the building.

SOLUTION

In solving problems of this type, it is advisable to draw a diagram such as Figure 1.5. We let

$$x = \text{the frontage of the building, in feet.}$$

Then $3x = \text{the depth of the building, in feet,}$

and $3x - 45 = \text{the depth of the rear portion, in feet.}$

The area of the rear portion of the building is $x(3x - 45)$ square feet. This area is 2100 square feet, so

$x(3x - 45) = 2100$	
$3x^2 - 45x = 2100$	Distributive property
$3x^2 - 45x - 2100 = 0$	Subtract 2100 from both sides.
$x^2 - 15x - 700 = 0$	Divide both sides by 3.
$(x - 35)(x + 20) = 0$	Factor.
$x - 35 = 0 \quad \text{or} \quad x + 20 = 0$	Zero-product property
$x = 35 \quad \text{or} \qquad x = -20$	Solve for x.

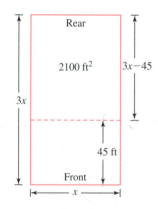

Rear

2100 ft² ┤ $3x - 45$

$3x$

45 ft

Front

x

FIGURE 1.5

Because the dimensions of the building are positive numbers, we reject $x = -20$. Thus,

$$x = 35 \qquad \text{Frontage in feet}$$
$$3x = 105 \qquad \text{Depth in feet}$$

Check: $3x - 45 = 3(35) - 45 = 105 - 45 = 60$ is the depth of the rear portion, and the area of the rear portion is $60 \cdot 35 = 2100$ square feet.   ■ ■ ■

Practice Problem 10   In Example 10, find approximate dimensions of the building assuming that its depth is five times its frontage.   ■

STUDY TIP

Recall that two rectangles are similar if the lengths of corresponding sides are proportional.

Golden Rectangle

A rectangle of length p and width q, with $p > q$, is called a *golden rectangle* if you can divide the rectangle into a square with side of length q and a smaller rectangle that is similar to the original one. See Figure 1.6. The ratio of the longer side to the shorter side of a golden rectangle is called the *golden ratio*. In Figure 1.6, the golden ratio is $\dfrac{p}{q}$.

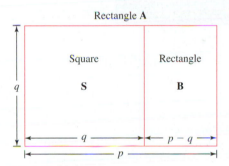

FIGURE 1.6 Golden Rectangle

EXAMPLE 11   **Calculating the Golden Ratio**

Use Figure 1.6 of a golden rectangle to calculate the golden ratio.

SOLUTION

The rectangle **A** in Figure 1.6 is a golden rectangle. This means that we can divide rectangle **A** into a square **S** and a smaller rectangle **B** such that

$$\frac{\text{Longer side of } \mathbf{A}}{\text{Shorter side of } \mathbf{A}} = \frac{\text{Longer side of } \mathbf{B}}{\text{Shorter side of } \mathbf{B}}.$$

Thus, because rectangle **B** is the smaller rectangle, $p - q < q$, and we have

$$\frac{p}{q} = \frac{q}{p - q}$$
$$p(p - q) = q \cdot q \qquad \text{Multiply both sides by } q(p - q) \text{; simplify.}$$
$$p^2 - pq = q^2 \qquad \text{Distributive property}$$
$$\left(\frac{p}{q}\right)^2 - \frac{p}{q} = 1 \qquad \text{Divide both sides by } q^2 \text{; simplify.}$$
$$\Phi^2 - \Phi = 1 \qquad \text{Replace } \frac{p}{q} \text{ with } \Phi \text{, the golden ratio.}$$

continued on the next page

$$\Phi^2 - \Phi - 1 = 0 \qquad \text{Write in standard form, treating } \Phi \text{ as the variable.}$$

$$\Phi = \frac{-(-1) \pm \sqrt{(-1)^2 - 4(1)(-1)}}{2(1)} \qquad \text{Substitute 1 for } a \text{ and } -1 \text{ for both } b \text{ and } c.$$

$$= \frac{1 \pm \sqrt{5}}{2} \qquad \text{Simplify.}$$

Because the ratio of two positive numbers is positive, we reject the negative root $\frac{1 - \sqrt{5}}{2}$. Therefore, the golden ratio is

$$\Phi = \frac{1 + \sqrt{5}}{2} \approx 1.618.$$

Practice Problem 11   Find the length of a golden rectangle whose width is 36 feet.

SECTION 1.4 ■ Exercises

A EXERCISES Basic Skills and Concepts

1. Any equation of the form $ax^2 + bx + c = 0$ with $a \neq 0$ is called a(n) _____ equation.

2. If $P(x)$, $D(x)$, and $Q(x)$ are polynomials and $P(x) = D(x)Q(x)$, then the solutions of $P(x) = 0$ are the solutions of $Q(x) = 0$ together with the solutions of _____ = 0.

3. If you complete the square in the quadratic equation $ax^2 + bx + c = 0$, you get the quadratic formula for the solutions $x = $ _____.

4. If $b^2 - 4ac < 0$, the quadratic equation has _____ solutions; if $b^2 - 4ac = 0$, the equation has _____ solution; $b^2 - 4ac > 0$, the equation has _____ real solutions.

5. *True or False* Every quadratic equation has, at most, two real solutions.

6. *True or False* If we are given $x^2 + kx$, we form a perfect square by adding k^2.

In Exercises 7–14, show by substitution whether the number r is a solution of the corresponding quadratic equation.

7. $x^2 + 4x - 12 = 0$; $r = -6$

8. $x^2 - 8x - 9 = 0$; $r = 9$

9. $3x^2 + 7x - 6 = 0$; $r = \dfrac{2}{3}$

10. $2x^2 - 5x - 3 = 0$; $r = -\dfrac{1}{2}$

11. $x^2 - 4x + 1 = 0$; $r = 2 - \sqrt{3}$

12. $x^2 - 6x + 1 = 0$; $r = 3 + 2\sqrt{2}$

13. $4x^2 - 8x + 13 = 0$; $r = 2 + 3i$

14. $x^2 - 6x + 13 = 5$; $r = 5 - 2i$

15. Find k assuming that $x = 1$ is a solution of the equation $kx^2 + x - 3 = 0$.

16. Find k assuming that $x = \sqrt{7}$ is a solution of the equation $kx^2 + x - 3 = 0$.

In Exercises 17–36, solve each equation by factoring.

17. $x^2 - 5x = 0$

18. $x^2 - 5x + 4 = 0$

19. $x^2 + 5x = 14$

20. $x^2 - 11x = 12$

21. $x^2 = 5x + 6$

22. $x = x^2 - 12$

23. $2x^2 + 5x - 3 = 0$

24. $2x^2 - 9x + 10 = 0$

25. $3y^2 + 5y + 2 = 0$

26. $6x^2 + 11x + 4 = 0$

27. $5x^2 + 12x + 4 = 0$

28. $3x^2 - 2x - 5 = 0$

29. $2x^2 + x = 15$

30. $6x^2 = 1 - x$

31. $12x^2 - 10x = 12$

32. $-x^2 + 10x + 1200 = 0$

33. $18x^2 - 45x = -7$

34. $18x^2 + 57x + 45 = 0$

35. $4x^2 - 10x - 750 = 0$

36. $12x^2 + 43x + 36 = 0$

In Exercises 37–46, solve each equation by the square root method.

37. $3x^2 = 48$

38. $2x^2 = 50$

39. $x^2 + 1 = 5$

40. $2x^2 - 1 = 17$

41. $x^2 + 5 = 1$

42. $4x^2 + 9 = 0$

43. $(x - 1)^2 = 16$

44. $(2x - 3)^2 = 25$

45. $(3x - 2)^2 + 16 = 0$

46. $(2x + 3)^2 + 25 = 0$

In Exercises 47–56, add a constant term to the expression to make it a perfect square.

47. $x^2 + 4x$ **48.** $y^2 + 10y$

49. $x^2 + 6x$ **50.** $y^2 - 8y$

51. $x^2 - 7x$ **52.** $x^2 - 3x$

53. $x^2 + \dfrac{1}{3}x$ **54.** $x^2 - \dfrac{3}{2}x$

55. $x^2 + ax$ **56.** $x^2 - \dfrac{2a}{3}x$

In Exercises 57–70, solve each equation by completing the square.

57. $x^2 + 2x - 5 = 0$ **58.** $x^2 + 6x = -7$

59. $x^2 - 3x - 1 = 0$ **60.** $x^2 - x - 3 = 0$

61. $2r^2 + 3r = 9$ **62.** $3k^2 - 5k + 1 = 0$

63. $z^2 - 2z + 2 = 0$ **64.** $x^2 - 6x + 11 = 0$

65. $2x^2 - 20x + 49 = -7$ **66.** $4y^2 + 4y + 5 = 0$

67. $5x^2 - 6x = 4x^2 + 6x - 3$

68. $x^2 + 7x - 5 = x - x^2$

69. $5y^2 + 10y + 4 = 2y^2 + 3y + 1$

70. $3x^2 - 1 = 5x^2 - 3x - 5$

In Exercises 71–86, solve each equation by using the quadratic formula.

71. $x^2 + 2x - 4 = 0$ **72.** $m^2 + 3m + 2 = 0$

73. $6x^2 = 7x + 5$ **74.** $t^2 + 7 = 4t$

75. $3z^2 - 2z = 7$ **76.** $6y^2 + 11y = 10$

77. $3p^2 + 8p + 4 = 0$ **78.** $8(x^2 - x) = x^2 - 3$

79. $x^2 = 5(x - 1)$ **80.** $(x + 13)(x + 5) = -2$

81. $t(t + 1) = 3t^2 + 1$

82. $3(x^2 + 1) = 2x^2 + 4x + 1$

83. $2t^2 - 5 = 0$ **84.** $3k^2 - 48 = 0$

85. $9k^2 + 25 = 0$ **86.** $3k^2 + 4 = 0$

In Exercises 87–94, find the discriminant and determine the number and type of roots of each equation.

87. $4x^2 - 12x + 9 = 0$ **88.** $3x^2 - 4x + 3 = 0$

89. $2y^2 = 6 - y$ **90.** $9y^2 + 24y = -16$

91. $17x - 12 = 6x^2$ **92.** $5x^2 = 7x - 3$

93. $-3x^2 + 21 = 0$ **94.** $x^2 + 2x + \dfrac{1}{2} = 0$

In Exercises 95–98, write your answer rounded to two decimal places.

95. Find the length of the golden rectangle whose width is 14.72 in.

96. Find the length of the golden rectangle whose width is 18.63 ft.

97. Find the width of the golden rectangle whose length is 8.46 cm.

98. Find the width of the golden rectangle whose length is 4.68 m.

B EXERCISES Applying the Concepts

99. Dimensions. The length of a rectangular plot is three times its width. Assuming that the area of the plot is 10,800 square feet, find the dimensions of the plot.

100. Dimensions. The diagonal of a square tile is 4 inches longer than its side. Find the length of the side.

101. Finding integers. Find two numbers whose sum is 28 and whose product is 147.

102. Finding integers. Find an integer such that the sum of twice the square of the integer and the integer itself is 55.

103. Geometry. The length of a rectangle is 5 centimeters greater than its width. The area of the rectangle is 500 square centimeters. Find the dimensions of the rectangle.

104. Geometry. The sides of a rectangle are in the ratio $3:2$. The area of the rectangle is 216 square centimeters. Find the dimensions of the rectangle.

105. Cutting a wire. A wire of length 16 inches is to be cut into two pieces; then each piece will be bent to form a square. Find the length of the two pieces assuming that the sum of the areas of the two squares is 10 square inches.

106. Cutting a wire. A piece of wire is 38 inches long. The wire is cut into two pieces; then each piece is bent to form a square. Find the length of each piece if three times the area of the larger square exceeds the area of the smaller square by 95.75 square inches.

107. Manufacturing. The surface area A of a cylinder with height h and radius r is given by the equation $A = 2\pi rh + 2\pi r^2$. A company makes soup cans by using 32π square inches of aluminum sheet for each can. Assuming that the height of the can is 6 inches, find the radius of the can.

108. Concrete patio. To make a rectangular concrete patio, a homeowner used 70 feet of forming, into which she poured 138 cubic feet of concrete to form a slab 6 inches thick. What were the dimensions of the patio?

109. Making a box. The volume of a box is given by the equation $V = \text{length} \cdot \text{width} \cdot \text{height}$. A topless box is to be made from a square sheet of tin by cutting 5-inch squares from each of the four corners and then shaping the tin into an open box by turning up the sides. Assuming that the box is to hold 480 cubic inches, find the size of the piece of tin to be used.

110. Making a box. A 2-inch square is cut from each corner of a rectangular piece of cardboard whose length exceeds the width by 4 inches. The sides are then turned up to form an open box. Assuming that the volume of the box is 64 cubic inches, find the dimensions of the box.

111. Bus travel. Two buses leave Atlanta at the same time, one traveling west at 40 miles per hour and the other traveling north at 30 miles per hour. When will they be 400 miles apart?

112. Plane travel. Two planes left San Francisco at 2 P.M. One flew east at a certain speed, and the other flew south at a speed that was 100 kilometers per hour more than the speed of the plane heading east. The planes were 2100 kilometers apart at 5 P.M. How fast was each plane flying?

113. Dimensions of a family room. The length of a rectangular rug is $\frac{7}{6}$ times its width. The rug is placed on a rectangular family room floor, with a margin of 1 foot all around. The area of the floor is four times that of the margin. Find the dimensions of the family room.

114. Gardening. You want to expand your present 25-foot-by-15-foot garden by planting a border of flowers. The border is to be of the same width around the entire garden. The flowers you bought will fill an area of 624 square feet. How wide should the border be?

115. Sky divers. Sky divers are in free fall from the time they jump out of a plane until they open their parachutes. A sky diver jumps from 5000 feet. The diver's height h above the ground t seconds after the jump is described by the equation $h = -16t^2 + 5000$. Find the time during which the diver is in free fall assuming that the parachute opens at 1000 feet.

116. Physics. Suppose you throw a ball straight up from the ground with a velocity of 112 feet per second. As the ball moves up, gravity slows it. Eventually, the ball begins to fall back to the ground. The height h of the ball after t seconds in the air is given by the equation $h = -16t^2 + 112t$.
  a. Find the height of the ball after two seconds.
  b. How long will it take the ball to reach a height of 96 feet?
  c. How long after you throw the ball will it return to the ground?

Height of a projectile. The height, h, of a projectile thrown straight up from the top of a 480-foot hill t seconds after being thrown upward with initial velocity v_0 ft/sec is given by $h = -16t^2 + v_0 t + 480$. In Exercises 117–120, for a given value of v_0, find the time(s) when the projectile will (a) be at a height of 592 ft above the ground and (b) crash on the ground. Round your answers to two decimal places.

117. $v_0 = 96$ **118.** $v_0 = 112$

119. $v_0 = 256$ **120.** $v_0 = 64$

C EXERCISES Beyond the Basics

In Exercises 121–126, find the values of k for which the given equation has equal roots.

121. $x^2 - kx + 3 = 0$ **122.** $x^2 + 3kx + 8 = 0$

123. $2x^2 + kx + k = 0$ **124.** $kx^2 + 2x + 6 = 0$

125. $x^2 + k^2 = 2(k + 1)x$ **126.** $kx^2 + (k + 3)x + 4 = 0$

127. Assuming that r and s are the roots of the quadratic equation $ax^2 + bx + c = 0$, show that
$$r + s = -\frac{b}{a} \quad \text{and} \quad r \cdot s = \frac{c}{a}.$$

128. Find the sum and the product of the roots of the following equations without solving the equation.
  a. $3x^2 + 5x - 5 = 0$
  b. $3x^2 - 7x = 1$
  c. $\sqrt{3}x^2 = 3x + 4$
  d. $(1 + \sqrt{2})x^2 - \sqrt{2}x + 5 = 0$

In Exercises 129 and 130, determine k so that the sum and the product of the roots are equal.

129. $2x^2 + (k - 3)x + 3k - 5 = 0$

130. $5x^2 + (2k - 3)x - 2k + 3 = 0$

131. Suppose r and s are the roots of the quadratic equation $ax^2 + bx + c = 0$. Show that you can write $ax^2 + bx + c = a(x - r)(x - s)$. Thus, the quadratic trinomial can be written in factored form.
[*Hint:* From Exercise 127, $b = -a(r + s)$ and $c = ars$. Substitute in $ax^2 + bx + c$ and factor.]

132. Use Exercise 131 to factor each trinomial.
 a. $4x^2 + 4x - 5$
 b. $25x^2 + 40x + 11$
 c. $25x^2 - 30x + 34$
 d. $72x^2 + 95x - 1000$

133. If we know the roots r and s in advance, then by Exercise 131, a quadratic equation with roots r and s can be expressed in the form

$$a(x - r)(x - s) = 0.$$

Form a quadratic equation that has the listed numbers as roots.
 a. $-3, 4$
 b. $5, 5$
 c. $3 + \sqrt{2}, 3 - \sqrt{2}$

134. Solve the equation $3x^2 - 4xy + 5 - y^2 = 0$
 a. for x in terms of y.
 b. for y in terms of x.

135. **How to construct a golden rectangle.** Draw a square $ABCD$ with side of length x, as in the accompanying diagram. Let M be the midpoint of DC. The segment MB sweeps out the arc BF so that F lies on the line through the segment DC. Show that the rectangle $AEFD$ is a golden rectangle.

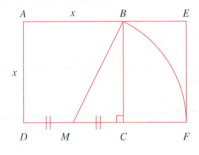

136. If the two roots of the quadratic equation

$$ax^2 + bx + c = 0$$

are reciprocals of each other, which of the following must be true?
 i. $b = c$ **ii.** $a = c$ **iii.** $a = 0$ **iv.** $b = 0$

137. Use Exercise 127 to show that a quadratic equation cannot have three distinct roots.

138. Suppose p is a positive integer greater than 1. Which of the following describe the roots of the equation $x^2 - (p + 1)x + p = 0$?
 i. real and equal   **ii.** real and irrational
 iii. complex   **iv.** integers and unequal

139. Amar, Akbar, and Anthony attempt to solve a barely legible quadratic equation written on an old paper. Amar misreads the constant term and finds the solutions 8 and 2. Akbar misreads the coefficient of x and finds the roots -9 and -1. Anthony solves the correct equation. What are Anthony's solutions of the equation?

Critical Thinking

140. If $a, b,$ and k are real numbers, show that the solutions of the equation $(x - a)(x - b) = k^2$ are real.

141. Show that for $0 < a < 4$, the equation $ax(1 - x) = 1$ has no real solutions.

Solving Other Types of Equations

Objectives

1. Solve equations by factoring.
2. Solve rational equations.
3. Solve equations involving radicals.
4. Solve equations with rational exponents.
5. Solve equations that are quadratic in form.

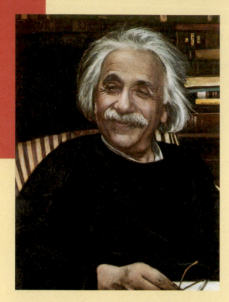

Albert Einstein 1879–1955

Einstein began his schooling in Munich, Germany. Einstein renounced his German citizenship in 1896 and entered the Swiss Federal Polytechnic School in Zurich to be trained as a teacher in physics and mathematics. Unable to find a teaching job, he accepted a position as a technical assistant in the Swiss patent office. In 1905, Einstein earned his doctorate from the University of Zurich.

While at the patent office, he produced much of his remarkable work on the special theory of relativity. In 1909, he was appointed Professor Extraordinary at Zurich. In 1914, he was appointed professor at the University of Berlin. He immigrated to America to take the position of professor of physics at Princeton University. By 1949, Einstein was unwell. He drew up his will in 1950.

He left his scientific papers to the Hebrew University in Jerusalem. In 1952, the Israeli government offered the post of second president to Einstein. He politely declined the offer.

One week before his death, Einstein agreed to let Bertrand Russell place his name on a manifesto urging all nations to give up nuclear weapons.

TIME DILATION

Do you ever feel like time moves really quickly or really slowly? For example, don't the hours fly by when you are hanging out with your best friend or the seconds drag on endlessly when a dentist is working on your teeth? But you cannot really speed up or slow time down, right? It always "flows" at the same rate.

Einstein said "no." He asserted in his "Special Theory of Relativity" that time is relative. It speeds up or slows down depending on how fast one object is moving relative to another. He concluded that the closer we come to traveling at the speed of light, the more time will slow down for us relative to someone not moving. He called this slowing of time due to motion *time dilation*. Time dilation can be quantified by the equation

$$t_0 = t\sqrt{1 - \frac{v^2}{c^2}},$$

where t_0 = the "proper time" measured in the moving frame of reference,

t = the time observed in the *fixed* frame of reference,

v = the speed of the moving frame of reference, and

c = the speed of light.

We investigate this effect in Example 12. ■

1 Solve equations by factoring.

Solving Equations by Factoring

The method of factoring used in solving quadratic equations can also be used to solve certain nonquadratic equations. Specifically, we use the following strategy for solving those equations that can be expressed in factored form with 0 *on one side*.

FINDING THE SOLUTION: A PROCEDURE

EXAMPLE 1 **Solving Equations by Factoring**

OBJECTIVE
Use factoring techniques to solve an equation.

Step 1 **Make one side zero.** Move all nonzero terms in the equation to one side (say the left side) so that the other side (the right side) is 0.

Step 2 **Factor the left side.**

Step 3 **Use the zero-product property.** Set each factor in Step 2 equal to 0 and then solve the resulting equations.

Step 4 **Check your solutions.**

EXAMPLE
Solve $x^3 + 3x^2 = 4x + 12$.

$$x^3 + 3x^2 - 4x - 12 = 0 \qquad \text{Subtract } 4x + 12 \text{ from both sides.}$$

$$x^2(x + 3) - 4(x + 3) = 0 \qquad \text{Group for factoring.}$$
$$(x + 3)(x^2 - 4) = 0 \qquad \text{Factor out } x + 3.$$
$$(x + 3)(x + 2)(x - 2) = 0 \qquad \text{Factor } x^2 - 4.$$

$$x + 3 = 0 \quad \text{or} \quad x + 2 = 0 \quad \text{or} \quad x - 2 = 0$$
$$x = -3 \quad \text{or} \quad x = -2 \quad \text{or} \quad x = 2$$

Check:
Check the solutions in the original equation.

$$\text{For } x = -3, (-3)^3 + 3(-3)^2 \overset{?}{=} 4(-3) + 12$$
$$-27 + 27 = -12 + 12 \checkmark$$

You should also verify that $x = -2$ and $x = 2$ are solutions of $x^3 + 3x^2 = 4x + 12$. The solution set is $\{-3, -2, 2\}$. ■ ■ ■

Practice Problem 1 Solve: $x^3 + 2x^2 - x - 2 = 0$ ■

EXAMPLE 2 **Solving an Equation by Factoring**

Solve by factoring: $x^4 = 9x^2$

SOLUTION
Step 1 **Move all terms to the left side.**

$$x^4 = 9x^2 \qquad \text{Original equation}$$
$$x^4 - 9x^2 = 0 \qquad \text{Subtract } 9x^2 \text{ from both sides.}$$

RECALL

$$A^2 - B^2 = (A + B)(A - B)$$

Difference of two squares, page 35

Step 2 **Factor.**

$$x^2(x^2 - 9) = 0 \qquad x^4 - 9x^2 = x^2(x^2 - 9)$$
$$x^2(x + 3)(x - 3) = 0 \qquad x^2 - 9 = (x + 3)(x - 3)$$

Step 3 **Set each factor equal to zero.**

$$x^2 = 0 \quad \text{or} \quad x + 3 = 0 \quad \text{or} \quad x - 3 = 0 \qquad \text{Zero-product property}$$
$$x = 0 \quad \text{or} \quad x = -3 \quad \text{or} \quad x = 3 \qquad \text{Solve each equation for } x.$$

Step 4 **Check:**

Let $x = 0$	Let $x = -3$	Let $x = 3$
$(0)^4 \overset{?}{=} 9(0)^2$	$(-3)^4 \overset{?}{=} 9(-3)^2$	$(3)^4 \overset{?}{=} 9(3)^2$
$0 = 0 \checkmark$	$81 = 81 \checkmark$	$81 = 81 \checkmark$

Thus, the solution set of the equation is $\{-3, 0, 3\}$. Note that 0 is a root of multiplicity 2.

Practice Problem 2   Solve: $x^4 = 4x^2$

◆ **WARNING**

A common error when solving an equation such as $x^4 = 9x^2$ in Example 2 is to divide both sides by x^2, obtaining $x^2 = 9$ so that ± 3 are the two solutions. This leads to the loss of the solution 0. Remember, we can divide only by quantities that are not 0!

EXAMPLE 3   **Solving an Equation by Factoring**

Solve by factoring: $x^3 - 2x^2 = -x + 2$

SOLUTION

Step 1   **Move all terms to the left side.**

$$
\begin{array}{ll}
x^3 - 2x^2 = -x + 2 & \text{Original equation} \\
x^3 - 2x^2 + x - 2 = -x + 2 + x - 2 & \text{Add } x - 2 \text{ to both sides.} \\
x^3 - 2x^2 + x - 2 = 0 & \text{Simplify.}
\end{array}
$$

Step 2   **Factor.** Because there are four terms, we factor by grouping.

$$
\begin{array}{ll}
(x^3 - 2x^2) + (x - 2) = 0 & \\
x^2(x - 2) + 1 \cdot (x - 2) = 0 & \text{Factor } x^2 \text{ from the first two terms} \\
& \text{and rewrite } x - 2 \text{ as } 1 \cdot (x - 2). \\
(x - 2)(x^2 + 1) = 0 & \text{Factor out the common factor } (x - 2).
\end{array}
$$

Step 3   **Set each factor equal to zero.**

$$
\begin{array}{c|c}
x - 2 = 0 & x^2 + 1 = 0 \\
x = 2 & x^2 = -1 \\
& x = \pm i
\end{array}
$$

Step 4   **Check:**

$$
\begin{array}{c|c|c}
\text{Let } x = 2. & \text{Let } x = i. & \text{Let } x = -i. \\
(2)^3 - 2(2)^2 \overset{?}{=} -2 + 2 & (i)^3 - 2(i)^2 \overset{?}{=} -(i) + 2 & (-i)^3 - 2(-i)^2 \overset{?}{=} -(-i) + 2 \\
8 - 8 \overset{?}{=} 0 & -i + 2 = -i + 2 \checkmark & i + 2 = i + 2 \checkmark \\
0 = 0 \checkmark & &
\end{array}
$$

Thus, the solution set of the equation is $\{2, i, -i\}$.

Practice Problem 3   Solve by factoring: $x^3 - 5x^2 = 4x - 20$

2 Solve rational equations.

Rational Equations

Recall that if at least one algebraic expression with the variable in the denominator appears in an equation, then the equation is a *rational equation*.

When we multiply a rational equation by an expression containing the variable, we may introduce a solution that satisfies the new equation but does not satisfy the original equation. Such a solution is called an **extraneous solution** or **extraneous root**.

EXAMPLE 4   **Solving a Rational Equation**

Solve: $\dfrac{1}{6} + \dfrac{1}{x+1} = \dfrac{1}{x}$

SOLUTION

Step 1   **Find the LCD.** The LCD from the denominators

$$6, x+1, \text{ and } x \text{ is } 6x(x+1).$$

Step 2   **Multiply both sides of the equation by the LCD and make the right side 0.**

$$6x(x+1)\left[\dfrac{1}{6} + \dfrac{1}{x+1}\right] = 6x(x+1)\left[\dfrac{1}{x}\right] \qquad \text{Multiply both sides by the LCD.}$$

$$\dfrac{6x(x+1)}{6} + \dfrac{6x(x+1)}{x+1} = \dfrac{6x(x+1)}{x} \qquad \text{Distributive property}$$

$$x(x+1) + 6x = 6(x+1) \qquad \text{Simplify.}$$

$$x^2 + x + 6x = 6x + 6 \qquad \text{Distributive property}$$

$$x^2 + x - 6 = 0 \qquad \text{Subtract } 6x + 6 \text{ from both sides.}$$

Step 3   **Factor.**

$$x^2 + x - 6 = 0$$

$$(x+3)(x-2) = 0 \qquad \text{Factor.}$$

Step 4   **Set each factor equal to zero.**

$$x + 3 = 0 \qquad \text{or} \qquad x - 2 = 0$$

$$x = -3 \qquad \text{or} \qquad x = 2$$

Step 5   **Check:** You should check the solutions, -3 and 2, in the original equation. The solution set is $\{-3, 2\}$. ■ ■ ■

Practice Problem 4   Solve: $\dfrac{1}{x} - \dfrac{12}{5x+10} = \dfrac{1}{5}$ ■

EXAMPLE 5   **Solving a Rational Equation with an Extraneous Solution**

Solve: $\dfrac{x}{x-1} - \dfrac{1}{x+1} = \dfrac{2x}{x^2-1}$

SOLUTION

Step 1   The LCD of the denominators $(x-1), (x+1), (x^2-1) = (x-1)(x+1)$ is $(x-1)(x+1)$.

Step 2 $(x-1)(x+1)\left[\dfrac{x}{x-1} - \dfrac{1}{x+1}\right]$ Multiply both sides by the LCD.

$$= (x-1)(x+1)\left[\dfrac{2x}{x^2-1}\right]$$

$$(x-1)(x+1)\dfrac{x}{x-1} - (x-1)(x+1)\dfrac{1}{x+1} \qquad \text{Distributive property;}$$
$$(x-1)(x+1)$$
$$= (x^2-1)\cdot\dfrac{2x}{x^2-1} \qquad = x^2 - 1$$

$$(x+1)x - (x-1) = 2x \qquad \text{Simplify.}$$

$$x^2 + x - x + 1 - 2x = 2x - 2x \qquad \text{Distributive property; add } -2x \text{ to both sides.}$$

$$x^2 - 2x + 1 = 0 \qquad \text{Simplify.}$$

Step 3 $(x - 1)(x - 1) = 0$ Factor.

Step 4 $x - 1 = 0$ or $x = 1$ Solve for x.

Step 5 **Check:** Substitute $x = 1$ in the original equation.

$$\frac{x}{x - 1} - \frac{1}{x + 1} = \frac{2x}{x^2 - 1}$$

$$\frac{1}{1 - 1} - \frac{1}{1 + 1} \stackrel{?}{=} \frac{2(1)}{1^2 - 1}$$

The denominator of the first term is 0, so the equation is undefined for $x = 1$. Therefore, $x = 1$ is not a solution of the original equation; it is extraneous. The original equation has no solution. ■ ■ ■

Practice Problem 5   Solve: $\dfrac{x}{x - 2} - \dfrac{2}{x + 2} = \dfrac{4x}{x^2 - 4}$ ■

3 Solve equations involving radicals.

Equations Involving Radicals

If an equation involves radicals or rational exponents, the method of raising both sides to a positive integer power is often used to remove them. When this is done, the solution set of the new equation always contains the solutions of the original equation. However, in some cases, the new equation has *more* solutions than the original equation. For instance, consider the equation $x = 4$. If we square both sides, we get $x^2 = 16$. Notice that the given equation, $x = 4$, has only one solution (namely, 4), whereas the new equation, $x^2 = 16$, has two solutions: 4 and -4. Extraneous solutions may be introduced whenever both sides of an equation are raised to an *even* power, Thus, it is *essential that we check* all solutions obtained after raising both sides of an equation to any even power.

EXAMPLE 6   **Solving Equations Involving Radicals**

Solve: $x = \sqrt{6x - x^3}$

SOLUTION

Because $(\sqrt{a})^2 = a$, we raise both sides of the equation to the second power to eliminate the radical sign. The solutions of the original equation are among the solutions of the equation

$$x^2 = (\sqrt{6x - x^3})^2$$ Square both sides of the equation.

$$x^2 = 6x - x^3$$ $(\sqrt{6x - x^3})^2 = 6x - x^3$

$$x^3 + x^2 - 6x = 0$$ Add $x^3 - 6x$ to both sides.

$$x(x^2 + x - 6) = 0$$ Factor out x from each term.

$$x(x + 3)(x - 2) = 0$$ Factor $x^2 + x - 6$.

$$x = 0 \quad \text{or} \quad x + 3 = 0 \quad \text{or} \quad x - 2 = 0$$ Zero-product property

$$x = 0 \quad \text{or} \quad x = -3 \quad \text{or} \quad x = 2$$ Solve each equation.

Hence, the only possible solutions of $x = \sqrt{6x - x^3}$ are $-3, 0$, and 2.

Check:

Let $x = -3$.	Let $x = 0$.	Let $x = 2$.
$-3 \stackrel{?}{=} \sqrt{6(-3) - (-3)^3}$	$0 \stackrel{?}{=} \sqrt{6(0) - (0)^3}$	$2 \stackrel{?}{=} \sqrt{6(2) - (2)^3}$
$-3 \stackrel{?}{=} \sqrt{-18 + 27}$	$0 \stackrel{?}{=} \sqrt{0}$	$2 \stackrel{?}{=} \sqrt{12 - 8}$
$-3 \stackrel{?}{=} \sqrt{9}$	$0 \stackrel{?}{=} 0 ✓$	$2 \stackrel{?}{=} \sqrt{4}$
$-3 \neq 3 ✗$		$2 = 2 ✓$

Thus, -3 is an extraneous solution; 0 and 2 are the solutions of the equation $x = \sqrt{6x - x^3}$. The solution set is $\{0, 2\}$.   ■ ■ ■

Practice Problem 6   Solve: $x = \sqrt{x^3 - 6x}$ ■

EXAMPLE 7   **Solving an Equation Involving a Radical**

Solve: $\sqrt{2x + 1} + 1 = x$

SOLUTION

Step 1   **We begin by isolating the radical to one side of the equation.**

$$\sqrt{2x + 1} + 1 = x \qquad \text{Original equation}$$
$$\sqrt{2x + 1} = x - 1 \qquad \text{Subtract 1 from both sides.}$$

Step 2   **Next, we square both sides and simplify.**

$$(\sqrt{2x + 1})^2 = (x - 1)^2 \qquad \text{Square both sides.}$$
$$2x + 1 = x^2 - 2x + 1 \qquad (\sqrt{a})^2 = a$$
$$2x + 1 - (2x + 1) = x^2 - 2x + 1 - (2x + 1) \qquad \text{Subtract } 2x + 1 \text{ from both sides.}$$
$$0 = x^2 - 4x \qquad \text{Simplify.}$$
$$0 = x(x - 4) \qquad \text{Factor.}$$

Step 3   **Set each factor equal to zero.**

$$x = 0 \quad \text{or} \quad x - 4 = 0$$
$$x = 0 \quad \text{or} \qquad x = 4 \qquad \text{Solve for } x.$$

Step 4   **Check:**

Let $x = 0$.	Let $x = 4$.
$\sqrt{2(0) + 1} + 1 \overset{?}{=} 0$	$\sqrt{2(4) + 1} + 1 \overset{?}{=} 4$
$\sqrt{1} + 1 \overset{?}{=} 0$	$\sqrt{9} + 1 \overset{?}{=} 4$
$1 + 1 \overset{?}{=} 0$	$3 + 1 \overset{?}{=} 4$
$2 \neq 0$	$4 = 4 \checkmark$

Thus, 0 is an extraneous root; the only solution of the given equation is 4, and the solution set is $\{4\}$.   ■ ■ ■

Practice Problem 7   Solve: $\sqrt{6x + 4} + 2 = x$ ■

Next, we illustrate a method for solving radical equations that contain two or more square root expressions.

EXAMPLE 8   **Solving an Equation Involving Two Radicals**

Solve: $\sqrt{2x - 1} - \sqrt{x - 1} = 1$

SOLUTION

Step 1   **We first isolate one of the radicals.**

$$\sqrt{2x - 1} - \sqrt{x - 1} = 1 \qquad \text{Original equation}$$
$$\sqrt{2x - 1} = 1 + \sqrt{x - 1} \qquad \text{Add } \sqrt{x - 1} \text{ to both sides to isolate the radical } \sqrt{2x - 1}.$$

Step 2   **Square both sides of the last equation.**

$$(\sqrt{2x-1})^2 = (1 + \sqrt{x-1})^2 \qquad \text{Square both sides.}$$
$$2x - 1 = 1^2 + 2 \cdot 1 \cdot \sqrt{x-1} + (\sqrt{x-1})^2 \qquad (A+B)^2 = A^2 + 2AB + B^2$$
$$2x - 1 = 1 + 2\sqrt{x-1} + x - 1 \qquad \text{Simplify; use } (\sqrt{a})^2 = a.$$
$$2x - 1 = 2\sqrt{x-1} + x \qquad \text{Simplify.}$$

Step 3   **The last equation still contains a radical. We repeat the process of isolating the radical expression.**

$$x - 1 = 2\sqrt{x-1} \qquad \text{Subtract } x \text{ from both sides and}$$
$$\text{simplify to isolate } 2\sqrt{x-1}.$$
$$(x-1)^2 = (2\sqrt{x-1})^2 \qquad \text{Square both sides.}$$
$$x^2 - 2x + 1 = 4(x-1) \qquad \text{Use } (A-B)^2 = A^2 - 2AB + B^2$$
$$\text{and } (AB)^2 = A^2 B^2.$$
$$x^2 - 2x + 1 = 4x - 4 \qquad \text{Distributive property}$$
$$x^2 - 6x + 5 = 0 \qquad \text{Add } -4x + 4 \text{ to both sides; simplify.}$$
$$(x-5)(x-1) = 0 \qquad \text{Factor.}$$

Step 4   **Set each factor equal to 0.**

$$x - 5 = 0 \quad \text{or} \quad x - 1 = 0$$
$$x = 5 \quad \text{or} \quad x = 1 \qquad \text{Solve for } x.$$

Step 5   **Check:** You should check that the solutions of the original equation are 1 and 5. The solution set is $\{1, 5\}$.   ■ ■ ■

Practice Problem 8   Solve: $\sqrt{x-5} + \sqrt{x} = 5$ ■

SOLVING EQUATIONS CONTAINING SQUARE ROOTS

Step 1   Isolate one radical to one side of the equation.

Step 2   Square both sides of the equation in Step 1 and simplify.

Step 3   If the equation obtained in Step 2 contains a radical, repeat Steps 1 and 2 to get an equation that is free of radicals.

Step 4   Solve the equation obtained in Steps 1–3.

Step 5   Check the solutions in the original equation.

---

4   Solve equations with rational exponents.

Equations with Rational Exponents

If a given equation can be expressed in the form $u^{m/n} = k$, then we isolate u by raising both sides to the power $\dfrac{n}{m}$ $\left(\text{the reciprocal of } \dfrac{m}{n} \right)$.

SOLVING EQUATIONS OF THE FORM $u^{m/n} = k$

Let m and n be positive integers, k a real number, and $\dfrac{m}{n}$ in lowest terms. Then if

     m is odd   **m is even**

    and $u^{m/n} = k$,   and $u^{m/n} = k$,

       $u = k^{n/m}$ $u = \pm k^{n/m}$

EXAMPLE 9   **Solving Equations with Rational Exponents**

Solve.

a. $2(2x - 1)^{3/2} - 30 = 24$ **b.** $(3x - 1)^{2/3} + 5 = 9$

SOLUTION

a. $2(2x - 1)^{3/2} - 30 = 24$ Original equation

$2(2x - 1)^{3/2} = 24 + 30 = 54$ Add 30 to both sides and simplify.

$(2x - 1)^{3/2} = \dfrac{54}{2} = 27$ Divide both sides by 2.

$((2x - 1)^{3/2})^{2/3} = (27)^{2/3}$ Raise both sides to the power $\dfrac{2}{3}$

with odd $m = 3$.

$2x - 1 = 9$ $(27)^{2/3} = \left(\sqrt[3]{27}\right)^2 = 3^2 = 9$

$2x = 10$ Isolate x.

$x = 5$ Solve for x.

Check: $2(2x - 1)^{3/2} - 30 = 24$

$2(2 \cdot 5 - 1)^{3/2} - 30 \overset{?}{=} 24$

$2(9)^{3/2} - 30 \overset{?}{=} 24$ $2 \cdot 5 - 1 = 9$

$2(27) - 30 \overset{?}{=} 24$ $(9)^{3/2} = (\sqrt{9})^3 = (3)^3 = 27$

$54 - 30 \overset{?}{=} 24$ ✓ Yes

b. $(3x - 1)^{2/3} + 5 = 9$ Original equation

$(3x - 1)^{2/3} = 9 - 5 = 4$ Subtract 5 from both sides and simplify.

$\left[(3x - 1)^{2/3}\right]^{3/2} = \pm(4)^{3/2}$ Because $m = 2$ is even, we use the $\pm$ sign.

$3x - 1 = \pm 8$ $(4)^{3/2} = (\sqrt{4})^3 = (2)^3 = 8$

$3x = 1 \pm 8$ Add 1 to both sides.

Then

or

$3x = 1 + 8$	$3x = 1 - 8$
$3x = 9$	$3x = -7$
$x = 3$	$x = -\dfrac{7}{3}$

Verify that both $x = 3$ and $x = -\dfrac{7}{3}$ satisfy the original equation. The solution set is $\left\{-\dfrac{7}{3}, 3\right\}$.   ■ ■ ■

Practice Problem 9

Solve: **a.** $3(x - 2)^{3/5} + 4 = 7$ **b.** $(2x + 1)^{4/3} - 7 = 9$ ■

5 Solve equations that are quadratic in form.

Equations That Are Quadratic in Form

An equation in a variable x is **quadratic in form** if it can be written as

$$au^2 + bu + c = 0 \qquad (a \neq 0),$$

where u is an expression in the variable x. We solve the equation $au^2 + bu + c = 0$ for u. Then the solutions of the original equation can be obtained by replacing u with the expression in x that u represents.

EXAMPLE 10   **Solving an Equation Quadratic in Form**

Solve: $x^{2/3} - 5x^{1/3} + 6 = 0$

SOLUTION

The given equation is not a quadratic equation. However, if we let $u = x^{1/3}$, then $u^2 = (x^{1/3})^2 = x^{2/3}$. The original equation becomes a quadratic equation if we replace $x^{1/3}$ with u.

$$
\begin{array}{ll}
x^{2/3} - 5x^{1/3} + 6 = 0 & \text{Original equation} \\
u^2 - 5u + 6 = 0 & \text{Replace } x^{1/3} \text{ with } u. \\
(u - 2)(u - 3) = 0 & \text{Factor.} \\
u - 2 = 0 \quad \text{or} \quad u - 3 = 0 & \text{Zero-product property} \\
u = 2 \quad \text{or} \quad u = 3 & \text{Solve for } u.
\end{array}
$$

Because $u = x^{1/3}$, we find x from the equations $u = 2$ and $u = 3$.

$$
\begin{array}{ll}
x^{1/3} = 2 \quad \text{or} \quad x^{1/3} = 3 & \text{Replace } u \text{ with } x^{1/3}. \\
x = 2^3 \quad \text{or} \quad x = 3^3 & \text{Here } \dfrac{m}{n} = \dfrac{1}{3}, \text{ so } m = 1 \text{ is odd.} \\
x = 8 \quad \text{or} \quad x = 27 & \text{are two apparent solutions.}
\end{array}
$$

Check: You should check that both $x = 8$ and $x = 27$ satisfy the equation. The solution set is $\{8, 27\}$.   ■ ■ ■

Practice Problem 10   Solve: $x^{2/3} - 7x^{1/3} + 6 = 0$ ■

◆ **WARNING**

When u replaces an expression in the variable x in an equation that is quadratic in form, you are not finished once you have found values for u. You then must replace u with the expression in x and solve for x.

EXAMPLE 11   **Solving an Equation That Is Quadratic in Form**

Solve: $\left(x + \dfrac{1}{x}\right)^2 - 6\left(x + \dfrac{1}{x}\right) + 8 = 0$

SOLUTION

We let $u = x + \dfrac{1}{x}$. Then $u^2 = \left(x + \dfrac{1}{x}\right)^2$. With this substitution, the original equation becomes a quadratic equation in u.

$$
\begin{array}{ll}
\left(x + \dfrac{1}{x}\right)^2 - 6\left(x + \dfrac{1}{x}\right) + 8 = 0 & \text{Original equation} \\
u^2 - 6u + 8 = 0 & \text{Replace } x + \dfrac{1}{x} \text{ with } u. \\
(u - 2)(u - 4) = 0 & \text{Factor.} \\
u - 2 = 0 \quad \text{or} \quad u - 4 = 0 & \text{Zero-product property} \\
u = 2 \quad \text{or} \quad u = 4 & \text{Solve for } u.
\end{array}
$$

Replacing u with $x + \dfrac{1}{x}$ in these equations, we have

$$
x + \frac{1}{x} = 2 \quad \text{or} \quad x + \frac{1}{x} = 4.
$$

Next, we solve each of these equations for x.

$$x + \frac{1}{x} = 2$$

$x^2 + 1 = 2x$	Multiply both sides by x (the LCD).
$x^2 - 2x + 1 = 0$	Subtract $2x$ from both sides; simplify.
$(x - 1)^2 = 0$	Factor the perfect-square trinomial.
$x - 1 = 0$	Square root property
$x = 1$	Solve for x.

Thus, $x = 1$ is a possible solution of the original equation.

$$x + \frac{1}{x} = 4$$

$x^2 + 1 = 4x$	Multiply both sides by x (the LCD).
$x^2 - 4x + 1 = 0$	Subtract $4x$ from both sides.
$x = \dfrac{-(-4) \pm \sqrt{(-4)^2 - 4(1)(1)}}{2(1)}$	Quadratic formula: $a = 1, b = -4, c = 1$
$x = \dfrac{4 \pm \sqrt{12}}{2}$	Simplify.
$x = \dfrac{4 \pm 2\sqrt{3}}{2}$	$\sqrt{12} = \sqrt{4 \cdot 3} = \sqrt{4}\sqrt{3} = 2\sqrt{3}$
$x = \dfrac{2(2 \pm \sqrt{3})}{2}$	Factor the numerator.
$x = 2 \pm \sqrt{3}$	Remove the common factor.

Thus, $2 + \sqrt{3}$ and $2 - \sqrt{3}$ are also possible solutions of the original equation. Consequently, the possible solutions of the original equation are 1, $2 + \sqrt{3}$, and $2 - \sqrt{3}$.

You should verify that 1, $2 + \sqrt{3}$, and $2 - \sqrt{3}$ are solutions of the original equation. The solution set is $\{1, 2 - \sqrt{3}, 2 + \sqrt{3}\}$. ■ ■ ■

Practice Problem 11 Solve: $\left(1 + \dfrac{1}{x}\right)^2 - 6\left(1 + \dfrac{1}{x}\right) + 8 = 0$ ■

The next example illustrates the paradoxical effect of time dilation (page 132) when one travels at speeds that are close to the speed of light.

EXAMPLE 12 **Investigating Space Travel (Your Older Sister Is Younger Than You Are)**

Your sister is five years older than you are. She decides that she has had enough of Earth and needs a vacation. She takes a trip to the Omega-One star system. Her trip to Omega-One and back in a spacecraft traveling at an average speed v took 15 years according to the clock and calendar on the spacecraft. But upon landing back on Earth, she discovers that her voyage took 25 years according to the time on Earth. This means that although you were five years younger than your sister before her vacation, you are five years older than her after her vacation! Use the time-dilation equation $t_0 = t\sqrt{1 - \dfrac{v^2}{c^2}}$ from the introduction to this section to calculate the speed of the spacecraft.

SOLUTION

Substitute $t_0 = 15$ (moving-frame time) and $t = 25$ (fixed-frame time) to obtain

$$15 = 25\sqrt{1 - \frac{v^2}{c^2}}$$

$$\frac{3}{5} = \sqrt{1 - \frac{v^2}{c^2}} \qquad \text{Divide both sides by 25; } \frac{15}{25} = \frac{3}{5}.$$

$$\frac{9}{25} = 1 - \frac{v^2}{c^2} \qquad \text{Square both sides.}$$

$$\frac{v^2}{c^2} = 1 - \frac{9}{25} \qquad \text{Add } \frac{v^2}{c^2} - \frac{9}{25} \text{ to both sides.}$$

$$\left(\frac{v}{c}\right)^2 = \frac{16}{25} \qquad \text{Simplify.}$$

$$\frac{v}{c} = \pm\frac{4}{5} \qquad \text{Square root property}$$

$$\frac{v}{c} = \frac{4}{5} \qquad \text{Reject the negative value.}$$

$$v = \frac{4}{5}c = 0.8c \qquad \text{Multiply both sides by } c.$$

Thus, the spacecraft must have been traveling at 80% of the speed of light $(0.8c)$ when your sister found the "trip of youth." ■ ■ ■

Practice Problem 12 Suppose your sister's trip took 20 years according to the clock and calendar on the spacecraft, but 25 years judging by time on Earth. Find the speed of the spacecraft. ■

SECTION 1.5 ■ Exercises

A EXERCISES Basic Skills and Concepts

1. If an apparent solution does not satisfy the equation, it is called a(n) _____ solution.

2. When solving a rational equation, we multiply both sides by the _____.

3. If $x^{3/4} = 8$, then $x =$ _____.

4. If $x^{4/3} = 16$, then $x =$ _____ or $x =$ _____.

5. *True or False* Rational equations always have extraneous solutions.

6. *True or False* If $x^{2/3} = k$, then $x = \pm\sqrt{k^3}$.

In Exercises 7–16, find the real solutions of each equation by factoring.

7. $x^3 = 2x^2$

8. $3x^4 - 27x^2 = 0$

9. $(\sqrt{x})^3 = \sqrt{x}$

10. $(\sqrt{x})^5 = 16\sqrt{x}$

11. $x^3 + x = 0$

12. $x^3 - 1 = 0$

13. $x^4 - x^3 = x^2 - x$

14. $x^3 - 36x = 16(x - 6)$

15. $x^4 = 27x$

16. $3x^4 = 24x$

In Exercises 17–26, solve each equation by multiplying both sides by the LCD.

17. $\dfrac{x + 1}{3x - 2} = \dfrac{5x - 4}{3x + 2}$

18. $\dfrac{x}{2x + 1} = \dfrac{3x + 2}{4x + 3}$

19. $\dfrac{1}{x} + \dfrac{2}{x + 1} = 1$

20. $\dfrac{x}{x + 1} + \dfrac{x + 1}{x + 2} = 1$

21. $\dfrac{6x - 7}{x} - \dfrac{1}{x^2} = 5$

22. $\dfrac{1}{x} + \dfrac{2}{x + 1} + \dfrac{3}{x + 2} = 0$

23. $\dfrac{1}{x} + \dfrac{1}{x - 3} = \dfrac{7}{3x - 5}$

24. $\dfrac{x + 3}{x - 1} + \dfrac{x + 4}{x + 1} = \dfrac{8x + 5}{x^2 - 1}$

25. $\dfrac{5}{x + 1} - \dfrac{4}{2x + 2} + \dfrac{2}{2x - 1} = \dfrac{13}{18}$

26. $\dfrac{6}{2x - 2} - \dfrac{1}{x + 1} = \dfrac{2}{2x + 2} + 1$

In Exercises 27–34, solve each equation. Check your answers.

27. $\dfrac{x}{x-5} - \dfrac{5}{x+5} = \dfrac{10x}{x^2-25}$

28. $\dfrac{x}{x-4} - \dfrac{4}{x+4} = \dfrac{8x}{x^2-16}$

29. $\dfrac{x}{x-3} + \dfrac{3}{x+3} = \dfrac{6x}{x^2-9}$

30. $\dfrac{x}{x+7} + \dfrac{7}{x-7} = \dfrac{14x}{x^2-49}$

31. $\dfrac{1}{x-1} + \dfrac{x}{x+3} = \dfrac{4}{x^2+2x-3}$

32. $\dfrac{x}{x-2} + \dfrac{2}{x+3} = -\dfrac{10}{x^2+x-6}$

33. $\dfrac{2x}{x+3} - \dfrac{x}{x-1} = \dfrac{14}{x^2+2x-3}$

34. $\dfrac{2(x+1)}{x-2} - \dfrac{x}{x+1} = \dfrac{9}{x^2-x-2}$

In Exercises 35–56, solve each equation. Check your answers.

35. $\sqrt[3]{3x-1} = 2$ **36.** $\sqrt[3]{2x+3} = 3$

37. $\sqrt{x-1} = -2$ **38.** $\sqrt{3x+4} = -1$

39. $x + \sqrt{x+6} = 0$ **40.** $x - \sqrt{6x+7} = 0$

41. $\sqrt{y+6} = y$ **42.** $r + 11 = 6\sqrt{r+3}$

43. $\sqrt{6y-11} = 2y-7$ **44.** $\sqrt{3y+1} = y-1$

45. $t - \sqrt{3t+6} = -2$ **46.** $\sqrt{5x^2-10x+9} = 2x-1$

47. $\sqrt{x-3} = \sqrt{2x-5} - 1$

48. $x + \sqrt{x+1} = 5$ **49.** $\sqrt{2y+9} = 2 + \sqrt{y+1}$

50. $\sqrt{m-1} = \sqrt{m-5}$ **51.** $\sqrt{7z+1} - \sqrt{5z+4} = 1$

52. $\sqrt{3q+1} - \sqrt{q-1} = 2$

53. $\sqrt{2x+5} + \sqrt{x+6} = 3$

54. $\sqrt{5x-9} - \sqrt{x+4} = 1$

55. $\sqrt{2x-5} - \sqrt{x-3} = 1$

56. $\sqrt{3x+5} + \sqrt{6x+3} = 3$

In Exercises 57–60, solve each equation. Check your answers.

57. $(x-4)^{3/2} = 27$ **58.** $(x+7)^{3/2} = 64$

59. $(5x-2)^{2/3} - 5 = 4$ **60.** $(2x-6)^{2/3} + 9 = 13$

In Exercises 61–84, solve each equation by using an appropriate substitution. Check your answers.

61. $x - 5\sqrt{x} + 6 = 0$ **62.** $x - 3\sqrt{x} + 2 = 0$

63. $2y - 15\sqrt{y} = -7$ **64.** $y + 44 = 15\sqrt{y}$

65. $x^{-2} - x^{-1} - 42 = 0$ **66.** $x^{1/2} + 3 - 4x^{-1/2} = 0$

67. $x^{2/3} - 6x^{1/3} + 8 = 0$ **68.** $x^{2/5} + x^{1/5} - 2 = 0$

69. $2x^{1/2} + 3x^{1/4} - 2 = 0$ **70.** $2x^{1/2} - x^{1/4} - 1 = 0$

71. $x^4 - 13x^2 + 36 = 0$ **72.** $x^4 - 7x^2 + 12 = 0$

73. $2t^4 + t^2 - 1 = 0$ **74.** $81y^4 + 1 = 18y^2$

75. $p^2 - 3 + 4\sqrt{p^2-3} - 5 = 0$

76. $x^2 - 3 - 4\sqrt{x^2-3} - 12 = 0$

77. $(3t+1)^2 - 3(3t+1) + 2 = 0$

78. $(7z+5)^2 + 2(7z+5) - 15 = 0$

79. $(\sqrt{y}+5)^2 - 9(\sqrt{y}+5) + 20 = 0$

80. $(2\sqrt{t}+1)^2 - 2(2\sqrt{t}+1) - 3 = 0$

81. $(x^2-4)^2 - 3(x^2-4) - 4 = 0$

82. $(x^2+2)^2 - 5(x^2+2) - 6 = 0$

83. $(x^2-3x)^2 - 2(x^2-3x) - 8 = 0$

84. $(x^2-4x)^2 + 7(x^2-4x) + 12 = 0$

B EXERCISES Applying the Concepts

85. **Fractions.** The numerator of a fraction is two less than the denominator. The sum of the fraction and its reciprocal is $\dfrac{25}{12}$. Find the numerator and denominator assuming that each is a positive integer.

86. **Fractions.** The numerator of a fraction is five less than the denominator. The sum of the fraction and six times its reciprocal is $\dfrac{25}{2}$. Find the numerator and denominator assuming that each is an integer.

87. **Investment.** Latasha bought some stock for $1800. If the price of each share of the stock had been $18 less, she could have bought five more shares for the same $1800.
a. How many shares of the stock did she buy?
b. How much did she pay for each share?

88. **Bus charter.** A civic club charters a bus for one day at a cost of $575. When two more people join the group, each person's cost decreases by $2. How many people were in the original group?

89. **Depth of a well.** A stone is dropped into a well, and the time it takes to hear the splash is 4 seconds. Find the depth of the well (to the nearest foot). Assume that the speed of sound is 1100 feet per second and that $d = 16t^2$ is the distance d an object falls in t seconds.

90. **Estimating the horizon.** You are in a hot air balloon 2 miles above the ocean. (See the accompanying diagram.) As far as you can see, there is only water. Given that the radius of the earth is 3960 miles, how far away is the horizon?

91. **Rate of current.** A motorboat is capable of traveling at a speed of 10 miles per hour in still water. On a particular day, the boat took 30 minutes longer to travel a distance of 12 miles upstream than it took to travel the same distance downstream. What was the rate of the current in the stream on that day?

92. **Train speed.** A freight train requires $2\frac{1}{2}$ hours longer to make a 300-mile journey than an express train does. If the express train averages 20 miles per hour faster than the freight train, how long does it take the express train to make the trip?

93. **Washing the family car.** A couple washed the family car in 24 minutes. Previously, when they each had washed the car alone, it took the husband 20 minutes longer to wash the car than it took the wife. How long did it take the wife to wash the car?

94. **Filling a swimming pool.** A small swimming pool can be filled by two hoses together in 1 hour and 12 minutes. The larger hose alone fills the pool in 1 hour less than the smaller hose. How long does it take the smaller hose to fill the pool?

95. **The area of a rectangular plot.** Together the diagonal and the longer side of a rectangular plot are three times the length of the shorter side. If the longer side exceeds the shorter side by 100 feet, what is the area of the plot?

96. **Emergency at sea.** You are at sea in your motorboat at a point A located 40 km from a point B that lies 130 kilometers from a hospital on a long, straight road. (See the accompanying figure.) While at point A, your companion feels sick and has to be taken to the hospital. You call the hospital on your cell phone to send an ambulance. Your motorboat can travel at 40 kilometers per hour, and the ambulance can travel at 80 kilometers per hour. How far from point B will you be when you and the ambulance arrive simultaneously at point C?

97. **Planning road construction.** Two towns A and B are located 8 and 5 miles, respectively, from a long, straight highway. The points C and D on the highway closest to the two towns are 18 miles apart. (Remember that the shortest distance from a point to a line is the length of the perpendicular segment from the point to the line.) Where should point E be located on the highway so that the sum of the lengths of the roads from A to E and E to B is 23 miles? (See the accompanying figure.) [*Hint:* $205x^2 - 4392x + 18972 = (x - 6)(205x - 3162)$.]

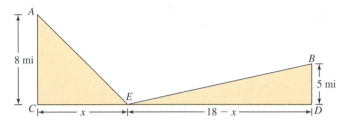

98. **Department store purchases.** A department store buyer purchased some shirts for $540. Then she sold all but eight of them. On each shirt sold, a profit of $6 (over the purchase price) was made. The shirts sold brought in revenue of $780.
   a. How many shirts were bought initially?
   b. What was the purchase price of each shirt?
   c. What was the selling price of each shirt?

C EXERCISES Beyond the Basics

In Exercise 99–108, solve each equation by any method.

99. $\left(\dfrac{x^2 - 3}{x}\right)^2 - 6\left(\dfrac{x^2 - 3}{x}\right) + 8 = 0$

100. $\left(\dfrac{x + 2}{3x + 1}\right)^2 - 5\left(\dfrac{x + 2}{3x + 1}\right) + 6 = 0$

101. $\sqrt{\sqrt{3x - 2} + 2\sqrt{4x + 1}} = 2\sqrt{2}$

102. $\dfrac{5}{x - 1} - \dfrac{3x}{x + 1} = \dfrac{9}{x^2 - 1}$

103. $\dfrac{x + 2}{x} + \dfrac{2x + 6}{x^2 + 4x} = -\dfrac{1}{x + 4}$

104. $\left(x^2 + \dfrac{1}{x^2}\right) - 7\left(x - \dfrac{1}{x}\right) + 8 = 0$

$\left[Hint: u = x - \dfrac{1}{x}, u^2 + 2 = x^2 + \dfrac{1}{x^2}.\right]$

105. $\left(\dfrac{t}{t + 2}\right)^2 + \dfrac{2t}{t + 2} - 15 = 0$

106. $6t^{-2/5} - 17t^{-1/5} + 5 = 0$

107. $2x^{1/3} + 2x^{-1/3} - 5 = 0$

108. $x^{4/3} - 4x^{2/3} + 3 = 0$

109. $x^{4/3} - 5x^{2/3} + 6 = 0$

110. $8\sqrt{\dfrac{x}{x + 3}} - \sqrt{\dfrac{x + 3}{x}} = 2$

111. Solve $x + \sqrt{x} - 42 = 0$ using these two methods.
   a. Rationalize the equation and then solve it.
   b. Use the substitution method.

112. Solve $\left(\dfrac{x}{x+1}\right)^2 - \dfrac{2x}{x+1} - 8 = 0$ using these two methods.
   a. Use the substitution method.
   b. Multiply by the LCD.

In Exercises 111–114, solve for x in terms of the other variables. (All letters denote positive real numbers.)

113. $\dfrac{x+b}{x-b} = \dfrac{x-5b}{2x-5b}$

114. $\dfrac{1}{x+a} + \dfrac{1}{x-2a} = \dfrac{1}{x-3a}$

115. $3x - 2a = \sqrt{a(3x - 2a)}$

116. $3a - \sqrt{ax} = \sqrt{a(3x + a)}$

Critical Thinking

In Exercises 115–119, find the value of x.

117. $x = \sqrt{1 + \sqrt{1 + \sqrt{1 + \cdots}}}$
   [*Hint:* $x = \sqrt{1 + x}.$]

118. $x = \sqrt{20 + \sqrt{20 + \sqrt{20 + \cdots}}}$

119. $x = \sqrt{n + \sqrt{n + \sqrt{n + \cdots}}}$, where n is a positive integer.

120. $x = 2 + \dfrac{1}{2 + \dfrac{1}{2 + \dfrac{1}{2 + \cdots}}}$

121. Assume that $x > 1$. Solve for x.

$$\sqrt{x + 2\sqrt{x - 1}} + \sqrt{x - 2\sqrt{x - 1}} = 4$$

$$[\textit{Hint:} \; x + 2\sqrt{x - 1} = (x - 1) + 1 + 2\sqrt{x - 1}$$
$$= (\sqrt{x - 1} + 1)^2.]$$

Linear Inequalities

Before Starting this Section, Review

1. Inequalities (Section P.1, page 5)
2. Intervals (Section P.1, page 7)
3. Linear equations (Section 1.1, page 85)

Objectives

1. Learn the vocabulary for discussing inequalities.
2. Solve and graph linear inequalities.
3. Solve and graph a compound inequality.
4. Solve and graph an inequality involving the reciprocal of a linear expression.

THE BERMUDA TRIANGLE

The "Bermuda Triangle," or "Devil's Triangle," is the expanse of the Atlantic Ocean between Florida, Bermuda, and Puerto Rico covering approximately 500,000 square miles of sea. This mysterious stretch of sea has an unusually high occurrence of disappearing ships and planes.

For example, on Halloween 1991, pilot John Verdi and his copilot were flying a Grumman Cougar jet over the triangle. They radioed the nearest tower to get permission to increase their altitude. The tower agreed and watched as the jet began the ascent and then disappeared off the radar. The jet did not fly out of range of the radar, did not descend, and did not radio a mayday (distress call). It just vanished, and the plane and crew were never recovered.

One explanation for such disappearances is based on the theory that strange compass readings occur in crossing the Atlantic due to confusion caused by the three north poles: magnetic (toward which compasses point), grid (the real North Pole, at 90 degrees latitude), and true or celestial north (determined by Polaris, the North Star). To get data to test this theory, a plane set out from Miami to Bermuda, a distance of 1035 miles, with the intention of setting the plane on automatic pilot once it reached a cruising speed of 300 miles per hour.

The plane was 150 miles along its path from Miami to Bermuda when it reached the 300-mile-per-hour cruising speed and was set on automatic pilot. After what length of time would you be confident in saying that the plane had encountered trouble? The answer to this question is found in Example 2, which uses a linear inequality, the topic discussed in this section. ■

---

1 Learn the vocabulary for discussing inequalities.

Inequalities

When we replace the equal sign (=) in an equation with any of the four inequality symbols $<$, $\leq$, $>$, or $\geq$, the resulting expression is an *inequality*. Here are some examples.

Equation	Replace = with	Inequality
$x = 5$	$<$	$x < 5$
$3x + 2 = 14$	$\leq$	$3x + 2 \leq 14$
$5x + 7 = 3x + 23$	$>$	$5x + 7 > 3x + 23$
$x^2 = 0$	$\geq$	$x^2 \geq 0$

In general, an **inequality** is a statement that one algebraic expression is less than or is less than or equal to another algebraic expression. The vocabulary for discussing inequalities is quite similar to that used for equations. The **domain** of a variable in an inequality is the set of all real numbers for which both sides of the inequality are defined. The real numbers that result in a true statement when those numbers are substituted for the variable in the inequality are called **solutions** of the inequality. Because replacing x with 1 in the inequality $3x + 2 \leq 14$ results in the true statement

$$3(1) + 2 \leq 14,$$

the number 1 is a solution of the inequality $3x + 2 \leq 14$. We also say that 1 *satisfies* the inequality $3x + 2 \leq 14$. To **solve** an inequality means to find all solutions of the inequality—that is, its solution set. The most elementary equations, such as $x = 5$, have only one solution. However, even the most elementary inequalities, such as $x < 5$, have infinitely many solutions. In fact, their solution sets are *intervals*, and we frequently graph the solution sets for inequalities in one variable on a number line. The graph of the inequality $x < 5$ is the interval $(-\infty, 5)$, shown in Figure 1.7.

$$x < 5, \text{ or } (-\infty, 5)$$

FIGURE 1.7 Graph of an inequality

Inequalities are classified the same way as equations: *conditional, inconsistent,* or *identities*. A **conditional inequality** such as $x < 5$ has in its domain at least one solution and at least one number that is not a solution. An inequality that *no* real number satisfies is called an **inconsistent inequality**, and an inequality that is satisfied by *every* real number in the domain of the variable is called an **identity**.

Because $x^2 = x \cdot x$ is the product of (1) two positive factors, (2) two negative factors, or (3) two zero factors, x^2 is either a positive number or zero. That is, x^2 is never negative, or is **nonnegative**. We call this fact the nonnegative identity.

NONNEGATIVE IDENTITY

$$x^2 \geq 0$$

for any real number x.

Two inequalities that have exactly the same solution set are called **equivalent inequalities**. The basic method of solving inequalities is similar to the method for solving equations: We replace a given inequality by a series of equivalent inequalities until we arrive at an equivalent inequality, such as $x < 5$, whose solution set we already know.

The following operations produce equivalent inequalities.

1. Simplifying one or both sides of an inequality by combining like terms and eliminating parentheses
2. Adding or subtracting the same expression on both sides of the inequality

---

Caution is required in multiplying or dividing both sides of an inequality by a real number or an expression representing a real number. Notice what happens when we multiply an inequality by -1.

We know that $2 < 3$, but how does $-2 = (-1)(2)$ compare with $-3 = (-1)(3)$? We have $-2 > -3$. See Figure 1.8.

-3 is left of -2   2 is left of 3

FIGURE 1.8

If we multiply (or divide) both sides of the inequality $2 < 3$ by -1, to get a correct result, we must exchange the $<$ symbol for the $>$ symbol. This exchange of symbols is called reversing the **sense** or the **direction** of the inequality.

The following chart describes the way multiplication and division affect inequalities.

If C represents a real number, then the following inequalities are all equivalent.			
Sign of C	**Inequality**	**Sense**	**Example**
	$A < B$	$<$	$3x < 12$
C positive	$A \cdot C < B \cdot C$	Unchanged	$\frac{1}{3}(3x) < \frac{1}{3}(12)$
C positive	$\frac{A}{C} < \frac{B}{C}$	Unchanged	$\frac{3x}{3} < \frac{12}{3}$
C negative	$A \cdot C > B \cdot C$	Reversed	$-\frac{1}{3}(3x) > -\frac{1}{3}(12)$
C negative	$\frac{A}{C} > \frac{B}{C}$	Reversed	$\frac{3x}{-3} > \frac{12}{-3}$

Similar results apply when $<$ is replaced throughout with any of the symbols $\leq$, $>$, or $\geq$.

2 Solve and graph linear inequalities.

Linear Inequalities

A **linear inequality in one variable** is an inequality that is equivalent to one of the forms

$$ax + b < 0 \quad \text{or} \quad ax + b \leq 0,$$

where a and b represent real numbers and $a \neq 0$.

Inequalities such as $2x - 1 > 0$ and $2x - 1 \geq 0$ are linear inequalities because they are equivalent to $-2x + 1 < 0$ and $-2x + 1 \leq 0$, respectively.

EXAMPLE 1 **Solving and Graphing Linear Inequalities**

Solve each inequality and graph its solution set.

a. $7x - 11 < 2(x - 3)$ **b.** $8 - 3x \leq 2$

SOLUTION

a. $7x - 11 < 2(x - 3)$

$7x - 11 < 2x - 6$	Distributive property
$7x - 11 + 11 < 2x - 6 + 11$	Add 11 to both sides.
$7x < 2x + 5$	Simplify.
$7x - 2x < 2x + 5 - 2x$	Subtract $2x$ from both sides.
$5x < 5$	Simplify.
$\dfrac{5x}{5} < \dfrac{5}{5}$	Divide both sides by 5.
$x < 1$	Simplify.

STUDY TIP

A linear inequality becomes a linear equation when the inequality symbol is replaced with the equal sign ($=$).

FIGURE 1.9

FIGURE 1.10

The solution set is $\{x \mid x < 1\}$, or in interval notation, $(-\infty, 1)$. The graph is shown in Figure 1.9.

b. $8 - 3x \leq 2$

$$8 - 3x - 8 \leq 2 - 8 \qquad \text{Subtract 8 from both sides.}$$
$$-3x \leq -6 \qquad \text{Simplify.}$$
$$\frac{-3x}{-3} \geq \frac{-6}{-3} \qquad \text{Divide both sides by } -3. \text{ (Reverse the direction of the inequality symbol.)}$$
$$x \geq 2 \qquad \text{Simplify.}$$

The solution set is $\{x \mid x \geq 2\}$, or in interval notation, $[2, \infty)$. The graph is shown in Figure 1.10. ■ ■ ■

Practice Problem 1 Solve each inequality and graph the solution set.

a. $4x + 9 > 2(x + 6) + 1$ **b.** $7 - 2x \geq -3$ ■

EXAMPLE 2 **Calculating the Results of the Bermuda Triangle Experiment**

In the introduction to this section, we discussed an experiment to test the reliability of compass settings and flight by automatic pilot along one edge of the Bermuda Triangle. The plane is 150 miles along its path from Miami to Bermuda, cruising at 300 miles per hour, when it notifies the tower that it is now set on automatic pilot.

The entire trip is 1035 miles, and we want to determine how much time we should let pass before we become concerned that the plane has encountered trouble.

SOLUTION

Let t = time elapsed since the plane went on autopilot. Then

$$300t = \text{distance the plane has flown in } t \text{ hours}$$
$$150 + 300t = \text{plane's distance from Miami after } t \text{ hours.}$$

Concern is not warranted unless the Bermuda tower does not see the plane, which means that the following is true.

| Plane's distance from Miami | $\geq$ | Distance from Miami to Bermuda |

$$150 + 300t \geq 1035 \qquad \text{Replace the verbal description with an inequality.}$$
$$150 + 300t - 150 \geq 1035 - 150 \qquad \text{Subtract 150 from both sides.}$$
$$300t \geq 885 \qquad \text{Simplify.}$$
$$\frac{300t}{300} \geq \frac{885}{300} \qquad \text{Divide both sides by 300.}$$
$$t \geq 2.95 \qquad \text{Simplify.}$$

Because 2.95 is roughly three hours, the tower will suspect trouble if the plane has not arrived in three hours. ■ ■ ■

Practice Problem 2 How much time should pass in Example 2 if the plane was set on automatic pilot at 340 miles per hour when it was 185 miles from Miami? ■

If an inequality is true for all real numbers, its solution set is $(-\infty, \infty)$. If an inequality has no solution, its solution set is $\varnothing$.

EXAMPLE 3 **Solving an Inequality**

Write the solution set of each inequality.

a. $7(x + 2) - 20 - 4x < 3(x - 1)$ **b.** $2(x + 5) + 3x < 5(x - 1) + 3$

SOLUTION

a.

$$7(x + 2) - 20 - 4x < 3(x - 1) \qquad \text{Original inequality}$$
$$7x + 14 - 20 - 4x < 3x - 3 \qquad \text{Distributive property}$$
$$3x - 6 < 3x - 3 \qquad \text{Combine like terms.}$$
$$-6 < -3 \qquad \text{Add } -3x \text{ to both sides and simplify.}$$

The last inequality is equivalent to the original inequality and is always true. So the solution set of the original inequality is $(-\infty, \infty)$.

b.

$$2(x + 5) + 3x < 5(x - 1) + 3 \qquad \text{Original inequality}$$
$$2x + 10 + 3x < 5x - 5 + 3 \qquad \text{Distributive property}$$
$$5x + 10 < 5x - 2 \qquad \text{Combine like terms.}$$
$$10 < -2 \qquad \text{Add } -5x \text{ and simplify.}$$

The resulting inequality is equivalent to the original inequality and is always false. So the solution set of the original inequality is $\varnothing$. ■■■

Practice Problem 3 Write the solution set of each inequality.

a. $2(4 - x) + 5x < 4(x + 1) + 7$ **b.** $3(x - 2) + 5 \geq 7(x - 1) - 4(x - 2)$

■

3 Solve and graph a compound inequality.

Combining Two Inequalities

Sometimes we are interested in the solution set of two or more inequalities. The combination of two or more inequalities is called a **compound inequality**.

Suppose E_1 is an inequality with solution set (interval) I_1 and E_2 is another inequality with solution set I_2. Then the solution set of the compound inequality "E_1 and E_2" is $I_1 \cap I_2$ and the solution set of the compound inequality "E_1 or E_2" is $I_1 \cup I_2$.

EXAMPLE 4 **Solving and Graphing a Compound Inequality**

Graph and write the solution set of the compound inequality.

$$2x + 7 \leq 1 \quad \text{or} \quad 3x - 2 < 4(x - 1)$$

SOLUTION

$$2x + 7 \leq 1 \quad \text{or} \quad 3x - 2 < 4(x - 1) \qquad \text{Original inequalities}$$
$$2x + 7 \leq 1 \quad \text{or} \quad 3x - 2 < 4x - 4 \qquad \text{Distributive property}$$
$$2x \leq -6 \quad \text{or} \qquad -x < -2 \qquad \text{Isolate variable term.}$$
$$x \leq -3 \quad \text{or} \qquad x > 2 \qquad \text{Solve for } x; \text{ reverse second inequality symbol.}$$

Graph each inequality and select the *union* of the two intervals.

$x \leq -3$

$x > 2$

$x \leq -3 \quad \text{or} \quad x > 2$

The solution set of the compound inequality is $(-\infty, -3] \cup (2, \infty)$. ■■■

Practice Problem 4 Write the solution set of the compound inequality.

$$3x - 5 \geq 7 \quad \text{or} \quad 5 - 2x \geq 1$$

■

EXAMPLE 5   **Solving and Graphing a Compound Inequality**

Solve the compound inequality and graph the solution set.

$$2(x - 3) + 5 < 9 \quad \text{and} \quad 3(1 - x) - 2 \le 7$$

SOLUTION

$2(x - 3) + 5 < 9$ and $3(1 - x) - 2 \le 7$		Original inequalities
$2x - 6 + 5 < 9$ and $3 - 3x - 2 \le 7$		Distributive property
$2x < 10$ and $-3x \le 6$		Isolate variable term.
$x < 5$ and $x \ge -2$		Solve for x; reverse second inequality symbol.

Graph each inequality and select the interval common to both inequalities.

$x < 5$

$x \ge -2$

$x < 5 \quad \text{and} \quad x \ge -2$

The solution set of the compound inequality is the interval $[-2, 5)$. ■ ■ ■

Practice Problem 5   Solve and graph the compound inequality.

$$2(3 - x) - 3 < 5 \quad \text{and} \quad 2(x - 5) + 7 \le 3 \qquad ■$$

Sometimes we are interested in a joint inequality such as $-5 < 2x + 3 \le 9$. This use of two inequality symbols in a single expression is shorthand for $-5 < 2x + 3$ *and* $2x + 3 \le 9$. Fortunately, solving such inequalities requires no new principles, as we see in the next example.

EXAMPLE 6   **Solving and Graphing a Compound Inequality**

Solve the inequality $-5 < 2x + 3 \le 9$ and graph its solution set.

SOLUTION

We must find all real numbers that are solutions of *both* inequalities

$$-5 < 2x + 3 \quad \text{and} \quad 2x + 3 \le 9.$$

Let's see what happens when we solve these inequalities separately.

$-5 < 2x + 3$	$2x + 3 \le 9$	Original inequalities
$-5 - 3 < 2x + 3 - 3$	$2x + 3 - 3 \le 9 - 3$	Subtract 3 from both sides.
$-8 < 2x$	$2x \le 6$	Simplify.
$\dfrac{-8}{2} < \dfrac{2x}{2}$	$\dfrac{2x}{2} \le \dfrac{6}{2}$	Divide both sides by 2.
$-4 < x$	$x \le 3$	Simplify.

The solution of the original pair of inequalities consists of all real numbers x such that $-4 < x$ *and* $x \le 3$. This solution may be written more compactly as $\{x \mid -4 < x \le 3\}$. In interval notation, we write $(-4, 3]$. The graph is shown in Figure 1.11.

$-4 < x \le 3$, or $(-4, 3]$

FIGURE 1.11

Notice that we did the same thing to both inequalities in each step of the solution process. We can accomplish this simultaneous solution more efficiently by working on both inequalities at the same time as follows.

$$-5 < 2x + 3 \le 9 \qquad \text{Original inequality}$$

$$-5 - 3 < 2x + 3 - 3 \le 9 - 3 \qquad \text{Subtract 3 from each part.}$$

$$-8 < 2x \le 6 \qquad \text{Simplify each part.}$$

$$\frac{-8}{2} < \frac{2x}{2} \le \frac{6}{2} \qquad \text{Divide each part by 2.}$$

$$-4 < x \le 3 \qquad \text{Simplify each part.}$$

The solution set is $(-4, 3]$, the same solution set we found previously.   ■ ■ ■

Practice Problem 6 Solve and graph $-6 \le 4x - 2 < 4$.   ■

4 Solve and graph an inequality involving the reciprocal of a linear expression.

Inequalities Involving the Reciprocal of a Linear Expression

In Section 3.7, we solve nonlinear inequalities. However, a "sign property" of reciprocals allows us to exchange certain nonlinear inequalities for equivalent linear inequalities. The **reciprocal sign property** says that any number and its reciprocal have the same sign.

RECIPROCAL SIGN PROPERTY

If $x \ne 0$, x and $\dfrac{1}{x}$ are both positive or both negative. In symbols,

if $x > 0$, then $\dfrac{1}{x} > 0$ and if $x < 0$, then $\dfrac{1}{x} < 0$.

EXAMPLE 7   **Solving and Graphing an Inequality by Using the Reciprocal Sign Property**

Solve and graph $(3x - 12)^{-1} > 0$.

SOLUTION

$$(3x - 12)^{-1} > 0 \qquad \text{Original inequality}$$

$$\frac{1}{3x - 12} > 0 \qquad a^{-1} = \frac{1}{a}; \text{ here } a = 3x - 12.$$

$$3x - 12 > 0 \qquad \text{Reciprocal sign property}$$

$$3x > 12 \qquad \text{Add 12 to both sides.}$$

$$\frac{3x}{3} > \frac{12}{3} \qquad \text{Divide both sides by 3.}$$

$$x > 4$$

The solution set is $\{x \,|\, x > 4\}$, or in interval notation, $(4, \infty)$. The graph is shown in Figure 1.12.   ■ ■ ■

$x > 4$, or $(4, \infty)$

FIGURE 1.12

Practice Problem 7 Solve and graph $(2x - 8)^{-1} > 0$.   ■

If you know that a particular quantity, represented by x, can vary between two values, then the procedures for working with inequalities can be used to find the values between which a linear expression $mx + k$ will vary. We show how to do this in Example 8.

EXAMPLE 8   **Finding the Interval of Values for a Linear Expression**

If $-2 < x < 5$, find real numbers a and b so that $a < 3x - 1 < b$.

SOLUTION

We start with the interval for x.

$$-2 < x < 5$$
$$3(-2) < 3x < 3(5) \qquad \text{Multiply each part by 3 to get } 3x \text{ in the middle.}$$
$$-6 < 3x < 15 \qquad \text{Simplify.}$$
$$-6 - 1 < 3x - 1 < 15 - 1 \qquad \text{Subtract 1 from each part to get } 3x - 1 \text{ in the middle.}$$
$$-7 < 3x - 1 < 14 \qquad \text{Simplify.}$$

We have $a = -7$ and $b = 14$.   ■ ■ ■

Practice Problem 8   Assuming that $-3 \leq x \leq 2$, find real numbers a and b so that $a \leq 3x + 5 \leq b$.   ■

Our next example demonstrates a practical use of the method shown in Example 8.

EXAMPLE 9   **Finding a Fahrenheit Temperature from a Celsius Range**

The weather in London is predicted to range between 10° and 20° Celsius during the three-week period you will be working there. To decide what kind of clothes to take, you want to convert the temperature range to Fahrenheit temperatures. The formula for converting Celsius temperature C to Fahrenheit temperature F is $F = \dfrac{9}{5}C + 32$. What range of Fahrenheit temperatures might you find in London during your stay there?

SOLUTION

First, we express the Celsius temperature range as an inequality. Let C = temperature in Celsius degrees.

For the three weeks under consideration, $10 \leq C \leq 20$.

We want to know the range of $F = \dfrac{9}{5}C + 32$ when $10 \leq C \leq 20$.

$$10 \leq C \leq 20$$
$$\left(\frac{9}{5}\right)(10) \leq \frac{9}{5}C \leq \left(\frac{9}{5}\right)(20) \qquad \text{Multiply each part by } \frac{9}{5}.$$
$$18 \leq \frac{9}{5}C \leq 36 \qquad \text{Simplify.}$$
$$18 + 32 \leq \frac{9}{5}C + 32 \leq 36 + 32 \qquad \text{Add 32 to each part.}$$
$$50 \leq \frac{9}{5}C + 32 \leq 68 \qquad \text{Simplify.}$$
$$50 \leq F \leq 68 \qquad F = \frac{9}{5}C + 32$$

So the temperature range from 10° to 20° Celsius corresponds to a range from 50° to 68° Fahrenheit.   ■ ■ ■

Practice Problem 9   What range in Fahrenheit degrees corresponds to the range of 15° to 25° Celsius?   ■

SECTION 1.6 ■ Exercises

A EXERCISES Basic Skills and Concepts

In Exercises 1–10, fill in the blank with the correct inequality symbol using the rules for producing equivalent inequalities.

1. If $x < 8$, then $x - 8$ _____ 0.

2. If $x \leq -3$, then $x + 3$ _____ 0.

3. If $x \geq 5$, then $x - 5$ _____ 0.

4. If $x > 7$, then $x - 7,$ _____ 0.

5. If $\dfrac{x}{2} < 6$, then x _____ 12.

6. If $\dfrac{x}{3} < 2$, then x _____ 6.

7. If $-2x \leq 4$, then x _____ -2.

8. If $-3x > 12$, then x _____ -4.

9. If $x + 3 < 2$, then x _____ -1.

10. If $x - 5 \leq 3$, then x _____ 8.

In Exercises 11–18, graph the solution set of each inequality and write it in interval notation.

11. $-2 < x < 5$
12. $-5 \leq x \leq 0$
13. $0 < x \leq 4$
14. $1 \leq x < 7$
15. $x \geq -1$
16. $x > 2$
17. $-5x \geq 10$
18. $-2x < 2$

In Exercises 19–40, solve each inequality. Write the solution in interval notation and graph the solution set.

19. $x + 3 < 6$
20. $x - 2 < 3$
21. $1 - x \leq 4$
22. $7 - x > 3$
23. $2x + 5 < 9$
24. $3x + 2 \geq 7$
25. $3 - 3x > 15$
26. $8 - 4x \geq 12$
27. $3(x + 2) < 2x + 5$
28. $4(x - 1) \geq 3x - 1$
29. $3(x - 3) \leq 3 - x$
30. $-x - 2 \geq x - 10$
31. $6x + 4 > 3x + 10$
32. $4(x - 4) > 3(x - 5)$
33. $8(x - 1) - x \leq 7x - 12$
34. $3(x + 2) + 2x \geq 5x + 18$
35. $5(x + 2) \leq 3(x + 1) + 10$
36. $x - 4 > 2(x + 8)$
37. $2(x + 1) + 3 \geq 2(x + 2) - 1$
38. $5(2 - x) + 4x \leq 12 - x$
39. $2(x + 1) - 2 \leq 3(2 - x) + 9$
40. $4(1 - x) + 2x > 5(2 - x) + 4x$

In Exercises 41–50, solve each rational inequality.

41. $9x - 6 \geq \dfrac{3}{2}x + 9$

42. $\dfrac{7x - 3}{2} < 3x - 4$

43. $\dfrac{x - 3}{3} \leq 2 + \dfrac{x}{2}$

44. $\dfrac{2x - 3}{4} \geq 3 - \dfrac{x}{2}$

45. $\dfrac{3x + 1}{2} < x - 1 + \dfrac{x}{2}$

46. $\dfrac{2x - 1}{3} \geq \dfrac{x + 1}{4} + \dfrac{x}{12}$

47. $\dfrac{x - 3}{2} \geq \dfrac{x}{3} + 1$

48. $\dfrac{2x + 1}{3} < \dfrac{x - 1}{2} + \dfrac{1}{6}$

49. $\dfrac{3x + 1}{3} - \dfrac{x}{2} \leq \dfrac{x + 2}{2}$

50. $\dfrac{x - 1}{3} + \dfrac{x + 1}{4} \leq \dfrac{x}{2} + \dfrac{x}{12}$

In Exercises 51–58, solve each compound *or* inequality.

51. $2x + 5 < 1$ or $2 + x > 4$

52. $3x - 2 > 7$ or $2(1 - x) > 1$

53. $\dfrac{2x - 3}{4} \leq 2$ or $\dfrac{4 - 3x}{2} \geq 2$

54. $\dfrac{5 - 3x}{3} \geq \dfrac{1}{6}$ or $\dfrac{x - 1}{3} \leq 1$

55. $\dfrac{2x + 1}{3} \geq x + 1$ or $\dfrac{x}{2} - 1 > \dfrac{x}{3}$

56. $\dfrac{x + 2}{2} < \dfrac{x}{3} + 1$ or $\dfrac{x - 1}{3} > \dfrac{x + 1}{5}$

57. $\dfrac{x - 1}{2} > \dfrac{x}{3} - 1$ or $\dfrac{2x + 5}{3} \leq \dfrac{x + 1}{6}$

58. $\dfrac{2x + 1}{3} \leq \dfrac{x}{4} + 1$ or $\dfrac{3 - x}{2} > \dfrac{x}{3} - 1$

In Exercises 59–66, solve each compound *and* inequality.

59. $3 - 2x \leq 7$ and $2x - 3 \leq 7$

60. $6 - x \leq 3x + 10$ and $7x - 14 \leq 3x + 14$

61. $2(x + 1) + 3 \geq 1$ and $2(2 - x) > -6$

62. $3(x + 1) - 2 \geq 4$ and $3(1 - x) + 13 > 4$

63. $2(x + 1) - 3 > 7$ and $3(2x + 1) + 1 < 10$

64. $5(x + 2) + 7 < 2$ and $2(5 - 3x) + 1 < 17$

65. $5 + 3(x - 1) < 3 + 3(x + 1)$ and $3x - 7 \leq 8$

66. $2x - 3 > 11$ and $5(2x + 1) < 3(4x + 1) - 2(x - 1)$

In Exercises 67–78, solve each combined inequality.

67. $3 < x + 5 < 4$
68. $9 \leq x + 7 \leq 12$
69. $-4 \leq x - 2 < 2$
70. $-3 < x + 5 < 4$
71. $-9 \leq 2x + 3 \leq 5$
72. $-2 \leq 3x + 1 \leq 7$

73. $0 \le 1 - \dfrac{x}{3} < 2$ **74.** $0 < 5 - \dfrac{x}{2} \le 3$

75. $-1 < \dfrac{2x - 3}{5} \le 0$ **76.** $-4 \le \dfrac{5x - 2}{3} \le 0$

77. $5x \le 3x + 1 < 4x - 2$

78. $3x + 2 < 2x + 3 < 4x - 1$

In Exercises 79–82, solve each reciprocal inequality.

79. $(3x + 6)^{-1} < 0$ **80.** $(2x - 8)^{-1} < 0$

81. $(2 - 4x)^{-1} > 0$ **82.** $(10 - 5x)^{-1} > 0$

In Exercises 83–88, find a and b.

83. If $-2 < x < 1$, then $a < x + 7 < b$.

84. If $1 < x < 5$, then $a < 2x + 3 < b$.

85. If $-1 < x < 1$, then $a < 2 - x < b$.

86. If $3 < x < 7$, then $a < 1 - 3x < b$.

87. If $0 < x < 4$, then $a < 5x - 1 < b$.

88. If $-4 < x < 0$, then $a < 3x + 4 < b$.

B EXERCISES   Applying the Concepts

89. Poster sales. A student club wants to produce a poster as a fund-raising device. The production cost is $1.80 per poster. There are additional expenses of $1200, independent of production or sales. How many posters must be sold at $2 each for the club to make a profit?

90. Telemarketing. A telemarketer is paid $25 plus 35% of the difference between the item's selling price and the item's cost. All items sell for at least $50 above cost and, at most, $125 above cost. For each sale made, over what range can a telemarketer's commission vary?

91. Appliance markup. The markup over the dealer's cost on a new refrigerator ranges from 15% to 20%. If the dealer's cost is $1750, over what range will the selling price vary?

92. Return on investment. An investor has $5000 to invest for a period of one year. Find the range of per annum simple interest rates required to generate interest between $200 and $275 inclusive.

93. Hybrid car trip. Sometime after passing a truck stop 300 miles from the start of her trip, Cora's hybrid car ran out of gas. Assuming that the tank could hold 12 gallons of gasoline and the hybrid car averaged 40 miles per gallon, find the range of gasoline (in gallons) that could have been in the tank at the start of the trip.

94. Average grade. Sean has taken three exams and earned scores of 85, 72, and 77 out of a possible 100 points. He needs an average of at least 80 to earn a B in the course. What range of scores on the fourth (and last) 100-point test will guarantee a B in the course?

95. Butterfat content. How much cream that is 30% butterfat must be added to milk that is 3% butterfat to have 270 quarts that are at least 4.5% butterfat?

96. Amplifier cost. The cost of an amplifier to a retailer is $340. At what price can the amplifier be sold if the retailer wants to make a profit of at least 20% of the selling price?

97. Pedometer cost. A company produces a pedometer at a cost of $3 each and sells the pedometer for $5 each. If the company has to recover an initial expense of $4000 before any profit is earned, how many pedometers must be sold to earn a profit in excess of $3000?

98. Car sales. A car dealer has three times as many SUVs and twice as many convertibles as four-door sedans. How many four-door sedans does the dealer have if she has at least 48 cars of these three types?

99. Temperature conversion. The formula for converting Fahrenheit temperature F to Celsius temperature C is $C = \dfrac{5}{9}(F - 32)$. What range in Celsius degrees corresponds to a range of 68° to 86° Fahrenheit?

100. Parking expense. The parking cost at the local airport (in dollars) is $C = 2 + 1.75(h - 1)$, where h is the number of hours the car is parked. For what range of hours is the parking cost between $37 and $51?

101. Car rental. Suppose a rental car costs $18 per day plus $0.25 per mile for each mile driven. What range of miles can be driven to stay within a $30-per-day car rental allowance?

102. Plumbing charges. A plumber charges $42 per hour. A repair estimate includes a fixed cost of $147 for parts and a labor cost that is determined by how long it takes to complete the job. If the total estimate is for at least $210 and, at most, $294, what time interval was estimated for the job?

C EXERCISES Beyond the Basics

103. For what value of k does $-2 \leq x \leq k$ have exactly one solution?

104. For what values of k is $k < x < k + 1$ a conditional inequality?

105. For what value of k is the interval $(-\infty, k]$ the solution set for the linear inequality $ax + b \leq 0$ if $a > 0$?

106. What interval is the solution set for the linear inequality $ax + b \geq 0$ if $a < 0$?

107. Solve $0 < (2x + 1)^{-1} < \dfrac{1}{2}$.

108. Solve $0 < (3x - 6)^{-1} < \dfrac{2}{3}$.

109. Find the domain of x in the expression $\sqrt{3x + 9}$.

110. Find the domain of x in the expression $\sqrt{3 - x}$.

111. Solve the inequality $-a \leq bx + c < a$ assuming that $b > 0$.

112. Solve the inequality $-a \leq bx + c < a$ assuming that $b < 0$.

Critical Thinking

113. The equation $x = x + 1$ is inconsistent. If possible, replace the equality sign in this equation with an appropriate inequality symbol so that the solution set of the resulting inequality is
  a. $\varnothing$.
  b. $(-\infty, \infty)$.
  c. neither (a) nor (b).

114. Repeat Exercise 113 by beginning with the identity $2(x + 3) = 2x + 6$.

115. Repeat Exercise 113 by beginning with the conditional equation $2x - 3 = 7$.

Equations and Inequalities Involving Absolute Value

Before Starting this Section, Review

1. Properties of absolute value (Section P.1, page 9)

2. Intervals (Section P.1, page 7)

3. Linear inequalities (Section 1.6, page 148)

Objectives

1 Solve equations involving absolute value.

2 Solve inequalities involving absolute value.

RESCUING A DOWNED AIRCRAFT

Rescue crews arrive at an airport in Maine to conduct an air search for a missing Cessna with two people aboard. If the rescue crews locate the missing aircraft, they are to mark the global positioning system (GPS) coordinates and notify the mission base. The search planes normally average 110 miles per hour, but weather conditions can affect the average speed by as much as 15 miles per hour (either slower or faster). If a search plane has 30 gallons of fuel and uses 10 gallons of fuel per hour, what is its possible search range (in miles)? In Example 5, we use an inequality involving absolute value to find the plane's possible search range. ∎

---

1 Solve equations involving absolute value.

Equations Involving Absolute Value

Recall that geometrically, the absolute value of a real number a is the distance from the origin on the number line to the point with coordinate a. The definition of absolute value is

$$|a| = a \text{ if } a \geq 0 \qquad \text{and} \qquad |a| = -a \text{ if } a < 0.$$

A summary of the most useful properties of absolute value can be found on page 9 in Section P.1.

Because the only two numbers on the number line that are exactly two units from the origin are 2 and -2, they are the only solutions of the equation $|x| = 2$. See Figure 1.13. This simple observation leads to the following rule for solving equations involving absolute value.

FIGURE 1.13 Distance from the origin

SOLUTIONS OF $|u| = a, a \geq 0$

If $a \geq 0$ and u is an algebraic expression, then

$$|u| = a \quad \text{is equivalent to} \quad u = a \text{ or } u = -a.$$
$$|u| = -a \quad \text{has no solution when } a > 0.$$

Note that if $a = 0$, then $u = 0$ is the only solution of $|u| = 0$.

> **EXAMPLE 1**   Solving an Equation Involving Absolute Value

Solve each equation.

a. $|x + 3| = 0$ **b.** $|2x - 3| - 5 = 8$

SOLUTION

a. Let $u = x + 3$. Then, $|u| = 0$ has only one solution: $u = 0$.

$$
\begin{array}{ll}
|x + 3| = 0 & |u| = 0 \\
x + 3 = 0 & u = 0 \\
x = -3 & \text{Solve for } x.
\end{array}
$$

Now we check the solution in the original equation.

Check:

$$
\begin{array}{ll}
|x + 3| = 0 & \text{Original equation} \\
|-3 + 3| \stackrel{?}{=} 0 & \text{Replace } x \text{ with } -3. \\
|0| = 0 \checkmark & \text{Simplify.}
\end{array}
$$

We have verified that -3 is a solution of the original equation.
The solution set is $\{-3\}$.

b. To use the rule for solving equations involving absolute value, we must first isolate the absolute value expression to one side of the equation.

$$
\begin{array}{ll}
|2x - 3| - 5 = 8 & \\
|2x - 3| = 13 & \text{Add 5 to both sides and simplify.}
\end{array}
$$

$|u| = 13$ is equivalent to $u = 13$ or $u = -13$. Let $u = 2x - 3$; then $|2x - 3| = 13$ is equivalent to

$$
\begin{array}{lll}
2x - 3 = 13 \quad \text{or} \quad 2x - 3 = -13. & u = 13 \text{ or } u = -13 \\
2x = 16 \qquad\qquad 2x = -10 & \text{Add 3 to both sides of each equation.} \\
x = 8 \qquad\qquad\quad x = -5 & \text{Divide both sides by 2.}
\end{array}
$$

We leave it to you to check these solutions.
The solution set is $\{-5, 8\}$.   ■ ■ ■

Practice Problem 1   Solve each equation.

a. $|x - 2| = 0$ **b.** $|6x - 3| - 8 = 1$ ■

With a little practice, you may find that you can solve these types of equations without actually writing u and the expression it represents. Then you can work with just the expression itself.

> **EXAMPLE 2**   Solving an Equation of the Form $|u| = |v|$

Solve $|x - 1| = |x + 5|$.

SOLUTION

If $|u| = |v|$, then u is equal to $|v|$ or u is equal to $-|v|$. Because $|v| = \pm v$ in every case, we have $u = v$ or $u = -v$. Thus,

$$|u| = |v| \text{ is equivalent to } u = v \text{ or } u = -v.$$
$$|x - 1| = |x + 5| \text{ is equivalent to}$$

$$
\begin{array}{lll}
x - 1 = x + 5 \quad \text{or} \quad x - 1 = -(x + 5) & u = x - 1, v = x + 5 \\
-1 = 5 \text{ (False)} \qquad\qquad x = -2 & \text{Solve for } x.
\end{array}
$$

The only solution of $|x - 1| = |x + 5|$ is -2, so the solution set is $\{-2\}$. We leave it to you to check this solution.   ■ ■ ■

Practice Problem 2   Solve $|x + 2| = |x - 3|$.   ■

EXAMPLE 3   **Solving an Equation of the Form $|u| = v$**

Solve $|2x - 1| = x + 5$.

SOLUTION

Letting $u = 2x - 1$ and $v = x + 5$, the equation $|u| = v$ is equivalent to

$u = v$	or	$u = -v$	
$2x - 1 = x + 5$		$2x - 1 = -(x + 5)$	
$2x - 1 = x + 5$		$2x - 1 = -x - 5$	Distributive property
$x = 6$		$3x = -4$	Isolate x term.
$x = 6$		$x = -\dfrac{4}{3}$	Solve for x.

Check:

$$x = 6 \qquad \text{or} \qquad x = -\frac{4}{3}$$

$$|2(6) - 1| \overset{?}{=} 6 + 5 \qquad \qquad \left|2\left(-\frac{4}{3}\right) - 1\right| \overset{?}{=} -\frac{4}{3} + 5$$

$$|11| \overset{?}{=} 11 \qquad \qquad \left|-\frac{8}{3} - 1\right| \overset{?}{=} \frac{-4 + 15}{3}$$

$$|11| \overset{?}{=} 11 \checkmark \text{ Yes} \qquad \qquad \left|-\frac{11}{3}\right| \overset{?}{=} \frac{11}{3}$$

$$\frac{11}{3} \overset{?}{=} \frac{11}{3} \checkmark \text{ Yes}$$

The solution set of the given equation is $\left\{-\dfrac{4}{3}, 6\right\}$.   ■ ■ ■

Practice Problem 3   Solve $|3x - 4| = 2(x - 1)$.   ■

2 Solve inequalities involving absolute value.

Inequalities Involving Absolute Value

The equation $|x| = 2$ has two solutions: 2 and -2. If x is any real number other than 2 or -2, then $|x| < 2$ or $|x| > 2$. Geometrically, this means that x is closer than two units to the origin if $|x| < 2$ or farther than two units from the origin if $|x| > 2$. Any number x that is closer than two units from the origin is in the interval $-2 < x < 2$. See Figure 1.14 (a). Any number x that is farther than two units from the origin is in the interval $(-\infty, -2)$ or the interval $(2, \infty)$, that is, either $x < -2$ or $x > 2$. See Figure 1.14 (b).

$-2 < x < 2$, or $(-2, 2)$	$x < -2$ or $x > 2$
	x in $(-\infty, -2)$ or x in $(2, \infty)$
(a)	(b)

FIGURE 1.14 Graphs of inequalities involving absolute value

This discussion is equally valid when we use any positive real number a instead of the number 2, and it leads to the following rules for replacing inequalities involving absolute value with equivalent inequalities that do *not* involve absolute value.

> ### RULES FOR SOLVING ABSOLUTE VALUE INEQUALITIES
>
> If $a > 0$ and u is an algebraic expression, then
>
> 1. $|u| < a$ is equivalent to $-a < u < a$.
> 2. $|u| \le a$ is equivalent to $-a \le u \le a$.
> 3. $|u| > a$ is equivalent to $u < -a$ or $u > a$.
> 4. $|u| \ge a$ is equivalent to $u \le -a$ or $u \ge a$.

EXAMPLE 4   Solving an Inequality Involving Absolute Value

Solve the inequality $|4x - 1| \le 9$ and graph the solution set.

SOLUTION

Rule 2 applies here with $u = 4x - 1$ and $a = 9$.

$$|4x - 1| \le 9 \text{ is equivalent to}$$

$$-9 \le 4x - 1 \le 9 \qquad \text{Rule 2, } -a \le u \le a$$
$$1 - 9 \le 4x - 1 + 1 \le 9 + 1 \qquad \text{Add 1 to each part.}$$
$$-8 \le 4x \le 10 \qquad \text{Simplify.}$$
$$-\frac{8}{4} \le \frac{4x}{4} \le \frac{10}{4} \qquad \text{Divide by 4; the sense of the inequality is unchanged.}$$
$$-2 \le x \le \frac{5}{2} \qquad \text{Simplify.}$$

The solution set is $\left\{ x \middle| -2 \le x \le \frac{5}{2} \right\}$; that is, the solution set is the closed interval $\left[-2, \frac{5}{2} \right]$. See Figure 1.15.   ▪ ▪ ▪

$-2 \le x \le \frac{5}{2}$, or $\left[-2, \frac{5}{2} \right]$

FIGURE 1.15

Practice Problem 4   Solve $|3x + 3| \le 6$ and graph the solution set.   ▪

Note the following techniques for solving inequalities similar to Example 4.

1. To solve $|4x - 1| < 9$, change $\le$ to $<$ throughout to obtain the solution $\left(-2, \frac{5}{2} \right)$.

2. To solve $|1 - 4x| \le 9$, note that $|1 - 4x| = |4x - 1|$ so the solution is the same as that for Example 4, $\left[-2, \frac{5}{2} \right]$.

EXAMPLE 5   Finding the Search Range of an Aircraft

In the introduction to this section, we wanted to find the possible search range (in miles) for a search plane that has 30 gallons of fuel and uses 10 gallons of fuel per hour. We were told that the search plane normally averages 110 miles per hour but that weather conditions can affect the average speed by as much as 15 miles per hour (either slower or faster). How do we find the possible search range?

SOLUTION

To find the distance a plane flies (in miles), we need to know how long (time, in hours) it flies and how fast (speed, in miles per hour) it flies.

Let $x =$ actual speed of the search plane, in miles per hour. We know that the actual speed is within 15 miles per hour of the average speed, 110 miles per hour. That is, |actual speed − average speed| ≤ 15 miles per hour.

$$|x - 110| \le 15 \qquad \text{Replace the verbal description with an inequality.}$$

Solve this inequality for x.

$$-15 \leq x - 110 \leq 15 \qquad \text{Rewrite the inequality using Rule 2.}$$
$$110 - 15 \leq x \leq 110 + 15 \qquad \text{Add 110 to each part.}$$
$$95 \leq x \leq 125 \qquad \text{Simplify.}$$

The actual speed of the search plane is between 95 and 125 miles per hour.
Because the plane uses 10 gallons of fuel per hour and has 30 gallons of fuel, it can fly $\frac{30}{10}$, or 3, hours. Thus, the actual number of miles the search plane can fly is $3x$.

From $\qquad\qquad\qquad 95 \leq x \leq 125,$

we have $\qquad\qquad 3(95) \leq 3x \leq 3(125) \qquad \text{Multiply each part by 3.}$
$$285 \leq 3x \leq 375 \qquad \text{Simplify.}$$

The search plane's range is between 285 and 375 miles. ■ ■ ■

Practice Problem 5 Repeat Example 5, but let the average speed of the plane be 115 miles per hour and suppose the wind speed can affect the average speed by 25 miles per hour. ■

EXAMPLE 6 **Solving an Inequality Involving Absolute Value**

Solve the inequality $|2x - 8| \geq 4$ and graph the solution set.

SOLUTION
Rule 4 applies here, with $u = 2x - 8$ and $a = 4$.

$$|2x - 8| \geq 4 \text{ is equivalent to}$$

$2x - 8 \leq -4$	or	$2x - 8 \geq 4$	Rule 4, $u \leq -a$ or $u \geq a$
$2x - 8 + 8 \leq -4 + 8$		$2x - 8 + 8 \geq 4 + 8$	Add 8 to each part.
$2x \leq 4$		$2x \geq 12$	Simplify.
$\dfrac{2x}{2} \leq \dfrac{4}{2}$		$\dfrac{2x}{2} \geq \dfrac{12}{2}$	Divide both sides of each part by 2.
$x \leq 2$		$x \geq 6$	Simplify.

$x \leq 2$ or $x \geq 6$
$(-\infty, 2] \cup [6, \infty)$

FIGURE 1.16

The solution set is $\{x \mid x \leq 2 \text{ or } x \geq 6\}$; that is, the solution set is the set of all real numbers x in either of the intervals $(-\infty, 2]$ or $[6, \infty)$. This set can also be written as $(-\infty, 2] \cup [6, \infty)$. See Figure 1.16. ■ ■ ■

Practice Problem 6 Solve $|2x + 3| \geq 6$ and graph the solution set. ■

EXAMPLE 7 **Solving Special Cases of Absolute Value Inequalities**

Solve each inequality.

a. $|3x - 2| > -5$ **b.** $|5x + 3| \leq -2$

SOLUTION

a. Because the absolute value is always nonnegative, $|3x - 2| > -5$ is true for all real numbers x. The solution set is the set of all real numbers; in interval notation, we write $(-\infty, \infty)$.

b. There is no real number with absolute value ≤ -2 because the absolute value of any number is nonnegative. The solution set for $|5x + 3| \leq -2$ is the empty set, $\varnothing$. ■ ■ ■

Practice Problem 7 Solve.

a. $|5 - 9x| > -3$ **b.** $|7x - 4| \leq -1$ ■

SECTION 1.7 ■ Exercises

A EXERCISES Basic Skills and Concepts

In Exercises 1–6, assume that $a > 0$.

1. The solution set of the equation $|x| = a$ is _____.

2. The solution set of the inequality $|x| < a$ is _____.

3. The solution set of the inequality $|x| \geq a$ is _____.

4. The equation $|u| = |v|$ is equivalent to $u =$ _____ or $u =$ _____.

5. *True or False* The solution set of $|3x - 2| < a$ is the same as the solution set of $|2 - 3x| < a$.

6. *True or False* If $a > 0$, the solution set of $|3x - 2| \leq -a$ is $\varnothing$.

In Exercises 7–34, solve each equation.

7. $|3x| = 9$
8. $|4x| = 24$

9. $|-2x| = 6$
10. $|-x| = 3$

11. $|x + 3| = 2$
12. $|x - 4| = 1$

13. $|6 - 2x| = 8$
14. $|6 - 3x| = 9$

15. $|6x - 2| = 9$
16. $|6x - 3| = 9$

17. $|2x + 3| - 1 = 0$
18. $|2x - 3| - 1 = 0$

19. $\dfrac{1}{2}|x| = 3$
20. $\dfrac{3}{5}|x| = 6$

21. $\left|\dfrac{1}{4}x + 2\right| = 3$
22. $\left|\dfrac{3}{2}x - 1\right| = 3$

23. $6|1 - 2x| - 8 = 10$
24. $5|1 - 4x| + 10 = 15$

25. $2|3x - 4| + 9 = 7$
26. $9|2x - 3| + 2 = -7$

27. $|2x + 1| = -1$
28. $|3x + 7| = -2$

29. $|x^2 - 4| = 0$
30. $|9 - x^2| = 0$

31. $|1 - 2x| = 3$
32. $|4 - 3x| = 5$

33. $\left|\dfrac{1}{3} - x\right| = \dfrac{2}{3}$
34. $\left|\dfrac{2}{5} - x\right| = \dfrac{1}{5}$

In Exercises 35–44, solve each equation.

35. $|x + 3| = |x + 5|$
36. $|x + 4| = |x - 8|$

37. $|3x - 2| = |6x + 7|$
38. $|2x - 4| = |4x + 6|$

39. $|2x - 1| = x + 1$
40. $|3x - 4| = 2(x - 1)$

41. $|4 - 3x| = x - 1$
42. $|2 - 3x| = 2x - 1$

43. $|3x + 2| = 2(x - 1)$
44. $|4x + 7| = x + 1$

In Exercises 45–60, solve each inequality.

45. $|3x| < 12$
46. $|2x| \leq 6$

47. $|4x| > 16$
48. $|3x| > 15$

49. $|x + 1| < 3$
50. $|x - 4| < 1$

51. $|x| + 2 \geq -1$
52. $|x| + 2 > -7$

53. $|2x - 3| < 4$
54. $|4x - 6| \leq 6$

55. $|5 - 2x| > 3$
56. $|3x - 3| \geq 15$

57. $|3x + 4| \leq 19$
58. $|9 - 7x| < 23$

59. $|2x - 15| < 0$
60. $|x + 5| \leq -3$

In Exercises 61–70, solve each inequality and graph the solution set.

61. $\left|1 - \dfrac{x}{2}\right| < 4$
62. $\left|2 - \dfrac{x}{2}\right| > 1$

63. $\left|\dfrac{x - 8}{2}\right| \leq 4$
64. $\left|\dfrac{2x + 15}{8}\right| > 4$

65. $\left|\dfrac{3}{8}x - 105\right| \geq 0$
66. $|27x - 43| \geq 0$

67. $\left|3x - \dfrac{7}{3}\right| - 2 < 3$
68. $\left|\dfrac{x}{2} - 1\right| - 1 \leq \dfrac{1}{2}$

69. $\left|\dfrac{3}{5}x - 2\right| \leq 4$
70. $\left|\dfrac{x}{2} - \dfrac{1}{3}\right| > \dfrac{2}{3}$

B EXERCISES Applying the Concepts

71. **Varying temperatures.** The inequality $|T - 75| \leq 20$, where T is in degrees Fahrenheit, describes the daily temperature in Tampa during December. Give an interpretation for this inequality assuming that the high and low temperatures in Tampa during December satisfy the equation $|T - 75| = 20$.

72. **Scale error.** A butcher's scale is accurate to within ±0.05 pound. A sirloin steak weighs 1.14 pounds on this scale. Let $x =$ actual weight of the steak. Write an absolute value inequality in x whose solution is the range of possible values for the actual weight of the steak.

73. **Company budget.** A company budgets $700 for office supplies. The actual expense for budget supplies must be within $\pm\$50$ of this figure. Let $x =$ actual expense for the office supplies. Write an absolute value inequality in x whose solution is the range of possible amounts for the expense of the office supplies.

74. **National achievement scores.** Suppose 68% of the scores on a national achievement exam will be within ±90 points of a score of 480. Let $x =$ a score among the 68% just described. Write an absolute value inequality in x whose solution is the range of possible scores within ±90 points of 480.

75. **Blood pressure.** Suppose 60% of Americans have a systolic blood pressure reading of 120, plus or minus 6.75. Let $x =$ a blood pressure reading among the 60% just described. Write an absolute value inequality in x whose solution set is the range of possible blood pressure readings within ±6.75 points of 120.

76. Gas mileage. A motorcycle has approximately 4 gallons of gas, with an error margin of $\pm\frac{1}{4}$ of a gallon. If the motorcycle gets 37 miles per gallon of gas, how many miles can the motorcycle travel?

77. Event planning. An event planner expects about 120 people at a private party. She knows that this estimate could be off by 15 people (more or fewer). Food for the event costs $48 per person. How much might the event planner's food expense be?

78. Company bonuses. Bonuses at a company are usually given to about 60 people each year. The bonuses are $1200 each, and the estimate of 60 recipients may be off by 7 people (more or fewer). How much might the company spend on bonuses this year?

79. Fishing revenue. Sarah sells the fish she catches to a local restaurant for 60 cents a pound. She can usually estimate how much her catch weighs to within $\pm\frac{1}{2}$ pound. She estimates that she has 32 pounds of fish. How much might she get paid for her catch?

80. Ticket sales. Ticket sales at an amusement park average about 460 on a Sunday. If actual Sunday sales are never more than 25 above or below the average and if each ticket costs $29.50, how much money might the park take in on a Sunday?

C EXERCISES Beyond the Basics

In Exercises 81–86, solve each equation or inequality.

81. $|x^2 + 4x - 3| = 5$ **82.** $|x^2 + 3x - 2| = 2$

83. $\dfrac{1}{|2x - 1|} < 2$ **84.** $\dfrac{3}{|x + 1|} \geq 1$

$\left[Hint: 0 < a < b \text{ implies that } \dfrac{1}{a} > \dfrac{1}{b}. \right]$

85. $|x|^2 - 4|x| - 7 = 5$ **86.** $2|x|^2 - |x| + 8 = 11$

87. Show that if $0 < a < b$ and $0 < c < d$, then $ac < bd$.

88. Show that if $0 < a < b < c$ and $0 < e < f < g$, then $ae < bf < cg$.

89. Show that if $x^2 < a$ and $a > 0$, the solution set of the inequality $x^2 < a$ is $\{x | -\sqrt{a} < x < \sqrt{a}\}$, or in interval notation, $(-\sqrt{a}, \sqrt{a})$.
[*Hint:* $a = (\sqrt{a})^2$; factor $x^2 - a$.]

90. Show that if $x^2 > a$ and $a > 0$, then the solution set of the inequality $x^2 > a$ is $(-\infty, -\sqrt{a}) \cup (\sqrt{a}, \infty)$.

In Exercises 91–102, write an absolute value inequality with variable x whose solution set is in the given interval(s).

91. $(1, 7)$ **92.** $[3, 8]$

93. $[-2, 10]$ **94.** $(-7, -1)$

95. $(-\infty, 3) \cup (11, \infty)$ **96.** $(-\infty, -1) \cup (5, \infty)$

97. $(-\infty, -5] \cup [10, \infty)$ **98.** $(-\infty, -3] \cup [-1, \infty)$

99. (a, b) **100.** $[c, d]$

101. $(-\infty, a) \cup (b, \infty)$ **102.** $(-\infty, c] \cup [d, \infty)$

103. Varying temperatures. The weather forecast predicts temperatures that will be between 8° of 70° Fahrenheit. Let x = temperature in degrees Fahrenheit. Write an absolute value inequality in x whose solution is the predicted range of temperatures.

In Exercises 104–109, solve each inequality for x. Use the fact that $|a| \leq |b|$ if and only if $a^2 \leq b^2$.

104. $|x + 1| < |x - 3|$ **105.** $|2x| \leq |x - 5|$

106. $\left|\dfrac{x - 1}{x + 1}\right| < 1$ **107.** $\left|\dfrac{3 - 2x}{1 + x}\right| \leq 4$

108. $\dfrac{1}{|x - 4|} - \dfrac{1}{|x + 7|} < 0$

109. $\dfrac{1}{|x - 3|} \geq \dfrac{1}{|x + 4|}$

Critical Thinking

110. For which values of x is $\sqrt{(x - 3)^2} = x - 3$?

111. For which values of x is $\sqrt{(x^2 - 6x + 8)^2} = x^2 - 6x + 8$?

112. Solve $|x - 3|^2 - 7|x - 3| + 10 = 0$.
[*Hint:* Let $u = |x - 3|$.]

Summary ■ Definitions, Concepts, and Formulas

1.1 Linear Equations in One Variable

i. An **equation** is a statement that two mathematical expressions are equal.

ii. The **solutions** or **roots** of an equation are those values (if any) of the variable that satisfy the equation.

iii. **Equivalent equations** are equations that have the same solution set.

iv. You can generate equivalent equations using the operations given on page 85.

v. The **standard form** of a conditional linear equation in x is $ax + b = 0, a \neq 0$.

vi. A **formula** is an equation that expresses a relationship between two or more variables.

1.2 Applications of Linear Equations

i. **Mathematical modeling** is the process of translating real-world problems into mathematical problems.

ii. **Procedure for solving applied problems:** See page 97.

iii. **Simple interest:** $I = $ Principal $\cdot$ Rate $\cdot$ Time

iv. **Uniform motion:** Distance $=$ Speed $\cdot$ Time

v. **Work rate:** If a job can be done in x units of time, the portion of the job completed in one unit of time is $\dfrac{1}{x}$.

1.3 Complex Numbers

i. Complex numbers are of the form $a + bi$, where a and b are real numbers and $i = \sqrt{-1}$; $i^2 = -1$.

ii. The number $a - bi$ is called the complex conjugate of $a + bi$.

iii. Operations with complex numbers can be performed as if they are binomials with the variable i. Set $i^2 = -1$ to simplify. Division is performed by first multiplying the numerator and the denominator by the complex conjugate of the denominator.

1.4 Quadratic Equations

i. The equation $ax^2 + bx + c = 0, a \neq 0$, is the **standard form** of the quadratic equation in x.

ii. **Zero-product property:** Let A and B be two algebraic expressions. Then $AB = 0$ if and only if $A = 0$ or $B = 0$.

iii. **Square root method:** If $u^2 = d$, then $u = \pm\sqrt{d}$.

iv. **Method for completing the square:** See page 121.

v. **Quadratic formula:** $x = \dfrac{-b \pm \sqrt{b^2 - 4ac}}{2a}$

vi. The quantity $b^2 - 4ac$ is called the **discriminant**.

1.5 Solving Other Types of Equations

Many types of equations can be solved by factoring. However, it is possible to introduce extraneous solutions when solving rational equations, equations involving radicals, or equations involving rational exponents. Remember to check the solutions. Extraneous solutions may be introduced when you

i. multiply both sides by an LCD that contains a variable.

ii. raise both sides to an even power.

1.6 Linear Inequalities

i. An **inequality** is a statement that one algebraic expression is less than or is less than or equal to another algebraic expression.

ii. The real numbers that result in a true statement when they are substituted for the variable in the inequality are called **solutions** of the inequality.

iii. A **linear inequality** is an inequality that can be written in the form $ax + b < 0$. The symbol $<$ can be replaced with $\leq$, $>$, or $\geq$.

1.7 Equations and Inequalities Involving Absolute Value

i. **Definition.** $|a| = a$ if $a \geq 0$, and $|a| = -a$ if $a < 0$.

ii. Let u be a variable or an algebraic expression and let $a > 0$. Then

a. $|u| < a$ if and only if $-a < u < a$.

b. $|u| > a$ if and only if $u < -a$ or $u > a$.

The properties in (a) and (b) are valid if $<$ and $>$ are replaced with $\leq$ and $\geq$, respectively.

REVIEW EXERCISES

Basic Skills and Concepts

In Exercises 1–18, solve each equation.

1. $5x - 4 = 11$
2. $12x + 7 = 31$
3. $3(2x - 4) = 9 - (x + 7)$
4. $4(x + 7) = 40 + 2x$
5. $3x + 8 = 3(x + 2) + 2$
6. $3x + 8 = 3(x + 1) + 4$
7. $x - (5x - 2) = 7(x - 1) - 2$
8. $7 + 4(3 + y) = 8(3y - 1) + 3$
9. $\dfrac{2}{x + 3} = \dfrac{5}{11x - 1}$
10. $\dfrac{7}{x + 2} = \dfrac{3}{x - 2}$
11. $\dfrac{y + 5}{2} + \dfrac{y - 1}{3} = \dfrac{7y + 3}{8} + \dfrac{4}{3}$
12. $\dfrac{y - 3}{6} - \dfrac{y - 4}{5} = -\dfrac{1}{6}$
13. $|2x - 3| = |4x + 5|$
14. $|5x - 3| = |x + 4|$
15. $|2x - 1| = |2x + 7|$
16. $|x - 1| = 2 - x$
17. $|3x - 2| = 2x + 1$
18. $|1 - 2x| = x + 5$

In Exercises 19–22, solve each equation for the indicated variable.

19. $p = k + gt$ for g
20. $RK = 4 + 3K$ for K
21. $T = \dfrac{2B}{B - 1}$ for B
22. $S = \dfrac{a}{1 - r}$ for r

In Exercises 23–46, solve each equation. (Include all complex solutions.)

23. $x^2 - 7x = 0$
24. $x^2 - 32x = 0$
25. $x^2 - 3x - 10 = 0$
26. $2x^2 - 9x - 18 = 0$
27. $(x - 1)^2 = 2x^2 + 3x - 5$
28. $(x + 2)^2 = x(3x + 2)$
29. $\dfrac{x^2}{4} + x = \dfrac{5}{4}$
30. $\dfrac{x^2}{4} + \dfrac{7}{16} = x$
31. $3x(x + 1) = 2x + 2$
32. $x^2 - x = 3(5 - x)$
33. $x^2 - 3x - 1 = 0$
34. $x^2 + 6x + 2 = 0$
35. $2x^2 + x - 1 = 0$
36. $x^2 + 4x + 1 = 0$
37. $3x^2 - 12x - 24 = 0$
38. $2x^2 + 4x - 3 = 0$
39. $2x^2 - x - 2 = 0$
40. $3x^2 - 5x + 1 = 0$
41. $x^2 - x + 1 = 0$
42. $3x^2 + 4x + 3 = 0$
43. $x^2 - 6x + 13 = 0$
44. $x^2 - 8x + 20 = 0$
45. $4x^2 - 8x + 13 = 0$
46. $3x^2 - 4x + 2 = 0$

In Exercises 47–50, find the discriminant and determine the number and type of roots.

47. $3x^2 - 11x + 6 = 0$
48. $x^2 - 14x + 49 = 0$
49. $5x^2 + 2x + 1 = 0$
50. $9x^2 = 25$

In Exercises 51–70, solve each equation.

51. $\sqrt{x^2 - 16} = 0$
52. $\sqrt{x + 6} = x$
53. $\sqrt{4 - 7x} = \sqrt{2}x$
54. $t - 2\sqrt{t} + 1 = 0$
55. $y - 2\sqrt{y} - 3 = 0$
56. $\sqrt{3x + 4} - \sqrt{x - 3} = 3$
57. $\sqrt{x - 1} = \sqrt{5 + \sqrt{x}}$
58. $\dfrac{1}{(1 - x)^2} - \dfrac{7}{1 - x} = -10$
59. $(7x + 5)^2 + 2(7x + 5) - 15 = 0$
60. $(x^2 - 1)^2 - 11(x^2 - 1) + 24 = 0$
61. $x^{2/3} + 3x^{1/3} - 4 = 0$
62. $x^{-2/3} + x^{-1/3} - 6 = 0$
63. $(\sqrt{t} + 5)^2 - 9(\sqrt{t} + 5) + 20 = 0$
64. $3\left(\dfrac{y - 1}{6}\right)^2 - 7\left(\dfrac{y - 1}{6}\right) = 0$
65. $4x^4 - 37x^2 + 9 = 0$
66. $\dfrac{1}{x} + \dfrac{1}{x - 1} = \dfrac{5}{6}$
67. $\dfrac{2x + 1}{2x - 1} = \dfrac{x - 1}{x + 1}$
68. $6 - \dfrac{2}{x} = \dfrac{4}{x - 1}$
69. $\left(\dfrac{7x}{x + 1}\right)^2 - 3\left(\dfrac{7x}{x + 1}\right) = 18$
70. $\left(\dfrac{4x^2 - 3}{x}\right)^2 = 1$

In Exercises 71–74, solve each equation for x in terms of other variables.

71. $x^2 + 2yx - 3y^2 = 0$
72. $x^2 + (y - 2z)x - 2yz = 0$
73. $x^2 + (3 - 2y)x + y^2 - 3y + 2 = 0$
74. $x^2 + (1 - 2y)x + y^2 - y - 2 = 0$

In Exercises 75–90, solve each inequality. Write the solution in interval notation.

75. $x + 5 < 3$
76. $2x + 1 < 9$
77. $3(x - 3) \le 8$
78. $x + 5 \le 19 + 3x$
79. $x + 2 \ge \dfrac{2}{3}x - 2x$
80. $2x + 1 \ge \dfrac{5x - 6}{3}$
81. $(5x + 15)^{-1} < 0$
82. $(3x - 1)^{-1} > 0$
83. $(2x - 5)^{-1} > 0$
84. $(3x + 4)^{-1} < 0$
85. $\dfrac{1}{6} > \dfrac{4 - 3x}{3}$
86. $\dfrac{2}{5} > \dfrac{3 - 2x}{2}$
87. $\dfrac{x - 3}{3} - 2 \le \dfrac{x}{6} + \dfrac{1}{2}$
88. $\dfrac{x + 1}{2} - 3 \le \dfrac{x}{4} + \dfrac{1}{2}$
89. $\dfrac{3 - 2x}{4} + 1 > \dfrac{x - 5}{3}$
90. $\dfrac{5 - 3x}{5} - 2 > \dfrac{2 - x}{3}$

In Exercises 91–98, solve each compound inequality. Write the solution in interval notation.

91. $3x - 1 < 2$ or $11 - 2x < 5$

92. $3 - 2x > 5$ or $15 - 3x < 6$

93. $4x - 5 < 7$ and $7 - 3x < 1$

94. $2x - 1 < 3$ and $4 - 3x > 1$

95. $-3 \le 2x + 1 < 7$ **96.** $\dfrac{1}{6} \le \dfrac{3x - 4}{3} \le 4$

97. $-3 < 3 - 2x \le 97$ **98.** $-\dfrac{1}{2} < \dfrac{4 - 3x}{3} \le \dfrac{1}{2}$

In Exercises 99–104, solve each inequality. Write the solution set in interval notation.

99. $|3x + 2| \le 7$ **100.** $|x - 4| \ge 2$

101. $4|x - 2| + 8 > 12$ **102.** $3|x - 1| + 4 < 10$

103. $\left|\dfrac{4 - x}{5}\right| \ge 1$ **104.** $\left|\dfrac{1 - x}{6}\right| < 1$

Applying the Concepts

105. A circular lens has a circumference of 22 centimeters. Find the radius of the lens.

106. A rectangle has a perimeter of 18 inches and a length of 5 inches. Find the width of the rectangle.

107. A trapezoid has an area of 32 square meters and a height of 8 meters. If one base is 5 meters, what is the length of the other base?

108. What principal must be deposited for four years at 7% annual simple interest to earn $354.20 in interest?

109. The volume of a box is 4212 cubic centimeters. If the box is 27 centimeters long and 12 centimeters wide, how high is it?

Box

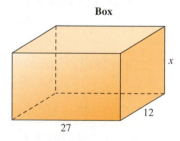

110. The ratio of the current assets of a business to its current liabilities is called the *current ratio* of the business. Assuming that the current ratio is 2.7 and the current assets total $256,500, find the current liabilities of the business.

111. A cylindrical shaft has a volume of 8750π cubic centimeters. If the radius is 5 centimeters, how long is the shaft?

112. A car dealer deducted 15% of the selling price of a car he had sold on consignment and gave the balance, $2210, to the owner. What was the selling price of the car?

113. The third angle in an isosceles triangle measures 40 degrees more than twice either base angle. Find the number of degrees in each angle of the triangle.

114. A 600-mile trip took 12 hours. Half of the time was spent traveling across flat terrain, and half was through hilly terrain. Assuming that the average rate over the hilly section was 20 miles per hour slower than the average rate over the flat section, find the two rates and the distance traveled at each rate.

115. A total of $30,000 is invested in two stocks. At the end of the year, one returned 6% and the other returned 8% on the original investment. How much was invested in each stock if the total profit was $2160?

116. The monthly note on a car that was leased for two years was $250 less than the monthly note on a car that was leased for a year and a half. The total income from the two leases was $21,300. Find the monthly note on each.

117. Two solutions, one containing $4\frac{1}{2}$% iodine and the other containing 12% iodine, are to be mixed to produce 10 liters of a 6% iodine solution. How many liters of each are required?

118. Two cars leave from the same place at the same time traveling in opposite directions. One travels 5 kilometers per hour faster than the other. After three hours, they are 495 kilometers apart. How fast is each car traveling?

119. A total of 28 handshakes were exchanged at the end of a party. Assuming that everyone shook hands with everyone else, how many people were at the party?

120. Find two positive numbers that differ by 7 and whose product is 408.

121. Lavina bought some shares of stock for $18,040. Later when the price had gone up $18 per share, she sold all but 20 of them for $20,000. How many shares had she bought?

122. Joann drives her car for 15 miles at a certain speed. She then increases her speed by 10 mph and drives for another 20 miles. If the total trip takes one hour, find her original speed.

123. One-fourth of a herd of wild horses is in a forest. Twice the square root of the number of horses in the herd has gone to the mountains, and the remaining 15 are on the bank of the river. What is the total number of horses in the herd?

124. A path of uniform width surrounds a rectangular swimming pool that is 30 ft long and 16 ft wide. Find the width of the path assuming that its area is 312 ft^2.

125. The Botany Club chartered a bus for $324 to go on a field trip, the cost to be divided equally among those attending. At the last minute, four more members decided to go, which reduced the cost by $0.90 per person. How many members went on the trip?

PRACTICE TEST A

1. Solve the equation $5x - 9 = 3x - 5$.

2. Solve the equation $\dfrac{7}{24} = \dfrac{x}{8} + \dfrac{1}{6}$.

3. Solve the equation $\dfrac{1}{x-2} - 5 = \dfrac{1}{x+2}$.

4. The width of a rectangle is 3 centimeters less than the length, and the area is 54 square centimeters. Find the dimensions of the rectangle.

5. Solve the equation $x^2 + 36 = -13x$.

6. If Fran invests $8200 at 6% per year, how much additional money must she invest at 8% to ensure that the interest she receives each year is 7% of the total investment?

7. A frame of uniform width borders a painting that is 25 inches long and 11 inches high. If the area of the framed picture is 351 square inches, find the width of the border.

8. Solve $-6x - 15 = (2x + 5)^2$.

9. Determine the constant that should be added to the binomial

$$x^2 + \frac{2}{3}x$$

so that it becomes a perfect-square trinomial. Then write and factor the trinomial.

10. Solve $3x^2 - 5x - 1 = 0$.

11. A box with a square base and no top is made from a square piece of cardboard by cutting 2-inch squares from each corner and folding up the sides. What length of side must the original cardboard square have if the volume of the box is 50 cubic inches?

12. Solve $x^2 + 12x + 33 = 0$.

13. Solve the polynomial equation

$$3x^4 - 75x^2 = 0$$

by factoring and then using the zero-product property.

14. Solve the equation

$$3x - 2 - 5\sqrt{x} = 0$$

by making an appropriate substitution.

15. Solve the absolute value equation

$$\left|\frac{1}{3}x + 5\right| = \left|\frac{2}{3}x + 7\right|.$$

In Problems 16 and 17, solve the inequality. Express the solution set in interval notation.

16. $\dfrac{x}{2} - 5 \geq \dfrac{4x}{9}$

17. $-4 < 2x - 3 < 4$

18. Solve the inequality

$$\left|\frac{2}{3}x - 1\right| - 2 > \frac{1}{3}$$

by first rewriting it as an equivalent inequality without absolute value bars. Express the solution set in interval notation.

19. Solve the linear inequality $\dfrac{2}{5}y - 13 \leq -\left(7 + \dfrac{13}{5}y\right)$.

20. Solve the inequality $0 \leq 5x - 2 \leq 8$.

PRACTICE TEST B

1. Solve the equation $2x - 2 = 5x + 34$.

   a. $\left\{\dfrac{17}{4}\right\}$ **b.** $\{-18\}$

   c. $\left\{\dfrac{15}{2}\right\}$ **d.** $\{-12\}$

2. Solve the equation $\dfrac{z}{2} = 2z + 35$.

   a. $\left\{\dfrac{47}{2}\right\}$ **b.** $\left\{-\dfrac{47}{2}\right\}$

   c. $\{23\}$ **d.** $\left\{-\dfrac{70}{3}\right\}$

3. Solve the equation $\dfrac{1}{t-2} - \dfrac{1}{2} = \dfrac{-2t}{4t-1}$.

   a. $\left\{\dfrac{4}{9}\right\}$ b. $\left\{\dfrac{1}{2}\right\}$

   c. $\left\{\dfrac{2}{3}\right\}$ d. $\{-2\}$

4. The length of a rectangle is 4 centimeters greater than the width, and the area is 77 square centimeters. Find the dimensions of the rectangle.
   a. 11 centimeters by 15 centimeters
   b. 5 centimeters by 9 centimeters
   c. 7 centimeters by 11 centimeters
   d. 9 centimeters by 13 centimeters

5. Solve the equation $x^2 + 12 = -7x$.
   a. $\{3, 4\}$ b. $\{-4, -3\}$
   c. $\{-3, 3, 4\}$ d. $\{4, 3\}$

6. If Rena invests \$7500 at 7% per year, how much additional money must she invest at 12% to ensure that the interest she receives each year is 10% of the total investment?
   a. \$11,250   b. \$12,375
   c. \$18,000   d. \$21,000

7. A frame of uniform width borders a painting that is 20 inches long and 13 inches high. Assuming that the area of the framed picture is 368 square inches, find the width of the border.
   a. 3.5 in.   b. 3 in.
   c. 4 in.   d. 1.5 in.

8. Solve $-6x - 2 = (3x + 1)^2$.

   a. $\varnothing$ b. $\left\{-\dfrac{1}{3}\right\}$

   c. $\left\{\dfrac{1}{3}, 1\right\}$ d. $\left\{-1, -\dfrac{1}{3}\right\}$

9. Determine the constant that should be added to the binomial

$$x^2 + \dfrac{1}{6}x$$

   so that it becomes a perfect-square trinomial. Then write and factor the trinomial.

   a. $\dfrac{1}{12}; x^2 + \dfrac{1}{6}x + \dfrac{1}{12} = \left(x + \dfrac{1}{6}\right)^2$

   b. $\dfrac{1}{144}; x^2 + \dfrac{1}{6}x + \dfrac{1}{144} = \left(x + \dfrac{1}{12}\right)^2$

   c. $\dfrac{1}{36}; x^2 + \dfrac{1}{6}x + \dfrac{1}{36} = \left(x + \dfrac{1}{6}\right)^2$

   d. $144; x^2 + \dfrac{1}{6}x + 144 = (x + 12)^2$

10. Solve $7x^2 + 10x + 2 = 0$.

   a. $\left\{\dfrac{-5 - \sqrt{11}}{7}, \dfrac{-5 + \sqrt{11}}{7}\right\}$

   b. $\left\{\dfrac{-5 - \sqrt{39}}{7}, \dfrac{-5 + \sqrt{39}}{7}\right\}$

   c. $\left\{\dfrac{-5 - \sqrt{11}}{14}, \dfrac{-5 + \sqrt{11}}{14}\right\}$

   d. $\left\{\dfrac{-10 - \sqrt{11}}{7}, \dfrac{-10 + \sqrt{11}}{7}\right\}$

11. A box with a square base and no top is made from a square piece of cardboard by cutting 3-inch squares from each corner and folding up the sides. What length of side must the original cardboard square have if the volume of the box is 675 cubic inches?
   a. 18 inches   b. 21 inches
   c. 15 inches   d. 14 inches

12. Solve $x^2 + 14x + 38 = 0$.
   a. $\{7 - \sqrt{38}, 7 + \sqrt{38}\}$
   b. $\{14 + \sqrt{38}\}$
   c. $\{7 + \sqrt{11}\}$
   d. $\{-7 - \sqrt{11}, -7 + \sqrt{11}\}$

13. Solve the polynomial equation

$$5x^4 - 45x^2 = 0$$

   by factoring and then using the zero-product property.
   a. $\{-3, 0, 3\}$ b. $\{-3, 3\}$
   c. $\{0\}$ d. $\{-3\sqrt{5}, 0, 3\sqrt{5}\}$

14. Solve the equation

$$x - 2048 - 32\sqrt{x} = 0$$

   by making an appropriate substitution.
   a. $\{4096\}$ b. $\{3072\}$
   c. $\{8192\}$ d. $\{2048\}$

15. Solve the absolute value equation

$$\left|\dfrac{1}{2}x + 2\right| = \left|\dfrac{3}{4}x - 2\right|.$$

   a. $\{12, 16\}$ b. $\varnothing$ c. $\{10\}$ d. $\{0, 16\}$

In Problems 16 and 17, solve the inequality. Express the solution set in interval notation.

16. $\dfrac{x}{6} - \dfrac{1}{3} \le \dfrac{x}{3} + 1$
   a. $(-\infty, -8)$ b. $(-\infty, -8)$
   c. $(-8, \infty)$ d. $[-8, \infty)$

17. $-13 \le -3x + 2 < -4$
   a. $[-5, -2)$ b. $(2, 5]$
   c. $(-5, -2]$ d. $[2, 5)$

18. Solve the inequality

$$8 + \left|1 - \dfrac{x}{2}\right| \ge 10$$

   by first rewriting it as an equivalent inequality without absolute value bars. Express the solution set in interval notation.
   a. $(-\infty, -6] \cup [2, \infty)$ b. $[-2, 6]$
   c. $(-\infty, -2] \cup [6, \infty)$ d. $[-6, 2]$

19. Solve the linear inequality $\dfrac{2}{3}x - 2 < \dfrac{5}{3}x$.
   a. $(-2, \infty)$ b. $(\infty, 2)$ c. $(-\infty, -2)$
   d. $\{-2\}$ e. $(2, \infty)$

20. Solve the inequality $0 \le 7x - 1 \le 13$.
   a. $\left[\dfrac{1}{7}, 2\right]$ b. $(-1, 2]$ c. $\dfrac{1}{7}$
   d. $[-1, 2]$ e. $\left(-\infty, \dfrac{1}{7}\right]$

Graphs and Functions

Hurricanes are tracked with the use of coordinate grids, and data from fields as diverse as medicine and sports are related and analyzed by means of functions. The material in this chapter introduces you to the versatile concepts that are the everyday tools of all dynamic industries.

The Coordinate Plane

Before Starting this Section, Review

1. The number line
   (Section P.1, page 8)

2. The Pythagorean Theorem
   (Section P.7, page 72)

Objectives

1. Plot points in the Cartesian coordinate plane.

1. Find the distance between two points.

3. Find the midpoint of a line segment.

A FLY ON THE CEILING

It is said that one day the French mathematician René Descartes noticed a fly buzzing around on a ceiling made of square tiles. He watched the fly for a long time and wondered how he could mathematically describe its location. Finally, he realized that he could describe the fly's position by its distance from the walls of the room. Descartes had just discovered the coordinate plane! In fact, the coordinate plane is sometimes called the Cartesian plane in his honor. The discovery led to the development of analytic geometry, the first mathematical blending of algebra and geometry.

Although the basic idea of graphing with coordinate axes dates all the way back to Apollonius in the second century B.C., Descartes, who lived in the 1600s, gets the credit for coming up with the two-axis system we use today. In Example 2, we see how the Cartesian plane helps visualize data on smoking. ■

1 Plot points in the Cartesian coordinate plane.

The Coordinate Plane

A visually powerful device for exploring relationships between numbers is the Cartesian plane. A pair of real numbers in which the order is specified is called an **ordered pair** of real numbers. The ordered pair (a, b) has **first component** a and **second component** b. Two ordered pairs (x, y) and (a, b) are **equal** if and only if $x = a$ and $y = b$.

Just as the real numbers are identified with points on a line, called the *number line* or the *coordinate line*, the sets of ordered pairs of real numbers are identified with points on a plane called the **coordinate plane** or the **Cartesian plane**. This identification is attributed to Descartes.

We begin with two coordinate lines, one horizontal and one vertical, that intersect at their zero points. The horizontal line (with positive numbers to the right) is usually called the **x-axis**, while the vertical line (with positive numbers up) is usually called the **y-axis**. The point of intersection is called the **origin**. The x-axis and y-axis are called **coordinate axes**, and the plane they determine is sometimes called the **xy-plane**. The axes divide the plane into four regions called **quadrants**, which are numbered as shown in Figure 2.1. The points on the axes do not belong to any of the quadrants.

Figure 2.1 shows how each ordered pair (a, b) of real numbers is associated with a unique point P in the plane and how each point in the plane is associated with a unique ordered pair of real numbers. The first component, a, is called the **x-coordinate** of P and the second component, b, is called the **y-coordinate** of P because we have called our horizontal axis the x-axis and our vertical axis the y-axis.

STUDY TIP

Although it is common to label the axes x and y, other letters are also used, especially in applications. In a *uv*-plane or an *st*-plane, the first letter in the name refers to the horizontal axis; the second, to the vertical axis.

FIGURE 2.1 **The Cartesian plane**

The signs of the x- and y-coordinates are shown in Figure 2.1 for each quadrant. We refer to the point corresponding to the ordered pair (a, b) as the **graph of the ordered pair** (a, b) in the coordinate system. However, we frequently ignore the distinction between an ordered pair and its graph. Either notation, $P(a, b)$ or $P = (a, b)$, designates the point P in the coordinate plane whose x-coordinate is a and whose y-coordinate is b.

EXAMPLE 1   **Graphing Points**

Graph the following points in the xy-plane.

$A(3, 1)$, $B(-2, 4)$, $C(-3, -4)$, $D(2, -3)$, and $E(-3, 0)$.

SOLUTION

Figure 2.2 shows a coordinate plane, along with the graph of the given points. These points are located by moving left, right, up, or down starting from the origin $(0, 0)$.

$A(3, 1)$	3 units right, 1 unit up	$D(2, -3)$	2 units right, 3 units down
$B(-2, 4)$	2 units left, 4 units up	$E(-3, 0)$	3 units left, 0 units up or down
$C(-3, -4)$	3 units left, 4 units down		

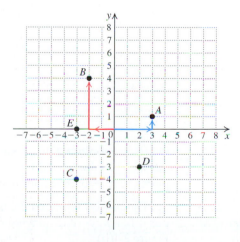

FIGURE 2.2 **Graphs of ordered pairs**

Practice Problem 1 Plot in the xy-plane the points

$$P(-2,2), Q(4,0), R(5,-3), S(0,-3), \text{ and } T\left(-2, \frac{1}{2}\right).$$

■

EXAMPLE 2 **Plotting Data on Adult Smokers in the United States**

The data in Table 2.1 show the prevalence of smoking among adults aged 18 years and older in the United States over the years 2000–2007.

TABLE 2.1

Year	2000	2001	2002	2003	2004	2005	2006	2007
Percent of Adult Smokers	23.2	22.7	22.4	21.6	20.9	21.1	20.8	20.8

Source: Center for Disease Control. www.cdc.gov

Plot the graph of the ordered pairs (year, percent of adult smokers) where the first coordinate represents a year and the second coordinate represents the percent of adult smokers in that year.

SOLUTION

We let x represent the years 2000–2007 and y represent the percent of adult smokers in those years. Because no data are given for the years 0–1999, it is customary to indicate these omissions graphically by showing a break in the x-axis. Another way to illustrate the data is to declare a year—say, 2000—as 0. Similar comments apply to the y-axis. The graph of the points (2000, 23.2), (2001, 22.7), (2002, 22.4), (2003, 21.6), (2004, 20.9), (2005, 21.1), (2006, 20.8), and (2007, 20.8) is shown in Figure 2.3. The graph shows that the percent of adult smokers has been declining almost every year since 2000.

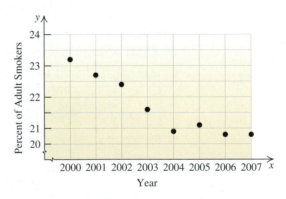

FIGURE 2.3 **Percent of adult smokers 2000–2007**   ■ ■ ■

Practice Problem 2 In Example 2, suppose the percent of adult smokers shown in Table 2.1 is decreased by 3 each year. Write and plot the corresponding ordered pairs (year, percent of adult smokers).   ■

The display in Figure 2.3 is called a **scatter diagram** of the data. There are numerous other ways of displaying the data. Two such ways are shown in Figure 2.4(a) and Figure 2.4(b).

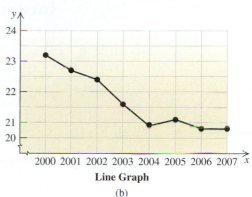

Bar Graph
(a)

Line Graph
(b)

FIGURE 2.4 Two methods for displaying data

Scales on a Graphing Utility

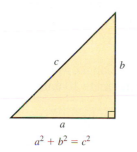

FIGURE 2.5 Viewing rectangle

When drawing a graph, you can use different scales for the x- and y-axes. Similarly, the scale can be set separately for each coordinate axis on a graphing utility. Once scales are set, you get a **viewing rectangle**, where your graphs are displayed. For example, in Figure 2.5, the scale for the x-axis (the distance between each tick mark) is 1, whereas the scale for the y-axis is 2. Read more about viewing rectangles in your graphing calculator manual.

Distance Formula

2 Find the distance between two points.

Suppose a Cartesian coordinate system has the same unit of measurement, such as inches or centimeters, on both axes. We can then calculate the distance between any two points in the coordinate plane in the given unit.

Recall that the Pythagorean Theorem states that in a **right triangle** with hypotenuse of length c and the other two sides of lengths a and b,

$$a^2 + b^2 = c^2, \quad \text{Pythagorean Theorem}$$

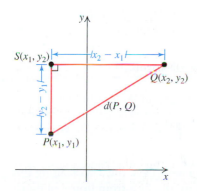

$$a^2 + b^2 = c^2$$

FIGURE 2.6 Pythagorean Theorem

as shown in Figure 2.6

Suppose we want to compute the distance between the two points $P(x_1, y_1)$ and $Q(x_2, y_2)$. We draw a horizontal line through the point Q and a vertical line through the point P to form the right triangle PQS, as shown in Figure 2.7.

The length of the horizontal side of the triangle is $|x_2 - x_1|$, and the length of the vertical side is $|y_2 - y_1|$. By the Pythagorean Theorem, we have

$$[d(P, Q)]^2 = |x_2 - x_1|^2 + |y_2 - y_1|^2$$
$$d(P, Q) = \sqrt{|x_2 - x_1|^2 + |y_2 - y_1|^2} \quad \text{Take the square root of both sides.}$$
$$d(P, Q) = \sqrt{(x_2 - x_1)^2 + (y_2 - y_1)^2} \quad |a - b|^2 = (a - b)^2$$

where $d(P, Q)$ is the distance between P and Q.

FIGURE 2.7 Visualizing the distance formula

DISTANCE FORMULA IN THE COORDINATE PLANE

Let $P = (x_1, y_1)$ and $Q = (x_2, y_2)$ be any two points in the coordinate plane. Then the distance between P and Q, denoted $d(P, Q)$, is given by the **distance formula**

$$d(P, Q) = \sqrt{(x_2 - x_1)^2 + (y_2 - y_1)^2}$$

EXAMPLE 3 Finding the Distance Between Two Points

Find the distance between the points $P(-2, 5)$ and $Q(3, -4)$.

SOLUTION

Let $(x_1, y_1) = (-2, 5)$ and $(x_2, y_2) = (3, -4)$. Then

$$x_1 = -2, \; y_1 = 5, \; x_2 = 3, \text{ and } y_2 = -4.$$

$$
\begin{aligned}
d(P, Q) &= \sqrt{(x_2 - x_1)^2 + (y_2 - y_1)^2} && \text{Distance formula} \\
&= \sqrt{[3 - (-2)]^2 + (-4 - 5)^2} && \text{Substitute the values for } x_1, x_2, y_1, y_2. \\
&= \sqrt{5^2 + (-9)^2} && \text{Simplify.} \\
&= \sqrt{25 + 81} = \sqrt{106} \approx 10.3 && \text{Simplify; use a calculator.} \quad ■■■
\end{aligned}
$$

Practice Problem 3   Find the distance between the points $(-5, 2)$ and $(-4, 1)$. ■

In the next example, we use the distance formula and the converse of the Pythagorean Theorem to show that the given triangle is a right triangle.

EXAMPLE 4   **Identifying a Right Triangle**

Let $A(4, 3)$, $B(1, 4)$, and $C(-2, -5)$ be three points in the plane.

a. Sketch the triangle ABC.

b. Find the length of each side of the triangle.

c. Show that ABC is a right triangle.

SOLUTION

a. A sketch of the triangle formed by the three points A, B, and C is shown in Figure 2.8.

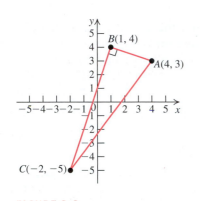

FIGURE 2.8

b. Using the distance formula, we have

$$d(A, B) = \sqrt{(4 - 1)^2 + (3 - 4)^2} = \sqrt{9 + 1} = \sqrt{10},$$

$$d(B, C) = \sqrt{[1 - (-2)]^2 + [4 - (-5)]^2} = \sqrt{9 + 81} = \sqrt{90} = 3\sqrt{10}$$

$$d(A, C) = \sqrt{[4 - (-2)]^2 + [3 - (-5)]^2} = \sqrt{36 + 64} = \sqrt{100} = 10.$$

c. We check whether the relationship $a^2 + b^2 = c^2$ holds in this triangle, where a, b, and c denote the lengths of its sides. The longest side, AC, has length 10 units.

$$
\begin{aligned}
[d(A, B)]^2 + [d(B, C)]^2 &= 10 + 90 && \text{Replace } [d(A, B)]^2 \text{ with } 10 \\
&&& \text{and } [d(B, C)]^2 \text{ with } 90. \\
&= 100 = (10)^2 = [d(A, C)]^2
\end{aligned}
$$

It follows from the converse of the Pythagorean Theorem that the triangle ABC is a right triangle. $\cdot$ ■■■

Practice Problem 4   Is the triangle with vertices $(6, 2)$, $(-2, 0)$, and $(1, 5)$ an isosceles triangle—that is, a triangle with two sides of equal length? ■

EXAMPLE 5   **Applying the Distance Formula to Baseball**

The baseball "diamond" is in fact a square with a distance of 90 feet between each of the consecutive bases. Use an appropriate coordinate system to calculate the distance the ball travels when the third baseman throws it from third base to first base.

SOLUTION

We can conveniently choose home plate as the origin and place the x-axis along the line from home plate to first base and the y-axis along the line from home plate to third base as shown in Figure 2.9. The coordinates of home plate (O), first base (A), second base (C), and third base (B) are shown in the figure.

C(90, 90)

2nd base

y

B(0, 90)   3rd base

1st base

A(90, 0)

x

Home plate   O(0, 0)

FIGURE 2.9

We are asked to find the distance between the points $A(90, 0)$ and $B(0, 90)$.

$$d(A, B) = \sqrt{(90 - 0)^2 + (0 - 90)^2}$$ Distance formula

$$= \sqrt{(90)^2 + (-90)^2}$$

$$= \sqrt{2(90)^2}$$ Simplify.

$$= 90\sqrt{2} \approx 127.28 \text{ feet}$$ Simplify; use a calculator.   ■ ■ ■

Practice Problem 5   Young players might play baseball in a square "diamond" with a distance of 60 feet between consecutive bases. Repeat Example 5 for a diamond with these dimensions.   ■

Midpoint Formula

3 Find the midpoint of a line segment.

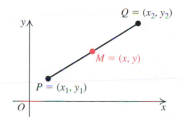

FIGURE 2.10 Midpoint of a segment

STUDY TIP

The coordinates of the midpoint of a line segment are found by taking the average of the x-coordinates and the average of the y-coordinates of the endpoints.

MIDPOINT FORMULA

The coordinates of the midpoint $M = (x, y)$ of the line segment joining $P = (x_1, y_1)$ and $Q = (x_2, y_2)$ are given by

$$M = (x, y) = \left(\frac{x_1 + x_2}{2}, \frac{y_1 + y_2}{2} \right).$$

We ask you to prove the midpoint formula in Exercise 68.

EXAMPLE 6   **Finding the Midpoint of a Line Segment**

Find the midpoint of the line segment joining the points $P(-3, 6)$ and $Q(1, 4)$.

SOLUTION

Let $(x_1, y_1) = (-3, 6)$ and $(x_2, y_2) = (1, 4)$. Then

$$x_1 = -3, y_1 = 6, x_2 = 1, \text{ and } y_2 = 4.$$

$$\text{Midpoint} = \left(\frac{x_1 + x_2}{2}, \frac{y_1 + y_2}{2} \right)$$ Midpoint formula

$$= \left(\frac{-3 + 1}{2}, \frac{6 + 4}{2} \right)$$ Substitute values for x_1, x_2, y_1, y_2.

$$= (-1, 5)$$ Simplify.

The midpoint of the line segment joining the points $P(-3, 6)$ and $Q(1, 4)$ is $M(-1, 5)$.   ■ ■ ■

Practice Problem 6   Find the midpoint of the line segment whose endpoints are $(5, -2)$ and $(6, -1)$.   ■

SECTION 2.1 ■ Exercises

A EXERCISES Basic Skills and Concepts

1. A point with a negative first coordinate and a positive second coordinate lies in the _____ quadrant.

2. Any point on the x-axis has second coordinate _____ .

3. The distance between the points $P = (x_1, y_1)$ and $Q = (x_2, y_2)$ is given by the formula $d(P, Q) =$ _____ .

4. The coordinates of the midpoint $M = (x, y)$ of the line segment joining $P = (x_1, y_1)$ and $Q = (x_2, y_2)$ are given by $(x, y) =$ _____ .

5. *True or False* For any points (x_1, y_1) and (x_2, y_2)
$\sqrt{(x_1 - x_2)^2 + (y_1 - y_2)^2} = \sqrt{(x_2 - x_1)^2 + (y_2 - y_1)^2}$.

6. *True or False* The point $(7, -4)$ is four units to the right and two units below the point $(3, 2)$.

7. Plot and label each of the given points in a Cartesian coordinate plane and state the quadrant, if any, in which each point is located. $(2, 2), (3, -1), (-1, 0), (-2, -5), (0, 0), (-7, 4), (0, 3), (-4, 2)$

8. **a.** Write the coordinates of any five points on the x-axis. What do these points have in common?
   b. Plot the points $(-2, 1), (0, 1), (0.5, 1), (1, 1)$, and $(2, 1)$. Describe the set of all points of the form $(x, 1)$, where x is a real number.

9. **a.** If the x-coordinate of a point is 0, where does that point lie?
   b. Plot the points $(-1, 1), (-1, 1.5), (-1, 2), (-1, 3)$, and $(-1, 4)$. Describe the set of all points of the form $(-1, y)$, where y is a real number.

10. What figure is formed by the set of all points in a Cartesian coordinate plane that have
    a. x-coordinate equal to -3?
    b. y-coordinate equal to 4?

11. Let $P(x, y)$ be a point in a coordinate plane.
    a. If the point $P(x, y)$ lies above the x-axis, what must be true of y?
    b. If the point $P(x, y)$ lies below the x-axis, what must be true of y?
    c. If the point $P(x, y)$ lies to the left of the y-axis, what must be true of x?
    d. If the point $P(x, y)$ lies to the right of the y-axis, what must be true of x?

12. Let $P(x, y)$ be a point in a coordinate plane. In which quadrant does P lie
    a. if x and y are both negative?
    b. if x and y are both positive?
    c. if x is positive and y is negative?
    d. if x is negative and y is positive?

In Exercises 13–22, find (a) the distance between P and Q and (b) the coordinates of the midpoint of the line segment PQ.

13. $P(2, 1), Q(2, 5)$

14. $P(3, 5), Q(-2, 5)$

15. $P(-1, -5), Q(2, -3)$

16. $P(-4, 1), Q(-7, -9)$

17. $P(-1, 1.5), Q(3, -6.5)$

18. $P(0.5, 0.5), Q(1, -1)$

19. $P(\sqrt{2}, 4), Q(\sqrt{2}, 5)$

20. $P(v - w, t), Q(v + w, t)$

21. $P(t, k), Q(k, t)$

22. $P(m, n), Q(-n, -m)$

In Exercises 23–30, determine whether the given points are collinear. Points are *collinear* if they can be labeled P, Q, and R so that $d(P, Q) + d(Q, R) = d(P, R)$.

23. $(0, 0), (1, 2), (-1, -2)$

24. $(3, 4), (0, 0), (-3, -4)$

25. $(4, -2), (-2, 8), (1, 3)$

26. $(9, 6), (0, -3), (3, 1)$

27. $(-1, 4), (3, 0), (11, -8)$

28. $(-2, 3), (3, 1), (2, -1)$

29. $(4, -4), (15, 1), (1, 2)$

30. $(1, 7), (-7, 8), (-3, 7.5)$

In Exercises 31–40, identify the triangle PQR as *isosceles* (two sides of equal length), *equilateral* (three sides of equal length), or a *scalene* triangle (three sides of different lengths).

31. $P(-5, 5), Q(-1, 4), R(-4, 1)$

32. $P(3, 2), Q(6, 6), R(-1, 5)$

33. $P(-4, 8), Q(0, 7), R(-3, 5)$

34. $P(-1, 4), Q(3, -2), R(7, 5)$

35. $P(6, 6), Q(-1, -1), R(-5, 3)$

36. $P(0, -1), Q(9, -9), R(5, 1)$

37. $P(1, 1), Q(-1, 4), R(5, 8)$

38. $P(-4, 4), Q(4, 5), R(0, -2)$

39. $P(1, -1), Q(-1, 1), R(-\sqrt{3}, -\sqrt{3})$

40. $P(-.5, -1), Q(-1.5, 1), R(\sqrt{3} - 1, \sqrt{3}/2)$

41. Show that the points $P(7, -12), Q(-1, 3), R(14, 11)$, and $S(22, -4)$ are the vertices of a square. Find the length of the diagonals.

42. Repeat Exercise 41 for the points $P(8, -10)$, $Q(9, -11), R(8, -12)$, and $S(7, -11)$.

43. Find x such that the point $(x, 2)$ is five units from $(2, -1)$.

44. Find y such that the point $(2, y)$ is 13 units from $(-10, -3)$.

45. Find the point on the *x*-axis that is equidistant from the points $(-5, 2)$ and $(2, 3)$.

46. Find the point on the *y*-axis that is equidistant from the points $(7, -4)$ and $(8, 3)$.

B EXERCISES Applying the Concepts

47. **Population.** The table shows the total population of the United States in millions. (181 represents 181,000,000.) The population is rounded to the nearest million. Plot the data in a Cartesian coordinate system.

Year	Total Population (millions)
1960	181
1965	194
1970	205
1975	216
1980	228
1985	238
1990	250
1995	267
2000	282
2005	296

Source: U.S. Census Bureau.

48. Use the midpoint formula for the years 1985 and 2005 to estimate the total population for 1995. What does this say about the line segment through the points $(1985, 238)$ and $(2005, 296)$?

In Exercises 49–52, use the following vital statistics table.

	Rate per 1000 Population			
Year	Births	Deaths	Marriages	Divorces
1950	24.1	9.6	11.1	2.6
1955	25.0	9.3	9.3	2.3
1960	23.7	9.5	8.5	2.2
1965	19.4	9.4	9.3	2.5
1970	18.4	9.5	10.6	3.5
1975	14.6	8.6	10.0	4.8
1980	15.9	8.8	10.6	5.2
1985	15.8	8.8	10.1	5.0
1990	16.7	8.6	9.4	4.7
1995	14.8	8.8	8.9	4.4
2000	14.4	8.7	8.5	4.2
2005	14.2	8.1	7.6	3.6

Source: U.S. Census Bureau.

The table shows the rate per 1000 population of live births, deaths, marriages, and divorces from 1950 to 2005 at five-year intervals. Plot the data in a Cartesian coordinate system and connect the points with line segments.

49. Plot (year, births).   **50.** Plot (year, deaths).

51. Plot (year, marriages).   **52.** Plot (year, divorces).

In Exercises 53–55, assume that the graph through all of the data points is a line segment. Use the midpoint formula to find the estimate.

53. **Murders in the United States.** In the United States, 16,740 murders were committed in 2005 and 16,930 in 2007. Estimate the number of murders committed in 2006. (*Source:* U.S. Census Bureau.)

54. **Spending on prescription drugs.** Americans spent $141 billion on prescription drugs in 2001 and $200 billion in 2005. Estimate the amount spent on prescription drugs during 2002, 2003, and 2004. (*Source:* U.S. Census Bureau.)

2001 2002 2003 2004

55. **Defense spending.** The government of the United States spent $350 billion on national defense in 2000 and $740 billion in 2008. Estimate the amount spent on defense in each of the years 2001–2007. (*Source:* www.usgovernmentspending.com)

56. **Length of a diagonal.** The application of the Pythagorean Theorem in three dimensions involves the relationship between the perpendicular edges of a rectangular block and the solid diagonal of the same block.

In the figure, show that $h^2 = a^2 + b^2 + c^2$.

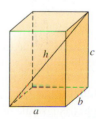

57. **Distance.** A pilot is flying from Dullsville to Middale to Pleasantville. With reference to an origin, Dullsville is located at $(2, 4)$, Middale at $(8, 12)$, and Pleasantville at $(20, 3)$, all numbers being in 100-mile units.
  a. Locate the positions of the three cities on a Cartesian coordinate plane.
  b. Compute the distance traveled by the pilot.
  c. Compute the direct distance between Dullsville and Pleasantville.

58. Docking distance. A rope is attached to the bow of a sailboat that is 24 feet from the dock. The rope is drawn in over a pulley 10 feet higher than the bow at the rate of 3 feet per second. Find the distance from the boat to the dock after t seconds.

10 feet

24 feet

C EXERCISES Beyond the Basics

59. Use coordinates to prove that the diagonals of a parallelogram bisect each other.
[*Hint:* Choose $(0, 0)$ $(a, 0)$, (b, c), and $(a + b, c)$ as the vertices of a parallelogram.]

60. Let $A(2, 3)$, $B(5, 4)$, and $C(3, 8)$ be three points in a coordinate plane. Find the coordinates of the point D such that the points A, B, C, and D form a parallelogram with
   a. AB as one of the diagonals.
   b. AC as one of the diagonals.
   c. BC as one of the diagonals.
   [*Hint:* Use Exercise 59.]

61. Prove that if the diagonals of a quadrilateral bisect each other, then the quadrilateral is a parallelogram.
[*Hint:* Choose $(0, 0)$, $(a, 0)$, (b, c), and (x, y) as the vertices of a quadrilateral and show that $x = a + b, y = c$.]

62. a. Show that $(1, 2)$, $(-2, 6)$, $(5, 8)$, and $(8, 4)$ are the vertices of a parallelogram. [*Hint:* Use Exercise 61.]
   b. Find (x, y) assuming that $(3, 2)$, $(6, 3)$, (x, y), and $(6, 5)$ are the vertices of a parallelogram.

63. Show that the sum of the squares of the lengths of the sides of a parallelogram is equal to the sum of the squares of the lengths of the diagonals. [*Hint:* Choose $(0, 0)$, $(a, 0)$, (b, c), and $(a + b, c)$ as the vertices of the parallelogram.]

64. Prove that in a right triangle, the midpoint of the hypotenuse is the same distance from each of the vertices.
[*Hint:* Let the vertices be $(0, 0)$, $(a, 0)$, and $(0, b)$.]

65. Show that the midpoints of the sides of any rectangle are the vertices of a rhombus (a quadrilateral with all sides of equal length). [*Hint:* Let the vertices of the rectangle be $(0, 0)$, $(a, 0)$, $(0, b)$, and (a, b).]

66. Show that in any triangle, the sum of the squares of the lengths of the medians is equal to three-fourths the sum of the squares of the lengths of the sides. [*Hint:* Choose $(0, 0)$, $(a, 0)$, and (b, c) as the vertices of a triangle.]

67. Trisecting a line segment. Let $A(x_1, y_1)$ and $B(x_2, y_2)$ be two points in a coordinate plane.
   a. Show that the point $C\left(\dfrac{2x_1 + x_2}{3}, \dfrac{2y_1 + y_2}{3}\right)$ is one-third of the way from A to B.
   [*Hint:* Show that $d(A, C) + d(C, B) = d(A, B)$ and that $d(A, C) = \dfrac{1}{3}d(A, B)$.]
   b. Show that the point $D\left(\dfrac{x_1 + 2x_2}{3}, \dfrac{y_1 + 2y_2}{3}\right)$ is two-thirds of the way from A to B. *The points C and D are called the points of trisection of segment AB.*
   c. Find the points of trisection of the line segment joining $(-1, 2)$ and $(4, 1)$.

68. Use the notation in the accompanying figure to prove the midpoint formula
$$M(x, y) = \left(\frac{x_1 + x_2}{2}, \frac{y_1 + y_2}{2}\right).$$

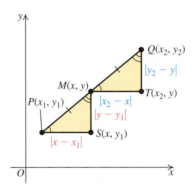

Critical Thinking

In Exercises 69–73, describe the set of points $P(x, y)$ in the xy-plane that satisfies the given condition.

69. a. $x = 0$ **b.** $y = 0$

70. a. $xy = 0$ **b.** $xy \neq 0$

71. a. $xy > 0$ **b.** $xy < 0$

72. a. $x^2 + y^2 = 0$ **b.** $x^2 + y^2 \neq 0$

73. Describe how to determine the quadrant in which a point lies from the signs of its coordinates.

Graphs of Equations

DOOMSANS DISCOVER DEER

A herd of 400 deer was introduced onto a small island called Dooms. The natives both liked and admired these beautiful creatures. However, the Doomsans soon discovered that deer meat is excellent food. Also, a rumor spread throughout Dooms that eating deer meat prolonged one's life. The natives then began to hunt the deer. The number of deer, y, after t years from the initial introduction of deer into Dooms is described by the equation $y = -t^4 + 96t^2 + 400$. When does the population of deer become extinct in Dooms? (See Example 6.) ■

---

1 Sketch a graph by plotting points.

Graph of an Equation

We say that a *relation* exists between the two quantities when one quantity changes with respect to the other quantity. The two changing (or varying) quantities in a relation are often represented by *variables*. A relation may sometimes be described by an equation, which may be a formula. Following are examples of equations showing relationships between two variables.

$$y = 2x + 1, \quad x^2 + y^2 = 4, \quad y = x^2, \quad x = y^2, \quad F = \frac{9}{5}C + 32, \quad q = -3p^2 + 30$$

An ordered pair (a, b) is said to **satisfy** an equation with variables x and y if, when a is substituted for x and b is substituted for y in the equation, the resulting statement is true. For example, the ordered pair $(2, 5)$ satisfies the equation $y = 2x + 1$ because replacing x with 2 and y with 5 yields $5 = 2(2) + 1$, a true statement. The ordered pair $(5, -2)$ does not satisfy the equation $y = 2x + 1$ because replacing x with 5 and y with -2 yields $-2 = 2(5) + 1$, which is false. An ordered pair that satisfies an equation is called a **solution** of the equation.

In an equation involving x and y, if the value of y can be found given the value of x, then we say that y is the **dependent variable** and x is the **independent variable**. In the equation $y = 2x + 1$, there are infinitely many choices for x, each resulting in a corresponding value of y. Hence, we have infinitely many solutions of the equation $y = 2x + 1$. When these solutions are plotted as points in the coordinate plane, they constitute the *graph of the equation*. The graph of an equation is thus a geometric picture of its solution set.

GRAPH OF AN EQUATION

The **graph of an equation** in two variables, such as x and y, is the set of all ordered pairs (a, b) in the coordinate plane that satisfies the equation.

TECHNOLOGY CONNECTION

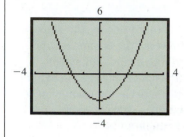 Calculator graph of $y = x^2 - 3$

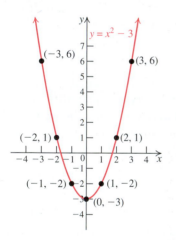

FIGURE 2.11 **Graph of an equation**

EXAMPLE 1　**Sketching a Graph by Plotting Points**

Sketch the graph of $y = x^2 - 3$.

SOLUTION

There are infinitely many solutions of the equation $y = x^2 - 3$. To find a few, we choose integer values of x between -3 and 3. Then we find the corresponding values of y as shown in Table 2.2.

TABLE 2.2

x	$y = x^2 - 3$	(x, y)
-3	$y = (-3)^2 - 3 = 9 - 3 = 6$	$(-3, 6)$
-2	$y = (-2)^2 - 3 = 4 - 3 = 1$	$(-2, 1)$
-1	$y = (-1)^2 - 3 = 1 - 3 = -2$	$(-1, -2)$
0	$y = 0^2 - 3 = 0 - 3 = -3$	$(0, -3)$
1	$y = 1^2 - 3 = 1 - 3 = -2$	$(1, -2)$
2	$y = 2^2 - 3 = 4 - 3 = 1$	$(2, 1)$
3	$y = 3^2 - 3 = 9 - 3 = 6$	$(3, 6)$

We plot the seven solutions (x, y) and join them with a smooth curve as shown in Figure 2.11. ▪ ▪ ▪

Practice Problem 1　Sketch the graph of $y = -x^2 + 1$. ▪

The bowl-shaped curve sketched in Figure 2.11 is called a *parabola*. It is easy to find parabolas in everyday settings. For example, when you throw a ball, the path it travels is a parabola. Also, the reflector behind a car's headlight is parabolic in shape.

Example 1 suggests the following three steps for sketching the graph of an equation by plotting points.

SKETCHING A GRAPH BY PLOTTING POINTS

Step 1　Make a representative table of solutions of the equation.

Step 2　Plot the solutions as ordered pairs in the coordinate plane.

Step 3　Connect the solutions in Step 2 with a smooth curve.

Comment. This point-plotting technique has obvious pitfalls. For instance, many different curves pass through the four points in Figure 2.12. Assume that these points are solutions of a given equation. There is no way to guarantee that any curve we pass through the plotted points is the actual graph of the equation. However, in general, the more solutions that are plotted, the more likely the resulting figure is to be a reasonably accurate graph of the equation.

FIGURE 2.12 Several graphs through four points

We now discuss some special features of the graph of an equation.

<table>
<tr><td>**2**</td><td>Find the intercepts of a graph.</td></tr>
</table>

Intercepts

Let's examine the points where a graph intersects (crosses or touches) the coordinate axes. Because all points on the x-axis have a y-coordinate of 0, any point where a graph intersects the x-axis has the form $(a, 0)$. See Figure 2.13. The number a is called an **x-intercept** of the graph. Similarly, any point where a graph intersects the y-axis has the form $(0, b)$, and the number b is called a **y-intercept** of the graph.

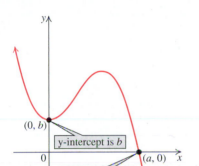

FIGURE 2.13 Intercepts of a graph

FINDING THE INTERCEPTS OF THE GRAPH OF AN EQUATION

Step 1   To find the x-intercepts of the graph of an equation, set $y = 0$ in the equation and solve for x.

Step 2   To find the y-intercepts of the graph of an equation, set $x = 0$ in the equation and solve for y.

STUDY TIP

Do not try to calculate the x-intercept by setting $x = 0$. The term x-*intercept* denotes the x-coordinate of the point where the graph touches or crosses the x-axis; so the y-coordinate must be 0.

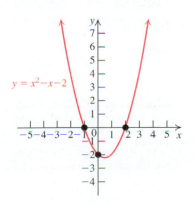

FIGURE 2.14 Intercepts of a graph

EXAMPLE 2   Finding Intercepts

Find the x- and y-intercepts of the graph of the equation $y = x^2 - x - 2$.

SOLUTION

Step 1   Set $y = 0$ in the equation and solve for x.

$$y = x^2 - x - 2 \qquad \text{Original equation}$$
$$0 = x^2 - x - 2 \qquad \text{Set } y = 0.$$
$$0 = (x + 1)(x - 2) \qquad \text{Factor.}$$
$$x + 1 = 0 \quad \text{or} \quad x - 2 = 0 \qquad \text{Zero-product property}$$
$$x = -1 \quad \text{or} \qquad x = 2 \qquad \text{Solve each equation for } x.$$

The x-intercepts are -1 and 2.

Step 2   Set $x = 0$ in the equation and solve for y.

$$y = x^2 - x - 2 \qquad \text{Original equation}$$
$$y = 0^2 - 0 - 2 \qquad \text{Set } x = 0.$$
$$y = -2 \qquad \text{Solve for } y.$$

The y-intercept is -2.

The graph of the equation $y = x^2 - x - 2$ is shown in Figure 2.14.   ■ ■ ■

Practice Problem 2   Find the intercepts of the graph of $y = 2x^2 + 3x - 2$.   ■

3 Find the symmetries in a graph.

Symmetry

The basic idea of symmetry can be seen by examining the points $(-3, 2)$ and $(3, 2)$. These points are the same distance from the y-axis and on a line segment perpendicular to the y-axis. We say that either point is the symmetric image, or mirror image, of the other in the y-axis.

The concept of **symmetry** helps us sketch graphs of equations. A graph has symmetry with respect to a line l, if one portion of the graph is a *mirror image* of another portion in l. As shown in Figure 2.15, suppose a line l is an **axis of symmetry**, or **line of symmetry** for some set of points. We can construct the mirror image of any point P not on l in this set by first drawing the perpendicular line segment from P to l. Then we extend this segment an equal distance on the other side to point P' so that the line l perpendicularly bisects the line segment $\overline{PP'}$. In Figure 2.15, we say that the point P' is the *symmetric image* of the point P with respect to the line l. Symmetry lets us use the information about part of the graph to draw the remainder of the graph.

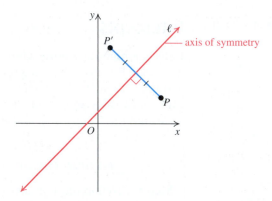

FIGURE 2.15 Symmetric points about a line

There are three basic types of symmetries.

TESTS FOR SYMMETRIES.

1. A graph is **symmetric with respect to** (or **in** or **about**) **the y-axis** if for every point (x, y) on the graph, the point $(-x, y)$ is also on the graph. See Figure 2.16(a).

2. A graph is **symmetric with respect to** (or **in** or **about**) **the x-axis** if for every point (x, y) on the graph, the point $(x, -y)$ is also on the graph. See Figure 2.16(b).

3. A graph is **symmetric with respect to** (or **in** or **about**) **the origin** if for every point (x, y) on the graph, the point $(-x, -y)$ is also on the graph. See Figure 2.16(c).

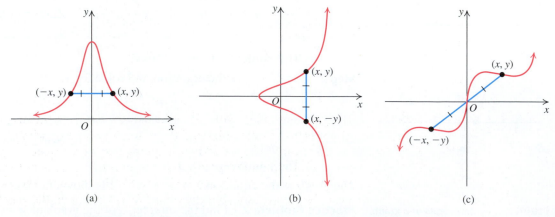

FIGURE 2.16 Three types of symmetry

TECHNOLOGY CONNECTION

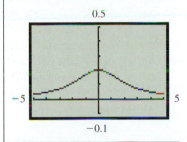

The calculator graph of $y = \dfrac{1}{x^2 + 5}$ reinforces the result of Example 3.

STUDY TIP

Note that if *only* even powers of x appear in an equation, the graph is automatically symmetric with respect to the y-axis because, for any integer n, $(-x)^{2n} = x^{2n}$. Consequently, we obtain the original equation when we replace x with $-x$.

EXAMPLE 3 **Checking for Symmetry**

Determine whether the graph of the equation $y = \dfrac{1}{x^2 + 5}$ is symmetric with respect to the y-axis.

SOLUTION

Replace x with $-x$ to see if $(-x, y)$ also satisfies the equation.

$$y = \frac{1}{x^2 + 5} \qquad \text{Original equation}$$

$$y = \frac{1}{(-x)^2 + 5} \qquad \text{Replace } x \text{ with } -x.$$

$$y = \frac{1}{x^2 + 5} \qquad \text{Simplify: } (-x)^2 = x^2.$$

Because replacing x with $-x$ gives us the original equation, the graph of $y = \dfrac{1}{x^2 + 5}$ is symmetric with respect to the y-axis. ■ ■ ■

Practice Problem 3 Check the graph of $x^2 - y^2 = 1$ for symmetry with respect to the y-axis. ■

EXAMPLE 4 **Checking for Symmetry**

Show that the graph of $x^3 + y^2 - xy^2 = 0$ is symmetric with respect to the x-axis.

SOLUTION

$$x^3 + y^2 - xy^2 = 0 \qquad \text{Original equation}$$
$$x^3 + (-y)^2 - x(-y)^2 = 0 \qquad \text{Replace } y \text{ with } -y.$$
$$x^3 + y^2 - xy^2 = 0 \qquad \text{Simplify: } (-y)^2 = y^2.$$

We obtain the original equation when y is replaced with $-y$. Thus, the graph of the equation is symmetric with respect to the x-axis. ■ ■ ■

Practice Problem 4 Show that the graph of $x^2 - y^3 = 2$ is symmetric with respect to the y-axis. ■

EXAMPLE 5 **Checking for Symmetry with Respect to the Origin**

Show that the graph of $y^5 = x^3$ is symmetric with respect to the origin, but not with respect to either axis.

SOLUTION

$$y^5 = x^3 \qquad \text{Original equation}$$
$$(-y)^5 = (-x)^3 \qquad \text{Replace } x \text{ with } -x \text{ and } y \text{ with } -y.$$
$$-y^5 = -x^3 \qquad \text{Simplify } (-a)^5 = (-1)^5 a^5 = -a^5.$$
$$y^5 = x^3 \qquad \text{Multiply both sides by } -1.$$

Because replacing x with $-x$ and y and $-y$ produces the original equation, $(-x, -y)$ also satisfies the equation and the graph is symmetric with respect to the origin.

The graph of $y^5 = x^3$ is not symmetric with respect to the y-axis because the point $(1, 1)$ is on the graph, but the point $(-1, 1)$ is not on the graph. Similarly, the graph of $y^5 = x^3$ is not symmetric with respect to the x-axis because the point $(1, 1)$ is on the graph, but the point $(1, -1)$ is not on the graph. ■ ■ ■

Practice Problem 5 Check the graph of $x^2 = y^3$ for symmetry in the x-axis, the y-axis, and the origin. ■

> ### EXAMPLE 6 Sketching a Graph by Using Intercepts and Symmetry

Initially, 400 deer are on Dooms island. The number y of deer on the island after t years is described by the equation

$$y = -t^4 + 96t^2 + 400.$$

a. Sketch the graph of the equation $y = -t^4 + 96t^2 + 400$.

b. Adjust the graph in part **a** to account for only the physical aspects of the problem.

c. When do deer become extinct on Dooms?

SOLUTION

a. (i) We find all intercepts. If we set $t = 0$ in the equation $y = -t^4 + 96t^2 + 400$, we obtain $y = 400$. Thus, the y-intercept is 400.

To find the t-intercepts, we set $y = 0$ in the given equation.

$y = -t^4 + 96t^2 + 400$	Original equation
$0 = -t^4 + 96t^2 + 400$	Set $y = 0$.
$t^4 - 96t^2 - 400 = 0$	Multiply both sides by -1 and interchange sides.
$(t^2 + 4)(t^2 - 100) = 0$	Factor.
$(t^2 + 4)(t + 10)(t - 10) = 0$	Factor $t^2 - 100$.
$t^2 + 4 = 0$ or $t + 10 = 0$ or $t - 10 = 0$	Zero-product property
$t = -10$ or $t = 10$	Solve for t; there is no real solution of $t^2 + 4 = 0$.

The t-intercepts are -10 and 10.

(ii) We check for symmetry. Note that t replaces x as the independent variable.
Symmetry in the t-axis:
Replacing y with $-y$ gives $-y = -t^4 + 96t^2 + 400$. The pair $(0, 400)$ is on the graph of $y = -t^4 + 96t^2 + 400$ but not of $-y = -t^4 + 96t^2 + 400$. Consequently, the graph is not symmetric in the t-axis.

Symmetry in the y-axis:
Replacing t with $-t$ in the equation $y = -t^4 + 96t^2 + 400$, we obtain $y = -(-t)^4 + 96(-t)^2 + 400 = -t^4 + 96t^2 + 400$, which is the original equation. Thus, $(-t, y)$ also satisfies the equation and the graph is symmetric in the y-axis.

Symmetry in the origin:
Replacing t with $-t$ and y with $-y$ in the equation $y = -t^4 + 96t^2 + 400$, we obtain $-y = -(-t)^4 + 96(-t)^2 + 400$, or $-y = -t^4 + 96t^2 + 400$. As we saw when we discussed symmetry in the t-axis, $(0, 400)$ is a solution but $(-0, -400)$ is not a solution of this equation; so the graph is not symmetric with respect to the origin.

(iii) We sketch the graph by plotting points for $t \geq 0$ (see Table 2.3) and then using symmetry in the y-axis (see Figure 2.17).

TABLE 2.3

t	$y = -t^4 + 96t^2 + 400$	(t, y)
0	400	$(0, 400)$
1	495	$(1, 495)$
5	2175	$(5, 2175)$
7	2703	$(7, 2703)$
9	1615	$(9, 1615)$
10	0	$(10, 0)$
11	−2625	$(11, -2625)$

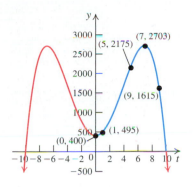

FIGURE 2.17

b. The graph pertaining to the physical aspects of the problem is the blue portion of the graph in Figure 2.17.

c. The positive t-intercept, 10, gives the time in years when the deer population of Dooms is 0, so deer are extinct after 10 years.   ■ ■ ■

Practice Problem 6 Repeat Example 6 assuming that the initial deer population is 324 and the number of deer on the island after t years is given by the equation $y = -t^4 + 77t^2 + 324$.   ■

4 Find the equation of a circle.

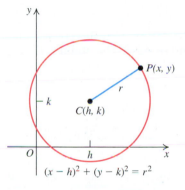

FIGURE 2.18 A circle with center (**h**, **k**) and radius **r**

Circles

So far, we have seen that an equation in x and y gives rise to a geometric figure that is its graph. Sometimes we can go the other way: A graph is described geometrically in terms of points in the plane, and these points, in turn, give rise to an equation of the graph in x and y. We illustrate the latter situation in the case of a circle.

Figure 2.18 is the graph of a circle with center $C(h, k)$ and radius r.

CIRCLE

A **circle** is a set of points in a Cartesian coordinate plane that are at a fixed distance r from a specified point (h, k). The fixed distance r is called the **radius** of the circle, and the specified point (h, k) is called the **center** of the circle.

A point $P(x, y)$ is on the circle if and only if its distance from the center C is r. Using the notation for the distance between the points P and C, $d(P, C)$, we have

$$d(P, C) = r$$
$$\sqrt{(x - h)^2 + (y - k)^2} = r \quad \text{Distance formula}$$
$$(x - h)^2 + (y - k)^2 = r^2 \quad \text{Square both sides.}$$

The equation $(x - h)^2 + (y - k)^2 = r^2$ is an equation of a circle with radius r and center (h, k). A point (x, y) is on the circle of radius r and center $C(h, k)$ if and only if it satisfies this equation.

STANDARD FORM FOR THE EQUATION OF A CIRCLE

The equation of a circle with center (h, k) and radius r is

(1) $$(x - h)^2 + (y - k)^2 = r^2,$$

and is called the **standard form** of an equation of a circle.

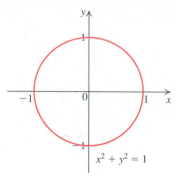

FIGURE 2.19 The unit circle

EXAMPLE 7   Finding the Equation of a Circle

Find the standard form of the equation of the circle with center $(-3, 4)$ and radius 7.

SOLUTION

$$(x - h)^2 + (y - k)^2 = r^2 \qquad \text{Standard form}$$
$$[x - (-3)]^2 + (y - 4)^2 = 7^2 \qquad \text{Replace } h \text{ with } -3, k \text{ with 4, and } r \text{ with 7.}$$
$$(x + 3)^2 + (y - 4)^2 = 49 \qquad \text{By convention, rewrite } [x - (-3)] = (x + 3).$$

▪▪▪

Practice Problem 7   Find the standard form of the equation of the circle with center $(3, -6)$ and radius 10. ▪

If an equation in two variables can be written in the standard form (1), with $r > 0$, then its graph is a circle with center (h, k) and radius r.

EXAMPLE 8   Graphing a Circle

Graph each equation.

a. $x^2 + y^2 = 1$ **b.** $(x + 2)^2 + (y - 3)^2 = 25$

SOLUTION

a. The equation $x^2 + y^2 = 1$ can be rewritten as

$$(x - 0)^2 + (y - 0)^2 = 1^2.$$

Comparing that equation with equation (1), we conclude that the given equation is an equation of a circle with center $(0, 0)$ and radius 1. The graph is shown in Figure 2.19. This circle is called the **unit circle**.

b. Rewriting the equation $(x + 2)^2 + (y - 3)^2 = 25$ as

$$[x - (-2)]^2 + (y - 3)^2 = 5^2, \qquad x + 2 = x - (-2)$$

we see that the graph of this equation is a circle with center $(-2, 3)$ and radius 5. The graph is shown in Figure 2.20.

▪▪▪

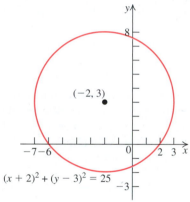

FIGURE 2.20 Circle with radius 5 and center $(-2, 3)$

Practice Problem 8   Graph the equation $(x - 2)^2 + (y + 1)^2 = 36$. ▪

TECHNOLOGY CONNECTION

To graph the equation $x^2 + y^2 = 1$ on a graphing calculator, we first solve for y.

$$y^2 = 1 - x^2 \qquad \text{Subtract } x^2 \text{ from both sides.}$$
$$y = \pm\sqrt{1 - x^2} \qquad \text{Square root property}$$

We then graph the two equations

$$Y_1 = \sqrt{1 - x^2} \quad \text{and} \quad Y_2 = -\sqrt{1 - x^2}$$

in the same viewing rectangle. The graph of $y = \sqrt{1 - x^2}$ is the upper semicircle ($y \geq 0$), and the graph of $y = -\sqrt{1 - x^2}$ is the lower semicircle ($y \leq 0$). The calculator graph in (a) does not look quite like a circle. The reason is that on most graphing calculators, the display screen is about two-thirds as high as it is wide. Graphing calculators have options on the ZOOM menu to square the viewing screen. We use the ZSquare option to make the display look like a circle, as shown in (b).

EQUATION ↔ CIRCLE

Note that stating that the equation

$$(x + 3)^2 + (y - 4)^2 = 25$$

represents the circle of radius 5 with center $(-3, 4)$ means two things:

(i) If the values of x and y are a pair of numbers that satisfy the equation, then they are the coordinates of a point on the circle with radius 5 and center $(-3, 4)$.

(ii) If a point is on the circle, then its coordinates satisfy the equation.

Let's expand and simplify the squared expressions in the standard form of the equation of the circle in the box above.

$$(x + 3)^2 + (y - 4)^2 = 25 \qquad \text{Given equation}$$
$$x^2 + 6x + 9 + y^2 - 8y + 16 = 25 \qquad \text{Expand squares.}$$
$$x^2 + y^2 + 6x - 8y = 0 \qquad \text{Simplify and rewrite.}$$

In general, if we expand the squared expressions in the standard form of the equation of a circle,

(1) $$(x - h)^2 + (y - k)^2 = r^2,$$

and then simplify, we obtain an equation of the form

(2) $$x^2 + y^2 + ax + by + c = 0.$$

Equation (2) is called the *general form* of the equation of a circle. The graph of the equation $Ax^2 + By^2 + Cx + Dy + E = 0$ is also a circle if $A \neq 0$ and $A = B$. You can convert this equation to the general form by dividing both sides by A.

GENERAL FORM OF THE EQUATION OF A CIRCLE

The **general form** of the equation of a circle is

$$x^2 + y^2 + ax + by + c = 0.$$

The graph of this equation is a circle, provided $d > 0$, where

$$d = \frac{1}{4}(a^2 + b^2 - 4c).$$

Conversely, suppose we are given an equation of a circle in general form. We can convert this equation to standard form by completing the squares on the x- and y-terms. We then obtain an equation of the form

(3) $$(x - h)^2 + (y - k)^2 = d.$$

If $d > 0$, the graph of equation (3) is a circle with center (h, k) and radius $\sqrt{d}$. If $d = 0$, the graph of equation (3) is the point (h, k). If $d < 0$, there is no graph.

EXAMPLE 9   Converting the General Form to Standard Form

Find the center and radius of the circle with equation

$$x^2 + y^2 - 6x + 8y + 10 = 0.$$

SOLUTION

We complete the squares (on both the terms involving x and the terms involving y) to write the given equation in standard form.

RECALL

A quadratic trinomial in the variable x with the coefficient of x^2 equal to 1 is a perfect square if the constant term is the square of one-half the coefficient of x.

$$x^2 + y^2 - 6x + 8y + 10 = 0 \qquad \text{Original equation}$$
$$(x^2 - 6x) + (y^2 + 8y) = -10 \qquad \text{Group the } x\text{-terms and } y\text{-terms separately; subtract 10 from both sides.}$$

$$(x^2 - 6x + 9) + (y^2 + 8y + 16) = -10 + 9 + 16 \qquad \text{Complete the square on both the } x\text{- and } y\text{-terms by adding 9 and 16 to both sides.}$$

$$(x - 3)^2 + (y + 4)^2 = 15 \qquad \text{Factor and simplify.}$$
$$(x - 3)^2 + [y - (-4)]^2 = (\sqrt{15})^2 \qquad \text{Rewrite in standard form.}$$

Comparing the last equation with the standard form, we have $h = 3$, $k = -4$, and $r = \sqrt{15}$. Thus, the given equation represents a circle with center $(3, -4)$ and radius $\sqrt{15} \approx 3.9$. ■ ■ ■

Practice Problem 9 Find the center and radius of the circle with equation $x^2 + y^2 + 4x - 6y - 12 = 0$. ■

SECTION 2.2 ■ Exercises

A EXERCISES Basic Skills and Concepts

1. The graph of an equation in two variables such as x and y is the set of all ordered pairs (a, b) _____.

2. If $(-2, 4)$ is a point on a graph that is symmetric with respect to the y-axis, then the point _____ is also on the graph.

3. If $(0, -5)$ is a point of a graph, then -5 is a(n) _____ intercept of the graph.

4. An equation in standard form of a circle with center $(1, 0)$ and radius 2 is _____.

5. *True or False* The graph of the equation $3x^2 - 2x + y + 3 = 0$ is a circle.

6. *True or False* If a graph is symmetric about the x-axis, then it must have at least one x-intercept.

In Exercises 7–14, determine whether the given points are on the graph of the equation.

Equation	Points
7. $y = x - 1$	$(-3, -4), (1, 0), (4, 3), (2, 3)$
8. $2y = 3x + 5$	$(-1, 1), (0, 2), \left(-\dfrac{5}{3}, 0\right), (1, 4)$
9. $y = \sqrt{x + 1}$	$(3, 2), (0, 1), (8, -3), (8, 3)$
10. $y = \dfrac{1}{x}$	$\left(-3, \dfrac{1}{3}\right), (1, 1), (0, 0), \left(2, \dfrac{1}{2}\right)$
11. $x^2 - y^2 = 1$	$(1, 0), (0, -1), (2, \sqrt{3}), (2, -\sqrt{3})$
12. $y^2 = x$	$(1, -1), (1, 1) (0, 0), (2, -\sqrt{2})$

13. Write the x- and y-intercepts of the graph.

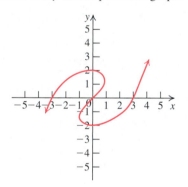

14. Write the x- and y-intercepts of the graph.

In Exercises 15–36, graph each equation by plotting points. Let $x = -3, -2, -1, 0, 1, 2,$ and 3 where applicable.

15. $y = x + 1$

16. $y = x - 1$

17. $y = 2x$

18. $y = \dfrac{1}{2}x$

19. $y = x^2$

20. $y = -x^2$

21. $y = |x|$

22. $y = |x + 1|$

23. $y = |x| + 1$

24. $y = -|x| + 1$

25. $y = 4 - x^2$

26. $y = x^2 - 4$

27. $y = \sqrt{9 - x^2}$

28. $y = -\sqrt{9 - x^2}$

29. $y = x^3$

30. $y = -x^3$

31. $y^3 = x$

32. $y^3 = -x$

33. $x = |y|$

34. $|x| = |y|$

35. $y = |2 - x|$

36. $|x| + |y| = 1$

In Exercises 37–46, find the *x*- and *y*-intercepts of the graph of the equation.

37. $3x + 4y = 12$

38. $\dfrac{x}{5} + \dfrac{y}{3} = 1$

39. $2x + 3y = 5$

40. $\dfrac{x}{2} - \dfrac{y}{3} = 1$

41. $y = x^2 - 6x + 8$

42. $x = y^2 - 5y + 6$

43. $x^2 + y^2 = 4$

44. $y = \sqrt{9 - x^2}$

45. $y = \sqrt{x^2 - 1}$

46. $xy = 1$

In Exercises 47–56, test each equation for symmetry with respect to the *x*-axis, the *y*-axis, and the origin.

47. $y = x^2 + 1$

48. $x = y^2 + 1$

49. $y = x^3 + x$

50. $y = 2x^3 - x$

51. $y = 5x^4 + 2x^2$

52. $y = -3x^6 + 2x^4 + x^2$

53. $y = -3x^5 + 2x^3$

54. $y = 2x^2 - |x|$

55. $x^2y^2 + 2xy = 1$

56. $x^2 + y^2 = 16$

In Exercises 57–60, specify the center and the radius of each circle.

57. $(x - 2)^2 + (y - 3)^2 = 36$

58. $(x + 1)^2 + (y - 3)^2 = 16$

59. $(x + 2)^2 + (y + 3)^2 = 11$

60. $\left(x - \dfrac{1}{2}\right)^2 + \left(y + \dfrac{3}{2}\right)^2 = \dfrac{3}{4}$

In Exercises 61–70, find the standard form of the equation of a circle that satisfies the given conditions. Graph each equation.

61. Center $(0, 1)$; radius 2

62. Center $(1, 0)$; radius 1

63. Center $(-1, 2)$; radius $\sqrt{2}$

64. Center $(-2, -3)$; radius $\sqrt{7}$

65. Center $(3, -4)$; passing through the point $(-1, 5)$

66. Center $(-1, 1)$; passing through the point $(2, 5)$

67. Center $(1, 2)$; touching the *x*-axis

68. Center $(1, 2)$; touching the *y*-axis

69. Diameter with endpoints $(7, 4)$ and $(-3, 6)$

70. Diameter with endpoints $(2, -3)$ and $(8, 5)$

In Exercises 71–76, find: a. The center and radius of each circle. b. The *x*- and *y*-intercepts of the graph of each circle.

71. $x^2 + y^2 - 2x - 2y - 4 = 0$

72. $x^2 + y^2 - 4x - 2y - 15 = 0$

73. $2x^2 + 2y^2 + 4y = 0$

74. $3x^2 + 3y^2 + 6x = 0$

75. $x^2 + y^2 - x = 0$

76. $x^2 + y^2 + 1 = 0$

B EXERCISES Applying the Concepts

In Exercises 77–80, a graph is described *geometrically* as the path of a point $P(x, y)$ on the graph. Find an equation for the graph described.

77. Geometry. $P(x, y)$ is on the graph if and only if the distance from $P(x, y)$ to the *x*-axis is equal to its distance from the *y*-axis.

78. Geometry. $P(x, y)$ is the same distance from the two points $(1, 2)$ and $(3, -4)$.

79. Geometry. $P(x, y)$ is the same distance from the point $(2, 0)$ and the *y*-axis.

80. Geometry. $P(x, y)$ is the same distance from the point $(0, 4)$ and the *x*-axis.

81. Corporate profits. The equation $P = -0.5t^2 - 3t + 8$ describes the monthly profits (in millions of dollars) of ABCD Corp. for the year 2004, with $t = 0$ representing July 2004.

  a. How much profit did the corporation make in March 2004?

  b. How much profit did the corporation make in October 2004?

  c. Sketch the graph of the equation.

  d. Find the *t*-intercept. What does it represent?

  e. Find the *P*-intercept. What does it represent?

82. Female students in colleges. The equation

$$P = -0.002t^2 + 0.093t + 8.18$$

models the approximate number (in millions) of female college students in the United States for the academic years 1995–2001, with $t = 0$ representing 1995.

  a. Sketch the graph of the equation.

  b. Find the positive *t*-intercept. What does it represent?

  c. Find the *P*-intercept. What does it represent?

(Source: Statistical Abstract of the United States.)

83. **Motion.** An object is thrown up from the top of a building that is 320 feet high. The equation $y = -16t^2 + 128t + 320$ gives the object's height (in feet) above the ground at any time t (in seconds) after the object is thrown.

a. What is the height of the object after 0, 1, 2, 3, 4, 5, and 6 seconds?

b. Sketch the graph of the equation $y = -16t^2 + 128t + 320$.

c. What part of the graph represents the physical aspects of the problem?

d. What are the intercepts of this graph, and what do they mean?

84. **Diving for treasure.** A treasure hunting team of divers is placed in a computer-controlled diving cage. The equation $d = \dfrac{40}{3}t - \dfrac{2}{9}t^2$ describes the depth d (in feet) the cage will descend in t minutes.

a. Sketch the graph of the equation $d = \dfrac{40}{3}t - \dfrac{2}{9}t^2$.

b. What part of the graph represents the physical aspects of the problem?

c. What is the total time of the entire diving experiment?

C EXERCISES Beyond the Basics

In Exercises 85–88, sketch the complete graph by using the given symmetry.

85.

y-axis symmetry

86.

x-axis symmetry

87.

origin symmetry

88.

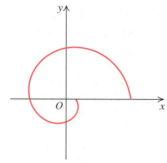

x-axis symmetry

In Exercises 89–94, graph each equation. [*Hint:* Use the zero-product property of numbers: $ab = 0$ if and only if $a = 0$ or $b = 0$.]

89. $(y - x)(y - 2x) = 0$

90. $y^2 = x^2$

91. $(y - x)(x^2 + y^2 - 4) = 0$

92. $(y - x + 1)(x^2 + y^2 - 6x + 8y + 24) = 0$

93. $x^2 - 9y^2 = 0$

94. $(y - x - 1)(y - x^2) = 0$

95. In the same coordinate system, sketch the graphs of the two circles with equations $x^2 + y^2 - 4x + 2y - 20 = 0$ and $x^2 + y^2 - 4x + 2y - 31 = 0$ and find the area of the region bounded by the two circles.

96. Sketch the graph (if possible) of each equation.

a. $x^2 + y^2 + 2x - 2y + 3 = 0$

b. $x^2 + y^2 + 2x - 2y + 2 = 0$

c. $x^2 + y^2 + 2x - 2y - 7 = 0$

Critical Thinking

97. Sketch the graph of $y^2 = 2x$ and explain how this graph is related to the graphs of $y = \sqrt{2x}$ and $y = -\sqrt{2x}$.

98. Show that a graph that is symmetric with respect to the x-axis and y-axis must be symmetric with respect to the origin. Give an example to show that the converse is not true.

99. *True or False* To find the x-intercepts of the graph of an equation, set $x = 0$ and solve for y. Explain your answer.

100. Write an equation whose graph has x-intercepts -2 and 3 and y-intercept 6. (Many answers are possible.)

101. Sketch a graph that has
  a. one x-intercept and one y-intercept.
  b. two x-intercepts and three y-intercepts.
  c. no intercepts.
  d. two x-intercepts and no y-intercept.

GROUP PROJECTS

1. a. Show that a circle with diameter having endpoints $A(0, 1)$ and $B(6, 8)$ intersects the x-axis at the roots of the equation $x^2 - 6x + 8 = 0$.
  b. Show that a circle with diameter having endpoints $A(0, 1)$ and $B(a, b)$ intersects the x-axis at the roots of the equation $x^2 - ax + b = 0$.
  c. Use graph paper, a ruler, and a compass to approximate the roots of the equation $x^2 - 3x + 1 = 0$.

2. a. Show that a circle with diameter having endpoints $A(1, 0)$ and $B(10, 7)$ intersects the y-axis at the roots of the equation $y^2 - 7y + 10 = 0$.
  b. Show that a circle with diameter having endpoints $A(1, 0)$ and $B(a, b)$ intersects the y-axis at the roots of the equation $y^2 - by + a = 0$.

Lines

Before Starting this Section, Review

1. Evaluating algebraic expressions (Section P.1, page 14)
2. Graphs of equations (Section 2.2, page 179)

Objectives

1. Find the slope of a line.
2. Write the point–slope form of the equation of a line.
3. Write the slope–intercept form of the equation of a line.
4. Recognize the equations of horizontal and vertical lines.
5. Recognize the general form of the equation of a line.
6. Find equations of parallel and perpendicular lines.
7. Use linear regression.

A TEXAN'S TALL TALE

Gunslinger Wild Bill Longley's *first* burial took place on October 11, 1878, after he was hanged before a crowd of thousands in Giddings, Texas.

Rumors persisted that Longley's hanging had been a hoax and that he had somehow faked his death and escaped execution. In 2001, Longley's descendants had his grave opened to determine whether the remains matched Wild Bill's description: a tall white male, age 27. Both the skeleton and some personal effects suggested that this was indeed Wild Bill. Modern science lent a hand too: The DNA of Helen Chapman, a descendant of Wild Bill's sister, was a perfect match.

Now the notorious gunman could be buried back in the Giddings cemetery—for the *second* time. How did the scientists conclude from the skeletal remains that Wild Bill was approximately 6 feet tall? See Example 8. ■

---

1 Find the slope of a line.

Slope of a Line

In this section, we study various forms of first-degree equations in two variables. Because the graphs of these equations are straight lines, they are called **linear equations**. Just as we measure weight and temperature by a number, we measure the "steepness" of a line by a number called its **slope**.

Consider two points $P(x_1, y_1)$ and $Q(x_2, y_2)$ on a nonvertical line, as shown in Figure 2.21. We say that the *rise* is the change in y-coordinates between the points (x_1, y_1) and (x_2, y_2) and that the *run* is the corresponding change in the x-coordinates. In moving from P to Q on the line, a positive run indicates movement to the right, as shown in Figure 2.21; a negative run indicates movement to the left. Similarly, a positive rise indicates upward movement; a negative rise indicates downward movement.

FIGURE 2.21 Slope of a line

SLOPE OF A LINE

The **slope** of a nonvertical line that passes through the points $P(x_1, y_1)$ and $Q(x_2, y_2)$ is denoted by m and is defined by

$$m = \frac{\text{rise}}{\text{run}} = \frac{y_2 - y_1}{x_2 - x_1}, \quad x_1 \neq x_2.$$

The slope of a vertical line is undefined.

Because $m = \dfrac{\text{rise}}{\text{run}} = \dfrac{\text{change in } y\text{-coordinates}}{\text{change in } x\text{-coordinates}}$, if the change in x is one unit to the right, the change in y equals the slope of the line. The slope of a line is therefore the **change in y per unit change in x**. In other words, the slope of a line measures the rate of change of y with respect to x. Roofs, staircases, graded landscapes, and mountainous roads all have slopes. For example, if the pitch (slope) of a section of a roof is 0.4, then for every horizontal distance of 10 feet in that section, the roof ascends 4 feet.

EXAMPLE 1 **Finding and Interpreting the Slope of a Line**

Sketch the graph of the line that passes through the points $P(1, -1)$ and $Q(3, 3)$. Find and interpret the slope of the line.

SOLUTION

The graph of the line is sketched in Figure 2.22. The slope m of this line is given by

$$
\begin{aligned}
m &= \frac{\text{rise}}{\text{run}} = \frac{\text{change in } y\text{-coordinates}}{\text{change in } x\text{-coordinates}} \\
&= \frac{(y\text{-coordinate of } Q) - (y\text{-coordinate of } P)}{(x\text{-coordinate of } Q) - (x\text{-coordinate of } P)} \\
&= \frac{(3) - (-1)}{(3) - (1)} = \frac{3 + 1}{3 - 1} = \frac{4}{2} = 2.
\end{aligned}
$$

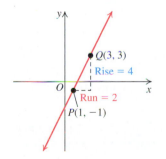

FIGURE 2.22 Interpreting slope

Interpretation

A slope of 2 means that the value of y increases two units for every one unit increase in the value of x. ■ ■ ■

Practice Problem 1 Find the slope of the line containing the points $(-7, 5)$ and $(6, -3)$. ■

Figure 2.23 shows several lines passing through the origin. The slope of each line is labeled.

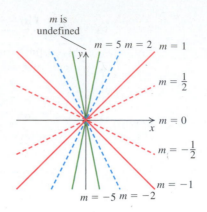

FIGURE 2.23 Slopes of lines

MAIN FACTS ABOUT SLOPES OF LINES

1. Scanning graphs from left to right, lines with positive slopes rise and lines with negative slopes fall.
2. The greater the absolute value of the slope, the steeper the line.
3. The slope of a vertical line is undefined.
4. The slope of a horizontal line is zero.

◆ **WARNING**   Be careful when using the formula for finding the slope of a line joining the points (x_1, y_1) and (x_2, y_2).

You can use either

$$m = \frac{y_2 - y_1}{x_2 - x_1} \qquad \text{or} \qquad m = \frac{y_1 - y_2}{x_1 - x_2},$$

but it is incorrect to use

$$m = \frac{y_1 - y_2}{x_2 - x_1} \qquad \text{or} \qquad m = \frac{y_2 - y_1}{x_1 - x_2}.$$

---

2 Write the point–slope form of the equation of a line.

Equation of a Line

We now find the equation of a line ℓ that passes through the point $A(x_1, y_1)$ and has slope m. Let $P(x, y)$, with $x \neq x_1$, be any point in the plane. Then $P(x, y)$ is on the line ℓ if and only if the slope of the line passing through $P(x, y)$ and $A(x_1, y_1)$ is m. This is true if and only if

$$\frac{y - y_1}{x - x_1} = m.$$

See Figure 2.24.

Multiplying both sides by $x - x_1$ gives $y - y_1 = m(x - x_1)$, which is also satisfied when $x = x_1$ and $y = y_1$.

FIGURE 2.24 Point–slope form

POINT–SLOPE FORM OF THE EQUATION OF A LINE

If a line has slope m and passes through the point (x_1, y_1), then the **point–slope form** of an equation of the line is

$$y - y_1 = m(x - x_1).$$

EXAMPLE 2   **Finding an Equation of a Line with Given Point and Slope**

Find the point–slope form of the equation of the line passing through the point $(1, -2)$ with slope $m = 3$. Then solve for y.

SOLUTION

$$y - y_1 = m(x - x_1) \qquad \text{Point–slope form}$$
$$y - (-2) = 3(x - 1) \qquad \text{Substitute } x_1 = 1, y_1 = -2, \text{ and } m = 3.$$
$$y + 2 = 3x - 3 \qquad \text{Simplify.}$$
$$y = 3x - 5 \qquad \text{Solve for } y. \qquad ■■■$$

Practice Problem 2   Find the point–slope form of the equation of the line passing through the point $(-2, -3)$ with slope $-\dfrac{2}{3}$. Then solve for y. ■

EXAMPLE 3   **Finding an Equation of a Line Passing Through Two Given Points**

Find the point–slope form of the equation of the line ℓ passing through the points $(3, 7)$ and $(-2, 1)$. Then solve for y.

SOLUTION

We first find the slope m of the line ℓ.

$$m = \frac{1 - 7}{(-2) - 3} = \frac{-6}{-2 - 3} = \frac{-6}{-5} = \frac{6}{5} \qquad m = \frac{y_2 - y_1}{x_2 - x_1}; \text{choose } (x_1, y_1) = (3, 7)$$

$$y - y_1 = m(x - x_1) \qquad\qquad\qquad \text{Point–slope form}$$

$$y - 7 = \frac{6}{5}(x - 3) \qquad\qquad\qquad \text{Substitute } x_1 = 3, y_1 = 7, \text{ and } m = \frac{6}{5}.$$

$$y = \frac{6}{5}x + \frac{17}{5} \qquad\qquad\qquad \text{Add 7 to both sides and simplify.} \quad ■■■$$

Practice Problem 3   Find the point–slope form of the equation of a line passing through the points $(-3, -4)$ and $(-1, 6)$. Then solve for y. ■

In Example 3, using the point $(-2, 1)$ for (x_1, y_1) in the point–slope form and solving for y would give the same answer.

3 Write the slope–intercept form of the equation of a line.

EXAMPLE 4   **Finding an Equation of a Line with a Given Slope and y-intercept**

Find the point–slope form of the equation of the line with slope m and y-intercept b. Then solve for y.

SOLUTION

Because the line has y-intercept b, the line passes through the point $(0, b)$. See Figure 2.25.

$$y - y_1 = m(x - x_1) \qquad \text{Point–slope form}$$
$$y - b = m(x - 0) \qquad \text{Substitute } x_1 = 0 \text{ and } y_1 = b.$$
$$y - b = mx \qquad \text{Simplify.}$$
$$y = mx + b \qquad \text{Solve for } y. \qquad ■■■$$

FIGURE 2.25 Slope–intercept form

Practice Problem 4   Find the point–slope form of the equation of the line with slope 2 and y-intercept -3. Then solve for y. ■

SLOPE–INTERCEPT FORM OF THE EQUATION OF A LINE

The **slope–intercept form** of the equation of the line with slope *m* and *y*-intercept *b* is

$$y = mx + b.$$

The linear equation in the form $y = mx + b$ displays the slope *m* (the coefficient of *x*) and the *y*-intercept *b* (the constant term). The number *m* tells which way and how much the line is tilted; the number *b* tells where the line intersects the *y*-axis.

EXAMPLE 5   **Graphing a Line Using its Slope and *y*-intercept**

Use the slope to find a point on the line whose equation is $y = \dfrac{2}{3}x + 2$ different from the *y*-intercept, then graph the line.

SOLUTION

The equation

$$y = \frac{2}{3}x + 2$$

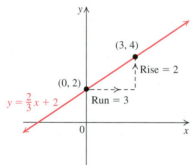

is in the slope–intercept form with slope $\dfrac{2}{3}$ and *y*-intercept 2. To sketch the graph, find two points on the line and draw a line through the two points. Use the *y*-intercept as one of the points; then use the slope to locate a second point. Because $m = \dfrac{2}{3}$, let 2 be the rise and 3 be the run. From the point $(0, 2)$, move three units to the right (run) and two units up (rise). This gives $(0 + \text{run}, 2 + \text{rise}) = (0 + 3, 2 + 2) = (3, 4)$ as the second point.

The line we want joins the points $(0, 2)$ and $(3, 4)$ and is shown in Figure 2.26. ■ ■ ■

FIGURE 2.26 Locating a second point on a line

Practice Problem 5   Repeat Example 5 for the line whose equation is $y = -\dfrac{2}{3}x + 4$.

■

4 Recognize the equations of horizontal and vertical lines.

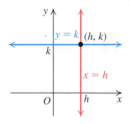

FIGURE 2.27 Vertical and horizontal lines

Equations of Horizontal and Vertical Lines

Consider the horizontal line through the point (h, k). The *y*-coordinate of every point on this line is *k*. So we can write its equation as $y = k$. This line has slope $m = 0$ and *y*-intercept *k*. Similarly, an equation of the vertical line through the point (h, k) is $x = h$. This line has undefined slope and *x*-intercept *h*. See Figure 2.27.

HORIZONTAL AND VERTICAL LINES

An equation of a horizontal line through (h, k) is $y = k$.

An equation of a vertical line through (h, k) is $x = h$.

EXAMPLE 6   **Recognizing Horizontal and Vertical Lines**

Discuss the graph of each equation in the *xy*-plane.

a. $y = 2$ **b.** $x = 4$

Calculator graph of $y = 2$

SOLUTION

a. The equation $y = 2$ may be considered as an equation in two variables x and y by writing

$$0 \cdot x + y = 2.$$

Any ordered pair of the form $(x, 2)$ is a solution of this equation. Some solutions are $(-1, 2)$, $(0, 2)$, $(2, 2)$, and $(7, 2)$. It follows that the graph is a line parallel to the x-axis and two units above it, as shown in Figure 2.28. Its slope is 0.

FIGURE 2.28 Horizontal line   **FIGURE 2.29** Vertical line

STUDY TIP

Students sometimes find it easy to remember that if a linear equation

(1) does not have y in it, the graph is parallel to the y-axis.
(2) does not have x in it, the graph is parallel to the x-axis.

b. The equation $x = 4$ may be written as

$$x + 0 \cdot y = 4.$$

Any ordered pair of the form $(4, y)$ is a solution of this equation. Some solutions are $(4, -5)$, $(4, 0)$, $(4, 2)$, and $(4, 6)$. The graph is a line parallel to the y-axis and four units to the right of it, as shown in Figure 2.29. Its slope is undefined.   ■ ■ ■

You cannot graph $x = 4$ by using the $\boxed{Y =}$ key on your calculator.

Practice Problem 6 Sketch the graphs of the lines $x = -3$ and $y = 7$.   ■

General Form of the Equation of a Line

5 Recognize the general form of the equation of a line.

An equation of the form

$$ax + by + c = 0,$$

where a, b, and c are constants and a and b are not both zero, is called the *general form* of the linear equation. Consider two possible cases: $b \neq 0$ and $b = 0$.

Suppose $b \neq 0$: We can isolate y on one side and rewrite the equation in slope–intercept form.

$$ax + by + c = 0 \qquad \text{Original equation}$$
$$by = -ax - c \qquad \text{Subtract } ax + c \text{ from both sides.}$$
$$y = -\frac{a}{b}x - \frac{c}{b} \qquad \text{Divide both sides by } b.$$

The result is an equation of a line in slope–intercept form with slope $-\dfrac{a}{b}$ and y-intercept $-\dfrac{c}{b}$.

Suppose $b = 0$: We know that $a \neq 0$ because not both a and b are zero, and we can solve for x.

$$ax + 0 \cdot y + c = 0 \qquad \text{Replace } b \text{ with 0.}$$
$$ax + c = 0 \qquad \text{Simplify.}$$
$$x = -\frac{c}{a} \qquad \text{Divide both sides by } a.$$

The graph of the equation $x = -\dfrac{c}{a}$ is a vertical line.

GENERAL FORM OF THE EQUATION OF A LINE

The graph of every linear equation

$$ax + by + c = 0,$$

where *a, b,* and *c* are constants and not both *a* and *b* are zero, is a line. The equation $ax + by + c = 0$ is called the **general form** of the equation of a line.

TECHNOLOGY CONNECTION

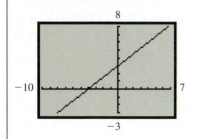 The calculator graph of $y = \dfrac{3}{4}x + 3$ reinforces the result of Example 7.

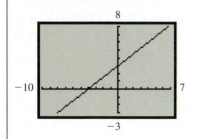

EXAMPLE 7   **Graphing a Linear Equation**

Find the slope, y-intercept, and x-intercept of the line with equation

$$3x - 4y + 12 = 0.$$

Then sketch the graph.

SOLUTION

First, solve for y to write the equation in slope–intercept form.

$$3x - 4y + 12 = 0 \qquad \text{Original equation}$$
$$4y = 3x + 12 \qquad \text{Subtract } 3x + 12; \text{ multiply by } -1.$$
$$y = \frac{3}{4}x + 3 \qquad \text{Divide by 4.}$$

This equation tells us that the slope m is $\dfrac{3}{4}$ and the y-intercept is 3. To find the x-intercept, set $y = 0$ in the original equation and obtain $3x + 12 = 0$, or $x = -4$. So the x-intercept is -4.

To sketch the graph of the equation we find two points on the graph. We use the intercepts $(-4, 0)$ and $(0, 3)$ and sketch the line joining these points. The graph is shown in Figure 2.30. To help prevent errors, find and graph a third point on the line.

FIGURE 2.30

Femur

Practice Problem 7 Sketch the graph of the equation $3x + 4y = 24$. ■

Forensic scientists use various clues to determine the identity of deceased individuals. Scientists know that certain bones serve as excellent sources of information about a person's height. The length of the *femur*, the long bone that stretches from the hip (pelvis) socket to the kneecap (patella), can be used to estimate a person's height. Knowing both the gender and race of the individual increases the accuracy of the relationship between the length of the bone and the height of the person.

EXAMPLE 8   Inferring Height from the Femur

The height H of a human male is related to the length x of his femur by the formula

$$H = 2.6x + 65,$$

where all measurements are in centimeters. The femur of Wild Bill Longley measured between 45 and 46 centimeters. Estimate the height of Wild Bill.

SOLUTION

Substituting $x = 45$ and $x = 46$ into the preceding formula, we have possible heights

$$H_1 = (2.6)(45) + 65 = 182$$
$$\text{and} \quad H_2 = (2.6)(46) + 65 = 184.6.$$

Therefore, Wild Bill's height was between 182 and 184.6 centimeters.

Converting to inches, we conclude that his height was between

$\dfrac{182}{2.54} \approx 71.7$ inches and $\dfrac{184.6}{2.54} \approx 72.7$ inches. He was approximately 6 feet tall. ■ ■ ■

Practice Problem 8 Suppose the femur of a person measures between 43 and 44 centimeters. Estimate the height of the person in centimeters. ■

Linear Depreciation

Linear depreciation occurs when an item loses a fixed value each year. In this situation, the value, V, of the item at any time, t, is given by a linear equation, $V = mt + b$.

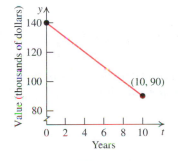

FIGURE 2.31

EXAMPLE 9   Linear Depreciation

The graph in Figure 2.31 represents the value of a small plane over a ten-year period. Each point (t, y) on the graph gives the plane's value y after t years.

a. What does the y-intercept represent?

b. What is the value of the plane after ten years?

c. How much does the plane depreciate in value each year?

d. Write a linear equation for the value of the plane after t years for $0 \le t \le 10$.

e. What does the slope found in part **d** represent?

SOLUTION

a. The y-intercept represents the purchase value of the plane, $140,000.

b. The graph point $(10, 90)$ gives the value of the plane after ten years as $90,000.

c. The value of the plane decreased from $140,000 to $90,000 over ten years.

This is $\dfrac{\$140{,}000 - \$90{,}000}{10} = \dfrac{\$50{,}000}{10} = \$5000$ per year.

d. Using the points $(0, 140)$ and $(10, 90)$, we find that the slope of the depreciation line is $\dfrac{140 - 90}{0 - 10} = \dfrac{50}{-10} = -5$. Then

$$y - y_1 = m(x - x_1) \qquad \text{Point–slope formula}$$
$$y - 140 = -5(x - 0) \qquad \text{Replace } y_1 \text{ with } 140, m \text{ with } -5, \text{ and } x_1 \text{ with } 0.$$
$$y = -5x + 140 \qquad \text{Simplify.}$$

e. The slope $m = -5$ gives the plane's yearly depreciation of $5000 per year. ■ ■ ■

Practice Problem 9 Repeat Example 9 assuming that the y-intercept is 210 and the point on the line is $(10, 110)$. ■

6 Find equations of parallel and perpendicular lines.

Parallel and Perpendicular Lines

We can use slope to decide whether two lines are parallel, perpendicular, or neither.

PARALLEL AND PERPENDICULAR LINES

Let ℓ_1 and ℓ_2 be two distinct lines with slopes m_1 and m_2, respectively. Then

ℓ_1 is parallel to ℓ_2 if and only if $m_1 = m_2$.

ℓ_1 is perpendicular to ℓ_2 if and only if $m_1 \cdot m_2 = -1$.

Any two vertical lines are parallel, and any horizontal line is perpendicular to any vertical line.

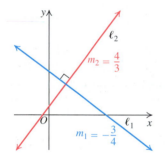

FIGURE 2.32 Perpendicular lines

In Exercises 123 and 124, you are asked to prove these assertions. The statement $m_1 \cdot m_2 = -1$ tells us that if the slope of a line ℓ is m, then the slope of a line perpendicular to ℓ is $-\dfrac{1}{m}$. In other words, the slopes of perpendicular lines are *negative reciprocals* of each other. For example, if the slope m_1 of a line ℓ_1 is $-\dfrac{3}{4}$, then the slope m_2 of any line ℓ_2 perpendicular to ℓ_1 is $-\dfrac{1}{-\dfrac{3}{4}} = \dfrac{4}{3}$. See Figure 2.32.

FINDING THE SOLUTION: A PROCEDURE

EXAMPLE 10   **Finding Equations of Parallel and Perpendicular Lines**

OBJECTIVE	EXAMPLE
Let $\ell : ax + by + c = 0$. Find equations of the lines ℓ_1 and ℓ_2 through the point (x_1, y_1) in general form with:	*Let $\ell: 2x - 3y + 6 = 0$. Find equations of the lines ℓ_1 and ℓ_2 line through the point $(2, 8)$ in general form with:*

OBJECTIVE

(a) ℓ_1 parallel to ℓ

(b) ℓ_2 perpendicular to ℓ

Step 1   Find the slope m of ℓ.

EXAMPLE

(a) ℓ_1 parallel to ℓ

(b) ℓ_2 perpendicular to ℓ

1. $\ell: 2x - 3y + 6 = 0$ \qquad Original equation

$$-3y = -2x - 6 \qquad \text{Subtract } 2x + 6.$$
$$y = \frac{2}{3}x + 2 \qquad \text{Divide by } -3.$$

The slope of ℓ is $m = \dfrac{2}{3}$.

Step 2   Write the slopes of ℓ_1 and ℓ_2.

The slope m_1 of ℓ_1 is m.

The slope m_2 of ℓ_2 is $-\dfrac{1}{m}$.

Step 3   Write the equations of ℓ_1 and ℓ_2.

Use the point–slope form to write the equations of ℓ_1 and ℓ_2.

Simplify to write the equations in the requested form.

2. $m_1 = \dfrac{2}{3}$ ℓ_1 is parallel to ℓ

$m_2 = -\dfrac{3}{2}$ ℓ_2 is perpendicular to ℓ

3. $\ell_1 : y - 8 = \dfrac{2}{3}(x - 2)$ Point–slope form

$2x - 3y + 20 = 0$ Simplify.

$\ell_2 : y - 8 = -\dfrac{3}{2}(x - 2)$ Point–slope form

$3x + 2y - 22 = 0$ Simplify.   ■ ■ ■

Practice Problem 10   Find (in general form) an equation of the line

a. through $(-2, 5)$ and parallel to the line containing $(2, 3)$ and $(5, 7)$.

b. through $(3, -4)$ and perpendicular to the line with equation $4x + 5y + 1 = 0$.

  7 Use linear regression.

Modeling Data Using Linear Regression

In many applications, the way in which one quantity depends on another can be modeled by a linear function. One technique for finding this linear function is called **linear regression**. To illustrate the method, we use Table 2.4, which gives the data for the number of registered motorcycles in the United States for the years 2000–2006.

TABLE 2.4

Year	2000	2001	2002	2003	2004	2005	2006
Registered motorcycles (millions)	4.3	4.9	5.0	5.4	5.8	6.2	6.7

The scatter diagram for the motorcycle data is shown in Figure 2.33(a), and the line used to model the data is added to the scatter diagram in Figure 2.33(b). The years 2000–2006 are represented by the numbers 0–6, respectively.

(a)

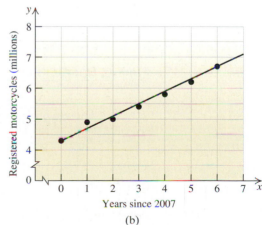

(b)

FIGURE 2.33

The line is called the **regression line** or the **least-squares line**. The slope–intercept equation for the line in Figure 2.33(b) is $y = 0.38x + 4.34$, where x is the number of years after 2000, and the slope and intercept values are rounded to the nearest hundredth. The equation is found by using either a graphing calculator or statistical formulas.

> **EXAMPLE 11** **Predictions Using Linear Regression**

Use the equation $y = 0.38x + 4.34$ to predict the number of registered motorcycles in the United States for 2007.

SOLUTION

Because 2007 is seven years after 2000, we set $x = 7$; then $y = 0.38(7) + 4.34 = 7$.

So according to the regression prediction, 7 million motorcycles were registered in the United States for 2007. The U.S. Department of Transportation reports that about 7.1 million motorcycles were registered in the United States that year. The quality of a prediction based on linear regression depends on how close the relationship between the data quantities is to being linear. ■ ■ ■

Practice Problem 11 Repeat Example 11 to predict the number of registered motorcycles in the United States for 2008. ■

The following Technology Connection shows how a graphing calculator displays the scatter diagram, regression line, and regression line equation.

TECHNOLOGY CONNECTION

You can use a graphing utility to investigate the linear regression model for data points that lie on or near a straight line. On the TI-84, the data for Example 11 are entered in five steps.

Step 1 Enter the data.

Press STAT 1 for STAT Edit. Enter the values of the independent variable x in L1 and the corresponding values of the dependent variable y in L2.

Step 2 Turn on STAT Plot

Press 2nd Y = 1 for Plot 1. Select *On*; then choose *Type*. ⌞ ⁚⁚ ⌝

Choose *X* List: L1; *Y* list: L2; and mark: ▫.

Step 3 Graph the scatterplot. Press Graph .

Step 4 Find the equation of the regression line.

Press STAT and highlight CALC. Press 4 for LinReg ($ax + b$). Press ENTER .

Step 5 Graph the regression line with the scatterplot.
Press [Y =].

Press [VARS] 5 for statistics. Highlight EQ and press [ENTER]. (You should see the regression equation in the $Y =$ window.)

Press [ZOOM] 9 for ZoomStat. (You should see the data graphed along with the regression line.)

Summary

1. We summarize different forms of equations whose graphs are lines.

Form	Equation	Slope	Other information
General	$ax + by + c = 0$ $a \neq 0$ or $b \neq 0$	$-\dfrac{a}{b}$ if $b \neq 0$ Undefined if $b = 0$	x-intercept is $-\dfrac{c}{a}, a \neq 0$. y-intercept is $-\dfrac{c}{b}, b \neq 0$.
Point–slope	$y - y_1 = m(x - x_1)$	m	Line passes through (x_1, y_1).
Slope–intercept	$y = mx + b$	m	x-intercept is $-\dfrac{b}{m}$ if $m \neq 0$; y-intercept is b.
Vertical line through (h, k)	$x = h$	Undefined	x-intercept is h; no y-intercept if $h \neq 0$. when $h = 0$, the graph is the y-axis.
Horizontal line through (h, k)	$y = k$	0	When $k \neq 0$, no x-intercept; y-intercept is k. When $k = 0$, the graph is the x-axis.

2. **Parallel and perpendicular lines:** Two lines with slopes m_1 and m_2 are
(a) parallel if $m_1 = m_2$, (b) perpendicular if $m_1 \cdot m_2 = -1$.

SECTION 2.3 ■ Exercises

A EXERCISES Basic Skills and Concepts

1. The number that measures the "steepness" of a line is called the _____.

2. In the slope formula $m = \dfrac{y_2 - y_1}{x_2 - x_1}$, the number $y_2 - y_1$ is called the _____.

3. If two lines have the same slope, they are _____.

4. The y-intercept of the line with equation $y = 3x - 5$ is _____.

5. A line perpendicular to a line with slope $-\dfrac{1}{5}$ has slope _____.

6. *True or False* A line with undefined slope is perpendicular to every line with slope 0.

7. *True or False* If $(1, 2)$ is a point on a line with slope 4, then so is the point $(2, 4)$.

8. *True or False* If the points $(1, 3)$ and $(2, y)$ are on a line with negative slope, then $y > 3$.

In Exercises 9–16, find the slope of the line through the given pair of points. Without plotting any points, state whether the line is rising, falling, horizontal, or vertical.

9. $(1, 3), (4, 7)$

10. $(0, 4), (2, 0)$

11. $(3, -2), (-2, 3)$

12. $(-3, 7), (-3, -4)$

13. $(3, -2), (2, -3)$

14. $(0.5, 2), (3, -3.5)$

15. $\left(\sqrt{2}, 1\right), \left(1 + \sqrt{2}, 5\right)$

16. $\left(1 - \sqrt{3}, 0\right), \left(1 + \sqrt{3}, 3\sqrt{3}\right)$

17. In the figure, identify the line with the given slope *m*.

   a. *m* = 1   **b.** *m* = −1

   c. *m* = 0   **d.** *m* is undefined.

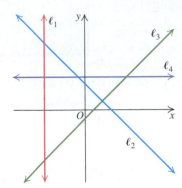

18. Find the slope of each line. (The scale is the same on both axes.)

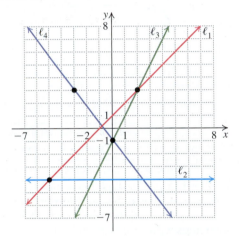

In Exercises 19 and 20, write an equation whose graph is described.

19. **a.** Graph is the *x*-axis.

     b. Graph is the *y*-axis.

20. **a.** Graph consists of all points with a *y*-coordinate of −4.

     b. Graph consists of all points with an *x*-coordinate of 5.

In Exercises 21–26, find an equation in slope–intercept form of the line that passes through the given point and has slope *m*. Also sketch the graph of the line by locating the second point with the rise-and-run method.

21. $(0, 4); m = \dfrac{1}{2}$ **22.** $(0, 4); m = -\dfrac{1}{2}$

23. $(2, 1); m = -\dfrac{3}{2}$ **24.** $(-1, 0); m = \dfrac{2}{5}$

25. $(5, -4); m = 0$ **26.** $(5, -4); m$ is undefined.

In Exercises 27–48, use the given conditions to find an equation in slope–intercept form of each nonvertical line. Write vertical lines in the form *x* = *h*.

27. Passing through $(0, 1)$ and $(1, 0)$

28. Passing through $(0, 1)$ and $(1, 3)$

29. Passing through $(-1, 3)$ and $(3, 3)$

30. Passing through $(-5, 1)$ and $(2, 7)$

31. Passing through $(-2, -1)$ and $(1, 1)$

32. Passing through $(-1, -3)$ and $(6, -9)$

33. Passing through $\left(\dfrac{1}{2}, \dfrac{1}{4}\right)$ and $(0, 2)$

34. Passing through $(4, -7)$ and $(4, 3)$

35. A vertical line through $(5, 1.7)$

36. A horizontal line through $(1.4, 1.5)$

37. A horizontal line through $(0, 0)$

38. A vertical line through $(0, 0)$

39. $m = 0$; *y*-intercept = 14

40. $m = 2$; *y*-intercept = 5

41. $m = -\dfrac{2}{3}$; *y*-intercept = −4

42. $m = -6$; *y*-intercept = −3

43. *x*-intercept = −3; *y*-intercept = 4

44. *x*-intercept = −5; *y*-intercept = −2

45. Parallel to $y = 5$; passing through $(4, 7)$

46. Parallel to $x = 5$; passing through $(4, 7)$

47. Perpendicular to $x = -4$; passing through $(-3, -5)$

48. Perpendicular to $y = -4$; passing through $(-3, -5)$

49. Let ℓ_1 be a line with slope $m = -2$. Determine whether the given line ℓ_2 is parallel to ℓ_1, perpendicular to ℓ_1, or neither.

   a. ℓ_2 is the line through the points $(1, 5)$ and $(3, 1)$.

   b. ℓ_2 is the line through the points $(7, 3)$ and $(5, 4)$.

   c. ℓ_2 is the line through the points $(2, 3)$ and $(4, 4)$.

50. A particle initially at $(-1, 3)$ moves along a line of slope $m = 4$ to a new position (x, y).

   a. Find *y* assuming that $x = 2$.

   b. Find *x* assuming that $y = 9$.

In Exercises 51–58, find the slope and intercepts from the equation of the line. Sketch the graph of each equation.

51. $x + 2y - 4 = 0$ **52.** $x = 3y - 9$

53. $3x - 2y + 6 = 0$ **54.** $2x = -4y + 15$

55. $x - 5 = 0$ **56.** $2y + 5 = 0$

57. $x = 0$ **58.** $y = 0$

59. Show that an equation of a line through the points $(a, 0)$ and $(0, b)$, with *a* and *b* nonzero, can be written in the form

$$\frac{x}{a} + \frac{y}{b} = 1.$$

This form is called the **two-intercept form** of the equation of a line.

In Exercises 60–64, use the two-intercept form of the equation of a line given in Exercise 59.

60. Find an equation of the line whose x-intercept is 4 and y-intercept is 3.

61. Find the x- and y-intercepts of the graph of the equation $2x + 3y = 6$.

62. Repeat Exercise 61 for the equation $3x - 4y + 12 = 0$.

63. Find an equation of the line that has equal intercepts and passes through the point $(3, -5)$.

64. Find an equation of the line making intercepts on the axes equal in magnitude but opposite in sign and passing through the point $(-5, -8)$.

65. Find an equation of the line passing through the points $(2, 4)$ and $(7, 9)$. Use the equation to show that the three points $(2, 4), (7, 9),$ and $(-1, 1)$ are on the same line.

66. Use the technique of Exercise 65 to check whether the points $(7, 2), (2, -3),$ and $(5, 1)$ are on the same line.

67. Write the slope-intercept form of the equation of the line shown in the figure perpendicular to the line with x-intercept -4.

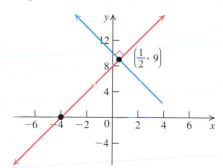

68. Write the point–slope form of the equation of the line shown in the figure parallel to the line through the origin.

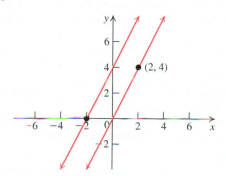

In Exercises 69–76, determine whether each pair of lines are parallel, perpendicular, or neither.

69. $x = -1$ and $2x + 7 = 9$

70. $x = 0$ and $y = 0$

71. $2x + 3y = 7$ and $y = 2$

72. $y = 3x + 1$ and $6y + 2x = 0$

73. $x = 4y + 8$ and $y = -4x + 1$

74. $4x + 3y = 1$ and $3 + y = 2x$

75. $3x + 8y = 7$ and $5x - 7y = 0$

76. $10x + 2y = 3$ and $y + 1 = -5x$

In Exercises 77–82, find the equation of the line in slope–intercept form satisfying the given conditions.

77. Parallel to $x + y = 1$; passing through $(1, 1)$

78. Parallel to $y = 6x + 5$; y-intercept of -2

79. Perpendicular to $3x - 9y = 18$; passing through $(-2, 4)$

80. Perpendicular to $-2x + y = 14$; passing through $(0, 0)$

81. Perpendicular to $y = 6x + 5$; y-intercept of 4

82. Parallel to $-2x + 3y - 7 = 0$; passing through $(1, 0)$

B EXERCISES Applying the Concepts

83. **Jet takeoff.** Upon takeoff, a jet climbs to 4 miles as it passes over 40 miles of land below it. Find the slope of the jet's climb.

84. **Road gradient.** Driving down the Smoky Mountains in Tennessee, Samantha descends 2000 feet in elevation while moving 4 miles horizontally away from a high point on the straight road. Find the slope (gradient) of the road (1 mile = 5280 feet).

In Exercises 85–92, write an equation of a line in slope–intercept form. To describe each situation, interpret (a) the variables x and y and (b) the meaning of the slope and the y-intercept.

85. **Christmas savings account.** You currently have $130 in a Christmas savings account. You deposit $7 per week into this account. (Ignore interest.)

86. **Golf club charges.** Your golf club charges a yearly membership fee of $1000 and $35 per session of golf.

87. **Wages.** Judy's job at a warehouse pays $11 per hour up to 40 hours per week. If she works more than 40 hours in a week, she is paid 1.5 times her usual wage. [*Hint:* You will need two equations.]

88. **Paying off a refrigerator.** You bought a new refrigerator for $700. You made a down payment of $100 and promised to pay $15 per month. (Ignore interest.)

89. **Converting currency.** In 2007, according to the International Monetary Fund, 1 U.S. dollar equaled 40.5 Indian rupees. Convert currency from rupees to dollars.

90. **Life expectancy.** In 2007, the life expectancy of a female born in the United States was 80.9 years and was increasing at a rate of 0.17 per year. (Assume that this rate of increase remains constant.)

91. Manufacturing. A manufacturer produces 50 TVs at a cost of $17,500 and 75 TVs at a cost of $21,250.

92. Demand. The demand for a product is zero units when the price per unit is $100 and 1000 units when the price per unit is zero dollars.

93. The graph represents the value of a piece of factory machinery over a ten-year period. Each point (t, y) gives the machine's value y after t years.
 a. What does the y-intercept represent?
 b. What is the value of the machine after ten years?
 c. How much does the machine depreciate in value each year?
 d. Write a linear equation for the value of the machine after t years, for $0 \le t \le 10$.
 e. What does the slope found in part d represent?

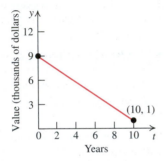

94. The graph represents the value of a piece of land over a ten-year period. Each point (t, y) gives the land's value y after t years.
 a. What does the y-intercept represent?
 b. What is the value of the land after five years?
 c. How much does the land depreciate in value each year?
 d. Write a linear equation for the value of the land after t years, for $0 \le t \le 5$.
 e. What does the slope found in part (d) represent?

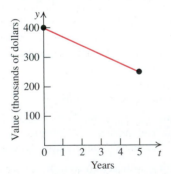

95. Depreciating a tractor. The value V of a tractor purchased for $14,000 and depreciated linearly at the rate of $1400 per year is given by $v = -1400t + 14,000$, where t represents the number of years since the purchase. Find the value of the tractor after (a) two years and (b) six years. When will the tractor have no value?

96. Depreciating a computer. A computer is purchased for $3000 and is to be depreciated linearly over four years. Assume that it has no value at the end of the four years.
 a. Find the amount depreciated per year.
 b. Find a linear equation relating the value V of the computer after t years of purchase.
 c. Graph the equation in part (b).

97. Playing for a concert. A famous band is considering playing a concert and charging $1000 plus $1.50 per person attending the concert. Write an equation relating the income y of the band to the number x of people attending the concert.

98. Car rental. The U Drive car rental agency charges $30.00 per day and $0.25 per mile.
 a. Find an equation relating the cost C of renting a car from U Drive for a one-day trip covering x miles.
 b. Draw the graph of the equation in part (a).
 c. How much does it cost to rent the car for one day covering 60 miles?
 d. How many miles were driven if the one-day rental cost was $47.75?

99. Female prisoners in Florida. The number of females in Florida's prisons rose from 2425 in 2000 to 4026 in 2006. (*Source:* Florida Department of Corrections.)
 a. Find a linear equation relating the number y of women prisoners to the year t. (Take $t = 0$ to represent 2000.)
 b. Draw the graph of the equation from part (a).
 c. How many women prisoners were there in 2003?
 d. Predict the number of women prisoners in 2010.

100. Fahrenheit and Celsius. In the Fahrenheit temperature scale, water boils at 212°F and freezes at 32°F. In the Celsius scale, water boils at 100°C and freezes at 0°C. Assume that the Fahrenheit temperature F and Celsius temperature C are linearly related.
 a. Find the equation in the slope–intercept form relating F and C, with C as the independent variable.
 b. What is the meaning of slope in part (a)?
 c. Find the Fahrenheit temperatures, to the nearest degree, corresponding to 40°C, 25°C, −5°C, and −10°C.
 d. Find the Celsius temperatures, to the nearest degree, corresponding to 100°F, 90°F, 75°F, −10°F, and −20°F.
 e. The normal body temperature in humans ranges from 97.6°F to 99.6°F. Convert this temperature range to degrees Celsius.
 f. When is the Celsius temperature the same numerical value as the Fahrenheit temperature?

101. Demand equation. A demand equation expresses a relationship between the demand q (the number of items demanded) and the price p per unit. A linear demand equation has the form $q = mp + b$. The slope m (usually negative) measures the change in demand per unit change in price. The y-intercept b gives the demand if the items were given away. A merchant can sell 480 T-shirts per week at a price of $4.00 each, but can sell only 400 per week if the price per T-shirt is increased to $4.50.

a. Find a linear demand equation for T-shirts.

b. Predict the demand for T-shirts assuming that the price per T-shirt is $5.

102. Supply equation. A supply equation expresses a relationship between the supply q (the number of units a supplier is willing to make available) and the unit price p (the price per item). A linear supply equation has the form $q = mp + b$. The slope m is usually positive. A manufacturer of ceiling fans can supply 70 fans per day at $50 apiece and will supply 100 fans a day at a price of $60 per fan.

a. Find a linear supply equation for the fans from this manufacturer.

b. Predict the supply at a price of $62 per fan.

103. Lake pollution. In 2002, tests showed that each 1000 liters of water from Lake Heron contained 7 milligrams of a polluting mercury compound. Two years later tests showed that 8 milligrams of the compound were contained in each 1000 liters of the lake's water. Assuming that the increase in pollutants is linear, find a linear equation relating the year in which the test was given to the number of milligrams of pollutant per 1000 liters of the lake's water. What does your equation predict the pollutant content per 1000 liters of water in Lake Heron to be in 2010?

104. Selling shoes. The Alpha shoe manufacturer has determined that the annual cost of making x pairs of its best-selling shoes, $\alpha 1$ (alpha one), is $25 per pair plus $75,000 in fixed overhead costs. Each pair of shoes that is manufactured is sold wholesale for $50.

a. Find the equations that model the cost, the revenue, and the profit (Profit = Revenue − Cost). Verify that all three equations are linear.

b. Graph the profit equation on the xy-coordinate system.

c. What is the slope of the line in part (b)? What is the practical meaning of this slope?

d. Find the intercepts of the line in part (b). What is the practical meaning of these intercepts?

105. Cost. The cost of producing four modems is $210.20, and the cost of producing ten modems is $348.80.

a. Write a linear equation in slope–intercept form relating cost to the number of modems produced. Sketch a graph of the equation.

b. What is the practical meaning of the slope and the intercepts of the equation in part (a)?

c. Predict the cost of producing 12 modems.

106. Viewers of a TV show. After five months, the number of viewers of *The Weekly Show* on TV was 5.73 million. After eight months, the number of viewers of the show rose to 6.27 million. Assume that the linear model applies.

a. Write an equation expressing the number of viewers V after x months.

b. Interpret the slope and the intercepts of the line in part (a).

c. Predict the number of viewers of the show after 11 months.

107. Health insurance coverage. In Michigan, 11.7% of the population was not covered by health insurance in 2000. In 2005, the percentage of uninsured people rose to 12.7%. Use the linear model to predict the percentage of people in Michigan that will not be covered in 2010. (*Source:* Michigan Department of Community Health.)

108. Oil consumption. The total world consumption of oil increased from 73.22 million barrels per day in 1999 to 82.59 million barrels per day in 2004. Create a linear model involving a relation between the year and the daily oil consumption in that year. Let $t = 0$ represent the year 1999. (*Source:* www.cia.gov)

109. Newspaper sales. The table shows the number of newspapers (in thousands) that sell at a given price (in nickels) per copy for the college newspaper *Midstate Oracle* at Midstate College.

Price (nickels)	1	2	3	4	5
Numbers of newspapers sold (thousands)	11	8	6	4	3

a. Use a graphing utility to find the equation of the line of regression relating the variables x (price in nickels) and y (number sold in thousands).

b. Use a graphing utility to plot the points and graph the regression line.

c. Approximately how many papers will be sold if the price per copy is 35¢?

110. Sales and advertising. Bildwell Computers keeps a careful record of its weekly advertising expenses (in thousands of dollars) and the number of computers (in thousands) it sells the following week. The table shows the data the company collected.

Weekly advertising expenses (in thousands of dollars)	100	200	300	400	500	600
Sales the following week (in thousands)	21	29	36	49	58	67

a. Use a graphing utility to find the equation of the line of regression equation relating the variables x (weekly advertising expenses in thousands of dollars) and y (number of computers sold, in thousands).

b. Use a graphing utility to plot the points and graph the regression line.

c. Approximately how many thousands of computers will be sold if the company spends $700,000 on advertising?

111. Height and weight of NBA players.

Cleveland Cavaliers Professional Basketball Players Height and Weight

Name of Player	Height (in inches)	Weight (in pounds)
Daniel Gibson	74	200
J.J. Hickson	81	242
Zydrunas Ilgauskas	87	260
Darnell Jackson	81	253
LeBron James	80	250
Tarence Kinsey	78	189
Aleksandar Pavlovic	79	235
Joe Smith	82	225
Wally Szczerbiak	79	240
Anderson Varejao	83	260
Ben Wallace	81	240
Delonte West	75	180
Jawad Williams	81	218
Mo Williams	73	190
Lorensen Wright	83	255

Source: http://www.nba.com/cavaliers/roster

a. Use a graphing utility to find the line of regression relating the variables x (height in inches) and y (weight in pounds).

b. Use a graphing utility to plot the points and graph the regression line.

c. Use the line in part (a) to predict the weight of an NBA player whose height is 76 inches.

112. Will anyone ever run a three-minute mile? In 1954, Roger Bannister of Britain cracked the four-minute mile mark. The following table shows the fastest record for running a mile in minutes up to a tenth of a second and the year the feat was accomplished.

Mile Records

Year	Time	Year	Time
1886	4:12.3	1958	3:54.5
1923	4:10.4	1966	3:51.3
1933	4:07.6	1979	3:48.9
1945	4:01.3	1985	3:46.3
1954	3:59.4	1999	3:43.1

a. Convert the table so that the times are given in seconds and are shown to the nearest tenth of a second.

b. Use a graphing utility to find the regression line for the data.

c. Determine to the nearest tenth of a second the time the regression line predicts for the years 1886, 1954, and 1999. Compare the actual and predicted times for those years.

d. Predict a time for 2020.

e. Predict the year in which a three-minute mile may occur.

C EXERCISES Beyond the Basics

113. The graph of a line has slope $m = 3$. If $P(-2, 3)$ and $Q(1, c)$ and are points on the graph, find c.

114. Find a real number c such that the line $3x - cy - 2 = 0$ has a y-intercept of -4.

In Exercises 115–118, show that the three given points are collinear by using (a) slopes and (b) the distance formula.

115. $(0, 1)$, $(1, 3)$, and $(-1, -1)$

116. $(1, 2)$, $(-1, 4)$, and $(2, 1)$

117. $(1, 2)$, $(0, -3)$, and $(-1, -8)$

118. $(1, .5)$, $(2, 0)$, and $(0.5, 0.75)$

119. Show that the triangle with vertices $A(1, 1)$, $B(-1, 4)$, and $C(5, 8)$ is a right triangle by using (a) slopes and (b) the converse of the Pythagorean Theorem.

120. Show that the four points $A(-4, -1)$, $B(1, 2)$, $C(3, 1)$, and $D(-2, -2)$ are the vertices of a parallelogram. [*Hint:* Opposite sides of a parallelogram are parallel.]

121. Show that the four points $A(-10, 9)$, $B(-2, 24)$, $C(5, 1)$, and $D(13, 16)$ are vertices of a square.

122. a. Find an equation of the line that is the perpendicular bisector of the line segment with endpoints $A(2, 3)$ and $B(-1, 5)$.

   b. Show that $y = x$ is the perpendicular bisector of the line segment with endpoints and $A(a, b)$ and $B(b, a)$.

123. Show that if two nonvertical lines are parallel, their slopes are equal. Start by letting the two lines be $\ell_1 : y = m_1 x + b_1$ and $\ell_2 : y = m_2 x + b_2$ $(b_1 > b_2)$. Show that if (x_1, y_1) and (x_2, y_2) are on ℓ_1, then $(x_1, y_1 - (b_1 - b_2))$ and $((x_2, y_2 - (b_1 - b_2))$ must be on ℓ_2. Then by computing m_1 and m_2, show that $m_1 = m_2$. See the accompanying figure.

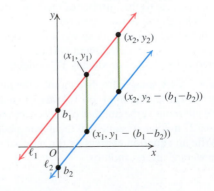

124. In the accompanying figure, note that lines ℓ_1 and ℓ_2 have slopes m_1 and m_2. Assume that ℓ_1 and ℓ_2 are perpendicular. Use the distance formula and the converse of the Pythagorean Theorem to show that $m_1 \cdot m_2 = -1$.

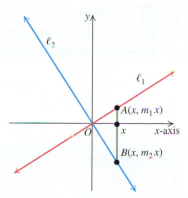

125. Geometry. Use a coordinate plane to prove that if two sides of a quadrilateral are congruent (equal in length) and parallel, then so are the other two sides.

126. Geometry. Use a coordinate plane to prove that the midpoints of the four sides of a quadrilateral are the vertices of a parallelogram.

127. Geometry. Find the coordinates of the point B of a line segment AB, given that $A = (2, 2)$ and $d(A, B) = 12.5$ and the slope of the line segment AB is $\dfrac{4}{3}$.

128. Geometry Show that an equation of a circle with (x_1, y_1) and (x_2, y_2) as endpoints of a diameter can be written in the form

$$(x - x_1)(x - x_2) + (y - y_1)(y - y_2) = 0.$$

[*Hint:* Use the fact that when the two endpoints of a diameter and any other point on the circle are used as vertices for a triangle, the angle opposite the diameter is a right angle.]

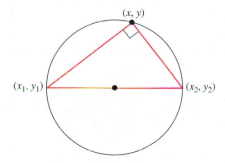

129. Geometry. Find an equation of the tangent line to the circle $x^2 + y^2 = 169$ at the point $(-5, 12)$. (See the accompanying figure.)

130. Geometry. Show that $xx_1 + yy_1 = a^2$ is an equation of the tangent line to the circle $x^2 + y^2 = a^2$ at the point (x_1, y_1) on the circle.

In Exercises 131–134, use a graphing device to graph the given lines on the same screen. Interpret your observations about the family of lines.

131. $y = 2x + b$ for $b = 0, \pm 2, \pm 4$

132. $y = mx + 2$ for $m = 0, \pm 2, \pm 4$

133. $y = m(x - 1)$ for $m = 0, \pm 3, \pm 5$

134. $y = m(x + 1) - 2$ for $m = 0, \pm 3, \pm 5$

Critical Thinking

135. By considering slopes, show that the points $(1, -1)$, $(-2, 5)$, and $(3, -5)$ lie on the same line.

136. By considering slopes, show that the points $(-9, 6)$, $(-2, 14)$, and $(-1, -1)$ are the vertices of a right triangle.

137. What can be said about the first coordinate of the point of intersection of two lines with equations $y = m_1 x + b_1$ and $y = m_2 x + b_2$ if $m_1 > m_2 > 0$ and
a. $b_1 > b_2$.
b. $b_1 < b_2$.

138. a. The equation $y + 2x + k = 0$ describes a family of lines for each value of k. Sketch the graphs of the members of this family for $k = 3, 0$, and -2. What is the common characteristic of the family?
b. Repeat part **a** for the family of lines $y - kx + 4 = 0$.

Relations and Functions

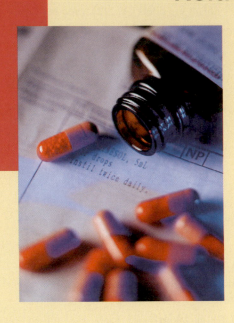

Before Starting this Section, Review

1. Graph of an equation (Section 2.2, page 179)
2. Equation of a circle (Section 2.2, page 185)

Objectives

1. Define relation.
2. Define function.
3. Find the domain of a function.
4. Get information about a function from its graph.
5. Solve applied problems by using functions.

THE UPS AND DOWNS OF DRUG LEVELS

Medicines in the form of a pill have a much harder time getting to work than you do. First, they are swallowed and dumped into a pool of acid in your stomach. Then they dissolve, leave the stomach, and start getting absorbed, mostly through the lining of the small intestine. Next, the blood from around the intestine carries the medicines to the liver, an organ designed to metabolize (break down) and remove foreign material from the blood. Because of this property of the liver, most drugs have a difficult time getting past it. The amount of drug that makes it into your bloodstream, compared with the amount that you put in your mouth, is called the drug's *bioavailability*. If a drug has low bioavailability, then much of the drug is destroyed before it reaches the bloodstream. The amount of drug you take in a pill is calculated to correct for this deficiency so that the amount you need actually ends up in your blood.

Once the drugs are past the liver, the bloodstream carries them throughout the rest of the body in about one minute. The study of the ways drugs are absorbed and move through the body is called *pharmacokinetics*, or PK for short. PK measures the ups and downs of drug levels in your body. In Example 9, we consider an application of *functions* to the levels of drugs in the bloodstream. (Adapted from *The ups and downs of drug levels*, by Bob Munk, www.thebody .com/content/treat/art1062.html) ■

---

1 Define relation.

Relations

Consider the following list of the six all-time top grossing movies in the United States through 2005.

Titanic	$600.8 million
Star Wars, Episode IV	$461.0 million
Shrek 2	$441.2 million
E.T.: The Extra-Terrestrial	$435.1 million
Star Wars, Episode I (*The Phantom Menace*)	$431.1 million
Spider-Man	$403.7 million

(*Source:* www.amazon.com)

For each of the movies listed, there is a corresponding amount of gross sales in millions of dollars. This correspondence can be written as a set of ordered pairs:

$$\{(\textit{Titanic}, \$600.8), (\textit{Star Wars IV}, \$461.0), (\textit{Shrek 2}, \$441.2),$$
$$(\textit{E.T.}, \$435.1), (\textit{Star Wars I}, \$431.1), (\textit{Spider-Man}, \$403.7)\}$$

Ordered pairs are an exceptionally useful method for displaying and organizing correspondences. Any set of ordered pairs is called a **relation**. When relations are written as a set of ordered pairs (x, y), we say that x corresponds to y.

DEFINITION OF A RELATION

Any set of ordered pairs is called a **relation**. The set of all first components is called the **domain** of the relation, and the set of all second components is called the **range** of the relation.

EXAMPLE 1   **Finding the Domain and Range of a Relation**

Find the domain and range of the relation

$$\{(\textit{Titanic}, \$600.8), (\textit{Star Wars IV}, \$461.0), (\textit{Shrek 2}, \$441.2),$$
$$(\textit{E.T.}, \$435.1), (\textit{Star Wars I}, \$431.1), (\textit{Spider-Man}, \$403.7)\}$$

SOLUTION

The domain is the set of all first components, or

$$\{\textit{Titanic}, \textit{Star Wars IV}, \textit{Shrek 2}, \textit{E.T.}, \textit{Star Wars I}, \textit{Spider-Man}\}.$$

The range is the set of all second components, or

$$\{\$600.8, \$461.0, \$441.2, \$435.1, \$431.1, \$403.7\}. \qquad ■ ■ ■$$

Practice Problem 1   Find the domain and range of the relation

$$\{(1, 2), (3, 0), (2, -2) (-2, -3), (-1, 1)\}. \qquad ■$$

2 Define function.

Functions

There are many ways of expressing a relationship between two quantities. For example, if you are paid \$10 per hour, then the relation between the number of hours, x, that you work and the amount of money, y, that you earn may be expressed by the equation $y = 10x$. Replacing x with 40 yields $y = 400$ and indicates that 40 hours of work corresponds to a \$400 paycheck. A special relationship such as $y = 10x$ in which to each element x in one set there corresponds a unique element y in another set is called a *function*. We say that your pay is a function of the number of hours you work. Because the value of y depends on the given value of x, y is called the **dependent variable** and x is called the **independent variable**. Any symbol may be used for the dependent or independent variables.

DEFINITION OF FUNCTION

A **function** is a relation in which each element of the domain corresponds to one and only one element of the range.

A relation may be defined by a **corespondence diagram**, in which an arrow points from each domain element to the element or elements in the range that correspond to it. To decide whether a relation defined by a correspondence diagram is a function, we must check whether *any* domain element is paired with more than one range element. If this happens, that domain element does not have a *unique* corresponding element of *Y* and the relation is not a function. See the correspondence diagrams in Figure 2.34.

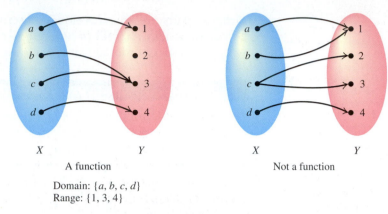

A function

Not a function

Domain: {*a*, *b*, *c*, *d*}
Range: {1, 3, 4}

FIGURE 2.34 **Correspondence diagrams**

BY THE WAY . . .

The functional notation $y = f(x)$ was first used by the great Swiss mathematician Leonhard Euler in the *Commentarii Academia Petropolitanae*, published in 1735.

We usually use single letters such as $f, F, g, G, h,$ and H as the name of a function. If we choose f as the name of a function, then for each x in the domain of f, there corresponds a unique y in its range. The number y is denoted by $f(x)$, read as "f of x" or "f at x." We call $f(x)$ the **value of f at the number x** and say that f *assigns* the value $f(x)$ to x.

Tables and graphs can be used to describe functions. The data in Table 2.5 (reproduced from Section 2.1) show the prevalence of smoking among adults aged 18 years and older in the United States over the years 2000–2007.

TABLE 2.5

Year	2000	2001	2002	2003	2004	2005	2006	2007
Percent of Adult Smokers	23.2	22.7	22.4	21.6	20.9	21.1	20.8	20.8

Source: Center for Disease Control. www.cdc.gov

The graph in Figure 2.35 is a **scatter diagram** that gives a visual representation of the same information. Scatter diagrams show the relationship between two variables by displaying data as points of a graph. The variable that might be considered the independent variable is plotted on the *x*-axis, and the dependent variable is plotted on the *y*-axis. Here the percent of adult smokers depends on the year and is plotted on the *y*-axis.

FIGURE 2.35 Scatter diagram

FIGURE 2.36 The function f as a machine

The function described by Table 2.5 and Figure 2.35 has for its domain the set {2000, 2001, 2002, 2003, 2004, 2005, 2006, 2007} and for its range the set {23.2, 22.7, 22.4, 21.6, 20.9, 21.1, 20.8}. The function assigns to each year in the domain the percent of adult smokers 18 years and older in the United States during that year.

Yet another nice way of picturing the concept of a function is as a "machine," as shown in Figure 2.36. If x is in the domain of a function, then the machine accepts it as an "input" and produces the value $f(x)$ as the "output."

◆ **WARNING**

In writing $y = f(x)$, do not confuse f, which is the name of the function, with $f(x)$, which is a number in the range of f that the function assigns to x. The symbol $f(x)$ does not mean "f times x."

STUDY TIP

The definition of a function is independent of the letters used to denote the function and the variables. For example,

$$s(x) = x^2, \ g(y) = y^2,$$
$$\text{and } h(t) = t^2$$

represent the same function.

Functions Defined by Equations When the relation defined by an equation in two variables is a function, we can often solve the equation for the dependent variable in terms of the independent variable. For example, the equation $y - x^2 = 0$ can be solved for y in terms of x.

$$y = x^2.$$

Now we can replace the dependent variable, in this case, y, with functional notation $f(x)$ and express the function as

$$f(x) = x^2 \quad (\text{read "}f \text{ of } x \text{ equals } x^2\text{"}).$$

Here $f(x)$ plays the role of y and is the value of the function f at x. For example, if $x = 3$ is an element (input value) in the domain of f, the corresponding element (output value) in the range is found by replacing x with 3 in the equation

$$f(x) = x^2$$
$$f(3) = 3^2 = 9 \qquad \text{Replace } x \text{ with 3.}$$

FIGURE 2.37 The function f as a machine

We say that the value of the function f at 3 is 9. See Figure 2.37. In other words, the number 9 in the range corresponds to the number 3 in the domain and the ordered pair $(3, 9)$ is an ordered pair of the function f.

If a function g is defined by an equation such as $y = x^2 - 6x + 8$, the notations

$$y = x^2 - 6x + 8 \quad \text{and} \quad g(x) = x^2 - 6x + 8$$

define the same function.

EXAMPLE 2 **Determining Whether an Equation Defines a Function**

Determine whether y is a function of x for each equation.

a. $6x^2 - 3y = 12$ **b.** $y^2 - x^2 = 4$

SOLUTION

Solve each equation for y in terms of x. If more than one value of y corresponds to the same value of x, then y is not a function of x.

a.
$6x^2 - 3y = 12$	Original equation
$6x^2 - 3y + 3y - 12 = 12 + 3y - 12$	Add $3y - 12$ to both sides.
$6x^2 - 12 = 3y$	Simplify.
$2x^2 - 4 = y$	Divide by 3.

The previous equation shows that only one value of y corresponds to each value of x. For example, if $x = 0$, then $y = 0 - 4 = -4$. So y is a function of x.

b.

$$y^2 - x^2 = 4 \qquad \text{Original equation}$$
$$y^2 - x^2 + x^2 = 4 + x^2 \qquad \text{Add } x^2 \text{ to both sides.}$$
$$y^2 = x^2 + 4 \qquad \text{Simplify.}$$
$$y = \pm\sqrt{x^2 + 4} \qquad \text{Square root property}$$

The previous equation shows that two values of y correspond to each value of x. For example, if $x = 0$, then $y = \pm\sqrt{0^2 + 4} = \pm\sqrt{4} = \pm 2$. Both $y = 2$ and $y = -2$ correspond to $x = 0$. Therefore, y is not a function of x.   ■■■

Practice Problem 2   Determine whether y is a function of x for each equation.

a. $2x^2 - y^2 = 1$ **b.** $x - 2y = 5$ ■

A graphing calculator allows you to store a number in a variable and then find the value of a function at that number. This screen shows the evaluation of the function $g(x) = x^2 - 6x + 8$ from Example 3(b) for $x = -2$.

```
-2→X
            -2
X^2-6X+8
            24
■
```

EXAMPLE 3   **Evaluating a Function**

Let g be the function defined by the equation

$$y = x^2 - 6x + 8.$$

Evaluate each function value.

a. $g(3)$ **b.** $g(-2)$ **c.** $g\left(\dfrac{1}{2}\right)$ **d.** $g(a + 2)$ **e.** $g(x + h)$

SOLUTION

Because the function is named g, we replace y with $g(x)$ and write

$$g(x) = x^2 - 6x + 8 \qquad \text{Replace } y \text{ with } g(x).$$

In this notation, the independent variable x is a placeholder. We can write $g(x) = x^2 - 6x + 8$ as

$$g(\ ) = (\ )^2 - 6(\ ) + 8.$$

This notation means that when you place any quantity within the parentheses on the left side of the equation, you must also place the same quantity into each set of parentheses on the right side of the equation.

a. $g(x) = x^2 - 6x + 8$ The given equation
    $g(3) = 3^2 - 6(3) + 8$ Replace x with 3 at each occurrence of x.
        $= 9 - 18 + 8 = -1$ Simplify.

The statement $g(3) = -1$ means that the value of the function g at 3 is -1.
As in part **a**, we can evaluate g at any value x in its domain.

    $g(x) = x^2 - 6x + 8$ Original equation

b. $g(-2) = (-2)^2 - 6(-2) + 8 = 24$ Replace x with -2 and simplify.

c. $g\left(\dfrac{1}{2}\right) = \left(\dfrac{1}{2}\right)^2 - 6\left(\dfrac{1}{2}\right) + 8 = \dfrac{21}{4}$ Replace x with $\dfrac{1}{2}$ and simplify.

d. $g(a + 2) = (a + 2)^2 - 6(a + 2) + 8$ Replace x with $(a + 2)$ in $g(x)$.
        $= a^2 + 4a + 4 - 6a - 12 + 8$ Recall: $(x + y)^2 = x^2 + 2xy + y^2$.
        $= a^2 - 2a$ Simplify.

e. $g(x + h) = (x + h)^2 - 6(x + h) + 8$ Replace x with $(x + h)$ in $g(x)$.
        $= x^2 + 2xh + h^2 - 6x - 6h + 8$ Simplify.   ■■■

Practice Problem 3   Let g be the function defined by the equation $y = -2x^2 + 5x$. Evaluate each function value.

a. $g(0)$ **b.** $g(-1)$ **c.** $g(x + h)$ ■

3 Find the domain of a function.

Domain of a Function

> **AGREEMENT ON DOMAIN**
>
> If the domain of a function f that is defined by an equation is not specified, then we agree that the domain of the function is the largest set of real numbers that results in real number outputs $f(x)$.

BY THE WAY . . .

The word *domain* comes from the Latin word *dominus*, or *master*, which in turn is related to *domus*, meaning "home." *Domus* is also the root for the words *domestic* and *domicile*.

When we use our agreement to find the domain of a function, first we usually find the values of the variable that do not result in real number outputs. Then we exclude those numbers from the domain. Remember that

1. division by zero is undefined.
2. the square root (or any even root) of a negative number is not a real number.

EXAMPLE 4   **Finding the Domain of a Function**

Find the domain of each function.

a. $f(x) = \dfrac{1}{1 - x^2}$ **b.** $g(x) = \sqrt{x}$ **c.** $h(x) = \dfrac{1}{\sqrt{x - 1}}$ **d.** $P(t) = 2t + 1$

SOLUTION

a. The function f is not defined when the denominator $1 - x^2$ is 0. Because $1 - x^2 = 0$ if $x = 1$ or $x = -1$, the domain of f is the set $\{x | x \neq -1$ and $x \neq 1\}$, or $(-\infty, -1) \cup (-1, 1) \cup (1, \infty)$ in interval notation.

b. Because the square root of a negative number is not a real number, negative numbers are excluded from the domain of g. The domain of $g(x) = \sqrt{x}$ is $\{x | x \geq 0\}$, or $[0, \infty)$ in interval notation.

c. The function $h(x) = \dfrac{1}{\sqrt{x - 1}}$ has *two* restrictions. The square root of a negative number is not a real number, so $\sqrt{x - 1}$ is a real number only if $x - 1 \geq 0$. However, we cannot allow $x - 1 = 0$ because $\sqrt{x - 1}$ is in the denominator. Therefore, we must have $x - 1 > 0$, or $x > 1$. The domain of h is $\{x | x > 1\}$, or in interval notation, $(1, \infty)$.

d. When any real number is substituted for t in $y = 2t + 1$, a unique real number is determined. The domain is the set of all real numbers: $\{t | t$ is a real number$\}$, or in interval notation, $(-\infty, \infty)$.   ■ ■ ■

Practice Problem 4   Find the domain of the function

$$f(x) = \frac{1}{\sqrt{1 - x}}.$$ ■

Graphs of Functions

The *graph of a function* f is the set of ordered pairs $(x, f(x))$ such that x is in the domain of f. That is, the graph of f is the graph of the equation $y = f(x)$. We sketch the graph of $y = f(x)$ by plotting points and joining them with a smooth curve. A filled in circle placed at the end of a graph indicates that the endpoint on the graph is *included* in the graph. An empty circle placed on the graph or its endpoint indicates that the point is *excluded* from the graph.

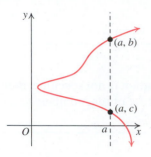

FIGURE 2.38 **Graph does not represent a function**

Figure 2.38 shows that not every curve in the plane is the graph of a function. In Figure 2.38, a vertical line intersects the curve at two distinct points (a, b) and (a, c). This curve cannot be the graph of $y = f(x)$ for any function f because having $f(a) = b$ and $f(a) = c$ means that f assigns two different range values (b and c) to the same domain element a. We express this statement in a slightly different way as follows:

VERTICAL-LINE TEST

If no vertical line intersects the graph of a curve (or scatterplot) at more than one point, then the curve (or scatterplot) is the graph of a function.

If we scan a graph from left to right, the vertical-line test tells us that the graph of a function cannot backtrack horizontally.

EXAMPLE 5   **Identifying the Graph of a Function**

Determine which graphs in Figure 2.39 are graphs of functions.

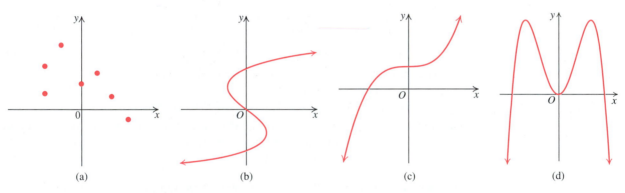

| (a) | (b) | (c) | (d) |

FIGURE 2.39 **The vertical-line test**

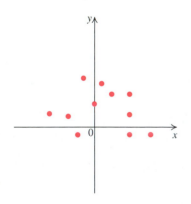

SOLUTION

The graphs in Figures 2.39(a) and 2.39(b) are not graphs of functions because a vertical line can be drawn through the two points farthest to the left in Figure 2.39(a) and the y-axis is one of many vertical lines that contains more than one point on the graph in Figure 2.39(b). The graphs in Figures 2.39(c) and 2.39(d) are the graphs of functions because no vertical line intersects either graph at more than one point. ■ ■ ■

Practice Problem 5   Decide whether the graph in the margin is the graph of a function.   ■

EXAMPLE 6   **Examining the Graph of a Function**

Let $y = f(x) = x^2 - 2x - 3$.

a. Is the point $(1, -3)$ on the graph of f?

b. Find all values of x so that $(x, 5)$ is on the graph of f.

c. Find the y-intercept of the graph of f.

d. Find all x-intercepts of the graph of f.

SOLUTION

a. We check whether $(1, -3)$ satisfies the equation $y = x^2 - 2x - 3$.

$$-3 \overset{?}{=} (1)^2 - 2(1) - 3 \qquad \text{Replace } x \text{ with 1 and } y \text{ with } -3.$$
$$-3 \overset{?}{=} -4 \qquad \text{No}$$

So $(1, -3)$ is *not* on the graph of f. See Figure 2.40.

b. Substitute 5 for y in $y = x^2 - 2x - 3$ and solve for x.

$$5 = x^2 - 2x - 3$$
$$0 = x^2 - 2x - 8 \qquad \text{Subtract 5 from both sides.}$$
$$0 = (x - 4)(x + 2) \qquad \text{Factor.}$$
$$x - 4 = 0 \quad \text{or} \quad x + 2 = 0 \qquad \text{Zero-product property}$$
$$x = 4 \quad \text{or} \quad x = -2 \qquad \text{Solve for } x.$$

The point $(x, 5)$ is on the graph of f only when $x = -2$ and $x = 4$. Both $(-2, 5)$ and $(4, 5)$ are on the graph of f. See Figure 2.40.

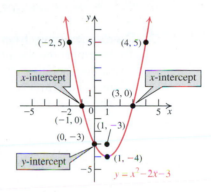

FIGURE 2.40

c. Find the point (x, y) with $x = 0$ in $y = x^2 - 2x - 3$.

$$y = 0^2 - 2(0) - 3 \qquad \text{Replace } x \text{ with 0 to find the } y\text{-intercept.}$$
$$y = -3 \qquad \text{Simplify.}$$

The only y-intercept is -3. See Figure 2.40.

d. Find all points (x, y) with $y = 0$ in $y = x^2 - 2x - 3$.

$$0 = x^2 - 2x - 3$$
$$0 = (x + 1)(x - 3) \qquad \text{Factor.}$$
$$x + 1 = 0 \quad \text{or} \quad x - 3 = 0 \qquad \text{Zero-product property}$$
$$x = -1 \quad \text{or} \quad x = 3 \qquad \text{Solve for } x.$$

The x-intercepts of the graph of f are -1 and 3. See Figure 2.40. ■■■

Practice Problem 6 Let $y = f(x) = x^2 + 4x - 5$.

a. Is the point $(2, 7)$ on the graph of f?

b. Find all values of x so that $(x, -8)$ is on the graph of f.

c. Find the y-intercept of the graph of f.

d. Find any x-intercepts of the graph of f. ■

4 Get information about a function from its graph.

Function Information from Its Graph

Given a graph in the xy-plane, we use the vertical line test to determine whether the graph is that of a function. We can obtain the following information from the graph of a function.

1. Point on a graph

A point (a, b) is on the graph of f means that a is in the domain of f and the value of f at a is b; that is, $f(a) = b$. We can visually determine whether a given points is on the graph of a function. See Figure 2.41.

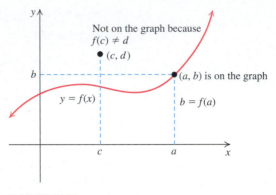

FIGURE 2.41

2. Domain and range from a graph

To determine the *domain* of a function, we look for the portion on the *x*-axis that is used in graphing f. We can find this portion by projecting (collapsing) the graph on the *x*-axis. This projection is the domain of f.

The *range* of f is the projection of its graph on the *y*-axis. See Figure 2.42.

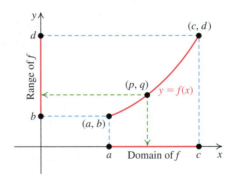

FIGURE 2.42 **Domain and range of *f***

EXAMPLE 7   **Finding the Domain and Range from a Graph**

Use the graphs in Figure 2.43 to find the domain and the range of each function.

FIGURE 2.43

SOLUTION

a. In Figure 2.43(a), the open circle at $(-2, 1)$ indicates that the point $(-2, 1)$ does not belong to the graph of f, while the full circle at the point $(3, 3)$ indicates that the point $(3, 3)$ is part of the graph. When we project the graph of $y = f(x)$ onto the x-axis, we obtain the interval $(-2, 3]$. The domain of f in interval notation is $(-2, 3]$. Similarly, the projection of the graph of f onto the y-axis gives the interval $(1, 4]$, which is the range of f.

b. The projection of the graph of g in Figure 2.43(b) onto the x-axis is made up of the two intervals $[-3, 1]$ and $[3, \infty)$. So the domain of g is $[-3, 1] \cup [3, \infty)$.

   To find the range of g, we project its graph onto the y-axis. The projection of the line segment joining $(-3, -1)$ and $(1, 1)$ onto the y-axis is the interval $[-1, 1]$. The projection of the horizontal ray starting at the point $(3, 4)$ onto the y-axis is just a single point at $y = 4$. Therefore, the range of g is $[-1, 1] \cup \{4\}$. ■ ■ ■

Practice Problem 7 Find the domain and the range of $y = h(x)$ in Figure 2.44.

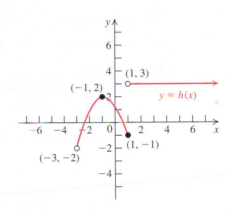

FIGURE 2.44 ■

3. Evaluations

a. **Finding $f(c)$** Given a number c in the domain of f, we find $f(c)$ from the graph of f.

   First, locate the number c on the x-axis. Draw a vertical line through c. It intersects the graph at the point $(c, f(c))$. Through this point, draw a horizontal line to intersect the y-axis. Read the value on the y-axis. See Figure 2.45.

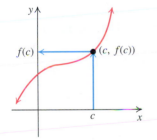

FIGURE 2.45 Locating $f(c)$ graphically

b. **Solving $f(x) = d$** Given d in the range of f, find values of x for which $f(x) = d$. Locate the number d on the y-axis. Draw a horizontal line through d. This line intersects the graph at point(s) (x, d). Through each of these points, draw vertical lines to intersect the x-axis. Read the values on the x-axis. These are the values of x for which $f(x) = d$. Figure 2.46 shows three solutions: x_1, x_2, and x_3 of the equation $f(x) = d$.

FIGURE 2.46 Graphically solving $f(x) = d$

| EXAMPLE 8 | **Obtaining Information from a Graph** |

Figure 2.47 shows the graph of a function f that relates the unemployment rate, y, of a country with its GDP in billions of dollars. Here unemployment rate = percentage of people unemployed and

$$GDP = \text{Gross domestic product, the monetary value of all goods and services produced in the country.}$$

FIGURE 2.47

Find and interpret.

a. The domain of f **b.** The range of f **c.** $f(30)$ **d.** $f(x) = 40$

SOLUTION

a. Because the independent variable x denotes the GDP, we have $x \geq 0$. The domain of f is the interval $[0, \infty)$. This interval is the projection of the graph onto the x-axis.

b. The unemployment rate, y, is the percentage of people unemployed; so $0 \leq y \leq 100$. The graph shows that as the GDP of the country increases, the unemployment rate drops but does not equal zero. So the range of f is $(0, 100]$.

c. From the graph, we find $f(30) = 80$. This means that if the country has $30 billion of GDP, its unemployment rate will be 80%.

d. From the graph, we find that $40 = f(60)$. This means that to have an unemployment rate of 40%, the GDP of the country must be $60 billion. ■ ■ ■

Practice Problem 8   In Figure 2.47, find and interpret $f(50)$ and $f(x) = 10$. ■

SUMMARY

A function is usually described in one or more of the following ways:

- A correspondence diagram
- A table of values
- An equation or a formula
- A scatter diagram or a graph
- A set of ordered pairs

Input	Output
x	y or $f(x)$
First coordinate	Second coordinate
Independent variable	Dependent variable
Domain is the set of all inputs.	Range is the set of all outputs.

---

5 Solve applied problems by using functions.

Applications

Level of Drugs in Bloodstream   When you take a dose of medication, the drug level in the blood goes up quickly and soon reaches its *peak*, called C_{max} (the maximum concentration). As your liver or kidneys remove the drug, your blood's drug levels drop until the next dose enters your bloodstream. The lowest drug level is called the *trough*, or C_{min} (the minimum concentration). The ideal dose should be strong enough to be effective without causing too many side effects. We start by setting upper and lower limits on the drug's blood levels, shown by the two horizontal

FIGURE 2.48 Drug levels

lines on a pharmacokinetics (PK) graph. See Figure 2.48. The upper line represents the drug level at which people start to develop serious side effects. The lower line represents the minimum drug level that provides the desired effect.

EXAMPLE 9   Cholesterol-Reducing Drugs

Many drugs used to lower high blood cholesterol levels are called *statins* and are very popular and widely prescribed. These drugs, along with proper diet and exercise, help prevent heart attacks and strokes. Recall from the introduction to this section that bioavailability is the amount of a drug you have ingested that makes it into your bloodstream. A statin with a bioavailability of 30% has been prescribed for Boris to treat his cholesterol levels. He is to take 20 milligrams daily. Every day his body filters out half the statin. Find the maximum concentration of the statin in the bloodstream on each of the first ten days of using the drug and graph the result.

SOLUTION

Because the statin has 30% bioavailability and Boris takes 20 milligrams per day, the maximum concentration in the bloodstream is 30% of 20 milligrams, or $20(0.3) = 6$ milligrams, from each day's prescription. Because half the statin is filtered out of the body each day, the daily maximum concentration is

$$\frac{1}{2}\,(\text{previous day's maximum concentration}) + 6.$$

Table 2.6 shows the maximum concentration of the drug for each of the first ten days. After the first day, each number in the second column is computed (to three decimal places) by adding 6 to one-half the number above it. The graph is shown in Figure 2.49.

TABLE 2.6

Day	Maximum Concentration
1	6.000
2	9.000
3	10.500
4	11.250
5	11.625
6	11.813
7	11.907
8	11.954
9	11.977
10	11.989

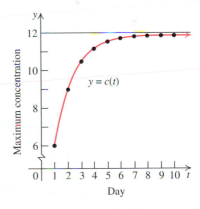

FIGURE 2.49 Maximum drug concentration

The graph shows that the maximum concentration of the statin in the bloodstream approaches 12 milligrams if Boris continues to take one pill every day.   ■ ■ ■

Practice Problem 9   In Example 9, use Figure 2.49 to find the domain and the range of the function. Also compute the function value at 6.   ■

Functions in Economics

The important concepts of *breaking even, earning a profit*, or *taking a loss* in economics will be discussed throughout this book. Here we define some of the elementary functions in economics.

Let x represent the number of units of an item manufactured and sold at a price of p dollars. Then the cost $C(x)$ of producing x items includes the **fixed cost** (such as rent, insurance, product design, setup, promotion, and overhead) and the **variable cost** (which depends on the number of items produced at a certain cost per item).

Linear Cost Function A linear cost function $C(x)$ is given by

$$C(x) = (\text{variable cost}) + (\text{fixed cost})$$
$$= ax + b,$$

where the fixed cost is b and the **marginal cost** (cost of producing each item) is a dollars per item.

Average cost, denoted by $\overline{C}(x)$, is defined by $\overline{C}(x) = \dfrac{C(x)}{x}$.

Linear Price–Demand Function Suppose x items can be sold (demanded) at a price of p dollars per item. Then a linear demand function usually has the form

$$p(x) = -mx + d \qquad \text{Expressing } p \text{ as a function of } x$$

or

$$x(p) = -np + k \qquad \text{Expressing } x \text{ as a function of } p$$

Both expressions indicate that a higher price will result in fewer items sold. The constants m, d, n, and k depend on the given situation, with $m > 0$ and $n > 0$. A **Price–Supply** function has the form $p = S(x)$ or $x = G(p)$, where a higher price results in a greater supply of items.

Revenue Function $R(x)$

$$\text{Revenue} = (\text{Price per item})(\text{Number of items sold})$$
$$R(x) = p \cdot x = (-mx + d)x \qquad\qquad p = p(x) = -mx + d$$

Profit Function $P(x)$

$$\text{Profit} = \text{Revenue} - \text{Cost}$$
$$P(x) = R(x) - C(x)$$

A manufacturing company will

(i) have a *profit* if $P(x) > 0$.

(ii) *break even* if $P(x) = 0$.

(iii) suffer a *loss* if $P(x) < 0$.

EXAMPLE 10   **Breaking Even**

Metro Entertainment Co. spent $100,000 on production costs for its off-Broadway play *Bride and Prejudice*. Once the play runs, each performance costs $1000 per show and the revenue from each show is $2400. Using x to represent the number of shows,

a. write the cost function $C(x)$.

b. write the revenue function $R(x)$.

c. write the profit function $P(x)$.

d. determine how many showings of *Bride and Prejudice* must be held for Metro to break even.

SOLUTION

a. $C(x) = $ (Variable cost) + (Fixed cost)

$\qquad = 1000x + 100{,}000$

b. $R(x) = $ (Revenue per show)(Number of shows)

$\qquad = 2400x$

c. $P(x) = R(x) - C(x)$

$\qquad = 2400x - (1000x + 100{,}000)$

$\qquad = 2400x - 1000x - 100{,}000 = 1400x - 100{,}000$

d. Metro will break even when $P(x) = 0$.

$$1400x - 100{,}000 = 0$$

$$1400x = 100{,}000$$

$$x = \frac{100{,}000}{1400} \approx 71.429 \qquad \text{Divide both sides by 1400.}$$

Only a whole number of shows is possible, so Metro needs 72 shows to break even.   ■■■

Practice Problem 10   Suppose in Example 10 that once the play runs, each performance costs $1200 and the revenue from each show is $2500. How do the results in Example 10 change?   ■

SECTION 2.4 ■ Exercises

A EXERCISES Basic Skills and Concepts

1. In the functional notation $y = f(x)$, x is the _____ variable.

2. If $f(-2) = 7$, then -2 is in the _____ of the function f and 7 is in the _____ of f.

3. If the point $(9, -14)$ is on the graph of a function f, then $f(9) = $ _____.

4. If $(3, 7)$ and $(3, 0)$ are points on a graph, then the graph cannot be the graph of a(n) _____.

5. To find the x-intercepts of the graph of an equation in x and y, we solve the equation _____.

6. *True or False* If a and b are in the domain of $f(x) = \dfrac{1}{x}$, then $a + b$ is also in the domain of f.

7. *True or False* If $x = -7$, then $-x$ is in the domain of $f(x) = \sqrt{x}$.

8. *True or False* The domain of $f(x) = \dfrac{1}{\sqrt{x + 2}}$ is all real x, $x \neq -2$.

In Exercises 9–14, determine the domain and range of each relation. Explain why each non-function is not a function.

9.

10.

11.

12.

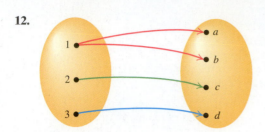

13.

x	0	3	8	0	3	8
y	−1	−2	−3	1	2	2

14.

x	−3	−1	0	1	2	3
y	−8	0	1	0	−3	−8

In Exercises 15–28, determine whether each equation defines *y* as a function of *x*.

15. $x + y = 2$

16. $x = y - 1$

17. $y = \dfrac{1}{x}$

18. $xy = -1$

19. $y^2 = x^2$

20. $x = |y|$

21. $y = \dfrac{1}{\sqrt{2x - 5}}$

22. $y = \dfrac{1}{\sqrt{x^2 - 1}}$

23. $2 - y = 3x$

24. $3x - 5y = 15$

25. $x + y^2 = 8$

26. $x = y^2$

27. $x^2 + y^3 = 5$

28. $x + y^3 = 8$

In Exercises 29–32, let $f(x) = x^2 - 3x + 1$, $g(x) = \dfrac{2}{\sqrt{x}}$, and $h(x) = \sqrt{2 - x}$.

29. Find $f(0), g(0), h(0), f(a)$, and $f(-x)$.

30. Find $f(1), g(1), h(1), g(a)$, and $g(x^2)$.

31. Find $f(-1), g(-1), h(-1), h(c)$, and $h(-x)$.

32. Find $f(4), g(4), h(4), g(2 + k)$, and $f(a + k)$.

33. Let $f(x) = \dfrac{2x}{\sqrt{4 - x^2}}$. Find each function value.

  a. $f(0)$ **b.** $f(1)$

  c. $f(2)$ **d.** $f(-2)$

  e. $f(-x)$

34. Let $g(x) = 2x + \sqrt{x^2 - 4}$. Find each function value.

  a. $g(0)$ **b.** $g(1)$

  c. $g(2)$ **d.** $g(-3)$

  e. $g(-x)$

In Exercises 35–48, find the domain of each function.

35. $f(x) = -8x + 7$

36. $f(x) = 2x^2 - 11$

37. $f(x) = \dfrac{1}{x - 9}$

38. $f(x) = \dfrac{1}{x + 9}$

39. $h(x) = \dfrac{2x}{x^2 - 1}$

40. $h(x) = \dfrac{x - 3}{x^2 - 4}$

41. $G(x) = \dfrac{\sqrt{x - 3}}{x + 2}$

42. $G(x) = \dfrac{\sqrt{x + 3}}{1 - x}$

43. $f(x) = \dfrac{3}{\sqrt{4 - x}}$

44. $f(x) = \dfrac{x}{\sqrt{2 - x}}$

45. $F(x) = \dfrac{x + 4}{x^2 + 3x + 2}$

46. $F(x) = \dfrac{1 - x}{x^2 + 5x + 6}$

47. $g(x) = \dfrac{\sqrt{x^2 + 1}}{x}$

48. $g(x) = \dfrac{1}{x^2 + 1}$

In Exercises 49–54, use the vertical-line test to determine whether the given graph represents a function.

49.

50.

51.

52.

53.

54.

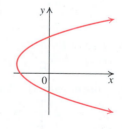

In Exercises 55–58, the graph of a function is given. Find the indicated function values.

55. $f(-4), f(-1), f(3), f(5)$

56. $g(-2), g(1), g(3), g(4)$

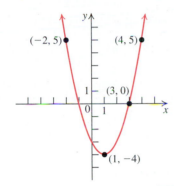

57. $h(-2), h(-1), h(0), h(1)$

58. $f(-1), f(0), f(1)$

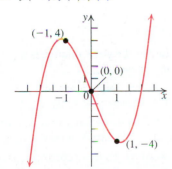

59. Let $h(x) = x^2 - x + 1$. Find x such that $(x, 7)$ is on the graph of h.

60. Let $H(x) = x^2 + x + 8$. Find x such that $(x, 7)$ is on the graph of H.

61. Let $f(x) = -2(x + 1)^2 + 7$.
   a. Is $(1, 1)$ a point of the graph of f?
   b. Find all x such that $(x, 1)$ is on the graph of f.
   c. Find all y-intercepts of the graph of f.
   d. Find all x-intercepts of the graph of f.

62. Let $f(x) = -3x^2 - 12x$.
   a. Is $(-2, 10)$ a point of the graph of f?
   b. Find x such that $(x, 12)$ is on the graph of g.
   c. Find all y-intercepts of the graph of f.
   d. Find all x-intercepts of the graph of f.

In Exercises 63–70, find the domain and the range of each function from its graph. The axes are marked in one-unit intervals.

63.

64.

65.

66.

67.

68.

69.

70.

For Exercises 71–78, refer to the graph of $y = f(x)$ **in the figure. The axes are marked off in one-unit intervals.**

71. Find the domain of f.

72. Find the range of f.

73. Find the x-intercepts of f.

74. Find the y-intercepts of f.

75. Find $f(-7)$, $f(1)$, and $f(5)$.

76. Find $f(-4)$, $f(-1)$, and $f(3)$.

77. Solve the equation $f(x) = 3$.

78. Solve the equation $f(x) = -2$.

In Exercises 79–84, refer to the graph of $y = g(x)$ **in the figure. The axes are marked off in one-unit intervals.**

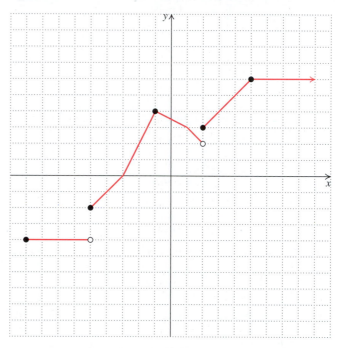

79. Find the domain of g.

80. Find the range of g.

81. Find $g(-4)$, $g(1)$, and $g(3)$.

82. Find $|g(-5) - g(5)|$.

83. Solve $g(x) = -4$.

84. Solve $g(x) = 6$.

B EXERCISES Applying the Concepts

In Exercises 85–88, state whether the given relation is a function and explain why or why not.

85. To each day of the year there corresponds the high temperature in your hometown on that day.

86. To each year since 1950 there corresponds the cost of a first-class postage stamp on January 1 of that year.

87. To each letter of the word *NUTS* there corresponds the states whose names start with that letter.

88. To each day of the week there corresponds the first names of people currently living in the United States born on that day of the week.

In Exercises 89 and 90, use the table listing the Super Bowl winners for the years 1995–2009.

Year	Super Bowl Winner
1995	San Francisco
1996	Dallas
1997	Green Bay
1998	Denver
1999	Denver
2000	St. Louis
2001	Baltimore
2002	New England
2003	Tampa Bay
2004	New England
2005	New England
2006	Pittsburgh
2007	Indianapolis
2008	New York Giants
2009	Pittsburgh

Source: www.superbowlhistory.net

89. Consider the correspondence whose domain values are the years 1995–2009 and whose range values are the corresponding winning teams. Is this a function? Explain your answer.

90. Consider the correspondence whose domain values are the names of the winning teams in the table and whose range values are the corresponding years. Is this a function? Explain your answer.

91. Square tiles. The area $A(x)$ of a square tile is a function of the length x of the square's side. Write a function rule for the area of a square tile. Find and interpret $A(4)$.

92. Cube. The volume $V(x)$ of a cube is a function of the length x of the cube's edge. Write a function rule for the volume of a cube. Evaluate the function for a cube with sides of length 3 inches.

93. Surface area. Is the total surface area S of a cube a function of the edge length x of the cube? If it is not a function, explain why not. If it is a function, write the function rule $S(x)$ and evaluate $S(3)$.

94. Measurement. One meter equals about 39.37 inches. Write a function rule for converting inches to meters. Evaluate the function for 59 inches.

95. Price–demand. Analysts for an electronics company determined that the price–demand function for its 30-inch LCD televisions is

$$p(x) = 1275 - 25x, \quad 1 \le x \le 30,$$

where p is the wholesale price per TV in dollars and x, in thousands, is the number of TVs that can be sold.
  a. Find and interpret $p(5)$, $p(15)$, and $p(30)$.
  b. Sketch the graph of $y = p(x)$.
  c. Solve the equation $p(x) = 650$ and interpret your result.

96. Revenue.
  a. Using the price–demand function $p(x) = 1275 - 25x, 1 \le x \le 30$, of Exercise 95, write the company's revenue function $R(x)$ and state its domain.
  b. Find and interpret $R(1)$, $R(5)$, $R(10)$, $R(15)$, $R(20)$, $R(25)$, and $R(30)$.
  c. Using the data from part (b), sketch the graph of $y = R(x)$.
  d. Solve the equation $R(x) = 4700$ and interpret your result.

97. Breaking even. Peerless Publishing Company intends to publish Harriet Henrita's next novel. The estimated cost is $75,000 plus $5.50 for each copy printed. The selling price per copy is $15. The bookstores retain 40% of the selling price as commission. Let x represent the number of copies printed and sold.
  a. Find the cost function $C(x)$.
  b. Find the revenue function $R(x)$.
  c. Find the profit function $P(x)$.
  d. How many copies of the novel must be sold for Peerless to break even?
  e. What is the company profit if 46,000 copies are sold?

98. Breaking even. Capital Records Company plans to produce a CD by the popular rapper Rapit. The fixed cost is $500,000, and the variable cost is $0.50 per CD. The company sells each CD to record stores for $5. Let x represent the number of CDs produced and sold.
  a. How many disks must be sold for the company to break even?
  b. How many CDs must be sold for the company to make a profit of $750,000?

99. Motion of a projectile. A stone thrown upward with an initial velocity of 128 feet per second will attain a height of h feet in t seconds, where

$$h(t) = 128t - 16t^2, \quad 0 \le t \le 8.$$

  a. What is the domain of h?
  b. Find $h(2)$, $h(4)$, and $h(6)$.
  c. How long will it take the stone to hit the ground?
  d. Sketch a graph of $y = h(t)$.

100. Drug prescription. A certain drug has been prescribed to treat an infection. The patient receives an injection of 16 milliliters of the drug every four hours. During this same four-hour period, the kidneys filter out one-fourth of the drug. Find the concentration of the drug after 4 hours, 8 hours, 12 hours, 16 hours, and 20 hours. Sketch the graph of the concentration of the drug in the bloodstream as a function of time.

101. **Drug prescription.** Every day Jane takes one 500-milligram aspirin tablet that has 80% bioavailability. During the same day, three-fourths of the aspirin is metabolized. Find the maximum concentration of the aspirin in the bloodstream on each of the first ten days of using the drug and graph the result.

C EXERCISES Beyond the Basics

102. Let $f(x) = 2x + 3$. Choose numerical values of a and b to verify each statement.

   i. $f(a + b) \neq f(a) + f(b)$

   ii. $f(ab) \neq f(a)f(b)$

   iii. $f\left(\dfrac{1}{a}\right) \neq \dfrac{1}{f(a)}$

   iv. $f\left(\dfrac{a}{b}\right) \neq \dfrac{f(a)}{f(b)}$

103. Explain whether $f(x)$ and $g(x)$ are equal.

   a. $f(x) = 2(\sqrt{x})^2 + 5,$ $g(x) = 2x + 5$

   b. $f(x) = 2(\sqrt[3]{x})^3 + 5,$ $g(x) = 2x + 5$

In Exercises 104–109, find an expression for $f(x)$ by solving for y and replacing y with $f(x)$. Then find the domain of f and compute $f(4)$.

104. $3x - 5y = 15$

105. $x = \dfrac{y}{y - 1}$

106. $x = \dfrac{2}{y - 4}$

107. $xy - 3 = 2y$

108. $(x^2 + 1)y + x = 2$

109. $yx^2 - \sqrt{x} = -2y$

In Exercises 110–115, state whether f and g represent the same function. Explain your reasons.

110. $f(x) = 3x - 4, 0 \leq x \leq 8$ $g(x) = 3x - 4$

111. $f(x) = 3x^2$ $g(x) = 3x^2, -5 \leq x \leq 5$

112. $f(x) = x - 1$ $g(x) = \dfrac{x^2 - 1}{x + 1}$

113. $f(x) = \dfrac{x + 2}{x^2 - x - 6}$ $g(x) = \dfrac{1}{x - 3}$

114. $f(x) = \dfrac{x^2 - 4}{x - 2}, 2 < x < \infty$

   $g(x) = x + 2, 2 < x < \infty$

115. $f(x) = \dfrac{x - 2}{x^2 - 6x + 8}, \quad 2 < x < \infty$

   $g(x) = \dfrac{1}{x - 4}, \quad 2 < x < \infty$

116. Let $f(x) = ax^2 + ax - 3$. Assuming that $f(2) = 15$, find a.

117. Let $g(x) = x^2 + bx + b^2$. Assuming that $g(6) = 28$, find b.

118. Let $h(x) = \dfrac{3x + 2a}{2x - b}$. Assuming that $h(6) = 0$ and $h(3)$ is undefined, find a and b.

119. Let $f(x) = 2x - 3$. Find $f(x^2)$ and $[f(x)]^2$.

120. Assuming that $g(x) = x^2 - \dfrac{1}{x^2}$, show that

$$g(x) + g\left(\dfrac{1}{x}\right) = 0.$$

121. Assuming that $f(x) = \dfrac{x - 1}{x + 1}$, show that

$$f\left(\dfrac{x - 1}{x + 1}\right) = -\dfrac{1}{x}.$$

122. Assuming that $f(x) = \dfrac{x + 3}{4x - 5}$ and $t = \dfrac{3 + 5x}{4x - 1}$, show that $f(t) = x$.

Critical Thinking

123. Write an equation of a function with each domain. Answers will vary.

   a. $[2, \infty)$

   b. $(2, \infty)$

   c. $(-\infty, 2]$

   d. $(-\infty, 2)$

124. Consider the graph of the function

$$y = f(x) = ax^2 + bx + c, a \neq 0.$$

   a. Write an equation whose solution yields the x-intercepts.

   b. Write an equation whose solution is the y-intercept.

   c. Write (if possible) a condition under which the graph of $y = f(x)$ has no x-intercepts.

   d. Write (if possible) a condition under which the graph of $y = f(x)$ has no y-intercepts.

Properties of Functions

Before Starting this Section, Review

1. Graph of a function (Section 2.4, page 215)

2. Evaluating a function (Section 2.4, page 214)

Objectives

1 Determine whether a function is increasing or decreasing on an interval.

2 Use a graph to locate relative maximum and minimum values.

3 Identify even and odd functions.

4 Find the average rate of change of a function.

ALGEBRA IN COUGHING

Coughing is the body's way of removing foreign material or mucus from the lungs and upper airway passages or reacting to an irritated airway. When you cough, the windpipe contracts to increase the velocity of the outgoing air.

The average flow velocity, v, can be modeled by the equation

$$v = c(r_0 - r)r^2 \text{ mm/sec,}$$

where r_0 is the rest radius of the windpipe in millimeters, r is its contracted radius, and c is a positive constant. In Example 3, we find the value of r that maximizes the velocity of the air going out. The diameter of the windpipe of a normal adult male varies between 25 mm and 27 mm, while that of a normal adult female varies between 21 mm and 23 mm. ■

---

1 Determine whether a function is increasing or decreasing on an interval.

Increasing and Decreasing Functions

Classifying functions according to how one variable changes with respect to the other variable can be very useful. Imagine a particle moving from left to right along the graph of a function. This means that the x-coordinate of the point on the graph is getting larger. If the corresponding y-coordinate of the point is getting larger, getting smaller, or staying the same, then the function is called *increasing, decreasing,* or *constant*, respectively.

INCREASING, DECREASING, AND CONSTANT FUNCTIONS

Let f be a function and let x_1 and x_2 be any two numbers in an open interval (a, b) contained in the domain of f. See Figure 2.50. The symbols a and b may represent real numbers, $-\infty$, or ∞. Then

(i) f is called an **increasing function** on (a, b) if $x_1 < x_2$ implies $f(x_1) < f(x_2)$.

(ii) f is called a **decreasing function** on (a, b) if $x_1 < x_2$ implies $f(x_1) > f(x_2)$.

(iii) f is **constant** on (a, b) if $x_1 < x_2$ implies $f(x_1) = f(x_2)$.

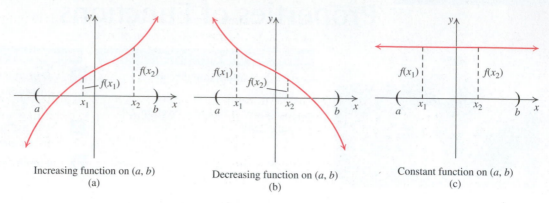

Increasing function on (a, b)
(a)

Decreasing function on (a, b)
(b)

Constant function on (a, b)
(c)

FIGURE 2.50 **Increasing, decreasing, and constant functions**

FIGURE 2.51 **The function $f(t)$ increases on $(0, 3)$ and decreases on $(3, 6)$.**

Geometrically, the graph of an increasing function on an open interval (a, b) rises as x-values increase on (a, b). This is because, by definition, as x-values increase from x_1 to x_2, the y-values also increase, from $f(x_1)$ to $f(x_2)$. Similarly, the graph of a decreasing function on (a, b) falls as x-values increase. The graph of a constant function is horizontal, or flat, over the interval (a, b).

Not every function always increases or decreases. The function that assigns the height of a ball tossed in the air to the length of time it is in the air is an example of a function that increases, but also decreases, over different intervals of its domain. See Figure 2.51.

EXAMPLE 1   **Tracking the Behavior of a Function**

From the graph of the function g in Figure 2.52(a), find the intervals over which g is increasing, is decreasing, or is constant.

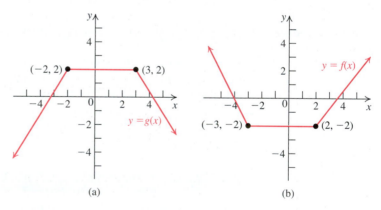

(a)

(b)

FIGURE 2.52

STUDY TIP

When specifying the intervals over which a function f is increasing, decreasing, or constant, you need to use the intervals in the *domain* of f, *not* in the range of f.

SOLUTION

The function g is increasing on the interval $(-\infty, -2)$. It is constant on the interval $(-2, 3)$. It is decreasing on the interval $(3, \infty)$. ■ ■ ■

Practice Problem 1   Using the graph in Figure 2.52(b), find the intervals over which f is increasing, is decreasing, or is constant.   ■

2 Use a graph to locate relative maximum and minimum values.

Relative Maximum and Minimum Values

The y-coordinate of a point that is higher (lower) than any nearby point on a graph is called a *relative maximum* (*relative minimum*) value of the function. These points are called the *relative maximum* (*relative minimum*) points of the graph. Relative maximum and minimum values are the y-coordinates of points corresponding to the peaks and troughs, respectively, of the graph.

DEFINITION OF RELATIVE MAXIMUM AND RELATIVE MINIMUM

If a is in the domain of a function f, we say that the value $f(a)$ is a **relative maximum of f** if there is an interval (x_1, x_2) containing a such that

$$f(a) \geq f(x) \text{ for every } x \text{ in the interval } (x_1, x_2).$$

We say that the value $f(a)$ is a **relative minimum of f** if there is an interval (x_1, x_2) containing a such that

$$f(a) \leq f(x) \text{ for every } x \text{ in the interval } (x_1, x_2).$$

The value $f(a)$ is called an **extreme value** of f if it is either a relative maximum value or a relative minimum value.

The graph of a function can help you find or approximate maximum or minimum values (if any) of the function. For example, the graph of a function f in Figure 2.53 shows that

Points whose y-coordinates are relative maximum values

Points whose y-coordinates are relative minimum values

FIGURE 2.53 Relative maximum and minimum values

- f has two relative maxima:
    (i) at $x = 1$ with relative maximum value $f(1) = 3$
    (ii) at $x = 4$ with relative maximum value $f(4) = 7$
- f has two relative minima:
    (i) at $x = 2$ with relative minimum value $f(2) = 1$
    (ii) at $x = 6$ with relative minimum value $f(6) = -2$

In Figure 2.53, each of the points $(1, 3)$, $(2, 1)$, $(4, 7)$, and $(6, -2)$ is called a **turning point**. At a turning point, a graph changes direction from increasing to decreasing or from decreasing to increasing. In calculus, you study techniques for finding the exact points at which a function has a relative maximum or a relative minimum value (if any). For the present, however, we settle for approximations of these points by using a graphing utility.

EXAMPLE 2   **Approximating Relative Extrema**

Use a graphing utility to approximate the relative maximum and the relative minimum point on the graph of the function $f(x) = x^3 - x^2$.

SOLUTION

The graph of f is shown in Figure 2.54. By using the TRACE and ZOOM features of a graphing utility, we see that the function f has a:

Relative minimum point estimated to be $(0.67, -0.15)$
Relative maximum point estimated to be $(0, 0)$. ■ ■ ■

FIGURE 2.54

Practice Problem 2   Repeat Example 2 for $g(x) = -2x^3 + 3x^2$. ■

EXAMPLE 3   **Algebra in Coughing**

The average flow velocity, v, of outgoing air through the windpipe is modeled by (see introduction to this section)

$$v = c(r_0 - r)r^2 \text{ mm/sec} \qquad \frac{r_0}{2} \leq r \leq r_0,$$

where r_0 is the rest radius of the windpipe, r is its contracted radius, and c is a positive constant. For Mr. Osborn, assume that $c = 1$ and $r_0 = 13$ mm. Use a graphing utility to estimate the value of r that will maximize the airflow v when Mr. Osborn coughs.

FIGURE 2.55 Flow velocity

3 Identify even and odd functions.

SOLUTION

The graph of v for $0 \leq r \leq 13$ is shown in Figure 2.55. By using the TRACE and ZOOM features, we see that the maximum point on the graph is estimated at $(8.67, 325)$. So Mr. Osborn's windpipe contracts to a radius of 8.67 mm to maximize the airflow velocity. ■■■

Practice Problem 3 Repeat Example 3 for Mrs. Osborn. Again, let $c = 1$ but $r_0 = 11$. ■

Even–Odd Functions and Symmetry

This topic, *even–odd functions*, uses ideas of symmetry discussed in Section 2.2.

EVEN FUNCTION

A function f is called an **even function** if for each x in the domain of f, $-x$ is also in the domain of f and

$$f(-x) = f(x).$$

The graph of an even function is symmetric about the y-axis.

EXAMPLE 4   **Graphing the Squaring Function**

Show that the squaring function $f(x) = x^2$ is an even function and sketch its graph.

SOLUTION

The function $f(x) = x^2$ is even because

$$
\begin{aligned}
f(-x) &= (-x)^2 & &\text{Replace } x \text{ with } -x. \\
&= x^2 & &(-x)^2 = x^2 \\
&= f(x) & &\text{Replace } x^2 \text{ with } f(x).
\end{aligned}
$$

To sketch the graph of $f(x) = x^2$, we make a table of values. See Table 2.7(a). We then plot the points that are in the first quadrant and join them with a smooth curve. Next, we use y-axis symmetry to find additional points on the graph of f in the second quadrant, which are listed in Table 2.7(b). See Figure 2.56.

TABLE 2.7(a)

x	0	$\dfrac{1}{2}$	1	$\dfrac{3}{2}$	2	$\dfrac{5}{2}$	3
$f(x) = x^2$	0	$\dfrac{1}{4}$	1	$\dfrac{9}{4}$	4	$\dfrac{25}{4}$	9

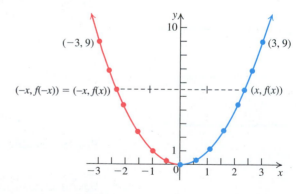

FIGURE 2.56 Graph of $f(x) = x^2$ ■■■

TABLE 2.7(b)

x	0	$-\dfrac{1}{2}$	-1	$-\dfrac{3}{2}$	-2	$-\dfrac{5}{2}$	-3
$f(x) = x^2$	0	$\dfrac{1}{4}$	1	$\dfrac{9}{4}$	4	$\dfrac{25}{4}$	9

■ ■ ■

Practice Problem 4   Show that the function $f(x) = -x^2$ is an even function and sketch its graph.   ■

ODD FUNCTION

A function f is an **odd function** if for each x in the domain of f, $-x$ is also in the domain of f and

$$f(-x) = -f(x).$$

The graph of an odd function is symmetric about the origin.

EXAMPLE 5   **Graphing the Cubing Function**

Show that the cubing function defined by $g(x) = x^3$ is an odd function and sketch its graph.

SOLUTION

The function $g(x) = x^3$ is an odd function because

$$\begin{aligned}
g(-x) &= (-x)^3 &&\text{Replace } x \text{ with } -x. \\
&= -x^3 &&(-x)^3 = -x^3 \\
&= -g(x) &&\text{Replace } x^3 \text{ with } g(x).
\end{aligned}$$

To sketch the graph of $g(x) = x^3$, we plot points in the first quadrant and then use symmetry about the origin to extend the graph to the third quadrant. The graph is shown in Figure 2.57.

FIGURE 2.57 Graph of $g(x) = x^3$ ■ ■ ■

Practice Problem 5   Show that the function $f(x) = -x^3$ is an odd function and sketch its graph.   ■

The function $h(x) = 2x + 3$, whose graph is sketched in Figure 2.58, is neither even nor odd, and its graph is neither symmetric with respect to the y-axis nor symmetric with respect to the origin. To see this algebraically, notice that $h(-x) = 2(-x) + 3 = -2x + 3$, which is not equivalent to $h(x)$ or $-h(x)$ (because $-h(x) = -2x - 3$).

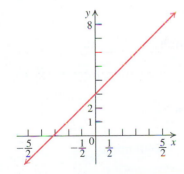

FIGURE 2.58 Graph of $h(x) = 2x + 3$

4 Find the average rate of change of a function.

Average Rate of Change

Suppose your salary increases from $25,000 a year to $40,000 a year over a five-year period. You then have

Change in salary: $40,000 − $25,000 = $15,000

Average rate of change in salary:

$$\frac{\$40,000 - \$25,000}{5} = \frac{\$15,000}{5} = \$3000 \text{ per year.}$$

Regardless of when you actually received the raises during the five-year period, the final salary is the same as if you received your average annual increase of $3000 *each* year.

The average rate of change can be defined in a more general setting.

AVERAGE RATE OF CHANGE OF A FUNCTION

Let $(a, f(a))$ and $(b, f(b))$ be two points on the graph of a function f. Then the **average rate of change** of $f(x)$ as x changes from a to b is defined by

$$\frac{f(b) - f(a)}{b - a}, a \neq b.$$

Note that the average rate of change of f as x changes from a to b is the slope of the line passing through the two points $(a, f(a))$ and $(b, f(b))$ on the graph of f.

FINDING THE SOLUTION: A PROCEDURE

EXAMPLE 6   Finding the Average Rate of Change

OBJECTIVE	EXAMPLE
Find the average rate of change of a function f as x changes from a to b.	*Find the average rate of change of $f(x) = 2 - 3x^2$ as x changes from $x = 1$ to $x = 3$.*
Step 1 Find $f(a)$ and $f(b)$.	1. $f(1) = 2 - 3(1)^2 = -1$ $a = 1$ $f(3) = 2 - 3(3)^2 = -25$ $b = 3$
Step 2 Use the values from Step 1 in the definition of average rate of change.	2. $\dfrac{f(b) - f(a)}{b - a} = \dfrac{-25 - (-1)}{3 - 1} = -\dfrac{24}{2}$ $= -12$ ■ ■ ■

Practice Problem 6 Find the average rate of change of $f(x) = 1 - x^2$ as x changes from 2 to 4. ■

EXAMPLE 7   Finding the Average Rate of Change

Find the average rate of change of $f(t) = 2t^2 - 3$ as t changes from $t = 5$ to $t = x, x \neq 5$.

SOLUTION

$$\text{Average rate of change} = \frac{f(b) - f(a)}{b - a} \qquad \text{Definition}$$

$$= \frac{f(x) - f(5)}{x - 5} \qquad b = x, a = 5$$

$$\text{Average rate of change} = \frac{(2x^2 - 3) - (2 \cdot 5^2 - 3)}{x - 5} \qquad \text{Substitute for } f(x) \text{ and } f(5).$$

$$= \frac{2x^2 - 3 - 2 \cdot 5^2 + 3}{x - 5}$$

$$= \frac{2(x^2 - 5^2)}{x - 5} \qquad \text{Simplify.}$$

$$= \frac{2(x + 5)(x - 5)}{x - 5} \qquad \text{Difference of squares}$$

$$= 2x + 10 \qquad \text{Simplify; } x - 5 \neq 0.$$

■ ■ ■

Practice Problem 7 Find the average rate of change of $f(t) = 1 - t$ as t changes from $t = 2$ to $t = x$, $x \neq 2$. ■

The average rate of change calculated in Example 7 is a *difference quotient* at 5. The difference quotient is an important concept in calculus.

DIFFERENCE QUOTIENT

For a function f, the **difference quotient** is

$$\frac{f(x + h) - f(x)}{h}, \quad h \neq 0.$$

EXAMPLE 8   Evaluating and Simplifying a Difference Quotient

Let $f(x) = 2x^2 - 3x + 5$. Find and simplify $\dfrac{f(x + h) - f(x)}{h}, h \neq 0$.

SOLUTION
First, we find $f(x + h)$.

$$f(x) = 2x^2 - 3x + 5 \qquad \text{Original equation}$$
$$f(x + h) = 2(x + h)^2 - 3(x + h) + 5 \qquad \text{Replace } x \text{ with } (x + h).$$
$$= 2(x^2 + 2xh + h^2) - 3(x + h) + 5 \qquad (x + h)^2 = x^2 + 2xh + h^2$$
$$= 2x^2 + 4xh + 2h^2 - 3x - 3h + 5 \qquad \text{Use the distributive property.}$$

Then we substitute into the difference quotient.

$$\frac{f(x + h) - f(x)}{h} = \frac{\overbrace{(2x^2 + 4xh + 2h^2 - 3x - 3h + 5)}^{f(x+h)} - \overbrace{(2x^2 - 3x + 5)}^{f(x)}}{h}$$

$$= \frac{2x^2 + 4xh + 2h^2 - 3x - 3h + 5 - 2x^2 + 3x - 5}{h} \qquad \text{Subtract.}$$

$$= \frac{4xh - 3h + 2h^2}{h} \qquad \text{Simplify.}$$

$$= \frac{h(4x - 3 + 2h)}{h} \qquad \text{Factor out } h.$$

$$= 4x - 3 + 2h \qquad \text{Remove the common factor } h. \qquad ■ ■ ■$$

Practice Problem 8 Repeat Example 8 for $f(x) = -x^2 + x - 3$. ■

SECTION 2.5 ■ Exercises

A EXERCISES Basic Skills and Concepts

1. A function f is decreasing if $x_1 < x_2$ implies that
   $f(x_1)$ _____ $f(x_2)$.

2. $f(a)$ is a relative maximum of f if there is an interval
   (x_1, x_2) containing a such that _____ for every
   x in the interval (x_1, x_2).

3. A function f is even if _____.

4. The average rate of change of f as x changes from
   $x = a$ to $x = b$ is _____, $a \neq b$.

5. *True or False* The average rate of change of f as x
   changes from a to b is the slope of the line through the
   points $(a, f(a))$ and $(b, f(b))$.

6. *True or False* A relative maximum or a relative
   minimum point on a graph is a turning point on the
   graph of a function.

**In Exercises 7–14, the graph of a function is given. For each
function, determine the intervals over which the function is
increasing, decreasing, or constant.**

7.

8.

9.

10.

11.

12.

13.

14.
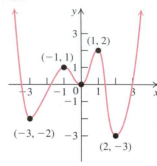

**In Exercises 15–22, locate relative maximum and relative
minimum points on the graph. State whether each relative
extremum point is a turning point.**

15. The graph of Exercise 7

16. The graph of Exercise 8

17. The graph of Exercise 9

18. The graph of Exercise 10

19. The graph of Exercise 11

20. The graph of Exercise 12

21. The graph of Exercise 13

22. The graph of Exercise 14

**In Exercises 23–32, the graph of a function is given. State
whether the function is odd, even, or neither.**

23.

24.

25.

26.

27.

28.

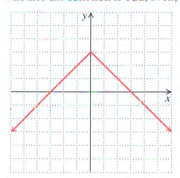

d. The relative extrema points, if any
e. Whether the function is odd, even, or neither

47.

29.

30.

48.

31.

32.

49.

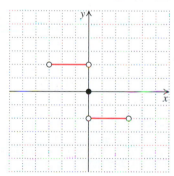

In Exercises 33–46, determine algebraically whether the given function is odd, even, or neither.

33. $f(x) = 2x^4 + 4$ **34.** $g(x) = 3x^4 - 5$

35. $f(x) = 5x^3 - 3x$ **36.** $g(x) = 2x^2 + 4x$

37. $f(x) = 2x + 4$ **38.** $g(x) = 3x + 7$

39. $f(x) = \dfrac{1}{x^2 + 4}$ **40.** $g(x) = \dfrac{x^2 + 2}{x^4 + 1}$

41. $f(x) = \dfrac{x^3}{x^2 + 1}$ **42.** $g(x) = \dfrac{x^4 + 3}{2x^3 - 3x}$

43. $f(x) = \dfrac{x}{x^5 - 3x^3}$ **44.** $g(x) = \dfrac{x^3 + 2x}{2x^5 - 3x}$

45. $f(x) = \dfrac{x^2 - 2x}{5x^4 + 7}$ **46.** $g(x) = \dfrac{3x^2 + 7}{x - 3}$

50.

In Exercises 47–54, the graph of a function is given; each grid line represents one unit. Use the graph to find each of the following.

a. The domain and the range of the function
b. The intercepts, if any
c. The intervals on which the function is increasing, decreasing, or constant

51.

52.

53.

54.

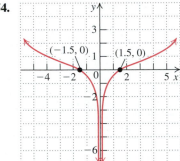

In Exercises 55–66, find the average rate of change of the function as x changes from a to b.

55. $f(x) = -2x + 7; a = -1, b = 3$

56. $f(x) = 4x - 9; a = -2, b = 2$

57. $g(x) = 2x^2; a = 0, b = 5$

58. $g(x) = -4x^2; a = -1, b = 4$

59. $h(x) = x^2 - 1; a = -2, b = 0$

60. $h(x) = 2 - x^2; a = 3, b = 4$

61. $f(x) = (3 - x)^2; a = 1, b = 3$

62. $f(x) = (x - 2)^2; a = -1, b = 5$

63. $g(x) = x^3; a = -1, b = 3$

64. $g(x) = -x^3; a = -1, b = 3$

65. $h(x) = \dfrac{1}{x}; a = 2, b = 6$

66. $h(x) = \dfrac{4}{x + 3}; a = -2, b = 4$

In Exercises 67–74, find and simplify the difference quotient of the form $\dfrac{f(x) - f(a)}{x - a}, x \neq a$.

67. $f(x) = 2x, a = 3$

68. $f(x) = 3x + 2, a = 2$

69. $f(x) = -x^2, a = 1$

70. $f(x) = 2x^2, a = -1$

71. $f(x) = 3x^2 + x, a = 2$

72. $f(x) = -2x^2 + x, a = 3$

73. $f(x) = \dfrac{4}{x}, a = 1$

74. $f(x) = -\dfrac{4}{x}, a = 1$

In Exercises 75–84, find and simplify the difference quotient of the form $\dfrac{f(x + h) - f(x)}{h}, h \neq 0$.

75. $f(x) = x$

76. $f(x) = 3x + 2$

77. $f(x) = x^2$

78. $f(x) = x^2 - x$

79. $f(x) = 2x^2 + 3x$

80. $f(x) = 3x^2 - 2x + 5$

81. $f(x) = 4$

82. $f(x) = -3$

83. $f(x) = \dfrac{1}{x}$

84. $f(x) = -\dfrac{1}{x}$

B EXERCISES   Applying the Concepts

Motion from a graph. For Exercises 85–94, use the accompanying graph of a particle moving on a coordinate line with velocity $v = f(t)$ in ft/sec at time t seconds. The axes are marked off at one-unit intervals. Use these terms to describe the motion state: moving forward/backward, increasing/decreasing speed, and resting. Recall that speed = |velocity|.

85. Find the domain of $v = f(t)$. Interpret.

86. Find the range of $v = f(t)$. Interpret.

87. Find the intervals over which the graph is above the t-axis. Interpret.

88. Find the intervals over which the graph is below the t-axis. Interpret.

89. Find the intervals over which $|v| = |f(t)|$ is increasing. Interpret for $v > 0$ and for $v < 0$.

90. Find the intervals over which $|v| = |f(t)|$ is decreasing. Interpret for $v > 0$ and for $v < 0$.

91. When does the particle have maximum speed?

92. When does the particle have minimum speed?

93. Describe the motion of the particle between $t = 5$ and $t = 6$.

94. Describe the motion of the particle between $t = 16$ and $t = 19$.

 95. Maximizing volume. An open cardboard box is to be made from a piece of 12-inch square cardboard by cutting equal squares from the four corners and turning up the sides. Let x be the length of the side of the square to be cut out.
  a. Show that the volume, V, of the box is given by

$$V = 4x^3 - 48x^2 + 144x.$$

  b. Find the domain of V.
  c. Use a graphing utility to graph the volume function and determine the range of V from its graph.
  d. Find the value of x for which V has a maximum value.

 96. Maximizing volume. Repeat Exercise 95 assuming that the box is to be made from a piece of cardboard with dimensions 9 in. by 15 in.

 97. Maximizing cost. The cost C of producing x toys, $2 \le x \le 10$ is given by

$$C = -x^2 + 12x + 20.$$

Use a graphing utility to graph the cost function and find the value of x for which C has a maximum value.

 98. Minimizing the sum. Find two positive numbers whose product is 32 and whose sum is as small as possible.
  a. Let one of the numbers be x, show that the sum

$$S = x + \frac{32}{x}.$$

  b. Use a graphing utility to graph the sum function and estimate the value of x for which S has a minimum value.

99. Cost. A computer notebook manufacturer has a daily fixed cost of $10,500 and a marginal cost of $210 per notebook.
  a. Find the daily cost $C(x)$ of manufacturing x notebooks per day.
  b. Find the cost of producing 50 notebooks per day.
  c. Find the average cost of a computer notebook assuming that 50 notebooks are manufactured each day.
  d. How many notebooks should be manufactured each day so that the average cost per notebook is $315?

100. Cost. The Just Juice Company has a monthly cost of $20,000 and a marginal cost of $4 per case of juice.
  a. Find the monthly cost $C(x)$ assuming that the company produces x cases of juice per month.
  b. Find the cost of producing 12,000 cases of juice per month.
  c. Find the average cost of a case of juice in part (b).
  d. How many cases of juice should be produced each month so that the average cost per case is $4.50?

C EXERCISES Beyond the Basics

101. Give an example (if possible) of a function matching each description. There are many different correct answers.
  a. Its graph is symmetric with respect to the y-axis.
  b. Its graph is symmetric with respect to the x-axis.
  c. Its graph is symmetric with respect to the origin.
  d. Its graph consists of a single point.
  e. Its graph is a horizontal line.
  f. Its graph is a vertical line.

102. What function is both odd and even?

103. Find algebraically the relative minimum point on the graph of $f(x) = (x - 2)^4 + 7$.

104. Find algebraically the relative minimum point on the graph of $g(x) = x^2 - 4x + 7$. [*Hint:* Complete the square.]

105. Find algebraically the relative maximum point on the graph of $f(x) = -(x + 1)^2 + 5$. Does its graph have a relative minimum point?

106. Give an example to illustrate that a relative maximum point on a graph need not be a turning point on the graph.

107. Give an example to illustrate that a relative minimum point on a graph need not be a turning point on the graph.

108. Show that every turning point on the graph of a function is either a relative maximum point or a relative minimum point.

Critical Thinking

Let f be a function with domain $[a, c]$ or $[c, b]$. Then a, b, and c are endpoints of this domain.

109. Define relative maximum at c.

110. Define relative minimum at c.

111. Give an example (if possible) of a function f matching each description.
  a. f has endpoint maximum and endpoint minimum.
  b. f has endpoint minimum but no endpoint maximum.
  c. f has endpoint maximum but no endpoint minimum.
  d. f has neither endpoint maximum nor endpoint minimum.

A Library of Functions

Before Starting this Section, Review

1. Equation of a line (Section 2.3, page 194)
2. Absolute value (Section P.1, page 8)
3. Graph of equation (Section 2.2, page 179)

Objectives

1. Relate linear functions to linear equations.
2. Graph square root and cube root functions.
3. Graph additional basic functions.
4. Evaluate and graph piecewise functions.

THE MEGATOOTH SHARK

The giant "Megatooth" shark (*Carcharodon megalodon*), once estimated to be 100 to 120 feet in length, is the largest meat-eating fish that ever lived. Its actual length has been the subject of scientific controversy. Sharks first appeared in the oceans more than 400 million years ago, almost 200 million years before the dinosaurs. A shark's skeleton is made of cartilage, which decomposes rather quickly, making complete shark fossils rare. Consequently, scientists rely on calcified vertebrae, fossilized teeth, and small skin scales to reconstruct the evolutionary record of sharks.

The great white shark is the closest living relative to the giant "Megatooth" shark and has been used as a model to reconstruct the "Megatooth." John Maisey, curator at the American Museum of Natural History in New York City, used a partial set of "Megatooth" teeth to make a comparison with the jaws of the great white shark. In Example 2, we use a formula to calculate the length of the "Megatooth" on the basis of the height of the tooth of the largest known "Megatooth" specimen. ■

---

1 Relate linear functions to linear equations.

Linear Functions

We know from Section 2.3 that the graph of a linear equation $y = mx + b$ is a straight line with slope m and y-intercept b. A *linear function* has a similar definition.

LINEAR FUNCTIONS

Let m and b be real numbers. The function $f(x) = mx + b$ is called a **linear function**. If $m = 0$, the function $f(x) = b$ is called a **constant function**. If $m = 1$ and $b = 0$, the resulting function $f(x) = x$ is called the **identity function**. See Figure 2.59.

The domain of a linear function is the interval $(-\infty, \infty)$ because the expression $mx + b$ is defined for any real number x. Figure 2.59 shows that the graph extends indefinitely to the left and right. The range of a nonconstant linear function also is the interval $(-\infty, \infty)$ because the graph extends indefinitely upward and downward. The range of a constant function $f(x) = b$ is the single real number b. See Figure 2.59(c).

GRAPHS OF $f(x) = mx + b$

(a)	(b)	(c)	(d)
$m > 0$	$m < 0$	$m = 0$ $f(x) = b$	$m = 1, b = 0$ $f(x) = x$

FIGURE 2.59 The graph of a linear function is a nonvertical line with slope m and y-intercept b.

EXAMPLE 1   Writing a Linear Function

Write a linear function g for which $g(1) = 4$ and $g(-3) = -2$.

SOLUTION

We need to find the equation of the line passing through the two points

$$(x_1, y_1) = (1, 4) \text{ and } (x_2, y_2) = (-3, -2).$$

The slope of the line is given by

$$m = \frac{y_2 - y_1}{x_2 - x_1} = \frac{-2 - 4}{-3 - 1} = \frac{-6}{-4} = \frac{3}{2}.$$

Next, we use the point–slope form of a line.

$$y - y_1 = m(x - x_1) \qquad \text{Point–slope form of a line}$$

$$y - 4 = \frac{3}{2}(x - 1) \qquad \text{Substitute } x_1 = 1, y_1 = 4, \text{ and } m = \frac{3}{2}.$$

$$y - 4 = \frac{3}{2}x - \frac{3}{2} \qquad \text{Distributive property}$$

$$y = \frac{3}{2}x + \frac{5}{2} \qquad \text{Add } 4 = \frac{8}{2} \text{ to both sides and simplify.}$$

$$g(x) = \frac{3}{2}x + \frac{5}{2} \qquad \text{Function notation}$$

The graph of the function $y = g(x)$ is shown in Figure 2.60.   ■ ■ ■

FIGURE 2.60 Point–slope form

Point on graph: $(1, 4)$, $g(x) = \frac{3}{2}x + \frac{5}{2}$, $(-3, -2)$

Practice Problem 1   Write a linear function g for which $g(-2) = 2$ and $g(1) = 8$.   ■

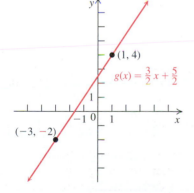

EXAMPLE 2   Determining the Length of the "Megatooth" Shark

The largest known "Megatooth" specimen is a tooth that has a total height of 15.6 centimeters. Calculate the length of the "Megatooth" shark from which it came by using the formula

$$\text{shark length} = (0.96)(\text{height of tooth}) - 0.22,$$

where shark length is measured in meters and tooth height is measured in centimeters.

SOLUTION

$$\text{Shark length} = (0.96)(\text{height of tooth}) - 0.22 \qquad \text{Formula}$$

$$\text{Shark length} = (0.96)(15.6) - 0.22 \qquad \text{Replace height of tooth with 15.6.}$$

$$= 14.756 \qquad \text{Simplify.}$$

The shark's length is approximately 14.8 meters ($\approx$48.6 feet).   ▪ ▪ ▪

Practice Problem 2   If the "Megatooth" specimen was a tooth measuring 16.4 cm, what was the length of the "Megatooth" shark from which it came?   ▪

2 Graph square root and cube root functions.

Square Root and Cube Root Functions

So far, we have learned to graph linear functions: $y = mx + b$; the squaring function: $y = x^2$; and the cubing function: $y = x^3$. We now add some common functions to this list.

Square Root Function   Recall (page 59) that for every nonnegative real number x, there is only one principal square root denoted by $\sqrt{x}$. Therefore, the equation $y = \sqrt{x}$ defines a function, called the **square root function**. The domain of the square root function is $[0, \infty)$.

EXAMPLE 3   **Graphing the Square Root Function**

Graph $f(x) = \sqrt{x}$.

SOLUTION

To sketch the graph of $f(x) = \sqrt{x}$, we make a table of values. For convenience, we have selected the values of x that are perfect squares. See Table 2.8. We then plot the ordered pairs (x, y) and draw a smooth curve through the plotted points. See Figure 2.61.

TABLE 2.8

x	$y = f(x)$	(x, y)
0	$\sqrt{0} = 0$	$(0, 0)$
1	$\sqrt{1} = 1$	$(1, 1)$
4	$\sqrt{4} = 2$	$(4, 2)$
9	$\sqrt{9} = 3$	$(9, 3)$
16	$\sqrt{16} = 4$	$(16, 4)$

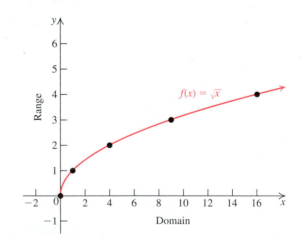

FIGURE 2.61 Graph of $y = \sqrt{x}$

The domain of $f(x) = \sqrt{x}$ is $[0, \infty)$, and its range also is $[0, \infty)$.   ▪ ▪ ▪

Practice Problem 3   Graph $g(x) = \sqrt{-x}$ and find its domain and range.   ▪

Cube Root Function

There is only one cube root for each real number x. Therefore, the equation $y = \sqrt[3]{x}$ defines a function, called the **cube root function**. The domain of the cube root function is $(-\infty, \infty)$.

EXAMPLE 4 **Graphing the Cube Root Function**

Graph $f(x) = \sqrt[3]{x}$.

SOLUTION

We use the point plotting method to graph $f(x) = \sqrt[3]{x}$. For convenience, we select values of x that are perfect cubes. See Table 2.9. The graph of $f(x) = \sqrt[3]{x}$ is shown in Figure 2.62.

TABLE 2.9

x	$y = f(x)$	(x, y)
-8	$\sqrt[3]{-8} = -2$	$(-8, -2)$
-1	$\sqrt[3]{-1} = -1$	$(-1, -1)$
0	$\sqrt[3]{0} = 0$	$(0, 0)$
1	$\sqrt[3]{1} = 1$	$(1, 1)$
8	$\sqrt[3]{8} = 2$	$(8, 2)$

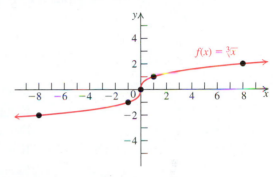

FIGURE 2.62 Graph of $y = \sqrt[3]{x}$

The domain of $f(x) = \sqrt[3]{x}$ is $(-\infty, \infty)$, and its range also is $(-\infty, \infty)$. ■ ■ ■

Practice Problem 4   Graph $g(x) = \sqrt[3]{-x}$ and find its domain and range. ■

3 Graph additional basic functions.

Basic Functions

As you progress through this course and future mathematics courses, you will repeatedly come across a small list of basic functions. The following box lists some of these common algebraic functions, along with their properties. You should try to produce these graphs by plotting points and using symmetries. The unit length in all of the graphs shown is the same on both axes.

A LIBRARY OF BASIC FUNCTIONS

Constant Function
$f(x) = c$

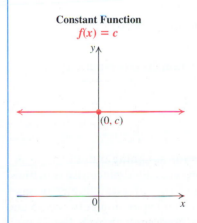

Domain: $(-\infty, \infty)$
Range: $\{c\}$
Constant on $(-\infty, \infty)$
Even function (y-axis symmetry)

Identity Function
$f(x) = x$

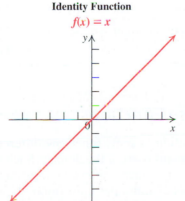

Domain: $(-\infty, \infty)$
Range: $(-\infty, \infty)$
Increasing on $(-\infty, \infty)$
Odd function (origin symmetry)

Squaring Function
$f(x) = x^2$

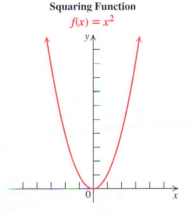

Domain: $(-\infty, \infty)$
Range: $[0, \infty)$
Decreasing on $(-\infty, 0)$
Increasing on $(0, \infty)$
Even function (y-axis symmetry)

Cubing Function

$$f(x) = x^3$$

Domain: $(-\infty, \infty)$
Range: $(-\infty, \infty)$
Increasing on $(-\infty, \infty)$
Odd function (origin symmetry)

Absolute Value Function

$$f(x) = |x|$$

Domain: $(-\infty, \infty)$
Range: $[0, \infty)$
Decreasing on $(-\infty, 0)$
Increasing on $(0, \infty)$
Even function (y-axis symmetry)

Square Root Function

$$f(x) = \sqrt{x} = x^{1/2}$$

Domain: $[0, \infty)$
Range: $[0, \infty)$
Increasing on $(0, \infty)$
Neither even nor odd (no symmetry)

Cube Root Function

$$f(x) = \sqrt[3]{x} = x^{1/3}$$

Domain: $(-\infty, \infty)$
Range: $(-\infty, \infty)$
Increasing on $(-\infty, \infty)$
Odd function (origin symmetry)

Reciprocal Function

$$f(x) = \frac{1}{x}$$

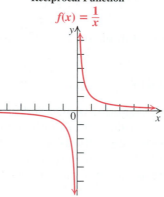

Domain: $(-\infty, 0) \cup (0, \infty)$
Range: $(-\infty, 0) \cup (0, \infty)$
Decreasing on $(-\infty, 0) \cup (0, \infty)$
Odd function (origin symmetry)

Reciprocal Square Function

$$f(x) = \frac{1}{x^2}$$

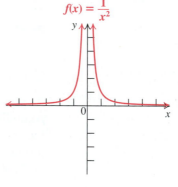

Domain: $(-\infty, 0) \cup (0, \infty)$
Range: $(0, \infty)$
Increasing on $(-\infty, 0)$
Decreasing on $(0, \infty)$
Even function (y-axis symmetry)

4 Evaluate and graph piecewise functions.

Piecewise Functions

In the definition of some functions, different rules for assigning output values are used on different parts of the domain. Such functions are called **piecewise functions**. For example, in Peach County, Georgia, a section of the interstate highway has a speed limit of 55 miles per hour (mph). If you are caught speeding between 56 and 74 mph, your fine is $50 plus $3 for every mile per hour over 55 mph. For 75 mph and higher, your fine is $150 plus $5 for every mile per hour over 75 mph.

Let $f(x)$ be the piecewise function that represents your fine for speeding at x miles per hour. We express $f(x)$ as a piecewise function.

$$f(x) = \begin{cases} 50 + 3(x - 55), & 56 \leq x < 75 \\ 150 + 5(x - 75), & x \geq 75 \end{cases}$$

The first line of the function means that if $56 \le x < 75$, your fine is $f(x) = 50 + 3(x - 55)$. Suppose you are caught driving 60 mph. Then

$$f(x) = 50 + 3(x - 55) \qquad \text{Expression used for } 56 \le x < 75$$
$$f(60) = 50 + 3(60 - 55) \qquad \text{Substitute 60 for } x.$$
$$= 65 \qquad \text{Simplify.}$$

Your fine for speeding at 60 mph is $65. The second line of the function means that if $x \ge 75$, your fine is $f(x) = 150 + 5(x - 75)$. Suppose you are caught driving 90 mph. Then

$$f(x) = 150 + 5(x - 75) \qquad \text{Expression used for } x \ge 75$$
$$f(90) = 150 + 5(90 - 75) \qquad \text{Substitute 90 for } x.$$
$$= 225 \qquad \text{Simplify.}$$

Your fine for speeding at 90 mph is $225.

FINDING THE SOLUTION: A PROCEDURE

EXAMPLE 5   **Evaluating a Piecewise Function**

OBJECTIVE	EXAMPLE
Evaluate $F(a)$ for piecewise function F.	Let $$F(x) = \begin{cases} x^2 & \text{if } x < 1 \\ 2x + 1 & \text{if } x \ge 1 \end{cases}$$ Find $F(0)$ and $F(2)$.
Step 1 **Determine which line of the function applies to the number a.**	**1.** Let $a = 0$. Because $0 < 1$, use the first line, $F(x) = x^2$. Let $a = 2$. Because $2 > 1$, use the second line, $F(x) = 2x + 1$.
Step 2 **Evaluate $F(a)$ using the line chosen in Step 1.**	**2.** $F(0) = (0)^2 = 0$ $F(2) = 2(2) + 1 = 5$ ■ ■ ■

Practice Problem 5

Let $f(x) = \begin{cases} x^2 & \text{if } x \le -1 \\ 2x & \text{if } x > -1 \end{cases}$

Find $f(-2)$ and $f(3)$.   ■

Graphing Piecewise Functions

Let's consider how to graph a piecewise function. The **absolute value function**, $f(x) = |x|$, can be expressed as a piecewise function by using the definition of absolute value.

$$f(x) = |x| = \begin{cases} -x & \text{if } x < 0 \\ x & \text{if } x \ge 0 \end{cases}$$

The first line in the function means that if $x < 0$, we use the equation $y = f(x) = -x$. If $x = -3$, then

$$y = f(-3) = -(-3) \qquad \text{Substitute } -3 \text{ for } x.$$
$$= 3 \qquad \text{Simplify.}$$

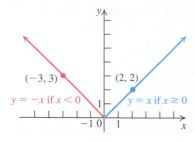

FIGURE 2.63 Graph of $y = |x|$

Thus, $(-3, 3)$ is a point on the graph of $y = |x|$. However, if $x \geq 0$, we use the second line in the function, which is $y = f(x) = x$. So if $x = 2$, then

$$y = f(2) = 2 \qquad \text{Substitute 2 for } x.$$

So $(2, 2)$ is a point on the graph of $y = |x|$. The two pieces $y = -x$ and $y = x$ are linear functions. We graph the appropriate parts of these lines ($y = -x$ for $x < 0$ and $y = x$ for $x \geq 0$) to form the graph of $y = |x|$. See Figure 2.63.

EXAMPLE 6   **Graphing a Piecewise Function**

Let

$$F(x) = \begin{cases} x^2 & \text{if } x < 1 \\ 3x + 1 & \text{if } x \geq 1 \end{cases}$$

Sketch the graph of $y = F(x)$.

SOLUTION

In the definition of F, the formula changes at $x = 1$. We call such numbers the **breakpoints** of the formula. For the function F, the only breakpoint is 1. Generally, to graph a piecewise function, we graph the function separately over the open intervals determined by the breakpoints and then graph the function at the breakpoints themselves. For the function $y = F(x)$, we graph the equation $y = x^2$ on the interval $(-\infty, 1)$. See Figure 2.64(a). Next, we graph the equation $y = 3x + 1$ on the interval $(1, \infty)$ and then at the breakpoint 1, where $y = F(1) = 3(1) + 1 = 4$. See Figure 2.64(b).

TECHNOLOGY CONNECTION

 To graph the greatest integer function, enter $Y_1 = \text{int}(x)$. If you have your calculator set to graph in **connected** mode, you will get the following incorrect graph.

For the graph to appear correctly, you must change from the **connected** to the **dot** mode.

Note that the step from $0 \leq x \leq 1$ is obscured by the x-axis.

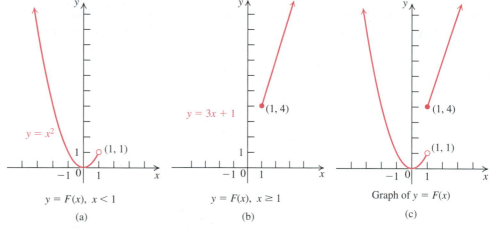

FIGURE 2.64 A piecewise function

Combining these portions, we obtain the graph of $y = F(x)$, shown in Figure 2.64(c). When we come to the edge of the part of the graph we are working with, we draw

(i) a filled in circle if that point is included.

(ii) an empty circle if that point is excluded.

You may find it helpful to think of this procedure as following the graph of $y = x^2$ when x is less than 1 and then jumping to the graph of $y = 3x + 1$ when x is equal to or greater than 1. ■ ■ ■

Practice Problem 6   Let $F(x) = \begin{cases} x^2 & \text{if } x \leq -1 \\ 2x & \text{if } x > -1 \end{cases}$

Sketch the graph of $y = F(x)$.

■

Some piecewise functions are called **step functions**. Their graphs look like the steps of a staircase.

The **greatest integer function** is denoted by $[x]$, or $\text{int}(x)$, where $[x]$ = the greatest integer less than or equal to x. For example,

$$[2] = 2, \quad [2.3] = 2, \quad [2.7] = 2, \quad [2.99] = 2$$

because 2 is the greatest integer less than or equal to $2, 2.3, 2.7,$ and 2.99. Similarly,

$$[-2] = -2, \quad [-1.9] = -2, \quad [-1.1] = -2, \quad [-1.001] = -2$$

because -2 is the greatest integer less than or equal to $-2, -1.9, -1.1,$ and -1.001.

In general, if m is an integer such that $m \le x < m + 1$, then $[x] = m$. In other words, if x is between two consecutive integers m and $m + 1$, then $[x]$ is assigned the smaller integer m.

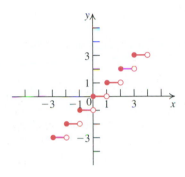

FIGURE 2.65 Graph of $y = [x]$

EXAMPLE 7   Graphing a Step Function

Graph the greatest integer function $f(x) = [x]$.

SOLUTION

Choose a typical closed interval between two consecutive integers—say, the interval $[2, 3]$. We know that between 2 and 3, the greatest integer function's value is 2. In symbols, if $2 \le x < 3$, then $[x] = 2$. Similarly, if $1 \le x < 2$, then $[x] = 1$. Therefore, the values of $[x]$ are constant between each pair of consecutive integers and jump by one unit at each integer. The graph of $f(x) = [x]$ is shown in Figure 2.65.   ■ ■ ■

Practice Problem 7   Find the values of $f(x) = [x]$ for $x = -3.4$ and $x = 4.7$.   ■

The greatest integer function f can be interpreted as a piecewise function.

$$f(x) = [x] = \begin{cases} \vdots \\ -2 \text{ if } -2 \le x < -1 \\ -1 \text{ if } -1 \le x < 0 \\ 0 \text{ if } 0 \le x < 1 \\ 1 \text{ if } 1 \le x < 2 \\ \vdots \end{cases}$$

SECTION 2.6 ■ Exercises

A EXERCISES Basic Skills and Concepts

1. The graph of the linear function $f(x) = b$ is a(n) _____ line.

2. The absolute value function can be expressed as a piecewise function by writing $f(x) = |x| = $ _____.

3. The graph of the function $f(x) = \begin{cases} x^2 + 2 \text{ if } x \le 1 \\ ax \text{ if } x > 1 \end{cases}$

   will have a break at $x = 1$ unless $a = $ _____.

4. If f is a function defined by $f(x) = \begin{cases} 6 \text{ if } x \le 3 \\ ax \text{ if } x \ge 3 \end{cases}$

   has no break point then $a = $ _____.

5. The function $f(x) = -x^4 - x^2 + 1$ is an example of a(n) _____ function.

6. *True or False* The function $f(x) = -x^{202} + x^{12} - x^2 + 3$ is an even function.

7. *True or False* The function $f(x) = x^3 + x + 1$ is an odd function.

8. *True or False* The function $f(x) = [x]$ is increasing on the interval $(0, 3)$.

In Exercises 9–18, write a linear function f that has the indicated values. Sketch the graph of f.

9. $f(0) = 1, f(-1) = 0$　　**10.** $f(1) = 0, f(2) = 1$

11. $f(-1) = 1, f(2) = 7$　　**12.** $f(-1) = -5, f(2) = 4$

13. $f(1) = 1, f(2) = -2$　　**14.** $f(1) = -1, f(3) = 5$

15. $f(-2) = 2, f(2) = 4$　　**16.** $f(2) = 2, f(4) = 5$

17. $f(0) = -1, f(3) = -3$　　**18.** $f(1) = \frac{1}{4}, f(4) = -2$

19. Let
$$f(x) = \begin{cases} x & \text{if } x \geq 2 \\ 2 & \text{if } x < 2 \end{cases}.$$
  a. Find $f(1), f(2)$, and $f(3)$.
  b. Sketch the graph of $y = f(x)$.

20. Let
$$g(x) = \begin{cases} 2x & \text{if } x < 0 \\ x & \text{if } x \geq 0 \end{cases}.$$
  a. Find $g(-1), g(0)$, and $g(1)$.
  b. Sketch the graph of $y = g(x)$.

21. Let
$$f(x) = \begin{cases} 1 & \text{if } x > 0 \\ -1 & \text{if } x < 0 \end{cases}.$$
  a. Find $f(-15)$ and $f(12)$.
  b. Sketch the graph of $y = f(x)$.
  c. Find the domain and the range of f.

22. Let
$$g(x) = \begin{cases} 2x + 4 & \text{if } x > 1 \\ x + 2 & \text{if } x \leq 1 \end{cases}.$$
  a. Find $g(-3), g(1)$, and $g(3)$.
  b. Sketch the graph of $y = g(x)$.
  c. Find the domain and the range of g.

In Exercises 23–34, sketch the graph of each function and from the graph, find its domain and its range.

23. $f(x) = \begin{cases} -2x & \text{if } x < 0 \\ x^2 & \text{if } x \geq 0 \end{cases}$　**24.** $g(x) = \begin{cases} |x| & \text{if } x < 1 \\ x^2 & \text{if } x \geq 1 \end{cases}$

25. $g(x) = \begin{cases} \dfrac{1}{x} & \text{if } x < 0 \\ \sqrt{x} & \text{if } x \geq 0 \end{cases}$　**26.** $h(x) = \begin{cases} \sqrt[3]{x} & \text{if } x < 1 \\ \sqrt{x} & \text{if } x \geq 1 \end{cases}$

27. $f(x) = \begin{cases} [x] & \text{if } x < 1 \\ \sqrt[3]{x} & \text{if } x \geq 1 \end{cases}$　**28.** $g(x) = \begin{cases} x^3 & \text{if } x < 0 \\ \sqrt{x} & \text{if } x \geq 0 \end{cases}$

29. $g(x) = \begin{cases} |x| & \text{if } x < 1 \\ x^3 & \text{if } x \geq 1 \end{cases}$　**30.** $f(x) = \begin{cases} \dfrac{1}{x} & \text{if } x < 0 \\ [x] & \text{if } x \geq 0 \end{cases}$

31. $f(x) = \begin{cases} |x| & \text{if } -2 \leq x < 1 \\ \sqrt{x} & \text{if } x \geq 1 \end{cases}$

32. $g(x) = \begin{cases} \sqrt[3]{x} & \text{if } -8 \leq x < 1 \\ 2x & \text{if } x \geq 1 \end{cases}$

33. $g(x) = \begin{cases} [x] & \text{if } x < 1 \\ -2x & \text{if } 1 \leq x \leq 4 \end{cases}$

34. $h(x) = \begin{cases} |x| & \text{if } x < 0 \\ \sqrt{x} & \text{if } 0 \leq x \leq 4 \end{cases}$

B EXERCISES Applying the Concepts

35. Converting volume. To convert the volume of a liquid measured in ounces to a volume measured in liters, we use the fact that 1 liter ≈ 33.81 ounces. Let x denote the volume measured in ounces and y denote the volume measured in liters.
  a. Write an equation for the linear function $y = f(x)$. What are the domain and range of f?
  b. Compute $f(3)$. What does it mean?
  c. How many liters of liquid are in a typical soda can containing 12 ounces of liquid?

36. Boiling point and elevation. The boiling point B of water (in degrees Fahrenheit) at elevation h (in thousands of feet) above sea level is given by the linear function $B(h) = -1.8h + 212$.
  a. Find and interpret the intercepts of the graph of the function $y = B(h)$.
  b. Find the domain of this function.
  c. Find the elevation at which water boils at $98.6°F$. Why is this height dangerous to humans?

37. Pressure at sea depth. The pressure P in atmospheres (atm) at a depth d feet is given by the linear function
$$P(d) = \frac{1}{33}d + 1.$$
  a. Find and interpret the intercepts of $y = P(d)$.
  b. Find $P(0), P(10), P(33)$, and $P(100)$.
  c. Find the depth at which the pressure is 5 atmospheres.

38. Speed of sound. The speed V of sound (in feet per second) in air at temperature T (in degrees Fahrenheit) is given by the linear function $V(T) = 1055 + 1.1T$.
  a. Find the speed of sound at $90°F$.
  b. Find the temperature at which the speed of sound is 1100 feet per second.

39. Manufacturer's cost. A manufacturer of printers has a total cost per day consisting of a fixed overhead of $6000 plus product costs of $50 per printer.
  a. Express the total cost C as a function of the number x of printers produced.
  b. Draw the graph of $y = C(x)$. Interpret the y-intercept.
  c. How many printers were manufactured on a day when the total cost was $11,500?

40. Supply function. When the price p of a commodity is $10 per unit, 750 units of it are sold. For every $1 increase in the unit price, the supply q increases by 100 units.
  a. Write the linear function $q = f(p)$.
  b. Find the supply when the price is $15 per unit.
  c. What is the price at which 1750 units can be supplied?

41. Apartment rental. Suppose a two-bedroom apartment in a neighborhood near your school rents for $900 per month. If you move in after the first of the month, the rent is prorated; that is, it is reduced linearly.
  a. Suppose you move into the apartment x days after the first of the month. (Assume that the month has 30 days.) Express the rent R as a function of x.
  b. Compute the rent if you move in six days after the first of the month.
  c. When did you move into the apartment if the landlord charged you a rent of $600?

42. College admissions. The average SAT scores (mathematics and critical reading) for incoming students at Central State College (CSC) have been rising linearly in recent years. In 2002, the average SAT score was 1120; in 2004, it was 1150. (The SAT testing format changed considerably in 2005.)
  a. Express the average SAT score at CSC as a function of time.
  b. If the trend continues, what will be the average SAT score of incoming students at CSC in 2010?
  c. If the trend continues, when will the average SAT score at CSC be 1300?

43. Breathing capacity. Suppose the average maximum breathing capacity for humans drops linearly from 100% at age 20 to 40% at age 80. What age corresponds to 50% capacity?

44. Drug dosage for children. If a is the adult dosage of a medicine and t is the child's age, then the *Friend's rule* for the child's dosage y is given by the formula $y = \dfrac{2}{25}ta$.
  a. Suppose the adult dosage is 60 milligrams. Find the dosage for a 5-year-old child.
  b. How old would a child have to be in order to be prescribed an adult dosage?

45. Air pollution. In Ballerenia, the average number y of deaths per month was observed to be linearly related to the concentration x of sulfur dioxide in the air. Suppose there are 30 deaths when $x = 150$ milligrams per cubic meter and 50 deaths when $x = 420$ milligrams per cubic meter.
  a. Write y as a function of x.
  b. Find the number of deaths when $x = 350$ milligrams per cubic meter.
  c. If the number of deaths per month is 45, what is the concentration of sulfur dioxide in the air?

46. Child shoe sizes. Children's shoe sizes start with size 0, having an insole length L of $3\frac{11}{12}$ inches and each full size being $\frac{1}{3}$ inch longer.
  a. Write the equation $y = L(S)$, where S is the shoe size.

  b. Find the length of the insole of a child's shoe of size 4.
  c. What size (to the nearest half size) shoe will fit a child whose insole length is 6.1 inches?

47. State income tax. Suppose a state's income tax code states that the tax liability T on x dollars of taxable income is as follows:

$$T(x) = \begin{cases} 0.04x & \text{if } 0 \le x < 20{,}000 \\ 800 + 0.06x & \text{if } x \ge 20{,}000 \end{cases}$$

  a. Graph the function $y = T(x)$.
  b. Find the tax liability on each taxable income.
    (i) $12,000
    (ii) $20,000
    (iii) $50,000
  c. Find your taxable income if you had each tax liability.
    (i) $600
    (ii) $1200
    (iii) $2300

48. Federal income tax. The tax table for the 2009 U.S. Income Tax for a single taxpayer is as follows.

If taxable income is over	But not over	The tax is
$0	$8350	10% of the amount over $0.
$8350	$33,950	$835 plus 15% of the amount over $8350.
$33,950	$82,250	$4675 plus 25% of the amount over $33,950.
$82,250	$171,550	$16,750 plus 28% of the amount over $82,250.
$171,550	$372,950	$41,754 plus 33% of the amount over $171,550.
$372,950	No limit	$108,216 plus 35% of the amount over $372,950.

Source: Internal Revenue Service.

  a. Express the information given in the table as a six-part piecewise function f, using x as a variable representing taxable income.
  b. Find the tax liability on each taxable income.
    (i) $35,000
    (ii) $100,000
    (iii) $500,000
  c. Find your taxable income if you had each tax liability.
    (i) $3500
    (ii) $12,700
    (iii) $35,000

C EXERCISES Beyond the Basics

49.

$$\text{Let } f(x) = \begin{cases} 3x + 5, & -3 \leq x < -1 \\ 2x + 1, & -1 \leq x < 2 \\ 2 - x, & 2 \leq x \leq 4 \end{cases}.$$

  a. Find the following.
   (i) $f(-2)$
   (ii) $f(-1)$
   (iii) $f(3)$
  b. Find x when $f(x) = 2$.
  c. Sketch a graph of $y = f(x)$.

50.

$$\text{Let } g(x) = \begin{cases} 3, & -3 \leq x < -1 \\ -6x - 3, & -1 \leq x < 0 \\ 3x - 3, & 0 \leq x \leq 1 \end{cases}.$$

  a. Find the following.
   (i) $g(-2)$
   (ii) $g\left(\dfrac{1}{2}\right)$
   (iii) $g(0)$
   (iv) $g(-1)$
  b. Find x when **(i)** $g(x) = 0$ and **(ii)** $2g(x) + 3 = 0$.
  c. Sketch a graph of $y = g(x)$.

In Exercises 51 and 52, a function is given. For each function,
  a. find the domain and range.
  b. find the intervals over which the function is increasing, decreasing, or constant.
  c. state whether the function is odd, even, or neither.

51. $f(x) = x - [\![x]\!]$

52. $f(x) = \dfrac{1}{[\![x]\!]}$

53. Let $f(x) = \dfrac{|x|}{x}, x \neq 0$. Find $|f(x) - f(-x)|$.

54. **Windchill Index (WCI).** Suppose the outside air temperature is T degrees Fahrenheit and the wind speed is v miles per hour. A formula for the WCI based on observations is as follows:

$$WCI = \begin{cases} T, & 0 \leq v \leq 4 \\ 91.4 + (91.4 - T)(0.0203v - 0.304\sqrt{v} - 0.474), & 4 < v < 45 \\ 1.6T - 55, & v \geq 45 \end{cases}$$

  a. Find the WCI to the nearest degree if the outside air temperature is 40°F and
   (i) $v = 2$ miles per hour.
   (ii) $v = 16$ miles per hour.
   (iii) $v = 50$ miles per hour.

  b. Find the air temperature to the nearest degree if
   (i) the WCI is $-58°$F and $v = 36$ miles per hour.
   (ii) the WCI is $-10°$F and $v = 49$ miles per hour.

Critical Thinking

55. **Postage rate function.** The U.S. Postal Service uses the function $f(x) = -[\![-x]\!]$, called the **ceiling integer function**, to determine postal charges. For instance, if you have an envelope that weighs 3.2 ounces, you will be charged the rate for $f(3.2) = -[\![-3.2]\!] = -(-4) = 4$ ounces. In 2009, it cost 44¢ for the first ounce (or a fraction of it) and 17¢ for each additional ounce (or a fraction of it) to mail a first-class letter in the United States.
  a. Express the cost C (in cents) of mailing a letter weighing x ounces.
  b. Graph the function $y = C(x)$.
  c. What are the domain and range of $y = C(x)$?

56. The cost C of parking a car at the metropolitan airport is \$4 for the first hour and \$2 for each additional hour or fraction thereof. Write the cost function $C(x)$, where x is the number of parking hours, in terms of the greatest integer function.

57. **Car rentals.** The weekly cost of renting a compact car from U-Rent is \$150. There is no charge for driving the first 100 miles, but there is a 20¢ charge for each mile (or fraction thereof) driven over 100 miles.
  a. Express the cost C of renting a car for a week and driving it for x miles.
  b. Sketch the graph of $y = C(x)$.
  c. How many miles were driven if the weekly rental cost was \$190?

Transformations of Functions

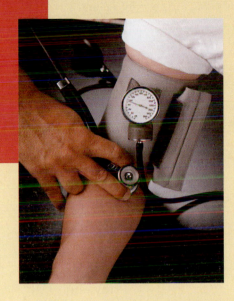

Before Starting this Section, Review

1. Graphs of basic functions (Section 2.6, page 243)
2. Symmetry (Section 2.2, page 182)
3. Completing the square (Section 1.4, page 121)

Objectives

1 Learn the meaning of transformations.

2 Use vertical or horizontal shifts to graph functions.

3 Use reflections to graph functions.

4 Use stretching or compressing to graph functions.

MEASURING BLOOD PRESSURE

Blood pressure is the force of blood per unit area against the walls of the arteries. Blood pressure is recorded as two numbers: the systolic pressure (as the heart beats) and the diastolic pressure (as the heart relaxes between beats). If your blood pressure is "120 over 80," it is rising to a maximum of 120 millimeters of mercury as the heart beats and is falling to a minimum of 80 millimeters of mercury as the heart relaxes.

The most precise way to measure blood pressure is to place a small glass tube in an artery and let the blood flow out and up as high as the heart can pump it. That was the way Stephen Hales (1677–1761), the first investigator of blood pressure, learned that a horse's heart could pump blood 8 feet 3 inches up a tall tube. Such a test, however, consumed a great deal of blood. Jean Louis Marie Poiseuille (1797–1869) greatly improved the process by using a mercury-filled manometer, which allowed a smaller, narrower tube. Even so, a blood pressure check in the 1840s was a very different experience than what people routinely undergo today.

In Example 8, we investigate Poiseuille's Law for arterial blood flow. ■

---

1 Learn the meaning of transformations.

Transformations

If a new function is formed by performing certain operations on a given function f, then the graph of the new function is called a **transformation** of the graph of f. For example, the graphs of $y = |x| + 2$ and $y = |x| - 3$ are transformations of the graph of $y = |x|$; that is, the graph of each is a special modification of the graph of $y = |x|$.

2 Use vertical or horizontal shifts to graph functions.

Vertical and Horizontal Shifts

EXAMPLE 1 **Graphing Vertical Shifts**

Let $f(x) = |x|$, $g(x) = |x| + 2$, and $h(x) = |x| - 3$. Sketch the graphs of these functions on the same coordinate plane. Describe how the graphs of g and h relate to the graph of f.

TECHNOLOGY CONNECTION

 Many graphing calculators use abs(x) for $|x|$. The graphs of $y = |x|$, $y = |x| - 3$, and $y = |x| + 2$ illustrate that these graphs are identical in shape. They differ only in vertical placement.

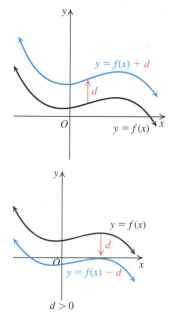

SOLUTION

Make a table of values and graph the equations $y = f(x)$, $y = g(x)$, and $y = h(x)$.

TABLE 2.10

x	$y = |x|$	$y = |x| + 2$	$y = |x| - 3$
-5	5	7	2
-3	3	5	0
-1	1	3	-2
0	0	2	-3
1	1	3	-2
3	3	5	0
5	5	7	2

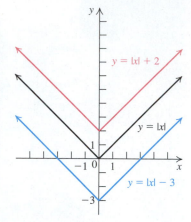

FIGURE 2.66 Vertical shifts of $y = |x|$

Notice that in Table 2.10 and Figure 2.66, for each value of x, the value of $y = |x| + 2$ is 2 more than the value of $y = |x|$. So the graph of $y = |x| + 2$ is the graph of $y = |x|$ shifted two units up.

We also note that for each value of x, the value of $y = |x| - 3$ is 3 less than the value of $y = |x|$. So the graph of $y = |x| - 3$ is the graph of $y = |x|$ shifted three units down. ■ ■ ■

Practice Problem 1 Let

$$f(x) = x^3, g(x) = x^3 + 1, \text{ and } h(x) = x^3 - 2.$$

Sketch the graphs of all three functions on the same coordinate plane. Describe how the graphs of g and h relate to the graph of f. ■

Example 1 illustrates the concept of the vertical (up or down) shift of a graph.

VERTICAL SHIFT

Let $d > 0$. The graph of $y = f(x) + d$ is the graph of $y = f(x)$ shifted d units *up*, and the graph of $y = f(x) - d$ is the graph of $y = f(x)$ shifted d units *down*. See Figure 2.67.

Next, we consider the operation that shifts a graph horizontally.

FIGURE 2.67 Vertical shifts

EXAMPLE 2 **Writing Functions for Horizontal Shifts**

Let $f(x) = x^2, g(x) = (x - 2)^2$, and $h(x) = (x + 3)^2$. A table of values for f, g, and h is given in Table 2.11. The three functions f, g, and h are graphed on the same coordinate plane in Figure 2.68. Describe how the graphs of g and h relate to the graph of f.

The graphs of $y = x^2$, $y = (x + 3)^2$, and $y = (x - 2)^2$ illustrate that these graphs are identical in shape. They differ only in horizontal placement.

TABLE 2.11

x	$y = x^2$	$y = (x - 2)^2$	x	$y = x^2$	$y = (x + 3)^2$
-4	**16**	36	-4	16	1
-3	**9**	25	-3	9	0
-2	**4**	16	-2	4	1
-1	**1**	9	-1	1	4
0	**0**	4	0	**0**	9
1	**1**	1	1	**1**	16
2	4	0	2	**4**	25
3	9	1	3	**9**	36
4	16	4	4	**16**	49
	(a)			**(b)**	

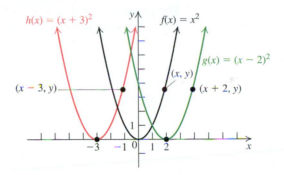

FIGURE 2.68 Horizontal shifts

SOLUTION

First, notice that all three functions are squaring functions.

a. Replacing x with $x - 2$ on the right side of the defining equation for f gives the equation for g.

$$f(x) = \quad x^2 \qquad \text{Defining equation for } f$$
$$\downarrow$$
$$g(x) = (x - 2)^2 \qquad \text{Replace } x \text{ with } x - 2 \text{ to obtain the defining equation for } g.$$

The x-intercept of f is 0. We can find the x-intercept of g by solving $g(x) = (x - 2)^2 = 0$ for x. We get $x - 2 = 0$, or $x = 2$.

This means that $(2, 0)$ is a point on the graph of g, whereas $(0, 0)$ is a point on the graph of f. In general, each point (x, y) on the graph of f has a corresponding point $(x + 2, y)$ on the graph of g. So the graph of $g(x) = (x - 2)^2$ is just the graph of $f(x) = x^2$ shifted two units to the *right*. Noticing that $f(0) = 0$ and $g(2) = 0$ will help you remember which way to shift the graph. Table 2.11(a) and Figure 2.68 illustrate these ideas.

b. Replacing x with $x + 3$ on the right-hand side of the defining equation for f gives the equation for h.

$$f(x) = \quad x^2 \qquad \text{Defining equation for } f$$
$$\downarrow$$
$$h(x) = (x + 3)^2 \qquad \text{Replace } x \text{ with } x + 3 \text{ to obtain the defining equation for } h.$$

Shift right, $c > 0$

To find the x-intercept of h, we solve $h(x) = (x + 3)^2 = 0$. We get $x + 3 = 0$ and find that $x = -3$. This means that $(-3, 0)$ is a point on the graph of h, whereas $(0, 0)$ is a point on the graph of f. In general, each point (x, y) on the graph of f has a corresponding point $(x - 3, y)$ on the graph of h. So the graph of $h(x) = (x + 3)^2$ is just the graph of $f(x) = x^2$ shifted three units to the *left*. Noticing that $f(0) = 0$ and $h(-3) = 0$ will help you remember which way to shift the graph.

Table 2.11(b) and Figure 2.68 confirm these considerations. ■ ■ ■

Practice Problem 2   Let

$$f(x) = x^3, g(x) = (x - 1)^3, \text{ and } h(x) = (x + 2)^3.$$

Sketch the graphs of all three functions on the same coordinate plane. Describe how the graphs of g and h relate to the graph of f. ■

y = f(x)

y = f(x − c)

Shift left, $c < 0$

FIGURE 2.69 Horizontal shifts

HORIZONTAL SHIFTS
The graph of $y = f(x - c)$ is the graph of $y = f(x)$ shifted $\lvert c \rvert$ units to the right if $c > 0$, to the left if $c < 0$. See Figure 2.69.

Replacing x with $x - c$ in the equation $y = f(x)$ results in a function whose graph is the graph of f shifted $\lvert c \rvert$ units to the right $(c > 0)$ or to the left $(c < 0)$.

FINDING THE SOLUTION: A PROCEDURE

EXAMPLE 3   **Graphing Combined Vertical and Horizontal Shifts**

OBJECTIVE	**EXAMPLE**
Sketch the graph of $g(x) = f(x - c) + d$, where f is a function whose graph is known.	*Sketch the graph of $g(x) = \sqrt{x + 2} - 3$.*
Step 1 Identify and graph the known function f.	**1.** Choose $f(x) = \sqrt{x}$. The graph of $y = \sqrt{x}$ is shown in Step 4.
Step 2 Identify the constants d and c.	**2.** $g(x) = \sqrt{x - (-2)} + (-3)$, so $c = -2$ and $d = -3$.
Step 3 Graph $y = f(x - c)$ by shifting the graph of f horizontally to the **(i)** right if $c > 0$. **(ii)** left if $c < 0$.	**3.** Because $c = -2 < 0$, the graph of $y = \sqrt{x + 2}$ is the graph of f shifted horizontally two units to the left. See the blue graph in Step 4.

Step 4 Graph $y = f(x - c) + d$ by shifting the graph of $y = f(x - c)$ vertically

(i) up if $d > 0$.
(ii) down if $d < 0$.

4. Because $d = -3 < 0$, graph $y = \sqrt{x + 2} - 3$ by shifting the graph of $y = \sqrt{x + 2}$ three units down.

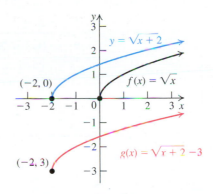

Horizontal and vertical shifts

■ ■ ■

Practice Problem 3 Sketch the graph of

$$f(x) = \sqrt{x - 2} + 3.$$

■

3 Use reflections to graph functions.

Reflections

Comparing the Graphs of $y = f(x)$ and $y = -f(x)$ Consider the graph of $f(x) = x^2$, shown in Figure 2.70. Table 2.12 gives some values for $f(x)$ and $g(x) = -f(x)$. Note that the y-coordinate of each point in the graph of $g(x) = -f(x)$ is the opposite of the y-coordinate of the corresponding point on the graph of $f(x)$. So the graph of $y = -x^2$ is the reflection of the graph of $y = x^2$ in the x-axis. This means that the points (x, x^2) and $(x, -x^2)$ are the same distance from but on opposite sides of the x-axis. See Figure 2.70.

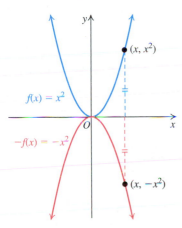

FIGURE 2.70 Reflection in the *x*-axis

TABLE 2.12

x	$f(x) = x^2$	$g(x) = -f(x) = -x^2$
-3	9	-9
-2	4	-4
-1	1	-1
0	0	0
1	1	-1
2	4	-4
3	9	-9

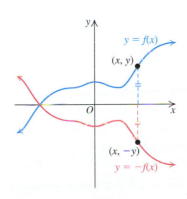

FIGURE 2.71 Reflection in the *x*-axis

REFLECTION IN THE *x*-AXIS

The graph of $y = -f(x)$ is a reflection of the graph of $y = f(x)$ in the x-axis. If a point (x, y) is on the graph of $y = f(x)$, then the point $(x, -y)$ is on the graph of $y = -f(x)$. See Figure 2.71.

Comparing the Graphs of $y = f(x)$ and $y = f(-x)$ To compare the graphs of $f(x)$ and $g(x) = f(-x)$, consider the graph of $f(x) = \sqrt{x}$. Then $f(-x) = \sqrt{-x}$. See Figure 2.72. Table 2.13 gives some values for $f(x)$ and $g(x) = f(-x)$. The domain of f is $[0, \infty)$, and the domain of g is $(-\infty, 0]$. Each point (x, y) on the graph of f has a corresponding point $(-x, y)$ on the graph of g. So the graph of $y = f(-x)$ is the reflection of the graph of $y = f(x)$ in the y-axis. This means that the points $(x, \sqrt{x})$ and $(-x, \sqrt{x})$ are the same distance from but on opposite sides of the y-axis. See Figure 2.72.

TABLE 2.13

x	$f(x) = \sqrt{x}$	$-x$	$g(x) = \sqrt{-x}$
-4	Undefined	4	2
-1	Undefined	1	1
0	0	0	0
1	1	-1	Undefined
4	2	-4	Undefined

FIGURE 2.72 Graphing $y = \sqrt{-x}$

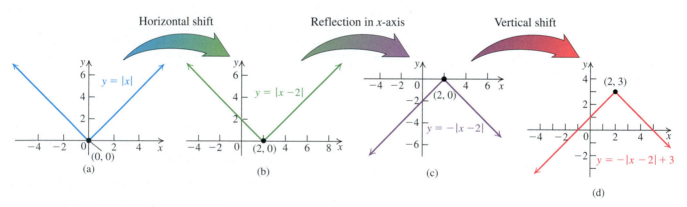

FIGURE 2.73 Reflection in the y-axis

REFLECTION IN THE y-AXIS

The graph of $y = f(-x)$ is a reflection of the graph of $y = f(x)$ in the y-axis. If a point (x, y) is on the graph of $y = f(x)$, then the point $(-x, y)$ is on the graph of $y = f(-x)$. See Figure 2.73.

EXAMPLE 4 **Combining Transformations**

Explain how the graph of $y = -|x - 2| + 3$ can be obtained from the graph of $y = |x|$.

SOLUTION

Start with the graph of $y = |x|$. Follow the point $(0, 0)$ on $y = |x|$. See Figure 2.74(a).

Step 1 Shift the graph of $y = |x|$ two units to the right to obtain the graph of $y = |x - 2|$. The point $(0, 0)$ moves to $(2, 0)$. See Figure 2.74(b).

Horizontal shift   Reflection in x-axis   Vertical shift

FIGURE 2.74 Transformations of $y = |x|$

Step 2 Reflect the graph of $y = |x - 2|$ in the x-axis to obtain the graph of $y = -|x - 2|$. The point $(2, 0)$ remains $(2, 0)$. See Figure 2.74(c).

Step 3 Finally, shift the graph of $y = -|x - 2|$ three units up to obtain the graph of $y = -|x - 2| + 3$. The point $(2, 0)$ moves to $(2, 3)$. See Figure 2.74(d). Algebraically, replace x with 2 to get $y = -|2 - 2| + 3 = 3$. ■■■

Practice Problem 4 Explain how the graph of $y = -(x - 1)^2 + 2$ can be obtained from the graph of $y = x^2$. ■

4 Use stretching or compressing to graph functions.

Stretching or Compressing

The transformations that shift or reflect a graph change only the graph's position, not its shape. Such transformations are called **rigid transformations**. We now look at transformations that distort the shape of a graph, called **nonrigid transformations**. We consider the relationship of the graphs of $y = af(x)$ and $y = f(bx)$ to the graph of $y = f(x)$.

Comparing the Graphs of $y = f(x)$ and $y = af(x)$

> **EXAMPLE 5**   **Stretching or Compressing a Function Vertically**
>
> Let $f(x) = |x|$, $g(x) = 2|x|$, and $h(x) = \dfrac{1}{2}|x|$. Sketch the graphs of f, g, and h on the same coordinate plane and describe how the graphs of g and h are related to the graph of f.

SOLUTION

The graphs of $y = |x|$, $y = 2|x|$, and $y = \dfrac{1}{2}|x|$ are sketched in Figure 2.75. Table 2.14 gives some typical function values.

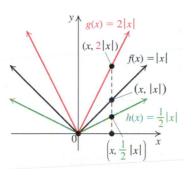

FIGURE 2.75 Vertical stretch and compression

TABLE 2.14

x	$f(x) = |x|$	$g(x) = 2|x|$	$h(x) = \dfrac{1}{2}|x|$
-2	2	4	1
-1	1	2	$\dfrac{1}{2}$
0	0	0	0
1	1	2	$\dfrac{1}{2}$
2	2	4	1

The graph of $y = 2|x|$ is the graph of $y = |x|$ vertically stretched (expanded) by multiplying each of its y-coordinates by 2. It is twice as high as the graph of $|x|$ at every real number x. The result is a taller V-shaped curve. See Figure 2.75.

The graph $y = \dfrac{1}{2}|x|$ is the graph of $y = |x|$ vertically compressed (shrunk) by multiplying each of its y-coordinates by $\dfrac{1}{2}$. It is half as high as the graph of $|x|$ at every real number x. The result is a flatter V-shaped curve. See Figure 2.75.   ■ ■ ■

Practice Problem 5   Let $f(x) = \sqrt{x}$ and $g(x) = 2\sqrt{x}$. Sketch the graphs of f and g on the same coordinate plane and describe how the graph of g is related to the graph of f.   ■

$a > 1$: stretch

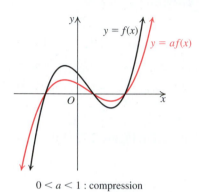

$0 < a < 1$: compression

FIGURE 2.76 Vertical stretch and compression

VERTICAL STRETCHING OR COMPRESSING

The graph of $y = af(x)$ is obtained from the graph of $y = f(x)$ by multiplying the y-coordinate of each point on the graph of $y = f(x)$ by a and leaving the x-coordinate unchanged. The result (see Figure 2.76) is as follows:

1. A **vertical stretch** away from the x-axis if $a > 1$

2. A **vertical compression** toward the x-axis if $0 < a < 1$

If $a < 0$, first graph $y = |a|f(x)$ by stretching or compressing the graph of $y = f(x)$ vertically. Then reflect the resulting graph in the x-axis.

Comparing the Graphs of $y = f(x)$ and $y = f(bx)$ Given a function $f(x)$, let's explore the effect of the constant b in graphing the function $y = f(bx)$. Consider the graphs of $y = f(x)$ and $y = f(2x)$ in Figure 2.77(a). Multiplying the *independent* variable x by 2 compresses the graph $y = f(x)$ horizontally toward the y-axis. Because the value $\frac{1}{2}x$ is half the value of x, and $f\left(2\left[\frac{1}{2}x\right]\right) = f(x)$, a point on the x-axis will be only half as far from the origin when $y = f(2x)$ has the same y value as $y = f(x)$. See Figure 2.77(a).

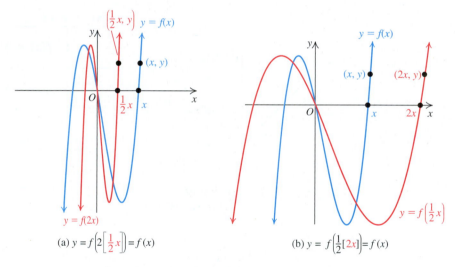

(a) $y = f\left(2\left[\frac{1}{2}x\right]\right) = f(x)$ (b) $y = f\left(\frac{1}{2}[2x]\right) = f(x)$

FIGURE 2.77 Horizontal stretch and compression

Now consider the graphs of $y = f(x)$ and $y = f\left(\frac{1}{2}x\right)$ in Figure 2.77(b). Multiplying the *independent* variable x by $\frac{1}{2}$ stretches the graph of $y = f(x)$ horizontally away from the y-axis. Because the value $2x$ is twice the value of x, and $f\left(\frac{1}{2}[2x]\right) = f(x)$, a point on the x-axis will be twice as far from the origin when $y = f\left(\frac{1}{2}x\right)$ has the same y value as $y = f(x)$. See Figure 2.77(b).

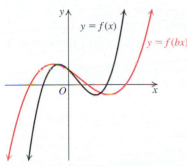

$y = f(x)$

$y = f(bx)$

$0 < b < 1$: stretch horizontally

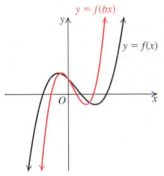

$y = f(bx)$

$y = f(x)$

$b > 1$: compress horizontally

FIGURE 2.78 Horizontal stretch and compression

$(1, 3)$

$y = f(x)$

$(-2, -1)$

FIGURE 2.79

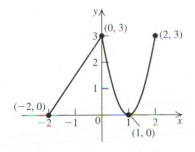

$(0, 3)$

$(2, 3)$

$(-2, 0)$

$(1, 0)$

HORIZONTAL STRETCHING OR COMPRESSING

The graph of $y = f(bx)$ is obtained from the graph of $y = f(x)$ by multiplying the x-coordinate of each point on the graph of $y = f(x)$ by $\dfrac{1}{b}$ and leaving the y-coordinate unchanged. The result (see Figure 2.78) is:

1. A **horizontal stretch** away from the y-axis if $0 < b < 1$
2. A **horizontal compression** toward the y-axis if $b > 1$

If $b < 0$, first graph $f(|b|x)$ by stretching or compressing the graph of $y = f(x)$ horizontally. Then reflect the graph of $y = f(|b|x)$ in the y-axis.

EXAMPLE 6   Stretching or Compressing a Function Horizontally

Using the graph of the function $y = f(x)$ in Figure 2.79, whose formula is not given, sketch the following graphs.

a. $f\left(\dfrac{1}{2}x\right)$ **b.** $f(2x)$ **c.** $f(-2x)$

SOLUTION

a. To graph $y = f\left(\dfrac{1}{2}x\right)$, we stretch the graph of $y = f(x)$ horizontally. In other words, we transform each point (x, y) in Figure 2.79 to the point $(2x, y)$ in Figure 2.80(a). For example, the points $(-2, -1)$ and $(1, 3)$ transform to $(-4, -1)$ and $(2, 3)$, respectively.

b. To graph $y = f(2x)$, we compress the graph of $y = f(x)$ horizontally. So, we transform each point (x, y) in Figure 2.79 to $\left(\dfrac{1}{2}x, y\right)$ in Figure 2.80(b). Here the points $(-2, -1)$ and $(1, 3)$ transform to $(-1, -1)$ and $\left(\dfrac{1}{2}, 3\right)$, respectively.

$(2, 3)$

$y = f\left(\tfrac{1}{2}x\right)$

$(-4, -1)$

(a)

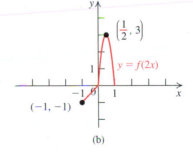

$\left(\tfrac{1}{2}, 3\right)$

$y = f(2x)$

$(-1, -1)$

(b)

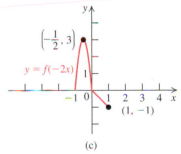

$\left(-\tfrac{1}{2}, 3\right)$

$y = f(-2x)$

$(1, -1)$

(c)

FIGURE 2.80

c. To graph $y = f(-2x)$, we reflect the graph of $y = f(2x)$ in Figure 2.80(b) in the y-axis. So we transform each point (x, y) in Figure 2.80(b) to the point $(-x, y)$ in Figure 2.80(c). For example, the points $(-1, -1)$ and $\left(\dfrac{1}{2}, 3\right)$ transform to $(1, -1)$ and $\left(-\dfrac{1}{2}, 3\right)$, respectively.   ■ ■ ■

Practice Problem 6 The graph of a function $y = f(x)$ is given in the margin. Sketch the graphs of the following functions.

a. $f\left(\dfrac{1}{2}x\right)$ **b.** $f(2x)$ ■

Multiple Transformations in Sequence

When graphing requires more than one transformation of a basic function, it is helpful to perform transformations in the following order:

1. Horizontal shifts   **2.** Stretch or compress   **3.** Reflections   **4.** Vertical shifts

EXAMPLE 7   Combining Transformations

Sketch the graph of the function $f(x) = 3 - 2(x - 1)^2$.

SOLUTION

Begin with the basic function $y = x^2$. Then apply the necessary transformations in a sequence of steps. You should follow the transformation of the point $(2, 4)$ through each of the five steps. The result of each step is shown in Figure 2.81.

Step 1 $y = x^2$ Identify a related function whose graph is familiar. In this case, use $y = x^2$. See Figure 2.81(a).

Step 2 $y = (x - 1)^2$ Replace x with $x - 1$; shift the graph of $y = x^2$ one unit to the right. See Figure 2.81(b).

Step 3 $y = 2(x - 1)^2$ Multiply by 2; stretch the graph of $y = (x - 1)^2$ vertically by a factor of 2. See Figure 2.81(c).

Step 4 $y = -2(x - 1)^2$ Multiply by -1. Reflect the graph of $y = 2(x - 1)^2$ in the x-axis. See Figure 2.81(d).

Step 5 $y = 3 - 2(x - 1)^2$ Add 3. Shift the graph of $y = -2(x - 1)^2$ three units up. See Figure 2.81(e).

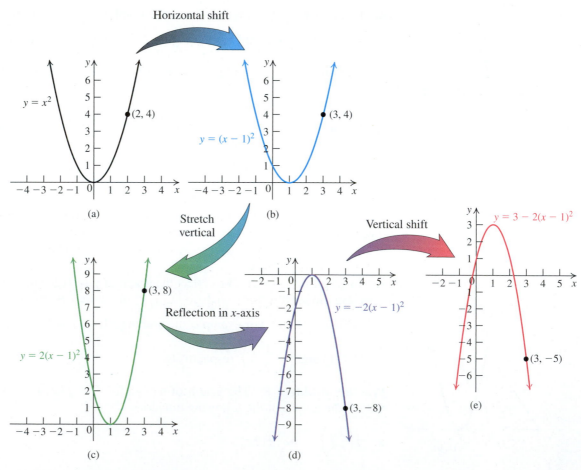

FIGURE 2.81 Multiple transformations

Practice Problem 7   Sketch the graph of the function

$$f(x) = 3\sqrt{x + 1} - 2.$$ ■

FIGURE 2.82

EXAMPLE 8   Using Poiseuille's Law for Arterial Blood Flow

For an artery with radius R, the velocity v of the blood flow at a distance r from the center of the artery is given by

$$v = c(R^2 - r^2),$$

where c is a constant that is determined for a particular artery. See Figure 2.82.

To emphasize that the velocity v depends on the distance r from the center of the artery, we write $v = v(r)$, or $v(r) = c(R^2 - r^2)$. Starting with the graph of $y = r^2$, sketch the graph of $y = v(r)$, where the artery has radius $R = 3$ and $c = 10^4$.

SOLUTION

$$v(r) = c(R^2 - r^2) \qquad \text{Original equation}$$
$$v(r) = 10^4(9 - r^2) \qquad \text{Substitute } c = 10^4 \text{ and } R = 3.$$
$$v(r) = 9 \cdot 10^4 - 10^4 r^2 \qquad \text{Distributive property}$$

Sketch the graph of $y = v(r)$ in Figure 2.83 through a sequence of transformations: Stretch the graph of $y = r^2$ vertically by a factor of 10^4, reflect the resulting graph in the x-axis, and shift this graph $9 \cdot 10^4$ units up. Figure 2.83 shows the graph of $y = v(r) = 10^4(9 - r^2)$.

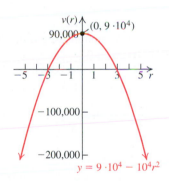

FIGURE 2.83 Velocity of flow, *r* units from the artery's center

The velocity of the blood decreases from the center of the artery ($r = 0$) to the artery wall ($r = 3$), where it ceases to flow. The portions of the graph in Figure 2.83 corresponding to negative values of r, or values of r greater than 3, do not have any physical significance.   ■ ■ ■

Practice Problem 8   Suppose in Example 8 that the artery has radius $R = 3$ and that $c = 10^3$. Sketch the graph of $y = v(r)$.   ■

Summary of Transformations of $y = f(x)$

To graph	Draw the graph of f and	Make these changes to the equation $y = f(x)$	Change the graph point (x, y) to
Vertical shifts (for $d > 0$)	Shift the graph of f		
$y = f(x) + d$	up d units.	Add d to $f(x)$.	$(x, y + d)$
$y = f(x) - d$	down d units.	Subtract d from $f(x)$.	$(x, y - d)$
Horizontal shifts	Shift the graph of f		
$y = f(x - c)$	• to the right c units if $c > 0$.	Replace x with $x - c$.	$(x - c, y)$
	• to the left $|c|$ units if $c < 0$.		
Reflection in the x-axis			
$y = -f(x)$	Reflect the graph of f in the x-axis.	Multiply $f(x)$ by -1.	$(x, -y)$
Reflection in the y-axis			
$y = f(-x)$	Reflect the graph of f in the y-axis.	Replace x with $-x$.	$(-x, y)$
Vertical stretching or compressing			
$y = af(x)$	Multiply each y-coordinate of $y = f(x)$ by a. The graph of $y = f(x)$ is stretched vertically away from the x-axis if $a > 1$ and is compressed vertically toward the x-axis if $0 < a < 1$. If $a < 0$, first sketch the graph of $y = |a|f(x)$, and then reflect it in the x-axis.	Multiply $f(x)$ by a.	(x, ay)
Horizontal stretching or compressing			
$y = f(bx)$	Multiply each x-coordinate of $y = f(x)$ by $\dfrac{1}{b}$. The graph of $y = f(x)$ is stretched away from the y-axis if $0 < b < 1$ and is compressed toward the y-axis if $b > 1$. If $b < 0$, first sketch the graph of $y = f(|b|x)$, and then reflect it in the y-axis.	Replace x with bx.	$\left(\dfrac{x}{b}, y\right)$

Summary of Combining Transformations in Sequence

To graph $y = af(bx - c) + d$, with $b > 0$

• Rewrite the equation: $y = af(bx - c) + d$ as $y = af\left[b\left(x - \dfrac{c}{b}\right)\right] + d$

(continued)

• Starting with the graph of $y = f(x)$, perform the indicated sequence of tranformations. Here is one possible sequence:

Here $\frac{c}{b} < 0$; graph shifts left.

Here $b > 1$; graph compresses horizontally.

Here $|a| > 1$; graph stretches vertically.

Here $a < 0$; graph reflects about the x-axis.

Here $d > 0$; graph shifts up.

Note: If $b < 0$, graph $y = af(|b|x - c) + d$, but reflect this graph in the y-axis before completing the last step, the vertical shift, d.

SECTION 2.7 ■ Exercises

A EXERCISES Basic Skills and Concepts

1. The graph of $y = f(x) - 3$ is found by vertically shifting the graph of $y = f(x)$ three units ———.

2. The graph of $y = f(x + 5)$ is found by horizontally shifting the graph of $y = f(x)$ five units to the ———.

3. The graph of $y = f(bx)$ is a horizontal compression of the graph of $y = f(x)$ if b ———.

4. The graph of $y = f(-x)$ is found by reflecting the graph of $y = f(x)$ in the ———.

5. *True or False* The graph of $y = f(x)$ and $y = f(-x)$ cannot be the same.

6. *True or False* You get the same graph by shifting the graph of $y = x^2$ two units up, reflecting the shifted graph in the x-axis or reflecting the graph of $y = x^2$ in the x-axis, and then shifting the reflected graph up two units.

In Exercises 7–20, describe the transformations that produce the graphs of g and h from the graph of f.

7. $f(x) = \sqrt{x}$
   a. $g(x) = \sqrt{x} + 2$ b. $h(x) = \sqrt{x} - 1$

8. $f(x) = |x|$
   a. $g(x) = |x| + 1$ b. $h(x) = |x| - 2$

9. $f(x) = x^2$
   a. $g(x) = (x + 1)^2$ b. $h(x) = (x - 2)^2$

10. $f(x) = \dfrac{1}{x}$
    a. $g(x) = \dfrac{1}{x + 2}$ b. $h(x) = \dfrac{1}{x - 3}$

11. $f(x) = \sqrt{x}$
    a. $g(x) = \sqrt{x + 1} - 2$
    b. $h(x) = \sqrt{x - 1} + 3$

12. $f(x) = x^2$
    a. $g(x) = -x^2$ b. $h(x) = (-x)^2$

13. $f(x) = |x|$
  a. $g(x) = -|x|$ **b.** $h(x) = |-x|$
14. $f(x) = \sqrt{x}$
  a. $g(x) = 2\sqrt{x}$ **b.** $h(x) = \sqrt{2x}$
15. $f(x) = \dfrac{1}{x}$

  a. $g(x) = \dfrac{2}{x}$ **b.** $h(x) = \dfrac{1}{2x}$
16. $f(x) = x^3$
  a. $g(x) = (x-2)^3 + 1$
  b. $h(x) = -(x+1)^3 + 2$
17. $f(x) = \sqrt{x}$
  a. $g(x) = -\sqrt{x} + 1$ **b.** $h(x) = \sqrt{-x} + 1$
18. $f(x) = [\![x]\!]$
  a. $g(x) = [\![x-1]\!] + 2$ **b.** $h(x) = 3[\![x]\!] - 1$
19. $f(x) = \sqrt[3]{x}$
  a. $g(x) = \sqrt[3]{x} + 1$ **b.** $h(x) = \sqrt[3]{x+1}$
20. $f(x) = \sqrt[3]{x}$
  a. $g(x) = 2\sqrt[3]{1-x} + 4$
  b. $h(x) = -\sqrt[3]{x-1} + 3$

In Exercises 21–32, match each function with its graph (a)–(l).

21. $y = -|x| + 1$ **22.** $y = -\sqrt{-x}$

23. $y = \sqrt{x^2}$ **24.** $y = \dfrac{1}{2}|x|$

25. $y = \sqrt{x+1}$ **26.** $y = 2|x| - 3$
27. $y = 1 - 2\sqrt{x}$ **28.** $y = -|x-1| + 1$
29. $y = (x-1)^2$ **30.** $y = -x^2 + 3$
31. $y = -2(x-3)^2 - 1$ **32.** $y = 3 - \sqrt{1-x}$

(a)

(b)

(c)

(d)

(e)

(f)

(g)

(h)

(i)

(j)

(k)

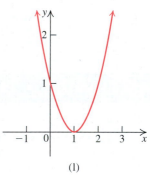
(l)

In Exercises 33–62, graph each function by starting with a function from the library of functions and then using the techniques of shifting, compressing, stretching, and/or reflecting.

33. $f(x) = x^2 - 2$ **34.** $f(x) = x^2 + 3$
35. $g(x) = \sqrt{x} + 1$ **36.** $g(x) = \sqrt{x} - 4$
37. $h(x) = |x+1|$ **38.** $h(x) = |x-2|$
39. $f(x) = (x-3)^3$ **40.** $f(x) = (x+2)^3$
41. $g(x) = (x-2)^2 + 1$ **42.** $g(x) = (x+3)^2 - 5$
43. $h(x) = -\sqrt{x}$ **44.** $h(x) = \sqrt{-x}$

45. $f(x) = -\dfrac{1}{x}$ **46.** $f(x) = -\dfrac{1}{2x}$

47. $g(x) = \dfrac{1}{2}|x|$ **48.** $g(x) = 4|x|$

49. $h(x) = -x^3 + 1$ **50.** $h(x) = -(x + 1)^3$

51. $f(x) = 2(x + 1)^2 - 1$ **52.** $f(x) = -(x - 1)^2$

53. $g(x) = 5 - x^2$ **54.** $g(x) = 2 - (x + 3)^2$

55. $h(x) = |1 - x|$ **56.** $h(x) = -2\sqrt{x - 1}$

57. $f(x) = -|x + 3| + 1$ **58.** $f(x) = 2 - \sqrt{x}$

59. $g(x) = -\sqrt{-x} + 2$ **60.** $g(x) = 3\sqrt{2 - x}$

61. $h(x) = 2[\![x + 1]\!]$ **62.** $h(x) = [\![-x]\!] + 1$

In Exercises 63–74, write an equation for a function whose graph fits the given description.

63. The graph of $f(x) = x^3$ is shifted two units up.

64. The graph of $f(x) = \sqrt{x}$ is shifted three units left.

65. The graph of $f(x) = |x|$ is reflected in the x-axis.

66. $f(x) = \sqrt{x}$ is reflected in the y-axis.

67. The graph of $f(x) = x^2$ is shifted three units right and two units up.

68. The graph of $f(x) = x^2$ is shifted two units left and reflected in the x-axis.

69. The graph of $f(x) = \sqrt{x}$ is shifted three units left, reflected in the x-axis, and shifted two units down.

70. The graph of $f(x) = \sqrt{x}$ is shifted two units down, reflected in the x-axis, and compressed vertically by a factor of $\frac{1}{2}$.

71. The graph of $f(x) = x^3$ is shifted four units left, stretched vertically by a factor of 3, reflected in the y-axis, and shifted two units up.

72. The graph of $f(x) = x^3$ is reflected in the x-axis, shifted one unit up, shifted one unit left, and reflected in the y-axis.

73. The graph of $f(x) = |x|$ is shifted four units right, compressed horizontally by a factor of 2, reflected in the x-axis, and shifted three units down.

74. The graph of $f(x) = |x|$ is shifted two units right, reflected in the y-axis, stretched horizontally by a factor of 2 and shifted three units down.

In Exercises 75–84, graph the function $y = g(x)$ given the following graph of $y = f(x)$.

75. $g(x) = f(x) - 1$ **76.** $g(x) = f(x) + 3$

77. $g(x) = -f(x)$ **78.** $g(x) = f(-x)$

79. $g(x) = f(2x)$ **80.** $g(x) = f\left(\frac{1}{2}x\right)$

81. $g(x) = f(x + 1)$ **82.** $g(x) = f(x - 2)$

83. $g(x) = 2f(x)$ **84.** $g(x) = \frac{1}{2}f(x)$

In Exercises 85–92, graph the function $y = g(x)$ given the following graph of $y = f(x)$.

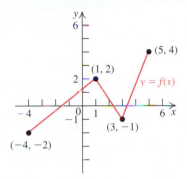

85. $g(x) = f(x) + 1$ **86.** $g(x) = -2f(x)$

87. $g(x) = f\left(\frac{1}{2}x\right)$ **88.** $g(x) = f(-2x)$

89. $g(x) = f(x - 1)$ **90.** $g(x) = f(2 - x)$

91. $g(x) = -2f(x + 1) + 3$

92. $g(x) = -f(-x + 1) - 2$

B EXERCISES Applying the Concepts

In Exercises 93–96, let f be the function that associates the employee number x of each employee of the ABC Corporation with his or her annual salary $f(x)$ in dollars.

93. **Across-the-board raise.** Each employee was awarded an across-the-board raise of $800 per year. Write a function $g(x)$ to describe the new salary.

94. **Percentage raise.** Suppose each employee was awarded a 5% raise. Write a function $h(x)$ to describe the new salary.

95. **Across-the-board and percentage raises.** Suppose each employee was awarded a $500 across-the-board raise and an additional 2% of his or her increased salary. Write a function $p(x)$ to describe the new salary.

96. **Across-the-board percentage raise.** Suppose the employees making $30,000 or more received a 2% raise, while those making less than $30,000 received a 10% raise. Write a piecewise function to describe these new salaries.

97. **Health plan.** The ABC Corporation pays for its employees' health insurance at an annual cost (in dollars) given by

$$C(x) = 5,000 + 10\sqrt{x - 1},$$

where x is the number of employees covered.
  a. Use transformations of the graph of $y = \sqrt{x}$ to sketch the graph of $y = C(x)$.
  b. Assuming that the company has 400 employees, find its annual outlay for the health coverage.

98. In Exercise 97, assume that the insurance company increases the annual cost of health insurance for the ABC Corporation by 10%. Write a function that reflects the new annual cost of health coverage.

99. **Demand.** The weekly demand for paper hats produced by Mythical Manufacturers is given by

$$x(p) = 109{,}561 - (p + 1)^2,$$

where x represents the number of hats that can be sold at a price of p cents each.
  a. Use transformations on the graph of $y = p^2$ to sketch the graph of $y = x(p)$.
  b. Find the price at which 69,160 hats can be sold.
  c. Find the price at which no hats can be sold.

100. **Revenue.** The weekly demand for cashmere sweaters produced by Wool Shop, Inc., is $x(p) = -3p + 600$. The revenue is given by $R(p) = -3p^2 + 600p$. Describe how to sketch the graph of $y = R(p)$ by applying transformations to the graph of $y = p^2$. [*Hint:* First, write $R(p)$ in the form $-3(p - h)^2 + k$.]

101. **Daylight.** At 60° north latitude, the graph of $y = f(t)$ gives the number of hours of daylight. (On the t-axis, note that $1 =$ January and $12 =$ December.)

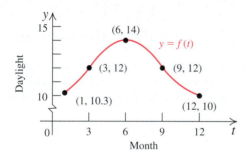

Sketch the graph of $y = f(t) - 12$.

102. Use the graph of $y = f(t)$ of Exercise 101 to sketch the graph of $y = 24 - f(t)$. Interpret the result.

C EXERCISES Beyond the Basics

103. Use transformations on $y = [\![x]\!]$ to sketch the graph of $y = -2[\![x - 1]\!] + 3$.

104. Use transformations on $y = [\![x]\!]$ to sketch the graph of $y = 3[\![x + 2]\!] - 1$.

In Exercises 105–112, by completing the square on each quadratic expression, use transformations on $y = x^2$ to sketch the graph of $y = f(x)$.

105. $f(x) = x^2 + 4x$
     [*Hint:* $x^2 + 4x = (x^2 + 4x + 4) - 4 = (x + 2)^2 - 4$.]

106. $f(x) = x^2 - 6x$

107. $f(x) = -x^2 + 2x$

108. $f(x) = -x^2 - 2x$

109. $f(x) = 2x^2 - 4x$

110. $f(x) = 2x^2 + 6x + 3.5$

111. $f(x) = -2x^2 - 8x + 3$

112. $f(x) = -2x^2 + 2x - 1$

In Exercises 113–118, sketch the graph of each function.

113. $y = |2x + 3|$
114. $y = |[\![x]\!]|$
115. $y = |4 - x^2|$
116. $y = \left|\dfrac{1}{x}\right|$
117. $y = \sqrt{|x|}$
118. $y = [\![|x|]\!]$

Critical Thinking

119. If f is a function with x-intercept -2 and y-intercept 3, find the corresponding x-intercept and y-intercept, if possible, for
  a. $y = f(x - 3)$.   b. $y = 5f(x)$.
  c. $y = -f(x)$.   d. $y = f(-x)$.

120. If f is a function with x-intercept 4 and y-intercept -1, find the corresponding x-intercept and y-intercept, if possible, for
  a. $y = f(x + 2)$.   b. $y = 2f(x)$.
  c. $y = -f(x)$.   d. $y = f(-x)$.

121. Suppose the graph of $y = f(x)$ is given. Using transformations, explain how the graph of $y = g(x)$ differs from the graph of $y = h(x)$.
  a. $g(x) = f(x) + 3$, $h(x) = f(x + 3)$
  b. $g(x) = f(x) - 1$, $h(x) = f(x - 1)$
  c. $g(x) = 2f(x)$, $h(x) = f(2x)$
  d. $g(x) = -3f(x)$, $h(x) = f(-3x)$

122. Suppose the graph of $y = f(-4x)$ is given. Explain how to obtain the graph of $y = f(x)$.

Combining Functions; Composite Functions

Before Starting this Section, Review

1. Domain of function (Section 2.4, page 215)
2. Area of a circle (Section P.7, page 74)
3. Rational inequalities (Section 1.6, page 148)

Objectives

1 Learn basic operations on functions.

2 Form composite functions.

3 Find the domain of a composite function.

4 Decompose a function.

5 Apply composition to practical problems.

EXXON VALDEZ OIL SPILL DISASTER

The oil tanker *Exxon Valdez* departed from the Trans-Alaska Pipeline terminal at 9:12 P.M., March 23, 1989. After passing through the Valdez Narrows, the tanker encountered icebergs in the shipping lanes, and Captain Hazelwood ordered the vessel to be taken out of the shipping lanes to go around the icebergs. For reasons that remain unclear, the ship failed to make the turn back into the shipping lanes, and the tanker ran aground on Bligh Reef at 12:04 A.M., March 24, 1989.

At the time of the accident, the *Exxon Valdez* was carrying 53 million gallons of oil, and approximately 11 million gallons were spilled. The amount of spilled oil is roughly equivalent to 125 Olympic-sized swimming pools. This spill was one of the largest ever to occur in the United States and is widely considered the worst worldwide in terms of damage to the environment. The spill covered a large area, stretching for 460 miles from Bligh Reef to the tiny village of Chigmik on the Alaska Peninsula. In Example 6, we calculate the area covered by an oil spill. ■

1 Learn basic operations on functions.

Combining Functions

Just as numbers can be added, subtracted, multiplied, and divided to produce new numbers, so functions can be added, subtracted, multiplied, and divided to produce other functions. To add, subtract, multiply, or divide functions, we simply add, subtract, multiply, or divide their output values.

For example, if $f(x) = x^2 - 2x + 4$ and $g(x) = x + 2$, then

$$f(x) + g(x) = (x^2 - 2x + 4) + (x + 2) = x^2 - x + 6.$$

This new function $y = x^2 - x + 6$ is called the sum function $f + g$.

SUM, DIFFERENCE, PRODUCT, AND QUOTIENT OF FUNCTIONS

Let f and g be two functions. The **sum** $f + g$, the **difference** $f - g$, the **product** fg, and the **quotient** $\dfrac{f}{g}$ are functions whose domains consist of those values of x that are common to the domains of f and g, except that for $\dfrac{f}{g}$, all x for which $g(x) = 0$ must also be excluded. These functions are defined as follows:

- Sum $\qquad\qquad (f + g)(x) = f(x) + g(x)$
- Difference $\qquad (f - g)(x) = f(x) - g(x)$
- Product $\qquad\; (fg)(x) = f(x) \cdot g(x)$
- Quotient $\qquad\; \left(\dfrac{f}{g}\right)(x) = \dfrac{f(x)}{g(x)}, \quad$ provided that $g(x) \neq 0$

EXAMPLE 1   **Combining Functions**

Let $f(x) = x^2 - 6x + 8$ and $g(x) = x - 2$. Find each of the following functions.

a. $(f - g)(4)$ **b.** $\left(\dfrac{f}{g}\right)(3)$

c. $(f + g)(x)$ **d.** $(f - g)(x)$

e. $(fg)(x)$ **f.** $\left(\dfrac{f}{g}\right)(x)$

SOLUTION

a.
$$f(4) = 4^2 - 6(4) + 8 = 0 \qquad \text{Replace } x \text{ with 4 in } f(x)$$
$$g(4) = 4 - 2 = 2 \qquad \text{Replace } x \text{ with 4 in } g(x)$$
$$(f - g)(4) = f(4) - g(4) \qquad \text{Definition of difference}$$
$$= 0 - 2 = -2 \qquad \text{Replace } f(4) \text{ with 0, } g(4) \text{ with 2}$$

b.
$$f(3) = 3^2 - 6(3) + 8 = -1 \qquad \text{Replace } x \text{ with 3 in } f(x)$$
$$g(3) = 3 - 2 = 1 \qquad \text{Replace } x \text{ with 3 in } g(x)$$
$$\left(\dfrac{f}{g}\right)(3) = \dfrac{f(3)}{g(3)} = \dfrac{-1}{1} = -1 \qquad \text{Replace } f(3) \text{ with } -1, g(3) \text{ with 1}$$

c.
$$(f + g)(x) = f(x) + g(x) \qquad\qquad \text{Definition of sum}$$
$$= (x^2 - 6x + 8) + (x - 2) \qquad \text{Add } f(x) \text{ and } g(x).$$
$$= x^2 - 5x + 6 \qquad\qquad \text{Combine terms.}$$

d.
$$(f - g)(x) = f(x) - g(x) \qquad\qquad \text{Definition of difference}$$
$$= (x^2 - 6x + 8) - (x - 2) \qquad \text{Subtract } g(x) \text{ from } f(x).$$
$$= x^2 - 7x + 10 \qquad\qquad \text{Combine terms.}$$

e.
$$(fg)(x) = f(x) \cdot g(x) \qquad\qquad\qquad \text{Definition of product}$$
$$= (x^2 - 6x + 8)(x - 2) \qquad\qquad \text{Multiply } f(x) \text{ and } g(x).$$
$$= x^2(x - 2) - 6x(x - 2) + 8(x - 2) \qquad \text{Distributive property}$$
$$= x^3 - 2x^2 - 6x^2 + 12x + 8x - 16 \qquad \text{Distributive property}$$
$$= x^3 - 8x^2 + 20x - 16 \qquad\qquad \text{Combine terms.}$$

f. $\left(\dfrac{f}{g}\right)(x) = \dfrac{f(x)}{g(x)}, \quad g(x) \neq 0$ Definition of quotient

$\qquad = \dfrac{x^2 - 6x + 8}{x - 2}, \quad x - 2 \neq 0$ Replace $f(x)$ with $x^2 - 6x + 8$ and $g(x)$ with $x - 2$.

$\qquad = \dfrac{(x - 2)(x - 4)}{x - 2}, \quad x \neq 2$ Factor the numerator.

$\qquad = x - 4, \quad x \neq 2$

Because f and g are polynomials, the domain of both f and g is the set of all real numbers, or in interval notation, $(-\infty, \infty)$. So the domain for $f + g, f - g$, and fg is $(-\infty, \infty)$. However, for $\dfrac{f}{g}$, we must exclude 2 because $g(2) = 0$. The domain for $\dfrac{f}{g}$ is the set of all real numbers $x, x \neq 2$; in interval notation, $(-\infty, 2) \cup (2, \infty)$.   ■■■

Practice Problem 1   Let $f(x) = 3x - 1$ and $g(x) = x^2 + 2$. Find $(f + g)(x)$, $(f - g)(x), (fg)(x)$, and $\left(\dfrac{f}{g}\right)(x)$.   ■

◆ **WARNING**   When rewriting a function such as

$$\left(\frac{f}{g}\right)(x) = \frac{(x - 2)(x - 4)}{x - 2}, \quad x \neq 2,$$

by removing the common factor, remember to specify "$x \neq 2$."
We identify the domain before we simplify and write

$$\left(\frac{f}{g}\right)(x) = x - 4, \quad x \neq 2.$$

---

2 Form composite functions.

Composition of Functions

There is yet another way to construct a new function from two functions. Figure 2.84 shows what happens when you apply one function $g(x) = x^2 - 1$ to an input (domain) value x and then apply a second function $f(x) = \sqrt{x}$ to the output value, $g(x)$, from the first function. In this case, the output from the first function becomes the input for the second function.

FIGURE 2.84   Composition of functions

If f and g are two functions, the composition of the function f with the function g is written as $f \circ g$ and is defined by the equation

$$(f \circ g)(x) = f(g(x)) \quad \text{(read ``}f\text{ of }g\text{ of }x\text{''),}$$

where the domain of $f \circ g$ consists of those values x in the domain of g for which $g(x)$ is in the domain of f.

Figure 2.85 may help clarify the definition of $f \circ g$.

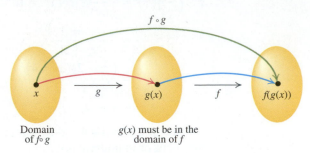

FIGURE 2.85 Composition of the function f with g

Evaluating $(f \circ g)(x)$ By definition,

$$(f \circ g)(x) = f(g(x))$$

with *inner function* pointing to $g(x)$ and *outer function* pointing to f.

Thus, to evaluate $(f \circ g)(x)$, we must

(i) evaluate the inner function g at x.

(ii) use the result $g(x)$ as the input for the outer function f.

EXAMPLE 2   Evaluating a Composite Function

Let $f(x) = x^3$ and $g(x) = x + 1$. Find each composite function.

a. $(f \circ g)(1)$ **b.** $(g \circ f)(1)$ **c.** $(f \circ f)(-1)$

SOLUTION

a. $(f \circ g)(1) = f(g(1))$ Definition of $f \circ g$
$\qquad\qquad = f(2)$ $g(1) = 1 + 1 = 2$
$\qquad\qquad = 2^3 = 8$ Evaluate $f(2)$ and simplify.

b. $(g \circ f)(1) = g(f(1))$ Definition of $g \circ f$
$\qquad\qquad = g(1)$ $f(1) = 1^3 = 1$
$\qquad\qquad = 1 + 1 = 2$ Evaluate $g(1)$ and simplify.

c. $(f \circ f)(-1) = f(f(-1))$ Definition of $f \circ f$
$\qquad\qquad = f(-1)$ $f(-1) = (-1)^3 = -1$
$\qquad\qquad = (-1)^3 = -1$ Evaluate $f(-1)$ and simplify.   ■ ■ ■

Practice Problem 2   Let $f(x) = -5x$ and $g(x) = x^2 + 1$. Find each composite function.

a. $(f \circ g)(0)$ **b.** $(g \circ f)(0)$ ■

EXAMPLE 3   Finding Composite Functions

Let $f(x) = 2x + 1$ and $g(x) = x^2 - 3$. Find each composite function.

a. $(f \circ g)(x)$ **b.** $(g \circ f)(x)$ **c.** $(f \circ f)(x)$

SOLUTION

a. $(f \circ g)(x) = f(g(x))$ Definition of $f \circ g$
$\qquad\qquad = f(x^2 - 3)$ Replace $g(x)$ with $x^2 - 3$.
$\qquad\qquad = 2(x^2 - 3) + 1$ Replace x with $x^2 - 3$ in $f(x) = 2x - 1$.
$\qquad\qquad = 2x^2 - 5$ Simplify.

b. $(g \circ f)(x) = g(f(x))$ Definition of $g \circ f$

$\qquad = g(2x + 1)$ Replace $f(x)$ with $2x + 1$.

$\qquad = (2x + 1)^2 - 3$ Replace x with $2x + 1$ in $g(x) = x^2 - 3$.

$\qquad = 4x^2 + 4x + 1 - 3$ $(A + B)^2 = A^2 + 2AB + B^2$

$\qquad = 4x^2 + 4x - 2$ Simplify.

c. $(f \circ f)(x) = f(f(x))$ Definition of $f \circ f$

$\qquad = f(2x + 1)$ Replace $f(x)$ with $2x + 1$.

$\qquad = 2(2x + 1) + 1$ Replace x with $2x + 1$ in $f(x) = 2x + 1$.

$\qquad = 4x + 3$ Simplify.   ■ ■ ■

Practice Problem 3 Let $f(x) = 2 - x$ and $g(x) = 2x^2 + 1$. Find $(g \circ f)(x)$ and $(f \circ g)(x)$.   ■

Parts **a** and **b** of Example 3 illustrate that in general, $f \circ g \neq g \circ f$. In other words, the composition of functions is not commutative.

◆ **WARNING**   Note that $f(g(x))$ is not $f(x) \cdot g(x)$. In $f(g(x))$, the output of g is *used as an input* for f; in contrast, in $f(x) \cdot g(x)$, the output $f(x)$ is *multiplied* by the output $g(x)$.

3 Find the domain of a composite function.

Domain of Composite Functions

Let f and g be two functions. The domain of the composite function $f \circ g$ consists of those values of x in the domain of g for which $g(x)$ is in the domain of f. So to find the domain of $f \circ g$, we first find the domain of g. Then from the domain of g, we exclude those values of x such that $g(x)$ is not in the domain of f.

EXAMPLE 4   **Finding the Domain of a Composite Function**

Let $f(x) = x + 1$ and $g(x) = \dfrac{1}{x}$.

a. Find $(f \circ g)(x)$ and its domain.   **b.** Find $(g \circ f)(x)$ and its domain.

SOLUTION

a. $(f \circ g)(x) = f(g(x))$ Definition of $f \circ g$

$\qquad = f\left(\dfrac{1}{x}\right)$ Replace $g(x)$ with $\dfrac{1}{x}$.

$\qquad = \dfrac{1}{x} + 1$ Replace x with $\dfrac{1}{x}$ in $f(x) = x + 1$.

Because 0 is not in the domain of $g(x) = \dfrac{1}{x}$, we must exclude 0 from the domain of $f \circ g$. This is the only restriction on the domain of $f \circ g$ because the domain of f is $(-\infty, \infty)$. So the domain of $f \circ g$ is all real x, $x \neq 0$. In interval notation, the domain of $f \circ g$ is $(-\infty, 0) \cup (0, \infty)$.

b. $(g \circ f)(x) = g(f(x))$ Definition of $g \circ f$

$\qquad = g(x + 1)$ Replace $f(x)$ with $x + 1$.

$\qquad = \dfrac{1}{x + 1}$ Replace x with $x + 1$ in $g(x) = \dfrac{1}{x}$.

The domain of f is $(-\infty, \infty)$, so we need to exclude from the domain of $g \circ f$ only values of x for which $f(x)$ is not in the domain of g. That is, values of x such that $f(x) = 0$. If $f(x) = 0$, then $x + 1 = 0$ and $x = -1$. Therefore, the domain of $g \circ f$ is all real numbers x, $x \neq -1$, or $(-\infty, -1) \cup (-1, \infty)$.   ■ ■ ■

STUDY TIP

If the function $f \circ g$ can be simplified, determine the domain *before* simplifying. For example, if $f(x) = x^2$ and $g(x) = \sqrt{x}$,

$(f \circ g)(x) = f(g(x)) = f(\sqrt{x})$.

Before simplifying:

$(f \circ g)(x) = (\sqrt{x})^2, x \geq 0$

After simplifying:

$(f \circ g)(x) = x, x \geq 0$

Practice Problem 4 Let $f(x) = \sqrt{x}$ and $g(x) = 2x + 5$. Find $(g \circ f)(x)$ and its domain. ■

4 Decompose a function.

Decomposition of a Function

In the composition of two functions, we combine two functions and create a new function. However, sometimes it is more useful to use the concept of composition to *decompose* a function into simpler functions. For example, consider the function $H(x) = \sqrt{x^2 - 2}$. A natural way to write $H(x)$ as the composition of two functions is to let $f(x) = \sqrt{x}$ and $g(x) = x^2 - 2$ so that $f(g(x)) = f(x^2 - 2) = \sqrt{x^2 - 2} = H(x)$.

A function may be decomposed into simpler functions in several different ways by choosing various "inner" functions.

FINDING THE SOLUTION: A PROCEDURE

EXAMPLE 5 Decomposing a Function

OBJECTIVE
Write a function H as a composite of simpler functions f and g so that $H = f \circ g$.

EXAMPLE
Write

$$H(x) = \frac{1}{\sqrt{2x^2 + 1}}$$

as $f \circ g$ for some functions f and g.

Step 1 Define $g(x)$ as any expression in the defining equation for H.

1. Let $g(x) = 2x^2 + 1$.

Step 2 To get $f(x)$ from the defining equation for H, (1) replace the letter H with f and (2) replace the expression chosen for $g(x)$ with x.

2. $H(x) = \dfrac{1}{\sqrt{2x^2 + 1}}$

becomes

$$f(x) = \frac{1}{\sqrt{x}}$$

Replace the expression $2x^2 + 1$ from Step 1 with x and replace H with f.

Step 3 Now we have

$$H(x) = f(g(x))$$
$$= f \circ g(x).$$

3. $f(g(x)) = f(2x^2 + 1)$

$$= \frac{1}{\sqrt{2x^2 + 1}}$$
$$= H(x)$$ ■■■

Practice Problem 5 Repeat Example 5, letting $g(x) = \sqrt{2x^2 + 1}$. Find $f(x)$ and write $H(x) = f(g(x))$. ■

5 Apply composition to practical problems.

Applications

EXAMPLE 6 Calculating the Area of an Oil Spill from a Tanker

Suppose oil spills from a tanker into the Pacific Ocean and the area of the spill is a perfect circle. (Winds, tides, and coastlines prevent real spills from forming circles.) The radius of this oil slick increases (because oil continues to spill) at the rate of 2 miles per hour.

a. Express the area of the oil slick as a function of time.

b. Calculate the area covered by the oil slick in six hours.

SOLUTION

The area A of the oil slick is a function of its radius r,

$$A = f(r) = \pi r^2$$

The radius is a function of time. We know that r increases at the rate of 2 miles per hour. So in t hours, the radius r of the oil slick is given by the function

$$r = g(t) = 2t \qquad \text{Distance} = \text{Rate} \times \text{Time}$$

a. Therefore, the area A of the oil slick is a function of the radius, which is itself a function of time. It is a composite function given by

$$A = f(g(t)) = f(2t) = \pi(2t)^2$$
$$= 4\pi t^2.$$

b. Substitute $t = 6$ into A to find the area covered by the oil in six hours.

$$A = 4\pi(6)^2 = 4\pi(36)$$
$$= 144\pi \approx 452 \text{ square miles} \qquad \blacksquare\blacksquare\blacksquare$$

Practice Problem 6   What would the result of Example 6 be if the radius of the oil slick increased at a rate of 3 miles per hour?   ■

EXAMPLE 7   **Applying Composition to Sales**

A car dealer offers an 8% discount off the manufacturer's suggested retail price (MSRP) of x dollars for any new car on his lot. At the same time, the manufacturer offers a $4000 rebate for each purchase of a new car.

a. Write a function $r(x)$ that represents the price after only the rebate.
b. Write a function $d(x)$ that represents the price after only the dealer's discount.
c. Write the functions $(r \circ d)(x)$ and $(d \circ r)(x)$. What do they represent?
d. Calculate $(d \circ r)(x) - (r \circ d)(x)$. Interpret this expression.

SOLUTION

a. The MSRP is x dollars, and the manufacturer's rebate is $4000 for each car; so

$$r(x) = x - 4000$$

represents the price of a car after the rebate.

b. The dealer's discount is 8% of x, or $0.08x$; therefore,

$$d(x) = x - 0.08x = 0.92x$$

represents the price of the car after the dealer's discount.

c. **(i)** $(r \circ d)(x) = r(d(x))$ Apply the dealer's discount first.

$$= r(0.92x) \qquad d(x) = 0.92x$$
$$= 0.92x - 4000 \qquad \text{Evaluate } r(0.92x) \text{ using } r(x) = x - 4000.$$

Then $(r \circ d)(x) = 0.92x - 4000$ represents the price when the dealer's discount is applied first.

(ii) $(d \circ r)(x) = d(r(x))$ Apply the manufacturer's rebate first.

$$= d(x - 4000) \qquad r(x) = x - 4000$$
$$= 0.92(x - 4000) \qquad \text{Evaluate } d(x - 4000) \text{ using } d(x) = 0.92x.$$
$$= 0.92x - 3680$$

Thus, $(d \circ r)(x) = 0.92x - 3680$ represents the price when the manufacturer's rebate is applied first.

d. $(d \circ r)(x) - (r \circ d)(x) = (0.92x - 3680) - (0.92x - 4000)$ Subtract function values.

$$= \$320$$

This equation shows that any car, regardless of its price, will cost $320 more if
you apply the rebate first and then the discount. ■ ■ ■

Practice Problem 7   Repeat Example 7 if the dealer offers a 6% discount and the
manufacturer offers a $4500 rebate?   ■

SECTION 2.8 ■ Exercises

A EXERCISES Basic Skills and Concepts

1. If $f(2) = 12$ and $g(2) = 4$, then $(g - f)(2) = $

   _____.

2. If $f(x) = 2x - 1$ and $f(x) + g(x) = 0$, then
   $g(x) = $ _____.

3. If $f(x) = x$ and $g(x) = 1$, then $(f \circ g)(x) = $

   _____.

4. If $f(1) = 4$ and $g(4) = 7$, then $(g \circ f)(1) = $

   _____.

5. *True or False* If $f(x) = 2x$ and $g(x) = \dfrac{1}{2x}$, then

   $(f \circ g)(x) = x$.

6. *True or False* It cannot happen that $f \circ g = g \circ f$.

**In Exercises 7–10, functions f and g are given. Find each of
the given values.**

a. $(f + g)(-1)$ b. $(f - g)(0)$

c. $(f \cdot g)(2)$ d. $\left(\dfrac{f}{g}\right)(1)$

7. $f(x) = 2x; g(x) = -x$

8. $f(x) = 1 - x^2; g(x) = x + 1$

9. $f(x) = \dfrac{1}{\sqrt{x + 2}}; g(x) = 2x + 1$

10. $f(x) = \dfrac{x}{x^2 - 6x + 8}; g(x) = 3 - x$

**In Exercises 11–22, functions f and g are given. Find each of
the following functions and state its domain.**

a. $f + g$ b. $f - g$ c. $f \cdot g$

d. $\dfrac{f}{g}$ e. $\dfrac{g}{f}$

11. $f(x) = x - 3; g(x) = x^2$

12. $f(x) = 2x - 1; g(x) = x^2$

13. $f(x) = x^3 - 1; g(x) = 2x^2 + 5$

14. $f(x) = x^2 - 4; g(x) = x^2 - 6x + 8$

15. $f(x) = 2x - 1; g(x) = \sqrt{x}$

16. $f(x) = 1 - \dfrac{1}{x}; g(x) = \dfrac{1}{x}$

17. $f(x) = \dfrac{2}{x + 1}, g(x) = \dfrac{x}{x + 1}$

18. $f(x) = \dfrac{5x - 1}{x + 1}; g(x) = \dfrac{4x + 10}{x + 1}$

19. $f(x) = \dfrac{x^2}{x + 1}; g(x) = \dfrac{2x}{x^2 - 1}$

20. $f(x) = \dfrac{x - 3}{x^2 - 25}; g(x) = \dfrac{x - 3}{x^2 + 9x + 20}$

21. $f(x) = \dfrac{2x}{x^2 - 16}; g(x) = \dfrac{2x - 7}{x^2 - 7x + 12}$

22. $f(x) = \dfrac{x^2 + 3x + 2}{x^3 + 4x}; g(x) = \dfrac{2x^3 + 8x}{x^2 + x - 2}$

**In Exercises 23 and 24, use each diagram to evaluate
$(g \circ f)(x)$. Then evaluate $(g \circ f)(2)$ and $(g \circ f)(-3)$.**

23. $x \rightarrow \boxed{f(x) = x^2 - 1} \rightarrow \rightarrow \boxed{g(x) = 2x + 3} \rightarrow$

24. $x \rightarrow \boxed{f(x) = |x + 1|} \rightarrow \rightarrow \boxed{g(x) = 3x^2 - 1} \rightarrow$

**In Exercises 25–36, let $f(x) = 2x + 1$ and $g(x) = 2x^2 - 3$.
Evaluate each expression.**

25. $(f \circ g)(2)$ 26. $(g \circ f)(2)$

27. $(f \circ g)(-3)$ 28. $(g \circ f)(-5)$

29. $(f \circ g)(0)$ 30. $(g \circ f)\left(\dfrac{1}{2}\right)$

31. $(f \circ g)(-c)$ 32. $(f \circ g)(c)$

33. $(g \circ f)(a)$ 34. $(g \circ f)(-a)$

35. $(f \circ f)(1)$ 36. $(g \circ g)(-1)$

In Exercises 37–42, the functions *f* and *g* are given. Find *f* ∘ *g* and its domain.

37. $f(x) = \dfrac{2}{x+1}; g(x) = \dfrac{1}{x}$

38. $f(x) = \dfrac{1}{x-1}; g(x) = \dfrac{2}{x+3}$

39. $f(x) = \sqrt{x-3}; g(x) = 2 - 3x$

40. $f(x) = \dfrac{x}{x-1}; g(x) = 2 + 5x$

41. $f(x) = |x|; g(x) = x^2 - 1$

42. $f(x) = 3x - 2; g(x) = |x - 1|$

In Exercises 43–56, the functions *f* and *g* are given. Find each composite function and describe its domain.

a. $f \circ g$ b. $g \circ f$
c. $f \circ f$ d. $g \circ g$

43. $f(x) = 2x - 3; g(x) = x + 4$

44. $f(x) = x - 3; g(x) = 3x - 5$

45. $f(x) = 1 - 2x; g(x) = 1 + x^2$

46. $f(x) = 2x - 3; g(x) = 2x^2$

47. $f(x) = 2x^2 + 3x; g(x) = 2x - 1$

48. $f(x) = x^2 + 3x; g(x) = 2x$

49. $f(x) = x^2; g(x) = \sqrt{x}$

50. $f(x) = x^2 + 2x; g(x) = \sqrt{x+2}$

51. $f(x) = \dfrac{1}{2x-1}; g(x) = \dfrac{1}{x^2}$

52. $f(x) = x - 1; g(x) = \dfrac{x}{x+1}$

53. $f(x) = |x|; g(x) = -2$

54. $f(x) = 3; g(x) = 5$

55. $f(x) = 1 + \dfrac{1}{x}; g(x) = \dfrac{1+x}{1-x}$

56. $f(x) = \sqrt[3]{x+1}; g(x) = x^3 + 1$

In Exercises 57–66, express the given function *H* as a composition of two functions *f* and *g* such that $H(x) = (f \circ g)(x)$.

57. $H(x) = \sqrt{x+2}$

58. $H(x) = |3x + 2|$

59. $H(x) = (x^2 - 3)^{10}$

60. $H(x) = \sqrt{3x^2 + 5}$

61. $H(x) = \dfrac{1}{3x-5}$

62. $H(x) = \dfrac{5}{2x+3}$

63. $H(x) = \sqrt[3]{x^2 - 7}$

64. $H(x) = \sqrt[4]{x^2 + x + 1}$

65. $H(x) = \dfrac{1}{|x^3 - 1|}$

66. $H(x) = \sqrt[3]{1 + \sqrt{x}}$

B EXERCISES Applying the Concepts

67. **Cost, revenue, and profit.** A retailer purchases *x* shirts from a wholesaler at a price of $12 per shirt. Her selling price for each shirt is $22. The state has 7% sales tax. Interpret each of the following functions.
   a. $f(x) = 12x$
   b. $g(x) = 22x$
   c. $h(x) = g(x) + 0.07g(x)$
   d. $P(x) = g(x) - f(x)$

68. **Cost, revenue, and profit.** The demand equation for a product is given by $x = 5000 - 5p$, where *x* is the number of units produced and sold at price *p* (in dollars) per unit. The cost (in dollars) of producing *x* units is given by $C(x) = 4x + 12{,}000$. Express each of the following as a function of price.
   a. Cost
   b. Revenue
   c. Profit

69. **Cost and revenue.** A manufacturer of radios estimates that his daily cost of producing *x* radios is given by the equation $C = 350 + 5x$. The equation $R = 25x$ represents the revenue in dollars from selling *x* radios.
   a. Write and simplify the profit function

$$P(x) = R(x) - C(x).$$

   b. Find $P(20)$. What does the number $P(20)$ represent?
   c. How many radios should the manufacturer produce and sell to have a daily profit of $500?
   d. Find the composite function $(R \circ x)(C)$. What does this function represent?
   [*Hint:* To find $x(C)$, solve $C = 350 + 5x$ for *x*.]

70. **Mail order.** You order merchandise worth *x* dollars from D-Bay Manufacturers. The company charges you sales tax of 4% of the purchase price plus a shipping-and-handling fee of $3 plus 2% of the after-tax purchase price.
   a. Write a function $g(x)$ that represents the sales tax.
   b. What does the function $h(x) = x + g(x)$ represent?
   c. Write a function $f(x)$ that represents the shipping-and-handling fee.
   d. What does the function $T(x) = h(x) + f(x)$ represent?

71. **Consumer issues.** You work in a department store in which employees are entitled to a 30% discount on their purchases. You also have a coupon worth $5 off any item.
   a. Write a function $f(x)$ that models discounting an item by 30%.
   b. Write a function $g(x)$ that models applying the coupon.
   c. Use a composition of your two functions from (a) and (b) to model your cost for an item assuming that the clerk applies the discount first and then the coupon.

d. Use a composition of your two functions from (a) and (b) to model your cost for an item assuming that the clerk applies the coupon first and then the discount.

e. Use the composite functions from (c) and (d) to find how much more an item costs assuming that the clerk applies the coupon first.

72. Consumer issues. You work in a department store in which employees are entitled to a 20% discount on their purchases. In addition, the store is offering a 10% discount on all items.
a. Write a function $f(x)$ that models discounting an item by 20%.
b. Write a function $g(x)$ that models discounting an item by 10%.
c. Use a composition of your two functions from (a) and (b) to model taking the 20% discount first.
d. Use a composition of your two functions from (a) and (b) to model taking the 10% discount first.
e. Use the composite functions from (c) and (d) to decide which discount you would prefer to take first.

73. Test grades. Professor Harsh gave a test to his college algebra class, and nobody got more than 80 points (out of 100) on the test. One problem worth 8 points had insufficient data, so nobody could solve that problem. The professor adjusted the grades for the class by (1) increasing everyone's score by 10% and (2) giving everyone 8 bonus points. Let x represent the original score of a student.
a. Write statements (1) and (2) as functions $f(x)$ and $g(x)$, respectively.
b. Find $(f \circ g)(x)$ and explain what it means.
c. Find $(g \circ f)(x)$ and explain what it means.
d. Evaluate $(f \circ g)(70)$ and $(g \circ f)(70)$.
e. Does $(f \circ g)(x) = (g \circ f)(x)$?
f. Suppose a score of 90 or better results in an A. What is the lowest original score that will result in an A if the professor uses
(i) $(f \circ g)(x)$? **(ii)** $(g \circ f)(x)$?

74. Sales commissions. Henrita works as a salesperson in a department store. Her weekly salary is $200 plus a 3% bonus on weekly sales over $8000. Suppose her sales in a week are x dollars.
a. Let $f(x) = 0.03x$. What does this mean?
b. Let $g(x) = x - 8000$. What does this mean?
c. Which composite function, $(f \circ g)(x)$ or $(g \circ f)(x)$, represents Henrita's bonus?
d. What was Henrita's salary for the week in which her sales were $17,500?
e. What were Henrita's sales for the week in which her salary was $521?

75. Enhancing a fountain. A circular fountain has a radius of x feet. A circular fence is installed around the fountain at a distance of 30 feet from its edge.
a. Write the function $f(x)$ that represents the area of the fountain.

b. Write the function $g(x)$ that represents the entire area enclosed by the fence.
c. What area does the function $g(x) - f(x)$ represent?
d. The cost of the fence was $4200, installed at the rate of $10.50 per running foot (per perimeter foot). You are to prepare an estimate for paving the area between the fence and the fountain at $1.75 per square foot. To the nearest dollar, what should your estimate be?

76. Running track design. An outdoor running track with semicircular ends and parallel sides is constructed. The length of each straight portion of the sides is 180 meters. The track has a uniform width of 4 meters throughout. The inner radius of each semicircular end is x meters.
a. Write the function $f(x)$ that represents the area enclosed within the *outer* edge of the running track.
b. Write the function $g(x)$ that represents the area of the field enclosed within the *inner* edge of the running track.
c. What does the function $f(x) - g(x)$ represent?
d. Suppose the inner perimeter of the track is 900 meters.
(i) Find the area of the track.
(ii) Find the outer perimeter of the track.

77. Area of a disk. The area A of a circular disk of radius r units is given by $A = f(r) = \pi r^2$. Suppose a metal disk is being heated and its radius r is increasing according to the equation $r = g(t) = 2t + 1$, where t is time in hours.
a. Find $(f \circ g)(t)$.
b. Determine A as a function of time.
c. Compare parts (a) and (b).

78. Volume of a balloon. The volume V of a spherical balloon of radius r inches is given by the formula $V = f(r) = \dfrac{4}{3}\pi r^3$. Suppose the balloon is being inflated and its radius is increasing at the rate of 2 inches per second. That is, $r = g(t) = 2t$, where t is time in seconds.
a. Find $(f \circ g)(t)$.
b. Determine V as a function of time.
c. Compare parts (a) and (b).

79. Currency exchange. On July 17, 2009, each British pound (symbol £) was worth \$1.64 and each Mexican peso was worth £0.04587 (*Source:* www.xe.com/ucc)

  a. Write a function f that converts pounds to dollars.
  b. Write a function g that converts pesos to pounds.
  c. Which function, $f \circ g$ or $g \circ f$, converts pesos to dollars?
  d. How much was 1000 pesos worth in dollars on July 17, 2009?

C EXERCISES Beyond the Basics

80. Odd and even functions. State whether each of the following functions is an odd function, an even function, or neither. Justify your statement.

  a. The sum of two even functions
  b. The sum of two odd functions
  c. The sum of an even function and an odd function
  d. The product of two even functions
  e. The product of two odd functions
  f. The product of an even function and an odd function

81. Odd and even functions. State whether the composite function $f \circ g$ is an odd function, an even function, or neither in the following situations.

  a. f and g are odd functions.
  b. f and g are even functions.
  c. f is odd and g is even.
  d. f is even and g is odd.

82. Let $h(x)$ be any function whose domain contains $-x$ whenever it contains x. Show that

  a. $f(x) = h(x) + h(-x)$ is an even function.
  b. $g(x) = h(x) - h(-x)$ is an odd function.
  c. $h(x)$ can always be expressed as the sum of an even function and an odd function.

83. Use Exercise 82 to write each of the following functions as the sum of an even function and an odd function.

  a. $h(x) = x^2 - 2x + 3$
  b. $h(x) = [\![x]\!] + x$

Critical Thinking

84. Let $f(x) = \dfrac{1}{[\![x]\!]}$ and $g(x) = \sqrt{x(2 - x)}$. Find the domain of each of the following functions.

  a. $f(x)$ **b.** $g(x)$
  c. $f(x) + g(x)$ **d.** $\dfrac{f(x)}{g(x)}$

85. For each of the following functions, find the domain of $f \circ f$.

  a. $f(x) = \dfrac{1}{\sqrt{-x}}$ **b.** $f(x) = \dfrac{1}{\sqrt{1 - x}}$

Inverse Functions

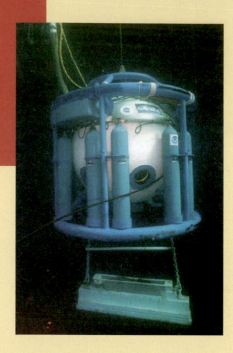

Before Starting this Section, Review

1. Vertical-line test (Section 2.4, page 216)

2. Domain and range of a function (Section 2.4, page 215)

3. Composition of functions (Section 2.8, page 269)

4. Line of symmetry (Section 2.2, page 182)

5. Solving linear and quadratic equations (Sections 1.1 and 1.4)

Objectives

1. Define an inverse function.

2. Find the inverse function.

3. Use inverse functions to find the range of a function.

4. Apply inverse functions in the real world.

WATER PRESSURE ON UNDERWATER DEVICES

The pressure at any depth in the ocean is due to the weight of the overlying water: the deeper you go, the greater the pressure. Pressure is a vital consideration in the design of all tools and vehicles that work beneath the water's surface. Every 33 feet of depth increases the pressure by approximately 15 pounds per square inch (1 atmosphere). Computing the pressure precisely requires a somewhat complicated equation that takes into account the water's salinity, temperature, and slight compressibility. However, a basic formula is given by "pressure equals depth times 15, divided by 33." So, we have pressure (psi) = depth (ft) × 15(psi per atm)/33(ft per atm), where *psi* means "pounds per square inch" and *atm* abbreviates "atmosphere."

For example, at 99 feet below the surface, the pressure is 45 psi. At 1 mile (5280 feet) below the surface, the pressure is (5280 × 15)/33 = 2400 psi. Suppose the pressure gauge on a diving bell breaks and shows a reading of 1800 psi. We can determine the bell's depth when the gauge failed by using an inverse function. We return to this problem in Example 9. ■

---

1 Define an inverse function.

Inverses

DEFINITION OF A ONE-TO-ONE FUNCTION

A function is a **one-to-one function** if each y-value in its range corresponds to only one x-value in its domain.

Let f be a one-to-one function. Then the preceding definition says that for any two numbers x_1 and x_2 in the domain of f, if $x_1 \neq x_2$, then $f(x_1) \neq f(x_2)$. That is, f *is a one-to-one function if different x-values correspond to different y-values.*

Figure 2.86(a) represents a one-to-one function, and Figure 2.86(b) represents a function that is not one-to-one. The relation shown in Figure 2.86(c) is not a function.

Domain   Range   Domain   Range   Domain   Range

One-to-one function: Each   Not a one-to-one function: One   Not a function: One x-value, x_2,
y-value corresponds to only y-value, y_2, corresponds to two   corresponds to two different
one x-value.   different x-values. y-values.
(a)   (b)   (c)

FIGURE 2.86 Deciding whether a diagram represents a function, a one-to-one function, or neither

With a one-to-one function, different x-values correspond to different y-values. Consequently, if a function f is *not* one-to-one, then there are at least two numbers x_1 and x_2 in the domain of f such that $x_1 \neq x_2$ and $f(x_1) = f(x_2)$. Geometrically, this means that the horizontal line passing through the two points $(x_1, f(x_1))$ and $(x_2, f(x_2))$ contains these *two* points on the graph of f. See Figure 2.87. The geometric interpretation of a one-to-one function is called the *horizontal-line test*.

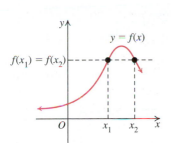

FIGURE 2.87 The function f is *not* one-to-one

HORIZONTAL-LINE TEST

A function f is one-to-one if no horizontal line intersects the graph of f in more than one point.

EXAMPLE 1   Using the Horizontal-Line Test

Use the horizontal-line test to determine which of the following functions are one-to-one.

a. $f(x) = 2x + 5$ **b.** $g(x) = x^2 - 1$ **c.** $h(x) = 2\sqrt{x}$

SOLUTION

We see that no horizontal line intersects the graphs of f (Figure 2.88(a)) or h (Figure 2.88(c)) in more than one point; therefore, the functions f and h are one-to-one. The function g is not one-to-one because the horizontal line $y = 3$ (among others) in Figure 2.88(b) intersects the graph of g at more than one point.

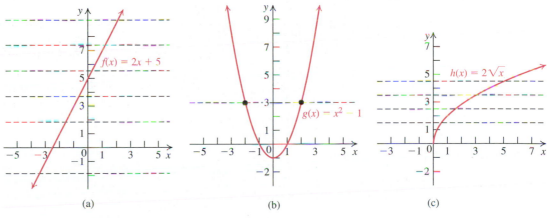

(a)   (b)   (c)

FIGURE 2.88 Horizontal-line test

Practice Problem 1 Use the horizontal-line test to determine whether $f(x) = (x - 1)^2$ is a one-to-one function.   ■

STUDY TIP

The reason that only a one-to-one function can have an inverse is that if $x_1 \neq x_2$ but $f(x_1) = y$ and $f(x_2) = y$, then $f^{-1}(y)$ must be x_1 as well as x_2. This is not possible because $x_1 \neq x_2$.

DEFINITION OF AN INVERSE FUNCTION f

Let f represent a one-to-one function. Then if y is in the range of f, there is only one value of x in the domain of f such that $f(x) = y$. We define the inverse of f, called the **inverse function of f**, denoted f^{-1}, by $f^{-1}(y) = x$ if and only if $y = f(x)$.

From this definition we have the following:

Domain of f = Range of f^{-1} and   Range of f = Domain of f^{-1}

◆ **WARNING**   The notation $f^{-1}(x)$ does not mean $\dfrac{1}{f(x)}$. The expression $\dfrac{1}{f(x)}$ represents the reciprocal of $f(x)$ and is sometimes written as $[f(x)]^{-1}$.

EXAMPLE 2   **Relating the Values of a Function and Its Inverse**

Assume that f is a one-to-one function.

a. If $f(3) = 5$, find $f^{-1}(5)$.   **b.** If $f^{-1}(-1) = 7$, find $f(7)$.

SOLUTION
By definition, $f^{-1}(y) = x$ if and only if $y = f(x)$.

a. Let $x = 3$ and $y = 5$. Now reading the definition from right to left, $5 = f(3)$ if and only if $f^{-1}(5) = 3$. So, $f^{-1}(5) = 3$.

b. Let $y = -1$ and $x = 7$. Now $f^{-1}(-1) = 7$ if and only if $f(7) = -1$. So, $f(7) = -1$.   ■ ■ ■

Practice Problem 2   Assume that f is a one-to-one function.

a. If $f(-3) = 12$, find $f^{-1}(12)$.   **b.** If $f^{-1}(4) = 9$, find $f(9)$.   ■

Consider the following input–output diagram for $f^{-1} \circ f$.

$$
\begin{array}{ccc}
y = f(x) & & f^{-1}(y) = x \\
\text{input} \xrightarrow{\ \ f\ \ } \text{output} & \text{input} \xrightarrow{\ \ f^{-1}\ \ } \text{output} \\
x \qquad\qquad y & y \qquad\qquad x
\end{array}
$$

$$
\text{input} \xrightarrow{\qquad\qquad f^{-1} \circ f \qquad\qquad} \text{output}
$$
$$
x \qquad\qquad\qquad\qquad\qquad\qquad\qquad x
$$

The diagram on page 280 suggests the following:

> ### INVERSE FUNCTION PROPERTY
>
> Let f denote a one-to-one function. Then
>
> **1.** $(f \circ f^{-1})(x) = f(f^{-1}(x)) = x$ for every x in the domain of f^{-1}.
> **2.** $(f^{-1} \circ f)(x) = f^{-1}(f(x)) = x$ for every x in the domain of f.
>
> Further, if g is any function such that
>
> $$f(g(x)) = x \text{ for all } x \text{ in the domain of } g$$
> $$\text{and} \quad g(f(x)) = x, \text{ for all } x \text{ in the domain of } f$$
> $$\text{then} \quad g = f^{-1}.$$

One interpretation of the equation $(f^{-1} \circ f)(x) = x$ is that f^{-1} undoes anything that f does to x. For example, let

$$f(x) = x + 2 \qquad f \text{ adds 2 to any input } x.$$

To undo what f does to x, we should subtract 2 from x. That is, the inverse of f should be

$$g(x) = x - 2 \qquad g \text{ subtracts 2 from } x.$$

Let's verify that $g(x) = x - 2$ is indeed the inverse function of x.

$$
\begin{aligned}
(f \circ g)(x) &= f(g(x)) && \text{Definition of } f \circ g \\
&= f(x - 2) && \text{Replace } g(x) \text{ with } x - 2. \\
&= (x - 2) + 2 && \text{Replace } x \text{ with } x - 2 \text{ in } f(x) = x + 2. \\
&= x && \text{Simplify.}
\end{aligned}
$$

We leave it for you to check that $(g \circ f)(x) = x$.

EXAMPLE 3   Verifying Inverse Functions

Verify that the following pairs of functions are inverses of each other.

$$f(x) = 2x + 3 \quad \text{and} \quad g(x) = \frac{x - 3}{2}$$

SOLUTION

$$
\begin{aligned}
(f \circ g)(x) &= f(g(x)) && \text{Definition of } f \circ g \\
&= f\!\left(\frac{x - 3}{2}\right) && \text{Replace } g(x) \text{ with } \frac{x - 3}{2}. \\
&= 2\!\left(\frac{x - 3}{2}\right) + 3 && \text{Replace } x \text{ with } \frac{x - 3}{2} \text{ in } f(x) = 2x + 3. \\
&= x && \text{Simplify.}
\end{aligned}
$$

So, $f(g(x)) = x$.

$$
\begin{aligned}
(g \circ f)(x) &= g(f(x)) && \text{Definition of } g \circ f \\
&= g(2x + 3) && \text{Replace } f(x) \text{ with } 2x + 3. \\
&= \frac{(2x + 3) - 3}{2} && \text{Replace } x \text{ with } 2x + 3 \text{ in } g(x) = \frac{x - 3}{2}. \\
&= x && \text{Simplify.}
\end{aligned}
$$

So, $g(f(x)) = x$. Because $f(g(x)) = g(f(x)) = x$, f and g are inverses of each other.

■ ■ ■

Practice Problem 3 Verify that $f(x) = 3x - 1$ and $g(x) = \dfrac{x + 1}{3}$ are inverses of each other.

■

In Example 3, notice how the functions f and g neutralize (undo) the effect of each other. The function f takes an input x, *multiplies* it by 2, and *adds* 3; g neutralizes (or undoes) this effect by *subtracting* 3 and *dividing* by 2. This process is illustrated in Figure 2.89. Notice that g reverses the operations performed by f *and* the order in which they are done.

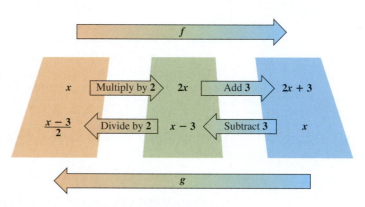

FIGURE 2.89 Function g undoes f

2 Find the inverse function.

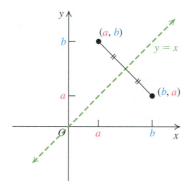

FIGURE 2.90 Relationship between points (a, b) and (b, a)

Finding the Inverse Function

Let $y = f(x)$ be a one-to-one function; then f has an inverse function. Suppose (a, b) is a point on the graph of f. Then $b = f(a)$. This means that $a = f^{-1}(b)$; so (b, a) is a point on the graph of f^{-1}. The points (a, b) and (b, a) are symmetric with respect to the line $y = x$, as shown in Figure 2.90. (See Exercises 85 and 86.) That is, if the graph paper is folded along the line $y = x$, the points (a, b) and (b, a) will coincide. Therefore, we have the following property.

> **SYMMETRY PROPERTY OF THE GRAPHS OF f AND f^{-1}**
>
> The graph of a one-to-one function f and the graph of f^{-1} are symmetric with respect to the line $y = x$.

EXAMPLE 4 **Finding the Graph of f^{-1} from the Graph of f**

The graph of a function f is shown in Figure 2.91. Sketch the graph of f^{-1}.

SOLUTION

By the horizontal-line test, f is a one-to-one function; so it has an inverse which is also a function.

The graph of f consists of two line segments: one joining the points $(-3, -5)$ and $(-1, 2)$ and the other joining the points $(-1, 2)$ and $(4, 3)$.

The graph of f^{-1} is the reflection of the graph of f in the line $y = x$. The reflections of the points $(-3, -5)$, $(-1, 2)$, and $(4, 3)$ in the line $y = x$ are $(-5, -3)$, $(2, -1)$, and $(3, 4)$, respectively.

FIGURE 2.91 Graph of f   **FIGURE 2.92 Graph of f^{-1}**

The graph of f^{-1} consists of two line segments: one joining the points $(-5, -3)$ and $(2, -1)$ and the other joining the points $(2, -1)$ and $(3, 4)$. See Figure 2.92. ■ ■ ■

Practice Problem 4   Use the graph of a function f in Figure 2.93 to sketch the graph of f^{-1}.

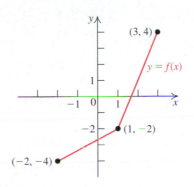

FIGURE 2.93 Graphing f^{-1} from the graph of f   ■

The symmetry between the graphs of f and f^{-1} about the line $y = x$ tells us that we can find an equation for the inverse function $y = f^{-1}(x)$ from the equation of a one-to-one function $y = f(x)$ by interchanging the roles of x and y in the equation $y = f(x)$. This results in the equation $x = f(y)$. Then we solve the equation $x = f(y)$ for y in terms of x to get $y = f^{-1}(x)$.

FINDING THE SOLUTION: A PROCEDURE

EXAMPLE 5   **Finding an Equation for f^{-1}**

OBJECTIVE	**EXAMPLE**
Find the inverse of a one-to-one function f.	*Find the inverse of $f(x) = 3x - 4$.*
Step 1 Replace $f(x)$ with y in the equation defining $f(x)$.	$y = 3x - 4$
Step 2 Interchange x and y.	$x = 3y - 4$
Step 3 Solve the equation in Step 2 for y.	$x + 4 = 3y$ Add 4 to both sides.
	$\dfrac{x + 4}{3} = y$ Divide both sides by 3.
Step 4 Replace y with $f^{-1}(x)$.	$f^{-1}(x) = \dfrac{x + 4}{3}$ We usually with $f^{-1}(x)$ on the left ■ ■ ■

Practice Problem 5   Find the inverse of $f(x) = -2x + 3$.   ■

EXAMPLE 6   **Finding the Inverse Function**

Find the inverse of the one-to-one function

$$f(x) = \frac{x + 1}{x - 2}, \quad x \neq 2.$$

SOLUTION

Step 1 $y = \dfrac{x + 1}{x - 2}$ Replace $f(x)$ with y.

Step 2 $x = \dfrac{y + 1}{y - 2}$ Interchange x and y.

Step 3   Solve $x = \dfrac{y + 1}{y - 2}$ for y.   This is the most challenging step.

$x(y - 2) = y + 1$ Multiply both sides by $y - 2$.

$xy - 2x = y + 1$ Distributive property

$xy - 2x + 2x - y = y + 1 + 2x - y$ Add $2x - y$ to both sides.

$xy - y = 2x + 1$ Simplify.

$y(x - 1) = 2x + 1$ Factor out y.

$y = \dfrac{2x + 1}{x - 1}$ Divide both sides by $x - 1$ assuming that $x \neq 1$.

Step 4 $f^{-1}(x) = \dfrac{2x + 1}{x - 1}, x \neq 1$ Replace y with $f^{-1}(x)$.

To see if our calculations are accurate, we compute $f(f^{-1}(x))$ and $f^{-1}(f(x))$.

$$f^{-1}(f(x)) = f^{-1}\left(\frac{x + 1}{x - 2}\right) = \frac{2\left(\dfrac{x + 1}{x - 2}\right) + 1}{\dfrac{x + 1}{x - 2} - 1} = \frac{2x + 2 + x - 2}{x + 1 - x + 2} = \frac{3x}{3} = x$$

You should also check that $f(f^{-1}(x)) = x$.   ■ ■ ■

Practice Problem 6   Find the inverse of the one-to-one function $f(x) = \dfrac{x}{x + 3}$, $x \neq -3$.   ■

3 Use inverse functions to find the range of a function.

Finding the Range of a One-to-One Function

In Section 2.6, we mentioned that it is not always easy to determine the range of a function that is defined by an equation. However, suppose a one-to-one function f has inverse f^{-1}. Then the range of f is the domain of f^{-1}.

EXAMPLE 7   **Finding the Domain and Range of a One-to-One Function**

Find the domain and the range of the function $f(x) = \dfrac{x + 1}{x - 2}$ of Example 6.

SOLUTION

The domain of $f(x) = \dfrac{x + 1}{x - 2}$ is the set of all real numbers x such that $x \neq 2$.

In interval notation, the domain of f is $(-\infty, 2) \cup (2, \infty)$.

From Example 6, $f^{-1}(x) = \dfrac{2x + 1}{x - 1}$, $x \neq 1$; therefore,

$$\text{Range of } f = \text{Domain of } f^{-1} = \{x | x \neq 1\}.$$

In interval notation, the range of f is $(-\infty, 1) \cup (1, \infty)$.

Practice Problem 7   Find the domain and the range of the function $f(x) = \dfrac{x}{x + 3}$.

TECHNOLOGY CONNECTION

With a graphing calculator, you can also see that the graphs of $g(x) = x^2 - 1$, $x \geq 0$ and $g^{-1}(x) = \sqrt{x - 1}$ are symmetric in the line $y = x$.

Enter Y_1, Y_2, and Y_3 as $g(x)$, $g^{-1}(x)$, and $f(x) = x$, respectively.

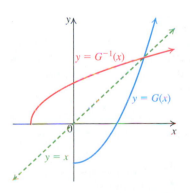

If a function f is not one-to-one, then it does not have an inverse function. Sometimes by changing its domain, we can produce an interesting function that does have an inverse. (This technique is frequently used in trigonometry.) We saw in Example 1(b) that $g(x) = x^2 - 1$ is not a one-to-one function; so g does not have an inverse function. However, the horizontal-line test shows that the function

$$G(x) = x^2 - 1, \quad x \geq 0$$

with domain $[0, \infty)$ is one-to-one. See Figure 2.94. Therefore, G has an inverse function G^{-1}.

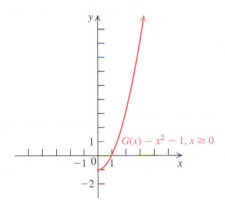

FIGURE 2.94 The function G has an inverse.

EXAMPLE 8   **Finding an Inverse Function**

Find the inverse of $G(x) = x^2 - 1$, $x \geq 0$.

SOLUTION

Step 1	$y = x^2 - 1, \quad x \geq 0$	Replace $G(x)$ with y.
Step 2	$x = y^2 - 1, \quad y \geq 0$	Interchange x and y.
Step 3	$x + 1 = y^2, \quad y \geq 0$	Add 1 to both sides.
	$y = \pm\sqrt{x + 1}, \quad y \geq 0$	Solve for y.

Because $y \geq 0$ from Step 2, we reject $y = -\sqrt{x + 1}$. So we choose

$$y = \sqrt{x + 1}.$$

Step 4	$G^{-1}(x) = \sqrt{x + 1}$	Replace y with $G^{-1}(x)$.

The graphs of G and G^{-1} are shown in Figure 2.95.

Practice Problem 8 · Find the inverse of $G(x) = x^2 - 1$, $x \leq 0$.

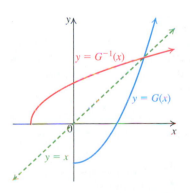

FIGURE 2.95 Graphs of G and G^{-1}

4 Apply inverse functions in the real world.

Applications

Functions that model real-life situations are frequently expressed as formulas with letters that remind you of the variable they represent. In finding the inverse of a function expressed by a formula, interchanging the letters could be very confusing. Accordingly, we omit Step 2, keep the letters the same, and just solve the formula for the other variable.

EXAMPLE 9　　**Water Pressure on Underwater Devices**

At the beginning of this section, we gave the formula for finding the water pressure p (in pounds per square inch) at a depth d (in feet) below the surface. This formula can be written as $p = \dfrac{15d}{33}$. Suppose the pressure gauge on a diving bell breaks and shows a reading of 1800 psi. How far below the surface was the bell when the gauge failed?

SOLUTION

We want to find the unknown depth in terms of the known pressure. This depth is given by the inverse of the function $p = \dfrac{15d}{33}$. To find the inverse, we solve the given equation for d.

$$p = \frac{15d}{33} \qquad \text{Original equation}$$

$$33p = 15d \qquad \text{Multiply both sides by 33.}$$

$$d = \frac{33p}{15} \qquad \text{Solve for } d.$$

Now we can use this formula to find the depth when the gauge read 1800 psi. We let $p = 1800$.

$$d = \frac{33p}{15} \qquad \text{Depth from pressure equation}$$

$$d = \frac{33(1800)}{15} \qquad \text{Replace } p \text{ with 1800.}$$

$$d = 3960$$

The device was 3960 feet below the surface when the gauge failed. ■ ■ ■

Practice Problem 9　In Example 9, suppose the pressure gauge showed a reading of 1650 psi. Determine the depth of the bell when the gauge failed. ■

SECTION 2.9 ■ Exercises

A EXERCISES Basic Skills and Concepts

1. If no horizontal line intersects the graph of a function f in more than one point, then f is a(n) _____ function.

2. A function f is one-to-one when different x-values correspond to _____.

3. If $f(x) = 3x$, then $f^{-1}(x) = $ _____.

4. The graphs of a function f and its inverse f^{-1} are symmetric in the line _____.

5. *True or False* If a function f has an inverse, then the domain of the inverse function is the range of f.

6. *True or False* It is possible for a function to be its own inverse, that is, for $f = f^{-1}$.

In Exercises 7–14, the graph of a function is given. Use the horizontal-line test to determine whether the function is one-to-one.

7.

8.

9.

10.

11.

12.

13.

14.
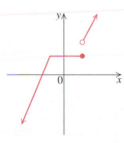

In Exercises 15–24, assume that the function f is one-to-one.

15. If $f(2) = 7$, find $f^{-1}(7)$.

16. If $f^{-1}(4) = -7$, find $f(-7)$.

17. If $f(-1) = 2$, find $f^{-1}(2)$.

18. If $f^{-1}(-3) = 5$, find $f(5)$.

19. If $f(a) = b$, find $f^{-1}(b)$.

20. If $f^{-1}(c) = d$, find $f(d)$.

21. Find $(f^{-1} \circ f)(337)$.

22. Find $(f \circ f^{-1})(25\pi)$.

23. Find $(f \circ f^{-1})(-1580)$.

24. Find $(f^{-1} \circ f)(9728)$.

25. For $f(x) = 2x - 3$, find each of the following.
   a. $f(3)$ **b.** $f^{-1}(3)$
   c. $(f \circ f^{-1})(19)$ **d.** $(f \circ f^{-1})(5)$

26. For $f(x) = x^3$, find each of the following.
   a. $f(2)$ **b.** $f^{-1}(8)$
   c. $(f \circ f^{-1})(15)$ **d.** $(f^{-1} \circ f)(27)$

27. For $f(x) = x^3 + 1$, find each of the following.
   a. $f(1)$ **b.** $f^{-1}(2)$ **c.** $(f \circ f^{-1})(269)$

28. For $g(x) = \sqrt[3]{2x^3 - 1}$, find each of the following.
   a. $g(1)$ **b.** $g^{-1}(1)$ **c.** $(g^{-1} \circ g)(135)$

In Exercises 29–34, show that f and g are inverses of each other by verifying that $f(g(x)) = x = g(f(x))$.

29. $f(x) = 3x + 1;\quad g(x) = \dfrac{x - 1}{3}$

30. $f(x) = 2 - 3x;\quad g(x) = \dfrac{2 - x}{3}$

31. $f(x) = x^3;\quad g(x) = \sqrt[3]{x}$

32. $f(x) = \dfrac{1}{x};\quad g(x) = \dfrac{1}{x}$

33. $f(x) = \dfrac{x - 1}{x + 2};\quad g(x) = \dfrac{1 + 2x}{1 - x}$

34. $f(x) = \dfrac{3x + 2}{x - 1};\quad g(x) = \dfrac{x + 2}{x - 3}$

In Exercises 35–40, the graph of a function f is given. Sketch the graph of f^{-1}.

35.

36.

37.

38.

39.

40.

In Exercises 41–52,
a. Determine whether the given function is a one-to-one function.
b. If the function is one-to-one, find its inverse.
c. Sketch the graph of the function and its inverse on the same coordinate axes.
d. Give the domain and intercepts of each one-to-one function and its inverse function.

41. $f(x) = 15 - 3x$ **42.** $g(x) = 2x + 5$

43. $f(x) = \sqrt{4 - x^2}$ **44.** $f(x) = -\sqrt{9 - x^2}$

45. $f(x) = \sqrt{x} + 3$ **46.** $f(x) = 4 - \sqrt{x}$

47. $g(x) = \sqrt[3]{x + 1}$ **48.** $h(x) = \sqrt[3]{1 - x}$

49. $f(x) = \dfrac{1}{x - 1}, x \neq 1$ **50.** $g(x) = 1 - \dfrac{1}{x}, x \neq 0$

51. $f(x) = 2 + \sqrt{x + 1}$ **52.** $f(x) = -1 + \sqrt{x + 2}$

53. Find the domain and range of the function f of Exercise 33.

54. Find the domain and range of the function f of Exercise 34.

In Exercises 55–58, assume that the given function is one-to-one. Find the inverse of the function. Also find the domain and the range of the given function.

55. $f(x) = \dfrac{x + 1}{x - 2}, x \neq 2$ **56.** $g(x) = \dfrac{x + 2}{x + 1}, x \neq -1$

57. $f(x) = \dfrac{1 - 2x}{1 + x}, x \neq -1$ **58.** $h(x) = \dfrac{x - 1}{x - 3}, x \neq 3$

In Exercises 59–66, find the inverse of each function and sketch the graph of the function and its inverse on the same coordinate axes.

59. $f(x) = -x^2, x \geq 0$ **60.** $g(x) = -x^2, x \leq 0$

61. $f(x) = |x|, x \geq 0$ **62.** $g(x) = |x|, x \leq 0$

63. $f(x) = x^2 + 1, x \leq 0$ **64.** $g(x) = x^2 + 5, x \geq 0$

65. $f(x) = -x^2 + 2, x \leq 0$ **66.** $g(x) = -x^2 - 1, x \geq 0$

B EXERCISES Applying the Concepts

67. Temperature scales. Scientists use the **Kelvin** temperature scale, in which the lowest possible temperature (called absolute zero) is 0 K. (K denotes degrees Kelvin.)
   The function $K(C) = C + 273$ gives the relationship between the Kelvin temperature (K) and Celsius temperature (C).
a. Find the inverse function of $K(C) = C + 273$. What does it represent?
b. Use the inverse function from (a) to find the Celsius temperature corresponding to 300 K.
c. A comfortable room temperature is 22°C. What is the corresponding Kelvin temperature?

68. Temperature scales. The boiling point of water is 373 K, or 212°F; the freezing point of water is 273 K, or 32°F. The relationship between Kelvin and Fahrenheit temperatures is linear.
a. Write a linear function expressing $K(F)$ in terms of F.
b. Find the inverse of the function in part (a). What does it mean?
c. A normal human body temperature is 98.6°F. What is the corresponding Kelvin temperature?

69. Composition of functions. Use Exercises 67 and 68 and the composition of functions to
a. write a function that expresses F in terms of C.
b. write a function that expresses C in terms of F.

70. Celsius and Fahrenheit temperatures. Show that the functions in (a) and (b) of Exercise 69 are inverse functions.

71. Currency exchange. Alisha went to Europe last summer. She discovered that when she exchanged her U.S. dollars for euros, she received 25% fewer euros than the number of dollars she exchanged. (She got 75 euros for every 100 U.S. dollars.) When she returned to the United States, she got 25% more dollars than the number of euros she exchanged.
a. Write each conversion function.
b. Show that in part (a), the two functions are not inverse functions.
c. Does Alisha gain or lose money after converting both ways?

72. Hourly wages. Anwar is a short-order cook in a diner. He is paid $4 per hour plus 5% of all food sales per hour. His average hourly wage w in terms of the food sales of x dollars is $w = 4 + 0.05x$.
a. Write the inverse function. What does it mean?
b. Use the inverse function to estimate the hourly sales at the diner if Anwar averages $12 per hour.

73. Hourly wages. In Exercise 72, suppose in addition that Anwar is guaranteed a minimum wage of $7 per hour.
a. Write a function expressing his hourly wage w in terms of food sales per hour. [*Hint:* Use a piecewise function.]
b. Does the function in part (a) have an inverse? Explain.
c. If the answer in part (b) is yes, find the inverse function. If the answer is no, restrict the domain so that the new function has an inverse.

74. Simple pendulum. If a pendulum is released at a certain point, the **period** is the time the pendulum takes to swing along its path and return to the point from which it was released. The period T (in seconds) of a simple pendulum is a function of its length l (in feet) and is given by $T = (1.11)l$.
a. Find the inverse function. What does it mean?
b. Use the inverse function to calculate the length of the pendulum assuming that its period is two seconds.
c. The convention center in Portland, Oregon, has the longest pendulum in the United States. The pendulum's length is 90 feet. Find the period.

75. Water supply. Suppose x is the height of the water above the opening at the base of a water tank. The velocity V of water that flows from the opening at the base is a function of x and is given by $V(x) = 8\sqrt{x}$.

  a. Find the inverse function. What does it mean?

  b. Use the inverse function to calculate the height of the water in the tank when the flow is **(i)** 30 feet per second and **(ii)** 20 feet per second.

76. Physics. A projectile is fired from the origin over horizontal ground. Its altitude y (in feet) is a function of its horizontal distance x (in feet) and is given by

$$y = 64x - 2x^2.$$

  a. Find the inverse function where the function is increasing.

  b. Use the inverse function to compute the horizontal distance when the altitude of the projectile is **(i)** 32 feet, **(ii)** 256 feet, and **(iii)** 512 feet.

77. Loan repayment. Chris purchased a car at 0% interest for five years. After making a down payment, she agreed to pay the remaining $36,000 in monthly payments of $600 per month for 60 months.

  a. What does the function $f(x) = 36{,}000 - 600x$ represent?

  b. Find the inverse of the function in part (a). What does the inverse function represent?

  c. Use the inverse function to find the number of months remaining to make payments if the balance due is $22,000.

78. Demand function. A marketing survey finds that the number x (in millions) of computer chips the market will purchase is a function of its price p (in dollars) and is estimated by $x = 8p^2 - 32p + 1200, 0 < p \le 2$.

  a. Find the inverse function. What does it mean?

  b. Use the inverse function to estimate the price of a chip if the demand is 1180.5 million chips.

C EXERCISES Beyond the Basics

In Exercises 79 and 80, show that f and g are inverses of each other by verifying that $f(g(x)) = x = g(f(x))$.

79.

x	1	3	4
$f(x)$	3	5	2

x	3	5	2
$g(x)$	1	3	4

80.

x	1	2	3	4
$f(x)$	-2	0	-3	1

x	-2	0	-3	1
$g(x)$	1	2	3	4

81. Let $f(x) = \sqrt{4 - x^2}$.

  a. Sketch the graph of $y = f(x)$.

  b. Is f one-to-one?

  c. Find the domain and the range of f.

82. Let $f(x) = \dfrac{[x] + 2}{[x] - 2}$, where $[x]$ is the greatest integer function.

  a. Find the domain of f.

  b. Is f one-to-one?

83. Let $f(x) = \dfrac{x}{x + 1} = 1 - \dfrac{1}{x + 1}$.

  a. Show that f is one-to-one.

  b. Find the inverse of f.

  c. Find the domain and the range of f.

84. Let $g(x) = \sqrt{1 - x^2}, 0 \le x \le 1$.

  a. Show that g is one-to-one.

  b. Find the inverse of g.

  c. Find the domain and the range of g.

85. Let $P(3, 7)$ and $Q(7, 3)$ be two points in the plane.

  a. Show that the midpoint M of the line segment PQ lies on the line $y = x$.

  b. Show that the line $y = x$ is the perpendicular bisector of the line segment $\overline{PQ}$; that is, show that the line $y = x$ contains the midpoint found in (a) and makes a right angle with $\overline{PQ}$. You will have shown that P and Q are symmetric about the line $y = x$.

86. Use the procedure outlined in Exercise 85 to show that the points (a, b) and (b, a) are symmetric about the line $y = x$. (See Figure 2.90.)

87. The graph of $f(x) = x^3$ is given in Section 2.6.

  a. Sketch the graph of $g(x) = (x - 1)^3 + 2$ by using transformations of the graph of f.

  b. Find $g^{-1}(x)$.

  c. Sketch the graph of $g^{-1}(x)$ by reflecting the graph of g in part (a) about the line $y = x$.

88. Inverse of composition of functions. We show that if f and g have inverses, then

$$(f \circ g)^{-1} = g^{-1} \circ f^{-1}. \text{ (Note the order.)}$$

Let $f(x) = 2x - 1$ and $g(x) = 3x + 4$.
a. Find the following:
   (i) $f^{-1}(x)$ (ii) $g^{-1}(x)$
  (iii) $(f \circ g)(x)$ (iv) $(g \circ f)(x)$
   (v) $(f \circ g)^{-1}(x)$ (vi) $(g \circ f)^{-1}(x)$
 (vii) $(f^{-1} \circ g^{-1})(x)$ (viii) $(g^{-1} \circ f^{-1})(x)$
b. From part (a), conclude that
   (i) $(f \circ g)^{-1} = g^{-1} \circ f^{-1}$.
  (ii) $(g \circ f)^{-1} = f^{-1} \circ g^{-1}$.

89. Repeat Exercise 88 for $f(x) = 2x + 3$ and $g(x) = x^3 - 1$.

90. Show that $f(x) = x^{2/3}, x \leq 0$ is a one-to-one function. Find $f^{-1}(x)$ and verify that $f^{-1}(f(x)) = x$ for every x in the domain of f.

Critical Thinking

91. Does every odd function have an inverse? Explain.

92. Is there an even function that has an inverse? Explain.

93. Does every increasing or decreasing function have an inverse? Explain.

94. A relation R is a set of ordered pairs (x, y). The inverse of R is the set of ordered pairs (y, x).
a. Give an example of a function whose inverse relation is not a function.
b. Give an example of a relation R whose inverse is a function.

SUMMARY ■ Definitions, Concepts, and Formulas

2.1 The Coordinate Plane

i. **Ordered pair.** A pair of numbers in which the order is specified is called an ordered pair of numbers.

ii. **Distance formula.** The distance between two points $P(x_1, y_1)$ and $Q(x_2, y_2)$, denoted by $d(P, Q)$, is given by

$$d(P, Q) = \sqrt{(x_2 - x_1)^2 + (y_2 - y_1)^2}.$$

iii. **Midpoint formula.** The coordinates of the midpoint $M(x, y)$ of the line segment joining $P(x_1, y_1)$ and $Q(x_2, y_2)$ are given by

$$(x, y) = \left(\frac{x_1 + x_2}{2}, \frac{y_1 + y_2}{2} \right).$$

2.2 Graphs of Equations

i. The graph of an equation in two variables, say, x and y, is the set of all ordered pairs (a, b) in the coordinate plane that satisfies the given equation. A graph of an equation, then, is a picture of its solution set.

ii. **Sketching the graph of an equation by plotting points**

Step 1 Make a representative table of solutions of the equation.

Step 2 Plot the solutions in Step 1 as ordered pairs in the coordinate plane.

Step 3 Connect the representative solutions in Step 2 with a smooth curve.

iii. Intercepts

1. The x-intercept is the x-coordinate of a point on the graph where the graph touches or crosses the x-axis. To find the x-intercepts, set $y = 0$ in the equation and solve for x.

2. The y-intercept is the y-coordinate of a point on the graph where the graph touches or crosses the y-axis. To find the y-intercepts, set $x = 0$ in the equation and solve for y.

iv. Symmetry A graph is symmetric with respect to

a. the y-axis if for every point (x, y) on the graph, the point $(-x, y)$ is also on the graph. That is, replacing x with $-x$ in the equation produces an equivalent equation.

b. the x-axis if for every point (x, y) on the graph, the point $(x, -y)$ is also on the graph. That is, replacing y with $-y$ in the equation produces an equivalent equation.

c. the *origin* if for every point (x, y) on the graph, the point $(-x, -y)$ is also on the graph. That is, replacing x with $-x$ and y with $-y$ in the equation produces an equivalent equation.

v. Circle A circle is the set of points in the coordinate plane that are at a fixed distance r from a fixed point (h, k). The fixed distance r is called the radius of the circle, and the fixed point (h, k) is called the center of the circle. The equation

$$(x - h)^2 + (y - k)^2 = r^2$$

is called the *standard equation* of a circle.

2.3 Lines

i. The slope m of a nonvertical line through the points $P(x_1, y_1)$ and $Q(x_2, y_2)$ is

$$m = \frac{\text{rise}}{\text{run}} = \frac{(y\text{-coordinate of } Q) - (y\text{-coordinate of } P)}{(x\text{-coordinate of } Q) - (x\text{-coordinate of } P)}$$

$$= \frac{y_2 - y_1}{x_2 - x_1}.$$

The slope of a vertical line is undefined. The slope of a horizontal line is 0.

ii. Equation of a line

$y - y_1 = m(x - x_1)$	Point–slope form
$y = mx + b$	Slope–intercept form
$y = k$	Horizontal line
$x = k$	Vertical line
$ax + by + c = 0$	General form

iii. Parallel and perpendicular lines
Two distinct lines with respective slopes m_1 and m_2 are

1. Parallel if $m_1 = m_2$.

2. Perpendicular if $m_1 m_2 = -1$.

iv. Linear regression A method for modeling data using linear functions.

2.4 Relations and Functions

i. Any set of ordered pairs is a **relation**. The set of all first components is the **domain**, and the set of all second components is the **range**. The graph of all of the ordered pairs is the graph of the relation.

ii. A **function** is a relation in which each element of the domain corresponds to one and only one element in the range.

iii. If a function f is defined by an equation, then its domain is the largest set of real numbers for which $f(x)$ is a real number.

iv. **Vertical-line test.** If each vertical line intersects a graph at no more than one point, the graph is the graph of a function.

2.5 Properties of Functions

i. Let x_1 and x_2 be any two numbers in the interval (a, b).

a. f is **increasing** on (a, b) if $x_1 < x_2$ implies $f(x_1) < f(x_2)$.

b. f is **decreasing** on (a, b) if $x_1 < x_2$ implies $f(x_1) > f(x_2)$.

c. f is **constant** on (a, b) if $x_1 < x_2$ implies $f(x_1) = f(x_2)$.

ii. Assume (x_1, x_2) is an interval in the domain of f containing a number a.

a. $f(a)$ is **relative minimum of f** if $f(a) \le f(x)$ for every x in (x_1, x_2).

b. $f(a)$ is **relative maximum of f** if $f(a) \ge f(x)$ for every x in (x_1, x_2).

c. A point (a, b) on the graph of f is a **turning point** if the graph of f changes direction at (a, b) from increasing to decreasing or from decreasing to increasing.

iii. Suppose that for each x in the domain of f, $-x$ is also in the domain of f.

a. f is called an **even function** if $f(-x) = f(x)$. The graph of an even function is symmetric about the y-axis.

b. f is called an **odd function** if $f(-x) = -f(x)$. The graph of an odd function is symmetric about the origin.

iv. The **average rate of change** of $f(x)$ as x changes from a to b is defined by $\dfrac{f(b) - f(a)}{b - a}, b \neq a.$

v. The expression $\dfrac{f(x) - f(a)}{x - a}, x \neq a$ is called a difference quotient at a.

The expression $\dfrac{f(x + h) - f(x)}{h}, h \neq 0$ is also called a difference quotient.

2.6 A Library of Functions

i. Basic Functions

Identity function	$f(x) = x$
Constant function	$f(x) = c$
Squaring function	$f(x) = x^2$
Cubing function	$f(x) = x^3$
Absolute value function	$f(x) = |x|$
Square root function	$f(x) = \sqrt{x}$
Cube root function	$f(x) = \sqrt[3]{x}$
Reciprocal function	$f(x) = \dfrac{1}{x}$
Reciprocal square function	$f(x) = \dfrac{1}{x^2}$
Greatest integer function	$f(x) = [\![x]\!]$

ii. For piecewise functions, different rules are used in different parts of the domain.

2.7 Transformations of Functions

i. Vertical and horizontal shifts. Let f be a function and c and d be positive numbers.

To Graph	Shift the graph of f
$f(x) + d$	up d units
$f(x) - d$	down d units
$f(x - c)$	right c units
$f(x + c)$	left c units

ii. Reflections:

To Graph	Reflect the graph of f in the
$-f(x)$	x-axis
$f(-x)$	y-axis

iii. Stretching and compressing. The graph of $g(x) = af(x)$ for $a > 0$ has the same shape as the graph of $f(x)$ and is taller if $a > 1$ and flatter if $0 < a < 1$. If a is negative, the graph of $|a|f(x)$ is reflected in the x-axis.

The graph of $g(x) = f(bx)$ for $b > 0$ is obtained from the graph of f by stretching away from the y-axis if $0 < b < 1$ and compressing horizontally toward the y-axis if $b > 1$. If b is negative, the graph of $f(|b|x)$ is reflected in the y-axis.

2.8 Combining Functions; Composite Functions

i. Given two functions f and g, for all values of x for which both $f(x)$ and $g(x)$ are defined, the functions can be combined to form **sum**, **difference**, **product**, and **quotient** functions.

ii. The **composition** of f and g is defined by $(f \circ g)(x) = f(g(x)).$

The input of f is the output of g. The domain of $f \circ g$ is the set of all x in the domain of g such that $g(x)$ is in the domain of f.

In general $f \circ g \neq g \circ f$.

2.9 Inverse Functions

i. A function f is **one-to-one** if for any x_1 and x_2 in the domain of f, $f(x_1) = f(x_2)$ implies $x_1 = x_2$.

ii. Horizontal-line test. If each horizontal line intersects the graph of a function f in, at most, one point, then f is a one-to-one function.

iii. Inverse function. Let f be a one-to-one function. Then g is the inverse of f, and we write $g = f^{-1}$, if $(f \circ g)(x) = x$ for every x in the domain of g and $(g \circ f)(x) = x$ for every x in the domain of f. The graph of f^{-1} is the reflection of the graph of f in the line $y = x$.

REVIEW EXERCISES

Basic Skills and Concepts

In Exercises 1–8, state whether the given statement is true or false.

1. $(0, 5)$ is the midpoint of the line segment joining $(-3, 1)$ and $(3, 11)$.

2. The equation $(x + 2)^2 + (y + 3)^2 = 5$ is the equation of a circle with center $(2, 3)$ and radius 5.

3. If a graph is symmetric with respect to the x-axis and the y-axis, then it must be symmetric with respect to the origin.

4. If a graph is symmetric with respect to the origin, then it must be symmetric with respect to the x-axis and the y-axis.

5. In the graph of the equation of the line $3y = 4x + 9$, the slope is 4 and the y-intercept is 9.

6. If the slope of a line is 2, then the slope of any line perpendicular to it is $\dfrac{1}{2}$.

7. The slope of a vertical line is undefined.

8. The equation $(x - 2)^2 + (y + 3)^2 = -25$ is the equation of a circle with center $(2, -3)$ and radius 5.

In Exercises 9–14, find
a. the distance between the point P and Q.
b. the coordinates of the midpoint of the line segment PQ.
c. the slope of the line containing the points P and Q.

9. $P(3, 5), Q(-1, 3)$ 10. $P(-3, 5), Q(3, -1)$

11. $P(4, -3), Q(9, -8)$ 12. $P(2, 3), Q(-7, -8)$

13. $P(2, -7), Q(5, -2)$ 14. $P(-5, 4), Q(10, -3)$

15. Show that the points $A(0, 5), B(-2, -3)$, and $C(3, 0)$ are the vertices of a right triangle.

16. Show that the points $A(1, 2), B(4, 8), C(7, -1)$, and $D(10, 5)$ are the vertices of a rhombus.

17. Which of the points $(-6, 3)$ and $(4, 5)$ is closer to the origin?

18. Which of the points $(-6, 4)$ and $(5, 10)$ is closer to the point $(2, 3)$?

19. Find a point on the x-axis that is equidistant from the points $(-5, 3)$ and $(4, 7)$.

20. Find a point on the y-axis that is equidistant from the points $(-3, -2)$ and $(2, -1)$.

In Exercises 21–24, specify whether the given graph has axis or origin symmetry (or neither).

21.

22.

23.

24.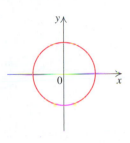

In Exercises 25–34, sketch the graph of the given equation. List all intercepts and describe any symmetry of the graph.

25. $x + 2y = 4$ 26. $3x - 4y = 12$

27. $y = -2x^2$ 28. $x = -y^2$

29. $y = x^3$ 30. $x = -y^3$

31. $y = x^2 + 2$ 32. $y = 1 - x^2$

33. $x^2 + y^2 = 16$ 34. $y = x^4 - 4$

In Exercises 35–37, find the standard form of the equation of the circle that satisfies the given conditions.

35. Center $(2, -3)$, radius 5

36. Diameter with endpoints $(5, 2)$ and $(-5, 4)$

37. Center $(-2, -5)$, touching the y-axis

In Exercises 38–42, describe and sketch the graph of the given equation.

38. $2x - 5y = 10$

39. $\dfrac{x}{2} - \dfrac{y}{5} = 1$

40. $(x + 1)^2 + (y - 3)^2 = 16$

41. $x^2 + y^2 - 2x + 4y - 4 = 0$

42. $3x^2 + 3y^2 - 6x - 6 = 0$

In Exercises 43–47, find the slope–intercept form of the equation of the line that satisfies the given conditions.

43. Passing through $(1, 2)$ with slope -2

44. x-intercept 2, y-intercept 5

45. Passing through $(1, 3)$ and $(-1, 7)$

46. Passing through $(1, 3)$, perpendicular to the x-axis

47. Determine whether the lines in each pair are parallel, perpendicular, or neither.
    a. $y = 3x - 2$ and $y = 3x + 2$
    b. $3x - 5y + 7 = 0$ and $5x - 3y + 2 = 0$
    c. $ax + by + c = 0$ and $bx - ay + d = 0$
    d. $y + 2 = \dfrac{1}{3}(x - 3)$ and $y - 5 = 3(x - 3)$

48. Write the point–slope form of the equation of the line shown in the figure that is
   a. parallel to the line in the figure with x-intercept 4.
   b. perpendicular to the line in the figure with y-intercept 1.

In Exercises 49–58, graph each relation. State the domain and range of each. Determine which relation determines y as a function of x.

49. $\{(0, 1), (-1, 2), (1, 0), (2, -1)\}$

50. $\{(0, 1), (0, -1), (3, 2), (3, -2)\}$

51. $x - y = 1$ **52.** $y = \sqrt{x - 2}$

53. $x^2 + y^2 = 0.04$ **54.** $x = -\sqrt{y}$

55. $x = 1$ **56.** $y = 2$

57. $y = |x + 1|$ **58.** $x = y^2 + 1$

In Exercises 59–76, let $f(x) = 3x + 1$ and $g(x) = x^2 - 2$. Find each of the following.

59. $f(-2)$ **60.** $g(-2)$

61. x if $f(x) = 4$ **62.** x if $g(x) = 2$

63. $(f + g)(1)$ **64.** $(f - g)(-1)$

65. $(f \cdot g)(-2)$ **66.** $(g \cdot f)(0)$

67. $(f \circ g)(3)$ **68.** $(g \circ f)(-2)$

69. $(f \circ g)(x)$ **70.** $(g \circ f)(x)$

71. $(f \circ f)(x)$ **72.** $(g \circ g)(x)$

73. $f(a + h)$ **74.** $g(a - h)$

75. $\dfrac{f(x + h) - f(x)}{h}$ **76.** $\dfrac{g(x + h) - g(x)}{h}$

In Exercises 77–82, graph each function and state its domain and range. Determine the intervals over which the function is increasing, decreasing, or constant.

77. $f(x) = -3$ **78.** $f(x) = x^2 - 2$

79. $g(x) = |3x - 2|$ **80.** $h(x) = \sqrt{36 - x^2}$

81. $f(x) = \begin{cases} x + 1 & \text{if } x \geq 0 \\ -x + 1 & \text{if } x < 0 \end{cases}$

82. $g(x) = \begin{cases} x & \text{if } x \geq 0 \\ x^2 & \text{if } x < 0 \end{cases}$

In Exercises 83–86, use transformations to graph each pair of functions on the same coordinate axes.

83. $f(x) = \sqrt{x},\ g(x) = \sqrt{x + 1}$

84. $f(x) = |x|,\ g(x) = 2|x - 1| + 3$

85. $f(x) = \dfrac{1}{x},\ g(x) = -\dfrac{1}{x - 2}$

86. $f(x) = x^2,\ g(x) = (x + 1)^2 - 2$

In Exercises 87–92, state whether each function is odd, even, or neither. Discuss the symmetry of each graph.

87. $f(x) = x^2 - x^4$ **88.** $f(x) = x^3 + x$

89. $f(x) = |x| + 3$ **90.** $f(x) = 3x + 5$

91. $f(x) = |x$ **92.** $f(x) = \dfrac{2}{x}$

In Exercises 93–96, express each function as a composition of two functions.

93. $f(x) = \sqrt{x^2 - 4}$ **94.** $g(x) = (x^2 - x + 2)^{50}$

95. $h(x) = \sqrt{\dfrac{x - 3}{2x + 5}}$ **96.** $H(x) = (2x - 1)^3 + 5$

In Exercises 97–100, determine whether the given function is one-to-one. If the function is one-to-one, find its inverse and sketch the graph of the function and its inverse on the same coordinate axes.

97. $f(x) = x + 2$ **98.** $f(x) = -2x + 3$

99. $f(x) = \sqrt[3]{x - 2}$ **100.** $f(x) = 8x^3 - 1$

In Exercises 101 and 102, assume that f is a one-to-one function. Find f^{-1} and find the domain and range of f.

101. $f(x) = \dfrac{x - 1}{x + 2}, x \neq -2$ **102.** $f(x) = \dfrac{2x + 3}{x - 1}, x \neq 1$

103. For the graph of a function f shown in the figure,
   a. Write a formula for f as a piecewise function.
   b. Find the domain and range of f.
   c. Find the intercepts of the graph of f.
   d. Draw the graph of $y = f(-x)$.
   e. Draw the graph of $y = -f(x)$.
   f. Draw the graph of $y = f(x) + 1$.
   g. Draw the graph of $y = f(x + 1)$.
   h. Draw the graph of $y = 2f(x)$.
   i. Draw the graph of $y = f(2x)$.
   j. Draw the graph of $y = f\left(\dfrac{1}{2}x\right)$.
   k. Explain why f is one-to-one.
   l. Draw the graph of $y = f^{-1}(x)$.

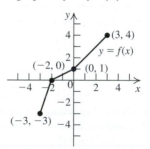

Applying the Concepts

104. Scuba diving. The pressure P on the body of a scuba diver increases linearly as she descends to greater depths d in seawater. The pressure at a depth of 10 feet is 19.2 pounds per square inch, and the pressure at a depth of 25 feet is 25.95 pounds per square inch.
   a. Write the equation relating P and d (with d as independent variable) in slope–intercept form.

b. What is the meaning of the slope and the y-intercept in part (a)?

c. Determine the pressure on a scuba diver who is at a depth of 160 feet.

d. Find the depth at which the pressure on the body of the scuba diver is 104.7 pounds per square inch.

105. Waste disposal. The Sioux Falls, South Dakota, council considered the following data regarding the cost of disposing of waste material.

Year	1996	2000
Waste (in pounds)	87,000	223,000
Cost	$54,000	$173,000

Source: Sioux Falls Health Department.

Assume that the cost C (in dollars) is linearly related to the waste w (in pounds).

a. Write the linear equation relating C and w in slope–intercept form.

b. Explain the meaning of the slope and the intercepts of the equation in part (a).

c. Suppose the city has projected 609,000 pounds of waste for the year 2008. Calculate the projected cost of disposing of the waste for 2008.

d. Suppose the city's projected budget for waste collection in 2012 is $1 million. How many pounds of waste can be handled?

106. Checking your speedometer. A common way to check the accuracy of your speedometer is to drive one or more measured (not odometer) miles, holding the speedometer at a constant 60 miles per hour and checking your watch at the start and end of the measured distance.

a. How does your watch show you whether the speedometer is accurate? What mathematics is involved?

b. Why shouldn't you use your odometer to measure the number of miles?

107. Playing blackjack. Chloe, a bright mathematician, read a book on counting cards in blackjack and went to Las Vegas to try her luck. She started with $100 and discovered that the amount of money she had at time t hours after the start of the game could be expressed by the function $f(t) = 100 + 55t - 3t^2$.

a. What amount of money had Chloe won or lost in the first two hours?

b. What was the average rate at which Chloe was winning or losing money during the first two hours?

c. Did she lose all of her money? If so, when?

d. If she played until she lost all of her money, what was the average rate at which she was losing her money?

108. Volume discounting. Major cola distributors sell a case (containing 24 cans) of cola to a retailer at a price of $4. They offer a discount of 20% for purchases over 100 cases and a discount of 25% for purchases over 500 cases. Write a piecewise function that describes this pricing scheme. Use x as the number of cases purchased and $f(x)$ as the price paid.

109. Air pollution. The daily level L of carbon monoxide in a city is a function of the number of automobiles in the city. Suppose $L(x) = 0.5\sqrt{x^2 + 4}$, where x is the number of automobiles (in hundred thousands). Suppose further that the number of automobiles in a given city is growing according to the formula $x(t) = 1 + 0.002t^2$, where t is time in years measured from now.

a. Form a composite function describing the daily pollution level as a function of time.

b. What pollution level is expected in five years?

110. Toy manufacturing. After being in business for t years, a toy manufacturer is making $x = 5000 + 50t + 10t^2$ GI Jimmy toys per year. The sale price p in dollars per toy has risen according to the formula $p = 10 + 0.5t$. Write a formula for the manufacturer's yearly revenue as

a. a function of time t.

b. a function of price p.

111. Height and weight of NBA players The table shows the roster for the National Basketball Association (NBA) team Boston Celtics for 2009.

Name	Height (in inches)	Weight (in pounds)
Ray Allen	77	205
Tony Allen	76	213
Marquis Daniels	78	200
Glen Davis	81	289
Kevin Garnett	83	253
J.R. Giddens	77	215
Eddie House	73	175
Lester Hudson	75	190
Kendrick Perkins	82	280
Paul Pierce	79	235
Rajon Rondo	73	171
Brian Scalabrine	81	235
Bill Walker	78	220
Rasheed Wallace	83	230
Shelden Williams	81	250

Source: http://www.nba.com/celtics/roster/2009-2010-roster.html

a. Use a graphing utility to find the line of regression relating the variables x (height in inches) and y (weight in pounds).

b. Use a graphing utility to plot the points and graph the regression line.

c. Use the line in part (a) to predict the weight of an NBA player whose height is 77 inches.

PRACTICE TEST A

1. Find the equation of a circle (in standard form) having a diameter with endpoints $(-2, 3)$ and $(-4, 5)$.

2. Determine which symmetries the graph of the equation $3x + 2xy^2 = 1$ has.

3. Find the x- and y-intercepts of the graph of $y = x^2(x - 3)(x + 1)$.

4. Graph the equation $x^2 + y^2 - 2x - 2y = 2$.

5. Write the slope–intercept form of the equation of the line with slope -1 passing through the point $(2, 7)$.

6. Write an equation of the line parallel to the line $8x - 2y = 7$ passing through $(2, -1)$.

7. Use $f(x) = -2x + 1$ and $g(x) = x^2 + 3x + 2$ to find $(fg)(2)$.

8. Use $f(x) = 2x - 3$ and $g(x) = 1 - 2x^2$ to evaluate $g(f(2))$.

9. Use $f(x) = x^2 - 2x$ to find $(f \circ f)(x)$.

10. If $f(x) = \begin{cases} x^3 - 2 & \text{if } x \le 0 \\ 1 - 2x^2 & \text{if } > 0 \end{cases}$, find (a) $f(-1)$, (b) $f(0)$, and (c) $f(1)$.

11. Find the domain of the function $f(x) = \dfrac{\sqrt{x}}{\sqrt{1 - x}}$.

12. Determine the average rate of change of the function $f(x) = 2x + 7$ between $x = 1$ and $x = 4$.

13. Determine whether the function $f(x) = 2x^4 - \dfrac{3}{x^2}$ is even, odd, or neither.

14. Find the intervals where the function whose graph is shown in the second column is increasing or decreasing.

15. Starting with the graph of $y = \sqrt{x}$, describe the sequence of transformations (in order) required to sketch the graph of $f(x) = 2\sqrt{x - 3} + 4$.

16. A ball is thrown upward from the ground. After t seconds, the height h (in feet) above the ground is given by $h = 25 - (2t - 5)^2$. How many seconds does the ball take to reach a height of 25 feet?

17. Assuming that f is a one-to-one function and $f(2) = 7$, find $f^{-1}(7)$.

18. Find the inverse function $f^{-1}(x)$ of the one-to-one function $f(x) = \dfrac{2x}{x - 1}$.

19. Now that Jo has saved $1000 for a car, her dad takes over and deposits $100 into her account each month. Write an equation that relates the total amount of money, A, in Jo's account to the number of months, x, that Jo's dad has been putting money in her account.

20. The cost C in dollars for renting a car for one day is a function of the number of miles traveled, m. For a car renting for $30.00 per day and $0.25 per mile, this function is given by

$$C(m) = 0.25m + 30.$$

   a. Find the cost of renting the car for one day and driving 230 miles.
   b. If the charge for renting the car for one day is $57.50, how many miles were driven?

PRACTICE TEST B

1. The graph of the equation $|x| + 2|y| = 2$ is symmetric with respect to which of the following?

   I. the origin **II.** the x-axis **III.** the y-axis
   a. I only **b.** II only
   c. III only **d.** I, II, and III

2. Which are the x- and y-intercepts of the graph of $y = x^2 - 9$?

   a. x-intercept 3, y-intercept -9
   b. x-intercepts ± 3, y-intercept -9
   c. x-intercept 3, y-intercept 9
   d. x-intercepts ± 3, y-intercepts ± 9

3. The slope of a line is undefined if the line is parallel to
   a. the x-axis **b.** the line $x - y = 0$
   c. the line $x + y = 0$ **d.** the y-axis

4. Which is an equation of the circle with center $(0, -5)$ and radius 7?
   a. $x^2 + y^2 - 10y = 0$ **b.** $x^2 + (y - 5)^2 = 7$
   c. $x^2 + (y - 5)^2 = 49$ **d.** $x^2 + y^2 + 10y = 24$

5. Which is an equation of the line passing through $(2, -3)$ and having slope -1?
a. $y + 3 = -(x + 2)$ **b.** $y - 3 = -(x - 2)$
c. $y + 3 = -(x - 2)$ **d.** $y + 3 = -(x + 2)$

6. What is a second point on the line through $(3, 2)$ having slope $-\dfrac{1}{2}$?
a. $(2, 4)$ **b.** $(5, 1)$
c. $(7, 0)$ **d.** $(4, 4)$

7. Which is an equation of the line parallel to the line $6x - 3y = 5$ passing through $(-1, 2)$?
a. $y - 2 = -2(x + 1)$ **b.** $y - 2 = 6(x + 1)$
c. $y - 2 = -6(x + 1)$ **d.** $y - 2 = 2(x + 1)$

8. Use $f(x) = 3x - 5$ and $g(x) = 2 - x^2$ to find $(f \circ g)(x)$.
a. $-9x^2 + 30x - 23$ **b.** $-3x^2 + 1$
c. $3x^2 + 1$ **d.** $9x^2 - 30x + 23$

9. Use $f(x) = 2x^2 - x$ to find $(f \circ f)(x)$.
a. $8x^4 - 8x^3 + x$ **b.** $8x^4 + x$
c. $4x^4 - 4x^3 + x^2 - x$ **d.** $4x^4 - x^3 + x^2$
e. $-x^4 + 6x^2 - x$

10. Assuming that $g(t) = \dfrac{1 - t}{1 + t}$, find $g(a - 1)$.
a. $\dfrac{2 - a}{2 + a}$ **b.** $\dfrac{1 - a}{1 + a}$
c. $\dfrac{2 - a}{a}$ **d.** -1

11. The domain of the function $f(x) = \sqrt{x} + \sqrt{1 - x}$ is
a. $[0, 1]$ **b.** $(-\infty, -1] \cup [0, +\infty)$
c. $(-\infty, 1]$ **d.** $[0, +\infty)$

12. Let $f(x) = x^2 + 3x - 4$. Find all values of x for which $(x, 6)$ is on the graph of f.
a. $-5, 1$ **b.** $-2, 5$
c. $2, -4$ **d.** $2, -5$

13. Find the intervals where the function whose graph is shown here is increasing or decreasing.
a. increasing on $(0, 3)$, decreasing on $(-3, 0) \cup (3, 4)$
b. increasing on $(-1, 3)$, decreasing on $(2, -1) \cup (3, 1)$
c. increasing on $(-3, 3)$, decreasing on $(2, -1) \cup (3, 1)$
d. increasing on $(1.2, 4)$, decreasing on $(-3, -1.4)$

14. Suppose the graph of f is given. Describe how the graph of $y = f(x + 4)$ can be obtained from the graph of f.
a. By shifting the graph of f four units to the left
b. By shifting the graph of f four units down
c. By shifting the graph of f four units up
d. By shifting the graph of f four units to the right
e. By shifting the graph of f in the x-axis

15. Suppose the graph of f is given. Describe how the graph of $y = -2f(x - 4)$ can be obtained from the graph of f.
a. By shifting the graph of f four units to the left, shrinking vertically by a factor of $\dfrac{1}{2}$, and reflecting in the x-axis
b. By shifting the graph of f four units to the right, stretching vertically by a factor of 2, and reflecting in the x-axis
c. By shifting the graph of f four units to the left, shrinking vertically by a factor of $\dfrac{1}{2}$, and reflecting in the x-axis
d. By shifting the graph of f four units to the right, stretching vertically by a factor of 2, and reflecting in the y-axis

16. Which of the following is a one-to-one function?
a. $f(x) = |x + 3|$ **b.** $f(x) = x^2$
c. $f(x) = \sqrt{x^2 + 9}$ **d.** $f(x) = x^3 + 2$

17. If f is a one-to-one function and $f(3) = 5$, then $f^{-1}(5)$ equals
a. $\dfrac{1}{5}$ **b.** $\dfrac{1}{3}$
c. 3 **d.** -5

18. Find the inverse function $f^{-1}(x)$ of $f(x) = \dfrac{x}{3x + 2}$.
a. $\dfrac{3x}{x - 2}$ **b.** $\dfrac{2x}{3x - 1}$
c. $\dfrac{2x}{1 - 3x}$ **d.** $\dfrac{x}{x - 3}$

19. The ideal weight w (in pounds) for a man of height x inches is found by subtracting 190 from five times his height. Write a linear equation that gives the ideal weight of a man of height x inches. Then find the ideal weight of a man whose height is 70 inches.
a. $w = \dfrac{x + 190}{5}$, 160 lb   **b.** $w = 5x - 190$, 160 lb
c. $w = 190 - \dfrac{x}{5}$, 176 lb   **d.** $w = \dfrac{x + 190}{5}$, 52 lb

20. Cassie rented an intermediate sedan at $25.00 per day plus $0.20 per mile. How many miles can Cassie travel in one day for $50.00?
a. 125   **b.** 1025
c. 250   **d.** 175

CUMULATIVE REVIEW EXERCISES ■ Chapters P–2

1. Simplify.
   a. $\left(\dfrac{x^3}{y^2}\right)^2 \left(\dfrac{y^2}{x^3}\right)^3$ b. $\dfrac{x^{-1}y^{-1}}{x^{-1}+y^{-1}}$

2. Factor.
   a. $2x^2 + x - 15$ b. $x^3 - 2x^2 + 4x - 8$

3. Combine and simplify.
   a. $\sqrt{75} + \sqrt{108} - \sqrt{192}$
   b. $\dfrac{x-1}{x+1} - \dfrac{x-2}{x+2}$

4. Rationalize the denominator.
   a. $\dfrac{1}{2+\sqrt{3}}$ b. $\dfrac{1}{\sqrt{5}-2}$

In Exercises 5–9, solve each equation for x.

5. a. $3x - 7 = 5$ b. $\dfrac{1}{x-1} = \dfrac{3}{x-1}$

6. a. $x^2 - 3x = 0$ b. $x^2 + 3x - 10 = 0$

7. a. $2x^2 - x + 3 = 0$ b. $4x^2 - 12x + 9 = 0$

8. a. $x - 6\sqrt{x} + 8 = 0$
   b. $\left(x - \dfrac{1}{x}\right)^2 - 10\left(x - \dfrac{1}{x}\right) + 21 = 0$

9. a. $\sqrt{3x-1} = 2x - 1$
   b. $\sqrt{1-x} = 2 - \sqrt{2x+1}$

In Exercises 10–12, solve each inequality for x and write the solution in interval notation.

10. a. $2x - 5 < 11$ b. $-3x + 4 > -5$

11. a. $-3 < 2x - 3 < 5$ b. $5 \le 1 - 2x \le 7$

12. a. $|2x - 1| \le 7$ b. $|2x - 3| \ge 5$

13. Show that the triangle with vertices $A(5, -2)$, $B(6, 5)$, and $C(2, 2)$ is isosceles.

14. Sketch the graph of $4x^2 - 9y^2 = 0$. [*Hint:* $4x^2 - 9y^2 = (2x + 3y)(2x - 3y)$.]

15. Show that the points $(-3, -1)$, $(2, 4)$, $(5, 3)$, and $(6, 2)$ lie on a circle with center $(2, -1)$.

16. Find the center and radius of the circle with equation $x^2 + y^2 - 6x + 4y + 9 = 0$.

In Exercises 17–22, find the slope–intercept form of the equation of the line satisfying the given condition.

17. Slope $= -3$; y-intercept 5

18. Slope $= 2$; x-intercept 4

19. The line is perpendicular to $y = 2x + 3$ and passes through $(2, -1)$.

20. The line is parallel to $y = 2x + 3$ and passes through $(2, -1)$.

21. The line is the perpendicular bisector of the line segment joining $(3, -1)$ and $(5, 7)$.

22. The line is parallel to $x = -2$ and passes through $(5, 7)$.

23. Find x assuming that the line through $(x, 5)$ and $(5, 11)$ is parallel to a line with slope 2.

24. Find x assuming that the line through $(x, 3)$ and $(3, 7)$ is perpendicular to a line with slope $\dfrac{1}{2}$.

In Exercise 25–28, graph each equation.

25. $12x = 4y - 6$

26. $x^2 - 2x + y^2 - 4y - 4 = 0$

27. $f(x) = \sqrt{x + 2} + 3$

28. $f(x) = -(x - 1)^2 + 4$

29. Ms. Gutiérrez bought some used books for \$1650. She kept 16 of the books and sold the rest at a profit of \$10 each. If she recovered her original \$1650 from this sale, how many books did she purchase initially?

30. The monthly note on a car that was leased for two years was \$250 less than the monthly note on a car that was leased for a year and a half. The total expense for the two leases was \$21,000. Find the monthly note for each lease.

31. Let $f(x) = \sqrt{x + 1} - 3$.
    a. Find the domain of f.
    b. Find the intercepts.
    c. Find $f(-1)$.
    d. Solve $f(x) > 0$.

32. Let $f(x) = \begin{cases} -x & \text{if } x < 0 \\ x^2 & \text{if } x \ge 0 \end{cases}$.
    a. Find $f(-2)$, $f(0)$, and $f(2)$.
    b. Find the intervals on which f is increasing, decreasing, or constant.

33. Let $f(x) = \dfrac{1}{x-2}$ and $g(x) = \dfrac{2}{x}$.
    a. Find $(f \circ g)(x)$ and state its domain.
    b. Find $(g \circ f)(x)$ and state its domain.

Polynomial and Rational Functions

Polynomials and rational functions are used in navigation, in computer-aided geometric design, in the study of supply and demand in business, in the design of suspension bridges, in computations determining the force of gravity on planets, and in virtually every area of inquiry requiring numerical approximations. Their applicability is so widespread that it is hard to imagine an area in which they have not made an impact.

Quadratic Functions

Before Starting this Section, Review

1. Linear functions (Section 2.6, page 240)
2. Completing the square (Section 1.4, page 121)
3. Quadratic formula (Section 1.4, page 123)
4. Transformations (Section 2.7, page 251)

Objectives

1. Graph a quadratic function in standard form.
2. Graph a quadratic function.
3. Solve problems modeled by quadratic functions.

YOURS, MINE, AND OURS

The movie *Yours, Mine, and Ours,* released in 1968, is a comedy about the ultimate blended family. The movie was based on the true story of Helen North, a widow with eight children, who married Frank Beardsley, a widower with ten children. The May 1964 issue of *Time* magazine had the following announcement:

May. 1, 1964

Born. To Francis Beardsley, 48, chief warrant officer at the Navy postgraduate school at Monterey, Calif., and Helen North Beardsley, 34: their second child, second daughter; in Carmel, Calif. The couple's 19 other children (he had ten, she eight from previous marriages) all voted on a name for the new Beardsley, came up "unanimously," with Helen Monica.

The film *Yours, Mine, and Ours* details the nuances of everyday life in a house full of children. The poster shown in the margin is from a remake of this popular story. In Example 5, we discuss the maximum possible number of fights among the children in such a family. ■

---

1 Graph a quadratic function in standard form.

Quadratic Function

A function whose defining equation is a polynomial is called a *polynomial function*. In this section, you will study **quadratic functions**, which are polynomial functions of degree 2.

QUADRATIC FUNCTION

A function of the form

$$f(x) = ax^2 + bx + c,$$

where a, b, and c, are real numbers with $a \neq 0$, is called a **quadratic function.**

Note that the squaring function $s(x) = x^2$ (whose graph is a parabola) is the quadratic function $f(x) = ax^2 + bx + c$ with $a = 1, b = 0,$ and $c = 0$. We will show that we can graph any quadratic function by transforming the graph of $s(x) = x^2$. Therefore, because the graph of $y = x^2$ is a parabola, the graph of any quadratic function is also a parabola.

First, we compare the graphs of $f(x) = ax^2$ for $a > 0$ and $g(x) = ax^2$ for $a < 0$. These are quadratic functions with $b = 0$ and $c = 0$.

$f(x) = ax^2, a > 0$	$g(x) = ax^2, a < 0$
The graph of $f(x) = ax^2$ is obtained from the graph of the squaring function $s(x) = x^2$ by vertically stretching or compressing it. Here are three such graphs on the same axes for $a = 2, a = 1,$ and $a = \frac{1}{2}$.	The graph of $g(x) = ax^2$ is the reflection of the graph of $f(x) = |a|x^2$ in the x-axis. Here are three graphs for $a = -2, a = -1,$ and $a = -\frac{1}{2}$.

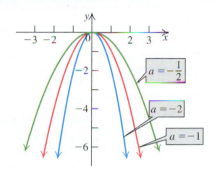

Standard Form of a Quadratic Function

Any quadratic function $f(x) = ax^2 + bx + c$ with $a \neq 0$ can be rewritten in the form $f(x) = a(x - h)^2 + k$, called the **standard form** of the quadratic function f. By using transformations (see Section 2.7), we can obtain the graph of $f(x) = a(x - h)^2 + k$ from the graph of the squaring function $s(x) = x^2$ as follows:

$s(x) = x^2$ The squaring function

$\downarrow$

$g(x) = (x - h)^2$ Horizontal translation to the right if $h > 0$ and to the left if $h < 0$

$\downarrow$

$G(x) = a(x - h)^2$ Vertical stretching or compressing. If $a < 0$, the graph is also reflected in the x-axis.

$\downarrow$

$f(x) = a(x - h)^2 + k$ Vertical translation, up if $k > 0$ and down if $k < 0$

These transformations show that the graph of a quadratic function $f(x) = a(x - h)^2 + k$ is a parabola that opens up if $a > 0$ and down if $a < 0$. Its domain is $(-\infty, \infty)$. The parabola is symmetric with respect to the vertical line $x = h$. The line of symmetry is called the **axis** (or **axis of symmetry**) of the parabola. The point (h, k) where the axis meets the parabola is called the **vertex** of the parabola. The vertex of any parabola that opens up is the *lowest point* of the graph, and the vertex of any parabola that opens down is the *highest point* of the graph. Therefore, k is the **minimum value** for any parabola that opens up or k is the **maximum value** for any parabola that opens down. The standard form is particularly useful because it immediately identifies the vertex (h, k).

$f(x) = a(x - h)^2 + k$

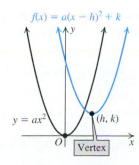

FIGURE 3.1 Graphs of quadratic functions

STANDARD FORM OF A QUADRATIC FUNCTION

The quadratic function

$$f(x) = a(x - h)^2 + k, a \neq 0$$

is in **standard form**. The graph of f is a parabola with vertex (h, k). See Figure 3.1. The parabola is symmetric with respect to the line $x = h$, called the **axis** of the parabola. If $a > 0$, the parabola opens up, and if $a < 0$, the parabola opens down. If $a > 0$, k is the **minimum value** of f, and if $a < 0$, k is the **maximum value** of f.

EXAMPLE 1   **Finding a Quadratic Function**

Find the standard form of the quadratic function f whose graph has vertex $(-3, 4)$ and passes through the point $(-4, 7)$. Does f have a maximum or a minimum value?

SOLUTION

$$f(x) = a(x - h)^2 + k \qquad \text{Standard form of a quadratic function}$$

$$f(x) = a[x - (-3)]^2 + 4 \qquad \text{Replace } h \text{ with } -3 \text{ and } k \text{ with } 4.$$

$$y = a(x + 3)^2 + 4 \qquad \text{Replace } f(x) \text{ with } y \text{ and simplify.}$$

$$7 = a(-4 + 3)^2 + 4 \qquad \text{Replace } x \text{ with } -4 \text{ and } y \text{ with } 7 \text{ because } (-4, 7)$$
$$\text{lies on the graph of } y = f(x).$$

$$7 = a + 4 \qquad \text{Simplify.}$$

$$a = 3 \qquad \text{Solve for } a.$$

The standard form is

$$f(x) = 3(x + 3)^2 + 4 \qquad a = 3, h = -3, k = 4$$

Because $a = 3 > 0, f$ has a minimum value of 4 at $x = -3$. ■ ■ ■

Practice Problem 1 Find the standard form of the quadratic function f whose graph has vertex $(1, -5)$ and passes through the point $(3, 7)$. Does f have a maximum or a minimum value? ■

FINDING THE SOLUTION: A PROCEDURE

EXAMPLE 2   **Graphing a Quadratic Function in Standard Form**

OBJECTIVE	**EXAMPLE**
Sketch the graph of $f(x) = a(x - h)^2 + k$.	*Sketch the graph of $f(x) = -3(x + 2)^2 + 12$.*
Step 1 **The graph is a parabola** because it has the form $f(x) = a(x - h)^2 + k$. Identify a, h, and k.	The graph of $f(x) = -3(x + 2)^2 + 12$ $$= (-3)[x - (-2)]^2 + 12$$ $$\qquad \uparrow \qquad\qquad \uparrow \qquad\quad \uparrow$$ $$\qquad a \qquad\qquad h \qquad\quad k$$ is a parabola; $a = -3$, $h = -2$, and $k = 12$.
Step 2 **Determine how the parabola opens**. If $a > 0$, the parabola opens *up*. If $a < 0$, it opens *down*.	Because $a = -3 < 0$, the parabola opens down.
Step 3 **Find the vertex (h, k)**. If $a > 0$ (or $a < 0$), the function f has a minimum (or a maximum) value k at $x = h$.	The vertex $(h, k) = (-2, 12)$. Because the parabola opens down, the function f has a maximum value of 12 at $x = -2$.
Step 4 **Find the x-intercepts (if any)**. Set $f(x) = 0$ and solve the equation $a(x - h)^2 + k = 0$ for x. If the solutions are real numbers, they are the x-intercepts. If not, the parabola lies above the x-axis, when $a > 0$ and below the x-axis, when $a < 0$.	$$0 = -3(x + 2)^2 + 12 \qquad \text{Set } f(x) = 0.$$ $$3(x + 2)^2 = 12 \qquad \text{Add } 3(x + 2)^2 \text{ to both sides.}$$ $$(x + 2)^2 = 4 \qquad \text{Divide both sides by 3.}$$ $$x + 2 = \pm 2 \qquad \text{Square root property}$$ $$x = -2 \pm 2 \qquad \text{Subtract 2 from both sides.}$$ $$x = 0 \quad \text{or} \quad x = -4 \qquad \text{Solve for } x.$$ The x-intercepts are 0 and -4. The parabola passes through the points $(0, 0)$ and $(-4, 0)$.

Step 5 **Find the *y*-intercept**. Replace *x* with 0. Then $f(0) = ah^2 + k$ is the *y*-intercept.

$f(0) = -3(0 + 2)^2 + 12 = -3(4) + 12 = 0$. The *y*-intercept is 0.

Step 6 **Sketch the graph**. Plot the points found in Steps 3–5 and join them, forming a parabola. Show the axis $x = h$ of the parabola by drawing a dashed vertical line.

If there are no *x*-intercepts, draw the half of the parabola that passes through the vertex and the *y*-intercept. Then use the axis of symmetry to draw the other half.

The axis of the parabola is the vertical line $x = -2$. The graph of the parabola is shown in the figure.

$(-2, 12)$

$f(x) = -3(x + 2)^2 + 12$

$x = -2$

Practice Problem 2   Graph the quadratic function $f(x) = -2(x + 1)^2 + 3$.   ■

2 Graph a quadratic function.

Graphing a Quadratic Function $f(x) = ax^2 + bx + c$

A quadratic function $f(x) = ax^2 + bx + c$ can be changed to its standard form $f(x) = a(x - h)^2 + k$ by *completing the square*.

Converting $f(x) = ax^2 + bx + c$ to Standard Form

$f(x) = ax^2 + bx + c$ Original function; $a \neq 0$

$\displaystyle = a\left(x^2 + \frac{b}{a}x\right) + c$ Factor out *a* from $ax^2 + bx$.

$\displaystyle = a\left(x^2 + \frac{b}{a}x + \frac{b^2}{4a^2} - \frac{b^2}{4a^2}\right) + c$ To complete the square, add and subtract $\left(\dfrac{1}{2} \cdot \dfrac{b}{a}\right)^2 = \dfrac{b^2}{4a^2}$ within the parentheses.

$\displaystyle = a\left(x^2 + \frac{b}{a}x + \frac{b^2}{4a^2}\right) - a \cdot \frac{b^2}{4a^2} + c$ Distributive property

$\displaystyle = a\left(x + \frac{b}{2a}\right)^2 - \frac{b^2}{4a} + c$ $\left(x + \dfrac{b}{2a}\right)^2 = x^2 + \dfrac{b}{a}x + \dfrac{b^2}{4a^2}$

Comparing this form with the standard form $f(x) = a(x - h)^2 + k$, we have:

$$h = -\frac{b}{2a} \quad \text{*x*-coordinate of the vertex}$$

$$\text{and} \quad k = c - \frac{b^2}{4a} \quad \text{*y*-coordinate of the vertex}$$

Notice that $f\left(-\dfrac{b}{2a}\right) = a\left(-\dfrac{b}{2a} + \dfrac{b}{2a}\right)^2 - \dfrac{b^2}{4a} + c$ Replace *x* with $-\dfrac{b}{2a}$ in the last equation for $f(x)$.

$$= -\frac{b^2}{4a} + c = k.$$

So to find the vertex $(h, k) = \left(-\dfrac{b}{2a}, f\left(-\dfrac{b}{2a}\right)\right)$, you find the x-coordinate $h = -\dfrac{b}{2a}$

of the vertex and then calculate $k = f\left(-\dfrac{b}{2a}\right)$ to find its y-coordinate.

FINDING THE SOLUTION: A PROCEDURE

| EXAMPLE 3 | **Graphing a Quadratic Function** |

OBJECTIVE

Graph $f(x) = ax^2 + bx + c, a \neq 0$.

Step 1 **Identify a, b, and c.**

Step 2 **Determine how the parabola opens.**
If $a > 0$, the parabola opens *up*; if $a < 0$, the parabola opens *down*.

Step 3 **Find the vertex (h, k).**
Use the formula

$$(h, k) = \left(-\frac{b}{2a}, f\left(-\frac{b}{2a}\right)\right).$$

k is a minimum value of f if $a > 0$
k is a maximum value of f if $a < 0$

Step 4 **Find the x-intercepts (if any).**
Set $f(x) = 0$ and solve
$ax^2 + bx + c = 0$. If the solutions are real numbers, they are the x-intercepts. If not, the parabola lies above the x-axis, when $a > 0$ and below the x-axis, when $a < 0$.

Step 5 **Find the y-intercept. Set $x = 0$.** The result, $f(0) = c$, is the y-intercept.

Step 6 The parabola is symmetric with respect to its axis, $x = -\dfrac{b}{2a}$. Use this symmetry to find additional points.

Step 7 Draw a parabola through the points found in Steps 3–6.

If there are no x-intercepts, draw the half of the parabola that passes through the vertex and the y-intercept. Then use the axis of symmetry to draw the other half.

EXAMPLE

Sketch the graph of $f(x) = 2x^2 + 8x - 10$.

In the equation $y = f(x) = 2x^2 + 8x - 10$,
$a = 2$, $b = 8$, and $c = -10$.

Because $a = 2 > 0$, the parabola opens up.

$$h = -\frac{b}{2a} = -\frac{8}{2(2)} = -2$$
$$k = f(h) = f(-2)$$
$$= 2(-2)^2 + 8(-2) - 10 \qquad \text{Replace } h \text{ with } -2 \text{ in } f(x).$$
$$= -18 \qquad \text{Simplify.}$$

The function f has a minimum value of -18 at $x = -2$.

$$2x^2 + 8x - 10 = 0 \qquad \text{Set } f(x) = 0.$$
$$2(x^2 + 4x - 5) = 0 \qquad \text{Factor out 2.}$$
$$2(x + 5)(x - 1) = 0 \qquad \text{Factor.}$$
$$x + 5 = 0 \quad \text{or} \quad x - 1 = 0 \qquad \text{Zero-product property}$$
$$x = -5 \quad \text{or} \quad x = 1 \qquad \text{Solve for } x.$$

The x-intercepts are at the points $(-5, 0)$ and $(1, 0)$.

Set $x = 0$ to obtain $f(0) = 2(0)^2 + 8(0) - 10 = -10$. The y-intercept is at the point $(0, -10)$.

The axis of symmetry is $x = -2$. The symmetric image of $(0, -10)$ with respect to the axis $x = -2$ is $(-4, -10)$.

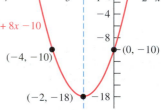

The parabola passing through the points in Steps 3–6 is sketched in the figure.

Practice Problem 3 Graph the quadratic function $f(x) = 3x^2 - 3x - 6$. ■

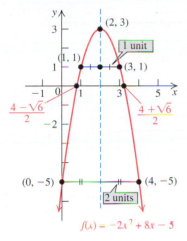

$$f(x) = -2x^2 + 8x - 5$$

FIGURE 3.2

3 Solve problems modeled by quadratic functions.

RECALL

For a quadratic function
$f(x) = ax^2 + bx + c$, the vertex
is $(h, k) = \left(-\dfrac{b}{2a}, f\left(-\dfrac{b}{2a}\right)\right)$ and
k is the function's minimum
value if $a > 0$ and its maximum
value if $a < 0$.

BY THE WAY . . .

This example illustrates a
counting principle that we will
study in Section 8.6

Note: The domain of a
function that models a real-
life problem may consist of
only integer values. However,
in solving these problems we
frequently find it convenient
to extend the domain to a
suitable interval of real numbers containing the relevant
integers.

EXAMPLE 4   **Identifying the Characteristics of a Quadratic Function from Its Graph**

The graph of the function $f(x) = -2x^2 + 8x - 5$ is shown in Figure 3.2.
Find the domain and range of f.

SOLUTION

The graph in Figure 3.2 is a parabola that opens down and whose vertex is $(2, 3)$.
   The domain of f is $(-\infty, \infty)$. Because the parabola opens down, the maximum
value of f is the y-coordinate 3 of its vertex $(2, 3)$. Every y value with $y < 3$ corresponds to two graph points, so the range of f is $(-\infty, 3]$.   ■ ■ ■

Practice Problem 4   Graph the function $f(x) = 3x^2 - 6x - 1$ and find the domain
and range of f.   ■

Applications

Many applications of quadratic functions involve finding the maximum or minimum value of the function.

EXAMPLE 5   *Yours, Mine, and Ours*

A widower with ten children marries a widow who also has children. After their
marriage, they have their own children. If the total number of children is 24 and we
assume that the children of the same parents do not fight, find the maximum
possible number of fights among the children. (In this example, if Sean and Misty
fight, no matter how many times they fight, we count it as one fight.)

SOLUTION

Suppose the widow had x children from her previous marriages. Then the couple has
$24 - 10 - x = 14 - x$ additional children after their marriage. Because the
children of the same parents do not fight, there are no fights among the ten children
the widower brought into the marriage, among the x children the widow brought into
the marriage, or among the $14 - x$ children of the couple (widower and widow).
   The possible number of fights among the children of

(i) the widower (10 children) and the widow (x children) is $10x$.

(ii) the widower (10 children) and the couple ($14 - x$ children) is $10(14 - x)$.

(iii) the widow (x children) and the couple ($14 - x$ children) is $x(14 - x)$.

   The possible number y of all fights is given by

$$y = 10x + 10(14 - x) + x(14 - x) \qquad \text{The sum of i, ii, and iii.}$$
$$= 10x + 140 - 10x + 14x - x^2 \qquad \text{Distributive property}$$
$$= 140 + 14x - x^2 \qquad \text{Simplify.}$$

In the quadratic function $y = f(x) = -x^2 + 14x + 140$, we have $a = -1, b = 14$,
and $c = 140$.
   The vertex (h, k) is given by

$$h = -\frac{b}{2a} = -\frac{14}{2(-1)} = 7,$$
$$k = f(7) = -(7)^2 + 14(7) + 140 = 189.$$

Because $a < 0$, the function f has a maximum value. So the maximum possible
number of fights among the children is 189.   ■ ■ ■

Practice Problem 5   Repeat Example 5 assuming that the widower had eight
children from his previous marriage.   ■

SECTION 3.1 ■ Exercises

A EXERCISES Basic Skills and Concepts

1. A point where the axis of the parabola meets the parabola is called the _____.

2. The vertex of the graph of $f(x) = -2(x + 3)^2 - 5$ is _____.

3. *True or False* The graph of the function in Exercise 2 opens down.

4. *True or False* The graph of $f(x) = -2 - x + x^2$ opens down.

5. The x-coordinate of the vertex of the parabola of Exercise 4 is _____.

6. *True or False* The y-coordinate of the vertex of the parabola $f(x) = x^2 - 2x + 5$ is $f(1)$.

7. *True or False* If $a > 0$, then the minimum value of $f(x) = ax^2 + bx + c$ is $f\left(-\dfrac{b}{2a}\right)$.

8. *True or False* If $a < 0$, then $f(x) = ax^2 + bx + c$ has no minimum value.

In Exercises 9–16, match each quadratic function with its graph.

9. $y = -\dfrac{1}{3}x^2$

10. $y = -3x^2$

11. $y = -3(x + 1)^2$

12. $y = 2(x + 1)^2$

13. $y = (x - 1)^2 + 2$

14. $y = (x - 1)^2 - 3$

15. $y = 2(x + 1)^2 - 3$

16. $y = -3(x + 1)^2 + 2$

(a)

(b)

(c)

(d)

(e)

(f)

(g)

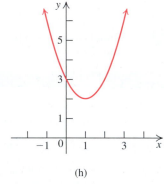

(h)

In Exercises 17–20, find a quadratic function of the form $y = ax^2$ that passes through the given point.

17. $(2, -8)$

18. $(-3, 3)$

19. $(2, 20)$

20. $(-3, -6)$

In Exercises 21–30, find the quadratic function $y = f(x)$ that has the given vertex and whose graph passes through the given point. Write the function in standard form.

21. Vertex $(0, 0)$; passing through $(-2, 8)$

22. Vertex $(2, 0)$; passing through $(1, 3)$

23. Vertex $(-3, 0)$; passing through $(-5, -4)$

24. Vertex $(0, 1)$; passing through $(-1, 0)$

25. Vertex $(2, 5)$; passing through $(3, 7)$

26. Vertex $(-3, 4)$; passing through $(0, 0)$

27. Vertex $(2, -3)$; passing through $(-5, 8)$

28. Vertex $(-3, -2)$; passing through $(0, -8)$

29. Vertex $\left(\dfrac{1}{2}, \dfrac{1}{2}\right)$; passing through $\left(\dfrac{3}{4}, -\dfrac{1}{4}\right)$

30. Vertex $\left(-\dfrac{3}{2}, -\dfrac{5}{2}\right)$; passing through $\left(1, \dfrac{55}{8}\right)$

In Exercises 31–34, the graph of a quadratic function $y = f(x)$ is given. Find the standard form of the function.

31.

32.

33.

34.

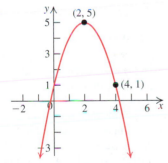

In Exercises 35–44, graph each function by starting with the graph of $y = x^2$ and using transformations.

35. $f(x) = 3x^2$

36. $f(x) = \dfrac{1}{3}x^2$

37. $g(x) = (x - 4)^2$

38. $g(x) = (x + 3)^2$

39. $f(x) = -2x^2 - 4$

40. $f(x) = -x^2 + 3$

41. $g(x) = (x - 3)^2 + 2$

42. $g(x) = (x + 1)^2 - 3$

43. $f(x) = -3(x - 2)^2 + 4$

44. $f(x) = -2(x + 1)^2 + 3$

In Exercises 45–52, graph the given function by writing it in the standard form $y = a(x - h)^2 + k$ and then using transformations on $y = x^2$. Find the vertex, the axis, and the intercepts of the parabola.

45. $y = x^2 + 4x$

46. $y = x^2 - 2x + 2$

47. $y = 6x - 10 - x^2$

48. $y = 8 + 3x - x^2$

49. $y = 2x^2 - 8x + 9$

50. $y = 3x^2 + 12x - 7$

51. $y = -3x^2 + 18x - 11$

52. $y = -5x^2 - 20x + 13$

In Exercises 53–60, (a) determine whether the graph of the given quadratic function opens up or down, (b) find the vertex $\left(-\dfrac{b}{2a}, f\left(-\dfrac{b}{2a}\right)\right)$, (c) find the axis of symmetry, (d) find the x- and y-intercepts, and (e) sketch the graph of the function.

53. $y = x^2 - 8x + 15$

54. $y = x^2 + 8x + 13$

55. $y = x^2 - x - 6$

56. $y = x^2 + x - 2$

57. $y = x^2 - 2x + 4$

58. $y = x^2 - 4x + 5$

59. $y = 6 - 2x - x^2$

60. $y = 2 + 5x - 3x^2$

In Exercises 61–68, a quadratic function f is given. (a) Determine whether the given quadratic function has a maximum value or a minimum value. Then find this value. (b) Find the range of f.

61. $f(x) = x^2 - 4x + 3$

62. $f(x) = -x^2 + 6x - 8$

63. $f(x) = -4 + 4x - x^2$

64. $f(x) = x^2 - 6x + 9$

65. $f(x) = 2x^2 - 8x + 3$

66. $f(x) = 3x^2 + 12x - 5$

67. $f(x) = -4x^2 + 12x + 7$

68. $f(x) = 8x - 5 - 2x^2$

B EXERCISES Applying the Concepts

69. Ignoring air resistance, the height of a baseball that is hit 3 feet above the ground at an angle of 45° with an initial velocity of 132 ft/sec is given by the equation

$$h(x) = -\frac{32}{(132)^2}x^2 + x + 3,$$

where x is the horizontal distance (in feet) the ball traveled after being hit. Baseball fans will notice that air resistance severely shortens the actual distance a baseball travels. Use the graph to answer the following questions.

a. Approximately how many feet did the ball travel horizontally before hitting the ground?

b. Approximately how high did the ball go?

c. Use the function h to find answers to (a) and (b) rounded to the nearest integer.

70. The number of apartments that can be rented in a 100-unit apartment complex decreases by one for each increase of $25 in rent. The revenue generated from rent after x increases of $25 is given by the equation $R(x) = -25x^2 + 1700x + 80,000$. Use the graph to answer the following questions.

   a. Approximately what was the maximum revenue from the apartments?

   b. Approximately how many $25 increases generated the maximum revenue?

   c. Use the function R to find answers to (a) and (b).

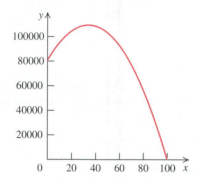

71. Maximizing revenue. A revenue function is given by $R(x) = 15 + 114x - 3x^2$, where x is the number of units produced and sold. Find the value of x for which the revenue is maximum.

72. Maximizing revenue. A demand function $p = 200 - 4x$, where p is the price per unit and x is the number of units produced and sold (the demand). Find x for which the revenue is maximum. [Recall that $R(x) = x \cdot p$.]

73. Minimizing cost. The total cost of a product is $C = 200 - 50x + x^2$, where x is the number of units produced. Find the value of x for which the total cost is minimum.

74. Maximizing profit. A manufacturer produces x items of a product at a total cost of $(50 + 2x)$ dollars. The demand function for the product is $p = 100 - x$. Find the value of x for which the profit is maximum. How much is the maximum profit? [Recall that profit = revenue − cost.]

75. Geometry. Find the dimensions of a rectangle of maximum area assuming that the perimeter of the rectangle is 80 units. What is the maximum area?

76. Enclosing area. A rancher with 120 meters of fence intends to enclose a rectangular region along a river (which serves as a natural boundary requiring no fence). Find the maximum area that can be enclosed.

77. Enclosing playing fields. You have 600 meters of fencing, and you plan to lay out two identical playing fields. The fields can be side-by-side and separated by a fence, as shown in the figure. Find the dimensions and the maximum area of each field.

78. Enclosing area on a budget. You budget $2400 for constructing a rectangular enclosure that consists of a high surrounding fence and a lower inside fence that divides the enclosure in half. The high fence costs $8 per foot, and the low fence costs $4 per foot. Find the dimensions and the maximum area of each half of the enclosure.

79. Maximizing apple yield. From past surveys and records, an orchard owner in northern Michigan has determined that if 26 apple trees per acre are planted, each tree yields 500 apples, on average. The yield decreases by ten apples per tree for each additional tree planted. How many trees should be planted per acre to achieve the maximum total yield?

80. Buying and selling beef. A steer weighing 300 pounds gains 8 pounds per day and costs $1.00 a day to keep. You bought this steer today at a market price of $1.50 per pound. However, the market price is falling $0.02 per pound per day. When should you sell the steer to maximize your profit?

81. Going on a tour. A group of students plans a tour. The charge per student is $72 if 20 students go on the trip. If more than 20 students participate, the charge per student is reduced by $2 times the number of students over 20. Find the number of students that will furnish the maximum revenue. What is the maximum revenue?

82. Computer disks. A 5-inch-radius computer disk contains information in units called *bytes* that are arranged in concentric tracks on the disk. Assume that m bytes per inch can be put on each track. The tracks are uniformly spread (that is, there is a fixed number, say, p, of tracks per inch measured radially across the disk). If the number of bytes on each track must be the same, where should the innermost track be located to get the maximum number of bytes on the disk?

83. Motion of a projectile. A projectile is fired straight up with a velocity of 64 ft/s. Its altitude (height) h after t seconds is given by

$$h(t) = -16t^2 + 64t.$$

   a. What is the maximum height of the projectile?
   b. When does the projectile hit the ground?

84. Motion of a projectile. From the top of a building, a projectile is shot 400 feet high with a velocity of 64 feet per second. Its altitude (height) h after t seconds is given by

$$h(t) = -16t^2 + 64t + 400.$$

   a. What is the maximum height of the projectile?
   b. When does the projectile hit the ground?

85. Architecture. A window is to be constructed in the shape of a rectangle surmounted by a semicircle. Assuming that the perimeter of the window is 18 feet, find its dimensions for the maximum area.

86. Architecture. The shape of the Gateway Arch in St. Louis, Missouri, is a *catenary* curve, which closely resembles a parabola. The function

$$y = f(x) = -0.00635x^2 + 4x$$

models the shape of the arch, where y is the height in feet and x is the horizontal distance from the base of the left side of the arch in feet.
   a. Graph the function $y = f(x)$.
   b. What is the width of the arch at the base?
   c. What is the maximum height of the arch?

87. Football. The altitude or height h (in feet) of a punted football can be modeled by the quadratic function

$$h = -0.01x^2 + 1.18x + 2,$$

where x (in feet) is the horizontal distance from the point of impact with the punter's foot. (See accompanying figure.)

   a. Find the vertex of the parabola.
   b. If the opposing team fails to catch the ball, how far away from the punter does the ball hit the ground (to the nearest foot)?
   c. What is the maximum height of the punted ball (to the nearest foot)?
   d. The nearest defensive player is 6 feet from the point of impact. How high must the player reach to block the punt (to the nearest foot)?
   e. How far downfield has the ball traveled when it reaches a height of 7 feet for the second time?

88. Field hockey. A *scoop* in field hockey is a pass that propels the ball from the ground into the air. Suppose a player makes a scoop with an upward velocity of 30 feet per second. The function

$$h = -16t^2 + 30t$$

models the altitude or height h in feet at time t in seconds. Will the ball ever reach a height of 16 feet?

C EXERCISES Beyond the Basics

In Exercises 89–94, find a quadratic function of the form $f(x) = ax^2 + bx + c$ that satisfies the given conditions.

89. The graph of f is obtained by shifting the graph of $y = 3x^2$ two units horizontally to the left and three units vertically up.

90. The graph of f passes through the point $(1, 5)$ and has vertex $(-3, -2)$.

91. The graph of f has vertex $(1, -2)$ and y-intercept 4.

92. The graph of f has x-intercepts 2 and 6 and y-intercept 24.

93. The graph of f has x-intercept 7 and y-intercept 14, and $x = 3$ is the axis of symmetry.

94. The graph of f is obtained by shifting the graph of $y = x^2 + 2x + 2$ three units horizontally to the right and two units vertically down.

In Exercises 95–98, find two quadratic functions, one opening up and the other down, whose graphs have the given x-intercepts.

95. $-2, 6$

96. $-3, 5$

97. $-7, -1$

98. $2, 10$

99. Let $f(x) = 4x - x^2$. Solve $f(a + 1) - f(a - 1) = 0$ for a.

100. Let $f(x) = x^2 + 1$. Find $(f \circ f)(x)$.

Critical Thinking

101. **Symmetry about the line $x = h$.** The graph of a polynomial function $y = f(x)$ is symmetric in the line $x = h$ if $f(h + p) = f(h - p)$ for every number p. For $f(x) = ax^2 + bx + c, a \neq 0$, set $f(h + p) = f(h - p)$ and show that $h = -\dfrac{b}{2a}$. This exercise proves that the graph of a quadratic function is symmetric in its axis $x = -\dfrac{b}{2a}$.

102. In the graph of $y = 2x^2 - 8x + 9$, find the coordinates of the point symmetric to the point $(-1, 19)$ in the axis of symmetry.

Polynomial Functions

Johannes Kepler (1571–1630)

Kepler was born in Weil in southern Germany and studied at the University of Tübingen. He studied with Michael Mastlin (professor of mathematics), who privately taught Kepler the Copernican theory of the sun-centered universe while hardly daring to recognize it openly in his lectures. Kepler knew that to work out a detailed version of Copernicus's theory, he had to have access to the observations of the Danish astronomer Tycho Brahe (1546–1601). Kepler's correspondence with Brahe resulted in his appointment as Brahe's assistant. Brahe died about 18 months after Kepler's arrival. During those 18 months, Kepler learned enough about Brahe's work to use the material in working out his own major project. He was appointed as Imperial Mathematician by Emperor Rudolf to succeed Tycho Brahe and spent the next 11 years in Prague. He produced the famous three laws of planetary motion. His investigations were published in 1609 in his book *Nova Astronomia*.

Before Starting this Section, Review

1. Graphs of equations (Section 2.2, page 180)
2. Solving linear equations (Section 1.1, page 85)
3. Solving quadratic equations (Section 1.4, page 124)
4. Transformations (Section 2.7, page 262)
5. Relative maximum and minimum values (Section 2.5, page 231)

Objectives

1. Learn properties of the graphs of polynomial functions.
2. Determine the end behavior of polynomial functions.
3. Find the zeros of a polynomial function by factoring.
4. Identify the relationship between degrees, real zeros, and turning points.
5. Graph polynomial functions.

KEPLER'S WEDDING RECEPTION

Kepler is best known as an astronomer who discovered the three laws of planetary motion. However, Kepler's primary field was not astronomy, but mathematics. He served as a mathematician in Austrian Emperor Mathias's court. After the death of his wife Barbara, Kepler remarried in 1613. His new wife, Susanna, had a crash course in Kepler's character. Legend has it that at his own wedding reception, Kepler observed that the Austrian vintners could quickly and mysteriously compute the capacities (volumes) of a variety of wine barrels. Each barrel had a hole, called a *bunghole*, in the middle of its side. The vintner would insert a rod, called the *bungrod*, into the hole until it hit the far corner. Then he would announce the volume. During the reception, Kepler occupied his mind with the problem of finding the volume of a wine barrel with a bungrod.

In Example 10, we show you how he solved the problem for barrels that are perfect cylinders. But barrels are not perfect cylinders: They are wider in the middle and narrower at the top and bottom, and in between, the sides make a graceful curve that appears to be an arc of a circle. What could be the formula for the volume of such a shape? Kepler tackled this problem in his important book *Nova Stereometria Doliorum Vinariorum* (1615) and developed a complete mathematical theory in relation to it. ∎

---

1 Learn properties of the graphs of polynomial functions.

Polynomial Functions

We begin with some terminology. A **polynomial function of degree n** is a function of the form

$$f(x) = a_n x^n + a_{n-1} x^{n-1} + \cdots + a_2 x^2 + a_1 x + a_0,$$

where n is a nonnegative integer and the *coefficients* $a_n, a_{n-1}, \ldots, a_2, a_1, a_0$ are real numbers with $a_n \neq 0$. The term $a_n x^n$ is called the **leading term**, the number a_n (the coefficient of x^n) is called the **leading coefficient**, and a_0 is the **constant term**. A constant function $f(x) = a(a \neq 0)$, which may be written as $f(x) = ax^0$, is a polynomial of degree 0. Because the zero function $f(x) = 0$ can be written in several ways, such as $0 = 0x^6 + 0x = 0x^5 + 0 = 0x^4 + 0x^3 + 0$, no degree is assigned to it.

In Section 2.4, we discussed polynomial functions of degree 0 and 1. In Section 3.1, we studied polynomial functions of degree 2. Polynomials of degree 3, 4, and 5 are also called **cubic, quartic**, and **quintic polynomials**, respectively. In this section, we concentrate on graphing polynomial functions of degree 3 or more.

Here are some common properties shared by all polynomial functions.

COMMON PROPERTIES OF POLYNOMIAL FUNCTIONS

1. The domain of a polynomial function is the set of all real numbers.

2. The graph of a polynomial function is a **continuous curve**. This means that the graph has no holes or gaps and can be drawn on a sheet of paper without lifting the pencil. See Figure 3.3.

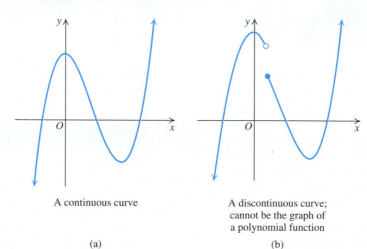

A continuous curve

(a)

A discontinuous curve; cannot be the graph of a polynomial function

(b)

FIGURE 3.3

3. The graph of a polynomial function is a **smooth curve**. This means that the graph of a polynomial function does not contain any sharp corners. See Figure 3.4.

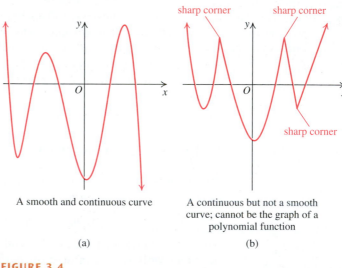

A smooth and continuous curve

(a)

A continuous but not a smooth curve; cannot be the graph of a polynomial function

(b)

FIGURE 3.4

EXAMPLE 1   **Polynomial Functions**

State which functions are polynomial functions. For each polynomial function, find its degree, the leading term, and the leading coefficient.

a. $f(x) = 5x^4 - 2x + 7$

b. $g(x) = 7x^2 - x + 1, \ 1 \le x \le 5$

SOLUTION

a. $f(x) = 5x^4 - 2x + 7$ is a polynomial function. Its degree is 4, the leading term is $5x^4$, and the leading coefficient is 5.

b. $g(x) = 7x^2 - x + 1, 1 \leq x \leq 5$ is not a polynomial function because its domain is not $(-\infty, \infty)$. ■ ■ ■

Practice Problem 1   Repeat Example 1 for each function.

a. $f(x) = \dfrac{x^2 + 1}{x - 1}$ **b.** $g(x) = 2x^7 + 5x^2 - 17$ ■

Power Functions

The simplest nth-degree polynomial function is a function of the form $f(x) = ax^n$ and is called a power function.

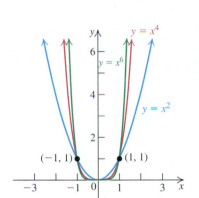

FIGURE 3.5 Power functions of even degree

POWER FUNCTION

A function of the form

$$f(x) = ax^n$$

is called a **power function of degree n**, where a is a nonzero real number and n is a positive integer.

The graph of a power function can be obtained by stretching or shrinking the graph of $y = x^n$ (and reflecting it about the x-axis if $a < 0$). So the basic shape of the graph of a power function $f(x) = ax^n$ is similar to the shape of the graphs of the equations $y = \pm x^n$.

Power Functions of Even Degree Let $f(x) = ax^n$. If n is even, then $(-x)^n = x^n$. So in this case, $f(-x) = a(-x)^n = ax^n = f(x)$. Therefore, a power function is an even function when n is even; so its graph is symmetric with respect to the y-axis.

The graph of $y = x^n$ (when n is even) is similar to the graph of $y = x^2$; the graphs of $y = x^2$, $y = x^4$, and $y = x^6$ are shown in Figure 3.5.

Notice that each of these graphs passes through the points $(-1, 1)$, $(0, 0)$, and $(1, 1)$. On the interval $-1 < x < 1$, the larger the exponent, the flatter the graph. (See Exercise 95.) However, on the intervals $(-\infty, -1)$ and $(1, \infty)$, the larger the exponent is, the more rapidly the graph rises. (See Exercise 96.)

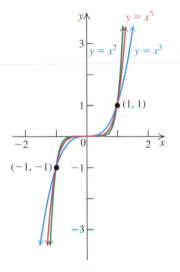

FIGURE 3.6 Power functions of odd degree

Power Functions of Odd Degree Again, let $f(x) = ax^n$. If n is odd, then $(-x)^n = -x^n$. In this case, we have $f(-x) = a(-x)^n = -ax^n = -f(x)$. So the power function $f(x) = ax^n$ is an odd function when n is odd; its graph is symmetric with respect to the origin. When n is odd, the graph of $f(x) = x^n$ is similar to the graph of $y = x^3$. The graphs of $y = x^3$, $y = x^5$, and $y = x^7$ are shown in Figure 3.6. If n is an odd positive integer, then the graph of $y = x^n$ passes through the points $(-1, -1)$, $(0, 0)$, and $(1, 1)$. The larger the value of n is, the flatter the graph is near the origin (in the interval $-1 < x < 1$) and the more rapidly it rises (or falls) on the intervals $(-\infty, -1)$ and $(1, \infty)$.

2 Determine the end behavior of polynomial functions.

End Behavior of Polynomial Functions

We next study the behavior of the function $y = f(x)$ when the independent variable x is large in absolute value. First, we consider the expression x *approaches infinity*. The notation is $x \to \infty$, meaning that x gets larger and larger without bound: x can

STUDY TIP

Here is a summary for the axis directions associated with the symbols $-\infty$ and ∞.

$x \to -\infty$	Left
$x \to \infty$	Right
$y \to -\infty$	Down
$y \to \infty$	Up

assume values greater than 1, 10, 100, 1000, ..., without end. Similarly, $x \to -\infty$ (read "x approaches negative infinity") means that x can assume values less than $-1, -10, -100, -1000, \ldots$, without end. Of course, x may be replaced by any other independent variable.

The behavior of a function $y = f(x)$ as $x \to \infty$ or $x \to -\infty$ is called the **end behavior** of the function. In the accompanying box, we describe the behavior of the power function $y = f(x) = ax^n$ as $x \to \infty$ or as $x \to -\infty$. Every polynomial function exhibits end behavior similar to one of the four functions $y = x^2$, $y = -x^2$, $y = x^3$, or $y = -x^3$.

End Behavior for the Graph of $y = f(x) = ax^n$

n Even	**n Odd**

Case 1

$a > 0$

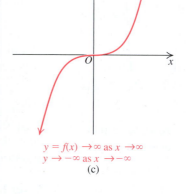

$y = f(x) \to \infty$ as $x \to \infty$ or $x \to -\infty$

(a)

The graph rises to the left and right, similar to $y = x^2$.

Case 3

$a > 0$

$y = f(x) \to \infty$ as $x \to \infty$
$y \to -\infty$ as $x \to -\infty$

(c)

The graph rises to the right and falls to the left, similar to $y = x^3$.

Case 2

$a < 0$

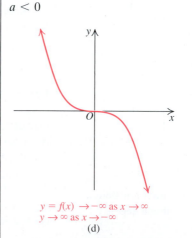

$y = f(x) \to -\infty$ as $x \to \infty$ or $x \to -\infty$

(b)

The graph falls to the left and right, similar to $y = -x^2$.

Case 4

$a < 0$

$y = f(x) \to -\infty$ as $x \to \infty$
$y \to \infty$ as $x \to -\infty$

(d)

The graph rises to the left and falls to the right, similar to $y = -x^3$.

In the next example, we show that the end behavior of a polynomial function is determined by its leading term. If a polynomial $P(x)$ has approximately the same values as $f(x) = ax^n$ when $|x|$ is very large, we write $P(x) \approx ax^n$ when $|x|$ is very large.

Calculator graph of
$P(x) = 2x^3 + 5x^2 - 7x + 11$

TABLE 3.1

x	$\dfrac{5}{x}$
10	0.5
10^2	0.05
10^3	0.005
10^4	0.0005

EXAMPLE 2   **Understanding the End Behavior of a Polynomial Function**

Let $P(x) = 2x^3 + 5x^2 - 7x + 11$ be a polynomial function of degree 3. Show that $P(x) \approx 2x^3$ when $|x|$ is very large.

SOLUTION

$$P(x) = 2x^3 + 5x^2 - 7x + 11 \qquad \text{Given polynomial}$$

$$= x^3\left(2 + \frac{5}{x} - \frac{7}{x^2} + \frac{11}{x^3}\right) \qquad \text{Distributive property}$$

When $|x|$ is very large, the terms $\dfrac{5}{x}$, $-\dfrac{7}{x^2}$, and $\dfrac{11}{x^3}$ are close to 0. Table 3.1 shows some values of $\dfrac{5}{x}$ as x increases. (You should compute the values of $-\dfrac{7}{x^2}$ and $\dfrac{11}{x^3}$ for $x = 10, 10^2, 10^3$, and 10^4.) When $|x|$ is very large, we then have

$$P(x) = x^3\left(2 + \frac{5}{x} - \frac{7}{x^2} + \frac{11}{x^3}\right) \approx x^3(2 + 0 - 0 + 0)$$

$$= 2x^3.$$

So the polynomial $P(x) \approx 2x^3$ when $|x|$ is very large. ■ ■ ■

Practice Problem 2   Let $P(x) = 4x^3 + 2x^2 + 5x - 17$. Show that $P(x) \approx 4x^3$ when $|x|$ is very large. ■

We can use the technique of Example 2 to show that the end behavior of any polynomial function is determined by its leading term.

The end behaviors of polynomial functions are shown in the following graphs.

THE LEADING-TERM TEST

Let $f(x) = a_nx^n + a_{n-1}x^{n-1} + \cdots + a_1x + a_0 (a_n \neq 0)$ be a polynomial function. Its leading term is a_nx^n. The behavior of the graph of f as $x \to \infty$ or $x \to -\infty$ is similar to one of the following four graphs and is described as shown in each case.

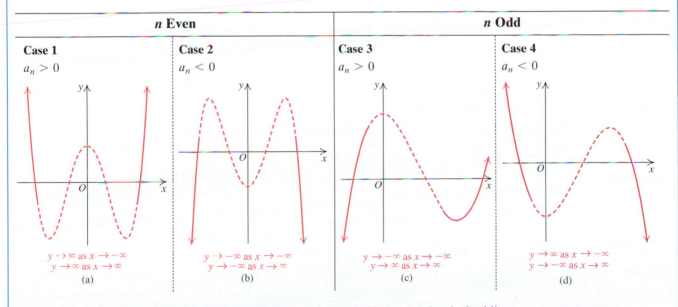

n **Even**		n **Odd**	
Case 1 $a_n > 0$	**Case 2** $a_n < 0$	**Case 3** $a_n > 0$	**Case 4** $a_n < 0$
$y \to \infty$ as $x \to -\infty$ $y \to \infty$ as $x \to \infty$ (a)	$y \to -\infty$ as $x \to -\infty$ $y \to -\infty$ as $x \to \infty$ (b)	$y \to -\infty$ as $x \to -\infty$ $y \to \infty$ as $x \to \infty$ (c)	$y \to \infty$ as $x \to -\infty$ $y \to -\infty$ as $x \to \infty$ (d)

This test does not describe the middle portion of each graph, shown by the dashed lines.

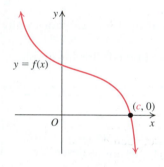

$f(x) = -2x^3 + 3x + 4$

FIGURE 3.7 End behavior of
$f(x) = -2x^3 + 3x + 4$

3 Find the zeros of a polynomial function by factoring.

EXAMPLE 3 **Using the Leading-Term Test**

Use the leading-term test to determine the end behavior of the graph of

$$y = f(x) = -2x^3 + 3x + 4.$$

SOLUTION

Here $n = 3$ (odd) and $a_n = -2 < 0$. Case 4 applies. The graph of $f(x)$ rises to the left and falls to the right. See Figure 3.7. This behavior is described as $y \to \infty$ as $x \to -\infty$ and $y \to -\infty$ as $x \to \infty$. ■ ■ ■

Practice Problem 3 What is the end behavior of $f(x) = -2x^4 + 5x^2 + 3$? ■

Zeros of a Function

Let f be a function. An input c in the domain of f that produces output 0 is called a *zero* of the function f. For example, if $f(x) = (x - 3)^2$, then 3 is a zero of f because $f(3) = (3 - 3)^2 = 0$. In other words, a number c is a **zero** of f if $f(c) = 0$. The zeros of a function f can be obtained by solving the equation $f(x) = 0$. One way to solve a polynomial equation $f(x) = 0$ is to factor $f(x)$ and then use the zero-product property.

If c is a real number and $f(c) = 0$, then c is called a **real zero** of f. Geometrically, this means that the graph of f has an x-intercept at $x = c$.

FIGURE 3.8 A zero of f

REAL ZEROS OF POLYNOMIAL FUNCTIONS

If f is a polynomial function and c is a real number, then the following statements are equivalent:

1. c is a **zero** of f.
2. c is a **solution** (or **root**) of the equation $f(x) = 0$.
3. c is an **x-intercept** of the graph of f. The point $(c, 0)$ is on the graph of f. See Figure 3.8.

EXAMPLE 4 **Finding the Zeros of a Polynomial Function**

Find all real zeros of each polynomial function.

a. $f(x) = x^3 + 2x^2 - x - 2$ **b.** $g(x) = x^3 - 2x^2 + x - 2$

SOLUTION

a. We factor $f(x)$ and then solve the equation $f(x) = 0$.

$$
\begin{aligned}
f(x) &= x^3 + 2x^2 - x - 2 && \text{Given function} \\
&= (x^3 + 2x^2) - (x + 2) && \text{Group terms.} \\
&= x^2(x + 2) - 1(x + 2) && \text{Distributive property} \\
&= (x + 2)(x^2 - 1) && \text{Distributive property} \\
&= (x + 2)(x + 1)(x - 1) && \text{Factor } x^2 - 1.
\end{aligned}
$$

$$(x + 2)(x + 1)(x - 1) = 0 \qquad \text{Set } f(x) = 0.$$
$$x + 2 = 0, \quad x + 1 = 0, \quad \text{or} \quad x - 1 = 0 \qquad \text{Zero-product property}$$
$$x = -2, \quad x = -1, \quad \text{or} \quad x = 1 \qquad \text{Solve each equation.}$$

The zeros of $f(x)$ are -2, -1, and 1. (See the calculator graph of f on page 317.)

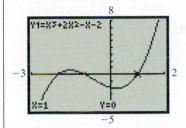

$f(x) = x^3 + 2x^2 - x - 2$

We can verify that the zeros of f are -2, -1, and 1 using the TRACE function.

$g(x) = x^3 - 2x^2 + x - 2$

b. By grouping terms, we factor $g(x)$ to obtain

$$g(x) = x^3 - 2x^2 + x - 2$$
$$= x^2(x - 2) + 1(x - 2) \qquad \text{Group terms.}$$
$$= (x - 2)(x^2 + 1) \qquad \text{Factor out } x - 2.$$
$$0 = (x - 2)(x^2 + 1) \qquad \text{Set } g(x) = 0.$$
$$x - 2 = 0 \quad \text{or} \quad x^2 + 1 = 0 \qquad \text{Zero-product property}$$

Solving for x, we obtain 2 as the only real zero of $g(x)$ because $x^2 + 1 > 0$ for all real numbers x. The calculator graph of g supports our conclusion. ■■■

Practice Problem 4 Find all real zeros of $f(x) = 2x^3 - 3x^2 + 4x - 6$. ■

Finding the zeros of an unfactorable polynomial of degree 3 or more is one of the most important problems of mathematics. You will learn more about this topic in Sections 3.3 and 3.4. For now, we will approximate the zeros by using the *Intermediate Value Theorem*. Recall that the graph of a polynomial function $f(x)$ is a continuous curve. Suppose you locate two numbers a and b such that the function values $f(a)$ and $f(b)$ have opposite signs (one positive and the other negative). Then the continuous graph of f connecting the points $(a, f(a))$ and $(b, f(b))$ must cross the x-axis (at least once) somewhere between a and b. In other words, the function f has a zero between a and b. See Figure 3.9.

AN INTERMEDIATE VALUE THEOREM

Let a and b be two numbers with $a < b$. If f is a polynomial function such that $f(a)$ and $f(b)$ have opposite signs, then there is at least one number c with $a < c < b$, for which $f(c) = 0$.

Notice that Figure 3.9(c) shows more than one zero of f. That is why the Intermediate Value Theorem says that f has *at least one* zero between a and b.

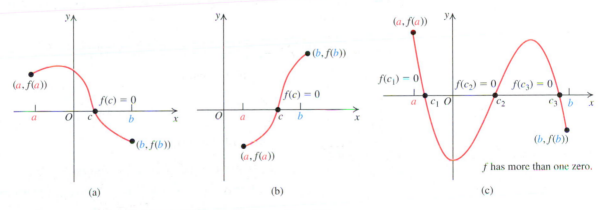

FIGURE 3.9 Each function has at least one zero between a and b.

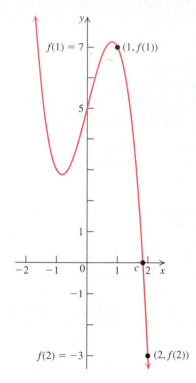

FIGURE 3.10 A zero lies between 1 and 2

4 Identify the relationship between degrees, real zeros, and turning points.

EXAMPLE 5 **Using the Intermediate Value Theorem**

Show that the function $f(x) = -2x^3 + 4x + 5$ has a real zero between 1 and 2.

SOLUTION

$$f(1) = -2(1)^3 + 4(1) + 5 \qquad \text{Replace } x \text{ with 1 in } f(x).$$
$$= -2 + 4 + 5 = 7 \qquad \text{Simplify.}$$
$$f(2) = -2(2)^3 + 4(2) + 5 \qquad \text{Replace } x \text{ with 2.}$$
$$= -2(8) + 8 + 5 = -3 \qquad \text{Simplify.}$$

Because $f(1)$ is positive and $f(2)$ is negative, they have opposite signs. So by the Intermediate Value Theorem, the polynomial function $f(x) = -2x^3 + 4x + 5$ has a real zero between 1 and 2. See Figure 3.10. ■■■

Practice Problem 5 Show that $f(x) = 2x^3 - 3x - 6$ has a real zero between 1 and 2. ■

In Example 5, if you want to find the zero between 1 and 2 more accurately, you can divide the interval $[1, 2]$ into tenths and find the value of the function at each of the points 1, 1.1, 1.2,..., 1.9, and 2. You will find that

$$f(1.8) = 0.536 \quad \text{and} \quad f(1.9) = -1.118.$$

So by the Intermediate Value Theorem, f has a zero between 1.8 and 1.9. By continuing this process, you can find the zero of f between 1.8 and 1.9 to any degree of accuracy.

Zeros and Turning Points

In Section 3.5, you will study the Fundamental Theorem of Algebra. One of its consequences is:

REAL ZEROS OF A POLYNOMIAL FUNCTION

A polynomial function of degree n with real coefficients has, *at most*, n real zeros.

EXAMPLE 6 **Finding the Number of Real Zeros**

Find the number of distinct real zeros of the following polynomial functions.

a. $f(x) = (x - 1)(x + 2)(x - 3)$ **b.** $g(x) = (x + 1)(x^2 + 1)$
c. $h(x) = (x - 3)^2(x + 1)^3$

SOLUTION

a. $f(x) = (x - 1)(x + 2)(x - 3) = 0$ $\qquad$ Set $f(x) = 0$.
$x - 1 = 0$ or $x + 2 = 0$ or $x - 3 = 0$ $\qquad$ Zero-product property
$x = 1$ or $x = -2$ or $x = 3$ $\qquad$ Solve for x.

So $f(x)$ has three real zeros: 1, -2, and 3.

b. $g(x) = (x + 1)(x^2 + 1) = 0$ $\qquad$ Set $g(x) = 0$.
$x + 1 = 0$ or $x^2 + 1 = 0$ $\qquad$ Zero-product property
$x = -1$ $\qquad$ Solve for x.

Because $x^2 + 1 > 0$ for all real x, the equation $x^2 + 1 = 0$ has no real solution. So -1 is the only real zero of $g(x)$.

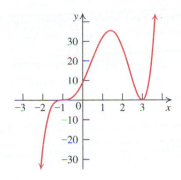

FIGURE 3.11 Graph of
$h(x) = (x - 3)^2 (x + 1)^3$

c.
$$h(x) = (x - 3)^2(x + 1)^3 = 0 \quad \text{Set } h(x) = 0.$$
$$(x - 3)^2 = 0 \quad \text{or} \quad (x + 1)^3 = 0 \quad \text{Zero-product property}$$
$$x = 3 \quad \text{or} \quad x = -1 \quad \text{Solve for } x.$$

The real zeros of $h(x)$ are 3 and -1. Thus, $h(x)$ has two distinct real zeros. See Figure 3.11 ■ ■ ■

Practice Problem 6 Find the distinct real zeros of
$$f(x) = (x + 1)^2(x - 3)(x + 5).$$ ■

In Example 6(c), the factor $(x - 3)$ appears twice in $h(x)$. We say that 3 is a zero repeated *twice*, or that 3 is a *zero of multiplicity* 2. Notice that the graph of $y = h(x)$ in Figure 3.11 *touches* the x-axis at $x = 3$, where h has a zero of multiplicity 2 (similar to $y = x^2$ at $x = 0$). In contrast, the factor $(x + 1)$ appears thrice in $h(x)$ and the graph of h *crosses* the x-axis at $x = -1$, where it has a zero of multiplicity 3 (similar to the graph of $y = x^3$ at $x = 0$). We next state the geometric significance of the multiplicity of a zero.

MULTIPLICITY OF A ZERO

If c is a zero of a polynomial function $f(x)$ and the corresponding factor $(x - c)$ occurs exactly m times when $f(x)$ is factored completely, then c is called a **zero of multiplicity m.**

1. If m is odd, the graph of f crosses the x-axis at $x = c$.

2. If m is even, the graph of f touches but does not cross the x-axis at $x = c$.

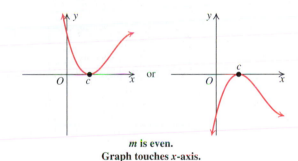

m is odd.
Graph crosses x-axis.

m is even.
Graph touches x-axis.

EXAMPLE 7 **Finding the Zeros and Their Multiplicity**

Find the zeros of the polynomial function $f(x) = x^2(x + 1)(x - 2)$ and give the multiplicity of each zero.

SOLUTION
The polynomial $f(x)$ is already in factored form.
$$f(x) = x^2(x + 1)(x - 2) = 0 \quad \text{Set } f(x) = 0.$$
$$x^2 = 0 \quad \text{or} \quad x + 1 = 0 \quad \text{or} \quad x - 2 = 0 \quad \text{Zero-product property}$$
$$x = 0 \quad \text{or} \quad x = -1 \quad \text{or} \quad x = 2 \quad \text{Solve for } x.$$

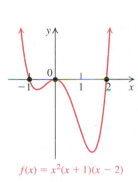

$f(x) = x^2(x + 1)(x - 2)$

FIGURE 3.12

The function has three distinct zeros: 0, -1, and 2. The zero $x = 0$ has multiplicity 2, and each of the zeros -1 and 2 has multiplicity 1. The graph of f is given in Figure 3.12.

Notice that the graph of f in Figure 3.12 crosses the x-axis at -1 and 2 (the zeros of odd multiplicity), while it touches but does not cross the x-axis at $x = 0$ (the zero of even multiplicity). ■ ■ ■

Practice Problem 7 Write the zeros of $f(x) = (x - 1)^2(x + 3)(x + 5)$ and give the multiplicity of each. ■

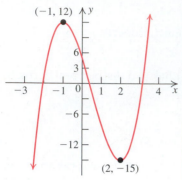

The graph of $f(x) = 2x^3 - 3x^2 - 12x + 5$ has turning points at $(-1, 12)$ and $(2, -15)$.

FIGURE 3.13 Turning points

Turning Points   The graph of $f(x) = 2x^3 - 3x^2 - 12x + 5$ is shown in Figure 3.13. The value $f(-1) = 12$ is a local (or relative) maximum value of f. Similarly, the value $f(2) = -15$ is a local (or relative) minimum value of f. (See Section 2.5 for definitions.) The graph points $(-1, 12)$ and $(2, -15)$ are the turning points. Recall that at each turning point, the graph changes direction from increasing to decreasing or from decreasing to increasing.

NUMBER OF TURNING POINTS

If $f(x)$ is a polynomial of degree n, then the graph of f has, *at most*, $(n - 1)$ turning points.

EXAMPLE 8   **Finding the Number of Turning Points**

Use a graphing calculator and the window $-10 \le x \le 10$; $-30 \le y \le 30$ to find the number of turning points of the graph of each polynomial function.

a. $f(x) = x^4 - 7x^2 - 18$

b. $g(x) = x^3 + x^2 - 12x$

c. $h(x) = x^3 - 3x^2 + 3x - 1$

SOLUTION

The graphs of f, g, and h are shown in Figure 3.14.

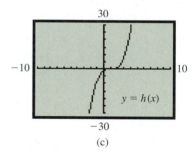

(a)   (b)   (c)

FIGURE 3.14 Turning points on a calculator graph

a. Function f has three turning points: two local minimum and one local maximum.

b. Function g has two turning points: one local maximum and one local minimum.

c. Function h has no turning points; it is increasing on the interval $(-\infty, \infty)$. ■ ■ ■

Practice Problem 8   Find the number of turning points of the graph of $f(x) = -x^4 + 3x^2 - 2$.   ■

5 Graph polynomial functions.

Graphing a Polynomial Function

You can graph a polynomial function by constructing a table of values for a large number of points, plotting the points, and connecting the points with a smooth curve. (In fact, a graphing calculator uses this method.) Generally, however, this process is a poor way to graph a function by hand because it is tedious, is error-prone, and may give misleading results. Similarly, a graphing calculator may give misleading results if the viewing window is not chosen appropriately. Next, we discuss a better strategy for graphing a polynomial function.

FINDING THE SOLUTION: A PROCEDURE

EXAMPLE 9   **Graphing a Polynomial Function**

OBJECTIVE	**EXAMPLE**

OBJECTIVE

Sketch the graph of a polynomial function.

EXAMPLE

Sketch the graph of $f(x) = -x^3 - 4x^2 + 4x + 16$.

Step 1 **Determine the end behavior.** Apply the leading-term test.

Because the degree 3 of f is odd and the leading coefficient -1 is negative, the end behavior (Case 4, page 315) is as shown in this sketch.

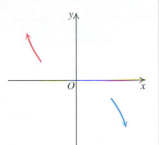

Step 2 **Find the real zeros of the polynomial function.** Set $f(x) = 0$. Solve the equation $f(x) = 0$ by factoring or by using a graphing calculator. The real zeros will give you the x-intercepts of the graph.

$$\begin{aligned}
-x^3 - 4x^2 + 4x + 16 &= 0 \\
-x^2(x + 4) + 4(x + 4) &= 0 \\
-(x + 4)(x^2 - 4) &= 0 \\
-(x + 4)(x + 2)(x - 2) &= 0 \\
x = -4, x = -2, \text{ or } x &= 2
\end{aligned}$$

There are three zeros, each of multiplicity 1; so the graph crosses the x-axis at each zero.

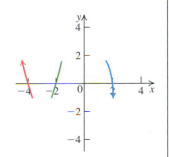

Step 3 **Find the y-intercept** by computing $f(0)$.

The y-intercept is $f(0) = 16$. The graph passes through the point $(0, 16)$.

Step 4 **Use symmetry** (if any). Check whether the function is odd, even, or neither.

$$\begin{aligned}
f(-x) &= -(-x)^3 - 4(-x)^2 + 4(-x) + 16 \\
&= x^3 - 4x^2 - 4x + 16
\end{aligned}$$

Because $f(-x) \neq f(x)$, there is no symmetry in the y-axis. Also, because $f(-x) \neq -f(x)$, there is no symmetry with respect to the origin.

Step 5 **Determine the value of $f(x)$** at any number less than the smallest zero or greater than the largest zero as a starting point for sketching the graph.

The three zeros of f are -4, -2, and 2. We choose -5 as a convenient number less than the smallest zero of f. Then we find $f(-5) = -(-5)^3 - 4(-5)^2 + 4(-5) + 16 = 21$.

Step 6 **Draw the graph.** Use the multiplicities of each zero to decide whether the graph crosses the x-axis. The end behavior determines the graph over the extremes of the x-axis.

To check whether the graph is drawn correctly, use the fact that the number of turning points is less than the degree of the polynomial.

Use the graph point $(-5, 21)$ as a starting point. The graph crosses the x-axis at each zero: -4, -2, and 2.

The number of turning points is 2, which is less than 3, the degree of f.

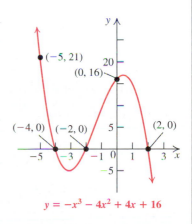

$$y = -x^3 - 4x^2 + 4x + 16$$

■ ■ ■

Practice Problem 9   Graph $f(x) = -x^4 + 5x^2 - 4$.

■

FIGURE 3.15 A wine barrel

Volume of a Wine Barrel Kepler first analyzed the problem of finding the volume of a wine barrel for a cylindrical barrel. (Review the information about this problem given in the introduction to this section and see Figure 3.15.) The volume V of a cylinder with radius r and height h is given by the formula $V = \pi r^2 h$. In Figure 3.15, $h = 2z$; so $V = 2\pi r^2 z$. Let x represent the value measured by the rod. Then by the Pythagorean Theorem,

$$x^2 = 4r^2 + z^2.$$

Kepler's challenge was to calculate V given only x; in other words, to express V as a function of x. To solve this problem, Kepler observed the key fact that Austrian wine barrels were made with the same height-to-diameter ratio. Kepler assumed that the winemakers would have made a smart choice for this ratio, one that would maximize the volume for a fixed value of r. He calculated the ratio to be $\sqrt{2}$.

EXAMPLE 10   **Volume of a Wine Barrel**

Express the volume V of the wine barrel in Figure 3.15 as a function of x. Assume that $\dfrac{\text{height}}{\text{diameter}} = \dfrac{2z}{2r} = \dfrac{z}{r} = \sqrt{2}$.

SOLUTION

We have the following relationships:

$$V = 2\pi r^2 z \qquad \text{Volume with radius } r \text{ and height } h = 2z$$

$$\frac{z}{r} = \sqrt{2} \qquad \text{Given height-to-diameter ratio}$$

$$x^2 = 4r^2 + z^2 \qquad \text{Pythagorean Theorem}$$

$$\frac{x^2}{r^2} = 4 + \frac{z^2}{r^2} \qquad \text{Divide both sides by } r^2.$$

$$\frac{x^2}{r^2} = 4 + 2 = 6 \qquad \text{Because } \frac{z}{r} = \sqrt{2}, \frac{z^2}{r^2} = \left(\frac{z}{r}\right)^2 = (\sqrt{2})^2 = 2.$$

From $\dfrac{x^2}{r^2} = 6$, $r^2 = \dfrac{x^2}{6}$; so $r = \dfrac{x}{\sqrt{6}}$. Using $\dfrac{z}{r} = \sqrt{2}$, we obtain

$$z = r\sqrt{2} = \frac{x}{\sqrt{6}}\sqrt{2} \qquad \text{Replace } r \text{ with } \frac{x}{\sqrt{6}}.$$

$$= \frac{x}{\sqrt{3}} \qquad \frac{\sqrt{2}}{\sqrt{6}} = \frac{\sqrt{2}}{\sqrt{2}\cdot\sqrt{3}} = \frac{1}{\sqrt{3}}$$

Substituting $z = \dfrac{x}{\sqrt{3}}$ and $r^2 = \dfrac{x^2}{6}$ in the expression for V, we have

$$V = 2\pi r^2 z$$

$$= 2\pi\left(\frac{x^2}{6}\right)\left(\frac{x}{\sqrt{3}}\right) \qquad \text{Replace } z \text{ with } \frac{x}{\sqrt{3}} \text{ and } r^2 \text{ with } \frac{x^2}{6}.$$

$$= \frac{\pi}{3\sqrt{3}}x^3 \qquad \text{Simplify.}$$

$$\approx 0.6046x^3 \qquad \text{Use a calculator.}$$

Consequently, Austrian vintners could easily calculate the volume of a wine barrel by using the formula $V \approx 0.6x^3$, where x was the length measured by the rod.   ■ ■ ■

Practice Problem 10 Use Kepler's formula to compute the volume of wine (in liters) in a barrel with $x = 70$ cm. Note: 1 decimeter (dm) $= 10$ cm and 1 liter $= (1 \text{ dm})^3$.   ■

SECTION 3.2 ▪ Exercises

A EXERCISES Basic Skills and Concepts

1. Consider the polynomial $2x^5 - 3x^4 + x - 6$. The degree of this polynomial is _____, its leading term is _____, its leading coefficient is _____, and its constant term is _____ .

2. The domain of a polynomial function is the _____.

3. The behavior of the function $y = f(x)$ as $x \to \infty$ or $x \to -\infty$ is called the _____ of the function.

4. The end behavior of any polynomial function is determined by its _____ term.

5. A number c for which $f(c) = 0$ is called a(n) _____ of the function f.

6. If c is a zero of even multiplicity for a function f, then the graph of f _____ the x-axis at c.

7. The graph of a polynomial function of degree n has, at most, _____ turning points.

8. The zeros of a polynomial function $y = f(x)$ are determined by solving the equation _____.

In Exercises 9–14, for each polynomial function, find the degree, the leading term, and the leading coefficient.

9. $f(x) = 2x^5 - 5x^2$

10. $f(x) = 3 - 5x - 7x^4$

11. $f(x) = \dfrac{2x^3 + 7x}{3}$

12. $f(x) = -7x + 11 + \sqrt{2}x^3$

13. $f(x) = \pi x^4 + 1 - x^2$

14. $f(x) = 5$

In Exercises 15–22, explain why the given function is not a polynomial function.

15. $f(x) = x^2 + 3|x| - 7$

16. $f(x) = 2x^3 - 5x, \quad 0 \le x \le 2$

17. $f(x) = \dfrac{x^2 - 1}{x + 5}$

18. $f(x) = \begin{cases} 2x + 3, & x \ne 0 \\ 1, & x = 0 \end{cases}$

19. $f(x) = 5x^2 - 7\sqrt{x}$

20. $f(x) = x^{4.5} - 3x^2 + 7$

21. $f(x) = \dfrac{x^2 - 1}{x - 1}, \quad x \ne 1$

22. $f(x) = 3x^2 - 4x + 7x^{-2}$

In Exercises 23–28, explain why the given graph cannot be the graph of a polynomial function.

23.

24.

25.

26.

27.

28.

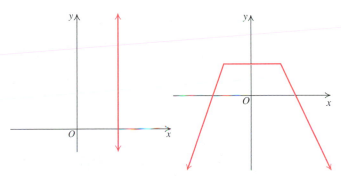

In Exercises 29–34, match the polynomial function with its graph. Use the leading-term test and the y-intercept.

29. $f(x) = -x^4 + 3x^2 + 4$

30. $f(x) = x^6 - 7x^4 + 7x^2 + 15$

31. $f(x) = x^4 - 10x^2 + 9$

32. $f(x) = x^3 + x^2 - 17x + 15$

33. $f(x) = x^3 + 6x^2 + 12x + 8$

34. $f(x) = -x^3 - 2x^2 + 11x + 12$

(a)

(b)

(c)   (d)

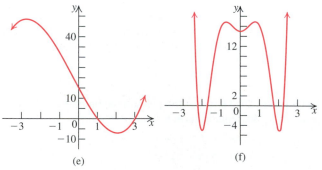

(e)   (f)

In Exercises 35–38, the graph of a polynomial function f is given. Find the unique equation for $f(x)$ from its graph.

35.   **36.**

37.   **38.**

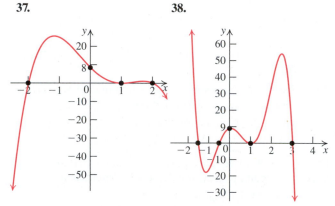

In Exercises 39–46, find the real zeros of each polynomial function and state the multiplicity for each zero and state whether the graph crosses or touches but does not cross the x-axis at each x-intercept. What is the maximum possible number of turning points?

39. $f(x) = 5(x - 1)(x + 5)$

40. $f(x) = 3(x + 2)^2(x - 3)$

41. $f(x) = (x + 1)^2(x - 1)^3$

42. $f(x) = 2(x + 3)^2(x + 1)(x - 6)$

43. $f(x) = -5\left(x + \dfrac{2}{3}\right)\left(x - \dfrac{1}{2}\right)^2$

44. $f(x) = x^3 - 6x^2 + 8x$

45. $f(x) = -x^4 + 6x^3 - 9x^2$

46. $f(x) = x^4 - 5x^2 - 36$

In Exercises 47–50, use the Intermediate Value Theorem to show that each polynomial $P(x)$ has a real zero in the specified interval. Approximate this zero to two decimal places.

47. $P(x) = x^4 - x^3 - 10; [2, 3]$

48. $P(x) = x^4 - x^2 - 2x - 5; [1, 2]$

49. $P(x) = x^5 - 9x^2 - 15; [2, 3]$

50. $P(x) = x^5 + 5x^4 + 8x^3 + 4x^2 - x - 5; [0, 1]$

In Exercises 51–64, for each polynomial function f,
  a. Describe the end behavior of f.
  b. Sketch the graph of f.

51. $f(x) = x^2 + 3$

52. $f(x) = x^2 - 4x + 4$

53. $f(x) = x^2 + 4x - 21$

54. $f(x) = -x^2 + 4x + 12$

55. $f(x) = -2x^2(x + 1)$

56. $f(x) = x - x^3$

57. $f(x) = x^2(x - 1)^2$

58. $f(x) = x^2(x + 1)(x - 2)$

59. $f(x) = (x - 1)^2(x + 3)(x - 4)$

60. $f(x) = (x + 1)^2(x^2 + 1)$

61. $f(x) = -x^2(x^2 - 1)(x + 1)$

62. $f(x) = -x^2(x^2 - 4)(x + 2)$

63. $f(x) = x(x + 1)(x - 1)(x + 2)$

64. $f(x) = x^2(x^2 + 1)(x - 2)$

B EXERCISES Applying the Concepts

65. Drug reaction. Let $f(x) = 3x^2(4 - x)$.
  a. Find the zeros of f and their multiplicities.
  b. Sketch the graph of $y = f(x)$.
  c. How many turning points are in the graph of f?
  d. A patient's reaction $R(x)$ ($R(x) \geq 0$) to a certain drug, the size of whose dose is x, is observed to be $R(x) = 3x^2(4 - x)$. What is the domain of the function $R(x)$? What portion of the graph of $f(x)$ constitutes the graph of $R(x)$?

 66. In Exercise 65, use a graphing calculator to estimate the value of x for which $R(x)$ is a maximum.

67. Cough and air velocity. The normal radius of the windpipe of a human adult at rest is approximately 1 cm. When you cough, your windpipe contracts to, say, a radius r. The velocity v at which the air is expelled depends on the radius r and is given by the formula $v = a(1 - r)r^2$, where a is a positive constant.
  a. What is the domain of v?
  b. Sketch the graph of $v(r)$.

 68. In Exercise 67, use a graphing calculator to estimate the value of r for which v is a maximum.

69. Maximizing revenue. A manufacturer of slacks has discovered that the number x of khakis sold to retailers per week at a price p dollars per pair is given by the formula

$$p = 27 - \left(\frac{x}{300}\right)^2.$$

  a. Write the weekly revenue $R(x)$ as a function of x.
  b. What is the domain of $R(x)$?
  c. Graph $y = R(x)$.

 70. a. In Exercise 69, use a graphing calculator to estimate the number of slacks sold to maximize the revenue.
  b. What is the maximum revenue from part (a)?

71. Maximizing production. An orange grower in California hires migrant workers to pick oranges during the season. He has 12 employees, and each can pick 400 oranges per hour. He has discovered that if he adds more workers, the production per worker decreases due to lack of supervision. When x new workers (above the 12) are hired, each worker picks $400 - 2x^2$ oranges per hour.
  a. Write the number $N(x)$ of oranges picked per hour as a function of x.
  b. Find the domain of $N(x)$.
  c. Graph the function $y = N(x)$.

 72. a. In Exercise 71, estimate the maximum number of oranges that can be picked per hour.
  b. Find the number of employees involved in part (a).

73. Making a box. From a rectangular 8×15 piece of cardboard, four congruent squares with sides of length x are cut out, one at each corner. (See the accompanying figure.) The resulting crosslike piece is then folded to form a box.
  a. Find the volume V of the box as a function of x.
  b. Sketch the graph of $y = V(x)$.

 74. In Exercise 73, use a graphing calculator to estimate the largest possible value of the volume of the box.

75. Mailing a box. According to Postal Service regulations, the girth plus the length of a parcel sent by Priority Mail may not exceed 108 inches.
  a. As a function of x, express the volume V of a rectangular parcel with two square faces (with sides of length x inches) that can be sent by mail.
  b. Sketch a graph of $V(x)$.

76. In Exercise 75, use a graphing calculator to estimate the largest possible volume of the box that can be sent by Priority Mail.

77. Luggage design. AAA Airlines requires that the total outside dimensions (length + width + height) of a piece of checked-in luggage not exceed 62 inches. You have designed a suitcase in the shape of a box whose height (x inches) equals its width. Your suitcase meets AAA's requirements.
  a. Write the volume V of your suitcase as a function of x.
  b. Graph the function $y = V(x)$.

 78. In Exercise 77, use a graphing calculator to find the approximate dimensions of the piece of luggage with the largest volume that you can check in on this airline.

79. Luggage dimensions. American Airlines requires that the total dimensions (length + width + height) of a carry-on bag not exceed 45 inches. You have designed a rectangular boxlike bag whose length is twice its width x. Your bag meets American's requirements.
  a. Write the volume V of your bag as a function of x, where x is measured in inches.
  b. Graph the function $y = V(x)$.

 80. In Exercise 79, use a graphing calculator to estimate the dimensions of the bag with the largest volume that you can carry on an American flight. (*Source:* www.aa.com)

81. Worker productivity. An efficiency study of the morning shift at an assembly plant indicates that the number y of units produced by an average worker t hours after 6 A.M. may be modeled by the formula $y = -t^3 + 6t^2 + 16t$. Sketch the graph of $y = f(t)$.

 82. In Exercise 81, estimate the time in the morning when a worker is most efficient.

83. Death rates from AIDS. According to the *Mortality and Morbidity* report of the U.S. Centers for Disease Control and Prevention, the following table gives the death rates per 100,000 population from AIDS in the United States for the years 1993–2000.

Year	Death Rate	Year	Death Rate
1993	14.5	1997	6.1
1994	16.2	1998	4.9
1995	16.3	1999	5.4
1996	11.7	2000	5.2

To model the death rate, a data analysis program produced the cubic polynomial

$$P(t) = 0.21594t^3 - 2.226t^2 + 3.7583t + 14.6788,$$

where $t = 0$ represents 1993.

Sketch the graph of $y = P(t)$.

 84. In Exercise 83, when does the highest point occur for $0 \le t \le 20$?

C EXERCISES Beyond the Basics

In Exercises 85–94, use transformations of the graph of $y = x^4$ or $y = x^5$ to graph each function $f(x)$. Find the zeros of $f(x)$ and their multiplicity.

85. $f(x) = (x - 1)^4$

86. $f(x) = -(x + 1)^4$

87. $f(x) = x^4 + 2$

88. $f(x) = x^4 + 81$

89. $f(x) = -(x - 1)^4$

90. $f(x) = \frac{1}{2}(x - 1)^4 - 8$

91. $f(x) = x^5 + 1$

92. $f(x) = (x - 1)^5$

93. $f(x) = 8 - \frac{(x + 1)^5}{4}$

94. $f(x) = 81 + \frac{(x + 1)^5}{3}$

95. Prove that if $0 < x < 1$, then $0 < x^4 < x^2$. This shows that the graph of $y = x^4$ is flatter than the graph of $y = x^2$ on the interval $(0, 1)$.

96. Prove that if $x > 1$, then $x^4 > x^2$. This shows that the graph of $y = x^4$ is higher than the graph of $y = x^2$ on the interval $(1, \infty)$.

 97. Sketch the graph of $y = x^2 - x^4$. Use a graphing calculator to estimate the maximum vertical distance between the graphs $y = x^2$ and $y = x^4$ on the interval $(0, 1)$. Also estimate the value of x at which this maximum distance occurs.

 98. Repeat Exercise 97 by sketching the graph of $y = x^3 - x^5$.

In Exercises 99–102, write a polynomial with the given characteristics.

99. A polynomial of degree 3 that has exactly two x-intercepts.

100. A polynomial of degree 3 that has three distinct x-intercepts and whose graph rises to the left and falls to the right.

101. A polynomial of degree 4 that has exactly two distinct x-intercepts and whose graph falls to the left and right.

102. A polynomial of degree 4 that has exactly three distinct x-intercepts and whose graph rises to the left and right.

In Exercises 103–106, find the smallest possible degree of a polynomial with the given graph. Explain your reasoning.

103.

104.

105.

106.

Critical Thinking

107. Is it possible for the graph of a polynomial function to have no y-intercepts? Explain.

108. Is it possible for the graph of a polynomial function to have no x-intercepts? Explain.

109. Is it possible for the graph of a polynomial function of degree 3 to have exactly one local maximum and no local minimum? Explain.

110. Is it possible for the graph of a polynomial function of degree 4 to have exactly one local maximum and exactly one local minimum? Explain.

Dividing Polynomials

Before Starting this Section, Review

1. Multiplying and dividing polynomials (Section P.3, page 31)

2. Evaluating a function (Section 2.4, page 213)

3. Factoring trinomials (Section P.4, page 44)

Objectives

1   Learn the Division Algorithm.

2   Use synthetic division.

3   Use the Remainder and Factor Theorems.

GREENHOUSE EFFECT AND GLOBAL WARMING

The **greenhouse effect** refers to the fact that Earth is surrounded by an atmosphere that acts like the transparent cover of a greenhouse, allowing sunlight to filter through while trapping heat. It is becoming increasingly clear that we are experiencing *global warming*. This means that Earth's atmosphere is becoming warmer near its surface. Scientists who study climate agree that global warming is due largely to the emission of "greenhouse gases" such as carbon dioxide (CO_2), chlorofluorocarbons (CFCs), methane (CH_4), ozone (O_3), and nitrous oxide (N_2O). The temperature of Earth's surface increased by 1°F over the twentieth century. The late 1990s and early 2000s were some of the hottest years ever recorded. Projections of future warming suggest a global temperature increase of between 2.5°F and 10.4°F by 2100.

Global warming will result in long-term changes in climate, including a rise in average temperatures, unusual rainfalls, and more storms and floods. The sea level may rise so much that people will have to move away from coastal areas. Some regions of the world may become too dry for farming.

The production and consumption of energy is one of the major causes of greenhouse gas emissions. In Example 8, we look at the pattern for the consumption of petroleum in the United States. ■

---

1   Learn the division algorithm.

The Division Algorithm

Dividing polynomials is similar to dividing integers. Both division processes are closely connected to multiplication. The fact that $10 = 5 \cdot 2$ is also expressed as $\dfrac{10}{5} = 2$. By comparison, in equation

RECALL

A polynomial $P(x)$ is an expression of the form

$$a_n x^n + \cdots + a_1 x + a_0.$$

(1) $x^3 - 1 = (x - 1)(x^2 + x + 1)$,   Factor using difference of cubes.

dividing both sides by $x - 1$, we obtain

$$\frac{x^3 - 1}{x - 1} = x^2 + x + 1, \quad x \neq 1.$$

POLYNOMIAL FACTOR

A polynomial $D(x)$ is a **factor** of a polynomial $F(x)$ if there is a polynomial $Q(x)$ such that $F(x) = D(x) \cdot Q(x)$.

It is helpful to compare the division process for integers with that of polynomials. Notice that

$$
\begin{array}{r}
2 \\
3\overline{)7} \\
6 \\
\hline
1
\end{array}
$$

can also be written as

$$7 = 3 \cdot 2 + 1$$

or

$$\frac{7}{3} = 2 + \frac{1}{3}.$$

An *algorithm* is a finite sequence of precise steps that leads to a solution of a problem.

Next, consider the equation

(2) $x^3 - 2 = (x - 1)(x^2 + x + 1) - 1$ Subtract 1 from both sides of equation (1).

Dividing both sides by $(x - 1)$, we can rewrite equation (2) in the form

$$\frac{x^3 - 2}{x - 1} = x^2 + x + 1 - \frac{1}{x - 1}, \quad x \neq 1.$$

Here we say that when the **dividend** $x^3 - 2$ is divided by the **divisor** $x - 1$, the **quotient** is $x^2 + x + 1$ and the **remainder** is -1. This result is an example of the *division algorithm*.

THE DIVISION ALGORITHM

If a polynomial $F(x)$ is divided by a polynomial $D(x)$ and $D(x)$ is not the zero polynomial, then there are unique polynomials $Q(x)$ and $R(x)$ such that

$$\frac{F(x)}{D(x)} = Q(x) + \frac{R(x)}{D(x)} \qquad \text{or}$$

$$\underset{\text{Dividend}}{F(x)} = \underset{\text{Divisor}}{D(x)} \cdot \underset{\text{Quotient}}{Q(x)} + \underset{\text{Remainder}}{R(x)}$$

Either $R(x)$ is the *zero polynomial* or the degree of $R(x)$ is less than the degree of $D(x)$.

The division algorithm says that the dividend is equal to the divisor times the quotient plus the remainder.

The expression $\dfrac{F(x)}{D(x)}$ is **improper** if: degree of $F(x) \geq$ degree of $D(x)$. However, the expression $\dfrac{R(x)}{D(x)}$ is always **proper** because by the division algorithm, degree of $R(x) <$ degree of $D(x)$.

For an improper expression $\dfrac{F(x)}{D(x)}$, you may recall the **long division** and the **synthetic division** processes shown in the next few examples.

FINDING THE SOLUTION: A PROCEDURE

EXAMPLE 1   **Long Division**

OBJECTIVE	EXAMPLE
Find the quotient and remainder when one polynomial is divided by another.	*Find the quotient and remainder when $2x^2 + x^5 + 7 + 4x^3$ is divided by $x^2 + 1 - x$.*
Step 1 Write the terms in the dividend and the divisor in descending powers of the variable.	*Dividend:* $x^5 + 4x^3 + 2x^2 + 7$ *Divisor:* $x^2 - x + 1$

continued on the next page

Step 2 Insert terms with zero coefficients in the dividend for any missing powers of the variable.

$$x^5 + 0x^4 + 4x^3 + 2x^2 + 0x + 7$$

$$\frac{x^5}{x^2} = x^3 \quad \text{Divide first terms.}$$

Step 3 Divide the first term in the dividend by the first term in the divisor to obtain the first term in the quotient.

$$x^3$$
$$x^2 - x + 1\overline{)x^5 + 0x^4 + 4x^3 + 2x^2 + 0x + 7}$$

Step 4 Multiply the divisor by the first term in the quotient and subtract the product from the dividend.

$$x^3$$
$$x^2 - x + 1\overline{)x^5 + 0x^4 + 4x^3 + 2x^2 + 0x + 7}$$
$$\underline{(-)(x^5 - x^4 + x^3)} \quad\longleftarrow\quad x^3(x^2 - x + 1)$$
$$x^4 + 3x^3 + 2x^2 + 0x + 7 \quad \text{Remainder}$$

Subtract.

$$\frac{x^4}{x^2} = x^2 \qquad \frac{4x^3}{x^2} = 4x$$

Step 5 Treat the remainder obtained in Step 4 as a new dividend and repeat Steps 3 and 4. Continue this process until a remainder is obtained that is of lower degree than the divisor.

$$x^3 + x^2 + 4x + 5 \quad\longleftarrow\quad \frac{5x^2}{x^2} = 5$$
$$x^2 - x + 1\overline{)x^5 + 0x^4 + 4x^3 + 2x^2 + 0x + 7}$$
$$\underline{(-)(x^5 - x^4 + x^3)}$$
$$x^4 + 3x^3 + 2x^2 + 0x + 7 \quad\longleftarrow\quad \text{New dividend}$$
$$\underline{(-)(x^4 - x^3 + x^2)} \quad\longleftarrow\quad x^2(x^2 - x + 1)$$
$$4x^3 + x^2 + 0x + 7 \quad \text{Remainder}$$
$$\underline{(-)(4x^3 - 4x^2 + 4x)} \quad\longleftarrow\quad 4x(x^2 - x + 1)$$
$$5x^2 - 4x + 7 \quad \text{Remainder}$$
$$\underline{(-)(5x^2 - 5x + 5)} \quad\longleftarrow\quad 5(x^2 - x + 1)$$
$$x + 2 \quad \text{Remainder}$$

Step 6 Write the quotient and the remainder.

$$\text{Quotient} = x^3 + x^2 + 4x + 5 \qquad \text{Remainder} = x + 2$$

■ ■ ■

Practice Problem 1 Find the quotient and remainder when $3x^3 + 4x^2 + x + 7$ is divided by $x^2 + 1$.

■

EXAMPLE 2 **Using Long Division**

Divide $x^4 - 13x^2 + x + 35$ by $x^2 - x - 6$.

SOLUTION

Because the dividend does not contain an x^3 term, we use a zero coefficient for the missing term.

$$x^2 + x - 6 \quad\longleftarrow\quad \text{Quotient}$$
$$x^2 - x - 6\overline{)x^4 + 0x^3 - 13x^2 + x + 35}$$
$$\underline{x^4 - x^3 - 6x^2}$$
$$x^3 - 7x^2 + x + 35$$
$$\underline{x^3 - x^2 - 6x}$$
$$-6x^2 + 7x + 35$$
$$\underline{-6x^2 + 6x + 36}$$
$$x - 1 \quad\longleftarrow\quad \text{Remainder}$$

The quotient is $x^2 + x - 6$, and the remainder is $x - 1$.

We can write this result in the form

$$\frac{x^4 - 13x^2 + x + 35}{x^2 - x - 6} = x^2 + x - 6 + \frac{x - 1}{x^2 - x - 6}.$$

■ ■ ■

Practice Problem 2   Divide $x^4 + 5x^2 + 2x + 6$ by $x^2 - x + 3$.

■

2 Use synthetic division.

Synthetic Division

You can decrease the work involved in finding the quotient and remainder when the divisor $D(x)$ is of the form $x - a$ by using the process known as *synthetic division*. The procedure is best explained by considering an example.

Long Division	Synthetic Division
$$\begin{array}{r} 2x^2 + 1x + 6 \\ x - 3\overline{)2x^3 - 5x^2 + 3x - 14} \\ \underline{2x^3 - 6x^2} \\ 1x^2 + 3x - 14 \\ \underline{1x^2 - 3x} \\ 6x - 14 \\ \underline{6x - 18} \\ 4 \end{array}$$	$$\begin{array}{r|rrrr} 3 & 2 & -5 & 3 & -14 \\ & & 6 & 3 & 18 \\ \hline & 2 & 1 & 6 & 4 \end{array}$$ ← Remainder

Each circled term in the long division process is the same as the term above it. Furthermore, the boxed terms are terms in the dividend written in a new position. If these two sets of terms are eliminated in the synthetic division process, a more efficient format results.

The essential steps in the process can be carried out as follows:

(i) Bring down 2 from the first line to the third line.

$$\begin{array}{r|rrrr} 3 & 2 & -5 & 3 & -14 \\ & \downarrow \\ \hline & 2 \end{array}$$

(ii) Multiply 2 in the third line by the 3 in the divisor position and place the product, 6, under -5; then add vertically to obtain 1.

$$\begin{array}{r|rrrr} 3 & 2 & -5 & 3 & -14 \\ & & 6 \\ \hline & 2 & 1 \end{array}$$

(iii) Multiply 1 by 3 and place the product 3 under 3; then add vertically to obtain 6.

$$\begin{array}{r|rrrr} 3 & 2 & -5 & 3 & -14 \\ & & 6 & 3 \\ \hline & 2 & 1 & 6 \end{array}$$

(iv) Finally, multiply 6 by 3 and place the product, 18, under -14; then add to obtain 4.

$$\begin{array}{r|rrrr} 3 & 2 & -5 & 3 & -14 \\ & & 6 & 3 & 18 \\ \hline & 2 & 1 & 6 & 4 \end{array}$$

The first three terms, namely, 2, 1, and 6, are the coefficients of the quotient, $2x^2 + x + 6$, and the last number, 4, is the remainder. The procedure for synthetic division is given next.

TECHNOLOGY CONNECTION

Synthetic division can be done on a graphing calculator. In this problem, enter the first coefficient, 2. Then repeatedly multiply the answer by 3 and add the next coefficient.

```
                    2
Ans*3+ -5
                    1
Ans*3+3
                    6
Ans*3+ -14
                    4
```

FINDING THE SOLUTION: A PROCEDURE

EXAMPLE 3   Synthetic Division

OBJECTIVE

Divide a polynomial $F(x)$ by $x - a$.

EXAMPLE

Divide $2x^4 - 3x^2 + 5x - 63$ by $x + 3$.

$$\underline{}\,\big|\ 2 \quad 0 \quad -3 \quad 5 \quad -63$$

Step 1   Arrange the coefficients of $F(x)$ in order of *descending powers of x*, supplying zero as the coefficient of each missing power.

Step 2   Replace the divisor $x - a$ with a.

Rewrite $x + 3 = x - (-3); a = -3$

$$-3\,\big|\ 2 \quad 0 \quad -3 \quad 5 \quad -63$$

Step 3   Bring the first (leftmost) coefficient down below the line. Multiply it by a and write the resulting product one column to the right and above the line.

$$-3\,\big|\ 2 \quad 0 \quad -3 \quad 5 \quad -63$$
$$-6$$
$$2$$

⊗ means "multiply by −3."

Step 4   Add the product obtained in the previous step to the coefficient directly above it and write the resulting sum directly below it and below the line. This sum is the "newest" number below the line.

$$-3\,\big|\ 2 \quad 0 \quad -3 \quad 5 \quad -63$$
$$-6$$
$$2 \quad -6$$

Step 5   Multiply the newest number below the line by a, write the resulting product one column to the right and above the line, and repeat Step 4.

$$-3\,\big|\ 2 \quad 0 \quad -3 \quad 5 \quad -63$$
$$-6 \quad 18$$
$$2 \quad -6 \quad 15$$

Step 6   Repeat Step 5 until a product is added to the constant term. Separate the last number below the line with a short vertical line.

$$-3\,\big|\ 2 \quad 0 \quad -3 \quad 5 \quad -63$$
$$-6 \quad 18 \quad -45 \quad 120$$
$$2 \quad -6 \quad 15 \quad -40 \quad 57$$

Step 7   The last number below the line is the remainder, and the other numbers, reading from left to right, are the coefficients of the quotient, which has degree one less than the dividend $F(x)$.

$$-3\,\big|\ 2 \quad 0 \quad -3 \quad 5 \quad -63 \qquad \text{Dividend: degree 4}$$
$$-6 \quad 18 \quad -45 \quad 120$$
$$2 \quad -6 \quad 15 \quad -40 \;\big|\; 57 \qquad \text{Quotient: degree 3}$$

$$2x^3 - 6x^2 + 15x - 40 \qquad \text{Remainder}$$

Quotient

Practice Problem 3   Divide $2x^3 - 7x^2 + 5$ by $x - 3$.

STUDY TIP

A quick way to see if you remembered to put 0 coefficients in place of missing terms when dividing a polynomial of degree n by $x - a$ is to check that there are $n + 1$ numbers in the row next to $\underline{a}\,\big|$.

EXAMPLE 4   Using Synthetic Division

Use synthetic division to divide $2x^4 + x^3 - 16x^2 + 18$ by $x + 2$.

SOLUTION

Because $x - a = x + 2 = x-(-2)$, we have $a = -2$. Write the coefficients of the dividend in a line, supplying 0 as the coefficient of the missing x-term. Then carry out the steps of synthetic division.

$$\begin{array}{r|rrrrr} -2| & 2 & 1 & -16 & 0 & 18 \\ & & -4 & 6 & 20 & -40 \\ \hline & 2 & -3 & -10 & 20 & |-22 \end{array}$$

The quotient is $2x^3 - 3x^2 - 10x + 20$, and the remainder is -22; so the result is

$$\frac{2x^4 + x^3 - 16x^2 + 18}{x + 2} = 2x^3 - 3x^2 - 10x + 20 + \frac{-22}{x + 2}$$

$$= 2x^3 - 3x^2 - 10x + 20 - \frac{22}{x + 2}.$$ ■ ■ ■

Practice Problem 4   Use synthetic division to divide $2x^3 + x^2 - 18x - 7$ by $x - 3$.   ■

3 Use the Remainder and Factor Theorems.

The Remainder and Factor Theorems

In Example 4, we discovered that when the polynomial

$$F(x) = 2x^4 + x^3 - 16x^2 + 18 \text{ is divided by } x + 2,$$

the remainder is -22, a constant.

Now evaluate F at -2. We have

$$F(-2) = 2(-2)^4 + (-2)^3 - 16(-2)^2 + 18 \qquad \text{Replace } x \text{ with } -2 \text{ in } F(x).$$
$$= 2(16) - 8 - 16(4) + 18 = -22 \qquad \text{Simplify.}$$

We see that $F(-2) = -22$, the remainder we found when we divided $F(x)$ by $x + 2$. Is this result a coincidence? No, the following theorem asserts that such a result is always true.

REMAINDER THEOREM

If a polynomial $F(x)$ is divided by $x - a$, then the remainder R is given by

$$R = F(a).$$

Note that in the statement of the Remainder Theorem, $F(x)$ plays the dual role of representing a polynomial and a function defined by the same polynomial.

The Remainder Theorem has a very simple proof: The division algorithm says that if a polynomial $F(x)$ is divided by a polynomial $D(x) \neq 0$, then

$$F(x) = D(x) \cdot Q(x) + R(x),$$

where $Q(x)$ is the quotient and $R(x)$ is the remainder. Suppose $D(x)$ is a linear polynomial of the form $x - a$. Because the degree of $R(x)$ is less than the degree of $D(x)$, $R(x)$ is a constant. Call this constant R. In this case, by the division algorithm, we have

$$F(x) = (x - a)Q(x) + R.$$

Replacing x with a in $F(x)$, we obtain

$$F(a) = (a - a)Q(a) + R = 0 + R = R.$$

This proves the Remainder Theorem.

EXAMPLE 5 **Using the Remainder Theorem**

Find the remainder when the polynomial

$$F(x) = 2x^5 - 4x^3 + 5x^2 - 7x + 2$$

is divided by $x - 1$.

SOLUTION

We could find the remainder by using long division or synthetic division, but a quicker way is to evaluate $F(x)$ when $x = 1$. By the Remainder Theorem, $F(1)$ is the remainder.

$$F(1) = 2(1)^5 - 4(1)^3 + 5(1)^2 - 7(1) + 2 \quad \text{Replace } x \text{ with 1 in } F(x).$$
$$= 2 - 4 + 5 - 7 + 2 = -2$$

The remainder is -2. ■■■

Practice Problem 5 Use the Remainder Theorem to find the remainder when

$$F(x) = x^{110} - 2x^{57} + 5 \text{ is divided by } x - 1. \quad ■$$

Sometimes the Remainder Theorem is used in the other direction: to evaluate a polynomial at a specific value of x. This use is illustrated in the next example.

EXAMPLE 6 **Using the Remainder Theorem**

Let $f(x) = x^4 + 3x^3 - 5x^2 + 8x + 75$. Find $f(-3)$.

SOLUTION

One way of solving this problem is to evaluate $f(x)$ when $x = -3$.

$$f(-3) = (-3)^4 + 3(-3)^3 - 5(-3)^2 + 8(-3) + 75 \quad \text{Replace } x \text{ with } -3.$$
$$= 6 \qquad\qquad\qquad\qquad\qquad\qquad\qquad \text{Simplify.}$$

Another way is to use synthetic division to find the remainder when the polynomial $f(x)$ is divided by $(x + 3)$.

$$
\begin{array}{r|rrrrr}
-3 & 1 & 3 & -5 & 8 & 75 \\
   &   & -3 & 0 & 15 & -69 \\
\hline
   & 1 & 0 & -5 & 23 & 6 \\
\end{array}
$$

The remainder is 6. By the Remainder Theorem, $f(-3) = 6$. ■■■

Practice Problem 6 Use synthetic division to find $f(-2)$, where

$$f(x) = x^4 + 10x^2 + 2x - 20. \quad ■$$

Note that if we divide any polynomial $F(x)$ by $x - a$, we have

$$F(x) = (x - a)Q(x) + R \qquad \text{Division algorithm}$$
$$F(x) = (x - a)Q(x) + F(a) \qquad R = F(a), \text{ by the Remainder Theorem}$$

The last equation says that if $F(a) = 0$, then

$$F(x) = (x - a)Q(x)$$

and $(x - a)$ is a factor of $F(x)$. Conversely, if $(x - a)$ is a factor of $F(x)$, then dividing $F(x)$ by $(x - a)$ gives a remainder of 0. Therefore, by the Remainder Theorem, $F(a) = 0$. This argument proves the Factor Theorem.

> **FACTOR THEOREM**
>
> A polynomial $F(x)$ has $(x - a)$ as a factor if and only if $F(a) = 0$.

The Factor Theorem shows that if $F(x)$ is a polynomial, then the following problems are equivalent:

1. Factoring $F(x)$
2. Finding the zeros of the function $F(x)$ defined by the polynomial expression
3. Solving (or finding the roots of) the polynomial equation $F(x) = 0$

Note that if a is a zero of the polynomial function $F(x)$, then $F(x) = (x - a)Q(x)$. Any solution of the equation $Q(x) = 0$ is also a zero of $F(x)$. Because the degree of $Q(x)$ is less than the degree of $F(x)$, it is simpler to solve the equation $Q(x) = 0$ to find other possible zeros of $F(x)$. The equation $Q(x) = 0$ is called the **depressed equation** of $F(x)$.

The next example illustrates how to use the Factor Theorem and the depressed equation to solve a polynomial equation.

EXAMPLE 7   Using the Factor Theorem

Given that 2 is a zero of the function $f(x) = 3x^3 + 2x^2 - 19x + 6$, solve the polynomial equation $3x^3 + 2x^2 - 19x + 6 = 0$.

SOLUTION

Because 2 is a zero of $f(x)$, we have $f(2) = 0$. The Factor Theorem tells us that $(x - 2)$ is a factor of $f(x)$. Next, we use synthetic division to divide $f(x)$ by $(x - 2)$.

$$
\begin{array}{r|rrrr}
2 & 3 & 2 & -19 & 6 \\
  &   & 6 & 16 & -6 \\
\hline
  & 3 & 8 & -3 & 0 \leftarrow \text{Remainder} \\
\end{array}
$$

$$\underbrace{\uparrow \qquad \uparrow \qquad \uparrow}_{\text{coefficients of the quotient}}$$

We now have the coefficients of the quotient $Q(x)$, with $f(x) = (x - 2)Q(x)$.

$$f(x) = 3x^3 + 2x^2 - 19x + 6 = (x - 2)\underset{\uparrow}{(3x^2 + 8x - 3)}.$$

$$\text{Quotient}$$

Any solution of the depressed equation $3x^2 + 8x - 3 = 0$ is a zero of f. Because this equation is of degree 2, any method of solving a quadratic equation may be used to solve it.

$$
\begin{array}{ll}
3x^2 + 8x - 3 = 0 & \text{Depressed equation} \\
(3x - 1)(x + 3) = 0 & \text{Factor} \\
3x - 1 = 0 \quad \text{or} \quad x + 3 = 0 & \text{Zero-product property} \\
x = \dfrac{1}{3} \quad \text{or} \quad x = -3 & \text{Solve each equation.}
\end{array}
$$

The solution set is $\left\{-3, \dfrac{1}{3}, 2\right\}$.

■ ■ ■

Practice Problem 7   Solve the equation $3x^3 - x^2 - 20x - 12 = 0$ given that one solution is -2.

■

Application Energy is measured in British thermal units (BTU). One BTU is the amount of energy required to raise the temperature of 1 pound of water 1°F at or near 39.2°F; 1 million BTU is equivalent to the energy in 8 gallons of gasoline. Fuels such as coal, petroleum, and natural gas are measured in quadrillion BTU (1 quadrillion = 10^{15}), abbreviated "quads."

EXAMPLE 8   Petroleum Consumption

The petroleum consumption C (in quads) in the United States from 1970–1995 can be modeled by the function

$$C(x) = 0.003x^3 - 0.139x^2 + 1.621x + 28.995,$$

where $x = 0$ represents 1970, $x = 1$ represents 1971, and so on.

The model indicates that $C(5) = 34$ quads of petroleum were consumed in 1975. Find another year between 1975 and 1995 when the model indicates that 34 quads of petroleum were consumed. (The model was slightly modified from the data obtained from the *Statistical Abstracts of the United States*.)

SOLUTION

We are given that

$$C(5) = 34$$
$$C(x) = (x - 5)Q(x) + 34 \qquad \text{Division algorithm and Remainder Theorem}$$
$$C(x) - 34 = (x - 5)Q(x) \qquad \text{Subtract 34 from both sides.}$$

Therefore, 5 is a zero of $F(x) = C(x) - 34$

$$= 0.003x^3 - 0.139x^2 + 1.621x - 5.005.$$

Finding another year between 1975 and 1995 when the petroleum consumption was 34 quads requires us to find another zero of $F(x) = C(x) - 34$ that is between 5 and 25. Because 5 is a zero of $F(x)$, we use synthetic division to find the quotient $Q(x)$.

$$
\begin{array}{r|rrrr}
5 & 0.003 & -0.139 & 1.621 & -5.005 \\
  &   & 0.015 & -0.62 & 5.005 \\
\hline
  & 0.003 & -0.124 & 1.001 & 0
\end{array}
$$

We solve the depressed equation $Q(x) = 0$.

$$0.003x^2 - 0.124x + 1.001 = 0 \qquad Q(x) = 0.003x^2 - 0.124x + 1.001$$

$$x = \frac{0.124 \pm \sqrt{(-0.124)^2 - 4(0.003)(1.001)}}{2(0.003)} \qquad \text{Quadratic formula}$$

$$x = 11 \text{ or } x = 30.\overline{3} \qquad \text{Use a calculator.}$$

Because x needs to be between 5 and 25, this model tells us that in 1981 ($x = 11$), the petroleum consumption was also 34 quads. ▪▪▪

Practice Problem 8 The value of $C(x) = 0.23x^3 - 4.255x^2 + 0.345x + 41.05$ is 10 when $x = 3$. Find another positive number x for which $C(x) = 10$. ▪

SECTION 3.3 ■ Exercises

A EXERCISES Basic Skills and Concepts

1. In the division $\dfrac{x^4 - 2x^3 + 5x^2 - 2x + 1}{x^2 - 2x + 3} = x^2 + 2 +$

   $\dfrac{2x - 5}{x^2 - 2x + 3}$, the dividend is _____, the divisor

   is _____, the quotient is _____, and the
   remainder is _____.

2. If $P(x), Q(x)$, and $F(x)$ are polynomials and
   $F(x) = P(x) \cdot Q(x)$, then the factors of $F(x)$ are
   _____ and _____.

3. The Remainder Theorem states that if a polynomial
   $F(x)$ is divided by $(x - a)$, then the remainder
   $R =$ _____.

4. The Factor Theorem states that $(x - a)$ is a factor of a
   polynomial $F(x)$ if and only if _____ = 0.

5. *True* or *False* You cannot use synthetic division to divide
   a polynomial by $x^2 - 1$.

6. *True* or *False* When a polynomial of degree $n + 1$ is
   divided by a polynomial of degree n, the remainder is a
   constant polynomial.

**In Exercises 7–14, use long division to find the quotient and
the remainder.**

7. $\dfrac{6x^2 - x - 2}{2x + 1}$

8. $\dfrac{4x^3 - 2x^2 + x - 3}{2x - 3}$

9. $\dfrac{3x^4 - 6x^2 + 3x - 7}{x + 1}$

10. $\dfrac{x^6 + 5x^3 + 7x + 3}{x^2 + 2}$

11. $\dfrac{4x^3 - 4x^2 - 9x + 5}{2x^2 - x - 5}$

12. $\dfrac{y^5 + 3y^4 - 6y^2 + 2y - 7}{y^2 + 2y - 3}$

13. $\dfrac{z^4 - 2z^2 + 1}{z^2 - 2z + 1}$

14. $\dfrac{6x^4 + 13x - 11x^3 - 10 - x^2}{3x^2 - 5 - x}$

**In Exercises 15–30, use synthetic division to find the quotient
and the remainder when the first polynomial is divided by the
second polynomial.**

15. $x^3 - x^2 - 7x + 2$; $x - 1$

16. $2x^3 - 3x^2 - x + 2$; $x + 2$

17. $x^3 + 4x^2 - 7x - 10$; $x + 2$

18. $x^3 + x^2 - 13x + 2$; $x - 3$

19. $x^4 - 3x^3 + 2x^2 + 4x + 5$; $x - 2$

20. $x^4 - 5x^3 - 3x^2 + 10$; $x - 1$

21. $2x^3 + 4x^2 - 3x + 1$; $x - \dfrac{1}{2}$

22. $3x^3 + 8x^2 + x + 1$; $x - \dfrac{1}{3}$

23. $2x^3 - 5x^2 + 3x + 2$; $x + \dfrac{1}{2}$

24. $3x^3 - 2x^2 + 8x + 2$; $x + \dfrac{1}{3}$

25. $x^5 + x^4 - 7x^3 + 2x^2 + x - 1$; $x - 1$

26. $2x^5 + 4x^4 - 3x^3 - 7x^2 + 3x - 2$; $x + 2$

27. $x^5 + 1$; $x + 1$

28. $x^5 + 1$; $x - 1$

29. $x^6 + 2x^4 - x^3 + 5$; $x + 1$

30. $3x^5$; $x - 1$

**In Exercises 31–34, use synthetic division and the Remainder
Theorem to find each function value. Check your answer by
evaluating the function at the given x-value.**

31. $f(x) = x^3 + 3x^2 + 1$
    a. $f(1)$ b. $f(-1)$
    c. $f\left(\dfrac{1}{2}\right)$ d. $f(10)$

32. $g(x) = 2x^3 - 3x^2 + 1$
    a. $g(-2)$ b. $g(-1)$
    c. $g\left(-\dfrac{1}{2}\right)$ d. $g(7)$

33. $h(x) = x^4 + 5x^3 - 3x^2 - 20$
    a. $h(1)$ b. $h(-1)$
    c. $h(-2)$ d. $h(2)$

34. $f(x) = x^4 + 0.5x^3 - 0.3x^2 - 20$
    a. $f(0.1)$ b. $f(0.5)$
    c. $f(1.7)$ d. $f(-2.3)$

**In Exercises 35–42, use the Factor Theorem to show that the
linear polynomial is a factor of the second polynomial. Check
your answer by synthetic division.**

35. $x - 1$; $2x^3 + 3x^2 - 6x + 1$

36. $x - 3$; $3x^3 - 9x^2 - 4x + 12$

37. $x + 1$; $5x^4 + 8x^3 + x^2 + 2x + 4$

38. $x + 3$; $3x^4 + 9x^3 - 4x^2 - 9x + 9$

39. $x - 2$; $x^4 + x^3 - x^2 - x - 18$

40. $x + 3$; $x^5 + 3x^4 + x^2 + 8x + 15$

41. $x + 2$; $x^6 - x^5 - 7x^4 + x^3 + 8x^2 + 5x + 2$

42. $x - 2$; $2x^6 - 5x^5 + 4x^4 + x^3 - 7x^2 - 7x + 2$

**In Exercises 43–46, find the value of k for which the first
polynomial is a factor of the second polynomial.**

43. $x + 1$; $x^3 + 3x^2 + x + k$ [*Hint:* Let
    $f(x) = x^3 + 3x^2 + x + k$. Then $f(-1) = 0$.]

44. $x - 1$; $-x^3 + 4x^2 + kx - 2$

45. $x - 2$; $2x^3 + kx^2 - kx - 2$

46. $x - 1$; $k^2 - 3kx^2 - 2kx + 6$

In Exercises 47–50, use the Factor Theorem to show that the first polynomial *is not* a factor of the second polynomial. [*Hint:* Show that the remainder is not zero.]

47. $x - 2; -2x^3 + 4x^2 - 4x + 9$

48. $x + 3; -3x^3 - 9x^2 + 5x + 12$

49. $x + 2; 4x^4 + 9x^3 + 3x^2 + x + 4$

50. $x - 3; 3x^4 - 8x^3 + 5x^2 + 7x - 3$

B EXERCISES Applying the Concepts

51. **Geometry.** The area of a rectangle is $(2x^4 - 2x^3 + 5x^2 - x + 2)$ square centimeters. Its length is $(x^2 - x + 2)$ cm. Find its width.

52. **Geometry.** The volume of a rectangular solid is $(x^4 + 3x^3 + x + 3)$ cubic inches. Its length and width are $(x + 3)$ and $(x + 1)$ inches, respectively. Find its height.

53. **Cell phone sales.** The weekly revenue from one type of cell phone sold at a mall outlet is a quadratic function, R, of its price, x, in dollars. The weekly revenue is $3000 if the phone is priced at $40 or $60 (fewer phones are sold at $60) and is $2400 if the phone is priced at $30.
    a. Find an equation for $R(x) - 3000$.
    b. Find an equation for $R(x)$.
    c. Find the maximum weekly revenue and the price that generates this revenue.

54. **Ticket sales.** The evening revenue from a theater is a quadratic function, R, of its ticket price, x, in dollars. The evening revenue is $4725 if the tickets are priced at $8 or $12 (fewer tickets are sold at $12) and is $4125 if the tickets are priced at $6.
    a. Find an equation for $R(x) - 4725$.
    b. Find an equation for $R(x)$.
    c. Find the maximum weekly revenue and the price that generates this revenue.

55. **Total energy consumption.** The total energy consumption (in quads) in the United States from 1970 to 1995 is approximated by the function

$$C(x) = 0.0439x^3 - 1.4911x^2 + 14.224x + 41.76,$$

where $x = 0$ represents 1970. The model indicates that 81 quads of energy were consumed in 1979. Find another year after 1979 when the model indicates that 81 quads of energy were consumed. (The model was modified from the data obtained from the *Statistical Abstracts of the United States.*)

56. **Lottery sales.** In the lottery game called *Lotto*, players typically select six digits from a large field of numbers. Varying prizes are offered for matching three through six numbers drawn by the lottery. Lotto sales (in billions of dollars) in the United States from 1983 through 2003 are approximated by the function

$$s(x) = 0.0039x^3 - 0.2095x^2 + 3.2201x - 2.7505,$$

where $x = 0$ represents 1980. The model indicates that $8.6 billion in Lotto sales occurred in 1985. Find another year after 1985 when the model indicates that $8.6 billion in Lotto sales occurred. (The model was modified from the data obtained from the *Statistical Abstracts of the United States.*)

57. **Marine Corps reserves.** The number of U.S. Marine Corps reserves, M (in thousands), from 2000–2007 can be modeled by the function

$$M(x) = -0.525x^3 + 5.25x^2 - 11.025x + 99.1,$$

where $x = 0$ represents 2000, $x = 1$ represents 2001, and so on.
The model indicates that there were 99.1 thousand Marine Corps reserves in 2003. Find two other years between 2000–2007 during which time the model shows that the Marine Corps had 99.1 thousand reserves. (The model was slightly modified from data obtained from the *U.S. Census Bureau.*)

58. **Unemployment.** U.S. unemployment, U (in percent), in the United States from 2000–2009 can be modeled by the function

$$U(x) = 0.0427x^3 - 0.5678x^2 + 2.058x + 3.4883,$$

where $x = 0$ represents 2000, $x = 1$ represents 2001, and so on.
The model indicates a 5.7% level of unemployment in 2003. Find two other years between 2000–2009 when the model indicates that a 5.7% level of unemployment occurred.
(The model was slightly modified from data obtained from the *U.S. Bureau of Labor Statistics.*)

C EXERCISES Beyond the Basics

59. Show that $\dfrac{1}{2}$ is a root of multiplicity 2 of the equation

$$4x^3 + 8x^2 - 11x + 3 = 0.$$

60. Show that $-\dfrac{2}{3}$ is a root of multiplicity 2 of the equation

$$9x^3 + 3x^2 - 8x - 4 = 0.$$

61. Assuming that n is a positive integer, find the values of n for which each of the following is true.
    a. $x + a$ is a factor of $x^n + a^n$.
    b. $x + a$ is a factor of $x^n - a^n$.
    c. $x - a$ is a factor of $x^n + a^n$.
    d. $x - a$ is a factor of $x^n - a^n$.

62. Use Exercise 61 to prove the following:
    a. $7^{11} - 2^{11}$ is exactly divisible by 5.
    b. $(19)^{20} - (10)^{20}$ is exactly divisible by 261.

63. Synthetic division is a process of dividing a polynomial $F(x)$ by $x - a$. Modify the process to use synthetic

division so that it finds the remainder when the first polynomial is divided by the second.

a. $2x^3 + 3x^2 + 6x - 2$; $2x - 1$

$$\left[Hint: 2x - 1 = 2\left(x - \frac{1}{2} \right). \right]$$

b. $2x^3 - x^2 - 4x + 1; 2x + 3$

64. Let $g(x) = 3x^3 - 5cx^2 - 3c^2x + 3c^3$. Use synthetic division to find each function value.

a. $g(-c)$ **b.** $g(2c)$

In Exercises 65–69, let $f(x) = ax^3 + bx^2 + cx + d, a \neq 0$ **and** $g(x) = kx^4 + lx^3 + mx^2 + nx + p, k \neq 0$. **[*Hint:* Use the end behavior of polynomials and the Intermediate Value Theorem.]**

65. Show that f has exactly one real zero or three real zeros.

66. Show that if $kp < 0$, then g has exactly two real zeros or exactly four real zeros.

67. Given an example of a fourth degree polynomial that has no real zeros.

68. Suppose $dp < 0$ and $kp < 0$. Show that the graphs of f and g must intersect.

69. Give an example to show that the results of Exercise 68 may not hold if $dp < 0$ and $kp > 0$.

Critical Thinking

70. **True or False.**

a. $\dfrac{1}{3}$ is a possible zero of $P(x) = x^5 - 3x^2 + x + 3$.

b. $\dfrac{2}{5}$ is a possible zero of $P(x) = 2x^7 - 5x^4 - 17x + 25$.

The Real Zeros of a Polynomial Function

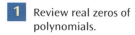

Before Starting this Section, Review

1. Synthetic division (Section 3.3, page 331)

2. Zero-product property (Section 1.4, page 118)

3. Multiplicity of a zero (Section 3.2, page 319)

Objectives

1 Review real zeros of polynomials.

2 Use the Rational Zeros Theorem.

3 Find the possible number of positive and negative zeros of polynomials.

4 Find the bounds on the real zeros of polynomials.

5 Learn a procedure for finding the real zeros of a polynomial function.

DOW JONES INDUSTRIAL AVERAGE

The Dow Jones Industrial Average (DJIA) reflects the average cost per share of stock from 30 companies chosen by the board of the New York Stock Exchange as the most outstanding companies to provide a quantitative view of the stock market and, by extension, the U.S. economy. To calculate the DJIA on any day, we do not simply add the prices of the component stocks and divide by 30. The reason is that during the last 100-plus years, there have been many stock splits, spinoffs, and stock substitutions. Without making an adjustment to the divisor (30), the value of the DJIA would be distorted.

To understand this, consider a stock split. Suppose two stocks are trading at $10 and $40; the average of the two is $25. If the company with $40 stock has a two-for-one split, its shares are now priced at $20 each. The average of the two stocks falls to $15 $\left(\dfrac{10 + 20}{2} = 15 \right)$. But due to the two-for-one split, the value of investment is unchanged. The $40 stock simply sells for $20 with twice as many shares available. So to keep the DJIA at $25, we use the divisor x so that $\dfrac{10 + 20}{x} = 25$. Solving for x, we have $x = \dfrac{30}{25} = 1.2$.

Over time, the divisor 30 for the DJIA has been adjusted several times, mostly downward. It stood at 0.125552709 at the end of 2008. This explains why the DJIA can be reported as 10,000 even though the sum of the prices of all of its component stocks is a fraction of this number. In Exercise 77, we model an investment between 2000–2009 by a polynomial function. ■

---

1 Review real zeros of polynomials.

Real Zeros of a Polynomial Function

We first review what we have learned about the real zeros of a polynomial function.

1. **Definition** If c is a real number in the domain of a function f and $f(c) = 0$, then c is called a real zero of f.
   a. Geometrically, this means that the graph of f has an x-intercept at $x = c$.
   b. Algebraically, this means that $(x - c)$ is a factor of $f(x)$. In other words, we can write $f(x) = (x - c)g(x)$ for some polynomial function $g(x)$.

2. **Number of real zeros** A polynomial function of degree n has, at most, n real zeros.

3. **Polynomial in factored form** If a polynomial $F(x)$ can be written in factored form, we find its zeros by solving $F(x) = 0$, using the zero product property of real numbers.

4. **Intermediate Value Theorem** If a polynomial $F(x)$ cannot be factored, we find approximate values of the real zeros (if any) of $F(x)$ by using the Intermediate Value Theorem. (See page 317.)

In this section, we investigate other useful tools for finding the real zeros of a polynomial function.

2 Use the Rational Zeros Theorem.

Rational Zeros Theorem

We use the following test (see Exercise 95 for a proof) to find *possible* rational zeros of a polynomial function with integer coefficients. We then use the Factor Theorem to determine whether any of these numbers is, in fact, a zero of a polynomial function $F(x)$.

RECALL

The ratio $\frac{p}{q}$ of integers p and q is in lowest terms if p and q have no common factor other than 1.

RATIONAL ZEROS THEOREM

If $F(x) = a_n x^n + a_{n-1} x^{n-1} + \cdots + a_2 x^2 + a_1 x + a_0$ is a polynomial function with integer coefficients ($a_n \neq 0, a_0 \neq 0$) and $\frac{p}{q}$ is a rational number in lowest terms that is a zero of $F(x)$, then

1. p is a factor of the constant term a_0.

2. q is a factor of the leading coefficient a_n.

EXAMPLE 1   Using the Rational Zeros Theorem

Find all rational zeros of $F(x) = 2x^3 + 5x^2 - 4x - 3$.

SOLUTION

First, we list all *possible* rational zeros of $F(x)$.

$$\text{Possible rational zeros} = \frac{p}{q} = \frac{\text{Factors of the constant term, } -3}{\text{Factors of the leading coefficient, } 2}$$

Factors of -3: $\pm 1, \pm 3$

Factors of 2: $\pm 1, \pm 2$

All combinations for the rational zeros can be found by using just the *positive* factors of the leading coefficient. The possible rational zeros are

$$\frac{\pm 1}{1}, \frac{\pm 1}{2}, \frac{\pm 3}{1}, \frac{\pm 3}{2}$$

or

$$\pm 1, \pm \frac{1}{2}, \pm \frac{3}{2}, \pm 3.$$

We use synthetic division to determine whether a rational zero exists among these eight candidates. We start by testing 1. If 1 is not a rational zero, we will test other possibilities.

$$
\begin{array}{r|rrrr}
1 & 2 & 5 & -4 & -3 \\
  &   & 2 & 7 & 3 \\
\hline
  & 2 & 7 & 3 & 0
\end{array}
$$

The zero remainder tells us that $(x - 1)$ is a factor of $F(x)$, with the other factor being $2x^2 + 7x + 3$. So

$$F(x) = (x - 1)(2x^2 + 7x + 3)$$
$$= (x - 1)(2x + 1)(x + 3) \qquad \text{Factor } 2x^2 + 7x + 3.$$

We find the zeros of $F(x)$ by solving the equation $(x - 1)(2x + 1)(x + 3) = 0$.

$$x - 1 = 0 \quad \text{or} \quad 2x + 1 = 0 \quad \text{or} \quad x + 3 = 0 \qquad \text{Zero-product property}$$

$$x = 1 \quad \text{or} \quad x = -\frac{1}{2} \quad \text{or} \quad x = -3 \qquad \text{Solve each equation.}$$

The solution set is $\left\{ 1, -\frac{1}{2}, -3 \right\}$. The rational zeros of F are -3, $-\frac{1}{2}$, and 1. ■ ■ ■

Practice Problem 1 Find all rational zeros of $F(x) = 2x^3 + 3x^2 - 6x - 8$. ■

Recall that the *zeros* of a polynomial function $y = F(x)$ are also the *roots* or the *solutions* of the polynomial equation $F(x) = 0$.

EXAMPLE 2 **Solving a Polynomial Equation**

Solve $3x^3 - 8x^2 - 8x + 8 = 0$.

SOLUTION

$$\text{Possible rational roots} = \frac{\text{Factors of the constant term, 8}}{\text{Factors of the leading coefficient, 3}}$$

$$= \frac{\pm 1, \pm 2, \pm 4, \pm 8}{\pm 1, \pm 3}$$

$$= \pm 1, \pm 2, \pm 4, \pm 8, \pm\frac{1}{3}, \pm\frac{2}{3}, \pm\frac{4}{3}, \pm\frac{8}{3}$$

The graph of $y = 3x^3 - 8x^2 - 8x + 8$ is shown in Figure 3.16. The calculator graph is shown in the margin.

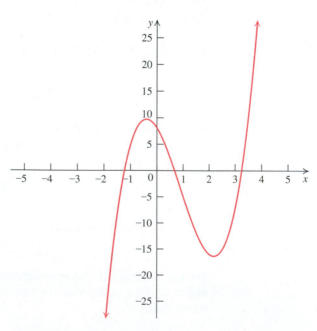

FIGURE 3.16 Graph of $y = 3x^3 - 8x^2 - 8x + 8$

From the graph in Figure 3.16, it appears that an x-intercept is between 0.5 and 1. We check whether $x = \dfrac{2}{3}$ is a root by synthetic division.

$$
\begin{array}{r|rrrr}
\frac{2}{3} & 3 & -8 & -8 & 8 \\
 & & 2 & -4 & -8 \\
\hline
 & 3 & -6 & -12 & 0 \checkmark \text{ Yes}
\end{array}
$$

Because the remainder in the synthetic division is 0, $\left(x - \dfrac{2}{3} \right)$ is a factor of the original equation with the depressed equation $3x^2 - 6x - 12 = 0$.

So

$$3x^3 - 8x^2 - 8x + 8 = 0 \qquad \text{Original equation}$$

$$\left(x - \frac{2}{3} \right)(3x^2 - 6x - 12) = 0 \qquad \text{Synthetic division}$$

$$\left(x - \frac{2}{3} \right) = 0 \quad \text{or} \quad 3x^2 - 6x - 12 = 0 \qquad \text{Zero-product property}$$

$$x = \frac{2}{3} \quad\bigg|\quad x = \frac{-(-6) \pm \sqrt{(-6)^2 - 4(-12)(3)}}{2(3)} \qquad \begin{array}{l}\text{Solve for } x \text{ using the} \\ \text{quadratic formula}\end{array}$$

$$= \frac{6 \pm \sqrt{36 + 144}}{6}$$

$$= \frac{6 \pm \sqrt{180}}{6} = \frac{6 \pm \sqrt{(36)(5)}}{6}$$

$$= \frac{6 \pm 6\sqrt{5}}{6} = 1 \pm \sqrt{5}$$

You can see from the graph in Figure 3.16 that $1 \pm \sqrt{5}$ are also the x-intercepts of the graph.

The three real roots of the given equation are

$$\frac{2}{3}, 1 + \sqrt{5}, \text{ and } 1 - \sqrt{5}. \qquad \blacksquare\,\blacksquare\,\blacksquare$$

Practice Problem 2   Solve $2x^3 - 9x^2 + 6x - 1 = 0$.   ■

3 Find the possible number of positive and negative zeros of polynomials.

Descartes's Rule of Signs

Descartes's Rule of Signs is used to find the possible number of positive and negative zeros of a polynomial function.

If the terms of a polynomial are written in descending order, we say that a **variation of sign** occurs when the signs of two consecutive terms differ. For example, in the polynomial $2x^5 - 5x^3 - 6x^2 + 7x + 3$, the signs of the terms are $+\ -\ -\ +\ +$. Therefore, there are two variations of sign, as follows:

There are three variations of sign in

$$x^5 - 3x^3 + 2x^2 + 5x - 7.$$

DESCARTES'S RULE OF SIGNS

Let $F(x)$ be a polynomial function with real coefficients and with terms written in descending order.

1. The number of positive zeros of F is equal to the number of variations of sign of $F(x)$ or is less than that number by an even integer.

2. The number of negative zeros of F is equal to the number of variations of sign of $F(-x)$ or is less than that number by an even integer.

When using Descartes's Rule, a zero of multiplicity m should be counted as m zeros.

EXAMPLE 3   **Using Descartes's Rule of Signs**

Find the possible number of positive and negative zeros of

$$f(x) = x^5 - 3x^3 - 5x^2 + 9x - 7.$$

SOLUTION

There are three variations of sign in $f(x)$.

$$f(x) = x^5 - 3x^3 - 5x^2 + 9x - 7$$

The number of positive zeros of $f(x)$ must be 3 or $3 - 2 = 1$. Furthermore,

$$f(-x) = (-x)^5 - 3(-x)^3 - 5(-x)^2 + 9(-x) - 7$$
$$= -x^5 + 3x^3 - 5x^2 - 9x - 7.$$

The polynomial $f(-x)$ has two variations of sign; therefore, the number of negative zeros of $f(x)$ is 2 or $2 - 2 = 0$.   ■ ■ ■

Practice Problem 3   Find the possible number of positive and negative zeros of

$$f(x) = 2x^5 + 3x^2 + 5x - 1.$$ ■

4 Find the bounds on the real zeros of polynomials.

Bounds on the Real Zeros

If a polynomial function $F(x)$ has no zeros greater than a number k, then k is called an **upper bound** on the zeros of $F(x)$. Similarly, if $F(x)$ has no zeros less than a number k, then k is called a **lower bound** on the zeros of $F(x)$. These bounds have the following rules.

RULES FOR BOUNDS

Let $F(x)$ be a polynomial function with real coefficients and a positive leading coefficient. Suppose $F(x)$ is synthetically divided by $x - k$.

1. If $k > 0$ and each number in the last row is zero or positive, then k is an upper bound on the zeros of $F(x)$.

2. If $k < 0$ and numbers in the last row alternate in sign, then k is a lower bound on the zeros of $F(x)$.

Zeros in the last row can be regarded as positive or negative.

EXAMPLE 4 **Finding the Bounds on the Zeros**

Find upper and lower bounds on the zeros of

$$F(x) = x^4 - x^3 - 5x^2 - x - 6.$$

SOLUTION

By the Rational Zeros Theorem, the possible rational zeros of $F(x)$ are $\pm 1, \pm 2, \pm 3,$ and ± 6. There is only one variation of sign in $F(x) = x^4 - x^3 - 5x^2 - x - 6$, so $F(x)$ has only one positive zero. First, we try synthetic division by $x - k$ with $k = 1, 2, 3, \ldots$. The first integer that makes each number in the last row a 0 or a positive number is an upper bound on the zeros of $F(x)$.

$$
\begin{array}{r|rrrr}
1 & 1 & -1 & -5 & -1 & -6 \\
  &   & 1 & 0 & -5 & -6 \\
\hline
  & 1 & 0 & -5 & -6 & -12
\end{array}
\quad \text{Synthetic division with } k = 1
$$

$$
\begin{array}{r|rrrr}
2 & 1 & -1 & -5 & -1 & -6 \\
  &   & 2 & 2 & -6 & -14 \\
\hline
  & 1 & 1 & -3 & -7 & -20
\end{array}
\quad \text{Synthetic division with } k = 2
$$

$$
\begin{array}{r|rrrr}
3 & 1 & -1 & -5 & -1 & -6 \\
  &   & 3 & 6 & 3 & 6 \\
\hline
  & 1 & 2 & 1 & 2 & 0
\end{array}
\quad \text{Synthetic division with } k = 3
$$

$\leftarrow$ All numbers are 0 or positive.

By the rule for bounds, 3 is an upper bound on the zeros of $F(x)$.

We now try synthetic division by $x - k$ with $k = -1, -2, -3, \ldots$. The first negative integer for which the numbers in the last row alternate in sign is a lower bound on the zeros of $F(x)$.

$$
\begin{array}{r|rrrr}
-1 & 1 & -1 & -5 & -1 & -6 \\
   &   & -1 & 2 & 3 & -2 \\
\hline
   & 1 & -2 & -3 & 2 & -8
\end{array}
\quad \text{Synthetic division with } k = -1
$$

$$
\begin{array}{r|rrrr}
-2 & 1 & -1 & -5 & -1 & -6 \\
   &   & -2 & 6 & -2 & 6 \\
\hline
   & 1 & -3 & 1 & -3 & 0
\end{array}
\quad \text{Synthetic division with } k = -2
$$

These numbers alternate in sign
$\leftarrow$ if we count 0 as positive.

By the rule of bounds, -2 is a lower bound on the zeros of $F(x)$. ■■■

Practice Problem 4   Find upper and lower bounds on the zeros of

$$f(x) = 2x^3 + 5x^2 + x - 2.$$ ■

An upper bound or a lower bound on the zeros of a polynomial function is not unique. For example, the zeros of $P(x) = x^3 + 2x^2 - x - 2$ are $-2, -1$, and 1. Any number greater than or equal to 1 is an upper bound on these zeros, and any number less than or equal to -2 is a lower bound on these zeros.

5 Learn a procedure for finding the real zeros of a polynomial function.

Find the Real Zeros of a Polynomial Function

We outline steps for gathering information and locating the real zeros of a polynomial function.

Step 1   Find the maximum number of real zeros by using the degree of the polynomial function.

Step 2   Find the possible number of positive and negative zeros by using Descartes's Rule of Signs.

Step 3   Write the set of possible rational zeros.

Step 4 Test the smallest positive integer in the set in Step 3, the next larger, and so on, until an integer zero or an upper bound of the zeros is found.

a. If a zero is found, use the function represented by the depressed equation in further calculations.

b. If an upper bound is found, discard all larger numbers in the set of possible rational zeros.

Step 5 Test the positive fractions that remain in the set in Step 3 after considering any bound that has been found.

Step 6 Use modified Steps 4 and 5 for negative numbers in the set in Step 3.

Step 7 If a depressed equation is quadratic, use any method (including the quadratic formula) to solve this equation.

EXAMPLE 5   Finding the Real Zeros of a Polynomial Function

Find the real zeros of $f(x) = 2x^5 - x^4 - 14x^3 + 7x^2 + 24x - 12$.

SOLUTION

Step 1 f has, at most, 5 real zeros.

Step 2 $f(x) = 2x^5 - x^4 - 14x^3 + 7x^2 + 24x - 12$

There are three variations of sign in $f(x)$. So f has 3 or 1 positive zeros.

$$f(-x) = -2x^5 - x^4 + 14x^3 + 7x^2 - 24x - 12$$

There are two variations of signs in $f(-x)$. So f has 2 or 0 negative zeros.

Step 3 Possible rational zeros are

$$\left\{\pm 1, \pm 2, \pm 3, \pm 4, \pm 6, \pm 12, \pm\frac{1}{2}, \pm\frac{3}{2}\right\}.$$

Step 4 We test for zeros, 1, 2, 3, 4, 6, and 12, until a zero or an upper bound is found.

Because 2 is a zero, $(x - 2)$ is a factor of f. We have

$$f(x) = (x - 2)(2x^4 + 3x^3 - 8x^2 - 9x + 6).$$

We consider the zeros of $Q_1(x) = 2x^4 + 3x^3 - 8x^2 - 9x + 6$.

The possible positive rational zeros of $Q_1(x)$ are $\left\{1, 2, 3, 6, \frac{1}{2}, \frac{3}{2}\right\}$. We discard 1 because 1 is not a zero of the original function. We test 2 because it may be a repeated zero of f.

The last row shows that 2 is not a zero of $Q_1(x)$ but is an upper bound (all entries in the last row are positive) for the zeros of $Q_1(x)$, hence, also on the zeros of f.

Step 5 We test the remaining positive entries $\dfrac{1}{2}$ and $\dfrac{3}{2}$ in the list.

$$
\begin{array}{r|rrrrr}
\dfrac{1}{2} & 2 & 3 & -8 & -9 & 6 \\
 & & 1 & 2 & -3 & -6 \\
\hline
 & 2 & 4 & -6 & -12 & \;0 \;\checkmark
\end{array}
\qquad
\begin{array}{r|rrrrr}
\dfrac{3}{2} & 2 & 3 & -8 & -9 & 6 \\
 & & 3 & 9 & \dfrac{3}{2} & -\dfrac{45}{4} \\
\hline
 & 2 & 6 & 1 & -\dfrac{15}{2} & \;-\dfrac{21}{4}\;\text{✗}
\end{array}
$$

Note that $\dfrac{1}{2}$ is a zero of $Q_1(x)$. Using the last row in synthetic division, we have

$$f(x) = (x - 2)\left(x - \frac{1}{2}\right)(2x^3 + 4x^2 - 6x - 12) = 0$$

or $(x - 2)(2x - 1)(x^3 + 2x^2 - 3x - 6) = 0$ Simplify.

Step 6 We now find the negative zeros of $Q_2(x) = x^3 + 2x^2 - 3x - 6$. We test $-1, -2, -3,$ and -6.

$$
\begin{array}{r|rrrr}
-1 & 1 & 2 & -3 & -6 \\
 & & -1 & -1 & 4 \\
\hline
 & 1 & 1 & -4 & \;-2\;\text{✗}
\end{array}
\qquad
\begin{array}{r|rrrr}
-2 & 1 & 2 & -3 & -6 \\
 & & -2 & 0 & 6 \\
\hline
 & 1 & 0 & -3 & \;0\;\checkmark
\end{array}
$$

Step 7 We can solve the depressed equation $x^2 - 3 = 0$ to obtain $x = \pm\sqrt{3}$. The real zeros of f are $-2, -\sqrt{3}, \dfrac{1}{2}, \sqrt{3},$ and 2. Because $f(x)$ has, at most, five real zeros, we have found all of them.   ■ ■ ■

Practice Problem 5 Find the real zeros of

$$f(x) = 3x^4 - 11x^3 + 22x - 12.$$ ■

SECTION 3.4 ■ Exercises

A EXERCISES Basic Skills and Concepts

1. Let $P(x) = a_n x^n + \cdots + a_0$ with integer coefficients. If $\dfrac{p}{q}$ is a rational zero of $P(x)$, then

   $$\dfrac{p}{q} = \dfrac{\text{possible factors of } \underline{\hspace{1cm}}}{\text{possible factors of } \underline{\hspace{1cm}}}.$$

2. If the terms of a polynomial are written in descending order, then a variation of sign occurs when the signs of two _____ terms differ.

3. The number of positive zeros of a polynomial function $P(x)$ is equal to the number of _____ of sign of $P(x)$ or is less than that number by (a)n _____.

4. If a polynomial $P(x)$ has no zero greater than a number k, then k is called a(n) _____ on the zeros of $P(x)$. If $P(x)$ has no zero less than a number m, then m is called a(n) _____ on the zeros of $P(x)$.

5. *True or False* Possible rational zeros of $P(x) = 9x^3 - 9x^2 - x + 1$ are $\pm 1, \pm 3, \pm 9$.

6. *True or False* The polynomial function $P(x) = x^3 + x + 1$ has exactly one real zero.

In Exercises 7–10, find the set of possible rational zeros of the given function.

7. $f(x) = 3x^3 - 4x^2 + 5$

8. $g(x) = 2x^4 - 5x^2 - 2x + 1$

9. $h(x) = 4x^4 - 9x^2 + x + 6$

10. $F(x) = 6x^6 + 5x^5 + x - 35$

In Exercises 11–26, find all rational zeros of the given polynomial function.

11. $f(x) = x^3 - x^2 - 4x + 4$

12. $f(x) = x^3 - 4x^2 + x + 6$

13. $f(x) = x^3 + x^2 + 2x + 2$

14. $f(x) = x^3 + 3x^2 + 2x + 6$

15. $g(x) = 2x^3 + x^2 - 13x + 6$

16. $g(x) = 6x^3 + 13x^2 + x - 2$

17. $g(x) = 3x^3 - 2x^2 + 3x - 2$

18. $g(x) = 2x^3 + 3x^2 + 4x + 6$

19. $h(x) = 3x^3 + 7x^2 + 8x + 2$

20. $h(x) = 2x^3 + x^2 + 8x + 4$

21. $h(x) = x^4 - x^3 - x^2 - x - 2$

22. $h(x) = 2x^4 + 3x^3 + 8x^2 + 3x - 4$

23. $F(x) = x^4 - x^3 - 13x^2 + x + 12$

24. $G(x) = 3x^4 + 5x^3 + x^2 + 5x - 2$

25. $F(x) = x^4 - 2x^3 + 10x^2 - x + 1$

26. $G(x) = x^6 + 2x^4 + x^2 + 2$

In Exercises 27–40, determine the possible number of positive and negative zeros of the given function by using Descartes's Rule of Signs.

27. $f(x) = 5x^3 - 2x^2 - 3x + 4$

28. $g(x) = 3x^3 + x^2 - 9x - 3$

29. $f(x) = 2x^3 + 5x^2 - x + 2$

30. $g(x) = 3x^4 + 8x^3 - 5x^2 + 2x - 3$

31. $h(x) = 2x^5 - 5x^3 + 3x^2 + 2x - 1$

32. $F(x) = 5x^6 - 7x^4 + 2x^3 - 1$

33. $G(x) = -3x^4 - 4x^3 + 5x^2 - 3x + 7$

34. $H(x) = -5x^5 + 3x^3 - 2x^2 - 7x + 4$

35. $f(x) = x^4 + 2x^2 + 4$ 36. $f(x) = 3x^4 + 5x^2 + 6$

37. $g(x) = 2x^5 + x^3 + 3x$ 38. $g(x) = 2x^5 + 4x^3 + 5x$

39. $h(x) = -x^5 - 2x^3 + 4$ 40. $h(x) = 2x^5 + 3x^3 + 1$

In Exercises 41–48, determine upper and lower bounds on the zeros of the given function.

41. $f(x) = 3x^3 - x^2 + 9x - 3$

42. $g(x) = 2x^3 - 3x^2 - 14x + 21$

43. $F(x) = 3x^3 + 2x^2 + 5x + 7$

44. $G(x) = x^3 + 3x^2 + x - 4$

45. $h(x) = x^4 + 3x^3 - 15x^2 - 9x + 31$

46. $H(x) = 3x^4 - 20x^3 + 28x^2 + 19x - 13$

47. $f(x) = 6x^4 + x^3 - 43x^2 - 7x + 7$

48. $g(x) = 6x^4 + 23x^3 + 25x^2 - 9x - 5$

In Exercises 49–68, find all rational zeros of f. Then (if necessary) use the depressed equation to find all roots of the equation $f(x) = 0$.

49. $f(x) = x^3 + 5x^2 - 8x + 2$

50. $f(x) = x^3 - 7x^2 - 5x + 3$

51. $f(x) = 2x^3 - x^2 - 6x + 3$

52. $f(x) = 3x^3 + 2x^2 - 15x - 10$

53. $f(x) = 2x^3 - 9x^2 + 6x - 1$

54. $f(x) = 2x^3 - 3x^2 - 4x - 1$

55. $f(x) = x^4 + x^3 - 5x^2 - 3x + 6$

56. $f(x) = x^4 - 2x^3 - 5x^2 + 4x + 6$

57. $f(x) = x^4 - 3x^3 + 3x - 1$

58. $f(x) = x^4 - 6x^3 - 7x^2 + 54x - 18$

59. $f(x) = 2x^4 - 5x^3 - 4x^2 + 15x - 6$

60. $f(x) = 3x^4 - 8x^3 - 18x^2 + 40x + 15$

61. $f(x) = 6x^4 - x^3 - 13x^2 + 2x + 2$

62. $f(x) = 6x^4 + x^3 - 19x^2 - 3x + 3$

63. $f(x) = x^5 - 2x^4 - 4x^3 + 8x^2 + 3x - 6$

64. $f(x) = x^5 + x^4 - 6x^3 - 6x^2 + 8x + 8$

65. $f(x) = 2x^5 + x^4 - 11x^3 - x^2 + 15x - 6$

66. $f(x) = 2x^5 - x^4 - 9x^3 + 10x + 4$

67. $f(x) = 2x^5 - 13x^4 + 27x^3 - 17x^2 - 5x + 6$

68. $f(x) = 3x^5 + x^4 - 9x^3 - 3x^2 + 6x + 2$

B EXERCISES Applying the Concepts

69. **Geometry.** The length of a rectangle is $x^2 - 2x + 3$, and its width is $x - 2$. Find its dimensions assuming that its area is 306 square units.

70. **Making a box.** A square piece of tin 18 inches on each side is to be made into a box without a top by cutting a square from each corner and folding up the flaps to form the sides. What size corners should be cut so that the volume of the box is 432 cubic inches?

71. **Cost function.** The total cost of a product (in dollars) is given by $C(x) = 3x^3 - 6x^2 + 108x + 100$, where x is the demand for the product. Find the value of x that gives a total cost of $628.

72. **Profit.** For the product in Exercise 71, the demand function is $p(x) = 330 + 10x - x^2$. Find the value of x that gives the profit of $910.

73. **Production cost.** The total cost per month, in thousands of dollars, of producing x printers (with x measured in hundreds) is given by $C(x) = x^3 - 15x^2 + 5x + 50$. How many units must be produced so that the total monthly cost is $125,000?

74. **Break even.** If the demand function for the printers in Exercise 73 is $p(x) = -3x + 3 + \dfrac{74}{x}$, find the number of printers that must be produced and sold to break even.

75. Oil import-export. A country maintains its normal oil needs by importing the deficit and exporting its surplus oil. The accompanying graph shows the country's balance of import and export (in million barrels) for 2000–2009, where $x = 0$ represents 2000.
   a. How many years did the country export oil?
   b. Construct a polynomial function of minimum degree that models this graph.
   c. Use part (b) to estimate the oil import or export in 2005.

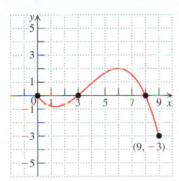

76. Number of tourists. The accompanying graph shows the number of tourists (in thousands) above or below normal during 2008, where $x = 1$ represents January.
   a. Construct a polynomial function of minimum degree that models this graph.
   b. Use part (a) to estimate the level of tourism in September 2008.

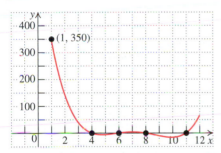

77. Investment. The accompanying graph shows the profit/loss (in thousand dollars) of Ms. Sharpy's investment in the stock market during 2000–2009, where $x = 0$ represents 2000.
   a. Construct a polynomial of minimum degree that models the graph.
   b. Use part (a) to estimate the profit/loss in 2004.

   c. Use part (a) to determine the year(s) when Ms. Sharpy realized a profit of $600.
   d. Use part (a) to determine the maximum profit and the year it occurred.

78. County government. The accompanying graph shows the surplus/deficit (in million dollars) for a certain county government during 2000–2009, where $x = 0$ represents 2000.
   a. Construct a polynomial function of minimum degree that models the graph.
   b. Use part (a) to estimate the surplus/deficit in 2003 and 2007.
   c. Use part (a) to determine the maximum amount of deficit.

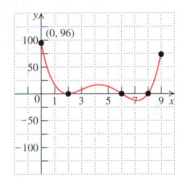

C EXERCISES Beyond the Basics

In Exercises 79–82, show that the given number is irrational. [*Hint:* To show that $x = 1 + \sqrt{3}$ is irrational, form the equivalent equations $x - 1 = \sqrt{3}$, $(x - 1)^2 = 3$, and $x^2 - 2x - 2 = 0$ and show that the last equation has no rational roots.]

79. $\sqrt{3}$ **80.** $\sqrt[3]{4}$

81. $3 - \sqrt{2}$ **82.** $(9)^{\frac{2}{3}}$

In Exercises 83–94, find a polynomial equation of least possible degree with integer coefficient whose roots include the given numbers. Use the result: If an equation with integer coefficients has a root $(a + \sqrt{b})$ where a and b are rational but $\sqrt{b}$ is irrational, then the equation also has the root $(a - \sqrt{b})$.

83. $1, -1, 0$ **84.** $2, 1, -2$

85. $-\dfrac{1}{2}, 2, \dfrac{7}{3}$ **86.** $-\dfrac{1}{3}, -\dfrac{1}{2}, \dfrac{1}{6}$

87. $1 + \sqrt{2}, 3$ **88.** $3 - \sqrt{5}, 2$

89. $1 + \sqrt{3}, 3 - \sqrt{2}$ **90.** $2 - \sqrt{5}, 2 + \sqrt{3}$

91. $\sqrt{2} + \sqrt{3}$ **92.** $2 + \sqrt{3} + \sqrt{5}$

93. $3 - \sqrt{2} + \sqrt{5}$ **94.** $\sqrt{2} + \sqrt{3} + \sqrt{5}$

95. To prove the rational zeros test, we assume the **Fundamental Theorem of Arithmetic**, which states that every integer has a unique prime factorization. Supply reasons for each step in the following proof.

(i) $F\left(\dfrac{p}{q}\right) = 0$, and $\dfrac{p}{q}$ is in lowest terms.

(ii) $a_n\left(\dfrac{p}{q}\right)^n + a_{n-1}\left(\dfrac{p}{q}\right)^{n-1} + \cdots + a_1\left(\dfrac{p}{q}\right) + a_0 = 0$

(iii) $a_n p^n + a_{n-1} p^{n-1} q + \cdots + a_1 p q^{n-1} + a_0 q^n = 0$

(iv) $a_n p^n + a_{n-1} p^{n-1} q + \cdots + a_1 p q^{n-1} = -a_0 q^n$

(v) p is a factor of the left side of (iv).

(vi) p is a factor of $-a_0 q^n$.

(vii) p is a factor of a_0.

(viii) $a_{n-1} p^{n-1} q + \cdots + a_1 p q^{n-1} + a_0 q^n = -a_n p^n$

(ix) q is a factor of the left side of the equation in (viii).

(x) q is a factor of $-a_n p^n$.

(xi) q is a factor of a_n.

96. Find all rational roots of the equation

$$2x^4 + 2x^3 + \frac{1}{2}x^2 + 2x - \frac{3}{2} = 0.$$

[*Hint:* Multiply both sides of the equation by the lowest common denominator.]

Critical Thinking

97. True or False.

a. $\dfrac{1}{3}$ is a possible zero of $P(x) = x^5 - 3x^2 + x + 3$.

b. $\dfrac{2}{5}$ is a possible zero of $P(x) = 2x^7 - 5x^4 - 17x + 25$.

The Complex Zeros of a Polynomial Function

Before Starting this Section, Review

1. Long division (Section 3.4, page 329)
2. Synthetic division (Section 3.4, page 331)
3. Rational zeros theorem (Section 3.4, page 341)
4. Complex numbers (Section 1.3, page 108)
5. Conjugate of a complex number (Section 1.3, page 112)

Objectives

1. Learn basic facts about the complex zeros of polynomials.
2. Use the Conjugate Pairs Theorem to find zeros of polynomials.

Gerolamo Cardano (1501–1576)

Cardano was trained as a physician, but his range of interests included philosophy, music, gambling, mathematics, astrology, and the occult sciences. Cardano was a colorful man, full of contradictions. His lifestyle was deplorable even by the standards of the times, and his personal life was filled with tragedies. His wife died in 1546. He saw one son executed for wife poisoning. He cropped the ears of a second son who attempted the same crime. In 1570, he was imprisoned for six months for heresy when he published the horoscope of Jesus. Cardano spent the last few years in Rome, where he wrote his autobiography, *De Propria Vita*, in which he did not spare the most shameful revelations.

CARDANO ACCUSED OF STEALING FORMULA

In sixteenth-century Italy, getting and keeping a teaching job at a university was difficult. University appointments were, for the most part, temporary. One way a professor could convince the administration that he was worthy of continuing in his position was by winning public challenges. Two contenders for a position would present each other with a list of problems, and later in a public forum, each person would present his solutions to the other's problems. As a result, scholars often kept secret their new methods of solving problems.

In 1535, in a public contest primarily about solving the cubic equation, Nicolo Tartaglia (1499–1557) won over his opponent Antonio Maria Fiore. Gerolamo Cardano (1501–1576), in the hope of learning the secret of solving a cubic equation, invited Tartaglia to visit him. After many promises and much flattery, Tartaglia revealed his method on the promise, probably given under oath, that Cardano would keep it confidential. When Cardano's book *Ars Magna* (The Great Art) appeared in 1545, the formula and the method of proof were fully disclosed. Angered at this apparent breach of a solemn oath and feeling cheated out of the rewards of his monumental work, Tartaglia accused Cardano of being a liar and a thief for stealing his formula. Thus began the bitterest feud in the history of mathematics, carried on with name-calling and mudslinging of the lowest order. As Tartaglia feared, the formula (see Group Project 1) has forever since been known as Cardano's formula for the solution of the cubic equation. ∎

---

1. Learn basic facts about the complex zeros of polynomials.

The Factor Theorem connects the concept of *factors* and *zeros* of a polynomial, and we continue to investigate this connection.

Because the equation $x^2 + 1 = 0$ has no real roots, the polynomial function $P(x) = x^2 + 1$ has no real zeros. However, if we replace x with the complex number i, we have

$$P(i) = i^2 + 1 = (-1) + 1 = 0.$$

Consequently, i is a complex number for which $P(i) = 0$; that is, i is a *complex zero* of $P(x)$. In Section 3.2, we stated that a polynomial of degree n can have, at most, n real zeros. We now extend our number system to allow the variables and coefficients of polynomials to represent complex numbers. When we want to emphasize that complex numbers may be used in this way, we call the polynomial $P(x)$ a **complex**

RECALL

A number z of the form $z = a + bi$ is a **complex number**, where a and b are real numbers and $i = \sqrt{-1}$.

polynomial. If $P(z) = 0$ for a complex number z, we say that z is a zero or a **complex zero** of $P(x)$. In the complex number system, every nth-degree polynomial equation has exactly n roots and every nth-degree polynomial can be factored into *exactly n* linear factors. This fact follows from the Fundamental Theorem of Algebra.

FUNDAMENTAL THEOREM OF ALGEBRA

Every polynomial function

$$P(x) = a_n x^n + a_{n-1} x^{n-1} + \cdots + a_1 x + a_0 \quad (n \geq 1, a_n \neq 0)$$

with complex coefficients $a_n, a_{n-1}, \ldots, a_1, a_0$ has at least one complex zero.

RECALL

If $b = 0$, the complex number $a + bi$ is a real number. So the set of real numbers is a subset of the set of complex numbers.

This theorem was proved by the mathematician Karl Friedrich Gauss at age 20. The proof is beyond the scope of this book, but if we are allowed to use complex numbers, we can use the theorem to prove that every polynomial has a complete factorization.

FACTORIZATION THEOREM FOR POLYNOMIALS

If $P(x)$ is a complex polynomial of degree $n \geq 1$ with leading coefficient a, it can be factored into n (not necessarily distinct) linear factors of the form

$$P(x) = a(x - r_1)(x - r_2) \cdots (x - r_n),$$

where $a, r_1, r_2, \ldots, r_n$ are complex numbers.

To prove the Factorization Theorem for Polynomials, we notice that if $P(x)$ is a polynomial of degree 1 or higher with leading coefficient a, then the Fundamental Theorem of Algebra states that there is a zero r_1 for which $P(r_1) = 0$. By the Factor Theorem, $(x - r_1)$ is a factor of $P(x)$ and

$$P(x) = (x - r_1)Q_1(x),$$

where the degree of $Q_1(x)$ is $n - 1$ and its leading coefficient is a. If $n - 1$ is positive, then we apply the Fundamental Theorem to $Q_1(x)$; this gives us a zero, say, r_2, of $Q_1(x)$. By the Factor Theorem, $(x - r_2)$ is a factor of $Q_1(x)$ and

$$Q_1(x) = (x - r_2)Q_2(x),$$

where the degree of $Q_2(x)$ is $n - 2$ and its leading coefficient is a. Then

$$P(x) = (x - r_1)(x - r_2)Q_2(x).$$

This process can be continued until $P(x)$ is completely factored as

$$P(x) = a(x - r_1)(x - r_2) \cdots (x - r_n),$$

where a is the leading coefficient and $r_1, r_2, \ldots, r_n$ are zeros of $P(x)$. The proof is now complete.

In general, the numbers $r_1, r_2, \ldots, r_n$ may not be distinct. As with the polynomials we encountered previously, if the factor $(x - r)$ appears m times in the complete factorization of $P(x)$, then we say that r is a zero of *multiplicity m*. The polynomial $P(x) = (x - 2)^3(x - i)^2$ has zeros

$$2 \text{ (of multiplicity 3) and } i \text{ (of multiplicity 2)}.$$

The polynomial $P(x) = (x - 2)^3(x - i)^2$ has only 2 and i as its distinct zeros, although it is of degree 5.

We have the following theorem.

NUMBER OF ZEROS THEOREM

Any polynomial of degree n has exactly n zeros, provided a zero of multiplicity k is counted k times.

EXAMPLE 1   **Constructing a Polynomial Whose Zeros Are Given**

Find a polynomial $P(x)$ of degree 4 with a leading coefficient of 2 and zeros $-1, 3, i,$ and $-i$. Write $P(x)$

a. in completely factored form.   **b.** by expanding the product found in part **a.**

SOLUTION

a. Because $P(x)$ has degree 4, we write

$$P(x) = a(x - r_1)(x - r_2)(x - r_3)(x - r_4). \qquad \text{Completely factored form}$$

Now replace the leading coefficient a with 2 and the zeros $r_1, r_2, r_3,$ and r_4 with $-1, 3, i,$ and $-i$ (in any order). Then

$$P(x) = 2[x - (-1)](x - 3)(x - i)[x - (-i)]$$
$$= 2(x + 1)(x - 3)(x - i)(x + i) \qquad \text{Simplify.}$$

b. $P(x) = 2(x + 1)(x - 3)(x - i)(x + i)$ From part **a**

$\qquad = 2(x + 1)(x - 3)(x^2 + 1)$ Expand $(x - i)(x + i)$.

$\qquad = 2(x + 1)(x^3 - 3x^2 + x - 3)$ Expand $(x - 3)(x^2 + 1)$.

$\qquad = 2(x^4 - 2x^3 - 2x^2 - 2x - 3)$ Expand $(x + 1)(x^3 - 3x^2 + x - 3)$.

$\qquad = 2x^4 - 4x^3 - 4x^2 - 4x - 6$ Distributive property   ■ ■ ■

Practice Problem 1   Find a polynomial $P(x)$ of degree 4 with a leading coefficient of 3 and zeros $-2, 1, 1 + i, 1 - i$. Write $P(x)$

a. in completely factored form.   **b.** by expanding the product found in part **a.**   ■

---

2 Use the Conjugate Pairs Theorem to find zeros of polynomials.

Conjugate Pairs Theorem

In Example 1, we constructed a polynomial that had both i and its conjugate, $-i$, among its zeros. Our next result says that for all polynomials with real coefficients, nonreal zeros occur in conjugate pairs.

CONJUGATE PAIRS THEOREM

If $P(x)$ is a polynomial function whose coefficients are real numbers and if $z = a + bi$ is a zero of P, then its conjugate, $\bar{z} = a - bi$, is also a zero of P.

RECALL

The conjugate of a complex number $z = a + bi$ is $\bar{z} = a - bi$. For two complex numbers z_1 and z_2, we have (Section 1.3)

$$\overline{z_1 + z_2} = \overline{z_1} + \overline{z_2}$$
$$\overline{z_1 z_2} = \overline{z_1}\,\overline{z_2}.$$

To prove the Conjugate Pairs Theorem, let

$$P(x) = a_n x^n + a_{n-1} x^{n-1} + \cdots + a_1 x + a_0,$$

where $a_n, a_{n-1}, \ldots, a_1, a_0$ are real numbers. Now suppose that $P(z) = 0$. Then we must show that $P(\bar{z}) = 0$ also. We will use the fact that the conjugate of a real number is identical to the real number a (because $\bar{a} = a - 0i = a$) and then use the conjugate sum and product properties $\overline{z_1 + z_2} = \overline{z_1} + \overline{z_2}$ and $\overline{z_1 z_2} = \overline{z_1}\ \overline{z_2}$.

$$P(z) = a_n z^n + a_{n-1} z^{n-1} + \cdots + a_1 z + a_0 \qquad \text{Replace } x \text{ with } z.$$

$$P(\bar{z}) = a_n (\bar{z})^n + a_{n-1}(\bar{z})^{n-1} + \cdots + a_1 \bar{z} + a_0 \qquad \text{Replace } z \text{ with } \bar{z}.$$

$$= \overline{a_n}\overline{z^n} + \overline{a_{n-1}}\overline{z^{n-1}} + \cdots + \overline{a_1}\,\bar{z} + \overline{a_0} \qquad (\bar{z})^k = \overline{z^k}, \overline{a_k} = a_k$$

$$= \overline{a_n z^n} + \overline{a_{n-1}z^{n-1}} + \cdots + \overline{a_1 z} + \overline{a_0} \qquad \overline{z_1}\ \overline{z_2} = \overline{z_1 z_2}$$

$$= \overline{a_n z^n + a_{n-1}z^{n-1} + \cdots + a_1 z + a_0} \qquad \overline{z_1} + \overline{z_2} = \overline{z_1 + z_2}$$

$$= \overline{P(z)} = \bar{0} = 0 \qquad P(z) = 0$$

We see that if $P(z) = 0$, then $P(\bar{z}) = 0$, and the theorem is proved.

This theorem has several uses. If, for example, we know that $2 - 5i$ is a zero of a polynomial with real coefficients, then we know that $2 + 5i$ is also a zero. Because nonreal zeros occur in conjugate pairs, we know that there will always be an *even* number of nonreal zeros. Therefore, any polynomial of odd degree with real coefficients has at least one zero that is a real number.

ODD-DEGREE POLYNOMIALS WITH REAL ZEROS

Any polynomial $P(x)$ of odd degree with real coefficients must have at least one real zero.

For example, the polynomial $P(x) = 5x^7 + 2x^4 + 9x - 12$ has at least one zero that is a real number because $P(x)$ has degree 7 (odd degree) and has real numbers as coefficients.

STUDY TIP

In problems such as Example 2, once you determine all of the zeros, you can use the Factorization Theorem to write a polynomial with those zeros.

EXAMPLE 2   **Using the Conjugate Pairs Theorem**

A polynomial $P(x)$ of degree 9 with real coefficients has the following zeros: 2, of multiplicity 3; $4 + 5i$, of multiplicity 2; and $3 - 7i$. Write all nine zeros of $P(x)$.

SOLUTION

Because complex zeros occur in conjugate pairs, the conjugate $4 - 5i$ of $4 + 5i$ is a zero of multiplicity 2 and the conjugate $3 + 7i$ of $3 - 7i$ is a zero of $P(x)$. The nine zeros of $P(x)$, including the repetitions, are

$$2, 2, 2, 4 + 5i, 4 + 5i, 4 - 5i, 4 - 5i, 3 + 7i, \text{ and } 3 - 7i. \qquad \blacksquare\blacksquare\blacksquare$$

Practice Problem 2   A polynomial $P(x)$ of degree 8 with real coefficients has the following zeros: -3 and $2 - 3i$, each of multiplicity 2, and i. Write all eight zeros of $P(x)$. ■

Every polynomial of degree n has exactly n zeros and can be factored into a product of n linear factors. If the polynomial has real coefficients, then, by the Conjugate Pairs Theorem, its nonreal zeros occur as conjugate pairs. Consequently, if $r = a + bi$ is a zero, then so is $\bar{r} = a - bi$, and both $x - r$ and $x - \bar{r}$ are linear factors of the polynomial. Multiplying these factors, we have

$$(x - r)(x - \bar{r}) = x^2 - (r + \bar{r})x + r\bar{r} \qquad \text{Use FOIL.}$$

$$= x^2 - 2ax + (a^2 + b^2) \qquad r + \bar{r} = 2a; r\bar{r} = a^2 + b^2$$

The quadratic polynomial $x^2 - 2ax + (a^2 + b^2)$ has real coefficients and is *irreducible* (cannot be factored any further) *over the real numbers*. This result shows that each pair of nonreal conjugate zeros can be combined into one *irreducible* quadratic factor.

> ### FACTORIZATION THEOREM FOR A POLYNOMIAL WITH REAL COEFFICIENTS
>
> Every polynomial with real coefficients can be uniquely factored over the real numbers as a product of linear factors and/or irreducible quadratic factors.

EXAMPLE 3   **Finding the Complex Zeros of a Polynomial Function**

Given that $2 - i$ is a zero of $P(x) = x^4 - 6x^3 + 14x^2 - 14x + 5$, find the remaining zeros.

SOLUTION

Because $P(x)$ has real coefficients, the conjugate $\overline{2 - i} = 2 + i$ is also a zero. By the Factorization Theorem, the linear factors $[x - (2 - i)]$ and $[x - (2 + i)]$ appear in the factorization of $P(x)$. Consequently, their product

$$
\begin{aligned}
[x - (2 - i)][x - (2 + i)] &= (x - 2 + i)(x - 2 - i) \\
&= [(x - 2) + i][(x - 2) - i] &&\text{Regroup.} \\
&= (x - 2)^2 - i^2 &&(A + B)(A - B) = A^2 - B^2 \\
&= x^2 - 4x + 4 + 1 &&\text{Expand } (x - 2)^2, i^2 = -1. \\
&= x^2 - 4x + 5 &&\text{Simplify.}
\end{aligned}
$$

is also a factor of $P(x)$. We divide $P(x)$ by $x^2 - 4x + 5$ to find the other factor.

> **RECALL**
>
> Dividend =
> Quotient · Divisor + Remainder

$$
\begin{array}{r}
x^2 - 2x + 1 \quad \leftarrow \text{Quotient} \\
\text{Divisor} \rightarrow x^2 - 4x + 5 \overline{)x^4 - 6x^3 + 14x^2 - 14x + 5} \quad \leftarrow \text{Dividend} \\
\underline{x^4 - 4x^3 + 5x^2} \\
-2x^3 + 9x^2 - 14x \\
\underline{-2x^3 + 8x^2 - 10x} \\
x^2 - 4x + 5 \\
\underline{x^2 - 4x + 5} \\
0 \quad \leftarrow \text{Remainder}
\end{array}
$$

$$
\begin{aligned}
P(x) &= (x^2 - 2x + 1)(x^2 - 4x + 5) &&P(x) = \text{Quotient} \cdot \text{Divisor} \\
&= (x - 1)(x - 1)(x^2 - 4x + 5) &&\text{Factor } x^2 - 2x + 1. \\
&= (x - 1)(x - 1)[x - (2 - i)][x - (2 + i)] &&\text{Factor } x^2 - 4x + 5.
\end{aligned}
$$

The zeros of $P(x)$ are 1 (of multiplicity 2), $2 - i$, and $2 + i$.   ■ ■ ■

Practice Problem 3   Given that $2i$ is a zero of $P(x) = x^4 - 3x^3 + 6x^2 - 12x + 8$, find the remaining zeros.   ■

In Example 3, because $x^2 - 4x + 5$ is irreducible over the real number system, the form $P(x) = (x - 1)(x - 1)(x^2 - 4x + 5)$ is a factorization of $P(x)$ into linear and irreducible quadratic factors. However, the form $P(x) = (x - 1)(x - 1)$ $[x - (2 - i)][x - (2 + i)]$ is a factorization of $P(x)$ into linear factors over the complex numbers.

EXAMPLE 4   **Finding the Zeros of a Polynomial Function**

Find all zeros of the polynomial function $P(x) = x^4 - x^3 + 7x^2 - 9x - 18$.

SOLUTION

Because the degree of $P(x)$ is 4, $P(x)$ has four zeros. The Rational Zeros Theorem tells us that the possible rational zeros are

$$\pm 1, \pm 2, \pm 3, \pm 6, \pm 9, \pm 18.$$

Testing these possible zeros by synthetic division, we find that 2 is a zero.

$$\begin{array}{r|rrrr}
2 & 1 & -1 & 7 & -9 & -18 \\
  &   & 2 & 2 & 18 & 18 \\
\hline
  & 1 & 1 & 9 & 9 & 0
\end{array}$$

Because $P(2) = 0$, $(x - 2)$ is a factor of $P(x)$ and the quotient is $x^3 + x^2 + 9x + 9$. We can solve the depressed equation $x^3 + x^2 + 9x + 9 = 0$ by factoring by grouping.

$x^2(x + 1) + 9(x + 1) = 0$	Group terms, distributive property
$(x^2 + 9)(x + 1) = 0$	Distributive property
$x^2 + 9 = 0$ or $x + 1 = 0$	Zero-product property
$x^2 = -9$ or $x = -1$	Solve each equation.
$x = \pm\sqrt{-9} = \pm 3i$	Solve $x^2 = -9$ for x.

The four zeros of $P(x)$ are $-1, 2, 3i$, and $-3i$. The complete factorization of $P(x)$ is

$$P(x) = x^4 - x^3 + 7x^2 - 9x - 18 = (x + 1)(x - 2)(x - 3i)(x + 3i).$$

■ ■ ■

Practice Problem 4 Find all zeros of the polynomial

$$P(x) = x^4 - 8x^3 + 22x^2 - 28x + 16.$$

■

SECTION 3.5 ■ Exercises

A EXERCISES Basic Skills and Concepts

1. The Fundamental Theorem of Algebra states that a polynomial function of degree $n \geq 1$ has at least one _____ zero.

2. The Number of Zeros Theorem states that a polynomial function of degree n has exactly _____ zeros, provided a zero of multiplicity k is counted _____ times.

3. If P is a polynomial function with real coefficients and if $z = a + bi$ is a zero of P, then _____ is also a zero of P.

4. If $2 - 3i$ is a zero of a polynomial function with real coefficients, then so is _____.

5. *True or False* A polynomial function of degree $n \geq 1$ with real coefficients has at least one real zero and, at most, n real zeros.

6. *True or False* A polynomial function $P(x)$ of degree $n \geq 1$ with real coefficients can be factored into n factors (not necessarily distinct) as

$$P(x) = a(x - r_1)(x - r_2)\ldots(x - r_n),$$

where $a, r_1, r_2, \ldots r_n$ are real numbers.

In Exercises 7–12, find all solutions of the equation in the complex number system.

7. $x^2 + 25 = 0$

8. $(x - 2)^2 + 9 = 0$

9. $x^2 + 4x + 4 = -9$

10. $x^3 - 8 = 0$

11. $(x - 2)(x - 3i)(x + 3i) = 0$

12. $(x - 1)(x - 2i)(2x - 6i)(2x + 6i) = 0$

In Exercises 13–18, find the remaining zeros of a polynomial $P(x)$ with real coefficients and with the specified degree and zeros.

13. Degree 3; zeros: $2, 3 + i$

14. Degree 3; zeros: $2, 2 - i$

15. Degree 4; zeros: $0, 1, 2 - i$

16. Degree 4; zeros: $-i, 1 - i$

17. Degree 6; zeros: $0, 5, i, 3i$

18. Degree 6; zeros: $2i, 4 + i, i - 1$

In Exercises 19–22, find the polynomial $P(x)$ with real coefficients having the specified degree, leading coefficient, and zeros.

19. Degree 4; leading coefficient 2; zeros: $5 - i, \ 3i$

20. Degree 4; leading coefficient -3; zeros: $2 + 3i, \ 1 - 4i$

21. Degree 5; leading coefficient 7; zeros: 5 (multiplicity 2), $1, 3 - i$.

22. Degree 6; leading coefficient 4; zeros: 3, 0 (multiplicity 3), $2 - 3i$.

In Exercises 23–26, use the given zero to write $P(x)$ as a product of linear and irreducible quadratic factors.

23. $P(x) = x^4 + x^3 + 9x^2 + 9x$, zero: $3i$

24. $P(x) = x^4 - 2x^3 + x^2 + 2x - 2$, zero: $1 - i$

25. $P(x) = x^5 - 5x^4 + 2x^3 + 22x^2 - 20x$, zero: $3 - i$

26. $P(x) = 2x^5 - 11x^4 + 19x^3 - 17x^2 + 17x - 6$, zero: i

In Exercises 27–36, find all zeros of each polynomial function.

27. $P(x) = x^3 - 9x^2 + 25x - 17$

28. $P(x) = x^3 - 5x^2 + 7x + 13$

29. $P(x) = 3x^3 - 2x^2 + 22x + 40$

30. $P(x) = 3x^3 - x^2 + 12x - 4$

31. $P(x) = 2x^4 - 10x^3 + 23x^2 - 24x + 9$

32. $P(x) = 9x^4 + 30x^3 + 14x^2 - 16x + 8$

33. $P(x) = x^4 - 4x^3 - 5x^2 + 38x - 30$

34. $P(x) = x^4 + x^3 + 7x^2 + 9x - 18$

35. $P(x) = 2x^5 - 11x^4 + 19x^3 - 17x^2 + 17x - 6$

36. $P(x) = x^5 - 2x^4 - x^3 + 8x^2 - 10x + 4$

In Exercises 37–40, find an equation of a polynomial function of least degree having the given complex zeros, intercepts, and graph.

37. f has complex zero $3i$

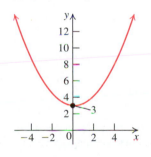

38. f has complex zero $-i$

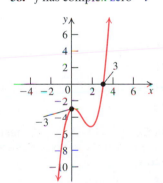

39. f has complex zeros i and $2i$

40. f has complex zero $-2i$

C EXERCISES Beyond the Basics

41. The solutions -1 and 1 of the equation $x^2 = 1$ are called the square roots of 1. The solutions of the equation $x^3 = 1$ are called the cube roots of 1. Find the cube roots of 1. How many are there?

42. The solutions of the equation $x^n = 1$, where n is a positive integer, are called the nth roots of 1 or "nth roots of unity." Explain the relationship between the solutions of the equation $x^n = 1$ and the zeros of the complex polynomial $P(x) = x^n - 1$. How many nth roots of 1 are there?

In Exercises 43–46, use the given zero and synthetic division to determine all of the zeros of $P(x)$. Then factor the depressed equation to write $P(x)$ as a product of linear factors.

43. $P(x) = x^2 + (i - 2)x - 2i$, $x = -i$

44. $P(x) = x^2 + 3ix - 2$, $x = -2i$

45. $P(x) = x^3 - (3 + i)x^2 - (4 - 3i)x + 4i$, $x = i$

46. $P(x) = x^3 - (4 + 2i)x^2 + (7 + 8i)x - 14i$, $x = 2i$

In Exercises 47–50, find an equation with real coefficients of a polynomial function f that has the given characteristics. Then write the end behavior of the graph of $y = f(x)$.

47. Degree: 3; zeros: $2, 1 + 2i$; y-intercept: 40

48. Degree: 3; zeros: $1, 2 - 3i$; y-intercept: -26

49. Degree: 4; zeros: $1, -1, 3 + i$; y-intercept: 20

50. Degree: 4; zeros: $1 - 2i, 3 - 2i$; y-intercept: 130

Critical Thinking

51. Show that if $r_1, r_2, \ldots, r_n$ are the roots of the equation

$$a_n x^n + a_{n-1} x^{n-1} + \cdots + a_1 x + a_0 = 0 \, (a_n \neq 0),$$

then the sum of the roots satisfies

$$r_1 + r_2 + \cdots + r_n = -\frac{a_{n-1}}{a_n}$$

and the product of the roots satisfies

$$r_1 \cdot r_2 \cdot \cdots \cdot r_n = (-1)^n \frac{a_0}{a_n}.$$

[*Hint:* Factor the polynomial; then multiply it out using $r_1, r_2, \ldots, r_n$ and compare coefficients with those of the original polynomial.]

GROUP PROJECTS

1. In his book *Ars Magna*, Cardano explained how to solve cubic equations. He considered the following example:

$$x^3 + 6x = 20$$

   a. Explain why this equation has exactly one real solution.

   b. Cardano explained the method as follows: "I take two cubes v^3 and u^3 whose difference is 20 and whose product is 2, that is, a third of the coefficient of x. Then, I say that $x = v - u$ is a solution of the equation." Show that if $v^3 - u^3 = 20$ and $vu = 2$, then $x = v - u$ is indeed the solution of the equation $x^3 + 6x = 20$.

   c. Solve the system

$$v^3 - u^3 = 20$$
$$uv = 2$$

   to find u and v.

   d. Consider the equation $x^3 + px = q$, where p is a positive number. Using your work in parts (a), (b), and (c) as a guide, show that the unique solution of this equation is

$$x = \sqrt[3]{\frac{q}{2} + \sqrt{\left(\frac{q}{2}\right)^2 + \left(\frac{p}{3}\right)^3}} - \sqrt[3]{-\frac{q}{2} + \sqrt{\left(\frac{q}{2}\right)^2 + \left(\frac{p}{3}\right)^3}}.$$

   e. Consider an arbitrary cubic equation

$$x^3 + ax^2 + bx + c = 0.$$

   Show that the substitution $x = y - \dfrac{a}{3}$ allows you to write the cubic equation as

$$y^3 + py = q.$$

2. Use Cardano's method to solve the equation $x^3 + 6x^2 + 10x + 8 = 0$.

Rational Functions

Before Starting this Section, Review

1. Domain of a function (Section 2.4, page 215)
2. Zeros of a function (Section 3.2, page 316)
3. Symmetry (Section 2.2, page 182)
4. Quadratic formula (Section 1.4, page 124)

Objectives

1. Define a rational function.
2. Find vertical asymptotes (if any).
3. Find horizontal asymptotes (if any).
4. Graph rational functions.
5. Graph rational functions with oblique asymptotes.
6. Graph a revenue curve.

FEDERAL TAXES AND REVENUES

Federal governments levy income taxes on their citizens and corporations to pay for defense, infrastructure, and social services. The relationship between tax rates and tax revenues shown in Figure 3.17 is a "revenue curve."

All economists agree that a zero tax rate produces no tax revenues and that no one would bother to work with a 100% tax rate; so tax revenue would be zero. It then follows that in a given economy, there is some optimal tax rate T between zero and 100% at which people are willing to maximize their output and still pay their taxes. This optimal rate brings in the most revenue for the government. However, because no one knows the actual shape of the revenue curve, it is impossible to find the exact value of T.

Notice in Figure 3.17 that between the two extreme rates are two rates (such as a and b in the figure) that will collect the same amount of revenue. In a democracy, politicians can argue that taxes are currently too high (at some point b in Figure 3.17) and should therefore be reduced to encourage incentives and harder work (this is *supply-side economics*); at the same time, the government will generate more revenue. Others can argue that we are well to the left of T (at some point a in Figure 3.17), in which case the tax rate should be raised to generate more revenue. In Example 9, we sketch the graph of a revenue curve for a hypothetical economy. ■

FIGURE 3.17 **A revenue curve**

---

1 Define a rational function.

Rational Functions

Recall that a sum, difference, or product of two polynomial functions is also a polynomial function. However, the quotient of two polynomial functions is generally *not* a polynomial function. We call the quotient of two polynomial functions a *rational function*.

RATIONAL FUNCTION

A function f that can be expressed in the form

$$f(x) = \frac{N(x)}{D(x)}$$

where the numerator $N(x)$ and the denominator $D(x)$ are polynomials and $D(x)$ is not the zero polynomial, is called a **rational function**. The domain of f consists of all real numbers for which $D(x) \neq 0$.

Examples of rational functions are

$$f(x) = \frac{x-1}{2x+1}, \quad g(y) = \frac{5y}{y^2+1}, \quad h(t) = \frac{3t^2 - 4t + 1}{t-2}, \quad \text{and} \quad F(z) = 3z^4 - 5z^2 + 6z + 1.$$

($F(z)$ has the constant polynomial function $D(z) = 1$ as its denominator.)

By contrast, $S(x) = \dfrac{\sqrt{1-x^2}}{x+1}$ and $G(x) = \dfrac{|x|}{x}$ are not rational functions.

EXAMPLE 1   Finding the Domain of a Rational Function

Find the domain of each rational function.

a. $f(x) = \dfrac{3x^2 - 12}{x-1}$ **b.** $g(x) = \dfrac{x}{x^2 - 6x + 8}$ **c.** $h(x) = \dfrac{x^2 - 4}{x-2}$

SOLUTION

a. The domain of $f(x) = \dfrac{3x^2 - 12}{x-1}$ is the set of all real numbers x for which $x - 1 \neq 0$ (that is, $x \neq 1$), or in interval notation, $(-\infty, 1) \cup (1, \infty)$.

b. The domain of g is the set of all real numbers x for which the denominator $D(x)$ is nonzero.

$$
\begin{array}{ll}
D(x) = x^2 - 6x + 8 = 0 & \text{Set } D(x) = 0. \\
(x-2)(x-4) = 0 & \text{Factor } D(x). \\
x - 2 = 0 \quad \text{or} \quad x - 4 = 0 & \text{Zero-product property} \\
x = 2 \quad \text{or} \quad x = 4 & \text{Solve for } x.
\end{array}
$$

The domain of g is the set of all real numbers x except 2 and 4, or in interval notation, $(-\infty, 2) \cup (2, 4) \cup (4, \infty)$.

c. The domain of $h(x) = \dfrac{x^2 - 4}{x-2}$ is the set of all real numbers x except 2, or in interval notation, $(-\infty, 2) \cup (2, \infty)$.   ■ ■ ■

Practice Problem 1   Find the domain of the rational function $f(x) = \dfrac{x-3}{x^2 - 4x - 5}$.   ■

RECALL

The graph of a polynomial function has no holes, gaps, or sharp corners.

In Example 1(**c**), note that the functions $h(x) = \dfrac{x^2 - 4}{x-2} = \dfrac{(x-2)(x+2)}{x-2}$ and $H(x) = x + 2$ are not equal: The domain of $H(x)$ is the set of all real numbers, but the domain of $h(x)$ is the set of all real numbers x except 2. The graph of $y = H(x)$ is a line with slope 1 and y-intercept 2. The graph of $y = h(x)$ is the same line as $y = H(x)$ except that the point $(2, 4)$ is missing. See Figure 3.18.

FIGURE 3.18 Functions $H(x)$ and $h(x)$ have different graphs.

As with the quotient of integers, if the polynomials $N(x)$ and $D(x)$ have no common factors, then the rational function $f(x) = \dfrac{N(x)}{D(x)}$ is said to be in **lowest terms**. The zeros of $N(x)$ and $D(x)$ will play an important role in the graphing of the function f.

2 Find vertical asymptotes (if any).

Vertical Asymptotes

Consider the function $f(x) = \dfrac{1}{x}$. We know that $f(0)$ is undefined, meaning that f is not assigned any value at $x = 0$. How does $f(x)$ behave "near" the excluded value $x = 0$? Consider the following table.

x	1	0.1	0.01	0.001	0.0002	10^{-9}	10^{-50}
$f(x) = \dfrac{1}{x}$	1	10	100	1000	5000	10^{9}	10^{50}

We see that as x approaches 0 (with $x > 0$), $f(x)$ increases without bound. In symbols, we write "as $x \to 0^{+}$, $f(x) \to \infty$." This statement is read "As x approaches 0 *from the right*, $f(x)$ approaches infinity."

The symbol $x \to a^{+}$, read "x approaches a from the right," refers only to values of x that are greater than a. The symbol $x \to a^{-}$ is read "x approaches a from the left" and refers only to values of x that are less than a.

For the function $f(x) = \dfrac{1}{x}$, we can make another table of values that includes negative values of x near zero, such as $x = -1, -0.1, -0.001, \ldots, -10^{-9}$, and so on. We can conclude that as $x \to 0^{-}$, $f(x) \to -\infty$ and say that as x approaches 0 *from the left*, $f(x)$ approaches negative infinity.

The graph of $f(x) = \dfrac{1}{x}$ is the *reciprocal function* from Chapter 2 (see Figure 3.19).

The vertical line $x = 0$ (the y-axis) is called a *vertical asymptote*.

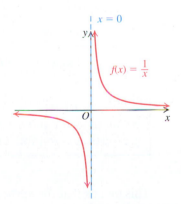

FIGURE 3.19 Reciprocal function

VERTICAL ASYMPTOTE

The line $x = a$ is called a **vertical asymptote** of the graph of a function f if $|f(x)| \to \infty$ as $x \to a^{+}$ or as $x \to a^{-}$.

This definition says that if the line $x = a$ is a vertical asymptote of the graph of a rational function f, then the graph of f *near* $x = a$ behaves like one of the graphs in Figure 3.20.

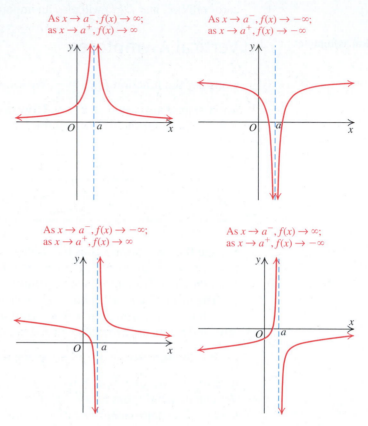

As $x \to a^-, f(x) \to \infty$;
as $x \to a^+, f(x) \to \infty$

As $x \to a^-, f(x) \to -\infty$;
as $x \to a^+, f(x) \to -\infty$

As $x \to a^-, f(x) \to -\infty$;
as $x \to a^+, f(x) \to \infty$

As $x \to a^-, f(x) \to \infty$;
as $x \to a^+, f(x) \to -\infty$

FIGURE 3.20 **Behavior of a rational function near a vertical asymptote**

If $x = a$ is a vertical asymptote for a rational function f, then a is not in the domain; so the graph of f *does not cross* the line $x = a$.

LOCATING VERTICAL ASYMPTOTES OF RATIONAL FUNCTIONS

If $f(x) = \dfrac{N(x)}{D(x)}$ is a rational function, where $N(x)$ and $D(x)$ do not have a common factor and a is a real zero of $D(x)$, then the line with equation $x = a$ is a vertical asymptote of the graph of f.

This means that the vertical asymptotes (if any) are found by locating the real zeros of the denominator.

Graphing calculators give very different graphs for

$$g(x) = \frac{1}{x^2 - 9}$$

depending on whether the mode is set to *connected* or *dot*.

In connected mode, many calculators connect the dots between plotted points, producing vertical lines at $x = -3$ and $x = 3$. Dot mode avoids graphing these vertical lines by plotting the same points but not connecting them. However, dot mode fails to display continuous sections of the graph.

Connected mode

Dot mode

EXAMPLE 2 **Finding Vertical Asymptotes**

Find all vertical asymptotes of the graph of each rational function.

a. $f(x) = \dfrac{1}{x - 1}$ **b.** $g(x) = \dfrac{1}{x^2 - 9}$ **c.** $h(x) = \dfrac{1}{x^2 + 1}$

SOLUTION

a. There are no common factors in the numerator and denominator of $f(x) = \dfrac{1}{x - 1}$, and the only *zero* of the denominator is 1. Therefore, $x = 1$ is a vertical asymptote of $f(x)$.

b. The rational function $g(x)$ is in lowest terms. Factoring $x^2 - 9 = (x + 3)(x - 3)$, we see that the *zeros* of the denominator are -3 and 3. Therefore, the lines $x = -3$ and $x = 3$ are the two vertical asymptotes of $g(x)$.

c. Because the denominator $x^2 + 1$ has no real zeros, the graph of $h(x)$ has *no* vertical asymptotes. ■ ■ ■

Practice Problem 2 Find the vertical asymptotes of the graph of

$$f(x) = \frac{x + 1}{x^2 + 3x - 10}.$$ ■

The next example illustrates that the graph of a rational function may have gaps (missing points) with or without vertical asymptotes.

EXAMPLE 3 **Rational Function Whose Graph Has a Hole**

Find all vertical asymptotes of the graph of each rational function.

a. $h(x) = \dfrac{x^2 - 9}{x - 3}$ **b.** $g(x) = \dfrac{x + 2}{x^2 - 4}$

 Calculator graph for

$$h(x) = \frac{x^2 - 9}{x - 3}$$

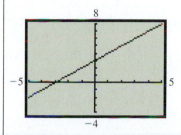

A calculator graph does not show the hole in the graph, but a calculator table explains the situation.

FIGURE 3.21 Graph with a hole

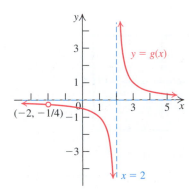

FIGURE 3.22 Graph with a hole and an asymptote

3 Find horizontal asymptotes (if any).

SOLUTION

a. $h(x) = \dfrac{x^2 - 9}{x - 3}$ Given function

$ = \dfrac{(x + 3)(x - 3)}{x - 3}$ Factor $x^2 - 9 = (x + 3)(x - 3)$ in the numerator.

$ = x + 3, \quad \text{if } x \neq 3$ Simplify.

The graph of $h(x)$ is the line $y = x + 3$ with a gap (or hole) at $x = 3$. See Figure 3.21.

    The graph of $h(x)$ has no vertical asymptote at $x = 3$, the zero of the denominator, because the numerator and the denominator have the factor $(x - 3)$ in common.

b. $g(x) = \dfrac{x + 2}{x^2 - 4}$ Given function

$ = \dfrac{x + 2}{(x + 2)(x - 2)}$ Factor the denominator.

$ = \dfrac{1}{x - 2}, x \neq -2$ Simplify.

The graph of $g(x)$ has a hole at $x = -2$ and a vertical asymptote at $x = 2$. See Figure 3.22.   ■ ■ ■

Practice Problem 3   Find all vertical asymptotes of the graph of $f(x) = \dfrac{3 - x}{x^2 - 9}$.   ■

Horizontal Asymptotes

A vertical asymptote identifies the behavior of the graph of a function near a specific real number but says nothing about the function's end behavior. Consider again the function $f(x) = \dfrac{1}{x}$. The larger we make x, the closer $\dfrac{1}{x}$ is to 0: As $x \to \infty$, $f(x) = \dfrac{1}{x} \to 0$. Similarly, as $x \to -\infty$, $f(x) = \dfrac{1}{x} \to 0$. The x-axis, with equation $y = 0$, is a *horizontal asymptote* of the graph of $f(x) = \dfrac{1}{x}$. In fact, for any real number a and any positive integer n, as $|x| \to \infty$, $g(x) = \dfrac{a}{x^n} \to 0$. So the x-axis is a horizontal asymptote of the graph of $g(x) = \dfrac{a}{x^n}$.

HORIZONTAL ASYMPTOTE

The line $y = k$ is a **horizontal asymptote** of the graph of a function f if $f(x) \to k$ as $x \to \infty$ or as $x \to -\infty$.

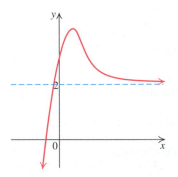

FIGURE 3.23 Horizontal asymptote

    In general, a line $y = k$ is a horizontal asymptote if either $f(x) \to k$ as $x \to \infty$ or $f(x) \to k$ as $x \to -\infty$. For example, in Figure 3.23, the line with equation $y = 2$ is a horizontal asymptote of the graph shown because $f(x) \to 2$ as $x \to \infty$. However, it can be shown that this situation is not possible for a rational function.

The graph of a function $y = f(x)$ can cross its horizontal asymptote (in contrast to a vertical asymptote). The graph of $g(x) = \dfrac{x + 2}{x^2 - 4}$, shown in Figure 3.22, has a horizontal asymptote: $y = 0$. This can be verified algebraically as follows:

$$g(x) = \frac{x + 2}{x^2 - 4}$$

$$= \frac{\dfrac{x + 2}{x^2}}{\dfrac{x^2 - 4}{x^2}} \qquad \text{Divide numerator and denominator by } x^2.$$

$$= \frac{\dfrac{x}{x^2} + \dfrac{2}{x^2}}{\dfrac{x^2}{x^2} - \dfrac{4}{x^2}} \qquad \frac{a \pm b}{c} = \frac{a}{c} \pm \frac{b}{c}$$

$$= \frac{\dfrac{1}{x} + \dfrac{2}{x^2}}{1 - \dfrac{4}{x^2}} \qquad \text{Simplify each expression.}$$

As $|x| \to \infty$, the expressions $\dfrac{1}{x}, \dfrac{2}{x^2}$, and $\dfrac{4}{x^2}$ all approach 0; so

$$g(x) \to \frac{0 + 0}{1 - 0} = \frac{0}{1} = 0.$$

Because $g(x) \to 0$ as $|x| \to \infty$, the line $y = 0$ (the x-axis) is a horizontal asymptote.

Note that although a rational function can have more than one vertical asymptote (see Example 2(b)), it can have, at most, one horizontal asymptote. We can find the horizontal asymptote (if any) of a rational function by dividing the numerator and denominator by the highest power of x that appears in the denominator and investigating the resulting expression as $|x| \to \infty$. Alternatively, we may use the following rules.

RULES FOR LOCATING HORIZONTAL ASYMPTOTES

Let f be a rational function given by

$$f(x) = \frac{N(x)}{D(x)} = \frac{a_n x^n + a_{n-1} x^{n-1} + \cdots + a_2 x^2 + a_1 x + a_0}{b_m x^m + b_{m-1} x^{m-1} + \cdots + b_2 x^2 + b_1 x + b_0}, a_n \neq 0, b_m \neq 0,$$

in lowest terms. To find whether the graph of f has one horizontal asymptote or no horizontal asymptote, we compare the degree of the numerator, n, with that of the denominator, m:

1. If $n < m$, then the x-axis ($y = 0$) is the horizontal asymptote.

2. If $n = m$, then the line with equation $y = \dfrac{a_n}{b_m}$ is the horizontal asymptote.

3. If $n > m$, then the graph of f has *no* horizontal asymptote.

EXAMPLE 4   **Finding the Horizontal Asymptote**

Find the horizontal asymptote (if any) of the graph of each rational function.

a. $f(x) = \dfrac{5x + 2}{1 - 3x}$ **b.** $g(x) = \dfrac{2x}{x^2 + 1}$ **c.** $h(x) = \dfrac{3x^2 - 1}{x + 2}$

SOLUTION

a. The numerator and denominator of $f(x) = \dfrac{5x + 2}{1 - 3x}$ are both of degree 1. The leading coefficient of the numerator is 5, and that of the denominator is -3. By Rule 2, the line $y = \dfrac{5}{-3} = -\dfrac{5}{3}$ is the horizontal asymptote of the graph of f.

b. For the function $g(x) = \dfrac{2x}{x^2 + 1}$, the degree of the numerator is 1 and that of the denominator is 2. By Rule 1, the line $y = 0$ (the x-axis) is the horizontal asymptote.

c. The degree of the numerator, 2, of $h(x)$ is greater than the degree of its denominator, 1. By Rule 3, the graph of h has no horizontal asymptote. ■ ■ ■

Practice Problem 4   Find the horizontal asymptote (if any) of the graph of each function.

a. $f(x) = \dfrac{2x - 5}{3x + 4}$ **b.** $g(x) = \dfrac{x^2 + 3}{x - 1}$ ■

---

4   Graph rational functions.

Graphing Rational Functions

FINDING THE SOLUTION: A PROCEDURE

EXAMPLE 5   **Graphing a Rational Function**

OBJECTIVE	**EXAMPLE**
Graph $f(x) = \dfrac{N(x)}{D(x)}$, where $f(x)$ is in lowest terms.	Sketch the graph of $f(x) = \dfrac{2x^2 - 2}{x^2 - 9}$.

Step 1   **Find the intercepts.** The x-intercepts are found by solving the equation $N(x) = 0$. The y-intercept is $f(0)$.

$$
\begin{aligned}
2x^2 - 2 &= 0 && \text{Set } N(x) = 0. \\
2(x - 1)(x + 1) &= 0 && \text{Factor.} \\
x - 1 = 0 \quad \text{or} \quad x + 1 &= 0 && \text{Zero-product property} \\
x = 1 \quad \text{or} \quad x &= -1 && \text{Solve for } x.
\end{aligned}
$$

The x-intercepts are -1 and 1, so the graph passes through the points $(-1, 0)$ and $(1, 0)$.

$$
\begin{aligned}
f(0) &= \frac{2(0)^2 - 2}{(0)^2 - 9} && \text{Replace } x \text{ with 0 in } f(x). \\
&= \frac{2}{9} && \text{Simplify.}
\end{aligned}
$$

The y-intercept is $\dfrac{2}{9}$, so the graph of f passes through the point $\left(0, \dfrac{2}{9}\right)$.

Step 2   **Find the vertical asymptotes (if any).**
Solve $D(x) = 0$ to find the vertical
asymptotes of the graph. Sketch the
vertical asymptotes.

$$x^2 - 9 = 0 \qquad \text{Set } D(x) = 0.$$
$$(x - 3)(x + 3) = 0 \qquad \text{Factor.}$$
$$x - 3 = 0 \quad \text{or} \quad x + 3 = 0 \qquad \text{Zero-product property}$$
$$x = 3 \quad \text{or} \qquad x = -3 \qquad \text{Solve for } x.$$

The vertical asymptotes of the graph of f are the lines
$x = -3$ and $x = 3$.

Step 3   **Find the horizontal asymptote (if any).**
Use the rules on page 365 for finding
the horizontal asymptote of a rational
function.

Because $n = m = 2$, by Rule 2, the horizontal asymptote is

$$y = \frac{\text{Leading coefficient of } N(x)}{\text{Leading coefficient of } D(x)} = \frac{2}{1} = 2.$$

Step 4   **Test for symmetry.** If $f(-x) = f(x)$,
then f is symmetric with respect to the
y-axis. If $f(-x) = -f(x)$, then f is
symmetric with respect to the origin.

$$f(-x) = \frac{2(-x)^2 - 2}{(-x)^2 - 9} = \frac{2x^2 - 2}{x^2 - 9} = f(x). \text{ The graph of}$$
f is symmetric in the y-axis. This is the only symmetry.

Step 5   **Find additional graph points.** Find
x-values for additional graph points
by using the zeros of $N(x)$ and $D(x)$.
Choose one x-value less than the
smallest zero, one x-value between
each two consecutive zeros, and one
x-value larger than the largest zero.
Calculate the corresponding y-values
to determine where the graph of f is
above the x-axis and where it is below
the x-axis.

The zeros of $N(x)$ and $D(x)$ are $-3, -1, 1,$ and 3. These zeros
divide the x-axis into five intervals. (See the figure.)

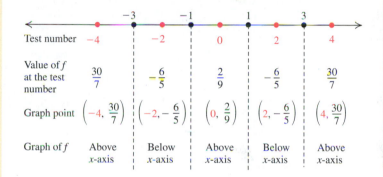

Test number	-4	-2	0	2	4
Value of f at the test number	$\dfrac{30}{7}$	$-\dfrac{6}{5}$	$\dfrac{2}{9}$	$-\dfrac{6}{5}$	$\dfrac{30}{7}$
Graph point	$\left(-4, \dfrac{30}{7}\right)$	$\left(-2, -\dfrac{6}{5}\right)$	$\left(0, \dfrac{2}{9}\right)$	$\left(2, -\dfrac{6}{5}\right)$	$\left(4, \dfrac{30}{7}\right)$
Graph of f	Above x-axis	Below x-axis	Above x-axis	Below x-axis	Above x-axis

Step 6   **Sketch the graph.** Plot the points and
asymptotes found in Steps 1–5 and use
symmetry to sketch the graph of f.

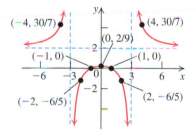

■ ■ ■

Practice Problem 5   Sketch the graph of $f(x) = \dfrac{2x}{x^2 - 1}$.

EXAMPLE 6   **Graphing a Rational Function**

Sketch the graph of $f(x) = \dfrac{x^2 + 2}{(x + 2)(x - 1)}$.

SOLUTION

Step 1   Because $x^2 + 2 > 0$, the graph has no x-intercepts.

$$f(0) = \frac{0^2 + 2}{(0 + 2)(0 - 1)} \qquad \text{Replace } x \text{ with } 0 \text{ in } f(x).$$

$$= -1 \qquad \text{Simplify.}$$

The y-intercept is -1.

Step 2   Set $(x + 2)(x - 1) = 0$. Solving for x, we have $x = -2$ or $x = 1$. The vertical asymptotes are the lines $x = -2$ and $x = 1$.

Step 3 $f(x) = \dfrac{x^2 + 2}{x^2 + x - 2}$. By Rule 2, page 365, the horizontal asymptote is $y = \dfrac{1}{1} = 1$ because the leading coefficient for both the numerator and denominator is 1.

Step 4   **Symmetry.** None

Step 5   **Additional points.** The zeros of the denominator, -2 and 1, yield the following table:

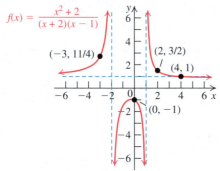

$$f(x) = \frac{x^2 + 2}{(x + 2)(x - 1)}$$

Test number	-3	0	2
Value of f at test number	$\dfrac{11}{4}$	-1	$\dfrac{3}{2}$
Graph point	$\left(-3, \dfrac{11}{4}\right)$	$(0, -1)$	$\left(2, \dfrac{3}{2}\right)$
Graph	Above x-axis	Below x-axis	Above x-axis

FIGURE 3.24 Graph crossing horizontal asymptote

Step 6   The graph of f is shown in Figure 3.24.   ■ ■ ■

Practice Problem 6   Sketch the graph of $f(x) = \dfrac{2x^2 - 1}{2x^2 + x - 3}$.   ■

Notice in Figure 3.24 that the graph of f crosses the horizontal asymptote $y = 1$. To find the point of intersection, set $f(x) = 1$ and solve for x.

$$\frac{x^2 + 2}{x^2 + x - 2} = 1 \qquad \text{Set } f(x) = 1.$$

$$x^2 + 2 = x^2 + x - 2 \qquad \text{Multiply both sides by } x^2 + x - 2.$$

$$x = 4 \qquad \text{Solve for } x.$$

The graph of f crosses the horizontal asymptote at the point $(4, 1)$.

We know that the graphs of polynomial functions are continuous (all in one piece). The next example shows that the graph of a rational function (other than a polynomial function) can also be continuous.

EXAMPLE 7   **Graphing a Rational Function**

Sketch a graph of $f(x) = \dfrac{x^2}{x^2 + 1}$.

SOLUTION

Step 1 Because $f(0) = 0$ and setting $f(x) = 0$, we have $x = 0$. Therefore, 0 is both the x-intercept and the y-intercept for the graph of f.

Step 2 Because the denominator $x^2 + 1 > 0$ for all real numbers x, there are no vertical asymptotes.

Step 3 By Rule 2 for locating horizontal asymptotes, the horizontal asymptote is $y = 1$.

Step 4 $f(-x) = \dfrac{(-x)^2}{(-x)^2 + 1} = \dfrac{x^2}{x^2 + 1} = f(x)$. The graph is symmetric with respect to the y-axis.

Step 5 The graph of $y = f(x)$ is above the x-axis except at $x = 0$.

Step 6 The graph of $y = f(x)$ is shown in Figure 3.25.   ■ ■ ■

FIGURE 3.25 A continuous rational function

Practice Problem 7   Sketch the graph of $f(x) = \dfrac{x^2 + 1}{x^2 + 2}$.   ■

5 Graph rational functions with oblique asymptotes.

Oblique Asymptotes

Suppose

$$f(x) = \frac{N(x)}{D(x)}$$

and the degree of $N(x)$ is greater than the degree of $D(x)$. We know from Rule 3 on page 365 that the graph of f has no horizontal asymptote. Use either long division or synthetic division to obtain

$$f(x) = \frac{N(x)}{D(x)} = Q(x) + \frac{R(x)}{D(x)},$$

where the degree of $R(x)$ is less than the degree of $D(x)$. Rule 1 on page 365 tells us that the x-axis is a horizontal asymptote for $\dfrac{R(x)}{D(x)}$; that is, as $x \to \infty$ or as $x \to -\infty$, the expression $\dfrac{R(x)}{D(x)} \to 0$.

Therefore, as $x \to \pm\infty$,

$$f(x) \to Q(x) + 0 = Q(x).$$

This means that as $|x|$ gets very large, the graph of $f(x)$ behaves like the graph of the polynomial $Q(x)$. If the degree of $N(x)$ is exactly one more than the degree of $D(x)$, then $Q(x)$ will have the linear form $mx + b$. In this case, $f(x)$ is said to have an **oblique** (or **slanted**) asymptote. Consider, for example, the function

$$f(x) = \frac{x^2 + x}{x - 1} = x + 2 + \frac{2}{x - 1}.$$

FIGURE 3.26 Graph with vertical and oblique asymptotes

Now as $x \to \pm\infty$, the expression $\dfrac{2}{x - 1} \to 0$ so that the graph of f approaches the graph of the oblique asymptote—the line $y = x + 2$, as shown in Figure 3.26.

EXAMPLE 8   **Graphing a Rational Function with an Oblique Asymptote**

Sketch the graph of $f(x) = \dfrac{x^2 - 4}{x + 1}$.

SOLUTION

Step 1   **Intercepts.** $f(0) = \dfrac{0 - 4}{0 + 1} = -4$, so the y-intercept is -4. Set $x^2 - 4 = 0$.

We have $x = -2$ and $x = 2$, so the x-intercepts are -2 and 2. So the graph of f passes through the points $(0, -4)$, $(-2, 0)$, and $(2, 0)$.

Step 2   **Vertical asymptotes.** Set $x + 1 = 0$. We get $x = -1$, so the line $x = -1$ is a vertical asymptote.

Step 3   **Asymptotes.** Because the degree of the numerator is greater than the degree of the denominator, $f(x)$ has no horizontal asymptote. However, by long division,

$$f(x) = \frac{x^2 - 4}{x + 1} = x - 1 + \frac{-3}{x + 1} = x - 1 - \frac{3}{x + 1}.$$

As $x \to \pm\infty$, the expression $\dfrac{3}{x + 1} \to 0$; so the line $y = x - 1$ is an oblique asymptote—the graph gets close to the line $y = x - 1$ as $x \to \infty$ and as $x \to -\infty$.

Step 4   **Symmetry.** You should check that there is no y-axis or origin symmetry.

Step 5   **Additional points.** Use the intervals determined by the zeros of the numerator and denominator.

$$f(x) = \frac{x^2 - 4}{x + 1} = \frac{(x + 2)(x - 2)}{x + 1}$$

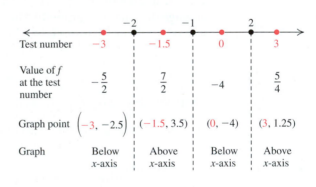

		-2		-1		2	
Test number	-3		-1.5		0		3
Value of f at the test number	$-\dfrac{5}{2}$		$\dfrac{7}{2}$		-4		$\dfrac{5}{4}$
Graph point	$(-3, -2.5)$		$(-1.5, 3.5)$		$(0, -4)$		$(3, 1.25)$
Graph	Below x-axis		Above x-axis		Below x-axis		Above x-axis

Vertical asymptote
$(-1.5, 3.5)$
Oblique asymptote
$y = x - 1$
$(3, 1.25)$
$(-3, -2.5)$
$f(x) = \dfrac{x^4 - 4}{x + 1}$
$x = -1$

FIGURE 3.27

Step 6   The graph of f is shown in Figure 3.27.   ■ ■ ■

Practice Problem 8   Sketch the graph of $f(x) = \dfrac{x^2 + 2}{x - 1}$.   ■

6   Graph a revenue curve.

Graph of a Revenue Curve

EXAMPLE 9   **Graphing a Revenue Curve**

The revenue curve for an economy of a country is given by

$$R(x) = \frac{x(100 - x)}{x + 10},$$

where x is the tax rate in percent and $R(x)$ is the tax revenue in billions of dollars.

a. Find and interpret $R(10)$, $R(20)$, $R(30)$, $R(40)$, $R(50)$, and $R(60)$.

b. Sketch the graph of $y = R(x)$ for $0 \le x \le 100$.

c. Use a graphing calculator to estimate the tax rate that yields the maximum revenue.

SOLUTION

a. $R(10) = \dfrac{10(100 - 10)}{10 + 10} = \$45$ billion. This means that if the income is taxed at the rate of 10%, the total revenue for the government will be \$45 billion.

$$\text{Similarly, } R(20) \approx \$53.3 \text{ billion,}$$
$$R(30) = \$52.5 \text{ billion,}$$
$$R(40) = \$48 \quad \text{billion,}$$
$$R(50) \approx \$41.7 \text{ billion, and}$$
$$R(60) \approx \$34.3 \text{ billion.}$$

b. The graph of the function $y = R(x)$ for $0 \le x \le 100$ is shown in Figure 3.28.

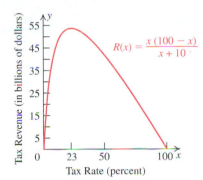

FIGURE 3.28

c. From the calculator graph of

$$y = \frac{100x - x^2}{x + 10},$$

by using the TRACE feature, you can see that the tax rate of about 23% produces the maximum tax revenue of about \$53.7 billion for the government. (See the Group Project for an algebraic proof.) ■ ■ ■

Practice Problem 9 Repeat Example 9 for $R(x) = \dfrac{x(100 - x)}{x + 20}$. ■

TECHNOLOGY CONNECTION

TRACE

$$y = \frac{100x - x^2}{x + 10}$$

to approximate the maximum revenue.

SECTION 3.6 ■ Exercises

A EXERCISES Basic Skills and Concepts

1. A rational function can be expressed in the form _____.

2. The line $x = a$ is a vertical asymptote of f if $|f(x)| \to \infty$ as $x \to$ _____ or as $x \to$ _____.

3. The line $y = k$ is a horizontal asymptote of f if $f(x) \to k$ as $x \to$ _____ or as $x \to$ _____.

4. If an asymptote is neither horizontal nor vertical, then it is called a(n) _____.

5. *True or False* Every rational function has at least one vertical asymptote.

6. *True or False* Every rational function has, at most, one horizontal asymptote.

In Exercises 7–14, find the domain of each rational function.

7. $f(x) = \dfrac{x - 3}{x + 4}$

8. $f(x) = \dfrac{x + 1}{x - 1}$

9. $g(x) = \dfrac{x - 1}{x^2 + 1}$

10. $g(x) = \dfrac{x + 2}{x^2 + 4}$

11. $h(x) = \dfrac{x - 3}{x^2 - x - 6}$

12. $h(x) = \dfrac{x - 7}{x^2 - 6x - 7}$

13. $F(x) = \dfrac{2x + 3}{x^2 - 6x + 8}$

14. $F(x) = \dfrac{3x - 2}{x^2 - 3x + 2}$

In Exercises 15–24, use the graph of the rational function $f(x)$ to complete each statement.

15. As $x \to 1^+$, $f(x) \to$ _____ .

16. As $x \to 1^-$, $f(x) \to$ _____ .

17. As $x \to -2^+$, $f(x) \to$ _____ .

18. As $x \to -2^-$, $f(x) \to$ _____ .

19. As $x \to \infty$, $f(x) \to$ _____ .

20. As $x \to -\infty$, $f(x) \to$ _____ .

21. The domain of $f(x)$ is _____ .

22. There are _____ vertical asymptotes.

23. The equations of the vertical asymptotes of the graph are _____ and _____ .

24. The equation of the horizontal asymptote of the graph is _____ .

In Exercises 25–34, find the vertical asymptotes, if any, of the graph of each rational function.

25. $f(x) = \dfrac{x}{x - 1}$

26. $f(x) = \dfrac{x + 3}{x - 2}$

27. $g(x) = \dfrac{(x + 1)(2x - 2)}{(x - 3)(x + 4)}$

28. $g(x) = \dfrac{(2x - 1)(x + 2)}{(2x + 3)(3x - 4)}$

29. $h(x) = \dfrac{x^2 - 1}{x^2 + x - 6}$

30. $h(x) = \dfrac{x^2 - 4}{3x^2 + x - 4}$

31. $f(x) = \dfrac{x^2 - 6x + 8}{x^2 - x - 12}$

32. $f(x) = \dfrac{x^2 - 9}{x^3 - 4x}$

33. $g(x) = \dfrac{2x + 1}{x^2 + x + 1}$

34. $g(x) = \dfrac{x^2 - 36}{x^2 + 5x + 9}$

In Exercises 35–42, find the horizontal asymptote, if any, of the graph of each rational function.

35. $f(x) = \dfrac{x + 1}{x^2 + 5}$

36. $f(x) = \dfrac{2x - 1}{x^2 - 4}$

37. $g(x) = \dfrac{2x - 3}{3x + 5}$

38. $g(x) = \dfrac{3x + 4}{-4x + 5}$

39. $h(x) = \dfrac{x^2 - 49}{x + 7}$

40. $f(x) = \dfrac{3x^3 + 2}{x^2 + 5x + 11}$

41. $f(x) = \dfrac{2x^2 - 3x + 7}{3x^3 + 5x + 11}$

42. $h(x) = \dfrac{x + 3}{x^2 - 9}$

In Exercises 43–48, match the rational function with its graph.

43. $f(x) = \dfrac{2}{x - 3}$

44. $f(x) = \dfrac{x - 2}{x + 3}$

45. $f(x) = \dfrac{1}{x^2 - 2x}$

46. $f(x) = \dfrac{x}{x^2 + 1}$

47. $f(x) = \dfrac{x^2 + 2x}{x - 3}$

48. $f(x) = \dfrac{x^2}{x^2 - 4}$

a.

b.

c.

d.

e.

f.

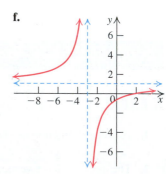

In Exercises 49–64, use the six-step procedure on pages 366–367 to graph each rational function.

49. $f(x) = \dfrac{2x}{x-3}$

50. $f(x) = \dfrac{-x}{x-1}$

51. $f(x) = \dfrac{x}{x^2-4}$

52. $f(x) = \dfrac{x}{1-x^2}$

53. $h(x) = \dfrac{-2x^2}{x^2-9}$

54. $h(x) = \dfrac{4-x^2}{x^2}$

55. $f(x) = \dfrac{2}{x^2-2}$

56. $f(x) = \dfrac{-2}{x^2-3}$

57. $g(x) = \dfrac{x+1}{(x-2)(x+3)}$

58. $g(x) = \dfrac{x-1}{(x+1)(x-2)}$

59. $h(x) = \dfrac{x^2}{x^2+1}$

60. $h(x) = \dfrac{2x^2}{x^2+4}$

61. $f(x) = \dfrac{x^2-4}{x^2-9}$

62. $f(x) = \dfrac{x^2-1}{x^2+2x-8}$

63. $g(x) = \dfrac{(x-2)^2}{x-2}$

64. $g(x) = \dfrac{(x-1)^2}{x-1}$

In Exercises 65–68, find an equation of a rational function having the given asymptotes, intercepts, and graph.

65.

66.

67.

68.

In Exercises 69–76, find the oblique asymptote and sketch the graph of each rational function.

69. $f(x) = \dfrac{2x^2+1}{x}$

70. $f(x) = \dfrac{x^2-1}{x}$

71. $g(x) = \dfrac{x^3-1}{x^2}$

72. $g(x) = \dfrac{2x^3+x^2+1}{x^2}$

73. $h(x) = \dfrac{x^2-x+1}{x+1}$

74. $h(x) = \dfrac{2x^2-3x+2}{x-1}$

75. $f(x) = \dfrac{x^3-2x^2+1}{x^2-1}$

76. $h(x) = \dfrac{x^3-1}{x^2-4}$

B EXERCISES Applying the Concepts

For Exercises 77 and 78, use the following definition: Given a cost function C, the **average cost** of the first x items is found from the formula

$$\overline{C}(x) = \frac{C(x)}{x}, x > 0.$$

77. Average cost. The Genuine Trinket Co. manufactures "authentic" trinkets for gullible tourists. Fixed daily costs are $2000, and it costs $0.50 to produce each trinket.
   a. Write the cost function C for producing x trinkets.
   b. Write the average cost $\overline{C}$ of producing x trinkets.
   c. Find and interpret $\overline{C}(100), \overline{C}(500)$, and $\overline{C}(1000)$.
   d. Find and interpret the horizontal asymptote of the rational function $\overline{C}(x)$.

78. Average cost. The monthly cost C of producing x portable CD players is given by

$$C(x) = -0.002x^2 + 6x + 7000.$$

   a. Write the average cost function $\overline{C}(x)$.
   b. Find and interpret $\overline{C}(100), \overline{C}(500)$, and $\overline{C}(1000)$.
   c. Find and interpret the oblique asymptote of $\overline{C}(x)$.

79. Biology: birds collecting seeds. A bird is collecting seed from a field that contains 100 grams of seed. The bird collects x grams of seed in t minutes, where

$$t = f(x) = \frac{4x+1}{100-x}, 0 < x < 100.$$

   a. Sketch the graph of $t = f(x)$.
   b. How long does it take the bird to collect
      (i) 50 grams?
      (ii) 75 grams?
      (iii) 95 grams?
      (iv) 99 grams?
   c. Complete the following statements if applicable:
      (i) As $x \to 100^-, f(x) \to$ _____.
      (ii) As $x \to 100^+, f(x) \to$ _____.
   d. Does the bird ever collect all of the seed from the field?

80. Criminology. Suppose the city of Las Vegas decides to be crime-free. The estimated cost of catching and convicting $x\%$ of the criminals is given by

$$C(x) = \frac{1000}{100-x} \text{ million dollars.}$$

   a. Find and interpret $C(50), C(75), C(90)$, and $C(99)$.
   b. Sketch the graph of $C(x), 0 \le x < 100$.
   c. What happens to $C(x)$ as $x \to 100^-$?
   d. What percentage of the criminals can be caught and convicted for $30 million?

81. Environment. Suppose an environmental agency decides to get rid of the impurities from the water of a polluted river. The estimated cost of removing $x\%$ of the impurities is given by

$$C(x) = \frac{3x^2 + 50}{x(100 - x)} \text{ billion dollars.}$$

a. How much will it cost to remove 50% of the impurities?
b. Sketch the graph of $y = C(x)$.
c. Estimate the percentage of the impurities that can be removed at a cost of $30 billion.

82. Biology. The growth function $g(x)$ describes the growth rate of organisms as a function of some nutrient concentration x. Suppose

$$g(x) = a\frac{x}{k + x}, \quad x \geq 0,$$

where a and k are positive constants.
a. Find the horizontal asymptote of the graph of $g(x)$. Use the asymptote to explain why a is called the *saturation level* of the nutrient.
b. Show that k is the half-saturation constant; that is, show that if $x = k$, then $g(x) = \frac{a}{2}$.

83. Population of bacteria. The population P (in thousands) of a colony of bacteria at time t (in hours) is given by

$$P(t) = \frac{8t + 16}{2t + 1}, \quad t \geq 0.$$

a. Find the initial population of the colony. (Find the population at $t = 0$ hours.)
b. What is the long-term behavior of the population? (What happens when $t \to \infty$?)

84. Drug concentration. The concentration c (in milligrams per liter) of a drug in the bloodstream of a patient at time $t \geq 0$ (in hours since the drug was injected) is given by

$$c(t) = \frac{5t}{t^2 + 1}, \quad t \geq 0.$$

a. By plotting points or using a graphing calculator, sketch the graph of $y = c(t)$.
b. Find and interpret the horizontal asymptote of the graph of $y = c(t)$.
c. Find the approximate time when the concentration of drug in the bloodstream is maximal.
d. At what time is the concentration level equal to 2 milligrams per liter?

85. Book publishing. The printing and binding cost for a college algebra book is $10. The editorial cost is $200,000. The first 2500 books are samples and are given free to professors. Let x be the number of college algebra books produced.
a. Write a function f describing the average cost of saleable books.
b. Find the average cost of a saleable book assuming that 10,000 books are produced.
c. How many books must be produced to bring the average cost of a saleable book under $20?

d. Find the vertical and horizontal asymptotes of the graph of $y = f(x)$. How are they related to this situation?

86. A "phony" sale. A jewelry merchant at a local mall wants to have a " p percent off" sale. She marks up the stock by q percent to break even with respect to the prices before the sale.
a. Show that $q = \frac{p}{1 - p}$, $0 < p < 1$.
b. Sketch the graph of q in part (a).
c. How much should she mark up the stock to break even if she wants to have a "25% off" sale?

C EXERCISES Beyond the Basics

In Exercises 87–96, use transformations of the graph of $y = \frac{1}{x}$ or $y = \frac{1}{x^2}$ to graph each rational function $f(x)$.

87. $f(x) = -\frac{2}{x}$

88. $f(x) = \frac{1}{2 - x}$

89. $f(x) = \frac{1}{(x - 2)^2}$

90. $f(x) = \frac{1}{x^2 + 2x + 1}$

91. $f(x) = \frac{1}{(x - 1)^2} - 2$

92. $f(x) = \frac{1}{(x + 2)^2} + 3$

93. $f(x) = \frac{1}{x^2 + 12x + 36}$

94. $f(x) = \frac{3x^2 + 18x + 28}{x^2 + 6x + 9}$

95. $f(x) = \frac{x^2 - 2x + 2}{x^2 - 2x + 1}$

96. $f(x) = \frac{2x^2 + 4x - 3}{x^2 + 2x + 1}$

97. Sketch and discuss the graphs of the rational functions of the form $y = \frac{a}{x^n}$, where a is a nonzero real number and n is a positive integer, in the following cases:
a. $a > 0$ and n is odd.
b. $a < 0$ and n is odd.
c. $a > 0$ and n is even.
d. $a < 0$ and n is even.

98. Graphing the reciprocal of a polynomial function. Let $f(x)$ be a polynomial function and $g(x) = \frac{1}{f(x)}$. Justify the following statements about the graphs of $f(x)$ and $g(x)$.
a. If c is a zero of $f(x)$, then $x = c$ is a vertical asymptote of the graph of $g(x)$.
b. If $f(x) > 0$, then $g(x) > 0$, and if $f(x) < 0$, then $g(x) < 0$.
c. The graphs of f and g intersect for those values (and only those values) of x for which $f(x) = g(x) = \pm 1$.
d. When f is increasing (decreasing, constant) on an interval of its domain, g is decreasing (increasing, constant) on that interval.

99. Use Exercise 98 to sketch the graphs of $f(x) = x^2 - 4$ and $g(x) = \frac{1}{x^2 - 4}$ on the same coordinate axes.

100. Use Exercise 98 to sketch the graphs of $f(x) = x^2 + 1$ and $g(x) = \frac{1}{x^2 + 1}$ on the same coordinate axes.

101. Recall that the inverse of a function $f(x)$ is written as $f^{-1}(x)$, while the reciprocal of $f(x)$ can be written as either $\dfrac{1}{f(x)}$ or $[f(x)]^{-1}$. The point of this exercise is to make clear that the reciprocal of a function has nothing to do with the inverse of a function. Let $f(x) = 2x + 3$. Find both $[f(x)]^{-1}$ and $f^{-1}(x)$. Compare the two functions. Graph all three functions on the same coordinate axes.

102. Let $f(x) = \dfrac{x-1}{x+2}$. Find $[f(x)]^{-1}$ and $f^{-1}(x)$. Graph all three functions on the same coordinate axes.

103. Sketch the graph of $g(x) = \dfrac{2x^3 + 3x^2 + 2x - 4}{x^2 - 1}$. Discuss the end behavior of $g(x)$.

104. Sketch the graph of $f(x) = \dfrac{x^3 - 2x^2 + 1}{x - 2}$. Discuss the end behavior of $f(x)$.

In Exercises 105 and 106, the graph of the rational function $f(x)$ crosses the horizontal asymptote. Graph $f(x)$ and find the point of intersection of the curve with its horizontal asymptote.

105. $f(x) = \dfrac{x^2 + x - 2}{x^2 - 2x - 3}$ **106.** $f(x) = \dfrac{4 - x^2}{2x^2 - 5x - 3}$

In Exercises 107–110, find an equation of a rational function f satisfying the given conditions.

107. Vertical asymptote: $x = 3$
Horizontal asymptote: $y = -1$
x-intercept: 2

108. Vertical asymptotes: $x = -1$, $x = 1$
Horizontal asymptote: $y = 1$
x-intercept: 0

109. Vertical asymptote: $x = 0$
Slant asymptote: $y = -x$
x-intercepts: -1, 1

110. Vertical asymptote: $x = 3$
Slant asymptote: $y = x + 4$
y-intercept: 2; $f(4) = 14$

Critical Thinking

In Exercises 111–116, make up a rational function $f(x)$ that has all of the characteristics given in the exercise.

111. Has a vertical asymptote at $x = 2$, a horizontal asymptote at $y = 1$, and a y-intercept at $(0, -2)$.

112. Has vertical asymptotes at $x = 2$ and $x = -1$, a horizontal asymptote at $y = 0$, a y-intercept at $(0, 2)$, and an x-intercept at 4.

113. Has $f\left(\dfrac{1}{2}\right) = 0$; $f(x) \to 4$ as $x \to \pm\infty$,
$f(x) \to \infty$ as $x \to 1^-$, and $f(x) \to \infty$ as $x \to 1^+$.

114. Has $f(0) = 0$; $f(x) \to 2$ as $x \to \pm\infty$; has no vertical asymptotes and is symmetric about the y-axis.

115. Has $y = 3x + 2$ as an oblique asymptote; has a vertical asymptote at $x = 1$.

116. Is it possible for the graph of a rational function to have both a horizontal and oblique asymptote? Explain.

GROUP PROJECTS Finding the Range and Turning Points

Find the range of the rational function $y = \dfrac{1}{x^2 + x - 2}$. First, solve this equation for x as follows:

$y = \dfrac{1}{x^2 + x - 2}$ Original equation

$(x^2 + x - 2)y = 1$ Multiply both sides by $(x^2 + x - 2)$.

$x^2 y + xy - 2y = 1$

$yx^2 + yx - (2y + 1) = 0$ This is a quadratic equation in x.

Use the quadratic formula with $a = y$, $b = y$, and $c = -(2y + 1)$.

$$x = \frac{-y \pm \sqrt{y^2 + 4y(2y + 1)}}{2y}$$

(1) $$x = \frac{-y \pm \sqrt{9y^2 + 4y}}{2y}.$$

Finally, for x to be a real number, the **discriminant** must be nonnegative. So $9y^2 + 4y \geq 0$, or $y(9y + 4) \geq 0$.

Sign graph

Test numbers -1 $-\dfrac{4}{9}$ $-\dfrac{1}{4}$ 0 1

Using the sign graph, we have $y \leq \dfrac{-4}{9}$ or $y \geq 0$. However, equation (1) is not defined for $y = 0$. Thus, the range of $f(x) = \dfrac{1}{x^2 + x - 2}$ is $\left(-\infty, \dfrac{-4}{9}\right] \cup (0, \infty)$. The value $y = -\dfrac{4}{9}$ is a local maximum value. Substitute $y = -\dfrac{4}{9}$ in (1) to obtain $x = -\dfrac{1}{2}$. Hence, $\left(-\dfrac{1}{2}, -\dfrac{4}{9}\right)$ is a local maximum point on the graph of f.

Use the preceding technique to find the range and turning points (if any) of the graph of each function.

1. $f(x) = \dfrac{x}{x^2 - 4}$

2. $f(x) = \dfrac{x^2 - 4}{x + 1}$ (See Example 8.)

3. $R(x) = \dfrac{x(100 - x)}{x + 10}$ (See Example 9.)

4. $f(x) = \dfrac{x^2 + 2}{(x + 2)(x - 1)}$ (See Example 6.)

Polynomial and Rational Inequalities

Before Starting this Section, Review

1. Linear equations (Section 1.1, page 85)
2. Quadratic equations (Section 1.4, page 118)
3. Zero-product property (Section 1.4, page 118)
4. Inequalities (Section 1.6, page 146)

Objectives

1 Solve quadratic inequalities.

2 Solve polynomial inequalities.

3 Solve rational inequalities.

ACCIDENT INVESTIGATION

Long before the famous traffic fatality in 1997 involving Princess Diana and Dodi Al Fayed, insurance companies and police have investigated the scenes of traffic accidents for evidence. You may have seen investigators measuring skid marks at the scene of an accident. These measurements yield information about the speed at which the car making the marks was traveling at the time of the accident. Under certain road conditions such as a wet blacktop surface, the distance (in feet) it takes a car traveling v miles per hour to stop is given by the equation

$$d = 0.05v^2 + v.$$

The skid marks at one accident where those conditions were met were over 75 feet long. The accident occurred in a 25-mile-per-hour speed zone. Was the driver going over the speed limit? We answer that question in Example 2, using the methods of this section. ■

Polynomial Inequalities

When you want to test the chlorine level in a swimming pool, you test the water at one place and are confident that the chlorine level is the same throughout the pool. Somewhat like the swimming pool, because the polynomial functions are continuous, they have a property that allows you to test *any point* in any interval of numbers that does *not* contain a root of the polynomial. If the value of the polynomial at that point is positive, the value at *every* point in the interval is also positive. Similarly, if the value at the point tested is negative, the value of the polynomial at *every* point in the interval is also negative. We learn to use this property to solve inequalities involving polynomials.

◆ **WARNING**

Before attempting to solve any polynomial or rational inequality, rearrange the inequality so that a polynomial or rational expression is on the left-hand side of the inequality symbol and 0 is on the right-hand side.

1 Solve quadratic inequalities.

Using Test Points to Solve Inequalities

The **test-point** method we describe next works for all quadratic polynomials. We first consider quadratic polynomials that have two distinct real roots. In this case, we find the two roots (using the quadratic formula if necessary) and then make **sign graphs** of the quadratic polynomial. The two roots divide the number line into three intervals, and the sign of the quadratic polynomial is determined in each interval by testing *any* number in the interval.

EXAMPLE 1 **Using the Test-Point Method to Solve a Quadratic Inequality**

Solve $x^2 > x + 6$. Write the solution in interval notation and graph the solution set.

SOLUTION

We first rearrange the inequality so that 0 is on the right side.

$$x^2 > x + 6 \qquad \text{Original inequality}$$
$$x^2 - x - 6 > x + 6 - x - 6 \qquad \text{Subtract } x + 6 \text{ from both sides.}$$
$$x^2 - x - 6 > 0 \qquad \text{Simplify.}$$

To determine where $x^2 - x - 6$ is positive, we first solve the *associated equation*.

$$x^2 - x - 6 = 0$$
$$(x + 2)(x - 3) = 0 \qquad \text{Factor.}$$
$$x + 2 = 0 \quad \text{or} \quad x - 3 = 0 \qquad \text{Zero-product property}$$
$$x = -2 \quad \text{or} \qquad x = 3 \qquad \text{Solve for } x.$$

The two roots of the equation $x^2 - x - 6 = 0$ are -2 and 3. These roots divide the number line into three intervals as shown in Figure 3.29. We select a convenient "test point" in each of the intervals $(-\infty, -2)$, $(-2, 3)$, and $(3, \infty)$. The points we selected, namely, $-3, 0$, and 4, are shown on the number line, but any other three points, one from each respective interval, would work as well. We then evaluate $x^2 - x - 6$ at each test point.

Test Interval	Test Point	Value of $x^2 - x - 6$	Result
$(-\infty, -2)$	-3	$(-3)^2 - (-3) - 6 = 6$	Positive
$(-2, 3)$	0	$(0)^2 - (0) - 6 = -6$	Negative
$(3, \infty)$	4	$(4)^2 - (4) - 6 = 6$	Positive

The sign of $x^2 - x - 6$ at each test point gives the sign of $x^2 - x - 6$ for every point in the interval containing that test point. These results are shown in the sign graph in Figure 3.29.

TECHNOLOGY CONNECTION

FIGURE 3.29 Sign of $x^2 - x - 6$

📈 In the calculator graph of $y = x^2 - x - 6$, to solve $x^2 - x - 6 > 0$, we are looking for the values of x where y values are positive. That means finding the intervals on the x-axis over which the graph is above the x-axis

To find where $x^2 - x - 6$ is positive, we locate the + signs on the sign graph. We observe that $x^2 - x - 6$ is positive (> 0) for each number x in the interval $(-\infty, -2)$ or in the interval $(3, \infty)$. The solution set of the inequality $x^2 - x - 6 > 0$ can be written as $(-\infty, -2) \cup (3, \infty)$. The graph of the inequality is shown in Figure 3.30.

FIGURE 3.30 Solution set of $x^2 - x - 6 > 0$

The calculator graph of $y = x^2 - x - 6$ in the margin confirms our solution. ■ ■ ■

Practice Problem 1 Solve $x^2 - 5x + 4 < 0$. Write the solution set in interval notation and graph the solution set. ■

EXAMPLE 2   **Calculating Speeds from Telltale Skid Marks**

In the introduction to this section, a car involved in an accident left skid marks over 75 feet long. Under the road conditions at the accident, the distance d (in feet) it takes a car traveling v miles per hour to stop is given by the equation

$$d = 0.05v^2 + v.$$

The accident occurred in a 25-mile-per-hour speed zone. Was the driver going over the speed limit?

SOLUTION

To answer the question, we solve the inequality

$$(\text{stopping distance}) > 75 \text{ feet}, \quad \text{or} \quad 0.05v^2 + v > 75.$$

To find out what speeds correspond to a stopping distance of more than 75 feet, we first rearrange the inequality so that only 0 appears on the right side.

$0.05v^2 + v > 75$	Stopping distance $d > 75$ feet
$0.05v^2 + v - 75 > 0$	Subtract 75 from both sides.
$0.05v^2 + v - 75 = 0$	Write the associated equation.
$5v^2 + 100v - 7500 = 0$	Multiply both sides by 100 to eliminate decimals.
$v^2 + 20v - 1500 = 0$	Divide both sides by 5.
$(v + 50)(v - 30) = 0$	Factor.
$v + 50 = 0 \quad \text{or} \quad v - 30 = 0$	Zero-product property
$v = -50 \quad \text{or} \quad v = 30$	Solve for v.

The two roots, -50 and 30, divide the number line into three test intervals: $(-\infty, -50)$, $(-50, 30)$, and $(30, \infty)$. Now select a convenient test point in each of these intervals. The points we selected, namely, -60, 0, and 40, are shown on the number line in Figure 3.31. Next, we evaluate $0.05v^2 + v - 75$ for each test point.

Test Interval	Test Point	Value of $0.05v^2 + v - 75$	Result
$(-\infty, -50)$	-60	$0.05(-60)^2 + (-60) - 75 = 45$	Positive
$(-50, -30)$	0	$0.05(0)^2 + 0 - 75 = -75$	Negative
$(30, \infty)$	40	$0.05(40)^2 + 40 - 75 = 45$	Positive

The sign of $0.05v^2 + v - 75$ at each test point gives the sign of $0.05v^2 + v - 75$ for every point in the interval containing the test point. These results are shown in the sign graph in Figure 3.31.

FIGURE 3.31

For this situation, we look at only the positive values of v. Note that the numbers corresponding to speeds between 0 and 30 miles per hour (that is, $0 \le v \le 30$) are not solutions of $0.05v^2 + v > 75$. Thus, the car was traveling *more* than 30 miles per hour. The driver was going over the speed limit.   ■ ■ ■

Practice Problem 2   Would the driver in Example 2 have been speeding if the skid marks had been 35 feet long and the accident had taken place in a 20-mile-per-hour speed zone?   ■

Once we rearrange a quadratic inequality so that 0 is on the right side, we set the left side equal to 0 and solve the resulting equation. What if the resulting equation has only one real solution or no real solution?

If the resulting equation has only one real solution, this root divides the number line into two sections, and the sign graph will consist of "0" at the one solution and all "+" signs or all "−" signs everywhere else. This sign graph can then be used to solve the inequality.

If the resulting equation has no real solution, we use the following result, which is true for all polynomial equations (not just quadratic equations).

ONE-SIGN THEOREM

If a polynomial equation has no real solution, then the polynomial is always positive or always negative.

EXAMPLE 3   **Using the One-Sign Theorem to Solve a Quadratic Inequality**

Solve $x^2 - 2x + 2 > 0$.

SOLUTION

This quadratic inequality already has 0 on the right side, so we set the left side equal to 0 to obtain the associated equation. Because there are no obvious factors, we evaluate the discriminant to see if there are any real roots.

$$x^2 - 2x + 2 = 1 \cdot x^2 - 2x + 2 = 0 \qquad a = 1, b = -2, c = 2$$
$$b^2 - 4ac = (-2)^2 - 4(1)(2) \qquad \text{Evaluate the discriminant.}$$
$$= -4 \qquad \text{The discriminant is negative.}$$

Because the discriminant, −4, is negative, $x^2 - 2x + 2 = 0$ has no real roots and $x^2 - 2x + 2$ is always positive (> 0) or always negative (< 0). To find out which, we can use any value of x as a test point. Calculations with $x = 0$ are simple, so we pick 0 as a test point.

Let $x = 0$. Then $(0)^2 - 2(0) + 2 = 2$, which is positive; so $x^2 - 2x + 2$ is *always* positive. The solution set is the set of all real numbers, which is written in interval notation as $(-\infty, \infty)$. ■ ■ ■

Practice Problem 3   Solve $x^2 + 2x - 2 > 0$. ■

2 Solve polynomial inequalities.

Other Polynomial Inequalities

Regardless of the degree of the polynomial in an inequality, if we can factor the polynomial, we can use test points and a sign graph to solve it.

FINDING THE SOLUTION: A PROCEDURE

EXAMPLE 4   **Solving a Polynomial Inequality**

OBJECTIVE	**EXAMPLE**
Solve a polynomial inequality $P(x) > 0, P(x) \geq 0, P(x) < 0$ or $P(x) \leq 0$.	*Solve:* $x^3 + 2x + 7 \leq 3x^2 + 6x - 5$
Step 1 Arrange the inequality so that 0 is on the right-hand side.	$x^3 + 2x + 7 \leq 3x^2 + 6x - 5$ Original inequality $x^3 - 3x^2 - 4x + 12 \leq 0$ Subtract $3x^2 + 6x - 5$ from both sides simplify.

continued on the next page

Step 2 Solve the associated equation $P(x) = 0$. The real solutions of this equation form the boundary points.

Solve: $x^3 - 3x^2 - 4x + 12 = 0$

$$x^2(x - 3) - 4(x - 3) = 0 \qquad \text{Factor by grouping.}$$

$$(x - 3)(x^2 - 4) = 0 \qquad \text{Distributive property}$$

$$(x - 3)(x - 2)(x + 2) = 0 \qquad \text{Difference of two squares}$$

$$x - 3 = 0 \text{ or } x - 2 = 0 \text{ or } x + 2 = 0 \qquad \text{Zero-product property}$$

$$x = 3 \text{ or } \qquad x = 2 \text{ or } \qquad x = -2 \qquad \text{Solve for } x.$$

The boundary points are $-2, 2,$ and 3.

Step 3 Locate the boundary points on the number line. These boundary points divide the number line into intervals. Write these intervals.

The boundary points divide the number line into four intervals: $(-\infty, -2), (-2, 2), (2, 3),$ and $(3, \infty)$

Step 4 Select a test point in each interval from Step 3 and evaluate $P(x)$ at each test point to determine the sign of $P(x)$ on that interval.

Interval	Test Point	Value of $P(x) = x^3 - 3x^2 - 4x + 12$	Result
$(-\infty, -2)$	-3	$(-3)^3 - 3(-3)^2 - 4(-3) + 12 = -30$	Negative
$(-2, 2)$	0	$(0)^3 - 3(0)^2 - 4(0) + 12 = 12$	Positive
$(2, 3)$	2.5	$(2.5)^3 - 3(2.5)^2 - 4(2.5) + 12 = -1.125$	Negative
$(3, \infty)$	4	$(4)^3 - 3(4)^2 - 4(4) + 12 = 12$	Positive

Step 5 Write the solution set by selecting the intervals in Step 4 that satisfy the given inequality. If the inequality is $P(x) \leq 0$ (or ≥ 0), include the corresponding boundary points. Graph the solution set.

The Solution set is $(-\infty, -2] \cup [2, 3]$.

Practice Problem 4 Solve: $x^3 + 2x^2 - 5x - 21 \geq 4x - 2x^2 + 15$

EXAMPLE 5   **Solving a Polynomial Inequality**

Solve: $(x - 1)(x + 1)(x^2 + 1) \leq 0$

SOLUTION

Step 1   The inequality has a 0 on one side, and $P(x)$ is in factored form. We proceed to the next step.

Step 2   We find the real solutions of the associated equation.

$$(x - 1)(x + 1)(x^2 + 1) = 0 \qquad \text{Replace} \leq \text{with} = \text{in } (x - 1)(x + 1)(x^2 + 1) \leq 0.$$

$$x - 1 = 0, \text{ or } x + 1 = 0 \text{ or } x^2 + 1 = 0 \qquad \text{Zero-product property}$$

Thus, $x = 1$ or $x = -1$. (The equation $x^2 + 1 = 0$ has no real solution.)

Step 3 The solutions -1 and 1 of $(x - 1)(x + 1)(x^2 + 1) = 0$ divide the number line into three intervals: $(-\infty, -1)$, $(-1, 1)$, and $(1, \infty)$. See Figure 3.32.

Step 4 We select convenient test points in each of these three intervals. We pick -2, 0, and 2. Next, we compute the value of $(x - 1)(x + 1)(x^2 + 1)$ at each test point to determine the sign at each point.

Interval	Test Point	Value of $(x - 1)(x + 1)(x^2 + 1)$	Sign
$(-\infty, -1)$	-2	$(-2 - 1)(-2 + 1)((-2)^2 + 1) = 15$	$+$
$(-1, 1)$	0	$(0 - 1)(0 + 1)((0)^2 + 1) = -1$	$-$
$(1, \infty)$	2	$(2 - 1)(2 + 1)((2)^2 + 1) = 15$	$+$

FIGURE 3.32

Step 5 We see from Figure 3.32 that the solution set consists of all x between -1 and 1, including both -1 and 1. The solution set is shown in Figure 3.33.

$-1 \leq x \leq 1$, or $[-1, 1]$

FIGURE 3.33 ■ ■ ■

Practice Problem 5 Solve $x^4 \leq 3x^2 + 4$. Write the solution in interval notation and graph the solution set. ■

3 Solve rational inequalities.

Rational Inequalities

An inequality involving a rational expression rather than a polynomial is called a **rational inequality**. The additional consideration in a rational inequality is the polynomial in the denominator. As with polynomial inequalities, our first step is to rearrange the inequality (if necessary) so that the right side is 0.

EXAMPLE 6 **Solving a Rational Inequality**

Solve $\dfrac{3}{x - 1} \geq 1$. Write the solution in interval notation and graph the solution set.

SOLUTION

Step 1 We first rewrite $\dfrac{3}{x - 1} \geq 1$ to get 0 on the right-hand side.

$$\frac{3}{x - 1} \geq 1 \qquad \text{Original inequality}$$

$$\frac{3}{x - 1} - 1 \geq 0 \qquad \text{Subtract 1 from both sides.}$$

$$\frac{3 - (x - 1)}{x - 1} \geq 0 \qquad \text{Use } x - 1 \text{ as a common denominator.}$$

$$R(x) = \frac{4 - x}{x - 1} \geq 0 \qquad \text{Simplify the numerator.}$$

Step 2 **Set the numerator and the denominator of the left side equal to 0.** The real solutions of these equations are the boundary points.

$$\text{numerator} = 0 \qquad \text{and} \qquad \text{denominator} = 0$$
$$4 - x = 0 \qquad\qquad\qquad x - 1 = 0$$
$$4 = x \qquad\qquad\qquad\qquad x = 1$$

Step 3 **Write the intervals using the boundary points from Step 2.** The numbers 1 and 4 divide the number line into three intervals: $(-\infty, 1)$, $(1, 4)$, and $(4, \infty)$.

Step 4 **Select a test point in each interval from step 3 and determine the sign of $R(x)$.**

We choose $0, 2$, and 5 as convenient test points in the intervals of step 3. The test points are shown on the sign graph in Figure 3.34.

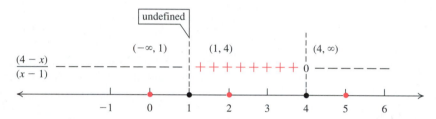

FIGURE 3.34

Step 5 **Write the solution set of the inequality.**

From the sign graph in Figure 3.34, we determine the real numbers for which $\frac{4 - x}{x - 1}$ is positive (> 0); these are the numbers in the interval $(1, 4)$, as the $+$ signs indicate. However, we want to know where $\frac{4 - x}{x - 1}$ is *either* positive *or* 0. The expression $\frac{4 - x}{x - 1}$ is undefined for $x = 1$ and is 0 only if $x = 4$. The solution set is then $\{x \mid 1 < x \leq 4\}$, or in interval notation, $(1, 4]$. The graph is shown in Figure 3.35. ▪ ▪ ▪

$1 < x \leq 4$, or $(1, 4]$

FIGURE 3.35

Practice Problem 6 Solve $\frac{2}{1 - 3x} > 1$. Write the solution in interval notation and graph the solution set. ▪

◆ **WARNING**   You cannot solve a rational inequality by multiplying both sides by the LCD, as you would with a rational equation. Remember that you reverse the sense of an inequality when multiplying by a negative expression. However, when an expression contains a variable, you do not know whether the expression is positive or negative.

EXAMPLE 7   **Solving a Rational Inequality**

Solve: $\dfrac{(x-1)(x+2)}{(x+4)(x-3)} \le 0$

SOLUTION

Step 1   There is no need to rewrite the inequality because there is a zero on the right side.

Step 2   Set numerator = 0   and   denominator = 0.

$$(x-1)(x+2)=0 \quad \text{and} \quad (x+4)(x-3)=0$$

$x-1=0$ or $x+2=0$ and $x+4=0$ or $x-3=0$ Zero-product property

$x=1$ or $x=-2$ and $x=-4$ or $x=3$ Solve for x.

Step 3   The numbers $-4, -2, 1$, and 3 divide the number line into five intervals. See Figure 3.36.

Step 4   We choose test points $-5, -3, 0, 2$, and 4 in the intervals of Step 3.

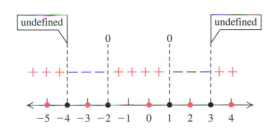

FIGURE 3.36

Interval	Test Value	Sign of $R(x) = \dfrac{(x-1)(x+2)}{(x+4)(x-3)}$	Result
$(-\infty, -4)$	-5	$\dfrac{(-5-1)(-5+2)}{(-5+4)(-5-3)} = \dfrac{(-)(-)}{(-)(-)}$	Positive
$(-4, -2)$	-3	$\dfrac{(-3-1)(-3+2)}{(-3+4)(-3-3)} = \dfrac{(-)(-)}{(+)(-)}$	Negative
$(-2, 1)$	0	$\dfrac{(0-1)(0+2)}{(0+4)(0-3)} = \dfrac{(-)(+)}{(+)(-)}$	Positive
$(1, 3)$	2	$\dfrac{(2-1)(2+2)}{(2+4)(2-3)} = \dfrac{(+)(+)}{(+)(-)}$	Negative
$(3, \infty)$	4	$\dfrac{(4-1)(4+2)}{(4+4)(4-3)} = \dfrac{(+)(+)}{(+)(+)}$	Positive

FIGURE 3.37 Solution set of
$\dfrac{(x-1)(x+2)}{(x+4)(x-3)} \le 0$

Step 5   From Step 4, we conclude that $R(x) < 0$ on the intervals $(-4, -2)$ and $(1, 3)$. Because $R(x) = 0$ when the numerator is 0, we must include the values $x = -2$ and $x = 1$ in the solution of the inequality $R(x) \le 0$.

The solution set of the given inequality in interval notation is $(-4, -2] \cup [1, 3)$, as shown in Figure 3.37. ■ ■ ■

Practice Problem 7   Solve: $\dfrac{(x+3)(x-2)}{(x+5)(x-3)} \le 0$ ■

SECTION 3.7 ■ Exercises

A EXERCISES Basic Skills and Concepts

In Exercises 1–50, solve each inequality by the test-point method. Write the solution in interval notation.

1. $(x-3)(x+2) < 0$ **2.** $(x-2)(x+3) < 0$

3. $(x+5)(x-2) < 0$ **4.** $(x-5)(x+2) > 0$

5. $(x+4)(x-1) \le 0$ **6.** $(x-4)(x+1) \le 0$

7. $(x-6)(x+1) \ge 0$ **8.** $(x+6)(x-1) \ge 0$

9. $x(x+3) < 0$ **10.** $x(x-3) > 0$

11. $x^2 \le 9$ **12.** $x^2 \ge 1$

13. $25 - x^2 < 0$ **14.** $4 - x^2 > 0$

15. $x^2 + 4x - 12 \leq 0$ 16. $x^2 - 8x + 7 > 0$

17. $6x^2 + 7x - 3 \geq 0$ 18. $4x^2 - 2x - 2 < 0$

19. $25x^2 + 5x - 6 \leq 0$ 20. $12x^2 - 5x - 7 < 0$

21. $(x - 2)^2 \leq 1$ 22. $(x + 3)^2 \geq 9$

23. $2x^2 + 5x < 12$ 24. $6x^2 + 11x \geq 7$

25. $2x(x + 3)(x - 3) \leq 0$ 26. $3x(x + 2)(x - 2) \geq 0$

27. $(x + 3)(x + 1)(x - 1) \geq 0$

28. $(x + 4)(x - 1)(x + 2) \leq 0$

29. $x^3 - 4x^2 - 12x > 0$ 30. $x^3 + 8x^2 + 15x > 0$

31. $x^3 - x^2 - 4x + 4 < 0$ 32. $x^3 + x^2 - 9x - 9 < 0$

33. $x^3 + x^2 + 2x + 2 \geq 0$ 34. $x^3 - x^2 + 3x - 3 \geq 0$

35. $x^3 - 2x^2 - x + 2 \leq 0$ 36. $x^3 + 4x^2 - 11x - 30 \leq 0$

37. $x^2 - 2x - 2 < 0$ 38. $x^2 - 4x - 1 \leq 0$

39. $x^2 \geq -1$ 40. $2x^2 \geq -5$

41. $x^2 + 2x < -1$ 42. $4x^2 + 12x < -9$

43. $x^3 - x^2 \geq 0$ 44. $x^3 - 9x^2 \geq 0$

45. $x^4 \leq 2x^2$ 46. $x^4 \geq 3x^2$

47. $x^2 \geq 1$ 48. $x^3 < -8$

49. $x^4 \leq 16$ 50. $x^4 > 9$

In Exercises 51–82, solve each rational inequality. Write the solution in interval notation.

51. $\dfrac{x + 2}{x - 5} < 0$ 52. $\dfrac{x - 3}{x + 1} > 0$

53. $\dfrac{3x + 2}{x - 3} \leq 0$ 54. $\dfrac{3 - 2x}{4x + 5} \geq 0$

55. $\dfrac{x + 4}{x} < 0$ 56. $\dfrac{x}{x - 2} > 0$

57. $\dfrac{x + 1}{x + 2} \leq 3$ 58. $\dfrac{x - 1}{x - 2} \geq 3$

59. $\dfrac{(x - 2)(x + 2)}{x} > 0$ 60. $\dfrac{(x - 1)(x + 3)}{x - 2} < 0$

61. $\dfrac{(x + 1)(x - 3)}{(x + 2)(x - 4)} \leq 0$ 62. $\dfrac{x(x - 4)}{(x - 2)(x + 3)} \leq 0$

63. $\dfrac{(x - 2)(x + 1)}{(x - 3)(x + 5)} \geq 0$ 64. $\dfrac{(x - 1)(x - 3)}{(x + 2)(x + 4)} \geq 0$

65. $\dfrac{x^2 - 1}{x^2 - 4} \leq 0$ 66. $\dfrac{x^2 - 9}{x^2 - 64} \leq 0$

67. $\dfrac{4x^2 - 9}{x^2 - 36} \geq 0$ 68. $\dfrac{x^2 - 25}{9x^2 - 49} \geq 0$

69. $x + \dfrac{8}{x} \leq 6$ 70. $x - \dfrac{12}{x} > 1$

71. $2 - \dfrac{2x}{3x - 4} > 0$ 72. $1 + \dfrac{1}{2 - x} \geq 0$

73. $\dfrac{x + 4}{3x - 2} \geq 1$ 74. $\dfrac{2x - 3}{x + 3} \leq 1$

75. $3 \leq \dfrac{2x + 6}{2x + 1}$ 76. $\dfrac{x - 2}{2x + 1} < -1$

77. $\dfrac{x + 5}{x - 2} > \dfrac{x}{x + 2}$ 78. $\dfrac{x - 1}{x + 1} > \dfrac{x}{x - 1}$

79. $\dfrac{x + 2}{x - 3} \geq \dfrac{x - 1}{x + 3}$ 80. $\dfrac{x + 1}{x - 2} \geq \dfrac{x}{x - 1}$

81. $\dfrac{x - 1}{x + 1} \leq \dfrac{x + 2}{x - 3}$ 82. $\dfrac{x + 3}{x + 1} \leq \dfrac{x - 1}{x - 2}$

B EXERCISES Applying the Concepts

83. **Free-falling object.** The distance d traveled in t seconds by an object dropped from a height h is given by $d = 16t^2$. If a water bottle is dropped from a hot air balloon when the balloon is between 64 and 100 feet above the ground, how long (in seconds) does it take for the bottle to hit the ground?

84. **Interest rate.** The amount of money A received at the end of two years when P dollars is invested at a compound rate r is $A = P(1 + r)^2$. A firm is investing $100,000 for a two-year period and expects to get a return of no less than $110,000 and no more than $115,000. What range of interest rates (to the nearest percent) will provide the expected return?

85. **Temperature.** The number N of water mites in a water sample depends on the temperature t in degrees Fahrenheit and is given by $N = 110t - t^2$. At what temperature will the number of mites exceed 1000?

86. **Falling object.** The height h of an object thrown from the top of a ski lift 1584 feet high after t seconds is $h = -16t^2 + 32t + 1584$. For what times is the height of the object at least 1200 feet?

87. **Company profit.** The profit P (in millions of dollars) of a company for next year is estimated to satisfy the inequality $P(P - 3) < 4(P - 3)$. What is the estimated profit for this company for next year?

88. **Area of a rectangle.** The length of a rectangle is 5 meters larger than the width. Find the range of widths that result in such a rectangle whose area is at least 204 square meters.

89. **Blackjack.** A gambler playing blackjack in a casino using four decks noticed that 20% of the cards that had been dealt were jacks, queens, kings, or aces.

After x cards had been dealt, he knew that the likelihood that the next card dealt would be a jack, a queen, a king, or an ace was $\dfrac{64 - 0.2x}{208 - x}$. For what values of x is this likelihood greater than 50%?

90. Area of a triangle. The base of a triangle is 3 centimeters greater than the height. Find the possible heights h so that the area of such a triangle will be at least 5 square centimeters.

C EXERCISES Beyond the Basics

91. Find the numbers k for which the quadratic equation $2x^2 + kx + 2 = 0$ has two real solutions.

92. Find the numbers k for which the quadratic equation $2x^2 + kx + 2 = 0$ has no real solutions.

93. Find the numbers k for which the quadratic equation $x^2 + kx + k = 0$ has two real solutions.

94. Find the numbers k for which the quadratic equation $x^2 + kx + k = 0$ has no real solutions.

In Exercises 95–100, find the domain of each function.

95. $f(x) = \sqrt{x^2 - 1}$ **96.** $f(x) = \sqrt{x^2 - x - 2}$

97. $f(x) = \dfrac{1}{\sqrt{1 - x^2}}$ **98.** $f(x) = \dfrac{\sqrt{-x}}{x^2 - 9}$

99. $g(x) = \sqrt{\dfrac{x - 1}{x + 3}}$ **100.** $g(x) = \sqrt{\dfrac{x - 2}{1 - x}}$

101. Find all values of c for which it is possible to find two numbers whose product is 36 and whose sum is c.

102. Find all values of c for which it is possible to find two numbers whose sum is 12 and the sum of whose squares is c.

103. An import firm pays a tax of $10 on each radio it imports. In addition to this tax, a penalty tax must be paid if more than 1000 radios are imported. The penalty tax is computed by multiplying 5 cents by the number of radios imported in excess of 1000. This tax must be paid on each radio imported. If 1006 radios are imported, the penalty tax is $6 \cdot 5 = 30$ cents and is paid on each of the 1006 radios. If the firm wants to spend no more than a total of $640,000 on import taxes, how many radios can it import?

104. A TV quiz program pays a contestant $100 for each correct answer for ten questions. If all ten questions are answered correctly, bonus questions are asked. The reward for every correct answer is increased by $50 for each bonus question that is correctly answered. Any incorrect answer ends the game. If two bonus questions are answered correctly, the contestant receives $200 *for each of the 12 questions*. If a contestant won more than $3500, how many questions must have been answered correctly?

Critical Thinking

105. Give an example of a quadratic inequality with each of the following solution sets.
   a. $(-4, 5)$
   b. $[-2, 6]$
   c. $(-\infty, \infty)$
   d. $\varnothing$
   e. $\{3\}$
   f. $(-\infty, 2) \cup (2, \infty)$

106. Give an example of an inequality with each of the following solution sets.
   a. $(-2, 4]$
   b. $[3, 5)$
   c. Is there a quadratic inequality whose solution set is $(2, 5]$?

Variation

Before Starting this Section, Review

1. Equation of a line (Section 2.3, page 194)

2. Circumference and area of a circle (Section P.7, page 74)

Objectives

1 Solve direct variation problems.

2 Solve inverse variation problems.

3 Solve joint and combined variation problems.

Sir Isaac Newton 1642–1727

Issac Newton was a mathematician and physicist, one of the foremost scientific intellects of all time. He entered Cambridge University in 1661 and was elected a Fellow of Trinity College in 1667 and Lucasian Professor of Mathematics in 1669. He remained at the university, lecturing most years, until 1696.

As a firm opponent of the attempt by King James II to make the universities into Catholic institutions, Newton was elected Member of Parliament for the University of Cambridge to the Convention Parliament of 1689, and he sat again from 1701–1702.

Meanwhile, in 1696, he moved to London as warden of the Royal Mint. He became master of the Mint in 1699, an office he retained until his death. He was elected a Fellow of the Royal Society of London in 1671, and in 1703, he became president, being annually reelected for the rest of his life. One of his major works, *Opticks*, appeared the next year; he was knighted in Cambridge in 1705.

NEWTON AND THE APPLE

According to legend, Isaac Newton was sitting under an apple tree, and when an apple fell on his head, he suddenly thought of the Law of Universal Gravitation. As in many such legends, the story may not be true in detail, but it contains elements of the truth.

Before we tell you a more realistic version of this story, let's recall some terminology associated with the motion of objects.

$$\text{Recall that speed} = \frac{\text{distance}}{\text{time}}$$

(i) The **velocity** of an object is the speed of an object in a specific direction.

(ii) **Acceleration:** If the velocity of an object is changing, we say that the object is accelerating. Acceleration is the rate of change of velocity.

(iii) **Force** = Mass × Acceleration.

What really happened with the apple? Perhaps the correct version of the story is that upon observing an apple falling from a tree, Newton concluded that the apple accelerated because the velocity of the apple changed from when it began hanging on the tree (zero velocity) and then moved to the ground. He decided that a force must act on the apple to cause this acceleration. He called this force "gravity" and the associated acceleration (Force = Mass × Acceleration) the "acceleration due to gravity." He conjectured further that if the force of gravity reaches the top of the apple tree, it might reach even farther; might it reach all the way to the moon? In that case, to keep the moon moving in a circular orbit, rather than wandering into outer space, Earth must exert a gravitational force on the moon. Newton realized that the force that brought the apple to the ground and the force that holds the moon in orbit were the same. He concluded that any two objects in the universe exert gravitational attraction on each other, whereupon he gave us the *Law of Universal Gravitation*. (See Example 6.)

Some scientists believe that it was Kepler's Third Law of Planetary Motion (see Exercise 53), not an apple, that led Newton to his Law of Universal Gravitation. ■

1 Solve direct variation problems.

Direct Variation

Two types of relationships between quantities (variables) occur so frequently in mathematics that they are given special names: *direct variation* and *inverse variation*.

Direct variation describes any relationship between two quantities in which any increase (or decrease) in one causes a proportional increase (or decrease) in the other.

STUDY TIP

The statements

"*y* varies directly as *x*,"
"*y* varies as *x*,"
"*y* is directly proportional to *x*,"
"*y* is proportional to *x*," and
"*y* = *kx*, *k* ≠ 0"

have the same meaning.

For example, the sales tax in Tampa, Florida, in 2008 was 7%; so the sales tax on a $1000 purchase was $70. If we *double* the purchase to $2000, the sales tax also *doubles*, to $140. If we *reduce* the purchase by *half* to $500, the sales tax is also *reduced* by half, to $35. We say that the sales tax *varies directly* as the purchase price. The equation

$$\text{Sales tax} = (0.07)(\text{Purchase price})$$

expresses the fact that the sales tax was a constant (0.07) multiple of the purchase price.

DIRECT VARIATION

A quantity *y* is said to **vary directly** as the quantity *x*, or *y* is **directly proportional** to *x*, if there is a constant *k* such that $y = kx$. This constant $k \neq 0$ is called the **constant of variation** or the **constant of proportionality**.

FINDING THE SOLUTION: A PROCEDURE

EXAMPLE 1   Solving Variation Problems

OBJECTIVE

Solve variation problems.

Step 1 Write the equation with the constant of variation, *k*.

Step 2 Substitute the given values of the variables into the equation in Step 1 to find the value of the constant *k*.

Step 3 Rewrite the equation in Step 1 with the value of *k* from Step 2.

Step 4 Use the equation from Step 3 to answer the question posed in the problem.

EXAMPLE

Suppose *y* varies as *x* and $x = \frac{4}{3}$ when *y* = 20. Find *y* when *x* = 8.

$$y = kx \qquad \text{\textit{y} varies as \textit{x}.}$$

$$20 = k\left(\frac{4}{3}\right) \qquad \text{Replace \textit{y} with 20 and \textit{x} with } \frac{4}{3}.$$

$$20\left(\frac{3}{4}\right) = k\left(\frac{4}{3}\right)\left(\frac{3}{4}\right) \qquad \text{Multiply both sides by } \frac{3}{4}.$$

$$15 = k \qquad \text{Simplify to solve for \textit{k}.}$$

$$y = 15x \qquad \text{Replace \textit{k} with 15.}$$

$$y = 15(8) \qquad \text{Replace \textit{x} with 8.}$$

$$y = 120 \qquad \text{Simplify.}$$

So *y* = 120 when *x* = 8. ■■■

Practice Problem 1   Suppose *y* varies directly as *x*. If *y* is 6 when *x* is 30, find *y* when *x* = 120. ■

EXAMPLE 2   Direct Variation in Electrical Circuits

The current in a circuit connected to a 220-volt battery is 50 amperes. If the current is directly proportional to the voltage of the attached battery, what voltage battery is needed to produce a current of 75 amperes?

SOLUTION

Let *I* = current in amperes and *V* = voltage in volts of the battery.

Step 1 $\quad I = kV \qquad$ *I* varies directly as *V*.

Step 2 $\quad 50 = k(220) \qquad$ Substitute *I* = 50, *V* = 220.

$$\frac{50}{220} = k \qquad \text{Solve for \textit{k}.}$$

$$\frac{5}{22} = k \qquad \text{Simplify.}$$

Step 3 $I = \dfrac{5}{22}V$ Replace k with $\dfrac{5}{22}$ in $I = kV$.

Step 4 $75 = \dfrac{5}{22}V$ Substitute $I = 75$ in Step 3.

       $\dfrac{22}{5} \cdot 75 = V$ Solve for V.

       $330 = V$ Simplify.

A battery of 330 volts is needed to produce 75 amperes of current.   ■ ■ ■

Practice Problem 2   In Example 2, if the current in the circuit is 60 amperes, what voltage battery is needed to produce a current of 75 amperes?   ■

We can generalize the concept of direct variation to variation with powers.

DIRECT VARIATION WITH POWERS

"A quantity y varies directly as the nth power of x" means that

$$y = kx^n,$$

where k is a nonzero constant and $n > 0$.

Note that when $n = 1$, y varies directly as x. In the formula for the area of a circle, $A = \pi r^2$, A varies directly as the square (or second power) of the radius r, with π as the constant of variation.

EXAMPLE 3   **Solving a Problem Involving Direct Variation with Powers**

Suppose you had forgotten the formula for the volume of a sphere but were told that the volume V of a sphere varies directly as the cube of its radius r. In addition, you are given the fact that $V = 972\pi$ when $r = 9$. Find V when $r = 6$.

SOLUTION

Step 1 $V = kr^3$ V varies directly as r^3.

Step 2 $972\pi = k(9)^3$ Replace V with 972π and r with 9.

         $972\pi = k(729)$ $9^3 = 729$

            $k = \dfrac{972\,\pi}{729}$ Solve for k.

              $= \dfrac{4}{3}\pi$ Simplify.

Step 3 $V = \dfrac{4}{3}\pi r^3$ Substitute $k = \dfrac{4}{3}\pi$ in Step 1.

Step 4 $V = \dfrac{4}{3}\pi(6)^3$ Replace r with 6.

         $= 288\,\pi$ cubic units   ■ ■ ■

Practice Problem 3   Assuming that y varies directly as the square of x and $y = 48$ when $x = 2$, find y when $x = 5$.   ■

2 Solve inverse variation problems.

Inverse Variation

INVERSE VARIATION

A quantity y **varies inversely** as the quantity x, or y is **inversely proportional** to x, if

$$y = \frac{k}{x},$$

where k is a nonzero constant.

Suppose a plane takes four hours to fly from Atlanta to Denver at an average speed of 300 miles per hour. If we *increase* the average speed to 600 miles per hour (twice the previous speed), the flight time *decreases* to two hours (half the previous time). For the Atlanta–Denver trip, the time of flight varies inversely as the speed of the plane:

$$\text{Time} = \frac{\text{Distance}}{\text{Speed}}$$

$$= \frac{k}{\text{Speed}}, \text{ where } k = \text{distance between Atlanta and Denver.}$$

We see that the distance is the constant of variation when time and speed are the variables.

EXAMPLE 4   **Solving an Inverse Variation Problem**

Suppose y varies inversely as x and $y = 35$ when $x = 11$. Find y when $x = 55$.

SOLUTION

Step 1 $y = \dfrac{k}{x}$ y varies inversely as x.

Step 2 $35 = \dfrac{k}{11}$ Substitute $y = 35$, $x = 11$.

$35 \cdot 11 = k$ Solve for k.

$385 = k$ Simplify.

Step 3 $y = \dfrac{385}{x}$ Replace k with 385 in $y = \dfrac{k}{x}$.

Step 4 $y = \dfrac{385}{55}$ Replace x with 55.

$y = 7$ Simplify.   ■ ■ ■

Practice Problem 4 A varies inversely as B, and $A = 12$ when $B = 5$. Find A when $B = 3$.   ■

INVERSE VARIATION WITH POWERS

A quantity y varies inversely as the nth power of x means that $y = \dfrac{k}{x^n}$, where k is a nonzero constant and $n > 0$.

---

3 Solve joint and combined variation problems.

EXAMPLE 5 **Solving a Problem Involving Inverse Variation with Powers**

The intensity of light varies inversely as the square of the distance from the light source. If Rita doubles her distance from a lamp, what happens to the intensity of light at her new location?

SOLUTION

Let I be the intensity of light at a distance d from the light source. Because I is inversely proportional to the square of d, we have

$$I = \frac{k}{d^2}.$$

If we replace d with $2d$, the intensity I_1 at the new location is given by

$$I_1 = \frac{k}{(2d)^2} \qquad \text{Replace } d \text{ with } 2d \text{ in } I = \frac{k}{d^2}.$$

$$I_1 = \frac{k}{4d^2} \qquad (2d)^2 = 2^2 d^2 = 4d^2$$

Therefore,

$$I_1 = \frac{I}{4} \qquad \text{Replace } \frac{k}{d^2} \text{ with } I.$$

This equation tells us that if Rita doubles her distance from the light source, the intensity of light at the new location will be one-fourth the intensity at the original location. ■ ■ ■

Practice Problem 5 Assuming that y varies inversely as the square root of x and $y = \frac{3}{4}$ when $x = 16$, find x when $y = 2$. ■

Joint and Combined Variation

Sometimes different types of variations occur in a problem that has more than one independent variable.

JOINT AND COMBINED VARIATION

The expression *z varies jointly as x and y* means that $z = kxy$ for some nonzero constant k. Similarly, if n and m are positive numbers, then the expression *z varies jointly as the nth power of x and mth power of y* means that $z = kx^n y^m$ for some nonzero constant k.

For example, the volume V of a right circular cylinder with radius r and height h is given by

$$V = \pi r^2 h.$$

Using the vocabulary of variation, we say, "The volume V of a right circular cylinder varies jointly as its height h and the square of its radius r." The constant of variation is π.

We can combine the joint variation and inverse variation. For example, the relationship

$$w = \frac{kx^2 y^3}{\sqrt{z}}$$

can be described as "w varies jointly as the square of x and cube of y and inversely as the square root of z." The same relationship also can be stated as "w varies directly as $x^2 y^3$ and inversely as $\sqrt{z}$."

EXAMPLE 6 Newton's Law of Universal Gravitation

Newton's Law of Universal Gravitation says that every object in the universe attracts every other object with a force acting along the line of the centers of the two objects. Also, this attracting force is directly proportional to the product of the two masses and inversely proportional to the square of the distance between the two objects.

a. Write the law symbolically.

b. Estimate the value of g (the acceleration due to gravity) near the surface of Earth. Use the following estimates: Radius of Earth $R_E = 6.38 \times 10^6$ meters, and mass of Earth $M_E = 5.98 \times 10^{24}$ kilograms.

SOLUTION

a. Let m_1 and m_2 be the masses of the two objects and r be the distance between their centers. See Figure 3.38. Let F denote the gravitational force between the objects.

FIGURE 3.38 Gravitational attraction

Then according to Newton's Law of Universal Gravitation,

$$F = G \cdot \frac{m_1 m_2}{r^2}.$$

The constant of proportionality G is called the *universal gravitational constant*. It is "universal" because it is thought to be the same at all places and all times. If the masses m_1 and m_2 are measured in kilograms, r is measured in meters, and the force F is measured in newtons, then the value of G is approximately 6.67×10^{-11} cubic meters per kilogram per second squared. This value was experimentally verified by Henry Cavendish in 1795.

FIGURE 3.39 Finding g on Earth

b. *Estimating the value of g* (acceleration due to gravity)

We use Newton's Laws: Force = Mass × Acceleration = $m \cdot g$. See Figure 3.39.

For an object of mass m near the surface of Earth.

$$\text{Force} = G \cdot \frac{m_1 m_2}{r^2} \qquad \text{Law of Universal Gravitation}$$

$$m \cdot g = G \cdot \frac{m M_E}{R_E^2} \qquad \text{Force} = m \cdot g; \text{ replace } m_1 \text{ with } m \text{ and } m_2 \text{ with } M_E.$$

$$g = G \cdot \frac{M_E}{R_E^2} \qquad \text{Divide both sides by } m.$$

$$= \frac{(6.67 \times 10^{-11} \text{ m}^3/\text{kg/sec}^2) \cdot (5.98 \times 10^{24} \text{ kg})}{(6.38 \times 10^6)^2 \text{ m}^2} \qquad \text{Substitute appropriate values for } G, M_E, \text{ and } R_E.$$

$$\approx 9.8 \text{ m/sec}^2 \qquad \text{Use a calculator.} \quad ■ ■ ■$$

Practice Problem 6 The mass of Mars is about 6.42×10^{23} kilograms, and its radius is about 3397 kilometers. What is the acceleration due to gravity near the surface of Mars? ■

We summarize the steps involved in solving most variation problems.

SOLVING VARIATION PROBLEMS

Step 1 Write the equation that models the problem.

    a. y varies directly with x. $y = kx$

    b. y varies with the nth power of x. $y = kx^n$

    c. y varies inversely with x. $y = \dfrac{k}{x}$

    d. y varies inversely with the nth power of x. $y = \dfrac{k}{x^n}$

    e. z varies jointly with the nth power of x and the mth power of y. $z = kx^n y^m$

    f. z varies directly with the nth power of x and inversely with the mth power of y. $z = \dfrac{kx^n}{y^m}$

Step 2 Substitute the given values in the equation in Step 1 and solve for k.

Step 3 Rewrite the equation in Step 1 with the value of k from Step 2.

Step 4 Use the equation from Step 3 to answer the question posed in the problem.

If a variation problem uses a more complicated relationship than that in Step 1, simply write the appropriate equation involving k and the variables and proceed with Steps 2–4.

SECTION 3.8 ■ Exercises

A EXERCISES Basic Skills and Concepts

In Exercises 1–8, write each statement as an equation.

1. P varies directly as T.

2. P varies inversely as V.

3. y varies as the square root of x.

4. F varies jointly as m_1 and m_2.

5. V varies jointly as the cube of x and fourth power of y.

6. w varies directly as the square root of x and inversely as y.

7. z varies jointly as x and u and inversely as the square of v.

8. S varies jointly as the square root of x, the square of y, and the cube of z and inversely as the square of u.

In Exercises 9–24, use the four-step procedure to solve for the variable requested.

9. x varies directly as y, and $x = 15$ when $y = 30$. Find x when $y = 28$.

10. y varies directly as x, and $y = 3$ when $x = 2$. Find y when $x = 7$.

11. s varies directly as the square of t, and $s = 64$ when $t = 2$. Find s when $t = 5$.

12. y varies directly as the cube of x, and $y = 270$ when $x = 3$. Find x when $y = 80$.

13. r varies inversely as u, and $r = 3$ when $u = 11$. Find r when $u = \dfrac{1}{3}$.

14. y varies inversely as z, and $y = 24$ when $z = \dfrac{1}{6}$. Find y when $z = 1$.

15. B varies inversely with the cube of A. If $B = 1$ when $A = 2$, find B when $A = 4$.

16. y varies inversely with the cube root of x, and $y = 10$ when $x = 2$. Find x when $y = 40$.

17. z varies jointly as x and y, and $z = 42$ when $x = 2$ and $y = 3$. Find y when $z = 56$ and $x = 2$.

18. m varies directly as q and inversely as p, and $m = \dfrac{1}{2}$ when $p = 26$ and $q = 13$. Find m when $p = 14$ and $q = 7$.

19. z varies directly as the square of x, and $z = 32$ when $x = 4$. Find z when $x = 5$.

20. u varies inversely as the cube of t, and $u = 9$ when $t = 2$. Find u when $t = 6$.

21. P varies jointly as T and the square of Q, and $P = 36$ when $T = 17$ and $Q = 6$. Find P when $T = 4$ and $Q = 9$.

22. a varies jointly as b and the square root of c, and $a = 9$ when $b = 13$ and $c = 81$. Find a when $b = 5$ and $c = 9$.

23. z varies directly as the square root of x and inversely as the square of y, and $z = 24$ when $x = 16$ and $y = 3$. Find x when $z = 27$ and $y = 2$.

24. z varies jointly as u and the cube of v and inversely as the square of w; $z = 9$ when $u = 4$, $v = 3$, and $w = 2$. Find w when $u = 27$, $v = 2$, and $z = 8$.

In Exercises 25–28, solve for the variable requested **without determining the constant of variation, k.** Use the fact that if

$$x_1 = ky_1 \text{ and } x_2 = ky_2, \text{ then } \frac{x_1}{y_1} = k = \frac{x_2}{y_2} \text{ so that } \frac{x_1}{y_1} = \frac{x_2}{y_2}.$$

25. Assuming that y is proportional to x and $y = 12$ when $x = 16$, find y when $x = 8$.

26. Assuming that z varies directly as w and $z = 17$ when $w = 22$, find z when $w = 110$.

27. Assuming that y is directly proportional to x and $y = 100$ for the value x_0 of x, find y when x_0 is doubled—that is, when $x = 2x_0$.

28. Assuming that x varies directly as the square root of y and $x = 2$ when $y = 9$, find y when $x = 3$.

B EXERCISES Applying the Concepts

29. Hubble constant. The American astronomer Edwin Powell Hubble (1889–1953) is renowned for having determined that there are other galaxies in the universe beyond the Milky Way. In 1929, he stated that the galaxies observed at a particular time recede from each other at a speed that is directly proportional to the distance between them. The constant of proportionality is denoted by the letter H. Write Hubble's statement in the form of an equation.

30. Malthusian doctrine. The British scientist Thomas Robert Malthus (1766–1834) was a pioneer of population science and economics. In 1798, he stated that populations grow faster than the means that can sustain them. One of his arguments was that the rate of change, R, of a given population is directly proportional to the size P of the population. Write the equation that describes Malthus's argument.

31. Converting units of length. Use the fact that 1 foot $\approx$ 30.5 centimeters.
   a. Write an equation that expresses the fact that a length measured in centimeters is directly proportional to the length measured in feet.

b. Convert the following measurements into centimeters.
   (i) 8 feet   **(ii)** 5 feet 4 inches
c. Convert the following measurements into feet.
   (i) 57 centimeters   **(ii)** 1 meter 24 centimeters

32. Converting weight. Use the fact that 1 kilogram $\approx$ 2.20 pounds.
   a. Write an equation that expresses the fact that a weight measured in pounds is directly proportional to the weight measured in kilograms.
   b. Convert the following measurements to pounds.
      (i) 125 grams
      (ii) 4 kilograms
      (iii) 2.4 kilograms
   c. Convert the following measurements to kilograms.
      (i) 27 pounds   **(ii)** 160 pounds

33. Protein from soybeans. The quantity P of protein obtained from dried soybeans is directly proportional to the quantity Q of dried soybeans used. If 20 grams of dried soybeans produces 7 grams of protein, how much protein can be produced from 100 grams of soybeans?

34. Wages. A person's weekly wages W are directly proportional to the number of hours h worked per week. If Samantha earned $600 for a 40-hour week, how much would she earn if she worked only 25 hours a particular week? What does the constant of proportionality mean here?

35. Physics. The distance d that an object falls varies directly as the square of the time t during which it is falling. If an object falls 64 feet in 2 seconds, how long will it take the object to fall 9 feet?

36. Physics. *Hooke's Law* states that the force F required to stretch a spring by x units is directly proportional to x. Assuming that a force of 10 pounds stretches a spring by 4 inches, find the force required to stretch a spring by 6 inches.

37. Chemistry. *Boyle's Law* states that at a constant temperature, the pressure P of a compressed gas is inversely proportional to its volume V. If the pressure is 20 pounds per square inch when the volume of the gas is 300 cubic inches, what is the pressure when the gas is compressed to 100 cubic inches?

38. Chemistry. In the Kelvin temperature scale, the lowest possible temperature (called absolute zero) is 0 K, where K denotes degrees Kelvin. The relationship between Kelvin temperature (T_K) and Celsius temperature (T_C) is given by $T_K = T_C + 273$.

The pressure P exerted by a gas varies directly as its temperature T_K and inversely as its volume V. Assume that at a temperature of 260 K, a gas occupies 13 cubic inches at a pressure of 36 pounds per square inch.

a. Find the volume of the gas when the temperature is 300 K and the pressure is 40 pounds per square inch.

b. Find the pressure when the temperature is 280 K and the volume is 39 cubic inches.

39. Weight. The weight of an object varies inversely as the square of the object's distance from the center of Earth. The radius of Earth is 3960 miles.

a. If an astronaut weighs 120 pounds on the surface of Earth, how much does she weigh 6000 miles above the surface of Earth?

b. If a miner weighs 200 pounds on the surface of Earth, how much does he weigh 10 miles below the surface of Earth?

40. Making a profit. Suppose you wanted to make a profit buying gold by weight at one altitude and selling at another altitude for the same price per unit weight. Should you buy or sell at the higher altitude? (Use the information from Exercise 39.)

In Exercises 41 and 42, use Newton's Law of Universal Gravitation. (See Example 6.)

41. Gravity on the moon. The mass of the moon is about 7.4×10^{22} kilograms, and its radius is about 1740 kilometers. How much is the acceleration due to gravity on the surface of the moon?

42. Gravity on the sun. The mass of the sun is about 2×10^{30} kilograms, and its radius is 696,000 kilometers. How much is the acceleration due to gravity on the surface of the sun?

43. Illumination. The intensity I of illumination from a light source is inversely proportional to the square of the distance d from the source. Suppose the intensity is 320 candlepower at a distance of 10 feet from a light source.

a. What is the intensity at 5 feet from the source?

b. How far away from the source will the intensity be 400 candlepower?

44. Speed and skid marks. Police estimate that the speed s of a car in miles per hour varies directly as the square root of d, where d in feet is the length of the skid marks left by a car traveling on a dry concrete pavement. A car traveling 48 miles per hour leaves skid marks of 96 feet.

a. Write an equation relating s and d.

b. Use the equation in part (a) to estimate the speed of a car whose skid marks stretched (i) 60 feet, (ii) 150 feet, and (iii) 200 feet.

c. Suppose you are driving 70 miles per hour and slam on your brakes. How long will your skid marks be?

45. Simple pendulum. *Periodic motion* is motion that repeats itself over successive equal intervals of time. The time required for one complete repetition of the motion is called the *period*. The period of a simple pendulum varies directly as the square root of its length. What is the effect on the length if the period is doubled?

(*Note:* The amplitude of a pendulum has no effect on its period. This is what makes pendulums such good timekeepers. Because they invariably lose energy due to friction, their amplitude decreases but their period remains constant.)

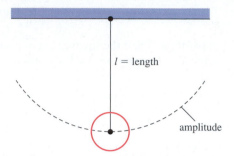

46. Biology. The volume V of a lung is directly proportional to its internal surface area A. A lung with volume 400 cubic centimeters from a certain species has an average internal surface area of 100 square centimeters. Find the volume of a lung of a member of this species assuming that the lung's internal surface area is 120 square centimeters.

47. Horsepower. The horsepower H of an automobile engine varies directly as the square of the piston radius R and the number N of pistons.

a. Write the given information in equation form.

b. What is the effect on the horsepower if the piston radius is doubled?

c. What is the effect on the horsepower if the number of pistons is doubled?

d. What is the effect on the horsepower if the radius of the pistons is cut in half and the number of pistons is doubled?

48. Safe load. The safe load that a rectangular beam can support varies jointly as the width and square of the depth of the beam and inversely as its length. A beam 4 inches wide, 6 inches deep, and 25 feet long can support a safe load of 576 pounds. Find the safe load for a beam that is of the same material but is 6 inches wide, 10 inches deep, and 20 feet long.

C EXERCISES Beyond the Basics

49. Energy from a windmill. The energy E from a windmill varies jointly as the square of the length l of the blades and the cube of the wind velocity v.

a. Express the given information as an equation with k as the constant of variation.

b. Assuming that blades of length 10 feet and wind velocity 8 miles per hour generate 1920 watts of electric power, find k.

c. How much electric power would be generated if the blades were 8 feet long and the wind velocity was 25 miles per hour?

d. If the velocity of the wind doubles, what happens to E?

e. If the length of blades doubles, what happens to E?

f. What happens to E if both the length of the blades and the wind velocity are doubled?

50. Intensity of light. The intensity of light, I_d, at a distance d from the source of light varies directly as the intensity I of the source and inversely as d^2. It is known that at a distance of 2 meters, a 100-watt bulb produces an intensity of approximately 2 watts per square meter.
  a. Find the constant of variation, k.
  b. Suppose a 200-watt bulb is located on a wall 2 meters above the floor. What is the intensity of light at a point A that is 3 meters from the wall?

  c. What is the intensity of illumination at the point A in the figure if the bulb is raised by 1 meter?

51. Metabolic rate. Metabolism is the sum total of all physical and chemical changes that take place within an organism. According to the laws of thermodynamics, all of these changes will ultimately release heat; so the metabolic rate is a measure of heat production by an animal. Biologists have found that the normal resting metabolic rate of a mammal is directly proportional to the $\frac{3}{4}$ power of its body weight.

  The resting metabolic rate of a person weighing 75 kilograms is 75 watts.
  a. Find the constant of proportionality, k.
  b. Estimate the resting metabolic rate of a brown bear weighing 450 kilograms.
  c. What is the effect on metabolic rate if the body weight is multiplied by 4?
  d. What is the approximate weight of an animal with a metabolic rate of 250 watts?

52. Comparing gravitational forces. The masses of the sun, earth, and moon are 2×10^{30} kilograms, 6×10^{24} kilograms, and 7.4×10^{22} kilograms, respectively. The earth–sun distance is about 400 times the earth–moon distance. Use Newton's Law of Universal Gravitation to compare the gravitational attraction between the sun and earth with that between the earth and moon.

53. Kepler's Third Law. Suppose an object of mass M_1 orbits around an object of mass M_2. Let r be the average distance in meters between the centers of the two objects and let T be the *orbital period* (the time in seconds the object completes one orbit). Kepler's Third Law states that T^2 is directly proportional to r^3 and inversely proportional to $M_1 + M_2$. The constant of proportionality is $\frac{4\pi^2}{G}$.
  a. Write the equation that expresses Kepler's Third Law.
  b. Earth orbits the sun once each year at a distance of about 1.5×10^8 kilometers. Find the mass of the sun. Use these estimates:
  Mass of Earth + Mass of sun $\approx$ Mass of sun,
  $G = 6.67 \times 10^{-11}$, 1 year = 3.15×10^7 seconds

54. Use Kepler's Third Law. The moon orbits Earth in about 27.3 days at an average distance of about 384,000 kilometers. Find the mass of Earth. [*Hint:* Convert the orbital period, defined in Exercise 53, to seconds; also use mass of earth + mass of moon $\approx$ mass of earth.]

55. Spreading a rumor. Let P be the population of a community. The rate R (per day) at which a rumor spreads in the community is jointly proportional to the number N of people who have already heard the rumor and the number $(P - N)$ of people who have not yet heard the rumor.
  A rumor about the president of a college is spread in his college community of 10,000 people. Five days after the rumor started, 1000 people had heard it and it was spreading at the rate of 45 additional people per day.
  a. Write an equation relating R, P, and N.
  b. Find the constant k of variation.
  c. Find the rate at which the rumor was spreading when one-half of the college community had heard it.
  d. How many people had heard the rumor when it was spreading at the rate of 100 people per day?

Critical Thinking

56. Electricity. The current I in an electric circuit varies directly as the voltage V and inversely as the resistance R. If the resistance is increased by 20%, what percent increase must occur in the voltage to increase the current by 30%?

57. Precious stones. The value of a precious stone is proportional to the square of its weight.
  a. Calculate the loss incurred by cutting a diamond worth $1000 into two pieces whose weights are in the ratio 2:3.
  b. A precious stone worth $25,000 is accidentally dropped and broken into three pieces, the weights of which are in the ratio 5:9:11. Calculate the loss incurred due to breakage.
  c. A diamond breaks into five pieces, the weights of which are in the ratio 1:2:3:4:5. Assuming that the resulting loss is $85,000, find the value of the original diamond. Also calculate the value of a diamond whose weight is twice that of the original diamond.

58. Bus service. The profit earned in running a bus service is jointly proportional to the distance and the number of passengers over a certain fixed number. The profit is $80 when 30 passengers are carried a distance of 40 km and $180 when 35 passengers are carried 60 km. What is the minimum number of passengers that results in no loss?

59. Weight of a sphere. The weight of a sphere is directly proportional to the cube of its radius. A metal sphere has a hollow space about its center in the form of a concentric sphere, and its weight is $\frac{7}{8}$ times the weight of a solid sphere of the same radius and material. Find the ratio of the inner to the outer radius of the hollow sphere.

60. Train speed. A locomotive engine can go 24 miles per hour, and its speed is reduced by a quantity that varies directly as the square root of the number of cars it pulls. Pulling four cars, its speed is 20 miles per hour. Find the greatest number of cars the engine can pull.

SUMMARY ■ Definitions, Concepts, and Formulas

3.1 Quadratic Functions

i. A **quadratic function** f is a function of the form

$$f(x) = ax^2 + bx + c, a \neq 0.$$

ii. The **standard form** of a quadratic function is

$$f(x) = a(x - h)^2 + k, a \neq 0.$$

iii. The graph of a quadratic function is a transformation of the graph of $y = x^2$.

iv. The graph of a quadratic function is a parabola with vertex

$$(h, k) = \left(-\frac{b}{2a}, f\left(-\frac{b}{2a} \right) \right).$$

v. The maximum (if $a < 0$) or minimum (if $a > 0$) value of a quadratic function $f(x) = ax^2 + bx + c$ occurs at the y-coordinate of the vertex of the parabola.

3.2 Polynomial Functions

i. A function f of the form $f(x) = a_n x^n + a_{n-1} x^{n-1} + \cdots + a_1 x + a_0, a_n \neq 0$ is a polynomial function of degree n.

ii. The graph of a polynomial function is smooth and continuous.

iii. The end behavior of the graph of a polynomial function depends upon the sign of the leading coefficient and the degree (even-odd) of the polynomial.

iv. A real number c is a **zero** of a function f if $f(c) = 0$. Geometrically, c is an x-intercept of the graph of $y = f(x)$.

v. If in the factorization of a polynomial function $f(x)$ the factor $(x - a)$ occurs exactly m times, then a is a zero of **multiplicity** m. If m is odd, the graph of $y = f(x)$ crosses the x-axis at a; if m is even, the graph touches but does not cross the x-axis at a.

vi. If the degree of a polynomial function $f(x)$ is n, then $f(x)$ has, at most, n real zeros and the graph of $f(x)$ has at most $(n - 1)$ turning points.

vii. **Intermediate Value Theorem:** Let $f(x)$ be a polynomial function and a and b be two numbers such that $a < b$. If $f(a)$ and $f(b)$ have opposite signs, then there is at least one number c, with $a < c < b$, for which $f(c) = 0$.

viii. See page 321 for graphing a polynomial function.

3.3 Dividing Polynomials

i. **Division Algorithm:** If a polynomial $F(x)$ is divided by a polynomial $D(x) \neq 0$, there are unique polynomials $Q(x)$ and $R(x)$ such that $F(x) = D(x)Q(x) + R(x)$, where either $R(x) = 0$ or deg $R(x) <$ deg $D(x)$. In words, "The dividend equals the product of the divisor and the quotient plus the remainder."

ii. Synthetic division is a shortcut for dividing a polynomial $F(x)$ by $x - a$.

iii. **Remainder Theorem:** If a polynomial $F(x)$ is divided by $(x - a)$, the remainder is $F(a)$.

iv. **Factor Theorem:** A polynomial function $F(x)$ has $(x - a)$ as a factor if and only if $F(a) = 0$.

3.4 The Real Zeros of a Polynomial Function

i. **Rational Zeros Theorem:** If $\dfrac{p}{q}$ is a rational zero in lowest terms for a polynomial function with integer coefficients, then p is a factor of the constant term and q is a factor of the leading coefficient.

ii. **Descartes's Rule of Signs.** Let $F(x)$ be a polynomial function with real coefficients.

    a. The number of positive zeros of F is equal to the number of variations of signs of $F(x)$ or is less than that number by an even integer.

    b. The number of negative zeros of F is equal to the number of variations of sign of $F(-x)$ or is less than that number by an even integer.

iii. **Rules for Bounds on the Zeros.** Suppose a polynomial $F(x)$ is synthetically divided by $x - k$.

    a. If $k > 0$ and each number in the last row is zero or positive, then k is an upper bound on the zeros of $F(x)$.

    b. If $k < 0$ and the numbers in the last row alternate in sign, then k is a lower bound on the zeros of $F(x)$.

3.5 The Complex Zeros of a Polynomial Function

i. **Fundamental Theorem of Algebra.** An nth-degree polynomial equation has at least one complex zero.

ii. **Factorization Theorem for Polynomials.** If $P(x)$ is a polynomial of degree $n \geq 1$, it can be factored into n (not necessarily distinct) linear factors of the form $P(x) = a(x - r_1)(x - r_2) \cdots (x - r_n)$, where a, $r_1, r_2, \ldots, r_n$ are complex numbers.

iii. **Number of Zeros Theorem.** A polynomial of degree n has exactly n complex zeros provided that a zero of multiplicity k is counted k times.

iv. **Conjugate Pairs Theorem.** If $a + bi$ is a zero of the polynomial function P (with real coefficients), then $a - bi$ is also a zero of P.

3.6 Rational Functions

i. A function $F(x) = \dfrac{N(x)}{D(x)}$, where $N(x)$ and $D(x)$ are polynomials and $D(x) \neq 0$, is called a rational function. The domain of $F(x)$ is the set of all real numbers except the real zeros of $D(x)$.

ii. The line $x = a$ is a vertical asymptote of the graph of f if $|f(x)| \to \infty$ as $x \to a^+$ or as $x \to a^-$.

iii. If $\dfrac{N(x)}{D(x)}$ is in lowest terms, then the graph of $F(x)$ has vertical asymptotes at the real zeros of $D(x)$.

iv. The line $y = k$ is a horizontal asymptote of the graph of f if $f(x) \to k$ as $x \to \infty$ or as $x \to -\infty$.

v. A procedure for graphing rational functions is given on page 366.

3.7 Polynomial and Rational Inequalities

Test numbers can be used to solve polynomial or rational inequalities. The inequality is rearranged so that a polynomial or rational expression is on the left side and the number 0 is on the right side.

3.8 Variation

k is a nonzero constant called the constant of variation.

Variation	Equation
y varies directly with x.	$y = kx$
y varies with the nth power of x.	$y = kx^n$
y varies inversely with x.	$y = \dfrac{k}{x}$
y varies inversely with the nth power of x.	$y = \dfrac{k}{x^n}$
z varies jointly with the nth power of x and the mth power of y.	$z = kx^n y^m$
z varies directly with the nth power of x and inversely with the mth power of y.	$z = \dfrac{kx^n}{y^m}$

REVIEW EXERCISES

Basic Skills and Concepts

In Exercises 1–10, graph each quadratic function by finding (i) whether the parabola opens up or down and by finding (ii) its vertex, (iii) its axis, (iv) its x-intercepts, (v) its y-intercept, and (vi) the intervals over which the function is increasing and decreasing.

1. $y = (x - 1)^2 + 2$

2. $y = (x + 2)^2 - 3$

3. $y = -2(x - 3)^2 + 4$

4. $y = -\dfrac{1}{2}(x + 1)^2 + 2$

5. $y = -2x^2 + 3$

6. $y = 2x^2 + 4x - 1$

7. $y = 2x^2 - 4x + 3$

8. $y = -2x^2 - x + 3$

9. $y = 3x^2 - 2x + 1$

10. $y = 3x^2 - 5x + 4$

In Exercises 11–14, determine whether the given quadratic function has a maximum or a minimum value and then find that value.

11. $f(x) = 3 - 4x + x^2$

12. $f(x) = 8x - 4x^2 - 3$

13. $f(x) = -2x^2 - 3x + 2$

14. $f(x) = \dfrac{1}{2}x^2 - \dfrac{3}{4}x + 2$

In Exercises 15–18, graph each polynomial function by using transformations on the appropriate function $y = x^n$.

15. $f(x) = (x + 1)^3 - 2$

16. $f(x) = (x + 1)^4 + 2$

17. $f(x) = (1 - x)^3 + 1$

18. $f(x) = x^4 + 3$

In Exercises 19–24, for each polynomial function f,

(i) determine the end behavior of f.

(ii) determine the zeros of f. State the multiplicity of each zero. Determine whether the graph of f crosses or only touches the axis at each x-intercept.

(iii) find the x- and y-intercepts of the graph of f.

(iv) use test numbers to find the intervals over which the graph of f is above or below the x-axis.

(v) find any symmetry.

(vi) sketch the graph of $y = f(x)$.

19. $f(x) = x(x - 1)(x + 2)$

20. $f(x) = x^3 - x$

21. $f(x) = -x^2(x - 1)^2$

22. $f(x) = -x^3(x - 2)^2$

23. $f(x) = -x^2(x^2 - 1)$

24. $f(x) = -(x - 1)^2(x^2 + 1)$

In Exercises 25–28, divide by using long division.

25. $\dfrac{6x^2 + 5x - 13}{3x - 2}$ 26. $\dfrac{8x^2 - 14x + 15}{2x - 3}$

27. $\dfrac{8x^4 - 4x^3 + 2x^2 - 7x + 165}{x + 1}$

28. $\dfrac{x^3 - 3x^2 + 4x + 7}{x^2 - 2x + 6}$

In Exercises 29–32, divide by using synthetic division.

29. $\dfrac{x^3 - 12x + 3}{x - 3}$ 30. $\dfrac{-4x^3 + 3x^2 - 5x}{x - 6}$

31. $\dfrac{2x^4 - 3x^3 + 5x^2 - 7x + 165}{x + 1}$

32. $\dfrac{3x^5 - 2x^4 + x^2 - 16x - 132}{x + 2}$

In Exercises 33–36, a polynomial function $f(x)$ and a constant c are given. Find $f(c)$ by (i) evaluating the function and (ii) using synthetic division and the Remainder Theorem.

33. $f(x) = x^3 - 3x^2 + 11x - 29; c = 2$

34. $f(x) = 2x^3 + x^2 - 15x - 2; c = -2$

35. $f(x) = x^4 - 2x^2 - 5x + 10; c = -3$

36. $f(x) = x^5 + 2; c = 1$

In Exercises 37–40, a polynomial function $f(x)$ and a constant c are given. Use synthetic division to show that c is a zero of $f(x)$. Use the result to final all zeros of $f(x)$.

37. $f(x) = x^3 - 7x^2 + 14x - 8; c = 2$

38. $f(x) = 2x^3 - 3x^2 - 12x + 4; c = -2$

39. $f(x) = 3x^3 + 14x^2 + 13x - 6; c = \dfrac{1}{3}$

40. $f(x) = 4x^3 + 19x^2 - 13x + 2; c = \dfrac{1}{4}$

In Exercises 41 and 42, use the Rational Zeros Theorem to list all possible rational zeros of $f(x)$.

41. $f(x) = x^4 + 3x^3 - x^2 - 9x - 6$

42. $f(x) = 9x^3 - 36x^2 - 4x + 16$

In Exercises 43–46, use Descartes's Rule of Signs and the Rational Zeros Theorem to find all real zeros of each polynomial function.

43. $f(x) = 5x^3 + 11x^2 + 2x$

44. $f(x) = x^3 + 2x^2 - 5x - 6$

45. $f(x) = x^3 + 3x^2 - 4x - 12$

46. $f(x) = 2x^3 - 9x^2 + 12x - 5$

47. $x^3 - 4x^2 - 5x + 14$

48. $x^3 - 5x^2 + 3x + 1$

49. $2x^3 - 5x^2 - 2x + 2$

50. $3x^3 - 29x^2 + 29x + 13$

In Exercises 51–56, find all of the zeros of $f(x)$, real and nonreal.

51. $f(x) = x^3 - 7x + 6$; one zero is 2.

52. $f(x) = x^4 + x^3 - 3x^2 - x + 2$; 1 is a zero of multiplicity 2.

53. $f(x) = x^4 - 2x^3 + 6x^2 - 18x - 27$; two zeros are -1 and 3.

54. $f(x) = 4x^3 - 19x^2 + 32x - 15$; one zero is $2 - i$.

55. $f(x) = x^4 + 2x^3 + 9x^2 + 8x + 20$; one zero is $-1 + 2i$.

56. $f(x) = x^5 - 7x^4 + 24x^3 - 32x^2 + 64$; $2 + 2i$ is a zero of multiplicity 2.

In Exercises 57–68, solve each equation in the complex number system.

57. $x^3 - x^2 - 4x + 4 = 0$

58. $2x^3 + x^2 - 12x - 6 = 0$

59. $4x^3 - 7x - 3 = 0$

60. $x^3 + x^2 - 8x - 6 = 0$

61. $x^3 - 8x^2 + 23x - 22 = 0$

62. $x^3 - 3x^2 + 8x + 12 = 0$

63. $3x^3 - 5x^2 + 16x + 6 = 0$

64. $2x^3 - 9x^2 + 18x - 7 = 0$

65. $x^4 - x^3 - x^2 - x - 2 = 0$

66. $x^4 - x^3 - 13x^2 + x + 12 = 0$

67. $2x^4 - x^3 - 2x^2 + 13x - 6 = 0$

68. $3x^4 - 14x^3 + 28x^2 - 10x - 7 = 0$

In Exercises 69 and 70, show that the given equation has no rational roots.

69. $x^3 + 13x^2 - 6x - 2 = 0$

70. $3x^4 - 9x^3 - 2x^2 - 15x - 5 = 0$

In Exercises 71 and 72, use the Intermediate Value Theorem to find the value of the real root between 1 and 2 of each equation to two decimal places.

71. $x^3 + 6x^2 - 28 = 0$ 72. $x^3 + 3x^2 - 3x - 7 = 0$

In Exercises 73–80, graph each rational function by following the six-step procedure outlined in Section 3.6.

73. $f(x) = 1 + \dfrac{1}{x}$ 74. $f(x) = \dfrac{2 - x}{x}$

75. $f(x) = \dfrac{x}{x^2 - 1}$ 76. $f(x) = \dfrac{x^2 - 9}{x^2 - 4}$

77. $f(x) = \dfrac{x^3}{x^2 - 9}$ 78. $f(x) = \dfrac{x + 1}{x^2 - 2x - 8}$

79. $f(x) = \dfrac{x^4}{x^2 - 4}$ 80. $f(x) = \dfrac{x^2 + x - 6}{x^2 - x - 12}$

In Exercises 81–96, solve each inequality.

81. $(x - 1)(x + 2) \geq 0$ 82. $(x + 3)(x + 5) \leq 0$

83. $2x^2 - 5x - 3 \leq 0$ 84. $3x^2 + 5x - 2 \geq 0$

85. $(x + 1)(x - 1)(x + 3) > 0$

86. $(x + 2)(2x - 1)(x - 2) > 0$

87. $x^3 - x^2 - 8x + 8 < 0$ 88. $x^3 + x^2 - 5x - 5 < 0$

89. $x^3 - x^2 - 4x - 6 > 0$ 90. $x^3 + 6x^2 + 14x + 12 > 0$

91. $\dfrac{3x}{x - 1} \geq 1$ 92. $\dfrac{2x}{x + 3} \geq 3$

93. $\dfrac{(x - 5)(x + 2)}{x(x + 4)} \geq 0$ 94. $\dfrac{(x - 1)(x + 3)}{(x + 2)(x + 5)} \geq 0$

95. $\dfrac{x^2 - 4x + 3}{x^2 + 6x + 8} \leq 0$ 96. $\dfrac{x^2 + 6x - 7}{x^2 + 2x - 15} \leq 0$

Applying the Concepts

97. **Variation.** Assuming that y varies directly as x and $y = 12$ when $x = 4$, find y when $x = 5$.

98. **Variation.** Assuming that p varies inversely as q and $p = 4$ when $q = 3$, find p when $q = 4$.

99. **Variation.** Assuming that s varies directly as the square of t and $s = 20$ when $t = 2$, find s when $t = 3$.

100. **Variation.** Assuming that y varies inversely as x^2 and $y = 3$ when $x = 8$, find x when $y = 12$.

101. **Missile path.** A missile fired from the origin of a coordinate system follows a path described by the equation $y = -\dfrac{1}{10}x^2 + 20x$, where the x-axis is at ground level. Sketch the missile's path and determine what its maximum altitude is and where it hits the ground.

102. **Minimizing area.** Suppose a wire 20 centimeters long is to be cut into two pieces, each of which will be formed into a square. Find the size of each piece that minimizes the total area.

103. A farmer wants to fence off three identical adjoining rectangular pens, each 400 square feet in area. (See the figure.) What should the width and length of each pen be so that the least amount of fence is used?

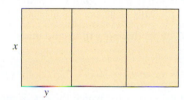

104. Suppose the outer boundary of the pens in Exercise 103 requires heavy fence that costs $5 per foot and two internal partitions cost $3 per foot. What dimensions x and y will minimize the cost?

105. **Electric circuit.** In the circuit shown in the figure, the voltage $V = 100$ volts and the resistance $R = 50$ ohms. We want to determine the size of the remaining resistor (x ohms). The power absorbed by the circuit is given by

$$p(x) = \dfrac{V^2 x}{(R + x)^2}.$$

a. Graph the function $y = p(x)$.

b. Use a graphing calculator to find the value of x that maximizes the power absorbed.

106. **Maximizing area.** A sheet of paper for a poster is 18 square feet in area. The margins at the top and bottom are 9 inches each, and the margin on each side is 6 inches. What should the dimensions of the paper be if the printed area is to be a maximum?

107. **Maximizing profit.** A manufacturer makes and sells printers to retailers at $24 per unit. The total daily cost C in dollars of producing x printers is given by

$$C(x) = 150 + 3.9x + \dfrac{3}{1000}x^2.$$

a. Write the profit P as a function of x.

b. Find the number of printers the manufacturer should produce and sell to achieve maximum profit.

c. Find the average cost $\overline{C}(x) = \dfrac{C(x)}{x}$. Graph $y = \overline{C}(x)$.

108. **Meteorology.** The function $p = \dfrac{69.1}{a + 2.3}$ relates the atmospheric pressure p in inches of mercury to the altitude a in miles from the surface of Earth.

a. Find the pressure on Mount Kilimanjaro at an altitude of 19,340 feet.

b. Is there an altitude at which the pressure is 0?

109. **Wages.** An employee's wages are directly proportional to the time he or she has worked. Sam earned $280 for 40 hours. How much would Sam earn if he worked 35 hours?

110. **Car's stopping distance.** The distance required for a car to come to a stop after its brakes are applied is directly proportional to the square of its speed. If the stopping distance for a car traveling 30 miles per hour is 25 feet, what is the stopping distance for a car traveling 66 miles per hour?

111. **Illumination.** The amount of illumination from a source of light varies directly as the intensity of the source and inversely as the square of the distance from the source. At what distance from a light source of intensity 300 candlepower will the illumination be one-half the illumination 6 inches from the source?

112. **Chemistry.** *Charles's Law* states that at a constant pressure, the volume V of a gas is directly proportional to its temperature T (in Kelvin degrees). If a bicycle tube is filled with 1.2 cubic feet of air at a temperature of 295 K, what will the volume of the air in the tube be if the temperature rises to 310 K while the pressure stays the same?

113. Electric circuits. The current I (measured in amperes) in an electrical circuit varies inversely as the resistance R (measured in ohms) when the voltage is held constant. The current in a certain circuit is 30 amperes when the resistance is 300 ohms.
   a. Find the current in the circuit if the resistance is decreased to 250 ohms.
   b. What resistance will yield a current of 60 amperes?

114. Safe load. The safe load that a circular column can support varies directly as the square root of its radius and inversely as the square of its length. A pillar with radius 4 inches and length 12 feet can safely support a 20-ton load. Find the load that a pillar of the same material with diameter 6 inches and length 10 feet can safely support.

115. Spread of disease. An infectious cold virus spreads in a community at a rate R (per day) that is jointly proportional to the number of people who are infected with the virus and the number of people in the community who are not infected yet. After the tenth day of the start of a certain infection, 15% of the total population of 20,000 people of Pollutville had been infected and the virus was spreading at the rate of 255 people per day.
   a. Find the constant of proportionality, k.
   b. At what rate is the disease spreading when one-half of the population is infected?
   c. Find the number of people infected when the rate of infection reached 95 people per day.

116. Coulomb's Law. The electric charge is measured in coulombs. (The charges on an electron and a proton, which are equal and opposite, are approximately 1.602×10^{-19} coulomb.) Coulomb's Law states that the force F between two particles is jointly proportional to their charges q_1 and q_2 and inversely proportional to the square of the distance between the two particles. Two charges are acted upon by a repulsive force of 96 units. What is the force if the distance between the particles is quadrupled?

PRACTICE TEST A

1. Find the x-intercepts of the graph of $f(x) = x^2 - 6x + 2$.

2. Graph $f(x) = 3 - (x + 2)^2$.

3. Find the vertex of the parabola described by $y = -7x^2 + 14x + 3$.

4. Find the domain of the function $f(x) = \dfrac{x^2 - 1}{x^2 + 3x - 4}$.

5. Find the quotient and remainder of $\dfrac{x^3 - 2x^2 - 5x + 6}{x + 2}$.

6. Graph the polynomial function $P(x) = x^5 - 4x^3$.

7. Find all of the zeros of $f(x) = 2x^3 - 2x^2 - 8x + 8$ given that 2 is one of the zeros.

8. Use the Remainder Theorem to find the value $P(-2)$ of the polynomial $P(x) = x^4 + 5x^3 - 7x^2 + 9x + 17$.

9. Find all rational roots of the equation $x^3 - 5x^2 - 4x + 20 = 0$ and then find the irrational roots if there are any.

10. Find the zeros of the polynomial function $f(x) = x^4 + x^3 - 15x^2$.

11. For $P(x) = 2x^{18} - 5x^{13} + 6x^3 - 5x + 9$, list all possible rational zeros found by the Rational Zeros Test, but do not check to see which values are actually zeros.

12. Describe the end behavior of $f(x) = (x + 3)^3(x - 5)^2$.

13. Find the zeros and the multiplicity of each zero for $f(x) = (x^2 - 4)(x + 2)^2$.

14. Determine how many positive and how many negative real zeros the polynomial function $P(x) = 3x^6 + 2x^3 - 7x^2 + 8x$ can have.

15. Find the horizontal and the vertical asymptotes of the graph of
$$f(x) = \frac{2x^2 + 3}{x^2 - x - 20}.$$

16. Solve $x^4 - 5x^3 + 6x^2 > 0$.

17. Solve $\dfrac{(x - 3)(x + 5)}{(x - 5)(x - 1)} \le 0$.

18. Suppose that y is directly proportional to x and inversely proportional to the square of t.
   Assuming that $y = 6$ when $x = 8$ and $t = 2$, find y when $x = 12$ and $t = 3$.

19. The cost C of producing x thousand units of a product is given by
$$C = x^2 - 30x + 335 \text{ (dollars)}.$$
   Find the value of x for which the cost is minimum.

20. From a rectangular 8×17 piece of cardboard, four congruent squares with sides of length x are cut out, one at each corner. The sides can then be folded to form a box. Find the volume V of the box as a function of x.

PRACTICE TEST B

1. Find the x-intercepts of the graph of
$f(x) = x^2 + 5x + 3$.

   a. $\dfrac{-5 \pm i\sqrt{13}}{2}$ **b.** $\dfrac{-5 \pm \sqrt{13}}{2}$

   c. $\dfrac{-5 \pm i\sqrt{37}}{2}$ **d.** $\dfrac{-5 \pm \sqrt{37}}{2}$

2. Which is the graph of $f(x) = 4 - (x - 2)^2$?

     (a)   (b)

     (c)   (d)

3. Find the vertex of the parabola described by
$y = 6x^2 + 12x - 5$.

   a. $(-1, -11)$ **b.** $(1, -5)$

   c. $(-1, 13)$ **d.** $(1, 13)$

4. Which of the following is *not* in the domain of the
function $f(x) = \dfrac{x^2}{x^2 + x - 6}$?

   I. -3

   II. 0

   III. 2

   a. I and II   **b.** I and III

   c. II and III   **d.** I only

5. Find the quotient and remainder when $x^3 - 8x + 6$ is
divided by $x + 3$.

   a. $x^2 - 8; 2$ **b.** $x^2 - 8; 0$

   c. $x^2 - 3x + 1; x + 3$ **d.** $x^2 - 3x + 1; 3$

6. Which is the graph of the polynomial $P(x) = x^4 + 2x^3$?

     (a)   (b)

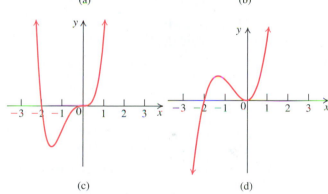

     (c)   (d)

7. Find all of the zeros of $f(x) = 3x^3 - 26x^2 + 61x - 30$
given that 3 is one of the zeros (that is, $f(3) = 0$).

   a. $3, -5, -\dfrac{2}{3}$ **b.** $3, 2, \dfrac{5}{3}$

   c. $3, 5, \dfrac{2}{3}$ **d.** $3, -2, \dfrac{5}{3}$

8. Use the Remainder Theorem to find the value $P(-3)$ of
the polynomial $P(x) = x^4 + 4x^3 + 7x^2 + 10x + 15$.

   a. 13   **b.** 15

   c. 21   **d.** 6

9. Find all rational roots of the equation
$-x^3 + x^2 + 8x - 12 = 0$ and then find the irrational
roots if there are any.

   a. -3 and 2   **b.** -3 and $\sqrt{2}$

   c. $-1, -2$, and 3   **d.** -3 and $\sqrt{3}$

10. Find the zeros of the polynomial function
$f(x) = x^3 + x^2 - 30x$.

   a. $x = -6, x = 5, x = 0$ **b.** $x = 0, x = -6$

   c. $x = 4, x = 5$ **d.** $x = 0, x = 4, x = 5$

11. For $P(x) = x^{30} - 4x^{25} + 6x^2 + 60$, list all possible
rational zeros found by the Rational Zeros Theorem, but
do not check to see which values are actually zeros.

   a. $\pm 2, \pm 3, \pm 4, \pm 5, \pm 6, \pm 8, \pm 10, \pm 12, \pm 15, \pm 20,$
      ± 30, and ± 60

b. $\pm 1, \pm 2, \pm 3, \pm 4, \pm 5, \pm 6, \pm 10, \pm 12, \pm 15, \pm 18, \pm 20,$
$\pm 24, \pm 30,$ and ± 60

c. $\pm 1, \pm 3, \pm 4, \pm 5, \pm 6, \pm 12, \pm 15, \pm 20, \pm 30,$ and ± 60

d. $\pm 1, \pm 2, \pm 3, \pm 4, \pm 5, \pm 6, \pm 10, \pm 12, \pm 15, \pm 20,$
$\pm 30,$ and ± 60

12. Which of the following correctly describes the end behavior of $f(x) = (x + 1)^2 (x - 2)^2$?

a. $\begin{cases} y \to \infty \text{ as } x \to -\infty \\ y \to -\infty \text{ as } x \to \infty \end{cases}$ b. $\begin{cases} y \to -\infty \text{ as } x \to -\infty \\ y \to -\infty \text{ as } x \to \infty \end{cases}$

c. $\begin{cases} y \to \infty \text{ as } x \to -\infty \\ y \to \infty \text{ as } x \to \infty \end{cases}$ d. $\begin{cases} y \to -\infty \text{ as } x \to -\infty \\ y \to \infty \text{ as } x \to \infty \end{cases}$

13. Find the zeros and the multiplicity of each zero for $f(x) = (x^2 - 1)(x + 1)^2$.

a. $\begin{cases} \text{zero } 1, \quad \text{multiplicity } 1 \\ \text{zero } -1, \quad \text{multiplicity } 2 \end{cases}$

b. $\begin{cases} \text{zero } 1, \quad \text{multiplicity } 1 \\ \text{zero } -1, \quad \text{multiplicity } 3 \end{cases}$

c. $\begin{cases} \text{zero } 1, \quad \text{multiplicity } 2 \\ \text{zero } -1, \quad \text{multiplicity } 2 \end{cases}$

d. $\begin{cases} \text{zero } i, \quad \text{multiplicity } 2 \\ \text{zero } -1, \quad \text{multiplicity } 2 \end{cases}$

14. Determine how many positive and how many negative real zeros the polynomial $P(x) = x^5 - 4x^3 - x^2 + 6x - 3$ can have.

a. positive 3, negative 2
b. positive 2 or 0, negative 2 or 0
c. positive 3 or 1, negative 2 or 0
d. positive 2 or 0, negative 3 or 1

15. The horizontal and the vertical asymptotes of the graph of $f(x) = \dfrac{x^2 + x - 2}{x^2 + x - 12}$ are

a. $x = -2, y = 3, y = -4.$
b. $x = 1, y = 3, y = -4.$
c. $y = 1, x = 3, x = -4.$
d. $y = -2, x = 3, x = -4.$

16. Solve $x^3 - 4x^2 - 12x \le 0$.
a. $[-2, 0] \cup [6, \infty)$ b. $(-\infty, -2] \cup [0, 6]$
c. $(-\infty, -2) \cup (0, 6)$ d. $(-\infty, -2] \cup [0, 6)$

17. Solve $\dfrac{2x - 1}{x - 3} \ge 1$.

a. $(-2, 3)$ b. $(-\infty, -2) \cup (3, \infty)$
c. $(-\infty, -2] \cup (3, \infty)$ d. $(-\infty, -2] \cup [3, \infty)$

18. Suppose S is directly proportional to the square of t and inversely proportional to the cube of x.
If $S = 27$ when $t = 3$ and $x = 1$, find S when $t = 6$ and $x = 3$.

a. 54 b. 4 c. $\dfrac{4}{3}$ d. 9

19. The cost C of producing x thousand units of a product is given by

$$C = x^2 - 24x + 319 \text{ (dollars)}.$$

Find the value of x for which the cost is minimum.
a. 319 b. 12 c. 175 d. 24

20. From a rectangular 10×12 piece of cardboard, four congruent squares with sides of length x are cut out, one at each corner. The sides can then be folded to form a box. Find the volume V of the box as a function of x.
a. $V = 2x(10 - x)(12 - x)$
b. $V = 2x(10 - 2x)(12 - 2x)$
c. $V = x(10 - x)(12 - x)$
d. $V = x(10 - 2x)(12 - 2x)$

CUMULATIVE REVIEW EXERCISES ■ Chapters P–3

1. Find the distance between the points $P(-1, 3)$ and $Q(2, 5)$.

2. Find the midpoint of the line segment joining $P(2, -5)$ and $Q(-8, -3)$.

3. Find the x- and y-intercepts of the graph of the equation $y = x^2 - 2x - 8$ and sketch the graph.

In Exercises 4 and 5, find the slope and intercepts of the line and sketch the graph.

4. $x + 3y - 6 = 0$ 5. $x = 2y - 6$

6. Find an equation in standard form of the circle with center $(2, -3)$ and radius 4.

7. Find the center and radius of the circle with equation

$$x^2 + y^2 + 2x - 4y - 4 = 0.$$

In Exercises 8 and 9, find the slope–intercept form of the line satisfying the given condition.

8. The line has slope 3 and passes through $(1, -2)$.

9. The line is parallel to $2x + 3y = 5$ and passes through $(1, 3)$.

In Exercises 10 and 11, find the domain of each function.

10. $f(x) = \dfrac{1}{2x + 3}$ 11. $p(x) = \dfrac{1}{\sqrt{4 - 2x}}$

12. Let $f(x) = x^2 - 2x + 3$. Find $f(-2), f(3), f(x + h)$, and $\dfrac{f(x + h) - f(x)}{h}$.

13. Let $f(x) = \sqrt{x}$ and $g(x) = x^2 + 1$. Find each of the following.
   a. $f(g(x))$ b. $g(f(x))$
   c. $f(f(x))$ d. $g(g(x))$

14. Let $f(x) = \begin{cases} 3x + 2 & \text{if } x \le 2 \\ 4x - 1 & \text{if } 2 < x \le 3. \\ 6 & \text{if } x > 3 \end{cases}$
   a. Find $f(1)$, $f(3)$, and $f(4)$.
   b. Sketch the graph of $y = f(x)$.

15. Let $f(x) = 2x - 3$. Find $f^{-1}(x)$.

16. Use transformations on $y = \sqrt{x}$ to sketch the graph of each function.
   a. $f(x) = \sqrt{x + 2}$ b. $g(x) = -2\sqrt{x + 1} + 3$

17. Use the Rational Zeros Theorem to list all possible rational zeros of $f(x) = 2x^4 - 3x^2 + 5x - 6$.

In Exercises 18–21, graph each function.

18. $f(x) = 2x^2 - 4x + 1$ 19. $f(x) = -x^2 + 2x + 3$

20. $f(x) = (x - 1)^2(x + 2)$ 21. $f(x) = \dfrac{x^2 - 1}{x^2 - 4}$

22. Let $f(x) = x^4 - 3x^3 + 2x^2 + 2x - 4$. Given that $1 + i$ is a zero of f, find all zeros of f.

23. Suppose y varies as the square root of x and $y = 6$ when $x = 4$. Find y when $x = 9$.

24. A drug manufactured by a pharmaceutical company is sold in bulk at a price of \$150 per unit. The total production cost (in dollars) for x units in one week is

$$C(x) = 0.02x^2 + 100x + 3000.$$

How many units of the drug must be manufactured and sold in a week to maximize the profit? What is the maximum profit?

25. The profit (in dollars) for a product is given by

$$P(x) = 0.02x^3 + 48.8x^2 - 2990x + 25{,}000,$$

where x is the number of units produced and sold. One break-even point occurs when $x = 10$. Use synthetic division to find another break-even point for the product.

Exponential and Logarithmic Functions

The world of finance, the growth of many populations, the molecular activity that results in an atomic explosion, and countless everyday events are modeled with exponential and logarithmic functions.

Exponential Functions

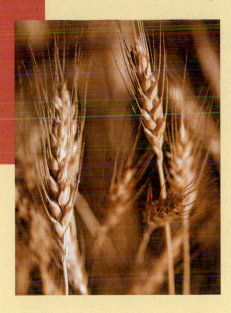

Before Starting this Section, Review

1. Integer exponents (Section P.2, page 20)

2. Rational exponents (Section P.6, page 64)

3. Graphing and transformations (Section 2.7, page 260)

4. One–to–one functions (Section 2.9, page 280)

5. Increasing and decreasing functions (Section 2.5, page 229)

Objectives

1. Define an exponential function.
2. Graph exponential functions.
3. Solve exponential equations.
4. Use transformations on exponential functions.

FOOLING A KING

One night in northwest India, a wise man named Shashi invented a wonderful new game called "Shatranj" (chess). The next morning he took it to the king Rai Bhalit, who marveled at it and said to Shashi, "Name your reward." Shashi merely requested that 1 grain of wheat be placed on the first square of the chessboard, 2 grains on the second, 4 on the third, 8 on the fourth, and so on, for all 64 squares. The king agreed to his request, thinking the man was an eccentric fool for asking for only a few grains of wheat when he could have had gold, jewels, or even his daughter's hand in marriage.

You may know the end of this story. Much more wheat was needed to satisfy Shashi's request. The king could never fulfill his promise to Shashi, so instead the king had him beheaded.

The details of this rapidly growing wheat phenomenon are given in the margin. Table 4.1 shows the number of grains of wheat that need to be stacked on each square of the chessboard.

These data can be modeled by the function

$$g(n) = 2^{n-1},$$

where $g(n)$ is the number of grains of wheat and n is the number of the square (or the nth square) on the chessboard. Each square has twice the number of grains as the previous square. The function g is an example of an *exponential function* with base 2, with n restricted to $1, 2, 3, \ldots, 64$—the number of the chessboard squares. The term *exponential function* comes from the fact that the variable n occurs in the exponent. The base 2 is the *growth factor*.

When the growth rate of a quantity is directly proportional to the existing amount, the growth can be modeled by an exponential function. For example, exponential functions are often used to model the growth of investments and populations, the cell division of living organisms, and the decay of radioactive material. ■

TABLE 4.1 Grains of Wheat on a Chessboard

Square Number	Grains Placed on This Square
1	$1 \, (= 2^0)$
2	$2 \, (= 2^1)$
3	$4 \, (= 2^2)$
4	$8 \, (= 2^3)$
5	$16 \, (= 2^4)$
6	$32 \, (= 2^5)$
⋮	⋮
63	2^{62}
64	2^{63}

1 Define an exponential function.

Exponential Functions

EXPONENTIAL FUNCTION

A function f of the form

$$f(x) = a^x, a > 0, a \neq 1,$$

is called an **exponential function with base a.** Its domain is $(-\infty, \infty)$.

In the definition of the exponential function, we rule out the base $a = 1$ because in this case, the function is simply the constant function $f(x) = 1^x$, or $f(x) = 1$. We exclude negative bases so that the domain can include all real numbers. For example, a cannot be -2 because if $a = -2$ then $f\left(\frac{1}{2}\right) = (-2)^{1/2} = \sqrt{-2}$ is not a real number.

Other functions that have variables in the exponent, such as $f(x) = 4 \cdot 3^x$, $g(x) = -5 \cdot 4^{3x-2}$, and $h(x) = c \cdot a^x$ ($a > 0$ and $a \neq 1$), are also called exponential functions.

Evaluate Exponential Functions

We can evaluate exponential functions by using the laws of exponents and/or calculators, as in the next example.

EXAMPLE 1   **Evaluating Exponential Functions**

a. Let $f(x) = 3^{x-2}$. Find $f(4)$.

b. Let $g(x) = -2 \cdot 10^x$. Find $g(-2)$.

c. Let $h(x) = \left(\frac{1}{9}\right)^x$. Find $h\left(-\frac{3}{2}\right)$.

d. Let $F(x) = 4^x$. Find $F(3.2)$.

SOLUTION

a. $f(4) = 3^{4-2} = 3^2 = 9$

b. $g(-2) = -2 \cdot 10^{-2} = -2 \cdot \frac{1}{10^2} = -2 \cdot \frac{1}{100} = -0.02$

c. $h\left(-\frac{3}{2}\right) = \left(\frac{1}{9}\right)^{-\frac{3}{2}} = (9^{-1})^{-\frac{3}{2}} = 9^{\frac{3}{2}} = (\sqrt{9})^3 = 27$

d. $F(3.2) = 4^{3.2} \approx 84.44850629$ Use a calculator.   ■ ■ ■

Practice Problem 1   Let $f(x) = \left(\frac{1}{4}\right)^x$. Find $f(2), f(0), f(-1), f\left(\frac{5}{2}\right)$, and $f\left(-\frac{3}{2}\right)$.

■

Because the domain of an exponential function is $(-\infty, \infty)$, in Example 1, we evaluated exponential functions at some rational numbers. But what is the meaning of a^x when x is an irrational number? For example, what do expressions such as $3^{\sqrt{2}}$ and 2^π mean? It turns out that the definition of a^x (with $a > 0$ and x irrational) requires methods discussed in calculus. However, we can understand the basis for the definition from the following discussion. Suppose we want to define the number 2^π. We use several numbers that approximate π. A calculator shows that $\pi \approx 3.14159265\ldots$. We successively approximate 2^π by using the rational powers shown in Table 4.2.

TABLE 4.2

Values of $f(x) = 2^x$ for rational values of x that approach π				
x	3	3.1	3.14	3.141	3.1415
2^x	$2^3 = 8$	$2^{3.1} = 8.5\ldots$	$2^{3.14} = 8.81\ldots$	$2^{3.141} = 8.821\ldots$	$2^{3.1415} = 8.8244\ldots$

It can be shown that the powers $2^3, 2^{3.1}, 2^{3.14}, 2^{3.141}, 2^{3.1415}, \ldots$ approach exactly one number. We define 2^π as that number. Table 4.2 shows that $2^\pi \approx 8.82$ (correct to two decimal places).

For our work with exponential functions, we need the following fundamental facts:

1. An exponential function $f(x) = a^x$ is defined for all real numbers x.
2. The graph of an exponential function is a continuous (unbroken) curve.
3. The rules of exponents hold for all real numbers.

RULES OF EXPONENTS

Let a, b, x, and y be real numbers with $a > 0$ and $b > 0$.

$$a^x \cdot a^y = a^{x+y} \qquad (a^x)^y = a^{xy}$$

$$\frac{a^x}{a^y} = a^{x-y} \qquad a^0 = 1$$

$$(ab)^x = a^x b^x \qquad a^{-x} = \frac{1}{a^x} = \left(\frac{1}{a}\right)^x$$

---

2 Graph exponential functions.

Graphing Exponential Functions

Let's see how to sketch the graph of an exponential function. Although the domain is the set of all real numbers, we usually evaluate the functions only for the integer values of x (for ease of computation). To evaluate a^x for noninteger values of x, use your calculator. Recall that the graph of a function $f(x)$ is the graph of the equation $y = f(x)$. We can use either $f(x) = a^x$ or $y = a^x$ to represent a given function.

EXAMPLE 2   **Graphing an Exponential Function with Base $a > 1$**

Graph the exponential function $f(x) = 3^x$.

SOLUTION

First, make a table of a few values of x and the corresponding values of y.

x	-3	-2	-1	0	1	2	3
$y = 3^x$	$\dfrac{1}{27}$	$\dfrac{1}{9}$	$\dfrac{1}{3}$	1	3	9	27

Next, plot the points (see Figure 4.1(a)) and draw a smooth curve through them. See Figure 4.1(b).

FIGURE 4.1 An exponential function $y = a^x$ with $a > 1$

The graph in Figure 4.1(b) is typical of the graphs of exponential functions $f(x) = a^x$ when $a > 1$. Note that the x-axis ($y = 0$) is the horizontal asymptote of the graph of $y = a^x$. ■ ■ ■

Practice Problem 2   Sketch the graph of

$$f(x) = 2^x.$$ ■

Let's now sketch the graph of an exponential function $f(x) = a^x$ when $0 < a < 1$.

EXAMPLE 3   **Graphing an Exponential Function $f(x) = a^x$, with $0 < a < 1$**

Sketch the graph of $y = \left(\dfrac{1}{2}\right)^x$.

SOLUTION
Make a table similar to the one in Example 2.

x	-3	-2	-1	0	1	2	3
$y = \left(\dfrac{1}{2}\right)^x$	8	4	2	1	$\dfrac{1}{2}$	$\dfrac{1}{4}$	$\dfrac{1}{8}$

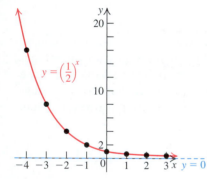

FIGURE 4.2 An exponential function $y = a^x$ with $0 < a < 1$

Plotting these points and drawing a smooth curve through them, we get the graph of $y = \left(\dfrac{1}{2}\right)^x$, shown in Figure 4.2. In this graph, as x increases in the positive direction, $y = \left(\dfrac{1}{2}\right)^x$ decreases toward 0, and $y = 0$ is the horizontal asymptote. This graph is typical of the graph of an exponential function $y = a^x$ with $0 < a < 1$. ■ ■ ■

Practice Problem 3   Sketch the graph of

$$f(x) = \left(\frac{1}{3}\right)^x.$$ ■

In general, exponential functions have two basic shapes determined by the base a. If $a > 1$, the graph of $y = a^x$ is rising as in Figure 4.3(a). If $0 < a < 1$, the graph is falling as in Figure 4.3(b). To get a correct shape for the graph of $f(x) = a^x$, it is helpful to graph the three points: $\left(-1, \dfrac{1}{a}\right)$, $(0, 1)$, and $(1, a)$.

Recall (page 256) that the graph of $y = f(-x)$ is the reflection of the graph of $y = f(x)$ about the y–axis. Because $y = \left(\dfrac{1}{a}\right)^x = (a^{-1})^x = a^{-x}$, the graph of

TECHNOLOGY CONNECTION

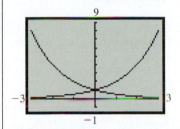

Graph $y_1 = 2^x$ and $y_2 = \left(\dfrac{1}{2}\right)^x$ on the same screen.

Notice that the graphs of y_1 and y_2 are reflections of each other in the y-axis.

FIGURE 4.3 Basic shapes of $y = a^x$

$y = \left(\dfrac{1}{a}\right)^x$ can be obtained by reflecting the graph of $y = a^x$ about the y–axis. The calculator graphs of $y = 2^x$ and $y = \left(\dfrac{1}{2}\right)^x$ in the margin support this observation.

In Figure 4.4, we sketch the graphs of four exponential functions on the same set of axes. These graphs illustrate some general properties of exponential functions.

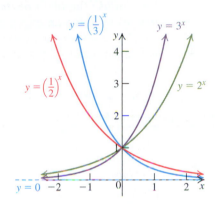

FIGURE 4.4 Graphs of some exponential functions

Properties of Exponential Functions Let $y = f(x) = a^x$, $a > 0$, $a \neq 1$.

1. The domain of $f(x) = a^x$ is $(-\infty, \infty)$.
2. The range of $f(x) = a^x$ is $(0, \infty)$: The entire graph lies above the x-axis.
3. For $a > 1$,
   (i) f is an increasing function; so the graph rises to the right.
   (ii) as $x \to \infty$, $y \to \infty$.
   (iii) as $x \to -\infty$, $y \to 0$.
4. For $0 < a < 1$,
   (i) f is a decreasing function; so the graph falls to the right.
   (ii) as $x \to -\infty$, $y \to \infty$.
   (iii) as $x \to \infty$, $y \to 0$.

RECALL

A function f is one-to-one if $f(x_1) = f(x_2)$ implies that $x_1 = x_2$.

5. Each exponential function f is one-to-one. So
   (i) if $a^u = a^v$, then $u = v$.
   (ii) f has an inverse.
6. The graph of $f(x) = a^x$ has no x-intercepts, so it never crosses the x-axis. No value of x will cause $f(x) = a^x$ to equal 0.
7. The graph of f is a smooth and continuous curve, and it passes through the points $\left(-1, \dfrac{1}{a}\right)$, $(0, 1)$, and $(1, a)$.

REMINDER

Recall that a line $y = b$ is called a horizontal asymptote if $f(x) \to b$ as $x \to \infty$ or as $x \to -\infty$.

8. From 3(iii) and 4(iii), we conclude that the x-axis is a horizontal asymptote for every exponential function of the form $f(x) = a^x$.

> **EXAMPLE 4** **Finding a Base Value for an Exponential Function**

Find a if the graph of the exponential function $f(x) = a^x$ contains the point $(2, 49)$.

SOLUTION

Write the function in the form $y = f(x)$.

$$y = a^x$$
$$49 = a^2 \qquad \text{The point } (2, 49) \text{ is on the graph.}$$
$$a = \pm\sqrt{49} = \pm 7 \qquad \text{Solve for } a.$$

We discard the -7 because the base for an exponential function must be positive. Therefore, $a = 7$ and the exponential function is $f(x) = 7^x$. ■ ■ ■

Practice Problem 4 Find a assuming that the exponential function $f(x) = a^x$ contains the point $(-1, 5)$. ■

3 Solve exponential equations.

Exponential Equations

We can use the one–to–one property of the exponential function (if $a^u = a^v$, then $u = v$) to solve an exponential equation. This technique works for equations in which both sides can be written as powers of the same base. Later in the chapter you will learn how to solve more general exponential equations.

FINDING THE SOLUTION: A PROCEDURE

> **EXAMPLE 5** **Solving Exponential Equations**

OBJECTIVE	**EXAMPLE**
Solve an exponential equation when both sides can be written as powers of the same base.	*Solve for x:* $5^{x(x-3)} = \dfrac{1}{25}$
Step 1 Write (if necessary) both sides of the exponential equation as powers of the same base.	**1.** $5^{x(x-3)} = \dfrac{1}{25}$ — Given equation $\qquad 5^{x(x-3)} = 5^{-2} \qquad \dfrac{1}{25} = \dfrac{1}{5^2} = 5^{-2}$
Step 2 Use the one-to-one property of exponential functions to equate exponents.	**2.** $x(x-3) = -2 \qquad$ If $a^u = a^v$, then $u = v$.
Step 3 Solve for the variable.	**3.** $\quad x^2 - 3x = -2 \qquad$ Distributive property $\quad x^2 - 3x + 2 = 0 \qquad$ Write in standard form. $\quad (x-1)(x-2) = 0 \qquad$ Factor. $\quad x - 1 = 0$ or $x - 2 = 0 \qquad$ Zero-product property $\quad x = 1$ or $x = 2 \qquad$ Solve for x.
Step 4 Check your solutions.	**4.** We leave to you the check that $x = 1$ and $x = 2$ satisfy the equation. The solution set is $\{1, 2\}$. ■ ■ ■

Practice Problem 5 Solve for x: $3^{2x+1} = 243$ ■

4 Use transformations on exponential functions.

Transformations on Exponential Functions

We can apply the transformations discussed in Section 2.7 to graph functions related to the basic exponential function $f(x) = a^x$.

REMINDER

The graph of the function $f(x) = a^x$ is the graph of the equation $y = a^x$.

Transformations on Exponential Function $f(x) = a^x$		
Transformation	**Equation**	**Effect on Graph**
Horizontal Shift	$y = a^{x-b}$ $= f(x - b)$	Shift the graph of $y = a^x$ **(i)** $\lvert b \rvert$ units left if $b < 0$. **(ii)** b units right if $b > 0$.
Vertical Shift	$y = a^x + b$ $= f(x) + b$	Shift the graph of $y = a^x$ **(i)** b units up if $b > 0$. **(ii)** $\lvert b \rvert$ units down if $b < 0$.
Stretching or Compressing (vertically)	$y = ca^x = cf(x)$	Multiply the y coordinates by c. The graph of $y = a^x$ is vertically **(i)** stretched if $c > 1$. **(ii)** compressed if $0 < c < 1$.
Stretching or Compressing (horizontally)	$y = a^{cx} = f(cx)$	Multiply the x coordinates by c. The graph of $y = a^x$ is horizontally **(i)** compressed if $c > 1$. **(ii)** stretched if $0 < c < 1$.
Reflection $(c > 0)$	$y = -ca^x = -cf(x)$ $y = a^{-cx} = f(-cx)$	Reflect the graph of $y = ca^x$ in the x-axis. Reflect the graph of $y = a^{cx}$ in the y-axis.

EXAMPLE 6   **Sketching Graphs**

Use transformations to sketch the graph of each function.

a. $f(x) = 3^x - 4$ **b.** $f(x) = 3^{x+1}$

c. $f(x) = -3^x$ **d.** $f(x) = -3^x + 2$

State the domain and the range of each function and the horizontal asymptote of its graph.

SOLUTION

Figure 4.5 describes the transformations of the exponential function $f(x) = 3^x$.

a. Subtract 4 from 3^x: Shift the graph of $y = 3^x$ four units down.

b. Replace x in 3^x with $x + 1$: Shift the graph of $y = 3^x$ one unit left.

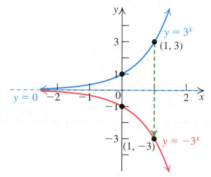

c. Multiply 3^x by -1: Reflect the graph of $y = 3^x$ in the x-axis.

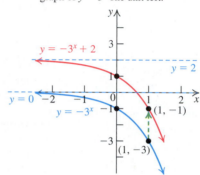

d. Add 2 to -3^x: Shift the graph of $y = -3^x$ from part **c** two units up.

FIGURE 4.5

a. Domain: $(-\infty, \infty)$; range: $(-4, \infty)$; horizontal asymptote: $y = -4$

b. Domain: $(-\infty, \infty)$; range: $(0, \infty)$; horizontal asymptote: x-axis $(y = 0)$

c. Domain: $(-\infty, \infty)$; range: $(-\infty, 0)$; horizontal asymptote: x-axis $(y = 0)$

d. Domain: $(-\infty, \infty)$; range: $(-\infty, 2)$; horizontal asymptote: $y = 2$ ■ ■ ■

Practice Problem 6 Sketch the graph of $y = 3^{x-1} + 4$. ■

EXAMPLE 7 **Comparing Exponential and Power Functions**

Compare the graphs of $f(x) = x^2$ and $g(x) = 2^x$ for $x \geq 0$.

SOLUTION

Figure 4.6 shows the graphs of f and g over the interval $[0, 5]$. We notice the following:

(i) The graph of $g(x) = 2^x$ is higher than the graph of $f(x) = x^2$ in the interval $[0, 2)$; so $2^x > x^2$ for $0 \leq x < 2$.

(ii) The graph of f intersects the graph of g at $x = 2$; so $2^x = x^2$ at $x = 2$. Both function values are 4, and $(2, 4)$ is a point on both graphs.

(iii) In the interval $(2, 4)$, the graph of f is higher than the graph of g; that is, $x^2 > 2^x$ for $2 < x < 4$.

(iv) The graph of f intersects the graph of g again at $x = 4$; that is, $2^x = x^2$ at $x = 4$. Both function values are 16, and $(4, 16)$ is a point on both graphs.

(v) For $x > 4$, the graph of g is higher than the graph of f; that is, $2^x > x^2$ for $x > 4$.

■ ■ ■

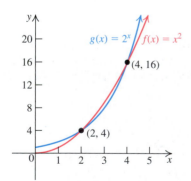

FIGURE 4.6 Graphs of
$f(x) = x^2$ and $g(x) = 2^x$

Practice Problem 7 Compare the graphs of $f(x) = x^4$ and $g(x) = 4^x$ for $x \geq 0$. ■

The result of Example 7 can be generalized. That is, for large values of x, the exponential function with base $a > 1$ is greater than any power function. In other words, for large values of x, we have

$$a^x > x^n \quad (a > 1 \text{ and } n \text{ any positive integer})$$

In fact, for large values of x, any exponential function $y = a^x (a > 1)$ eventually becomes greater than any polynomial function.

EXAMPLE 8   **Bacterial Growth**

A technician to the French microbiologist Louis Pasteur noticed that a certain culture of bacteria in milk doubles every hour. If the bacteria count $B(t)$ is modeled by the equation

$$B(t) = 2000 \cdot 2^t,$$

with t in hours, find

a. the initial number of bacteria.

b. the number of bacteria after ten hours.

c. the time when the number of the bacteria will be 32,000.

SOLUTION

a. Initial size $B_0 = B(0) = 2000 \cdot 2^0 = 2000 \cdot 1 = 2000$ Substitute $t = 0$.

b. $B(10) = 2000 \cdot 2^{10} = 2,048,000$ Substitute $t = 10$.

c. We have to find t when $B(t) = 32,000$ in the equation $B(t) = 2000 \cdot 2^t$.

$$32,000 = 2000 \cdot 2^t$$
$$16 = 2^t \qquad \text{Divide both sides by 2000.}$$
$$2^4 = 2^t \qquad \text{Rewrite 16 as } 2^4.$$
$$4 = t \qquad \text{If } a^x = a^y, \text{ then } x = y.$$

After four hours, the number of bacteria will be 32,000. ■ ■ ■

Practice Problem 8   In Example 8, find the time when the bacteria will be 128,000. ■

SECTION 4.1 ■ Exercises

A EXERCISES Basic Skills and Concepts

1. For the exponential function $f(x) = a^x, a > 0, a \neq 1$, the domain is _____ and the range is _____.

2. The graph of $f(x) = 3^x$ has y-intercept _____ and has _____ x-intercept.

3. The horizontal asymptote of the graph of $y = \left(\dfrac{1}{3}\right)^x$ is the _____.

4. The exponential function $f(x) = a^x$ is increasing if a is _____ and is decreasing if a is _____.

5. *True or False* The graphs of $y = 2^x$ and $y = \left(\dfrac{1}{2}\right)^x$ are symmetric with respect to the x-axis.

6. *True or False* For the equation $y = a^x (a > 0, a \neq 1)$, $y \to \infty$ as $x \to \infty$.

In Exercises 7–14, explain whether the given equation defines an exponential function. Give reasons for your answers. Write the base for each exponential function.

7. $y = x^3$

8. $y = 4^x$

9. $y = 2^{-x}$

10. $y = (-5)^x$

11. $y = x^x$

12. $y = 4^2$

13. $y = 0^x$

14. $y = (1.8)^x$

In Exercises 15–24, evaluate each exponential function for the given value. (Use a calculator if necessary.)

15. $f(x) = 5^{x-1}, f(3)$

16. $f(x) = -2^{x+1}, f(3)$

17. $f(x) = 3^{1-x}, f(3.2)$

18. $f(x) = 6^{x-1}, f(2.8)$

19. $g(x) = -\left(\dfrac{1}{2}\right)^{x+1}, g(-3.5)$

20. $g(x) = -\left(\dfrac{2}{3}\right)^{2x-1}, g(2.5)$

21. $g(x) = \left(\dfrac{1}{4}\right)^{1-2x}, g\left(\dfrac{3}{2}\right)$

22. $g(x) = -\left(\dfrac{2}{5}\right)^{3x-1}, g\left(\dfrac{4}{3}\right)$

23. $h(x) = 1 - 3^{-x}, h(1)$

24. $h(x) = 3 - 5^{-x}, h(2)$

In Exercises 25–30, find the exponential function of the form $f(x) = a^x$ that contains the given graph point.

25. $(2, 16)$

26. $\left(-2, \dfrac{1}{9}\right)$

27. $(3, 216)$

28. $\left(3, \dfrac{1}{125}\right)$

29. $(-2, 16)$

30. $\left(\dfrac{1}{3}, \dfrac{1}{2}\right)$

In Exercises 31 and 32, find the function of the form $f(x) = c \cdot a^x$ that contains the two given points.

31. $(0, 3)$ and $(2, 12)$

32. $(0, 5)$ and $(1, 15)$

In Exercises 33–40, sketch the graph of the given function by making a table of values. (Use a calculator if necessary.)

33. $f(x) = 4^x$

34. $g(x) = 10^x$

35. $g(x) = \left(\dfrac{3}{2}\right)^{-x}$

36. $h(x) = 7^{-x}$

37. $h(x) = \left(\dfrac{1}{4}\right)^x$

38. $f(x) = \left(\dfrac{1}{10}\right)^x$

39. $f(x) = (1.3)^{-x}$

40. $g(x) = (0.7)^{-x}$

41. How are the graphs in Exercises 33 and 37 related? Can we obtain the graph of Exercise 37 from that of Exercise 33? If so, how?

42. Repeat Exercise 41 for the graphs in Exercises 34 and 38.

Match each exponential function given in Exercises 43–46 with one of the graphs labeled (a), (b), (c), and (d).

43. $f(x) = 5^x$

44. $f(x) = -5^x$

45. $f(x) = 5^{-x}$

46. $f(x) = 5^{-x} + 1$

(a) (b) (c) (d)

In Exercises 47–54, start with the basic exponential function $f(x) = 3^x$ and use transformations to sketch the graph of the function $g(x)$. State the domain and range of $g(x)$ and the horizontal asymptote of its graph.

47. $g(x) = 3^{x-1}$

48. $g(x) = 3^{x-1} - 2$

49. $g(x) = \dfrac{1}{2} \cdot 3^{x+1}$

50. $g(x) = \dfrac{1}{2} \cdot 3^{x+1} - 2$

51. $g(x) = -2 \cdot 3^{x+1} + 1$

52. $g(x) = 2 \cdot 3^{x+1} - 1$

53. $g(x) = -2 \cdot 3^{x-1} + 4$

54. $g(x) = \dfrac{1}{2} \cdot 3^{-x+1} - 2$

In Exercises 55–64, start with the graph of $y = 3^x$. (a) Describe a sequence of transformations that results in the graph of $g(x)$; (b) Find the range of $g(x)$; (c) Find the horizontal asymptote of the graph of $g(x)$.

55. $g(x) = 3^{x/2}$

56. $g(x) = 3^{-x/3}$

57. $g(x) = \left(\sqrt[3]{3}\right)^x - 1$

58. $g(x) = (\sqrt{3})^x + 2$

59. $g(x) = 3^{2x-1} + 2$

60. $g(x) = 3^{2x+1} - 2$

61. $g(x) = 5(3^{2x+1}) + 4$

62. $g(x) = -2(3^{2x-1}) + 4$

63. $g(x) = -2(3^{5-2x}) - 4$

64. $g(x) = 2(3^{2-3x}) - 4$

In Exercises 65–68, write an equation of each graph in the final position.

65. The graph of $y = 2^x$ is shifted two units left and five units up.

66. The graph of $y = 3^x$ is reflected in the y-axis and then shifted three units right.

67. The graph of $y = \left(\dfrac{1}{2}\right)^x$ is stretched vertically by a factor of 2 and then shifted five units down.

68. The graph of $y = 2^{-x}$ is reflected in the x-axis and then shifted three units up.

In Exercises 69–84, solve each equation for *x* by first rewriting both sides as powers of the same base.

69. $3^x = 81$ **70.** $9^x = 243$

71. $4^{|x|} = 128$ **72.** $3^{-x} = 27$

73. $2^{-x+1} = 16$ **74.** $4^{x-1} = 1$

75. $2^x = 4^{2x+1}$ **76.** $(\sqrt{2})^x = \dfrac{1}{16}$

77. $\left(\dfrac{3}{5}\right)^x = \dfrac{125}{27}$ **78.** $\left(\dfrac{2}{3}\right)^x = \left(\dfrac{9}{4}\right)^3$

79. $2^{|x|} = 16$ **80.** $4^{-|x|} = \dfrac{1}{128}$

81. $3^x = (9)^{x-1} \cdot (27)^{1-3x}$ **82.** $5^{x^2} = \left(\dfrac{1}{25}\right)^{-8}$

83. $9^x - 3^x = 0$ **84.** $2^{x^2} = (8)^{2x} \cdot \dfrac{1}{256}$

B EXERCISES Applying the Concepts

85. Suppose a metal block is cooling so that its temperature T (in °C) is given by $T = 200 \cdot 4^{-0.1t}$, where t is in hours.
   a. Find the temperature after
       (i) 2 hours.   **(ii)** 3.5 hours.
   b. How long has the cooling been taking place if the block now has a temperature of 100°C?
   c. Find the eventual temperature $(t \to \infty)$.

86. Suppose the worldwide sales (in millions) of a computer chip called pentium p are approximated by

$$s(t) = 50 - 30 \cdot 2^{-t},$$

   where t represents the number of years the pentium p has been on the market.
   a. Find each function value.
       (i) $s(0)$ **(ii)** $s(1)$
       (iii) $s(5)$ **(iv)** $s(10)$
   b. Graph $y = s(x)$. Find the horizontal asymptote. What is the meaning of this horizontal asymptote?

87. The population (in thousands) of people of East Indian origin in the United States is approximated by the function

$$p(t) = 1600(2)^{0.1047t},$$

   where t is the number of years since 2000.
   a. Find the population of this group in
       (i) 2000.   **(ii)** 2008.
   b. Predict the population in 2015.
   c. When will the population reach 6.4 million? (The function $p(t)$ is modeled by the authors from the data from the weekly magazine *India Abroad*.)

88. Suppose we have a large sheet of paper 0.015 centimeter thick and we tear the paper in half and put the pieces on top of each other. We keep tearing and stacking in this manner, always tearing each piece in half. How high will the resulting pile of paper be if we continue the process of tearing and stacking
   a. 30 times?
   b. 40 times?
   c. 50 times? [*Hint:* Use a calculator.]

89. A jar with a volume of 1000 cubic centimeters contains bacteria that doubles in number every minute. If the jar is full in 60 minutes, how long did it take for the container to be half full?

90. A box contains radioactive plutonium. The number of kilograms $p(t)$ at time t years is given by the formula

$$p(t) = 2^{-0.0001754\,t}.$$

   Find the amount of plutonium at the following times:
   a. $t = 0$ **b.** $t = 800$
   c. $t = 1000$ **d.** $t = 10{,}000$

91. In Exercise 90, how long will it take until only one-half kilogram of plutonium is left?

92. The number of bacteria present after t minutes is given by the formula $n(t) = 1000(3)^{0.1t}$.
   a. Find the number of bacteria after
       (i) 5 minutes.
       (ii) 10 minutes.
   b. How long will it take for the bacteria population to reach 9000?

C EXERCISES Beyond the Basics

93. Sketch the graph of $f(x) = 2^{|x|} = \begin{cases} 2^x & \text{if } x \ge 0 \\ 2^{-x} & \text{if } x < 0 \end{cases}$.
   Check for symmetry. Find the range of this function.

94. Sketch the graph of $g(x) = 2^{-|x|}$.

95. Sketch the graph of $g(x) = 2^{x^2}$. Check for symmetry. Find the domain and range of this function.

96. **a.** Sketch the graph of $h(x) = 2^{-x^2}$.
   b. Find the domain and range of $h(x)$.
   c. Why is it not possible to obtain the graph of $h(x)$ by reflecting the graph of $g(x)$ of Exercise 95 in the y-axis?

97. Let $f(x) = 3^x + 3^{-x}$ and $g(x) = 3^x - 3^{-x}$. Find each of the following:
   a. $f(x) + g(x)$
   b. $f(x) - g(x)$
   c. $[f(x)]^2 - [g(x)]^2$
   d. $[f(x)]^2 + [g(x)]^2$

98. Repeat Exercise 97 for the functions $f(x) = a^x + a^{-x}$ and $g(x) = a^x - a^{-x}$.

In Exercises 99–102, solve each equation for *x*.

99. $\left(\dfrac{3}{5}\right)^{2x-1} = \left(\dfrac{25}{9}\right)^{3x+1}$ **100.** $\left(\dfrac{3}{4}\right)^{x^2+8} = \left(\dfrac{64}{27}\right)^{2x}$

101. $\left(\dfrac{1}{2}\right)^{|x-3|} = \left(\dfrac{1}{8}\right)^5$ **102.** $\left(\dfrac{2}{3}\right)^{|x+2|} = \left(\dfrac{81}{16}\right)^{-3}$

In Exercises 103–106, solve the given equation for x. [*Hint*: These equations are quadratic in form.]

103. $2^{2x} - 12 \cdot 2^x + 32 = 0$

104. $3^{2x} - 4 \cdot 3^{x+1} + 27 = 0$

105. $4(2^x + 2^{-x}) = 17$

106. $4^{1+x} + 4^{1-x} = 10$

107. Solve: $p^{2x}(p^2 + 1) = p(p^{3x} + p^x)$, $p > 0$.
[*Hint*: Simplify to $(p^{x+1} - 1)(p^{x-1} - 1) = 0$.]

108. Growth of bacteria. The bacterium *Escherichia coli*, commonly known as *E. coli*, is found in the human digestive tract. Suppose a colony of *E. coli* starts with a single cell and each *E. coli* cell divides into two cells one-third hour after its birth. Suppose also that no *E. coli* cells die.
   a. Write a function that models the number of *E. coli* cells after t hours.
   b. Given that the mass of a single *E. coli* cell is approximately $5 \cdot 2^{-43}$ grams, find the mass of the colony after ten hours.
   c. When will the mass of the colony be 20 grams?

109. Internet users. The number of Internet users worldwide was estimated to be 0.4 billion in 2000 and 1.1 billion in 2005. The growth can be approximated by an exponential function.
   a. Find an exponential function of the form $y = c \cdot 2^{kx}$ that models the number of Internet users, where $x = 0$ corresponds to 2000 and y corresponds to the number of users in billions.

   b. Graph the function and estimate the number of users in 2008.
   c. Use the graph to estimate the year when there are 2.5 billion users.

110. Spread of AIDS. In 1987, the number of AIDS cases in the United States was estimated at 50,000. In 2005, this number was estimated at 1 million. Assume that the trend continued.
   a. Set up a function of the form

   $$f(t) = c \cdot 2^{kt}$$

   to describe the number of cases of AIDS, where t is time measured in years from 1987.
   b. According to the model in (a), how many cases of AIDS occurred in 2000?
   c. Use the model to predict the number of cases of AIDS in 2012.
   d. According to the model, when was the number of cases of AIDS 1.6 million?

Critical Thinking

111. Give a convincing argument to show that the equation $2^x = k$ has exactly one solution for every $k > 0$. Support your argument with graphs.

112. Find all solutions of the equation $2^x = 2x$. How do you know you have found all solutions?

The Natural Exponential Function

Before Starting this Section, Review

1. Exponential function (Section 4.1, page 406)

2. Graphing and transformations (Section 2.7, page 260)

Objectives

1. Define simple interest.

2. Develop a compound interest formula.

3. Understand the number *e*.

4. Graph exponential functions.

5. Evaluate exponential functions.

DESCENDANT SEEKS LOAN REPAYMENT

George Washington could not tell a lie, but apparently he could leave behind a whopping debt.

In 1777, Jacob DeHaven, a wealthy Pennsylvania trader, responded to a desperate plea from Washington when it seemed that the Revolutionary War was about to be lost. Mr. DeHaven loaned $450,000 worth of gold and supplies to the Continental Congress to rescue the troops at Valley Forge. His loan provided badly needed supplies, provisions, and salaries for the Continental Army. George Washington went on to defeat the British in the war. After the war, the government offered to repay DeHaven's loan in continental dollars, the then worthless currency of the fledgling country. But DeHaven refused and died penniless in 1812.

In 1990, DeHaven's descendants sued the U.S. government for repayment of the loan. "If I owed the government this money, I would be expected to pay the principal, interest, and any penalties that might have been incurred," said Carolyn Cokerham, one of nearly 2000 living descendants of DeHaven. Many of his descendants have tried unsuccessfully to get the money repaid through the years. How much money did DeHaven's descendants claim the government owed them on the 1990 anniversary of the loan at the then prevailing interest rate of 6%? (See Example 5.) ■

To answer the DeHaven and related financial questions, let's first review some terminology concerning interest.

1 Define simple interest.

SIMPLE INTEREST

A fee charged for borrowing a lender's money is the **interest**, denoted by *I*.

The original, or initial, amount of money borrowed is the **principal**, denoted by *P*.

The period of time during which the borrower pays back the principal plus the interest is the **time**, denoted by *t*.

The **interest rate** is the percent charged for the use of the principal for the given period. The interest rate, denoted by *r*, is expressed as a decimal. Unless stated otherwise, the period is assumed to be one year; that is, *r* is an *annual* rate.

The amount of interest computed only on the principal is called **simple interest**.

RECALL

8% = 8 percent

= 8 per hundred

$= \dfrac{8}{100}$

= 0.08

When money is deposited with a bank, the bank becomes the borrower. For example, depositing $1000 in an account at 8% interest means that the principal P is $1000 and the interest rate r is 0.08 for the bank as the borrower.

SIMPLE INTEREST FORMULA

The simple interest I on a principal P at a rate r (expressed as a decimal) per year for t years is

$$I = Prt. \quad (1)$$

EXAMPLE 1   **Calculating Simple Interest**

Juanita has deposited $8000 in a bank for five years at a simple interest rate of 6%.

a. How much interest will she receive?

b. How much money will be in her account at the end of five years?

SOLUTION

a. $P = \$8000, r = 0.06,$ and $t = 5$

$\quad\quad\quad I = Prt$ Equation (1)

$\quad\quad\quad\quad = \$8000\,(0.06)(5)$ Substitute values.

$\quad\quad\quad\quad = \$2400$ Simplify.

b. In five years, the amount A she will receive is the original principal plus the interest earned:

$$A = P + I$$
$$= \$8000 + \$2400$$
$$= \$10{,}400$$

■ ■ ■

Practice Problem 1   Find the amount that will be in a bank account assuming that $10,000 is deposited at a simple interest rate of 7.5% for two years.   ■

Given P and r, the amount $A(t)$ calculated at simple interest and due in t years is found by using the formula

$$A(t) = P + Prt. \quad (2)$$

Equation (2) is a linear function of t. Simple interest problems are examples of **linear growth**, which take place when the growth of a quantity occurs at a constant rate and so can be modeled by a linear function.

2   Develop a compound interest formula.

Compound Interest

In the real world, simple interest is rarely used for periods of more than one year. Instead, **compound interest** is used—the interest paid on both the principal and the accrued (previously earned) interest.

To illustrate compound interest, suppose $1000 is deposited in a bank account paying 4% annual interest. At the end of one year, the account will contain the original $1000 plus the 4% interest earned on the $1000.

$$\$1000 + (0.04)(\$1000) = \$1040$$

Similarly, if P represents the initial amount deposited at an interest rate r (expressed as a decimal) per year, then the amount A_1 in the account after one year is

$$A_1 = P + rP \qquad \text{Principal } P \text{ plus interest earned}$$
$$A_1 = P(1 + r) \qquad \text{Factor out } P.$$

During the second year, the account earns interest on the new principal A_1. The amount A_2 in the account after the second year will be equal to A_1 plus the interest on A_1.

$$A_2 = A_1 + rA_1 \qquad A_1 \text{ plus interest earned on } A_1$$
$$= A_1(1 + r) \qquad \text{Factor out } A_1.$$
$$= P(1 + r)(1 + r) \qquad A_1 = P(1 + r)$$
$$A_2 = P(1 + r)^2$$

The amount A_3 in the account after the third year is

$$A_3 = A_2 + rA_2 \qquad A_2 \text{ plus interest earned on } A_2$$
$$= A_2(1 + r) \qquad \text{Factor out } A_2.$$
$$= P(1 + r)^2(1 + r) \qquad A_2 = P(1 + r)^2$$
$$A_3 = P(1 + r)^3$$

In general, the amount A in the account after t years is given by

$$A = P(1 + r)^t. \qquad (3)$$

We say that this type of interest is **compounded annually** because it is paid once a year.

EXAMPLE 2   **Calculating Compound Interest**

Juanita deposits $8000 in a bank at the interest rate of 6% compounded annually for five years.

a. How much money will she have in her account after five years?

b. How much interest will she receive?

SOLUTION

a. Here $P = \$8000$, $r = 0.06$, and $t = 5$; so

$$A = P(1 + r)^t$$
$$= \$8000(1 + 0.06)^5 \qquad P = \$8000, r = 0.06, t = 5$$
$$= \$8000(1.06)^5$$
$$= \$10,705.80 \qquad \text{Use a calculator.}$$

b. Interest $= A - P = \$10,705.80 - \$8000 = \$2705.80$. ■ ■ ■

Practice Problem 2 Repeat Example 2 assuming that the bank pays 7.5% interest compounded annually. ■

Comparing Examples 1 and 2, we see that compounding Juanita's interest made her money grow faster. We expect this because the function $A(t) = P(1 + r)^t$ is an exponential function with base $(1 + r)$.

Banks or financial institutions usually pay savings account interest more than once a year. They pay a smaller amount of interest more frequently. Suppose the annual interest rate r quoted (also called the **nominal rate**) is compounded n times per year (at equal intervals) instead of annually. Then for each period, the interest rate is $\dfrac{r}{n}$, and there are $n \cdot t$ periods in t years. Accordingly, we can restate the formula $A(t) = P(1 + r)^t$ as follows.

COMPOUND INTEREST FORMULA

$$A = P\left(1 + \frac{r}{n}\right)^{nt} \quad (4)$$

A = amount after t years

P = principal

r = annual interest rate (expressed as a decimal number)

n = number of times interest is compounded each year

t = number of years

The total amount accumulated after t years, denoted by A, is also called the **future value** of the investment.

EXAMPLE 3   **Using Different Compounding Periods to Compare Future Values**

BY THE WAY . . .

Compounding that occurs 1, 2, 4, 12, and 365 times a year is known as compounding annually, semiannually, quarterly, monthly, and daily, respectively.

Assuming that $100 is deposited in a bank that pays 5% annual interest, find the future value A after one year if the interest is compounded

(i) annually.

(ii) semiannually.

(iii) quarterly.

(iv) monthly.

(v) daily.

SOLUTION

In the following computations, $P = 100, r = 0.05$, and $t = 1$. Only n, the number of times interest is compounded each year, changes. Because $t = 1, nt = n(1) = n$.

(i) Annual Compounding: $\qquad A = P\left(1 + \dfrac{r}{n}\right)^{nt}$ $\qquad\qquad n = 1; t = 1$

$A = 100(1 + 0.05) = \$105.00$

(ii) Semiannual Compounding: $\qquad A = P\left(1 + \dfrac{r}{2}\right)^{2}$ $\qquad\qquad n = 2; t = 1$

$A = 100\left(1 + \dfrac{0.05}{2}\right)^{2}$

$\approx \$105.06$ $\qquad\qquad$ Use a calculator.

(iii) Quarterly Compounding: $\qquad A = P\left(1 + \dfrac{r}{4}\right)^{4}$ $\qquad\qquad n = 4; t = 1$

$A = 100\left(1 + \dfrac{0.05}{4}\right)^{4}$

$\approx \$105.09$ $\qquad\qquad$ Use a calculator.

(iv) Monthly Compounding: $\qquad A = P\left(1 + \dfrac{r}{12}\right)^{12}$ $\qquad\qquad n = 12; t = 1$

$A = 100\left(1 + \dfrac{0.05}{12}\right)^{12}$

$\approx \$105.12$ $\qquad\qquad$ Use a calculator.

3 Understand the number *e*.

(v) Daily Compounding:

$$A = P\left(1 + \frac{r}{365}\right)^{365} \qquad n = 365; t = 1$$

$$A = 100\left(1 + \frac{0.05}{365}\right)^{365}$$

$$\approx \$105.13 \qquad \text{Use a calculator.}$$

■ ■ ■

Practice Problem 3   Repeat Example 3 assuming that \$5000 is deposited at a 6.5% annual rate.   ■

Compound Interest Formula

Notice in Example 3 that the future value A increases with n, the number of compounding periods. (Of course P, r, and t are fixed.) The question is, "If n increases indefinitely (100, 1000, 10,000 times, and so on), does the amount A also increase indefinitely?" Let's see why the answer is no. Let $h = \frac{n}{r}$. Then we have

$$A = P\left(1 + \frac{r}{n}\right)^{nt} \qquad \text{Equation (4)}$$

$$= P\left[\left(1 + \frac{r}{n}\right)^{\frac{n}{r}}\right]^{rt} \qquad nt = \frac{n}{r} \cdot rt$$

$$= P\left[\left(1 + \frac{1}{h}\right)^{h}\right]^{rt} \qquad h = \frac{n}{r}, \text{so} \frac{1}{h} = \frac{r}{n} \qquad (5)$$

Table 4.3 shows values of the expression $\left(1 + \frac{1}{h}\right)^{h}$ as h takes on increasingly larger values.

TABLE 4.3

h	$\left(1 + \dfrac{1}{h}\right)^{h}$
1	2
2	2.25
10	2.59374
100	2.70481
1000	2.71692
10,000	2.71815
100,000	2.71827
1,000,000	2.71828

Table 4.3 suggests that as h gets larger and larger, $\left(1 + \frac{1}{h}\right)^{h}$ gets closer and closer to a fixed number. This observation can be proven, and the fixed number is denoted by e in honor of the famous mathematician Leonhard Euler (pronounced "oiler"). The number e, an irrational number, is sometimes called the **Euler number**.

The value of e to 15 places is

$$e = 2.718281828459045.$$

We sometimes write

$$e = \lim_{h \to \infty} \left(1 + \frac{1}{h}\right)^{h},$$

which means that when h is very large, $\left(1 + \frac{1}{h}\right)^{h}$ has a value very close to e. Note that in equation (5), as n gets very large, the quantity $h = \frac{n}{r}$ also gets very large because r is fixed. Therefore, the expression inside the brackets approaches e; so the compounded amount $A = P\left(1 + \frac{r}{n}\right)^{nt}$ approaches $A = Pe^{rt}$.

Continuous Compounding   When interest is compounded continuously, the amount A after one year, with principal P and interest rate r (written as a decimal number), is Pe^{r}; the amount A after t years is given by the following formula.

> **CONTINUOUS COMPOUND INTEREST FORMULA**
>
> $$A = Pe^{rt} \qquad (6)$$
>
> A = amount after t years
> P = principal
> r = annual interest rate (expressed as a decimal number)
> t = number of years

TECHNOLOGY CONNECTION

 Graphs of $y_1 = e$ and $y_2 = \left(1 + \frac{1}{x}\right)^{x}$ in the window $0 < x < 20$, $0 < y < 3$ are shown.

As $x \to \infty$, $\left(1 + \frac{1}{x}\right)^{x} \to e$.

TECHNOLOGY CONNECTION

 Most calculators have the $\boxed{e^x}$ or $\boxed{\text{EXP}}$ key for calculating e^x for a given value of x.

On a TI-83 calculator, to evaluate a number such as $\boxed{e^5}$, you press $\boxed{e^x}$ 5 $\boxed{\text{ENTER}}$ and read the display 148.4131591.

EXAMPLE 4   **Calculating Continuous Compound Interest**

Find the amount when a principal of $8300 is invested at a 7.5% annual rate of interest compounded continuously for eight years and three months.

SOLUTION

We use formula (6), with $P = \$8300$ and $r = 0.075$. We convert eight years and three months to 8.25 years.

$$A = \$8300e^{(0.075)(8.25)} \qquad \text{Use the formula } A = Pe^{rt}.$$
$$\approx \$15{,}409.83 \qquad \text{Use a calculator.} \qquad ■\,■\,■$$

Practice Problem 4   Repeat Example 4 assuming that $9000 is invested at a 6% annual rate.   ■

EXAMPLE 5   **Calculating the Amount of Repaying a Loan**

How much money did the government owe DeHaven's descendants for 213 years on the $450,000 loan at the interest rate of 6%?

SOLUTION

a. With simple interest,

$$A = P + Prt = P(1 + rt)$$
$$= \$450{,}000\,[1 + (0.06)(213)] = \$6.201 \text{ million.}$$

b. With interest compounded yearly,

$$A = P(1 + r)^t = \$450{,}000\,(1 + 0.06)^{213}$$
$$\approx \$1.105 \times 10^{11} = \$110.500 \text{ billion.}$$

c. With interest compounded quarterly,

$$A = P\left(1 + \frac{r}{4}\right)^{4t} = \$450{,}000 \left(1 + \frac{0.06}{4}\right)^{4(213)}$$
$$\approx \$1.45305 \times 10^{11} = \$145.305 \text{ billion.}$$

d. With interest compounded continuously,

$$A = Pe^{rt} = \$450{,}000e^{0.06\,(213)}$$
$$\approx \$1.5977 \times 10^{11} = \$159.770 \text{ billion.}$$

Notice the dramatic difference between quarterly and continuous compounding and the dramatic difference between simple and compound interest.   ■ ■ ■

Practice Problem 5   Repeat Example 5 assuming that the interest rate is 4%.   ■

4 Graph exponential functions.

The Natural Exponential Function

The exponential function

$$f(x) = e^x$$

with base e is so prevalent in the sciences that it is often referred to as *the* exponential function or the natural exponential function. We use a calculator to find e^x to two decimal places for $x = -2, -1, 0, 1,$ and 2 in Table 4.4.

TABLE 4.4

x	e^x
-2	0.14
-1	0.37
0	1
1	2.72
2	7.39

The graph of $f(x) = e^x$ is sketched in Figure 4.7 by using the ordered pairs in Table 4.4.

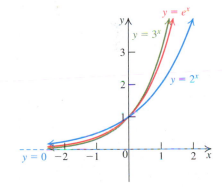

FIGURE 4.7 The graph of $y = e^x$ is between the graphs $y = 2^x$ and $y = 3^x$.

Because $2 < e < 3$, the graph of $y = e^x$ lies between the graphs $y = 2^x$ and $y = 3^x$. The function $f(x) = e^x$ has all of the properties of exponential functions with base $a > 1$ listed in Section 4.1.

We can apply the transformations from Section 2.7 to the natural exponential function.

EXAMPLE 6   **Sketching a Graph**

Use transformations to sketch the graph of
$$g(x) = e^{x-1} + 2.$$

SOLUTION

We start with the natural exponential function $f(x) = e^x$ and shift its graph one unit right to obtain the graph of $y = e^{x-1}$.

We then shift the graph of $y = e^{x-1}$ up two units to obtain the graph of $g(x) = e^{x-1} + 2$. See Figure 4.8.

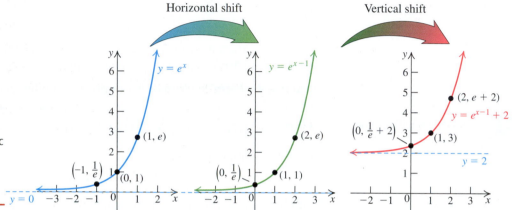

FIGURE 4.8 **Graphing $g(x) = e^{x-1} + 2$**

Practice Problem 6   Sketch the graph of $g(x) = -e^{x-1} - 2$.

Numerous applications of the exponential function are based on the fact that many different quantities have a growth (or decay) rate proportional to their size. Using calculus, we can show that in such cases, we obtain the following mathematical model.

MODEL FOR EXPONENTIAL GROWTH OR DECAY

$$A(t) = A_0 e^{kt} \quad (7)$$

$A(t) =$ the amount at time t

$A_0 = A(0)$, the initial amount

$k =$ relative rate of growth $(k > 0)$ or decay $(k < 0)$

$t =$ time

---

5 Evaluate exponential functions.

EXAMPLE 7

In 2000, the human population of the world was approximately 6 billion and the annual rate of growth was about 2.1%. Using the model given in equation (7), estimate the population of the world in the following years.

a. 2030   **b.** 1990

SOLUTION

a. The year 2000 corresponds to $t = 0$. So $A_0 = 6$ (billion), $k = 0.021$, and 2030 corresponds to $t = 30$.

$$A(30) = 6e^{(0.021)(30)} \qquad A(t) = A_0e^{kt}$$
$$A(30) \approx 11.265663 \qquad \text{Use a calculator.}$$

Thus, the model predicts that if the rate of growth is 2.1% per year, over 11.26 billion people will be in the world in 2030.

b. The year 1990 corresponds to $t = -10$ (because 1990 is ten years prior to 2000). We have

$$A(-10) = 6e^{(0.021)(-10)} \qquad A(t) = A_0e^{kt}$$
$$= 6e^{(-0.21)} \qquad \text{Simplify.}$$
$$\approx 4.8635055 \qquad \text{Use a calculator.}$$

Thus, the model predicts that the world had over 4.86 billion people in 1990. (The actual population in 1990 was 5.28 billion.)   ■ ■ ■

Practice Problem 7   Repeat Example 7 assuming that the annual rate of growth was 2.3%.   ■

SECTION 4.2 ■ Exercises

A EXERCISES Basic Skills and Concepts

1. The formula for simple interest is $I =$ _____.

2. The value of e to two decimal places is _____.

3. The formula for continuous compound interest is $A =$ _____.

4. As x gets larger and larger, the value of $\left(1 + \dfrac{1}{x}\right)^x$ gets closer to _____.

5. *True or False* Let $y = e^{-x}$. As $x \to \infty$, $y \to 0$.

6. *True or False* The graph of $y = e^x - 1$ is obtained by shifting the graph of $y = e^x$ horizontally one unit to the right.

In Exercises 7–12, find the simple interest and amount for each value of principal P, rate r per year, and time t.

7. $P = \$5000, r = 10\%, t = 5$ years

8. $P = \$10,000, r = 5\%, t = 10$ years

9. $P = \$6750, r = 4.6\%, t = 4$ years and 6 months

10. $P = \$8620, r = 5.7\%, t = 6$ years and 3 months

11. $P = \$7800, r = 6\dfrac{7}{8}\%, t = 10$ years and 9 months

12. $P = \$8670, r = 4\dfrac{1}{8}\%, t = 6$ years and 8 months

In Exercises 13–18, find (a) the future value of the given principal P and (b) the interest earned in the given period.

13. $P = \$5250$ at 8% compounded quarterly for 10 years

14. $P = \$3500$ at 6.5% compounded annually for 13 years

15. $P = \$6240$ at 7.5% compounded monthly for 12 years

16. $P = \$8000$ at 6.5% compounded daily for 15 years

17. $P = \$7500$ at 5% compounded continuously for 10 years

18. $P = \$8500$ at $4\dfrac{3}{4}\%$ compounded continuously for 8 years

In Exercises 19–22, find the principal P that will generate the given future value A.

19. $A = \$10,000$ at 8% compounded annually for 10 years

20. $A = \$10,000$ at 8% compounded quarterly for 10 years

21. $A = \$10,000$ at 8% compounded daily for 10 years

22. $A = \$10,000$ at 8% compounded continuously for 10 years

In Exercises 23–30, starting with the graph of $y = e^x$, use transformations to sketch the graph of each function and state its horizontal asymptote.

23. $f(x) = e^{-x}$

24. $f(x) = -e^x$

25. $f(x) = e^{x-2}$

26. $f(x) = e^{2-x}$

27. $f(x) = 1 + e^x$

28. $f(x) = 2 - e^{-x}$

29. $f(x) = -e^{x-2} + 3$

30. $f(x) = 3 + e^{2-x}$

In Exercises 31–42, start with the graph of $y = e^x$ (a) describe a sequence of transformations that results in the graph of $y = f(x)$, (b) find the range of $f(x)$, and (c) find the horizontal asymptote for the graph of f.

31. $f(x) = e^{x/2} - 1$

32. $f(x) = e^{x/3} + 1$

33. $f(x) = e^{3x} + 2$

34. $f(x) = e^{2x} - 2$

35. $f(x) = e^{2x-1} - 3$

36. $f(x) = e^{2x+1} + 3$

37. $f(x) = e^{3-2x} + 4$

38. $f(x) = e^{5-2x} - 4$

39. $f(x) = -2e^{3x+1} + 3$

40. $f(x) = 3e^{3x-1} - 2$

41. $f(x) = 3e^{2-3x} - 4$

42. $f(x) = -5e^{3-2x} + 4$

B EXERCISES Applying the Concepts

43. Price appreciation. In 2008, the median price of a house in Miami was $190,000. Assuming a rate of increase of 3% per year, what can we expect the price of such a house to be in 2013?

44. Price appreciation. In 2008, the median price of a compact car was $16,800. If the rate of increase is assumed to be 4% per year, what can we expect the price of such a car to be in 2010?

45. Investment. How much should a mother invest at the time her son is born to provide him with $80,000 at age 21? Assume that the interest is 7% compounded quarterly.

46. Investment. Sabrina is planning to retire in 20 years. She calculates that she will need $100,000 in addition to the money she will get from her retirement plans (including Social Security). How much should she invest today at 10% compounded daily to accomplish her objective?

47. Depreciation. Trans Trucking Co. purchased a truck for $80,000. The company depreciated the truck at the end of each year at the rate of 15% of its current value. What is the value of the truck at the end of the fifth year?

48. Depreciation. Suppose you bought a car for $21,600. After using the car for three years, you wreck it. The insurance company will pay you the depreciated value of the car at the rate of 20% of its previous year's value. How much would you receive from the insurance company if your car was a total loss?

49. Manhattan Island purchase. In 1626, Peter Du Minuit purchased Manhattan Island from the Native Americans for 60 Dutch guilders (about $24). Suppose the $24 was invested in 1626 at a 6% rate. How much money would that investment be worth in 2006 if the interest was
  a. simple interest.
  b. compounded annually.
  c. compounded monthly.
  d. compounded continuously.

50. Investment. Ms. Ann Scheiber retired from government service in 1941 with a monthly pension of $83 and $5000 in savings. At the time of her death in January 1995 at the age of 101, Ms. Scheiber had turned the $5000 into $22 million through shrewd investments in the stock market. She bequeathed all of it to Yeshiva University in New York City. What annual rate of return compounded annually would turn $5000 to a whopping $22 million in 54 years?

51. Population. The population of Sometown, USA, was 12,000 in 1990 and grew to 15,000 in 2000. Assume that the population will continue to grow exponentially at the same constant rate. What will the population of Sometown be in 2010? $\left[\textit{Hint: Show that } e^k = \left(\dfrac{5}{4}\right)^{1/10} . \right]$

52. Population. The population of the United States was about 280 million in 2000 and 300 million in 2006. Assume that the population will continue to grow exponentially at the same rate. What is the predicted population of the United States in 2020?

53. Medicine. Tests show that a new ointment X helps heal wounds. If A_0 square millimeters is the area of the original wound, then the area A of the wound after n days of application of the ointment X is given by $A = A_0 e^{-0.43n}$. Assuming that the area of the original wound was 10 square millimeters, find the area of the wound after ten days of application of the new ointment.

54. Medicine. In the model of Exercise 53, suppose the area of the wound after 20 days of application of the ointment is 0.1 square millimeter. What was the area of the original wound?

55. Cooling. The temperature T (in °C) of coffee at time t minutes after its removal from the microwave is given by the equation

$$T = 25 + 73e^{-0.28t}.$$

Find the temperature of the coffee at each time listed.
  a. $t = 0$
  b. $t = 10$
  c. $t = 20$
  d. after a long time

56. Medicine. The radioactive substance iodine-131 is used in measuring heart, liver, and thyroid activity. The quantity Q (in grams) remaining t days after the element is purchased is given by the equation

$$Q = 40e^{-0.086643397t}.$$

Calculate the amount at each time listed.
  a. initially purchased
  b. 8 days after purchase
  c. 16 days after purchase
  d. 24 days after purchase

57. The accompanying graph shows the dollar amount (in thousands) in Sherry's bank account, with $x = 0$ representing the year 2010.
  a. Write an exponential function that models the graph.
  b. Find the percent interest rate assuming that it is compounded annually. Round your answer to two decimal places.

c. Find the interest rate assuming that it is compounded monthly. Round your answer to two decimal places.

d. Assuming that the pattern continues, find the amount in Sherry's account in 2025.

58. The accompanying graph shows the dollar value (in thousands) of Mike's car, with $x = 0$ representing 2010.

a. Write an exponential function that models this graph.

b. Find the annual percentage rate of depreciation assuming that the car's value decays exponentially. Round your answer to two decimal places.

c. Find the anticipated value of Mike's car in 2015.

C EXERCISES Beyond the Basics

In Exercises 59–64, sketch the graph of the given function.

59. $f(x) = \begin{cases} e^{-x} & \text{if } x > 0 \\ e^{x} & \text{if } x \le 0 \end{cases}$

60. $f(x) = \begin{cases} e^{x} & \text{if } x > 0 \\ -e^{x} & \text{if } x \le 0 \end{cases}$

61. $f(x) = \begin{cases} 1 + x & \text{if } x > 0 \\ e^{x} & \text{if } x \le 0 \end{cases}$

62. $g(x) = \begin{cases} 1 + x^2 & \text{if } x > 0 \\ e^{-x} & \text{if } x \le 0 \end{cases}$

63. $f(x) = \begin{cases} e^{-x} & \text{if } x > 0 \\ 1 & \text{if } x \le 0 \end{cases}$

64. $g(x) = \begin{cases} e^{-x} & \text{if } x \le 0 \\ 1 & \text{if } 0 < x < 1 \\ x^2 & \text{if } x \ge 1 \end{cases}$

65. Let $f(x) = e^x$. Show that

a. $\dfrac{f(x + h) - f(x)}{h} = e^x \left(\dfrac{e^h - 1}{h} \right)$.

b. $f(x + y) = f(x)f(y)$.

c. $f(-x) = \dfrac{1}{f(x)}$.

66. You need the following definition for this exercise: Let $0! = 1; 1! = 1; 2! = 2 \cdot 1$; and, in general, $n! = n(n - 1)(n - 2) \cdots 3 \cdot 2 \cdot 1$. The symbol $n!$ is read "n factorial." Now let

$$S_n = 2 + \frac{1}{2!} + \frac{1}{3!} + \frac{1}{4!} + \cdots + \frac{1}{n!}.$$

Compute S_n for $n = 5, 10,$ and 15. Compare the resulting values with the value of e on page 421.

67. The hyperbolic cosine function (abbreviated as cosh x, often pronounced as "kosh x or coshine x") is defined by $\cosh x = \dfrac{e^x + e^{-x}}{2}$.

a. What is the domain of cosh x?

b. Is cosh x odd, even, or neither?

c. Sketch the graph of $y = \cosh x$ by plotting points.

68. The hyperbolic sine function (abbreviated as sinh x, often pronounced as "shine x") is defined by

$$\sinh x = \frac{e^x - e^{-x}}{2}.$$

a. What is the domain of sinh x?

b. Is sinh x odd, even, or neither?

c. Sketch the graph of $y = \sinh x$ by plotting points.

In Exercises 69 and 70, prove that the given equations are true for all real numbers.

69. $\cosh x + \sinh x = e^x$

70. $(\cosh x)^2 - (\sinh x)^2 = 1$

71. Define the **tangent hyperbolic function** (tanh x) by $\tanh x = \dfrac{e^x - e^{-x}}{e^x + e^{-x}}$ and the **secant hyperbolic function** (sech x) by sech $x = \dfrac{1}{\cosh x}$. Then show that $1 - (\tanh x)^2 = (\text{sech } x)^2$.

72. If an investment of P dollars returns A dollars after one year, the **effective annual interest rate** or **annual yield** y is defined by the equation

$$A = P(1 + y).$$

Show that if the sum of P dollars is invested at a nominal rate r per year, compounded m times per year, the effective annual yield y is given by the equation

$$y = \left(1 + \frac{r}{m} \right)^m - 1.$$

73. Calculate the effective annual yield if money is invested at 8% annual rate compounded

a. quarterly.

b. monthly.

c. daily.

d. continuously.

[*Hint:* Use the formula from Exercise 72.]

74. Sumeet's investment in a mutual fund increased from $30,000 to $45,000 in five years. What was his effective annual yield?

[*Hint:* Use $A = P(1 + y)^t$, where t is the number of years and y is the effective annual yield.]

75. Fidelity Federal offers three types of investments paying, respectively,

9.5% continuous interest, and
10% simple interest, and
9.6% interest compounded monthly.

Which investment will yield the greatest return?

76. The relative rate of growth of $A(t) = Pe^{kt}$ in the interval $[t, t + h]$ is defined by $\dfrac{A(t + h) - A(t)}{A(t)}$.

   a. Show that $\dfrac{A(t + 1) - A(t)}{A(t)} = e^k - 1$.

   b. Show that $\dfrac{A(t + h) - A(t)}{A(t)} = e^{kh} - 1$.

Critical Thinking

77. Let $f(x) = e^{-x^2}$.

   a. Calculate $f(0), f(1), f(2),$ and $f(3)$. (Use a calculator.)

   b. Find the axis of symmetry of f.

   c. Find $\lim f(x)$ as $x \to \infty$ and as $x \to -\infty$.

   d. Sketch the graph of f.

78. An important function in statistics is the probability density function, defined by

$$f(x) = \frac{1}{\sigma\sqrt{2\pi}} e^{-\frac{(x-\mu)^2}{2\sigma^2}},$$

where σ (the Greek letter sigma) and μ (the Greek letter mu) are constants with $\sigma > 0$ and μ a real number.

   a. Find the axis of symmetry of f.

   b. Find $\lim f(x)$ as $x \to \infty$ and as $x \to -\infty$.

   c. Describe how to sketch the graph of f from the graph of $y = e^{-x^2}$.

Logarithmic Functions

Before Starting this Section, Review

1. Finding the inverse of a one-to-one function (Section 2.9, page 283)
2. Polynomial and rational inequalities (Section 3.7, page 376)
3. Graphing transformations (Section 2.7, page 260)

Objectives

1. Define logarithmic functions.
2. Evaluate logarithms.
3. Find the domains of logarithmic functions.
4. Graph logarithmic functions.
5. Use logarithms to evaluate exponential equations.

THE McDONALD'S COFFEE CASE

Stella Liebeck of Albuquerque, New Mexico, was in the passenger seat of her grandson's car when she was severely burned by McDonald's coffee in February 1992. Liebeck ordered coffee that was served in a foam cup at the McDonald's drive-through window. While attempting to add cream and sugar to her coffee, Liebeck spilled the entire cup into her lap. A vascular surgeon determined that Liebeck suffered full-thickness burns (third-degree burns) over 6% of her body. She sued McDonald's for damages.

During the preliminary phase of the trial, McDonald's said that on the basis of a consultant's advice, the company brewed its coffee at 195° to 205°F and held it at 180°F to 190°F to maintain optimal taste.

Coffee served at home is generally between 135°F and 140°F. Prior to the Liebeck case, the prestigious Shriner's Burn Institute of Cincinnati had published warnings to the franchise food industry that beverages above 130°F were causing serious scald burns. Liebeck's lawsuit was settled out of court, and both parties agreed to keep the award for damages confidential.

Avoiding Further Lawsuits Corporate lawyers informed McDonald's management that further lawsuits from customers who spill their coffee can be avoided if the temperature of the coffee is 125°F or less when it is delivered to the customers. In Example 11, we learn how long to wait after brewing coffee before delivering it to customers. ∎

---

1 Define logarithmic functions.

Logarithmic Functions

Recall from Section 4.1 that every exponential function

$$f(x) = a^x, a > 0, a \neq 1,$$

is a *one-to-one* function and therefore has an inverse function. Let's find the inverse of an exponential function. First, we consider the exponential function $f(x) = 3^x$. If we think of an exponential function f as "putting an exponent on" a particular base, then the inverse function f^{-1} must "lift the exponent off" to undo the effect of f, as illustrated below.

$$x \xrightarrow{\ f\ } 3^x \xrightarrow{\ f^{-1}\ } x$$

We try to find the inverse of $f(x) = 3^x$ by the procedure outlined in Section 2.9.

Step 1 Replace $f(x)$ in the equation $f(x) = 3^x$ with y: $y = 3^x$.

Step 2 Interchange x and y in the equation in Step 1: $x = 3^y$.

Step 3 Solve the equation for y in Step 2.

In Step 3, we need to solve the equation $x = 3^y$ for y. Finding y is straightforward in equations such as $3 = 3^y$ or $27 = 3^y$, but how do you solve the equation $5 = 3^y$ for y? There actually is a real number y that solves the equation $5 = 3^y$, but we have not introduced the name for this number yet. (We were in a similar position concerning the solution of $y^3 = 7$ before we introduced the name for the solution, namely, $\sqrt[3]{7}$). We could say that y is "the exponent on 3 that gives 5." Instead, we use the word *logarithm* (or *log* for short) to describe this exponent. We use the notation $\log_3 5$ (read "log of 5 with base 3") to express the solution of the equation $5 = 3^y$. This log notation is a new way to write information about exponents: $y = \log_3 x$ means $x = 3^y$. In other words, $\log_3 x$ *is the exponent to which 3 must be raised to obtain x.* Therefore, the inverse function of the exponential function $f(x) = 3^x$ is written

$$f^{-1}(x) = \log_3 x \quad \text{or} \quad y = \log_3 x, \text{where } y = f^{-1}(x).$$

The previous discussion could be repeated for any base a instead of 3 provided that $a > 0$ and $a \neq 1$.

DEFINITION OF THE LOGARITHMIC FUNCTION

For $x > 0$, $a > 0$, and $a \neq 1$,

$$y = \log_a x \quad \text{if and only if} \quad x = a^y.$$

The function $f(x) = \log_a x$ is called the **logarithmic function with base *a*.**

The definition of the logarithm says that the two equations

$$y = \log_a x \text{ (\textbf{logarithmic form})}$$
$$x = a^y \text{ (\textbf{exponential form})}$$

are equivalent. For instance, $\log_3 81 = 4$ because 3 must be raised to the fourth power to obtain 81.

EXAMPLE 1   **Converting from Exponential to Logarithmic Form**

Write each exponential equation in logarithmic form.

a. $4^3 = 64$ **b.** $\left(\dfrac{1}{2}\right)^4 = \dfrac{1}{16}$ **c.** $a^{-2} = 7$

SOLUTION

a. $4^3 = 64$ is equivalent to $\log_4 64 = 3$.

b. $\left(\dfrac{1}{2}\right)^4 = \dfrac{1}{16}$ is equivalent to $\log_{1/2} \dfrac{1}{16} = 4$.

c. $a^{-2} = 7$ is equivalent to $\log_a 7 = -2$.   ■ ■ ■

Practice Problem 1   Write each exponential equation in logarithmic form.

a. $2^{10} = 1024$ **b.** $9^{-1/2} = \dfrac{1}{3}$ **c.** $p = a^q$ ■

EXAMPLE 2   **Converting from Logarithmic Form to Exponential Form**

Write each logarithmic equation in exponential form.

a. $\log_3 243 = 5$ **b.** $\log_2 5 = x$ **c.** $\log_a N = x$

SOLUTION

a. $\log_3 243 = 5$ is equivalent to $243 = 3^5$.

b. $\log_2 5 = x$ is equivalent to $5 = 2^x$.

c. $\log_a N = x$ is equivalent to $N = a^x$.

■ ■ ■

Practice Problem 2   Write each logarithmic equation in exponential form.

a. $\log_2 64 = 6$ b. $\log_v u = w$

■

Evaluating Logarithms

2 Evaluate logarithms.

The technique of converting from logarithmic form to exponential form can be used to evaluate some logarithms by inspection.

> **EXAMPLE 3**   **Evaluating Logarithms**

Find the value of each of the following logarithms.

a. $\log_5 25$ b. $\log_2 16$ c. $\log_{1/3} 9$ d. $\log_7 7$ e. $\log_6 1$ f. $\log_4 \dfrac{1}{2}$

SOLUTION

Logarithmic Form	Exponential Form	Value (from the one-to-one property)
a. $\log_5 25 = y$	$25 = 5^y$ or $5^2 = 5^y$	$y = 2$
b. $\log_2 16 = y$	$16 = 2^y$ or $2^4 = 2^y$	$y = 4$
c. $\log_{1/3} 9 = y$	$9 = \left(\dfrac{1}{3}\right)^y$ or $3^2 = 3^{-y}$	$y = -2$
d. $\log_7 7 = y$	$7 = 7^y$ or $7^1 = 7^y$	$y = 1$
e. $\log_6 1 = y$	$1 = 6^y$ or $6^0 = 6^y$	$y = 0$
f. $\log_4 \dfrac{1}{2} = y$	$\dfrac{1}{2} = 4^y$ or $2^{-1} = 2^{2y}$	$y = -\dfrac{1}{2}$

■ ■ ■

Practice Problem 3   Evaluate.

a. $\log_3 9$ b. $\log_9 \dfrac{1}{3}$ c. $\log_{1/2} 32$

■

> **EXAMPLE 4**   **Using the Definition of Logarithm**

Solve each equation.

a. $\log_5 x = -3$ b. $\log_3 \dfrac{1}{27} = y$

c. $\log_z 1000 = 3$ d. $\log_2 (x^2 - 6x + 10) = 1$

SOLUTION

a. $\log_5 x = -3$

$\quad\quad x = 5^{-3}$ Exponential form

$\quad\quad x = \dfrac{1}{5^3} = \dfrac{1}{125}$ $\quad a^{-n} = \dfrac{1}{a^n}$

b. $\log_3 \dfrac{1}{27} = y$

$\dfrac{1}{27} = 3^y$ Exponential form

$3^{-3} = 3^y$ $\dfrac{1}{27} = \dfrac{1}{3^3} = 3^{-3}$

$-3 = y$ If $a^u = a^v$, then $u = v$.

c. $\log_z 1000 = 3$

$1000 = z^3$ Exponential form

$10^3 = z^3$ $1000 = 10^3$

$10 = z$ Take the cube root of both sides.

d. $\log_2 (x^2 - 6x + 10) = 1$

$x^2 - 6x + 10 = 2^1$ Exponential form

$x^2 - 6x + 8 = 0$ $2^1 = 2$; subtract 2 from both sides.

$(x - 2)(x - 4) = 0$ Factor.

$x - 2 = 0$ or $x - 4 = 0$ Zero-product property

$x = 2$ or $x = 4$ Solve for x.   ▪▪▪

Practice Problem 4   Solve each equation.

a. $\log_2 x = 3$ **b.** $\log_z 125 = 3$ **c.** $\log_3(x^2 - x - 5) = 0$ ▪

3   Find the domains of logarithmic functions.

Domain of Logarithmic Functions

Because the exponential function $f(x) = a^x$ has domain $(-\infty, \infty)$ and range $(0, \infty)$, its inverse function $y = \log_a x$ has **domain $(0, \infty)$** and **range $(-\infty, \infty)$**. Therefore, the logarithms of 0 and of negative numbers are not defined; so expressions such as $\log_a(-2)$ and $\log_a(0)$ are meaningless.

EXAMPLE 5   **Finding the Domain**

Find the domain of $f(x) = \log_3(2 - x)$.

SOLUTION

Because the domain of the basic logarithmic function $y = \log_3 x$ is $(0, \infty)$, the expression $2 - x$ must be positive. The domain of f is the set of all real numbers x, where

$$2 - x > 0$$
$$2 > x \quad \text{or} \quad x < 2 \quad \text{Solve for } x.$$

The domain of f is $(-\infty, 2)$.   ▪▪▪

Practice Problem 5   Find the domain of $f(x) = \log_{10} \sqrt{1 - x}$.   ▪

Because $a^1 = a$ and $a^0 = 1$ for any base a, expressing these equations in logarithmic notation gives

$$\log_a a = 1 \quad \text{and} \quad \log_a 1 = 0.$$

Also, letting $f(x) = a^x$ and $f^{-1}(x) = \log_a x$, the equation

$$f^{-1}(f(x)) = x \text{ can be written as } f^{-1}(a^x) = \log_a a^x = x$$

and the equation

$$f(f^{-1}(x)) = x \text{ can be written as } f(\log_a x) = a^{\log_a x} = x.$$

We summarize these relationships as basic properties of logarithms.

> ### BASIC PROPERTIES OF LOGARITHMS
>
> For any base $a > 0$, with $a \neq 1$,
>
> **1.** $\log_a a = 1$.
> **2.** $\log_a 1 = 0$.
> **3.** $\log_a a^x = x$ for any real number x.
> **4.** $a^{\log_a x} = x$ for any $x > 0$.

4 Graph logarithmic functions.

Graphs of Logarithmic Functions

We now look at other properties of logarithmic functions and begin with an example showing how to graph a logarithmic function.

EXAMPLE 6   Sketching a Graph

Sketch the graph of $y = \log_3 x$.

RECALL

Remember that "$\log_3 x$" means the exponent on 3 that gives x.

Plotting points (Method 1) To find selected ordered pairs on the graph of $y = \log_3 x$, we choose the x-values to be powers of 3. We can easily compute the logarithms of these values by using Property 3, namely, $\log_3 3^x = x$. We compute the y values in Table 4.5.

FIGURE 4.9 **Graph by plotting points**

TABLE 4.5

x	$y = \log_3 x$	(x, y)
$\dfrac{1}{27}$	Because $3^{-3} = \dfrac{1}{27}$, $y = \log_3 \dfrac{1}{27} = -3$.	$\left(\dfrac{1}{27}, -3\right)$
$\dfrac{1}{9}$	Because $3^{-2} = \dfrac{1}{9}$, $y = \log_3 \dfrac{1}{9} = -2$.	$\left(\dfrac{1}{9}, -2\right)$
$\dfrac{1}{3}$	Because $3^{-1} = \dfrac{1}{3}$, $y = \log_3 \dfrac{1}{3} = -1$.	$\left(\dfrac{1}{3}, -1\right)$
1	Because $3^0 = 1$, $y = \log_3 1 = 0$.	$(1, 0)$
3	Because $3^1 = 3$, $y = \log_3 3 = 1$.	$(3, 1)$
9	Because $3^2 = 9$, $y = \log_3 9 = 2$.	$(9, 2)$

Plotting these ordered pairs and connecting them with a smooth curve gives us the graph of $y = \log_3 x$, shown in Figure 4.9.

Using the inverse function (Method 2) Because $y = \log_3 x$ is the inverse of the function $y = 3^x$, we first graph the exponential function $y = 3^x$, then we reflect the graph of $y = 3^x$ in the line $y = x$. Both graphs are shown in Figure 4.10. ■ ■ ■

Practice Problem 6 Sketch the graph of

$$y = \log_2 x.$$

FIGURE 4.10 **Graph by using the inverse function**

We can graph $y = \log_{1/3} x$ by plotting points or by using the graph of its inverse. See Figure 4.11. The graph of $y = \log_3 x$ shown in Figure 4.10 is increasing. The graph of $y = \log_{1/3} x$ shown in Figure 4.11, on the other hand, is decreasing.

In general, logarithmic functions have two basic shapes, determined by the base a. If $a > 1$, the graph of $y = \log_a x$ is increasing, as in Figure 4.12(a). If $0 < a < 1$, the graph is decreasing, as in Figure 4.12(b).

FIGURE 4.11

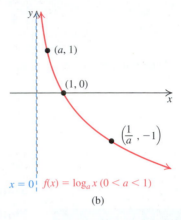

FIGURE 4.12 Basic shapes of $y = \log_a x$

PROPERTIES OF EXPONENTIAL AND LOGARITHMIC FUNCTIONS	
Exponential Function $y = a^x$	**Logarithmic Function $y = \log_a x$**
1. The domain is $(-\infty, \infty)$, and the range is $(0, \infty)$.	The domain is $(0, \infty)$, and the range is $(-\infty, \infty)$.
2. The y-intercept is 1, and there is no x-intercept.	The x-intercept is 1, and there is no y-intercept.
3. The x-axis ($y = 0$) is the horizontal asymptote.	The y-axis ($x = 0$) is the vertical asymptote.
4. The graph is a continuous smooth curve that passes through the points $\left(-1, \frac{1}{a}\right)$, $(0, 1)$, and $(1, a)$.	The graph is a continuous smooth curve that passes through the points $\left(\frac{1}{a}, -1\right)$, $(1, 0)$, and $(a, 1)$.
5. The function is one-to-one; that is, $a^u = a^v$ if and only if $u = v$.	The function is one-to-one; that is, $\log_a u = \log_a v$ if and only if $u = v$.
6. The function is increasing if $a > 1$ and decreasing if $0 < a < 1$.	The function is increasing if $a > 1$ and decreasing if $0 < a < 1$.

Once we know the shape of a logarithmic graph, we can shift it vertically or horizontally, stretch it, compress it, check answers with it, and interpret the graph.

EXAMPLE 7 **Using Transformations**

Start with the graph of $f(x) = \log_3 x$ and use transformations to sketch the graph of each function.

a. $f(x) = \log_3 x + 2$ **b.** $f(x) = \log_3 (x - 1)$

c. $f(x) = -\log_3 x$ **d.** $f(x) = \log_3 (-x)$

State the domain and range and the vertical asymptote for the graph of each function.

SOLUTION

We start with the graph of $f(x) = \log_3 x$ and use the transformations shown in Figure 4.13.

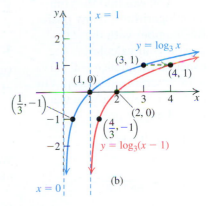

a. $f(x) = \log_3 x + 2$

Adding 2 to $\log_3 x$ shifts the graph two units up. Domain: $(0, \infty)$; range: $(-\infty, \infty)$; vertical asymptote: $x = 0$ (y-axis)

b. $f(x) = \log_3 (x - 1)$

Replacing x with $x - 1$ in $\log_3 x$ shifts the graph one unit right. Domain: $(1, \infty)$; range $(-\infty, \infty)$; vertical asymptote: $x = 1$

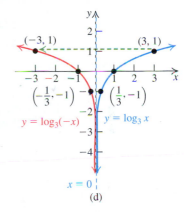

c. $f(x) = -\log_3 (x)$

Multiplying $\log_3 x$ by -1 reflects the graph in the x-axis. Domain: $(0, \infty)$; range: $(-\infty, \infty)$; vertical asymptote: $x = 0$

d. $f(x) = \log_3 (-x)$

Replacing x with $-x$ in $\log_3 x$ reflects the graph in the y-axis. Domain: $(-\infty, 0)$; range: $(-\infty, \infty)$; vertical asymptote: $x = 0$

FIGURE 4.13 Transformations on $y = \log_3 x$ ■ ■ ■

Practice Problem 7   Use transformations to sketch the graph of
$$y = -\log_2 (x - 3).$$
■

Common Logarithms

The logarithm with base 10 is called the **common logarithm** and is denoted by omitting the base; so
$$\log x = \log_{10} x.$$

$$y = \log x\ (x > 0) \quad \text{if and only if} \quad x = 10^y.$$

TECHNOLOGY CONNECTION

You can find the common logarithms of any positive number with the $\boxed{\text{LOG}}$ key on your calculator. The exact key sequence will vary with different calculators. On most calculators, if you want to find log 3.7, press

3.7 $\boxed{\text{LOG}}$ $\boxed{\text{ENTER}}$

or

$\boxed{\text{LOG}}$ 3.7 $\boxed{\text{ENTER}}$

and see the display
0.5682017.
Although the display reads 0.5682017, this number is an approximate value; so you should write
$\log_{10} 3.7 \approx 0.5682017$.

Applying the basic properties of logarithms (see page 433) to common logarithms, we have the following:

1. $\log 10 = 1$
2. $\log 1 = 0$
3. $\log 10^x = x$
4. $10^{\log x} = x$

We use Property 3 to evaluate the common logarithms of numbers that are powers of 10. For example,

$$\log 1000 = \log 10^3 = 3$$
$$\text{and} \quad \log 0.01 = \log 10^{-2} = -2.$$

We use a calculator by pressing the LOG key to find the common logarithms of numbers that are not powers of 10.

The graph of $y = \log x$ is similar to the graph in Figure 4.12(a) because the base for $\log x$ is understood to be 10.

EXAMPLE 8 **Using Transformations to Sketch a Graph**

Sketch the graph of

$$y = 2 - \log(x - 2).$$

SOLUTION

We start with the graph of the function $f(x) = \log x$ and use the order of transformations (see page 260) indicated in Figure 4.14 to sketch the graph.

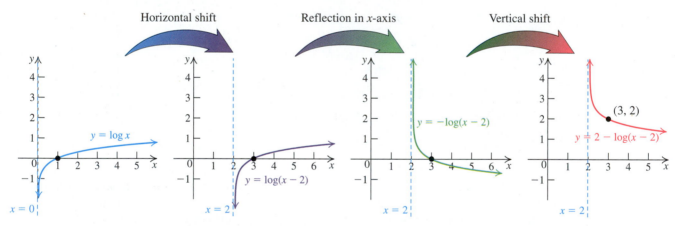

Step 1 Replacing x with $x - 2$ in $\log x$ shifts the graph of $y = \log x$ two units right.

Step 2 Multiplying $\log(x - 2)$ by -1 reflects the graph of $y = \log(x - 2)$ in the x-axis.

Step 3 Adding 2 to $-\log(x - 2)$ shifts the graph of $y = -\log(x - 2)$ two units up.

FIGURE 4.14 Transformations on $y = \log x$

The domain of $f(x) = 2 - \log(x - 2)$ is $(2, \infty)$, the range is $(-\infty, \infty)$, and the vertical asymptote is the line $x = 2$. ■ ■ ■

Practice Problem 8 Sketch the graph of

$$y = \log(x - 3) - 2.$$

■

TECHNOLOGY CONNECTION

To evaluate natural logarithms, we use the $\boxed{\text{LN}}$ key on a calculator. To find $\ln(2.3)$ on most calculators, press

2.3 $\boxed{\text{LN}}$ $\boxed{\text{ENTER}}$

or

$\boxed{\text{LN}}$ 2.3 $\boxed{\text{ENTER}}$

and see the display 0.832909123.

Natural Logarithms

In most applications in calculus and the sciences, the convenient base for logarithms is the number e. The logarithms with base e are called **natural logarithms** and are denoted by $\ln x$ (read "ell en x" or "lawn x") so that

$$\ln x = \log_e x.$$

$$y = \ln x \,(x > 0) \quad \text{if and only if} \quad x = e^y.$$

Applying the basic properties of logarithms (see page 433) to natural logarithms, we have the following:

1. $\ln e = 1$
2. $\ln 1 = 0$
3. $\ln e^x = x$
4. $e^{\ln x} = x$

We can use Property 3 to evaluate the natural logarithms of powers of e.

EXAMPLE 9   **Evaluating the Natural Logarithm Function**

Evaluate each expression.

a. $\ln e^4$ **b.** $\ln \dfrac{1}{e^{2.5}}$ **c.** $\ln 3$

SOLUTION

a. $\ln e^4 = 4$ — Property (3)

b. $\ln \dfrac{1}{e^{2.5}} = \ln e^{-2.5} = -2.5$ — Property (3)

c. $\ln 3 \approx 1.0986123$ — Use a calculator.   ■ ■ ■

Practice Problem 9   Evaluate each expression.

a. $\ln \dfrac{1}{e}$ **b.** $\ln 2$ ■

We summarize the basic properties of the three logarithms with bases a, 10, and e.

Basic Properties of Logarithms			
(base a)	$\log_a a = 1$	$\log_a 1 = 0$	$\log_a a^x = x$	$a^{\log_a x} = x$
(base 10)	$\log 10 = 1$	$\log 1 = 0$	$\log 10^x = x$	$10^{\log x} = x$
(base e)	$\ln e = 1$	$\ln 1 = 0$	$\ln e^x = x$	$e^{\ln x} = x$

5 Use logarithms to solve exponential equations.

Investments

EXAMPLE 10   **Doubling Your Money**

a. How long will it take to double your money if it earns 6.5% compounded continuously?

b. At what rate of return, compounded continuously, would your money double in five years?

SOLUTION

If P dollars is invested and you want to double it, the final amount $A = 2P$.

a.

$A = Pe^{rt}$	Continuous compounding formula
$2P = Pe^{0.065t}$	$A = 2P, r = 0.065$
$2 = e^{0.065t}$	Divide both sides by P.
$\ln 2 = 0.065t$	Logarithmic form of $x = e^y$
$\dfrac{\ln 2}{0.065} = t$	Divide both sides by 0.065.
$t \approx 10.66$	Use a calculator.

It will take about 11 years to double your money.

b.

$A = Pe^{rt}$	Continuous compounding formula
$2P = Pe^{5r}$	$A = 2P, t = 5$
$2 = e^{5r}$	Divide both sides by P.
$\ln 2 = 5r$	Logarithmic form of $x = e^y$
$\dfrac{\ln 2}{5} = r$	Divide both sides by 5.
$r \approx 0.1386$	Use a calculator.

Your investment will double in five years at the rate of 13.86%. ■ ■ ■

Practice Problem 10 Repeat Example 10 for tripling $(A = 3P)$ your money. ■

Newton's Law of Cooling

When a cool drink is removed from a refrigerator and placed in a room at normal temperature, the drink warms to the temperature of the room. When removed from the oven, a pizza baked at a high temperature cools to the temperature of the room.

In situations such as these, the rate at which an object's temperature changes at any given time is proportional to the difference between its temperature and the temperature of the surrounding medium. This observation is called *Newton's Law of Cooling*, although, as in the case of a cool drink, it applies to warming as well.

NEWTON'S LAW OF COOLING

Newton's Law of Cooling states that

$$T = T_S + (T_0 - T_S)e^{-kt},$$

where T is the temperature of the object at time t, T_S is the surrounding temperature, and T_0 is the value of T at $t = 0$.

EXAMPLE 11 **McDonald's Hot Coffee**

The local McDonald's franchise has discovered that when coffee is poured from a coffeemaker whose contents are 180°F into a noninsulated pot, after one minute, the coffee cools to 165°F if the room temperature is 72°F. How long should the employees wait before pouring the coffee from this noninsulated pot into cups to deliver it to customers at 125°F?

SOLUTION

We use Newton's Law of Cooling with $T_0 = 180$ and $T_S = 72$.

$$T = 72 + (180 - 72)e^{-kt}$$
$$T = 72 + 108e^{-kt} \qquad (1) \qquad \text{Simplify.}$$

From the given data, we have $T = 165$ when $t = 1$.

$$165 = 72 + 108e^{-k} \qquad \text{Substitute data into equation (1).}$$
$$93 = 108e^{-k} \qquad \text{Subtract 72 from both sides.}$$
$$\frac{93}{108} = e^{-k} \qquad \text{Solve for } e^{-k}.$$
$$\ln\left(\frac{93}{108}\right) = -k \qquad \text{Write in logarithmic form.}$$
$$-0.1495317 \approx -k \qquad \text{Use a calculator.}$$
$$k \approx 0.1495317$$

With this value of k, equation (1) becomes

$$T = 72 + 108e^{-0.1495317t} \qquad (2)$$
$$125 = 72 + 108e^{-0.1495317t} \qquad \text{Replace } T \text{ with 125 in equation (2).}$$
$$\frac{125 - 72}{108} = e^{-0.1495317t} \qquad \text{Solve for } e^{-0.1495317t}.$$
$$\frac{53}{108} = e^{-0.1495317t} \qquad \text{Simplify.}$$
$$\ln\left(\frac{53}{108}\right) = -0.1495317t \qquad \text{Write in logarithmic form.}$$
$$t = \frac{-1}{0.1495317}\ln\left(\frac{53}{108}\right) \qquad \text{Solve for } t.$$
$$\approx 4.76 \qquad \text{Use a calculator.}$$

The employees should wait approximately 4.76 minutes (realistically, about 5 minutes) to deliver the coffee at 125°F to the customers. ■ ■ ■

Practice Problem 11   Repeat Example 11 assuming that the coffee is to be delivered to the customers at 120°F. ■

In the next example, we use the mathematical model

$$A = A_0 e^{rt}, \qquad (3)$$

where A is the quantity after t units of time, A_0 is the initial (original) quantity when the experiment started ($t = 0$), and r is the growth or decay rate per period.

EXAMPLE 12   **Chemical Toxins in a Lake**

A chemical spill deposits 60,000 cubic meters of soluble toxic waste into a large lake. If 20% of the waste is removed every year, how many years will it take to reduce the toxin to 1000 cubic meters?

SOLUTION

In the equation $A = A_0 e^{rt}$, we need to find A_0, r, and the time t when $A = 1000$.

1. **Find A_0.** Initially ($t = 0$), we are given $A_0 = 60,000$. So our equation is

$$A = 60,000e^{rt}. \qquad (4)$$

2. Find *r*. When $t = 1$ year, the amount of toxin will be 80% of its initial value, or $\dfrac{80}{100}(60{,}000) = 48{,}000$. So

$$48{,}000 = 60{,}000e^{r(1)} \qquad \text{Substitute in equation (4).}$$

$$\frac{48{,}000}{60{,}000} = e^r \qquad \text{Divide both sides by 60,000.}$$

$$\frac{4}{5} = e^r \qquad \text{Simplify.}$$

$$\ln\left(\frac{4}{5}\right) = r \qquad \text{Write in logarithmic form.}$$

3. Find *t* when *A* = 1000.

$$1000 = 60{,}000e^{t\ln\left(\frac{4}{5}\right)} \qquad \text{Substitute the value of } r \text{ in equation (4).}$$

$$\frac{1}{60} = e^{t\ln\left(\frac{4}{5}\right)} \qquad \text{Divide by 60,000 and simplify.}$$

$$\ln\left(\frac{1}{60}\right) = t\ln\left(\frac{4}{5}\right) \qquad \text{Write in logarithmic form.}$$

$$t = \frac{\ln\left(\dfrac{1}{60}\right)}{\ln\left(\dfrac{4}{5}\right)} \qquad \text{Solve for } t.$$

$$t \approx 18.3 \qquad \text{Use a calculator.}$$

It will take approximately 18 years to reduce toxin to 1000 m^3. ■ ■ ■

Practice Problem 12 Repeat Example 12 assuming that 25% of the waste is removed every year. ■

SECTION 4.3 ■ Exercises

A EXERCISES Basic Skills and Concepts

1. The domain of the function $y = \log_a x$ is _____, and its range is _____.

2. The logarithmic form $y = \log_a x$ is equivalent to the exponential form _____.

3. The logarithm with base 10 is called the _____ logarithm, and the logarithm with base e is called the _____ logarithm.

4. $a^{\log_a x} = $ _____, and $\log_a a^x = $ _____.

5. *True or False* The graph of $\log_a x, a > 0, a \neq 1$, is an increasing function.

6. *True or False* The graph of $y = \log_a x, a > 0, a \neq 1$, has no horizontal asymptote.

In Exercises 7–18, write each exponential equation in logarithmic form.

7. $5^2 = 25$

8. $81^{1/2} = 9$

9. $(49)^{-\frac{1}{2}} = \frac{1}{7}$

10. $\left(\frac{1}{16}\right)^{1/2} = \frac{1}{4}$

11. $\left(\frac{1}{16}\right)^{-\frac{1}{2}} = 4$

12. $(a^2)^2 = a^4$

13. $10^0 = 1$

14. $10^4 = 10{,}000$

15. $(10)^{-1} = 0.1$

16. $3^x = 5$

17. $a^2 = 5$

18. $a^e = \pi$

In Exercises 19–30, write each logarithmic equation in exponential form.

19. $\log_2 32 = 5$

20. $\log_7 49 = 2$

21. $\log_{10} 100 = 2$

22. $\log_{10} 10 = 1$

23. $\log_{10} 1 = 0$

24. $\log_a 1 = 0$

25. $\log_{10} 0.01 = -2$

26. $\log_{1/5} 5 = -1$

27. $\log_8 2 = \frac{1}{3}$

28. $\log 1000 = 3$

29. $\ln 2 = x$

30. $\ln \pi = a$

In Exercises 31–40, evaluate each expression without using a calculator.

31. $\log_5 125$

32. $\log_9 81$

33. $\log 10{,}000$

34. $\log_3 \dfrac{1}{3}$

35. $\log_2 \dfrac{1}{8}$

36. $\log_4 \dfrac{1}{64}$

37. $\log_3 \sqrt{27}$

38. $\log_{27} 3$

39. $\log_{16} 2$

40. $\log_5 \sqrt{125}$

In Exercises 41–52, solve each equation.

41. $\log_5 x = 2$

42. $\log_5 x = -2$

43. $\log_5 (x - 2) = 3$

44. $\log_5 (x + 1) = -2$

45. $\log_2 \left(\dfrac{1}{16} \right) = x$

46. $\log_x \left(\dfrac{1}{5} \right) = -1$

47. $\log_{16} \sqrt{x - 1} = \dfrac{1}{4}$

48. $\log_{27} \sqrt[3]{1 - x} = \dfrac{1}{3}$

49. $\log_a (x - 1) = 0$

50. $\log_a (x^2 + 5x + 7) = 0$

51. $\log_2 (x^2 - 7x + 14) = 1$

52. $\log (x^2 + 5x + 16) = 1$

In Exercises 53–60, find the domain of each function.

53. $f(x) = \log_2 (x + 1)$ **54.** $f(x) = \log_2 (x + 3)$

55. $g(x) = \log_3 (x - 5)$ **56.** $g(x) = \log_3 (x - 8)$

57. $f(x) = \log_2 \sqrt{x - 1}$ **58.** $f(x) = \log_2 \sqrt{x - 3}$

59. $g(x) = \log_4 (-x)$ **60.** $g(x) = \log_4 \sqrt{-x}$

61. Match each logarithmic function with one of the graphs labeled (a)–(f).
  a. $f(x) = \log x$
  b. $f(x) = -\log |x|$
  c. $f(x) = -\log (-x)$
  d. $f(x) = \log (x - 1)$
  e. $y = (\log x) - 1$
  f. $f(x) = \log (-1 - x)$

(a)

(b)

(c)   (d)

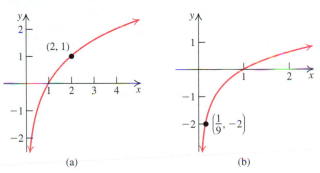

(e)   (f)

62. For each graph, find a function of the form $f(x) = \log_a x$ that represents the graph.

(a)   (b)

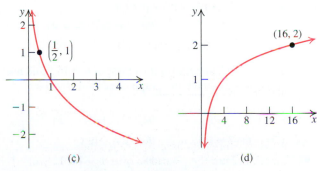

(c)   (d)

In Exercises 63–66, find an equation of the form $y = d + \log \left[\pm(x - c) \right]$ for the given graph.

63.   **64.**

65.

66.

In Exercises 67–74, graph the given function by using transformations on the appropriate basic graph of the form $y = \log_a x$. State the domain and range of the function and the vertical asymptote of the graph.

67. $f(x) = \log_4 (x + 3)$ 68. $g(x) = \log_{1/2} (x - 1)$

69. $y = -\log_5 x$ 70. $y = 2 \log_7 (x)$

71. $y = \log_{1/5} (-x)$ 72. $y = 1 + \log_{1/5} (-x)$

73. $y = |\log_3 x|$ 74. $y = \log_3 |x|$

In Exercises 75–82, begin with the graph of $f(x) = \log_2 x$ and use transformations to sketch the graph of each function. Find the domain and range of the function and the vertical asymptote of the graph.

75. $y = \log_2 (x - 1)$ 76. $y = \log_2 (-x)$

77. $y = \log_2 (3 - x)$ 78. $y = \log_2 x$

79. $y = 2 + \log_2 (3 - x)$ 80. $y = 4 - \log_2 (3 - x)$

81. $y = \log_2 |x|$ 82. $y = \log_2 x^2$

In Exercises 83–88, evaluate each expression.

83. $\log_4 (\log_3 81)$ 84. $\log_4 [\log_3 (\log_2 8)]$

85. $5^{\log_5 7}$ 86. $10^{\log 13}$

87. $e^{\ln 5}$ 88. $e^{\ln e^2}$

In Exercises 89–92, solve each equation.

89. $\ln e^{x^2 - 3x} = 10$ 90. $\log 10^{21 - x^2} = 4x$

91. $10^{\log(2x+3)} = 11$ 92. $e^{\ln (x^2 - 2x)} = 8$

In Exercises 93–98, begin with the graph of $y = \ln x$ and use transformations to sketch the graph of each of the given functions.

93. $y = \ln (x + 2)$ 94. $y = \ln (2 - x)$

95. $y = -\ln (2 - x)$ 96. $y = 2 \ln x$

97. $y = 3 - 2 \ln x$ 98. $y = 1 - \ln (1 - x)$

B EXERCISES Applying the Concepts

In Exercises 99–106, use the model $A = A_0 e^{kt}$.

99. **Doubling your money.** How long would it take to double your money if you invested P dollars at the rate of 8% compounded continuously?

100. **Investment goal.** How long would it take to grow your investment from $10,000 to $120,000 at the rate of 10% compounded continuously?

101. **Rate for doubling your money.** At what annual rate of return, compounded continuously, would your investment double in six years?

102. **Rate-of-investment goal.** At what annual rate of return, compounded continuously, would your investment grow from $8000 to $50,000 in 25 years?

103. **Population of Canada.** The population of Canada was 31.3 million in 2000 and 33.4 million in 2007. Determine the time from 2000 until Canada's population (a) doubles and (b) triples.

104. **Population of Canada.** In Exercise 103, find the time from 2000 until Canada's population reaches 50 million.

105. **Infectious disease.** Suppose during any month, the number of cases of a disease is reduced by 5%. If at present there are 20,000 cases, how long will it take to reduce the number to 1000?

106. **Toxic chemicals in a lake.** Sixteen thousand cubic meters of soluble toxic chemical spill takes place in a lake. Fifteen percent of the toxin is removed every year. Let $T(n)$ represent the amount of toxin left after n years.
   a. Find a formula for $T(n)$.
   b. How much toxin will be left after 12 years?
   c. When will 80% of the toxin be eliminated?

107. **Newton's Law of Cooling.** A thermometer is taken from a room at 75°F to the outdoors, where the temperature is 20°F. The reading on the thermometer drops to 50°F after one minute.
   a. Find the reading on the thermometer after
      (i) five minutes.
      (ii) ten minutes.
      (iii) one hour.
   b. How long will it take for the reading to drop to 22°F?

108. **Newton's Law of Cooling.** The last bit of ice in a picnic cooler has melted ($T_0 = 32$°F). The temperature in the park is 85°F. After 30 minutes, the temperature in the cooler is 40°F. How long will it take for the temperature inside the cooler to reach 50°F?

109. **Cooking salmon.** A salmon filet initially at 50°F is cooked in an oven at a constant temperature of 400°F. After ten minutes, the temperature of the filet rises to 160°F. How long does it take until the salmon is medium rare at 220°F?

110. **The time of murder.** A forensic specialist took the temperature of a victim's body lying in a street at 2:10 A.M. and found it to be 85.7°F. At 2:40 A.M., the temperature of the body was 84.8°F. When was the murder committed if the air temperature during the night was 55°F? [Remember, normal body temperature is 98.6°F.]

C EXERCISES Beyond the Basics

111. Sketch the graph of $f(x) = \begin{cases} \ln x & \text{if } x > 0 \\ \ln(-x) & \text{if } x < 0 \end{cases}$

112. Sketch the graph of $f(x) = \begin{cases} \ln(1-x) & \text{if } x < 1 \\ e^{-x} & \text{if } x \geq 1 \end{cases}$

113. Find the domain of $f(x) = \log_3\left(\dfrac{x-2}{x+1}\right)$.

114. Find the domain of $g(x) = \log_2\left(\dfrac{x+3}{x-2}\right)$.

115. **Finding the domain.** Find the domain of each function.
   a. $f(x) = \log_2(\log_3 x)$
   b. $f(x) = \log(\ln(x-1))$
   c. $f(x) = \ln(\log(x-1))$
   d. $f(x) = \log(\log(\log(x-1)))$

116. **Finding the inverse.** Find the inverse of each function in Exercise 115.

In Exercises 117–120, describe the transformations on the graph of $y = \ln x$ to sketch the graph of $y = f(x)$.

117. $f(x) = 4 + \ln(2x - 3)$

118. $f(x) = 3 - \ln\left(\dfrac{x}{2} - 1\right)$

119. $f(x) = -4 + \ln\left(3 - \dfrac{x}{2}\right)$

120. $f(x) = -4 - \ln(3 - 2x)$

121. **Present value of an investment.** Recall that if P dollars is invested in an account at an interest rate r compounded continuously, then the amount A (called the *future value of P*) in the account t years from now will be $A = Pe^{rt}$. Solving the equation for P, we get $P = Ae^{-rt}$. In this formulation, P is called the present value of the investment.
   a. The future value of an investment is $100,000. Find its present value at 7% compounded continuously for 20 years.
   b. Find the interest rate r compounded continuously that is needed to have $50,000 be the present value of $75,000 in ten years.

122. Your uncle is 40 years old, and he wants to have an annual interest income of $50,000 each year at age 65. What is the present value of his pension if the money can be invested at all times at a continuously compounded interest rate of
   a. 5%?
   b. 8%?
   c. 10%?

In Exercises 123–126, solve for x in terms of y.

123. $y = \ln\left(x + \sqrt{x^2 + 1}\right)$

124. $y = \ln\left(x - \sqrt{x^2 - 1}\right), x \geq 1$

125. $y = \dfrac{e^x - e^{-x}}{2}$

126. $y = \ln\left(\dfrac{e^x + e^{-x}}{2}\right)$

Critical Thinking

In Exercises 127 and 128 evaluate each expression without using a calculator

127. $2^{\log_2 3} - 3^{\log_3 2}$

128. $(\log_3 4 + \log_2 9)^2 - (\log_3 4 - \log_2 9)^2$

129. Solve for x: $\log_3[\log_4(\log_2 x)] = 0$

Rules of Logarithms

Before Starting this Section, Review

1. Definition of logarithm (Section 4.3, page 430)
2. Basic properties of logarithms (Section 4.3, page 433)
3. Rules of exponents (Section P.6, page 64)

Objectives

1. Learn the rules of logarithms.
2. Change the base of a logarithm.
3. Apply logarithms in growth and decay.

King Tut's Golden Mask

THE BOY KING TUT

Today the most famous pharaoh of Ancient Egypt is King Nebkheperure Tutankhamun, popularly called "King Tut." He was only 9 years old when he became a pharaoh. In 2005, a team of Egyptian scientists headed by Dr. Zahi Hawass determined that he was 19 years old when he died (around 1346 B.C.).

Tutankhamun was a short-lived boy king who, unlike the great Egyptian Kings Khufu (builder of the Great Pyramid), Amenhotep III (builder of temples through-out Egypt), and Ramses II (prolific builder and usurper), accomplished nothing sig-nificant during his reign. In fact, little was known about him prior to Howard Carter's discovery of his tomb (and the amazing treasures it held) in the Valley of the Kings on November 4, 1922. Carter, with his benefactor Lord Carnarvon at his side, entered the tomb's burial chamber on November 26, 1922. Lord Carnarvon died seven weeks after entering the burial chamber, giving rise to the theory of the "curse" of King Tut.

The work on the tomb continued until 1933. The tomb contained a pristine mummy of an Egyptian king lying intact in his original burial furniture. He was accompanied by a small slice of the royal world of the pharaohs: golden chariots, statues of gold and ebony, a fleet of miniature ships to accommodate his trip to the hereafter, his throne of gold, bottles of perfume, precious jewelry, and more. The "Treasures of Tutankhamun" exhibition, first shown at the British Museum in London in 1972, traveled to many countries, including the United States. In 2005, a new exhibition, "Tutankhamun and the Golden Age of the Pharaohs," visited the United States.

In Example 7, we discuss the age of a work of art found in King Tut's tomb. ■

---

1 Learn the rules of logarithms.

Rules of Logarithms

Recall the basic properties of logarithms from Section 4.3 for $a > 0, a \neq 1$.

$$\log_a a = 1 \qquad (1)$$
$$\log_a 1 = 0 \qquad (2)$$
$$\log_a a^x = x, \ x \text{ real} \qquad (3)$$
$$a^{\log_a x} = x, \ x > 0 \qquad (4)$$

In this section, we discuss some important rules of logarithms that are helpful in calculations, simplifications, and applications.

Rules of Logarithms

Let M, N, and a be positive real numbers with $a \neq 1$ and let r be any real number.

Rule	Description	Examples
1. Product Rule: $\log_a(MN) = \log_a M + \log_a N$	The logarithm of the product of two (or more) numbers is the sum of the logarithms of the numbers.	$\log_2(5 \cdot 17) = \log_2 5 + \log_2 17$ $\log(3x) = \log 3 + \log x$ $\ln(5 \cdot 7) = \ln 5 + \ln 7$
2. Quotient Rule: $\log_a\left(\dfrac{M}{N}\right) = \log_a M - \log_a N$	The logarithm of the quotient of two numbers is the difference of the logarithms of the numbers.	$\ln \dfrac{5}{7} = \ln 5 - \ln 7$ $\log\left(\dfrac{5}{x}\right) = \log 5 - \log x$ $\log_2\left(\dfrac{y}{11}\right) = \log_2 y - \log_2 11$
3. Power Rule: $\log_a M^r = r \log_a M$	The logarithm of a number to the power r is r times the logarithm of the number.	$\ln 5^7 = 7 \ln 5$ $\log(5)^{3/2} = \dfrac{3}{2} \log 5$ $\log_2(7)^{-3} = -3 \log_2 7$

Rules 1, 2, and 3 follow from the corresponding rules of the exponents:

$$a^u \cdot a^v = a^{u+v} \qquad \text{Product rule}$$

$$\frac{a^u}{a^v} = a^{u-v} \qquad \text{Quotient rule}$$

$$(a^u)^r = a^{u\,r} \qquad \text{Power rule}$$

For example, to prove the product rule for logarithms, we let

$$\log_a M = u \qquad \text{and} \qquad \log_a N = v.$$

The corresponding exponential forms of these equations are

$$M = a^u \qquad \text{and} \qquad N = a^v.$$

Then

$$MN = a^u \cdot a^v$$
$$MN = a^{u+v} \qquad \qquad \text{Product rule of exponents}$$
$$\log_a MN = u + v \qquad \qquad \text{Logarithmic form}$$
$$\log_a MN = \log_a M + \log_a N \qquad \text{Replace } u \text{ with } \log_a M \text{ and } v \text{ with } \log_a N.$$

This proves the Product rule of logarithms. In Exercise 104, you are asked to prove the Quotient rule and the Power rule similarly by using the corresponding rules for exponents.

REMINDER

Radicals can also be written as exponents. Recall that

$$\sqrt{x} = x^{1/2}$$

and

$$\sqrt[3]{x} = x^{1/3}.$$

In general,

$$\sqrt[m]{x} = x^{1/m}.$$

EXAMPLE 1   **Using Rules of Logarithms to Evaluate Expressions**

Given that $\log_5 z = 3$ and $\log_5 y = 2$, evaluate each expression.

a. $\log_5(yz)$ **b.** $\log_5(125y^7)$ **c.** $\log_5 \sqrt{\dfrac{z}{y}}$ **d.** $\log_5(z^{1/30}y^5)$

SOLUTION

a. $\log_5(yz) = \log_5 y + \log_5 z \qquad$ Product rule

$\qquad\qquad\quad = 2 + 3 = 5 \qquad$ Use the given values and simplify.

b. $\log_5 (125y^7) = \log_5 125 + \log_5 y^7$ Product rule

$\qquad\qquad\quad = \log_5 5^3 + \log_5 y^7$ $125 = 5^3$

$\qquad\qquad\quad = 3 + 7 \log_5 y$ $\log_a a^x = x$ and power rule

$\qquad\qquad\quad = 3 + 7(2) = 17$ Use the given value and simplify.

c. $\log_5 \sqrt{\dfrac{z}{y}} = \log_5 \left(\dfrac{z}{y}\right)^{1/2}$ Rewrite radical as exponent.

$\qquad\qquad = \dfrac{1}{2} \log_5 \dfrac{z}{y}$ Power rule

$\qquad\qquad = \dfrac{1}{2} \left(\log_5 z - \log_5 y\right)$ Quotient rule

$\qquad\qquad = \dfrac{1}{2} (3 - 2) = \dfrac{1}{2}$ Use the given values and simplify.

d. $\log_5 (z^{1/30} y^5) = \log_5 z^{1/30} + \log_5 y^5$ Product rule

$\qquad\qquad\quad = \dfrac{1}{30} \log_5 z + 5 \log_5 y$ Power rule

$\qquad\qquad\quad = \dfrac{1}{30}(3) + 5(2)$ Use the given values.

$\qquad\qquad\quad = 0.1 + 10 = 10.1$ Simplify.   ■ ■ ■

Practice Problem 1 Evaluate each expression for the given values in Example 1.

a. $\log_5 (y/z)$ **b.** $\log_5 (y^2 z^3)$ ■

In many applications in more advanced mathematics courses, the rules of logarithms are used in both directions; that is, the rules are read from left to right and from right to left. For example, by the product rule of logarithms, we have

$$\log_2 3x = \log_2 3 + \log_2 x.$$

The expression $\log_2 3 + \log_2 x$ is the *expanded form* of $\log_2 3x$, while $\log_2 3x$ is the *condensed form*, or the *single logarithmic form*, of $\log_2 3 + \log_2 x$.

The notation used for the rules of logarithms depends on the choice of base.

Rules of Logarithms

	Product Rule	Quotient Rule	Power Rule
(base a)	$\log_a (MN) = \log_a M + \log_a N$	$\log_a \left(\dfrac{M}{N}\right) = \log_a M - \log_a N$	$\log_a M^r = r \log_a M$
(base 10)	$\log (MN) = \log M + \log N$	$\log \left(\dfrac{M}{N}\right) = \log M - \log N$	$\log M^r = r \log M$
(base e)	$\ln (MN) = \ln M + \ln N$	$\ln \left(\dfrac{M}{N}\right) = \ln M - \ln N$	$\ln M^r = r \ln M$

EXAMPLE 2   Writing Expressions in Expanded Form

Write each expression in expanded form.

a. $\log_2 \dfrac{x^2 (x - 1)^3}{(2x + 1)^4}$ **b.** $\ln \sqrt{x^3 y^2 z^5}$

SOLUTION

a. $\log_2 \dfrac{x^2(x-1)^3}{(2x+1)^4} = \log_2 x^2(x-1)^3 - \log_2 (2x+1)^4$ Quotient rule

$$= \log_2 x^2 + \log_2 (x-1)^3 - \log_2 (2x+1)^4 \qquad \text{Product rule}$$

$$= 2\log_2 x + 3\log_2 (x-1) - 4\log_2 (2x+1) \qquad \text{Power rule}$$

b. $\ln \sqrt{x^3 y^2 z^5} = \ln (x^3 y^2 z^5)^{1/2}$ $\sqrt{a} = a^{1/2}$

$$= \frac{1}{2} \ln x^3 y^2 z^5 \qquad \text{Power rule}$$

$$= \frac{1}{2} [\ln x^3 + \ln y^2 + \ln z^5] \qquad \text{Product rule}$$

$$= \frac{1}{2} [3 \ln x + 2 \ln y + 5 \ln z] \qquad \text{Power rule}$$

$$= \frac{3}{2} \ln x + \ln y + \frac{5}{2} \ln z \qquad \text{Distributive property} \qquad ■\ ■\ ■$$

Practice Problem 2   Write each expression in expanded form.

a. $\ln \dfrac{2x-1}{x+4}$ **b.** $\log \sqrt{\dfrac{4xy}{z}}$ ■

EXAMPLE 3	**Writing Expressions in Condensed Form**

Write each expression in condensed form.

a. $\log 3x - \log 4y$ **b.** $2 \ln x + \dfrac{1}{2} \ln (x^2 + 1)$

c. $2 \log_2 5 + \log_2 9 - \log_2 75$ **d.** $\dfrac{1}{3} [\ln x + \ln (x+1) - \ln (x^2 + 1)]$

SOLUTION

a. $\log 3x - \log 4y = \log \left(\dfrac{3x}{4y} \right)$ Quotient rule

b. $2 \ln x + \dfrac{1}{2} \ln (x^2 + 1) = \ln x^2 + \ln (x^2 + 1)^{1/2}$ Power rule

$$= \ln (x^2 \sqrt{x^2 + 1}) \qquad \text{Product rule;}\ \sqrt{a} = a^{1/2}$$

c. $2 \log_2 5 + \log_2 9 - \log_2 75 = \log_2 5^2 + \log_2 9 - \log_2 75$ Power rule

$$= \log_2 (25 \cdot 9) - \log_2 75 \qquad 5^2 = 25;\ \text{Product rule}$$

$$= \log_2 \frac{25 \cdot 9}{75} \qquad \text{Quotient rule}$$

$$= \log_2 3 \qquad \frac{25 \cdot 9}{75} = \frac{9}{3} = 3$$

d. $\dfrac{1}{3} [\ln x + \ln (x+1) - \ln (x^2 + 1)] = \dfrac{1}{3} [\ln x(x+1) - \ln (x^2 + 1)]$ Product rule

$$= \frac{1}{3} \ln \left[\frac{x(x+1)}{x^2+1} \right] \qquad \text{Quotient rule}$$

$$= \ln \sqrt[3]{\frac{x(x+1)}{x^2+1}} \qquad \text{Power rule;}\ a^{1/3} = \sqrt[3]{a} \qquad ■\ ■\ ■$$

Practice Problem 3   Write in condensed form: $\dfrac{1}{2} [\log (x+1) + \log (x-1)]$ ■

Be careful when using the rules of logarithms to simplify expressions. For example, there is no property that allows you to rewrite $\log_a (x + y)$.

In general,

$$\log_a (x + y) \neq \log_a x + \log_a y. \quad (5)$$

To illustrate this inequality, let $x = 100$, $y = 10$, and $a = 10$. Then the value of the left side of (5) is

$$\log (100 + 10) = \log (110) \approx 2.0414 \qquad \text{Use a calculator}$$

The value of the right side of (5) is

$$\begin{aligned}
\log 100 + \log 10 &= \log 10^2 + \log 10 \\
&= 2 \log 10 + \log 10 \qquad \text{Power rule} \\
&= 2 (1) + 1 = 3.
\end{aligned}$$

Therefore, $\log (100 + 10) \neq \log 100 + \log 10$.

When graphing logarithmic functions, you must also be aware of the fact that changing the form of the expression in the defining equation may affect the domain. For example, $\log x^2 = 2 \log x$ for all values of x for which both sides are defined. However, the domain of $f(x) = \log x^2$ is $(-\infty, 0) \cup (0, \infty)$, whereas the domain of $g(x) = 2 \log x$ is $(0, \infty)$. See Figure 4.15.

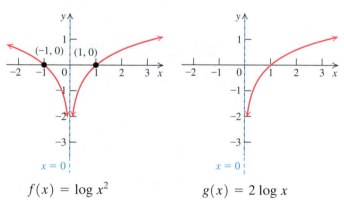

$$f(x) = \log x^2 \qquad\qquad g(x) = 2 \log x$$

FIGURE 4.15

◆ **WARNING**   Avoid the following common errors.

$$\log_a (M + N) = \log_a M + \log_a N.$$
$$\log_a (M - N) = \log_a M - \log_a N.$$
$$(\log_a M)(\log_a N) = \log_a MN.$$
$$\log_a \left(\frac{M}{N} \right) = \frac{\log_a M}{\log_a N}.$$
$$\frac{\log_a M}{\log_a N} = \log_a M - \log_a N.$$
$$(\log_a M)^r = r \log_a M.$$

2 Change the base of a logarithm.

Change of Base

Calculators usually come with two types of log keys: a key for natural logarithms (base e, $\boxed{\text{LN}}$) and a key for common logarithms (base 10, $\boxed{\text{LOG}}$). These two types of logarithms are frequently used in applications. Sometimes, however, we need to evaluate logarithms for bases other than e and 10. The *change-of-base formula* helps us evaluate these logarithms.

Suppose we are given $\log_b x$ and we want to find an equivalent expression in terms of logarithms with base a. We let

$$u = \log_b x. \quad (6)$$

In exponential form, we have

$$x = b^u. \quad (7)$$

Then

$$\log_a x = \log_a b^u \qquad \text{Take } \log_a \text{ of each side of (7).}$$
$$\log_a x = u \log_a b \qquad \text{Power rule}$$
$$u = \frac{\log_a x}{\log_a b} \qquad \text{Solve for } u.$$
$$\log_b x = \frac{\log_a x}{\log_a b} \quad (8) \qquad \text{Substitute for } u \text{ from (6).}$$

Equation (8) is the change-of-base formula. By choosing $a = 10$ and $a = e$, we can state the change-of-base formula as follows:

CHANGE-OF-BASE FORMULA

Let a, b, and x be positive real numbers with $a \neq 1$ and $b \neq 1$. Then $\log_b x$ can be converted to a different base as follows:

$$\log_b x = \underset{\text{(base } a\text{)}}{\frac{\log_a x}{\log_a b}} = \underset{\text{(base 10)}}{\frac{\log x}{\log b}} = \underset{\text{(base } e\text{)}}{\frac{\ln x}{\ln b}}$$

TECHNOLOGY CONNECTION

 To graph a logarithmic function whose base is not e or 10, use the change-of-base formula. For example, graph $y = \log_5 x$ by entering $y_1 = \text{LN}(X)/\text{LN}(5)$.

EXAMPLE 4 **Using a Change of Base to Compute Logarithms**

Compute $\log_5 13$ by changing to

a. common logarithms. **b.** natural logarithms.

SOLUTION

a. $\log_5 13 = \dfrac{\log 13}{\log 5}$ Change to base 10.

      ≈ 1.59369 Use a calculator.

b. $\log_5 13 = \dfrac{\ln 13}{\ln 5}$ Change to base e.

      ≈ 1.59369 Use a calculator. ■ ■ ■

Practice Problem 4 Find $\log_3 15$. ■

EXAMPLE 5   **Matching Data to an Exponential Curve**

Find the exponential function of the form $f(x) = ae^{bx}$ that passes through the points $(0, 2)$ and $(3, 8)$.

SOLUTION

We need to find the values of a and b. Because the graph passes through the point $(0, 2)$, we have

$$2 = f(0) = ae^{b(0)} = ae^0 = a \cdot 1 = a;$$

so $a = 2$. Next, for the graph to pass through the point $(3, 8)$, we must have

$$8 = f(3) = ae^{b(3)} = 2e^{3b} \qquad \text{Replace } a \text{ with 2.}$$

Now solve the equation $8 = 2e^{3b}$ for b.

$$4 = e^{3b} \qquad \text{Divide both sides by 2.}$$
$$\ln 4 = 3b \qquad \text{Logarithmic form}$$
$$b = \frac{1}{3}\ln 4 \qquad \text{Divide both sides by 3.}$$

Therefore, the desired exponential function is

$$f(x) = 2e^{(\frac{1}{3}\ln 4)x}. \qquad \blacksquare\blacksquare\blacksquare$$

Practice Problem 5   Find an exponential function $f(x) = ae^{bx}$ that passes through the points $(0, 5)$ and $(1, 20)$.   ■

3 | Apply logarithms in growth and decay.

Growth and Decay

Exponential growth (or decay) occurs when a quantity grows (or decreases) at a rate proportional to its size. The standard growth formula introduced in Section 4.2 is

$$A(t) = A_0 e^{kt}, \qquad (9)$$

where $A(t)$ = the amount of substance (or population) at time t,
$A_0 = A(0)$ is the initial amount,
k = relative rate of growth (if $k > 0$) or decay (if $k < 0$), and
t = time.

The natural logarithm function provides a new tool for working with this model. The next diagram represents growth and decay changes over time.

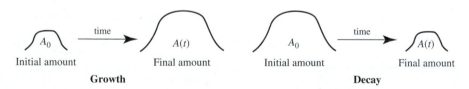

Initial amount   Final amount   Initial amount   Final amount

Growth   **Decay**

Half-Life

The **half-life** of any quantity whose value decreases with time is the time required for the quantity to decay to half its initial value.

Radioactive substances undergo exponential decay. Krypton-85, for example, has a half-life of about ten years; this means that any initial quantity of krypton-85 will dissipate or decay to half that amount in about ten years.

If h is the half-life of a substance undergoing exponential decay at the rate k, then any mass A_0 decays to $\frac{1}{2}A_0$ in time h. This fact leads to a half-life formula.

$$\frac{1}{2}A_0 = A_0e^{kh} \qquad A(t) = A_0e^{kt} \text{ with } t = h, \text{ and } A(h) = \frac{1}{2}A_0$$

$$\frac{1}{2} = e^{kh} \qquad \text{Divide both sides by } A_0.$$

$$\ln\left(\frac{1}{2}\right) = \ln\left(e^{kh}\right) \qquad \text{Take the natural log of both sides.}$$

$$-\ln(2) = kh \qquad \ln\left(\frac{1}{2}\right) = \ln\left(2^{-1}\right) = -\ln(2); \ln(e^x) = x$$

$$h = -\frac{\ln(2)}{k} \qquad \text{Solve for } h.$$

HALF-LIFE FORMULA

The half-life h of a substance undergoing exponential decay at a rate k ($k < 0$) is given by the formula

$$h = -\frac{\ln(2)}{k}.$$

EXAMPLE 6   **Finding the Half-Life of a Substance**

In an experiment, 18 grams of the radioactive element sodium-24 decayed to 6 grams in 24 hours. Find its half-life to the nearest hour.

SOLUTION

$$A(t) = A_0e^{kt} \qquad \text{Exponential growth and decay model}$$

$$6 = 18e^{k(24)} \qquad A_0 = 18, A(24) = 6$$

$$\frac{1}{3} = e^{k(24)} \qquad \text{Divide both sides by 18.}$$

$$\ln\left(\frac{1}{3}\right) = \ln\left(e^{k(24)}\right) \qquad \text{Take the natural log of both sides.}$$

$$-\ln(3) = 24k \qquad \ln\left(\frac{1}{3}\right) = \ln\left(3^{-1}\right) = -\ln(3); \ln(e^x) = x$$

$$k = -\frac{\ln(3)}{24} \qquad \text{Solve for } k.$$

So $\qquad A(t) = 18e^{-\frac{\ln(3)}{24}t}.$

To find the half-life of sodium-24, we use the formula

$$h = -\frac{\ln(2)}{k} \qquad \text{Half-life formula}$$

$$h = -\frac{\ln(2)}{-\dfrac{\ln(3)}{24}} \qquad \text{Replace } k \text{ with } -\frac{\ln(3)}{24}.$$

$$h = \frac{24\ln 2}{\ln 3} \approx 15 \text{ hours} \qquad \text{Simplify; use a calculator.} \qquad ■■■$$

Practice Problem 6 In an experiment, 100 grams of the radioactive element strontium-90 decayed to 66 grams in 15 years. Find its half-life to the nearest year. ■

Radiocarbon Dating

Ordinary carbon, called carbon-12 (^{12}C), is stable and does not decay. However, carbon-14 (^{14}C), is a form of carbon that decays radioactively with a half-life of 5700 years. Sunlight constantly produces carbon-14 in Earth's atmosphere. When a living organism breathes or eats, it absorbs carbon-12 and carbon-14. After the organism dies, no more carbon-14 is absorbed; so the age of its remains can be calculated by determining how much carbon-14 has decayed. The method of radiocarbon dating was developed by the American scientist W. F. Libby.

EXAMPLE 7   King Tut's Treasure

In 1960, a group of specialists from the British Museum in London investigated whether a piece of art containing organic material found in Tutankhamun's tomb had been made during his reign or (as some historians claimed) whether it belonged to an earlier period. We know that King Tut died in 1346 B.C. and ruled Egypt for ten years. What percent of the amount of carbon-14 originally contained in the object should be present in 1960 if the object was made during Tutankhamun's reign?

SOLUTION

Because the half-life of carbon-14 is approximately 5700 years, we can rewrite the equation $A(t) = A_0 e^{kt}$ as

$$A(5700) = A_0 e^{5700k}.$$

Now we can find the value of k.

$$\frac{1}{2} A_0 = A_0 e^{5700k} \qquad \text{Replace } A(5700) \text{ with } \frac{1}{2} A_0.$$

$$\frac{1}{2} = e^{5700k} \qquad \text{Divide both sides by } A_0.$$

$$\ln \left(\frac{1}{2} \right) = 5700k \qquad \text{Logarithmic form}$$

$$k = \frac{\ln \left(\frac{1}{2} \right)}{5700} \qquad \text{Solve for } k.$$

$$\approx -0.0001216 \qquad \text{Use a calculator.}$$

Substituting this value of k into equation (9), we obtain

$$A(t) = A_0 e^{-0.0001216t} \qquad (10)$$

The time t that elapsed between King Tut's death and 1960 is

$$t = 1960 + 1346 \qquad \text{1346 B.C. to A.D. 1960}$$

$$= 3306.$$

$$A(3306) = A_0 e^{-0.0001216(3306)} \qquad \text{Replace } t \text{ with 3306 in (10).}$$

$$\approx 0.66897 A_0.$$

Therefore, the percent of the original amount of carbon-14 remaining in the object (after 3306 years) is 66.897%.

Because King Tut began ruling Egypt ten years before he died, the time t_1 that elapsed from the beginning of his reign to 1960 is 3316 years.

$$A(3316) = A_0 e^{-0.0001216(3316)}$$ Replace t with 3316 in (10).
$$\approx 0.66816 A_0.$$

Therefore, a piece of art made during King Tut's reign would have between 66.816% and 66.897% of carbon-14 remaining in 1960.   ■ ■ ■

Practice Problem 7 Repeat Example 7 assuming that the art object was made 194 years before King Tut's death.   ■

SECTION 4.4 ■ Exercises

A EXERCISES Basic Skills and Concepts

1. $\log_a MN = \underline{\hspace{1cm}} + \underline{\hspace{1cm}}.$

2. $\underline{\hspace{1cm}} = \log_a M - \log_a N.$

3. $\log_a M^r = \underline{\hspace{1cm}}.$

4. The change-of-base formula with base e is $\log_a M = \underline{\hspace{1cm}}.$

5. *True or False* $\log_a (u + v) = \log_a u + \log_a v.$

6. *True or False* $\log \dfrac{x}{10} = \log x - 1.$

In Exercises 7–18, given that $\log x = 2, \log y = 3, \log 2 \approx 0.3,$ and $\log 3 \approx 0.48,$ evaluate each expression without using a calculator.

7. $\log 6$
8. $\log 4$
9. $\log 5 \left[Hint: 5 = \dfrac{10}{2} \right]$
10. $\log (3x)$
11. $\log \left(\dfrac{2}{x} \right)$
12. $\log x^2$
13. $\log (2x^2 y)$
14. $\log xy^3$
15. $\log \sqrt[3]{x^2 y^4}$
16. $\log (\log x^2)$
17. $\log \sqrt[3]{48}$
18. $\log_2 3$

In Exercises 19–38, write each expression in expanded form.

19. $\ln [x(x - 1)]$
20. $\ln \dfrac{x(x + 1)}{(x - 1)^2}$
21. $\log \dfrac{\sqrt{x^2 + 1}}{x + 3}$
22. $\log \left(\dfrac{x - 1}{x + 3} \right)$
23. $\log_3 \dfrac{(x - 1)^2}{(x + 1)^5}$
24. $\log_4 \left(\dfrac{x^2 - 9}{x^2 - 6x + 8} \right)^{\frac{2}{3}}$
25. $\log_b x^2 y^3 z$
26. $\log_b \sqrt{xyz}$
27. $\ln \dfrac{e^2}{x^3 y}$
28. $\ln \dfrac{ey^2}{xz^3}$

29. $\log \sqrt{x \sqrt{yz}}$
30. $\log_b \left(x \log_b b^{\sqrt{yz}} \right)$
31. $\ln \left[\dfrac{x \sqrt{x - 1}}{x^2 + 2} \right]$
32. $\ln \left[\dfrac{\sqrt{x - 2} \sqrt[3]{x + 1}}{x^2 + 3} \right]$
33. $\ln \left[\dfrac{(x + 1)^2}{(x - 3)\sqrt{x + 4}} \right]$
34. $\ln \left[\dfrac{2x + 3}{(x + 4)^2 (x - 3)^4} \right]$
35. $\ln \left[(x + 1)\sqrt{\dfrac{x^2 + 2}{x^2 + 5}} \right]$
36. $\ln \left[\dfrac{\sqrt[3]{2x + 1}\,(x + 1)}{(x - 1)^2 (3x + 2)} \right]$
37. $\ln \left[\dfrac{x^3 (3x + 1)^4}{\sqrt{x^2 + 1}(x + 2)^{-5}(x - 3)^2} \right]$
38. $\ln \left[\dfrac{(x + 1)^{\frac{1}{2}}(x^2 - 2)^{\frac{2}{5}}}{(2x - 1)^{\frac{3}{2}}(x^2 + 2)^{-\frac{4}{5}}} \right]$

In Exercises 39–50, write each expression in condensed form.

39. $\log_2 x + \log_2 7$
40. $\log_2 x - \log_2 3$
41. $\log \dfrac{5}{2} - \log \dfrac{7}{9}$
42. $2 \log (x - 1)$
43. $\dfrac{1}{2} \log x - \log y + \log z$
44. $\dfrac{1}{2} (\log x + \log y)$
45. $\dfrac{1}{5} (\log_2 z + 2 \log_2 y)$
46. $\dfrac{1}{3} (\log x - 2 \log y + 3 \log z)$
47. $\ln x + 2 \ln y + 3 \ln z$
48. $2 \ln x - 3 \ln y + 4 \ln z$
49. $2 \ln x - \dfrac{1}{2} \ln (x^2 + 1)$
50. $2 \ln x + \dfrac{1}{2} \ln (x^2 - 1) - \dfrac{1}{2} \ln (x^2 + 1)$

In Exercises 51–58, use the change-of-base formula and a calculator to evaluate each logarithm.

51. $\log_2 5$

52. $\log_4 11$

53. $\log_{1/2} 3$

54. $\log_{\sqrt{3}} 12.5$

55. $\log_{\sqrt{5}} \sqrt{17}$

56. $\log_{15} 123$

57. $\log_2 7 + \log_4 3$

58. $\log_2 9 - \log_{\sqrt{2}} 5$

 In Exercises 59–62, use a graphing utility and the change-of-base formula to graph each function.

59. $y = \log_3 x$

60. $y = \log_7 x$

61. $y = \log_5 (x + 2)$

62. $y = \log_{\frac{1}{2}} (x - 1)$

In Exercises 63–74, find the value of each expression without using a calculator.

63. $\log_3 \sqrt{3}$

64. $\log_{1/4} 4$

65. $\log_3 (\log_2 8)$

66. $2^{\log_2 2}$

67. $5^{2\log_5 3 + \log_5 2}$

68. $e^{3\ln 2 - 2\ln 3}$

69. $10^{3\log 2 - \log 4}$

70. $10^{\log_{1/10} 3}$

71. $10^{\log_{100} 3}$

72. $\log 4 + 2 \log 5$

73. $\log_2 160 - \log_2 5$

74. $e^{\log_{e^2} 25}$

In Exercises 75–78, find the exponential function of the form $f(x) = ae^{bx}$ that passes through the given points.

75. $(0, 2)$ and $(2, 8)$

76. $(0, 8)$ and $(2, 5)$

77. $(0, 100)$ and $(5, 1000)$

78. $(0, 10)$ and $(10, 2)$

B EXERCISES Applying the Concepts

In Exercises 79–82, assume the exponential growth model $A(t) = A_0 e^{kt}$ and a world population of 6.5 billion in 2006.

79. **Rate of population growth.** If the world added about 90 million people in 2007, what was the rate of growth?

80. **Population.** Use the rate of growth from Exercise 79 to estimate the year in which the world population will be
a. 12 billion.
b. 20 billion.

81. **Population growth.** If the population must stay below 20 billion during the next 100 years, what is the maximum acceptable annual rate of growth?

82. **Population decline.** If world population must decline below 5 billion during the next 25 years, what must the annual rate of decline be?

83. **Continuous compounding.** If $1000 is deposited in a bank at 10% interest compounded continuously, how many years will it take for the money to grow to $3500?

84. **Continuous compounding.** At what rate of interest compounded continuously will an investment double in six years?

85. **Half-life.** Iodine-131, a radioactive substance that is effective in locating brain tumors, has a half-life of only eight days. A hospital purchased 20 grams of the substance but had to wait five days before it could be used. How much of the substance was left after five days?

86. **Half-life.** Plutonium-241 has a half-life of 13 years. A laboratory purchased 10 grams of the substance but did not use it for two years. How much of the substance was left after two years?

87. **Half-life.** Tritium is used in nuclear weapons to increase their power. Fifty grams of tritium decayed to 34 grams after seven years. Calculate the half-life of tritium to the nearest year.

88. **Half-life.** Sixty grams of magnesium-28 decayed to 7.5 grams after 63 hours. What is the half-life of magnesium-28 to the nearest hour?

Effective medicine dosage. Use the following model in Exercises 89–92: The concentration $C(t)$ of a drug administered to a patient intravenously jumps to its highest level almost immediately. The concentration subsequently decays exponentially according to the law

$$C(t) = C_0 e^{-kt},$$

where $C(0) = C_0$. The physician administering the drug would like to have

$$m < C(t) < M,$$

where m is the concentration below which the drug is ineffective and M is the concentration above which the drug is dangerous.

89. **Drug concentration.** Suppose immediately after a certain drug is administered, the concentration is 5 milligrams per milliliter. Ten hours later the concentration drops to 1.5 milligrams per milliliter. Determine the value of k for this drug.

90. **Drug concentration.** For the drug of Exercise 89, suppose the maximum safe level is $M = 12$ milligrams per milliliter and the minimum effective level is $m = 3$ milligrams per milliliter. If the initial concentration is M, find the maximum possible time between doses of this drug.

91. **Half-life.** The half-life of Valium is 36 hours. Suppose a patient receives 16 milligrams of the drug at 8 A.M. How much Valium is in the patient's blood at 4 P.M. the same day?

92. **Half-life.** The half-life of aspirin in the blood is 12 hours. Estimate the time required for the aspirin to decay to 90% of the original amount ingested.

C EXERCISES Beyond the Basics

93. Show that
$$\log_b (\sqrt{x^2 + 1} - x) + \log_b (\sqrt{x^2 + 1} + x) = 0.$$

94. Show that
$$\log_b(\sqrt{x+1} + \sqrt{x}) = -\log_b(\sqrt{x+1} - \sqrt{x}).$$

95. Show that $(\log_b a)(\log_a b) = 1$.

96. Show that $\log\dfrac{a}{b} + \log\dfrac{b}{a} = 0$.

97. By the power rule for logarithms, $\log x^2 = 2\log x$. However, the graphs of $f(x) = \log x^2$ and $g(x) = 2\log x$ are not identical. Explain why.

98. Simplify each expression.

a. $\log\left(\dfrac{a}{b}\right) + \log\left(\dfrac{b}{a}\right) + \log\left(\dfrac{c}{a}\right) + \log\left(\dfrac{a}{c}\right)$

b. $\log\left(\dfrac{a^2}{bc}\right) + \log\left(\dfrac{b^2}{ca}\right) + \log\left(\dfrac{c^2}{ab}\right)$

c. $\log_2 3 \cdot \log_3 4$

d. $\log_a b \cdot \log_b c \cdot \log_c a$

99. Find the number of digits in N assuming that $\log_2(\log_2 N) = 4$.

100. Let $f(x) = \log_a x$. Show that

a. $f\left(\dfrac{1}{x}\right) = -f(x) = \log_{1/a} x$.

b. $\dfrac{f(x+h) - f(x)}{h} = \log_a\left(1 + \dfrac{h}{x}\right)^{\frac{1}{h}}, h \neq 0.$

101. State whether each of the following is true or false for all permissible values of the variable.

a. $\log(x+2) = \log x + \log 2$

b. $\log 2x = \log x + \log 2$

c. $\log_2 3x = 3\log_2 x$

d. $(\log_3 2x)^4 = 4\log_3 2x$

e. $\log_5 x^2 = 2\log_5 x$

f. $\log\dfrac{x}{3} = \log x - \log 3$

g. $\log\dfrac{x}{4} = \dfrac{\log x}{\log 4}$

h. $\ln\left(\dfrac{1}{x}\right) = -\ln x$

i. $\log_2 x^2 = (\log_2 x)(\log_2 x)$

j. $\log|10x| = 1 + \log|x|$

102. **a.** Prove the quotient rule for logarithms:
$$\log_a \frac{M}{N} = \log_a M - \log_a N$$

b. Prove the power rule for logarithms:
$$\log_a M^r = r\log_a M$$

103. Assuming that $\log\left(\dfrac{a+b}{3}\right) = \dfrac{1}{2}(\log a + \log b)$, show that $a^2 + b^2 - 7ab = 0$.

104. Evaluate $\dfrac{1}{1+\log_c ab} + \dfrac{1}{1+\log_a bc} + \dfrac{1}{1+\log_b ca}$.

[*Hint:* Convert to common logarithms and then simplify.]

Critical Thinking

105. **What went wrong?** Find the error in the following argument:

$$3 < 4$$
$$3\log\frac{1}{2} < 4\log\frac{1}{2}$$
$$\log\left(\frac{1}{2}\right)^3 < \log\left(\frac{1}{2}\right)^4$$
$$\log\frac{1}{8} < \log\frac{1}{16}$$
$$\frac{1}{8} < \frac{1}{16}$$
$$2 < 1$$

Exponential and Logarithmic Equations and Inequalities

Before Starting this Section, Review

1. One-to-one property of exponential functions (Section 4.1, page, 409)

2. One-to-one property of logarithmic functions (Section 4.3, page, 434)

3. Rules of logarithms (Section 4.4, page, 445).

Objectives

1 Solve exponential equations.

2 Solve applied problems involving exponential equations.

3 Solve logarithmic equations.

4 Use the logistic growth model.

5 Use logarithmic and exponential inequalities.

LOGISTIC GROWTH MODEL

In Section 4.4, we discussed the population growth equation

$$A(t) = A_0 e^{kt}. \qquad (1)$$

Equation (1) is a realistic model for a biological population that has plenty of food, enough space to grow, and no threat from predators.

However, most populations are constrained by limitations on resources such as the plants that populations eat. In the 1830s, the Belgian scientist P. F. Verhulst added another constraint to the population growth model: *There is a limited sustainable maximum population M of a species that the habitat can support.* In population biology, *M* is called the *carrying capacity*.

Verhulst replaced equation (1) with

$$P(t) = \frac{M}{1 + ae^{-kt}}, \qquad (2)$$

where *k* is the rate of growth and *a* is a positive constant. In 1840, using the data from the first five U.S. censuses, he made a prediction of the U.S. population for the year 1940. As things turned out, it was off by less than 1%.

The mathematical model represented by equation (2) is called the *logistic growth model* or the *Verhulst model*. Verhulst introduced the term *logistique* to refer to the "loglike" qualities of the *S*-shaped graph of equation (2). See Figure 4.16.

To sketch the graph of equation (2), we notice that the *y*-intercept is

$$P(0) = \frac{M}{1 + ae^0} = \frac{M}{1 + a}.$$

As $t \to \infty$, the quantity $ae^{-kt} \to 0$ (because *k* is positive). Consequently, the denominator $1 + ae^{-kt} \to 1$; therefore, $P(t) \to M$ and the line $y = M$ is a horizontal asymptote. Similarly, as $t \to -\infty$, $e^{-kt} \to \infty$. In this case, the denominator $1 + ae^{-kt} \to \infty$ and $P(t) \to 0$. The *x*-axis, or $y = 0$, is another horizontal asymptote. See the graph

of $y = \frac{6}{1 + 2e^{-1.5t}}$ in Figure 4.16.

In Example 10 we use the logistic function defined by equation (2). ■

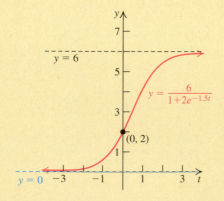

FIGURE 4.16 A logistic curve

1 Solve exponential equations.

Solving Exponential Equations

An equation containing terms of the form $a^x (a > 0, a \neq 1)$ is called an **exponential equation**. Here are some examples of simple exponential equations:

$$2^x = 15$$
$$9^x = 3^{x+1}$$
$$7^x = 13^{2x}$$
$$5 \cdot 2^{x-3} = 17$$

RECALL

The one-to-one property of exponential and logarithmic functions states that

- if $a^u = a^v$, then $u = v$.
- if $\log_a u = \log_a v$, then $u = v$.

If both sides of an equation can be expressed as a power of the same base, then we can use the one-to-one property of exponential functions to solve the equation. We used this strategy in Section 4.1.

EXAMPLE 1   **Solving an Exponential Equation**

Solve each equation.

a. $25^x = 125$ **b.** $9^x = 3^{x+1}$

SOLUTION

a.
$$\begin{aligned}
(5^2)^x &= 5^3 &&\text{Rewrite 25 as } 5^2 \text{ and 125 as } 5^3. \\
5^{2x} &= 5^3 &&\text{Power rule for exponents} \\
2x &= 3 &&\text{One-to-one property} \\
x &= \frac{3}{2} &&\text{Solve for } x.
\end{aligned}$$

b.
$$\begin{aligned}
(3^2)^x &= 3^{x+1} &&\text{Rewrite 9 as } 3^2. \\
3^{2x} &= 3^{x+1} &&\text{Power rule of exponents} \\
2x &= x + 1 &&\text{One-to-one property} \\
x &= 1 &&\text{Solve for } x. \quad \blacksquare\blacksquare\blacksquare
\end{aligned}$$

Practice Problem 1 Solve each equation.

a. $3^x = 243$ **b.** $8^x = 4$ ■

We can use logarithms to solve those equations for which both sides cannot be readily expressed with the same base. For example, the fact that $\log 2^x = x \log 2$ allows us to solve the equation $2^x = 9$ by first taking the log of both sides.

$$2^x = 9$$
$$\log 2^x = \log 9$$
$$x \log 2 = \log 9$$
$$x = \frac{\log 9}{\log 2}$$

An approximation $x \approx 3.1699$ is found by using a calculator.

The general procedure for solving exponential equations using logarithms is given next.

FINDING THE SOLUTION: A PROCEDURE

EXAMPLE 2   **Solving Exponential Equations Using the Logarithms**

OBJECTIVE	**EXAMPLE**
Solve exponential equations when both sides are not expressed with the same base.	*Solve for x:* $5 \cdot 2^{x-3} = 17$

Step 1 Isolate the exponential expression on one side of the equation.

1. $2^{x-3} = \dfrac{17}{5} = 3.4$ $\qquad$ Solve for 2^{x-3}

Step 2 Take the common or natural logarithm of both sides.

2. $\ln 2^{x-3} = \ln(3.4)$ $\qquad$ Take ln of both sides.

Step 3 Use the power rule, $\log_a M^r = r \log_a M$.

3. $(x-3)\ln 2 = \ln(3.4)$ $\qquad$ Power Rule of logarithms

Step 4 Solve for the variable.

4. $x - 3 = \dfrac{\ln(3.4)}{\ln 2}$ $\qquad$ Divide both sides by ln 2.

$x = \dfrac{\ln(3.4)}{\ln 2} + 3 \approx 4.766$ $\qquad$ Solve for x; use a calculator.

Practice Problem 2   Solve for x: $7 \cdot 3^{x+1} = 11$.

EXAMPLE 3   **Solving an Exponential Equation with Different Bases**

Solve the equation $5^{2x-3} = 3^{x+1}$ and approximate the answer to three decimal places.

SOLUTION

When different bases are involved, we begin with Step 2 of the procedure above.

$\ln 5^{2x-3} = \ln 3^{x+1}$ $\qquad$ Take the natural log of both sides.

$(2x - 3)\ln 5 = (x + 1)\ln 3$ $\qquad$ Power rule of logarithms

$2x \ln 5 - 3 \ln 5 = x \ln 3 + \ln 3$ $\qquad$ Distributive property

$2x \ln 5 - x \ln 3 = \ln 3 + 3 \ln 5$ $\qquad$ Collect terms containing x on one side.

$x(2 \ln 5 - \ln 3) = \ln 3 + 3 \ln 5$ $\qquad$ Factor out x on the left side.

$x = \dfrac{\ln 3 + 3 \ln 5}{2 \ln 5 - \ln 3}$ $\qquad$ Solve for x to obtain an *exact* solution.

≈ 2.795 $\qquad$ Use a calculator.

Practice Problem 3   Solve the equation $3^{x+1} = 2^{2x}$.

EXAMPLE 4   **An Exponential Equation of Quadratic Form**

Solve the equation $3^x - 8 \cdot 3^{-x} = 2$.

SOLUTION

$3^x(3^x - 8 \cdot 3^{-x}) = 2(3^x)$ $\qquad$ Multiply both sides by 3^x.

$3^{2x} - 8 \cdot 3^0 = 2 \cdot 3^x$ $\qquad$ Distributive property; $3^x \cdot 3^{-x} = 3^0$

$3^{2x} - 8 = 2 \cdot 3^x$ $\qquad$ $3^0 = 1$

$3^{2x} - 2 \cdot 3^x - 8 = 0$ $\qquad$ Subtract $2 \cdot 3^x$ from both sides.

The last equation is quadratic in form. To see this, we let $y = 3^x$; then $y^2 = (3^x)^2 = 3^{2x}$. So

$$3^{2x} - 2 \cdot 3^x - 8 = 0$$

$$y^2 - 2y - 8 = 0 \qquad \text{Substitute: } y = 3^x, y^2 = 3^{2x}.$$

$$(y + 2)(y - 4) = 0 \qquad \text{Factor.}$$

$$y + 2 = 0 \quad \text{or} \quad y - 4 = 0 \qquad \text{Zero-product property}$$

$$y = -2 \quad \text{or} \qquad y = 4 \qquad \text{Solve for } y.$$

$$3^x = -2 \quad \text{or} \qquad 3^x = 4 \qquad \text{Replace } y \text{ with } 3^x.$$

But $3^x = -2$ is not possible because $3^x > 0$ for all real numbers x. Solve $3^x = 4$.

$$\ln 3^x = \ln 4 \qquad \text{Take the natural log of both sides.}$$

$$x \ln 3 = \ln 4 \qquad \text{Power rule of logarithms}$$

$$x = \frac{\ln 4}{\ln 3} \qquad \text{An exact solution}$$

$$\approx 1.262 \qquad \text{Use a calculator.}$$

■ ■ ■

Practice Problem 4 Solve the equation $e^{2x} - 4e^x - 5 = 0$. ■

2 Solve applied problems involving exponential equations.

Applications of Exponential Equations

EXAMPLE 5 **Finding an Interest Rate**

What interest rate is required to double your investment of P dollars in three years if interest is compounded

a. annually? **b.** continuously?

SOLUTION

a. The amount accumulated after t years from an investment of P dollars when interest is compounded annually is

$$A = P(1 + r)^t$$

$$\text{So} \qquad 2P = P(1 + r)^3 \qquad P \text{ must double to } 2P \text{ when } t = 3.$$

$$2 = (1 + r)^3 \qquad \text{Divide both sides by } P.$$

$$\sqrt[3]{2} = 1 + r \qquad \text{Take the cube root of both sides.}$$

$$r = \sqrt[3]{2} - 1 \approx 0.259921 \qquad \text{Solve for } r; \text{ use a calculator.}$$

The annual interest rate required to double the principal in three years is 25.99%.

b. For continuous interest, we use the formula

$$A = Pe^{rt}$$

$$2P = Pe^{r \cdot 3} \qquad P \text{ must double to } 2P \text{ when } t = 3.$$

$$2 = e^{3r} \qquad \text{Divide both sides by } P.$$

$$\ln 2 = 3r \qquad \text{Take the natural log of both sides.}$$

$$r = \frac{\ln 2}{3} \approx 0.231049 \qquad \text{Solve for } r; \text{ use a calculator.}$$

The continuous interest rate required to double the principle in three years is 23.10%.

■ ■ ■

Practice Problem 5 What interest rate is required to triple your investment of P dollars in five years if interest is compounded

a. annually? **b.** continuously? ■

EXAMPLE 6   **Solving a Population Growth Problem**

The following table shows the approximate population and annual growth rate of the United States and Pakistan in 2005.

Country	Population	Annual Population Growth Rate
United States	295 million	1.0%
Pakistan	162 million	3.1%

Source: www.cia.gov

Use the alternate population model $P(t) = P_0(1 + r)^t$, where P_0 is the initial population and t is the time in years since 2005. Assume that the growth rate for each country stays the same.

a. Estimate the population of each country in 2015.

b. In what year will the population of the United States be 350 million?

c. In what year will the population of Pakistan be the same as the population of the United States?

SOLUTION

The given model is

$$P(t) = P_0(1 + r)^t.$$

a. In 2005, the initial year, the U.S. population was $P_0 = 295$. In ten years, 2015, the population of the United States will be

$$295(1 + 0.01)^{10} \approx 325.86 \text{ million} \qquad P_0 = 295, r = 0.01, t = 10$$

In the same way, the population of Pakistan in 2015 will be

$$162(1 + 0.031)^{10} \approx 219.84 \text{ million} \qquad P_0 = 162, r = 0.031, t = 10$$

b. To find the year when the population of the United States will be 350 million, we solve for t in the equation

$$350 = 295(1 + 0.01)^t \qquad P(t) = 350, P_0 = 295, r = 0.01$$

$$\frac{350}{295} = (1.01)^t \qquad \text{Divide both sides by 295.}$$

$$\ln\left(\frac{350}{295}\right) = \ln(1.01)^t \qquad \text{Take natural logs of both sides.}$$

$$\ln\left(\frac{350}{295}\right) = t \ln(1.01) \qquad \text{Power rule for logarithms}$$

$$t = \frac{\ln\left(\dfrac{350}{295}\right)}{\ln(1.01)} \approx 17.18 \qquad \text{Solve for } t \text{ and use a calculator.}$$

The population of the United States will be 350 million approximately 17.18 years after 2005—that is, sometime in 2022.

c. To find when the population of the two countries will be the same, we solve for t in the equation

$$295(1 + 0.01)^t = 162(1 + 0.031)^t \qquad \text{Equate populations for both countries.}$$

$$295(1.01)^t = 162(1.031)^t$$

$$\frac{295}{162} = \left(\frac{1.031}{1.01}\right)^t \qquad \text{Divide both sides by } 162(1.01)^t \text{ and rewrite.}$$

$$\ln\left(\frac{295}{162}\right) = \ln\left(\frac{1.031}{1.01}\right)^t \qquad \text{Take natural logs of both sides.}$$

$$\ln\left(\frac{295}{162}\right) = t \ln\left(\frac{1.031}{1.01}\right) \qquad \text{Power rule for logarithms}$$

$$t = \frac{\ln\left(\dfrac{295}{162}\right)}{\ln\left(\dfrac{1.031}{1.01}\right)} \qquad \text{Solve for } t.$$

$$\approx 29.13 \text{ years} \qquad \text{Use a calculator.}$$

Therefore, the two populations will be equal in about 29.13 years, that is, during 2034. ■ ■ ■

Practice Problem 6 Repeat Example 6 assuming that the rates of growth for the United States and Pakistan are 1.1% and 3.3%, respectively. ■

3 Solve logarithmic equations.

Solving Logarithmic Equations

Equations that contain terms of the form $\log_a x$ are called **logarithmic equations**. Here are some examples of logarithmic equations.

$$\log_2 x = 4$$

$$\log_3 (2x - 1) = \log_3 (x + 2)$$

$$\log_2 (x - 3) + \log_2 (x - 4) = 1$$

To solve an equation such as $\log_2 x = 4$, we rewrite it in the equivalent exponential form:

$$\log_2 x = 4 \text{ means } x = 2^4 = 16$$

Thus, $x = 16$ is a solution of the equation $\log_2 x = 4$. Because the domain of logarithmic functions consists of positive numbers, we must check apparent solutions of logarithmic equations in the original equation.

EXAMPLE 7 **Solving a Logarithmic Equation**

Solve $4 + 3 \log_2 x = 1$.

SOLUTION

$$4 + 3 \log_2 x = 1 \qquad \text{Original equation}$$

$$3 \log_2 x = 1 - 4 \qquad \text{Subtract 4 from both sides.}$$

$$3 \log_2 x = -3 \qquad \text{Simplify.}$$

$$\log_2 x = -1 \qquad \text{Divide both sides by 3.}$$

$$x = 2^{-1} \qquad \text{Exponential form}$$

$$x = \frac{1}{2} \qquad a^{-n} = \frac{1}{a^n}$$

Check: Substitute $x = \dfrac{1}{2} = 2^{-1}$ into $4 + 3\log_2 x = 1$.

$$4 + 3\log_2 2^{-1} \overset{?}{=} 1$$
$$4 + 3(-1)\log_2 2 \overset{?}{=} 1 \qquad \text{Power rule for logarithms}$$
$$4 - 3 \overset{?}{=} 1 \qquad \log_2 2 = 1$$
$$1 \overset{?}{=} 1 \qquad \text{Yes}$$

This check shows that the solution set is $\left\{\dfrac{1}{2}\right\}$.   ■ ■ ■

Practice Problem 7   Solve $1 + 2\ln x = 4$.   ■

If each side of an equation can be expressed as a single logarithm with the same base, then we can use the one-to-one property of logarithms to solve the equation.

REMINDER

The Product Rule says that
$\log_a M + \log_a N = \log_a MN$.

EXAMPLE 8   **Using the One-to-One Property of Logarithms**

Solve $\log_4 x + \log_4 (x + 1) = \log_4 (x - 1) + \log_4 6$.

SOLUTION

$$\log_4 x + \log_4 (x + 1) = \log_4 (x - 1) + \log_4 6 \qquad \text{Original equation}$$
$$\log_4 [x(x + 1)] = \log_4 [6(x - 1)] \qquad \text{Product rule for logarithms}$$
$$x(x + 1) = 6(x - 1) \qquad \text{If } \log_a u = \log_a v, \text{ then } u = v.$$
$$x^2 + x = 6x - 6 \qquad \text{Distributive property}$$
$$x^2 - 5x + 6 = 0 \qquad \text{Add } -6x + 6 \text{ to both sides.}$$
$$(x - 2)(x - 3) = 0 \qquad \text{Factor.}$$
$$x - 2 = 0 \quad \text{or} \quad x - 3 = 0 \qquad \text{Zero-product property}$$
$$x = 2 \quad \text{or} \quad x = 3 \qquad \text{Solve for } x.$$

Now we check these possible solutions in the original equation.

Check $x = 2$

$$\log_4 2 + \log_4 (2 + 1) \overset{?}{=} \log_4 (2 - 1) + \log_4 6$$
$$\log_4 2 + \log_4 3 \overset{?}{=} \log_4 1 + \log_4 6$$
$$\log_4 (2 \cdot 3) \overset{?}{=} \log_4 (1 \cdot 6) \qquad \text{Yes}$$

Check $x = 3$

$$\log_4 3 + \log_4 (3 + 1) \overset{?}{=} \log_4 (3 - 1) + \log_4 6$$
$$\log_4 3 + \log_4 4 \overset{?}{=} \log_4 2 + \log_4 6$$
$$\log_4 (3 \cdot 4) \overset{?}{=} \log_4 (2 \cdot 6) \qquad \text{Yes}$$

The solution set is $\{2, 3\}$.   ■ ■ ■

Practice Problem 8   Solve $\ln (x + 5) + \ln (x + 1) = \ln (x - 1)$.   ■

EXAMPLE 9   **Using the Product and Quotient Rules**

Solve the following equations.

a. $\log_2 (x - 3) + \log_2 (x - 4) = 1$ **b.** $\log_2 (x + 4) + \log_2 (x + 3) = 1$

SOLUTION

a.
$$\log_2 (x - 3) + \log_2 (x - 4) = 1 \qquad \text{Original equation}$$
$$\log_2 [(x - 3)(x - 4)] = 1 \qquad \text{Product rule for logarithms}$$
$$(x - 3)(x - 4) = 2^1 \qquad \text{Exponential form}$$
$$x^2 - 7x + 10 = 0 \qquad \text{Write in standard form.}$$
$$(x - 2)(x - 5) = 0 \qquad \text{Factor.}$$
$$x - 2 = 0 \quad \text{or} \quad x - 5 = 0 \qquad \text{Zero-product property}$$
$$x = 2 \quad \text{or} \quad x = 5 \qquad \text{Solve for } x.$$

We check the possible solutions in the original equation.

Check $x = 2$	**Check $x = 5$**
$\log_2(2 - 3) + \log_2(2 - 4) \overset{?}{=} 1$	$\log_2(5 - 3) + \log_2(5 - 4) \overset{?}{=} 1$
$\log_2(-1) + \log_2(-2) \overset{?}{=} 1$ No	$\log_2 2 + \log_2 1 \overset{?}{=} 1$
Because logarithms are not defined	$1 + 0 \overset{?}{=} 1$
for negative numbers, $x = 2$	$1 \overset{?}{=} 1$ Yes
is an extraneous solution of the	
original equation.	

The solution set is $\{5\}$.

b.
$$\log_2(x + 4) + \log_2(x + 3) = 1 \qquad \text{Original equation}$$
$$\log_2[(x + 4)(x + 3)] = 1 \qquad \text{Product rule}$$
$$(x + 4)(x + 3) = 2^1 \qquad \text{Exponential form}$$
$$x^2 + 7x + 10 = 0 \qquad \text{Write in standard form.}$$
$$(x + 2)(x + 5) = 0 \qquad \text{Factor.}$$
$$x = -2 \text{ or } x = -5 \qquad \text{Solve for } x.$$

We check the possible solutions in the original equation.

Check $x = -2$	**Check $x = -5$**
$\log_2(-2 + 4) + \log_2(-2 + 3) \overset{?}{=} 1$	$\log_2(-5 + 4) + \log_2(-5 + 3) \overset{?}{=} 1$
$\log_2 2 + \log_2 1 \overset{?}{=} 1$	$\log_2(-1) + \log_2(-2) \overset{?}{=} 1$
$1 + 0 \overset{?}{=} 1$ Yes	$\log_2(-1)$ and $\log_2(-2)$ are undefined. No

The solution set is $\{-2\}$.   ■ ■ ■

Practice Problem 9   Solve $\log_3(x - 8) + \log_3 x = 2$.   ■

4 Use the logistic growth model.

EXAMPLE 10   **Using Logarithms in the Logistic Growth Model**

Suppose the carrying capacity M of the human population on Earth is 35 billion. In 1987, the world population was about 5 billion. Use the logistic growth model of P. F. Verhulst to calculate the average rate, k, of growth of the population given that the population was about 6 billion in 2003.

$$P(t) = \frac{M}{1 + ae^{-kt}} \qquad (2) \qquad \text{From the section introduction}$$

SOLUTION

If 1987 represents $t = 0$, then $P(0) = 5$ and $M = 35$; so equation (2) becomes

$$5 = \frac{35}{1 + ae^{-k(0)}}$$

$$5 = \frac{35}{1 + a}$$

$$5(1 + a) = 35 \qquad \text{Multiply both sides by } 1 + a.$$

$$a = 6 \qquad \text{Solve for } a.$$

Equation (2), with $M = 35$ and $a = 6$, becomes

$$P(t) = \frac{35}{1 + 6e^{-kt}}. \qquad (3)$$

We now solve equation (3) for k given that $t = 16$ (for 2003) and $P(16) = 6$.

$$6 = \frac{35}{1 + 6e^{-16k}}$$

$$6 + 36e^{-16k} = 35 \qquad \text{Multiply both sides by } 1 + 6e^{-16k} \text{ and distribute.}$$

$$e^{-16k} = \frac{29}{36} \qquad \text{Isolate } e^{-16k}.$$

$$-16k = \ln\left(\frac{29}{36}\right) \qquad \text{Logarithmic form}$$

$$k = -\frac{1}{16}\ln\left(\frac{29}{36}\right) \qquad \text{Solve for } k.$$

$$k \approx 0.0135 = 1.35\% \qquad \text{Use a calculator.}$$

The average growth rate of the world population was approximately 1.35%. ■ ■ ■

Practice Problem 10   Repeat Example 10 assuming that the estimated world population in 2005 was about 6.5 billion.   ■

The magnitude M of an earthquake is defined in the Richter scale by

$$M = \log\left(\frac{I}{I_0}\right), \qquad (4)$$

where I is the intensity of the earthquake and I_0 is a zero-level intensity earthquake below which we would not be aware of Earth's movement.

EXAMPLE 11   **Comparing Two Earthquakes**

Compare the intensity of the Mexico City earthquake of 1985, which registered 8.1 on the Richter scale, to that of the San Francisco earthquake of 1989, which measured 7.1 on the Richter scale.

SOLUTION

Let I_M and I_S denote the intensities of the Mexico City and San Francisco earthquakes, respectively. By equation (4), we have

$$8.1 = \log\left(\frac{I_M}{I_0}\right) \qquad 7.1 = \log\left(\frac{I_S}{I_0}\right) \qquad \text{Use given values in equation (4).}$$

$$\frac{I_M}{I_0} = 10^{8.1} \qquad \frac{I_S}{I_0} = 10^{7.1} \qquad \text{Exponential form of } y = \log x$$

$$I_M = 10^{8.1} I_0 \qquad I_S = 10^{7.1} I_0 \qquad \text{Solve for } I_M \text{ and } I_S.$$

Dividing I_M by I_S and simplifying, we get

$$\frac{I_M}{I_S} = \frac{10^{8.1} I_0}{10^{7.1} I_0} = \frac{10^{8.1}}{10^{7.1}} = 10^{8.1-7.1} = 10^1 = 10$$

$$I_M = 10 I_S.$$

This equation shows that the intensity of the Mexico City earthquake was ten times that of the San Francisco earthquake. ■ ■ ■

Practice Problem 11   Compare the intensity of the Assam (India) earthquake of 1950, I_A, which registered 8.4 on the Richter scale, to that of the Tangshan (China) earthquake of 1976, I_T, which registered 7.9 on the Richter scale.   ■

5 Use logarithmic and exponential inequalities.

Logarithmic and Exponential Inequalities

The properties of logarithms and exponents, together with the same techniques we used to solve polynomials, are used to solve inequalities involving logarithmic and exponential expressions.

The exponential and logarithmic functions with base $a > 1$ are increasing functions. That is, if $x_1 < x_2$, then $a^{x_1} < a^{x_2}$, and if $0 < x_1 < x_2$, then $\log_a x_1 < \log_a x_2$. As their graphs suggest, it is also true for base $a > 1$ that if $a^{x_1} < a^{x_2}$, then $x_1 < x_2$ and if $\log_a x_1 < \log_a x_2$, then $x_1 < x_2$.

$$y = a^x;\ a > 1 \qquad\qquad y = \log_a x;\ a > 1$$

When working with inequalities, it is important to remember that $y = \log_a x$, $a > 1$, is negative for $0 < x < 1$.

EXAMPLE 12 **Solving an Inequality Involving an Exponential Expression**

Solve: $5(0.7)^x + 3 < 18$

SOLUTION

$5(0.7)^x + 3 < 18$	Given inequality
$(0.7)^x < 3$	Isolate $(0.7)^x$ on one side.
$\ln(0.7)^x < \ln 3$	Take the natural log of both sides.
$x \ln(0.7) < \ln 3$	Power rule for logarithms
$x > \dfrac{\ln 3}{\ln(0.7)} \approx -3.080$	Divide both sides by $\ln(0.7)$; reverse the sense of the inequality because $\ln(0.7) < 0$. ■ ■ ■

RECALL

If $A < B$ and $C < 0$ then $AC > BC$.

Practice Problem 12 Solve: $3(0.5)^x + 7 > 19$ ■

EXAMPLE 13 **Calculating Carbon Emissions in the United States and China**

In 2000, the United States emitted about 1400 million tons of carbon into the atmosphere (about one-fourth of worldwide emissions). In the same year, China emitted about 850 million tons. Suppose the annual rate of growth of the carbon emissions in the United States and China are 1.5% and 4.5%, respectively. After how many years will China be emitting more carbon into Earth's atmosphere than the United States will?

SOLUTION

We let $t = 0$ correspond to 2000, the year for which we have the data. After t years, the amount of carbon emitted by the United States will be

$$A_1(t) = 1400e^{0.015t} \qquad \left(1.5\% = \frac{1.5}{100} = 0.015\right)$$

and that by China will be $A_2(t) = 850e^{0.045t}$.

We need to find t so that $A_2(t) > A_1(t)$. That is,

$$850e^{0.045t} > 1400e^{0.015t}.$$

We have

$850e^{0.045t} \cdot e^{-0.015t} > 1400e^{0.015t} \cdot e^{-0.015t}$	Multiply both sides by $e^{-0.015t}$.
$850e^{0.03t} > 1400$	Product rule for exponents; $e^0 = 1$
$e^{0.03t} > \dfrac{28}{17}$	Divide both sides by 850 and simplify.
$\ln e^{0.03t} > \ln\left(\dfrac{28}{17}\right)$	$y = \ln x$ is an increasing function; so if $a > b$, then $\ln a > \ln b$.
$0.03t > \ln\left(\dfrac{28}{17}\right)$	$\ln e^x = x$
$t > \dfrac{\ln\left(\dfrac{28}{17}\right)}{0.03}$	Divide both sides by 0.03.
≈ 16.63	Use a calculator.

So in less than 17 years from 2000 (around the year 2017), under the present assumptions, China will be emitting more carbon into Earth's atmosphere than the United States will. ■ ■ ■

Practice Problem 13 Repeat Example 13 assuming that the rate of growth of carbon emissions is 1.3% in the United States and 4.7% in China. ■

EXAMPLE 14 **Solving an Inequality Involving a Logarithmic Expression**

Solve: $\log(2x - 5) \leq 1$

SOLUTION

We first identify the domain for $\log(2x - 5)$. Because $2x - 5$ must be positive, we solve $2x - 5 > 0$; so $2x > 5$ or $x > \dfrac{5}{2}$.

Then	$\log(2x - 5) \leq 1$	Given inequality
	$10^{\log(2x-5)} \leq 10^1$	If $u < v$, then $10^u < 10^v$.
	$2x - 5 \leq 10$	$a^{\log_a x} = x$
	$x \leq \dfrac{15}{2}$	Solve for x.

For $\log(2x - 5)$ to be a real number, we must have $x > \dfrac{5}{2}$. Combining this with the fact that $x \leq \dfrac{15}{2}$, we find that the solution set for $\log(2x - 5) \leq 1$ is $\left(\dfrac{5}{2}, \dfrac{15}{2}\right]$. ■ ■ ■

Practice Problem 14 Solve: $\ln(1 - 3x) > 2$ ■

SECTION 4.5 ■ Exercises

A EXERCISES Basic Skills and Concepts

1. An equation that contains terms of the form a^x is called a(n) _____ equation.

2. An equation that contains terms of the form $\log_a x$ is called a(n) _____ equation.

3. The equation $y = \dfrac{M}{1 + ae^{-bx}}$ represents a(n) _____ model.

4. *True or False* A logistic curve always has two horizontal asymptotes.

5. *True or False* Because the domain of $f(x) = \log_a x$ is the interval $(0, \infty)$, a logarithmic equation cannot have a negative solution.

6. *True or False* The equation $8^{2x} = 4^{3x}$ is true for all real values of x.

In Exercises 7–22, solve each equation.

7. $2^x = 16$
8. $3^x = 243$

9. $8^x = 32$
10. $5^{x-1} = 1$

11. $4^{|x|} = 128$
12. $9^{|x|} = 243$

13. $5^{-|x|} = 625$
14. $3^{-|x|} = 81$

15. $\ln x = 0$
16. $\ln (x - 1) = 1$

17. $\log_2 x = -1$
18. $\log_2 (x + 1) = 3$

19. $\log_3 |x| = 2$
20. $\log_2 |x + 1| = 3$

21. $\dfrac{1}{2} \log x - 2 = 0$

22. $\dfrac{1}{3} \log (x + 1) - 1 = 0$

In Exercises 23–52, solve each exponential equation and approximate the result correct to three decimal places.

23. $2^x = 3$
24. $3^x = 5$

25. $2^{2x+3} = 15$
26. $3^{2x+5} = 17$

27. $5 \cdot 2^x - 7 = 10$
28. $3 \cdot 5^x + 4 = 11$

29. $3 \cdot 4^{2x-1} + 4 = 14$

30. $2 \cdot 3^{4x-5} - 7 = 10$

31. $5^{1-x} = 2^x$
32. $3^{2x-1} = 2^{x+1}$

33. $2^{1-x} = 3^{4x+6}$
34. $5^{2x+1} = 3^{x-1}$

35. $2 \cdot 3^{x-1} = 5^{x+1}$
36. $5 \cdot 2^{2x+1} = 7 \cdot 3^{x-1}$

37. $(1.065)^t = 2$

38. $(1.0725)^t = 2$

39. $2^{2x} - 4 \cdot 2^x = 21$

40. $4^x - 4^{-x} = 2$

41. $9^x - 6 \cdot 3^x + 8 = 0$

42. $\dfrac{3^x + 5 \cdot 3^{-x}}{3} = 2$

43. $3^{3x} - 4 \cdot 3^{2x} + 2 \cdot 3^x = 8$

44. $2^{3x} + 3 \cdot 2^{2x} - 2^x = 3$

45. $\dfrac{3^x - 3^{-x}}{3^x + 3^{-x}} = \dfrac{1}{4}$

46. $\dfrac{e^x - e^{-x}}{e^x + e^{-x}} = \dfrac{1}{3}$

47. $\dfrac{4}{2 + 3^x} = 1$

48. $\dfrac{7}{2^x - 1} = 3$

49. $\dfrac{17}{5 - 3^x} = 7$

50. $\dfrac{15}{3 + 2 \cdot 5^x} = 4$

51. $\dfrac{5}{2 + 3^x} = 4$
52. $\dfrac{7}{3 + 5 \cdot 2^x} = 4$

In Exercises 53–70, solve each logarithmic equation.

53. $3 + \log (2x + 5) = 2$

54. $1 + \log (3x - 4) = 0$

55. $\log (x^2 - x - 5) = 0$

56. $\log (x^2 - 6x + 9) = 0$

57. $\log_4 (x^2 - 7x + 14) = 1$

58. $\log_4 (x^2 + 5x + 10) = 1$

59. $\ln (2x - 3) - \ln (x + 5) = 0$

60. $\log (x + 8) + \log (x - 1) = 1$

61. $\log x + \log (x + 9) = 1$

62. $\log_5 (3x - 1) - \log_5 (2x + 7) = 0$

63. $\log_a (5x - 2) - \log_a (3x + 4) = 0$

64. $\log (x - 1) + \log (x + 2) = 1$

65. $\log_6 (x + 2) + \log_6 (x - 3) = 1$

66. $\log_2 (3x - 2) - \log_2 (5x + 1) = 3$

67. $\log_3 (2x - 7) - \log_3 (4x - 1) = 2$

68. $\log_4 \sqrt{x + 3} - \log_4 \sqrt{2x - 1} = \dfrac{1}{4}$

69. $\log_7 3x + \log_7 (2x - 1) = \log_7 (16x - 10)$

70. $\log_3 (x + 1) + \log_3 2x = \log_3 (3x + 1)$

In Exercises 71–78, find a and k and then evaluate the function. Round your answer to three decimal places.

71. Let $f(x) = 20 + a2^{kx}$ with $f(0) = 50$ and $f(1) = 140$. Find $f(2)$.

72. $f(x) = 40 + a(4^{kx})$ with $f(0) = -216$ and $f(2) = 39$. Find $f(1)$.

73. $f(x) = 16 + a(3^{kx})$ with $f(0) = 21$ and $f(4) = 61$. Find $f(2)$.

74. $f(x) = 50 + a(2^{kx})$ with $f(0) = 34$ and $f(4) = 46$. Find $f(2)$.

75. Let $f(x) = \dfrac{10}{3 + ae^{kx}}$ with $f(0) = 2$ and $f(1) = \dfrac{1}{2}$. Find $f(2)$.

76. Let $f(x) = \dfrac{6}{a + 2e^{kx}}$ with $f(0) = 1$ and $f(1) = 0.8$. Find $f(2)$.

77. Let $f(x) = \dfrac{4}{a + 4e^{kx}}$ with $f(0) = 2$ and $f(1) = 9$. Find $f(2)$.

78. Let $f(x) = \dfrac{a}{1 + 3e^{kx}}$ with $f(0) = -1$ and $f(2) = -2$. Find $f(1)$.

In Exercises 79–88, solve each inequality.

79. $5(0.3)^x + 1 \le 11$

80. $(0.1)^x - 4 > 15$

81. $-3(1.2)^x + 11 \ge 8$

82. $-7(0.4)^x + 19 < 5$

83. $\log(5x + 15) < 2$

84. $\log(2x + 0.9) > -1$

85. $\ln(x - 5) \ge 1$

86. $\ln(4x + 10) \le 2$

87. $\log_2(3x - 7) < 3$

88. $\log_2(5x - 4) > 4$

B EXERCISES Applying the Concepts

89. **Investment.** Find the time required for an investment of $10,000 to grow to $18,000 at an annual interest rate of 6% if the interest is compounded
  a. yearly.
  b. quarterly.
  c. monthly.
  d. daily.
  e. continuously.

90. **Investment.** How long will it take for an investment of $100 to double in value if the annual rate of interest is 7.2% compounded
  a. yearly.
  b. quarterly.
  c. monthly.
  d. continuously.

In Exercises 91–96, use the following table, which shows various statistics for Canada, Mexico, and the United Kingdom in 2000. Assume the same annual growth rate to continue for each country and use the continuous compounding model.

Item	Canada	Mexico	United Kingdom
Population	33 million	108.7 million	60.8 million
Annual growth rate	0.9%	1.2%	0.3%
Price/liter of milk	$1.97	$0.92	$0.70
Price/kilogram of bread	$2.05	$1.48	$1.20
Annual inflation rate	2%	3.4%	3%

91. **Annual growth rate.** Find all of the data for each country for each year.
  a. 2012
  b. 2018

92. **Annual growth rate.** When will the population of Mexico reach 250 million people?

93. **Annual growth rate.** In what year will the population of Canada be 60 million people?

94. **Annual growth rate.** In Mexico in 2000, the average person spent approximately $500 on food. Use the 2000 inflation rate to calculate the cost of this food in
  a. 2012.
  b. 2018.

95. **Annual growth rate.** In each country, when will the price of a kilogram of bread be $10?

96. **Annual growth rate.** In what year will the price of a liter of milk be the same in Mexico and Canada?

97. **Inflation.** In the fictional land of Sardonia, the price of a car costing $20,000 in U.S. currency will double in eight years. What will the car cost in five years? What is the annual rate of inflation?

98. **Doubling time.** An amount P dollars is deposited in a regular bank account. How long does it take to double the initial investment (called the **doubling time**) if
  a. the interest rate is r compounded annually.
  b. the interest rate is r (per year) compounded continuously.

99. **Population growth.** It is estimated that in t years from 2000, the population of Sardonia will be

$$P(t) = \frac{32}{5 + 3e^{-kt}} \text{ million.}$$

  a. Find the average rate of growth if the population of Sardonia in 2004 was 4.2 million.
  b. What was the population of Sardonia in 2002?
  c. Sketch the graph of $y = P(t)$.
  d. What will the population of Sardonia be in 2020?
  e. What will happen to the population in the long run?

100. Epidemic outbreak. The number of people in a community who became infected during an epidemic t weeks after its outbreak is given by the function

$$f(t) = \frac{20,000}{1 + ae^{-kt}},$$

where 20,000 people of the community are susceptible to the disease. Assuming that 1000 people were infected initially and 8999 had been infected by the end of the fourth week,
a. find the number of people infected after 8 weeks.
b. after how many weeks will 12,400 people be infected?

101. Spreading a rumor. A jealous student started a malicious rumor on an isolated college campus of 5000 students. The number of people who heard the rumor within t days after it was started is given by the function

$$R(t) = \frac{5000}{1 + ae^{-kt}}.$$

Half the students had heard the rumor within ten days.
a. Find a and k and graph the function $y = R(t)$.
b. How many students would have heard the rumor within 15 days after it started?
c. How many students altogether will hear the rumor?

102. Biological growth. In his laboratory experiment in 1934, G. F. Gause placed paramecia (unicellular microorganisms) in 5 cubic centimeters of a saline (salt) solution with a constant amount of food and measured their growth on a daily basis. He found that the population $P(t)$ after t days was approximated by

$$P(t) = \frac{4490}{1 + e^{5.4094 - 1.0255t}}, \quad t \geq 0.$$

a. What was the initial size of the paramecia?
b. What was the carrying capacity of the medium?
c. Graph the equation $y = P(t)$.

103. Marriage rate. The marriage rate in the United States in 2006 was 0.73%, and about 2,164,000 marriages took place that year. (*Source*: Department of Health and Human Services.)
    Use the model $M(t) = M_0(1 + r)^t$ with a constant marriage rate and $t = 0$ corresponding to the year 2006 to estimate
a. the number of marriages in 2026.
b. the year in which the number of marriages will be 4 million.

104. Divorce rate. The divorce rate in the United States in 2006 was 0.36%, and 1,125,000 divorces occurred that year. Use the model $D(t) = D_0(1 + r)^t$ with a constant divorce rate and $t = 0$ corresponding to the year 2006 to estimate
a. the number of divorces in 2026.
b. the year in which the number of divorces will be 2 million.

105. Light absorption in physics. The light intensity I (in lumens) at a depth of x feet in Lake Elizabeth is given by

$$\log\left(\frac{I}{12}\right) = -0.025x.$$

a. Find the light intensity at a depth of 30 feet.
b. At what depth is the light intensity 4 lumens?

106. Rule of 70. Bankers use the **rule of 70** to estimate the doubling time for money invested at different rates. The rule of 70 states that

$$\text{Doubling time} \approx \frac{70}{100r} \text{ years,}$$

where r is the annual interest rate (expressed as a decimal number). Explain why this formula works.

Richter scale. In Exercises 107–110, use the following information. The energy E (measured in joules) released by an earthquake of magnitude M on the Richter scale $m = \log\left(\dfrac{I}{I_0}\right)$ is given by the equation

$$\log E = 4.4 + 1.5M.$$

107. The Great China Earthquake of 1920 registered 8.6 on the Richter scale. Let $I_0 = 1$.
a. What was the intensity of this earthquake?
b. How many joules of energy were released?

108. Suppose one earthquake registers one more point on the Richter scale than another earthquake.
a. How are their corresponding intensities related?
b. How are their released energies related?

109. If one earthquake is 150 times as intense as another, what is the difference in the Richter scale readings of the two earthquakes?

110. If the energy released by one earthquake is 150 times that of another, what is the difference in the Richter scale readings of the two earthquakes?

Sound. In Exercises 111–114, use the following information. The *loudness* (level of sound) of a sound is measured in *decibels* (dB). It is one-tenth of a *bel* (named after Alexander Graham Bell). The loudness L of a sound is related to its intensity I by the equation

$$L = 10\log\left(\frac{I}{I_0}\right),$$

where $I_0 = 10^{-12}$ watts per square meter (W/m²) is the intensity of the faintest sound the human ear can detect.

111. Find the dB level L of a TV at average volume from 12 ft away when it has an intensity of 250×10^{-7} (W/m²).

112. Compare the intensity of a 65 dB sound to that of a 42 dB sound.

113. Find the intensity of a 73 dB sound.

114. The intensity of one sound is 5000 times that of another sound. Find the difference in the dB levels of the two sounds.

C EXERCISES Beyond the Basics

115. Solve for t: $P = \dfrac{M}{1 + e^{-kt}}$

116. Find k so that
   a. $2^x = e^{kx}$.
   b. $e^x = 2^{kx}$.

117. If $x = \dfrac{e^y - e^{-y}}{e^y + e^{-y}}$, show that $y = \dfrac{1}{2}\ln\left(\dfrac{1+x}{1-x}\right)$.

118. If $x = \dfrac{a^y + a^{-y}}{a^y - a^{-y}}$, show that $y = \dfrac{1}{2}\log_a\left(\dfrac{x+1}{x-1}\right)$.

119. If $\dfrac{\log x}{2} = \dfrac{\log y}{3} = \dfrac{\log z}{5}(=k)$, show that
   a. $xy = z$.
   b. $y^2 z^2 = x^8$.
   [*Hint:* First, express x, y, and z in terms of k.]

120. The pressure P and the volume V of a gas are related by the formula

$$PV^\alpha = k,$$

where α and k are constants that depend on the gas. Find α for a given gas with $P_1 = 2$ atmospheres (atm), $V_1 = 3.2$ cubic feet, $P_2 = 3$ atm, and $V_2 = 2.3$ cubic feet.

121. **Annuity.** The future value A of an ordinary annuity with payment size R, periodic rate i, and a term of n payments is given by the formula $A = R\dfrac{(1+i)^n - 1}{i}$.

Solve this equation for n.

122. The present value P of an annuity with payment size R, periodic rate i, and a term of n payments is given by the formula

$$P(1+i)^n = R\dfrac{(1+i)^n - 1}{i}.$$

Solve this equation for n.

In Exercises 123–126, solve each equation for x.

123. $(\log x)^2 = \log x$

124. $\log_2 x + \log_4 x = 6$

125. $\log_4 x^2(x-1)^2 - \log_2(x-1) = 1$

126. $\log(\log x^{10}) = 1$

127. Find the value of $\log_2(\log_2(\log_3(\log_3 27^3)))$.

128. If $f(x) = \log_a\left(\dfrac{1+x}{1-x}\right)$, show that $f\left(\dfrac{2x}{1+x^2}\right) = 2f(x)$.

129. If $\log 3 = 0.477$ and $(1000)^x = 3$, then the value of x is
   a. 0.159.   **b.** 10.   **c.** 0.0477.   **d.** 0.0159.

130. The simplified value of $\log(\log x^2) - \log(\log x)$ is
   a. 2.   **b.** $\dfrac{1}{2}$.   **c.** $\log 2$.   **d.** $2\log x$.

Critical Thinking

In Exercises 131 and 132, x, y, and z represent positive numbers.

131. Evaluate: $\dfrac{x^{\log y}}{x^{\log z}} \cdot \dfrac{y^{\log z}}{y^{\log x}} \cdot \dfrac{z^{\log x}}{z^{\log y}}$

132. Evaluate: $\dfrac{1}{\log_{xy} xyz} + \dfrac{1}{\log_{yz} xyz} + \dfrac{1}{\log_{zx} xyz}$

133. **Logistic function.** For $f(t) = \dfrac{P}{1 + ae^{-kt}}$, it can be shown that the maximum rate of growth occurs when $f(t) = \dfrac{P}{2}$.

Find the time when the rate of growth is maximum.

SUMMARY ■ Definitions, Concepts, and Formulas

4.1 Exponential Functions

i. A function $f(x) = a^x$, with $a > 0$ and $a \neq 1$, is called an exponential function with base a.

Rules of exponents: $a^x a^y = a^{x+y}, \dfrac{a^x}{a^y} = a^{x-y}, (a^x)^y = a^{xy}$,

$$a^0 = 1, a^{-x} = \dfrac{1}{a^x}$$

ii. Exponential functions are one-to-one: If $a^u = a^v$, then $u = v$.

iii. If $a > 1$, then $f(x) = a^x$ is an increasing function; $f(x) \to \infty$ as $x \to \infty$ and $f(x) \to 0$ as $x \to -\infty$.

iv. If $0 < a < 1$, then $f(x) = a^x$ is a decreasing function; $f(x) \to 0$ as $x \to \infty$ and $f(x) \to \infty$ as $x \to -\infty$.

v. The graph of $f(x) = a^x$ has y-intercept 1, and the x-axis is a horizontal asymptote.

4.2 The Natural Exponential Function

i. *Simple interest formula.* If P dollars is invested at an interest rate r per year for t years, then the simple interest is given by the formula $I = Prt$. The amount $A(t) = P + Prt$.

ii. *Compound interest.* P dollars invested at annual rate r compounded n times per year for t years amounts to
$$A_n = P\left(1 + \frac{r}{n}\right)^{nt}.$$

iii. The Euler constant $e = \lim\limits_{h \to \infty}\left(1 + \frac{1}{h}\right)^{h} \approx 2.718$.

iv. *Continuous compounding.* P dollars invested at an annual rate r compounded continuously for t years amounts to $A = Pe^{rt}$.

v. The function $f(x) = e^x$ is the natural exponential function.

4.3 Logarithmic Functions

i. For $a > 0$ and $a \neq 1$, $y = \log_a x$ if and only if $x = a^y$.

ii. Basic properties: $\log_a a = 1, \log_a 1 = 0$,
$$\log_a a^x = x, a^{\log_a x} = x \qquad \text{Inverse properties}$$

iii. The domain of $\log_a x$ is $(0, \infty)$, the range is $(-\infty, \infty)$, and the y-axis is a vertical asymptote. The x-intercept is 1.

iv. Logarithmic functions are one-to-one: If $\log_a x = \log_a y$, then $x = y$.

v. If $a > 1$, then $f(x) = \log_a x$ is an increasing function; $f(x) \to \infty$ as $x \to \infty$ and $f(x) \to -\infty$ as $x \to 0^+$.

vi. If $0 < a < 1$, then $f(x) = \log_a x$ is a decreasing function; $f(x) \to -\infty$ as $x \to \infty$ and $f(x) \to \infty$ as $x \to 0^+$.

vii. The common logarithmic function is $y = \log x$ (base 10); the natural logarithmic function is $y = \ln x$ (base e).

4.4 Rules of Logarithms

i. *Rules of logarithms:*
$$\log_a MN = \log_a M + \log_a N \qquad \text{Product rule}$$
$$\log_a \frac{M}{N} = \log_a M - \log_a N \qquad \text{Quotient rule}$$
$$\log_a M^r = r \log_a M \qquad \text{Power rule}$$

ii. *Change-of-base formula:*
$$\log_b x = \frac{\log_a x}{\log_a b} = \frac{\log x}{\log b} = \frac{\ln x}{\ln b}$$
$$\text{(base } a\text{) (base 10) (base } e\text{)}$$

4.5 Exponential and Logarithmic Equations and Inequalities

An *exponential equation* is an equation in which a variable occurs in one or more exponents.

A *logarithmic equation* is an equation that involves the logarithm of a function of the variable.

Exponential and logarithmic equations are solved by using some or all of the following techniques:

i. Using the one-to-one property of exponential and logarithmic functions

ii. Converting from exponential to logarithmic form or vice versa

iii. Using the Product, Quotient, and Power rules for exponents and logarithms

iv. Using similar techniques to solve logarithmic and exponential inequalities

REVIEW EXERCISES

Basic Skills

In Exercises 1–10, state whether the given statement is true or false.

1. The function $f(x) = a^x$ is exponential if $a > 0$.

2. The graph of $f(x) = 4^x$ approaches the x-axis as $x \to -\infty$.

3. The domain of $f(x) = \log(2 - x)$ is $(-\infty, 2]$.

4. The equation $u = 10^v$ means that $\log u = v$.

5. The inverse of $f(x) = \ln x$ is $g(x) = e^x$.

6. The graph of $y = a^x$ $(a > 0, a \neq 1)$ always contains the points $(0, 1)$ and $(1, a)$.

7. $\ln(M + N) = \ln M + \ln N$

8. $\log \sqrt{300} = 1 + \dfrac{1}{2}\log 3$

9. The functions $f(x) = 2^{-x}$ and $g(x) = \left(\dfrac{1}{2}\right)^x$ have the same graph.

10. $\ln u = \dfrac{\log u}{\log e}$

In Exercises 11–18, match the function with its graph in (a)–(h).

11. $f_1(x) = \log_2 x$

12. $f_2(x) = 2^x$

13. $f_3(x) = \log_2 (3 - x)$

14. $f_4(x) = \log_{1/2} x$

15. $f_5(x) = -\log_2 x$

16. $f_6(x) = \left(\dfrac{1}{2}\right)^x$

17. $f_7(x) = 3 - 2^{-x}$

18. $f_8(x) = \dfrac{6}{1 + 2e^{-x}}$

(a)

(b)

(c)

(d)

(e)

(f)

(g)

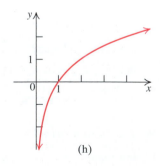

(h)

In Exercises 19–30, graph each function using transformations on an appropriate graph. Determine the domain, range, and asymptotes (if any).

19. $f(x) = 2^{-x}$

20. $g(x) = 2^{-0.5x}$

21. $h(x) = 3 + 2^{-x}$

22. $f(x) = e^{|x|}$

23. $g(x) = e^{-|x|}$

24. $h(x) = e^{-x+1}$

25. $f(x) = \ln(-x)$

26. $g(x) = 2\ln|x|$

27. $h(x) = 2\ln(x - 1)$

28. $f(x) = 2 - \left(\dfrac{1}{2}\right)^x$

29. $g(x) = 2 - \ln(-x)$

30. $h(x) = 3 + 2\ln(5 + x)$

In Exercises 31–34, sketch the graph of the given function using these two steps:

a. Find the intercepts.   **b.** Find the end behavior of f.

31. $f(x) = 3 - 2e^{-x}$

32. $f(x) = \dfrac{5}{2 + 3e^{-x}}$

33. $f(x) = e^{-x^2}$

34. $f(x) = 3 - \dfrac{6}{1 + 2e^{-x}}$

In Exercises 35–38, find a and k and then evaluate the function.

35. Let $f(x) = a(2^{kx})$ with $f(0) = 10$ and $f(3) = 640$. Find $f(2)$.

36. Let $f(x) = 50 - a(5^{kx})$ with $f(0) = 10$ and $f(2) = 0$. Find $f(1)$.

37. Let $f(x) = \dfrac{4}{1 + ae^{-kx}}$ with $f(0) = 1$ and $f(3) = \dfrac{1}{2}$. Find $f(4)$.

38. Let $f(x) = 6 - \dfrac{3}{1 + ae^{-kx}}$ with $f(0) = 5$ and $f(4) = 4$. Find $f(10)$.

In Exercises 39 and 40, find an exponential function of the form $f(x) = ca^x$ with the given graph.

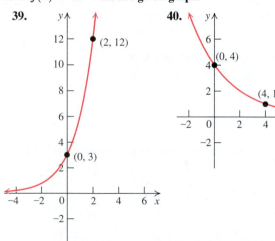

39.

40.

In Exercises 41 and 42, find a logarithmic function of the form $y = \log_a(x - c)$ with the given graph.

41.   **42.**

43. Start with the graph of $y = 2^x$. Find an equation of the graph that results when you
  a. shift the graph right one unit, shift up three units, and reflect about the x-axis.
  b. shift the graph right one unit, reflect about the x-axis, and shift up three units.

44. Start with the graph of $y = \ln x$. Find an equation of the graph that results when you
  a. shift the graph left one unit, stretch horizontally by a factor of 2, and compress vertically by a factor of 3.
  b. compress horizontally to $\frac{1}{3}$, shift left by one unit, and reflect about the y-axis.

In Exercises 45–48, write each logarithm in expanded form.

45. $\ln(xy^2z^3)$ **46.** $\log(x^3\sqrt{y - 1})$

47. $\ln\left[\dfrac{x\sqrt{x^2 + 1}}{(x^2 + 3)^2}\right]$ **48.** $\ln\sqrt{\dfrac{x^3 + 5}{x^3 - 7}}$

In Exercises 49–54, write y as a function of x.

49. $\ln y = \ln x + \ln 3$

50. $\ln y = \ln(C) + kx$; C and k are constants.

51. $\ln y = \ln(x - 3) - \ln(y + 2)$

52. $\ln y = \ln x - \ln(x^2y) - 2\ln y$

53. $\ln y = \dfrac{1}{2}\ln(x - 1) + \dfrac{1}{2}\ln(x + 1) - \ln(x^2 + 1)$

54. $\ln(y - 1) = \dfrac{1}{x} + \ln(y)$

In Exercises 55–78, solve each equation.

55. $3^x = 81$ **56.** $5^{x-1} = 625$

57. $2^{x^2+2x} = 16$ **58.** $3^{x^2-6x+8} = 1$

59. $3^x = 23$ **60.** $2^{x-1} = 5.2$

61. $273^x = 19$ **62.** $27 = 9^x \cdot 3^{x^2}$

63. $3^{2x} = 7^x$ **64.** $2^{x+4} = 3^{x+1}$

65. $(1.7)^{3x} = 3^{2x-1}$ **66.** $3(2^{x+5}) = 5(7^{2x-3})$

67. $\log_3(x + 2) - \log_3(x - 1) = 1$

68. $\log_3(x + 12) - \log_3(x + 4) = 2$

69. $\log_2(x + 2) + \log_2(x + 4) = 3$

70. $\log_5(3x + 7) + \log_5(x - 5) = 2$

71. $\log_5(x^2 - 5x + 6) - \log_5(x - 2) = 1$

72. $\log_3(x^2 - x - 6) - \log_3(x - 3) = 1$

73. $\log_6(x - 2) + \log_6(x + 1) = \log_6(x + 4) + \log_6(x - 3)$

74. $\ln(x - 2) - \ln(x + 2) = \ln(x - 1) - \ln(2x + 1)$

75. $2\ln 3x = 3\ln x$ **76.** $2\log x = \ln e$

77. $2^x - 8 \cdot 2^{-x} - 7 = 0$ **78.** $3^x - 24 \cdot 3^{-x} = 10$

In Exercises 79–82, solve each inequality.

79. $3(0.2)^x + 5 \leq 20$ **80.** $-4(0.2)^x + 15 < 6$

81. $\log(2x + 7) < 2$ **82.** $\ln(3x + 5) \leq 1$

Applying the Concepts

83. Doubling your money. How much time is required for a $1000 investment to double in value if interest is earned at the rate of 6.25% compounded annually?

84. Tripling your money. How much time (to the nearest month) is required to triple your money if the interest rate is 100% compounded continuously?

85. Half-life. The half-life of a certain radioactive substance is 20 hours. How long does it take for this substance to fall to 25% of its original value?

86. Plutonium-210. Find the half-life of plutonium-210 assuming that its decay equation is $Q = Q_0e^{-5 \cdot 10^{-3}t}$, where t is in days.

87. Comparing rates. You have $7000 to invest for seven years. Which investment will provide the greater return, 5% compounded yearly or 4.75% compounded monthly?

88. Investment growth. How long will it take $8000 to grow to $20,000 if the rate of interest is 7% compounded continuously?

89. Population. The formula $P(t) = 33e^{0.003t}$ models the population of Canada, in millions, t years after 2007.
  a. Estimate the population of Canada in 2017.
  b. According to this model, when will the population of Canada be 60 million?

90. House appreciation. The formula $C(t) = 100 + 25e^{.03t}$ models the average cost of a house in Sometown, USA, t years after 2000. The cost is expressed in thousands of dollars.
  a. Sketch the graph of $y = C(t)$.
  b. Estimate the average cost in 2010.
  c. According to this model, when will the average cost of a house in Sometown be $250,000?

91. Drug concentration. An experimental drug was injected into the bloodstream of a rat. The concentration $C(t)$ of the drug (in micrograms per milliliter of blood) after t hours was modeled by the function

$$C(t) = 0.3e^{-0.47t},\ 0.5 \leq t \leq 10.$$

a. Graph the function $y = C(t)$ for $0.5 \leq t \leq 10$.

b. When will the concentration of the drug be 0.029 micrograms per milliliter? (1 microgram $= 10^{-6}$ gram)

92. X-ray intensity. X-ray technicians are shielded by a wall insulated with lead. The equation $x = \dfrac{1}{152} \log\left(\dfrac{I_0}{I}\right)$ measures the thickness x (in centimeters) of the lead insulation required to reduce the initial intensity I_0 of X-rays to the desired intensity I.

a. What thickness of lead is required to reduce the intensity of X-rays to one-tenth their initial intensity?

b. What thickness of lead is required to reduce the intensity of X-rays to $\dfrac{1}{40}$ their initial intensity?

c. How much is the intensity reduced if the lead insulation is 10 centimeters thick?

93. Cooling tea. Chai (tea) is made by adding boiling water (212°F) to the chai mix. Suppose you make chai in a room with the air temperature at 75°F. According to Newton's Law of Cooling, the temperature of the chai t minutes after it is boiled is given by a function of the form $f(t) = 75 + ae^{-kt}$. After one minute, the temperature of the chai falls to 200°F. How long will it take for the chai to be drinkable at 150°F?

94. Spread of influenza. Approximately $P(t) = \dfrac{6}{1 + 5e^{-0.7t}}$ thousand people caught a new form of influenza within t weeks of its outbreak.

a. Sketch the graph of $y = P(t)$.

b. How many people had the disease initially?

c. How many people contracted the disease within four weeks?

d. If the trend continues, how many people in all will contract the disease?

95. Population density. The population density x miles from the center of a town called Greenville is approximated by the equation $D(x) = 5e^{0.08x}$, in thousands of people per square mile.

a. What is the population density at the center of Greenville?

b. What is the population density 5 miles from the center of Greenville?

c. Approximately how far from the center of Greenville would the density be 15,000 people per square mile?

96. Population. In 2000, the population of the United States was 280 million and the number of vehicles was 200 million. If the population of the United States is growing at the rate of 1% per year while the number of vehicles is growing at the rate of 3%, in what year will there be an average of one vehicle per person?

97. Light intensity. The *Bouguer–Lambert Law* states that the intensity I of sunlight filtering down through water at a depth x (in meters) decreases according to the exponential decay function $I = I_0e^{-kx}$, where I_0 is the intensity at the surface and $k > 0$ is an absorption constant that depends on the murkiness of the water.

Suppose the absorption constant of Carrollwood Lake was experimentally determined to be $k = 0.73$. How much light has been absorbed by the water at a depth of 2 meters?

98. Signal strength. The strength of a TV signal usually fades due to a damping effect of cable lines. If I_0 is the initial strength of the signal, then its strength I at a distance x miles is measured by the formula $I = I_0e^{-kx}$, where $k > 0$ is a damping constant that depends on the type of wire used. Suppose the damping constant has been measured experimentally to be $k = 0.003$. What fraction of the signal is lost at a distance of 10 miles? 20 miles?

99. Walking speed in a city. In 1976, Marc and Helen Bernstein discovered that in a city with population p, the average speed s (in feet per second) that a person walks on main streets can be approximated by the formula

$$s(p) = 0.04 + 0.86 \ln p.$$

a. What is the average walking speed of pedestrians in Tampa (population 470,000)?

b. What is the average walking speed of pedestrians in Bowman, Georgia (population 450)?

c. What is the estimated population of a town in which the estimated average walking speed is 4.6 feet per second?

100. Drinking and driving. Just after Eric had his last drink, the alcohol level in his bloodstream was 0.26 (milligram of alcohol per milliliter of blood). After one-half hour, his alcohol level was 0.18. The alcohol level $A(t)$ in a person follows the exponential decay law

$$A(t) = A_0e^{-kt},$$

where $k > 0$ depends on the individual.

a. What is the value of A_0 for Eric?

b. What is the value of k for Eric?

c. If the legal driving limit for alcohol level is 0.08, how long should Eric wait (after his last drink) before he will be able to drive legally?

101. Bacteria culture. The mass $m(t)$, in grams, of a bacteria culture grows according to the logistic growth model

$$m(t) = \dfrac{6}{1 + 5e^{-0.7t}},$$

where t is time measured in days.

a. What is the initial mass of the culture?

b. What happens to the mass in the long run?

c. When will the mass reach 5 grams?

102. A learning model. The number of units $n(t)$ produced per day after t days of training is given by

$$n(t) = 60(1 - e^{-kt}),$$

where $k > 0$ is a constant that depends on the individual.

a. Estimate k for Rita, who produced 20 units after one day of training.

b. How many units will Rita produce per day after ten days of training?

c. How many days should Rita be trained if she is expected to produce 40 units per day?

PRACTICE TEST A

1. Solve the equation $5^{-x} = 125$.

2. Solve the equation $\log_2 x = 5$.

3. State the range of $y = -e^x + 1$ and find the asymptote of its graph.

4. Evaluate $\log_2 \dfrac{1}{8}$.

5. Solve the exponential equation $\left(\dfrac{1}{4}\right)^{2-x} = 4$.

6. Evaluate $\log 0.001$ without using a calculator.

7. Rewrite the expression $\ln 3 + 5 \ln x$ in condensed form.

8. Solve the equation $2^{x+1} = 5$.

9. Solve the equation $e^{2x} + e^x - 6 = 0$.

10. Rewrite the expression $\ln \dfrac{2x^3}{(x+1)^5}$ in expanded logarithmic form.

11. Evaluate $\ln e^{-5}$.

12. Give the equation for the graph obtained by shifting the graph of $y = \ln x$ three units up and one unit right.

13. Sketch the graph of $y = 3^{x-1} + 2$.

14. State the domain of the function $f(x) = \ln(-x) + 4$.

15. Write $3 \ln x + \ln(x^3 + 2) - \dfrac{1}{2} \ln(3x^2 + 2)$ in condensed form.

16. Solve the equation $\log x = \log 6 - \log(x - 1)$.

17. Find x if $\log_x 9 = 2$.

18. Rewrite the expression $\ln \sqrt{2x^3 y^2}$ in expanded logarithmic form.

19. Suppose \$15,000 is invested in a savings account paying 7% interest per year. Write the formula for the amount in the account after t years if the interest is compounded quarterly.

20. Suppose the number of Hispanic people living in the United States is approximated by $H = 15,000e^{0.2t}$, where $t = 0$ represents 1960. According to this model, in which year did the Hispanic population in the United States reached 1.5 million?

PRACTICE TEST B

1. Solve the equation $3^{-x} = 9$.
   a. $\{2\}$ b. $\{-2\}$ c. $\left\{\dfrac{1}{2}\right\}$ d. $\left\{-\dfrac{1}{2}\right\}$ e. $\{\ln 2\}$

2. Solve the equation $\log_5 x = 2$.
   a. $\{10\}$ b. $\{25\}$ c. $\left\{\dfrac{5}{2}\right\}$ d. $\left\{\dfrac{2}{5}\right\}$ e. $\{2\}$

3. State the range and asymptote of $y = e^{-x} - 1$.
   a. $(0, \infty); y = -1$ b. $(-1, \infty); y = -1$
   c. $(-\infty, \infty); y = -1$ d. $(-1, \infty); y = 1$
   e. $(\infty, 1); y = 1$

4. Evaluate $\log_4 64$.
   a. 16   b. 8   c. 2   d. 3   e. 4

5. Find the solution of the exponential equation $\left(\dfrac{1}{3}\right)^{1-x} = 3$.
   a. $\left\{-\dfrac{1}{3}\right\}$ b. $\left\{\dfrac{1}{3}\right\}$ c. $\{1\}$ d. $\{2\}$

6. Evaluate $\log 0.01$.
   a. -1.99999999 b. -2 c. 2   d. 100

7. Which of the following expressions is equivalent to $\ln 7 + 2 \ln x$?
   a. $\ln(7 + 2x)$ b. $\ln(7x^2)$ c. $\ln(9x)$ d. $\ln(14x)$

8. Solve the equation $2^{x^2} = 3^x$.
   a. $\{0\}$ b. $\left\{\dfrac{\ln 3}{\ln 2}\right\}$
   c. $\left\{0, \dfrac{\ln 3}{\ln 2}\right\}$ d. $\{0, \ln 3 - \ln 2\}$

9. Solve the equation $e^{2x} - e^x - 6 = 0$.
   a. $\{-\ln 2\}$ b. $\{-\ln 3\}$ c. $\{\ln 6\}$ d. $\{\ln 3\}$

10. Rewrite the expression $\ln \dfrac{3x^2}{(x+1)^{10}}$ in expanded logarithmic form.
    a. $\ln 6x - 10 \ln x + 1$
    b. $2 \ln 3x - 10 \ln(x + 1)$
    c. $2 \ln 3 + 2 \ln x - 10 \ln(x + 1)$
    d. $\ln 3 + 2 \ln x - 10 \ln(x + 1)$

11. Find $\ln e^{3x}$.
    a. 3   b. $3x$ c. $3 + x$ d. x

12. The equation for the graph obtained by shifting the graph of $y = \log_3 x$ two units up and three units left is
    a. $y = \log_3(x - 3) + 2$.   b. $y = \log_3(x + 3) - 2$.
    c. $y = \log_3(x - 3) - 2$.   d. $y = \log_3(x + 3) + 2$.

13. Which of the following graphs is the graph of $y = 5^{x+1} - 4$?

(a)

(b)

(c)

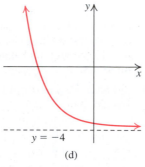

(d)

14. Find the domain of the function $f(x) = \ln(1 - x) + 3$.
  a. $(-\infty, 1)$ **b.** $(1, \infty)$ **c.** $(-\infty, -1)$
  d. $(-\infty, 3)$ **e.** $(3, \infty)$

15. Write $\ln x - 2\ln(x^2 + 1) + \dfrac{1}{2}\ln(x^4 + 1)$ in condensed form.

  a. $\ln \dfrac{x\sqrt{x^4 + 1}}{(x^2 + 1)^2}$ **b.** $\ln \dfrac{x}{x^2 + 1}$

  c. $\ln(x - (x^2 + 1)^2 + (x^4 + 1)^{1/2})$

  d. $2\ln \dfrac{x(1 + x^4)}{x^2 + 1}$

16. Solve the equation $\log x = \log 12 - \log(x + 1)$.
  a. $\left\{\dfrac{11}{2}\right\}$ **b.** $\left\{\dfrac{13}{2}\right\}$ **c.** $\{3, -4\}$

  d. $\{3\}$ **e.** $\left\{\dfrac{2}{12}\right\}$

17. Find x if $\log_x 16 = 4$.
  a. 4   **b.** 2   **c.** 64   **d.** $\dfrac{1}{4}$ **e.** 16

18. Rewrite the expression $\ln \sqrt[3]{5x^2y^3}$ in expanded logarithmic form.

  a. $\dfrac{1}{3}(\ln 5x^2 + \ln y^3)$

  b. $\dfrac{1}{3}(2\ln 5x + 3\ln y)$

  c. $\dfrac{1}{3}(\ln 5 + 2\ln x + 3\ln y)$

  d. $\dfrac{1}{3}\ln 5 + 2\ln x + 3\ln y$

  e. $\dfrac{1}{3}\ln 5 + 2 + \ln x + 3 + \ln y$

19. Suppose $12,000 is invested in a savings account paying 10.5% interest per year. Write the formula for the amount in the account after t years assuming that the interest is compounded monthly.
  a. $A = 12,000(1.105)^t$ **b.** $A = 12,000(1.525)^{2t}$
  c. $A = 12,000(1.2625)^{4t}$ **d.** $A = 12,000(1.00875)^{12t}$

20. The population of a certain city is growing according to the model $P = 10,000 \log_5(t + 5)$, where t is time in years. If $t = 0$ corresponds to the year 2000, what will the population of the city be in 2020?
  a. 30,000   **b.** 20,000
  c. 50,000   **d.** 10,000

CUMULATIVE REVIEW EXERCISES ■ Chapters P–4

1. Find the x- and y-intercepts of the graph of the equation $x^2 + (y - 1)^2 = 1$. Check for symmetry with respect to both axes and the origin.

2. Graph the equation $y = x^3$.

3. Find the center and radius of the circle $x^2 + 2x + y^2 - 4y - 20 = 0$. Graph the circle.

4. Write the slope–intercept form of the equation of the line that passes through $(-2, 4)$ and is perpendicular to the line $2x + 3y = 17$.

5. Find the domain of $f(x) = \log \dfrac{x - 2}{x + 3}$.

6. Evaluate the function $f(x) = \begin{cases} 2x - 3, & x < 1 \\ 2x^2 + 1, & x \geq 1 \end{cases}$ at each value specified.
  a. $f(-2)$ **b.** $f(0)$ **c.** $f(3)$

7. Identify the basic function and use transformations to sketch the graph of the function $f(x) = 3\sqrt{x + 1} - 2$.

8. Let $f(x) = 3\sqrt{x}$ and $g(x) = -\dfrac{1}{x}$.
  Find and write the domain of $(f \circ g)(x)$ in interval notation.

9. Determine whether the function has an inverse function; if so, find the inverse function.
   a. $f(x) = 2x^3 + 1$
   b. $f(x) = |x|$
   c. $f(x) = \ln x$

10. Divide
   a. $\dfrac{2x^3 + 3x + 1}{x^2 + 1}$ using long division.
   b. $\dfrac{2x^4 + 7x^3 - 2x + 3}{x + 3}$ using synthetic division.

11. Find a polynomial function of least degree with integer coefficients that has the given zeros.
   a. $1, 2, -3$
   b. $1, -1, i, -i$

12. Use rules of logarithms to rewrite the following expressions in expanded form.
   a. $\log_2 5x^3$
   b. $\log_a \sqrt[3]{\dfrac{xy^2}{z}}$
   c. $\ln \dfrac{3\sqrt{x}}{5y}$

13. Use rules of logarithms to rewrite the following expressions in condensed form.
   a. $\dfrac{1}{2}(\log x + \log y)$
   b. $3 \ln x - 2 \ln y$

14. Use a calculator to evaluate the expression. Round your answer to three decimal places.
   a. $\log_9 17$
   b. $\log_3 25$
   c. $\log_{1/2} 0.3$

15. Solve each equation.
   a. $\log_6 (x + 3) + \log_6 (x - 2) = 1$
   b. $5^{x^2 - 4x + 5} = 25$
   c. $3.1^{x-1} = 23$

16. Find all possible rational zeros of
   $$f(x) = x^4 - x^3 - 2x^2 - 2x + 4.$$
   Also find upper and lower bounds for the zeros of f.

17. Let $f(x) = (x - 1)^3(x + 2)^2(x - 3)$.
   a. Find the zeros of f along with their multiplicities.
   b. What is the behavior of f near each zero?
   c. What is the end behavior of f?
   d. Sketch the graph of f.

18. Let $f(x) = \dfrac{(x - 1)(x + 2)}{(x - 3)(x + 4)}$.
   a. Find the vertical asymptotes of f (if any).
   b. Find the horizontal asymptotes of f (if any).
   c. Find the intervals over which $f(x) > 0$ or $f(x) < 0$.
   d. Sketch the graph of f.

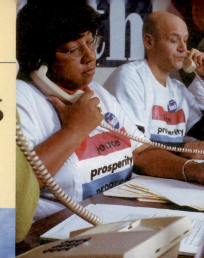

CHAPTER

5 Systems of Equations and Inequalities

TOPICS

Systems of equations are frequently the tools used to describe relationships among variables representing different interrelated quantities. These same systems of equations are used to predict results associated with changes in the quantities.

The quantities involved can come from the study of medicine, taxes, elections, business, agriculture, city planning, or any discipline in which multiple quantities interact.

Systems of Linear Equations in Two Variables

Before Starting this Section, Review

1. Graphs of equations (Section 2.2, page 179)

2. Graphs of linear equations (Section 2.3, page 196)

Objectives

1. Verify a solution to a system of equations.

2. Solve a system of equations by the graphical method.

3. Solve a system of equations by the substitution method.

4. Solve a system of equations by the elimination method.

5. Solve applied problems by solving systems of equations.

Al-Khwarizmi (780–850)

Al-Khwarizmi came to Baghdad from the town of Khwarizm, which is now called Khiva and is part of Uzbekistan. The term *algorithm* is a corruption of the name al-khwarizmi. Originally, the word *algorism* was used for the rules for performing arithmetic by using decimal notation. *Algorism* evolved into the word *algorithm* by the eighteenth century. The word *algorithm* now means a finite set of precise instructions for performing a computation or for solving a problem. (The picture shown here is a Soviet postage stamp from 1983 depicting Al-Khwarizmi.)

ALGEBRA–IRAQ CONNECTION

The most important contributions of medieval Islamic mathematicians lie in the area of algebra. One of the greatest Islamic scholars was Muhammad ibn Musa al-Khwarizmi (780–850). Al-Khwarizmi was one of the first scholars in the House of Wisdom, an academy of scientists, established by caliph al-Mamun in the city of Baghdad in what is now Iraq. Jews, Christians, and Muslims worked together in scholarly pursuits in the academy during this period. The eminent scholar al-Khwarizmi wrote several books on astronomy and mathematics. Western Europeans first learned about algebra from his books. The word *algebra* comes from the Arabic *al-jabr*, part of the title of his book *Kitab al-jabr wal-muqabala*.

This book was translated into Latin and was a widely used text. The Arabic word *al-jabr* means "restoration," as in restoring broken parts. (At one time, it was not unusual to see the sign "Algebrista y Sangrador," meaning "bone setter and blood letter," at the entrance of a Spanish barber's shop. The sign informed customers of the barber's side business.) Al-Khwarizmi used the term in the mathematical sense of removing a negative quantity on one side of an equation and restoring it as a positive quantity on the other side. ∎

---

1. Verify a solution to a system of equations.

System of Equations

A set of equations with common variables is called a **system of equations**. If each equation in a system of equations is linear, then the set of equations is called a **system of linear equations** or a **linear system of equations**. However, if at least one equation in a system of equations is nonlinear, then the set of equations is called a **system of nonlinear equations**. In this section, we solve systems of two linear equations in two variables, such as:

$$\begin{cases} 2x - y = 5 \\ x + 2y = 5 \end{cases}$$

Notice that we designate a system of equations by using a left brace. A system of equations is sometimes referred to as a set of *simultaneous equations*. A **solution of a system of equations in two variables** x and y is an ordered pair of numbers (a, b) such that when x is replaced with a and y is replaced with b, the resulting equations are true. The **solution set of a system of equations** is the set of all solutions of the system.

EXAMPLE 1   Verifying a Solution

Verify that the ordered pair $(3, 1)$ is a solution of the system of linear equations:

$$\begin{cases} 2x - y = 5 & (1) \\ x + 2y = 5 & (2) \end{cases}$$

SOLUTION

To verify that $(3, 1)$ is a solution, replace x with 3 and y with 1 in both equations.

Equation (1)	Equation (2)
$2x - y = 5$	$x + 2y = 5$
$2(3) - 1 \overset{?}{=} 5$	$3 + 2(1) \overset{?}{=} 5$
$5 = 5$ ✓	$5 = 5$ ✓

Because the ordered pair $(3, 1)$ satisfies both equations, it is a solution of the system.

■ ■ ■

Practice Problem 1   Verify that $(1, 3)$ is a solution of the system:

$$\begin{cases} x + y = 4 \\ 3x - y = 0 \end{cases}$$

■

In this section, you will learn three methods of solving a system of two equations in two variables: the *graphical method*, the *substitution method*, and the *elimination method*.

2 Solve a system of equations by the graphical method.

Graphical Method

Recall that the graph of a linear equation

$$ax + by = c \qquad (a \text{ and } b \text{ not both zero})$$

is a line. Consider the following system of linear equations:

$$\begin{cases} a_1x + b_1y = c_1 \\ a_2x + b_2y = c_2 \end{cases}$$

A solution of this system is a point (an ordered pair) that satisfies both equations. Therefore, to find or estimate the solution set of this system, we graph both equations on the same coordinate axes and find the coordinates of any points of intersection. This procedure is called the **graphical method**.

EXAMPLE 2   Solving a System by the Graphical Method

Use the graphical method to solve the system of equations.

$$\begin{cases} 2x - y = 4 & (1) \\ 2x + 3y = 12 & (2) \end{cases}$$

SOLUTION

Step 1   **Graph both equations on the same coordinate axes.**
   (i) We first graph equation (1) by finding its intercepts.
      a. To find the y-intercept, set $x = 0$ and solve for y.
         We have $2(0) - y = 4$, or $y = -4$; so the y-intercept is -4.

FIGURE 5.1 Graphical solution

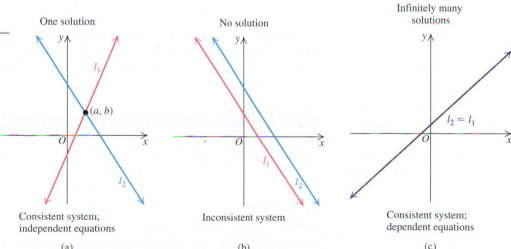
b. To find the x-intercept, set $y = 0$ and solve for x.
 We have $2x - 0 = 4$, or $x = 2$; so the x-intercept is 2.
The points $(0, -4)$ and $(2, 0)$ are on the graph of equation (1), which is sketched in red in Figure 5.1.

(ii) We now graph equation (2) using intercepts.
 Here the x-intercept is 6 and the y-intercept is 4; so joining the points $(0, 4)$ and $(6, 0)$ produces the line sketched in blue in Figure 5.1.

Step 2 **Find the point(s) of intersection of the two graphs.** We observe in Figure 5.1 that the point of intersection of the two graphs is $(3, 2)$.

Step 3 **Check your solution(s).** Replace x with 3 and y with 2 in equations (1) and (2).

$$
\begin{array}{c|c}
\text{Equation (1)} & \text{Equation (2)} \\
2x - y = 4 & 2x + 3y = 12 \\
2(3) - 2 \overset{?}{=} 4 & 2(3) + 3(2) \overset{?}{=} 12 \\
4 = 4 \checkmark & 12 = 12 \checkmark
\end{array}
$$

Step 4 **Write the solution set for the system.**
 The solution set is $\{(3, 2)\}$. ■ ■ ■

Practice Problem 2 Solve the system of equations graphically.

$$\begin{cases} x + y = 2 \\ 4x + y = -1 \end{cases}$$ ■

If a system of equations has at least one solution (as in Example 2), the system is **consistent**. A system of equations with no solution is **inconsistent**, and its solution set is the empty set, $\varnothing$.

A system of two linear equations in two variables must have one of the following types of solution sets.

1. One solution (the lines intersect; see Figure 5.2(a)): the system is consistent, and the equations in the system are said to be **independent**.

2. No solution (the lines are parallel; see Figure 5.2(b)): the system is inconsistent.

3. Infinitely many solutions (the lines coincide; see Figure 5.2(c)): the system is consistent, and the equations in the system are said to be **dependent**.

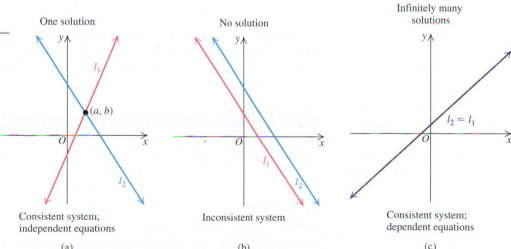

FIGURE 5.2 Possible solution sets of a system of two linear equations

3 Solve a system of equations by the substitution method.

Substitution Method

Another way to solve a linear system of equations is with the **substitution method**.

FINDING THE SOLUTION: A PROCEDURE

EXAMPLE 3   **The Substitution Method**

OBJECTIVE

Reduce the solution of the system to the solution of one equation in one variable by substitution.

Step 1 **Solve for one variable.** Choose one of the equations and express one of its variables in terms of the other variable.

Step 2 **Substitute.** Substitute the expression found in Step 1 into the other equation to obtain an equation in one variable.

Step 3 **Solve** the equation obtained in Step 2.

Step 4 **Back-substitute.** Substitute the value(s) you found in Step 3 back into the expression you found in Step 1. The result is the solution(s).

Step 5 **Check.** Check your answer(s) in the original equations.

EXAMPLE

Solve the system:

$$\begin{cases} 2x - 5y = 3 & (1) \\ y - 2x = 9 & (2) \end{cases}$$

We choose equation (2) and express y in terms of x.

$$y = 2x + 9 \quad (3) \qquad \text{Solve equation (2) for } y.$$

$$\begin{aligned} 2x - 5y &= 3 & \text{Equation (1)} \\ 2x - 5(2x + 9) &= 3 & \text{Substitute } 2x + 9 \text{ for } y. \\ 2x - 10x - 45 &= 3 & \text{Distributive property} \\ -8x - 45 &= 3 & \text{Combine like terms.} \\ -8x &= 3 + 45 & \text{Add 45 to both sides.} \\ x &= -6 & \text{Solve for } x. \end{aligned}$$

$$\begin{aligned} y &= 2x + 9 & \text{Equation (3) from Step 1} \\ y &= 2(-6) + 9 = -3 & \text{Substitute } x = -6. \end{aligned}$$

Because $x = -6$ and $y = -3$, the solution set is $\{(-6, -3)\}$.

Check: $x = -6$ and $y = -3$.

Equation (1)	Equation (2)
$2(-6) - 5(-3) \overset{?}{=} 3$	$-3 - 2(-6) \overset{?}{=} 9$
$-12 + 15 \quad = 3 \ ✓$	$-3 + 12 \quad = 9 \ ✓$ ■■■

Practice Problem 3   Solve: $\begin{cases} x - y = 5 \\ 2x + y = 7 \end{cases}$ ■

It is possible that in the process of solving the equation in Step 3, you obtain an equation of the form $0 = k$ (which can be written as $0x = k$ or $0y = k$), where k is a *nonzero* constant. In such cases, the false statement $0 = k$ indicates that the system is *inconsistent*.

EXAMPLE 4   **Attempting to Solve an Inconsistent System of Equations**

Solve the system of equations.

$$\begin{cases} x + y = 3 & (1) \\ 2x + 2y = 9 & (2) \end{cases}$$

SOLUTION

Step 1   Solve equation (1) for y in terms of x.

$$y = 3 - x \qquad \text{Add } -x \text{ to both sides.}$$

Step 2   Substitute this expression into equation (2).

$$\begin{aligned} 2x + 2y &= 9 & \text{Equation (2)} \\ 2x + 2(3 - x) &= 9 & \text{Replace } y \text{ with } 3 - x \text{ (from Step 1).} \end{aligned}$$

FIGURE 5.3 Inconsistent system

Step 3   Solve for x.

$$2x + 6 - 2x = 9 \qquad \text{Distributive property}$$

$$\underbrace{0 = 3}_{\text{False}} \qquad \text{Subtract 6 from both sides and simplify.}$$

Because the equation $0 = 3$ is false, the system is inconsistent. Figure 5.3 shows that the graphs of the two equations in this system are parallel lines. Because the lines do not intersect, the system has no solution. The solution set is $\varnothing$.   ■ ■ ■

Practice Problem 4   Solve the system of equations.

$$\begin{cases} x - 3y = 1 \\ -2x + 6y = 3 \end{cases}$$ ■

In Step 3, it is also possible to end up with an equation of the form $0 = 0$. In such cases, the equations are *dependent* and the system has infinitely many solutions.

EXAMPLE 5   **Solving a Dependent System**

Solve the system of equations.

$$\begin{cases} 4x + 2y = 12 & (1) \\ -2x - y = -6 & (2) \end{cases}$$

SOLUTION

Step 1   Solve equation (2) for y in terms of x.

$$-2x - y = -6 \qquad \text{Equation (2)}$$

$$-y = -6 + 2x \qquad \text{Add } 2x \text{ to both sides.}$$

$$y = 6 - 2x \qquad \text{Multiply both sides by } -1.$$

Step 2   Substitute $(6 - 2x)$ for y in equation (1).

$$4x + 2y = 12 \qquad \text{Equation (1)}$$

$$4x + 2(6 - 2x) = 12 \qquad \text{Replace } y \text{ with } 6 - 2x \text{ (from Step 1).}$$

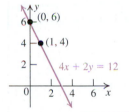

FIGURE 5.4 Dependent equations

Step 3   Solve for x.

$$4x + 12 - 4x = 12 \qquad \text{Distributive property}$$

$$\underbrace{0 = 0}_{\text{True}} \qquad \text{Subtract 12 from both sides and simplify.}$$

The equation $0 = 0$ is true for *every* value of x. Thus, *any* value of x can be used in the equation $y = 6 - 2x$ for back-substitution.

The solutions of the system are of the form $(x, 6 - 2x)$, and the solution set is

$$\{(x, 6 - 2x)\}.$$

BY THE WAY . . .

We use the informal notation

$$\{(x, 6 - 2x)\}$$

rather than

$$\{(x, 6 - 2x) | x \text{ is a real number}\}$$

In other words, the solution set consists of the infinite number of ordered pairs (x, y) lying on the line with equation $4x + 2y = 12$, as shown in Figure 5.4. You can find particular solutions by replacing x with any real number. For example, if we let $x = 0$ in $(x, 6 - 2x)$, we find that $(0, 6 - 2(0)) = (0, 6)$ is a solution of the system. Similarly, letting $x = 1$, we find that $(1, 4)$ is a solution.   ■ ■ ■

Practice Problem 5   Solve the system of equations.

$$\begin{cases} -2x + y = -3 \\ 4x - 2y = 6 \end{cases}$$ ■

4 Solve a system of equations by the elimination method.

Elimination Method

The **elimination method** of solving a system of equations is also called the **addition method**.

FINDING THE SOLUTION: A PROCEDURE

EXAMPLE 6   **The Elimination Method**

OBJECTIVE	**EXAMPLE**

OBJECTIVE
Solve a system of two linear equations by first eliminating one variable.

EXAMPLE
Solve the system.

$$\begin{cases} 2x + 3y = 21 & (1) \\ 3x - 4y = 23 & (2) \end{cases}$$

Step 1 **Adjust the coefficients.** If necessary, multiply both equations by appropriate numbers to get two new equations in which the coefficients of the variable to be eliminated are opposites.

We arbitrarily select y as the variable to be eliminated.

$$\begin{cases} 8x + 12y = 84 & \text{Multiply equation (1) by 4.} \\ 9x - 12y = 69 & \text{Multiply equation (2) by 3.} \end{cases}$$

Coefficients are opposites of each other.

Step 2 **Add the equations.** Add the resulting equations to get an equation in one variable.

$$\begin{array}{r} 8x + 12y = 84 \\ \underline{9x - 12y = 69} \\ 17x = 153 \end{array}$$

Adding the two equations from Step 1 eliminates the y-terms.

Step 3 **Solve** the resulting equation.

$$17x = 153 \quad \text{Equation from Step 2}$$
$$x = \frac{153}{17} \quad \text{Divide both sides by 17.}$$
$$x = 9 \quad \text{Simplify.}$$

Step 4 **Back-substitute** the value you found into one of the original equations to solve for the other variable.

$$2x + 3y = 21 \quad \text{Equation (1)}$$
$$2(9) + 3y = 21 \quad \text{Replace } x \text{ with 9 from Step 3.}$$
$$18 + 3y = 21 \quad \text{Simplify.}$$
$$3y = 3 \quad \text{Subtract 18 from both sides.}$$
$$y = 1 \quad \text{Solve for } y.$$

Step 5 **Write** the solution set from Steps 3 and 4.

The solution set is $\{(9, 1)\}$.

Step 6 **Check** your solution(s) in the original equations (1) and (2).

Check $x = 9$ and $y = 1$.

Equation (1)	Equation (2)
$2(9) + 3(1) \overset{?}{=} 21$	$3(9) - 4(1) = 23$
$18 + 3 = 21$ ✓	$27 - 4 = 23$ ✓

■ ■ ■

Practice Problem 6   Solve the system.

$$\begin{cases} 3x + 2y = 3 \\ 9x - 4y = 4 \end{cases}$$

■

As in the substitution method, if you add the equations in Step 2 and the resulting equation becomes $0 = k$, where $k \neq 0$, then the system is inconsistent: it has no solutions. As an illustration, if you solve the system of equations in Example 4 by the elimination method, you will obtain the equation $0 = 3$. You now conclude that the system is inconsistent. Similarly, in Step 2, if you obtain $0x + 0y = 0$ (or $0 = 0$), then the equations are dependent See Example 5.

STUDY TIP

You should use the elimination method when the terms involving one of the variables can easily be eliminated by adding multiples of the equations. Otherwise, use substitution to solve the system. The graphing method is usually used to confirm or visually interpret the result obtained from the other two methods.

The next example illustrates how some nonlinear systems can be solved by substituting variables to create a linear system and then solving the new system by the elimination method.

EXAMPLE 7   Using the Elimination Method

Solve the system.

$$\begin{cases} \dfrac{2}{x} + \dfrac{5}{y} = -5 & (1) \\ \dfrac{3}{x} - \dfrac{2}{y} = -17 & (2) \end{cases}$$

SOLUTION

Replace $\dfrac{1}{x}$ with u and $\dfrac{1}{y}$ with v. (Note that $\dfrac{2}{x} = 2\left(\dfrac{1}{x}\right) = 2u$, and so on.) Equations (1) and (2) become

$$\begin{cases} 2u + 5v = -5 & (3) \\ 3u - 2v = -17 & (4) \end{cases}$$

Now we solve the new system for u and v using the elimination method.

Step 1 We choose to eliminate the variable u.

$6u + 15v = -15$	(5)	Multiply equation (3) by 3.
$\underline{-6u + 4v = 34}$	(6)	Multiply equation (4) by -2.

Step 2 $19v = 19$ (7)   Add equations (5) and (6).

Step 3 $v = 1$ Solve equation (7) for v.

Step 4 $2u + 5v = -5$ Equation (3)

$2u + 5(1) = -5$ Back-substitute $v = 1$.

$u = -5$ Solve for u.

Step 5 Now solve for x and y, the variables in the original system.

$u = \dfrac{1}{x}$ and $v = \dfrac{1}{y}$ Original substitution

$-5 = \dfrac{1}{x}$ $1 = \dfrac{1}{y}$ Replace u with -5 and v with 1.

$x = -\dfrac{1}{5}$ $y = 1$ Solve for x and y.

The solution set of the original system is $\left\{\left(-\dfrac{1}{5}, 1\right)\right\}$.

Step 6 Verify that the ordered pair $\left(-\dfrac{1}{5}, 1\right)$ is the solution of the original system of equations (1) and (2). ■ ■ ■

Practice Problem 7 Solve the system.

$$\begin{cases} \dfrac{4}{x} + \dfrac{3}{y} = 1 \\ \dfrac{2}{x} - \dfrac{6}{y} = 3 \end{cases}$$ ■

5 Solve applied problems by solving systems of equations.

Applications

We can deduce from Section 2.4 that as the price of a product increases, demand for it decreases and as the price increases, the supply of the product also increases. The **equilibrium point** is the ordered pair (x, p) such that the number of units, x, and the price per unit, p, satisfy *both* the demand and supply equations.

STUDY TIP

The graph of a system of equations allows you to use geometric intuition to understand algebra. If the geometric and algebraic representations of a solution do not agree, you must go back and find the error(s).

EXAMPLE 8   Finding the Equilibrium Point

Find the equilibrium point if the supply and demand functions for a new brand of digital video recorder (DVR) are given by the system

$$p = 60 + 0.0012x \quad (1) \quad \text{Supply equation}$$
$$p = 80 - 0.0008x \quad (2) \quad \text{Demand equation}$$

where p is the price in dollars and x is the number of units.

SOLUTION

We substitute the value of p from equation (1) into equation (2) and solve the resulting equation.

$$p = 80 - 0.0008x \qquad \text{Demand equation (2)}$$
$$60 + 0.0012x = 80 - 0.0008x \qquad \text{Replace } p \text{ with } 60 + 0.0012x \text{ in (1).}$$
$$0.002x = 20 \qquad \text{Collect like terms and simplify.}$$
$$x = \frac{20}{0.002} = 10{,}000 \qquad \text{Solve for } x.$$

The equilibrium point occurs when the supply and demand for DVRs is 10,000 units. To find the price p, we back-substitute $x = 10{,}000$ into either of the original equations (1) or (2).

$$p = 60 + 0.0012x \qquad \text{Equation (1)}$$
$$= 60 + 0.0012(10{,}000) \qquad \text{Replace } x \text{ with } 10{,}000.$$
$$= 72 \qquad \text{Simplify.}$$

FIGURE 5.5 Price of DVRs

The equilibrium point is $(10{,}000, 72)$. See Figure 5.5. This means that at a price of $72 per unit, the supply and demand for DVRs is 10,000 units.

Check. Verify that the ordered pair $(10{,}000, 72)$ satisfies both equations (1) and (2).   ■ ■ ■

Practice Problem 8   Find the equilibrium point.

$$\begin{cases} p = 20 + 0.002x & \text{Supply equation} \\ p = 77 - 0.008x & \text{Demand equation} \end{cases}$$ ■

In general, to solve an applied problem involving two unknowns, we need to set up two equations using two variables to represent the two unknowns. The general rule is that *the number of equations formed must be equal to the number of unknown quantities involved.*

A EXERCISES Basic Skills and Concepts

1. The ordered pair (a, b) is a(n) _____ of a system of equations in x and y providing that when x is replaced with a and y is replaced with b, the resulting equations are true.

2. The two nongraphical methods for solving a system of equations are the _____ and _____ methods.

3. If in the process of solving a system of equations you get an equation of the form $0 = k$, where k is not zero, then the system is _____.

4. If in the process of solving a system of equations you get an equation of the form $0 = 0$, then the system has _____.

5. *True or False* A system consisting of two identical equations has no solution.

6. *True or False* If in the process of solving a system of two equations in x and y you get the equation $x = 5$, then the system has exactly one solution.

In Exercises 7–12, determine which ordered pairs are solutions of each system of equations.

7. $\begin{cases} 2x + 3y = 3 \\ 3x - 4y = 13 \end{cases}$ $(1, -3), (3, -1), (6, 3), \left(5, \dfrac{1}{2}\right)$

8. $\begin{cases} x + 2y = 6 \\ 3x + 6y = 18 \end{cases}$ $(2, 2), (-2, 4), (0, 3), (1, 2)$

9. $\begin{cases} 5x - 2y = 7 \\ -10x + 4y = 11 \end{cases}$ $\left(\dfrac{5}{4}, 1\right), \left(0, \dfrac{11}{4}\right), (1, -1), (3, 4)$

10. $\begin{cases} x - 2y = -5 \\ 3x - y = 5 \end{cases}$ $(1, 3), (-5, 0), (3, 4), (3, -4)$

11. $\begin{cases} x + y = 1 \\ \dfrac{1}{2}x + \dfrac{1}{3}y = 2 \end{cases}$ $(0, 1), (1, 0), \left(\dfrac{2}{3}, \dfrac{3}{2}\right), (10, -9)$

12. $\begin{cases} \dfrac{2}{x} + \dfrac{3}{y} = 2 \\ \dfrac{6}{x} + \dfrac{18}{y} = 9 \end{cases}$ $(3, 2), (2, 3), (4, 3), (3, 4)$

In Exercises 13–22, estimate the solution(s) (if any) of each system by the graphical method. Check your solution(s). For any dependent equations, write your answer with x being arbitrary.

13. $\begin{cases} x + y = 3 \\ x - y = 1 \end{cases}$

14. $\begin{cases} x + y = 10 \\ x - y = 2 \end{cases}$

15. $\begin{cases} x + 2y = 6 \\ 2x + y = 6 \end{cases}$

16. $\begin{cases} 2x - y = 4 \\ x - y = 3 \end{cases}$

17. $\begin{cases} 3x - y = -9 \\ y = 3x + 6 \end{cases}$

18. $\begin{cases} 5x + 2y = 10 \\ y = -\dfrac{5}{2}x - 5 \end{cases}$

19. $\begin{cases} x + y = 7 \\ y = 2x \end{cases}$

20. $\begin{cases} y - x = 2 \\ y + x = 9 \end{cases}$

21. $\begin{cases} 3x + y = 12 \\ y = -3x + 12 \end{cases}$

22. $\begin{cases} 2x + 3y = 6 \\ 6y = -4x + 12 \end{cases}$

In Exercises 23–36, determine whether each system is *consistent* or *inconsistent*. If the system is consistent, determine whether the equations are dependent or independent. Do not solve the system.

23. $\begin{cases} y = -2x + 3 \\ y = 3x + 5 \end{cases}$

24. $\begin{cases} 3x + y = 5 \\ 2x + y = 4 \end{cases}$

25. $\begin{cases} 2x + 3y = 5 \\ 3x + 2y = 7 \end{cases}$

26. $\begin{cases} 2x - 4y = 5 \\ 3x + 5y = -6 \end{cases}$

27. $\begin{cases} 3x + 5y = 7 \\ 6x + 10y = 14 \end{cases}$

28. $\begin{cases} 3x - y = 2 \\ 9x - 3y = 6 \end{cases}$

29. $\begin{cases} x + 2y = -5 \\ 2x - y = 4 \end{cases}$

30. $\begin{cases} x + 2y = -2 \\ 2x - 3y = 5 \end{cases}$

31. $\begin{cases} 2x - 3y = 5 \\ 6x - 9y = 10 \end{cases}$

32. $\begin{cases} 3x + y = 2 \\ 15x + 5y = 15 \end{cases}$

33. $\begin{cases} -3x + 4y = 5 \\ \dfrac{9}{2}x - 6y = \dfrac{15}{2} \end{cases}$

34. $\begin{cases} 6x + 5y = 11 \\ 9x + \dfrac{15}{2}y = 21 \end{cases}$

35. $\begin{cases} 7x - 2y = 3 \\ 11x - \dfrac{3}{2}y = 8 \end{cases}$

36. $\begin{cases} 4x + 7y = 10 \\ 10x + \dfrac{35}{2}y = 25 \end{cases}$

In Exercises 37–46, solve each system of equations by the substitution method. Check your solutions. For any dependent equations, write your answer in the ordered pair form given in Example 5.

37. $\begin{cases} y = 2x + 1 \\ 5x + 2y = 9 \end{cases}$

38. $\begin{cases} x = 3y - 1 \\ 2x - 3y = 7 \end{cases}$

39. $\begin{cases} 3x - y = 5 \\ x + y = 7 \end{cases}$

40. $\begin{cases} 2x + y = 2 \\ 3x - y = -7 \end{cases}$

41. $\begin{cases} 2x - y = 5 \\ -4x + 2y = 7 \end{cases}$

42. $\begin{cases} 3x + 2y = 5 \\ -9x - 6y = 15 \end{cases}$

43. $\begin{cases} \dfrac{2}{3}x + y = 3 \\ 3x + 2y = 1 \end{cases}$

44. $\begin{cases} x - 2y = 3 \\ 4x + 6y = 3 \end{cases}$

45. $\begin{cases} x - 2y = 5 \\ -3x + 6y = -15 \end{cases}$

46. $\begin{cases} x + y = 3 \\ 2x + 2y = 6 \end{cases}$

In Exercises 47–56, solve each system of equations by the elimination method. Check your solutions. For any dependent equations, write your answer as in Example 5.

47. $\begin{cases} x - y = 1 \\ x + y = 5 \end{cases}$

48. $\begin{cases} 2x - 3y = 5 \\ 3x + 2y = 14 \end{cases}$

49. $\begin{cases} x + y = 0 \\ 2x + 3y = 3 \end{cases}$

50. $\begin{cases} x + y = 3 \\ 3x + y = 1 \end{cases}$

51. $\begin{cases} 5x - y = 5 \\ 3x + 2y = -10 \end{cases}$ **52.** $\begin{cases} 3x - 2y = 1 \\ -7x + 3y = 1 \end{cases}$

53. $\begin{cases} x - y = 2 \\ -2x + 2y = 5 \end{cases}$ **54.** $\begin{cases} x + y = 5 \\ 2x + 2y = -10 \end{cases}$

55. $\begin{cases} 4x + 6y = 12 \\ 2x + 3y = 6 \end{cases}$ **56.** $\begin{cases} 4x + 7y = -3 \\ -8x - 14y = 6 \end{cases}$

In Exercises 57–76, use any method to solve each system of equations. For any dependent equations, write your answer as in Example 5.

57. $\begin{cases} 2x + y = 9 \\ 2x - 3y = 5 \end{cases}$ **58.** $\begin{cases} x + 2y = 10 \\ x - 2y = -6 \end{cases}$

59. $\begin{cases} 2x + 5y = 2 \\ x + 3y = 2 \end{cases}$ **60.** $\begin{cases} 4x - y = 6 \\ 3x - 4y = 11 \end{cases}$

61. $\begin{cases} 2x + 3y = 7 \\ 3x + y = 7 \end{cases}$ **62.** $\begin{cases} x = 3y + 4 \\ x = 5y + 10 \end{cases}$

63. $\begin{cases} 2x + 3y = 9 \\ 3x + 2y = 11 \end{cases}$ **64.** $\begin{cases} 3x - 4y = 0 \\ y = \dfrac{2x + 1}{3} \end{cases}$

65. $\begin{cases} \dfrac{x}{4} + \dfrac{y}{6} = 1 \\ x + 2(x - y) = 7 \end{cases}$ **66.** $\begin{cases} \dfrac{x}{3} + \dfrac{y}{5} = 12 \\ x - y = 4 \end{cases}$

67. $\begin{cases} 3x = 2(x + y) \\ 3x - 5y = 2 \end{cases}$ **68.** $\begin{cases} y + x + 2 = 0 \\ y + 2x + 1 = 0 \end{cases}$

69. $\begin{cases} 0.2x + 0.7y = 1.5 \\ 0.4x - 0.3y = 1.3 \end{cases}$ **70.** $\begin{cases} 0.6x + y = -1 \\ x - 0.5y = 7 \end{cases}$

71. $\begin{cases} \dfrac{2}{x} + \dfrac{5}{y} = -5 \\ \dfrac{3}{x} - \dfrac{2}{y} = -17 \end{cases}$ **72.** $\begin{cases} \dfrac{2}{x} + \dfrac{1}{y} = 3 \\ \dfrac{4}{x} - \dfrac{2}{y} = 0 \end{cases}$

73. $\begin{cases} \dfrac{3}{x} + \dfrac{1}{y} = 4 \\ \dfrac{6}{x} - \dfrac{1}{y} = 2 \end{cases}$ **74.** $\begin{cases} \dfrac{6}{x} + \dfrac{3}{y} = 0 \\ \dfrac{4}{x} + \dfrac{9}{y} = -1 \end{cases}$

75. $\begin{cases} \dfrac{5}{x} + \dfrac{10}{y} = 3 \\ \dfrac{2}{x} - \dfrac{12}{y} = -2 \end{cases}$ **76.** $\begin{cases} \dfrac{3}{x} + \dfrac{4}{y} = 1 \\ \dfrac{6}{x} + \dfrac{4}{y} = 3 \end{cases}$

In Exercises 77–82: a. Write a system of two linear equations in x and y whose graphs are given in the figure. b. Solve the system.

77.

78.

79.

80.

81.

82.

(3, 5)
(0, 2)

B EXERCISES Applying the Concepts

In Exercises 83–86, the demand and supply functions of a product are given. In each case, *p* represents the price in dollars per unit and *x* represents the number of units in hundreds. Find the equilibrium point.

83. $\begin{cases} 2p + x = 140 & \text{Demand equation} \\ 12p - x = 280 & \text{Supply equation} \end{cases}$

84. $\begin{cases} 7p + x = 150 & \text{Demand equation} \\ 10p - x = 20 & \text{Supply equation} \end{cases}$

85. $\begin{cases} 2p + x = 25 & \text{Demand equation} \\ x - p = 13 & \text{Supply equation} \end{cases}$

86. $\begin{cases} p + 2x = 96 & \text{Demand equation} \\ p - x = 39 & \text{Supply equation} \end{cases}$

In Exercises 87–102, use a system of equations to solve each problem.

87. Diameter of a pizza. The sum of the diameters of the largest and smallest pizzas sold at the Monster Pizza Shop is 29 inches. The difference in their diameters is 13 inches. Find the diameters of the largest and smallest pizzas.

88. Calories in hamburgers. The sum of the number of calories in a hamburger from Boston Burger and a hamburger from Carmen's Broiler is 1130. The difference in the number of calories in the hamburgers is 40. If the Boston Burger has the larger number of calories, how many calories are in each restaurant's hamburger?

89. Trash composition. Paper and plastic together account for 48% (by weight) of the total trash collected. If the weight of paper trash collected is five times the weight of plastic trash, what percent of the total trash collected is paper and what percent is plastic?

90. Cost of food and clothing. The average monthly combined cost of food and clothing for the Martínez family is $1000. If they spend four times as much for food as they do for clothes, what is the average monthly cost of each?

91. Mardi Gras parade "throws." Levon paid 40¢ for each string of beads and 30¢ for each doubloon (a special coin) he bought to throw from his Mardi Gras float. He paid a total of $265 for the two items. His doubloons and beads combined to a total of 770 items. How many doubloons and how many strings of beads did Levon buy?

92. Halloween candy. Janet bought 135 pieces of candy to give away on Halloween. She bought two kinds of candy, paying 24¢ apiece for one kind and 18¢ apiece for the other. If she spent $26.70 for the candy, how many pieces of each kind did she buy?

In Exercises 93 and 94, use the information in the following table.

Food Description McDonald's	Fat (gms)	Carb (gms)	Protein (gms)
Breakfast Burrito	20	21	13
Egg McMuffin	12	27	17

Source: www.shapefit.com/mcdonalds.html

Suppose you ate breakfast at McDonald's during one workweek (Monday–Friday).

93. McDonald's breakfast. If you consumed 123 grams of carbohydrates and 77 grams of protein, how many of each breakfast item did you eat during the workweek?

94. McDonald's breakfast. If you consumed 68 grams of fat and 129 grams of carbohydrates, how many of each breakfast item did you eat during the workweek?

95. Investment. Last year Mrs. García invested $50,000. She put part of the money in a real estate venture that paid 7.5% for the year and the rest in a small-business venture that returned 12% for the year. The combined income from the two investments for the year totaled $5190. How much did she invest at each rate?

96. Investment. Mr. Sharma invested a total of $30,000 in two ventures for a year. The annual return from one of them was 8%, and the other paid 10.5% for the year. He received a total income of $2550 from both investments. How much did he invest at each rate?

97. Tutoring other students. A student earns twice as much per hour for tutoring as she does working at McDougal's. If her average wage is $11.25 per hour, how much does she earn per hour at each job?

98. Plumber's earnings. A plumber earns $15 per hour more than her apprentice. For a 40-hour week, their combined earnings are $2200. What is the hourly rate for each?

99. Mixture problem. An herb that sells for $5.50 per pound is mixed with tea that sells for $3.20 per pound to produce a 100-pound mix that is worth $3.66 per pound. How many pounds of each ingredient does the mix contain?

100. Mixture problem. A chemist had a solution of 60% acid. She added some distilled water, reducing the acid concentration to 40%. She then added 1 more liter of water to further reduce the acid concentration to 30%. How much of the 30% solution did she then have?

101. Airplane speed. With the help of a tail wind, a plane travels 3000 kilometers in five hours. The return trip against the wind requires six hours. Assume that the direction and the wind speed are constant. Find both the speed of the plane in still air and the wind speed. [*Hint:* Remember that the wind helps the plane in one direction and hinders it in the other.]

102. Motorboat speed. A motorboat travels upstream a distance of 12 miles in two hours. If the current had been twice as strong, the trip would have taken three hours. How long should it take for the return trip down stream?

103. Break-even analysis. A local publishing company publishes *City Magazine*. The production and setup costs are $30,000, and the cost of producing each magazine is $2. Each magazine sells for $3.50. Assume that x magazines are published and sold.
   a. Write the total cost function $y = C(x)$ and the revenue function $y = R(x)$.
   b. Graph both functions from part (a) on the same coordinate plane.
   c. How many magazines must be sold to break even?

104. Break-even analysis. An electronic manufacturer plans to make digital portable music players (MP4s). Fixed costs will be $750,000, and it will cost $50 to manufacture each MP4, which will be sold for $125 to the retailers. Assume that x MP4s are manufactured and sold.
   a. Write the total cost function $y = C(x)$ and the revenue function $y = R(x)$.
   b. Graph both functions from part (a) on the same coordinate plane.
   c. How many MP4s must be sold to break even?

105. Making a job decision. Shanaysha has job offers from department stores A and B to sell major appliances. Store A offers her a fixed salary of $400 per week. Store B offers her $150 per week plus a commission of 4% of her weekly sales. How much should Shanaysha's weekly sales be for the offer from Store B to be better than that from Store A?

106. Making a job decision. Sheena has job offers from companies A and B to sell insurance. Company A offers her $25,000 per year plus 2% of her yearly sales premiums. Company B offers her $30,000 per year plus 1% of her yearly sales premiums. How much must Sheena earn in yearly sales premiums for the offer from Company B to be the better offer?

C EXERCISES Beyond the Basics

107. Solve for x and y.

$$\begin{cases} \dfrac{40}{x + y} + \dfrac{2}{x - y} = 5 \\ \dfrac{25}{x + y} - \dfrac{2}{x - y} = 1 \end{cases}$$

[*Hint:* Let $u = \dfrac{1}{x + y}$ and $v = \dfrac{1}{x - y}$, solve for u and v, and then solve for x and y.]

108. Solve for x and y.

$$\begin{cases} 6x + 3y = 7xy \\ 3x + 9y = 11xy \end{cases}$$

[*Hint:* Case (*i*), $xy = 0$; case (*ii*), $xy \neq 0$: divide by xy.]

109. For what value of the constant c does the following system have only one solution?

$$\begin{cases} x + 2y = 7 \\ 3x + 5y = 11 \\ cx + 3y = 4 \end{cases}$$

[*Hint:* Solve the first two equations for x and y and then substitute the solution into the third equation.]

110. Repeat Exercise 109 for the following system of equations.

$$\begin{cases} 5x + 7y = 11 \\ 13x + 17y = 19 \\ 3x + cy = 5 \end{cases}$$

111. Find the value of c for which the following system of equations represents the same line.

$$\begin{cases} 4x + 7y = 10 \\ 10x + cy = 25 \end{cases}$$

112. Find the value of c for which the following system of equations is inconsistent.

$$\begin{cases} 2x + cy = 11 \\ 5x - 7y = 5 \end{cases}$$

113. If $(x - 2)$ is a factor of both $f(x) = x^3 - 4x^2 + ax + b$ and $g(x) = x^3 - ax^2 + bx + 8$, find the values of a and b.

114. Let $l_1: a_1x + b_1y + c_1 = 0$ and $l_2: a_2x + b_2y + c_2 = 0$ be two nonparallel lines. Then for any constant k, show that the line $(a_1x + b_1y + c_1) + k(a_2x + b_2y + c_2) = 0$ is an equation of the line passing through the point of intersection of l_1 and l_2.

In Exercises 115–117, write your answer in slope–intercept form. [*Hint:* Use Exercise 114.]

115. Find an equation of the line that passes through the point of intersection of the lines with equations $2x + y = 3$ and $x - 3y = 12$ and that is parallel to the line with equation $3x + 2y = 8$.

116. Find an equation of the line that passes through the point of intersection of the lines with equations $5x + 2y = 7$ and $6x - 5y = 38$ and that passes through the point $(1, 3)$.

117. Find an equation of the line that passes through the point of intersection of the lines with equations $2x + 5y + 7 = 0$ and $13x - 10y + 3 = 0$ and that is perpendicular to the line with equation $7x + 13y = 8$.

Critical Thinking

118. Consider the system of equations

$$\begin{cases} a_1x + b_1y = c_1 \\ a_2x + b_2y = c_2 \end{cases}$$

where $a_1, b_1, c_1, a_2, b_2,$ and c_2 are constants. Find a relationship (an equation) among these constants such that the system of equations has

  a. only one solution. Find the solution in terms of the constants.

  b. no solution.

  c. infinitely many solutions.

119. Let $3x + 4y - 12 = 0$ be the equation of a line l_1 and let $P(5, 8)$ be a point.

  a. Find an equation of a line l_2 passing through the point P and perpendicular to l_1.

  b. Find the point Q of the intersection of the two lines l_1 and l_2.

  c. Find the distance $d(P, Q)$. This is the distance from the point P to the line l_1.

120. Use the procedure outlined in Exercise 115 to show that the (perpendicular) distance from a point $P(x_1, y_1)$ to the line $ax + by + c = 0$ is given by $\dfrac{|ax_1 + by_1 + c|}{\sqrt{a^2 + b^2}}$.

121. Use the formula from Exercise 120 to find the distance from the given point P to the given line l.

  a. $P(2, 3), l: x + y - 7 = 0$

  b. $P(-2, 5), l: 2x - y + 3 = 0$

  c. $P(3, 4), l: 5x - 2y - 7 = 0$

  d. $P(0, 0), l: ax + by + c = 0$

Systems of Linear Equations in Three Variables

Before Starting this Section, Review

1. Solving systems of linear equations (Section 5.1, page 480)

2. Solving a system of equations in two variables (Section 5.1, page 482)

Objectives

1   Solve a system of linear equations in three variables.

2   Classify systems as consistent and inconsistent.

3   Solve nonsquare systems.

4   Interpret linear systems in three variables geometrically.

5   Use linear systems in applications.

Cross section with tumor

FIGURE 5.6 **Cross section with tumor**

CAT SCANS

Computerized axial tomography imaging is more commonly known as a "CAT scan" or "CT scan." Tomography is from the Greek word *tomos*, meaning "slice or section," and *graphia*, meaning "describing." The CAT scan machine was invented in 1972 by British engineer Godfrey Hounsfield and, independently, by physicist Allan Cormack of Tufts University in Massachusetts. Hounsfield never attended a university, but he started experimenting with electrical and mechanical devices as a boy. He shared a Nobel Prize in 1979 with Cormack for inventing the scanner technology and was honored with knighthood in England for his contributions to medicine and science. Together with algebraic computations, CAT scanners use X-rays measured outside the patient's body to construct pictures of the patient's internal body organs, including the brain. The scanner does not "take" any pictures; instead, it *computes* pictures. The idea behind the CAT scanner is simple. When an X-ray passes through an object, it loses intensity. The denser the object, the more intensity the X-ray loses. Because tumors are denser than healthy tissues, an unexpected loss of intensity may indicate the presence of a tumor. To obtain sufficient information, a large number of properly arranged X-rays is needed. Imagine a cross section of body tissue with a superimposed mathematical grid. Each region bounded by the grid lines is called a *grid cell*. In an actual CAT scan, each grid cell is 2 square millimeters and contains 10,000 or more biological cells. The grid shown in Figure 5.6 would require many X-rays. Actual CAT scanners work with grids that have 512 × 512 grid cells and require hundreds of thousands of X-ray beams. In Example 6, we discuss a simple situation in which X-rays pass through three grid cells. ■

---

1   Solve a system of linear equations in three variables.

Systems of Linear Equations

A **linear equation in the variables** $x_1, x_2, \ldots, x_n$ is an equation that can be written in the form

$$a_1 x_1 + a_2 x_2 + \cdots + a_n x_n = b,$$

where b and the *coefficients* $a_1, a_2, \ldots, a_n$ are real numbers. The subscript n can be any positive integer. When n is small—say, between 2 and 5 (as in most textbook examples and exercises)—we normally use x, y, z, u, and v instead of x_1, x_2, x_3, x_4, and x_5. However, in real-life applications, n might be 100 or 1000 or even greater.

A **system of linear equations** (or a **linear system**) in three variables is a collection of two or more linear equations involving the same variables. For example,

$$\begin{cases} x + 3y + z = 0 \\ 2x - y + z = 5 \\ 3x - 3y + 2z = 10 \end{cases}$$

is a system of three linear equations in the three variables x, y, and z. An ordered triple (a, b, c) is a **solution** of a system of three linear equations in three variables x, y, and z if each equation in the system is a true statement when a, b, and c are substituted for x, y, and z, respectively.

EXAMPLE 1   **Verifying a Solution**

Determine whether the ordered triple $(2, -1, 3)$ is a solution of the following linear system.

$$\begin{cases} 5x + 3y - 2z = 1 & (1) \\ x - y + z = 6 & (2) \\ 2x + 2y - z = -1 & (3) \end{cases}$$

SOLUTION

We replace x with 2, y with -1, and z with 3 in all three equations.

Equation (1)	Equation (2)	Equation (3)
$5x + 3y - 2z = 1$	$x - y + z = 6$	$2x + 2y - z = -1$
$5(2) + 3(-1) - 2(3) \overset{?}{=} 1$	$2 - (-1) + 3 \overset{?}{=} 6$	$2(2) + 2(-1) - (3) \overset{?}{=} -1$
$10 - 3 - 6 \overset{?}{=} 1$	$2 + 1 + 3 \overset{?}{=} 6$	$4 - 2 - 3 \overset{?}{=} -1$
$1 = 1$ ✓	$6 = 6$ ✓	$-1 = -1$ ✓

Because the ordered triple $(2, -1, 3)$ satisfies all three equations, it is a solution of the system. ■ ■ ■

Practice Problem 1   Determine whether the ordered triple $(2, -3, 2)$ is a solution of the following linear system.

$$\begin{cases} x + y + z = 1 \\ 3x + 4y + z = -4 \\ 2x + y + 2z = 5 \end{cases}$$ ■

We now use a procedure called **Gaussian elimination** to convert a system of three equations in the variables x, y, and z into an equivalent system in *triangular form* with leading coefficient 1, as shown below.

$$\begin{cases} x + ay + bz = p & (1) \\ y + cz = q & (2) \\ z = z_0 & (3) \end{cases}$$

We can then solve the system by back-substitution. To get a system in triangular form, we use the following three operations to obtain equivalent systems (systems with the same solution set as the original system).

OPERATIONS THAT PRODUCE EQUIVALENT SYSTEMS

1. Interchange the position of any two equations.
2. Multiply any equation by a nonzero constant.
3. Add a nonzero multiple of one equation to another.

FINDING THE SOLUTION: A PROCEDURE

EXAMPLE 2   **The Gaussian Elimination Method**

OBJECTIVE

Solve a system of three equations in three variables by first converting the system to the triangular form

$$\begin{cases} x + ay + bz = p \\ \quad\quad y + cz = q \\ \quad\quad\quad\quad z = z_0 \end{cases}$$

Step 1 If necessary, rearrange the equations to obtain an x-term with a nonzero coefficient in the first equation. If the coefficient of the x-term is not 1, then multiply the first equation by the reciprocal of the coefficient of the x-term to get 1 as a leading coefficient.

Step 2 By adding appropriate multiples of the first equation, eliminate any x-terms from the second and third equations. Then multiply the resulting second equation by the reciprocal of the coefficient of the y-term to get 1 as a leading coefficient.

Step 3 Add an appropriate multiple of the second equation from Step 2 to eliminate any y-term from the third equation. Then solve the resulting equation for z; this gives the triangular form.

EXAMPLE

Solve the system of equations.

$$\begin{cases} 3x + 9y + 5z = 14 & (1) \\ 2x + \ y + \ z = 9 & (2) \\ \ x + 2y + 2z = 3 & (3) \end{cases}$$

Interchange equations (1) and (3).

$$\begin{cases} \ x + 2y + 2z = 3 & (3) \\ 2x + \ y + \ z = 9 & (2) \\ 3x + 9y + 5z = 14 & (1) \end{cases}$$

The first equation (equation (3)) already has a leading coefficient of 1.

To eliminate x from equation (2), add -2 times equation (3) to equation (2).

$$\begin{array}{rl} -2x - 4y - 4z = -6 & \quad -2(x + 2y + 3z = 3) \\ \underline{2x + \ y + \ z = \ \ 9} & \quad (2) \\ -3y - 3z = \ \ 3 & \quad (4) \quad \text{Add.} \end{array}$$

To eliminate x from equation (1), add -3 times equation (3) to equation (1).

$$\begin{array}{rl} -3x - 6y - 6z = -9 & \quad -3(x + 2y + 2z = 3) \\ \underline{3x + 9y + 5z = 14} & \quad (1) \\ 3y - \ z = \ \ 5 & \quad (5) \end{array}$$

$$\begin{cases} x + 2y + 2z = 3 & (3) \\ \quad\quad y + \ z = -1 & (6) \quad \text{Multiply equation (4) by } -\tfrac{1}{3}. \\ \quad 3y - \ z = 5 & (5) \end{cases}$$

To eliminate y from equation (5), add -3 times equation (6) to equation (5).

$$\begin{array}{rl} -3y - 3z = \ \ 3 & \quad -3(y + z = -1) \\ \underline{3y - \ z = \ \ 5} & \quad (5) \\ -4z = \ \ 8 & \quad \text{Add.} \\ z = -2 & \quad (7) \quad \text{Solve for } z. \end{array}$$

continued on the next page

The triangular form is:

$$\begin{cases} x + 2y + 2z = -3 & (3) \\ \quad\quad y + \;\; z = -1 & (6) \\ \quad\quad\quad\quad\quad z = -2 & (7) \end{cases}$$

Step 4 Back-substitute the value of z from Step 3 into one of the equations in Step 3 that contains only y and z and solve for y.

$$\begin{aligned} y + z &= -1 && \text{Equation (6) from Step 2} \\ y + (-2) &= -1 && \text{Substitute } z = -2. \\ y &= 1 && \text{Solve for } y. \end{aligned}$$

Step 5 Back-substitute the values of y and z from Steps 3 and 4 in any equation containing x, y, and z and solve for x.

$$\begin{aligned} x + 2y + 2z &= 3 && \text{Equation (3) from Step 1} \\ x + 2(1) + 2(-2) &= 3 && \text{Substitute } y = 1, z = -2. \\ x &= 5 && \text{Solve for } x. \end{aligned}$$

Step 6 Write the solution set.

The solution set is $\{(5, 1, -2)\}$.

Step 7 Check your answer in the original equations.

Check: Verify the solution by substituting $x = 5$, $y = 1$, and $z = -2$ in each of the original equations.

$$\begin{aligned} 3(5) + 9(1) + 5(-2) &= 14 \;\checkmark && (1) \\ 2(5) + (1) + (-2) &= 9 \;\checkmark && (2) \\ 5 + 2(1) + 2(-2) &= 3 \;\checkmark && (3) \end{aligned}$$

■ ■ ■

Practice Problem 2   Solve the system.

$$\begin{cases} 2x + 5y \quad\quad = 1 \\ \;\; x - 3y + 2z = 1 \\ -x + 2y + \;\; z = 7 \end{cases}$$

■

2 Classify systems as consistent and inconsistent.

Number of Solutions of a Linear System

As in the case of two variables, a system of linear equations in three variables may be **consistent** (having at least one solution) or **inconsistent** (having no solution). The equations in a consistent system may be **dependent** (the system has infinitely many solutions) or **independent** (the system has one solution).

INCONSISTENT SYSTEM

If in the process of converting a linear system to triangular form an equation of the form $0 = k$ occurs, where $k \neq 0$, then the system has no solution and is *inconsistent*.

EXAMPLE 3   **Attempting to Solve a Linear System with No Solution**

Solve the system of equations.

$$\begin{cases} \;\; x - \;\; y + 2z = 5 & (1) \\ 2x + \;\; y + \;\; z = 7 & (2) \\ 3x - 2y + 5z = 20 & (3) \end{cases}$$

SOLUTION

Steps 1–2 To eliminate x from equation (2), add -2 times equation (1) to equation (2).

$$
\begin{array}{rl}
-2x + 2y - 4z = -10 & \qquad -2(x - y + 2z = 5) \\
\underline{2x + y + z = 7} & \quad (2) \\
3y - 3z = -3 & \quad (4) \qquad \text{Add.}
\end{array}
$$

Next, eliminate x from equation (3) by using equation (1).

$$
\begin{array}{rl}
-3x + 3y - 6z = -15 & \qquad -3(x - y + 2z = 5) \\
\underline{3x - 2y + 5z = 20} & \quad (3) \\
y - z = 5 & \quad (5) \qquad \text{Add.}
\end{array}
$$

We now have the system:

$$
\begin{cases}
x - y + 2z = 5 & (1) \\
 3y - 3z = -3 & (4) \\
 y - z = 5 & (5)
\end{cases}
$$

Step 3 Multiply equation (4) by $\dfrac{1}{3}$ to obtain

$$
y - z = -1 \qquad (6)
$$

We can eliminate y from equation (5) by using equation (6).

$$
\begin{array}{rl}
-y + z = 1 & \qquad -1(y - z = 1) \\
\underline{y - z = 5} & \quad (5) \\
0 = 6 & \quad (7) \qquad \text{Add.}
\end{array}
$$

We now have the system in triangular form:

$$
\begin{cases}
x - y + 2z = 5 & (1) \\
y - z = -1 & (4) \\
\underbrace{0 = 6}_{\text{False}} & (7)
\end{cases}
$$

This system is equivalent to the original system. Because equation (7) is false, we conclude that the solution set of the system is $\varnothing$ and the system is inconsistent. ■ ■ ■

Practice Problem 3 Solve the system of equations.

$$
\begin{cases}
2x + 2y + 2z = 12 \\
-3x + y - 11z = -6 \\
2x + y + 4z = -8
\end{cases}
$$

■

DEPENDENT EQUATIONS

If in the process of converting a linear system to triangular form,

(i) an equation of the form $0 = k\,(k \neq 0)$ *does not* occur but

(ii) an equation of the form $0 = 0$ *does* occur, then the system of equations has infinitely many solutions and the equations are *dependent*.

EXAMPLE 4   **Solving a System with Infinitely Many Solutions**

Solve the system of equations.

$$\begin{cases} x - y + z = 7 & (1) \\ 3x + 2y - 12z = 11 & (2) \\ 4x + y - 11z = 18 & (3) \end{cases}$$

SOLUTION

Steps 1–2   Eliminate x from equation (2) by using equation (1).

$$\begin{array}{rl} -3x + 3y - 3z = -21 & \quad -3(x - y + z = 7) \\ \underline{3x + 2y - 12z = 11} & \quad (2) \\ 5y - 15z = -10 & \quad \text{Add.} \\ y - 3z = -2 & \quad (4) \quad \text{Divide both sides by 5.} \end{array}$$

Eliminate x from equation (3) by using equation (1).

$$\begin{array}{rl} -4x + 4y - 4z = -28 & \quad -4(x - y + z = 7) \\ \underline{4x + y - 11z = 18} & \quad (3) \\ 5y - 15z = -10 & \quad (5) \quad \text{Add.} \end{array}$$

We now have the following equivalent system.

$$\begin{cases} x - y + z = 7 & (1) \\ y - 3z = -2 & (4) \\ 5y - 15z = -10 & (5) \end{cases}$$

Step 3   We eliminate y from equation (5) by using equation (4).

$$\begin{array}{rl} -5y + 15z = 10 & \quad -5(y - 3z = -2) \\ \underline{5y - 15z = -10} & \quad (5) \\ 0 = 0 & \quad (6) \quad \text{Add.} \end{array}$$

We finally have the system in triangular form.

$$\begin{cases} x - y + z = 7 & (1) \\ y - 3z = -2 & (4) \\ \underbrace{0 = 0}_{\text{True}} & (6) \end{cases}$$

The equation $0 = 0$ is equivalent to $0z = 0$, which is true for every value of z. Solving equation (4) for y, we have $y = 3z - 2$. Substituting $y = 3z - 2$ into equation (1) and solving for x, we obtain

$$\begin{array}{ll} x = y - z + 7 & \quad \text{Solve equation (1) for } x. \\ x = (3z - 2) - z + 7 & \quad \text{Substitute } y = 3z - 2. \\ x = 2z + 5 & \quad \text{Simplify.} \end{array}$$

Thus, for every real number z, the triple $(x, y, z) = (2z + 5, 3z - 2, z)$ is a solution of the system. For example, when $z = 0$, the triple $(5, -2, 0)$ is a solution. When $z = 1$, the triple $(7, 1, 1)$ is a solution. Therefore, the given system has infinitely many solutions, one for each real number z, and the equations are dependent. The solution set for the system is $\{(2z + 5, 3z - 2, z)\}$.   ■■■

Practice Problem 4 Solve the system of equations.

$$\begin{cases} x + y + z = 5 \\ -4x - y - 8z = -29 \\ 2x + 5y - 2z = 1 \end{cases}$$ ∎

3 Solve nonsquare systems.

Nonsquare Systems

Sometimes in a linear system, the number of equations is not the same as the number of variables. Such systems are called **nonsquare linear systems**.

EXAMPLE 5 **Solving a Nonsquare Linear System**

Solve the system of linear equations.

$$\begin{cases} x + 5y + 3z = 7 & (1) \\ 2x + 11y - 4z = 6 & (2) \end{cases}$$

SOLUTION

Steps 1–2 We eliminate x from equation (2) by using equation (1).

$$\begin{array}{rl} -2x - 10y - 6z = -14 & \quad -2(x + 5y + 3z = 7) \\ \underline{2x + 11y - 4z = 6} & \quad (2) \\ y - 10z = -8 & \quad (3) \quad \text{Add.} \end{array}$$

We obtain the equivalent system.

$$\begin{cases} x + 5y + 3z = 7 & (1) \\ y - 10z = -8 & (3) \end{cases}$$

Steps 3–4 Because there is no third equation, Steps 3 and 4 are not needed.

Step 5 Solve equation (3) for y in terms of z to obtain

$$y = 10z - 8.$$

Step 6 Back-substitute $y = 10z - 8$ into equation (1) and solve for x.

$$\begin{array}{ll} x + 5y + 3z = 7 & \text{Equation (1)} \\ x + 5(10z - 8) + 3z = 7 & \text{Replace } y \text{ with } 10z - 8. \\ x = -53z + 47 & \text{Solve for } x. \end{array}$$

Steps 5 and 6 mean that z determines both the x and y values. The system has infinitely many solutions, one for each real number z. The solution set is $\{(-53z + 47, 10z - 8, z)\}$. ■■■

Practice Problem 5 Solve the system of linear equations.

$$\begin{cases} x + 3y + 2z = 4 \\ 2x + 7y - z = 5 \end{cases}$$ ∎

4 Interpret linear systems in three variables geometrically.

Geometric Interpretation

The graph of a linear equation in three variables, such as $ax + by + cz = d$ (where a, b, and c are not all zero), is a plane in three–dimensional space. Figure 5.7 on page 499 shows possible situations for a system of three linear equations in three variables.

a. Three planes intersect in a single point. The system has only one solution. The system is consistent and the equations are independent.

b. Three planes intersect in one line. The system has infinitely many solutions. The system is consistent and the equations are dependent.

FIGURE 5.7

c. Three planes coincide with each other. The system has infinitely many solutions. The system is consistent and the equations are dependent.

d. There are three parallel planes. The system has no solution. The system is inconsistent.

e. Two parallel planes are intersected by a third plane. The system has no solution. The system is inconsistent.

f. Three planes have no point in common. The system has no solution. The system is inconsistent.

5 Use linear systems in applications.

An Application to CAT Scans

CAT scanners report X-ray data in units called *linear attenuation units* (LAUs), also called *Hounsfield numbers*. The data are reported in such a way that if an X-ray passes through consecutive grid cells, the total weakening of the beam is the sum of the individual decreases. Table 5.1 gives the range of LAUs that determine whether the grid cell contains healthy tissue, tumorous tissue, bone, or metal.

TABLE 5.1 CAT Scanner Ranges

Type of Tissue	LAU Values
Healthy tissue	0.1625–0.2977
Tumorous tissue	0.2679–0.3930
Bone	0.38757–0.5108
Metal	1.54–∞

Reprinted with permission from *Mathematics Teacher,* May 1996. © 1996 by the National Council of Teachers of Mathematics.

Beam 2
Beam 1
A
B   C
Beam 3
X-ray detector
Three grid cells, three X-rays

FIGURE 5.8 Reprinted with permission from *Mathematics Teacher,* May 1996 ©, 1996 by the National Council of Teachers of Mathematics.

EXAMPLE 6   A CAT Scan with Three Grid cells

Let A, B, and C be three grid cells as shown in Figure 5.8. A CAT scanner reports the following data for a patient named Monica.

(i) Beam 1 is weakened by 0.80 unit as it passes through grid cells A and B.

(ii) Beam 2 is weakened by 0.55 unit as it passes through grid cells A and C.

(iii) Beam 3 is weakened by 0.65 unit as it passes through grid cells B and C.

Use Table 5.1 to determine which grid cells contain each type of tissue listed.

SOLUTION

Suppose grid cell A weakens the beam by x units, grid cell B weakens it by y units, and grid cell C weakens it by z units. Then we have the system

$$\begin{cases} x + y \phantom{{}+ z} = 0.80 & (1) \quad \text{(Beam 1)} \\ x \phantom{{}+ y} + z = 0.55 & (2) \quad \text{(Beam 2)} \\ \phantom{x + {}} y + z = 0.65 & (3) \quad \text{(Beam 3)} \end{cases}$$

To solve this system of equations, we use the elimination procedure. Adding -1 times equation (1) to equation (2), we obtain the equivalent system

$$\begin{cases} x + \ y \qquad = \ \ 0.80 & (1) \\ \qquad -y + z = -0.25 & (4) \\ \qquad \ \ y + z = \ \ 0.65 & (3) \end{cases} \quad -1 \text{ times equation (1) plus equation (2)}$$

Add equation (4) to equation (3) to get the system

$$\begin{cases} x + \ y \qquad = \ \ 0.80 & (1) \\ \qquad -y + \ z = -0.25 & (4) \\ \qquad \quad \ 2z = \ \ 0.40 & (5) \end{cases} \quad \text{or} \quad \begin{cases} x + y \qquad = 0.80 & (1) \\ \qquad y - z = 0.25 & (6) \\ \qquad \quad \ z = 0.20 & (7) \end{cases}$$

Back-substitute $z = 0.20$ in equation (6) to get $y - 0.20 = 0.25$, or $y = 0.45$. Next, back-substitute $y = 0.45$ in equation (1) to get $x + 0.45 = 0.80$. Solving for x gives $x = 0.35$. Now referring to Table 5.1, we conclude that cell A contains tumorous tissue (because $x = 0.35$), cell B contains bone (because $y = 0.45$), and cell C contains healthy tissue (because $z = 0.20$). ■ ■ ■

Practice Problem 6   In Example 6, suppose **(i)** beam 1 is weakened by 0.65 unit as it passes through grid cells A and B, **(ii)** beam 2 is weakened by 0.70 unit as it passes through grid cells A and C, and **(iii)** beam 3 is weakened by 0.55 unit as it passes through grid cells B and C. Use Table 5.1 to determine which grid cells contain each type of tissue listed. ■

Given three points in a coordinate plane that do not lie on a line, it can be shown that there is one and only one quadratic function whose graph contains these points.

We can use a system of linear equations to find the quadratic function whose graph passes through any three data points that are not collinear.

EXAMPLE 7   **Finding a Quadratic Function Using Graph Points**

Find the quadratic function whose graph contains the three points $(-2, 51)$, $(1, -9)$, and $(5, -33)$.

SOLUTION

Let $f(x) = ax^2 + bx + c$. Then $f(-2) = 51$, $f(1) = -9$, and $f(5) = -33$. So we have the system

$$\begin{cases} a(-2)^2 + b(-2) + c = \ \ 51 & (1) \\ a(1)^2 + \ \ b(1) + c = \ -9 & (2) \\ a(5)^2 + \ \ b(5) + c = -33 & (3) \end{cases} \quad \text{or} \quad \begin{cases} c - 2b + \ \ 4a = \ \ 51 & (1) \\ c + \ \ b + \quad a = -9 & (2) \\ c + 5b + 25a = -33 & (3) \end{cases}$$

Eliminate c from equation (2) by adding -1 times equation (1) to equation (2).

$$\begin{array}{rl} -c + 2b - 4a = -51 & \quad -1(c - 2b + 4a = 51) \\ \underline{\ \ c + \ \ b + \ \ a = \ -9} & \quad (2) \\ 3b - 3a = -60 & \quad (4) \qquad \text{Add.} \end{array}$$

Next, eliminate c from equation (3) by adding -1 times equation (1) to equation (3).

$$\begin{array}{rl} -c + 2b - \ \ 4a = -51 & \quad -1(c - 2b + 4a = 51) \\ \underline{\ \ c + 5b + 25a = -33} & \quad (3) \\ 7b + 21a = -84 & \quad (5) \qquad \text{Add.} \end{array}$$

We now have the system

$$\begin{cases} c - 2b + \ \ 4a = \ \ 51 & (1) \\ \quad \ \ 3b - \ \ 3a = -60 & (4) \\ \quad \ \ 7b + 21a = -84 & (5) \end{cases}$$

Multiply equation (4) by $\frac{1}{3}$ and equation (5) by $-\frac{1}{7}$ to get

$$
\begin{array}{rcl}
b - a &=& -20 \quad (6) \\
-b - 3a &=& 12 \quad (7) \\
\hline
- 4a &=& - 8 \quad (8) \quad \text{Add.}
\end{array}
$$

Multiplying equation (8) by $-\frac{1}{4}$ gives $a = 2$. Back-substituting into the system

$$
\begin{cases}
c - 2b + 4a = 51 \quad (1) \\
\quad\quad b - a = -20 \quad (6) \\
\quad\quad\quad\quad a = 2
\end{cases}
$$

gives $a = 2, b = -18, c = 7$. So $f(x) = 2x^2 - 18x + 7$.   ■ ■ ■

Practice Problem 7   Find the quadratic function whose graph contains the three points $(-3, -16), (0, 5)$, and $(2, 9)$.   ■

SECTION 5.2 ■ Exercises

A EXERCISES Basic Skills and Concepts

1. Systems of equations that have the same solution set are called _____ systems.

2. Adding a(n) _____ multiple of one equation to another equation in the system produces an equivalent system.

3. If any of the equations in a system has no solution, then the system is _____.

4. If the number of equations is different from the number of variables in a linear system, then the system is called a(n) _____ system.

5. *True or False* The graph of the equation $x - 2y + z = 1$ is a plane in three-dimensional space.

6. *True or False* If $\{(2z, 1 - z, z + 3)\}$ is the solution set of a linear system in three variables, then $(0, 1, 3)$ is a solution of this system.

In Exercises 7–10, determine whether the given ordered triple (x, y, z) is a solution of the given linear system.

7. $\begin{cases} 2x - 2y - 3z = 1 \\ \quad\quad 3y + 2z = -1 \quad (1, -1, 1) \\ \quad\quad y + z = 0 \end{cases}$

8. $\begin{cases} 3x - 2y + z = 2 \\ \quad\quad y + z = 5 \quad (2, 3, 2) \\ x \quad\quad + 3z = 8 \end{cases}$

9. $\begin{cases} x + 3y - 2z = 0 \\ 2x - y + 4z = 5 \quad (-10, 8, 7) \\ x - 11y + 14z = 0 \end{cases}$

10. $\begin{cases} x - 4y + 7z = 14 \\ 3x + 8y - 2z = 13 \quad (4, 1, 2) \\ 7x - 8y + 26z = 5 \end{cases}$

In Exercises 11–14, solve each triangular linear system.

11. $\begin{array}{l} x + y + z = 4 \\ \quad\quad y - 2z = 4 \\ \quad\quad\quad\quad z = -1 \end{array}$

12. $\begin{array}{l} x + 3y + z = 3 \\ \quad\quad y + 2z = 8 \\ \quad\quad\quad\quad z = 5 \end{array}$

13. $\begin{array}{l} x - 5y + 3z = -1 \\ \quad\quad y - 2z = -6 \\ \quad\quad\quad\quad z = 4 \end{array}$

14. $\begin{array}{l} x + 3y + 5z = 0 \\ \quad\quad y + 7z = 2 \\ \quad\quad\quad\quad z = \dfrac{1}{2} \end{array}$

15. In the system in Exercise 7, interchange equations (2) and (3). Write the new equivalent system. In the new system, eliminate y from the last equation. Convert the system to triangular form.

16. In Exercise 8, interchange equations (1) and (3). In the new equivalent system, eliminate x from the last equation. Convert the system to triangular form.

17. Convert the system of Exercise 9 to triangular form.

18. Convert the system of Exercise 10 to triangular form.

In Exercises 19–22, convert the system to triangular form and then use back-substitution to solve the system.

19. $\begin{cases} x - y + z = 6 \\ \quad\quad 2y + 3z = 5 \\ \quad\quad\quad\quad 2z = 6 \end{cases}$

20. $\begin{cases} 4x + 5y + 2z = -3 \\ \quad\quad 3y - z = 14 \\ \quad\quad\quad\quad - 3z = 15 \end{cases}$

21. $\begin{cases} 4x + 4y + 4z = 7 \\ 3x - 8y \quad\quad = 14 \\ \quad\quad\quad 4z = -1 \end{cases}$

22. $\begin{cases} 5x + 10y + 10z = 0 \\ \quad\quad 2y + 3z = -0.6 \\ \quad\quad\quad 4z = 1.6 \end{cases}$

In Exercises 23–44, find the solution set of each linear system. Identify inconsistent systems and dependent equations. For dependent equations, write your answer as in Example 4.

23. $\begin{cases} x + y + z = 6 \\ x - y + z = 2 \\ 2x + y - z = 1 \end{cases}$

24. $\begin{cases} x + y + z = 6 \\ 2x + 3y - z = 5 \\ 3x - 2y + 3z = 8 \end{cases}$

25. $\begin{cases} 2x + 3y + z = 9 \\ x + 2y + 3z = 6 \\ 3x + y + 2z = 8 \end{cases}$

26. $\begin{cases} 4x + 2y + 3z = 6 \\ x + 2y + 2z = 1 \\ 2x - y + z = -1 \end{cases}$

27. $\begin{cases} x - 4y + 7z = 14 \\ 3x + 8y - 2z = 13 \\ 7x - 8y + 26z = 5 \end{cases}$

28. $\begin{cases} x + y + 4z = 6 \\ x + 2y - 2z = 8 \\ 7x + 10y + 10z = 60 \end{cases}$

29. $\begin{cases} 2x + 3y + 2z = 7 \\ x + 3y - z = -2 \\ x - y + 2z = 8 \end{cases}$

30. $\begin{cases} x - y + 2z = 3 \\ 2x + 2y + z = 3 \\ x + y + 3z = 4 \end{cases}$

31. $\begin{cases} 4x - 2y + z = 5 \\ 2x + y - 2z = 4 \\ x + 3y - 2z = 6 \end{cases}$

32. $\begin{cases} x - 3y + 2z = 9 \\ 2x + 4y - 3z = -9 \\ 3x - 2y + 5z = 12 \end{cases}$

33. $\begin{cases} 2x + y + z = 6 \\ x + y - z = 1 \\ x + y + 2z = 4 \end{cases}$

34. $\begin{cases} 2x + y - 3z = 7 \\ x - y - 2z = 4 \\ 3x + 3y + 2z = 4 \end{cases}$

35. $\begin{cases} x - y - z = 3 \\ x + 9y + z = 3 \\ 2x + 3y - z = 6 \end{cases}$

36. $\begin{cases} x + y - z = 2 \\ 3x - y - z = 10 \\ 3x + y - 2z = 8 \end{cases}$

37. $\begin{cases} x + y \quad\quad = 0 \\ \quad\quad y + 2z = -4 \\ \quad\quad y + z = 4 - x \end{cases}$

38. $\begin{cases} 2x + 4y + 3z = 6 \\ x + \quad\quad 2z = -1 \\ x - 2y + z = -5 \end{cases}$

39. $\begin{cases} \quad\quad 2y - z = -4 \\ x + \quad\quad z = 3 \\ 2x + 3y \quad\quad = -1 \end{cases}$

40. $\begin{cases} x + y \quad\quad = 9 \\ \quad\quad 2y + 3z = 7 \\ x \quad\quad - 2z = 4 \end{cases}$

41. $\begin{cases} 3x - \quad\quad 2z = 11 \\ 2x + y \quad\quad = 8 \\ \quad\quad 2y + 3z = 1 \end{cases}$

42. $\begin{cases} 2x + y \quad\quad = 4 \\ x + \quad\quad 2z = 3 \\ \quad\quad 3y - z = 5 \end{cases}$

43. $\begin{cases} 2x + 6y \quad\quad + 11 = 0 \\ \quad\quad 6y - 18z + 1 = 0 \end{cases}$

44. $\begin{cases} 3x + 5y - 15 = 0 \\ 6x + 20y - 6z = 11 \end{cases}$

In Exercises 45–50, find the quadratic function whose graph contains the given points.

45. $(-1, 26), (1, -8), (5, -4)$

46. $(-3, 11), (-1, -1), (4, 4)$

47. $(-1, -3), (1, 1), (2, -3)$

48. $(-3, 40), (-2, 33), (1, -24)$

49. $(2, -11), (3, -13), (4, -11)$

50. $(-3, -4), (-1, 74), (1, 144)$

B EXERCISES Applying the Concepts

In Exercises 51–58, use a system of equations to solve each problem.

51. Investment. Miguel invested $20,000 in three different funds that paid 4%, 5%, and 6% for the year. The total income for the year from the three funds was $1060. The income from the 6% fund was twice the income from the 5% fund. What amount was invested in each fund?

52. Number problem. The sum of the digits in a three–digit number is 14. The sum of the hundreds digit and the units digit is equal to the tens digit. If the hundreds digit and the units digit are interchanged, the number is increased by 297. What is the original number?

53. Election campaign. Alex, Becky, and Courtney volunteered to stuff 741 envelopes with newsletters for Senator Douglas's reelection campaign. Alex could assemble 124 per hour; Becky, 118 per hour; and Courtney, 132 per hour. They worked a total of six hours. The sum of the number of hours that Becky and Courtney spent was twice what Alex spent. How long did each of them work?

54. Age of students. A college algebra class of 38 students at Central State College was made up of people who were 18, 19, and 20 years of age. The average of their ages was 18.5 years. How many of each age were in the class if the number of 18-year-olds was eight more than the combined number of 19- and 20-year-olds?

55. Coins in a machine. A vending machine's coin box contains nickels, dimes, and quarters. The total number of coins in the box is 300. The number of dimes is three times the number of nickels and quarters together. If the box contains $30.05, find the number of nickels, dimes, and quarters that it contains.

56. Sports. The Wildcats scored 46 points in a football game. Twice the number of points resulting from the sum of field goals and extra points equals two more than the number of points from touchdowns. Five times the number of points scored by field goals equals twice the number of points from touchdowns. Find the number of points resulting from touchdowns, field goals, and extra points. (A touchdown = 6 points, a field goal = 3 points, and an extra point = 1 point.)

57. Weekly wage. Amy worked 53 hours one week and was paid at three different rates. She earned $7.40 per hour for her normal daytime work, $9.20 per hour for night work, and $11.75 per hour for holiday work. If her total gross wages for the week were $452.20 and the number of regular daytime hours she worked exceeded the combined hours of night and holiday work by nine hours, how many hours of each category of work did Amy perform?

58. Components of a product. A manufacturer buys three components—A, B, and C—for use in making a toaster. She used as many units of A as she did of B and C combined. The cost of A, B, and C is \$4, \$5, and \$6 per unit, respectively. If she purchased 100 units of these components for a total of \$480, how many units of each did she purchase?

59. Coin puzzle. A box of coins contains \$100 in nickels, dimes, and quarters. If there are twice as many nickels as quarters and a total of 900 coins altogether, how many of each type of coin are in the box?

60. Coin puzzle. Tyler finds that his 308 coins (all dimes, quarters, and half-dollars) total \$54. If there are two and a half times as many dimes as quarters, how many of each type of coin are in the box?

61. Finding numbers. The sum of three numbers is 63. The second number is twice the first and four times the third. What are the numbers?

62. Finding numbers. The sum of three numbers is 14. The second number is the result of subtracting the third from the first, and the third number is the sum of the second and seven times the first. What are the numbers?

In Exercises 63–66, use Table 5.1 and Figure 5.8 to determine which grid cells of the patients contain healthy tissue, tumorous tissue, bone, or metal.

Beam decrease, in LAUs

Patient	Beam 1	Beam 2	Beam 3
63. Vicky	0.54	0.40	0.52
64. Yolanda	0.65	0.80	0.75
65. Nina	0.51	0.49	0.44
66. Srinivasan	0.44	2.21	2.23

67. Voting. Students identify themselves as Democrats, Republicans, or Independents for a simulated vote on government-sponsored health care. Ten percent more students register as Democrats than as Republicans, and the number of students that register as either Democrat or Republican is 9 times the number registering as Independents. If a total of 420 students register, how many registered in each category?

68. Marketing preserves. Kaitlyn needs 31 jars of strawberry preserves, 20 jars of peach preserves, and 26 jars of grape preserves. She can get a good price on packages containing one of the following:
Type 1: three jars of strawberry preserves, two jars of peach preserves, and two jars of grape preserves
Type 2: four jars of strawberry preserves, one jar of peach preserves, and two jars of grape preserves
Type 3: two jars of strawberry preserves, two jars of peach preserves, and three jars of grape preserves

How many packages of each type should she order?

69. Vacation rentals. Sunrise Corporation has the following rental units:
Type 1: four single rooms, three double rooms, and one double room with a kitchen
Type 2: five single rooms, four double rooms, and no double rooms with a kitchen
Type 3: eight single rooms, one double room, and two double rooms with a kitchen

It had 72 requests for single rooms, 40 requests for double rooms, and 13 requests for double rooms with a kitchen. How many rental units of each type were required?

70. Ice cream sales. Jon's Ice Cream Emporium finds that its sales for the warmer months of the year can be modeled by the quadratic function $S(x) = ax^2 + bx + c$, where $x = 1$ for January, $x = 2$ for February, and so on, and $S(x)$ is the number of gallons sold in month x. Assuming that 22 gallons of ice cream are sold in April, 62 gallons in June, and 38 gallons in August, find the coefficients a, b, c.

71. Crawfish prices. During crawfish season the demand for Swamp Critters' crawfish can be modeled by the quadratic function $D(x) = ax^2 + bx + c$, where x is the price per pound and $D(x)$ is the number of pounds sold at price x. If 6627.5 pounds of crawfish are sold at \$5.75 per pound, 6882.3 pounds of crawfish are sold at \$5.55 per pound, and 7130.7 pounds of crawfish are sold at \$5.35 per pound, find the coefficients a, b, c.

C EXERCISES Beyond the Basics

In Exercises 72–75, write a linear equation of the form $x + by + cz = d$ that is satisfied by all three of the given ordered triples.

72. $(1, 0, 0), (0, 1, 0), (0, 0, 1)$

73. $\left(\frac{1}{3}, 0, 0\right), (0, 4, 3), (1, 2, 2)$

74. $(3, -4, 0), \left(0, \frac{1}{4}, \frac{1}{2}\right), (1, 1, -4)$

75. $(0, 1, -10), \left(\frac{1}{8}, 0, \frac{1}{4}\right), \left(1, \frac{1}{3}, -2\right)$

In Exercises 76–79, find an equation of the parabola of the form $y = ax^2 + bx + c$ that passes through the three given points.

76. $(0, 1), (-1, 0), (1, 4)$

77. $(0, 2), (-1, 30), (2, 6)$

78. $(1, 2), (-1, 4), (2, 4)$

79. $(0, 3), (-1, 4), (1, 6)$

In Exercises 80–83, find an equation of the circle of the form $x^2 + y^2 + ax + by + c = 0$ that passes through the three given points.

80. $(0, 4), (2\sqrt{2}, 2\sqrt{2}), (-4, 0)$

81. $(0, 3), (0, -1), (\sqrt{3}, 2)$

82. $(1, 2), (6, -3), (4, 1)$

83. $(5, 6), (-1, 6), (3, 2)$

In Exercises 84 and 85, solve the system of equations.

84.
$$\begin{cases} \dfrac{1}{x} + \dfrac{3}{y} - \dfrac{1}{z} = 5 \\[2mm] \dfrac{2}{x} + \dfrac{4}{y} + \dfrac{6}{z} = 4 \\[2mm] \dfrac{2}{x} + \dfrac{3}{y} + \dfrac{1}{z} = 3 \end{cases}$$

85.
$$\begin{cases} \dfrac{1}{x} + \dfrac{2}{y} + \dfrac{3}{z} = 8 \\[2mm] \dfrac{2}{x} + \dfrac{5}{y} + \dfrac{9}{z} = 16 \\[2mm] \dfrac{3}{x} - \dfrac{4}{y} - \dfrac{5}{z} = 32 \end{cases}$$

$$\left[\textit{Hint: } \text{Let } \dfrac{1}{x} = u, \dfrac{1}{y} = v, \text{ and } \dfrac{1}{z} = w. \text{ Solve for } u, v, \text{ and } w. \right]$$

In Exercises 86 and 87, find the value of c for which the given system has a unique solution.

86.
$$\begin{cases} x + 2y - 5z - \ \ 9 = 0 \\ 3x - \ \ y + 2z - 14 = 0 \\ 2x + 3y - \ \ z - \ \ 3 = 0 \\ cx - 5y + \ \ z + \ \ 3 = 0 \end{cases}$$

87.
$$\begin{cases} x + 5y + \ \ z = 42 \\ 3x + \ \ y - 3z = c \\ x - \ \ y + 3z = 4 \\ x + \ \ y - \ \ z = 0 \end{cases}$$

88. Use Table 5.1 to investigate four grid cells, arranged in a square, as shown in the given figure. Figures (a) and (b) reprinted with permission from *Mathematics Teacher*, May 1996. © 1996 by National Council of Teachers of Mathematics.

a. For Figure (a), the following data were observed.

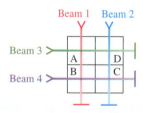

Four grid cells and four X-rays

(a)

1. Beam 1 decreased by 0.60 unit.
2. Beam 2 decreased by 0.75 unit.
3. Beam 3 decreased by 0.65 unit.
4. Beam 4 decreased by 0.70 unit.

Is there sufficient information in Figure (a) to determine which grid cells may contain tumorous tissue?

b. From two additional X-rays, as shown in Figure (b), the following data were observed.

Four grid cells and six X-rays

(b)

5. Beam 5 decreased by 0.85 unit.
6. Beam 6 decreased by 0.50 unit.

Show that the six X-rays represented in the two figures are sufficient to locate tumors in this situation but that, in fact, it is not necessary to use all six rays.

89. Efficiency in programming. Working together, database programmers Darren, Skylar, and Yolanda can complete an assignment in 10 hours; Darren and Skylar can complete the assignment in 15 hours; and Skylar and Yolanda can complete the assignment in 20 hours. How many hours would it take each programmer to complete the job working alone?

90. Filling a pool. Three pipes, labeled A, B, and C, can be used to fill a pool. If only A and B are used, the pool will be filled in four hours. If only B and C are used, the pool will be filled in three hours, and if only A and C are used, the pool will be filled in six hours. How many hours would it take each pipe to fill the pool alone?

Critical Thinking

91. Write a linear system of three equations in the variables x, y, and z that has $\{(1, -1, 2)\}$ as its solution set.

92. Write a linear system of three equations in the variables x, y, and z such that
a. the system has no solution.
b. the system has infinitely many solutions.

Systems of Nonlinear Equations

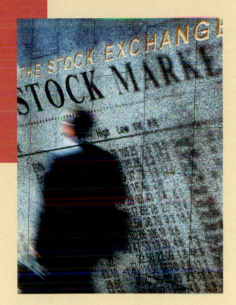

New York Stock Exchange

Before Starting this Section, Review

1. Solve systems of linear equations (Section 5.1, page 480)

Objectives

1 Solve systems of nonlinear equations by substitution.

2 Solve systems of nonlinear equations by elimination.

3 Solve applied problems using nonlinear systems.

INVESTING IN THE STOCK MARKET

A *stock* is a certificate that shows that you own a small fraction of a corporation. When you own stock in a company, you are referred to as a *shareholder*. When you buy stock, you are paying for a small percentage of everything the company owns, including a slice of its profits (or losses).

Suppose you inherit some money and want to invest $1000 in the stock market. Where do you invest your money? (This is a serious question. You may have to do a great deal of research on the companies that interest you, or you may have to get help from a broker.)

Suppose you get a tip from a friend who heard it from his brother-in-law (a taxi driver who overheard his customer's conversation) that Microsoft is coming out with a new version of Microsoft Windows that is going to double the company's business. Where do you buy Microsoft stock? You go to a broker. A brokerage house (such as Merrill Lynch) is a supermarket of stocks. You can tell any broker that you have $1000 and want to buy as much Microsoft stock as you can. The broker might reply, "A share of Microsoft at this time [the price may fluctuate every second] costs $31 [the actual closing price on October 17, 2007, was $31], and I am going to charge you $55 for my services. So you can buy $\frac{1000 - 55}{31} \approx 30.48$, or 30 shares of Microsoft." You then give the broker the money, and you become part owner of Microsoft. (Usually, the broker does not give you the paper stock certificate, but rather transfers ownership to you.)

You can make a profit on your investment if you sell your stock for more than the total amount you paid for it. You can also make money while you own certain stocks by receiving dividends. A *dividend* on a share of stock is a return on your investment and represents the portion of the company's profits allocated to that share. The dividend can be in the form of cash or additional shares of the company. In Example 3, we discuss such a scenario. ■

Systems of Nonlinear Equations

In Sections 5.1 and 5.2, we discussed systems of *linear* equations. We now consider systems of *nonlinear* equations involving two variables: in these systems, at least one equation is nonlinear.

To determine whether an ordered pair is a solution of a nonlinear system, you check whether it is a solution of each equation or inequality in the system. Thus, (2, 1) is a solution of the system

$$\begin{cases} x^2 + 3y = 7 & (1) \\ 5x^2 - 4y^2 = 16 & (2) \end{cases}$$

because $2^2 + 3(1) = 4 + 3 = 7$ and $5(2)^2 - 4(1)^2 = 20 - 4 = 16$ are both true statements.

1 Solve systems of nonlinear equations by substitution.

Solving Systems of Nonlinear Equations by Substitution

We follow the same steps used in the substitution method for solving a system of linear equations in Section 5.1. See page 482. This method is useful when one of the equations in the system is linear in one of the variables.

See page 482.

EXAMPLE 1   **Using Substitution to Solve a Nonlinear System**

Solve the system of equations by the substitution method.

$$\begin{cases} 4x + y = -3 & (1) \\ -x^2 + y = 1 & (2) \end{cases}$$

SOLUTION

Because both equations are linear in y, we solve either equation for y.

Step 1   **Solve for one variable.** In equation (2), express y in terms of x.

$$y = x^2 + 1 \quad (3) \qquad \text{Solve equation (2) for } y.$$

Step 2   **Substitute.** Substitute $x^2 + 1$ for y in equation (1).

$$\begin{aligned} 4x + y &= -3 & &\text{Equation (1)} \\ 4x + (x^2 + 1) &= -3 & &\text{Replace } y \text{ with } (x^2 + 1). \\ 4x + x^2 + 4 &= 0 & &\text{Add 3 to both sides.} \\ x^2 + 4x + 4 &= 0 & &\text{Rewrite in descending powers of } x. \end{aligned}$$

Step 3   **Solve** the resulting equation from Step 2.

$$\begin{aligned} (x + 2)(x + 2) &= 0 & &\text{Factor.} \\ x + 2 &= 0 & &\text{Zero-product property} \\ x &= -2 & &\text{Solve for } x. \end{aligned}$$

Step 4   **Back-substitute.** Substitute $x = -2$ in equation (3) to find the corresponding y-value.

$$\begin{aligned} y &= x^2 + 1 & &\text{Equation (3)} \\ y &= (-2)^2 + 1 & &\text{Replace } x \text{ with } -2. \\ y &= 5 & &\text{Simplify.} \end{aligned}$$

Step 5   From Steps 3 and 4, we have $x = -2$ and $y = 5$. The apparent solution set of the system is $\{(-2, 5)\}$.

Step 6   **Check:** Replace x with -2 and y with 5 in equation (1) and equation (2).

$$\begin{array}{ll} 4x + y = -3 & -x^2 + y = 1 \\ 4(-2) + 5 \overset{?}{=} -3 & -(-2)^2 + 5 \overset{?}{=} 1 \\ -8 + 5 = -3 \checkmark & -4 + 5 = 1 \checkmark \end{array}$$

The graphs of the line $4x + y = -3$ and the parabola $y = x^2 + 1$ confirm that the solution set is $\{(-2, 5)\}$. See Figure 5.9. ■ ■ ■

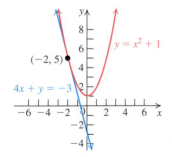

FIGURE 5.9

Practice Problem 1   Solve the system of equations by the substitution method.

$$\begin{cases} x^2 + y = 2 & (1) \\ 2x + y = 3 & (2) \end{cases}$$

2 Solve systems of nonlinear equations by elimination.

Solving Systems of Nonlinear Equations by Elimination

We follow the same steps outlined for solving systems of linear equations by the elimination method. See page 484. This method works well when both equations of the system are of the same degree in the same variable.

To use a graphing calculator to graph the system in Example 2, solve $x^2 + y^2 = 25$ for y and enter the two resulting equations.

$$y_1 = \sqrt{25 - x^2}$$
$$y_2 = -\sqrt{25 - x^2}$$

Then enter

$$y_3 = x^2 - 5$$

and use the ZSquare feature.

EXAMPLE 2 **Using Elimination to Solve a Nonlinear System**

Solve the system of equations by the elimination method.

$$\begin{cases} x^2 + y^2 = 25 & (1) \\ x^2 - y = 5 & (2) \end{cases}$$

SOLUTION

Step 1 **Adjust the coefficients.** Because x has the same power in both equations, we choose the variable x for elimination. We multiply equation (2) by -1 and obtain the equivalent system.

$$\begin{cases} x^2 + y^2 = 25 & (1) \\ -x^2 + y = -5 & (3) \end{cases}$$ Multiply equation (2) by -1.

Step 2 $\qquad\qquad y^2 + y = 20 \quad (4)$ Add equations (1) and (3).

Step 3 Solve the equation found in Step 2: $y^2 + y = 20$.

$$y^2 + y - 20 = 0 \qquad \text{Subtract 20 from both sides.}$$
$$(y + 5)(y - 4) = 0 \qquad \text{Factor.}$$
$$y + 5 = 0 \quad \text{or} \quad y - 4 = 0 \qquad \text{Zero-product property}$$
$$y = -5 \quad \text{or} \qquad y = 4 \qquad \text{Solve for } y.$$

Step 4 **Back-substitute** the values in one of the original equations to solve for the other variable.

(i) Substitute $y = -5$ in equation (2) and solve for x.

$$x^2 - y = 5 \qquad \text{Equation (2)}$$
$$x^2 - (-5) = 5 \qquad \text{Replace } y \text{ with } -5.$$
$$x^2 = 0 \qquad \text{Simplify.}$$
$$x = 0 \qquad \text{Solve for } x.$$

So $(0, -5)$ is a solution of the system.

(ii) Substitute $y = 4$ in equation (2) and solve for x.

$$x^2 - y = 5 \qquad \text{Equation (2)}$$
$$x^2 - 4 = 5 \qquad \text{Replace } y \text{ with } 4.$$
$$x^2 = 9 \qquad \text{Add 4 to both sides.}$$
$$x = \pm 3 \qquad \text{Square root property}$$

So $(3, 4)$ and $(-3, 4)$ are solutions of the system.
From **(i)** and **(ii)**, the apparent solution set for the system is $\{(0, -5), (3, 4), (-3, 4)\}$.

Step 5 **Check:**

	$(0, -5)$	$(3, 4)$	$(-3, 4)$
$x^2 + y^2 = 25$	$0^2 + (-5)^2 \overset{?}{=} 25$	$3^2 + 4^2 \overset{?}{=} 25$	$(-3)^2 + 4^2 \overset{?}{=} 25$
	$25 = 25$ ✓	$9 + 16 = 25$	$9 + 16 \overset{?}{=} 25$
		$25 = 25$ ✓	$25 = 25$ ✓
$x^2 - y = 5$	$0^2 - (-5) \overset{?}{=} 5$	$3^2 - 4 \overset{?}{=} 5$	$(-3)^2 - 4 \overset{?}{=} 5$
	$5 = 5$ ✓	$9 - 4 = 5$ ✓	$9 - 4 = 5$ ✓

FIGURE 5.10

The graphs of the circle $x^2 + y^2 = 25$ and the parabola $y = x^2 - 5$ sketched in Figure 5.10 confirm the three solutions.   ■ ■ ■

Practice Problem 2   Solve the system of equations.

$$\begin{cases} x^2 + 2y^2 = 34 \\ x^2 - y^2 = 7 \end{cases}$$ ■

Applications

In the next example, we solve an investment problem that involves a nonlinear system of equations.

3 Solve applied problems using nonlinear systems.

EXAMPLE 3   **Investing in Stocks**

Danielle bought 240 shares of Alpha Airlines stock at \$40 per share and paid \$100 in broker's fees. She kept these shares for three years, during which time she received a number of additional shares of the stock as dividends. At the end of three years, her dividends were worth \$2400 and she sold all of her stock and made a profit of \$7100 (after paying \$100 in broker's commission). How many shares of the stock did she receive as dividends? What was the selling price of each share of the stock?

SOLUTION

Let x = number of shares she received as dividends
   p = selling price per share

Then

$$xp = 2400 \qquad (1) \qquad \text{The total dividend is worth \$2400.}$$

The number of shares she sold at p dollars per share is $240 + x$.

$$\text{Revenue} = (240 + x)p$$
$$\text{Cost} = (240)(40) + 100 = 9700$$

number of shares bought   commission

price she paid per share

and

$$\text{Revenue} - \text{Cost} = \text{Profit}$$
$$(240 + x)p - 9700 = 7100$$
$$(240 + x)p = 16{,}800 \qquad \text{Add 9700 to both sides.}$$
$$240p + xp = 16{,}800 \qquad (2) \qquad \text{Distributive property}$$

We solve the system of equations:

$$\begin{cases} xp = 2400 & (1) \qquad \text{Revenue from the dividends} \\ 240p + xp = 16{,}800 & (2) \qquad \text{Revenue} - \text{Cost} = \text{Profit} \end{cases}$$

Substitute $xp = 2400$ from equation (1) into equation (2).

$$240p + xp = 16{,}800 \qquad (2)$$
$$240p + 2400 = 16{,}800 \qquad (3) \qquad \text{Substitute 2400 for } xp \text{ in (2).}$$
$$240p = 14{,}400 \qquad \text{Subtract 2400 from both sides.}$$
$$p = 60 \qquad \text{Solve for } p.$$

Back-substitute $p = 60$ in equation (1).

$$xp = 2400 \qquad \text{Equation (1)}$$
$$x(60) = 2400 \qquad \text{Replace } p \text{ with 60.}$$
$$x = \frac{2400}{60} = 40 \qquad \text{Solve for } x.$$

Danielle received $x = 40$ shares as dividends and sold her stock at $p = \$60$ per share. ■ ■ ■

Practice Problem 3 Rework Example 3 assuming that Danielle's dividends were worth \$1950 after three years and her profit was \$7850. ■

SECTION 5.3 ■ Exercises

A EXERCISES Basic Skills and Concepts

1. In a system of nonlinear equations, _____ equation must be nonlinear.

2. Both the _____ and _____ methods can also be used to solve systems of nonlinear equations.

3. The solutions of the system
$$\begin{cases} ax + by = c & (1) \\ (x - h)^2 + (y - k)^2 = r^2 & (2) \end{cases}$$
represent the points of _____ of the graphs of equations (1) and (2).

4. The system of equations in Exercise 3 has exactly one solution because the graph of equation (1) is a(n) _____ line to the graph of equation (2).

5. *True or False* The system of equations
$$\begin{cases} y = ax^2 \\ (x - h)^2 + (y - k)^2 = r^2 \end{cases}$$ may have 0, 1, 2, 3, or 4 real solutions.

6. *True or False* If $4x + 3y = 8$ is one of the equations in a system of equations, then the system is linear.

In Exercises 7–14, determine which of the given ordered pairs are solutions of each system of equations.

7. $\begin{cases} 2x + 3y = 3 \\ x - y^2 = 2 \end{cases}$ $(1, -3), (3, -1), (6, 3), \left(5, \dfrac{1}{2}\right)$

8. $\begin{cases} x + 2y = 6 \\ y = x^2 \end{cases}$ $(2, 2), (-2, 4), (0, 3), (1, 2)$

9. $\begin{cases} 5x - 2y = 7 \\ x^2 + y^2 = 2 \end{cases}$ $\left(\dfrac{5}{4}, 1\right), \left(0, \dfrac{11}{4}\right), (1, -1), (3, 4)$

10. $\begin{cases} x - 2y = -5 \\ x^2 + y^2 = 25 \end{cases}$ $(1, 3), (-5, 0), (3, 4), (3, -4)$

11. $\begin{cases} 4x^2 + 5y^2 = 180 \\ x^2 - y^2 = 9 \end{cases}$ $(5, 4), (-5, 4), (3, 0), (-5, -4)$

12. $\begin{cases} x^2 - y^2 = 3 \\ x^2 - 4x + y^2 = -3 \end{cases}$ $(2, 1), (2, -1), (-2, 1), (-2, -1)$

13. $\begin{cases} y = e^{x-1} \\ y = 2x - 1 \end{cases}$ $(1, 1), (0, -1), (2, e), (-1, -3)$

14. $\begin{cases} y = \ln(x + 1) \\ y = x \end{cases}$ $(1, 1), (0, 0), (-1, -1), (e - 1, 1)$

In Exercises 15–30, solve each system of equations by the substitution method. Check your solutions.

15. $\begin{cases} y = x^2 \\ y = x + 2 \end{cases}$

16. $\begin{cases} y = x^2 \\ x + y = 6 \end{cases}$

17. $\begin{cases} x^2 - y = 6 \\ x - y = 0 \end{cases}$

18. $\begin{cases} x^2 - y = 6 \\ 5x - y = 0 \end{cases}$

19. $\begin{cases} x^2 + y^2 = 9 \\ x = 3 \end{cases}$

20. $\begin{cases} x^2 + y^2 = 9 \\ y = 3 \end{cases}$

21. $\begin{cases} x^2 + y^2 = 5 \\ x - y = -3 \end{cases}$

22. $\begin{cases} x^2 + y^2 = 13 \\ 2x - 3y = 0 \end{cases}$

23. $\begin{cases} x^2 - 4x + y^2 = -2 \\ x - y = 2 \end{cases}$

24. $\begin{cases} x^2 - 8y + y^2 = -6 \\ 2x - y = 1 \end{cases}$

25. $\begin{cases} x - y = -2 \\ xy = 3 \end{cases}$

26. $\begin{cases} x - 2y = 4 \\ xy = 6 \end{cases}$

27. $\begin{cases} 4x^2 + y^2 = 25 \\ x + y = 5 \end{cases}$

28. $\begin{cases} x^2 + 4y^2 = 16 \\ x + 2y = 4 \end{cases}$

29. $\begin{cases} x^2 - y^2 = 24 \\ 5x - 7y = 0 \end{cases}$

30. $\begin{cases} x^2 - 7y^2 = 9 \\ x - y = 3 \end{cases}$

In Exercises 31–40, solve each system of equations (only real solutions) by the elimination method. Check your solutions.

31. $\begin{cases} x^2 + y^2 = 20 \\ x^2 - y^2 = 12 \end{cases}$

32. $\begin{cases} x^2 + 8y^2 = 9 \\ 3x^2 + y^2 = 4 \end{cases}$

33. $\begin{cases} x^2 + 2y^2 = 12 \\ 7y^2 - 5x^2 = 8 \end{cases}$

34. $\begin{cases} 3x^2 + 2y^2 = 77 \\ x^2 - 6y^2 = 19 \end{cases}$

35. $\begin{cases} x^2 - y = 2 \\ 2x - y = 4 \end{cases}$

36. $\begin{cases} x^2 + 3y = 0 \\ x - y = -12 \end{cases}$

37. $\begin{cases} x^2 + y^2 = 5 \\ 3x^2 - 2y^2 = -5 \end{cases}$　　**38.** $\begin{cases} x^2 + 4y^2 = 5 \\ 9x^2 - y^2 = 8 \end{cases}$

39. $\begin{cases} x^2 + y^2 + 2x = 9 \\ x^2 + 4y^2 + 3x = 14 \end{cases}$　　**40.** $\begin{cases} x^2 + y^2 = 9 \\ x^2 + y^2 - 18x = 0 \end{cases}$

In Exercises 41–54, use any method to solve each system of equations.

41. $\begin{cases} x + y = 8 \\ \quad xy = 15 \end{cases}$　　**42.** $\begin{cases} 2x + y = 8 \\ x^2 - y^2 = 5 \end{cases}$

43. $\begin{cases} x^2 + y^2 = 2 \\ 3x^2 + 3y^2 = 9 \end{cases}$　　**44.** $\begin{cases} x^2 + y^2 = 5 \\ 3x^2 - y^2 = -11 \end{cases}$

45. $\begin{cases} y^2 = 4x + 4 \\ y = 2x - 2 \end{cases}$　　**46.** $\begin{cases} xy = 125 \\ y = x^2 \end{cases}$

47. $\begin{cases} x^2 + 4y^2 = 25 \\ x - 2y + 1 = 0 \end{cases}$　　**48.** $\begin{cases} y = x^2 - 5x + 4 \\ 3x + y = 3 \end{cases}$

49. $\begin{cases} x^2 - 3y^2 = 1 \\ x^2 + 4y^2 = 8 \end{cases}$　　**50.** $\begin{cases} 4x^2 - y^2 = 12 \\ 4y^2 - x^2 = 12 \end{cases}$

51. $\begin{cases} x^2 - xy + 5x = 4 \\ 2x^2 - 3xy + 10x = -2 \end{cases}$

52. $\begin{cases} x^2 - xy + x = -4 \\ 3x^2 - 2xy - 2x = 4 \end{cases}$

53. $\begin{cases} x^2 + y^2 - 8x = -8 \\ x^2 - 4y^2 + 6x = 0 \end{cases}$　　**54.** $\begin{cases} 4x^2 + y^2 - 9y = -4 \\ 4x^2 - y^2 - 3y = 0 \end{cases}$

B EXERCISES Applying the Concepts

Exercises 55 and 56 show the demand and supply functions of a product. In each case, p represents the price per unit and x represents the number of units, in hundreds. Determine the price that gives market equilibrium and the number of units that are demanded and supplied at that price.

55. $\begin{cases} p + 2x^2 = 96 \\ p - 13x = 39 \end{cases}$　　**56.** $\begin{cases} p^2 + 6p + 3x = 75 \\ -p + x = 13 \end{cases}$

In Exercises 57–64, use systems of equations to solve the problem.

57. Numbers. Find two positive numbers whose sum is 24 and whose product is 143.

58. Numbers. Find two positive numbers whose difference is 13 and whose product is 114.

59. Commercial land. A commercial parcel of land is in the form of a trapezoid with two right angles, as shown in the figure. The oblique side is 100 meters in length, and two sides that meet at the vertex of one of the right angles are equal. The perimeter of the land is 360 meters. Find
　a. the lengths of the remaining sides.
　b. the area of the land.

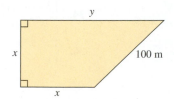

60. Dividing a pasture. The area of a rectangular pasture is 8750 square feet. The pasture is divided into three smaller pastures by two fences parallel to the shorter sides. The widths of two of the larger pastures are the same, and the width of the third is one-half that of the others. Find the possible dimensions of the original pasture if the perimeter of the smaller of the subdivisions is 190 feet.

61. Chartering a fishing boat. A group of students of a fraternity chartered a fishing boat that would cost $960. Before the date of the fishing trip, eight more students joined the group and agreed to pay their share of the cost of the charter. The cost per student was reduced by $6. Find the original number of students in the group and the original cost per student.

62. Carpet sale. A salesperson sold a square carpet and a rectangular carpet whose length was 6 feet more than its width. The combined area of the two carpets was 540 square feet. The price of the square carpet was $12 per square yard, and the price of the rectangular carpet was $10 per square yard. Assuming that the rectangular piece was $50 more than the square piece, find the dimensions of each carpet.

63. Buying and selling stocks. Flat Fee Brokerage, Inc., charges $100 commission for every transaction (purchase or sale) you make. Anju bought a block of stock containing over 400 shares that cost her $10,000, including the commission. After one year, she received a stock dividend of 30 shares. She then sold all of her stock for $3 more per share than it cost and made a profit of $1900 (after paying the commission). How many shares did she buy, and what was the original price of each share?

64. Trading stocks. Repeat Exercise 63 with the following details: Anju does not receive dividends and decides to keep 100 shares of the stock. She sells the remaining stock for $10 more per share than it cost and makes a profit of $3900 after paying the commission.

C EXERCISES Beyond the Basics

In Exercises 65–68, the given circle intersects the given line at the points A and B. Find the coordinates of A and B and compute the distance $d(A, B)$.

65. Circle: $x^2 + y^2 - 7x + 5y + 6 = 0$; line: x-axis

66. The circle of Exercise 65 and the y-axis

67. Circle: $x^2 + y^2 + 2x - 4y - 5 = 0$; line: $x - y + 1 = 0$

68. Circle: $x^2 + y^2 - 6x - 8y - 50 = 0$;
line: $2x + y - 5 = 0$

69. Show that the line with equation $x + 2y + 6 = 0$ is a tangent line to the circle with equation $x^2 + y^2 - 2x + 2y - 3 = 0$. (A line that intersects the circle at exactly one point is a **tangent line** to the circle.)

70. Repeat Exercise 69 for the line with equation $3x - 4y = 0$ and the circle with equation $x^2 + y^2 - 2x - 4y + 4 = 0$.

In Exercises 71–74, find the value(s) of c so that the graphs of the equations in the system are tangent.

71. $\begin{cases} x + y = 2 \\ x^2 + y^2 = c^2 \end{cases}$ **72.** $\begin{cases} cx - y = 5 \\ x^2 + y^2 = 16 \end{cases}$

73. $\begin{cases} 4x^2 + y^2 = 25 \\ 8x + 3y = c \end{cases}$ **74.** $\begin{cases} x^2 + 4y^2 = 9 \\ cx + 2y = -4 \end{cases}$

In Exercises 75–78, solve by letting $u = \dfrac{1}{x}$ and $v = \dfrac{1}{y}$.

75. $\begin{cases} \dfrac{1}{x} - \dfrac{1}{y} = 5 \\ \dfrac{1}{xy} - 24 = 0 \end{cases}$ **76.** $\begin{cases} \dfrac{1}{x} + \dfrac{1}{y} = 3 \\ \dfrac{1}{x^2} - \dfrac{1}{y^2} = -3 \end{cases}$

77. $\begin{cases} \dfrac{5}{x^2} - \dfrac{3}{y^2} = 2 \\ \dfrac{6}{x^2} + \dfrac{1}{y^2} = 7 \end{cases}$ **78.** $\begin{cases} \dfrac{3}{x^2} - \dfrac{2}{y^2} = 1 \\ \dfrac{6}{x^2} + \dfrac{5}{y^2} = 11 \end{cases}$

In Exercises 79–84, find square roots of each complex number w. [$z = a + bi$ is a square root of w if $z^2 = w$.]

79. $w = -5 + 12i$ **80.** $w = -16 - 30i$

81. $w = 7 - 24i$ **82.** $w = 3 - 4i$

83. $w = 13 + i8\sqrt{3}$ **84.** $w = -1 - i4\sqrt{5}$

In Exercises 85–90, solve each system of equations and verify the solutions graphically.

85. $\begin{cases} y = 3^x + 4 \\ y = 3^{2x} - 2 \end{cases}$ **86.** $\begin{cases} y = \ln x + 1 \\ y = \ln x - 2 \end{cases}$

87. $\begin{cases} y = 2^x + 3 \\ y = 2^{2x} + 1 \end{cases}$ **88.** $\begin{cases} y = 2^x + 3 \\ y = 4^x + 1 \end{cases}$

89. $\begin{cases} x = 3^y \\ 3^{2y} = 3x - 2 \end{cases}$ **90.** $\begin{cases} x = 3^y \\ x^2 = 3^{y+1} - 2 \end{cases}$

Critical Thinking

91. Consider a system of two equations in two variables.

$$\begin{cases} y = mx + b \\ y = Ax^3 + Bx^2 + Cx + D, \ A \neq 0 \end{cases}$$

Explain whether it is possible for this system to have
a. no real solutions.
b. one real solution.
c. two real solutions.
d. three real solutions.
e. four real solutions.
f. more than four real solutions.

92. Repeat Exercise 91 for the system.

$$\begin{cases} y = mx + b \\ y = Ax^4 + Bx^3 + Cx^2 + Dx + E, A \neq 0 \end{cases}$$

Systems of Inequalities

1 Horsepower
=
33,000 foot-pounds
per minute

Before Starting this Section, Review

1. Intersection of sets (Section P.1, page 6)
2. Graphing a line (Section 2.3, page 196)
3. Solving systems of equations (Sections 5.1 and 5.3)

Objectives

1 Graph a linear inequality in two variables.

2 Graph systems of linear inequalities in two variables.

3 Graph a nonlinear inequality in two variables.

4 Graph systems of nonlinear inequalities in two variables.

WHAT IS HORSEPOWER?

The definition of horsepower was originated by the inventor James Watt (1736–1819). To demonstrate the power (work capability) of the steam engine that he invented and patented in 1783, he rigged his favorite horse with a rope and some pulleys and showed that his horse could raise a 3300-pound weight 10 feet in the air in one minute. Watt called this amount of work 1 horsepower (hp). So he defined

$$1 \text{ hp} = 33{,}000 \text{ foot-pounds per minute } (3300 \times 10 = 33{,}000)$$

$$= 550 \text{ foot-pounds per second } \left(\frac{33{,}000}{60} = 550 \right).$$

Because horses were one of the main energy sources during that time, the public easily understood and accepted the horsepower measure. Watt labeled his steam engines in terms of equivalent horsepower. A 10 hp engine could do the work of 10 horses, or could lift 5500 pounds up 1 foot in one second. In the usual definition of power $= \dfrac{(\text{force})\,(\text{distance})}{\text{time}}$, the horse provides the force. In Exercise 83, we discuss engines with different amounts of horsepower. ■

---

1 Graph a linear inequality in two variables.

Graph of a Linear Inequality in Two Variables

The statements $x + y > 4$, $2x + 3y < 7$, $y \geq x$, and $x + y \leq 9$ are examples of linear inequalities in the variables x and y. As with equations, a **solution of an inequality in two variables** x and y is an ordered pair (a, b) that results in a true statement when x is replaced with a and y is replaced with b in the inequality. For example, the ordered pair $(3, 2)$ is a solution of the inequality $4x + 5y > 11$ because $4(3) + 5(2) = 22 > 11$ is a true statement. The ordered pair $(1, 1)$ is *not* a solution of the inequality $4x + 5y > 11$ because $4(1) + 5(1) = 9 > 11$ is a false statement. The set of all solutions of an inequality is called the **solution set of the inequality**. The **graph of an inequality in two variables** is the graph of the solution set of the inequality.

EXAMPLE 1 Graphing a Linear Inequality

Graph the inequality: $2x + y > 6$.

SOLUTION

First, graph the *equation* $2x + y = 6$. See Figure 5.11. Notice that if we solve this equation for y, we have $y = -2x + 6$. Any point $P(a, b)$ whose y-coordinate, b, is *greater* than $6 - 2a$ must be above this line. Therefore, the graph of $y > -2x + 6$, which is equivalent to $2x + y > 6$, consists of all points (x, y) in the plane that lie above the line $2x + y = 6$. The graph of the inequality $2x + y > 6$ is shaded in Figure 5.12.

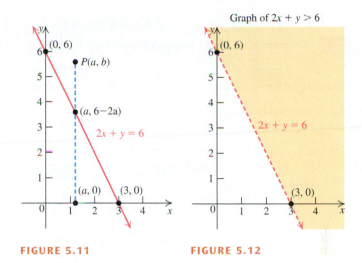

FIGURE 5.11   **FIGURE 5.12**   ■ ■ ■

Practice Problem 1   Graph the inequality $3x + y < 6$.   ■

In Figure 5.12, the graph of the equation $2x + y = 6$ is drawn as a dashed line to indicate that points on the line are *not* included in the solution set of the inequality $2x + y > 6$. In Figure 5.13(a), however, the graph of the inequality $2x + y \geq 6$ shows $2x + y = 6$ as a solid line: the solid line indicates that the points on the line *are* included in the solution set. Note that the region below the line $2x + y = 6$ in Figure 5.13(b) is the graph of the inequality $2x + y < 6$.

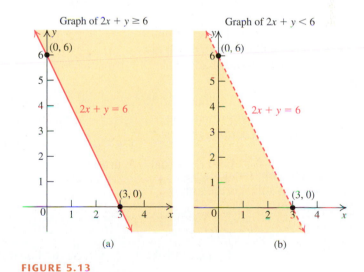

(a)   (b)

FIGURE 5.13

In general, the graph of a linear inequality's *corresponding equation* (found by changing the inequality symbol to an equal sign) is a line that separates the plane into two regions, each called a *half plane*. The line itself is called the **boundary** of each region. *All* of the points in one region satisfy the inequality, and *none* of the points in the other region are solutions. You can therefore test one point not on the boundary, called a **test point**, to determine which region represents the solution set.

FINDING THE SOLUTION: A PROCEDURE

EXAMPLE 2   **Graphing a Linear Inequality in Two Variables**

OBJECTIVE

Graph an inequality in two variables.

Step 1   Replace the inequality symbol with an equal ($=$) sign.

Step 2   Sketch the graph of the *corresponding equation* in Step 1. Use a dashed line for the boundary if the given inequality sign is $<$ or $>$ and a solid line if the inequality symbol is $\leq$ or $\geq$.

Step 3   The graph in Step 2 divides the plane into two regions. Select a *test point* in either region, but not on the graph of the equation in Step 1.

Step 4   **(i)** If the coordinates of the test point satisfy the inequality, then so do all of the points in that region. Shade that region.

(ii) If the test point's coordinates do not satisfy the inequality, shade the region that does not contain the test point.

The shaded region (including the boundary if it is solid) is the graph of the inequality.

EXAMPLE

Graph the linear inequality $y \geq -2x + 4$.

$$y = -2x + 4$$

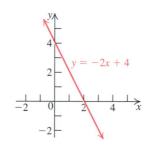

The graph of the equation $y = -2x + 4$ is shown in the figure.

The line drawn is solid because the given inequality sign is $\geq$.

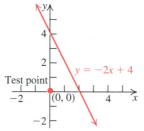

Select $(0, 0)$ as a test point.

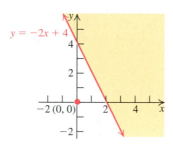

Because $0 \geq -2(0) + 4$ is false, the coordinates of the test point $(0, 0)$ do *not* satisfy the inequality $y \geq -2x + 4$.

The graph of the solution set of $y \geq -2x + 4$ is on the side of the line that does not contain $(0, 0)$. It is shaded in the figure.

■ ■ ■

Practice Problem 2   Graph the linear inequality $y \leq -2x + 4$.   ■

EXAMPLE 3   **Graphing Inequalities**

Sketch the graph of each inequality.

a. $x \geq 2$ **b.** $y < 3$ **c.** $x + y < 4$

SOLUTION

Steps	Inequality		
1. Change the inequality to an equality.	**a.** $x \geq 2$ $x = 2$	**b.** $y < 3$ $y = 3$	**c.** $x + y < 4$ $x + y = 4$
2. Graph the equation in Step 1 with a dashed line ($<$ or $>$) or a solid line ($\leq$ or $\geq$).			
3. Select a test point.	Test $(0, 0)$ in $x \geq 2$; $0 \geq 2$ is a false statement. The region not containing $(0, 0)$, together with the vertical line, is the solution set.	Test $(0, 0)$ in $y < 3$; $0 < 3$ is a true statement. The region containing $(0, 0)$ is the solution set.	Test $(0, 0)$ in $x + y < 4$; $0 + 0 < 4$ is a true statement. The region containing $(0, 0)$ is the solution set.
4. Shade the solution set.	(see figure)		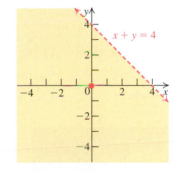

Practice Problem 3 Graph the inequality $x + y > 9$.

TECHNOLOGY CONNECTION

To shade a graph on a graphing calculator, first solve the corresponding equation for y. Enter the equation using the $\boxed{Y =}$ key. On some calculators, you can move the cursor to the symbol to the left of Y_1. Press $\boxed{\text{ENTER}}$ to cycle through the shading options: ◣ shades below the line, and ◥ shades above it.

2 Graph systems of linear inequalities in two variables.

Systems of Linear Inequalities in Two Variables

An ordered pair (a, b) is a **solution of a system of inequalities** involving two variables if it is a solution of each of its inequalities. For example, $(0, 1)$ is a solution of the system

$$\begin{cases} 2x + y \leq 1 \\ x - 3y < 0 \end{cases}$$

because both $2(0) + 1 \leq 1$ and $0 - 3(1) < 0$ are true statements. The **solution set of a system of inequalities** is the *intersection* of the solution sets of all inequalities in the system. To find the graph of this solution set, graph the solution set of each inequality in the system (with different-color shadings) on the same coordinate plane. The region common to all of the solution sets, which is where all of the shaded parts overlap, is the solution set.

A graphing calculator uses overlapping crosshatch shading to display the solution set of a system of inequalities. The solution set for the system

$$\begin{cases} 2x + 3y > 6 & (1) \\ y - x \leq 0 & (2) \end{cases}$$

is shown here.

EXAMPLE 4   **Graphing a System of Two Inequalities**

Graph the solution set of the system of inequalities.

$$\begin{cases} 2x + 3y > 6 & (1) \\ y - x \leq 0 & (2) \end{cases}$$

SOLUTION

Graph each inequality separately in the same coordinate plane.

Inequality 1	**Inequality 2**
$2x + 3y > 6$	$y - x \leq 0$
Step 1 $2x + 3y = 6$	**Step 1** $y - x = 0$
Step 2 Sketch $2x + 3y = 6$ as a dashed line by joining the points $(0, 2)$ and $(3, 0)$.	**Step 2** Sketch $y - x = 0$ as a solid line by joining the points $(0, 0)$ and $(1, 1)$.
Step 3 Test $(0, 0)$ in	**Step 3** Test $(1, 0)$ in

$$\begin{array}{ll} 2x + 3y > 6 & y - x \leq 0 \\ 2 \cdot 0 + 3 \cdot 0 \overset{?}{>} 6 & 0 - 1 \overset{?}{\leq} 0 \\ 0 > 6 \quad \text{False} & -1 \overset{?}{\leq} 0 \quad \text{True} \end{array}$$

Step 4 Shade as in Figure 5.14(a).   **Step 4** Shade as in Figure 5.14(b).

The graph of the solution set of the system of inequalities (1) and (2) (the region where the shadings overlap) is shown shaded in purple in Figure 5.14(c). ▪▪▪

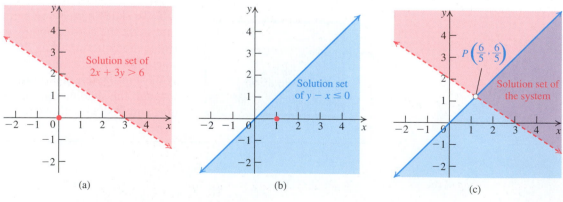

FIGURE 5.14 **Graph of a system of two inequalities**

Practice Problem 4   Graph the solution set of the system of inequalities.

$$\begin{cases} 3x - y > 3 \\ x - y \geq 1 \end{cases}$$ ■

The point $P\left(\dfrac{6}{5}, \dfrac{6}{5}\right)$ in Figure 5.14(c) is the point of intersection of the two lines $2x + 3y = 6$ and $y - x = 0$. Such a point is called a **corner point**, or **vertex**, of the solution set. It is found by solving the system of equations $\begin{cases} 2x + 3y = 6 \\ y - x = 0 \end{cases}$.

In this case, however, the point P is not in the solution set of the given system because the ordered pair $\left(\dfrac{6}{5}, \dfrac{6}{5}\right)$ satisfies only inequality (2), but not inequality (1), of Example 4. We show P as an open circle in Figure 5.14(c).

FIGURE 5.15

EXAMPLE 5   **Solving a System of Three Linear Inequalities**

Sketch the graph and label the vertices of the solution set of the system of linear inequalities.

$$\begin{cases} 3x + 2y \leq 11 & (1) \\ x - y \leq 2 & (2) \\ 7x - 2y \geq -1 & (3) \end{cases}$$

SOLUTION

First, on the same coordinate plane, sketch the graphs of the three linear equations that correspond to the three inequalities. Because either $\leq$ or $\geq$ occurs in all three inequalities, all of the equations are graphed as *solid* lines. See Figure 5.15.

Notice that $(0, 0)$ satisfies each of the inequalities but none of the equations. You can use $(0, 0)$ as the test point for each inequality.

Make the following conclusions:

(i)   Because $(0, 0)$ lies below the line $3x + 2y = 11$, the region on or below the line $3x + 2y = 11$ is in the solution set of inequality (1).

(ii) Because $(0, 0)$ is above the line $x - y = 2$, the region on or above the line $x - y = 2$ is in the solution set of inequality (2).

(iii) Finally, because $(0, 0)$ lies below the line $7x - 2y = -1$, the region on or below the line $7x - 2y = -1$ belongs to the solution set of inequality (3).

The region common to the regions in **(i)**, **(ii)**, and **(iii)** (including the boundaries) is the solution set of the given system of inequalities. The solution set is the shaded region in Figure 5.16, including the sides of the triangle.

Figure 5.16 also shows the vertices of the solution set. These vertices are obtained by solving each pair of equations in the system. Because all vertices are solutions of the given system, they are shown as (filled-in) points.

a.   To find the vertex $(3, 1)$, solve the system $\begin{cases} 3x + 2y = 11 & (1) \\ x - y = 2 & (2) \end{cases}$

Solve equation (2) for x to obtain $x = 2 + y$. Substitute $x = 2 + y$ in equation (1).

$$3(2 + y) + 2y = 11$$
$$6 + 3y + 2y = 11 \qquad \text{Distributive property}$$
$$y = 1 \qquad \text{Solve for } y.$$

Back-substitute $y = 1$ in equation (2) to find $x = 3$. The vertex is the point $(3, 1)$.

b.   Solve the system of equations $\begin{cases} x - y = 2 \\ 7x - 2y = -1 \end{cases}$

by the substitution method to find the vertex $(-1, -3)$.

FIGURE 5.16 **Graph of the solution set**

c. Solve the system of equations $\begin{cases} 3x + 2y = 11 \\ 7x - 2y = -1 \end{cases}$

by the elimination method to find the vertex $(1, 4)$.　■ ■ ■

Practice Problem 5　Sketch the graph (and label the corner points) of the solution set of the system of inequalities.

$$\begin{cases} 2x + 3y < 16 \\ 4x + 2y \geq 16 \\ 2x - y \leq 8 \end{cases}$$
　■

3 Graph a nonlinear inequality in two variables.

Nonlinear Inequality

We follow the same steps that we used in solving a linear inequality.

FINDING THE SOLUTION: A PROCEDURE

EXAMPLE 6　**Graphing a Nonlinear Inequality in Two Variables**

OBJECTIVE
Graph a nonlinear inequality in two variables.

EXAMPLE
Graph $y > x^2 - 2$.

Step 1　Replace the inequality symbol with an equal (=) sign.

$y = x^2 - 2$

Step 2　Sketch the graph of the *corresponding equation* in Step 1. Use a dashed curve if the given inequality sign is $<$ or $>$ and a solid curve if the inequality symbol is $\leq$ or $\geq$.

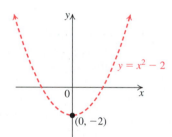

The graph of the parabola with vertex $(0, -2)$ is sketched as a dashed curve in the figure.

Step 3　The graph in Step 2 divides the plane into two regions. Select a *test point* in either region, but not on the graph.

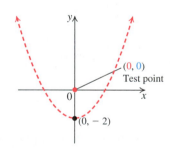

Step 4　**(i)** If the coordinates of the test point satisfy the inequality, then so do all of the points in that region. Shade that region.

(ii) If the test point's coordinates do not satisfy the inequality, shade the region that does not contain the test point.

The shaded region (including the boundary if it is a solid curve) is the graph of the inequality.

Because $0 > 0^2 - 2 = -2$, the coordinates of the test point, $(0, 0)$, satisfy the inequality $y > x^2 - 2$.

The graph of the solution set of $y > x^2 - 2$ is shaded.

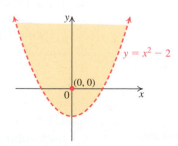

Practice Problem 6　Graph $y \leq x^2 - 2$.　■

4 Graph systems of nonlinear inequalities in two variables.

Nonlinear Systems

The procedure for solving a nonlinear system of inequalities is identical to the procedure for solving a linear system of inequalities.

EXAMPLE 7 **Solving a Nonlinear System of Inequalities**

Graph the solution set of the system of inequalities.

$$\begin{cases} y \le 4 - x^2 & (1) \\ y \ge \dfrac{3}{2}x - 3 & (2) \\ y \ge -6x - 3 & (3) \end{cases}$$

SOLUTION

Graph each inequality separately in the same coordinate plane. Because $(0, 0)$ is not a solution of any of the corresponding equations, use $(0, 0)$ as a test point for each inequality.

Inequality (1)	Inequality (2)	Inequality (3)
$y \le 4 - x^2$	$y \ge \dfrac{3}{2}x - 3$	$y \ge -6x - 3$
Step 1 $y = 4 - x^2$	$y = \dfrac{3}{2}x - 3$	$y = -6x - 3$
Step 2 The graph of $y = 4 - x^2$ is a parabola opening down with vertex $(0, 4)$. Sketch it as a solid curve.	The graph of $y = \dfrac{3}{2}x - 3$ is a line through the points $(0, -3)$ and $(2, 0)$. Sketch a solid line.	The graph of $y = -6x - 3$ is a line through the points $(0, -3)$ and $\left(-\dfrac{1}{2}, 0\right)$. Sketch a solid line.
Step 3 Testing $(0, 0)$ in $y \le 4 - x^2$ gives $0 \le 4$, a true statement. The solution set is below the parabola.	Testing $(0, 0)$ in $y \ge \dfrac{3}{2}x - 3$ gives $0 \ge -3$, a true statement. The solution set is above the line.	Testing $(0, 0)$ in $y \ge -6x - 3$ gives $0 \ge -3$, a true statement. The solution set is above the line.
Step 4 Shade the region as in Figure 5.17(a).	Shade the region as in Figure 5.17(b).	Shade the region as in Figure 5.17(c).

The region common to all three graphs is the graph of the solution set of the given system of inequalities. See Figure 5.17(d).

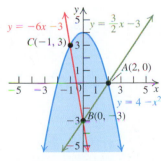

Solution for $y \le 4 - x^2$

(a)

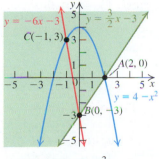

Solution for $y \ge \dfrac{3}{2}x - 3$

(b)

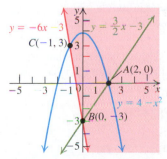

Solution for $y \ge -6x - 3$

(c)

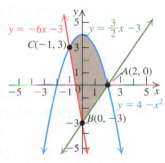

Solution of the system

(d)

FIGURE 5.17

Use the substitution method to locate the points of intersection A, B, and C of the corresponding equations. See Figure 5.17(d). Solve all pairs of corresponding equations.

The point $A(2, 0)$ is the solution of the system

$$\begin{cases} y = 4 - x^2 & \text{Boundary of inequality (1)} \\ y = \dfrac{3}{2}x - 3 & \text{Boundary of inequality (2)} \end{cases}$$

The point $B(0, -3)$ is the solution of the system

$$\begin{cases} y = \dfrac{3}{2}x - 3 & \text{Boundary of inequality (2)} \\ y = -6x - 3 & \text{Boundary of inequality (3)} \end{cases}$$

Finally, the point $C(-1, 3)$ is the solution of the system

$$\begin{cases} y = -6x - 3 & \text{Boundary of inequality (1)} \\ y = 4 - x^2 & \text{Boundary of inequality (2)} \end{cases}$$ ▪▪▪

Practice Problem 7 Graph the solution set of the system of inequalities.

$$\begin{cases} y \geq x^2 + 1 \\ y \leq -x + 13 \\ y < 4x + 13 \end{cases}$$ ▪

SECTION 5.4 ▪ Exercises

A EXERCISES Basic Skills and Concepts

1. In the graph of $x - 3y > 1$, the corresponding equation $x - 3y = 1$ is graphed as a _____ line.

2. If a test point from one of the two regions determined by an inequality's corresponding equation satisfies the inequality, then _____ in that region satisfy the inequality.

3. In a system of inequalities containing both $2x + y > 5$ and $2x - y \leq 3$, the point of intersection of the lines $2x + y = 5$ and $2x - y = 3$ is _____.

4. In a nonlinear system of equations or inequalities, _____ equation or inequality must be nonlinear.

5. *True or False* There are always infinitely many solutions of an inequality $ax + by < c$, where a, b, and c are real numbers and $a \neq 0$.

6. *True or False* If $x^2 + 2y < 5$ is one of the inequalities in a system of inequalities, then the point $(2, 1)$ cannot be used as a test point in graphing the system.

In Exercises 7–22, graph each inequality.

7. $x \geq 0$
8. $y \geq 0$
9. $x > -1$
10. $y > 2$
11. $x \geq 2$
12. $y \leq 3$
13. $y - x < 0$
14. $y + 2x \leq 0$
15. $x + 2y < 6$
16. $3x + 2y < 12$
17. $2x + 3y \geq 12$
18. $2x + 5y \geq 10$

19. $x - 2y \leq 4$
20. $3x - 4y \leq 12$
21. $3x + 5y < 15$
22. $5x + 7y < 35$

In Exercises 23–28, determine which ordered pairs are solutions of each system of inequalities.

23. $\begin{cases} x + y < 2 \\ 2x + y \geq 6 \end{cases}$
  a. $(0, 0)$ **b.** $(-4, -1)$ **c.** $(3, 0)$ **d.** $(0, 3)$

24. $\begin{cases} x - y < 2 \\ 3x + 4y \geq 12 \end{cases}$
  a. $(0, 0)$ **b.** $\left(\dfrac{16}{5}, \dfrac{3}{5}\right)$ **c.** $(4, 0)$ **d.** $\left(\dfrac{1}{2}, \dfrac{1}{2}\right)$

25. $\begin{cases} y < x + 2 \\ x + y \leq 4 \end{cases}$
  a. $(0, 0)$ **b.** $(1, 0)$ **c.** $(0, 1)$ **d.** $(1, 1)$

26. $\begin{cases} y \leq 2 - x \\ x + y \leq 1 \end{cases}$
  a. $(0, 0)$ **b.** $(0, 1)$ **c.** $(1, 2)$ **d.** $(-1, -1)$

27. $\begin{cases} 3x - 4y \leq 12 \\ x + y \leq 4 \\ 5x - 2y \geq 6 \end{cases}$
  a. $(0, 0)$ **b.** $(2, 0)$ **c.** $(3, 1)$ **d.** $(2, 2)$

28. $\begin{cases} 2 + y \leq 3 \\ x - 2y \leq 3 \\ 5x + 2y \geq 3 \end{cases}$
  a. $(0, 0)$ **b.** $(1, 0)$ **c.** $(1, 2)$ **d.** $(2, 1)$

In Exercises 29–44, graph the solution set of each system of linear inequalities and label the vertices (if any) of the solution set.

29. $\begin{cases} x + y \le 1 \\ x - y \le -1 \end{cases}$ **30.** $\begin{cases} x + y \ge 1 \\ y - x \ge 1 \end{cases}$

31. $\begin{cases} 3x + 5y \le 15 \\ 2x + 2y \le 8 \end{cases}$ **32.** $\begin{cases} -2x + y \le 2 \\ 3x + 2y \le 4 \end{cases}$

33. $\begin{cases} 2x + 3y \le 6 \\ 4x + 6y \ge 24 \end{cases}$ **34.** $\begin{cases} x + 2y \ge 6 \\ 2x + 4y \le 4 \end{cases}$

35. $\begin{cases} 3x + y \le 8 \\ 4x + 2y \ge 4 \end{cases}$ **36.** $\begin{cases} x + 2y \le 12 \\ 2x + 4y \ge 8 \end{cases}$

37. $\begin{cases} x \ge 0 \\ y \ge 0 \\ x + y \le 1 \end{cases}$ **38.** $\begin{cases} x \ge 0 \\ y \ge 0 \\ x + y > 2 \end{cases}$

39. $\begin{cases} x \ge 0 \\ y \ge 0 \\ 2x + 3y \ge 6 \end{cases}$ **40.** $\begin{cases} x \ge 0 \\ y \ge 0 \\ 3x + 4y \le 12 \end{cases}$

41. $\begin{cases} x + y \le 1 \\ x - y \ge 1 \\ 2x + y \le 1 \end{cases}$ **42.** $\begin{cases} x + y \le 1 \\ x - y \ge 1 \\ 2x + y \ge 1 \end{cases}$

43. $\begin{cases} x + y \le 1 \\ x - y < 2 \\ -x + y \le 3 \\ x + y \ge -4 \end{cases}$ **44.** $\begin{cases} x + y \le -1 \\ 2x - y > -2 \\ -2x + y \ge -3 \\ 2x + 2y \ge -4 \end{cases}$

In Exercises 45–50, use the following figure to indicate the regions (A–G) that correspond to the graph of the given system of inequalities.

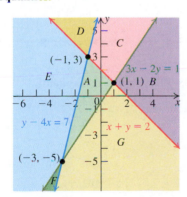

45. $\begin{cases} x + y \le 2 \\ 3x - 2y \le 1 \\ y - 4x \le 7 \end{cases}$ **46.** $\begin{cases} x + y \ge 2 \\ 3x - 2y \le 1 \\ y - 4x \le 7 \end{cases}$

47. $\begin{cases} x + y \ge 2 \\ 3x - 2y \ge 1 \\ y - 4x \le 7 \end{cases}$ **48.** $\begin{cases} x + y \ge 2 \\ 3x - 2y \le 1 \\ y - 4x \ge 7 \end{cases}$

49. $\begin{cases} x + y \le 2 \\ 3x - 2y \le 1 \\ y - 4x \ge 7 \end{cases}$ **50.** $\begin{cases} x + y \le 2 \\ 3x - 2y \ge 1 \\ y - 4x \ge 7 \end{cases}$

In Exercises 51–64, use the procedure for graphing a nonlinear inequality in two variables to graph each inequality.

51. $y \le x^2 + 2$ **52.** $y < x^2 - 3$

53. $y > x^2 - 1$ **54.** $y \ge x^2 + 1$

55. $y \le (x - 1)^2 + 3$ **56.** $y > -(x + 1)^2 + 4$

57. $x^2 + y^2 > 4$ **58.** $x^2 + y^2 \le 9$

59. $(x - 1)^2 + (y - 2)^2 \le 9$

60. $(x + 2)^2 + (y - 1)^2 > 4$

61. $y < 2^x$ **62.** $y \ge e^x$

63. $y < \log x$ **64.** $y \ge \ln x$

For Exercises 65–70, use the following figure to indicate the region or regions that correspond to the graph of each system of inequalities.

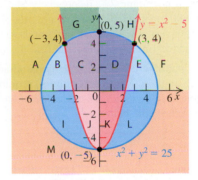

65. $\begin{cases} y \ge x^2 - 5 \\ x^2 + y^2 \le 25 \\ x \ge 0 \end{cases}$ **66.** $\begin{cases} y \ge x^2 - 5 \\ x^2 + y^2 \ge 25 \\ x \ge 0 \end{cases}$

67. $\begin{cases} y \le x^2 - 5 \\ x^2 + y^2 \le 25 \\ x \ge 0 \end{cases}$ **68.** $\begin{cases} y \ge x^2 - 5 \\ x^2 + y^2 \ge 25 \\ y \ge 0 \end{cases}$

69. $\begin{cases} y \le x^2 - 5 \\ x^2 + y^2 \le 25 \\ y \ge 0 \end{cases}$ **70.** $\begin{cases} y \ge x^2 - 5 \\ x^2 + y^2 \le 25 \\ y \ge 0 \end{cases}$

In Exercises 71–78, graph the region determined by each system of inequalities and label all points of intersection.

71. $\begin{cases} x + y \le 6 \\ y \ge x^2 \end{cases}$ **72.** $\begin{cases} x + y \ge 6 \\ y \ge x^2 \end{cases}$

73. $\begin{cases} y \ge 2x + 1 \\ y \le -x^2 + 2 \end{cases}$ **74.** $\begin{cases} y \le 2x + 1 \\ y \le -x^2 + 2 \end{cases}$

75. $\begin{cases} y \ge x \\ x^2 + y^2 \le 1 \end{cases}$ **76.** $\begin{cases} y \le x \\ x^2 + y^2 \le 1 \end{cases}$

77. $\begin{cases} x^2 + y^2 \le 25 \\ x^2 + y \le 5 \end{cases}$ **78.** $\begin{cases} x^2 + y^2 \le 25 \\ x^2 + y \ge 5 \end{cases}$

B EXERCISES Applying the Concepts

79. **Grocer's shelf space.** Your local grocery store has shelf space for, at most, 40 cases of Coke and Sprite. The grocer wants at least ten cases of Coke and five cases of Sprite. Graph the possible stock arrangements and find the vertices.

80. **DVD players.** An electronic store sells two types, A and B, of DVD players. Each player of type A and type B costs $60 and $80, respectively. The store wants at least 20 players in stock and does not want its investment in the players to exceed $6000. Graph the possible inventory of the players and find the vertices.

81. **Home builder.** A builder has 43 units of material and 25 units of labor available to build one- and/or two-story houses within one year. Assume that a two-story house requires seven units of material and four units of labor and a one-story house requires five units of material and three units of labor. Graph the possible inventory of houses and find the vertices.

82. **Financial planning.** Sabrina has $100,000 to invest in stocks and bonds. The annual rate of return for stocks and bonds is 4% and 7%, respectively. She wants at least $5600 income from her investments. She also wants to invest at least $60,000 in bonds. Graph her possible investment portfolio and find the vertices.

83. **Engine manufacturing.** A company manufactures 3 hp and 5 hp engines. The time (in hours) required to assemble, test, and package each type of engine is shown in the table below.

	Process		
Engine	**Assemble**	**Test**	**Package**
3 hp	3	2	0.5
5 hp	4.5	1	0.75

The maximum times (in hours) available for assembling, testing, and packaging are 360 hours, 200 hours, and 60 hours, respectively. Graph the possible processing activities and find the vertices.

84. **Pet food.** Vege Pet Food manufactures two types of dog food: Vegies and Yummies. Each bag of Vegies contains 3 pounds of vegetables and 2 pounds of cereal; each bag of Yummies contains 1.5 pounds of vegetables and 3.5 pounds of cereal. The total pounds of vegetables and cereal available per week are 20,000 pounds and 27,000 pounds, respectively. Graph the possible number of bags of each type of food and find the vertices.

C EXERCISES Beyond the Basics

In Exercises 85–92, write a system of linear inequalities that has the given graph.

85.

86.

87.

88.

89.

90.

91.

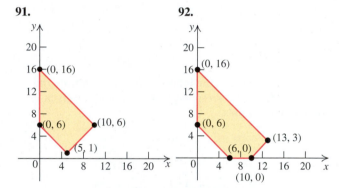

92.

In Exercises 93–102, graph each inequality.

93. $xy \geq 1$ 94. $xy < 3$

95. $y < 2^x$ 96. $y \leq 2^{x-1}$

97. $y \geq 3^{x-2}$ 98. $y > -3^x$

99. $y \geq \ln x$ 100. $y < -\ln x$

101. $y > \ln(x - 1)$ 102. $y \leq -\ln(x + 1)$

In Exercises 103 and 104, graph the solution set of each inequality.

103. $1 \leq x^2 + y^2 \leq 9$ 104. $x^2 \leq y \leq 4x^2$

Critical Thinking

In Exercises 105–106, graph the solution set of each inequality.

105. $|x| + |y| < 1$ 106. $|x + 2y| < 4$

Linear Programming

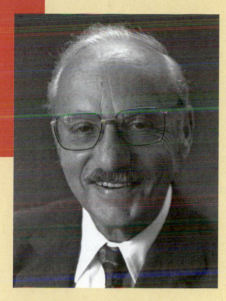

Before Starting this Section, Review

1. Intersection of sets (Section P.1, page 6)
2. Graphing a line (Section 2.3, page 196)
3. Solving systems of linear inequalities (Sections 5.4, page 516)

Objectives

1 State linear programming problems.

2 Solve linear programming problems.

3 Use linear programming in applications.

THE SUCCESS OF LINEAR PROGRAMMING

From 1941 to 1946, George Dantzig was head of the Combat Analysis Branch, U.S. Army Air Corps Headquarters Statistical Control. During this period, he became an expert on planning methods involving troop supply problems that arose during World War II. With a war on several fronts, the size of the problem of coordinating supplies was daunting. This problem was known as *programming*, a military term which at that time referred to plans or schedules for training, logistical supply, or the deployment of forces. Dantzig mechanized the planning process by introducing *linear programming*, where *programming* has the military meaning.

In recent years, linear programming has been applied to problems in almost all areas of human life. A quick check in the library at a university may uncover entire books on linear programming applied to business, agriculture, city planning, and many other areas. Even feedlot operators use linear programming to decide how much and what kinds of food they should give their livestock each day. In Example 3 and in the exercises, we consider several linear programming problems involving two variables. ■

George Dantzig 1914–2005

George Dantzig received numerous honors and awards for his work. Dantzig attributed his success in mathematics to his father, who encouraged him to develop his analytic power by making him solve thousands of geometry problems while George was still in high school.

During his first year of graduate studies, Dantzig arrived late in one of Professor Neyman's classes at the University of California at Berkeley. On the blackboard were two problems that Dantzig assumed had been assigned for homework. He submitted his homework a little late. He was surprised to learn that these were the two famous unsolved problems in statistics. The solutions of these problems became part of Dantzig's dissertation for his PhD degree from Berkeley.

1 State linear programming problems.

Linear Programming

Recall that if a function f with domain $[a, b]$ has a largest and a smallest value, the largest value is called the *maximum value* and the smallest value is called the *minimum value*. The process of finding the maximum or minimum value of a quantity is called **optimization**. For example, for the function

$$f(x) = x^2, \quad -2 \le x \le 3,$$

the maximum value of f occurs at $x = 3$ and is given by $f(3) = 3^2 = 9$, while the minimum value occurs at $x = 0$ and is given by $f(0) = 0^2 = 0$. See Figure 5.18. This is an example of an optimization problem involving one variable.

FIGURE 5.18 **Optimization**

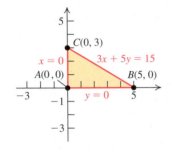

FIGURE 5.19 **Feasible solutions**

Now we will investigate optimization problems involving two variables, say, x and y. Assume the following conditions:

1. The quantity f to be maximized or minimized can be written as a linear expression in x and y; that is,

$$f = ax + by, \text{ where } a \neq 0, b \neq 0 \text{ are constants.}$$

2. The domain of the variables x and y is restricted to a region S that is determined by (is a solution set of) a system of linear inequalities.

A problem that satisfies conditions (1) and (2) is called a **linear programming problem**.

The inequalities that determine the region S are called **constraints**, the region S is called the **set of feasible solutions**, and $f = ax + by$ is called the **objective function**. A point in S at which f reaches its maximum (or minimum) value, together with the value of f at that point, is called an **optimal solution**.

Consider the following linear programming problem:

Find values of x and y that will make $f = 2x + 3y$ as large as possible (in other words, that will result in a maximum value for f), where x and y are restricted by the following constraints:

$$\begin{cases} 3x + 5y \leq 15 \\ \quad\quad x \geq 0 \\ \quad\quad y \geq 0 \end{cases}$$

To solve the problem, first sketch the graph of the permissible values of x and y; that is, find the set of feasible solutions determined by the constraints of the problem. Using the method explained in Section 5.4, you find that the set of feasible solutions is the shaded region shown in Figure 5.19.

Now you have to find the values of $f = 2x + 3y$ at each of the points in the shaded region of Figure 5.19 and then see where f takes on the maximum value. The shaded region, however, has infinitely many points. Because you cannot evaluate f at each point, you need another method to solve the problem.

Fortunately, it can be shown that if there is an optimal solution, it must occur at one of the vertices (corner points) of the feasible solution set. This means that when there is an optimal solution, *we can find the maximum (or minimum) value of f by computing the value of f at each vertex and then picking the largest (or smallest) value that results.*

SOLUTION OF A LINEAR PROGRAMMING PROBLEM

1. If a linear programming problem has a solution, it must occur at one of the vertices of the feasible solution set.

2. A linear programming problem may have many solutions, but at least one of them occurs at a vertex of the feasible solution set.

3. In any case, the value of the objective function is unique.

---

2 Solve linear programming problems.

Solving Linear Programming Problems

We next describe a procedure for solving linear programming problems that have a solution.

FINDING THE SOLUTION: A PROCEDURE

EXAMPLE 1   **Solving a Linear Programming Problem**

OBJECTIVE	EXAMPLE
Solve a linear programming problem.	*Maximize $f = 2x + 3y$ subject to the constraints given below.*

Step 1   Write an expression for the quantity to be maximized or minimized. This expression is the objective function.

Maximize the objective function

$$f = 2x + 3y$$

Step 2   Write all constraints as linear inequalities.

subject to the constraints

$$\begin{cases} 3x + 5y \le 15 \\ x \ge 0 \\ y \ge 0 \end{cases}$$

Step 3   Graph the solution set of the constraint inequalities. This set is the set of feasible solutions.

See Figure 5.19 for the solution set of the constraints.

Step 4   Find all vertices of the solution set in Step 3 by solving all pairs of equations corresponding to the constraint inequalities.

Find

$A(0,0)$ by solving $\begin{cases} x = 0 \\ y = 0 \end{cases}$

$B(5,0)$ by solving $\begin{cases} y = 0 \\ 3x + 5y = 15 \end{cases}$

and

$C(0,3)$ by solving $\begin{cases} x = 0 \\ 3x + 5y = 15 \end{cases}$

Step 5   Find the values of the objective function at each of the vertices in Step 4.

At $A(0,0), f = 2(0) + 3(0) = 0$.
At $B(5,0), f = 2(5) + 3(0) = 10$.
At $C(0,3), f = 2(0) + 3(3) = 9$.

Step 6   The largest of the values (if any) in Step 5 is the maximum value of the objective function, and the smallest of the values (if any) in Step 5 is the minimum value of the objective function.

At $B(5,0)$, the objective function f has a maximum value of 10. Although you are not asked for the minimum value, it is 0 at $A(0,0)$.

Practice Problem 1   Maximize $f = 4x + 5y$ subject to the constraints:

$$\begin{cases} 5x + 7y \le 35 \\ x \ge 0 \\ y \ge 0 \end{cases}$$

We stated in the box on page 524 that the optimal value of the objective function f is unique, but the solution (the point (x, y) at which f attains this optimal value) may not be unique. We next solve a linear programming problem that has infinitely many solutions.

EXAMPLE 2   **Solving a Linear Programming Problem Having a Nonunique Solution**

Minimize $f = 60x + 40y$ subject to the constraints $x + y \leq 8$, $x + 2y \geq 4$, $3x + 2y \geq 6$, $x \geq 0$, $y \geq 0$.

SOLUTION

Step 1   The objective function is $f = 60x + 40y$.

Step 2   We want to find the minimum value of f that can occur at the points (x, y) that satisfy the constraints

$$\begin{cases} x + y \leq 8 \\ x + 2y \geq 4 \\ 3x + 2y \geq 6 \\ x \geq 0 \\ y \geq 0 \end{cases}$$

Step 3   The graph of the solution of the inequalities in Step 2 is shaded in Figure 5.20.

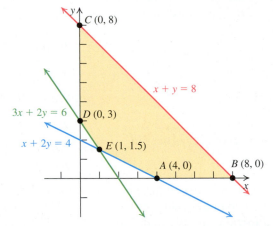

FIGURE 5.20 **Feasible solutions**

Step 4   Solving the pairs of equations corresponding to the constraint inequalities, we get the coordinates of the vertices $A = (4, 0)$, $B = (8, 0)$, $C = (0, 8)$, $D = (0, 3)$, and $E = (1, 1.5)$.

Step 5   Table 5.2 lists the vertices and the corresponding values of the objective function.

TABLE 5.2

Vertex (x, y)	Value of the Objective Function $f = 60x + 40y$
$A = (4, 0)$	$f = (60)(4) + (40)(0) = 240$
$B = (8, 0)$	$f = (60)(8) + (40)(0) = 480$
$C = (0, 8)$	$f = (60)(0) + (40)(8) = 320$
$D = (0, 3)$	$f = (60)(0) + (40)(3) = \boxed{120}$
$E = (1, 1.5)$	$f = (60)(1) + (40)(1.5) = \boxed{120}$

Step 6   We observe from Table 5.2 that the minimum value of f is 120 and that it occurs at two vertices, D and E. In fact, the unique minimum value occurs at every point on the line segment DE. So the linear programming problem has infinitely many solutions, each of which satisfies $3x + 2y = 6$.   ■ ■ ■

Practice Problem 2   Minimize $f = 60x + 40y$ subject to the constraints $2x + 3y \leq 30, 3x + 2y \leq 24, x + y \geq 3, x \geq 0, y \geq 0$.   ■

3 Use linear programming in applications.

Applications

The next example shows how linear programming is used to minimize the number of calories in a diet.

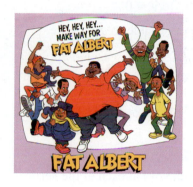

EXAMPLE 3   **Nutrition; Minimizing Calories**

Fat Albert wants to go on a crash diet and needs your help in designing a lunch menu. The menu is to include two items: soup and salad. The vitamin units (milligrams) and calorie counts in each ounce of soup and salad are given in Table 5.3.

TABLE 5.3

Item	Vitamin A	Vitamin C	Calories
Soup	1	3	50
Salad	1	2	40

The menu must provide at least

10 units of vitamin A.

24 units of vitamin C.

Find the number of ounces of each item in the menu needed to provide the required vitamins with the fewest number of calories.

SOLUTION

a. State the problem mathematically.

Step 1   **Write the objective function.** Let the lunch menu contain x ounces of soup and y ounces of salad. You need to minimize the total number of calories in the menu. Because the soup has 50 calories per ounce (see Table 5.3), x ounces of soup has $50x$ calories. Likewise, the salad has $40y$ calories. The number f of calories in the two items is therefore given by the objective function

$$f = 50x + 40y.$$

Step 2   **Write the constraints.** Because each ounce of soup provides one unit of vitamin A and each ounce of salad provides one unit of vitamin A, the two items together provide $1 \cdot x + 1 \cdot y = x + y$ units of vitamin A. The lunch must have at least ten units of vitamin A; this means that

$$x + y \geq 10.$$

Similarly, the two items provide $3x + 2y$ units of vitamin C. The menu must provide at least 24 units of vitamin C, so

$$3x + 2y \geq 24.$$

The fact that x and y cannot be negative means that $x \geq 0$ and $y \geq 0$. You now have four constraints.

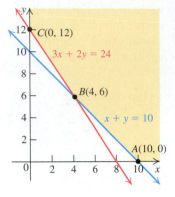

FIGURE 5.21 Feasible solutions

Summarize your information. Find x and y such that the value of

$$f = 50x + 40y \qquad \text{Objective function from Step 1}$$

is a *minimum*, with the restrictions

$$\begin{cases} x + y \geq 10 \\ 3x + 2y \geq 24 \\ x \geq 0 \\ y \geq 0 \end{cases} \qquad \text{Constraints from Step 2}$$

b. Solve the linear programming problem.

Step 3   **Graph the set of feasible solutions.** The set of feasible solutions is shaded in Figure 5.21. This set is bounded by the lines whose equations are

$$x + y = 10, 3x + 2y = 24, x = 0, \text{ and } y = 0.$$

Step 4   **Find the vertices.** The vertices of the set of feasible solutions are

$$A(10, 0), \text{found by solving } \begin{cases} y = 0 \\ x + y = 10 \end{cases}$$

$$B(4, 6), \text{found by solving } \begin{cases} x + y = 10 \\ 3x + 2y = 24 \end{cases}$$

$$C(0, 12), \text{found by solving } \begin{cases} x = 0 \\ 3x + 2y = 24 \end{cases}$$

Step 5   **Find the value of f at the vertices.** The value of f at each of the vertices is given in Table 5.4.

TABLE 5.4

Vertex (x, y)	Value of $f = 50x + 40y$
$(10, 0)$	$50(10) + 40(0) = 500$
$(4, 6)$	$50(4) + 40(6) = 440$
$(0, 12)$	$50(0) + 40(12) = 480$

Step 6   **Find the maximum or minimum value of f.** From Table 5.4, the smallest value of f is 440, which occurs when $x = 4$ and $y = 6$.

c. State the conclusion.

The lunch menu for Fat Albert should contain 4 ounces of soup and 6 ounces of salad. His intake of 440 calories will be as small as possible under the given constraints.   ■ ■ ■

Practice Problem 3   How does the answer to Example 3 change if 1 ounce of soup contains 30 calories and 1 ounce of salad contains 60 calories?   ■

Notice in Example 3 that the value of the objective function $f = 50x + 40y$ can be made as large as you want by choosing x or y (or both) to be very large. So f has a minimum value of 440 but no maximum value.

SECTION 5.5 ■ Exercises

A EXERCISES Basic Skills and Concepts

1. The process of finding the maximum or minimum value of a quantity is called _____.

2. In a linear programming problem, the linear expression f that is to be maximized or minimized is called a(n) _____.

3. The inequalities that determine the region S in a linear programming problem are called _____, and S is called the set of _____ solutions.

4. In linear programming, the expression for the objective function is _____, and all of the constraint inequalities are _____.

5. *True or False* In a linear programming problem, the optimal value of the objective function is unique.

6. *True or False* In a linear programming problem, the optimal value of the objective function occurs at a unique point.

In Exercises 7–12, find the maximum and the minimum value of each objective function f. The graph of the set of feasible solutions is as follows:

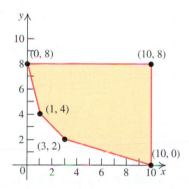

7. $f = x + y$
8. $f = 2x + y$
9. $f = x + 2y$
10. $f = 2x + 5y$
11. $f = 5x + 2y$
12. $f = 8x + 5y$

In Exercises 13–24, you are given an objective function f and a set of constraints. Find the set of feasible solutions determined by the given constraints and then maximize or minimize the objective function as directed.

13. Maximize $f = 9x + 13y$, subject to the constraints $x \geq 0, y \geq 0, 5x + 8y \leq 40, 3x + y \leq 12$.

14. Maximize $f = 7x + 6y$, subject to the constraints $x \geq 0, y \geq 0, 2x + 3y \leq 13, x + y \leq 5$.

15. Maximize $f = 5x + 7y$, subject to the constraints $x \geq 0, y \geq 0, x + y \leq 70, x + 2y \leq 100, 2x + y \leq 120$.

16. Maximize $f = 2x + y$, subject to the constraints $x \geq 0, y \geq 0, x + y \geq 5, 2x + 3y \leq 21, 4x + 3y \leq 24$.

17. Minimize $f = x + 4y$, subject to the constraints $x \geq 0, y \geq 0, x + 3y \geq 3, 2x + y \geq 2$.

18. Minimize $f = 5x + 2y$, subject to the constraints $x \geq 0, y \geq 0, 5x + y \geq 10, x + y \geq 6$.

19. Minimize $f = 13x + 15y$, subject to the constraints $x \geq 0, y \geq 0, 3x + 4y \geq 360, 2x + y \geq 100$.

20. Minimize $f = 40x + 37y$, subject to the constraints $x \geq 0, y \geq 0, 10x + 3y \geq 180, 2x + 3y \geq 60$.

21. Maximize $f = 15x + 7y$, subject to the constraints $3x + 8y \leq 24, 2x + 2y \leq 8, 5x + 3y \leq 18, x \geq 0, y \geq 0$.

22. Maximize $f = 17x + 10y$, subject to the constraints $2x + 3y \geq 12, 5y - 6x \leq 6, x \leq 4, x \geq 0, y \geq 2$.

23. Minimize $f = 8x + 16y$, subject to the constraints $2x + y \geq 40, x + 2y \geq 50, x + y \geq 35, x \geq 0, y \geq 0$.

24. Minimize $f = 12x + 12y$, subject to the constraints $2x + 3y \geq 4, 3x + y \leq 6, x \geq 0, y \geq 0$.

B EXERCISES Applying the Concepts

25. **Agriculture: crop planning.** A farmer owns 240 acres of cropland. He can use his land to produce corn and soybeans. A total of 320 labor hours is available to him during the production period of both of these commodities. Each acre for corn production requires two hours of labor, and each acre for soybean production requires one hour of labor. Assuming that the profit per acre in corn production is $50 and that in soybean production is $40, find the number of acres that he should allocate to the production of corn and soybeans so as to maximize his profit.

26. Repeat Exercise 25 assuming that the profit per acre in corn production is $30 and that in soybean production is $40.

27. **Manufacturing: production scheduling.** Two machines (I and II) each produce two grades of plywood: Grade A and Grade B. In one hour of operation, machine I produces 20 units of Grade A and 10 units of Grade B plywood. Also in one hour of operation, machine II produces 30 units of Grade A and 40 units of Grade B plywood. The machines are required to meet a production schedule of at least 1400 units of Grade A and 1200 units of Grade B plywood. Assuming that the cost of operating machine I is $50 per hour and the cost of operating machine II is $80 per hour, find the number of hours each machine should be operated so as to minimize the cost.

28. Repeat Exercise 27 assuming that the cost of operating machine I is $70 per hour and the cost of operating machine II is $90 per hour.

29. **Advertising.** The Sundial Cheese Company wants to allocate its $6000 advertising budget to two local media—television and newspaper—so that the exposure (number of viewers) to its advertisements is maximized.

There are 60,000 viewers exposed to each minute of television time and 20,000 readers exposed to each page of newspaper advertising. Each minute of television time costs $1000, and each page of newspaper advertising costs $500. How should the company allocate its budget if it has to buy at least one minute of television time and at least two pages of newspaper advertising?

30. Advertising. The Fantastic Bread Company allocates a maximum of $24,000 to advertise its product to television and newspaper. Each hour of television costs $4000, and each page of newspaper advertising costs $3000. The exposure from each hour of television time is assumed to be 120,000, and the exposure from each page of newspaper advertising is 80,000. Furthermore, the board of directors requires at least two hours of television time and one page of newspaper advertising. How should the advertising budget be divided to maximize exposure to the advertisements?

31. Agriculture. A Florida citrus company has 480 acres of land for growing oranges and grapefruit. Profits per acre are $40 for oranges and $30 for grapefruit. The total labor hours available during the production are 800. Each acre for oranges uses two hours of labor, and each acre for grapefruit uses one hour of labor. If the fixed costs are $3000, what is the maximum profit?

32. Agriculture. The Florida Juice Company makes two types of fruit punch—Fruity and Tangy—by blending orange juice and apple juice into a mixture. The fruit punch is sold in 5-gallon bottles. A bottle of Fruity earns a profit of $3, and a bottle of Tangy earns a $2 profit. A bottle of Fruity requires 3 gallons of orange juice and 2 gallons of apple juice, while a bottle of Tangy requires 4 gallons of orange juice and 1 gallon of apple juice. If 200 gallons of orange juice and 120 gallons of apple juice are available, find the maximum profit for the company.

33. Manufacturing. The Fantastic Furniture Company manufactures rectangular and circular tables. Each rectangular table requires one hour of assembling and one hour of finishing, and each circular table requires one hour of assembling and two hours of finishing. The company's 20 assemblers and 30 finishers each work 40 hours per week. Each rectangular table brings a profit of $3, and each circular table brings a profit of $4. If all of the tables manufactured are sold, how many of each type should be made to maximize profits?

34. Manufacturing. A company produces a portable global positioning system (GPS) and a portable digital video disc (DVD) player, each of which requires three stages of processing. The length of time for processing each unit is given in the following table.

Stage	GPS (hr/unit)	DVD (hr/unit)	Maximum Process Capacity (hr/day)
I	12	12	840
II	3	6	300
III	8	4	480
Profit per unit	$6	$4	

How many of each product should the company produce per day to maximize profit?

35. Building houses. A contractor has 100 units of concrete, 160 units of wood, and 400 units of glass. He builds terraced houses and cottages. A terraced house requires one unit of concrete, two units of wood, and two units of glass; a cottage requires one unit of concrete, one unit of wood, and five units of glass. A terraced house sells for $40,000, and a cottage sells for $45,000. How many houses of each type should the contractor build to obtain the maximum amount of money?

36. Hospitality. Mrs. Adams owns a motel consisting of 300 single rooms and an attached restaurant that serves breakfast. She knows that 30% of the male guests and 50% of the female guests will eat in the restaurant. Suppose she makes a profit of $18.50 per day from every guest who eats in her restaurant and $15 per day from every guest who does not eat in her restaurant. Assuming that Mrs. Adams never has more than 125 female guests and never more than 220 male guests, find the number of male and female guests that would provide the maximum profit.

37. Hospitality. In Exercise 36, find Mrs. Adams's maximum profit during a season in which she always has at least 100 guests assuming that she makes a profit of $15 per day from every guest who eats in her restaurant while she suffers a loss of $2 per day from every guest who does not eat in her restaurant.

38. Transportation. Major Motors, Inc., must produce at least 5000 luxury cars and 12,000 medium-priced cars. It must produce, at most, 30,000 units of compact cars. The company owns one factory in Michigan and one in North Carolina. The Michigan factory produces 20, 40, and 60 units of luxury, medium-priced, and compact cars, respectively, per day, while the North Carolina factory produces 10, 30, and 20 luxury, medium-priced, and compact cars, respectively, per day. If the Michigan factory costs $60,000 per day to operate and the North Carolina factory costs $40,000 per day to operate, find the number of days each factory should operate to minimize the cost and meet the requirements.

39. Nutrition. Elisa Epstein is buying two types of frozen meals: a meal of enchiladas with vegetables and a vegetable loaf meal. She wants to guarantee that she gets a minimum of 60 grams of carbohydrates, 40 grams of proteins, and 35 grams of fats. The enchilada meal contains 5, 3, and 5 grams of carbohydrates, proteins, and fats, respectively, per kilogram, while the vegetable loaf contains 2, 2, and 1 gram of carbohydrates, proteins, and fats, respectively, per kilogram. If the enchilada meal costs $3.50 per kilogram and the vegetable loaf costs $2.25 per kilogram, how many kilograms of each should Elisa buy to minimize the cost and still meet the minimum requirement?

40. Temporary help. The personnel director of the main post office plans to hire extra helpers during the holiday season. Because office space is limited, the number of temporary workers cannot exceed 10. She hires workers sent by Super Temps and by Ready Aid. From her past experience, she knows that a typical worker sent by Super Temps can handle 300 letters and 80 packages per day and that a typical worker from Ready Aid can handle 600 letters and 40 packages per day. The post office expects that the additional daily volume of letters and packages will be at least 3900 and 560, respectively. The daily wages for a typical worker from Super Temps and Ready Aid are $96 and $86, respectively. How many workers from each agency should be hired to keep the daily wages to a minimum?

C EXERCISES Beyond the Basics

41. Maximize $f = 8x + 7y$, subject to the constraints $x + 2y \geq 10, 8y - 7x \leq 40, x + y \leq 20, 4x - y \leq 40, x \geq 0, y \geq 0$.

42. Minimize $f = 3x + 2y$, subject to the constraints $6y - 5x \leq 80, 7x - 5y \leq 80, 5x + 8y \geq 115, 10x - y \leq 60, 0 \leq x \leq 20, y \geq 0$.

43. Minimize $f = 5x + 6y$, subject to the constraints $y \leq x + 5, 7x + 2y \leq 28, 0 \leq x \leq 3, 0 \leq y \leq 6$.

44. Maximize $f = 7x + 3y$, subject to the constraints $x + y \leq 100, x + y \geq 50, 2x + 5y \geq 150, 2x + y \geq 80, x \geq 0, y \geq 0$.

45. Suppose an objective function $f = ax + by$ has the same value M at two different points $P(x_1, y_1)$ and $Q(x_2, y_2)$. Prove that f has the value M at every point of the line segment PQ.

46. Suppose the feasible solution set of a linear programming problem is a quadrilateral. Is it possible that the optimal solution also occurs at an interior point of the quadrilateral?

Critical Thinking

47. Consider this problem: maximize $f = 2x + 3y$, subject to the constraints $y - x \leq 6, x \geq 0, 0 \leq y \leq 8$.
   a. Explain why the region S of feasible solutions is unbounded (its area is infinite).
   b. Show that the problem has no solution.

Partial-Fraction Decomposition

Georg Simon Ohm (1787–1854)

Georg Ohm was born in Erlangen, Germany. His father, Johann, a mechanic and self-educated man, was interested in philosophy and mathematics and gave his children an excellent education in physics, chemistry, mathematics, and philosophy through his own teachings. Ohm's Law, named in Georg Ohm's honor, appeared in his famous pamphlet *Die galvanische Kette, mathematisch bearbeitet* (1827). Ohm's work was not appreciated in Germany, where it was first published. Eventually, it was recognized, first by the British Royal Society, which awarded Ohm the Copley Medal in 1841. In 1849, Ohm became a curator of the Bavarian Academy's science museum and began to lecture at the University of Munich. Finally, in 1852 (two years before his death), Ohm achieved his lifelong ambition of becoming chair of physics at the University of Munich.

Before Starting this Section, Review

1. Division of polynomials (Section 3.3, page 329)
2. Factoring polynomials (Section P.4, page 39)
3. Irreducible polynomials (Section P.4, page 40)
4. Rational expressions (Section P.5, page 48)

Objectives

1. Become familiar with partial-fraction decomposition.

2. Decompose $\dfrac{P(x)}{Q(x)}$ when $Q(x)$ has only distinct linear factors.

3. Decompose $\dfrac{P(x)}{Q(x)}$ when $Q(x)$ has repeated linear factors.

4. Decompose $\dfrac{P(x)}{Q(x)}$ when $Q(x)$ has distinct irreducible quadratic factors.

5. Decompose $\dfrac{P(x)}{Q(x)}$ when $Q(x)$ has repeated irreducible quadratic factors.

OHM'S LAW

If a voltage is applied across a resistor (such as a wire), a current will flow. Ohm's Law states that in a circuit at constant temperature,

$$I = \frac{V}{R}$$

where V is the voltage (measured in volts), I is the current (measured in amperes), and R is the resistance (measured in ohms). A **parallel circuit** consists of two or more resistors connected as shown in Figure 5.22. In the figure, the current I from the source divides into two currents, I_1 and I_2, with I_1 going through one resistor and I_2 going through the other, so that $I = I_1 + I_2$. Suppose we want to find the total resistance R in the circuit due to the two resistors R_1 and R_2 connected in parallel. Ohm's Law can be used to show that in the parallel circuit in Figure 5.22, $\dfrac{1}{R} = \dfrac{1}{R_1} + \dfrac{1}{R_2}$. This result is called the *parallel property of resistors*. In Example 7, we examine the parallel property of resistors by using partial fractions.

FIGURE 5.22 Parallel resistors

1 Become familiar with partial-fraction decomposition.

Partial Fractions

Recall that to add two rational expressions such as $\dfrac{2}{x+3}$ and $\dfrac{3}{x-1}$ required you to find the least common denominator, rewrite expressions with the same denominator, and then add the corresponding numerators. This procedure gives

$$\frac{2}{x+3} + \frac{3}{x-1} = \frac{2(x-1)}{(x+3)(x-1)} + \frac{3(x+3)}{(x-1)(x+3)} = \frac{5x+7}{(x+3)(x-1)}.$$

In some applications of algebra to more advanced mathematics, you need a *reverse* procedure for *splitting* a fraction such as $\dfrac{5x+7}{(x-1)(x+3)}$ into the simpler fractions to obtain

$$\frac{5x+7}{(x+3)(x-1)} = \frac{2}{x+3} + \frac{3}{x-1}.$$

Each of the two fractions on the right is called a **partial fraction**. Their sum is called the **partial-fraction decomposition** of the rational expression on the left.

A rational expression $\dfrac{P(x)}{Q(x)}$ is called **improper** if the degree $P(x) \geq$ degree $Q(x)$ and is called **proper** if the degree $P(x) <$ degree $Q(x)$. Examples of improper rational expressions are

$$\frac{x^3}{x^2-1}, \quad \frac{(x+2)(x-1)}{(x-2)(x+3)}, \quad \text{and} \quad \frac{x^2+x+1}{2x+3}.$$

Using long division, we can express an improper rational expression $\dfrac{P(x)}{Q(x)}$ as the sum of a polynomial and a proper rational expression.

$$\underset{\substack{\uparrow \\ \text{Improper} \\ \text{Expression}}}{\frac{P(x)}{Q(x)}} \quad = \quad \underset{\substack{\uparrow \\ \text{Polynomial}}}{S(x)} \quad + \quad \underset{\substack{\uparrow \\ \text{Proper} \\ \text{Expression}}}{\frac{R(x)}{Q(x)}}$$

RECALL

Any polynomial $Q(x)$ with real coefficients can be factored so that each factor is linear or an irreducible quadratic.

We therefore restrict our discussion of the decomposition of $\dfrac{P(x)}{Q(x)}$ into partial fractions to cases involving proper rational expressions. We also assume that $P(x)$ and $Q(x)$ have no common factor.

The problem of decomposing a proper fraction $\dfrac{P(x)}{Q(x)}$ into partial fractions depends on the type of factors in the denominator $Q(x)$. In this section, we consider each of the four possible cases for $Q(x)$.

2 Decompose $\dfrac{P(x)}{Q(x)}$ when $Q(x)$ has only distinct linear factors.

$Q(x)$ Has Only Distinct Linear Factors

**CASE 1: THE DENOMINATOR IS THE PRODUCT
OF DISTINCT (NONREPEATED) LINEAR FACTORS**

Suppose $Q(x)$ can be factored as

$$Q(x) = (x - a_1)(x - a_2) \cdots (x - a_n)$$

with no factor repeated. The partial-fraction decomposition of $\dfrac{P(x)}{Q(x)}$ is of the form

$$\frac{P(x)}{Q(x)} = \frac{A_1}{x - a_1} + \frac{A_2}{x - a_2} + \cdots + \frac{A_n}{x - a_n},$$

where $A_1, A_2, \ldots, A_n$ are constants to be determined.

We can find the constants $A_1, A_2, \ldots, A_n$ by using the following procedure. Note that if the number of constants is small, we use the letters $A, B, C, \ldots$ instead of $A_1, A_2, A_3, \ldots$.

FINDING THE SOLUTION: A PROCEDURE

EXAMPLE 1 **Partial-Fraction Decomposition**

OBJECTIVE

Find the partial-fraction decomposition of a rational expression.

Step 1 Write the form of the partial-fraction decomposition with the unknown constants $A, B, C, \ldots$ in the numerators of the decomposition.

Step 2 Multiply both sides of the equation in Step 1 by the original denominator. Use the distributive property and eliminate common factors. Simplify.

Step 3 Write both sides of the equation in Step 2 in descending powers of x and equate the coefficients of like powers of x.

Step 4 Solve the linear system resulting from Step 3 for the constants $A, B, C, \ldots$.

Step 5 Substitute the values you found for $A, B, C, \ldots$ into the equation in Step 1 and write the partial-fraction decomposition.

EXAMPLE

Find the partial fraction decomposition of $\dfrac{3x + 26}{(x - 3)(x + 4)}$.

$$\frac{3x + 26}{(x - 3)(x + 4)} = \frac{A}{x - 3} + \frac{B}{x + 4}$$

Multiply both sides by $(x - 3)(x + 4)$ and simplify to obtain

$$3x + 26 = (x + 4)A + (x - 3)B$$

$$3x + 26 = (A + B)x + (4A - 3B)$$
$$\begin{cases} 3 = A + B & \text{Equate coefficients of } x. \\ 26 = 4A - 3B & \text{Equate constant coefficients.} \end{cases}$$

Solving the system of equations in Step 3, we obtain $A = 5$ and $B = -2$.

$$\frac{3x + 26}{(x - 3)(x + 4)} = \frac{5}{x - 3} + \frac{-2}{x + 4} \quad \text{Replace } A \text{ with 5 and } B \text{ with } -2 \text{ in Step 1.}$$
$$\frac{3x + 26}{(x - 3)(x + 4)} = \frac{5}{x - 3} - \frac{2}{x + 4} \quad \text{Rewrite}$$

■ ■ ■

Practice Problem 1 Find the partial fraction decomposition of $\dfrac{2x - 7}{(x + 1)(x - 2)}$. ■

EXAMPLE 2 **Finding the Partial-Fraction Decomposition When the Denominator Has Only Distinct Linear Factors**

Find the partial-fraction decomposition of the expression:

$$\frac{10x - 4}{x^3 - 4x}$$

SOLUTION

First, factor the denominator.

$$
\begin{aligned}
x^3 - 4x &= x(x^2 - 4) && \text{Distributive property} \\
&= x(x - 2)(x + 2) && x^2 - 4 = (x - 2)(x + 2)
\end{aligned}
$$

Each of the factors x, $x - 2$, and $x + 2$ becomes a denominator for a partial fraction.

Step 1 The partial-fraction decomposition is given by

$$\frac{10x - 4}{x(x - 2)(x + 2)} = \frac{A}{x} + \frac{B}{x - 2} + \frac{C}{x + 2}$$ Use A, B, and C instead of A_1, A_2, and A_3.

Step 2 Multiply both sides by the common denominator $x(x - 2)(x + 2)$.

Distribute

$$x(x - 2)(x + 2)\left[\frac{10x - 4}{x(x - 2)(x + 2)}\right] = x(x - 2)(x + 2)\left[\frac{A}{x} + \frac{B}{x - 2} + \frac{C}{x + 2}\right]$$

$$
\begin{aligned}
10x - 4 &= A(x - 2)(x + 2) + Bx(x + 2) + Cx(x - 2) && (1) && \text{Distributive property; simplify.} \\
&= A(x^2 - 4) + B(x^2 + 2x) + C(x^2 - 2x) && && \text{Multiply.} \\
&= Ax^2 - 4A + Bx^2 + 2Bx + Cx^2 - 2Cx && && \text{Distributive property} \\
&= (A + B + C)x^2 + (2B - 2C)x - 4A && && \text{Combine like terms.}
\end{aligned}
$$

Step 3 Now use the fact that *two equal polynomials have equal corresponding coefficients*. Writing $10x - 4 = 0x^2 + 10x - 4$, we have

$$0x^2 + 10x - 4 = (A + B + C)x^2 + (2B - 2C)x - 4A.$$

Equating corresponding coefficients leads to the system of equations.

$$
\begin{cases}
A + B + C = 0 && \text{Equate coefficients of } x^2. \\
2B - 2C = 10 && \text{Equate coefficients of } x. \\
-4A = -4 && \text{Equate constant coefficients.}
\end{cases}
$$

Step 4 Solve the system of equations in Step 3 to obtain $A = 1$, $B = 2$, and $C = -3$ (See Exercise 57).

Step 5 The partial-fraction decomposition is

$$\frac{10x - 4}{x^3 - 4x} = \frac{1}{x} + \frac{2}{x - 2} + \frac{-3}{x + 2} = \frac{1}{x} + \frac{2}{x - 2} - \frac{3}{x + 2}.$$

Alternative Solution of Example 2

In Example 2, to find the constants A, B, and C, we *equated coefficients* of like powers of x and solved the resulting system. An alternative (and sometimes quicker) method is to *substitute* well-chosen values for x in equation (1) found in Step 2.

$$10x - 4 = A(x - 2)(x + 2) + Bx(x + 2) + Cx(x - 2) \qquad (1)$$

You can reinforce your partial-fraction connection graphically:

The graph of

$$y_1 = \frac{10x - 4}{x^3 - 4x}$$

appears to be identical to the graph of

$$y_2 = \frac{1}{x} + \frac{2}{x - 2} - \frac{3}{x + 2}.$$

The graphs of y_1 and y_2 are the same.

3 Decompose $\dfrac{P(x)}{Q(x)}$ when $Q(x)$ has repeated linear factors.

Substitute $x = 2$ in equation (1) to cause the terms containing A and C to be 0.

$$10(2) - 4 = A(2 - 2)(2 + 2) + B(2)(2 + 2) + C(2)(2 - 2)$$
$$16 = 8B \qquad \text{Simplify.}$$
$$2 = B \qquad \text{Solve for } B.$$

Now substitute $x = -2$ in equation (1) to get

$$10(-2) - 4 = A(-2 - 2)(-2 + 2) + B(-2)(-2 + 2) + C(-2)(-2 - 2)$$
$$-24 = 8C \qquad \text{Simplify.}$$
$$-3 = C \qquad \text{Solve for } C.$$

Finally, substitute $x = 0$ in equation (1) to get

$$10(0) - 4 = A(0 - 2)(0 + 2) + B(0)(0 + 2) + C(0)(0 - 2)$$
$$-4 = -4A \qquad \text{Simplify.}$$
$$1 = A \qquad \text{Solve for } A.$$

Because we found the same values for A, B, and C, the partial-fraction decomposition will also be the same. ■ ■ ■

Practice Problem 2 Find the partial-fraction decomposition of

$$\frac{3x^2 + 4x + 3}{x^3 - x}.$$
■

$Q(x)$ Has Repeated Linear Factors

CASE 2: THE DENOMINATOR HAS A REPEATED LINEAR FACTOR

Let $(x - a)^m$ be the linear factor $(x - a)$ that is repeated m times in $Q(x)$. Then the portion of the partial-fraction decomposition of $\dfrac{P(x)}{Q(x)}$ that corresponds to the factor $(x - a)^m$ is

$$\frac{A_1}{x - a} + \frac{A_2}{(x - a)^2} + \cdots + \frac{A_m}{(x - a)^m},$$

where $A_1, A_2, \ldots, A_m$ are constants.

EXAMPLE 3 **Finding the Partial-Fraction Decomposition When the Denominator Has Repeated Linear Factors**

Find the partial-fraction decomposition of $\dfrac{x + 4}{(x + 3)(x - 1)^2}$.

SOLUTION

Step 1 The linear factor $(x - 1)$ is repeated twice, and the factor $(x + 3)$ is nonrepeating. So the partial-fraction decomposition has the form

$$\frac{x + 4}{(x + 3)(x - 1)^2} = \frac{A}{x + 3} + \frac{B}{x - 1} + \frac{C}{(x - 1)^2}.$$

Steps 2–4 Multiply both sides of the equation in Step 1 by the original denominator, $(x + 3)(x - 1)^2$; then use the distributive property on the right side and simplify to get

$$x + 4 = A(x - 1)^2 + B(x + 3)(x - 1) + C(x + 3).$$

Substitute $x = 1$ in the last equation to obtain

$$1 + 4 = A(1 - 1)^2 + B(1 + 3)(1 - 1) + C(1 + 3).$$

This simplifies to $5 = 4C$, or $C = \dfrac{5}{4}$.

To find A, we substitute $x = -3$ to get

$$-3 + 4 = A(-3 - 1)^2 + B(-3 + 3)(-3 - 1) + C(-3 + 3).$$

This simplifies to $1 = 16A$, or $A = \dfrac{1}{16}$.

To find B, we replace x with any convenient number, say, 0. We have

$$0 + 4 = A(0 - 1)^2 + B(0 + 3)(0 - 1) + C(0 + 3)$$

$$4 = A - 3B + 3C \qquad\qquad\text{Simplify.}$$

Now replace A with $\dfrac{1}{16}$ and C with $\dfrac{5}{4}$ in the last equation.

$$4 = \frac{1}{16} - 3B + 3\left(\frac{5}{4}\right)$$

$$64 = 1 - 48B + 60 \qquad\qquad\text{Multiply both sides by 16.}$$

$$B = -\frac{1}{16} \qquad\qquad\text{Solve for } B.$$

We have $A = \dfrac{1}{16}$, $B = -\dfrac{1}{16}$, and $C = \dfrac{5}{4}$.

Step 5   Substitute $A = \dfrac{1}{16}$, $B = -\dfrac{1}{16}$, and $C = \dfrac{5}{4}$ in the decomposition in Step 1 to get

$$\frac{x + 4}{(x + 3)(x - 1)^2} = \frac{\frac{1}{16}}{x + 3} + \frac{-\frac{1}{16}}{x - 1} + \frac{\frac{5}{4}}{(x - 1)^2}$$

or

$$\frac{x + 4}{(x + 3)(x - 1)^2} = \frac{1}{16(x + 3)} - \frac{1}{16(x - 1)} + \frac{5}{4(x - 1)^2}. \qquad ■■■$$

Practice Problem 3   Find the partial fraction decomposition of:

$$\frac{x + 5}{x(x - 1)^2} \qquad\qquad ■$$

4 Decompose $\dfrac{P(x)}{Q(x)}$ when $Q(x)$ has distinct irreducible quadratic factors.

$Q(x)$ Has Distinct Irreducible Quadratic Factors

> ### CASE 3: THE DENOMINATOR HAS A DISTINCT (NONREPEATED) IRREDUCIBLE QUADRATIC FACTOR
>
> Suppose $ax^2 + bx + c$ is an irreducible quadratic factor of $Q(x)$. Then the portion of the partial-fraction decomposition of $\dfrac{P(x)}{Q(x)}$ that corresponds to $ax^2 + bx + c$ has the form:
>
> $$\frac{Ax + B}{ax^2 + bx + c}$$

EXAMPLE 4   **Finding the Partial-Fraction Decomposition When the Denominator Has a Distinct Irreducible Quadratic Factor**

Find the partial-fraction decomposition of $\dfrac{3x^2 - 8x + 1}{(x - 4)(x^2 + 1)}$.

SOLUTION

Step 1   The factor $x - 4$ is linear, and $x^2 + 1$ is irreducible; so the partial fraction decomposition has the form:

$$\frac{3x^2 - 8x + 1}{(x - 4)(x^2 + 1)} = \frac{A}{x - 4} + \frac{Bx + C}{x^2 + 1}$$

Step 2   Multiply both sides of the decomposition in Step 1 by the original denominator, $(x - 4)(x^2 + 1)$, and simplify to get

$$3x^2 - 8x + 1 = A(x^2 + 1) + (Bx + C)(x - 4).$$

Substitute $x = 4$ to find $17 = 17A$, or $A = 1$.

Step 3   Collect like terms and write both sides of the equation in Step 2 in descending powers of x.

$$3x^2 - 8x + 1 = (A + B)x^2 + (-4B + C)x + (A - 4C)$$

Equate corresponding coefficients to get

$$\begin{cases} A + B = 3 & (1) & \text{Equate the coefficients of } x^2. \\ -4B + C = -8 & (2) & \text{Equate the coefficients of } x. \\ A - 4C = 1 & (3) & \text{Equate the constant coefficients.} \end{cases}$$

Step 4   Substitute $A = 1$ from Step 2 in equation (1) to get $1 + B = 3$, or $B = 2$. Substitute $A = 1$ in equation (3) to get $1 - 4C = 1$, or $C = 0$.

Step 5   Substitute $A = 1$, $B = 2$, and $C = 0$ into the decomposition in Step 1 to get

$$\frac{3x^2 - 8x + 1}{(x - 4)(x^2 + 1)} = \frac{1}{x - 4} + \frac{2x}{x^2 + 1}.$$

■ ■ ■

Practice Problem 4   Find the partial-fraction decomposition of:

$$\frac{3x^2 + 5x - 2}{x(x^2 + 2)}$$

■

5 Decompose $\dfrac{P(x)}{Q(x)}$ when $Q(x)$ has repeated irreducible quadratic factors.

$Q(x)$ Has Repeated Irreducible Quadratic Factors

CASE 4: THE DENOMINATOR HAS A REPEATED IRREDUCIBLE QUADRATIC FACTOR

Suppose the denominator $Q(x)$ has a factor $(ax^2 + bx + c)^m$, where $m \geq 2$ is an integer and $ax^2 + bx + c$ is irreducible. Then the portion of the partial-fraction decomposition of $\dfrac{P(x)}{Q(x)}$ that corresponds to the factor $(ax^2 + bx + c)^m$ has the form:

$$\frac{A_1x + B_1}{ax^2 + bx + c} + \frac{A_2x + B_2}{(ax^2 + bx + c)^2} + \cdots + \frac{A_mx + B_m}{(ax^2 + bx + c)^m}$$

| EXAMPLE 5 | **Finding the Partial-Fraction Decomposition When the Denominator Has a Repeated Irreducible Quadratic Factor** |

Find the partial-fraction decomposition of $\dfrac{2x^4 - x^3 + 13x^2 - 2x + 13}{(x - 1)(x^2 + 4)^2}$.

SOLUTION

Step 1   The denominator has a nonrepeating linear factor $(x - 1)$ and an irreducible quadratic factor $(x^2 + 4)$ that appears twice. Therefore, the decomposition has the form:

$$\frac{2x^4 - x^3 + 13x^2 - 2x + 13}{(x - 1)(x^2 + 4)^2} = \frac{A}{x - 1} + \frac{Bx + C}{x^2 + 4} + \frac{Dx + E}{(x^2 + 4)^2}$$

Step 2   Multiply both sides of the decomposition in Step 1 by the original denominator $(x - 1)(x^2 + 4)^2$, use the distributive property, and eliminate common factors to get:

$$2x^4 - x^3 + 13x^2 - 2x + 13 =$$
$$A(x^2 + 4)^2 + (Bx + C)(x - 1)(x^2 + 4) + (Dx + E)(x - 1)$$

Substitute $x = 1$ and simplify to obtain $25 = 25A$, or $A = 1$.

Steps 3–4   Multiply the factors on the right side of the equation in Step 2 and collect like terms to obtain

$$2x^4 - x^3 + 13x^2 - 2x + 13 =$$
$$(A + B)x^4 + (-B + C)x^3 + (8A + 4B - C + D)x^2 + (-4B + 4C - D + E)x + (16A - 4C - E).$$

Equate corresponding coefficients to get the system.

$$\begin{cases} A + B & = 2 & (1) & \text{Coefficient of } x^4 \\ -B + C & = -1 & (2) & \text{Coefficient of } x^3 \\ 8A + 4B - C + D & = 13 & (3) & \text{Coefficient of } x^2 \\ -4B + 4C - D + E & = -2 & (4) & \text{Coefficient of } x \\ 16A - 4C - E & = 13 & (5) & \text{Constant coefficient} \end{cases}$$

Now back-substitute $A = 1$ from Step 2 in equation (1) to get $B = 1$. Again, back-substitute $B = 1$ in equation (2) to get $C = 0$.

Back-substitute $A = 1$ and $C = 0$ in equation (5) to obtain $E = 3$. Again, back-substitute $A = 1$, $B = 1$, and $C = 0$ in equation (3) to get $D = 1$. Hence, $A = 1$, $B = 1$, $C = 0$, $D = 1$, and $E = 3$.

Step 5   Substitute $A = 1$, $B = 1$, $C = 0$, $D = 1$, and $E = 3$ in the equation in Step 1 to obtain the partial-fraction decomposition:

$$\frac{2x^4 - x^3 + 13x^2 - 2x + 13}{(x - 1)(x^2 + 4)^2} = \frac{1}{x - 1} + \frac{x}{x^2 + 4} + \frac{x + 3}{(x^2 + 4)^2} \quad ■ ■ ■$$

Practice Problem 5   Find the partial-fraction decomposition of:

$$\frac{x^2 + 3x + 1}{(x^2 + 1)^2} \qquad ■$$

| EXAMPLE 6 | **Learning a Clever Way to Add** |

Express the sum

$$\frac{1}{2 \cdot 3} + \frac{1}{3 \cdot 4} + \frac{1}{4 \cdot 5} + \cdots + \frac{1}{2009 \cdot 2010}$$

as a fraction of whole numbers in lowest terms.

SOLUTION

Each term in the sum is of the form $\dfrac{1}{k(k+1)}$. From the partial-fraction decomposition, we have

$$\frac{1}{k(k+1)} = \frac{1}{k} - \frac{1}{k+1}.$$

Therefore, by using this decomposition for each term, we get

$$\frac{1}{2 \cdot 3} + \frac{1}{3 \cdot 4} + \frac{1}{4 \cdot 5} + \cdots + \frac{1}{2009 \cdot 2010}$$

$$= \left(\frac{1}{2} - \frac{1}{3}\right) + \left(\frac{1}{3} - \frac{1}{4}\right) + \left(\frac{1}{4} - \frac{1}{5}\right) + \cdots + \left(\frac{1}{2009} - \frac{1}{2010}\right).$$

After removing parentheses, we see that most terms cancel in pairs $\Bigg($ such as $-\dfrac{1}{3}$ and $\dfrac{1}{3}\Bigg)$, leaving only the first and last terms. The original sum is therefore referred to as a *telescoping sum* that "collapses" to

$$\frac{1}{2} - \frac{1}{2010} = \frac{1005 - 1}{2010} = \frac{1004}{2010} = \frac{502}{1005}.$$ ■ ■ ■

Practice Problem 6 Express the sum $\dfrac{1}{4 \cdot 5} + \dfrac{1}{5 \cdot 6} + \dfrac{1}{6 \cdot 7} + \cdots + \dfrac{1}{3111 \cdot 3112}$ as a fraction of whole numbers in lowest terms. ■

EXAMPLE 7 **Calculating the Resistance in a Circuit**

The total resistance R due to two resistors connected in parallel is given by:

$$R = \frac{x(x+5)}{3x+10}$$

Use partial-fraction decomposition to write $\dfrac{1}{R}$ in terms of partial fractions. What does each term represent?

SOLUTION

From $R = \dfrac{x(x+5)}{3x+10}$, we get the following:

$$\frac{1}{R} = \frac{3x+10}{x(x+5)} \qquad \text{Take the reciprocal of both sides.}$$

$$\frac{3x+10}{x(x+5)} = \frac{A}{x} + \frac{B}{x+5} \qquad \text{Form of partial-fraction decomposition}$$

$$3x+10 = A(x+5) + Bx \qquad \text{Multiply both sides by } x(x+5) \text{ and simplify.}$$

Substitute $x = 0$ in the last equation to obtain $5A = 10$, or $A = 2$. Now substitute $x = -5$ in the same equation to get $-5 = -5B$, or $B = 1$. Thus, we have

$$\frac{1}{R} = \frac{3x+10}{x(x+5)} = \frac{2}{x} + \frac{1}{x+5}$$

$$\frac{1}{R} = \frac{1}{\dfrac{x}{2}} + \frac{1}{x+5} \qquad \text{Rewrite the right side.}$$

RECALL

The total resistance R in a circuit due to two resistors R_1 and R_2 connected in parallel satisfies the equation:

$$\frac{1}{R} = \frac{1}{R_1} + \frac{1}{R_2}$$

The last equation says that if two resistors with resistances $R_1 = \dfrac{x}{2}$ and $R_2 = x + 5$ are connected in parallel, they will produce a total resistance R, where

$$R = \frac{x(x + 5)}{3x + 10}.$$

■ ■ ■

Practice Problem 7   Repeat Example 7 assuming that $R = \dfrac{(x + 1)(x + 2)}{4x + 7}$.   ■

SECTION 5.6 ■ Exercises

A EXERCISES Basic Skills and Concepts

1. In a rational expression, if the degree of the numerator, $P(x)$, is less than the degree of the denominator, $Q(x)$, then the expression is _____.

2. The numerators in the partial-fraction decomposition of a rational expression are constants if the denominator, $Q(x)$, can be factored into _____.

3. In a rational expression, if $(x - 8)$ is a linear factor that is repeated three times in the denominator, then the portion of the expression's partial-fraction decomposition that corresponds to $(x - 8)^3$ has _____ terms.

4. *True or False* The form of the partial-fraction decomposition of a rational expression depends only on its denominator.

5. *True or False* Both $(x - 1)^2$ and $(x + 2)^2$ are irreducible quadratic factors of the polynomial $Q(x) = (x - 1)^2(x + 2)^2$.

6. *True or False* If the denominator of a rational expression is $Q(x) = x^3 - 9x$, then $x, x - 3$, and $x + 3$ become denominators in its partial-fraction decomposition.

In Exercises 7–16, write the form of the partial-fraction decomposition of each rational expression. You do not need to solve for the constants.

7. $\dfrac{1}{(x - 1)(x + 2)}$

8. $\dfrac{x}{(x + 1)(x - 3)}$

9. $\dfrac{1}{x^2 + 7x + 6}$

10. $\dfrac{3}{x^2 - 6x + 8}$

11. $\dfrac{2}{x^3 - x^2}$

12. $\dfrac{x - 1}{(x + 2)^2(x - 3)}$

13. $\dfrac{x^2 - 3x + 3}{(x + 1)(x^2 - x + 1)}$

14. $\dfrac{2x + 3}{(x - 1)^2(x^2 + x + 1)}$

15. $\dfrac{3x - 4}{(x^2 + 1)^2}$

16. $\dfrac{x - 2}{(2x + 3)^2(x^2 + 3)^2}$

In Exercises 17–48, find the partial-fraction decomposition of each rational expression.

17. $\dfrac{2x + 1}{(x + 1)(x + 2)}$

18. $\dfrac{7}{(x - 2)(x + 5)}$

19. $\dfrac{1}{x^2 + 4x + 3}$

20. $\dfrac{x}{x^2 + 5x + 6}$

21. $\dfrac{2}{x^2 + 2x}$

22. $\dfrac{x + 9}{x^2 - 9}$

23. $\dfrac{x}{(x + 1)(x + 2)(x + 3)}$

24. $\dfrac{x^2}{(x - 1)(x^2 + 5x + 4)}$

25. $\dfrac{x - 1}{(x + 1)^2}$

26. $\dfrac{3x + 2}{x^2(x + 1)}$

27. $\dfrac{2x^2 + x}{(x + 1)^3}$

28. $\dfrac{x}{(x - 1)(x + 1)}$

29. $\dfrac{-x^2 + 3x + 1}{x^3 + 2x^2 + x}$

30. $\dfrac{5x^2 - 8x + 2}{x^3 - 2x^2 + x}$

31. $\dfrac{1}{x^2(x + 1)^2}$

32. $\dfrac{2}{(x - 1)^2(x + 3)^2}$

33. $\dfrac{1}{(x^2 - 1)^2}$

34. $\dfrac{3}{(x^2 - 4)^2}$

35. $\dfrac{x - 1}{(2x - 3)^2}$

36. $\dfrac{6x + 1}{(3x + 1)^2}$

37. $\dfrac{6x + 7}{4x^2 + 12x + 9}$

38. $\dfrac{2x + 3}{9x^2 + 30x + 25}$

39. $\dfrac{x - 3}{x^3 + x^2}$

40. $\dfrac{1}{x^3 + x}$

41. $\dfrac{x^2 + 2x + 4}{x^3 + x^2}$

42. $\dfrac{x^2 + 2x - 1}{(x - 1)(x^2 + 1)}$

43. $\dfrac{x}{x^4 - 1}$

44. $\dfrac{3x^3 - 5x^2 + 12x + 4}{x^4 - 16}$

45. $\dfrac{1}{x(x^2 + 1)^2}$

46. $\dfrac{x^2}{(x^2 + 2)^2}$

47. $\dfrac{2x^2 + 3x}{(x^2 + 1)(x^2 + 2)}$

48. $\dfrac{x^2 - 2x}{(x^2 + 9)(x^2 + x + 7)}$

B EXERCISES Applying the Concepts

In Exercises 49–52, express each sum as a fraction of whole numbers in lowest terms.

49. $\dfrac{1}{1 \cdot 2} + \dfrac{1}{2 \cdot 3} + \dfrac{1}{3 \cdot 4} + \cdots + \dfrac{1}{n(n + 1)}$

50. $\dfrac{2}{1 \cdot 3} + \dfrac{2}{2 \cdot 4} + \dfrac{2}{3 \cdot 5} + \cdots + \dfrac{2}{100 \cdot 102}$

51. $\dfrac{2}{1\cdot 3} + \dfrac{2}{3\cdot 5} + \dfrac{2}{5\cdot 7} + \cdots + \dfrac{2}{(2n-1)(2n+1)}$

52. $\dfrac{2}{1\cdot 2\cdot 3} + \dfrac{2}{2\cdot 3\cdot 4} + \dfrac{2}{3\cdot 4\cdot 5} + \cdots + \dfrac{2}{100\cdot 101\cdot 102}$

$$\left[\textit{Hint: } \text{Write} \right.$$

$$\dfrac{2}{k(k+1)(k+2)} = \dfrac{1}{k} - \dfrac{2}{k+1} + \dfrac{1}{k+2}$$

$$\left. = \dfrac{1}{k} - \dfrac{1}{k+1} - \dfrac{1}{k+1} + \dfrac{1}{k+2}. \right]$$

In Exercises 53–56, the total resistance R in a circuit is given. Use partial-fraction decomposition to write $\dfrac{1}{R}$ in terms of partial fractions. Interpret each term in the partial fraction.

53. Circuit resistance. $R = \dfrac{(x+1)(x+3)}{2x+4}$

54. Circuit resistance. $R = \dfrac{(x+2)(x+4)}{7x+20}$

55. Circuit resistance. $R = \dfrac{R_1 R_2 R_3}{R_1 R_2 + R_2 R_3 + R_3 R_1}$

56. Circuit resistance. $R = \dfrac{x(x+2)(x+4)}{3x^2+12x+8}$

C EXERCISES   Beyond the Basics

57. Solve the system of equations in Example 2 to verify the values of the constants A, B, and C.

In Exercises 58–66, find the partial-fraction decomposition of each rational expression.

58. $\dfrac{1}{x^3+1}$

59. $\dfrac{4x}{(x^2-1)^2}$

60. $\dfrac{2x+3}{(x^2+1)(x+1)}$

61. $\dfrac{x+1}{(x^2+1)(x-1)^2}$

62. $\dfrac{x}{(x^2+1)^2(x-1)}$

63. $\dfrac{x^3}{(x+1)^2(x+2)^2}$

64. $\dfrac{2x^3+2x^2-1}{(x^2+x+1)^2}$

65. $\dfrac{15x}{x^3-27}$

66. $\dfrac{2x^2-9x+10}{x^3+8}$

67. Find the partial-fraction decomposition of

$$\dfrac{4x}{x^4+4}.$$

$[\textit{Hint: } x^4 + 4 = x^4 + 4 + 4x^2 - 4x^2 = (x^2+2)^2 - (2x)^2$
$= (x^2 - 2x + 2)(x^2 + 2x + 2).]$

68. Find the partial-fraction decomposition of

$$\dfrac{x^2-8x+18}{(x-5)^3}.$$

In Exercises 69–72, find the partial-fraction decomposition by first using long division.

69. $\dfrac{x^2+4x+5}{x^2+3x+2}$

70. $\dfrac{(x-1)(x+2)}{(x+3)(x-4)}$

71. $\dfrac{2x^4+x^3+2x^2-2x-1}{(x-1)(x^2+1)}$

72. $\dfrac{x^4}{(x-1)(x-2)(x-3)}$

5.1 Systems of Linear Equations in Two Variables

A **system of equations** is a set of equations with common variables.

A **solution** of a system of equations in two variables x and y is an ordered pair of numbers (a, b) that satisfies all equations in the system. The solution set of a system of two linear equations of the form $ax + by = c$ is the point(s) of intersection of the graphs of the equations.

Systems with no solution are called **inconsistent**, and systems with at least one solution are **consistent**. A system of two linear equations in two variables may have one solution (**independent equations**), no solution (**inconsistent system**), or infinitely many solutions (**dependent equations**).

Three methods of solving a system of equations are the graphical method, the substitution method, and the elimination or addition method.

5.2 Systems of Linear Equations in Three Variables

A **linear equation** in n variables $x_1, x_2, \ldots, x_n$ is an equation that can be written in the form

$$a_1x_1 + a_2x_2 + \cdots + a_nx_n = b,$$

where b and the coefficients $a_1, a_2, \ldots, a_n$ are real numbers.

A system of linear equations can be solved by transforming it into an *equivalent system* that is easier to solve.
The following operations produce equivalent systems:

1. Interchange the position of any two equations.

2. Multiply any equation by a nonzero constant.

3. Add a nonzero multiple of one equation to another.

The **Gaussian elimination method** is a procedure for converting a system of linear equations into an equivalent system in *triangular form*.

A linear system may have only one solution, infinitely many solutions, or no solution.

5.3 Systems of Nonlinear Equations

A system of equations in which at least one equation is nonlinear is called a **nonlinear system of equations**. The methods of substitution, elimination, or graphing is sometimes used to solve a nonlinear system.

5.4 Systems of Inequalities

A statement of the form $ax + by < c$ is a linear inequality in the variables x and y. The symbol $<$ may be replaced by $\leq$, $>$, or $\geq$. A procedure for graphing a linear inequality in two variables is described on page 514.

The graph of the solution set of a system of inequalities is obtained by graphing each inequality of the system in the same coordinate plane and then finding the region that is common to every graph in the system.

5.5 Linear Programming

In a linear programming model, a quantity f (to be maximized or minimized) that can be expressed in the form $f = ax + by$ is called an **objective function** of the variables x and y. The **constraints** are the restrictions placed on the variables x and y that can be expressed as a system of linear inequalities.

A procedure for solving a linear programming problem is given on page 515.

5.6 Partial-Fraction Decomposition

In partial–fraction decomposition, we undo the addition of rational expressions. A rational expression $\dfrac{P(x)}{Q(x)}$ is called a proper fraction if $\deg P(x) < \deg Q(x)$.

There are four cases to consider in decomposing $\dfrac{P(x)}{Q(x)}$ into partial fractions:

1. $Q(x)$ has only distinct linear factors. (See page 534.)

2. $Q(x)$ has repeated linear factors. (See page 536.)

3. $Q(x)$ has distinct irreducible quadratic factors. (See page 537.)

4. $Q(x)$ has repeated irreducible quadratic factors. (See page 538.)

REVIEW EXERCISES

Basic Skills and Concepts

In Exercises 1–18, solve each system of equations by using the method of your choice. Identify systems with no solution and systems with infinitely many solutions.

1. $\begin{cases} 3x - y = -5 \\ x + 2y = 3 \end{cases}$

2. $\begin{cases} x + 3y + 6 = 0 \\ y = 4x - 2 \end{cases}$

3. $\begin{cases} 2x + 4y = 3 \\ 3x + 6y = 10 \end{cases}$

4. $\begin{cases} x - y = 2 \\ 2x - 2y = 9 \end{cases}$

5. $\begin{cases} 3x - y = 3 \\ \dfrac{1}{2}x + \dfrac{1}{3}y = 2 \end{cases}$

6. $\begin{cases} 0.02y - 0.03x = -0.04 \\ 1.5x - 2y = 3 \end{cases}$

7. $\begin{cases} x + 3y + z = 0 \\ 2x - y + z = 5 \\ 3x - 3y + 2z = 10 \end{cases}$

8. $\begin{cases} 2x + y = 11 \\ 3y - z = 5 \\ x + 2z = 1 \end{cases}$

9. $\begin{cases} x + y = 1 \\ 3y + 2z = 0 \\ 2x - 3z = 7 \end{cases}$

10. $\begin{cases} 2x - 3y + z = 2 \\ x - 3y + 2z = -1 \\ 2x + 3y + 2z = 3 \end{cases}$

11. $\begin{cases} x + y + z = 1 \\ x + 5y + 5z = -1 \\ 3x - y - z = 4 \end{cases}$

12. $\begin{cases} x + 3y - 2z = -4 \\ 2x + 6y - 4z = 3 \\ x + y + z = 1 \end{cases}$

13. $\begin{cases} x + 4y + 3z = 1 \\ 2x + 5y + 4z = 4 \end{cases}$

14. $\begin{cases} 3x - y + 2z = 9 \\ x - 2y + 3z = 2 \end{cases}$

15. $\begin{cases} 3x - y = 2 \\ x + 2y = 9 \\ 3x + y = 10 \end{cases}$

16. $\begin{cases} x - 2y = 1 \\ 3x + 4y = 11 \\ 2x + 2y = 7 \end{cases}$

17. $\begin{cases} x + y + z = 3 \\ 2x - y + z = 4 \\ x + 4y + 2z = 5 \end{cases}$

18. $\begin{cases} x - y + z = 4 \\ x + 2y + 3z = 2 \\ 2x + y + 4z = 6 \end{cases}$

In Exercises 19–24, graph each system of inequalities.

19. $\begin{cases} x + y \le 1 \\ x - y \le 1 \\ x \ge 0 \end{cases}$

20. $\begin{cases} 2x + 3y \le 6 \\ 4x - 3y \le 12 \\ x \ge 0 \end{cases}$

21. $\begin{cases} 4x + y \le 7 \\ 2x + 5y \ge -1 \\ x - 2y \ge -5 \end{cases}$

22. $\begin{cases} 7x - 2y + 6 \le 0 \\ x + y + 14 \ge 0 \\ 2x - 3y + 9 \ge 0 \end{cases}$

23. $\begin{cases} x + y - 3 \le 0 \\ 3x - y - 9 \le 0 \\ y + 3 \ge 0 \\ 7x + 4y + 23 \ge 0 \end{cases}$

24. $\begin{cases} x + 5y - 11 \le 0 \\ 5x + y - 7 \le 0 \\ x - 5y - 17 \le 0 \\ 7x + y + 25 \ge 0 \end{cases}$

In Exercises 25–28, solve each linear programming problem.

25. Maximize $z = 2x + 3y$, subject to the constraints $x \ge 0, y \ge 0, 3x + 7y \le 21$.

26. Minimize $z = -5x + 2y$, subject to the constraints $x \ge -2, y \le 1, x - 2y + 2 \le 0$.

27. Minimize $z = x + 3y$, subject to the constraints $2x - 3y + 2 \ge 0, 4x + y - 10 \le 0, x - 3y - 9 \le 0, 3x + y + 3 \ge 0$.

28. Maximize $z = 2x + 5y$, subject to the constraints $x + 3y - 3 \le 0, 3x - y - 9 \le 0, x + 4y + 10 \ge 0, 3x - 2y + 2 \ge 0$.

In Exercises 29–34, solve each nonlinear system of equations.

29. $\begin{cases} x + 3y = 1 \\ x^2 - 3x = 7y + 3 \end{cases}$

30. $\begin{cases} 4x - y = 3 \\ y^2 - 2y = x - 2 \end{cases}$

31. $\begin{cases} x - y = 4 \\ 5x^2 + y^2 = 24 \end{cases}$

32. $\begin{cases} x - 2y = 7 \\ 2x^2 + 3y^2 = 29 \end{cases}$

33. $\begin{cases} xy = 2 \\ x^2 + 2y^2 = 9 \end{cases}$

34. $\begin{cases} xy = 3 \\ 2x^2 + 3y^2 = 21 \end{cases}$

In Exercises 35–40, graph each nonlinear system of inequalities.

35. $\begin{cases} x - y \le 1 \\ x^2 + y^2 \le 13 \end{cases}$

36. $\begin{cases} x - y \ge 1 \\ x^2 + y^2 \le 13 \end{cases}$

37. $\begin{cases} x - y \le 1 \\ x^2 + y^2 \le 13 \\ x \ge 0 \\ y \ge 0 \end{cases}$

38. $\begin{cases} x - y \ge 1 \\ x^2 + y^2 \ge 13 \\ x \le 4 \\ y \ge 0 \end{cases}$

39. $\begin{cases} y \le 3x \\ x^2 + y^2 \le 25 \\ x \ge 0 \\ y \ge 0 \end{cases}$

40. $\begin{cases} y \ge 3x \\ x^2 + y^2 \le 25 \\ x \ge 0 \\ y \ge 0 \end{cases}$

In Exercises 41–46, find the partial-fraction decomposition of each rational expression.

41. $\dfrac{x + 4}{x^2 + 5x + 6}$

42. $\dfrac{x + 14}{x^2 + 3x - 4}$

43. $\dfrac{3x^2 + x + 1}{x(x - 1)^2}$

44. $\dfrac{3x^2 + 2x + 3}{(x^2 - 1)(x + 1)}$

45. $\dfrac{x^2 + 2x + 3}{(x^2 + 4)^2}$

46. $\dfrac{2x}{x^4 - 1}$

Applying the Concepts

47. **Investment.** A speculator invested part of $15,000 in a high-risk venture and received a return of 12% at the end of the year. The rest of the $15,000 was invested at 4% annual interest. The combined annual income from the two sources was $1300. How much was invested at each rate?

48. **Agriculture.** A farmer earns a profit of $525 per acre of tomatoes and $475 per acre of soybeans. His soybean acreage is 5 acres more than twice his tomato acreage. If his total profit from the two crops is $24,500, how many acres of tomatoes and how many acres of soybeans does he have?

49. **Geometry.** The area of a rectangle is 63 square feet, and its perimeter is 33 feet. What are the dimensions of the rectangle?

50. Numbers. Twice the sum of the reciprocals of two numbers is 13, and the product of the numbers is $\frac{1}{9}$. Find the numbers.

51. Geometry. The hypotenuse of a right triangle is 17. If one leg of the triangle is increased by 1 and the other leg is increased by 4, the hypotenuse becomes 20. Find the sides of the triangle.

52. Paper route. Chris covers her paper route, which is 21 miles long, by 7:30 A.M. each day. If her average rate of travel were 1 mile faster each hour, she would cover the route by 7 A.M. What time does she start in the morning?

53. Agriculture. A rectangular pasture with an area of 6400 square meters is divided into three smaller pastures by two fences parallel to the shorter sides. The width of two of the smaller pastures is the same, and the width of the third is twice that of the others. Find the dimensions of the original pasture if the perimeter of the larger of the subdivisions is 240 m.

54. Leasing. Budget Rentals leases its compact cars for $23.00 per day plus $0.17 per mile. Dollar Rentals leases the same car for $24.00 per day plus $0.22 per mile.
   a. Find the cost functions describing the daily cost of leasing from each company.
   b. Graph the two functions in part (a) on the same coordinate axes.
   c. From which company should you lease the car if you plan to drive **(i)** 50 miles per day; **(ii)** 60 miles per day; **(iii)** 70 miles per day?

55. Break-even analysis. Auto-Sprinkler Corp. manufactures seven-day, 24-hour variable timers for lawn sprinklers. The corporation has a monthly fixed cost of $60,000 and a production cost of $12 for each timer manufactured. Each timer sells for $20.
   a. Write the cost function and the revenue function for selling x timers per month.
   b. Graph the two functions in part (a) on the same coordinate plane and find the break-even point graphically.
   c. Find the break-even point algebraically.
   d. How many timers should be sold in a month to realize a profit (before taxes) of 15% of the cost?

56. Equilibrium quantity and price. The demand equation for a product is $p = \dfrac{4000}{x}$, where p is the price per unit and x is the number of units of the product. The supply equation for the product is $p = \dfrac{x}{20} + 10$.
   a. Find the equilibrium quantity.
   b. Find the equilibrium price.
   c. Graph the supply and demand functions on the same coordinate plane and label the equilibrium point.

57. Apartment lease. Alisha and Sunita signed a lease on an apartment for nine months and split the rent evenly. At the end of six months, Alisha got married and moved out. She paid the landlady an amount equal to the difference between double-occupancy rental and single-occupancy rental for the remaining three months, and Sunita paid the single-occupancy rate for the same three months. If the nine-month rental cost Alisha $2250 and Sunita $3150, what were the single and double monthly rates?

58. Seating arrangement. The 600 graduating seniors at Central State College are seated in rows, each of which contains the same number of chairs, and every chair is occupied. If five more chairs were in each row, everyone could be seated in four fewer rows. How many chairs are in each row?

59. Passing a final exam. Twenty-six students in a college algebra class took a final exam on which the passing score was 70. The mean score of those who passed was 78, and the mean score of those who failed was 26. The mean of all scores was 72. How many students failed the exam?

60. Finding numbers. The average of a and b is 2.5, the average of b and c is 3.8, and the average of a and c is 3.1. Find the numbers a, b, and c.

61. May–December match. Steve and his wife Janet have the same birthday. When Steve was as old as Janet is now, he was twice as old as Janet was then. When Janet becomes as old as Steve is now, the sum of their ages will be 119. How old are Steve and Janet now?

62. Percentage increase. Let x and y be positive real numbers. Increasing x by y percent gives 46, while increasing y by x percent gives 21. Find x and y.

63. Criminals rob a bank. Three criminals—Butch, Sundance, and Billy—robbed a bank and divided the loot in the following manner: Because Butch planned the job and drove the getaway car, he got 75% as much as Sundance and Billy put together. Because Sundance was an expert safecracker, he got $500 more than 50% of Billy's and Butch's cut put together. Billy got $1000 less than three times the difference of Butch and Sundance's cut. How much did they steal, and what was each criminal's take?

64. Curve fitting. Find all of the curves of the form $y = ax^2 + bx + c$ that contain the points $(0, 1)$, $(1, 0)$, and $(-1, 6)$.

65. Using exponents. Given that $2^x = 8^y$ and $9^y = 3^{x-2}$, find x and y.

66. Area of an octagon. A square of side length 8 has its corners cut off to make it a regular octagon. Find the area of the octagon. (See the figure.)

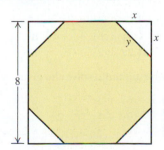

67. Building houses. A builder has 42 units of material and 32 units of labor available during a given period. She builds two-story houses or one-story houses or some of both. Suppose she makes a profit of $10,000 on each two-story house and $4000 on each one-story house. A two-story house requires seven units of material and one unit of labor, whereas a one-story house requires one unit of material and two units of labor. How many houses of each type should she build to make a maximum profit? What is the maximum profit?

68. Minimizing cost. An animal food is to be a mixture of two products X and Y. The content and cost of 1 pound of each product is given in the following table.

Product	Protein Grams	Fat Grams	Carbohydrates Grams	Cost
X	180	2	240	$0.75
Y	36	8	200	$0.56

How much of each product should be used to minimize the cost if each bag must contain at least 612 grams of protein; at least 22 grams of fat; and, at most, 1880 grams of carbohydrates?

69. Lobster prices. The Captain's Catch classifies its lobsters as small, medium, and large. A package of one large, two medium, and four small lobsters sells for $344; a package of two large and three small lobsters sells for $255; and a package of two large, two medium, and five small lobsters sells for $449. What is the individual price of each size lobster?

70. Ancient mathematics. The *Nine Chapters in the Mathematical Art*, a Chinese work written about 250 B.C. contains the following problem: Three sheafs of good crop, two sheaf of mediocre crop, and one sheaf of bad crop are sold for 39 dou. Two sheafs of good, three mediocre, and one bad are sold for 34 dou. And one good, two mediocre, and three bad are sold for 26 dou. What price is received for a sheaf of each of good crop, mediocre crop, and bad crop?

PRACTICE TEST A

In Problems 1–7, solve each system of equations.

1. $\begin{cases} 2x - y = 4 \\ 2x + y = 4 \end{cases}$
2. $\begin{cases} x + 2y = 8 \\ 3x + 6y = 24 \end{cases}$

3. $\begin{cases} -2x + y = 4 \\ 4x - 2y = 4 \end{cases}$
4. $\begin{cases} 3x + 3y = -15 \\ 2x - 2y = -10 \end{cases}$

5. $\begin{cases} \dfrac{5}{3}x + \dfrac{y}{2} = 14 \\ \dfrac{2}{3}x - \dfrac{y}{8} = 3 \end{cases}$
6. $\begin{cases} y = x^2 \\ 3x - y + 4 = 0 \end{cases}$

7. $\begin{cases} x - 3y = -4 \\ 2x^2 + 3x - 3y = 8 \end{cases}$

8. Two gold bars together weigh a total of 485 pounds. One bar weighs 15 pounds more than the other.
  (a) Write a system of equations that describes these relationships.
  (b) How much does each bar weigh?

9. Convert the given system to triangular form.

$$\begin{cases} 2x + y + 2z = 4 \\ 4x + 6y + z = 15 \\ 2x + 2y + 7z = -1 \end{cases}$$

10. Use back-substitution to solve the given linear system.

$$\begin{cases} x - 6y + 3z = -2 \\ 9y - 5z = 2 \\ 2z = 10 \end{cases}$$

In Exercises 11–13, solve the system of equations.

11. $\begin{cases} x + y + z = 8 \\ 2x - 2y + 2z = 4 \\ x + y - z = 12 \end{cases}$
12. $\begin{cases} x + z = -1 \\ 3y + 2z = 5 \\ 3x - 3y + z = -8 \end{cases}$

13. $\begin{cases} 2x - y + z = 2 \\ x + y - z = -1 \\ x - 5y + 5z = 7 \end{cases}$

14. A vending machine's coin box contains nickels, dimes, and quarters. The total number of coins in the box is 300. The number of dimes is three times the number of nickels and quarters together. Assuming that $30.65 is in the box, find the number of nickels, dimes, and quarters that it contains.

For Problems 15 and 16, write the form of the partial-fraction decomposition of the given rational expression. You do not need to solve for the constants.

15. $\dfrac{2x}{(x - 5)(x + 1)}$
16. $\dfrac{-5x^2 + x - 8}{(x - 2)(x^2 + 1)^2}$

17. Find the partial-fraction decomposition of the rational expression

$$\frac{x + 3}{(x + 4)^2(x - 7)}.$$

18. Graph the inequality $3x + y < 6$.

19. Graph the solution set of the system of inequalities

$$\begin{cases} y - x^2 \leq 3 \\ y - x > 0 \end{cases}$$

20. Maximize $z = 2x + y$, subject to the constraints $x \geq 0, y \geq 0, x + 3y \leq 3$.

PRACTICE TEST B

In Problems 1–7, solve each system of equations.

1. $\begin{cases} 2x - 3y = 16 \\ x - y = 7 \end{cases}$

    a. $\{(5, -2)\}$ **b.** $\{(8, 0)\}$

    c. $\{(0, -7)\}$ **d.** $\{(4, -3)\}$

2. $\begin{cases} 6x - 9y = -2 \\ 3x - 5y = -6 \end{cases}$

    a. $\{(-3, 0)\}$ **b.** $\left\{\left(0, \dfrac{5}{6}\right)\right\}$

    c. $\left\{\left(\dfrac{44}{3}, 10\right)\right\}$ **d.** $\{(1, 1)\}$

3. $\begin{cases} 2x - 5y = 9 \\ 4x - 10y = 18 \end{cases}$

    a. $\left\{\left(0, \dfrac{9}{2}\right)\right\}$ **b.** $\left\{\left(\dfrac{5}{2}y + \dfrac{9}{2}, y\right)\right\}$

    c. $\left\{\left(x, \dfrac{5}{2}x + \dfrac{9}{2}\right)\right\}$ **d.** $\varnothing$

4. $\begin{cases} 3x + 5y = 1 \\ -6x - 10y = 2 \end{cases}$

    a. $\left\{\left(0, -\dfrac{1}{5}\right)\right\}$ **b.** $\{(2, -1)\}$

    c $\varnothing$ **d.** $\left\{\left(-\dfrac{5}{3}y + \dfrac{1}{3}, y\right)\right\}$

5. $\begin{cases} \dfrac{1}{5}x + \dfrac{2}{5}y = 1 \\ \dfrac{1}{4}x - \dfrac{1}{3}y = \dfrac{-5}{12} \end{cases}$

    a. $\{(-15, 10)\}$ **b.** $\{(1, 2)\}$

    c. $\{(-10, -15)\}$ **d.** $\{(1, 0)\}$

6. $\begin{cases} x - 3y = \dfrac{1}{2} \\ -2x + 6y = -1 \end{cases}$

    a. $\left\{\left(\dfrac{1}{2}, 0\right)\right\}$ **b.** $\left\{\left(2, \dfrac{1}{2}\right)\right\}$

    c. $\varnothing$ **d.** $\left\{\left(3y + \dfrac{1}{2}, y\right)\right\}$

7. $\begin{cases} 3x + 4y = 12 \\ 3x^2 + 16y^2 = 48 \end{cases}$

    a. $\{(0, 4), (3, 0)\}$ **b.** $\{(0, -4)\}$

    c. $\{(0, -4), (3, 0)\}$ **d.** $\left\{(4, 0), \left(2, \dfrac{3}{2}\right)\right\}$

8. Student tickets for a dance cost \$2, and nonstudent tickets cost \$5. Three hundred tickets were sold, and the total ticket receipts were \$975. How many of each type of ticket were sold?

    a. 150 student tickets   **b.** 200 student tickets
        150 nonstudent tickets   100 nonstudent tickets

    c. 210 student tickets   **d.** 175 student tickets
        90 nonstudent tickets   125 nonstudent tickets

9. Convert the given system to triangular form.

$$\begin{cases} x + 3y + 3z = 4 \\ 2x + 5y + 4z = 5 \\ x + 2y + 2z = 6 \end{cases}$$

    a. $\begin{aligned} x + 3y + 3z &= 4 \\ y + 2z &= 8 \\ z &= 2 \end{aligned}$ **b.** $\begin{aligned} x + 3y + 3z &= 4 \\ 2y + z &= 1 \\ z &= 5 \end{aligned}$

    c. $\begin{aligned} x + 3y + 3z &= 4 \\ y + 2z &= 3 \\ z &= 5 \end{aligned}$ **d.** $\begin{aligned} x + 3y + 3z &= 4 \\ 2y + z &= -2 \\ z &= 5 \end{aligned}$

10. Use back-substitution to solve the linear system.

$$\begin{cases} 2x + y + z = 0 \\ 10y - 2z = 4 \\ 3z = 9 \end{cases}$$

    a. $\{(-2, 1, 3)\}$ **b.** $\{(0, -3, 3)\}$
    c. $\{(0, 2, 8)\}$ **d.** $\{(-1, 1, 1)\}$

11. Solve the given system of equations or state that the system is inconsistent.

$$\begin{cases} 2x + 13y + 6z = 1 \\ 3x + 10y + 11z = 15 \\ 2x + 10y + 8z = 8 \end{cases}$$

    a. $\{(3, -1, -4)\}$ **b.** $\{(1, -1, 2)\}$
    c. $\varnothing$ **d.** $(1, 6, 8)\}$

12. Solve the given system of equations or state that the system is inconsistent.

$$\begin{cases} 4x - y + 7z = -2 \\ 2x + y + 11z = 13 \\ 3x - y + 4z = -3 \end{cases}$$

    a. $\{(-3z + 1, -5z + 11, z)\}$
    b. $\{(-14, -14, 5)\}$
    c. $\left\{\left(-\dfrac{19}{5}, 3, \dfrac{8}{5}\right)\right\}$
    d. $\varnothing$

13. Solve the given system of equations or state that the system is inconsistent.

$$\begin{cases} x - 2y + 3z = 4 \\ 2x - y + z = 1 \\ x + y - 2z = -3 \end{cases}$$

    a. $\{(1, 6, 5)\}$ **b.** $\{(0, 1, 2)\}$
    c. $\{(x, 5x + 1, 3x + 2)\}$ **d.** $\varnothing$

14. Forty-six students will go to France, Italy, or Spain for six weeks during the summer. The number of students going to France or Italy is four more than the number going to Spain. The number of students going to France is two less than the number going to Spain. How many students are going to each country?

a. France: 23   b. France: 19
   Italy: 21   Italy: 6
   Spain: 2   Spain: 21
c. France: 21   d. France: 2
   Italy: 6   Italy: 21
   Spain: 19   Spain: 23

For Problems 15 and 16, write the form of the partial-fraction decomposition of the given rational functions. You do not need to solve for the constants.

15. $\dfrac{x}{(x+2)(x-7)}$

a. $\dfrac{A}{x+2} + \dfrac{Bx}{x-7}$ b. $\dfrac{Ax}{x+2} + \dfrac{Bx}{x-7}$

c. $\dfrac{A}{x+2} + \dfrac{B}{x-7}$ d. $\dfrac{A}{x+2} + \dfrac{Bx+C}{x-7}$

16. $\dfrac{7-x}{(x-3)(x+5)^2}$

a. $\dfrac{A}{x-3} + \dfrac{B}{x+5} + \dfrac{C}{(x+5)^2}$

b. $\dfrac{A}{x-3} + \dfrac{B}{x+5} + \dfrac{Cx+D}{(x+5)^2}$

c. $\dfrac{A}{x+3} + \dfrac{B}{(x+5)^2}$

d. $\dfrac{A}{x+3} + \dfrac{Bx+C}{(x+5)^2}$

17. Find the partial-fraction decomposition of the rational expression

$$\frac{x^2 + 15x + 18}{x^3 - 9x}.$$

a. $\dfrac{-2}{x} + \dfrac{26}{x-9}$ b. $\dfrac{-2}{x} + \dfrac{4}{x+3} - \dfrac{1}{x-3}$

c. $\dfrac{-2}{x} + \dfrac{4}{x-3} - \dfrac{1}{x+3}$ d. $\dfrac{-2}{x} + \dfrac{3x-15}{x^2-9}$

18. Which of the graphs at the right is the graph of $x + 3y > 3$?

19. Which of the graphs at the right is the graph of the solution set of the system of inequalities

$$\begin{cases} y - x^2 + 4 \geq 0 \\ 3x - y \leq 0 \end{cases}?$$

20. Maximize $z = 3x + 21y$, subject to the constraints

$x \geq 0, y \geq 0, 2x + y \leq 8, 2x + 3y \leq 16.$

a. 84   b. 112

c. $\dfrac{16}{3}$ d. 12

(a)   (b)

(c)   (d)

Figure for Exercise 18

(a)   (b)

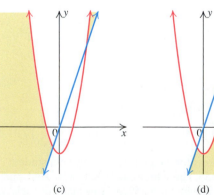

(c)   (d)

Figure for Exercise 19

CUMULATIVE REVIEW EXERCISES ▪ Chapters P–5

In Exercises 1–8, solve each equation or inequality.

1. $\dfrac{1}{x-1} + \dfrac{4}{x-4} = \dfrac{5}{x-5}$

2. $2\left(x + \dfrac{1}{x}\right)^2 - 7\left(x + \dfrac{1}{x}\right) + 5 = 0$

3. $\sqrt{3x-5} = x - 3$

4. $\dfrac{x-1}{x+3} \le 0$

5. $x^2 - 9x + 20 > 0$

6. $2^{x-1} = 5$

7. $\log_x 16 = 4$

8. $\log(x-3) + \log(x-1) = \log(2x-5)$

In Exercises 9–12, use transformations to graph each function.

9. $f(x) = |x+1| - 2$

10. $f(x) = -(x-2)^2 + 3$

11. $f(x) = 2^{x-1} - 3$

12. $f(x) = \sqrt{x-1} + 3$

13. Let $f(x) = 2x - 2$.
  a. Find $f^{-1}(x)$.
  b. Graph f and f^{-1} on the same coordinate plane.

14. Let $f(x) = x^3 - x^2 + x - 6$.
  a. List all possible rational zeros of f.
  b. Use synthetic division to test the possible rational zeros and find a real zero.
  c. Use the zero from part (b) to find all of the (real or complex) zeros of $f(x)$.

15. Expand and simplify: $\log_3 (9x^4)$

 16. A radioactive substance decays so that the amount present in t years is given by $A = A_0 e^{-kt}$, where A_0 is the initial amount. Find the half-life of the substance if
  a. $k = 0.05$.
  b. $k = 0.0002$.

17. If f is a one-to-one function and $f(2) = 9$, find $f^{-1}(9)$.

18. Find the time required to double your money if it is compounded continuously at 7.5%.

In Exercises 19 and 20, solve the system of equations.

19. $\begin{cases} 5x - 2y + 25 = 0 \\ 4y - 3x - 29 = 0 \end{cases}$

20. $\begin{cases} 2x - y + z = 3 \\ x + 3y - 2z = 11 \\ 3x - 2y + z = 4 \end{cases}$

The linear programming problems introduced in the previous chapter involving quantities from the study of such disciplines as medicine, traffic control, structural design, and city planning often involve an extremely large number of variables. To solve such problems successfully requires the use of powerful computers and new techinques. In this chapter, we study matrices and several methods of solving linear systems that are easily implemented on computers.

Matrices and Systems of Equations

Before Starting this Section, Review

1. Systems of equations (Section 5.1, page 479)

2. Equivalent systems (Section 5.2, page 494)

3. Gaussian elimination (Section 5.2, page 494)

Objectives

1 Define a matrix.

2 Use matrices to solve a system of linear equations.

3 Use Gaussian elimination to solve a system.

4 Use Gauss–Jordan elimination to solve a system.

Wassily Leontief 1906–1999

Leontief studied philosophy, sociology, and economics at the University of Leningrad and earned the degree of Learned Economist in 1925. He continued his studies at the University of Berlin, where he earned his PhD degree. In 1973, the Nobel Prize in Economic Sciences was awarded to Leontief for his work on input–output analysis of the U.S. economy.

LEONTIEF INPUT–OUTPUT MODEL

In 1949, the Russian-born Harvard Professor Wassily Leontief opened the door to a new era in mathematical modeling in economics. He divided the U.S. economy into 500 "sectors," such as the coal, automotive, and communications industries and agriculture. For each sector, he wrote a linear equation that described how that sector distributed its output to other sectors of the economy. In 1949, the largest computer was the Mark II, and it could not handle the resulting system of 500 equations in 500 variables. Leontief then combined some of the sectors and reduced the system to 42 equations in 42 variables to solve the problem of output distributions. As computers have become more powerful, scientists and engineers can now work on problems far more complex than they ever dreamed of a few decades ago. In Example 10, we explore Leontief's input–output model applied to a simple economy. ■

1 Define a matrix.

Definition of a Matrix

Great Builders Incorporated (GBI), builds ranch, colonial, and modern houses. GBI wants to compare the cost of labor (in millions of dollars) and the cost of material (in millions of dollars) involved in building each type of house during one year. These data could be represented as in Table 6.1.

TABLE 6.1

Type of House	Ranch	Colonial	Modern
Cost of labor (in millions)	12	13	14
Cost of material (in millions)	9	11	10

Or it could be represented more compactly by the following array:

$$\begin{bmatrix} 12 & 13 & 14 \\ 9 & 11 & 10 \end{bmatrix}$$

This rectangular array of numbers is called a *matrix* (plural, *matrices*). The matrix has two rows (the types of costs) and three columns (the types of houses).

DEFINITION OF A MATRIX

A **matrix** is a rectangular array of numbers.

$$A = \begin{bmatrix} a_{11} & a_{12} & \cdots & a_{1n} \\ a_{21} & a_{22} & \cdots & a_{2n} \\ \vdots & \vdots & & \vdots \\ a_{m1} & a_{m2} & \cdots & a_{mn} \end{bmatrix} \begin{matrix} \leftarrow \text{Row 1} \\ \leftarrow \text{Row 2} \\ \\ \leftarrow \text{Row } m \end{matrix}$$

Column 1   Column 2   Column n

If a matrix A has m rows and n columns, then we say that A is of **order m by n** (written $m \times n$).

The **entry** or **element** in the ith row and jth column is a real number and is denoted by the *double-subscript* notation a_{ij}. We call a_{ij} the (i, j)**th entry**. For example, a_{24} is the entry in the second row and fourth column of the matrix A. In general, we will use capital letters such as $A, B, \ldots$ to denote matrices and the corresponding lowercase letter $a_{ij}, b_{ij}, \ldots$ for their entries.

If A has n rows and n columns, then A is called a **square matrix of order n**. The entries $a_{11}, a_{22}, \ldots, a_{nn}$ form the **main diagonal** of A. A $1 \times n$ matrix is called a **row matrix**, and an $n \times 1$ matrix is called a **column matrix**.

EXAMPLE 1   **Determining the Order of Matrices**

Determine the order of each matrix. Identify square, row, and column matrices. Identify entries in the main diagonal of each square matrix.

a. $A = [3]$ **b.** $B = [3 \quad 5 \quad -7]$ **c.** $C = \begin{bmatrix} 0 & 1 \\ -3 & 4 \end{bmatrix}$ **d.** $D = \begin{bmatrix} 1 & 2 & 3 \\ 4 & 5 & 6 \\ 7 & 8 & 9 \end{bmatrix}$

SOLUTION

a. Matrix A with one row and one column is a 1×1 matrix. A is a square matrix of order 1. In A, $a_{11} = 3$. A is also a column and a row matrix.

b. Matrix B with one row and three columns is a 1×3 matrix. B is a row matrix.

c. Matrix C, a 2×2 matrix, is a square matrix of order 2. In C, the entries $c_{11} = 0$ and $c_{22} = 4$ form the main diagonal.

d. Matrix D, a 3×3 matrix, is a square matrix of order 3. In D, the entries $d_{11} = 1, d_{22} = 5$, and $d_{33} = 9$ form the main diagonal. ■ ■ ■

Practice Problem 1   Determine the order of each matrix.

a. $\begin{bmatrix} -1 & 3 \\ 7 & 4 \\ 0 & 0 \end{bmatrix}$ **b.** $[3 \quad -8]$ ■

2 Use matrices to solve a system of linear equations.

Using Matrices to Solve Linear Systems

To solve a system of linear equations by the elimination method, the particular symbols used for the variables do not matter; only the coefficients and the constants are important. We can display the constants and coefficients of a system in a matrix called the **augmented matrix** of the system. The matrix containing only the coefficients of the variables is the **coefficient matrix**. Consider the following system.

$$\begin{cases} x - y - z = 1 \\ 2x - 3y + z = 10 \\ x + y - 2z = 0 \end{cases}$$

Coefficients of x
Coefficients of y
Coefficients of z

Augmented Matrix:

$$\left[\begin{array}{ccc|c} 1 & -1 & -1 & 1 \\ 2 & -3 & 1 & 10 \\ 1 & 1 & -2 & 0 \end{array}\right]$$

Constants

Coefficient Matrix:

$$\begin{bmatrix} 1 & -1 & -1 \\ 2 & -3 & 1 \\ 1 & 1 & -2 \end{bmatrix}$$

Notice that the numbers in the first column of the augmented matrix are the coefficients of x, those in the second column are the coefficients of y, and those in the third column are the coefficients of z. The constants on the right side of the equations in the system are found in the fourth column. The vertical line in the augmented matrix is to remind you of the equal signs in the equations. If a variable does not appear in one of the equations, represent it with a zero in the matrix.

EXAMPLE 2   Writing the Augmented Matrix of a Linear System

Write the augmented matrix of the linear system.

$$\begin{cases} 2x + 3z = 1 \\ 2z + y = 5 \\ -4x + 5y = 7 \end{cases}$$

SOLUTION

First, write the system with the variables lined up in columns and insert zeros as coefficients of any missing variable.

$$\begin{cases} 2x + 0y + 3z = 1 \\ 0x + 1y + 2z = 5 \\ -4x + 5y + 0z = 7 \end{cases}$$

The augmented matrix of the given system is:

$$\left[\begin{array}{ccc|c} 2 & 0 & 3 & 1 \\ 0 & 1 & 2 & 5 \\ -4 & 5 & 0 & 7 \end{array}\right]$$ ■ ■ ■

Practice Problem 2   Write the augmented matrix of the linear system.

$$\begin{cases} 3y - z = 8 \\ x + 4y = 14 \\ -2y + 9z = 0 \end{cases}$$ ■

We can reverse this process and write a linear system from a given augmented matrix. For example, the augmented matrix

$$\left[\begin{array}{ccc|c} 9 & 2 & -1 & 6 \\ 3 & 4 & 7 & -8 \\ 0 & 1 & 5 & 11 \end{array}\right] \text{ corresponds to the system of equations } \begin{cases} 9x + 2y - z = 6 \\ 3x + 4y + 7z = -8. \\ y + 5z = 11 \end{cases}$$

TECHNOLOGY CONNECTION

Many graphing calculators let you name a matrix, specify the size, and then insert the entries for the matrix. No vertical line separates the column of constants, but this does not affect any of the matrix operations. The augmented matrix from Example 2 has been named A.

The basic strategy for solving a system of equations is to *replace the given system with an equivalent system* (one with the same solution set) *that is easier to solve*. In the previous chapter, we used three basic operations to solve a linear system:

1. Interchange two equations.
2. Multiply all of the terms in an equation by a nonzero constant.
3. Add a multiple of one equation to another equation. In other words, replace one equation with the sum of itself and a multiple of another equation.

In matrix terminology, the three corresponding operations are called the **elementary row operations**. Two matrices are **row-equivalent** if one can be obtained from the other by a sequence of elementary row operations, given next. The symbol R_i represents the ith row of a matrix.

Elementary Row Operations

Row Operation	In Symbols	Description
1. Interchange two rows.	$R_i \leftrightarrow R_j$	Interchange the ith and jth rows.
2. Multiply a row by a nonzero constant.	cR_j	Multiply the jth row by c, $c \neq 0$.
3. Add a multiple of one row to another row.	$cR_i + R_j \rightarrow R_j$	Replace the jth row by adding c times the ith row to it.

EXAMPLE 3 **Applying Elementary Row Operations**

Perform the row operations given in **a**, **b**, and **c**, in that order, on the matrix A:

$$A = \begin{bmatrix} 3 & -2 & -3 \\ 2 & 4 & 14 \end{bmatrix}$$

a. $R_1 \leftrightarrow R_2$,   **b.** $\dfrac{1}{2}R_1$,   **c.** $-3R_1 + R_2 \rightarrow R_2$

SOLUTION

a. $A = \begin{bmatrix} 3 & -2 & -3 \\ 2 & 4 & 14 \end{bmatrix} \xrightarrow{R_1 \leftrightarrow R_2} \begin{bmatrix} 2 & 4 & 14 \\ 3 & -2 & -3 \end{bmatrix} = B$ Interchange the 1st and 2nd rows of A.

b. $B = \begin{bmatrix} 2 & 4 & 14 \\ 3 & -2 & -3 \end{bmatrix} \xrightarrow{\frac{1}{2}R_1} \begin{bmatrix} 1 & 2 & 7 \\ 3 & -2 & -3 \end{bmatrix} = C$ Multiply the 1st row of B by $\frac{1}{2}$.

c. $C = \begin{bmatrix} 1 & 2 & 7 \\ 3 & -2 & -3 \end{bmatrix} \xrightarrow{-3R_1 + R_2 \rightarrow R_2} \begin{bmatrix} 1 & 2 & 7 \\ 0 & -8 & -24 \end{bmatrix}$ Add -3 times the 1st row of C to its 2nd row. ■■■

Practice Problem 3 Repeat Example 3 on the following matrix:

$$A = \begin{bmatrix} 3 & 4 & 5 \\ 2 & 4 & 6 \end{bmatrix}$$

■

In the next example, we solve a system of equations both with and without matrix notation and place the results side by side for comparison. We solve the system by converting it into triangular form by first using elimination and then back-substitution.

Many graphing calculators provide all three row operations. Consider the matrix A.

Here are some examples of row operations on the matrix A.

Swap rows 1 and 3.

Multiply row 1 by 4.

Add 3 times row 1 to row 2.

EXAMPLE 4   **Comparing Linear Systems and Matrices**

Solve the system of linear equations.

$$\begin{cases} x - y - z = 1 & (1) \\ 2x - 3y + z = 10 & (2) \\ x + y - 2z = 0 & (3) \end{cases}$$

SOLUTION

Linear System	**Augmented Matrix**

$$\begin{cases} x - y - z = 1 & (1) \\ 2x - 3y + z = 10 & (2) \\ x + y - 2z = 0 & (3) \end{cases}$$

$$A = \begin{bmatrix} 1 & -1 & -1 & | & 1 \\ 2 & -3 & 1 & | & 10 \\ 1 & 1 & -2 & | & 0 \end{bmatrix}$$

Add -2 times equation (1) to equation (2).
Add -1 times equation (1) to equation (3).

Use the first row to produce zeros at the (2, 1) and (3, 1) positions.

$$\begin{cases} x - y - z = 1 & (1) \\ -y + 3z = 8 & (4) \\ 2y - z = -1 & (5) \end{cases}$$

$$\xrightarrow[(-1)R_1+R_3 \to R_3]{-2R_1+R_2 \to R_2} \begin{bmatrix} 1 & -1 & -1 & | & 1 \\ 0 & -1 & 3 & | & 8 \\ 0 & 2 & -1 & | & -1 \end{bmatrix}$$

Multiply equation (4) by -1.

Produce a 1 at the (2, 2) position.

$$\begin{cases} x - y - z = 1 & (1) \\ y - 3z = -8 & (6) \\ 2y - z = -1 & (5) \end{cases}$$

$$\xrightarrow{(-1)R_2} \begin{bmatrix} 1 & -1 & -1 & | & 1 \\ 0 & 1 & -3 & | & -8 \\ 0 & 2 & -1 & | & -1 \end{bmatrix}$$

Add -2 times equation (6) to equation (5).

Use the second row to produce a zero at the (3, 2) position.

$$\begin{cases} x - y - z = 1 & (1) \\ y - 3z = -8 & (6) \\ 5z = 15 & (7) \end{cases}$$

$$\xrightarrow{-2R_2+R_3 \to R_3} \begin{bmatrix} 1 & -1 & -1 & | & 1 \\ 0 & 1 & -3 & | & -8 \\ 0 & 0 & 5 & | & 15 \end{bmatrix}$$

Equation (7), $5z = 15$, which corresponds to row 3 of the final matrix, gives the value $z = 3$. We find y by back-substitution into equation (6).

$y - 3z = -8$	Equation (6); or row 2 of the final matrix
$y - 3(3) = -8$	Replace z with 3.
$y = 1$	Solve for y.

We find x by back-substitution.

$x - y - z = 1$	Equation (1); or row 1 of the final matrix
$x - 1 - 3 = 1$	Replace y with 1 and z with 3.
$x = 5$	Solve for x.

The solution of the system is $x = 5$, $y = 1$, and $z = 3$. You should check this solution by substituting it into the original system of equations. The solution set is $\{(5, 1, 3)\}$. ■ ■ ■

Practice Problem 4   Solve the system of linear equations.

$$\begin{cases} x - 6y + 3z = -2 \\ 3x + 3y - 2z = -2 \\ 2x - 3y + z = -2 \end{cases}$$ ■

The last matrix in the solution of Example 4 is in *row-echelon form*, which is defined next. In the definition, a **nonzero row** means a row that contains at least one nonzero entry; a **leading entry** of a row is the leftmost nonzero entry in a nonzero row.

ROW-ECHELON FORM AND REDUCED ROW-ECHELON FORM

An $m \times n$ matrix is in **row-echelon form** if it has the following three properties:

1. The leading entry of each nonzero row is a 1.
2. The leading entry in any row is to the right of the leading entry in the row above it.
3. All nonzero rows are above the rows consisting entirely of zeros.

A matrix in row-echelon form having the following property is in **reduced row-echelon form:** Each leading 1 is the only nonzero entry in its column.

Property 2 says that the leading entries form an echelon (steplike) pattern that moves down and to the right.

The following matrices are in row-echelon form; the starred entries may be any value, including zero.

$$\begin{bmatrix} 1 & * & * \\ 0 & 1 & * \\ 0 & 0 & 1 \end{bmatrix} \qquad \begin{bmatrix} 1 & * & * & * \\ 0 & 1 & * & * \\ 0 & 0 & 1 & * \end{bmatrix} \qquad \begin{bmatrix} 1 & * & * & * \\ 0 & 0 & 1 & * \\ 0 & 0 & 0 & 1 \\ 0 & 0 & 0 & 0 \end{bmatrix}$$

The following matrices are in *reduced* row-echelon form because the entries above and below each leading 1 are zero. (Again, the starred entries may be any value, including zero.)

$$\begin{bmatrix} 0 & 1 & 0 & 0 \\ 0 & 0 & 1 & 0 \\ 0 & 0 & 0 & 1 \end{bmatrix} \qquad \begin{bmatrix} 1 & 0 & 0 & * \\ 0 & 1 & 0 & * \\ 0 & 0 & 1 & * \end{bmatrix} \qquad \begin{bmatrix} 1 & * & 0 & 0 \\ 0 & 0 & 1 & 0 \\ 0 & 0 & 0 & 1 \\ 0 & 0 & 0 & 0 \end{bmatrix}$$

An augmented matrix may be transformed into several different row-echelon form matrices by using different sequences of row operations. However, its *reduced* row-echelon form is *unique*. (See Exercise 103.)

3 Use Gaussian elimination to solve a system.

Gaussian Elimination

The method of solving a system of linear equations by transforming the augmented matrix of the system into row-echelon form and then using back-substitution to find the solution set is known as **Gaussian elimination**.

FINDING THE SOLUTION: A PROCEDURE

EXAMPLE 5   **Solving Linear Systems by Using Gaussian Elimination**

OBJECTIVE	EXAMPLE
Solve a system of linear equations by Gaussian elimination.	*Solve the system of equations.*

EXAMPLE

Solve the system of equations.

$$\begin{cases} x + 2y = -2 \\ 4x + 3y = 7 \end{cases}$$

Step 1 Write the augmented matrix of the system.

$$A = \begin{bmatrix} 1 & 2 & -2 \\ 4 & 3 & 7 \end{bmatrix} \qquad \text{Augmented matrix}$$

Step 2 Use elementary row operations to transform the augmented matrix into row-echelon form.

$$\xrightarrow{-4R_1 + R_2 \to R_2} \begin{bmatrix} 1 & 2 & -2 \\ 0 & -5 & 15 \end{bmatrix} \qquad \begin{array}{l} \text{This operation produces a 0} \\ \text{in the } (2, 1) \text{ position.} \end{array}$$

$$\xrightarrow{-\frac{1}{5}R_2} \begin{bmatrix} 1 & 2 & -2 \\ 0 & 1 & -3 \end{bmatrix} \qquad \begin{array}{l} \text{This operation produces a 1} \\ \text{in the } (2, 2) \text{ position.} \end{array}$$

Step 3 Write the system of linear equations that corresponds to the last matrix in Step 2.

$$\begin{cases} x + 2y = -2 & (1) \\ \quad\quad y = -3 & (2) \end{cases}$$

Step 4 Use the system of equations from Step 3, together with back-substitution, to find the system's solution set.

$$\begin{array}{ll} x + 2y = -2 & \text{Equation (1)} \\ x + 2(-3) = -2 & \text{Replace } y \text{ with } -3. \\ x = 4 & \text{Solve for } x. \end{array}$$

You should check the solution set, $\{(4, -3)\}$, by substituting $x = 4$ and $y = -3$ into the original system of equations. ■■■

Practice Problem 5   Solve by Gaussian elimination.

$$\begin{cases} x - 2y = 1 \\ 2x + 3y = 16 \end{cases}$$
■

EXAMPLE 6   **Solving a System by Using Gaussian Elimination**

Solve by Gaussian elimination.

$$\begin{cases} 2x + y + z = 6 \\ -3x - 4y + 2z = 4 \\ x + y - z = -2 \end{cases}$$

SOLUTION

STUDY TIP

Note that when we perform a row operation such as

$3R_1 + R_2 \to R_2$ on a matrix,

1. the entries in row 1 are *unchanged* and
2. we multiply each entry of row 1 by 3 and add this product to the corresponding entries of row 2 to obtain a *new row 2*.

Step 1 $A = \begin{bmatrix} 2 & 1 & 1 & 6 \\ -3 & -4 & 2 & 4 \\ 1 & 1 & -1 & -2 \end{bmatrix}$ The augmented matrix of the system

Step 2 $\xrightarrow{R_1 \leftrightarrow R_3} \begin{bmatrix} 1 & 1 & -1 & -2 \\ -3 & -4 & 2 & 4 \\ 2 & 1 & 1 & 6 \end{bmatrix}$

This operation produces a 1 in the $(1, 1)$ position.

$\xrightarrow[-2R_1 + R_3 \to R_3]{3R_1 + R_2 \to R_2} \begin{bmatrix} 1 & 1 & -1 & -2 \\ 0 & -1 & -1 & -2 \\ 0 & -1 & 3 & 10 \end{bmatrix}$

These operations produce zeros in the $(2, 1)$ and $(3, 1)$ positions.

$\xrightarrow{(-1)R_2} \begin{bmatrix} 1 & 1 & -1 & -2 \\ 0 & 1 & 1 & 2 \\ 0 & -1 & 3 & 10 \end{bmatrix}$

This operation produces a 1 in the $(2, 2)$ position.

Continued on the next page

$$\xrightarrow{R_2 + R_3 \rightarrow R_3} \begin{bmatrix} 1 & 1 & -1 & | & -2 \\ 0 & 1 & 1 & | & 2 \\ 0 & 0 & 4 & | & 12 \end{bmatrix}$$

This operation produces a zero in the $(3, 2)$ position.

$$\xrightarrow{\frac{1}{4}R_3} \begin{bmatrix} 1 & 1 & -1 & | & -2 \\ 0 & 1 & 1 & | & 2 \\ 0 & 0 & 1 & | & 3 \end{bmatrix}$$

This operation produces a 1 in the $(3, 3)$ position.

The last matrix is in row-echelon form.

Step 3 The system of equations corresponding to the last matrix in Step 2 is

$$\begin{cases} x + y - z = -2 & (1) & \text{From row (1)} \\ y + z = 2 & (2) & \text{From row (2)} \\ z = 3 & (3) & \text{From row (3)} \end{cases}$$

Step 4 Equation (3) in Step 3 gives the value $z = 3$. Back-substitute $z = 3$ in equation (2).

$$\begin{aligned} y + z &= 2 & \text{Equation (2)} \\ y + 3 &= 2 & \text{Replace } z \text{ with 3.} \\ y &= -1 & \text{Solve for } y. \end{aligned}$$

Now back-substitute $z = 3$ and $y = -1$ in equation (1).

$$\begin{aligned} x + y - z &= -2 & \text{Equation (1)} \\ x - 1 - 3 &= -2 \\ x &= 2 & \text{Solve for } x. \end{aligned}$$

You should check the solution set, $\{(2, -1, 3)\}$, by substituting those values into the original system of equations. ■ ■ ■

Practice Problem 6 Solve the system of equations.

$$\begin{cases} 2x + y - z = 7 \\ x - 3y - 3z = 4 \\ 4x + y + z = 3 \end{cases}$$ ■

EXAMPLE 7 **Attempting to Solve a System with No Solution**

Solve the system of equations by Gaussian elimination.

$$\begin{cases} y + 5z = -4 \\ x + 4y + 3z = -2 \\ 2x + 7y + z = 8 \end{cases}$$

SOLUTION

Step 1

$$A = \begin{bmatrix} 0 & 1 & 5 & | & -4 \\ 1 & 4 & 3 & | & -2 \\ 2 & 7 & 1 & | & 8 \end{bmatrix}$$

The augmented matrix of the system

Step 2 $\xrightarrow{R_1 \leftrightarrow R_2} \begin{bmatrix} 1 & 4 & 3 & | & -2 \\ 0 & 1 & 5 & | & -4 \\ 2 & 7 & 1 & | & 8 \end{bmatrix}$

This operation produces a 1 in the $(1, 1)$ position.

$$\xrightarrow{-2R_1 + R_3 \rightarrow R_3} \begin{bmatrix} 1 & 4 & 3 & | & -2 \\ 0 & 1 & 5 & | & -4 \\ 0 & -1 & -5 & | & 12 \end{bmatrix}$$

Row 2 already begins with a 0. The operation produces a 0 in $(3, 1)$ position.

$$\xrightarrow{R_2+R_3 \to R_3} \begin{bmatrix} 1 & 4 & 3 & -2 \\ 0 & 1 & 5 & -4 \\ 0 & 0 & 0 & 8 \end{bmatrix}$$

This operation produces a 0 in (3, 2) position.

$$\xrightarrow{\frac{1}{8}R_3} \begin{bmatrix} 1 & 4 & 3 & -2 \\ 0 & 1 & 5 & -4 \\ 0 & 0 & 0 & 1 \end{bmatrix}$$

The matrix is now in row-echelon form.

Step 3   The system of equations corresponding to the last matrix in Step 2 is

$$\begin{cases} x + 4y + 3z = -2 \\ \quad\quad y + 5z = -4 \\ \quad\quad\quad\quad 0 = 1 \end{cases}$$ A false statement

The equation $0 = 1$ can be rewritten as $0x + 0y + 0z = 1$. Because this equation is never true, we conclude that this system is inconsistent. Because this system is equivalent to the original system, the original system is also inconsistent.

RECALL

A system of equations is called *inconsistent* if it has no solution.

Step 4   The solution set for the system is $\varnothing$.   ■ ■ ■

Practice Problem 7   Solve the following system of equations by first transforming the augmented matrix into row-echelon form.

$$\begin{cases} 6x + 8y - 14z = 3 \\ 3x + 4y - 7z = 12 \\ 6x + 3y + z = 0 \end{cases}$$ ■

---

4   Use Gauss–Jordan elimination to solve a system.

Gauss–Jordan Elimination

If we continue the Gaussian elimination procedure until we have a *reduced* row-echelon form matrix, the procedure is called **Gauss–Jordan elimination**.

TECHNOLOGY CONNECTION

 Many graphing calculators can produce the unique reduced row-echelon form of an augmented matrix. Consider the initial augmented matrix A in Example 4.

```
[A]
[[1  -1  -1  1 ]
 [2  -3   1  10]
 [1   1  -2  0 ]]
```

We use the "rref" (reduced row-echelon form) option to obtain the reduced row-echelon form.

```
rref([A])
[[1  0  0  5]
 [0  1  0  1]
 [0  0  1  3]]
```

EXAMPLE 8   **Solving a System of Equations by Gauss–Jordan Elimination**

Solve the system given in Example 4 on page 555 by Gauss–Jordan elimination.

SOLUTION

$$\begin{cases} x - y - z = 1 \\ 2x - 3y + z = 10 \\ x + y - 2z = 0 \end{cases}$$ The given system

$$\begin{bmatrix} 1 & -1 & -1 & 1 \\ 0 & 1 & -3 & -8 \\ 0 & 0 & 5 & 15 \end{bmatrix}$$ The final augmented matrix of the system in Example 4

$$\xrightarrow{\frac{1}{5}R_3} \begin{bmatrix} 1 & -1 & -1 & 1 \\ 0 & 1 & -3 & -8 \\ 0 & 0 & 1 & 3 \end{bmatrix}$$ The matrix is now in row-echelon form.

$$\xrightarrow{R_2+R_1 \to R_1} \begin{bmatrix} 1 & 0 & -4 & -7 \\ 0 & 1 & -3 & -8 \\ 0 & 0 & 1 & 3 \end{bmatrix}$$ Produces only one nonzero value in column 2.

$$\xrightarrow[3R_3+R_2 \to R_2]{4R_3+R_1 \to R_1} \begin{bmatrix} 1 & 0 & 0 & 5 \\ 0 & 1 & 0 & 1 \\ 0 & 0 & 1 & 3 \end{bmatrix}$$ Produces only one nonzero value in column 3.

We now have an equivalent matrix in reduced row-echelon form. The corresponding system of equations for the last augmented matrix is

$$\begin{cases} x = 5 \\ y = 1 \\ z = 3 \end{cases}$$

Therefore, the solution set is $\{(5, 1, 3)\}$, as in Example 4. ■ ■ ■

Practice Problem 8 Solve the system by using Gauss–Jordan elimination.

$$\begin{cases} 2x - 3y - 2z = 0 \\ x + y - 2z = 7 \\ 3x - 5y - 5z = 3 \end{cases}$$ ■

EXAMPLE 9 **Solving a System with Infinitely Many Solutions**

Solve the system of equations.

$$\begin{cases} x + 2y + 5z = 4 \\ y + 4z = 4 \\ 2x + 4y + 10z = 8 \end{cases}$$

SOLUTION

The augmented matrix A of the system is given below. We want to find the equivalent system in reduced row-echelon form.

Because $a_{11} = 1$ and $a_{21} = 0$, we only need a zero at the $(3, 1)$ position of column (1).

$$A = \begin{bmatrix} 1 & 2 & 5 & | & 4 \\ 0 & 1 & 4 & | & 4 \\ 2 & 4 & 10 & | & 8 \end{bmatrix} \xrightarrow{-2R_1 + R_3 \rightarrow R_3} \begin{bmatrix} 1 & 2 & 5 & | & 4 \\ 0 & 1 & 4 & | & 4 \\ 0 & 0 & 0 & | & 0 \end{bmatrix}$$

Next, we need a zero at the $(1, 2)$ position.

$$\xrightarrow{-2R_2 + R_1 \rightarrow R_1} \begin{bmatrix} 1 & 0 & -3 & | & -4 \\ 0 & 1 & 4 & | & 4 \\ 0 & 0 & 0 & | & 0 \end{bmatrix}$$ The matrix is now in reduced row-echelon form.

The equivalent system is as follows:

$$x - 3z = -4$$
$$y + 4z = 4$$

Solving for x and y in terms of z, we obtain the following:

$$x = 3z - 4$$
$$y = -4z + 4$$

Each real number z results in a solution with $y = -4z + 4$ and $x = 3z - 4$, giving infinitely many solutions of the form $(3z - 4, -4z + 4, z)$. The solution set is $\{(3z - 4, -4z + 4, z)\}$ where z represents any real number. ■ ■ ■

Practice Problem 9 Solve the system of equations.

$$\begin{cases} x + z = -1 \\ 3y + 2z = 5 \\ 3x - 3y + z = -8 \end{cases}$$ ■

EXAMPLE 10   **Leontief Input–Output Model**

Consider an economy that has steel, coal, and transportation industries. There are two types of demands (measured in dollars) on the production of each industry: interindustry demand and external consumer demand. The outputs and requirements of the three industries are shown in Figure 6.1. For example, $1.00 of transportation output requires $0.10 from steel and $0.01 from coal.

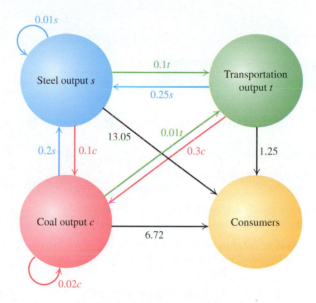

FIGURE 6.1 Output diagram of an economy

a. Write a system of equations that expresses the outputs of the three industries. Assume that all quantities are given in millions of dollars.

b. Verify that $s = 15$, $c = 10$, and $t = 8$ will meet both interindustry and consumer demand.

SOLUTION

a. To satisfy both consumer and interindustry demand, we obtain the following system of outputs. (s denotes total steel output; c, total coal output; t, total transportation output.)

$$\begin{cases} s = 0.01s + 0.1t + 0.1c + 13.05 & \text{Distribution of steel output} \\ c = 0.2s + 0.01t + 0.02c + 6.72 & \text{Distribution of coal output} \\ t = 0.25s + 0.3c + 1.25 & \text{Distribution of transportation output} \end{cases}$$

You can rewrite this system as:

$$\begin{cases} 0.99s - 0.1c - 0.1t = 13.05 & (1) \\ -0.2s + 0.98c - 0.01t = 6.72 & (2) \\ -0.25s - 0.3c + t = 1.25 & (3) \end{cases}$$

b. To verify that the given numbers (obtained by solving the system above using matrices) satisfy those equations, we substitute $s = 15$, $c = 10$, and $t = 8$ into each equation.

$$(0.99)(15) - (0.1)(10) - (0.1)(8) = 13.05 \checkmark \quad (1)$$
$$(-0.20)(15) + (0.98)(10) - (0.01)(8) = 6.72 \checkmark \quad (2)$$
$$(-0.25)(15) - (0.30)(10) + 8 = 1.25 \checkmark \quad (3)$$

In other words, output levels of $s = 15, c = 10$, and $t = 8$ million dollars will meet interindustry and consumer demands.   ■ ■ ■

Practice Problem 10   In Example 10, assume that consumer demand (in millions of dollars) is 12.46 for steel, 3 for coal, and 2.7 for transportation.

a. Write a system of equations to express the outputs for the three industries.

b. Verify that $s = 14, c = 6, t = 8$ will meet both interindustry and consumer demand.   ■

SECTION 6.1 ■ Exercises

A EXERCISES Basic Skills and Concepts

1. A matrix is any rectangular array of _____.

2. The array of coefficients and constants in a linear system is called the _____ of the system.

3. Two matrices are row-equivalent if one can be obtained from the other by a sequence of _____.

4. If a matrix is in row-echelon form and each leading entry 1 is the only nonzero entry in its column, then the matrix is in _____ form.

5. *True or False* If A is a 3×4 matrix, then each row of A has three entries.

6. *True or False* Every $m \times n$ ($n \geq 2$) matrix is the matrix of some linear system.

7. *True or False* The augmented matrix for the system of equations

$$\begin{cases} 2x + 3y & = 5 \\ 3y - 4z = -1 \\ x + 2z = 3 \end{cases} \quad \text{is} \quad \begin{bmatrix} 2 & 3 & 5 \\ 3 & -4 & -1 \\ 1 & 2 & 3 \end{bmatrix}.$$

8. *True or False* The augmented matrix for the system of equations

$$\begin{cases} 2x - y = 1 \\ 3x + y = 9 \\ 5x - 2y = 4 \end{cases} \quad \text{is} \quad \begin{bmatrix} 2x & -y & 1 \\ 3x & y & 9 \\ 5x & -2y & 4 \end{bmatrix}.$$

In Exercises 9–14, determine the order of each matrix.

9. $[7]$

10. $[1 \quad 3 \quad 5 \quad 9]$

11. $\begin{bmatrix} 3 & 4.3 & 7.5 & 8 \\ 2 & -1 & -3 & 0 \end{bmatrix}$

12. $\begin{bmatrix} -1 & 2 & 3 \\ 4 & -5 & 7 \\ 1 & 0 & 0 \end{bmatrix}$

13. $\begin{bmatrix} 4 & 5 & 1 \\ 0 & 0 & 0 \end{bmatrix}$

14. $\begin{bmatrix} -5 & 2 & 3 & 7 \\ 2.5 & e & -\pi & \frac{1}{2} \\ -e & \pi & 0 & 1 \end{bmatrix}$

15. Let $A = \begin{bmatrix} 1 & 2 & 3 & 4 \\ 5 & 6 & 7 & 8 \\ 9 & 10 & 11 & 12 \end{bmatrix}$. Identify the entries $a_{13}, a_{31}, a_{33},$ and a_{34}.

16. In matrix A from Exercise 15, identify the (i, j)th location of each entry.
a. 7   **b.** 10
c. 4   **d.** 12

17. Is the following array a matrix? Why or why not?

$$\begin{bmatrix} 2 & 0 & 4 \\ 1 & -3 & 5 \\ 0 & 0 \end{bmatrix}$$

18. Write the matrix A with the following entries:

$$a_{13} = 5, a_{22} = 2, a_{11} = 6, a_{12} = -5, a_{23} = 4, \text{ and } a_{21} = 7$$

In Exercises 19–24, write the augmented matrix for each system of linear equations.

19. $\begin{cases} 2x + 4y = 2 \\ x - 3y = 1 \end{cases}$ **20.** $\begin{cases} x_1 + 2x_2 = 7 \\ 3x_1 + 5x_2 = 11 \end{cases}$

21. $\begin{cases} 5x_1 - 11 = 7x_2 \\ 17x_2 - 19 = 13x_1 \end{cases}$ **22.** $\begin{cases} 2v - 10 = 3u \\ 5u + 7 = v \end{cases}$

23. $\begin{cases} -x + 2y + 3z = 8 \\ 2x - 3y + 9z = 16 \\ 4x - 5y - 6z = 32 \end{cases}$ **24.** $\begin{cases} x - y = 2 \\ 2x + 3z = -5 \\ y - 2z = 7 \end{cases}$

In Exercises 25–28, write the system of linear equations represented by each augmented matrix. Use $x, y,$ and z as the variables.

25. $\begin{bmatrix} 1 & 2 & -3 & 4 \\ -2 & -3 & 1 & 5 \\ 3 & -3 & 2 & 7 \end{bmatrix}$ **26.** $\begin{bmatrix} -1 & 2 & 3 & 6 \\ 2 & 3 & 1 & 2 \\ 4 & 3 & 2 & 1 \end{bmatrix}$

27. $\begin{bmatrix} 1 & -1 & 1 & 2 \\ 2 & 1 & -3 & 6 \end{bmatrix}$ **28.** $\begin{bmatrix} 1 & 1 & 1 & 2 \\ 1 & -1 & -1 & 4 \\ 2 & 3 & 1 & 6 \\ -1 & 1 & -1 & 8 \end{bmatrix}$

In Exercises 29–32, perform the indicated elementary row operations in the stated order.

29. $\begin{bmatrix} 2 & 3 & 5 \\ 1 & 2 & 3 \end{bmatrix}$; **(i)** $R_1 \leftrightarrow R_2$, **(ii)** $-2R_1 + R_2 \rightarrow R_2$, **(iii)** $-R_2$

30. $\begin{bmatrix} 2 & 4 & 2 \\ 1 & 5 & 7 \end{bmatrix}$; **(i)** $\frac{1}{2}R_1$, **(ii)** $(-1) R_1 + R_2 \rightarrow R_2$, **(iii)** $\frac{1}{3}R_2$

31. $\begin{bmatrix} 1 & 2 & 3 & 4 \\ 0 & 4 & -3 & 11 \\ 0 & 1 & 5 & -3 \end{bmatrix}$; **(i)** $R_2 \leftrightarrow R_3$, **(ii)** $-4R_2 + R_3 \to R_3$, **(iii)** $-\frac{1}{23}R_3$

32. $\begin{bmatrix} 1 & \frac{3}{2} & -2 & \frac{1}{2} \\ 0 & 2 & 3 & 4 \\ -2 & -1 & 7 & 3 \end{bmatrix}$; **(i)** $2R_1 + R_3 \to R_3$, **(ii)** $\frac{1}{2}R_2 + R_3 \to R_3$

In Exercises 33–36, identify the elementary row operation used and supply the missing entries in each row-equivalent matrix.

33. $\begin{bmatrix} 4 & 5 & -7 \\ 5 & 4 & -2 \end{bmatrix} \to \begin{bmatrix} 1 & ? & -\frac{7}{4} \\ 5 & 4 & -2 \end{bmatrix} \to \begin{bmatrix} 1 & ? & -\frac{7}{4} \\ 0 & -\frac{9}{4} & ? \end{bmatrix} \to$

$\begin{bmatrix} 1 & ? & -\frac{7}{4} \\ 0 & 1 & ? \end{bmatrix}$

34. $\begin{bmatrix} 2 & 6 & -8 \\ 3 & -1 & 2 \end{bmatrix} \to \begin{bmatrix} 1 & 3 & -? \\ 3 & -1 & 2 \end{bmatrix} \to \begin{bmatrix} 1 & 3 & -? \\ 0 & ? & 14 \end{bmatrix} \to$

$\begin{bmatrix} 1 & 3 & ? \\ 0 & 1 & ? \end{bmatrix}$

35. $\begin{bmatrix} 1 & 4 & 3 & 1 \\ 0 & -3 & -2 & 0 \\ 0 & 7 & 5 & -3 \end{bmatrix} \to \begin{bmatrix} 1 & 4 & 3 & 1 \\ 0 & 1 & ? & 0 \\ 0 & 7 & 5 & -3 \end{bmatrix} \to$

$\begin{bmatrix} 1 & 4 & 3 & 1 \\ 0 & 1 & ? & 0 \\ 0 & 0 & ? & -3 \end{bmatrix} \to \begin{bmatrix} 1 & 4 & 3 & 1 \\ 0 & 1 & ? & 0 \\ 0 & 0 & 1 & ? \end{bmatrix}$

36. $\begin{bmatrix} 1 & 1 & 1 & 3 \\ 2 & 3 & 3 & 8 \\ 1 & -3 & -2 & 5 \end{bmatrix} \to \begin{bmatrix} 1 & 1 & 1 & 3 \\ 0 & 1 & ? & 2 \\ 1 & -3 & -2 & 5 \end{bmatrix} \to$

$\begin{bmatrix} 1 & 1 & 1 & 3 \\ 0 & 1 & ? & 2 \\ 0 & -4 & ? & 2 \end{bmatrix} \to \begin{bmatrix} 1 & 1 & 1 & 3 \\ 0 & 1 & ? & 2 \\ 0 & 0 & ? & 10 \end{bmatrix}$

In Exercises 37–44, determine whether each matrix is in row-echelon form. If your answer is no, explain why. If your answer is yes, is the matrix in reduced row-echelon form?

37. $\begin{bmatrix} 0 & 1 & 2 \\ 1 & 0 & 3 \end{bmatrix}$

38. $\begin{bmatrix} 1 & 2 & 5 \\ 0 & 3 & 6 \end{bmatrix}$

39. $\begin{bmatrix} 1 & 0 & 0 & 2 \\ 0 & 1 & 0 & 3 \\ 0 & 0 & 1 & 4 \end{bmatrix}$

40. $\begin{bmatrix} 0 & 0 & 1 & 2 \\ 0 & 1 & 0 & 1 \\ 1 & 0 & 0 & 3 \end{bmatrix}$

41. $\begin{bmatrix} 1 & 2 & 0 & 2 \\ 0 & 0 & 1 & 5 \end{bmatrix}$

42. $\begin{bmatrix} 0 & 1 & -1 & -2 & 0 & 2 & 0 \\ 0 & 0 & 0 & 0 & 1 & 2 & 0 \end{bmatrix}$

43. $\begin{bmatrix} 1 & 0 & 0 & -2 \\ 0 & 1 & 0 & 3 \\ 0 & 0 & 1 & 2 \\ 0 & 0 & 0 & 0 \end{bmatrix}$

44. $\begin{bmatrix} 1 & 0 & 0 & 3 \\ 0 & 1 & 0 & 4 \\ 0 & 0 & 0 & 5 \\ 0 & 0 & 1 & 3 \end{bmatrix}$

In Exercises 45–54, the augmented matrix of a system of equations has been transformed to an equivalent matrix in row-echelon form or reduced row-echelon form. Using $x, y, z,$ and w as variables, write the system of equations corresponding to the matrix. If the system is consistent, solve it.

45. $\left[\begin{array}{cc|c} 1 & 2 & 1 \\ 0 & 1 & -2 \end{array}\right]$

46. $\left[\begin{array}{cc|c} 1 & 0 & 2 \\ 0 & 1 & 3 \end{array}\right]$

47. $\left[\begin{array}{ccc|c} 1 & 4 & 2 & 2 \\ 0 & 0 & 1 & 3 \end{array}\right]$

48. $\left[\begin{array}{ccc|c} 1 & 2 & 3 & 4 \\ 0 & 0 & 0 & 1 \end{array}\right]$

49. $\left[\begin{array}{ccc|c} 1 & 2 & 3 & 2 \\ 0 & 1 & -2 & 4 \\ 0 & 0 & 1 & -1 \end{array}\right]$

50. $\left[\begin{array}{ccc|c} 1 & 0 & 2 & 12 \\ 0 & 1 & 3 & 12 \\ 0 & 0 & 1 & 5 \end{array}\right]$

51. $\left[\begin{array}{cccc|c} 1 & 0 & 0 & 0 & 2 \\ 0 & 1 & 0 & 0 & -5 \\ 0 & 0 & 1 & 2 & 3 \\ 0 & 0 & 0 & 0 & 0 \end{array}\right]$

52. $\left[\begin{array}{cccc|c} 1 & 0 & 0 & 0 & 4 \\ 0 & 1 & 0 & 0 & 3 \\ 0 & 0 & 0 & 1 & 2 \end{array}\right]$

53. $\left[\begin{array}{cccc|c} 1 & 0 & 0 & 0 & -5 \\ 0 & 1 & 0 & 0 & 4 \\ 0 & 0 & 1 & 2 & 3 \\ 0 & 0 & 0 & 1 & 0 \end{array}\right]$

54. $\left[\begin{array}{cccc|c} 1 & 0 & 0 & 0 & 3 \\ 0 & 1 & 0 & 0 & 2 \\ 0 & 0 & 1 & 0 & 0 \\ 0 & 0 & 0 & 0 & 1 \end{array}\right]$

In Exercises 55–74, solve each system of equations by Gaussian elimination.

55. $\begin{cases} x - 2y = 11 \\ 2x - y = 13 \end{cases}$

56. $\begin{cases} 3x - 2y = 4 \\ 4x - 3y = 5 \end{cases}$

57. $\begin{cases} 2x - y = 5 \\ 6x - 3y = 9 \end{cases}$

58. $\begin{cases} 3x + 2y = 1 \\ 6x + 4y = 3 \end{cases}$

59. $\begin{cases} 2x - 3y = 3 \\ 4x - y = 11 \end{cases}$

60. $\begin{cases} 2x + 3y = 13 \\ 3x - 5y = -9 \end{cases}$

61. $\begin{cases} 3x - 5y = 4 \\ 4x - 15y = 13 \end{cases}$

62. $\begin{cases} -2x + 4y = 1 \\ 3x - 5y = -9 \end{cases}$

63. $\begin{cases} x - y = 1 \\ 2x + y = 5 \\ 3x - 4y = 2 \end{cases}$

64. $\begin{cases} y = 2x + 1 \\ 3x + 2y + 1.5 = 0 \\ 4x - 2y + 2 = 0 \end{cases}$

65. $\begin{cases} x + y + z = 6 \\ x - y + z = 2 \\ 2x + y - z = 1 \end{cases}$

66. $\begin{cases} 2x + 4y + z = 5 \\ x + y + z = 6 \\ 2x + 3y + z = 6 \end{cases}$

67. $\begin{cases} x - 2y + 3z = 1 \\ 3x - 6y + 9z = 6 \\ -x + 3y - 2z = 1 \end{cases}$

68. $\begin{cases} 5x + 8y + 6z = 14 \\ 2x + 5y + z = 12 \\ -3x + 6y - 12z = 20 \end{cases}$

69. $\begin{cases} 2x + 3y - z = 9 \\ x + y + z = 9 \\ 3x - y - z = -1 \end{cases}$

70. $\begin{cases} x + y + 2z = 4 \\ 2x - y + 3z = 9 \\ 3x - y - z = 2 \end{cases}$

71. $\begin{cases} x + y - 3z = -8 \\ 8x + 6y - 10z = -18 \\ 2x + y + z = 7 \end{cases}$ **72.** $\begin{cases} 2x + 3y - z = -2 \\ 5x + 5y - 3z = -1 \\ 3x + 2y - 2z = 1 \end{cases}$

73. $\begin{cases} 3x + 2y + 4z = 19 \\ 2x - y + z = 3 \\ 6x + 7y - z = 17 \end{cases}$ **74.** $\begin{cases} 4x + 3y + z = 8 \\ 2x + y + 4z = -4 \\ 3x + z = 1 \end{cases}$

In Exercises 75–82, solve each system of equations by Gauss–Jordan elimination.

75. $\begin{cases} x + 2y - z = 6 \\ 3x + y + 2z = 3 \\ 2x + 5y + 3z = 9 \end{cases}$ **76.** $\begin{cases} 2x + 4y - z = 9 \\ x + 3y - 3z = 4 \\ 3x + y + 2z = 7 \end{cases}$

77. $\begin{cases} 3x - 6y + 9z = 2 \\ 2x - 4y + 6z = 5 \\ -x + 3y - 2z = -1 \end{cases}$ **78.** $\begin{cases} 3x - 4y + z = 1 \\ 4x - 6y + z = 2 \\ 5x - 8y + z = 4 \end{cases}$

79. $\begin{cases} x + y - z = 4 \\ x + 3y + 5z = 10 \\ 3x + 5y + 3z = 18 \end{cases}$ **80.** $\begin{cases} x + y + z = -5 \\ 2x - y - z = -4 \\ y + z = -2 \end{cases}$

81. $\begin{cases} x - y = 1 \\ x - z = -1 \\ 2x + y - z = 3 \end{cases}$ **82.** $\begin{cases} 4x + 5z = 7 \\ y - 6z = 8 \\ 3x + 4y = 9 \end{cases}$

83. The reduced row-echelon forms of the augmented matrices of three systems of equations are given. How many solutions does each system have?

a. $\begin{bmatrix} 1 & 0 & 2 \\ 0 & 1 & 3 \end{bmatrix}$ **b.** $\begin{bmatrix} 1 & 2 & 3 & 8 \\ 0 & 0 & 1 & 2 \\ 0 & 0 & 0 & 0 \end{bmatrix}$

c. $\begin{bmatrix} 1 & 0 & 2 \\ 0 & 0 & 3 \end{bmatrix}$

In Exercises 84–86, determine whether the statement is true or false and justify your answer.

84. If A is a 5×7 matrix, then each column of A has seven entries.

85. The matrix $E = \begin{bmatrix} 1 & 3 & 0 \\ 0 & 0 & 1 \\ 0 & 0 & 0 \end{bmatrix}$ is in reduced row-echelon form.

86. If a matrix A is in reduced row-echelon form, then at least one of the entries in each column must be a 1.

B EXERCISES Applying the Concepts

In Exercises 87–90, Leontief's input–output model of a simplified economy is described. (See Example 10.) For each exercise, do the following.

a. Set up (without solving) a linear system whose solution will represent the required production schedule.

b. Write an augmented matrix for the economy.

c. Solve the system of equations by using part (b) to find the production schedule that will meet interindustry and consumer demand.

87. Two-sector economy. Consider an economy with only two industries: A and B. Suppose industry B needs

$0.10 worth of A's product for each $1.00 of output B produces and that industry A needs $0.20 worth of B's product for each $1.00 of output A produces. Consumer demand for A's product is $1000, and consumer demand for B's product is $780.

88. A company with two branches. A company has two interacting branches: A and B. Branch A consumes $0.50 of its own output and $0.20 of B's output for every $1.00 it produces. Branch B consumes $0.60 of A's output and $0.30 of its own output for every $1.00 of output. Consumer demand for A's product is $50,000, and consumer demand for B's product is $40,000.

89. Three-sector economy. An economy has three sectors: labor, transportation, and food industries. Suppose the demand on $1.00 in labor is $0.40 for transportation and $0.20 for food; the demand on $1.00 in transportation is $0.50 for labor and $0.30 for transportation; and the demand on $1.00 in food production is $0.50 for labor, $0.05 for transportation, and $0.35 for food. Outside consumer demand for the current production period is $10,000 for labor, $20,000 for transportation, and $10,000 for food.

90. Three-sector economy. Consider an economy with three industries A, B, and C, with outputs a, b, and c, respectively. Demand on the three industries is shown in the figure.

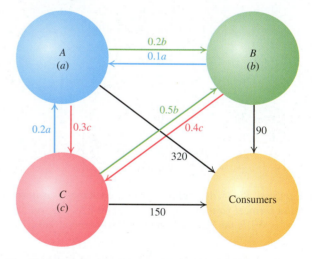

91. Heat transfer. In a study of heat transfer in a grid of wires, the temperature at an exterior node is maintained at a constant value (in °F) as shown in the accompanying figure. When the grid is in thermal equilibrium, the temperature at an interior node is the average of the temperatures at the four adjacent nodes. For instance

$$T_1 = \frac{0 + 0 + 300 + T_2}{4}, \text{ or } 4T_1 - T_2 = 300.$$ Find the temperatures $T_1, T_2,$ and T_3 when the grid is in thermal equilibrium.

92. Repeat Exercise 91 for the following grid of wires.

93. Traffic flow. The sketch shows the intersections of certain one-way streets. The traffic volume during one hour was observed. To keep the traffic moving, traffic controllers use the formula that the number of cars per hour entering an intersection equals the number of cars per hour exiting the intersection.

a. Set up a system of equations that keeps traffic moving.

b. Solve the system of equations in part (a).

94. Repeat Exercise 93 for the following pattern of one-way streets.

95. Modeling. A football was punted. Its height h above the ground at time t is given by the following table.

Time t (seconds)	Height h (feet)
$t = 0.5$	31
$t = 1$	51
$t = 2$	67

a. Use the system of equations to find the equation of the parabola of the form $h = at^2 + bt + c$.

b. What is the hang time (the time it takes for the ball to touch the ground) of the punt?

c. What is the maximum height of the ball?

96. Study Schedules A student spent a total of 18 hours one week studying for three courses: biology, English, and history. Twice the time spent studying English equals the time spent studying biology and history combined. Four times the time spent studying history equals the time spent studying biology and English combined. How much time (hours and minutes) did she spend studying each subject?

97. Coffee Mixture Chicory is to be combined with two types of coffee to fill an order for 14 lb of the resulting blend. There are 10 more pounds of coffee in the blend than there is chicory. The coffees sell for $8.50 per lb and $11.00 per lb, and chicory sells for $7.00 per lb. How many pounds of chicory and of each type of coffee must be combined if the price of the blend is $9.00 per lb?

C EXERCISES Beyond the Basics

98. Partial fractions. Find the partial-fraction decomposition of the rational expression

$$\frac{2x^2 + 22x}{(x - 1)^2(x + 2)(x + 3)}.$$

Use matrix methods to find the partial fractions.

In Exercises 99 and 100, solve each system of equations by Gauss–Jordan elimination.

99. $\begin{cases} x + y + z + w = 0 \\ x + 3y + 2z + 4w = 0 \\ 2x + z - w = 0 \end{cases}$

100. $\begin{cases} x - y + z - w = 4 \\ x + 2y + z + w = 2 \\ 2x + 3y + 4z + 5w = 5 \\ 3x + 4y + 2z - w = 8 \end{cases}$

101. $\begin{cases} -2x + 3y - 4z + 2w = 5 \\ 2x - y + z - w = 3 \\ x - 2y - 3z - 4w = -7 \\ x - y + 2z - 2w = -5 \end{cases}$

102. $\begin{cases} x + y + z + w = 4 \\ 2x - y + z - w = 2 \\ 3x - 6y - 3z - 3w = 9 \\ x - y + 2z - 2w = -7 \end{cases}$

103. Let $A = \begin{bmatrix} 1 & 2 & -3 & 1 \\ 1 & 0 & -3 & 2 \\ 0 & 1 & 1 & 0 \\ 2 & 3 & 0 & -2 \end{bmatrix}$.

    a. Find two different matrices B and C in row-echelon form that are row-equivalent to A.

    b. Show that the reduced row-echelon form of the matrices B and C of part (a) produce the same matrix.

104. Consider the following system of linear equations:

$$\begin{cases} x + y + z = 6 \\ x - y + z = 2 \\ 2x + y - z = 1 \end{cases}$$

    a. Find two different matrices B and C in row-echelon form that yield the same solution set.

    b. Show that the reduced row-echelon form of the matrices B and C of part (a) produces the same matrix.

105. a. Use the methods of this section to solve the general 2×2 system

$$\begin{cases} ax + by = m \\ cx + dy = n \end{cases}.$$

    b. Under what conditions on the coefficients does the system have **(i)** a unique solution, **(ii)** no solution, and **(iii)** infinitely many solutions?

106. Construct three different augmented matrices for linear systems whose solution set is $x = 1, y = 0, z = -2$. There are many different correct answers.

In Exercises 107 and 108, solve each system of equations by using row operations.

107. $\begin{cases} \log x + \log y + \log z = 6 \\ 3\log x - \log y + 3\log z = 10 \\ 5\log x + 5\log y - 4\log z = 3 \end{cases}$
[*Hint:* Let $u = \log x$, $v = \log y$, and $w = \log z$.]

108. $\begin{cases} 4 \cdot 2^x - 3 \cdot 3^y + 5^z = 1 \\ 2^x + 4 \cdot 3^y - 2 \cdot 5^z = 10 \\ 2 \cdot 2^x - 2 \cdot 3^y + 3 \cdot 5^z = 4 \end{cases}$
[*Hint:* Let $u = 2^x$, $v = 3^y$, and $w = 5^z$.]

109. Find the cubic function $y = ax^3 + bx^2 + cx + d$ whose graph passes through the points $(1, 5)$, $(-1, 1)$, $(2, 7)$, and $(-2, 11)$.

110. Find the cubic function $y = ax^3 + bx^2 + cx + d$ whose graph passes through the points $(1, 8)$, $(-1, 2)$, $(2, 8)$, and $(-2, 20)$.

Critical Thinking

111. Find the form of all 2×1 matrices in reduced row-echelon form.

112. Find the form of all 2×2 matrices in reduced row-echelon form.

113. Suppose a matrix A is transformed to a matrix B by an elementary row operation. Is there an elementary row operation that transforms B to A? Explain.

114. Are the following statements true or false? Justify your answers. Suppose a matrix A is in reduced row-echelon form.

    a. If we delete a row of A, then the remaining matrix is in reduced row-echelon form.

    b. Repeat part (a), replacing *row* with *column*.

Matrix Algebra

Before Starting this Section, Review

1. Properties of real numbers (Section P.1, page 10)

2. Solving systems of equations (Section 5.1, page 480)

Objectives

1. Define equality of two matrices.

2. Define matrix addition and scalar multiplication.

3. Define matrix multiplication.

4. Apply matrix multiplication to computer graphics.

COMPUTER GRAPHICS

The following fields illustrate the widespread and growing applications of computer graphics.

1. The *automobile industry*. Both computer-aided design (CAD) and computer-aided manufacturing (CAM) have revolutionized the automotive industry. Many months before a new car is built, engineers design and construct a *mathematical car*—a wire-frame model that exists only in computer memory and on graphics display terminals. The mathematical model organizes and influences each step of the design and manufacture of the car.

2. The *entertainment industry*. Computer graphics are used by the entertainment industry with spectacular success. They were used in *Star Wars* and other movies for special effects, and they are used in the design and creation of arcade games.

3. *Military*. From the beginning of the widespread use of computers (in the 1960s), computer graphics have been used by the military to prepare for or defend against war. Modern weaponry depends heavily on computer-generated images.

Engineers use matrix algebra techniques on graphic images to change their orientation of scale, to zoom in on them, or to switch between two- and three-dimensional views. In Example 10 and in the exercises, we examine how basic matrix algebra can be used to manipulate graphical images. ■

---

1 Define equality of two matrices.

Equality of Matrices

In the previous section, a matrix was defined as a rectangular array of numbers written within brackets. In this section and the next two, you will learn about the operations with and uses of matrices. A matrix may be represented in any of the following three ways:

1. A matrix may be denoted by capital letters such as A, B, or C.

2. A matrix may be denoted by its representative (i, j)th entry: $[a_{ij}]$, $[b_{ij}]$, $[c_{ij}]$, and so on.

3. A matrix may be written as a rectangular array of numbers:

$$A = [a_{ij}] = \begin{bmatrix} a_{11} & a_{12} & \cdots & a_{1n} \\ a_{21} & a_{22} & \cdots & a_{2n} \\ \vdots & \vdots & \cdots & \vdots \\ a_{m1} & a_{m2} & \cdots & a_{mn} \end{bmatrix}$$

We now examine some *algebraic* properties of matrices.

EQUALITY OF MATRICES

Two matrices $A = [a_{ij}]$ and $B = [b_{ij}]$ are said to be **equal,** written $A = B$, if

1. A and B have the same order $m \times n$ (that is, A and B have the same number m of rows and same number n of columns).

2. $a_{ij} = b_{ij}$ for all i and j; that is, each (i, j)th entry of A is equal to the corresponding (i, j)th entry of B.

EXAMPLE 1   **Determining Equality of Matrices**

Find x and y such that

$$\begin{bmatrix} x + y & 2 \\ 3 & x - y \end{bmatrix} = \begin{bmatrix} 4 & 2 \\ 3 & 1 \end{bmatrix}.$$

SOLUTION

Because equal matrices must have identical entries in corresponding positions, we have the following system of equations.

$$\begin{cases} x + y = 4 & (1) \quad \text{Equate the entries in the } (1,1) \text{ position.} \\ x - y = 1 & (2) \quad \text{Equate the entries in the } (2,2) \text{ position.} \end{cases}$$

Solve this system: Add the two equations to obtain $x = \dfrac{5}{2}$. Substitute $x = \dfrac{5}{2}$ in the equation $x - y = 1$ and solve for y to obtain $y = \dfrac{3}{2}$. So $x = \dfrac{5}{2}$ and $y = \dfrac{3}{2}$. ■ ■ ■

Practice Problem 1   Find x and y such that

$$\begin{bmatrix} 1 & 2x - y \\ x + y & 5 \end{bmatrix} = \begin{bmatrix} 1 & 1 \\ 2 & 5 \end{bmatrix}.$$

■

2 Define matrix addition and scalar multiplication.

Matrix Addition and Scalar Multiplication

Matrix addition differs somewhat from real number addition. You can add any two real numbers, but you can add two matrices only if they have the same order.

MATRIX ADDITION

If $A = [a_{ij}]$ and $B = [b_{ij}]$ are two $m \times n$ matrices, their **sum** $A + B$ is the $m \times n$ matrix defined by

$$A + B = [a_{ij} + b_{ij}],$$

for all i and j.

To find the sum of two matrices with a graphing calculator, name the matrices and then use the regular addition key with the names of the matrices.

Consider the matrices A and B in Example 2 as follows:

The sum is as follows:

The definition of matrix addition says that

(i) we can add two matrices of the same order simply by adding their corresponding entries, but that

(ii) the sum of two matrices of different orders is not defined.

EXAMPLE 2   Adding Matrices

Let $A = \begin{bmatrix} -2 & 3 \\ 4 & -1 \\ 0 & 2 \end{bmatrix}$, $B = \begin{bmatrix} 6 & 4 \\ -7 & 9 \\ 8 & -5 \end{bmatrix}$, and $C = \begin{bmatrix} 1 & 2 \\ 3 & 4 \end{bmatrix}$. Find each sum if possible.

a. $A + B$ **b.** $A + C$

SOLUTION

a. Because A and B have the same order, $A + B$ is defined. We have

$$A + B = \begin{bmatrix} -2 & 3 \\ 4 & -1 \\ 0 & 2 \end{bmatrix} + \begin{bmatrix} 6 & 4 \\ -7 & 9 \\ 8 & -5 \end{bmatrix} = \begin{bmatrix} -2+6 & 3+4 \\ 4-7 & -1+9 \\ 0+8 & 2-5 \end{bmatrix} = \begin{bmatrix} 4 & 7 \\ -3 & 8 \\ 8 & -3 \end{bmatrix}.$$

b. $A + C$ is not defined because A and C do not have the same order. ■ ■ ■

Practice Problem 2 Let $A = \begin{bmatrix} 2 & -1 & 4 \\ 5 & 0 & 9 \end{bmatrix}$ and $B = \begin{bmatrix} -8 & 2 & 9 \\ 7 & 3 & 6 \end{bmatrix}$. Find $A + B$. ■

You can also multiply any matrix A by any real number c. The real number c is called a **scalar**, and the multiplication process is called *scalar multiplication*.

SCALAR MULTIPLICATION

Let $A = [a_{ij}]$ be an $m \times n$ matrix and let c be a real number. Then the **scalar product** of c and A is denoted by cA and is defined by

$$cA = [ca_{ij}].$$

In other words, to multiply a matrix A by a real number c, we multiply each entry of A by c. When $c = -1$, we frequently write $-A$ instead of $(-1)A$. Further, the *difference* of two matrices, $A - B$, is defined to be the sum $A + (-1)B$.

MATRIX SUBTRACTION

If A and B are two $m \times n$ matrices, then their **difference** is defined by

$$A - B = A + (-1)B.$$

Subtraction $A - B$ is performed by subtracting the corresponding entries of B from those of A.

EXAMPLE 3   Performing Scalar Multiplication and Matrix Subtraction

Let $A = \begin{bmatrix} 1 & 2 & 0 \\ -1 & 3 & 1 \\ 2 & -1 & 4 \end{bmatrix}$ and $B = \begin{bmatrix} 2 & 1 & 3 \\ 1 & 0 & -2 \\ -3 & 4 & 5 \end{bmatrix}$. Find the following.

a. $3A$ **b.** $2B$ **c.** $3A - 2B$

SOLUTION

a. $3A = 3\begin{bmatrix} 1 & 2 & 0 \\ -1 & 3 & 1 \\ 2 & -1 & 4 \end{bmatrix} = \begin{bmatrix} 3(1) & 3(2) & 3(0) \\ 3(-1) & 3(3) & 3(1) \\ 3(2) & 3(-1) & 3(4) \end{bmatrix} = \begin{bmatrix} 3 & 6 & 0 \\ -3 & 9 & 3 \\ 6 & -3 & 12 \end{bmatrix}$

b. $2B = 2\begin{bmatrix} 2 & 1 & 3 \\ 1 & 0 & -2 \\ -3 & 4 & 5 \end{bmatrix} = \begin{bmatrix} 2(2) & 2(1) & 2(3) \\ 2(1) & 2(0) & 2(-2) \\ 2(-3) & 2(4) & 2(5) \end{bmatrix} = \begin{bmatrix} 4 & 2 & 6 \\ 2 & 0 & -4 \\ -6 & 8 & 10 \end{bmatrix}$

c. $3A - 2B = 3A + (-1)2B$

$= \begin{bmatrix} 3 & 6 & 0 \\ -3 & 9 & 3 \\ 6 & -3 & 12 \end{bmatrix} + \begin{bmatrix} -4 & -2 & -6 \\ -2 & 0 & 4 \\ 6 & -8 & -10 \end{bmatrix}$ Substitute from parts **a** and **b**, changing the sign of each entry in B.

$= \begin{bmatrix} 3 + (-4) & 6 + (-2) & 0 + (-6) \\ -3 + (-2) & 9 + 0 & 3 + 4 \\ 6 + 6 & -3 + (-8) & 12 + (-10) \end{bmatrix} = \begin{bmatrix} -1 & 4 & -6 \\ -5 & 9 & 7 \\ 12 & -11 & 2 \end{bmatrix}$ ■ ■ ■

Practice Problem 3 Let $A = \begin{bmatrix} 7 & -4 \\ 3 & 6 \\ 0 & -2 \end{bmatrix}$ and $B = \begin{bmatrix} 1 & -3 \\ 2 & 2 \\ 5 & 8 \end{bmatrix}$. Find $2A - 3B$. ■

The $m \times n$ matrix whose entries are all equal to 0 is called the $m \times n$ **zero matrix** and is denoted by $\mathbf{0}_{mn}$. When m and n are understood from the context, we simply write $\mathbf{0}$. The matrix $-A = (-1)A$ is called the **negative** of A. Many of the algebraic properties of matrix addition and scalar multiplication are similar to the familiar properties of real numbers.

MATRIX ADDITION AND SCALAR MULTIPLICATION PROPERTIES

Let A, B, and C be $m \times n$ matrices and c and d be scalars.

1.	$A + B = B + A$	Commutative property of addition
2.	$A + (B + C) = (A + B) + C$	Associative property of addition
3.	$A + \mathbf{0} = \mathbf{0} + A = A$	Additive identity property
4.	$A + (-A) = (-A) + A = \mathbf{0}$	Additive inverse property
5.	$(cd)A = c(dA)$	Associative property of scalar multiplication
6.	$1A = A$	Scalar identity property
7.	$c(A + B) = cA + cB$	Distributive property
8.	$(c + d)A = cA + dA$	Distributive property

In the next example, we use these properties to solve an equation involving matrices. Such an equation is called a **matrix equation**.

EXAMPLE 4 **Solving a Matrix Equation**

Solve the matrix equation $3A + 2X = 4B$ for X, where

$$A = \begin{bmatrix} 2 & 0 \\ 4 & 6 \end{bmatrix} \text{ and } B = \begin{bmatrix} 1 & 3 \\ 5 & 2 \end{bmatrix}.$$

SOLUTION

$3A + 2X = 4B$	The given equation
$2X = 4B - 3A$	Subtract the matrix $3A$ from both sides.
$X = \dfrac{1}{2}(4B - 3A)$	Multiply both sides by $\dfrac{1}{2}$.
$= \dfrac{1}{2}\left(4\begin{bmatrix} 1 & 3 \\ 5 & 2 \end{bmatrix} - 3\begin{bmatrix} 2 & 0 \\ 4 & 6 \end{bmatrix}\right)$	Substitute for A and B.
$= \dfrac{1}{2}\left(\begin{bmatrix} 4 & 12 \\ 20 & 8 \end{bmatrix} - \begin{bmatrix} 6 & 0 \\ 12 & 18 \end{bmatrix}\right)$	Scalar multiplication
$= \dfrac{1}{2}\begin{bmatrix} -2 & 12 \\ 8 & -10 \end{bmatrix}$	Subtract.
$= \begin{bmatrix} -1 & 6 \\ 4 & -5 \end{bmatrix}$	Scalar multiplication

You should check that the matrix $X = \begin{bmatrix} -1 & 6 \\ 4 & -5 \end{bmatrix}$ satisfies the given matrix equation.   ■ ■ ■

Practice Problem 4 Solve the matrix equation $5A + 3X = 2B$ for X, where

$$A = \begin{bmatrix} 1 & -1 \\ 3 & 5 \end{bmatrix} \text{ and } B = \begin{bmatrix} 2 & 7 \\ -3 & -5 \end{bmatrix}.$$ ■

3 Define matrix multiplication.

Matrix Multiplication

You know that the sum of two matrices is defined only if both matrices are of the same order. In the case of multiplication, however, the general rule is that if A is an $m \times p$ matrix and B is a $p \times n$ matrix, then the *product* AB is defined; otherwise, it is not defined.

order of A order of B
$m \times p$ $p \times n$
must be same
order of AB
$m \times n$

> **RULE FOR DEFINING THE PRODUCT AB**
>
> In order to define the product AB of two matrices A and B, the number of columns of A must be equal to the number of rows of B. If A is an $m \times p$ matrix and B is a $p \times n$ matrix, then the product AB is an $m \times n$ matrix.

EXAMPLE 5   **Determining the Order of the Product**

For each pair of matrices A and B, state whether the product matrix AB is defined. If it is, find the order of AB.

a. $A = \begin{bmatrix} 1 & 3 & 5 \end{bmatrix}, B = \begin{bmatrix} 2 \\ 3 \\ 7 \end{bmatrix}$ **b.** $A = \begin{bmatrix} 2 & 3 \\ -1 & 4 \\ 5 & 6 \end{bmatrix}, B = \begin{bmatrix} 2 & 1 & 3 & 5 \\ 4 & -1 & -2 & 6 \end{bmatrix}$

c. $A = \begin{bmatrix} 1 \\ -1 \\ 2 \end{bmatrix}, B = \begin{bmatrix} 2 & 1 & 4 \\ 0 & 2 & 2 \end{bmatrix}$

SOLUTION

a. Yes, the product AB is defined. A is a 1×3 matrix, and B is a 3×1 matrix; so the number of columns of A equals the number of rows of B. The product AB is a 1×1 matrix, which has a single entry.

b. Yes, AB is defined. A is a 3×2 matrix, and B is a 2×4 matrix; so the number of columns of A equals the number of rows of B. The product AB is a 3×4 matrix.

c. No, AB is not defined. A is a 3×1 matrix, and B is a 2×3 matrix; so the number of columns of A does not equal the number of rows of B. ■ ■ ■

Practice Problem 5 State whether the product matrix AB is defined. If AB is defined, find the order of the product matrix AB.

$$A = \begin{bmatrix} 1 & 4 & 7 \\ 0 & 2 & -2 \end{bmatrix}, \quad B = \begin{bmatrix} 4 \\ 3 \\ -1 \end{bmatrix}$$ ■

Before you learn to multiply two matrices A and B, consider the next example.

EXAMPLE 6 **Calculating the Total Revenue from Car Sales**

A car dealership sold 10 sports cars (S), 30 compacts (C), and 45 mini-compacts (M). The average price per car of type S, C, and M was 42, 24, and 17 (in thousands of dollars), respectively. Find the total revenue the dealership received from the sale of these cars.

SOLUTION

You can represent the number of cars of each type sold by a 1×3 matrix.

$$N = \begin{bmatrix} S & C & M \end{bmatrix} = \begin{bmatrix} 10 & 30 & 45 \end{bmatrix}$$

Using the price of each type of car, you can construct a 3×1 price matrix.

$$P = \begin{bmatrix} 42 \\ 24 \\ 17 \end{bmatrix}$$

Then the total revenue R received by the dealership from the sale of these cars last month is given by $R = 10(42) + 30(24) + 45(17) = 1905$ (or $1,905,000).

We define the right side of this equation to be the product NP of the two matrices N and P. In matrix notation,

$$NP = \begin{bmatrix} 10 & 30 & 45 \end{bmatrix} \begin{bmatrix} 42 \\ 24 \\ 17 \end{bmatrix} = 10(42) + 30(24) + 45(17) = 1905.$$ ■ ■ ■

Practice Problem 6 In Example 6, suppose the average price per car of types S, C, and M is 41, 26, and 19 (in thousands of dollars), respectively. Find the total revenue received by the dealership from the sale of these cars. ■

Considerations such as those in Example 6 suggest the following definition.

PRODUCT OF $1 \times n$ AND $n \times 1$ MATRICES

Suppose A is a $1 \times n$ matrix and B is an $n \times 1$ matrix.

$$A = \begin{bmatrix} a_1 & a_2 & a_3 \cdots a_n \end{bmatrix} \text{ and } B = \begin{bmatrix} b_1 \\ b_2 \\ \vdots \\ b_n \end{bmatrix}$$

We define the product AB by

$$AB = a_1 b_1 + a_2 b_2 + \cdots + a_n b_n.$$

We make the following observations about this definition.

1. Because A is a $1 \times n$ matrix and B is an $n \times 1$ matrix, the product AB is defined and is a 1×1 matrix.

2. To find the product AB, multiply the leftmost entry of A and the top entry of B; then move to the right in A and down in B while multiplying corresponding (first, second, third, and so on) entries; finally, add all of the resulting products.

EXAMPLE 7   **Finding the Product AB**

Find AB, where $A = \begin{bmatrix} 2 & 0 & 1 & -2 \end{bmatrix}$ and $B = \begin{bmatrix} 3 \\ 1 \\ -1 \\ 4 \end{bmatrix}$.

SOLUTION

Because A has order 1×4 and B has order 4×1, AB is defined. The product rule gives

$$AB = 2(3) + 0(1) + 1(-1) + (-2)(4) = -3.$$

Therefore, the product of the 1×4 matrix A and the 4×1 matrix B is the 1×1 matrix $[-3]$, or the number -3.   ■ ■ ■

Practice Problem 7   Find AB, where

$$A = \begin{bmatrix} 3 & -1 & 2 & 7 \end{bmatrix} \quad \text{and} \quad B = \begin{bmatrix} -2 \\ 0 \\ 1 \\ 5 \end{bmatrix}$$ ■

In Example 7, you may think of the answer -3 as the 1×1 matrix $[-3]$ or simply as the number -3. We will have occasion to make use of both points of view.

The rule for multiplying any two matrices can be reduced to the special case of multiplying a $1 \times p$ matrix by a $p \times 1$ matrix repeatedly.

MATRIX MULTIPLICATION

Let $A = [a_{ij}]$ be an $m \times p$ matrix and $B = [b_{ij}]$ be a $p \times n$ matrix. Then the **product** AB is the $m \times n$ matrix $C = [c_{ij}]$, where the entry c_{ij} of C is obtained by matrix multiplying the ith row of A by the jth column of B.

The definition of the product AB says that

$$c_{ij} = a_{i1}b_{1j} + a_{i2}b_{2j} + \cdots + a_{ip}b_{pj}.$$

Written in full, the matrix product $AB = C$ in the definition is as follows:

$$\begin{bmatrix} a_{11} & a_{12} & \cdots & a_{1p} \\ a_{21} & a_{22} & \cdots & a_{2p} \\ \vdots & \vdots & & \vdots \\ a_{i1} & a_{i2} & \cdots & a_{ip} \\ \vdots & \vdots & & \vdots \\ a_{m1} & a_{m2} & \cdots & a_{mp} \end{bmatrix} \begin{bmatrix} b_{11} & b_{12} & \cdots & b_{1j} & \cdots & b_{1n} \\ b_{21} & b_{22} & \cdots & b_{2j} & \cdots & b_{2n} \\ \vdots & \vdots & \cdots & \vdots & \cdots & \vdots \\ b_{p1} & b_{p2} & \cdots & b_{pj} & \cdots & b_{pn} \end{bmatrix} = \begin{bmatrix} c_{11} & c_{12} & \cdots & c_{1j} & \cdots & c_{1n} \\ \vdots & \vdots & \cdots & \vdots & & \vdots \\ c_{i1} & c_{i2} & \cdots & c_{ij} & \cdots & c_{in} \\ \vdots & \vdots & \vdots & \vdots & & \vdots \\ c_{m1} & c_{m2} & \cdots & c_{mj} & \cdots & c_{mn} \end{bmatrix}$$

TECHNOLOGY CONNECTION

 To find the product of two matrices using a graphing calculator, name the matrices and then use the regular multiplication key with the names of the matrices.

Consider the matrices A and B in Example 8.

We find the product AB.

```
[A]*[B]
        [[5 2 7]
         [0 8 8]]
```

EXAMPLE 8   **Finding the Product of Two Matrices**

Find the products AB and BA if possible.

$$A = \begin{bmatrix} 1 & 2 \\ -1 & 3 \end{bmatrix} \quad \text{and} \quad B = \begin{bmatrix} 3 & -2 & 1 \\ 1 & 2 & 3 \end{bmatrix}$$

SOLUTION

Because A is of order 2×2 and the order of B is 2×3, the product AB is defined and has order 2×3.

Let $AB = C = [c_{ij}]$. By the definition of the product AB, each entry c_{ij} of C is found by multiplying the ith row of A by the jth column of B. For example, c_{11} is found by multiplying the first row of A by the first column of B.

$$\begin{bmatrix} 1 & 2 \\ * & * \end{bmatrix}\begin{bmatrix} 3 & * & * \\ 1 & * & * \end{bmatrix} = \begin{bmatrix} 1(3) + 2(1) & * & * \\ * & * & * \end{bmatrix}$$

So

$$AB = \begin{bmatrix} 1 & 2 \\ -1 & 3 \end{bmatrix}\begin{bmatrix} 3 & -2 & 1 \\ 1 & 2 & 3 \end{bmatrix}$$

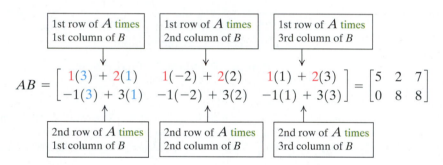

$$AB = \begin{bmatrix} 1(3) + 2(1) & 1(-2) + 2(2) & 1(1) + 2(3) \\ -1(3) + 3(1) & -1(-2) + 3(2) & -1(1) + 3(3) \end{bmatrix} = \begin{bmatrix} 5 & 2 & 7 \\ 0 & 8 & 8 \end{bmatrix}$$

The product BA is not defined because B is of order 2×3 and A is of order 2×2; that is, the number of columns of B is not the same as the number of rows of A. ■■■

Practice Problem 8   Find the products AB and BA.

$$A = \begin{bmatrix} 5 & 0 \\ 2 & -1 \end{bmatrix} \quad \text{and} \quad B = \begin{bmatrix} 8 & 1 \\ -2 & 6 \\ 0 & 4 \end{bmatrix}$$
■

In Example 8, we have $AB \neq BA$ because BA is undefined. Even if both AB and BA are defined, AB may not equal BA, as the next example illustrates.

EXAMPLE 9   **Finding the Product of Two Matrices**

Find the products AB and BA and show that $AB \neq BA$.

$$A = \begin{bmatrix} 2 & 3 \\ -1 & 0 \end{bmatrix} \quad \text{and} \quad B = \begin{bmatrix} 3 & 4 \\ 5 & 1 \end{bmatrix}$$

SOLUTION

$$AB = \begin{bmatrix} 2 & 3 \\ -1 & 0 \end{bmatrix}\begin{bmatrix} 3 & 4 \\ 5 & 1 \end{bmatrix} = \begin{bmatrix} 2(3) + 3(5) & 2(4) + 3(1) \\ -1(3) + 0(5) & -1(4) + 0(1) \end{bmatrix} = \begin{bmatrix} 21 & 11 \\ -3 & -4 \end{bmatrix}$$

$$BA = \begin{bmatrix} 3 & 4 \\ 5 & 1 \end{bmatrix}\begin{bmatrix} 2 & 3 \\ -1 & 0 \end{bmatrix} = \begin{bmatrix} 3(2) + 4(-1) & 3(3) + 4(0) \\ 5(2) + 1(-1) & 5(3) + 1(0) \end{bmatrix} = \begin{bmatrix} 2 & 9 \\ 9 & 15 \end{bmatrix}$$

Since the entry in row 1, column 1 of AB, 21, is not equal to the entry in row 1, column 1 of BA, 2, we have $AB \neq BA$. ■■■

Practice Problem 9 Find the products AB and BA.

$$A = \begin{bmatrix} 7 & 1 \\ 0 & 3 \end{bmatrix} \quad \text{and} \quad B = \begin{bmatrix} 2 & -1 \\ 4 & 4 \end{bmatrix}$$ ■

In Example 9, observe that $AB \neq BA$. Thus, in general, *matrix multiplication is not commutative.* Matrix multiplication, however, does share many of the properties of the multiplication of real numbers. Also, integral powers of square matrices are defined exactly as for real numbers. For a square matrix A, we write A^2 to mean AA, A^3 to mean AAA, and so on. We define A^0 by

$$A^0 = I_n = \begin{bmatrix} 1 & 0 & 0 & \cdots & 0 \\ 0 & 1 & 0 & \cdots & 0 \\ 0 & 0 & 1 & \cdots & 0 \\ \vdots & & \vdots & \vdots & \\ 0 & 0 & 0 & \cdots & 1 \end{bmatrix}.$$

The matrix I_n is called the **identity matrix** of order $n \times n$.

When the order of the matrix is clear from the context, we can drop the subscript n and write I for I_n. For any square matrix A, we have

$$\boxed{IA = A \quad \text{and} \quad AI = A}$$

So in matrix multiplication, the matrix I plays a role similar to that of the number 1 in the multiplication of real numbers.

PROPERTIES OF MATRIX MULTIPLICATION

Let $A, B,$ and C be matrices and let c be a scalar. Assume that each product and sum is defined. Then

1. $(AB)C = A(BC)$ Associative property of multiplication
2. **(i)** $A(B + C) = AB + AC$ Distributive property
    (ii) $(A + B)C = AC + BC$ Distributive property
3. $c(AB) = (cA)B = A(cB)$ Associative property of scalar multiplication

---

In exercises 43–46, you will be asked to verify these properties for given matricies.

4 Apply matrix multiplication to computer graphics.

Computer Graphics

Letters used for labels on a computer screen are very simple two-dimensional graphics. The next example illustrates how a letter (or a figure) is transformed when all of the coordinates of certain points on certain figure are multiplied by the same matrix.

FIGURE 6.2

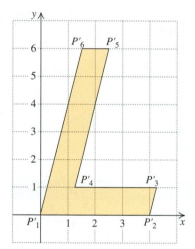

FIGURE 6.3

EXAMPLE 10   **Transforming a Letter**

The capital letter L in Figure 6.2 is determined by six points (or *vertices*) P_1 through P_6. The coordinates of the six points can be stored in a *data matrix D*. We draw line segments connecting these vertices in order.

$$\begin{array}{c} \\ x\text{-coordinate} \\ y\text{-coordinate} \end{array} \begin{array}{cccccc} P_1 & P_2 & P_3 & P_4 & P_5 & P_6 \end{array} \\ \begin{bmatrix} 0 & 4 & 4 & 1 & 1 & 0 \\ 0 & 0 & 1 & 1 & 6 & 6 \end{bmatrix} = D$$

If $A = \begin{bmatrix} 1 & 0.25 \\ 0 & 1 \end{bmatrix}$, compute AD and graph the figure it represents.

SOLUTION

The columns of the product matrix AD represent the transformed vertices of the letter L.

$$AD = \begin{bmatrix} 1 & 0.25 \\ 0 & 1 \end{bmatrix} \begin{bmatrix} 0 & 4 & 4 & 1 & 1 & 0 \\ 0 & 0 & 1 & 1 & 6 & 6 \end{bmatrix}$$

$$= \begin{bmatrix} 1\cdot 0 + 0.25\cdot 0 & 1\cdot 4 + 0.25\cdot 0 & 1\cdot 4 + 0.25\cdot 1 & 1\cdot 1 + 0.25\cdot 1 & 1\cdot 1 + 0.25\cdot 6 & 1\cdot 0 + 0.25\cdot 6 \\ 0\cdot 0 + 1\cdot 0 & 0\cdot 4 + 1\cdot 0 & 0\cdot 4 + 1\cdot 1 & 0\cdot 1 + 1\cdot 1 & 0\cdot 1 + 1\cdot 6 & 0\cdot 0 + 1\cdot 6 \end{bmatrix}$$

$$\begin{array}{cccccc} P'_1 & P'_2 & P'_3 & P'_4 & P'_5 & P'_6 \end{array}$$

$$= \begin{bmatrix} 0 & 4 & 4.25 & 1.25 & 2.5 & 1.5 \\ 0 & 0 & 1 & 1 & 6 & 6 \end{bmatrix}$$

Figure 6.3 shows the transformed vertices and the transformed figure formed by these vertices.   ■ ■ ■

Practice Problem 10   In Example 10, use the matrix $A = \begin{bmatrix} 0 & 1 \\ 1 & 0.25 \end{bmatrix}$. Graph the figure represented by the matrix AD.   ■

SECTION 6.2 ■ Exercises

A EXERCISES   Basic Skills and Concepts

1. Two $m \times n$ matrices $A = [a_{ij}]$ and $B = [b_{ij}]$ are equal if _____ for all i and j.

2. Let $A = [a_{ij}]$, $B = [b_{ij}]$, and $C = [c_{ij}]$. If $A + B = C$, then $c_{ij} =$ _____ for all i and for all j.

3. The product of a $1 \times n$ matrix A and an $n \times 1$ matrix B is a _____ matrix.

4. If A is an $m \times n$ matrix and B is an $n \times p$ matrix, then AB is defined and is a(n) _____ matrix.

5. *True or False* If AB and BA are defined, then $AB = BA$.

6. *True or False* Any two square matrices can be multiplied.

In Exercises 7–14, find the values of all variables.

7. $\begin{bmatrix} 2 \\ -3 \end{bmatrix} = \begin{bmatrix} x \\ u \end{bmatrix}$

8. $\begin{bmatrix} -\dfrac{y}{2} \\ x \end{bmatrix} = \begin{bmatrix} 4 \\ 3 \end{bmatrix}$

9. $\begin{bmatrix} 2 & x \\ y & -3 \end{bmatrix} = \begin{bmatrix} 2 & -1 \\ 3 & -3 \end{bmatrix}$

10. $\begin{bmatrix} 2 - x & 1 \\ -2 & 3 + y \end{bmatrix} = \begin{bmatrix} 3 & 1 \\ -2 & -3 \end{bmatrix}$

11. $\begin{bmatrix} 2x - 3y & -4 \\ 5 & 3x + y \end{bmatrix} = \begin{bmatrix} 1 & -4 \\ 5 & 7 \end{bmatrix}$

12. $\begin{bmatrix} 3x + y & -17 \\ 19 & 2 \end{bmatrix} = \begin{bmatrix} -\dfrac{1}{2} & -17 \\ 19 & 2x + 3y \end{bmatrix}$

13. $\begin{bmatrix} x - y & 1 & 2 \\ 4 & 3x - 2y & 3 \\ 5 & 6 & 5x - 10y \end{bmatrix} = \begin{bmatrix} -1 & 1 & 2 \\ 4 & -1 & 3 \\ 5 & 6 & 6 \end{bmatrix}$

14. $\begin{bmatrix} x + y & 2 & 3 \\ 2 & x - y & 4 \\ 3 & 4 & 2x + 3y \end{bmatrix} = \begin{bmatrix} -1 & 2 & 3 \\ 2 & 5 & 4 \\ 3 & 4 & -5 \end{bmatrix}$

In Exercises 15–22, find each of the following if possible.

a. $A + B$ b. $A - B$
c. $-3A$ d. $3A - 2B$
e. $(A + B)^2$ f. $A^2 - B^2$

15. $A = \begin{bmatrix} 1 & 2 \\ 3 & 4 \end{bmatrix}, B = \begin{bmatrix} -1 & 0 \\ 2 & -3 \end{bmatrix}$

16. $A = \begin{bmatrix} \frac{1}{3} & 1 \\ -1 & 2 \end{bmatrix}, B = \begin{bmatrix} 1 & 0 \\ 2 & -\frac{1}{2} \end{bmatrix}$

17. $A = \begin{bmatrix} 2 & 3 \\ -4 & 5 \end{bmatrix}, B = \begin{bmatrix} 1 & 0 & 2 \\ 3 & 1 & 4 \end{bmatrix}$

18. $A = \begin{bmatrix} 1 & 2 & 3 \\ -1 & -3 & 4 \end{bmatrix}, B = \begin{bmatrix} 1 & 0 \\ 2 & 3 \\ 1 & 4 \end{bmatrix}$

19. $A = \begin{bmatrix} 4 & 0 & -1 \\ -2 & 5 & 2 \\ 0 & 0 & 1 \end{bmatrix}, B = \begin{bmatrix} 3 & 1 & 0 \\ 1 & -4 & 2 \\ 2 & 1 & 3 \end{bmatrix}$

20. $A = \begin{bmatrix} 1 & 0 & 2 \\ 2 & 1 & 0 \\ 0 & 1 & 3 \end{bmatrix}, B = \begin{bmatrix} 3 & -1 & 2 \\ 0 & 2 & 1 \\ 1 & 4 & -1 \end{bmatrix}$

21. $A = \begin{bmatrix} 1 & 2 & -3 \\ 3 & 4 & 5 \\ 2 & -1 & 0 \end{bmatrix}, B = \begin{bmatrix} 3 & 1 & 0 \\ 1 & -4 & 2 \\ 2 & 1 & 3 \end{bmatrix}$

22. $A = \begin{bmatrix} 1 & 0 & 2 \\ 0 & 1 & 0 \\ 2 & 0 & 3 \end{bmatrix}, B = \begin{bmatrix} 3 & 0 & 0 \\ 0 & 2 & 1 \\ 0 & 1 & -1 \end{bmatrix}$

In Exercises 23–30, solve each matrix equation for X, where

$$A = \begin{bmatrix} 2 & 3 & -1 \\ 1 & -2 & 4 \end{bmatrix} \text{ and } B = \begin{bmatrix} -2 & 1 & 0 \\ 2 & 3 & 4 \end{bmatrix}.$$

23. $A + X = B$ **24.** $B + X = A$

25. $2X - A = B$ **26.** $2X - B = A$

27. $2X + 3A = B$ **28.** $3X + 2A = B$

29. $2A + 3B + 4X = 0$ **30.** $3X - 2A + 5B = 0$

In Exercises 31–40, find each product if possible.

a. AB b. BA

31. $A = \begin{bmatrix} 1 & 2 \\ 3 & 4 \end{bmatrix}, B = \begin{bmatrix} -2 & 1 \\ 3 & 5 \end{bmatrix}$

32. $A = \begin{bmatrix} 3 & 2 \\ 1 & 5 \\ 0 & 1 \end{bmatrix}, B = \begin{bmatrix} 1 & 3 & -2 \\ 2 & 5 & 0 \end{bmatrix}$

33. $A = \begin{bmatrix} 2 & -1 & 0 \\ -3 & 1 & 2 \end{bmatrix}, B = \begin{bmatrix} 1 & 5 \\ -2 & 3 \\ 4 & 0 \end{bmatrix}$

34. $A = \begin{bmatrix} 1 & 2 & 3 \\ 4 & 5 & 6 \end{bmatrix}, B = \begin{bmatrix} -1 & -3 & -6 \\ 2 & 5 & 7 \end{bmatrix}$

35. $A = [2 \ \ 3 \ \ 5], B = \begin{bmatrix} 1 \\ -2 \\ 4 \end{bmatrix}$

36. $A = \begin{bmatrix} 1 \\ 2 \\ -1 \end{bmatrix}, B = [-3 \ \ 0 \ \ 2]$

37. $A = [1 \ \ 2 \ \ 3], B = \begin{bmatrix} 1 & 2 & -1 \\ 0 & 3 & 1 \\ 2 & 0 & -3 \end{bmatrix}$

38. $A = \begin{bmatrix} 4 & -6 & 2 \\ 2 & 3 & 0 \\ 1 & 2 & -3 \end{bmatrix}, B = [-2 \ \ 0 \ \ 1]$

39. $A = \begin{bmatrix} 2 & 0 & 1 \\ 1 & 4 & 2 \\ 3 & -1 & 0 \end{bmatrix}, B = \begin{bmatrix} 3 & 1 & 0 \\ -1 & 2 & 0 \\ 4 & 5 & 2 \end{bmatrix}$

40. $A = \begin{bmatrix} 3 & -1 & 2 \\ 0 & 4 & -3 \\ 1 & -2 & 2 \end{bmatrix}, B = \begin{bmatrix} 2 & -5 & 0 \\ -1 & 2 & -1 \\ 3 & 0 & 2 \end{bmatrix}$

41. Let $A = \begin{bmatrix} 3 & 1 & 2 \\ 0 & 4 & 3 \\ 1 & -2 & 2 \end{bmatrix}$ and $B = \begin{bmatrix} 2 & 5 & 0 \\ 1 & 2 & -1 \\ 3 & 0 & 2 \end{bmatrix}$.

Verify that $AB \neq BA$.

42. Let $A = \begin{bmatrix} 5 & 2 & 4 \\ -6 & -3 & 10 \\ -2 & 6 & 5 \end{bmatrix}$ and $B = \begin{bmatrix} 4 & 1 & 2 \\ -3 & 0 & 5 \\ -1 & 3 & 4 \end{bmatrix}$.

Verify that $AB = BA$.

In Exercises 43–46, let

$$A = \begin{bmatrix} 1 & 2 \\ 3 & 4 \end{bmatrix}, B = \begin{bmatrix} 2 & -3 \\ 3 & 5 \end{bmatrix}, \text{ and } C = \begin{bmatrix} 0 & 1 \\ 2 & 4 \end{bmatrix}.$$

43. Verify the associative property for multiplication: $(AB)C = A(BC)$

44. Verify the distributive property: $A(B + C) = AB + AC$

45. Verify the distributive property: $(A + B)C = AC + BC$

46. For any real number c, verify that $c(AB) = (cA)(B) = A(cB)$.

B EXERCISES Applying the Concepts

47. **Cost matrix.** The Build-Rite Co. is building an apartment complex. The cost of purchasing and transporting specific amounts of steel, glass, and wood (in appropriate units) from two different locations is given by the following matrices.

$$
\begin{array}{ccc}
\text{Steel} & \text{Glass} & \text{Wood} \\
\end{array}
$$

$$A = \begin{bmatrix} 7 & 3 & 18 \\ 4 & 1 & 3 \end{bmatrix} \begin{array}{l} \text{Cost of Material} \\ \text{Transportation Cost} \end{array}$$

$$B = \begin{bmatrix} 6 & 2 & 20 \\ 3 & 1 & 4 \end{bmatrix} \begin{array}{l} \text{Cost of Material} \\ \text{Transportation Cost} \end{array}$$

Find the matrix representing the total cost of material and transportation for steel, glass, and wood from both locations.

48. Repeat Exercise 47 assuming that the order from location *A* is tripled and the order from location *B* is doubled.

49. Stock purchase. Ms. Goodbyer plans to buy 100 shares of computer stock, 300 shares of oil stock, and 400 shares of automobile stock. The computer stock is selling for $60 a share, oil stock is selling for $38 a share, and automobile stock is selling for $17 a share. Use matrix multiplication to calculate the total cost of Ms. Goodbyer's purchases.

50. Product export. The International Export Corporation received an export order for three of its products, say, *A*, *B*, and *C*. The export order is (in thousands) for 50 of product *A*, 75 of product *B*, and 150 of product *C*. The material cost, labor, and profit (each in appropriate units) required to produce each unit of the product are given in the following table.

	Material	Labor	Profit
Product A	20	2	20
Product B	15	3	25
Product C	25	1	15

Use matrix multiplication to compute the total material cost, total labor, and total profit assuming that the entire export order is filled.

51. Executive compensation. The Multinational Oil Corporation pays its top executives a salary, a cash bonus, and shares of its stock annually. In 2009, the chair of the board received $2.5 million in salary, a $1.5 million bonus, and 50,000 shares of stock; the president of the company received one-half the compensation of the chair; and each of the four vice presidents was paid $100,000 in salary, a $150,000 bonus, and 5000 shares of stock.
  a. Express payments to these executives in salary, bonus, and stock as a 3 × 3 matrix.
  b. Express the number of executives of each rank as a column matrix.
  c. Use matrix multiplication to compute the total amount the company paid to each executive in each category in 2009.

52. Calories and protein. The Browns (*B*) and Newgards (*N*) are neighboring families. The Brown family has two men, three women, and one child, while the Newgard family has one man, one woman, and two children. Both families are weight watchers. They have the same dietician, who recommends the following daily allowance of calories and proteins.

  Calories
    male adults: 2400 calories
    female adults: 1900 calories
    children: 1800 calories
  Protein (in grams)
    male adults: 55
    female adults: 45
    children: 33

Represent the preceding information in matrix form. Use matrix multiplication to compute the total requirements of calories and proteins for each of the two families.

In Exercise 53–56, a matrix *A* is given. Graph the figure represented by the matrix *AD*, where *D* is the matrix representation of the letter *L* of Example 10.

53. $A = \begin{bmatrix} 1 & 0 \\ 0 & -1 \end{bmatrix}$ **54.** $A = \begin{bmatrix} -1 & 0 \\ 0 & 1 \end{bmatrix}$

55. $A = \begin{bmatrix} -1 & 0 \\ 0.25 & 1 \end{bmatrix}$ **56.** $A = \begin{bmatrix} 1 & -0.25 \\ 0 & 1 \end{bmatrix}$

C EXERCISES Beyond the Basics

57. Let $A = \begin{bmatrix} 0 & 3 \\ 0 & 0 \end{bmatrix}$, $B = \begin{bmatrix} 2 & 1 \\ 3 & 0 \end{bmatrix}$, and $C = \begin{bmatrix} 5 & 4 \\ 3 & 0 \end{bmatrix}$.
Verify that $AB = AC$. This example shows that $AB = AC$ does not, in general, imply that $B = C$.

58. Let $A = \begin{bmatrix} 2 & -3 & -5 \\ -1 & 4 & 5 \\ 1 & -3 & -4 \end{bmatrix}$ and $B = \begin{bmatrix} -1 & 3 & 5 \\ 1 & -3 & -5 \\ -1 & 3 & 5 \end{bmatrix}$.
Verify that $AB = \mathbf{0}$. This example shows that $AB = \mathbf{0}$ does not imply that $A = \mathbf{0}$ or $B = \mathbf{0}$.

59. Select appropriate 2 × 2 matrices *A* and *B* to show that, in general, $(A + B)^2 \neq A^2 + 2AB + B^2$.

60. Repeat Exercise 59 to show that, in general, $A^2 - B^2 \neq (A - B)(A + B)$.

61. Let $A = \begin{bmatrix} 1 & -2 & 3 \\ 2 & -4 & 1 \\ 3 & -5 & 2 \end{bmatrix}$. Show that

$$3A^2 - 2A + I = \begin{bmatrix} 17 & -23 & 15 \\ -13 & 30 & 10 \\ -9 & 22 & 21 \end{bmatrix},$$

where $I = \begin{bmatrix} 1 & 0 & 0 \\ 0 & 1 & 0 \\ 0 & 0 & 1 \end{bmatrix}$.

62. Let $A = \begin{bmatrix} 1 & 3 & 2 \\ 2 & 0 & 3 \\ 1 & -1 & 1 \end{bmatrix}$. Show that

$$A^3 - 3A^2 + A - I = \begin{bmatrix} -3 & 14 & -3 \\ 5 & -2 & 8 \\ 5 & -7 & 6 \end{bmatrix},$$

where *I* is given in Exercise 61.

63. If possible, find a matrix *B* such that $\begin{bmatrix} 2 & 3 \\ 1 & 2 \end{bmatrix} B = \begin{bmatrix} 1 & 0 \\ 0 & 1 \end{bmatrix}$.

64. If possible find a matrix *B* such that

$$\begin{bmatrix} 1 & -2 \\ -2 & 4 \end{bmatrix} B = \begin{bmatrix} 1 & 0 \\ 0 & 1 \end{bmatrix}.$$

65. Find *x* and *y* assuming that

$$\left(4\begin{bmatrix} 2 & -1 & 3 \\ 1 & 0 & 2 \end{bmatrix} - \begin{bmatrix} 1 & 2 & -1 \\ 2 & -3 & 4 \end{bmatrix} \right) \begin{bmatrix} 2 \\ -1 \\ 1 \end{bmatrix} = \begin{bmatrix} x \\ y \end{bmatrix}.$$

66. Find x, y, and z assuming that

$$\begin{bmatrix} 3 & 2 & -1 \\ 4 & 9 & 2 \\ 5 & 0 & -2 \end{bmatrix} \begin{bmatrix} x \\ y \\ z \end{bmatrix} = \begin{bmatrix} 0 \\ 7 \\ 2 \end{bmatrix}.$$

Critical Thinking

67. Let A be a $3 \times n$ matrix and B be a $5 \times m$ matrix.
   a. Under what conditions is AB defined? What is the order of AB when this product is defined?
   b. Repeat part (a) for BA.

68. Let $A = \begin{bmatrix} 3 & -1 & 0 \\ -2 & 3 & 1 \end{bmatrix}$, $B = \begin{bmatrix} 2 \\ 3 \\ 4 \end{bmatrix}$, and $C = [-1 \quad 2]$. In

which order should all three matrices be multiplied to produce a number?

GROUP PROJECT

Market share. The Asma Corporation (A), Bronkial Brothers (B), and Coufmore Company (C) simultaneously decide to introduce a new cigarette at a time when each company has one-third of the market. During the year, the following occurs:

   (i) A retains 40% of its customers and loses 30% to B and 30% to C.
   (ii) B retains 30% of its customers and loses 60% to A and 10% to C.
   (iii) C retains 30% of its customers and loses 60% to A and 10% to B.

The foregoing information can be represented by the following *transition matrix*.

$$
\begin{array}{c c c c}
 & A & B & C \\
P = \begin{matrix} A \\ B \\ C \end{matrix} & \begin{bmatrix} 0.4 & 0.3 & 0.3 \\ 0.6 & 0.3 & 0.1 \\ 0.6 & 0.1 & 0.3 \end{bmatrix}
\end{array}
$$

Write the initial market share as $X = \begin{bmatrix} \dfrac{1}{3} & \dfrac{1}{3} & \dfrac{1}{3} \end{bmatrix}$.

   a. Compute XP^2 and interpret your result.
   b. Compute XP^3, XP^4, and XP^5; observe the pattern for the long run. Describe the long-term market share for each company.

The Matrix Inverse

Objectives

1 Verify the multiplicative inverse of a matrix.

2 Find the inverse of a matrix.

3 Find the inverse of a 2 × 2 matrix.

4 Use matrix inverses to solve systems of linear equations.

5 Use matrix inverses in applied problems.

CODING AND DECODING MESSAGES

Imagine you are a secret agent who is looking for spies in different parts of the world. Periodically, you communicate with headquarters via your laptop or by satellite. You expect that your communications will be intercepted. So before you broadcast any message, you must write it using a secret code. A message written with a secret code is called a **cryptogram**.

First, you might replace each letter in the message with a number and send the message as a sequence of numbers. The problem with this approach is that it is rather easy to crack such a code. For example, if letters such as *e* are always represented by the same symbol, then a code breaker (hacker) can guess which symbol represents each letter and eventually translate your message. In Example 8, we show you a better way to encode and decode a message using matrix multiplication. ∎

---

1 Verify the multiplicative inverse of a matrix.

Multiplicative Inverse of a Matrix

In arithmetic, you know that if a is a nonzero real number, then a has a unique reciprocal, denoted by $\dfrac{1}{a}$, or a^{-1}. The number a^{-1} is called the **multiplicative inverse** of a, and

$$aa^{-1} = a^{-1}a = 1.$$

Similarly, certain square matrices have *multiplicative inverses*, with an identity matrix acting as "1."

INVERSE OF A MATRIX

Let A be an $n \times n$ matrix and let I be the $n \times n$ identity matrix that has 1s on the main diagonal and 0s elsewhere. If there is an $n \times n$ matrix B such that

$$AB = I \quad \text{and} \quad BA = I,$$

then B is called the **inverse** of A and we write $B = A^{-1}$ (read "A inverse").

Notice that if B is the inverse of A, then A is the inverse of B.

Arthur Cayley

(1821–1895)

English mathematicians of the first third of the nineteenth century attempted to extend the basic ideas of algebra and to determine exactly how much one can generalize the properties of integers to other types of quantities. Although Sylvester (see page 568) coined the term *matrix*, it was Arthur Cayley who developed the algebra of matrices. Cayley studied mathematics at Trinity College, Cambridge, but because there was no suitable teaching job available, he decided to become a lawyer. Although he became skilled in legal work, he regarded the law as just a way to support himself. During his legal career, he was able to write more than 300 mathematical papers. In 1863, Cambridge University established a new post in mathematics and offered it to Cayley. He accepted the job eagerly even though it paid less money than he made as a lawyer.

EXAMPLE 1   Verifying the Inverse of a Matrix

Show that B is the inverse of A.

$$A = \begin{bmatrix} 1 & 3 \\ 2 & 7 \end{bmatrix} \text{ and } B = \begin{bmatrix} 7 & -3 \\ -2 & 1 \end{bmatrix}$$

SOLUTION
You need to verify that $AB = I$ and $BA = I$.

$$AB = \begin{bmatrix} 1 & 3 \\ 2 & 7 \end{bmatrix}\begin{bmatrix} 7 & -3 \\ -2 & 1 \end{bmatrix} = \begin{bmatrix} 1(7) + 3(-2) & 1(-3) + 3(1) \\ 2(7) + 7(-2) & 2(-3) + 7(1) \end{bmatrix} = \begin{bmatrix} 1 & 0 \\ 0 & 1 \end{bmatrix} = I$$

$$BA = \begin{bmatrix} 7 & -3 \\ -2 & 1 \end{bmatrix}\begin{bmatrix} 1 & 3 \\ 2 & 7 \end{bmatrix} = \begin{bmatrix} 7(1) + (-3)(2) & 7(3) + (-3)(7) \\ (-2)(1) + 1(2) & (-2)(3) + 1(7) \end{bmatrix} = \begin{bmatrix} 1 & 0 \\ 0 & 1 \end{bmatrix} = I$$

Because $AB = I$ and $BA = I$, it follows that $B = A^{-1}$.   ■ ■ ■

Practice Problem 1   Show that B is the inverse of A.

$$A = \begin{bmatrix} 3 & 2 \\ 2 & 1 \end{bmatrix} \quad \text{and} \quad B = \begin{bmatrix} -1 & 2 \\ 2 & -3 \end{bmatrix}$$ ■

Recall that in general (even for square matrices), $AB \neq BA$. However, if A and B are square matrices with $AB = I$, then it can be shown that $BA = I$. Therefore, to determine whether A and B are inverses, you need only check that either $AB = I$ or $BA = I$.

In working with real numbers, you know that if $a = 0$, then a^{-1} does not exist. So it should not come as a surprise that if A is a square *zero* matrix, then A^{-1} does not exist. Surprisingly, there are also nonzero matrices that do not have inverses.

EXAMPLE 2   Proving That a Particular Nonzero Matrix Has No Inverse

Show that the matrix A does not have an inverse.

$$A = \begin{bmatrix} 2 & 0 \\ 0 & 0 \end{bmatrix}$$

SOLUTION
Suppose A has an inverse B, where

$$B = \begin{bmatrix} x & y \\ z & w \end{bmatrix}.$$

Then you would have for some real numbers $x, y, z,$ and w

$$\begin{bmatrix} 2 & 0 \\ 0 & 0 \end{bmatrix}\begin{bmatrix} x & y \\ z & w \end{bmatrix} = \begin{bmatrix} 1 & 0 \\ 0 & 1 \end{bmatrix} \qquad AB = I$$

$$\begin{bmatrix} 2x & 2y \\ 0 & 0 \end{bmatrix} = \begin{bmatrix} 1 & 0 \\ 0 & 1 \end{bmatrix}. \qquad \text{Multiply the two matrices on the left side.}$$

Because these two matrices are equal, you must have

$$0 = 1 \qquad \text{Equate the entries in the } (2, 2) \text{ position.}$$

Because $0 = 1$ is a false statement, the matrix A does not have an inverse.   ■ ■ ■

Practice Problem 2   Show that the matrix $A = \begin{bmatrix} 3 & 1 \\ 3 & 1 \end{bmatrix}$ does not have an inverse.

■

2 Find the inverse of a matrix.

Finding the Inverse of a Matrix

It can be shown that if a square matrix A has an inverse, then the inverse is unique. This means that a square matrix cannot have more than one inverse. If a matrix A has an inverse, then A is called **invertible** or **nonsingular**. A square matrix that does not have an inverse is called a **singular** matrix. The following discussion derives a technique for finding the inverse of an invertible matrix.

Suppose we want to find the inverse of the matrix $A = \begin{bmatrix} 1 & 2 \\ 3 & 5 \end{bmatrix}$. We first let $B = \begin{bmatrix} x & y \\ z & w \end{bmatrix}$ be the inverse of the matrix A. Then

$$AB = I \qquad \text{Definition of inverse}$$

$$\begin{bmatrix} 1 & 2 \\ 3 & 5 \end{bmatrix}\begin{bmatrix} x & y \\ z & w \end{bmatrix} = \begin{bmatrix} 1 & 0 \\ 0 & 1 \end{bmatrix} \qquad \text{Substitute for } A, B, \text{ and } I.$$

$$\begin{bmatrix} x + 2z & y + 2w \\ 3x + 5z & 3y + 5w \end{bmatrix} = \begin{bmatrix} 1 & 0 \\ 0 & 1 \end{bmatrix} \qquad \text{Multiply matrices } A \text{ and } B.$$

Equating entries in the last matrix equation yields two systems of linear equations.

$$\begin{cases} x + 2z = 1 \\ 3x + 5z = 0 \end{cases} \qquad \begin{array}{l}\text{Equate the entries in the } (1, 1) \\ \text{and } (2, 1) \text{ positions.}\end{array}$$

$$\begin{cases} y + 2w = 0 \\ 3y + 5w = 1 \end{cases} \qquad \begin{array}{l}\text{Equate the entries in the } (1, 2) \\ \text{and } (2, 2) \text{ positions.}\end{array}$$

We solve the two systems by Gauss–Jordan elimination.

System of Equations	Matrix Form
$\begin{cases} x + 2z = 1 \\ 3x + 5z = 0 \end{cases}$	$\begin{bmatrix} 1 & 2 & \vert & 1 \\ 3 & 5 & \vert & 0 \end{bmatrix}$
$\begin{cases} y + 2w = 0 \\ 3y + 5w = 1 \end{cases}$	$\begin{bmatrix} 1 & 2 & \vert & 0 \\ 3 & 5 & \vert & 1 \end{bmatrix}$

Because the coefficient matrix for both systems of equations is the original matrix

$$A = \begin{bmatrix} 1 & 2 \\ 3 & 5 \end{bmatrix},$$

the calculations involved in the Gauss–Jordan elimination on the coefficient matrix A will be the same for both systems. Therefore, you can solve both systems at once by considering the "doubly augmented" matrix

$$C = \begin{bmatrix} 1 & 2 & \vert & 1 & 0 \\ 3 & 5 & \vert & 0 & 1 \end{bmatrix}.$$

Note that the matrix C is formed by adjoining the identity matrix I to matrix A:

$$C = [A\,\vert\,I]$$

Now apply Gauss–Jordan elimination to the matrix C.

$$C = \begin{bmatrix} 1 & 2 & \vert & 1 & 0 \\ 3 & 5 & \vert & 0 & 1 \end{bmatrix} \xrightarrow{-3R_1+R_2 \to R_2} \begin{bmatrix} 1 & 2 & \vert & 1 & 0 \\ 0 & -1 & \vert & -3 & 1 \end{bmatrix} \xrightarrow{(-1)R_2} \begin{bmatrix} 1 & 2 & \vert & 1 & 0 \\ 0 & 1 & \vert & 3 & -1 \end{bmatrix}$$

$$\xrightarrow{-2R_2+R_1 \to R_1} \begin{bmatrix} 1 & 0 & \vert & -5 & 2 \\ 0 & 1 & \vert & 3 & -1 \end{bmatrix}$$

TECHNOLOGY CONNECTION

To use a graphing calculator to find the inverse of an invertible matrix, name the matrix and then use the regular reciprocal key with the name of the matrix.

```
[A]
        [[1 2]
         [3 5]]
■
```

```
[A]⁻¹
        [[-5 2 ]
         [3  -1]]
■
```

Next convert the last doubly augmented matrix back into two systems of equations,

$$\begin{bmatrix} 1 & 0 & | & -5 \\ 0 & 1 & | & 3 \end{bmatrix} \text{ represents } \begin{cases} 1 \cdot x + 0 \cdot z = -5 \\ 0 \cdot x + 1 \cdot z = 3 \end{cases} \text{ or } \begin{cases} x = -5 \\ z = 3 \end{cases}.$$

Similarly,

$$\begin{bmatrix} 1 & 0 & | & 2 \\ 0 & 1 & | & -1 \end{bmatrix} \text{ represents } \begin{cases} 1 \cdot y + 0 \cdot w = 2 \\ 0 \cdot y + 1 \cdot w = -1 \end{cases} \text{ or } \begin{cases} y = 2 \\ w = -1 \end{cases}.$$

We conclude that $x = -5$, $y = 2$, $z = 3$, and $w = -1$.

Because $B = \begin{bmatrix} x & y \\ z & w \end{bmatrix}$, it follows that $B = \begin{bmatrix} -5 & 2 \\ 3 & -1 \end{bmatrix} = A^{-1}$.

Check

$$AA^{-1} = \begin{bmatrix} 1 & 2 \\ 3 & 5 \end{bmatrix}\begin{bmatrix} -5 & 2 \\ 3 & -1 \end{bmatrix} = \begin{bmatrix} 1 \cdot (-5) + 2 \cdot 3 & 1 \cdot 2 + 2 \cdot (-1) \\ 3 \cdot (-5) + 5 \cdot 3 & 3 \cdot 2 + 5 \cdot (-1) \end{bmatrix} = \begin{bmatrix} 1 & 0 \\ 0 & 1 \end{bmatrix} = I.$$

We have shown that to find the inverse of an invertible matrix A, we transform the matrix $[A \mid I]$ by a sequence of row operations into $[I \mid B]$, where $B = A^{-1}$.

PROCEDURE FOR FINDING THE INVERSE OF A MATRIX

Let A be an $n \times n$ matrix.

1. Form the $n \times 2n$ augmented matrix $[A \mid I]$, where I is the $n \times n$ identity matrix.

2. If there is a sequence of row operations that transforms A into I, then this same sequence of row operations will transform $[A \mid I]$ into $[I \mid B]$, where $B = A^{-1}$.

3. Check your work by showing that $AA^{-1} = I$.

If it is not possible to transform A into I by row operations, then A does not have an inverse. (This occurs if, at any step in the process, you obtain a matrix $[C \mid D]$ in which C has a row of zeros.)

EXAMPLE 3   **Finding the Inverse of a Matrix**

Find the inverse (if it exists) of the matrix $A = \begin{bmatrix} 1 & 2 & 3 \\ 2 & 5 & 7 \\ 2 & 4 & 6 \end{bmatrix}$.

SOLUTION

Step 1   Start by adjoining the identity 3×3 matrix $I = \begin{bmatrix} 1 & 0 & 0 \\ 0 & 1 & 0 \\ 0 & 0 & 1 \end{bmatrix}$ to the matrix A to form the augmented matrix

$$[A \mid I] = \begin{bmatrix} 1 & 2 & 3 & | & 1 & 0 & 0 \\ 2 & 5 & 7 & | & 0 & 1 & 0 \\ 2 & 4 & 6 & | & 0 & 0 & 1 \end{bmatrix}.$$

Step 2   Perform row operations on $[A \mid I]$ to transform it to a matrix of the form $[I \mid B]$.

$$\begin{bmatrix} 1 & 2 & 3 & | & 1 & 0 & 0 \\ 2 & 5 & 7 & | & 0 & 1 & 0 \\ 2 & 4 & 6 & | & 0 & 0 & 1 \end{bmatrix} \xrightarrow[-2R_1 + R_3 \to R_3]{-2R_1 + R_2 \to R_2} \begin{bmatrix} 1 & 2 & 3 & | & 1 & 0 & 0 \\ 0 & 1 & 1 & | & -2 & 1 & 0 \\ 0 & 0 & 0 & | & -2 & 0 & 1 \end{bmatrix} = [C \mid D]$$

Because the third row of matrix C consists of zeros, it is impossible to transform A into I by row operations. Hence, A is *not* invertible. ■ ■ ■

Practice Problem 3 Find the inverse (if it exists) of the matrix $A = \begin{bmatrix} 1 & 4 & -2 \\ -1 & 1 & 2 \\ 3 & 7 & -6 \end{bmatrix}$.

■

EXAMPLE 4 **Finding the Inverse of a Matrix**

Find the inverse (if it exists) of the matrix $A = \begin{bmatrix} 1 & 1 & 0 \\ 0 & 3 & 1 \\ 2 & 3 & 3 \end{bmatrix}$.

SOLUTION

Step 1 Start with the matrix

$$[A\,|\,I] = \begin{bmatrix} 1 & 1 & 0 & | & 1 & 0 & 0 \\ 0 & 3 & 1 & | & 0 & 1 & 0 \\ 2 & 3 & 3 & | & 0 & 0 & 1 \end{bmatrix} \qquad \text{Here } I = I_3$$

Step 2 Use row operations on $[A\,|\,I]$. Then $[A\,|\,I] \rightarrow$

$$\xrightarrow[\substack{-2R_1 + R_3 \rightarrow R_3}]{\frac{1}{3}R_2 \rightarrow R_2} \begin{bmatrix} 1 & 1 & 0 & | & 1 & 0 & 0 \\ 0 & 1 & \frac{1}{3} & | & 0 & \frac{1}{3} & 0 \\ 0 & 1 & 3 & | & -2 & 0 & 1 \end{bmatrix} \xrightarrow{(-1)R_2 + R_3 \rightarrow R_3} \begin{bmatrix} 1 & 1 & 0 & | & 1 & 0 & 0 \\ 0 & 1 & \frac{1}{3} & | & 0 & \frac{1}{3} & 0 \\ 0 & 0 & \frac{8}{3} & | & -2 & -\frac{1}{3} & 1 \end{bmatrix}$$

$$\xrightarrow{\frac{3}{8}R_3} \begin{bmatrix} 1 & 1 & 0 & | & 1 & 0 & 0 \\ 0 & 1 & \frac{1}{3} & | & 0 & \frac{1}{3} & 0 \\ 0 & 0 & 1 & | & -\frac{6}{8} & -\frac{1}{8} & \frac{3}{8} \end{bmatrix} \xrightarrow{-\frac{1}{3}R_3 + R_2 \rightarrow R_2} \begin{bmatrix} 1 & 1 & 0 & | & 1 & 0 & 0 \\ 0 & 1 & 0 & | & \frac{2}{8} & \frac{3}{8} & -\frac{1}{8} \\ 0 & 0 & 1 & | & -\frac{6}{8} & -\frac{1}{8} & \frac{3}{8} \end{bmatrix}$$

$$\xrightarrow{(-1)R_2 + R_1 \rightarrow R_1} \begin{bmatrix} 1 & 0 & 0 & | & \frac{6}{8} & -\frac{3}{8} & \frac{1}{8} \\ 0 & 1 & 0 & | & \frac{2}{8} & \frac{3}{8} & -\frac{1}{8} \\ 0 & 0 & 1 & | & -\frac{6}{8} & -\frac{1}{8} & \frac{3}{8} \end{bmatrix}.$$

Hence, A is invertible and

$$A^{-1} = \begin{bmatrix} \frac{6}{8} & -\frac{3}{8} & \frac{1}{8} \\ \frac{2}{8} & \frac{3}{8} & -\frac{1}{8} \\ -\frac{6}{8} & -\frac{1}{8} & \frac{3}{8} \end{bmatrix} = \frac{1}{8}\begin{bmatrix} 6 & -3 & 1 \\ 2 & 3 & -1 \\ -6 & -1 & 3 \end{bmatrix}.$$

Step 3 You should verify that $AA^{-1} = I$. ■ ■ ■

Practice Problem 4 Find the inverse (if it exists) of the matrix $A = \begin{bmatrix} 1 & 2 & 3 \\ -2 & 3 & 1 \\ 4 & 5 & -2 \end{bmatrix}$.

■

3 Find the inverse of a 2×2 matrix.

A Rule for Finding the Inverse of a 2×2 Matrix

You can quickly determine whether any 2×2 matrix is invertible and, if so, what its inverse is.

A RULE FOR FINDING THE INVERSE OF A 2×2 MATRIX

The matrix

$$A = \begin{bmatrix} a & b \\ c & d \end{bmatrix}$$

is invertible if and only if $ad - bc \neq 0$. Moreover, if $ad - bc \neq 0$, then

$$A^{-1} = \frac{1}{ad - bc} \begin{bmatrix} d & -b \\ -c & a \end{bmatrix}.$$

Note that if $ad - bc = 0$, the matrix A does not have an inverse.

EXAMPLE 5   **Finding the Inverse of a 2×2 Matrix**

Find the inverse (if it exists) of each matrix.

a. $A = \begin{bmatrix} 5 & 2 \\ 4 & 3 \end{bmatrix}$ **b.** $B = \begin{bmatrix} 4 & 6 \\ 2 & 3 \end{bmatrix}$

SOLUTION

a. For the matrix A, $a = 5$, $b = 2$, $c = 4$, and $d = 3$.
Here $ad - bc = (5)(3) - (2)(4) = 15 - 8 = 7 \neq 0$; so A is invertible and

$$A^{-1} = \frac{1}{7} \begin{bmatrix} 3 & -2 \\ -4 & 5 \end{bmatrix}$$ Substitute for $a, b, c,$ and d in $\dfrac{1}{ad - bc}\begin{bmatrix} d & -b \\ -c & a \end{bmatrix}$.

$$= \begin{bmatrix} \dfrac{3}{7} & -\dfrac{2}{7} \\ -\dfrac{4}{7} & \dfrac{5}{7} \end{bmatrix}$$ Scalar multiplication

You should verify that $AA^{-1} = I$.

b. For the matrix B, $a = 4$, $b = 6$, $c = 2$, and $d = 3$; so $ad - bc = (4)(3) - (6)(2) = 12 - 12 = 0$. Therefore, the matrix B does not have an inverse. ■ ■ ■

Practice Problem 5   Find the inverse (if it exists) of each matrix.

a. $A = \begin{bmatrix} 8 & 2 \\ 4 & 1 \end{bmatrix}$ **b.** $B = \begin{bmatrix} 8 & -2 \\ 3 & 1 \end{bmatrix}$ ■

4 Use matrix inverses to solve systems of linear equations.

Solving Systems of Linear Equations by Using Matrix Inverses

Matrix multiplication can be used to write a system of linear equations in matrix form.

System of Equations	Matrix Form
$\begin{cases} 3x - 2y = 4 \\ 4x - 3y = 5 \end{cases}$	$\begin{bmatrix} 3 & -2 \\ 4 & -3 \end{bmatrix}\begin{bmatrix} x \\ y \end{bmatrix} = \begin{bmatrix} 4 \\ 5 \end{bmatrix}$
$\begin{cases} 2x_1 + 4x_2 - x_3 = 9 \\ 3x_1 + x_2 + 2x_3 = 7 \\ x_1 + 3x_2 - 3x_3 = 4 \end{cases}$	$\begin{bmatrix} 2 & 4 & -1 \\ 3 & 1 & 2 \\ 1 & 3 & -3 \end{bmatrix}\begin{bmatrix} x_1 \\ x_2 \\ x_3 \end{bmatrix} = \begin{bmatrix} 9 \\ 7 \\ 4 \end{bmatrix}$

Solving a system of linear equations amounts to solving a corresponding matrix equation of the form

$$AX = B,$$

where A is the coefficient matrix of the system, X is the column matrix containing the variables, and B is the column matrix of the system's constants.

If A is a square matrix *and* A is invertible, then the matrix equation $AX = B$ has a unique solution, $X = A^{-1}B$, as the following derivation shows.

$AX = B$	Given equation
$A^{-1}(AX) = A^{-1}B$	Be careful to multiply by A^{-1} on the left on both sides of the equation because multiplication of matrices is not commutative.
$(A^{-1}A)X = A^{-1}B$	Associative property
$IX = A^{-1}B$	$A^{-1}A = I$
$X = A^{-1}B$	I is the identity matrix.

EXAMPLE 6   Solving a Linear System by Using an Inverse Matrix

Use a matrix inverse to solve the linear system:
$$\begin{cases} x + y \quad\quad = 4 \\ \quad\quad 3y + z = 7 \\ 2x + 3y + 3z = 21 \end{cases}$$

SOLUTION

Write the linear system in matrix form.

$$\underbrace{\begin{bmatrix} 1 & 1 & 0 \\ 0 & 3 & 1 \\ 2 & 3 & 3 \end{bmatrix}}_{A} \underbrace{\begin{bmatrix} x \\ y \\ z \end{bmatrix}}_{X} = \underbrace{\begin{bmatrix} 4 \\ 7 \\ 21 \end{bmatrix}}_{B}$$

Use zeros for coefficients of missing variables.

We need to solve the matrix equation $AX = B$ for X. Because the matrix A is invertible (see Example 4), the system has a unique solution $X = A^{-1}B$.

$$A^{-1} = \frac{1}{8}\begin{bmatrix} 6 & -3 & 1 \\ 2 & 3 & -1 \\ -6 & -1 & 3 \end{bmatrix}$$ Computed in Example 4 on page 584.

$$X = A^{-1}B$$

$$= \frac{1}{8}\begin{bmatrix} 6 & -3 & 1 \\ 2 & 3 & -1 \\ -6 & -1 & 3 \end{bmatrix}\begin{bmatrix} 4 \\ 7 \\ 21 \end{bmatrix}$$ Substitute for A^{-1} and B.

$$= \frac{1}{8}\begin{bmatrix} 6(4) - 3(7) + 1(21) \\ 2(4) + 3(7) - 1(21) \\ -6(4) - 1(7) + 3(21) \end{bmatrix}$$ Matrix multiplication

$$= \frac{1}{8}\begin{bmatrix} 24 \\ 8 \\ 32 \end{bmatrix} = \begin{bmatrix} 3 \\ 1 \\ 4 \end{bmatrix}$$ Scalar multiplication

The solution set is $\{(3, 1, 4)\}$, which you can check in the original system.   ■ ■ ■

Practice Problem 6   Repeat Example 6 for the linear system:

$$\begin{cases} 3x + 2y + 3z = 9 \\ 3x + y = 12 \\ x + z = 6 \end{cases}$$ ■

5 Use matrix inverses in applied problems.

Applications of Matrix Inverses

The Leontief Input–Output Model   We can use the inverse of a matrix to analyze input–output models that we studied in Section 6.1.

Suppose a simplified economy depends on two products: energy (E) and food (F). To produce 1 unit of E requires $\frac{1}{4}$ unit of E and $\frac{1}{2}$ unit of F. To produce 1 unit of F requires $\frac{1}{3}$ unit of E and $\frac{1}{4}$ unit of F. Then the interindustry consumption is given by the matrix

$$A = \begin{matrix} & \text{\textbf{Input}} \\ & \begin{matrix} E & F \end{matrix} \\ \begin{bmatrix} \frac{1}{4} & \frac{1}{2} \\ \frac{1}{3} & \frac{1}{4} \end{bmatrix} & \begin{matrix} E \\ F \end{matrix} \end{matrix} \quad \textbf{\textcolor{red}{Output}}$$

The matrix A is the **interindustry technology input–output matrix**, or simply the **technology matrix**, of the system. If the system is producing x_1 units of energy and x_2 units of food, then the column matrix $X = \begin{bmatrix} x_1 \\ x_2 \end{bmatrix}$ is called the **gross production matrix**.

Consequently,

$$AX = \begin{bmatrix} \frac{1}{4} & \frac{1}{2} \\ \frac{1}{3} & \frac{1}{4} \end{bmatrix} \begin{bmatrix} x_1 \\ x_2 \end{bmatrix} = \begin{bmatrix} \frac{1}{4}x_1 + \frac{1}{2}x_2 \\ \frac{1}{3}x_1 + \frac{1}{4}x_2 \end{bmatrix} \begin{matrix} \leftarrow \text{units consumed by } E \\ \leftarrow \text{units consumed by } F \end{matrix}$$

represents the interindustry consumptions. If the column matrix $D = \begin{bmatrix} d_1 \\ d_2 \end{bmatrix}$ represents consumer demand, then

$$\begin{aligned} D &= X - AX & &\text{Gross production minus interindustry consumption} \\ &= IX - AX & &I \text{ is the identity matrix.} \\ D &= (I - A)X & &\text{Distributive property} \\ (I - A)^{-1}D &= (I - A)^{-1}(I - A)X & &\text{Multiply both sides by } (I - A)^{-1} \text{ if } I - A \text{ is invertible.} \\ (I - A)^{-1}D &= IX & &(I - A)^{-1}(I - A) = I \\ &= X & &I \text{ is the identity matrix.} \end{aligned}$$

So if $(I - A)$ is invertible, the gross production matrix is $X = (I - A)^{-1}D$.

EXAMPLE 7   **Using the Leontief Input–Output Model**

In the preceding discussion, suppose the consumer demand for energy is 1000 units and for food is 3000 units. Find the level of production (X) that will meet interindustry and consumer demand.

SOLUTION

For the matrix A, we have

$$I - A = \begin{bmatrix} 1 & 0 \\ 0 & 1 \end{bmatrix} - \begin{bmatrix} \frac{1}{4} & \frac{1}{2} \\ \frac{1}{3} & \frac{1}{4} \end{bmatrix} = \begin{bmatrix} \frac{3}{4} & -\frac{1}{2} \\ -\frac{1}{3} & \frac{3}{4} \end{bmatrix}.$$

RECALL

If $ad - bc \neq 0$, then for $A = \begin{bmatrix} a & b \\ c & d \end{bmatrix}$, we have

$$A^{-1} = \frac{1}{ad - bc}\begin{bmatrix} d & -b \\ -c & a \end{bmatrix}.$$

Use the formula for the inverse of a 2×2 matrix.

$$(I - A)^{-1} = \frac{1}{\frac{3}{4} \cdot \frac{3}{4} - \left(-\frac{1}{2}\right)\left(-\frac{1}{3}\right)}\begin{bmatrix} \frac{3}{4} & \frac{1}{2} \\ \frac{1}{3} & \frac{3}{4} \end{bmatrix}$$

$$= \frac{48}{19}\begin{bmatrix} \frac{3}{4} & \frac{1}{2} \\ \frac{1}{3} & \frac{3}{4} \end{bmatrix} = \frac{1}{19}\begin{bmatrix} 36 & 24 \\ 16 & 36 \end{bmatrix} \qquad \text{Simplify.}$$

The matrix $D = \begin{bmatrix} 1000 \\ 3000 \end{bmatrix}$. Therefore, the gross production matrix X is

$$X = (I - A)^{-1}D = \frac{1}{19}\begin{bmatrix} 36 & 24 \\ 16 & 36 \end{bmatrix}\begin{bmatrix} 1000 \\ 3000 \end{bmatrix} \qquad \text{Substitute for } (I - A)^{-1} \text{ and } D.$$

$$= \frac{1}{19}\begin{bmatrix} 108{,}000 \\ 124{,}000 \end{bmatrix} = \begin{bmatrix} \frac{108{,}000}{19} \\ \frac{124{,}000}{19} \end{bmatrix}.$$

To meet the consumer demand for 1000 units of energy and 3000 units of food, the energy produced must be $\dfrac{108{,}000}{19}$ units and food production must be $\dfrac{124{,}000}{19}$ units.

■ ■ ■

Practice Problem 7 In Example 7, find the level of production (X) that will meet both the interindustry and consumer demand if consumer demand for energy is 800 units and consumer demand for food is 3400 units. ■

Cryptography

To encode or decode a message, associate each letter in the message with a number in the obvious manner. The blank corresponds to 0.

blank	A	B	C	D	E	F	G	H	I	J	K	L	M
0	1	2	3	4	5	6	7	8	9	10	11	12	13

N	O	P	Q	R	S	T	U	V	W	X	Y	Z
14	15	16	17	18	19	20	21	22	23	24	25	26

For example, the message

JOHNSON IS IN DANGER

becomes

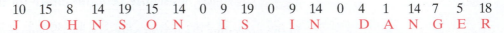

This message is easy to decode, so we will make the coding more complicated. First, partition these 20 numbers into groups of three, enclosed in brackets, and insert zeros at the end of the last bracket if necessary.

$$[10 \quad 15 \quad 8] [14 \quad 19 \quad 15] [14 \quad 0 \quad 9] [19 \quad 0 \quad 9] [14 \quad 0 \quad 4] [1 \quad 14 \quad 7] [5 \quad 18 \quad 0]$$
$$\text{J O H N S O N \quad I S \quad I N \quad D A N G E R}$$

Let's write these seven groups of three numbers as the seven columns of a 3×7 matrix.

$$M = \begin{bmatrix} 10 & 14 & 14 & 19 & 14 & 1 & 5 \\ 15 & 19 & 0 & 0 & 0 & 14 & 18 \\ 8 & 15 & 9 & 9 & 4 & 7 & 0 \end{bmatrix}$$

Now choose any invertible 3×3 matrix A, say,

$$A = \begin{bmatrix} 1 & 2 & 3 \\ 1 & 3 & 3 \\ 1 & 2 & 4 \end{bmatrix}.$$

We can use the matrix A to convert the message into code and then use A^{-1} to decode the message, as in the next example. The matrix A, called the **coding matrix**, is known to both the sender and the receiver of the message.

EXAMPLE 8   Encoding and Decoding a Message

Encode and decode the message

JOHNSON IS IN DANGER

by using the matrix A given above.

SOLUTION

Step 1   Express the message numerically and partition the numbers into groups of three.

$$[10 \quad 15 \quad 8] [14 \quad 19 \quad 15] [14 \quad 0 \quad 9] [19 \quad 0 \quad 9] [14 \quad 0 \quad 4] [1 \quad 14 \quad 7] [5 \quad 18 \quad 0]$$

Step 2   Write each group of three numbers in Step 1 as a column of a matrix M.

$$M = \begin{bmatrix} 10 & 14 & 14 & 19 & 14 & 1 & 5 \\ 15 & 19 & 0 & 0 & 0 & 14 & 18 \\ 8 & 15 & 9 & 9 & 4 & 7 & 0 \end{bmatrix}$$

Step 3   Select an invertible 3×3 matrix A such as

$$A = \begin{bmatrix} 1 & 2 & 3 \\ 1 & 3 & 3 \\ 1 & 2 & 4 \end{bmatrix}.$$

Step 4   Multiply the matrices in Steps 2 and 3 to form AM.

$$AM = \begin{bmatrix} 1 & 2 & 3 \\ 1 & 3 & 3 \\ 1 & 2 & 4 \end{bmatrix} \begin{bmatrix} 10 & 14 & 14 & 19 & 14 & 1 & 5 \\ 15 & 19 & 0 & 0 & 0 & 14 & 18 \\ 8 & 15 & 9 & 9 & 4 & 7 & 0 \end{bmatrix}$$
$$= \begin{bmatrix} 64 & 97 & 41 & 46 & 26 & 50 & 41 \\ 79 & 116 & 41 & 46 & 26 & 64 & 59 \\ 72 & 112 & 50 & 55 & 30 & 57 & 41 \end{bmatrix}$$

Matrix multiplication

The coded message sent will be the numbers from column 1 of AM, followed by the numbers from column 2, and so on.

64 79 72 97 116 112 41 41 50 46 46 55 26 26 30 50 64 57 41 59 41

Step 5   To decode the message in Step 4, write the numbers received in the message in groups of three and as columns of the matrix AM. Find the **decoding matrix** A^{-1} of the coding matrix A. You can verify that

$$A^{-1} = \begin{bmatrix} 6 & -2 & -3 \\ -1 & 1 & 0 \\ -1 & 0 & 1 \end{bmatrix}.$$

Step 6   Compute the message matrix M.

$$M = A^{-1}(AM) = \begin{bmatrix} 6 & -2 & -3 \\ -1 & 1 & 0 \\ -1 & 0 & 1 \end{bmatrix}\begin{bmatrix} 64 & 97 & 41 & 46 & 26 & 50 & 41 \\ 79 & 116 & 41 & 46 & 26 & 64 & 59 \\ 72 & 112 & 50 & 55 & 30 & 57 & 41 \end{bmatrix}$$

$$= \begin{bmatrix} 10 & 14 & 14 & 19 & 14 & 1 & 5 \\ 15 & 19 & 0 & 0 & 0 & 14 & 18 \\ 8 & 15 & 9 & 9 & 4 & 7 & 0 \end{bmatrix} \quad \text{Matrix multiplication}$$

Step 7   Finally, convert the numerical message from the matrix M in Step 6 back into English.

JOHNSON IS IN DANGER.   ■ ■ ■

Practice Problem 8   Rework Example 8, replacing the message JOHNSON IS IN DANGER with the message JACK IS NOW SAFE.   ■

SECTION 6.3 ■ Exercises

A EXERCISES Basic Skills and Concepts

1. For $n \times n$ matrices A and B, if $AB = I = BA$, then B is called the _____ of A.

2. An $n \times n$ matrix A is invertible if there is a matrix B such that _____ $= I$.

3. To find the inverse of an invertible matrix A, we transform $[A \mid I]$ by a sequence of row operations into $[I \mid B]$, where $B =$ _____.

4. $A = \begin{bmatrix} a & b \\ c & d \end{bmatrix}$ is invertible if and only if _____.

5. *True or False* Every square matrix is invertible.

6. *True or False* $A = \begin{bmatrix} 8 & 16 \\ 4 & 8 \end{bmatrix}$ is invertible.

In Exercises 7–16, determine whether B is the inverse of A by computing AB and BA.

7. $A = \begin{bmatrix} 1 & 2 \\ 1 & 3 \end{bmatrix}, B = \begin{bmatrix} 3 & -2 \\ -1 & 1 \end{bmatrix}$

8. $A = \begin{bmatrix} 3 & 2 \\ 4 & 3 \end{bmatrix}, B = \begin{bmatrix} 3 & -2 \\ -4 & 3 \end{bmatrix}$

9. $A = \begin{bmatrix} 3 & 2 \\ 1 & 4 \end{bmatrix}, B = \begin{bmatrix} \dfrac{2}{5} & -\dfrac{1}{5} \\ -\dfrac{1}{10} & \dfrac{3}{10} \end{bmatrix}$

10. $A = \begin{bmatrix} 2 & -3 \\ 4 & -3 \end{bmatrix}, B = \begin{bmatrix} -\dfrac{1}{2} & \dfrac{1}{2} \\ -\dfrac{2}{3} & \dfrac{1}{3} \end{bmatrix}$

11. $A = \begin{bmatrix} 1 & 0 & -1 \\ -1 & 1 & 1 \end{bmatrix}, B = \begin{bmatrix} 2 & 1 \\ 1 & 1 \\ 1 & 1 \end{bmatrix}$

12. $A = \begin{bmatrix} 1 & -2 & 1 \\ -8 & 6 & -2 \\ 5 & -3 & 1 \end{bmatrix}, B = \dfrac{1}{2}\begin{bmatrix} 0 & 1 & 2 \\ 1 & 2 & 3 \\ 3 & 1 & 1 \end{bmatrix}$

13. $A = \begin{bmatrix} -2 & 1 & 3 \\ 0 & -1 & 1 \\ 1 & 2 & 0 \end{bmatrix}, B = \dfrac{1}{8}\begin{bmatrix} -2 & 6 & 4 \\ 1 & -3 & 2 \\ 1 & 5 & 2 \end{bmatrix}$

14. $A = \begin{bmatrix} 2 & 3 & 1 \\ 1 & 2 & 3 \\ 3 & 1 & 2 \end{bmatrix}, B = \dfrac{1}{18}\begin{bmatrix} 1 & -5 & 7 \\ 7 & 1 & -5 \\ -5 & 7 & 1 \end{bmatrix}$

15. $A = \begin{bmatrix} 1 & 1 & 1 \\ 1 & 2 & 3 \\ -1 & 1 & -1 \end{bmatrix}, B = \dfrac{1}{4}\begin{bmatrix} 5 & -2 & -1 \\ 2 & 0 & 2 \\ -3 & 2 & -1 \end{bmatrix}$

16. $A = \begin{bmatrix} 1 & -1 & 0 \\ 0 & 1 & -1 \\ 1 & 0 & 1 \end{bmatrix}, B = \dfrac{1}{2}\begin{bmatrix} 1 & 1 & 1 \\ -1 & 1 & 1 \\ -1 & -1 & 1 \end{bmatrix}$

In Exercises 17–26, find A^{-1} (if it exists) by forming the matrix $[A\,|\,I]$ and using the row operation to obtain $[I\,|\,B]$, where $B = A^{-1}$.

17. $A = \begin{bmatrix} 2 & 0 \\ 1 & 3 \end{bmatrix}$

18. $A = \begin{bmatrix} 4 & 3 \\ 1 & 0 \end{bmatrix}$

19. $A = \begin{bmatrix} 2 & 4 \\ 3 & 6 \end{bmatrix}$

20. $A = \begin{bmatrix} 9 & 6 \\ 6 & 4 \end{bmatrix}$

21. $A = \begin{bmatrix} 1 & 6 & 4 \\ 0 & 2 & 3 \\ 0 & 1 & 2 \end{bmatrix}$

22. $A = \begin{bmatrix} -2 & 1 & 3 \\ 0 & -1 & 1 \\ 1 & 2 & 0 \end{bmatrix}$

23. $A = \begin{bmatrix} 2 & 4 & 3 \\ 0 & 1 & 1 \\ 2 & 2 & -1 \end{bmatrix}$

24. $A = \begin{bmatrix} 1 & 1 & 5 \\ 4 & 3 & -5 \\ 1 & 1 & 0 \end{bmatrix}$

25. $A = \begin{bmatrix} 1 & 2 & -1 \\ -1 & 1 & 2 \\ 2 & -1 & 1 \end{bmatrix}$

26. $A = \begin{bmatrix} 1 & 2 & 3 \\ 2 & 3 & 0 \\ 0 & 1 & 2 \end{bmatrix}$

In Exercises 27–32, find the inverse (if it exists) of $A = \begin{bmatrix} a & b \\ c & d \end{bmatrix}$ by using the formula $A^{-1} = \dfrac{1}{ad - bc}\begin{bmatrix} d & -b \\ -c & a \end{bmatrix}$. Check that $A^{-1}A = I$.

27. $A = \begin{bmatrix} 1 & 0 \\ 3 & 2 \end{bmatrix}$

28. $A = \begin{bmatrix} 3 & 4 \\ 5 & 6 \end{bmatrix}$

29. $A = \begin{bmatrix} 2 & -3 \\ -3 & 5 \end{bmatrix}$

30. $A = \begin{bmatrix} 3 & -4 \\ 2 & -2 \end{bmatrix}$

31. $A = \begin{bmatrix} a & -b \\ b & -a \end{bmatrix}$, $a^2 \neq b^2$

32. $A = \begin{bmatrix} 2 & -1 \\ 1 & -1 \end{bmatrix}$

In Exercises 33–36, write the given linear system of equations as a matrix equation $AX = B$.

33. $\begin{cases} 2x + 3y = -9 \\ x - 3y = 13 \end{cases}$

34. $\begin{cases} 5x - 4y = 7 \\ 4x - 3y = 5 \end{cases}$

35. $\begin{cases} 3x + 2y + z = 8 \\ 2x + y + 3z = 7 \\ x + 3y + 2z = 9 \end{cases}$

36. $\begin{cases} x + 3y + z = 4 \\ x - 5y + 2z = 7 \\ 3x + y - 4z = -9 \end{cases}$

In Exercises 37–40, write each matrix equation as a linear system of equations.

37. $\begin{bmatrix} 1 & -2 \\ 2 & 1 \end{bmatrix}\begin{bmatrix} x \\ y \end{bmatrix} = \begin{bmatrix} 0 \\ 5 \end{bmatrix}$

38. $\begin{bmatrix} 2 & 3 \\ 3 & -1 \end{bmatrix}\begin{bmatrix} x_1 \\ x_2 \end{bmatrix} = \begin{bmatrix} 0 \\ 11 \end{bmatrix}$

39. $\begin{bmatrix} 2 & 3 & 1 \\ 5 & 7 & -1 \\ 4 & 3 & 0 \end{bmatrix}\begin{bmatrix} x_1 \\ x_2 \\ x_3 \end{bmatrix} = \begin{bmatrix} -1 \\ 5 \\ 5 \end{bmatrix}$

40. $\begin{bmatrix} 3 & -2 & 3 \\ 5 & 0 & 4 \\ 2 & 7 & 0 \end{bmatrix}\begin{bmatrix} r \\ s \\ t \end{bmatrix} = \begin{bmatrix} 4 \\ 3 \\ -8 \end{bmatrix}$

41. Let $A = \begin{bmatrix} 1 & 2 & 5 \\ 2 & 3 & 8 \\ -1 & 1 & 2 \end{bmatrix}$. Show that $A^{-1} = \begin{bmatrix} 2 & -1 & -1 \\ 12 & -7 & -2 \\ -5 & 3 & 1 \end{bmatrix}$.

In Exercises 42–44, use the result of Exercise 41 to solve the system of equations.

42. $\begin{cases} x + 2y + 5z = 4 \\ 2x + 3y + 8z = 6 \\ -x + y + 2z = 3 \end{cases}$

43. $\begin{cases} x_1 + 2x_2 + 5x_3 = 1 \\ 2x_1 + 3x_2 + 8x_3 = 3 \\ -x_1 + x_2 + 2x_3 = -3 \end{cases}$

44. $\begin{cases} x + 2y + 5z = -4 \\ 2x + 3y + 8z = -6 \\ -x + y + 2z = -\dfrac{5}{2} \end{cases}$

45. a. Find the inverse of the matrix $A = \begin{bmatrix} 1 & 1 & 1 \\ 1 & 2 & 3 \\ 1 & 4 & 9 \end{bmatrix}$.

   b. Use the result of part (a) to solve the linear system.

   $\begin{cases} x + y + z = 6 \\ x + 2y + 3z = 14 \\ x + 4y + 9z = 36 \end{cases}$

46. a. Find the inverse of the matrix $A = \begin{bmatrix} 2 & 4 & -1 \\ 3 & 1 & 2 \\ 1 & 3 & -3 \end{bmatrix}$.

   b. Use the result of part (a) to solve the linear system.

   $\begin{cases} 2x + 4y - z = 9 \\ 3x + y + 2z = 7 \\ x + 3y - 3z = 4 \end{cases}$

B EXERCISES Applying the Concepts

In Exercises 47–52, solve each linear system by using an inverse matrix.

47. $\begin{cases} 3x + 7y = 11 \\ -5x + 4y = 13 \end{cases}$

48. $\begin{cases} x - 7y = 3 \\ 2x + 3y = 23 \end{cases}$

49. $\begin{cases} x + y + 2z = 7 \\ x - y - 3z = -6 \\ 2x + 3y + z = 4 \end{cases}$

50. $\begin{cases} x + y + z = 6 \\ 2x - 3y + 3z = 5 \\ 3x - 2y - z = -4 \end{cases}$

51. $\begin{cases} 2x + 2y + 3z = 7 \\ 5x + 3y + 5z = 3 \\ 3x + 5y + z = -5 \end{cases}$

52. $\begin{cases} 9x + 7y + 4z = 12 \\ 6x + 5y + 4z = 5 \\ 4x + 3y + z = 7 \end{cases}$

In Exercises 53–58, use the method presented in this section to solve the problem.

53. **Investment.** Liz inherited $90,000, and she split it into three investments. Part of the money she invested in a treasury bill that yields 3% annual interest, part she invested in bonds with an annual yield of 7%, and the rest she invested in a mutual fund. In 2008, when the mutual fund lost 11%, her net income from all three investments was $980. In 2009, when the mutual fund gained 8%, her net income from all three investments was $5920. How much money did Liz put in each investment?

54. **Investment.** Derrick invested a total of $40,000 in stocks A, B and C. At the beginning of each year, he kept the same level of investment for each stock. The following table depicts the performance of each stock.

Year	Percent Gain or Loss			Total Return
	Stock A	Stock B	Stock C	
2007	12%	8%	10%	$3880
2008	−15%	−12%	−16%	−$5560
2009	10%	6%	8%	$3080

How much did Derrick invest in each stock?

55. Sandwich sales. The Fresh Sandwich specializes in chicken, fish, and ham sandwiches. In April, May, and June 2008, they sold the following quantities.

	Sales Volume in Hundreds		
	April	May	June
Chicken	8	7	4
Fish	11	8	6
Ham	6	7	3
Monthly revenue, in hundreds, from the sales of these sandwiches	$164	$141	$86

Find the selling price of each of the three types of sandwiches.

56. Euler's formula. A plane graph connects points called vertices with nonintersecting lines called edges and divides the plane into regions. The accompanying figure has seven vertices, nine edges, and four regions. Euler's theorem states that in a plane graph,

$$V - E + R = 2,$$

where V = the number of vertices,

   E = the number of edges, and

   R = the number of regions.

In a certain plane graph, twice the number of edges is three times the number of vertices and twice the number of regions is one less than the number of edges. Find the number of vertices, edges, and regions for this graph. Sketch a plane graph that satisfies the conditions of this exercise.

57. Projectile. A projectile is fired up from the top of a hill. After t seconds, the height of the projectile above the ground is given by the equation $h = a + bt + ct^2$. A built-in sensor in the projectile sends the following measurements of its height above the ground.

Time t in seconds	1	2	3
Height h in feet	64	76	56

a. Find the height of the hill.
b. Find the height of the projectile after 4 seconds.

58. Projectile. On a planet the height h of a projectile fired up from an initial height h_0 with initial vertical velocity of v_0 ft/sec after t seconds is given by $h = \dfrac{1}{2} at^2 + v_0 t + h_0$, where a is the gravitational constant. The following measurements were recorded.

Time t in seconds	1	2	3
Height h in feet	100	92	76

a. Find the initial height.
b. Find the height of the projectile after 2.5 seconds.

In Exercises 59–62, technology matrix A, representing interindustry demand, and matrix D, representing consumer demand for the sectors in a two- or three-sector economy, are given. Use the matrix equation $X = (I - A)^{-1}D$ to find the production level X that will satisfy both demands.

59. Two-sector economy. In a two-sector economy,

$$A = \begin{bmatrix} 0.1 & 0.4 \\ 0.5 & 0.2 \end{bmatrix}, D = \begin{bmatrix} 50 \\ 30 \end{bmatrix}.$$

60. Two-sector economy. In a two-sector economy,

$$A = \begin{bmatrix} 0.2 & 0.5 \\ 0.1 & 0.6 \end{bmatrix}, D = \begin{bmatrix} 11 \\ 18 \end{bmatrix}.$$

61. Three-sector economy. In a three-sector economy,

$$A = \begin{bmatrix} 0.2 & 0.2 & 0 \\ 0.1 & 0.1 & 0.3 \\ 0.1 & 0 & 0.2 \end{bmatrix}, D = \begin{bmatrix} 400 \\ 600 \\ 800 \end{bmatrix}.$$

62. Three-sector economy. In a three-sector economy,

$$A = \begin{bmatrix} 0.5 & 0.4 & 0.2 \\ 0.2 & 0.3 & 0.1 \\ 0.1 & 0.1 & 0.3 \end{bmatrix}, D = \begin{bmatrix} 500 \\ 300 \\ 200 \end{bmatrix}.$$

In Exercises 63 and 64, use a 2 × 2 invertible coding matrix A of your choice. Remember to partition the numerical message into groups of two. (a) Write a cryptogram for each message. (b) Check by decoding.

63. Coding and decoding. CANNOT FIND COLOMBO.

64. Coding and decoding. FOUND HEAD OF TERRORIST NETWORK.

In Exercises 65 and 66, use a 3 × 3 invertible coding matrix A of your choice.

65. Code and then decode the message of Exercise 63.

66. Code and then decode the message of Exercise 64.

C EXERCISES Beyond the Basics

67. Suppose A is an invertible square matrix.
   a. Is A^{-1} invertible? Why or why not?
   b. If your answer to part (a) is yes, then what is $(A^{-1})^{-1}$?

68. Let $A^{-1} = \begin{bmatrix} 2 & 1 \\ 3 & -4 \end{bmatrix}$. Find A.

69. Let A be a square matrix. Show that if A^2 is invertible, then A must be invertible. [*Hint:* $A^2B = A(AB)$.]

70. Show that if A is invertible, then A^2 is invertible.

71. Show that if A and B are $n \times n$ invertible matrices, then AB is invertible and $(AB)^{-1} = B^{-1}A^{-1}$. (Note the order.)

72. Verify the result of Exercise 71 for

$$A = \begin{bmatrix} 1 & 2 \\ 3 & 4 \end{bmatrix}, B = \begin{bmatrix} -1 & 0 \\ 3 & -2 \end{bmatrix}.$$

73. Let A, B, and C be $n \times n$ matrices such that $AB = AC$. Show that if A is invertible, then $B = C$.

74. Is the result of Exercise 73 true if A is not invertible?

75. Let A and B be $n \times n$ matrices such that $AB = A$. Show that if A is invertible, then $B = I$.

76. Give an example to show that the result of Exercise 75 is not true if A is not invertible.

77. Let $A = \begin{bmatrix} 3 & 4 \\ 2 & 3 \end{bmatrix}$.

   a. Show that the matrix A satisfies the equation

$$A^2 - 6A + I = 0.$$

   b. Use the equation in part (a) to show that $A^{-1} = 6I - A$.

   c. Use the equation in part (b) to find A^{-1}.

78. Let $A = \begin{bmatrix} 2 & -1 & 1 \\ -1 & 2 & -1 \\ 1 & -1 & 2 \end{bmatrix}$. Assume that A is invertible.

   a. Show that A satisfies the equation
   $A^3 - 6A^2 + 9A - 4I = 0$.

   b. Use the equation in part (a) to show that

$$A^{-1} = \frac{1}{4}[A^2 - 6A + 9I].$$

   [*Hint:* Consider the product $\frac{1}{4}[A^2 - 6A + 9I]A$.]

   c. Use part (b) to find A^{-1}.

In Exercises 79–84, use your graphing utility to find the inverse of the coefficient matrix. Then use the inverse to solve each linear system.

79. $\begin{cases} x - y - z - w = -4 \\ x + y + z - w = 2 \\ 2x + y + z - w = 3 \\ x - y + z - w = -2 \end{cases}$

80. $\begin{cases} x + y + z - w = 1 \\ x - y - z + w = 1 \\ x - y + z + 3w = 1 \\ x + y + z + w = 3 \end{cases}$

81. $\begin{cases} x - y - z - w = -1 \\ x + y + z - w = -5 \\ x + y + z + w = 1 \\ x + y - z + w = 9 \end{cases}$

82. $\begin{cases} x + y + z + w = 2 \\ 3x - 4y + 2z + w = -9 \\ 4x + 3y + -2z - 5w = 15 \\ 2x - 3y + 4z + 2w = -12 \end{cases}$

83. $\begin{cases} x + y + z + u + v = 9 \\ x + y - z + 2u + 3v = 5 \\ 2x - y + 4z - 3u - v = 1 \\ x - y + z + 5u + 2v = 20 \\ 3x - y - z + u + 3v = -1 \end{cases}$

84. $\begin{cases} x + y + z + u + v = 3 \\ 2x + y - 2z + u - v = 2 \\ x + y + z + 2u + 3v = 2 \\ 3x + y + 4z - 2u - v = 6 \\ x - y - z + u + v = 1 \end{cases}$

Critical Thinking

85. Suppose A and B are 2×2 matrices and $B^{-1} = \begin{bmatrix} 1 & 2 \\ 3 & 4 \end{bmatrix}$.

   Find A

   a. if $AB = \begin{bmatrix} -1 & 3 \\ 0 & 2 \end{bmatrix}$.

   b. if $BA = \begin{bmatrix} -1 & 3 \\ 0 & 2 \end{bmatrix}$.

   c. if $B^{-1}AB = \begin{bmatrix} 1 & 1 \\ 1 & 2 \end{bmatrix}$.

   d. if $BAB^{-1} = \begin{bmatrix} 1 & 1 \\ 1 & 2 \end{bmatrix}$.

86. **a.** Is the matrix $\begin{bmatrix} x & -1 \\ 2 & x + 2 \end{bmatrix}$ invertible for all real numbers x?

   b. Is the matrix $\begin{bmatrix} x - 1 & -1 \\ 3 & x - 5 \end{bmatrix}$ invertible for all real numbers x?

87. *True or False* Justify your answers.
   a. If A and B are matrices for which $AB = BA$, then the formula $A^2B = BA^2$ must hold.
   b. If A and B are invertible matrices, then $A + B$ must be invertible.

88. Give an example of a 2×2 matrix A such that $A^2 = 0$. Show that the matrix $I + A$ is invertible and that $(I + A)^{-1} = I - A$.

89. Is it true that every nonsingular matrix is a square matrix? Why or why not?

Determinants and Cramer's Rule

Seki Köwa (1642–1708)

Seki Köwa has been called "the Japanese Newton." He was born the same year as Newton. Like Newton, he acquired much of his knowledge by self-study and was a brilliant problem solver. But the greatest similarity between the two men lies in the claim that Seki Kowa was the creator of *yenri* (calculus). Thus, Seki Köwa probably invented calculus in the East just as Newton and Leibniz invented calculus in the West.

A LITTLE HISTORY OF DETERMINANTS

It appears that determinants were first investigated in 1683 by the outstanding Japanese mathematician Seki Köwa in connection with systems of linear equations. Determinants were independently investigated in 1693 by the German mathematician Gottfried Leibniz (1646–1716). The French mathematician Alexandre-Théophile Vandermonde (1735–1796) was the first to give a coherent and systematic exposition of the theory of determinants. Determinants were used extensively in the nineteenth century by eminent mathematicians such as Cauchy, Jacobi, and Kronecker. Today determinants are of little numerical value in the large-scale matrix computations that occur so often. Nevertheless, a knowledge of determinants is useful in some applications of matrices. ■

---

1 Calculate the determinant of a 2 × 2 matrix.

Determinant of a 2 × 2 Matrix

Associated with each square matrix A is a number called the *determinant* of A.

STUDY TIP

A determinant can be thought of as a function whose domain is the set of all square matrices and whose range is the set of real numbers.

DETERMINANT OF A 2 × 2 MATRIX

The **determinant** of the matrix

$$A = \begin{bmatrix} a & b \\ c & d \end{bmatrix},$$

denoted by $\det(A)$, $|A|$, or $\begin{vmatrix} a & b \\ c & d \end{vmatrix}$, is called a **2 by 2 determinant** and is defined by

$$\det(A) = ad - bc.$$

To remember this definition, note that you form products along diagonal elements as shown by the arrows.

$$\begin{bmatrix} a & b \\ c & d \end{bmatrix}$$

$$-bc \qquad +ad$$

You attach a plus sign if you descend from left to right and a minus sign if you descend from right to left. Then you form the sum $ad + (-bc) = ad - bc$.

EXAMPLE 1   **Calculating the Determinant of a 2 × 2 Matrix**

Evaluate each determinant.

a. $\begin{vmatrix} 3 & -4 \\ 1 & 5 \end{vmatrix}$ **b.** $\begin{vmatrix} -2 & 3 \\ 4 & -5 \end{vmatrix}$

SOLUTION

a. $\begin{vmatrix} 3 & -4 \\ 1 & 5 \end{vmatrix} = (3)(5) - (-4)(1) = 15 - (-4) = 15 + 4 = 19$

b. $\begin{vmatrix} -2 & 3 \\ 4 & -5 \end{vmatrix} = (-2)(-5) - (3)(4) = 10 - 12 = -2$ ■ ■ ■

Practice Problem 1   Evaluate each determinant.

a. $\begin{vmatrix} 5 & 1 \\ 3 & -7 \end{vmatrix}$ **b.** $\begin{vmatrix} 2 & -9 \\ -4 & 18 \end{vmatrix}$ ■

RECALL

A matrix of order n is an $n \times n$ square matrix.

Recall (page 585) that the matrix $A = \begin{bmatrix} a & b \\ c & d \end{bmatrix}$ is invertible if and only if $ad - bc \neq 0$. You can rephrase this property by saying that a *matrix of order 2 is invertible if and only if its determinant is nonzero*. If the matrix A is invertible, then its inverse can be expressed in terms of its determinant as follows:

$$A^{-1} = \frac{1}{ad - bc} \begin{bmatrix} d & -b \\ -c & a \end{bmatrix} = \frac{1}{\det(A)} \begin{bmatrix} d & -b \\ -c & a \end{bmatrix}$$

◆ **WARNING**

Note carefully the notations used for matrices and determinants. Brackets, [ ], denote a matrix, while vertical bars, | |, denote the determinant of a matrix. Remember that *the determinant of a matrix is a number.*

---

2 Find minors and cofactors.

Minors and Cofactors

To define the determinants of square matrices of order 3 or higher, we need some additional terminology.

STUDY TIP

Signs of cofactors

$$\begin{bmatrix} + & - & + & - & \cdots \\ - & + & - & + & \cdots \\ + & - & + & - & \cdots \\ - & + & - & + & \cdots \\ \vdots & \vdots & \vdots & \vdots & \end{bmatrix}$$

Note how the positions where $i + j$ is even alternate with those where $i + j$ is odd.

MINORS AND COFACTORS IN AN $n \times n$ MATRIX

Let A be an $n \times n$ square matrix. The **minor** M_{ij} of the element a_{ij} is the determinant of the $(n - 1) \times (n - 1)$ matrix obtained by deleting the ith row and the jth column of A. The **cofactor** A_{ij} of the entry a_{ij} is given by

$$A_{ij} = (-1)^{i+j} M_{ij}.$$

When $i + j$ is an even integer, $(-1)^{i+j} = 1$. When $i + j$ is an odd integer, $(-1)^{i+j} = -1$. Therefore,

$$A_{ij} = \begin{cases} M_{ij} & \text{if } i + j \text{ is an even integer} \\ -M_{ij} & \text{if } i + j \text{ is an odd integer} \end{cases}.$$

Notice that to compute a minor in an $n \times n$ matrix, you must know how to compute the determinant of an $(n - 1) \times (n - 1)$ matrix. At this point in the text, because you have learned the definition of the determinant of a 2×2 matrix, you can find the minors and cofactors of entries in a 3×3 matrix.

EXAMPLE 2 **Finding Minors and Cofactors**

For the matrix $A = \begin{bmatrix} 2 & 3 & 4 \\ 5 & -3 & -6 \\ 0 & 1 & 7 \end{bmatrix}$, find

a. the minors M_{11}, M_{23}, and M_{32}.

b. the cofactors A_{11}, A_{23}, and A_{32}.

SOLUTION

a. **(i)** To find M_{11}, delete the first row and first column of the matrix A.

$$\begin{bmatrix} 2 & 3 & 4 \\ 5 & -3 & -6 \\ 0 & 1 & 7 \end{bmatrix}$$

Now find the determinant of the resulting matrix.

$$M_{11} = \begin{vmatrix} -3 & -6 \\ 1 & 7 \end{vmatrix} = (-3)(7) - (-6)(1) = -21 + 6 = -15$$

(ii) To find the minor M_{23}, delete the second row and third column of the matrix A.

$$\begin{bmatrix} 2 & 3 & 4 \\ 5 & -3 & -6 \\ 0 & 1 & 7 \end{bmatrix}$$

Now find the determinant of the resulting matrix.

$$M_{23} = \begin{vmatrix} 2 & 3 \\ 0 & 1 \end{vmatrix} = (2)(1) - (3)(0) = 2$$

(iii) To find M_{32}, delete the third row and second column of A.

$$\begin{bmatrix} 2 & 3 & 4 \\ 5 & -3 & -6 \\ 0 & 1 & 7 \end{bmatrix}$$

So $M_{32} = \begin{vmatrix} 2 & 4 \\ 5 & -6 \end{vmatrix} = (2)(-6) - (4)(5) = -12 - 20 = -32.$

b. To find the cofactors, use the formula $A_{ij} = (-1)^{i+j} M_{ij}$.

$A_{11} = (-1)^{1+1} M_{11} = (1)(-15) = -15$ $(-1)^{1+1} = (-1)^2 = 1; M_{11} = -15$

$A_{23} = (-1)^{2+3} M_{23} = (-1)(2) = -2$ $(-1)^{2+3} = (-1)^5 = -1; M_{23} = 2$

$A_{32} = (-1)^{3+2} M_{32} = (-1)(-32) = 32$ $(-1)^{3+2} = (-1)^5 = -1; M_{32} = -32$

■ ■ ■

Practice Problem 2 Repeat Example 2 for the matrix $A = \begin{bmatrix} 3 & -1 & 2 \\ 4 & 5 & 6 \\ 7 & 1 & 2 \end{bmatrix}$.

3 Evaluate the determinant of an $n \times n$ matrix.

Determinant of an $n \times n$ Matrix

There are several ways of defining the determinant of a square matrix. Here we give an *inductive definition*. This means that the definition of the determinant of a matrix of order n uses the determinants of the matrices of order $(n - 1)$.

---

n BY n DETERMINANT

Let A be a square matrix of order $n \geq 3$. The **determinant** of A is the sum of the entries in any row of A (or column of A), each multiplied by their respective cofactors.

Applying the preceding definition to find the determinant of a matrix is called **expanding by cofactors**. Expanding by the cofactors of the ith row gives

$$\det(A) = a_{i1}A_{i1} + a_{i2}A_{i2} + \cdots + a_{in}A_{in}.$$

Similarly, expanding by the cofactors of the jth column yields

$$\det(A) = a_{1j}A_{1j} + a_{2j}A_{2j} + \cdots + a_{nj}A_{nj}.$$

We usually say "expanding by the ith row (column)" instead of "expanding by the cofactors of the ith row (column)."

TECHNOLOGY CONNECTION

Use the Det option on a graphing calculator to find the determinant of a matrix.

EXAMPLE 3 Evaluating a 3 by 3 Determinant

Find the determinant of $A = \begin{bmatrix} 2 & 3 & 4 \\ 1 & -2 & 2 \\ 3 & 4 & -1 \end{bmatrix}$.

SOLUTION

Expanding by the first row, we have $|A| = a_{11}A_{11} + a_{12}A_{12} + a_{13}A_{13}$, where

$$A_{11} = (-1)^{1+1}M_{11} = (1)\begin{vmatrix} -2 & 2 \\ 4 & -1 \end{vmatrix}$$

$$A_{12} = (-1)^{1+2}M_{12} = (-1)\begin{vmatrix} 1 & 2 \\ 3 & -1 \end{vmatrix}$$

$$A_{13} = (-1)^{1+3}M_{13} = (1)\begin{vmatrix} 1 & -2 \\ 3 & 4 \end{vmatrix}$$

Thus,

$$\begin{vmatrix} 2 & 3 & 4 \\ 1 & -2 & 2 \\ 3 & 4 & -1 \end{vmatrix} = 2(1)\begin{vmatrix} -2 & 2 \\ 4 & -1 \end{vmatrix} + 3(-1)\begin{vmatrix} 1 & 2 \\ 3 & -1 \end{vmatrix} + 4(1)\begin{vmatrix} 1 & -2 \\ 3 & 4 \end{vmatrix}$$

$$= 2(2 - 8) - 3(-1 - 6) + 4(4 + 6) \qquad \text{Evaluate determinants of order 2.}$$

$$= -12 + 21 + 40 = 49 \qquad \text{Simplify.} \qquad \blacksquare\blacksquare\blacksquare$$

Practice Problem 3   Find the determinant of $A = \begin{bmatrix} 2 & -3 & 7 \\ -2 & -1 & 9 \\ 0 & 2 & -9 \end{bmatrix}$. ■

The definition says that we can expand the determinant by any row or column. If we expand by the third column in A in Example 3, we have

$$\begin{vmatrix} 2 & 3 & 4 \\ 1 & -2 & 2 \\ 3 & 4 & -1 \end{vmatrix} = 4(1)\begin{vmatrix} 1 & -2 \\ 3 & 4 \end{vmatrix} + 2(-1)\begin{vmatrix} 2 & 3 \\ 3 & 4 \end{vmatrix} + (-1)(1)\begin{vmatrix} 2 & 3 \\ 1 & -2 \end{vmatrix}$$

$$= 4(4 + 6) - 2(8 - 9) - 1(-4 - 3)$$

$$= 40 + 2 + 7 = 49$$

4 Apply Cramer's Rule

Cramer's Rule

You have already learned several methods for solving a system of linear equations. Another method, called **Cramer's Rule**, uses determinants. To understand the derivation of the rule, consider a simple system of two linear equations in two variables.

$$\begin{cases} a_1x + b_1y = c_1 & (1) \\ a_2x + b_2y = c_2 & (2) \end{cases}$$

We can solve this system by using the elimination method. Let's first eliminate y and solve for x.

$$\begin{array}{ll} a_1b_2x + b_1b_2y = c_1b_2 & \text{Multiply equation (1) by } b_2. \\ \underline{-a_2b_1x - b_1b_2y = -c_2b_1} & \text{Multiply equation (2) by } -b_1. \\ (a_1b_2 - a_2b_1)x = c_1b_2 - c_2b_1 & \text{Add to eliminate } y. \\ x = \dfrac{c_1b_2 - c_2b_1}{a_1b_2 - a_2b_1} & \begin{array}{l}\text{Divide both sides by} \\ a_1b_2 - a_2b_1 \text{ if } a_1b_2 - a_2b_1 \neq 0\end{array} \end{array}$$

Similarly, by eliminating x and solving for y, we have

$$y = \frac{a_1c_2 - a_2c_1}{a_1b_2 - a_2b_1} \qquad \text{If } a_1b_2 - a_2b_1 \neq 0$$

The solution of the system of equations (1) and (2) is therefore given by

$$x = \frac{c_1b_2 - c_2b_1}{a_1b_2 - a_2b_1} \quad y = \frac{a_1c_2 - a_2c_1}{a_1b_2 - a_2b_1} \quad \text{provided that } a_1b_2 - a_2b_1 \neq 0.$$

You can write the solution using determinants

$$x = \frac{\begin{vmatrix} c_1 & b_1 \\ c_2 & b_2 \end{vmatrix}}{\begin{vmatrix} a_1 & b_1 \\ a_2 & b_2 \end{vmatrix}} \qquad y = \frac{\begin{vmatrix} a_1 & c_1 \\ a_2 & c_2 \end{vmatrix}}{\begin{vmatrix} a_1 & b_1 \\ a_2 & b_2 \end{vmatrix}} \qquad \text{provided that } D = \begin{vmatrix} a_1 & b_1 \\ a_2 & b_2 \end{vmatrix} \neq 0.$$

Note that the denominator of both the x- and y-values in the solution is the determinant D of the coefficient matrix of the linear system.

Suppose we write $D_x = \begin{vmatrix} c_1 & b_1 \\ c_2 & b_2 \end{vmatrix}$ and $D_y = \begin{vmatrix} a_1 & c_1 \\ a_2 & c_2 \end{vmatrix}$, which are obtained by replacing the column containing the coefficients of x in D and the column containing the coefficients of y in D with the column of constants, respectively. Then the solution just found can be written as

$$x = \frac{D_x}{D} \qquad y = \frac{D_y}{D} \quad \text{provided that } D \neq 0.$$

This representation of the solution is Cramer's Rule for solving a system of linear equations.

CRAMER'S RULE FOR SOLVING TWO EQUATIONS IN TWO VARIABLES

The system

$$\begin{cases} a_1x + b_1y = c_1 \\ a_2x + b_2y = c_2 \end{cases}$$

of two equations in two variables has a unique solution (x, y) given by

$$x = \frac{D_x}{D} \quad \text{and} \quad y = \frac{D_y}{D}$$

provided that $D \neq 0$, where

$$D = \begin{vmatrix} a_1 & b_1 \\ a_2 & b_2 \end{vmatrix}, \quad D_x = \begin{vmatrix} c_1 & b_1 \\ c_2 & b_2 \end{vmatrix}, \quad \text{and} \quad D_y = \begin{vmatrix} a_1 & c_1 \\ a_2 & c_2 \end{vmatrix}.$$

EXAMPLE 4   **Using Cramer's Rule**

Use Cramer's Rule to solve the following system: $\begin{cases} 5x + 4y = 1 \\ 2x + 3y = 6 \end{cases}$

SOLUTION

Step 1   Form the determinant D of the coefficient matrix.

$$D = \begin{vmatrix} 5 & 4 \\ 2 & 3 \end{vmatrix} = 15 - 8 = 7$$

Because $D = 7 \neq 0$, the system has a unique solution.

Step 2   Replace the column of coefficients of x in D by the constant terms to get

$$D_x = \begin{vmatrix} 1 & 4 \\ 6 & 3 \end{vmatrix} = 3 - 24 = -21.$$

Step 3   Replace the column of coefficients of y in D by the constant terms to get

$$D_y = \begin{vmatrix} 5 & 1 \\ 2 & 6 \end{vmatrix} = 30 - 2 = 28.$$

Step 4   By Cramer's Rule,

$$x = \frac{D_x}{D} = \frac{-21}{7} = -3 \quad \text{and} \quad y = \frac{D_y}{D} = \frac{28}{7} = 4.$$

The solution is $x = -3$ and $y = 4$, or $\{(-3, 4)\}$.

Check: You should check the solution in the original system.   ■ ■ ■

Practice Problem 4   Use Cramer's Rule to solve the system:

$$\begin{cases} 2x + 3y = 7 \\ 5x + 9y = 4 \end{cases}$$

■

Cramer's Rule, as used in Example 4, can be generalized to a system of equations with more than two variables. We state the rule for three equations in three variables.

CRAMER'S RULE FOR SOLVING THREE EQUATIONS IN THREE VARIABLES

The system

$$\begin{cases} a_1x + b_1y + c_1z = k_1 \\ a_2x + b_2y + c_2z = k_2 \\ a_3x + b_3y + c_3z = k_3 \end{cases}$$

of three equations in three variables has a unique solution (x, y, z) given by

$$x = \frac{D_x}{D}, \quad y = \frac{D_y}{D}, \quad \text{and} \quad z = \frac{D_z}{D}$$

provided that $D \neq 0$, where

$$D = \begin{vmatrix} a_1 & b_1 & c_1 \\ a_2 & b_2 & c_2 \\ a_3 & b_3 & c_3 \end{vmatrix}, \quad D_x = \begin{vmatrix} k_1 & b_1 & c_1 \\ k_2 & b_2 & c_2 \\ k_3 & b_3 & c_3 \end{vmatrix}, \quad D_y = \begin{vmatrix} a_1 & k_1 & c_1 \\ a_2 & k_2 & c_2 \\ a_3 & k_3 & c_3 \end{vmatrix}, \quad \text{and} \quad D_z = \begin{vmatrix} a_1 & b_1 & k_1 \\ a_2 & b_2 & k_2 \\ a_3 & b_3 & k_3 \end{vmatrix}.$$

STUDY TIP

When you apply Cramer's Rule, make sure that all linear equations are written in the form

$$ax + by + cz = k.$$

EXAMPLE 5 **Using Cramer's Rule**

Use Cramer's Rule to solve the system of equations.

$$\begin{cases} 7x + y + z - 1 = 0 \\ 5x + 3z - 2y = 4 \\ 4x - z + 3y = 0 \end{cases}$$

SOLUTION

First, rewrite the system of equations so that the terms on the left side of the equal signs are in the proper order and only the constant terms appear on the right side.

$$\begin{cases} 7x + y + z = 1 \\ 5x - 2y + 3z = 4 \\ 4x + 3y - z = 0 \end{cases}$$

Step 1 $\quad D = \begin{vmatrix} 7 & 1 & 1 \\ 5 & -2 & 3 \\ 4 & 3 & -1 \end{vmatrix}$ D is the determinant of the coefficient matrix.

$$= 7\begin{vmatrix} -2 & 3 \\ 3 & -1 \end{vmatrix} - 5\begin{vmatrix} 1 & 1 \\ 3 & -1 \end{vmatrix} + 4\begin{vmatrix} 1 & 1 \\ -2 & 3 \end{vmatrix}$$ Expand by column 1.

$$= 7(2 - 9) - 5(-1 - 3) + 4(3 + 2)$$ Evaluate determinants of order 2.

$$= 7(-7) - 5(-4) + 4(5) = -9$$ Simplify.

Because $D \neq 0$, the system has a unique solution.

Step 2 $D_x = \begin{vmatrix} 1 & 1 & 1 \\ 4 & -2 & 3 \\ 0 & 3 & -1 \end{vmatrix}$ Replace the x-coefficients in the first column in D with the constants.

$= 1\begin{vmatrix} -2 & 3 \\ 3 & -1 \end{vmatrix} - 4\begin{vmatrix} 1 & 1 \\ 3 & -1 \end{vmatrix} + 0\begin{vmatrix} 1 & 1 \\ -2 & 3 \end{vmatrix}$ Expand by Column 1.

$= 1(2 - 9) - 4(-1 - 3)$ Evaluate determinants of order 2.

$= 1(-7) - 4(-4) = 9$ Simplify.

Step 3 $D_y = \begin{vmatrix} 7 & 1 & 1 \\ 5 & 4 & 3 \\ 4 & 0 & -1 \end{vmatrix}$ Replace column 2 in D with the constants.

$= -1\begin{vmatrix} 5 & 3 \\ 4 & -1 \end{vmatrix} + 4\begin{vmatrix} 7 & 1 \\ 4 & -1 \end{vmatrix} - 0\begin{vmatrix} 7 & 1 \\ 5 & 3 \end{vmatrix}$ Expand by column 2 because it contains a zero.

$= -1(-5 - 12) + 4(-7 - 4)$ Evaluate the determinants.

$= 17 - 44 = -27$ Simplify.

Step 4 $D_z = \begin{vmatrix} 7 & 1 & 1 \\ 5 & -2 & 4 \\ 4 & 3 & 0 \end{vmatrix}$ Replace column 3 in D with the constants.

$= 1\begin{vmatrix} 5 & -2 \\ 4 & 3 \end{vmatrix} - 4\begin{vmatrix} 7 & 1 \\ 4 & 3 \end{vmatrix} + 0\begin{vmatrix} 7 & 1 \\ 5 & -2 \end{vmatrix}$ Expand by column 3.

$= (15 + 8) - 4(21 - 4)$ Evaluate the determinants.

$= 23 - 4(17) = -45$ Simplify.

Step 5   Cramer's Rule gives the following values:

$$x = \frac{D_x}{D} = \frac{9}{-9} = -1 \qquad \text{From Steps 1 and 2}$$

$$y = \frac{D_y}{D} = \frac{-27}{-9} = 3 \qquad \text{From Steps 1 and 3}$$

$$z = \frac{D_z}{D} = \frac{-45}{-9} = 5 \qquad \text{From Steps 1 and 4}$$

Therefore, the solution set is $\{(-1, 3, 5)\}$.

Check: You should check the solution.   ■ ■ ■

Practice Problem 5   Use Cramer's Rule to solve the system.

$$\begin{cases} 3x + 2y + z = 4 \\ 4x + 3y + z = 5 \\ 5x + y + z = 9 \end{cases}$$ ■

◆ **WARNING**   It is important to note that Cramer's Rule does not apply if $D = 0$. In that case, the system is either consistent and has dependent equations (has infinitely many solutions) or inconsistent (has no solution). When $D = 0$, use a different method for solving systems of equations to find the solution set.

SECTION 6.4 ▪ Exercises

A EXERCISES Basic Skills and Concepts

1. The determinant of $\begin{bmatrix} a & b \\ c & d \end{bmatrix}$ is _____.

2. The minor of an element is the determinant you get by deleting the _____ and the _____ containing that element.

3. To expand an $n \times n$ determinant, you multiply each element of some row (or column) by its _____ and add the result.

4. A system of n linear equations in n variables has a unique solution provided the determinant of the _____ is not zero.

5. *True or False* A 2 by 2 determinant is always positive.

6. *True or False* To expand an $n \times n$ determinant, you multiply each element of some row by its minor and add the result.

In Exercises 7–16, evaluate each determinant.

7. $\begin{vmatrix} 2 & 3 \\ 4 & 5 \end{vmatrix}$

8. $\begin{vmatrix} 3 & -5 \\ 1 & 4 \end{vmatrix}$

9. $\begin{vmatrix} 4 & -2 \\ 3 & -3 \end{vmatrix}$

10. $\begin{vmatrix} -2 & \frac{1}{2} \\ 3 & 1 \end{vmatrix}$

11. $\begin{vmatrix} -1 & -3 \\ -4 & -5 \end{vmatrix}$

12. $\begin{vmatrix} \frac{1}{2} & \frac{1}{3} \\ \frac{1}{4} & \frac{1}{6} \end{vmatrix}$

13. $\begin{vmatrix} \frac{3}{8} & \frac{1}{2} \\ -\frac{1}{9} & 5 \end{vmatrix}$

14. $\begin{vmatrix} -\frac{1}{2} & \frac{1}{3} \\ \frac{1}{4} & -\frac{1}{3} \end{vmatrix}$

15. $\begin{vmatrix} \sqrt{a} & \sqrt{b} \\ \sqrt{b} & \sqrt{a} \end{vmatrix}$

16. $\begin{vmatrix} \sqrt{3} & 1 \\ -1 & \sqrt{3} \end{vmatrix}$

For Exercises 17–20, consider the following matrix:

$$A = \begin{bmatrix} 2 & -3 & 4 \\ 1 & -1 & 2 \\ 0 & 1 & 2 \end{bmatrix}$$

17. Find the minors.
   a. M_{21} **b.** M_{23} **c.** M_{32}

18. Find the cofactors.
   a. A_{21} **b.** A_{23} **c.** A_{32}

19. Find the minors.
   a. M_{11} **b.** M_{22} **c.** M_{31}

20. Find the cofactors.
   a. A_{11} **b.** A_{22} **c.** A_{31}

In Exercises 21–36, evaluate each determinant.

21. $\begin{vmatrix} 1 & 0 & -1 \\ 0 & 2 & 2 \\ -1 & 0 & 0 \end{vmatrix}$

22. $\begin{vmatrix} 2 & 0 & \frac{1}{2} \\ 1 & 0 & 2 \\ 4 & 0 & -5 \end{vmatrix}$

23. $\begin{vmatrix} 1 & 2 & 3 \\ 0 & 3 & 4 \\ 0 & 0 & 4 \end{vmatrix}$

24. $\begin{vmatrix} 2 & 3 & 4 \\ 0 & -4 & 6 \\ 0 & 0 & -5 \end{vmatrix}$

25. $\begin{vmatrix} 1 & 0 & 0 \\ 2 & 0 & 0 \\ 3 & 4 & 5 \end{vmatrix}$

26. $\begin{vmatrix} -1 & 0 & 0 \\ 3 & 2 & 0 \\ 4 & 5 & -3 \end{vmatrix}$

27. $\begin{vmatrix} 1 & 6 & 0 \\ 2 & 5 & 3 \\ 3 & 4 & 0 \end{vmatrix}$

28. $\begin{vmatrix} 1 & 0 & 2 \\ 0 & 5 & 7 \\ 3 & 1 & 0 \end{vmatrix}$

29. $\begin{vmatrix} 3 & 4 & 1 \\ 1 & 4 & 3 \\ 4 & 3 & 1 \end{vmatrix}$

30. $\begin{vmatrix} 3 & 1 & -2 \\ 4 & 0 & -4 \\ 2 & -1 & -3 \end{vmatrix}$

31. $\begin{vmatrix} 0 & 1 & 6 \\ 1 & 0 & 4 \\ 8 & 3 & 1 \end{vmatrix}$

32. $\begin{vmatrix} 0 & 2 & 3 \\ 1 & 0 & 1 \\ 3 & 2 & 0 \end{vmatrix}$

33. $\begin{vmatrix} a & b & 0 \\ 0 & a & b \\ b & 0 & a \end{vmatrix}$

34. $\begin{vmatrix} 0 & z & y \\ z & 0 & x \\ y & x & 0 \end{vmatrix}$

35. $\begin{vmatrix} a & b & c \\ c & a & b \\ b & c & a \end{vmatrix}$

36. $\begin{vmatrix} u & v & w \\ w & v & u \\ u & v & w \end{vmatrix}$

In Exercises 37–46, use Cramer's Rule (if applicable) to solve each system of equations.

37. $\begin{cases} x + y = 8 \\ x - y = -2 \end{cases}$

38. $\begin{cases} 4x + 3y = -1 \\ 2x - 5y = 9 \end{cases}$

39. $\begin{cases} 5x + 3y = 11 \\ 2x + y = 4 \end{cases}$

40. $\begin{cases} 2x - 7y = 13 \\ 5x + 6y = 9 \end{cases}$

41. $\begin{cases} 2x + 9y = 4 \\ 3x - 2y = 6 \end{cases}$

42. $\begin{cases} 5x + 3y = 1 \\ 2x - 5y = -12 \end{cases}$

43. $\begin{cases} 2x - 3y = 4 \\ 4x - 6y = 8 \end{cases}$

44. $\begin{cases} 3x + y = 2 \\ 6x + 2y = 4 \end{cases}$

45. $\begin{cases} \dfrac{2}{x} + \dfrac{3}{y} = 2 \\ \dfrac{5}{x} + \dfrac{8}{y} = \dfrac{31}{6} \end{cases}$

46. $\begin{cases} \dfrac{3}{x} - \dfrac{6}{y} = 2 \\ \dfrac{4}{x} + \dfrac{7}{y} = -3 \end{cases}$

$$\left[\text{\textit{Hint:} For Exercises 45 and 46, let } u = \frac{1}{x} \text{ and } v = \frac{1}{y}. \right]$$

In Exercises 47–56, solve each system of equations by means of Cramer's Rule. Use technology to compute the determinants involved.

47. $\begin{cases} x - 2y + z = -1 \\ 3x + y - z = 4 \\ y + z = 1 \end{cases}$

48. $\begin{cases} x + 3y = 4 \\ y + 3z = 7 \\ z + 4x = 6 \end{cases}$

49. $\begin{cases} x + y - z = -3 \\ 2x + 3y + z = 2 \\ 2y + z = 1 \end{cases}$ **50.** $\begin{cases} 3x + y + z = 10 \\ x + y - z = 0 \\ 5x - 9y = 1 \end{cases}$

51. $\begin{cases} x + y + z = 3 \\ 2x - 3y + 5z = 4 \\ x + 2y - 4z = -1 \end{cases}$ **52.** $\begin{cases} x + 2y - z = 3 \\ 3x - y + z = 8 \\ x + y + z = 0 \end{cases}$

53. $\begin{cases} 2x - 3y + 5z = 11 \\ 3x + 5y - 2z = 7 \\ x + 2y - 3z = -4 \end{cases}$ **54.** $\begin{cases} x - 3y - 1 = 0 \\ 2x - y - 4z = 2 \\ y + 2z - 4 = 0 \end{cases}$

55. $\begin{cases} 5x + 2y + z = 12 \\ 2x + y + 3z = 13 \\ 3x + 2y + 4z = 19 \end{cases}$ **56.** $\begin{cases} 2x + y + z = 7 \\ 3x - y - z = -2 \\ x + 2y - 3z = -4 \end{cases}$

B EXERCISES Applying the Concepts

Assume that the *area of a triangle* with vertices (x_1, y_1), (x_2, y_2), and (x_3, y_3) is equal to the absolute value of D, where

$$D = \frac{1}{2} \begin{vmatrix} x_1 & y_1 & 1 \\ x_2 & y_2 & 1 \\ x_3 & y_3 & 1 \end{vmatrix}.$$

In Exercises 57–60, use determinants to find the area of each triangle.

57. **Area of a triangle.** The triangle with vertices $(1, 2), (-3, 4)$, and $(4, 6)$.

58. **Area of a triangle.** The triangle with vertices $(3, 1), (4, 2)$, and $(5, 4)$.

59. **Area of a triangle.** The triangle with vertices $(-2, 1), (-3, -5)$, and $(2, 4)$.

60. **Area of a triangle.** The triangle with vertices $(-1, 2)$ and $(2, 4)$ and the origin.

Collinearity. Three points $A(x_1, y_1)$, $B(x_1, y_2)$, and $C(x_3, y_3)$ are *collinear* (lie on the same line) if and only if the area of the triangle ABC is 0. In terms of determinants, three points (x_1, y_1), (x_2, y_2), and (x_3, y_3) are collinear if and only if

$$\begin{vmatrix} x_1 & y_1 & 1 \\ x_2 & y_2 & 1 \\ x_3 & y_3 & 1 \end{vmatrix} = 0.$$

In Exercises 61–64, use determinants to find whether the three points are collinear.

61. $(0, 3), (-1, 1)$, and $(2, 7)$

62. $(2, 0), \left(1, \dfrac{1}{2}\right)$, and $(4, -1)$

63. $(0, -4), (3, -2)$, and $(1, -4)$

64. $\left(0, -\dfrac{1}{4}\right), (1, -1)$, and $(2, -2)$

Equation of a line. As explained in the discussion of collinearity, three points (x, y), (x_1, y_1), and (x_2, y_2) all lie on the same line if and only if

$$D = \begin{vmatrix} x & y & 1 \\ x_1 & y_1 & 1 \\ x_2 & y_2 & 1 \end{vmatrix} = 0.$$

In other words, a point $P(x, y)$ lies on the line passing through the points (x_1, y_1) and (x_2, y_2) if and only if the determinant $D = 0$. Therefore, the equation

$$\begin{vmatrix} x & y & 1 \\ x_1 & y_1 & 1 \\ x_2 & y_2 & 1 \end{vmatrix} = 0$$

is an equation of the line passing through the points (x_1, y_1) and (x_2, y_2).

In Exercises 65–68, use determinants to write an equation of the line passing through the given points. Then evaluate the determinant and write the equation of the line in slope–intercept form.

65. $(-1, -1), (1, 3)$ **66.** $(0, 4), (1, 1)$

67. $\left(0, \dfrac{1}{3}\right), (1, 1)$ **68.** $\left(1, -\dfrac{1}{3}\right), (2, 0)$

C EXERCISES Beyond the Basics

In Exercises 69–73, we list some properties of determinants. These properties are true for $n \times n$ matrices. Verify the properties for $n = 2$ or $n = 3$.

69. If each entry in any row or each entry in any column of a matrix A is zero, then $|A| = 0$. Verify each of the following.

   a. $\begin{vmatrix} 0 & 0 \\ 2 & 5 \end{vmatrix} = 0$ **b.** $\begin{vmatrix} 1 & 0 \\ 3 & 0 \end{vmatrix} = 0$

   c. $\begin{vmatrix} 1 & -2 & 3 \\ 0 & 0 & 0 \\ 4 & 5 & -7 \end{vmatrix} = 0$ **d.** $\begin{vmatrix} 4 & 5 & 0 \\ 6 & -7 & 0 \\ 8 & 15 & 0 \end{vmatrix} = 0$

70. If a matrix B is obtained from matrix A by interchanging two rows (or columns), then $|B| = -|A|$. Verify each of the following.

   a. $\begin{vmatrix} 2 & 3 \\ 4 & 5 \end{vmatrix} = -\begin{vmatrix} 4 & 5 \\ 2 & 3 \end{vmatrix}$

   b. $\begin{vmatrix} -5 & 3 \\ 2 & -4 \end{vmatrix} = -\begin{vmatrix} 3 & -5 \\ -4 & 2 \end{vmatrix}$

   c. $\begin{vmatrix} 1 & 3 & 5 \\ 0 & 1 & 2 \\ 3 & -1 & 4 \end{vmatrix} = -\begin{vmatrix} 1 & 3 & 5 \\ 3 & -1 & 4 \\ 0 & 1 & 2 \end{vmatrix}$

71. If two rows (or columns) of a matrix A have corresponding entries that are equal, then $|A| = 0$. Verify the following.

   a. $\begin{vmatrix} 2 & 3 & 5 \\ -1 & 4 & 8 \\ 2 & 3 & 5 \end{vmatrix} = 0$ **b.** $\begin{vmatrix} 3 & 4 & 3 \\ 1 & 2 & 1 \\ -1 & 6 & -1 \end{vmatrix} = 0$

72. If a matrix B is obtained from a matrix A by multiplying every entry of one row (or one column) by a real number c, then $|B| = c|A|$. Verify each of the following.

a. $\begin{vmatrix} -1 & 2 & 3 \\ 1 \cdot 5 & 4 \cdot 5 & 8 \cdot 5 \\ 2 & 3 & 6 \end{vmatrix} = 5 \begin{vmatrix} -1 & 2 & 3 \\ 1 & 4 & 8 \\ 2 & 3 & 6 \end{vmatrix}$

b. $\begin{vmatrix} 1 & 2 & 3 \\ -1 & 5 & 6 \\ 0 & 2 & 9 \end{vmatrix} = 3 \begin{vmatrix} 1 & 2 & 1 \\ -1 & 5 & 2 \\ 0 & 2 & 3 \end{vmatrix}$

73. If a matrix B is obtained from a matrix A by replacing any row (or column) of A by the sum of that row (or column) and c times another row (or column), then $|B| = |A|$.

Verify that the following two determinants are equal.

$$\begin{vmatrix} 2 & -3 & 4 \\ -4 & 7 & -8 \\ 5 & -1 & 3 \end{vmatrix} \xrightarrow{2R_1 + R_2 \to R_2} \begin{vmatrix} 2 & -3 & 4 \\ 0 & 1 & 0 \\ 5 & -1 & 3 \end{vmatrix}$$

74. Evaluate the determinant

$$D = \begin{vmatrix} 1 & 2 & 3 & 3 \\ 4 & 5 & 2 & 1 \\ 1 & 2 & 5 & 3 \\ 2 & 4 & 3 & 7 \end{vmatrix}$$

by following these steps:
(i) $(-1)R_1 + R_3 \to R_3$
(ii) Expand by the cofactors of row 3.
(iii) In **(ii)**, you will have a determinant of order 3. Evaluate it to obtain the value of D.

In Exercises 75–78, use the properties of determinants in Exercises 69–73 to obtain three zeros in one row (or column). Then evaluate the given determinant by evaluating the resulting determinant of order 3.

75. $\begin{vmatrix} 2 & 0 & -3 & 4 \\ 0 & 1 & 0 & 5 \\ 5 & 0 & -9 & 8 \\ 1 & 2 & 0 & 7 \end{vmatrix}$

76. $\begin{vmatrix} -3 & 1 & 2 & 3 \\ 0 & -2 & 4 & 7 \\ -9 & 3 & 5 & 2 \\ 2 & -4 & 3 & -12 \end{vmatrix}$

77. $\begin{vmatrix} 5 & 7 & 1 & 2 \\ 6 & 8 & 9 & 3 \\ 24 & 22 & 6 & 10 \\ 21 & 17 & 7 & 10 \end{vmatrix}$

78. $\begin{vmatrix} 1 & 2 & 3 & 0 \\ 2 & -3 & 1 & 0 \\ -1 & -3 & 4 & 5 \\ 1 & 3 & 7 & 4 \end{vmatrix}$

In Exercises 79–86, solve each equation for x.

79. $\begin{vmatrix} 3 & 2 \\ 6 & x \end{vmatrix} = 0$

80. $\begin{vmatrix} 3 & 2 \\ x & 12 \end{vmatrix} = 0$

81. $\begin{vmatrix} x & -2 \\ 1 & x-1 \end{vmatrix} = 0$

82. $\begin{vmatrix} 2x+7 & 4 \\ 1 & x \end{vmatrix} = 0$

83. $\begin{vmatrix} 1 & -3 & 1 \\ 4 & 7 & x \\ 0 & 2 & 2 \end{vmatrix} = 0$

84. $\begin{vmatrix} x & x+1 & x+2 \\ 2 & 3 & -1 \\ 3 & -2 & 4 \end{vmatrix} = 0$

85. $\begin{vmatrix} x & 0 & 1 \\ 0 & x & 0 \\ 1 & 0 & x \end{vmatrix} = 0$

86. $\begin{vmatrix} x & 2 & 3 \\ x & x & 1 \\ 2 & 0 & 1 \end{vmatrix} = -8$

In Exercises 87 and 88, use the formula on page 603 for the area of a triangle.

87. Assuming that the area of the triangle formed by the points $(-4, 2), (0, k)$, and $(-2, k)$ is 28 square units, find k.

88. Assuming that the area of the triangle formed by the points $(2, 3), (1, 2)$, and $(-2, k)$ is 3 square units, find k.

89. Solve the system by using Cramer's Rule.

$$\begin{cases} \dfrac{1}{x-2} + \dfrac{3}{y+1} = 13 \\ \dfrac{4}{x-2} - \dfrac{5}{y+1} = 1 \end{cases}$$

[*Hint:* Do not try to simplify the equations by clearing the denominators because the resulting equations will not be linear. Instead, let $u = \dfrac{1}{x-2}$ and $v = \dfrac{1}{y+1}$. Solve the system for u and v and then use this solution to solve for x and y.]

In Exercises 90 and 91, use Cramer's Rule to solve each system of equations.

90. $\begin{cases} \dfrac{6}{x+1} + \dfrac{4}{y-1} = 7 \\ \dfrac{8}{x+1} + \dfrac{5}{y-1} = 9 \end{cases}$

91. $\begin{cases} \dfrac{1}{3^x} + \dfrac{4}{4^y} = 25 \\ \dfrac{2}{3^x} - \dfrac{1}{4^y} = 14 \end{cases}$

Critical Thinking

92. Consider the system of equations.

$$\begin{cases} 2x + 3y - z = 5 \\ 3x - 2y + z = 2 \\ 7x + 4y - z = 12 \end{cases}$$

Verify that $(1, 2, 3)$ is a solution of the system. Explain whether Cramer's Rule can be used to solve the system.

6.1 Matrices and Systems of Equations

i. **Matrix.** An $m \times n$ matrix is a rectangular array of numbers with m rows and n columns.

ii. **Augmented matrix.** The augmented matrix of a system of linear equations is a matrix whose entries are the coefficients of the variables in the system together with the constants.

iii. **Elementary row operations.** For an augmented matrix of a system of linear equations, the following row operations will result in the matrix of an equivalent system:
1. Interchange any two rows.
2. Multiply any row by a nonzero constant.
3. Add a multiple of one row to another row.

iv. **Row-echelon form and reduced row-echelon form.** The elementary row operations are used to transform an augmented matrix to row-echelon form or reduced row-echelon form. (See page 556)

v. **Gaussian elimination** is the method of transforming the augmented matrix to row-echelon form and then using back-substitution to find the solution set of a system of linear equations.

vi. **Gauss–Jordan elimination.** If you continue the Gaussian elimination procedure until a reduced row-echelon form is obtained, the procedure is called Gauss–Jordan elimination.

6.2 Matrix Algebra

i. **Equality of matrices.** Two matrices $A = [a_{ij}]$ and $B = [b_{ij}]$ are equal if (i) A and B have the same order and (ii) all corresponding entries are equal.

ii. **Matrix addition.** The matrices A and B can be added or subtracted if they have the same order. Their sum or difference can be obtained by adding or subtracting the corresponding entries.

iii. **Scalar multiplication.** The product of a matrix by a real number (scalar) is obtained by multiplying every entry of the matrix by the real number.

iv. **Matrix multiplication.** The product AB of two matrices is defined only if the number of columns of A is equal to the number of rows of B. If A is an $m \times p$ matrix and B is a $p \times n$ matrix, then the product AB is an $m \times n$ matrix. The (i, j)th entry of AB is the sum of the products of the corresponding entries in the ith row of A and the jth column of B.

v. **The identity matrix.** The $n \times n$ identity matrix denoted by I_n or I is a matrix that has 1s on the main diagonal and 0s elsewhere. If A is an $n \times n$ matrix, then $AI = IA = A$.

6.3 The Matrix Inverse

i. **Inverse of a matrix.** Let A be an $n \times n$ matrix and I be the $n \times n$ identity matrix. If there is an $n \times n$ matrix B such that $AB = BA = I$, then A is **invertible** and B is called the **inverse** of A. We write $B = A^{-1}$.

ii. **Finding the inverse.** A procedure for finding A^{-1} (if it exists) is given on page 583.

iii. **Inverse of a 2 × 2 matrix.** The matrix $A = \begin{bmatrix} a & b \\ c & d \end{bmatrix}$ has an inverse if and only if $ad - bc \neq 0$. Moreover, if $ad - bc \neq 0$, then

$$A^{-1} = \frac{1}{ad - bc} \begin{bmatrix} d & -b \\ -c & a \end{bmatrix}.$$

iv. If A is invertible, then the solution of the matrix equation $AX = B$ is $X = A^{-1}B$.

6.4 Determinants and Cramer's Rule

i. **Determinant.** The determinant of a square matrix A is a real number denoted by $\det(A)$ or $|A|$.

ii. The determinant of a 2 × 2 matrix $A = \begin{bmatrix} a & b \\ c & d \end{bmatrix}$ is defined by $\det(A) = \begin{vmatrix} a & b \\ c & d \end{vmatrix} = ad - bc$.

iii. The determinant of a matrix of order $n \geq 3$ can be found by expanding by cofactors. (See page 597.)

iv. **Cramer's Rule** gives formulas for solving a square system of linear equations for which the determinant of the coefficient matrix is nonzero. (See page 599 for solving a system of two equations in two variables and page 600 for solving a system of three equations in three variables.)

REVIEW EXERCISES

Basic Skills and Concepts

In Exercises 1–4, determine the order of each matrix.

1. $\begin{bmatrix} -1 & 2 & 3 & 4 \end{bmatrix}$ **2.** $\begin{bmatrix} 5 \end{bmatrix}$

3. $\begin{bmatrix} 2 & 3 \\ -1 & 2 \\ 0 & 1 \end{bmatrix}$ **4.** $\begin{bmatrix} 1 & -1 & 2 & -4 \\ 5 & 4 & 3 & 2 \end{bmatrix}$

5. Let A be the matrix of Exercise 4. Identify the entries $a_{12}, a_{14}, a_{23},$ and a_{21}.

6. Write the 2×3 matrix A with entries $a_{22} = 3, a_{21} = 4,$ $a_{12} = 2, a_{13} = 5, a_{11} = -1,$ and $a_{23} = 1$.

In Exercises 7 and 8, write the augmented matrix for each system of linear equations.

7. $\begin{cases} 2x - 3y = 7 \\ 3x + y = 6 \end{cases}$

8. $\begin{cases} x + y - z = 6 \\ 2x - 3y - 2z = 2 \\ 5x - 3y + z = 8 \end{cases}$

In Exercises 9 and 10, convert each matrix to row-echelon form.

9. $\begin{bmatrix} 0 & 1 & 2 & 1 \\ 2 & 0 & 3 & 4 \\ 1 & -2 & 1 & 7 \end{bmatrix}$ **10.** $\begin{bmatrix} 3 & -1 & 2 & 12 \\ 1 & 1 & -1 & 2 \\ 1 & 2 & 1 & 7 \end{bmatrix}$

In Exercises 11 and 12, convert each matrix to reduced row-echelon form.

11. $\begin{bmatrix} 3 & 1 & 3 & 1 \\ 2 & 1 & 1 & 1 \\ 1 & -1 & -1 & 0 \end{bmatrix}$ **12.** $\begin{bmatrix} 3 & 4 & -4 & 2 \\ 2 & 1 & -2 & 1 \\ 1 & 1 & 2 & 1 \end{bmatrix}$

In Exercises 13–16, solve each system of equations by Gaussian elimination.

13. $\begin{cases} x + y - z = 0 \\ 2x + y - 2z = 3 \\ 3x - 2y + 3z = 9 \end{cases}$

14. $\begin{cases} x - y + 2z = 1 \\ x + 3y - z = 6 \\ 2x + y - 3z = 1 \end{cases}$

15. $\begin{cases} x - 2y + 3z = -2 \\ 2x - 3y + z = 9 \\ 3x - y + 2z = 5 \end{cases}$

16. $\begin{cases} 2x + y + z = 7 \\ x + 2y + z = 3 \\ x + 2y + 2z = 6 \end{cases}$

In Exercises 17–20, solve each system of equations by Gauss–Jordan elimination.

17. $\begin{cases} x - 2y - 2z = 11 \\ 3x + 4y - z = -2 \\ 4x + 5y + 7z = 7 \end{cases}$

18. $\begin{cases} x - y - 9z = 1 \\ 3x + 2y - z = 2 \\ 4x + 3y + 3z = 0 \end{cases}$

19. $\begin{cases} 2x - y + 3z = 4 \\ x + 3y + 3z = -2 \\ 3x + 2y - 6z = 6 \end{cases}$

20. $\begin{cases} x - y - 9z = 1 \\ 3x + 2y - z = 2 \\ 4x + 3y + 3z = 0 \end{cases}$

21. Find x and y if $\begin{bmatrix} x - y & 0 \\ 1 & x + y \end{bmatrix} = \begin{bmatrix} 1 & 0 \\ 1 & 3 \end{bmatrix}$.

22. Find x and y if
$$\begin{bmatrix} 2x + 3y & -1 & -2 \\ 0 & 1 & x - y \\ 2 & 3x + 4y & 5 \end{bmatrix} = \begin{bmatrix} 4 & -1 & -2 \\ 0 & 1 & -3 \\ 2 & 5 & 5 \end{bmatrix}.$$

23. Let $A = \begin{bmatrix} 1 & 2 \\ -3 & 4 \end{bmatrix}$ and $B = \begin{bmatrix} 2 & -3 \\ -5 & 6 \end{bmatrix}$. Find each of the following.
  a. $A + B$ **b.** $A - B$ **c.** $2A$
  d. $-3B$ **e.** $2A - 3B$

24. Let $A = \begin{bmatrix} 2 & 0 & -1 \\ 1 & 2 & 2 \\ -2 & 0 & 3 \end{bmatrix}$ and $B = \begin{bmatrix} 0 & 1 & -1 \\ 1 & -1 & 0 \\ -1 & 0 & 1 \end{bmatrix}$.

Find each of the following.
  a. $A + B$ **b.** $A - B$ **c.** $2A + 3B$

25. For the matrices A and B of Exercise 23, solve the matrix equation $3A + 2B - 3X = \mathbf{0}$ for X.

26. For the matrices A and B of Exercise 24, solve the matrix equation $A - 2X + 2B = \mathbf{0}$ for X.

In Exercises 27–30, find the products, if possible.
a. AB **b.** BA

27. $A = \begin{bmatrix} 0 & 1 \\ 2 & 3 \end{bmatrix}$, $B = \begin{bmatrix} -1 & -1 \\ -3 & 4 \end{bmatrix}$

28. $A = \begin{bmatrix} 0 & 1 & 2 \\ -1 & 0 & 1 \end{bmatrix}$, $B = \begin{bmatrix} 1 & 2 \\ 3 & -1 \\ 0 & 4 \end{bmatrix}$

29. $A = \begin{bmatrix} 1 & 2 & -1 \end{bmatrix}$, $B = \begin{bmatrix} 2 \\ 3 \\ 1 \end{bmatrix}$

30. $A = \begin{bmatrix} 1 & 0 \\ 2 & -1 \\ 3 & -2 \end{bmatrix}$, $B = \begin{bmatrix} 1 & 2 & 3 & 4 \\ 2 & -1 & 2 & 3 \end{bmatrix}$

31. Find a matrix $A = \begin{bmatrix} x & y \\ z & w \end{bmatrix}$ such that
$$\begin{bmatrix} 1 & 2 \\ 3 & -1 \end{bmatrix} A = \begin{bmatrix} 5 & 6 \\ 1 & -3 \end{bmatrix}.$$

32. Find a matrix $B = \begin{bmatrix} x & y \\ z & w \end{bmatrix}$ such that
$$B\begin{bmatrix} 1 & 2 \\ 3 & -1 \end{bmatrix} = \begin{bmatrix} 5 & 6 \\ 1 & -3 \end{bmatrix}.$$

Compare your answers for Exercises 31 and 32.

33. Find a matrix $A = \begin{bmatrix} x & y \\ z & w \end{bmatrix}$ such that
$$\begin{bmatrix} 5 & 3 \\ 4 & 2 \end{bmatrix} A = \begin{bmatrix} 1 & 0 \\ 0 & 1 \end{bmatrix}.$$

34. Find a matrix $A = \begin{bmatrix} x & y \\ z & w \end{bmatrix}$ such that

$$A\begin{bmatrix} 5 & 3 \\ 4 & 2 \end{bmatrix} = \begin{bmatrix} 1 & 0 \\ 0 & 1 \end{bmatrix}.$$

Compare your answers for Exercises 33 and 34.

35. Let $A = \begin{bmatrix} 0 & 1 \\ -1 & 3 \end{bmatrix}$.

   a. Find A^{-1}. (Use the formula on page 585.)
   b. Find $(A^2)^{-1}$.
   c. Find $(A^{-1})^2$.
   d. Use your answers in parts (b) and (c) to show that $(A^{-1})^2$ is the inverse of A^2.

36. Repeat Exercise 35 for $A = \begin{bmatrix} 3 & 4 \\ 2 & 3 \end{bmatrix}$.

In Exercises 37–40, show that B is the inverse of A.

37. $A = \begin{bmatrix} 7 & 6 \\ 6 & 5 \end{bmatrix}, \quad B = \begin{bmatrix} -5 & 6 \\ 6 & -7 \end{bmatrix}$

38. $A = \begin{bmatrix} -1 & 3 \\ 1 & -4 \end{bmatrix}, \quad B = \begin{bmatrix} -4 & -3 \\ -1 & -1 \end{bmatrix}$

39. $A = \begin{bmatrix} 2 & -1 & 0 \\ 1 & 0 & 4 \\ 1 & -1 & 1 \end{bmatrix}, \quad B = \frac{1}{5}\begin{bmatrix} 4 & 1 & -4 \\ 3 & 2 & -8 \\ -1 & 1 & 1 \end{bmatrix}$

40. $A = \begin{bmatrix} 2 & 0 & 1 \\ 0 & -1 & 2 \\ 1 & 0 & 1 \end{bmatrix}, \quad B = \begin{bmatrix} 1 & 0 & -1 \\ -2 & -1 & 4 \\ -1 & 0 & 2 \end{bmatrix}$

In Exercises 41–44, find the inverse (if it exists) of each matrix.

41. $\begin{bmatrix} 3 & 1 \\ 2 & 4 \end{bmatrix}$ **42.** $\begin{bmatrix} 3 & -4 \\ 1 & 2 \end{bmatrix}$

43. $\begin{bmatrix} 1 & 2 & -2 \\ -1 & 3 & 0 \\ 0 & -2 & 1 \end{bmatrix}$ **44.** $\begin{bmatrix} 1 & 2 & 1 \\ 2 & 2 & 4 \\ 0 & 0 & 3 \end{bmatrix}$

In Exercises 45–50, use an inverse matrix (if possible) to solve each system of linear equations.

45. $\begin{cases} x + 3y = 7 \\ 2x + 5y = 4 \end{cases}$ **46.** $\begin{cases} 3x + 5y = 4 \\ 2x + 4y = 5 \end{cases}$

47. $\begin{cases} x + 3y + 3z = 3 \\ x + 4y + 3z = 5 \\ x + 3y + 4z = 6 \end{cases}$ **48.** $\begin{cases} x + 2y + 3z = 6 \\ 2x + 4y + 5z = 8 \\ 3x + 5y + 6z = 10 \end{cases}$

49. $\begin{cases} x - y + z = 3 \\ 4x + 2y = 5 \\ 7x - y - z = 6 \end{cases}$ **50.** $\begin{cases} x + 2y + 4z = 7 \\ 4x + 3y - 2z = 6 \\ x - 3z = 4 \end{cases}$

In Exercises 51–54, find the determinant of each matrix.

51. $\begin{bmatrix} 2 & -5 \\ 3 & 4 \end{bmatrix}$ **52.** $\begin{bmatrix} -1 & 4 \\ 11 & -3 \end{bmatrix}$

53. $\begin{bmatrix} 12 & 21 \\ 4 & 7 \end{bmatrix}$ **54.** $\begin{bmatrix} -7 & 9 \\ -4 & 5 \end{bmatrix}$

In Exercises 55 and 56, find (a) the minors $M_{12}, M_{23},$ and M_{22} and (b) the cofactors $A_{12}, A_{23},$ and A_{22} of the given matrix A.

55. $A = \begin{bmatrix} 4 & -1 & 2 \\ -2 & -3 & 5 \\ 0 & 2 & -4 \end{bmatrix}$ **56.** $A = \begin{bmatrix} 1 & 2 & 3 \\ 4 & 5 & 6 \\ 7 & 8 & 9 \end{bmatrix}$

In Exercises 57 and 58, find the determinant of each matrix A by expanding (a) by the second row and (b) by the third column.

57. $A = \begin{bmatrix} 1 & -2 & 3 \\ 4 & -1 & -2 \\ -2 & 1 & 5 \end{bmatrix}$ **58.** $A = \begin{bmatrix} 1 & 4 & 3 \\ 6 & 8 & 10 \\ 2 & 5 & 4 \end{bmatrix}$

In Exercises 59 and 60, find the determinant of each matrix A.

59. $A = \begin{bmatrix} 1 & 2 & 0 \\ 0 & 1 & 2 \\ 1 & 0 & 2 \end{bmatrix}$ **60.** $A = \begin{bmatrix} 2 & 3 & 4 \\ 2 & 5 & 9 \\ 2 & 7 & 16 \end{bmatrix}$

In Exercises 61–64, use Cramer's Rule to solve each system of equations.

61. $\begin{cases} 5x + 3y = 11 \\ 2x + y = 4 \end{cases}$ **62.** $\begin{cases} 2x - 7y - 13 = 0 \\ 5x + 6y - 9 = 0 \end{cases}$

63. $\begin{cases} 3x + y - z = 14 \\ x + 3y - z = 16 \\ x + y - 3z = -10 \end{cases}$ **64.** $\begin{cases} 2x + 3y - 2z = 0 \\ 5y - 3x + 4z = 9 \\ 3x + 7y - 6z = 4 \end{cases}$

In Exercises 65 and 66, solve each equation for x.

65. $\begin{vmatrix} 1 & 2 & 4 \\ -3 & 5 & 7 \\ 1 & x & 4 \end{vmatrix} = 0$ **66.** $\begin{vmatrix} 1 & -1 & 2 \\ 0 & x & 1 \\ 3 & 2 & x-1 \end{vmatrix} = 14$

67. Show that the equation of the line passing through the points $(2, 3)$ and $(1, -4)$ is equivalent to the equation

$$\begin{vmatrix} x & y & 1 \\ 2 & 3 & 1 \\ 1 & -4 & 1 \end{vmatrix} = 0. \text{ (See page 603.)}$$

68. Use a determinant to find the area of the triangle with vertices $A(1, 1), B(4, 3),$ and $C(2, 7)$. (See page 603.)

Applying the Concepts

69. **Maximizing profit.** A company is considering which of three methods of production it should use in producing the three products $A, B,$ and C. The number of units of each product produced by each method is shown in this matrix:

$$\begin{array}{c} \\ Method\ 1 \\ Method\ 2 \\ Method\ 3 \end{array} \begin{array}{ccc} A & B & C \\ \begin{bmatrix} 4 & 8 & 2 \\ 5 & 7 & 1 \\ 5 & 4 & 8 \end{bmatrix} \end{array}$$

The profit per unit of the products $A, B,$ and C is \$10, \$4, and \$6, respectively. Use matrix multiplication to find which method maximizes total profit for the company.

70. Nutrition. Consider two families: Ardestanis (*A*) and Barkley (*B*). Family *A* consists of two adult males, three adult females, and one child. Family *B* consists of one adult male, one adult female, and two children. Their mutual dietician considers the age and weight of each member of the two families and recommends daily allowances for calories—adult males 2200, adult females 1700, and children 1500—and for protein—adult males 50 grams, adult females 40 grams, and children 30 grams.

Represent the information in the preceding paragraph by matrices. Use matrix multiplication to calculate the total requirements of calories and proteins for each family.

In Exercises 71–76, each problem produces a system of linear equations in two or three variables. Use the method of your choice from this chapter to solve the system of equations.

71. Airplane speed. An airplane traveled 1680 miles in 3 hours with the wind. It would have taken 3.5 hours to make the same trip against the wind. Find the speed of the plane and the velocity of the wind assuming that both are constant.

72. Walking speed. In 5 hours, Nertha walks 2 miles farther than Kristina walks in 4 hours; then in 12 hours, Kristina walks 10 miles farther than Nertha walks in 10 hours. How many miles per hour does each woman walk?

73. Friends. Andrew, Bonnie, and Chauncie have $320 between them. Andrew has twice as much as Chauncie, and Bonnie and Chauncie together have $20 less than Andrew. How much money does each person have?

74. Ratio. Steve and Nicole entered a double-decker bus in London on which there were $\frac{5}{8}$ as many men as women.

After they boarded the bus, there were $\frac{7}{11}$ as many men as women on the bus. How many people were on the bus before they entered it?

75. Hospital workers. Metropolitan Hospital employs three types of caregivers: registered nurses, licensed practical nurses, and nurse's aides. Each registered nurse is paid $75 per hour, each licensed practical nurse is paid $20 per hour, and each nurse's aide is paid $30 per hour. The hospital has budgeted $4850 per hour for the three categories of caregivers. The hospital requires that each caregiver spend some time in a class learning about advanced medical practices each week: registered nurses, 3 hours; licensed practical nurses, 5 hours; and nurse's aides, 4 hours. The advanced classes can accommodate a maximum of 530 person-hours each week. The hospital needs a total of 130 registered nurses, licensed practical nurses, and nurse's aides to meet the daily needs of the patients. Assume that the allowable payroll is met, the training classes are full, and the patients' needs are met. How many registered nurses, licensed practical nurses, and nurse's aides should the hospital employ?

76. Sales tax. A drugstore sells three categories of items:
  (i) Medicine, which is not taxed
  (ii) Nonmedical items, which are taxed at the rate of 8%
  (iii) Beer and cigarettes, which are taxed at the "sin tax" rate of 20%

Last Friday, the total sales (excluding taxes) of all three items were $18,500. The total tax collected was $1020. The sales of medicines exceeded the combined sales of the other items by $3500. Find the sales in each category.

PRACTICE TEST A

1. Give the order of the matrix and the value of the designated entry of the matrix.

$$\begin{bmatrix} 3 & -1 & -3 & 0 \\ 2 & 4 & 1 & 7 \\ 5 & 0 & 3 & 2 \\ 8 & 10 & 9 & 6 \\ -4 & 0 & 0 & 3 \end{bmatrix} \quad a_{43}$$

2. Write an augmented matrix for the system of equations.

$$\begin{cases} 7x - 3y + 9z = 5 \\ -2x + 4y + 3z = -12 \\ 8x - 5y + z = -9 \end{cases}$$

3. Write the system of linear equations represented by the augmented matrix

$$\begin{bmatrix} 4 & 0 & -1 & | & -3 \\ 1 & 3 & 0 & | & 9 \\ 2 & 7 & 5 & | & 8 \end{bmatrix}.$$

Use *x*, *y*, and *z* for the variables.

4. Find values for the variables so that the matrices

$$\begin{bmatrix} x - 5 & y + 3 \\ z & 9 \end{bmatrix} = \begin{bmatrix} 5 & 13 \\ 0 & 9 \end{bmatrix}$$

are equal.

In Problems 5 and 6, solve the system of equations by using matrices.

5. $\begin{cases} x + 2y + z = 6 \\ x + y - z = 7 \\ 2x - y + 2z = -3 \end{cases}$

6. $\begin{cases} 2x + y - 4z = 6 \\ -x + 3y - z = -2 \\ 2x - 6y + 2z = 4 \end{cases}$

7. Let $A = \begin{bmatrix} 5 & -2 \\ 4 & 0 \\ 7 & 6 \end{bmatrix}$ and $B = \begin{bmatrix} -3 & -1 \\ 0 & -8 \\ -4 & 6 \end{bmatrix}$. Find $A - B$.

In Problems 8 and 9, find the product AB or state that AB is not defined.

8. $A = \begin{bmatrix} 3 & -7 & 2 \end{bmatrix}$, $B = \begin{bmatrix} 0 \\ 1 \\ 4 \end{bmatrix}$

9. $A = \begin{bmatrix} -4 & 1 & 2 \\ -9 & 0 & 3 \end{bmatrix}$, $B = \begin{bmatrix} 2 & 8 \\ 7 & -5 \end{bmatrix}$

For Problems 10–13, let

$$A = \begin{bmatrix} -3 & 1 & 0 \\ 5 & 7 & 2 \end{bmatrix}, \quad B = \begin{bmatrix} -1 & 4 \\ 8 & 2 \end{bmatrix}, \quad \text{and} \quad C = \begin{bmatrix} 1 & 5 \\ 0 & 4 \end{bmatrix}.$$

10. Find $2A$.

11. Find $A + BA$.

12. Find C^2.

13. Find C^{-1}.

14. Find the inverse of the matrix

$$A = \begin{bmatrix} 2 & 1 & 3 \\ 1 & 2 & -1 \\ 3 & 1 & 5 \end{bmatrix}.$$

15. Write the linear system

$$\begin{cases} 5x + 2y = 32 \\ 3x + y = 18 \end{cases}$$

as a matrix equation in the form $AX = B$, where A is the coefficient matrix and B is the constant matrix.

16. Write the matrix equation

$$\begin{bmatrix} 12 & -3 \\ -2 & 7 \end{bmatrix} \begin{bmatrix} x \\ y \end{bmatrix} = \begin{bmatrix} 5 \\ -9 \end{bmatrix}$$

as a system of linear equations without matrices.

17. Solve the matrix equation $A - 5X = 2B$ for X, where

$$A = \begin{bmatrix} 1 & 5 & -2 \\ 4 & -2 & 7 \end{bmatrix} \text{ and } B = \begin{bmatrix} 2 & 5 & -11 \\ 18 & 8 & 11 \end{bmatrix}.$$

In Problems 18 and 19, evaluate the determinant.

18. $\begin{vmatrix} \dfrac{1}{2} & -\dfrac{1}{4} \\ \dfrac{1}{2} & \dfrac{3}{4} \end{vmatrix}$

19. $\begin{vmatrix} 1 & 3 & 5 \\ 2 & 0 & 10 \\ -3 & 1 & -15 \end{vmatrix}$

20. Use Cramer's Rule to write the solution of the system

$$\begin{cases} 2x - y + z = 3 \\ x + y + z = 6 \\ 4x + 3y - 2z = 4 \end{cases}$$

in the determinant form. (You do not need to find the solution.)

PRACTICE TEST B

1. Give the order of the matrix and the value of the designated entries of the matrix.

$$A = \begin{bmatrix} 0 & 3 & 0 & -10 & 6 \\ 14 & -5 & 7 & -7 & 8 \\ -15 & 9 & -6 & 13 & -2 \\ 0.4 & 11 & 4 & -3 & 0 \end{bmatrix} \quad a_{34}$$

   a. Order: 5×4; $a_{34} = 6$
   b. Order: 4×5; $a_{34} = 11$
   c. Order: 20; $a_{34} = -4$
   d. Order: 4×5; $a_{34} = 13$

2. Write an augmented matrix for the following system of equations:

$$\begin{cases} 8x + 4y + 5z = 10 \\ -2x + 3y + 6z = -9 \\ -2x + 8y + 6z = 4 \end{cases}$$

   a. $\begin{bmatrix} 8 & 4 & 5 \\ -2 & 3 & 6 \\ -2 & 8 & 6 \end{bmatrix}$ **b.** $\left[\begin{array}{ccc|c} 10 & 5 & 4 & 8 \\ -9 & 6 & 3 & -2 \\ 4 & 6 & 8 & -2 \end{array}\right]$

   c. $\left[\begin{array}{ccc|c} 8 & -2 & -2 & 10 \\ 4 & 3 & 8 & -9 \\ 5 & 6 & 6 & 4 \end{array}\right]$ **d.** $\left[\begin{array}{ccc|c} 8 & 4 & 5 & 10 \\ -2 & 3 & 6 & -9 \\ -2 & 8 & 6 & 4 \end{array}\right]$

3. Write the system of linear equations represented by the following augmented matrix.

$$\left[\begin{array}{ccc|c} -1 & 7 & 0 & -6 \\ 2 & 0 & 6 & 5 \\ 0 & 8 & -8 & 4 \end{array}\right]$$

Use x, y, and z for the variables.

   a. $\begin{cases} x + 7y + z = -6 \\ 2x + 6z = 5 \\ 8y + 8z = 4 \end{cases}$ **b.** $\begin{cases} -x + 7y + z = -6 \\ 2x + 6y = 5 \\ 8x - 8y = 4 \end{cases}$

   c. $\begin{cases} -x + 7y + z = -6 \\ 2x + y + 6z = 5 \\ x + 8y - 8z = 4 \end{cases}$ **d.** $\begin{cases} -x + 7y = -6 \\ 2x + 6z = 5 \\ 8y - 8z = 4 \end{cases}$

4. Find values for the variables so that the matrices

$$\begin{bmatrix} x + 3 & y + 4 \\ 7 & 3 \end{bmatrix} \text{ and } \begin{bmatrix} 9 & 2 \\ 7 & z \end{bmatrix}$$

are equal.

   a. $x = -6$; $y = 2$; $z = 3$ **b.** $x = 6$; $y = 9$; $z = 3$
   c. $x = 6$; $y = -2$; $z = 3$ **d.** $x = 9$; $y = 2$; $z = 3$

5. Find the value of z in the following system:

$$\begin{cases} 2x + y = 15 \\ 2y + z = 25 \\ 2z + x = 26 \end{cases}$$

   a. 4   **b.** 9   **c.** 11   **d.** 12

6. Find the value of $4x + 3y + z$ in the following system:

$$\begin{cases} 2x + y \quad\;\; = 17 \\ \quad\;\; y + 2z = 15 \\ x \quad\;\; + z = 9 \end{cases}$$

a. 41   **b.** 43   **c.** 55   **d.** 45

7. Let $A = \begin{bmatrix} -1 & 4 \\ 0 & 4 \\ 8 & -4 \end{bmatrix}$ and $B = \begin{bmatrix} 7 & 2 \\ 17 & 4 \\ 2 & 2 \end{bmatrix}$. Find $A - B$.

a. $\begin{bmatrix} 1 & 2 \\ 7 & 0 \\ 6 & -2 \end{bmatrix}$ **b.** $\begin{bmatrix} 3 & -3 \\ 7 & 0 \\ -6 & 6 \end{bmatrix}$

c. $\begin{bmatrix} -8 & 2 \\ -17 & 0 \\ 6 & -6 \end{bmatrix}$ **d.** $\begin{bmatrix} 1 & 5 \\ 7 & 8 \\ 10 & 0 \end{bmatrix}$

In Problems 8 and 9, find the product AB if possible.

8. $A = \begin{bmatrix} -8 & 2 & 9 \end{bmatrix}$, $B = \begin{bmatrix} 3 \\ 0 \\ -3 \end{bmatrix}$

a. $\begin{bmatrix} -24 & 0 & -27 \end{bmatrix}$ **b.** $\begin{bmatrix} -51 \end{bmatrix}$

c. $\begin{bmatrix} 210 \end{bmatrix}$ **d.** $\begin{bmatrix} -24 \\ 0 \\ -27 \end{bmatrix}$

9. $A = \begin{bmatrix} 3 & -2 & 1 \\ 0 & 4 & -2 \end{bmatrix}$, $B = \begin{bmatrix} 5 & 0 \\ -2 & 1 \end{bmatrix}$

a. AB is not defined.   **b.** $\begin{bmatrix} 15 & -10 & 5 \\ -6 & 8 & -4 \end{bmatrix}$

c. $\begin{bmatrix} 15 & -6 \\ -10 & 8 \\ 5 & -4 \end{bmatrix}$ **d.** $\begin{bmatrix} 15 & 0 \\ 0 & 4 \end{bmatrix}$

For Problems 10–13, let

$A = \begin{bmatrix} 2 & 1 & -3 \\ -5 & 2 & 1 \end{bmatrix}$, $B = \begin{bmatrix} -3 & 7 \\ 2 & 4 \end{bmatrix}$, and $C = \begin{bmatrix} 5 & 4 \\ 1 & 0 \end{bmatrix}$.

10. Find $2A$.

a. $\begin{bmatrix} 1 & \dfrac{1}{2} & -\dfrac{3}{2} \\ -\dfrac{5}{2} & 1 & \dfrac{1}{2} \end{bmatrix}$ **b.** $\begin{bmatrix} 4 & 2 & -6 \\ -10 & 4 & 2 \end{bmatrix}$

c. $\begin{bmatrix} 4 & 3 & -1 \\ -3 & 4 & 3 \end{bmatrix}$ **d.** $\begin{bmatrix} 0 & -1 & -5 \\ -7 & 0 & -1 \end{bmatrix}$

11. Find $A + BA$.

a. $\begin{bmatrix} 2 & -4 & 10 \\ -5 & 4 & 5 \end{bmatrix}$ **b.** $\begin{bmatrix} 4 & -2 & 6 \\ -10 & 4 & 2 \end{bmatrix}$

c. $\begin{bmatrix} -39 & 12 & 13 \\ -21 & 12 & -1 \end{bmatrix}$ **d.** $\begin{bmatrix} -41 & 17 & -2 \\ -16 & 6 & 10 \end{bmatrix}$

12. Find C^2.

a. $\begin{bmatrix} 10 & 8 \\ 2 & 0 \end{bmatrix}$ **b.** $\begin{bmatrix} 41 & 5 \\ 5 & 1 \end{bmatrix}$

c. $\begin{bmatrix} 29 & 20 \\ 5 & 4 \end{bmatrix}$ **d.** $\begin{bmatrix} 0 & 4 \\ 1 & 5 \end{bmatrix}$

13. Find C^{-1}.

a. $\begin{bmatrix} -\dfrac{5}{4} & 1 \\ \dfrac{1}{4} & 0 \end{bmatrix}$ **b.** $\begin{bmatrix} 0 & 1 \\ \dfrac{1}{4} & -\dfrac{5}{4} \end{bmatrix}$

c. $\begin{bmatrix} \dfrac{1}{4} & -\dfrac{5}{4} \\ 0 & 1 \end{bmatrix}$ **d.** $\begin{bmatrix} 0 & -1 \\ -\dfrac{1}{4} & -\dfrac{5}{4} \end{bmatrix}$

14. Find the inverse of the following matrix:

$$A = \begin{bmatrix} 1 & 0 & 0 \\ 2 & 1 & 0 \\ 3 & -4 & 1 \end{bmatrix}$$

a. $\begin{bmatrix} 1 & 0 & 0 \\ -2 & 1 & 0 \\ 3 & 4 & 1 \end{bmatrix}$ **b.** $\begin{bmatrix} 1 & 0 & 0 \\ -2 & 1 & 0 \\ -8 & 3 & 1 \end{bmatrix}$

c. $\begin{bmatrix} 1 & -2 & -11 \\ 0 & 1 & 4 \\ 0 & 0 & 1 \end{bmatrix}$ **d.** $\begin{bmatrix} 1 & 0 & 0 \\ -2 & 1 & 0 \\ -11 & 4 & 1 \end{bmatrix}$

15. Write the linear system as a matrix equation

$$\begin{cases} 4x + 2y = 15 \\ -2x + 4y = 9 \end{cases}$$

in the form $AX = B$, where A is the coefficient matrix and B is the constant matrix.

a. $\begin{bmatrix} 4 & 2 \\ -2 & 4 \end{bmatrix}\begin{bmatrix} x \\ y \end{bmatrix} = \begin{bmatrix} 15 \\ 9 \end{bmatrix}$ **b.** $\begin{bmatrix} 4 & 2 \\ 4 & -2 \end{bmatrix}\begin{bmatrix} x \\ y \end{bmatrix} = \begin{bmatrix} 15 \\ 9 \end{bmatrix}$

c. $\begin{bmatrix} 15 & 2 \\ 9 & -2 \end{bmatrix}\begin{bmatrix} x \\ y \end{bmatrix} = \begin{bmatrix} 4 \\ 4 \end{bmatrix}$ **d.** $\begin{bmatrix} 4 & -2 \\ 2 & 4 \end{bmatrix}\begin{bmatrix} x \\ y \end{bmatrix} = \begin{bmatrix} 15 \\ 9 \end{bmatrix}$

16. Write the matrix equation

$$\begin{bmatrix} -3 & 9 \\ -8 & -4 \end{bmatrix}\begin{bmatrix} x \\ y \end{bmatrix} = \begin{bmatrix} -4 \\ -5 \end{bmatrix}$$

as a system of linear equations.

a. $\begin{cases} -3x + 9y = -4 \\ -8x - 4y = -5 \end{cases}$ **b.** $\begin{cases} -3x + 9y = -4 \\ -4x - 8y = -5 \end{cases}$

c. $\begin{cases} 9x - 3y = -4 \\ -8x - 4y = -5 \end{cases}$ **d.** $\begin{cases} -3x + 9y = 4 \\ -8x - 4y = 5 \end{cases}$

17. Let $A = \begin{bmatrix} -2 & -3 & 1 \\ -5 & 3 & -2 \end{bmatrix}$ and $B = \begin{bmatrix} -1 & -2 & -1 \\ 1 & 0 & 1 \end{bmatrix}$.

Solve the matrix equation $A - 3X = -5B$ for X.

a. $X = \begin{bmatrix} -\dfrac{5}{2} & -\dfrac{9}{2} & -1 \\ -1 & \dfrac{3}{2} & \dfrac{1}{2} \end{bmatrix}$

b. $X = \begin{bmatrix} -\dfrac{7}{3} & -\dfrac{13}{3} & -\dfrac{4}{3} \\ 0 & 1 & 1 \end{bmatrix}$

c. $X = \begin{bmatrix} -1 & \dfrac{3}{2} & -1 \\[2mm] -\dfrac{5}{2} & -\dfrac{9}{2} & -1 \end{bmatrix}$

d. $X = \begin{bmatrix} -\dfrac{1}{3} & \dfrac{1}{3} & \dfrac{8}{3} \\[2mm] -\dfrac{20}{3} & 3 & -\dfrac{11}{3} \end{bmatrix}$

In Problems 18 and 19, evaluate the determinant.

18. $\begin{vmatrix} -8 & 5 \\ -4 & -1 \end{vmatrix}$

　　a. −28　　**b.** −44　　**c.** −12　　**d.** 28

19. $\begin{vmatrix} 2 & 3 & -2 \\ 3 & 0 & -3 \\ -3 & 0 & -5 \end{vmatrix}$

　　a. −72　　**b.** 18　　**c.** 72　　**d.** −18

20. Solve the system of equations

$$\begin{cases} x + y + z = -6 \\ x - y + 3z = -22 \\ 2x + y + z = -10 \end{cases}$$

by using Cramer's Rule.

　　a. $\varnothing$ 　　　　　　　**b.** $\{(-4, 3, -5)\}$
　　c. $\{(-5, -4, 3)\}$ 　　**d.** $\{(-5, 3, -4)\}$

CUMULATIVE REVIEW EXERCISES ▪ Chapters P–6

1. Assuming that the distance between the points $(2, -3)$ and $(-1, y)$ is five units, find y.

In Problems 2–8, solve each equation or inequality.

2. $\dfrac{4}{x - 1} - \dfrac{3}{x + 2} = \dfrac{18}{(x + 2)(x - 1)}$

3. $|2x - 5| = 3$

4. $4x^2 = 8x - 13$

5. $\left(\dfrac{3x - 1}{x + 5}\right)^2 - 3\left(\dfrac{3x - 1}{x + 5}\right) - 28 = 0$

6. $\log_2|x| + \log_2|x + 6| = 4$

7. $\dfrac{x + 2}{2x - 1} > 0$

8. $x^2 - 7x + 6 \le 0$

9. Write all possible rational zeros of

$$f(x) = 4x^3 + 8x^2 - 11x + 3.$$

10. Show that $\dfrac{1}{2}$ is a zero of multiplicity 2 of the function f in Exercise 9.

11. The horsepower required to propel a ship varies as the cube of the ship's speed. If the horsepower required for a speed of 15 miles per hour is 10,125, find the horsepower required for a speed of 20 miles per hour.

12. A city covers an area of 400 square miles. At present only 2% of its area is reserved for parks. In the unincorporated area outside the city limits, 20% of the area could be developed into parks. How many square miles of the incorporated area had to be annexed so that the city could develop 12 percent of its area as parks?

In Problems 13 and 14, solve the system of equations.

13. $\begin{cases} 3x + y = 2 \\ 4x + 5y = -1 \end{cases}$

14. $\begin{cases} 2x + y = 5 \\ y^2 - 2y = -3x + 5 \end{cases}$

15. Use transformations to sketch the graph of

$$f(x) = 3|x + 1| + 2.$$

16. Find the inverse of the matrix $\begin{bmatrix} 1 & 2 & -2 \\ -1 & 3 & 0 \\ 0 & -2 & 1 \end{bmatrix}$.

17. Use the result in Exercise 16 to solve the system of equations

$$\begin{cases} x + 2y - 2z = 5 \\ -x + 3y = 2 \\ -2y + z = -3 \end{cases}$$

18. Assuming that $f(x) = x^2 + 3x - 1$ and $g(x) = x + 2$, find **(a)** $F(x) = (f \circ g)(x)$ and **(b)** $F(4)$.

19. Let $f(x) = \dfrac{x}{x + 4}$. Find $f^{-1}(x)$.

20. In Exercise 19, find the domain and range of f.

Conic Sections

Greek mathematicians were fascinated by certain curves, called conic sections, now used to describe natural phenomena such as planetary orbits. Modern applications also include the construction of lenses, whispering galleries, navigation systems, and even medical devices. In this chapter, we investigate the conic sections and their many uses.

Conic Sections: Overview

Most of the sections in this chapter focus on the plane curves called **conics** or **conic sections**. As the name implies, these curves are the sections of a cone (similar to an ice cream cone) formed when a plane intersects the cone. ■

Euclid defined a cone as a surface generated by rotating a right triangle about one of its legs. However, the following description of a cone given by Apollonius is more appropriate.

RIGHT CIRCULAR CONE

Draw a circle on a plane. Draw a line *l*, called the **axis**, passing through the center of the circle and perpendicular to the plane. Choose a point *V* above the plane on this line. The surface consisting of all lines that simultaneously pass through the point *V* and the circle is called a **right circular cone** with **vertex** *V*. The vertex *V* separates the surface into two parts, called *nappes of the cone*. See Figure 7.1.

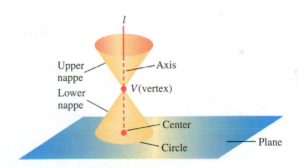

FIGURE 7.1 **Parts of a cone**

If we slice the cone with a plane, the intersections of the slicing plane and the cone are **conic sections**. If the slicing plane is horizontal (parallel to the first plane) and does not contain the vertex, the intersection is a **circle** (Figure 7.2(a)). If the slicing plane is inclined slightly from the horizontal, the intersection is an oval-shaped curve called an **ellipse** (Figure 7.2(b)). As the angle of the slicing plane increases, more elongated ellipses are formed (Figure 7.2(b)). If the angle of the slicing plane increases still further so that the slicing plane is parallel to one of the lines generating the cone, the intersection is called a **parabola** (Figure 7.2(c)). If the angle of the slicing plane increases yet further so that the slicing plane intersects both nappes of the cone, the resulting intersection is called a **hyperbola** (Figure 7.2(d)).

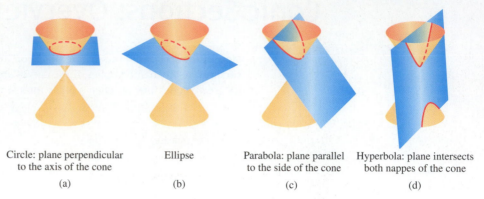

Circle: plane perpendicular to the axis of the cone

(a)

Ellipse

(b)

Parabola: plane parallel to the side of the cone

(c)

Hyperbola: plane intersects both nappes of the cone

(d)

FIGURE 7.2 **Conic sections**

Intersections of the slicing plane through the vertex results in points and lines called **degenerate conic sections**. See Figure 7.3.

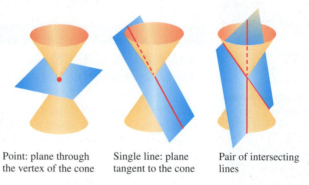

Point: plane through the vertex of the cone

Single line: plane tangent to the cone

Pair of intersecting lines

FIGURE 7.3 **Degenerate conic sections**

Just as a circle (see page 185) is defined as the set of points in the plane at a fixed distance r (the radius) from a fixed point (the center), we can define each of the conic sections just described as a set of points in the plane that satisfies certain geometric conditions. It can be shown that these alternate definitions give the same family of curves described earlier. This enables us to keep our work in a two-dimensional setting. Using these definitions, we will derive the equations of the conic sections and show that an equation of the form $Ax^2 + Cy^2 + Dx + Ey + F = 0$ is (except in degenerate cases represented in Figure 7.3.) the equation of a parabola, an ellipse, or a hyperbola. (A circle is a special case of an ellipse.)

We study these conic sections because of their practical applications. For example, a satellite dish, a flashlight lens, and a telescope lens have a parabolic shape; the planets travel in elliptical orbits, and comets travel in orbits that are either elliptical or hyperbolic. A comet with an elliptical orbit can be viewed from Earth more than once, while those with a hyperbolic orbit can be viewed only once!

The Parabola

Before Starting this Section, Review

1. Distance formula (Section 2.1, page 173)

2. Midpoint formula (Section 2.1, page 175)

3. Completing the square (Section 1.4, page 121)

4. Line of symmetry (Section 2.2, page 182)

5. Slopes of perpendicular lines (Section 2.2, page 200)

Objectives

1. Define a parabola geometrically.

2. Find an equation of a parabola.

3. Translate a parabola.

4. Use the reflecting property of parabolas.

Hubble Telescope

Edwin Hubble (1889–1953)

Edwin Hubble is renowned for discovering that there are other galaxies in the universe beyond the Milky Way. Hubble then wanted to classify the galaxies according to their content, distance, shape, and brightness patterns. He made the momentous discovery that the galaxies were moving away from each other at a rate proportional to the distance between them (Hubble's Law). This led to the calculation of the point where the expansion began and to confirmation of the Big Bang theory of the origin of the universe. Recent estimates place the age of the universe at about 20 billion years.

THE HUBBLE SPACE TELESCOPE

In 1918, the 2.5-meter (100-inch) Hooker Telescope was installed at Mount Wilson Observatory in Pasadena, California. With this telescope, astronomer Edwin Hubble measured the distances and velocities of the galaxies. His work led to today's concept of an expanding universe. In 1977, the National Aeronautics and Space Administration (NASA) named its largest, most complex, and most capable orbiting telescope in honor of Edwin Hubble.

On April 25, 1990, the Hubble Space Telescope was deployed into orbit from the space shuttle *Discovery*. However, due to a spherical aberration in the telescope's parabolic mirror, the Hubble had "blurred vision." (See Example 4.) In 1993, astronauts from the space shuttle *Endeavor* installed replacement instruments and supplemental optics to restore the telescope to full optimal performance.

Not since Galileo turned his telescope toward the heavens in 1610 has any event so changed our understanding of the universe as did the deployment of the Hubble Space Telescope. The first instrument of any kind that was specifically designed for routine servicing by spacewalking astronauts, the Hubble telescope brought stunning pictures from the end of the universe to the general public. ■

1 Define a parabola geometrically.

Geometric Definition of a Parabola

In Section 3.1, we named the graph of a quadratic function $y = ax^2 + bx + c$, $a \neq 0$, a parabola. In this section, we give a *geometric* definition of a parabola. Then we show that this definition leads to the graph of a quadratic function.

PARABOLA

Let l be a line and F a point not on the line l. See Figure 7.4(a). Let $\mathcal{P}$ be the plane determined by F and l. Then the set of all points P in the plane that are the same distance from F as they are from the line l is called a **parabola**. See Figure 7.4(b). Thus, a parabola is the set of points P for which $d(F, P) = d(P, l)$, where $d(P, l)$ denotes the distance between P and l.

(a)   (b)

FIGURE 7.4 Defining a parabola   **FIGURE 7.5**

The line l is the **directrix** of the parabola, and the point F is the **focus**. The line through the focus *perpendicular* to the directrix is called the **axis**, or the **axis of symmetry**, of the parabola. The point at which the axis intersects the parabola is called the **vertex**. See Figure 7.5. Note that the vertex of a parabola is the midpoint of the segment $\overline{FM}$.

2 Find an equation of a parabola.

Equation of a Parabola

A coordinate system allows us to transform the geometric description of a parabola into an algebraic *equation of the parabola*. This equation is particularly simple if we place the vertex V at the origin and the focus on one of the coordinate axes. Suppose for a moment that the focus F is on the positive x-axis. Then F has coordinates of the form $(a, 0)$ with $a > 0$, as shown Figure 7.6.

The vertex V lies on the parabola with focus F and directrix l; so we have the following:

$$d(V, l) = d(V, F) \qquad \text{Definition of parabola, where}$$
$$= a \qquad\qquad\quad V \text{ is a point on the parabola}$$

The vertex V at the origin is located midway between the focus and the directrix; so the equation of the directrix is $x = -a$.

Suppose a point $P(x, y)$ lies on the parabola shown in Figure 7.6. Then the distance from the point P to the directrix l is the distance between the points $P(x, y)$ and $A(-a, y)$. So a point $P(x, y)$ lies on this parabola if and only if

$$d(P, A) = d(P, F).$$

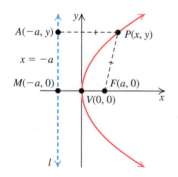

FIGURE 7.6 Deriving the equation of a parabola

Using the distance formula on each side gives

$$\sqrt{(x-a)^2 + y^2} = \sqrt{(x+a)^2}$$

$(x-a)^2 + y^2 = (x+a)^2$ Square both sides.

$x^2 - 2ax + a^2 + y^2 = x^2 + 2ax + a^2$ Expand squares.

$y^2 = 4ax$ Simplify.

The equation $y^2 = 4ax$ is called the **standard equation of a parabola with vertex (0, 0) and focus (a, 0)**. Similarly, if the focus of a parabola is placed on the negative x-axis, we obtain the equation $y^2 = -4ax$ as the standard equation of a parabola with vertex $(0, 0)$ and focus $(-a, 0)$.

By interchanging the roles of x and y, we find that $x^2 = 4ay$ is the standard equation of a parabola with vertex $(0, 0)$ and focus $(0, a)$. Similarly, the equation $x^2 = -4ay$ is the standard equation of a parabola with vertex $(0, 0)$ and focus $(0, -a)$.

Main Facts About a Parabola with $a > 0$

Standard Equation	$y^2 = 4ax$	$y^2 = -4ax$	$x^2 = 4ay$	$x^2 = -4ay$
Vertex	$(0, 0)$	$(0, 0)$	$(0, 0)$	$(0, 0)$
Description	Opens right	Opens left	Opens up	Opens down
Axis of Symmetry	$y = 0$ (x-axis)	$y = 0$ (x-axis)	$x = 0$ (y-axis)	$x = 0$ (y-axis)
Focus	$(a, 0)$	$(-a, 0)$	$(0, a)$	$(0, -a)$
Directrix	$x = -a$	$x = a$	$y = -a$	$y = a$
Graph				

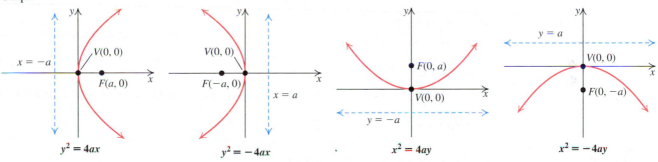

EXAMPLE 1   **Graphing a Parabola**

Graph each parabola and specify the vertex, focus, directrix, and axis.

a. $x^2 = -8y$ **b.** $y^2 = 5x$

SOLUTION

a. The equation $x^2 = -8y$ has the standard form $x^2 = -4ay$; so

$-4a = -8$ Equate coefficients of y.

$a = 2$ Divide both sides by -4.

The parabola opens down. The vertex is the origin, and the focus is $(0, -2)$. The directrix is the horizontal line $y = 2$; the axis of the parabola is the y-axis. You should plot some additional points to ensure the accuracy of your sketch.

TECHNOLOGY CONNECTION

You can use a graphing calculator to graph parabolas.

To graph $x^2 = 4ay$, just graph $Y_1 = \dfrac{1}{4a}x^2$.

For example, when $x^2 = -8y$, you have $4a = -8$; so you graph

$$Y_1 = -\dfrac{1}{8}x^2.$$

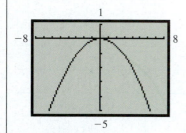

Because the focus is $(0, -2)$, substitute $y = -2$ in the equation $x^2 = -8y$ of the parabola to obtain

$$\begin{aligned} x^2 &= (-8)(-2) &&\text{Replace } y \text{ with } -2 \text{ in } x^2 = -8y. \\ x^2 &= 16 &&\text{Simplify.} \\ x &= \pm 4 &&\text{Solve for } x. \end{aligned}$$

Thus, the points $(4, -2)$ and $(-4, -2)$ are two symmetric points on the parabola to the right and left of the focus. The graph of this parabola is sketched in Figure 7.7.

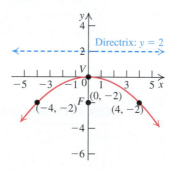

FIGURE 7.7 Graph of $x^2 = -8y$

b. The equation $y^2 = 5x$ has the standard form $y^2 = 4ax$.

$$\begin{aligned} 4a &= 5 &&\text{Equate coefficients of } x. \\ a &= \dfrac{5}{4} &&\text{Divide both sides by 4.} \end{aligned}$$

The parabola opens to the right. The vertex is the origin, and the focus is $\left(\dfrac{5}{4}, 0\right)$.

The directrix is the vertical line $x = -a = -\dfrac{5}{4}$; the axis of the parabola is the x-axis. To find two symmetric points on the parabola that are above and below the focus, substitute $x = \dfrac{5}{4}$ in the equation of the parabola.

$$\begin{aligned} y^2 &= 5x &&\text{Equation of the parabola} \\ y^2 &= 5\left(\dfrac{5}{4}\right) &&\text{Replace } x \text{ with } \dfrac{5}{4}. \\ y^2 &= \dfrac{25}{4} &&\text{Simplify.} \\ y &= \pm\dfrac{5}{2} &&\text{Solve for } y. \end{aligned}$$

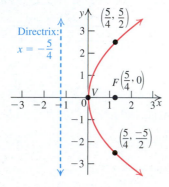

FIGURE 7.8 Graph of $y^2 = 5x$

Plot the two additional points $\left(\dfrac{5}{4}, \dfrac{5}{2}\right)$ and $\left(\dfrac{5}{4}, -\dfrac{5}{2}\right)$ on the parabola. The graph of $y^2 = 5x$ is sketched in Figure 7.8. ■ ■ ■

Practice Problem 1 Graph the parabola and specify the vertex, focus, directrix, and axis.

a. $x^2 = 12y$ **b.** $y^2 = -6x$ ■

The line segment joining the focus and two points symmetric with respect to the axis on a parabola is called the *latus rectum* of the parabola. Intuitively, the length of the latus rectum determines the "size of the opening" of the parabola.

To graph $y^2 = 4ax$, solve the equation for y to obtain

$$y = \pm\sqrt{4ax}.$$

Then graph $Y_1 = \sqrt{4ax}$ and $Y_2 = -\sqrt{4ax}$. For example, to graph $y^2 = 5x$, you graph

$$Y_1 = \sqrt{5x}, \quad Y_2 = -\sqrt{5x}.$$

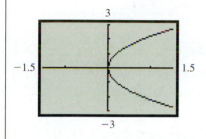

LATUS RECTUM

The line segment passing through the focus of a parabola, perpendicular to the axis, and having endpoints on the parabola is called the **latus rectum of the parabola**. Figure 7.9 shows that the length of the latus rectum for each of the graphs $y^2 = \pm 4ax$ and $x^2 = \pm 4ay$ for $a > 0$ is $4a$.

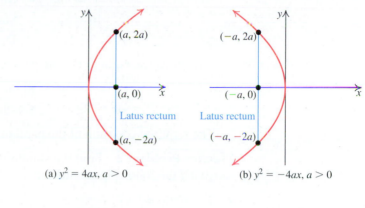

(a) $y^2 = 4ax$, $a > 0$ (b) $y^2 = -4ax$, $a > 0$

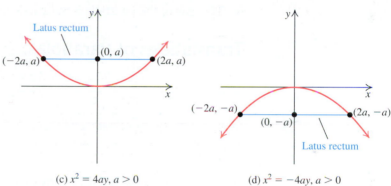

(c) $x^2 = 4ay$, $a > 0$ (d) $x^2 = -4ay$, $a > 0$

FIGURE 7.9 **The latus rectum**

EXAMPLE 2   **Finding the Equation of a Parabola**

Find the standard equation of a parabola with vertex $(0, 0)$ satisfying the given description.

a. The focus is $(-3, 0)$.

b. The axis of the parabola is the y-axis, and the graph passes through the point $(-4, 2)$.

SOLUTION

a. The vertex $(0, 0)$ of the parabola and its focus $(-3, 0)$ are on the x-axis; so the parabola opens to the left. Its equation must be of the form $y^2 = -4ax$ with $a = 3$.

$$
\begin{aligned}
y^2 &= -4ax &&\text{Form of the equation} \\
y^2 &= -4(3)x &&\text{Replace } a \text{ with 3.} \\
y^2 &= -12x &&\text{Simplify.}
\end{aligned}
$$

The equation of the parabola is $y^2 = -12x$.

b. Because the vertex of the parabola is $(0, 0)$, the axis is the y-axis, and the point $(-4, 2)$ is above the x-axis, the parabola opens up and the equation of the parabola is of the form

$$x^2 = 4ay.$$

This equation is satisfied by $x = -4$ and $y = 2$ because the parabola passes through the point $(-4, 2)$.

$$(-4)^2 = 4a(2) \quad \text{Replace } x \text{ with } -4 \text{ and } y \text{ with } 2.$$
$$16 = 8a \quad \text{Simplify.}$$
$$2 = a \quad \text{Divide both sides by 8.}$$

Therefore, the equation of the parabola is

$$x^2 = 4(2)y \quad \text{Replace } a \text{ with 2 in } x^2 = 4ay.$$
$$x^2 = 8y \quad \text{Simplify.}$$

The required equation of the parabola is $x^2 = 8y$. ■ ■ ■

Practice Problem 2 Find the standard equation of a parabola with vertex $(0, 0)$ satisfying the following conditions.

a. The focus is $(0, 2)$.

b. The axis of the parabola is the x-axis, and the graph passes through $(1, 2)$. ■

3 Translate a parabola.

Translations of Parabolas

The equation of a parabola with vertex $(0, 0)$ and axis on a coordinate axis is of the form $x^2 = \pm 4ay$ or $y^2 = \pm 4ax$. You can use translations of the graphs of these equations to find the equation of a parabola with vertex (h, k) and axis of symmetry parallel to a coordinate axis. We will obtain these translations by replacing x with $x - h$ (a horizontal shift) and y with $y - k$ (a vertical shift).

Main Facts About a Parabola with Vertex (h, k) and $a > 0$				
Standard Equation	$(y - k)^2 = 4a(x - h)$	$(y - k)^2 = -4a(x - h)$	$(x - h)^2 = 4a(y - k)$	$(x - h)^2 = -4a(y - k)$
Equation of axis	$y = k$	$y = k$	$x = h$	$x = h$
Description	Opens right	Opens left	Opens up	Opens down
Vertex	(h, k)	(h, k)	(h, k)	(h, k)
Focus	$(h + a, k)$	$(h - a, k)$	$(h, k + a)$	$(h, k - a)$
Directrix	$x = h - a$	$x = h + a$	$y = k - a$	$y = k + a$
Graph				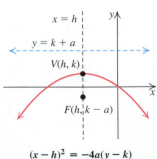

Expanding the standard equation of a parabola and collecting like terms results in an equation of the form

$$Ax^2 + Cy^2 + Dx + Ey + F = 0$$

with $A = 0$ or $C = 0$, but not both. Conversely, except in degenerate cases, any equation of this form can be put into one of the standard forms of a parabola by completing the squares on the x- or y-terms.

EXAMPLE 3   **Graphing a Parabola**

Find the vertex, focus, and directrix of the parabola $2y^2 - 8y - x + 7 = 0$. Sketch the graph of the parabola.

SOLUTION

First, complete the square on y. Then you can compare the resulting equation with one of the standard forms of the equation of a parabola.

$2y^2 - 8y - x + 7 = 0$	The given equation
$2y^2 - 8y = x - 7$	Isolate terms containing y.
$2(y^2 - 4y) = x - 7$	Factor out the common factor, 2.
$2(y^2 - 4y + 4) = x - 7 + 8$	Add $2\left(-\dfrac{4}{2}\right)^2 = 2(-2)^2 = 2 \cdot 4 = 8$ to both sides to complete the square.
$2(y - 2)^2 = x + 1$	$y^2 - 4y + 4 = (y - 2)^2$; simplify.
$(y - 2)^2 = \dfrac{1}{2}(x + 1)$	Divide both sides by 2.

The last equation is in one of the standard forms of the parabola. Comparing it with the form $(y - k)^2 = 4a(x - h)$, we have $h = -1$; $k = 2$; and $4a = \dfrac{1}{2}$, or $a = \dfrac{1}{8}$. The parabola opens to the right, the vertex is $(h, k) = (-1, 2)$, the focus is $(h + a, k) = \left(-1 + \dfrac{1}{8}, 2\right) = \left(-\dfrac{7}{8}, 2\right)$, the axis is the horizontal line $y = 2$, and the directrix is the vertical line $x = h - a = -1 - \dfrac{1}{8} = -\dfrac{9}{8}$. The graph is shown in Figure 7.10.   ■ ■ ■

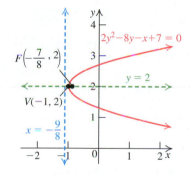

FIGURE 7.10 Converting
$2y^2 - 8y - x + 7 = 0$
to standard form

Practice Problem 3   Find the vertex, focus, and directrix of the parabola

$$2x^2 - 8x - y + 7 = 0.$$

Sketch the graph of the parabola.   ■

---

4 Use the reflecting property of parabolas.

Reflecting Property of Parabolas

A property of parabolas that is useful in applications is the **reflecting property**: if a reflecting surface has parabolic cross sections with a common focus, then all light rays entering the surface parallel to the axis will be reflected through the focus. See Figure 7.11(a) on the next page. This property is used in reflecting telescopes and satellite antennas because the light rays or radio waves bouncing off a parabolic surface are reflected to the focus, where they are collected and amplified.

Conversely, if a light source is located at the focus of a parabolic reflector, the reflected rays will form a beam parallel to the axis. See Figure 7.11(b). This principle is used in flashlights, searchlights, and other such devices.

(a)   (b)

FIGURE 7.11

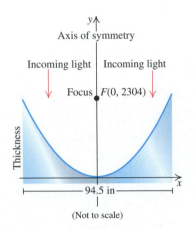

y

Axis of symmetry

Incoming light | Incoming light

Focus | $F(0, 2304)$

Thickness

94.5 in

(Not to scale)

FIGURE 7.12 Hubble Space Telescope

EXAMPLE 4 **Calculating Some Properties of the Hubble Space Telescope**

The parabolic mirror used in the Hubble Space Telescope has a diameter of 94.5 inches. See Figure 7.12. Find the equation of the parabola if its focus is 2304 inches from the vertex. What is the thickness of the mirror at the edges?

SOLUTION

Position the parabola so that its vertex is at the origin and its focus is on the positive y-axis. The equation of the parabola is of the form

$$x^2 = 4ay$$
$$x^2 = 4(2304)y \qquad a = 2304$$
$$x^2 = 9216y \qquad \text{Simplify.}$$

To find the thickness y of the mirror at the edge, substitute $x = 47.25$ (half the diameter) in the equation $x^2 = 9216y$ and solve for y.

$$(47.25)^2 = 9216y \qquad \text{Replace } x \text{ with } 47.25.$$

$$\frac{(47.25)^2}{9216} = y \qquad \text{Divide both sides by 9216.}$$

$$0.242248 \approx y \qquad \text{Use a calculator.}$$

Thus, the thickness of the mirror at the edges is approximately 0.242248 inch.

The Hubble telescope initially had "blurred vision" because the mirror was ground at the edges to a thickness of 0.24224 inch. In other words, it had been ground 0.000008 inch, or two-millionths of a meter, smaller than it should have been! The problem was fixed in 1993. ▪ ▪ ▪

Practice Problem 4 A small imitation of the Hubble's parabolic mirror has a diameter of 3 inches. Find the equation of the parabola assuming that its focus is 7.3 inches from the vertex. What is the thickness of the mirror at its edges, correct to four decimal places? ▪

SECTION 7.2 ▪ Exercises

A EXERCISES Basic Skills and Concepts

1. A parabola is the set of all points P in the plane that are equidistant from a fixed line called the _____ and a fixed point not on the line called the _____.

2. The line through the focus perpendicular to the directrix is called the _____ of the parabola.

3. The point at which the axis intersects the parabola is called the _____ of the parabola.

4. The graph of $y + 2 = 3(x - 4)^2$ is obtained by shifting the graph of $y = 3x^2$ to the right _____ units and shifting _____ two units.

5. *True or False* The vertex of a parabola is midway between its focus and its directrix.

6. *True or False* A parabola is symmetric about its axis.

In Exercises 7–14, find the focus and directrix of the parabola with the given equation. Then match each equation to one of the graphs labeled (a) through (h).

7. $x^2 = 2y$

8. $9x^2 = 4y$

9. $16x^2 = -9y$

10. $x^2 = -2y$

11. $y^2 = 2x$

12. $9y^2 = 16x$

13. $9y^2 = -16x$

14. $y^2 = -2x$

(a)

(b)

(c)

(d)

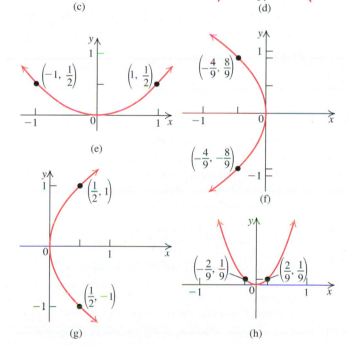

(e)

(f)

(g)

(h)

In Exercises 15–20, use a graphing device to graph each parabola.

15. $x^2 - 2y = 0$

16. $x^2 = 6y$

17. $y^2 = -3x$

18. $y^2 + 5x = 0$

19. $3x + 2y^2 = 0$

20. $y^2 - 5x = 0$

In Exercises 21–36, find the standard equation of the parabola that satisfies the given conditions. Also find the length of the latus rectum of each parabola.

21. Focus: $(0, 2)$; directrix: $y = 4$

22. Focus: $(0, 4)$; directrix: $y = -2$

23. Focus: $(-2, 0)$; directrix: $x = 3$

24. Focus: $(-1, 0)$; directrix: $x = -2$

25. Vertex: $(1, 1)$; directrix: $x = 3$

26. Vertex: $(1, 1)$ directrix: $y = 2$

27. Vertex: $(1, 1)$; directrix: $y = -3$

28. Vertex: $(1, 1)$; directrix: $x = -2$

29. Vertex: $(1, 0)$; focus: $(3, 0)$

30. Vertex: $(0, 1)$; focus: $(0, 2)$

31. Vertex: $(0, 1)$; focus: $(0, -2)$

32. Vertex: $(-1, 0)$; focus: $(-3, 0)$

33. Vertex: $(2, 3)$; directrix: $x = 4$

34. Vertex: $(2, 3)$; directrix: $y = 4$

35. Vertex: $(2, 3)$; directrix: $y = 1$

36. Vertex: $(2, 3)$; directrix: $x = 1$

In Exercises 37–40, use a graphing device to graph each parabola.

37. $(x + 1) = 4(y - 1)^2$

38. $(x - 2) = -4(y - 1)^2$

39. $(x - 1)^2 = -9(y + 1)$

40. $(x + 2)^2 = 6(y + 1)$

In Exercises 41–56, find the vertex, focus, and directrix of each parabola. Graph the equation.

41. $(y - 1)^2 = 2(x + 1)$

42. $(y + 2)^2 = 8(x - 3)$

43. $(x + 2)^2 = 3(y - 2)$

44. $(x - 3)^2 = 4(y + 1)$

45. $(y + 1)^2 = -6(x - 2)$

46. $(y - 2)^2 = -12(x + 3)$

47. $(x - 1)^2 = -10(y - 3)$

48. $(x + 2)^2 = -8(y + 3)$

49. $y = x^2 + 2x + 2$

50. $y = 3x^2 + 6x + 2$

51. $2y^2 + 4y - 2x + 1 = 0$

52. $3y^2 - 6y + x - 1 = 0$

53. $y + x^2 + x + \dfrac{5}{4} = 0$

54. $x = 8y^2 + 8y$

55. $x = 24y - 3y^2 - 2$

56. $y = 16x - 2x^2 + 3$

In Exercises 57–64, find two possible equations for the parabola with the given vertex V that passes through the given point P. The axis of symmetry of each parabola is parallel to a coordinate axis.

57. $V(0, 0), P(1, 2)$

58. $V(0, 0), P(-3, 2)$

59. $V(0, 1), P(2, 3)$

60. $V(1, 2), P(2, 1)$

61. $V(-2, 1), P(-3, 0)$

62. $V(1, -1), P(0, 0)$

63. $V(-1, 1), P(0, 2)$

64. $V(2, 3), P(3, -1)$

B EXERCISES Applying the Concepts

65. **Satellite dish.** A parabolic satellite dish is 40 inches across and 20 inches deep (from the vertex to the plane of the rim). How far from the vertex must the signal-receiving (receptor) unit be located to ensure that it is at the focus of the parabolic cross sections?

66. Repeat Exercise 65 assuming that the dish is 8 feet across and 3 feet deep.

67. **Solar heating.** A parabolic reflector mirror is to be used for the solar heating of water. The mirror is 18 feet across and 6 feet deep. Where should the heating element be placed to heat the water most rapidly?

68. **Flashlights.** A parabolic flashlight reflector is to be 4 inches across and 2 inches deep. Where should the lightbulb be placed?

69. **Flashlight.** The shape of a parabolic flashlight reflector can be described by the equation $x = 4y^2$. Where should the lightbulb be placed?

70. Repeat Exercise 69 assuming that the parabolic flashlight reflector has the shape described by the equation $x = \dfrac{1}{2}y^2$.

71. **Satellite dish.** The shape of a parabolic satellite dish can be described by the equation $y = 4x^2$. Where should the microphone be placed?

72. Repeat Exercise 71 assuming that the shape of the dish can be described by the equation $y = x^2$.

73. **Suspension bridge.** The ends of a suspension bridge cable are fastened to two towers that are 800 feet apart and 120 feet high. Assume that the cable hangs in the shape of a parabola and touches the roadbed midway between the towers. Find the length of the straight wire needed to suspend the roadbed from the cable at a distance of 250 feet from the tower.

74. **Suspension bridge.** Repeat Exercise 73 assuming that the suspension bridge cable is at a height of 8 feet above the midway between the towers.

75. **Kicking a football.** A football is kicked from the ground 70 yards along a parabolic path. The maximum height of

the ball is 30 yards. Find the height of the ball 5 yards before it hits the ground.

76. **Parabolic arch bridge.** A parabolic arch of a bridge has a 60-foot base and a height of 24 feet. Find the height of the arch at distances of 5, 10, and 20 feet from the center of the base.

77. **Variable cost.** The average variable cost y, in dollars, of a monthly output of x tons of a company producing a metal is given by $y = \dfrac{1}{10}x^2 - 3x + 50$. Find the output and cost at the vertex of the parabola.

78. Repeat Exercise 77 assuming that the variable cost is given by $y = \dfrac{1}{8}x^2 - 2x + 60$.

C EXERCISES Beyond the Basics

79. Find the coordinates of the points at which the line $2x - 3y + 16 = 0$ intersects the parabola $y^2 = 16x$.

80. Show that the line $y = 2x + 3$ intersects the parabola $y^2 = 24x$ at only one point. Find the point of intersection.

81. Find the equation of the directrix of a parabola with vertex $(-2, 2)$ and focus $(-6, 6)$.
[*Hint:* Note that the axis of this parabola is *not* parallel to a coordinate axis. Recall that the vertex is midway between the focus and the directrix.]

82. A parabola has focus $(-4, 5)$, and the equation of its directrix is $3x - 4y = 18$.
  a. Find the equation for the axis of the parabola.
  b. Find the point of intersection of the axis and the directrix.
  c. Use part (b) to find the vertex of this parabola.

83. Find an equation of two possible parabolas whose latus rectum is the line segment joining the points $(3, 5)$ and $(3, -3)$.

84. Repeat Exercise 83 assuming that the latus rectum is the line segment joining the points $(-5, 1)$ and $(3, 1)$.

85. Find an equation of the parabola with axis parallel to the y-axis, passing through the points $(0, 5)$, $(1, 4)$, and $(2, 7)$.

86. Find an equation of the parabola with axis parallel to the x-axis, passing through the points $(-1, -1)$, $(3, 1)$, and $(4, 0)$.

87. Find the vertex, focus, directrix, and axis of the parabola

$$x^2 - 8x + 2y + 4 = 0.$$

88. Repeat Exercise 87 for the parabola

$$Ax^2 + Dx + Ey + F = 0, \quad A \neq 0, E \neq 0.$$

89. Find the vertex, focus, directrix, and axis of the parabola

$$y^2 + 3x - 6y + 15 = 0.$$

90. Repeat Exercise 89 for the parabola

$$Cy^2 + Dx + Ey + F = 0, \quad C \neq 0, D \neq 0.$$

In Exercises 91–94, use the following definition: A *tangent line* to a parabola is a line that is not parallel to the axis of the parabola and that intersects the parabola at exactly one point.

91. Find an equation of the tangent line at the point $(1, 3)$ on the parabola $y = 3x^2$ by using the following steps.

Step 1 Let m be the slope of the tangent line. Then the equation of the tangent line is $y - 3 = m(x - 1)$. Solve this equation for y.

Step 2 Substitute the value of y from Step 1 in the equation $y = 3x^2$. You will obtain a quadratic equation in x.

Step 3 For the line in Step 1 to be the tangent line to the parabola, the roots of the quadratic equation in Step 2 must be equal. Set discriminant $= 0$ and solve for m.

Step 4 Use the value of m from Step 3 to write the equation of the tangent line to the parabola $y = 3x^2$.

92. Show that the equation of the tangent line at a point (x_1, y_1) on the parabola $y = 4ax^2$ is $y = 8ax_1x - y_1$. [*Hint:* Follow the steps in the previous exercise and note that $y_1 = 4ax_1^2$.]

93. Show that the equation of the tangent line at point (x_1, y_1) on the parabola $x = 4ay^2$ is $y = \dfrac{1}{8ay_1}x + \dfrac{y_1}{2}$.

94. **Reflection property of a parabola.** Let $P(x_1 \, y_1)$ be a point on the parabola $y = 4ax^2$ with focus $F = (0, a)$ and the directrix $y = -a$. Let AP be the tangent line to the parabola. (See the figure.)

a. Use Exercise 92 to show that $d(F, A) = a + y_1$. Then use the definition of the parabola to show that $d(P, F) = a + y_1$. Conclude that $\angle \beta = \angle \gamma$.

This will suggest the following construction of the tangent line: Let B be a point on the axis of the parabola such that BP is perpendicular to the axis. Let V be the vertex of the parabola. Find the point A on the axis of the parabola such that $d(B, V) = d(A, V)$. Then the line AP is the tangent line to the parabola at the point P.

b. Use geometry to show that the angle between the y-axis and the line AP equals the angle between the lines AP and the line $x = x_1$ (which is parallel to the axis of parabola).

This demonstrates the reflection property of a parabola, which says that the tangent line to a parabola at a point P makes equal angles with

1. the line through P and the focus.
2. the axis of the parabola.

Critical Thinking

95. a. Is a parabola always the graph of a function? Why or why not?

b. Find all possible values of the slope of the directrix of a parabola such that the parabola is always the graph of a function.

96. The equation $y^2 = 4kx$ describes a family of curves for each value of k. Sketch the graphs of the members of this family for $k = \pm 1, \pm 2, \pm 3$. What is the common characteristic of this family?

97. *True or False* The graph of $x = ay^2 + by + c \; (a \neq 0)$ is a parabola whose axis is parallel to the y-axis.

GROUP PROJECTS

98. It can be proved (see page 616) that the distance d from a point (x_1, y_1) to a line $ax + by + c = 0$ is given by

$$d = \frac{|ax_1 + by_1 + c|}{\sqrt{a^2 + b^2}}.$$

Use this fact and the definition of a parabola to find an equation of the parabola whose focus is $(4, 5)$ and whose directrix is the line $3x + 4y = 7$. Find an equation for the axis of this parabola.

99. Given that the vertex of a parabola is $(6, -3)$ and the directrix is the line $3x - 5y + 1 = 0$, find the coordinates of the focus.

The Ellipse

Lithotripter

Before Starting this Section, Review

1. Distance formula (Section 2.1, page 173)
2. Completing the square (Section 1.4, page 121)
3. Midpoint formula (Section 2.1, page 175)
4. Transformations of graphs (Section 2.7, page 262)
5. Symmetry (Section 2.2, page 183)

Objectives

1. Define an ellipse.
2. Find the equation of an ellipse.
3. Translate ellipses.
4. Use ellipses in applications.

WHAT IS LITHOTRIPSY?

Lithotripsy is a combination of two words: *litho* and *tripsy*. In Greek, the word *lith* denotes a stone and the word *tripsy* means "crushing." The complete medical term for lithotripsy is *extracorporeal shock-wave lithotripsy* (ESWL). Lithotripsy is a medical procedure in which a kidney stone is crushed into small sandlike pieces with the help of ultrasound high-energy shock waves, without breaking the skin.

The lithotripsy machine is called a *lithotripter*. The technology for the lithotripter was developed in Germany. The first successful treatment of a patient by a lithotripter was in February 1980 at Munich University. In the beginning, the patient had to lie in an elliptical tank filled with water and only small kidney stones were crushed. But as the medical research advanced, newer machines with better technology were manufactured. Now there is no need to keep the patient unconscious with an empty stomach, nor is the tub of water required.

In Example 5, you will see how the *reflecting property* of an ellipse is used in lithotripsy. ■

---

1 Define an ellipse.

Definition of Ellipse

ELLIPSE

An **ellipse** is the set of all points in the plane, the sum of whose distances from two fixed points is a constant. The fixed points are called the **foci** (the plural of *focus*) of the ellipse.

You can draw an ellipse with the use of thumbtacks, a fixed length of string, and a pencil. Stick a thumbtack at each of the two points (the foci) and tie one end of the string to each tack. Place a pencil inside the loop of the string and pull it taut. Move the pencil, keeping the string taut at all times. The pencil traces an ellipse, as shown in Figure 7.13.

The points of intersection of the ellipse with the line through the foci are called the **vertices** (plural of *vertex*). The line segment connecting the vertices is the **major axis** of the ellipse. The midpoint of the major axis is the **center** of the ellipse. The line segment that is perpendicular to the major axis at the center and with endpoints on the ellipse is called the **minor axis**. See Figure 7.14.

FIGURE 7.13 Drawing an ellipse

Points F_1 and F_2 are the foci.
Points V_1 and V_2 are the vertices.
Point C is the center.
Segment $\overline{V_1 V_2}$ is the major axis.
Segment $\overline{M_1 M_2}$ is the minor axis.
The sum of the distances
$PF_1 + PF_2$ from any point P on the
ellipse to the foci is constant.

FIGURE 7.14

2 Find the equation of an ellipse.

Equation of an Ellipse

To find a simple equation of an ellipse, we place the ellipse's major axis along one of the coordinate axes and the center of the ellipse at the origin. We will begin with the case in which the major axis lies along the x-axis. Let $F_1(-c, 0)$ and $F_2(c, 0)$ be the coordinates of the foci. See Figure 7.15. Note that the center at the origin is the midpoint of the line segment having the foci as endpoints. To simplify the equation, it is customary to let $2a$ be the constant distance referred to in the definition. Let $P(x, y)$ be a point on the ellipse. Then

$d(P, F_1) + d(P, F_2) = 2a$	Definition of ellipse
$\sqrt{(x + c)^2 + y^2} + \sqrt{(x - c)^2 + y^2} = 2a$	Distance formula
$\sqrt{(x + c)^2 + y^2} = 2a - \sqrt{(x - c)^2 + y^2}$	Isolate a radical.
$(x + c)^2 + y^2 = \left(2a - \sqrt{(x - c)^2 + y^2}\right)^2$	Square both sides.
$x^2 + 2cx + c^2 + y^2 = 4a^2 - 4a\sqrt{(x - c)^2 + y^2}$ $+ (x - c)^2 + y^2$	Expand squares.
$x^2 + 2cx + c^2 + y^2 = 4a^2 - 4a\sqrt{(x - c)^2 + y^2}$ $+ x^2 - 2cx + c^2 + y^2$	Expand squares.
$4cx - 4a^2 = -4a\sqrt{(x - c)^2 + y^2}$	Isolate the radical and simplify.
$cx - a^2 = -a\sqrt{(x - c)^2 + y^2}$	Divide both sides by 4.
$(cx - a^2)^2 = a^2[(x - c)^2 + y^2]$	Square both sides.
$c^2x^2 - 2a^2cx + a^4 = a^2(x^2 - 2cx + c^2 + y^2)$	Expand squares.
$(c^2 - a^2)x^2 - a^2y^2 = a^2c^2 - a^4$	Simplify and rearrange.
$(a^2 - c^2)x^2 + a^2y^2 = a^2(a^2 - c^2)$ (1)	Multiply both sides by -1 and factor $a^4 - a^2c^2$.

Now consider triangle F_1PF_2 in Figure 7.15. The *triangle inequality* from geometry states that the sum of the lengths of any two sides of a triangle is greater than the length of the third side; so for this triangle, we have

$d(P, F_1) + d(P, F_2) > d(F_1, F_2)$	
$2a > 2c$	$2a = d(P, F_1) + d(P, F_2); d(F_1, F_2) = 2c$
$a > c$	Divide both sides by 2.
$a^2 > c^2$	Square both sides.

Therefore, $a^2 - c^2 > 0$. Letting $b^2 = a^2 - c^2$, we obtain

$(a^2 - c^2)x^2 + a^2y^2 = a^2(a^2 - c^2)$	Equation (1)
$b^2x^2 + a^2y^2 = a^2b^2$	Replace $a^2 - c^2$ with b^2.
$\dfrac{x^2}{a^2} + \dfrac{y^2}{b^2} = 1$ (2)	Divide both sides by a^2b^2.

FIGURE 7.15

The last equation is the **standard form of the equation of an ellipse** with center $(0, 0)$ and foci $(-c, 0)$ and $(c, 0)$, where $b^2 = a^2 - c^2$.

To find the x-intercepts for the graph of this ellipse, set $y = 0$ in equation (2) to get $\dfrac{x^2}{a^2} = 1$. The solutions are $x = \pm a$, so the x-intercepts of the graph are $-a$ and a.

Therefore, the ellipse's vertices are $(-a, 0)$ and $(a, 0)$. The distance between the vertices $V_1(-a, 0)$ and $V_2(a, 0)$ is $2a$, so the length of the major axis is $2a$.

The y-intercepts of the ellipse (found by setting $x = 0$ in equation (2)) are $-b$ and b, so the endpoints of the minor axis are $(0, -b)$ and $(0, b)$. Consequently, the length of the minor axis is $2b$. (See the left-hand figure in the table below.)

Similarly, by reversing the roles of x and y, an equation of the ellipse with center $(0, 0)$ and foci $(0, -c)$ and $(0, c)$ on the y-axis is given by

$$\frac{x^2}{b^2} + \frac{y^2}{a^2} = 1, \text{where } b^2 = a^2 - c^2 \quad (3)$$

In equation (3), the major axis, of length $2a$, is along the y-axis and the minor axis, of length $2b$, is along the x-axis. (See the right-hand figure in the table below.)

If the major axis of an ellipse is along or parallel to the x-axis, the ellipse is called a **horizontal ellipse**, while an ellipse with major axis along or parallel to the y-axis is called a **vertical ellipse**.

We summarize the facts about horizontal and vertical ellipses with center $(0, 0)$. By the equation of a major or minor axis, we mean the equation of the line on which that axis lies.

Main Facts About an Ellipse with Center $(0, 0)$

Standard Equation	$\dfrac{x^2}{a^2} + \dfrac{y^2}{b^2} = 1; a > b > 0$ **(Horizontal ellipse)**	$\dfrac{x^2}{b^2} + \dfrac{y^2}{a^2} = 1; a > b > 0$ **(Vertical ellipse)**
Relationship between a, b, and c	$b^2 = a^2 - c^2$	$b^2 = a^2 - c^2$
Major axis along	x-axis	y-axis
Length of major axis	$2a$	$2a$
Minor axis along	y-axis	x-axis
Length of minor axis	$2b$	$2b$
Vertices	$(\pm a, 0)$	$(0, \pm a)$
Foci	$(\pm c, 0)$	$(0, \pm c)$
Endpoints of minor axis	$(0, \pm b)$	$(\pm b, 0)$
Symmetry	The graph is symmetric with respect to the x-axis, y-axis, and origin.	The graph is symmetric with respect to the x-axis, y-axis, and origin.
Graph	$$\frac{x^2}{a^2} + \frac{y^2}{b^2} = 1$$ **Horizontal ellipse**	$$\frac{x^2}{b^2} + \frac{y^2}{a^2} = 1$$ **Vertical ellipse**

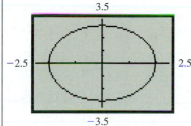 To graph the ellipse $\dfrac{x^2}{4} + \dfrac{y^2}{9} = 1$, first solve the equation for y:

$$\frac{y^2}{9} = 1 - \frac{x^2}{4}$$

or $y^2 = 9\left(1 - \dfrac{x^2}{4}\right)$

so that $y = \pm 3\sqrt{1 - \dfrac{x^2}{4}}$.

Then graph both functions

$$Y_1 = 3\sqrt{1 - \frac{x^2}{4}} \quad \text{and}$$

$$Y_2 = -3\sqrt{1 - \frac{x^2}{4}}.$$

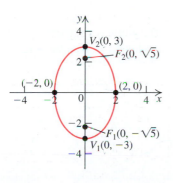

EXAMPLE 1   Finding an Equation of an Ellipse

Find the standard form of the equation of the ellipse that has vertex $(5, 0)$ and foci $(\pm 4, 0)$.

SOLUTION

Because the foci are $(-4, 0)$ and $(4, 0)$, the major axis is on the x-axis. We know that $c = 4$ and $a = 5$. Next find b^2:

$$b^2 = a^2 - c^2 \qquad \text{Relationship between } a, b, \text{ and } c$$
$$b^2 = (5)^2 - (4)^2 = 9 \qquad \text{Replace } c \text{ with 4 and } a \text{ with 5.}$$

Substituting 25 for a^2 and 9 for b^2 in the standard form for a horizontal ellipse, we get

$$\frac{x^2}{25} + \frac{y^2}{9} = 1. \qquad \blacksquare\,\blacksquare\,\blacksquare$$

Practice Problem 1 Find the standard form of the equation of the ellipse that has vertex $(0, 10)$ and foci $(0, -8)$ and $(0, 8)$. ■

EXAMPLE 2   Graphing an Ellipse

Sketch a graph of the ellipse whose equation is $9x^2 + 4y^2 = 36$. Find the foci of the ellipse.

SOLUTION

First, write the equation in standard form.

$$9x^2 + 4y^2 = 36 \qquad \text{Given equation}$$
$$\frac{9x^2}{36} + \frac{4y^2}{36} = 1 \qquad \text{Divide both sides by 36.}$$
$$\frac{x^2}{4} + \frac{y^2}{9} = 1 \qquad \text{Simplify.}$$

larger denominator

Because the denominator in the y^2-term is larger than the denominator in the x^2-term, the ellipse is a vertical ellipse. Here $a^2 = 9$ and $b^2 = 4$, so $c^2 = a^2 - b^2 = 9 - 4 = 5$. Thus, $a = 3$, $b = 2$, and $c = \sqrt{5}$. From the table on page 628, we know the following features of the vertical ellipse:

Vertices:	$(0, \pm a) = (0, \pm 3)$
Foci:	$(0, \pm c) = (0, \pm \sqrt{5})$
Length of major axis:	$2a = 2(3) = 6$
Length of minor axis:	$2b = 2(2) = 4$

The graph of the ellipse is shown in Figure 7.16. ■ ■ ■

Practice Problem 2 Sketch a graph of the ellipse whose equation is $4x^2 + y^2 = 16$. ■

FIGURE 7.16 Ellipse with equation $9x^2 + 4y^2 = 36$

3 Translate ellipses.

Translations of Ellipses

The graph of the equation

$$\frac{x^2}{a^2} + \frac{y^2}{b^2} = 1$$

is an ellipse with center $(0, 0)$ and major axis along a coordinate axis. As in the case of a parabola, horizontal and vertical shifts can be used to obtain the graph of an ellipse whose equation is

$$\frac{(x - h)^2}{a^2} + \frac{(y - k)^2}{b^2} = 1.$$

The center of such an ellipse is (h, k), and its major axis is parallel to a coordinate axis.

Main Facts About Horizontal and Vertical Ellipses with Center (h, k)		
Standard Equation	$\dfrac{(x - h)^2}{a^2} + \dfrac{(y - k)^2}{b^2} = 1;$ $\boldsymbol{a > b > 0}$ **(Horizontal ellipse)**	$\dfrac{(x - h)^2}{b^2} + \dfrac{(y - k)^2}{a^2} = 1;$ $\boldsymbol{a > b > 0}$ **(Vertical ellipse)**
Center	(h, k)	(h, k)
Major axis along the line	$y = k$	$x = h$
Length of major axis	$2a$	$2a$
Minor axis along the line	$x = h$	$y = k$
Length of minor axis	$2b$	$2b$
Vertices	$(h + a, k), (h - a, k)$	$(h, k + a), (h, k - a)$
Endpoints of minor axis	$(h, k - b), (h, k + b)$	$(h - b, k), (h + b, k),$
Foci	$(h + c, k), (h - c, k)$	$(h, k + c), (h, k - c)$
Equation involving a, b, and c	$c^2 = a^2 - b^2$	$c^2 = a^2 - b^2$
Symmetry	The graph is symmetric about the lines $x = h$ and $y = k$.	The graph is symmetric about the lines $x = h$ and $y = k$.
Graph		

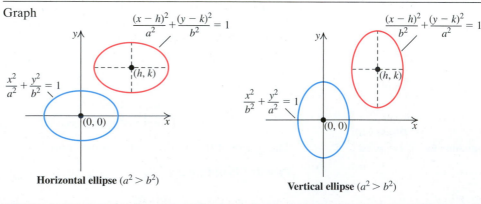

Horizontal ellipse $(a^2 > b^2)$ Vertical ellipse $(a^2 > b^2)$

The midpoint of the segment joining (x_1, y_1) and (x_2, y_2) is

$$\left(\frac{x_1 + x_2}{2}, \frac{y_1 + y_2}{2} \right).$$

Center (1, 2)

$F_1(-3, 2)$ $F_2(5, 2)$

FIGURE 7.17 Finding the center given the foci

FIGURE 7.18 Ellipse with foci $(-3, 2)$ and $(5, 2)$

| **EXAMPLE 3** | **Finding the Equation of an Ellipse** |

Find an equation of the ellipse that has foci $(-3, 2)$ and $(5, 2)$ and that has a major axis of length 10.

SOLUTION

Because the foci $(-3, 2)$ and $(5, 2)$ lie on the horizontal line $y = 2$, the ellipse is a horizontal ellipse. The center of the ellipse is the midpoint of the line segment joining the foci. Using the midpoint formula yields

$$h = \frac{-3 + 5}{2} = 1 \quad \text{and} \quad k = \frac{2 + 2}{2} = 2.$$

The center of the ellipse is $(1, 2)$. See Figure 7.17. Because the length of the major axis is 10, the vertices must be at a distance $a = 5$ units from the center.

In Figure 7.17, the foci are four units from the center. Thus, $c = 4$. Now find b^2:

$$b^2 = a^2 - c^2 = (5)^2 - (4)^2 = 9$$

Because the major axis is horizontal, the standard form of the equation of the ellipse is

$$\frac{(x - h)^2}{a^2} + \frac{(y - k)^2}{b^2} = 1 \qquad a > b > 0$$

$$\frac{(x - 1)^2}{25} + \frac{(y - 2)^2}{9} = 1 \qquad \begin{array}{l}\text{Replace } h \text{ with } 1, k \text{ with } 2,\\ a^2 \text{ with } 25, \text{ and } b^2 \text{ with } 9.\end{array}$$

For this horizontal ellipse with center $(1, 2)$, we have $a = 5$, $b = 3$, and $c = 4$.

Vertices:

$$(h \pm a, k) = (1 \pm 5, 2) = (-4, 2) \text{ and } (6, 2)$$

Endpoints of Minor Axis:

$$(h, k \pm b) = (1, 2 \pm 3) = (1, -1) \text{ and } (1, 5)$$

Using the center and these four points we sketch the graph shown in Figure 7.18. ■ ■ ■

Practice Problem 3   Find an equation of the ellipse that has foci at $(2, -3)$ and $(2, 5)$ and has a major axis of length 10. ■

Both standard forms of the equations of an ellipse with center (h, k) can be put into the form

$$Ax^2 + Cy^2 + Dx + Ey + F = 0 \qquad (4)$$

where neither A nor C is zero and both have the same sign. For example, consider the ellipse with equation

$$\frac{(x - 1)^2}{4} + \frac{(y + 2)^2}{9} = 1.$$

$$9(x - 1)^2 + 4(y + 2)^2 = 36 \qquad \text{Multiply both sides by 36.}$$
$$9(x^2 - 2x + 1) + 4(y^2 + 4y + 4) = 36 \qquad \text{Expand squares.}$$
$$9x^2 - 18x + 9 + 4y^2 + 16y + 16 = 36 \qquad \text{Distributive property}$$
$$9x^2 + 4y^2 - 18x + 16y - 11 = 0 \qquad \text{Simplify.}$$

Comparing this equation with equation (4), we have

$$A = 9, C = 4, D = -18, E = 16, \text{ and } F = -11.$$

Conversely, equation (4) can be put into a standard form for an ellipse by completing the squares in both of the variables x and y.

EXAMPLE 4   **Converting to Standard Form**

Find the center, vertices, and foci of the ellipse with equation

$$3x^2 + 4y^2 + 12x - 8y - 32 = 0.$$

SOLUTION
Complete squares on x and y.

$3x^2 + 4y^2 + 12x - 8y - 32 = 0$	Original equation
$(3x^2 + 12x) + (4y^2 - 8y) = 32$	Group like terms.
$3(x^2 + 4x) + 4(y^2 - 2y) = 32$	Factor out 3 and 4.
$3(x^2 + 4x + 4) + 4(y^2 - 2y + 1) = 32 + 12 + 4$	Complete squares on x and y: add $3 \cdot 4$ and $4 \cdot 1$ to each side.
$3(x + 2)^2 + 4(y - 1)^2 = 48$	Factor and simplify.
$\dfrac{(x + 2)^2}{16} + \dfrac{(y - 1)^2}{12} = 1$	Divide both sides by 48. (16 is the larger denominator.)

The last equation is the standard form for a horizontal ellipse with center $(-2, 1)$, $a^2 = 16$, $b^2 = 12$, and $c^2 = a^2 - b^2 = 16 - 12 = 4$. Thus, $a = 4$, $b = \sqrt{12} = 2\sqrt{3}$, and $c = 2$.

From the table on page 630, we have the following information:
The length of the major axis is $2a = 8$.
The length of the minor axis is $2b = 4\sqrt{3}$.

Center:	$(h, k) = (-2, 1)$
Foci:	$(h \pm c, k) = (-2 \pm 2, 1)$
	$= (-4, 1) \text{ and } (0, 1)$
Vertices:	$(h \pm a, k) = (-2 \pm 4, 1)$
	$= (-6, 1) \text{ and } (2, 1)$
Endpoints of minor axis:	$(h, k \pm b) = (-2, 1 \pm 2\sqrt{3})$
	$= (-2, 1 + 2\sqrt{3}) \text{ and } (-2, 1 - 2\sqrt{3})$
	$\approx (-2, 4.46) \text{ and } (-2, -2.46)$

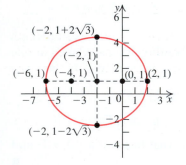

FIGURE 7.19 Ellipse with equation
$3x^2 + 4y^2 + 12x - 8y - 32 = 0$

The graph of the ellipse is shown in Figure 7.19. ■ ■ ■

Practice Problem 4   Find the center, vertices, and foci of the ellipse with equation

$$x^2 + 4y^2 - 6x + 8y - 29 = 0.$$ ■

A *circle* is considered an ellipse in which the two foci coincide at the center. Conversely, if the two foci of an ellipse coincide, then $2c = 0$ (remember, $2c$ is the distance between the two foci of an ellipse); so $c = 0$. Again, from the relationship $c^2 = a^2 - b^2 = 0$, we have $a = b$. Thus, the equation for the ellipse becomes the equation for a circle of radius $r = a$ having the same center as that of the ellipse.

4   Use ellipses in applications.

Applications

Ellipses have many applications. We mention just a few.

1. The orbits of the planets are ellipses with the sun at one focus. This fact alone is sufficient to explain the overwhelming importance of ellipses since the seventeenth century, when Kepler and Newton did their monumental work in astronomy.

2. Newton also reasoned that comets move in elliptical orbits about the sun. His friend Edmund Halley used this information to predict that a certain comet (now called Halley's comet) reappears about every 77 years. Halley saw the comet in 1682 and correctly predicted its return in 1759. The last sighting of the

Edmund Halley

(1656–1742)

Edmund Halley was a friend of Isaac Newton. Although Halley is known primarily for calculating the orbit of Halley's comet, he was also a biologist, a geologist, a sea captain, a spy, and the author of the first actuarial mortality tables.

FIGURE 7.20 Reflective property

comet was in 1986, and it is due to return in 2061. A recent study shows that Halley's comet has made approximately 2000 cycles so far and has about the same number to go before the sun erodes it away completely. (See Exercise 73.)

3. We can calculate the distance that a planet travels in one orbit around the sun.

4. **The reflecting property of ellipses.** The *reflecting property* for an ellipse says that a ray of light originating at one focus will be reflected to the other focus. See Figure 7.20. Sound waves also follow such paths. This property is used in the construction of "whispering galleries," such as the gallery at St. Paul's Cathedral in London. Such rooms have ceilings whose cross sections are elliptical with common foci. As a result, sounds emanating from one focus are reflected by the ceiling to the other focus. Thus, a whisper at one focus may not be audible at all at a nearby place, but may nevertheless be clearly heard far off at the other focus.

Example 5 illustrates how the *reflecting property* of an ellipse is used in lithotripsy. Recall from the introduction to this section that a *lithotripter* is a machine designed to crush kidney stones into small sandlike pieces with the help of high-energy shock waves. To focus these shock waves accurately, a lithotripter uses the reflective property of ellipses. A patient is carefully positioned so that the kidney stone is at one focus of the ellipsoid (a three-dimensional ellipse) and the shock-producing source is placed at the other focus. The high-energy shock waves generated at this focus are concentrated on the kidney stone at the other focus, thus pulverizing it without harming the patient.

EXAMPLE 5   Lithotripsy

An elliptical water tank has a major axis of length 6 feet and a minor axis of length 4 feet. The source of high-energy shock waves from a lithotripter is placed at one focus of the tank. To smash the kidney stone of a patient, how far should the stone be positioned from the source?

SOLUTION

Because the length of the major axis of the ellipse is 6 feet, we have $2a = 6$; so $a = 3$. Similarly, the minor axis of 4 feet gives $2b = 4$, or $b = 2$. To find c, we use the equation $c^2 = a^2 - b^2$. We have

$$c^2 = (3)^2 - (2)^2 = 5. \text{ Therefore, } c = \pm\sqrt{5}.$$

If we position the center of the ellipse at $(0, 0)$ and the major axis along the x-axis, then the foci of the ellipse are $(-\sqrt{5}, 0)$ and $(\sqrt{5}, 0)$. The distance between these foci is $2\sqrt{5} \approx 4.472$ feet. The kidney stone should be positioned 4.472 feet from the source of the shock waves.   ■ ■ ■

Practice Problem 5   In Example 5, if the tank has a major axis of 8 feet and a minor axis of 4 feet, how far should the stone be positioned from the source?   ■

SECTION 7.3 ■ Exercises

A EXERCISES Basic Skills and Concepts

1. An ellipse is the set of all points in the plane, the _____ of whose distances from two fixed points is a constant.

2. The points of intersection of the ellipse with the line through the foci are called _____ of the ellipse.

3. The standard equation of an ellipse with center $(0, 0)$, vertices $(\pm a, 0)$, and foci $(\pm c, 0)$ is $\dfrac{x^2}{a^2} + \dfrac{y^2}{b^2} = 1$, where $b^2 = $ _____.

4. In the equation $\dfrac{x^2}{m^2} + \dfrac{y^2}{n^2} = 1$, if **(i)** $m^2 > n^2$, the graph is a(n) _____ ellipse; **(ii)** if $m^2 < n^2$, the graph is a(n) _____ ellipse; **(iii)** if $m^2 = n^2$, the graph is a(n) _____.

5. *True or False* The ellipse $\dfrac{x^2}{a^2} + \dfrac{y^2}{b^2} = 1$ lies inside the box formed by the four lines $x = \pm a, y = \pm b$.

6. *True or False* The graph of $Ax^2 + Cy^2 + Dx + Ey + F = 0$ is a degenerate conic or an ellipse if $AC > 0$.

In Exercises 7–26, find the vertices and foci for each ellipse. Graph each equation.

7. $\dfrac{x^2}{16} + \dfrac{y^2}{4} = 1$

8. $\dfrac{x^2}{4} + \dfrac{y^2}{16} = 1$

9. $\dfrac{x^2}{9} + y^2 = 1$

10. $x^2 + \dfrac{y^2}{4} = 1$

11. $\dfrac{x^2}{25} + \dfrac{y^2}{16} = 1$

12. $\dfrac{x^2}{25} + \dfrac{y^2}{9} = 1$

13. $\dfrac{x^2}{16} + \dfrac{y^2}{36} = 1$

14. $\dfrac{x^2}{9} + \dfrac{y^2}{16} = 1$

15. $x^2 + y^2 = 4$

16. $x^2 + y^2 = 16$

17. $x^2 + 4y^2 = 4$

18. $9x^2 + y^2 = 9$

19. $9x^2 + 4y^2 = 36$

20. $4x^2 + 25y^2 = 100$

21. $3x^2 + 4y^2 = 12$

22. $4x^2 + 5y^2 = 20$

23. $5x^2 - 10 = -2y^2$

24. $3y^2 = 21 - 7x^2$

25. $2x^2 + 3y^2 = 7$

26. $3x^2 + 4y^2 = 11$

In Exercises 27–40, find the standard form of the equation for the ellipse satisfying the given conditions. Graph the equation.

27. Foci: $(\pm 1, 0)$; vertex $(3, 0)$

28. Foci: $(\pm 3, 0)$; vertex $(5, 0)$

29. Foci: $(0, \pm 2)$; vertex $(0, 4)$

30. Foci: $(0, \pm 3)$; vertex $(0, -6)$

31. Foci: $(\pm 4, 0)$; y-intercepts: ± 3

32. Foci: $(\pm 3, 0)$; y-intercepts: ± 4

33. Foci: $(0, \pm 2)$; x-intercepts: ± 4

34. Foci: $(0, \pm 3)$; x-intercepts: ± 5

35. Center: $(0, 0)$; vertical major axis of length 10; minor axis of length 6

36. Center: $(0, 0)$; vertical major axis of length 8; minor axis of length 4

37. Center: $(0, 0)$; vertices: $(\pm 6, 0)$; $c = 3$

38. Center: $(0, 0)$; vertices: $(0, \pm 5)$; $c = 3$

39. Center: $(0, 0)$; foci: $(0, \pm 2)$; $b = 3$

40. Center: $(0, 0)$; foci: $(\pm 3, 0)$; $b = 2$

In Exercises 41–56, find the center, foci, and vertices of each ellipse. Graph each equation.

41. $\dfrac{(x-1)^2}{4} + \dfrac{(y-1)^2}{9} = 1$

42. $\dfrac{(x-1)^2}{4} + \dfrac{(y-1)^2}{16} = 1$

43. $\dfrac{x^2}{16} + \dfrac{(y+3)^2}{4} = 1$

44. $\dfrac{(x+2)^2}{4} + \dfrac{y^2}{9} = 1$

45. $3(x-1)^2 + 4(y+2)^2 = 12$

46. $9(x+1)^2 + 4(y-2)^2 = 36$

47. $4(x+3)^2 + 5(y-1)^2 = 20$

48. $25(x-2)^2 + 4(y+5)^2 = 100$

49. $5x^2 + 9y^2 + 10x - 36y - 4 = 0$

50. $9x^2 + 4y^2 + 72x - 24y + 144 = 0$

51. $9x^2 + 5y^2 + 36x - 40y + 71 = 0$

52. $3x^2 + 4y^2 + 12x - 16y - 32 = 0$

53. $x^2 + 2y^2 - 2x + 4y + 1 = 0$

54. $2x^2 + 2y^2 - 12x - 8y + 3 = 0$

55. $2x^2 + 9y^2 - 4x + 18y + 12 = 0$

56. $3x^2 + 2y^2 + 12x - 4y + 15 = 0$

In Exercises 57–60, use a graphing utility to sketch the graph of each ellipse.

57. $\dfrac{x^2}{3} + \dfrac{y^2}{5} = 1$

58. $\dfrac{x^2}{6} + \dfrac{y^2}{4} = 1$

59. $x^2 + 3y^2 = 21$

60. $5x^2 + y^2 = 10$

B EXERCISES Applying the Concepts

61. **Semielliptical arch.** A portion of an arch in the shape of a *semiellipse* (half an ellipse) is 50 feet wide at the base and has a maximum height of 20 feet. Find the height of the arch at a distance of 10 feet from the center. (See the figure.)

62. In Exercise 61, find the distance between the two points at the base where the height of the arch is 10 feet.

63. **Semielliptical arch bridge.** A bridge is built in the shape of a semielliptical arch. The span of the bridge is 150 meters, and its maximum height is 45 meters. There are two vertical supports, each at a distance of 25 meters from the central position. Find their heights.

64. **Elliptical billiard table.** Some billiard tables (similar to pool tables, but without pockets) are manufactured in the shape of an ellipse. The foci of such tables are plainly marked for the convenience of the players. An elliptical billiard table is 6 feet long and 4 feet wide. Find the location of the foci.

65. "Ellipti-pool." In Exercise 64, suppose there is a pocket at one of the foci. A pool shark places the ball at the table and bets that he will be able to make the ball fall into the pocket 10 out of 10 times. Explain why you should or should not accept the bet.

66. Whispering gallery. The vertical cross section of the dome of Statuary Hall in Washington, D.C., is semi-elliptical. The hall is 96 feet long and 23 feet high. It is said that John C. Calhoun used the whispering-gallery phenomenon to eavesdrop on his adversaries. Where should Calhoun and his foes have been standing for the maximum whispering effect?

67. Whispering gallery. The vertical cross section of the dome of the Mormon Tabernacle in Salt Lake City, Utah, is semielliptical in shape. The cross section is 250 feet long with a maximum height of 80 feet. Where should two people stand to maximize the whispering effect?

In Exercises 68–71, use the fact (discovered by Johannes Kepler in 1609) that the planets move in elliptical orbits with the sun at one of the foci. Astronomers have measured the *perihelion* (the smallest distance from the planet to the sun) and *aphelion* (the largest distance from the planet to the sun) for each of the planets. The distances given in Table 7.1 are in millions of miles.

TABLE 7.1

Planet	Perihelion	Aphelion
Mercury	28.56	43.88
Venus	66.74	67.68
Earth	91.38	94.54
Mars	128.49	154.83
Jupiter	460.43	506.87
Saturn	837.05	936.37
Uranus	1699.45	1866.59
Neptune	2771.72	2816.42

68. Mars. Find an equation of Mars's orbit about the sun.
[*Hint:* Set up a coordinate system with the sun at one focus and the major axis lying on the x-axis. (See the figure.) Calculate a from the equation $2a =$ aphelion + perihelion. Calculate c from the equation $c = a -$ perihelion. Calculate b^2 from the equation $b^2 = a^2 - c^2$. Write the equation $\frac{x^2}{a^2} + \frac{y^2}{b^2} = 1$.]

69. Earth. Write an equation for the orbit of Earth about the sun.

70. Mercury. Write an equation for the orbit of the planet Mercury about the sun.

71. Saturn. Write an equation for the orbit of the planet Saturn about the sun.

72. Moon. The moon orbits Earth in an elliptical path with Earth at one focus. The major and minor axes of the orbit have lengths of 768,806 kilometers and 767,746 kilometers, respectively. Find the perihelion and aphelion of the orbit.

Moon's orbit

73. Halley's comet. In 1682, Edmund Halley (1656–1742) identified the orbit of a certain comet (now called Halley's comet) as an elliptical orbit about the sun, with the sun at one focus. He calculated the length of its major axis to be approximately 5.39×10^9 kilometers and its minor axis to be approximately 1.36×10^9 kilometers. Find the perihelion and aphelion of the orbit of Halley's comet.

74. Orbit of a satellite. Assume that Earth is a sphere with a radius of 3960 miles. A satellite has an elliptical orbit around Earth with the center of Earth as one of its foci. The maximum and minimum heights of the satellite from the surface of Earth are 164 miles and 110 miles, respectively. Find an equation for the orbit of the satellite.

C EXERCISES Beyond the Basics

75. Find an equation of the ellipse in standard form with center $(0, 0)$ and with major axis of length 10 passing through the point $\left(-3, \frac{16}{5}\right)$.

76. Find an equation of the ellipse in standard form with center $(0, 0)$ and with major axis of length 6 passing through the point $\left(1, \frac{3\sqrt{5}}{2}\right)$.

77. Find the lengths of the major and minor axes of the ellipse

$$\frac{x^2}{b^2} + \frac{y^2}{a^2} = 1,$$

given that the ellipse passes through the points $(2, 1)$ and $(1, -3)$.

78. Find an equation of the ellipse in standard form with center $(2, 1)$ passing through the points $(10, 1)$ and $(6, 2)$.

79. The **eccentricity** of an ellipse, denoted by e, is defined as
$$e = \frac{\text{distance between the foci}}{\text{distance between the vertices}} = \frac{2c}{2a} = \frac{c}{a}.$$ What happens when $e = 0$?

In Exercises 80–84, use the definition of e given in Exercise 79 to determine the eccentricity of each ellipse. (This e has nothing to do with the number e, discussed in Chapter 3, that is the base for natural logarithms.)

80. $\dfrac{x^2}{16} + \dfrac{y^2}{9} = 1$

81. $20x^2 + 36y^2 = 720$

82. $x^2 + 4y^2 = 1$

83. $\dfrac{(x + 1)^2}{25} + \dfrac{(y - 2)^2}{9} = 1$

84. $x^2 + 2y^2 - 2x + 4y + 1 = 0$

85. An ellipse has its major axis along the x-axis and its minor axis along the y-axis. Its eccentricity is $\dfrac{1}{2}$, and the distance between the foci is 4. Find an equation of the ellipse.

86. Find an equation of the ellipse whose major axis has end points $(-3, 0)$ and $(3, 0)$ and that has eccentricity $\dfrac{1}{3}$.

87. A point $P(x, y)$ moves so that its distance from the point $(4, 0)$ is always one-half its distance from the line $x = 16$. Show that the equation of the path of P is an ellipse of eccentricity $\dfrac{1}{2}$.

88. A point, $P(x, y)$ moves so that its distance from the point $(0, 2)$ is always one-third its distance from the line $y = 10$. Show that the equation of the path of P is an ellipse of eccentricity $\dfrac{1}{3}$.

89. The **latus rectum of an ellipse** is a line segment that passes through a focus and that is perpendicular to the major axis and has endpoints on the ellipse. Show that the length of the latus rectum of an ellipse is $\dfrac{2b^2}{a}$, where b is half the length of the minor axis and a is half the length of the major axis.

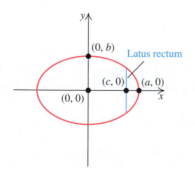

90. Find an equation of the ellipse with center $(0, 0)$, major axis along the x-axis, eccentricity $\dfrac{1}{\sqrt{2}}$, and latus rectum of length three units.

In Exercises 91–94, find all points of intersection of the given curves by solving systems of nonlinear equations. Sketch the graphs of the curves and show the points of intersection.

91. $\begin{cases} x + 3y = -2 \\ 4x^2 + 3y^2 = 7 \end{cases}$

92. $\begin{cases} x^2 = 2y \\ 2x^2 + y^2 = 12 \end{cases}$

93. $\begin{cases} x^2 + y^2 = 20 \\ 9x^2 + y^2 = 36 \end{cases}$

94. $\begin{cases} 9x^2 + 16y^2 = 36 \\ 18x^2 + 5y^2 = 45 \end{cases}$

In Exercises 95 and 96, use the following definition: A *tangent line* to an ellipse is a line that intersects the ellipse at exactly one point.

95. Find the equation of the tangent line to the ellipse with equation $\dfrac{x^2}{a^2} + \dfrac{y^2}{b^2} = 1$, with $a = \dfrac{5}{2}$ and $b = 5$ at the point $(2, 3)$.
[*Hint:* See the steps for Exercise 91 in Section 7.2 and (in Step 3) write the discriminant as a perfect square.]

96. Show that the equation of the tangent line to the ellipse with equation $\dfrac{x^2}{a^2} + \dfrac{y^2}{b^2} = 1$, at the point (x_1, y_1), is
$$y = -\frac{b^2 x_1}{a^2 y_1}x + \frac{b^2 x_1^2}{b^2 y_1^2} + y_1.$$
[*Hint:* See the steps for Exercise 91 in Section 7.2 and show that $(a^2 y_1 m + b^2 x_1)^2 = 0$.]

Critical Thinking

97. Find the number of circles with a radius of three units that touch both of the coordinate axes. Write the equation of each circle.

98. Find the number of ellipses with a major axis and a minor axis of lengths six units and four units, respectively, that touch both axes. Write the equation of each ellipse.

The Hyperbola

Before Starting this Section, Review

1. Distance formula (Section 2.1, page 173)
2. Midpoint formula (Section 2.1, page 175)
3. Completing the square (Section 1.4, page 121)
4. Oblique asymptotes (Section 3.6, page 369)
5. Transformations of graphs (Section 2.7, page 262)
6. Symmetry (Section 2.2, page 182)

Objectives

1. Define a hyperbola.
2. Find the asymptotes of a hyperbola.
3. Graph a hyperbola.
4. Translate hyperbolas.
5. Use hyperbolas in applications.

Alfred Lee Loomis (1887–1975)

Alfred Loomis was an American lawyer, investment banker, physicist, and patron of scientific research. Besides inventing LORAN, he made significant contributions to the ground-based technology for landing airplanes by instruments. President Roosevelt recognized the value of Loomis's work and described him as second perhaps only to Churchill as the civilian most responsible for the Allied victory in World War II.

WHAT IS LORAN?

LORAN stands for LOng-RAnge Navigation. The federal government runs the LORAN system, which uses land-based radio navigation transmitters to provide users with information on position and timing. During World War II, Alfred Lee Loomis was selected to chair the National Defense Research Committee. Much of his work focused on the problem of creating a light system for plane-carried radar. As a result of this effort, he invented LORAN, the long-range navigation system whose offshoot, LORAN-C, remains in widespread use. The navigational method provided by LORAN is based on the principle of determining the points of intersection of two hyperbolas to fix the two-dimensional position of the receiver. In Example 8, we illustrate how this is done.

The LORAN-C system is now being supplemented by the global positioning system (GPS), which works on the same principle as LORAN-C. The GPS system consists of 24 satellites that orbit 11,000 miles above Earth. The GPS was created by the U.S. Department of Defense to provide precise navigation information for targeting weapons systems anywhere in the world. The GPS is available to civilian users worldwide in a degraded mode. It is now used in aviation, navigation, and automobiles. The system can provide your exact position on Earth anywhere, anytime.

The GPS and LORAN-C work together, giving a highly accurate, duplicate navigation system for the Defense Department. ■

---

1. Define a hyperbola.

Definition of Hyperbola

HYPERBOLA

A **hyperbola** is the set of all points in the plane, the difference of whose distances from two fixed points is constant. The fixed points are called the **foci** of the hyperbola.

Figure 7.21 shows a hyperbola in *standard position*, with foci $F_1(-c, 0)$ and $F_2(c, 0)$ on the x-axis at equal distances from the origin. The two parts of the hyperbola are called **branches**.

FIGURE 7.21 $|PF_1 - PF_2|$ **is a constant for any P on the hyperbola**

As in the discussion of ellipses in Section 7.3 (see page 627), we take $2a$ as the positive constant referred to in the definition.

A point $P(x, y)$ lies on a hyperbola if and only if

$$|d(P, F_1) - d(P, F_2)| = 2a \qquad \text{Definition of hyperbola}$$
$$\sqrt{(x + c)^2 + y^2} - \sqrt{(x - c)^2 + y^2} = \pm 2a \qquad \text{Distance formula}$$

As in the case of an ellipse, we eliminate radicals and simplify (see Exercise 93) to obtain the equation.

$$(c^2 - a^2)x^2 - a^2y^2 = a^2(c^2 - a^2) \qquad (1)$$
$$b^2x^2 - a^2y^2 = a^2b^2 \qquad \text{Replace } (c^2 - a^2) \text{ with } b^2.$$
$$\frac{x^2}{a^2} - \frac{y^2}{b^2} = 1 \qquad (2) \qquad \text{Divide both sides by } a^2b^2.$$

Equation (2) is called the **standard form of the equation of a hyperbola** with center $(0, 0)$. The x-intercepts of the graph of equation (2) are $-a$ and a. The points corresponding to these x-intercepts are the **vertices** of the hyperbola. The distance between the vertices $V_1(-a, 0)$ and $V_2(a, 0)$ is $2a$. The line segment joining the two vertices is called the **transverse axis**. The midpoint of the transverse axis is the **center** of the hyperbola. The *center* is also the midpoint of the line segment joining the foci. See Figure 7.22. The line segment joining the points $(0, -b)$ and $(0, b)$ is called the **conjugate axis**.

Points F_1 and F_2 are foci.
Points V_1 and V_2 are vertices.
Point O is the center.
Segment $\overline{V_1V_2}$ is the transverse axis.
The difference of the distances $\left|PF_1 - PF_2\right|$ from any point P on the hyperbola is constant.

FIGURE 7.22 Parts of a hyperbola

Similarly, an equation of a hyperbola with center $(0, 0)$ and foci $(0, -c)$ and $(0, c)$ on the y-axis is given by:

$$\frac{y^2}{a^2} - \frac{x^2}{b^2} = 1, \text{ where } b^2 = c^2 - a^2 \qquad (3)$$

Here the vertices are $(0, -a)$ and $(0, a)$. The transverse axis of length $2a$ of the graph of equation (3) lies on the y-axis, and its conjugate axis is the segment joining the points $(-b, 0)$ and $(0, b)$ of length $2b$ that lies on the x-axis.

Main Facts About Hyperbolas Centered at (0, 0)

	$\dfrac{x^2}{a^2} - \dfrac{y^2}{b^2} = 1;\ a>0,\ b>0$	$\dfrac{y^2}{a^2} - \dfrac{x^2}{b^2} = 1;\ a>0,\ b>0$
Standard Equation		
Equation of transverse axis	$y = 0$ (x-axis)	$x = 0$ (y-axis)
Length of transverse axis	$2a$	$2a$
Equation of conjugate axis	$x = 0$ (y-axis)	$y = 0$ (x-axis)
Length of conjugate axis	$2b$	$2b$
Vertices	$(\pm a, 0)$	$(0, \pm a)$
Endpoints of conjugate axis	$(0, \pm b)$	$(\pm b, 0)$
Foci	$(\pm c, 0)$, where $c^2 = a^2 + b^2$	$(0, \pm c)$, where $c^2 = a^2 + b^2$
Description	Hyperbola has a *left branch* and a *right branch*. (Hyperbola opens left and right.)	Hyperbola has an *upper branch* and a *lower branch*. (Hyperbola opens up and down.)
Graph	Horizontal transverse axis	Vertical transverse axis

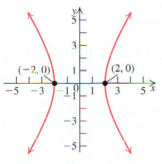

EXAMPLE 1   Determining the Orientation of a Hyperbola

Does the hyperbola

$$\frac{x^2}{4} - \frac{y^2}{10} = 1$$

have its transverse axis on the x-axis or y-axis?

SOLUTION

The orientation (left–right branches or up–down branches) of a hyperbola is determined *by noting where the minus sign occurs in the standard equation.* From the table above, we see that the transverse axis is on the x-axis when the minus sign precedes the y^2-term and is on the y-axis when the minus sign precedes the x^2-term. In the equation in this example, the minus sign precedes the y^2-term; so the transverse axis is on the x-axis. See Figure 7.23. ■ ■ ■

Practice Problem 1   Does the hyperbola $\dfrac{y^2}{8} - \dfrac{x^2}{5} = 1$ have its transverse axis on the x-axis or the y-axis?   ■

FIGURE 7.23 $\dfrac{x^2}{4} - \dfrac{y^2}{10} = 1$

Find the vertices and foci for the hyperbola

$$28y^2 - 36x^2 = 63.$$

SOLUTION

First, convert the equation to the standard form.

$$28y^2 - 36x^2 = 63 \qquad \text{Given equation}$$

$$\frac{28y^2}{63} - \frac{36x^2}{63} = 1 \qquad \text{Divide both sides by 63.}$$

$$\frac{4y^2}{9} - \frac{4x^2}{7} = 1 \qquad \text{Simplify.}$$

$$\frac{y^2}{\dfrac{9}{4}} - \frac{x^2}{\dfrac{7}{4}} = 1 \qquad \text{Divide the numerators and denominators of each term by 4.}$$

The last equation is the equation of a hyperbola in standard form with center $(0, 0)$. Because the coefficient of x^2 is negative, the transverse axis lies on the y-axis.

Here $a^2 = \dfrac{9}{4}$ and $b^2 = \dfrac{7}{4}$. The equation $c^2 = a^2 + b^2$ gives the value of c^2.

$$c^2 = a^2 + b^2$$

$$= \frac{9}{4} + \frac{7}{4} \qquad \text{Replace } a^2 \text{ with } \frac{9}{4} \text{ and } b^2 \text{ with } \frac{7}{4}.$$

$$= \frac{16}{4} = 4 \qquad \text{Simplify.}$$

Now $a^2 = \dfrac{9}{4}$ gives $a = \dfrac{3}{2}$ and $c^2 = 4$ gives $c = 2$ because we require $a > 0$ and $c > 0$.

The vertices of the hyperbola are $\left(0, -\dfrac{3}{2}\right)$ and $\left(0, \dfrac{3}{2}\right)$. The foci of the hyperbola are $(0, -2)$ and $(0, 2)$. ■ ■ ■

Practice Problem 2 Find the vertices and foci for the hyperbola

$$x^2 - 4y^2 = 8. \qquad ■$$

EXAMPLE 3 **Finding the Equation of a Hyperbola**

Find the standard form of the equation of a hyperbola with vertices $(\pm 4, 0)$ and foci $(\pm 5, 0)$.

SOLUTION

Because the foci of the hyperbola $(-5, 0)$ and $(5, 0)$ are on the x-axis, the transverse axis lies on the x-axis. The center of the hyperbola is midway between the foci, at $(0, 0)$. The standard form of such a hyperbola is

$$\frac{x^2}{a^2} - \frac{y^2}{b^2} = 1.$$

We need to find a^2 and b^2.

The distance a between the center $(0,0)$ to either vertex, $(-4,0)$ or $(4,0)$, is 4; so $a = 4$ and $a^2 = 16$. The distance c between the center $(0,0)$ to either focus, $(-5,0)$ or $(5,0)$, is 5; so $c = 5$ and $c^2 = 25$. Now use the equation $b^2 = c^2 - a^2$ to find $b^2 = c^2 - a^2 = 25 - 16 = 9$. Substitute $a^2 = 16$ and $b^2 = 9$ in $\dfrac{x^2}{a^2} - \dfrac{y^2}{b^2} = 1$ to get the standard form of the equation of the hyperbola

$$\frac{x^2}{16} - \frac{y^2}{9} = 1.$$

Practice Problem 3   Find the standard form of the equation of a hyperbola with vertices at $(0, \pm 4)$ and foci at $(0, \pm 6)$.

2 Find the asymptotes of a hyperbola.

The Asymptotes of a Hyperbola

Recall from Section 3.6 that a line $y = mx + b$ is a horizontal asymptote (if $m = 0$) or an oblique asymptote (if $m \neq 0$) of the graph of $y = f(x)$ if the distance from the line to the points on the graph of f approaches 0 as $x \to \infty$ or as $x \to -\infty$. When horizontal or oblique asymptotes exist, they provide us with information about the end behavior of the graph of f. This is especially helpful in graphing a hyperbola.

To find the asymptotes of the hyperbola with equation $\dfrac{x^2}{a^2} - \dfrac{y^2}{b^2} = 1$, we first solve this equation for y.

$\dfrac{x^2}{a^2} - \dfrac{y^2}{b^2} = 1$	Given equation of hyperbola
$-\dfrac{y^2}{b^2} = -\dfrac{x^2}{a^2} + 1$	Subtract $\dfrac{x^2}{a^2}$ from both sides.
$\dfrac{y^2}{b^2} = \dfrac{x^2}{a^2} - 1$	Multiply both sides by -1.
$y^2 = b^2\left(\dfrac{x^2}{a^2} - 1\right)$	Multiply both sides by b^2.
$y^2 = b^2\left(\dfrac{x^2}{a^2} - \dfrac{x^2}{a^2} \cdot \dfrac{a^2}{x^2}\right)$	Write $1 = \dfrac{x^2}{a^2} \cdot \dfrac{a^2}{x^2}$.
$y^2 = \dfrac{b^2 x^2}{a^2}\left(1 - \dfrac{a^2}{x^2}\right)$	Factor out $\dfrac{x^2}{a^2}$.
$y = \pm\dfrac{bx}{a}\sqrt{1 - \dfrac{a^2}{x^2}}$	Square root property

As $x \to \infty$ or as $x \to -\infty$, the quantity $\dfrac{a^2}{x^2}$ approaches 0. Thus, $\sqrt{1 - \dfrac{a^2}{x^2}}$ approaches 1 and the value of y approaches $\pm\dfrac{bx}{a}$. Therefore, the lines $y = \dfrac{b}{a}x$ and $y = -\dfrac{b}{a}x$ are the oblique asymptotes for the graph of the hyperbola $\dfrac{x^2}{a^2} - \dfrac{y^2}{b^2} = 1$.

STUDY TIP

A convenient device can be used to obtain equations of the asymptotes of a hyperbola. Substitute 0 for the 1 on the right side of the equation of the hyperbola and then solve for y in terms of x. For example, for the hyperbola $\dfrac{x^2}{a^2} - \dfrac{y^2}{b^2} = 1$, replace the 1 on the right side with 0 to obtain $\dfrac{x^2}{a^2} - \dfrac{y^2}{b^2} = 0$. Solve this equation for y.

$$\frac{y^2}{b^2} = \frac{x^2}{a^2}$$

$$y^2 = \frac{b^2}{a^2}x^2$$

The lines $y = \pm\dfrac{b}{a}x$ are the asymptotes of the hyperbola.

Similarly, you can show that the lines $y = \dfrac{a}{b}x$ and $y = -\dfrac{a}{b}x$ are the oblique asymptotes for the graph of the hyperbola $\dfrac{y^2}{a^2} - \dfrac{x^2}{b^2} = 1$.

THE ASYMPTOTES OF A HYPERBOLA WITH CENTER (0, 0)

1. The graph of the hyperbola $\dfrac{x^2}{a^2} - \dfrac{y^2}{b^2} = 1$ has transverse axis along the x-axis and has two asymptotes:

$$y = \frac{b}{a}x \quad \text{and} \quad y = -\frac{b}{a}x$$

2. The graph of the hyperbola $\dfrac{y^2}{a^2} - \dfrac{x^2}{b^2} = 1$ has transverse axis along the y-axis and has two asymptotes:

$$y = \frac{a}{b}x \quad \text{and} \quad y = -\frac{a}{b}x$$

EXAMPLE 4   **Finding the Asymptotes of a Hyperbola**

Determine the asymptotes of each hyperbola.

a. $\dfrac{x^2}{4} - \dfrac{y^2}{9} = 1$ **b.** $\dfrac{y^2}{9} - \dfrac{x^2}{16} = 1$

SOLUTION

a. The hyperbola $\dfrac{x^2}{4} - \dfrac{y^2}{9} = 1$ is of the form $\dfrac{x^2}{a^2} - \dfrac{y^2}{b^2} = 1$; so $a = 2$ and $b = 3$.

Substituting these values into $y = \dfrac{b}{a}x$ and $y = -\dfrac{b}{a}x$, we get the asymptotes

$$y = \frac{3}{2}x \quad \text{and} \quad y = -\frac{3}{2}x.$$

b. The hyperbola $\dfrac{y^2}{9} - \dfrac{x^2}{16} = 1$ has the form $\dfrac{y^2}{a^2} - \dfrac{x^2}{b^2} = 1$; so $a = 3$ and $b = 4$.

Substituting these values into $y = \dfrac{a}{b}x$ and $y = -\dfrac{a}{b}x$, we get the asymptotes

$$y = \frac{3}{4}x \quad \text{and} \quad y = -\frac{3}{4}x.$$

The hyperbola of parts **a** and **b** and their asymptotes are shown in Figures 7.24(a) and 7.24(b), respectively.   ■ ■ ■

Practice Problem 4   Determine the asymptotes of the hyperbola $\dfrac{y^2}{4} - \dfrac{x^2}{9} = 1$. ■

(a)

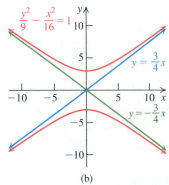

(b)

FIGURE 7.24 Asymptotes

3 Graph a hyperbola.

Graphing a Hyperbola with Center (0, 0)

Consider the hyperbola with equation

$$\frac{x^2}{a^2} - \frac{y^2}{b^2} = 1.$$

The vertices of this hyperbola are $(-a, 0)$ and $(a, 0)$. The endpoints of the conjugate axis are $(0, -b)$ and $(0, b)$. The rectangle with vertices (a, b), $(-a, b)$, $(-a, -b)$, and $(a, -b)$ is called the **fundamental rectangle** of the hyperbola. See Figure 7.25 on page 644. The diagonals of the fundamental rectangle have slopes $\frac{b}{a}$ and $-\frac{b}{a}$. Thus, the extensions of these diagonals are the asymptotes of the hyperbola.

FINDING THE SOLUTION: A PROCEDURE

EXAMPLE 5   **Graphing a Hyperbola Centered at (0,0)**

OBJECTIVE	EXAMPLE

OBJECTIVE

Sketch the graph of a hyperbola in the form

(i) $\dfrac{x^2}{a^2} - \dfrac{y^2}{b^2} = 1$ *or* **(ii)** $\dfrac{y^2}{a^2} - \dfrac{x^2}{b^2} = 1.$

EXAMPLE

Sketch the graph of

a. $16x^2 - 9y^2 = 144.$ **b.** $25y^2 - 4x^2 = 100.$

Step 1 **Write the equation in standard form.** Determine transverse axis and the orientation of the hyperbola.

$$16x^2 - 9y^2 = 144$$
$$\frac{16x^2}{144} - \frac{9y^2}{144} = \frac{144}{144}$$
$$\frac{x^2}{9} - \frac{y^2}{16} = 1$$

Minus sign precedes y^2-term, and transverse axis is along the x-axis; hyperbola opens left and right.

$$25y^2 - 4x^2 = 100$$
$$\frac{25y^2}{100} - \frac{4x^2}{100} = \frac{100}{100}$$
$$\frac{y^2}{4} - \frac{x^2}{25} = 1$$

Minus sign precedes x^2-term, and transverse axis is along the y-axis; hyperbola opens up and down.

Step 2 **Locate vertices** and the endpoints of the conjugate axis.

Because $a^2 = 9$, $a = 3$. Also, $b^2 = 16$; so $b = 4$. The vertices are $(\pm 3, 0)$, and the endpoints of the conjugate axis are $(0, \pm 4)$.

Because $a^2 = 4$, $a = 2$. Also, $b^2 = 25$; so $b = 5$. The vertices are $(0, \pm 2)$, and the endpoints of the conjugate axis are $(\pm 5, 0)$.

Step 3 Lightly sketch the fundamental rectangle by drawing dashed lines parallel to the coordinate axes through the points in Step 2.

Draw dashed lines $x = 3$, $x = -3, y = 4$, and $y = -4$ to form the fundamental rectangle with vertices $(3, 4)$, $(-3, 4)$, $(-3, -4)$, and $(3, -4)$.

Draw dashed lines $y = 2$, $y = -2, x = 5$, and $x = -5$ to form the fundamental rectangle with vertices $(5, 2)$, $(-5, 2)$, $(-5, -2)$, and $(5, -2)$.

continued on the next page

Step 4 Sketch the asymptotes. Extend the diagonals of the fundamental rectangle. These are the asymptotes.

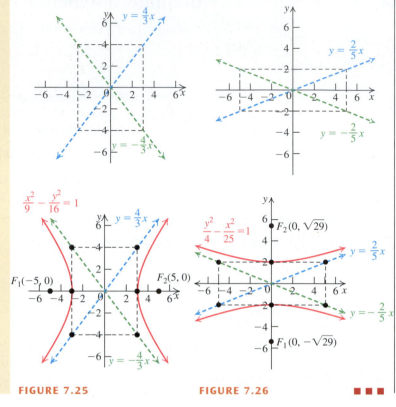

Step 5 Sketch the graph. Draw both branches of the hyperbola through the vertices, approaching the asymptotes. See Figures 7.25 and 7.26. The foci are located on the transverse axis, c units from the center, where $c^2 = a^2 + b^2$.

FIGURE 7.25   **FIGURE 7.26**   ■ ■ ■

Practice Problem 5   Sketch the graph of the equations.

a. $25x^2 - 4y^2 = 100$ **b.** $9y^2 - x^2 = 1$

4 Translate hyperbolas.

Translations of Hyperbolas

We can use horizontal and vertical shifts to find the standard form of the equations of hyperbolas centered at (h, k).

Main Properties of Hyperbolas Centered at (h, k)		
Standard Equation	$\dfrac{(x - h)^2}{a^2} - \dfrac{(y - k)^2}{b^2} = 1;$ $a > 0, b > 0$	$\dfrac{(y - k)^2}{a^2} - \dfrac{(x - h)^2}{b^2} = 1;$ $a > 0, b > 0$
Equation of transverse axis	$y = k$	$x = h$
Length of transverse axis	$2a$	$2a$
Equation of conjugate axis	$x = h$	$y = k$
Length of conjugate axis	$2b$	$2b$
Center	(h, k)	(h, k)
Vertices	$(h - a, k)$ and $(h + a, k)$	$(h, k - a)$ and $(h, k + a)$
Endpoints of conjugate axis	$(h, k - b)$ and $(h, k + b)$	$(h - b, k)$ and $(h + b, k)$
Foci	$(h - c, k)$ and $(h + c, k)$; $c^2 = a^2 + b^2$	$(h, k - c)$ and $(h, k + c)$; $c^2 = a^2 + b^2$
Asymptotes	$y - k = \pm\dfrac{b}{a}(x - h)$	$y - k = \pm\dfrac{a}{b}(x - h)$

We now describe a procedure for sketching the graph of a hyperbola centered at (h, k).

FINDING THE SOLUTION: A PROCEDURE

EXAMPLE 6   **Graphing a Hyperbola Centered at (h, k)**

OBJECTIVE

Sketch the graph of either

$$\frac{(x - h)^2}{a^2} - \frac{(y - k)^2}{b^2} = 1 \text{ or}$$

$$\frac{(y - k)^2}{a^2} - \frac{(x - h)^2}{b^2} = 1.$$

EXAMPLE

Sketch the graph of each equation.

a. $\dfrac{(x - 1)^2}{4} - \dfrac{(y + 2)^2}{9} = 1.$ **b.** $\dfrac{(y + 2)^2}{9} - \dfrac{(x - 1)^2}{4} = 1.$

Step 1 Plot the center (h, k) and draw horizontal and vertical dashed lines through the center.

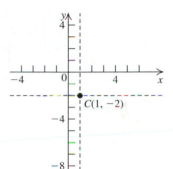

Step 2 Locate the vertices and the endpoints of the conjugate axis. Lightly sketch the fundamental rectangle, with sides parallel to the coordinate axes, through these points.

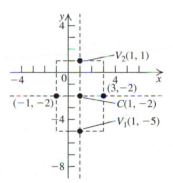

Step 3 Sketch dashed lines through opposite vertices of the fundamental rectangle. These are the asymptotes.

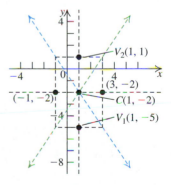

Step 4 Draw both branches of the hyperbola through the vertices, approaching the asymptotes.

continued on the next page

Step 5 The foci are located on the transverse axis, c units from the center, where $c^2 = a^2 + b^2$.

Practice Problem 6 Sketch the graph of each equation.

a. $\dfrac{(x + 1)^2}{4} - \dfrac{(y - 1)^2}{16} = 1$ **b.** $\dfrac{(y - 1)^2}{4} - \dfrac{(x + 1)^2}{9} = 1$ ■

Expanding the binomials in the equation of a hyperbola in standard form and collecting like terms results in an equation of the form

$$Ax^2 + Cy^2 + Dx + Ey + F = 0,$$

with A and C of opposite sign (that is, $AC < 0$.) Conversely, except in degenerate cases, any equation of this form can be put into one of the standard forms of a hyperbola by completing the squares on the x- and y-terms.

EXAMPLE 7   **Graphing a Hyperbola**

Show that $9x^2 - 16y^2 + 18x + 64y - 199 = 0$ is an equation of a hyperbola; then graph the hyperbola.

SOLUTION
Because we plan on completing squares, first group the x- and y-terms on the left side and the constants on the right side.

$9x^2 - 16y^2 + 18x + 64y - 199 = 0$	Given equation
$(9x^2 + 18x) + (-16y^2 + 64y) = 199$	Group terms.
$9(x^2 + 2x) - 16(y^2 - 4y) = 199$	Factor out 9 and -16.
$9(x^2 + 2x + 1) - 16(y^2 - 4y + 4) = 199 + 9 - 64$	Complete the squares: add $9 \cdot 1$ and $-16 \cdot 4$.
$9(x + 1)^2 - 16(y - 2)^2 = 144$	Factor and simplify.
$\dfrac{9(x + 1)^2}{144} - \dfrac{16(y - 2)^2}{144} = 1$	Divide both sides by 144 to obtain 1 on the right side.
$\dfrac{(x + 1)^2}{16} - \dfrac{(y - 2)^2}{9} = 1$	Simplify.

The last equation is the standard form of the equation of a hyperbola with center $(-1, 2)$.

Now we sketch the graph of this hyperbola.

Steps 1–2   **Locate the vertices.** For this hyperbola with center $(-1, 2)$, $a^2 = 16$ and $b^2 = 9$. Therefore, $a = 4$ and $b = 3$. From the table on page 644, with $h = -1$ and $k = 2$, the vertices are $(h - a, k) = (-1 - 4, 2) = (-5, 2)$ and $(h + a, k) = (-1 + 4, 2) = (3, 2)$.

Draw the fundamental rectangle. The vertices of the fundamental rectangle are $(3, -1)$, $(3, 5)$, $(-5, 5)$, and $(-5, -1)$.

Step 3   **Sketch the asymptotes.** Extend the diagonals of the fundamental rectangle to sketch the asymptotes.

$$y - 2 = \frac{3}{4}(x + 1) \quad \text{and} \quad y - 2 = -\frac{3}{4}(x + 1)$$

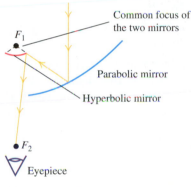

FIGURE 7.27 **The hyperbola** $9x^2 - 16y^2 + 18x + 64y - 199 = 0$

Step 4   **Sketch the graph.** Draw two branches of the hyperbola opening to the left and right, starting from the vertices $(-5, 2)$ and $(3, 2)$ and approaching the asymptotes. See Figure 7.27.   ■ ■ ■

Practice Problem 7   Show that $x^2 - 4y^2 - 2x + 16y - 20 = 0$ is an equation of a hyperbola and graph the hyperbola.   ■

5 Use hyperbolas in applications.

Applications

Hyperbolas have many applications. We list a few of them here.

1. Comets that do not move in elliptical orbits around the sun almost always move in a hyperbolic orbit. (In theory, they also can move in parabolic orbits.)

2. Boyle's Law states that if a perfect gas is kept at a constant temperature, then its pressure P and volume V are related by the equations $PV = c$, where c is a constant. The graph of this equation is a hyperbola. In Exercise 94(b), you are asked to explore the case where $c = 8$. In this case, the transverse axis is not parallel to a coordinate axis.

3. The hyperbola has the reflecting property that a ray of light from a source at one focus of a hyperbolic mirror (a mirror with hyperbolic cross sections) is reflected along the line through the other focus.

   The reflecting properties of the parabola and hyperbola are combined into one design for a reflecting telescope. See Figure 7.28. The parallel rays from a star are finally focused at the eyepiece at F_2.

4. The definition of a hyperbola forms the basis of several important navigational systems.

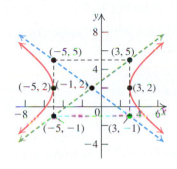

FIGURE 7.28

How Does LORAN Work?

Suppose two stations A and B (several miles apart) transmit synchronized radio signals. The LORAN receiver in your ship measures the difference in reception times of these synchronized signals. The radio signals travel at a speed of 186,000 miles per second. Using this information, you can determine the difference $2a$ in the distance of your ship's receiver from the two transmitters. By the definition of a hyperbola, this information places your ship somewhere on a hyperbola with foci A and B. With two pairs of transmitters, the position of your ship can be determined at a point of intersection of the two hyperbolas. (See Exercises 87 and 88.)

EXAMPLE 8	**Using LORAN**

LORAN navigational transmitters A and B are located at $(-130, 0)$ and $(130, 0)$, respectively. A receiver P on a fishing boat somewhere in the first quadrant listens to the pair (A, B) of transmissions and computes the difference of the distances from the boat to A and B as 240 miles. Find the equation of the hyperbola on which P is located.

SOLUTION

With reference to the transmitters A and B, P is located on a hyperbola with $2a = 240$, or $a = 120$, and the hyperbola has foci $(-130, 0)$ and $(130, 0)$.

$$\frac{x^2}{a^2} - \frac{y^2}{b^2} = 1 \qquad \text{Standard from of the equation of a hyperbola}$$

$$a^2 = (120)^2 = 14{,}400 \qquad \text{Replace } a \text{ with } 120.$$

$$b^2 = c^2 - a^2 = (130)^2 - (120)^2 = 2500 \qquad \text{Replace } c \text{ with } 130 \text{ and } a \text{ with } 120.$$

$$\frac{x^2}{14{,}400} - \frac{y^2}{2500} = 1 \qquad \text{Replace } a^2 \text{ with } 14{,}400 \text{ and } b^2 \text{ with } 2500.$$

Therefore, the location of the ship lies on a hyperbola (see Figure 7.29) with equation

$$\frac{x^2}{14{,}400} - \frac{y^2}{2500} = 1.$$

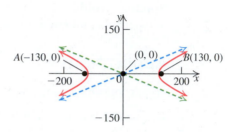

FIGURE 7.29

Practice Problem 8 In Example 8, the transmitters A and B are located at $(-150, 0)$ and $(150, 0)$, respectively, and the difference of the distances from the boat to A and B is 260 miles. Find the equation of the hyperbola on which P is located.

SECTION 7.4 ▪ Exercises

A EXERCISES Basic Skills and Concepts

1. A hyperbola is a set of all points in the plane, the _____ of whose distances from two fixed points is constant.

2. The line segment joining the two vertices of a hyperbola is called the _____ axis.

3. The standard equation of the hyperbola with center $(0, 0)$, vertices $(\pm a, 0)$, and foci $(\pm c, 0)$ is

$$\frac{x^2}{a^2} - \frac{y^2}{b^2} = 1, \text{ where } b^2 = \text{_____}.$$

4. For the hyperbola $\dfrac{y^2}{a^2} - \dfrac{x^2}{b^2} = 1$, the vertices are _____ and _____ and the foci are $(0, \pm c)$, where $c^2 = $ _____.

5. *True or False* The graph of $\dfrac{x^2}{a^2} - \dfrac{y^2}{b^2} = -1$ is a hyperbola.

6. *True or False* The graph of $Ax^2 + Cy^2 + Dx + Ey + F = 0$ is a degenerate conic or a hyperbola if $AC < 0$.

In Exercises 7–14, the equation of a hyperbola is given. Match each equation with its graph shown in (a)–(h).

7. $\dfrac{x^2}{9} - \dfrac{y^2}{4} = 1$

8. $\dfrac{x^2}{4} - \dfrac{y^2}{25} = 1$

9. $\dfrac{y^2}{25} - \dfrac{x^2}{4} = 1$

10. $\dfrac{y^2}{4} - \dfrac{x^2}{9} = 1$

11. $4x^2 - 25y^2 = 100$

12. $16x^2 - 49y^2 = 196$

13. $16y^2 - 9x^2 = 100$

14. $9y^2 - 16x^2 = 144$

(a)

(b)

(c)   (d)

(e)   (f)

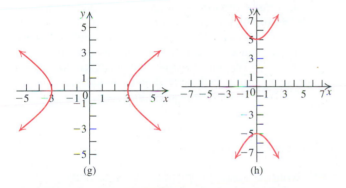

(g)   (h)

In Exercises 15–26, the equation of a hyperbola is given.
  a. Find and plot the vertices, the foci, and the transverse axis of the hyperbola.
  b. State how the hyperbola opens.
  c. Find and plot the vertices of the fundamental rectangle.
  d. Sketch the asymptotes and write their equations.
  e. Graph the hyperbola by using the vertices and the asymptotes.

15. $x^2 - \dfrac{y^2}{4} = 1$ **16.** $y^2 - \dfrac{x^2}{4} = 1$

17. $x^2 - y^2 = -1$ **18.** $y^2 - x^2 = -1$

19. $9y^2 - x^2 = 36$ **20.** $9x^2 - y^2 = 36$

21. $4x^2 - 9y^2 - 36 = 0$ **22.** $4x^2 - 9y^2 + 36 = 0$

23. $y = \pm\sqrt{4x^2 + 1}$ **24.** $y = \pm 2\sqrt{x^2 + 1}$

25. $y = \pm\sqrt{9x^2 - 1}$ **26.** $y = \pm 3\sqrt{x^2 - 4}$

In Exercises 27–40, find an equation of the hyperbola satisfying the given conditions. Graph the hyperbola.

27. Vertices: $(\pm 2, 0)$; foci: $(\pm 3, 0)$

28. Vertices: $(\pm 3, 0)$; foci: $(\pm 5, 0)$

29. Vertices: $(0, \pm 4)$; foci: $(0, \pm 6)$

30. Vertices: $(0, \pm 5)$; foci: $(0, \pm 8)$

31. Center: $(0, 0)$; vertex: $(0, 2)$; focus: $(0, 5)$

32. Center: $(0, 0)$; vertex: $(0, -1)$; focus: $(0, -4)$

33. Center: $(0, 0)$; vertex: $(1, 0)$; focus: $(-5, 0)$

34. Center: $(0, 0)$; vertex: $(-3, 0)$; focus: $(6, 0)$

35. Foci: $(0, \pm 5)$; the length of the transverse axis is 6.

36. Foci: $(\pm 2, 0)$; the length of the transverse axis is 2.

37. Foci: $(\pm\sqrt{5}, 0)$; asymptotes: $y = \pm 2x$

38. Foci: $(0, \pm 10)$; asymptotes: $y = \pm 3x$

39. Vertices: $(0, \pm 4)$; asymptotes: $y = \pm x$

40. Vertices: $(\pm 3, 0)$; asymptotes: $y = \pm 2x$

In Exercises 41–48, use a graphing device to graph each hyperbola.

41. $\dfrac{x^2}{2} - \dfrac{y^2}{4} = 1$ **42.** $\dfrac{x^2}{6} - \dfrac{y^2}{2} = 1$

43. $2y^2 - x^2 = 6$ **44.** $y^2 - 3x^2 = 12$

45. $(x + 1)^2 - 2y^2 = 4$ **46.** $3x^2 - (y - 2)^2 = 6$

47. $2x^2 - y^2 - 4x - 2y - 7 = 0$

48. $4y^2 - 3x^2 - 16y - 6x + 1 = 0$

In Exercises 49–68, the equation of a hyperbola is given.
  a. Find and plot the center, vertices, transverse axis, and asymptotes of the hyperbola.
  b. Use the vertices and the asymptotes to graph the hyperbola.

49. $\dfrac{(x - 1)^2}{9} - \dfrac{(y + 1)^2}{16} = 1$

50. $\dfrac{(y - 1)^2}{16} - \dfrac{(x + 1)^2}{4} = 1$

51. $\dfrac{(x + 2)^2}{25} - \dfrac{y^2}{49} = 1$

52. $\dfrac{(x + 1)^2}{9} - \dfrac{(y + 2)^2}{36} = 1$

53. $\dfrac{(x + 4)^2}{25} - \dfrac{(y + 3)^2}{49} = 1$

54. $\dfrac{(x - 3)^2}{9} - \dfrac{(y + 1)^2}{9} = 1$

55. $4x^2 - (y + 1)^2 = 25$ **56.** $9(x - 1)^2 - y^2 = 144$

57. $(y + 1)^2 - 9(x - 2)^2 = 25$

58. $6(x - 4)^2 - 3(y + 3)^2 = 4$

59. $x^2 - y^2 + 6x = 36$ **60.** $x^2 + 4x - 4y^2 = 12$

61. $x^2 - 4y^2 - 4x = 0$ **62.** $x^2 - 2y^2 - 8y = 12$

63. $2x^2 - y^2 + 12x - 8y + 3 = 0$

64. $y^2 - 9x^2 - 4y - 30x = 33$

65. $3x^2 - 18x - 2y^2 - 8y + 1 = 0$

66. $4y^2 - 9x^2 - 8y - 36x = 68$

67. $y^2 + 2\sqrt{2} - x^2 + 2\sqrt{2}x = 1$

68. $x^2 + x = \dfrac{y^2 - 1}{4}$

In Exercises 69–78, identify the conic section represented by each equation and sketch the graph.

69. $x^2 - 6x + 12y + 33 = 0$

70. $9y^2 = x^2 - 4x$

71. $y^2 - 9x^2 = -1$

72. $4x^2 + 9y^2 - 8x + 36y + 4 = 0$

73. $x^2 + y^2 - 4x + 8y = 16$

74. $2y^2 + 3x - 4y - 7 = 0$

75. $2x^2 - 4x + 3y + 8 = 0$

76. $y^2 - 9x^2 + 18x - 4y = 14$

77. $4x^2 + 9y^2 + 8x - 54y + 49 = 0$

78. $x^2 - 4y^2 + 6x + 12y = 0$

B EXERCISES Applying the Concepts

79. **Sound of an explosion.** Points A and B are 1000 meters apart, and it is determined from the sound of an explosion heard at these points at different times that the location of the explosion is 600 meters closer to A than to B. Show that the location of the explosion is restricted to points on a hyperbola and find the equation of the hyperbola. [*Hint:* Let the coordinates of A be $(-500, 0)$ and those of B be $(500, 0)$.]

80. **Sound of an explosion.** Points A and B are 2 miles apart. The sound of an explosion at A was heard three seconds before it was heard at B. Assume that the sound travels at 1100 feet/second. Show that the location of the explosion is restricted to a hyperbola and find the equation of the hyperbola. [*Hint:* Recall that 1 mile = 5280 feet.]

81. **Thunder and lightning.** Nicole and Juan, who are 8000 feet apart, hear thunder. Nicole hears the thunder three seconds before Juan does. Find the equation of the hyperbola whose points are locations where the lightning might have struck. Assume that the sound travels at 1100 feet/second.

82. **Locating source of thunder.** In Exercise 81, Valerie is at a location midway between Nicole and Juan, and she hears the thunder two seconds after Juan does. Determine the location where the lightning strikes in relation to the three people involved.

83. **LORAN.** Two LORAN stations, A and B, are situated 300 kilometers apart along a straight coastline. Simultaneous radio signals are sent from each station to a ship. The ship receives the signal from A 0.0005 second before the signal from B. Assume that the radio signals travel 300,000 kilometers per second. Find the equation of the hyperbola on which the ship is located.

84. **LORAN.** In Exercise 83, assuming that the ship were to follow the graph of the hyperbola, determine the location of the ship when it reaches the coastline.

85. **Target practice.** A gun at G and a target at T are 1600 feet apart. The muzzle velocity of the bullet is 2000 feet per second. A person at P hears the crack of the gun and the thud of the bullet at the same time. Show that the

possible locations of the person are on a hyperbola. Find the equation of the hyperbola with G at $(-800, 0)$ and T at $(800, 0)$. Assume that the speed of sound is 1100 feet/second. [*Hint:* Show that $|d(P, G) - d(P, T)|$ is a constant.]

86. **LORAN.** Two LORAN stations, A and B, lie on an east–west line with A 250 miles west of B. A plane is flying west on a line 50 miles north of the line AB. Radio signals are sent (traveling at 980 feet per microsecond) simultaneously from A and B, and the one sent from B arrives at the plane 500 microseconds before the one from A. Where is the plane?

87. **Using LORAN.** LORAN navigational transmitters A, B, C, and D are located at $(-100, 0)$, $(100, 0)$, $(0, -150)$, and $(0, 150)$, respectively. A navigator P on a ship somewhere in the second quadrant listens to the pair (A, B) of transmitters and computes the difference of the distances from the ship to A and B as 120 miles. Similarly, by listening to the pair (C, D) of transmitters, navigator P computes the difference of the distances from the ship to C and D as 80 miles. Use a calculator to find the coordinates of the ship.

88. **Using LORAN.** LORAN navigational transmitters A, B, and C are located at $(200, 0)$, $(-200, 0)$, and $(200, 1000)$, respectively. A navigator on a fishing boat listens to the pair (A, B) of transmitters and finds that the difference of the distances from the boat to A and B is 300 miles. She also finds that the difference of the distances from the boat to A and C is 400 miles. Use a calculator to find the possible locations of the boat.

C EXERCISES Beyond the Basics

89. Write the standard form of the equation of the hyperbola for which the difference of the distances from any point on the hyperbola to $(-5, 0)$ and $(5, 0)$ is equal to (a) two units; (b) four units; (c) eight units.

90. Sketch the graphs of all three hyperbolas in Exercise 89 on the same coordinate plane.

91. Write the standard form of the equation of the hyperbola for which the difference of the distances from any point on the hyperbola to the points F_1 and F_2 is two units, where
   a. F_1 is $(0, 6)$ and F_2 is $(0, -6)$.
   b. F_1 is $(0, 4)$ and F_2 is $(0, -4)$.
   c. F_1 is $(0, 3)$ and F_2 is $(0, -3)$.

92. Sketch the graphs of all three hyperbolas in Exercise 91 on the same coordinate plane.

93. Rewrite equation

$$\sqrt{(x + c)^2 + y^2} - \sqrt{(x - c)^2 + y^2} = \pm 2a$$

from page 627 as

$$\sqrt{(x + c)^2 + y^2} = \pm 2a + \sqrt{(x - c)^2 + y^2}.$$

Square both sides to obtain

$$(x + c)^2 + y^2 = 4a^2 \pm 4a\sqrt{(x - c)^2 + y^2} + (x - c)^2 + y^2.$$

Simplify and isolate the radical; then square both sides again to show that the equation can be simplified to

$$(c^2 - a^2)x^2 - a^2y^2 = a^2(c^2 - a^2).$$

94. a. Let $p > 0$. Prove that the graph of the equation

$$xy = \frac{p^2}{2}$$ is a hyperbola by showing that this equation

is the equation of the hyperbola with foci (p, p) and $(-p, -p)$ and with difference $2p$ in distance.

[*Hint:* Show that the equation

$$\sqrt{(x + p)^2 + (y + p)^2} - \sqrt{(x - p)^2 + (y - p)^2} = 2p$$

simplifies to $xy = \dfrac{p^2}{2}$.]

b. Use part (a) to sketch the graph of the hyperbola $xy = 8$. Find the coordinates of the foci.

95. A hyperbola for which the lengths of the transverse and conjugate axes are equal is called an **equilateral hyperbola**. Show that the asymptotes of an equilateral hyperbola are perpendicular to one another.

96. The **latus rectum of a hyperbola** is a line segment that passes through a focus, is perpendicular to the transverse axis, and has endpoints on the hyperbola. Show that the length of the latus rectum of the hyperbola

$$\frac{x^2}{a^2} - \frac{y^2}{b^2} = 1 \text{ is } \frac{2b^2}{a}.$$

97. The *eccentricity* **of a hyperbola**, denoted by e, is defined by

$$e = \frac{\text{Distance between the foci}}{\text{Distance between the vertices}} = \frac{2c}{2a} = \frac{c}{a}.$$

Show that for every hyperbola, $e > 1$. What happens when $e = 1$?

In Exercises 98–102, find the eccentricity and the length of the latus rectum of each hyperbola.

98. $\dfrac{x^2}{16} - \dfrac{y^2}{9} = 1$ **99.** $36x^2 - 25y^2 = 900$

100. $x^2 - y^2 = 49$ **101.** $8(x - 1)^2 - (y + 2)^2 = 2$

102. $5x^2 - 4y^2 - 10x - 8y - 19 = 0$

103. For the hyperbola $\dfrac{x^2}{a^2} - \dfrac{y^2}{b^2} = 1$ with eccentricity e, show that $b^2 = a^2(e^2 - 1)$.

104. Find an equation of a hyperbola assuming that its foci are $(\pm 6, 0)$ and the length of its latus rectum is 10.

105. A point $P(x, y)$ moves in the plane so that its distance from the point $(3, 0)$ is always twice its distance from the line $x = -1$. Show that the equation of the path of P is a hyperbola of eccentricity 2.

106. A point $P(x, y)$ moves in the plane so that its distance from the point $(0, -3)$ is always three times its distance from the line $y = 1$. Show that the equation of the path of P is a hyperbola of eccentricity 3.

In Exercises 107–112, find all points of intersection of the given curves. Make a sketch of the curves that shows the points of intersection.

107. $y - 2x - 20 = 0$ and $y^2 - 4x^2 = 36$

108. $x - 2y^2 = 0$ and $x^2 = 8y^2 + 5$

109. $x^2 + y^2 = 15$ and $x^2 - y^2 = 1$

110. $x^2 + 9y^2 = 9$ and $4x^2 - 25y^2 = 36$

111. $9x^2 + 4y^2 = 72$ and $x^2 - 9y^2 = -77$

112. $3y^2 - 7x^2 = 5$ and $9x^2 - 2y^2 = 1$

In Exercises 113 and 114, use the following definition. A *tangent line* **to a hyperbola is a line that is not parallel to either asymptote of the hyperbola and that intersects the hyperbola at only one point.**

113. Find the equation of the tangent line to the hyperbola with equation $\dfrac{x^2}{a^2} - \dfrac{y^2}{b^2} = 1$, with $a = \sqrt{\dfrac{5}{3}}$ and $b = \sqrt{\dfrac{5}{7}}$ at the point $(2, 1)$.
[*Hint:* See the steps for Exercise 91 in Section 7.2 and (in Step 3) write the discriminant as a perfect square.]

114. Show that the equation of the tangent line to the hyperbola with equation $\dfrac{x^2}{a^2} - \dfrac{y^2}{b^2} = 1$, at the point (x_1, y_1), is $y = \dfrac{b^2x_1}{a^2y_1}x - \dfrac{b^2x_1^2}{a^2y_1^2} + y_1$.
[*Hint:* See the steps for Exercise 92 in Section 7.2 and show that $(a^2y_1m - b^2x_1)^2 = 0$.]

Critical Thinking

115. Explain why each of the following represents a degenerate conic section. Where possible, sketch the graph.
a. $4x^2 - 9y^2 = 0$
b. $x^2 + 4y^2 + 6 = 0$
c. $8x^2 + 5y^2 = 0$
d. $x^2 + y^2 - 4x + 8y = -20$
e. $y^2 - 2x^2 = 0$

GROUP PROJECT

Discuss the graph of a quadratic equation in x and y of the form

$$Ax^2 + Cy^2 + Dx + Ey + F = 0, \quad AC \neq 0.$$

Rewrite that equation in the form

$$A\left(x^2 + \frac{D}{A}x + \frac{D^2}{4A^2}\right) + C\left(y^2 + \frac{E}{C}y + \frac{E^2}{4C^2}\right)$$
$$= \frac{D^2}{4A} + \frac{E^2}{4C} - F.$$

Include in your discussion the types of graphs you will obtain in each of the following cases:
i. $A > 0, C > 0$ **ii.** $A > 0, C < 0$
iii. $A < 0, C > 0$ **iv.** $A < 0, C < 0$

In each case, discuss how the classification is affected by the sign of

$$\frac{D^2}{4A} + \frac{E^2}{4C} - F.$$

SUMMARY ■ Definitions, Concepts, and Formulas

7.1 Conic Sections: Overview

Conic sections are the curves formed when a plane intersects the surface of a right circular cone. Except in some degenerate cases, the curves formed are the circle, the ellipse, the parabola, and the hyperbola.

7.2 The Parabola

i. **Definition.** A **parabola** is the set of all points in the plane that are the same distance from a fixed line and a fixed point not on the line. The fixed line is called the **directrix**, and the fixed point is called the **focus**.

ii. **Axis** or **axis of symmetry.** This is the line that passes through the focus and is perpendicular to the directrix.

iii. **Vertex.** This is the point at which the parabola intersects its axis. The vertex is halfway between the focus and the directrix.

iv. **Standard forms of equations of parabolas with vertex at $(0, 0)$ and $a > 0$** are

$$y^2 = \pm 4ax \text{ and } x^2 = \pm 4ay. \text{ (See the table on page 617.)}$$

v. The **latus rectum** of a parabola is the line segment that passes through the focus, is perpendicular to the axis of the parabola, and has endpoints on the parabola.

vi. **Standard forms of equations of parabolas with vertex (h, k) and $a > 0$** are $(y - k)^2 = \pm 4a(x - h)$ and $(x - h)^2 = \pm 4a(y - k)$. (See the table on page 620.)

7.3 The Ellipse

i. **Definition.** An **ellipse** is the set of all points in the plane, the sum of whose distances from two fixed points is a constant. The fixed points are called the **foci** of the ellipse.

ii. **Vertices.** These are the two points where the line through the foci intersects the ellipse.

iii. **Major axis.** This is the line segment joining the two vertices of the ellipse.

iv. **Center.** This is the midpoint of the line segment joining the foci. It is also the midpoint of the major axis.

v. **Minor axis.** This is the line segment that passes through the center of the ellipse, is perpendicular to the major axis, and has endpoints on the ellipse.

vi. **Standard forms of equations of ellipses with center $(0, 0)$, $a > b > 0$, $c < a$, and $b^2 = a^2 - c^2$** are

$$\frac{x^2}{a^2} + \frac{y^2}{b^2} = 1, \text{ with foci } (\pm c, 0) \text{ and vertices } (\pm a, 0),$$

and

$$\frac{x^2}{b^2} + \frac{y^2}{a^2} = 1, \text{ with foci } (0, \pm c) \text{ and vertices } (0, \pm a).$$

(See the table on page 628.)

vii. **Standard forms of equations of ellipses with center (h, k), $a > b > 0$, and $b^2 = a^2 - c^2$** are

$$\frac{(x - h)^2}{a^2} + \frac{(y - k)^2}{b^2} = 1 \text{ and } \frac{(x - h)^2}{b^2} + \frac{(y - k)^2}{a^2} = 1.$$

(See the table on page 630.)

7.4 The Hyperbola

i. **Definition.** A **hyperbola** is the set of all points in the plane, the difference of whose distances from two fixed points is a constant. The fixed points are called the **foci** of the hyperbola.

ii. **Vertices.** These are the two points where the line through the foci intersects the hyperbola.

iii. **Transverse axis.** This is the line segment joining the two vertices of the hyperbola.

iv. **Center.** This is the midpoint of the line segment joining the foci. It is also the midpoint of the transverse axis.

v. **Standard forms of equations of hyperbolas with center $(0, 0)$, $c > a$, and $b^2 = c^2 - a^2$** are

$$\frac{x^2}{a^2} - \frac{y^2}{b^2} = 1, \text{ with foci } (\pm c, 0) \text{ and vertices } (\pm a, 0),$$

and

$$\frac{y^2}{a^2} - \frac{x^2}{b^2} = 1, \text{ with foci } (0, \pm c) \text{ and vertices } (0, \pm a).$$

(See the table on page 639.)

vi. **Conjugate axis.** This is a line segment of length $2b$ that passes through the center of the hyperbola and is perpendicularly bisected by the transverse axis.

vii. **Asymptotes.** The asymptotes of the hyperbola $\frac{x^2}{a^2} - \frac{y^2}{b^2} = 1$ are the two lines $y = \pm \frac{b}{a} x$; the asymptotes of the hyperbola $\frac{y^2}{a^2} - \frac{x^2}{b^2} = 1$ are the lines $y = \pm \frac{a}{b} x$.

viii. **Graphing.** A procedure for graphing a hyperbola centered at $(0, 0)$ is given on page 643.

ix. **Standard forms of equations of hyperbolas centered at (h, k)** are

$$\frac{(x - h)^2}{a^2} - \frac{(y - k)^2}{b^2} = 1 \text{ and } \frac{(y - k)^2}{a^2} - \frac{(x - h)^2}{b^2} = 1.$$

See the table on page 644. Also see page 645 for a procedure for graphing such hyperbolas.

REVIEW EXERCISES

In Exercises 1–8, sketch the graph of each parabola. Determine the vertex, focus, axis, and directrix of each parabola.

1. $y^2 = -6x$ 2. $y^2 = 12x$

3. $x^2 = 7y$ 4. $x^2 = -3y$

5. $(x - 2)^2 = -(y + 3)$ 6. $(y + 1)^2 = 5(x + 2)$

7. $y^2 = -4y + 2x + 1$ 8. $-x^2 + 2x + y = 0$

In Exercises 9–12, find an equation of the parabola satisfying the given conditions.

9. Vertex: $(0, 0)$; focus: $(-3, 0)$

10. Vertex: $(0, 0)$; focus: $(0, 4)$

11. Focus: $(0, 4)$; directrix: $y = -4$

12. Focus: $(-3, 0)$; directrix: $x = 3$

In Exercises 13–20, sketch the graph of each ellipse. Determine the foci, vertices, and endpoints of the minor axis of the ellipse.

13. $\dfrac{x^2}{25} + \dfrac{y^2}{4} = 1$ 14. $\dfrac{x^2}{9} + \dfrac{y^2}{36} = 1$

15. $4x^2 + y^2 = 4$ 16. $16x^2 + y^2 = 64$

17. $16(x + 1)^2 + 9(y + 4)^2 = 144$

18. $4(x - 1)^2 + 3(y + 2)^2 = 12$

19. $x^2 + 9y^2 + 2x - 18y + 1 = 0$

20. $4x^2 + y^2 + 8x - 10y + 13 = 0$

In Exercises 21–24, find an equation of the ellipse satisfying the given conditions.

21. Vertices: $(\pm 4, 0)$; endpoints of minor axis: $(0, \pm 2)$

22. Vertices: $(0, \pm 6)$; endpoints of minor axis: $(\pm 2, 0)$

23. Length of major axis: 20; foci: $(\pm 5, 0)$

24. Length of minor axis: 16; foci: $(0, \pm 6)$

In Exercises 25–32, sketch the graph of each hyperbola. Determine the vertices, the foci, and the asymptotes of each hyperbola.

25. $\dfrac{y^2}{16} - \dfrac{x^2}{4} = 1$ 26. $\dfrac{x^2}{16} - \dfrac{y^2}{9} = 1$

27. $8x^2 - y^2 = 8$ 28. $4y^2 - 4x^2 = 1$

29. $\dfrac{(x + 2)^2}{9} - \dfrac{(y - 3)^2}{4} = 1$

30. $\dfrac{(y + 1)^2}{6} - \dfrac{(x - 2)^2}{8} = 1$

31. $4y^2 - x^2 + 40y - 4x + 60 = 0$

32. $4x^2 - 9y^2 + 16x - 54y - 29 = 0$

In Exercises 33–36, find an equation of the hyperbola satisfying the given condition.

33. Vertices: $(\pm 1, 0)$; foci: $(\pm 2, 0)$

34. Vertices: $(0, \pm 2)$; foci: $(0, \pm 4)$

35. Vertices: $(\pm 2, 0)$; asymptotes: $y = \pm 3x$

36. Vertices: $(0, \pm 3)$; asymptotes: $y = \pm x$

In Exercises 37–48, identify each equation as representing a circle, a parabola, an ellipse, or a hyperbola. Sketch the graph of the conic.

37. $5x^2 - 4y^2 = 20$ 38. $x^2 - x + y = 1$

39. $3x^2 + 4y^2 + 8y - 12x - 6 = 0$

40. $2y - x^2 = 0$ 41. $x^2 + y^2 + 2x - 3 = 0$

42. $2x + y^2 = 0$ 43. $y^2 = x^2 + 3$

44. $x^2 = 10 - 3y^2$

45. $3x^2 + 3y^2 - 6x + 12y + 5 = 0$

46. $y + 2x - x^2 + 2y^2 = 0$

47. $9x^2 + 8y^2 = 36$

48. $2y^2 + 4y = 3x^2 - 6x + 9$

49. Find an equation of the hyperbola whose foci are the vertices of the ellipse $4x^2 + 9y^2 = 36$ and whose vertices are the foci of this ellipse.

50. Find an equation of the ellipse whose foci are the vertices of the hyperbola $9x^2 - 16y^2 = 144$ and whose vertices are the foci of this hyperbola.

In Exercises 51–54, find all points of intersection of the given curves and make a sketch.

51. $x^2 - 4y^2 = 36$ and $x - 2y - 20 = 0$

52. $y^2 - 8x^2 = 5$ and $y - 2x^2 = 0$

53. $3x^2 - 7y^2 = 5$ and $9y^2 - 2x^2 = 1$

54. $x^2 - y^2 = 1$ and $x^2 + y^2 = 7$

55. **Parabolic curve.** Water flowing from the end of a horizontal pipe 20 feet above the ground describes a parabolic curve whose vertex is at the end of the pipe. If at a point 6 feet below the line of the pipe the flow of water has curved outward 8 feet beyond a vertical line through the end of the pipe, how far beyond this vertical line will the water strike the ground?

20 ft   6 ft   8 ft   ? ft

56. **Parabolic arch.** A parabolic arch has a height of 20 meters and a width of 36 meters at the base. Assuming that the vertex of the parabola is at the top of the arch, find the height of the arch at a distance of 9 meters from the center of the base.

57. Football. A football is 12 inches long, and a cross section containing a seam is an ellipse with a minor axis of length 7 inches. Suppose every cross section of the ball formed by a plane perpendicular to the major axis of the ellipse is a circle. Find the circumference of such a circular cross section located 2 inches from an end of the ball.

58. Production cost. A company assembles computers at two locations A and B that are 1000 miles apart. The per unit cost of production at location A is \$20 less than at location B. Assume that the route of delivery of the computers is along a straight line and that the delivery cost is 25¢ per unit per mile. Find the equation of the curve at any point the computers can be supplied from either location at the same cost.
[*Hint*: Take A at $(-500, 0)$ and B at $(500, 0)$.]

PRACTICE TEST A

1. Find the standard form of the parabola with focus $(0, 12)$ and directrix with equation $x = -12$.

2. Convert the equation $y^2 - 2y + 8x + 25 = 0$ to the standard form for a parabola by completing the square.

3. Find the focus and directrix of the parabola with equation $x^2 = -9y$.

4. Graph the parabola with equation $y^2 = -5x$.

5. Find the vertex, focus, and directrix of the parabola with equation $(x + 2)^2 = -8(y - 1)$.

6. The base of a water slide is parabolic in shape and is 6 feet wide and 2 feet deep. Find the height of the slide 1 foot from its center.

7. Find an equation of the parabola with vertex $(3, -1)$ and directrix $x = -3$.

In Problems 8–10, find the standard form of the equation of the ellipse satisfying the given conditions.

8. Foci: $(0, -2), (0, 2)$; vertices: $(0, -4), (0, 4)$

9. Foci: $(-2, 0), (2, 0)$; y-intercepts: $-5, 5$

10. Major axis horizontal with length 18; minor axis of length 4; center $(0, 0)$

11. Graph the ellipse with equation $9x^2 + y^2 = 9$.

12. Find the vertices and foci for the hyperbola with equation $49y^2 - x^2 = 49$.

13. Graph the hyperbola with equation $\dfrac{x^2}{49} - \dfrac{y^2}{4} = 1$.

14. Find the standard form of the equation of the hyperbola with foci $(0, -\sqrt{45}), (0, \sqrt{45})$ and vertices $(0, -6), (0, 6)$.

In Problems 15 and 16, convert the equation to the standard form for a hyperbola by completing the squares on x and y.

15. $y^2 - x^2 + 2x = 2$

16. $x^2 - 2x - 4y^2 - 16y = 19$

In Problems 17–20, identify the conic section and sketch its graph.

17. $x^2 - 9y^2 = -16$

18. $x^2 - 2y + 4y + 1 = 0$

19. $x^2 + y^2 + 2x - 6y = 3$

20. $25x^2 + 4y^2 = 100$

PRACTICE TEST B

1. Find the standard form of the equation of the parabola with focus $(-10, 0)$ and directrix $x = 10$.
   a. $y^2 = -10x$ b. $y^2 = -40x$
   c. $x^2 = -40y$ d. $y^2 = 40x$

2. Convert the equation $y^2 - 4y - 5x + 24 = 0$ to the standard form for a parabola by completing the square.
   a. $(y + 2)^2 = 5(x - 4)$ b. $(y - 2)^2 = 5(x - 4)$
   c. $(y + 2)^2 = -5(x - 4)$ d. $(y - 2)^2 = 5(x + 4)$

3. Find the focus and directrix of the parabola with equation $x = 7y^2$.

   a. focus: $\left(\dfrac{1}{28}, 0\right)$; directrix: $x = -\dfrac{1}{28}$

   b. focus: $\left(0, \dfrac{1}{28}\right)$; directrix: $y = -\dfrac{1}{28}$

   c. focus: $\left(\dfrac{1}{28}, 0\right)$; directrix: $x = \dfrac{1}{28}$

   d. focus: $\left(\dfrac{1}{7}, 0\right)$; directrix: $x = -\dfrac{1}{7}$

4. Graph the parabola with equation $y^2 = -7x$.

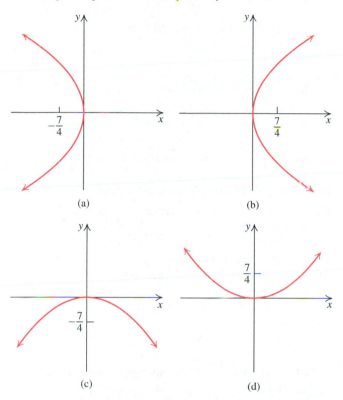

    (a)   (b)

    (c)   (d)

5. Find the vertex, focus, and directrix of the parabola with equation $(x - 3)^2 = 12(y - 1)$.
   a. vertex: $(3, 1)$ focus: $(3, 4)$ directrix: $y = -2$
   b. vertex: $(-3, -1)$ focus: $(-3, 2)$ directrix: $y = -4$
   c. vertex: $(3, 1)$ focus: $(3, -2)$ directrix: $x = 4$
   d. vertex: $(1, 3)$ focus: $(1, 6)$ directrix: $y = 0$

6. The parabolic arch of a bridge has a 180-foot base and a height of 25 feet. Find the height of the arch 45 feet from the center of the base.
   a. 12.5 ft   b. 6.25 ft
   c. 16.7 ft   d. 18.75 ft

7. Find an equation of the parabola with vertex $(-2, 1)$ and directrix $x = 2$.
   a. $(y - 1)^2 = -16(x + 2)$
   b. $(y - 1)^2 = 16(x + 2)$
   c. $(y + 2)^2 = -16(x - 1)$
   d. $(y + 2)^2 = 16(x - 1)$

In Problems 8–10, find the standard form of the equation of the ellipse satisfying the given conditions.

8. Foci: $(-3, 0), (3, 0)$; vertices: $(-5, 0), (5, 0)$

   a. $\dfrac{x^2}{9} + \dfrac{y^2}{16} = 1$ b. $\dfrac{x^2}{25} + \dfrac{y^2}{16} = 1$

   c. $\dfrac{x^2}{9} + \dfrac{y^2}{25} = 1$ d. $\dfrac{x^2}{16} + \dfrac{y^2}{25} = 1$

9. Foci: $(0, -3), (0, 3)$; y-intercepts: $-7, 7$

   a. $\dfrac{x^2}{49} + \dfrac{y^2}{40} = 1$ b. $\dfrac{x^2}{9} + \dfrac{y^2}{49} = 1$

   c. $\dfrac{x^2}{9} + \dfrac{y^2}{40} = 1$ d. $\dfrac{x^2}{40} + \dfrac{y^2}{49} = 1$

10. Major axis vertical with length 16; length of minor axis $= 8$; center: $(0, 0)$

   a. $\dfrac{x^2}{16} + \dfrac{y^2}{64} = 1$ b. $\dfrac{x^2}{64} + \dfrac{y^2}{256} = 1$

   c. $\dfrac{x^2}{8} + \dfrac{y^2}{64} = 1$ d. $\dfrac{x^2}{64} + \dfrac{y^2}{16} = 1$

11. Graph the ellipse with equation $9(x - 1)^2 + 4(y - 2)^2 = 36$.

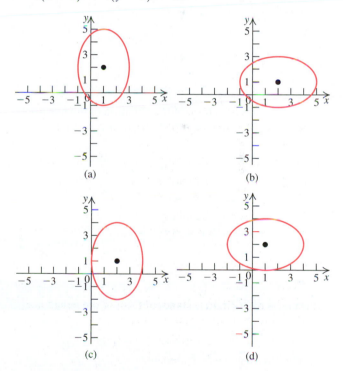

    (a)   (b)

    (c)   (d)

12. Find the vertices and foci of the hyperbola with equation $\dfrac{x^2}{121} - \dfrac{y^2}{4} = 1$.

   a. vertices: $(-11, 0), (11, 0)$ foci: $(-2, 0), (2, 0)$
   b. vertices: $(0, -1), (0, 11)$ foci: $(-5\sqrt{5}, 0), (5\sqrt{5}, 0)$
   c. vertices: $(-11, 0), (11, 0)$ foci: $(-5\sqrt{5}, 0), (5\sqrt{5}, 0)$
   d. vertices: $(-2, 0), (2, 0)$ foci: $(-5\sqrt{5}, 0), (5\sqrt{5}, 0)$

13. Graph the hyperbola with equation $\dfrac{x^2}{25} - \dfrac{y^2}{4} = 1$.

(a) (b)

(c) (d)

14. Find the standard form of the equation of the hyperbola with foci $(0, -10)$, $(0, 10)$ and vertices $(0, -5)$, $(0, 5)$.

a. $\dfrac{y^2}{25} - \dfrac{x^2}{100} = 1$ **b.** $\dfrac{y^2}{25} - \dfrac{x^2}{75} = 1$

c. $\dfrac{x^2}{25} - \dfrac{y^2}{100} = 1$ **d.** $\dfrac{x^2}{25} - \dfrac{y^2}{75} = 1$

In Problems 15 and 16, convert the equation to the standard form for a hyperbola by completing the square on x and y.

15. $y^2 - 4x^2 - 2y - 16x - 19 = 0$

a. $\dfrac{(y-2)^2}{4} - (x+4)^2 = 1$

b. $\dfrac{(x-1)^2}{4} - (y+2)^2 = 1$

c. $\dfrac{(y-1)^2}{4} - (x+2)^2 = 1$

d. $(x+2)^2 - \dfrac{(y-1)^2}{4} = 1$

16. $4y^2 - 9x^2 - 16y - 36x - 56 = 0$

a. $\dfrac{(y-2)^2}{4} - \dfrac{(x+2)^2}{9} = 1$

b. $\dfrac{(x-2)^2}{4} - \dfrac{(y+2)^2}{9} = 1$

c. $\dfrac{(y-2)^2}{9} - \dfrac{(x+2)^2}{4} = 1$

d. $\dfrac{(y+2)^2}{9} - \dfrac{(x-2)^2}{4} = 1$

In Problems 17–20, identify the conic section.

17. $3x^2 + 2y^2 - 6x + 4y - 1 = 0$
a. parabola **b.** circle
c. ellipse **d.** hyperbola

18. $x^2 + 2x + 4y + 5 = 0$
a. parabola **b.** circle
c. ellipse **d.** hyperbola

19. $4x^2 - y^2 + 16x + 2y + 11 = 0$
a. parabola **b.** circle
c. ellipse **d.** hyperbola

20. $5y^2 - 3x^2 + 20y - 6x + 2 = 0$
a. parabola **b.** circle
c. ellipse **d.** hyperbola

CUMULATIVE REVIEW EXERCISES ■ Chapters P–7

1. Let $f(x) = x^2 - 3x + 2$. Find $\dfrac{f(x+h) - f(x)}{h}$.

2. Sketch the graph of $f(x) = \sqrt{x+1} - 2$.

3. Let $f(x) = 2x - 3$. Find the inverse function f^{-1}. Verify that $f(f^{-1}(x)) = x$.

4. Sketch the graph of $f(x) = \left(\dfrac{1}{2}\right)^{x+2}$.

5. Solve the equation $\log_5(x-1) + \log_5(x-2) = 3\log_5\sqrt[3]{6}$.

6. Express

$$\log_a \sqrt[3]{x\sqrt{yz}}$$

in terms of logarithms of x, y, and z.

7. Solve the inequality

$$\frac{x}{x-2} \geq 1.$$

8. The current I in an electric circuit is given by

$$I = \frac{V}{R}(1 - e^{-0.3t}).$$

Use natural logarithms to solve for t.

In Problems 9–12, solve each system of equations.

9. $\begin{cases} 1.4x - 0.5y = 1.3 \\ 0.4x + 1.1y = 4.1 \end{cases}$

10. $\begin{cases} 2x + y - 4z = 3 \\ x - 2y + 3z = 4 \\ -3x + 4y - z = -2 \end{cases}$

11. $\begin{cases} y = 2 - \log x \\ y - \log(x + 3) = 1 \end{cases}$

12. $\begin{cases} y = x^2 - 1 \\ 3x^2 + 8y^2 = 8 \end{cases}$

13. Find the determinant of the matrix $\begin{bmatrix} 1 & 4 & 7 \\ 2 & 5 & 8 \\ 3 & 6 & 9 \end{bmatrix}$.

14. Use Cramer's Rule to solve the system of equations.

$$\begin{cases} 2x - 3y = -4 \\ 5x + 7y = 1 \end{cases}$$

15. Find the inverse of the matrix $A = \begin{bmatrix} 3 & -2 \\ -5 & 4 \end{bmatrix}$.

16. Find an equation of the line that passes through the point of intersection of the lines $x + 2y - 3 = 0$ and $3x + 4y - 5 = 0$ and that is perpendicular to the line $x - 3y + 5 = 0$. Write your answer in slope–intercept form.

17. Solve the equation $2x^4 - 5x^2 + 3 = 0$.

In Problems 18–20, an equation of a conic section is given. Identify the conic and sketch its graph.

18. $x^2 - y^2 = -4$

19. $9x^2 + 9y^2 = 144$

20. Sketch the graph of the rational function $f(x) = \dfrac{x}{x^2 - 16}$.

Probability was studied in ancient Eastern civilizations, including those in Babylonia and Egypt. Probability is most famously used in gambling, and Egyptians even used objects made from the heel bones of animals, called *tali*, as a type of dice. Today probability is used not only in gambling, but also in polling, insurance, business, the life sciences, and many other areas. In this chapter, we learn to understand and compute the probability of an event.

Sequences and Series

Before Starting this Section, Review

1. Simplifying algebraic expressions (Section P.1, page 14)

2. Finding functional values (Section 2.4, page 214)

Objectives

1. Use sequence notation and find specific and general terms in a sequence.

2. Use factorial notation.

3. Use summation notation to write partial sums of a series.

THE FAMILY TREE OF THE HONEYBEE

The bee that most of us know best is the honeybee, an insect that lives in a colony and has an unusual family tree. Perhaps the most surprising fact about honeybees is that not all of them have two parents. This phenomenon begins with a special female in the colony called the *queen*. Many other female bees, called *worker bees*, live in the colony, but unlike the queen bee, they produce no eggs. Male bees do no work and are produced from the queen's unfertilized eggs; so male bees have no father, only a mother. The female bees are produced as a result of the queen bee mating with a male bee. Consequently, all female bees have two parents—a male and a female—whereas male bees have just one parent—a female. In this section, we study sequences, and in Example 5, we see how a famous sequence accurately counts a honeybee's ancestors. ■

---

1 Use sequence notation and find specific and general terms in a sequence.

Sequences

The word *sequence* is used in mathematics in much the same way it is used in ordinary English. If someone saw a *sequence* of bad movies, you know that the person could list the first bad movie he or she saw, the second bad movie, and so on.

DEFINITION OF A SEQUENCE

An **infinite sequence** is a function whose domain is the set of positive integers. The function values, written as

$$a_1, a_2, a_3, a_4, \ldots, a_n, \ldots,$$

are called the **terms** of the sequence. The **nth term**, a_n, is called the **general term** of the sequence.

If the domain of a function consists of only the first n positive integers, the sequence is called a **finite sequence**. In computer science, finite sequences of the form $a_1, a_2, \ldots, a_n$ are called *strings*. Sometimes for convenience, we use 0 instead of 1 as the first subscript, and we write the sequence terms as $a_0, a_1, a_2, a_3, \ldots$. The nth term of this sequence is a_{n-1}.

Notice in Example 1 that converting $a_n = 5n - 1$ to standard function notation gives us $f(n) = 5n - 1$, a linear function.

TECHNOLOGY CONNECTION

 A graphing calculator can write the terms of a sequence and graph the sequence. You must first select Sequence mode.

The first four terms of

the sequence $a_n = \dfrac{1}{n + 1}$

and the graph of this sequence are shown here. Use the right arrow key ▶ to scroll to the right to display the sequence values corresponding to $n = 3$ and $n = 4$.

Now press GRAPH.

EXAMPLE 1   Writing the Terms of a Sequence from the General Term

Write the first four terms of each sequence.

a. $a_n = 5n - 1$ **b.** $a_n = \dfrac{1}{n + 1}$

SOLUTION

The first four terms of each sequence are found by replacing n with the integers 1, 2, 3, and 4 in the equation defining a_n.

a. $a_n = 5n - 1$ General term of the sequence
$a_1 = 5(1) - 1 = 4$ 1st term: Replace n with 1.
$a_2 = 5(2) - 1 = 9$ 2nd term: Replace n with 2.
$a_3 = 5(3) - 1 = 14$ 3rd term: Replace n with 3.
$a_4 = 5(4) - 1 = 19$ 4th term: Replace n with 4.

b. $a_n = \dfrac{1}{n + 1}$ General term of the sequence

$a_1 = \dfrac{1}{1 + 1} = \dfrac{1}{2}$ 1st term: Replace n with 1.

$a_2 = \dfrac{1}{2 + 1} = \dfrac{1}{3}$ 2nd term: Replace n with 2.

$a_3 = \dfrac{1}{3 + 1} = \dfrac{1}{4}$ 3rd term: Replace n with 3.

$a_4 = \dfrac{1}{4 + 1} = \dfrac{1}{5}$ 4th term: Replace n with 4.   ■ ■ ■

Practice Problem 1   Write the first four terms of the sequence with general term $a_n = -2^n$.   ■

We can use subscripts on variables other than a to represent the terms of a sequence. In Example 2, we use the variable b.

EXAMPLE 2   Writing the First Several Terms of a Sequence

Write the first six terms of the sequence defined by

$$b_n = (-1)^{n+1}\left(\dfrac{1}{n}\right).$$

SOLUTION

Recall that all even powers of -1 equal 1 and all odd powers of -1 equal -1.

$b_1 = (-1)^{1+1}\left(\dfrac{1}{1}\right) = (-1)^2(1) = 1$ Replace n with 1

$b_2 = (-1)^{2+1}\left(\dfrac{1}{2}\right) = (-1)^3\left(\dfrac{1}{2}\right) = -\dfrac{1}{2}$ Replace n with 2

$b_3 = (-1)^{3+1}\left(\dfrac{1}{3}\right) = (-1)^4\left(\dfrac{1}{3}\right) = \dfrac{1}{3}$ Replace n with 3

$b_4 = (-1)^{4+1}\left(\dfrac{1}{4}\right) = (-1)^5\left(\dfrac{1}{4}\right) = -\dfrac{1}{4}$ Replace n with 4

$b_5 = (-1)^{5+1}\left(\dfrac{1}{5}\right) = (-1)^6\left(\dfrac{1}{5}\right) = \dfrac{1}{5}$ Replace n with 5

$b_6 = (-1)^{6+1}\left(\dfrac{1}{6}\right) = (-1)^7\left(\dfrac{1}{6}\right) = -\dfrac{1}{6}$ Replace n with 6   ■ ■ ■

Practice Problem 2   Write the first six terms of the sequence defined by $a_n = \dfrac{(-1)^n}{n^2}$.   ∎

Frequently, the first few terms of a sequence exhibit a pattern that *suggests* a natural choice for the general term of the sequence.

> **EXAMPLE 3**   **Finding a General Term of a Sequence from a Pattern**

Write the general term a_n for a sequence whose first five terms are given.

a. $1, 4, 9, 16, 25, \ldots$ **b.** $0, \dfrac{1}{2}, -\dfrac{2}{3}, \dfrac{3}{4}, -\dfrac{4}{5}, \ldots$

SOLUTION

Write the position number of the term above each term of the sequence and look for a pattern that connects the term to the position number of the term.

a.

n:	1	2	3	4	5	$\ldots$	n
term:	1,	4,	9,	16,	25,	$\ldots$,	a_n

Apparent pattern: Here $1 = 1^2, 4 = 2^2, 9 = 3^2, 16 = 4^2$, and $25 = 5^2$. Each term is the square of the position number of that term. This suggests $a_n = n^2$.

b.

n:	1	2	3	4	5	$\ldots$	n
term:	0,	$\dfrac{1}{2}$,	$-\dfrac{2}{3}$,	$\dfrac{3}{4}$,	$-\dfrac{4}{5}$,	$\ldots$,	a_n

Apparent pattern: When the terms alternate in sign and $n = 1$, we use factors such as $(-1)^n$ if we want to begin with the factor -1 or we use factors such as $(-1)^{n+1}$ if we want to begin with the factor 1. Notice that each term can be written as a quotient with denominator equal to the position number and numerator equal to one less than the position number, suggesting the general term $a_n = (-1)^n \left(\dfrac{n-1}{n} \right)$.   ■ ■ ■

Practice Problem 3   Write a general term for a sequence whose first five terms are $0, -\dfrac{1}{2}, \dfrac{2}{3}, -\dfrac{3}{4}, \dfrac{4}{5}, \ldots$.   ∎

In Example 3, we found the general term of a sequence by finding a pattern in the first few terms of the sequence. However, when only a finite number of successive terms are given without a rule that defines the general term, a *unique* general term cannot be found.

To indicate why this is so, consider the two sequences:

$$a_n = n^2 \text{ and } b_n = n^2 + (n-1)(n-2)(n-3)(n-4).$$

The first four terms of these two sequences are identical: $1, 4, 9, 16$. However, the fifth term is different: $a_5 = 25$, but $b_5 = 49$. Thus, either a_n or b_n could correctly describe a sequence whose first four terms are $1, 4, 9,$ and 16. So you can *never* find a unique general term from a finite number of terms in a sequence.

Recursive Formulas

Up to this point, we have expressed the general term as a function of the position of the term in the sequence. A second approach is to define a sequence *recursively*. A **recursive formula** requires that one or more of the first few terms of the sequence be specified and all other terms be defined in relation to previously defined terms. An example will help clarify how this works.

The **Fibonacci sequence** is a famous sequence that is defined recursively and shows up often in nature. In this sequence, we specify the first two terms as $a_0 = 1$ and $a_1 = 1$; each subsequent term is the sum of the two terms immediately preceding it. So we have

$a_0 = 1, a_1 = 1, a_2 = a_0 + a_1 = 1 + 1 = 2$ Add the two given terms.

$a_1 = 1, a_2 = 2, a_3 = a_1 + a_2 = 1 + 2 = 3$ Add the two previous terms.

$a_2 = 2, a_3 = 3, a_4 = a_2 + a_3 = 2 + 3 = 5$ Add the two previous terms.

$a_3 = 3, a_4 = 5, a_5 = a_3 + a_4 = 3 + 5 = 8$ Add the two previous terms.

The first six terms of the sequence are then

$$1, 1, 2, 3, 5, 8.$$

This sequence can also be defined in subscript notation.

$$a_0 = 1, a_1 = 1, a_k = a_{k-2} + a_{k-1}, \text{for all } k \geq 2$$

Replacing k with 2, 3, 4, and 5, respectively, in the equation $a_k = a_{k-2} + a_{k-1}$ will again produce the values $a_2 = 2, a_3 = 3, a_4 = 5$, and $a_5 = 8$.

EXAMPLE 4   **Finding Terms of a Recursively Defined Sequence**

Write the first five terms of the recursively defined sequence

$$a_1 = 4, a_{n+1} = 2a_n - 9.$$

SOLUTION

We are given the first term of the sequence: $a_1 = 4$.

$a_2 = 2a_1 - 9$ Replace n with 1 in $a_{n+1} = 2a_n - 9$.

$a_2 = 2(4) - 9 = -1$ Replace a_1 with 4.

$a_3 = 2a_2 - 9$ Replace n with 2 in $a_{n+1} = 2a_n - 9$.

$a_3 = 2(-1) - 9 = -11$ Replace a_2 with -1.

$a_4 = 2a_3 - 9$ Replace n with 3 in $a_{n+1} = 2a_n - 9$.

$a_4 = 2(-11) - 9 = -31$ Replace a_3 with -11.

$a_5 = 2a_4 - 9$ Replace n with 4 in $a_{n+1} = 2a_n - 9$.

$a_5 = 2(-31) - 9 = -71$ Replace a_4 with -31.

Thus, the first five terms of the sequence are

$$4, -1, -11, -31, -71.$$ ■ ■ ■

Practice Problem 4 Write the first five terms of the recursively defined sequence $a_1 = -3, a_{n+1} = 2a_n + 5$. ■

EXAMPLE 5   **Diagramming the Family Tree of a Honeybee**

Find a sequence that accurately counts the ancestors of a male honeybee. The first term of the sequence should give the number of parents of a male honeybee; the second term, the number of grandparents; the third term, the number of great-grandparents; and so on.

SOLUTION

Recall from the introduction to this section that female honeybees have two parents—a male and a female—but that male honeybees have just one parent—a female.

First, we consider the family tree of a male honeybee. A male bee has one parent: a female. Because his mother had two parents, he has two grandparents. Because his grandmother had two parents and his grandfather had only one parent, he has three great-grandparents.

Now to count the bee's ancestors, we count backward using Figure 8.1. You should recognize the first five terms of the Fibonacci sequence: $1, 2, 3, 5, 8$. This sequence is defined by $a_0 = 1$, $a_1 = 1$, $a_2 = 2$, and $a_k = a_{k-2} + a_{k-1}$, for all $k \geq 3$. ■ ■ ■

Practice Problem 5 Find a sequence that accurately counts the ancestors of a female honeybee. ■

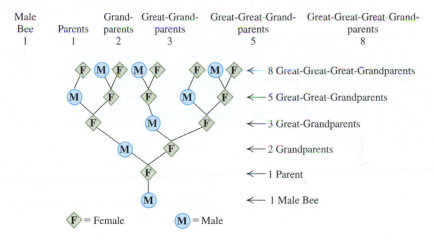

FIGURE 8.1

2 Use factorial notation.

Factorial Notation

Special types of products, called factorials, appear in some very important sequences.

BY THE WAY . . .

The symbol $n!$ was first introduced by Christian Kremp of Strasbourg in 1808 as a convenience for the printer. Alternative symbols $\lfloor n$ and $\pi(n)$ were used until late in the nineteenth century.

A typical calculator will compute approximate values of $n!$ up to $n = 69$.

$$69! \approx 1.7112 \times 10^{98}$$

DEFINITION OF FACTORIAL

For any positive integer n, **n factorial** (written $n!$) is defined as

$$n! = n \cdot (n - 1) \cdots 4 \cdot 3 \cdot 2 \cdot 1.$$

As a special case, **zero factorial** (written $0!$) is defined as

$$0! = 1.$$

Here are the first eight values of $n!$

$0! = 1$	$4! = 4 \cdot 3 \cdot 2 \cdot 1 = 24$
$1! = 1$	$5! = 5 \cdot 4 \cdot 3 \cdot 2 \cdot 1 = 120$
$2! = 2 \cdot 1 = 2$	$6! = 6 \cdot 5 \cdot 4 \cdot 3 \cdot 2 \cdot 1 = 720$
$3! = 3 \cdot 2 \cdot 1 = 6$	$7! = 7 \cdot 6 \cdot 5 \cdot 4 \cdot 3 \cdot 2 \cdot 1 = 5040$

The values of $n!$ get large very quickly. For example, $10! = 3,628,800$ and $12! = 479,001,600$.

In simplifying some expressions involving factorials, it is helpful to note that

$$\boxed{n! = n \cdot (n - 1)!}$$

TECHNOLOGY CONNECTION

EXAMPLE 6   **Simplifying a Factorial Expression**

Simplify.

a. $\dfrac{16!}{14!}$ **b.** $\dfrac{(n + 1)!}{(n - 1)!}$

SOLUTION

a. $\dfrac{16!}{14!} = \dfrac{16 \cdot 15 \cdot \cancel{14!}}{\cancel{14!}}$ $16! = 16 \cdot 15! = 16 \cdot 15 \cdot 14!$

        $= 16 \cdot 15 = 240$ Divide out the factor 14!

b. $\dfrac{(n + 1)!}{(n - 1)!} = \dfrac{(n + 1) \cdot n \cdot \cancel{(n - 1)!}}{\cancel{(n - 1)!}}$ $(n + 1)! = (n + 1) \cdot n! = (n + 1) \cdot n \cdot (n - 1)!$

        $= (n + 1)n$ Divide out the common factor $(n - 1)!$

                                                            ▪ ▪ ▪

Practice Problem 6   Simplify.

a. $\dfrac{13!}{12!}$ **b.** $\dfrac{n!}{(n - 3)!}$ ▪

EXAMPLE 7   **Writing Terms of a Sequence Involving Factorials**

Write the first five terms of the sequence whose general term is $a_n = \dfrac{(-1)^{n+1}}{n!}$.

SOLUTION

Replace n in the formula for the general term with each positive integer from 1 through 5.

$$a_1 = \frac{(-1)^{1+1}}{1!} = \frac{(-1)^2}{1} = 1 \qquad a_4 = \frac{(-1)^{4+1}}{4!} = \frac{(-1)^5}{24} = -\frac{1}{24}$$

$$a_2 = \frac{(-1)^{2+1}}{2!} = \frac{(-1)^3}{2} = -\frac{1}{2} \qquad a_5 = \frac{(-1)^{5+1}}{5!} = \frac{(-1)^6}{120} = \frac{1}{120}$$

$$a_3 = \frac{(-1)^{3+1}}{3!} = \frac{(-1)^4}{6} = \frac{1}{6}$$

Using these five terms, we could write this sequence as

$$1, -\frac{1}{2}, \frac{1}{6}, -\frac{1}{24}, \frac{1}{120}, \ldots .$$

                                                        ▪ ▪ ▪

Practice Problem 7   Write the first five terms of the sequence whose general term is

$$a_n = \frac{(-1)^n 2^n}{n!}.$$

                                                        ▪

3 Use summation notation to write partial sums of a series.

Summation Notation

If we know the general term of a sequence, we can represent a *sum* of terms of the sequence by using *summation* (or *sigma*) *notation*. In this notation, the Greek letter Σ (capital sigma) indicates that we are to *add* the given terms. The letter Σ corresponds to the English letter *S*, the first letter of the word *sum*.

> ### SUMMATION NOTATION
>
> The sum of the first n terms of sequence $a_1, a_2, a_3, \ldots, a_n, \ldots$ is denoted by
> $$\sum_{i=1}^{n} a_i = a_1 + a_2 + a_3 + \cdots + a_n.$$
> The letter i in the summation notation is called the **index of summation**, n is called the **upper limit**, and 1 is called the **lower limit** of the summation.

In summation notation, the index need not start at $i = 1$. Also, any letter may be used in place of the index i.

TECHNOLOGY CONNECTION

 A graphing calculator in Sequence mode can sum terms of a series.

The viewing window shows $\sum_{j=4}^{7} (2j^2 - 1)$. Note that the calculator replaces the variable j with n.

```
sum(seq(2*n^2-1,
n,4,7,1)
            248
■
```

EXAMPLE 8   **Evaluating Sums Given in Summation Notation**

Find each sum.   **a.** $\displaystyle\sum_{i=1}^{9} i$ **b.** $\displaystyle\sum_{j=4}^{7} (2j^2 - 1)$ **c.** $\displaystyle\sum_{k=0}^{4} \frac{2^k}{k!}$

SOLUTION

a. Replace i with successive integers from 1 to 9 inclusive; then add.
$$\sum_{i=1}^{9} i = 1 + 2 + 3 + 4 + 5 + 6 + 7 + 8 + 9 = 45$$

b. The index of summation is j. Evaluate $(2j^2 - 1)$ for all consecutive integers from 4 through 7 inclusive; then add.
$$\sum_{j=4}^{7} (2j^2 - 1) = [2(4)^2 - 1] + [2(5)^2 - 1] + [2(6)^2 - 1] + [2(7)^2 - 1]$$
$$= 31 + 49 + 71 + 97 = 248$$

c. The index of summation is k. Evaluate $\dfrac{2^k}{k!}$ for consecutive integers 0 through 4, inclusive; then add.
$$\sum_{k=0}^{4} \frac{2^k}{k!} = \frac{2^0}{0!} + \frac{2^1}{1!} + \frac{2^2}{2!} + \frac{2^3}{3!} + \frac{2^4}{4!}$$
$$= \frac{1}{1} + \frac{2}{1} + \frac{4}{2} + \frac{8}{6} + \frac{16}{24} = 1 + 2 + 2 + \frac{4}{3} + \frac{2}{3} = 7 \qquad ∎∎∎$$

Practice Problem 8   Find the following sum: $\displaystyle\sum_{k=0}^{3}(-1)^k k!$ ∎

The familiar properties of real numbers, such as the distributive, commutative, and associative properties, can be used to prove the summation properties listed next.

SUMMATION PROPERTIES

Let a_k and b_k represent the general terms of two sequences and let c represent any real number. Then

1. $\displaystyle\sum_{k=1}^{n} c = c \cdot n$
2. $\displaystyle\sum_{k=1}^{n} ca_k = c\sum_{k=1}^{n} a_k$

3. $\displaystyle\sum_{k=1}^{n} (a_k + b_k) = \sum_{k=1}^{n} a_k + \sum_{k=1}^{n} b_k$
4. $\displaystyle\sum_{k=1}^{n} (a_k - b_k) = \sum_{k=1}^{n} a_k - \sum_{k=1}^{n} b_k$

5. $\displaystyle\sum_{k=1}^{n} a_k = \sum_{k=1}^{j} a_k + \sum_{k=j+1}^{n} a_k,$ for $1 \le j < n$

Series

We all know how to add numbers, but do you think it is possible to add infinitely many numbers? It *is* sometimes possible to add infinitely many numbers, although a complete explanation of this kind of addition must be put off until a calculus course. For now, we need some definitions.

DEFINITION OF SERIES

Let $a_1, a_2, a_3, \ldots, a_k, \ldots$ be an infinite sequence. Then

1. The sum of the first n terms of the sequence is called the **nth partial sum** of the sequence and is denoted by

$$a_1 + a_2 + a_3 + \cdots + a_n = \sum_{i=1}^{n} a_i.$$

This sum is a **finite series**.

2. The sum of all terms of the infinite sequence is called an **infinite series** and is denoted by

$$a_1 + a_2 + a_3 + \cdots = \sum_{i=1}^{\infty} a_i.$$

EXAMPLE 9   **Writing a Partial Sum in Summation Notation**

Write each sum in summation notation.

a. $3 + 5 + 7 + \cdots + 21$ **b.** $\dfrac{1}{4} + \dfrac{1}{9} + \cdots + \dfrac{1}{49}$

SOLUTION

a. The finite series $3 + 5 + 7 + \cdots + 21$ is a sum of consecutive odd numbers from 3 to 21. Each of these numbers can be expressed in the form $2k + 1$, starting with $k = 1$ and ending with $k = 10$. With k as the index of summation, 1 as the lower limit, and 10 as the upper limit, we write

$$3 + 5 + 7 + \cdots + 21 = \sum_{k=1}^{10} (2k + 1).$$

b. The finite series $\frac{1}{4} + \frac{1}{9} + \cdots + \frac{1}{49}$ is a sum of fractions, each of which has numerator 1 and denominator k^2, starting with $k = 2$ and ending with $k = 7$. We write

$$\frac{1}{4} + \frac{1}{9} + \cdots + \frac{1}{49} = \sum_{k=2}^{7} \frac{1}{k^2}.$$

■ ■ ■

Practice Problem 9 Write the following in summation notation: $2 - 4 + 6 - 8 + 10 - 12 + 14$

■

SECTION 8.1 ■ Exercises

A EXERCISES Basic Skills and Concepts

1. An infinite sequence is a function whose domain is the set of _____.

2. If a sequence is defined by $a_n = 2n - \dfrac{10}{n}$, then $a_5 =$ _____.

3. By definition, $0! =$ _____.

4. Evaluate $\displaystyle\sum_{k=0}^{5} \dfrac{2k + 1}{2}$.

5. *True or False* If $a_1 = 3$ and $a_{n+1} = a_n^2$, then $a_3 = 27$.

6. *True or False* $\dfrac{(2n)!}{n!} = 2$

In Exercises 7–26, write the first four terms of each sequence.

7. $a_n = 3n - 2$

8. $a_n = 2n + 1$

9. $a_n = 1 - \dfrac{1}{n}$

10. $a_n = 1 + \dfrac{1}{n}$

11. $a_n = -n^2$

12. $a_n = n^3$

13. $a_n = \dfrac{2n}{n + 1}$

14. $a_n = \dfrac{3n}{n^2 + 1}$

15. $a_n = (-1)^{n+1}$

16. $a_n = (-3)^{n-1}$

17. $a_n = 3 - \dfrac{1}{2^n}$

18. $a_n = \left(\dfrac{3}{2}\right)^n$

19. $a_n = 0.6$

20. $a_n = -0.4$

21. $a_n = \dfrac{(-1)^n}{n!}$

22. $a_n = \dfrac{n}{n!}$

23. $a_n = (-1)^n 3^{-n}$

24. $a_n = (-1)^n 3^{n-1}$

25. $a_n = \dfrac{e^n}{2n}$

26. $a_n = \dfrac{2^n}{e^n}$

In Exercises 27–40, write a general term a_n for each sequence. Assume that n begins with 1.

27. $1, 4, 7, 10, \ldots$

28. $7, 9, 11, 13, \ldots$

29. $\dfrac{1}{2}, \dfrac{1}{3}, \dfrac{1}{4}, \dfrac{1}{5}, \ldots$

30. $\dfrac{2}{3}, \dfrac{3}{4}, \dfrac{4}{5}, \dfrac{5}{6}, \ldots$

31. $2, -2, 2, -2, \ldots$

32. $-3, 6, -9, 12, \ldots$

33. $\dfrac{1}{2}, \dfrac{3}{4}, \dfrac{9}{8}, \dfrac{27}{16}, \ldots$

34. $\dfrac{-1}{2}, \dfrac{1}{4}, \dfrac{-1}{8}, \dfrac{1}{16}, \ldots$

35. $1 \cdot 2, 2 \cdot 3, 3 \cdot 4, 4 \cdot 5, \ldots$

36. $\dfrac{-1}{1 \cdot 2}, \dfrac{1}{2 \cdot 3}, \dfrac{-1}{3 \cdot 4}, \dfrac{1}{4 \cdot 5}, \ldots$

37. $2 + \dfrac{1}{2}, 2 - \dfrac{1}{3}, 2 + \dfrac{1}{4}, 2 - \dfrac{1}{5}, \ldots$

38. $1 + \dfrac{1}{2}, 1 + \dfrac{1}{3}, 1 + \dfrac{1}{4}, 1 + \dfrac{1}{5}, \ldots$

39. $\dfrac{3^2}{2}, \dfrac{3^3}{3}, \dfrac{3^4}{4}, \dfrac{3^5}{5}, \ldots$

40. $\dfrac{e}{2}, \dfrac{e^2}{4}, \dfrac{e^3}{8}, \dfrac{e^4}{16}, \ldots$

In Exercises 41–50, write the first five terms of each recursively defined sequence.

41. $a_1 = 2, a_{n+1} = a_n + 3$

42. $a_1 = 5, a_{n+1} = a_n - 1$

43. $a_1 = 3, a_{n+1} = 2a_n$

44. $a_1 = 1, a_{n+1} = \dfrac{1}{2}a_n$

45. $a_1 = 7, a_{n+1} = -2a_n + 3$

46. $a_1 = -4, a_{n+1} = -3a_n - 5$

47. $a_1 = 2, a_{n+1} = \dfrac{1}{a_n}$

48. $a_1 = -1, a_{n+1} = \dfrac{-1}{a_n}$

49. $a_1 = 25, a_{n+1} = \dfrac{(-1)^n}{5a_n}$

50. $a_1 = 12, a_{n+1} = \dfrac{(-1)^n}{3a_n}$

In Exercises 51–56, use a graphing calculator to (a) find the first ten terms of the sequence and (b) graph the first ten terms of the sequence.

51. $a_n = 3n^2 - 1$

52. $a_n = 4 - \dfrac{3}{n}$

53. $a_n = n\left(1 - \dfrac{1}{n}\right)$

54. $a_n = n^3 - n^2$

55. $a_n = 1 - \dfrac{1}{a_{n-1}}, a_1 = \dfrac{1}{2}$

56. $a_n = a_{n-1}^2, a_1 = 1$

In Exercises 57–64, simplify the factorial expression.

57. $\dfrac{3!}{5!}$

58. $\dfrac{8!}{10!}$

59. $\dfrac{12!}{11!}$

60. $\dfrac{20!}{18!}$

61. $\dfrac{n!}{(n+1)!}$

62. $\dfrac{(n-1)!}{(n-2)!}$

63. $\dfrac{(2n+1)!}{(2n)!}$

64. $\dfrac{(2n+1)!}{(2n-1)!}$

In Exercises 65–76, find each sum.

65. $\displaystyle\sum_{k=1}^{7} 5$

66. $\displaystyle\sum_{j=1}^{4} 12$

67. $\displaystyle\sum_{j=0}^{5} j^2$

68. $\displaystyle\sum_{k=0}^{4} k^3$

69. $\displaystyle\sum_{i=1}^{5} (2i-1)$

70. $\displaystyle\sum_{k=0}^{6} (1-3k)$

71. $\displaystyle\sum_{j=1}^{7} \dfrac{j+1}{j}$

72. $\displaystyle\sum_{k=0}^{5} \dfrac{1}{k+1}$

73. $\displaystyle\sum_{i=1}^{6} (-1)^i 3^{i-1}$

74. $\displaystyle\sum_{k=0}^{4} (-1)^{k+1} k$

75. $\displaystyle\sum_{k=1}^{3} (2-k^2)$

76. $\displaystyle\sum_{j=0}^{4} (j^3+1)$

In Exercises 77–84, write each sum in summation notation.

77. $1 + 3 + 5 + 7 + \cdots + 101$

78. $2 + 4 + 6 + 8 + \cdots + 102$

79. $\dfrac{1}{5(1)} + \dfrac{1}{5(2)} + \dfrac{1}{5(3)} + \dfrac{1}{5(4)} + \cdots + \dfrac{1}{5(11)}$

80. $1 + \dfrac{2}{2\cdot3} + \dfrac{2}{3\cdot4} + \dfrac{2}{4\cdot5} + \cdots + \dfrac{2}{9\cdot10}$

81. $1 - \dfrac{1}{2} + \dfrac{1}{3} - \dfrac{1}{4} + \cdots - \dfrac{1}{50}$

82. $1 - 2 + 3 - 4 + \cdots + (-50)$

83. $\dfrac{1}{2} + \dfrac{2}{3} + \dfrac{3}{4} + \dfrac{4}{5} + \cdots + \dfrac{10}{11}$

84. $2 + \dfrac{2^2}{2} + \dfrac{2^3}{3} + \dfrac{2^4}{4} + \cdots + \dfrac{2^{10}}{10}$

 In Exercises 85–90, use a graphing calculator to find each sum.

85. $\displaystyle\sum_{i=1}^{10} 12i^2$

86. $\displaystyle\sum_{i=1}^{50} (2i+7)$

87. $\displaystyle\sum_{k=5}^{30} \dfrac{7}{1-k^2}$

88. $\displaystyle\sum_{k=10}^{25} \dfrac{(k+1)^2}{k}$

89. $\displaystyle\sum_{j=1}^{100} (-1)^j j$

90. $\displaystyle\sum_{j=8}^{50} \left(7 + \dfrac{(-1)^j}{j}\right)$

B EXERCISES Applying the Concepts

91. **Free fall.** In the absence of friction, a freely falling body will fall about 16 feet the first second, 48 feet the next second, 80 feet the third second, 112 feet the fourth second, and so on. How far has it fallen during
  a. the seventh second?
  b. the nth second?

92. **Bacterial growth.** A colony of 1000 bacteria doubles in size every hour. How many bacteria will there be in
  a. two hours?
  b. five hours?
  c. n hours?

93. **Workplace fines.** A contractor who quits work on a house was told she would be fined $50 if she did not resume work on Monday, $75 if she failed to resume work on Tuesday, $100 if she did not resume work on Wednesday, and so on (including weekends). How much will her fine be on the ninth day she fails to show up for work?

94. **Appreciating value.** A painting valued at $30,000 is expected to appreciate $1280 the first year, $1240 the second year, $1200 the third year, and so on. How much will the painting appreciate in
  a. the seventh year?
  b. the tenth year?

95. **Cell phone use.** At the end of the first six months after a company began providing cell phones to its sales force, it averaged 600 cell minutes per month. For the next four years, its monthly cell phone use doubled every six months. How many cell minutes per month were being used at the end of three years?

96. **Motorcycle acceleration.** A motorcycle travels 10 yards the first second and then increases its speed by 20 yards per second in each succeeding second. At this rate, how far will the motorcycle travel during
  a. the eighth second?
  b. the nth second?

97. **Compound interest.** Suppose $10,000 is deposited into an account that earns 6% interest compounded semiannually. The balance in the account after n compounding periods is given by the sequence

$$A_n = 10{,}000\left(1 + \dfrac{0.06}{2}\right)^n, n = 1, 2, 3, \ldots.$$

  a. Find the first six terms of this sequence.
  b. Find the balance in the account after eight years.

98. **Compound interest.** Suppose $100 is deposited at the beginning of a year into an account that earns 8% interest compounded quarterly. The balance in the account after n compounding periods is

$$A_n = 100\left(1 + \dfrac{0.08}{4}\right)^n, n = 1, 2, 3, \ldots.$$

  a. Find the first six terms of this sequence.
  b. Find the balance in the account after ten years.

99. **Real estate value.** Analysts estimate that a $100,000 condominium will increase in value 5% each year for the next seven years. Find the value of the condominium in each of those years. Write a formula for a sequence whose first seven terms give these values.

100. **Diminishing savings.** Laura withdraws 10% of her $50,000 savings each year. Find the amount remaining in her account for each of the next ten years. Write a formula for a sequence whose first ten terms give these values.

C EXERCISES Beyond the Basics

101. Find a formula for the nth term of the sequence defined recursively by $a_1 = \sqrt{2}$, $a_{n+1} = \sqrt{2a_n}$. [*Hint:* Write each of the first five terms as a power of 2.]

102. The triangular tiles used in the figures shown have white interiors and gold edges. A sequence of figures is obtained by adding one triangle to the previous figure.

One tile   Two tiles   Three tiles

    a. Write a recursive sequence whose nth term, a_n, gives the number of gold edges in the nth figure.

    b. Write a recursive sequence whose nth term, b_n, gives the number of gold edges that lie on the perimeter of the nth figure.

103. The figure shows a sequence of successively smaller squares formed by connecting the midpoints of the sides of the preceding larger square. The area of the largest square is 1.

    a. Write the first five terms of the sequence whose general term, a_n, gives the area of the nth square.

    b. Write the general term, a_n, of the sequence.

104. A sequence of concentric circles is designed so that the radius of the first circle is 1 and each successive circle has a radius that is twice the length of the radius of the preceding circle. Write a recursive sequence whose nth term, a_n, gives the area of the nth circle.

105. Find the first ten terms of the sequence defined recursively by $a_1 = a_2 = 1$, $a_n = a_{a_{n-1}} + a_{n-a_{n-1}}$. In 1989, a prize of $1000 was offered to whoever first discovered (for this sequence) a value of n for which
$$\left|\frac{a_i}{i} - \frac{1}{2}\right| < \frac{1}{20} \text{ for all } i > n.$$
A mathematician named Colin L. Mallows of AT&T/Bell Laboratories found that $n = 1489$ and claimed the prize. (*Source:* http://el.media.mit.edu/logo-foundation/pubs/papers/easy_as_11223.html)

106. A sequence of numbers $a_0, a_1, a_2, a_3, \ldots$ satisfies the equation
$$a_n^2 = (-1)^n a_{n-1} + a_{n+1}.$$
Assuming that $a_0 = 1$ and $a_1 = 3$, find a_3.

107. Find a sequence with general term a_m so that
$$\sum_{n=1}^{20} n^2 = \sum_{m=0}^{19} a_m.$$

108. Find a real number c such that
$$\sum_{n=1}^{50} (n-3)^2 = \sum_{n=1}^{50} (n^2 - 6n) + \sum_{n=1}^{50} c.$$

109. Find a lower limit p and an upper limit q such that
$$\sum_{n=0}^{10} n^3 = \sum_{m=p}^{q} (m+2)^3.$$

110. Give an example of two sequences a_k and b_k such that
$$\sum_{k=1}^{20} a_k b_k \neq \left(\sum_{k=1}^{20} a_k\right)\left(\sum_{k=1}^{20} b_k\right).$$

111. Show that $7! - 2(5!) = 40(5!)$.

112. Show that $\dfrac{10!}{6!\,4!} + \dfrac{10!}{5!\,5!} = \dfrac{11!}{6!\,5!}$.

Critical Thinking

113. Find the largest integer value for the upper limit k if
$$\sum_{n=1}^{k} (n^2 + n)$$
$$= \sum_{n=1}^{k} [(n-1)(n-2)(n-3)(n-4)(n-5) + n^2 + n].$$

114. **The Ulam conjecture.** Define the sequence a_n recursively by
$$a_n = \begin{cases} \dfrac{a_{n-1}}{2} & \text{if } a_{n-1} \text{ is even.} \\[2mm] 3a_{n-1} + 1 & \text{if } a_{n-1} \text{ is odd} \end{cases}$$

The Ulam conjecture says that if your first term a_0 is any positive integer, the sequence a_n will eventually reach the integer 1. Verify the conjecture for $a_0 = 13$.

Arithmetic Sequences; Partial Sums

Before Starting this Section, Review

1. The general term of a sequence (Section 8.1, page 659)

2. Partial sums (Section 8.1, page 666)

Objectives

1. Identify an arithmetic sequence and find its common difference.

2. Find the sum of the first *n* terms of an arithmetic sequence.

FALLING SPACE JUNK

On April 27, 2000, more than 300 kilograms (700 pounds) of "space junk" crashed to the ground in South Africa. The material was eventually identified as part of a Delta 2 rocket used to launch a global positioning satellite in 1996. Once an object begins a free fall toward Earth, it falls (in the absence of friction) 16 feet in the first second, 48 feet in the next second, 80 feet in the third second, and so on. The number of feet traveled in succeeding seconds is the sequence

$$16, 48, 80, 112, \ldots.$$

This is an example of an *arithmetic sequence*. We investigate such a sequence in Example 5. ■

---

1 Identify an arithmetic sequence and find its common difference.

When the difference between any two consecutive terms of a sequence is always the same number, the sequence is called an *arithmetic sequence*. In other words, the sequence $a_1, a_2, a_3, a_4, \ldots$ is an arithmetic sequence if $a_2 - a_1 = a_3 - a_2 = a_4 - a_3 = \ldots$.

DEFINITION OF ARITHMETIC SEQUENCE

The sequence

$$a_1, a_2, a_3, a_4, \ldots, a_n, \ldots$$

is an **arithmetic sequence**, or an **arithmetic progression**, if there is a number d such that each term in the sequence except the first is obtained from the preceding term by adding d to it. The number d is called the **common difference** of the arithmetic sequence. We have

$$d = a_{n+1} - a_n, \quad n \geq 1.$$

EXAMPLE 1 **Finding the Common Difference**

Find the common difference of each arithmetic sequence.

a. $3, 23, 43, 63, 83, \ldots$ **b.** $29, 19, 9, -1, -11, \ldots$

SOLUTION

a. The common difference is $23 - 3 = 20$ (or $43 - 23, 63 - 43,$ or $83 - 63$).

b. The common difference is $19 - 29 = -10$ (or $9 - 19, -1 - 9,$ or $-11 - (-1)$).

■ ■ ■

Practice Problem 1 Find the common difference of the arithmetic sequence

$$3, -2, -7, -12, -17, \ldots.$$

■

An arithmetic sequence can be completely specified by giving the first term a_1 and the common difference d. Let's see why. Suppose we are told that the sequence $a_1, a_2, a_3, a_4, \ldots$ is an arithmetic sequence and that $a_1 = 5$ and $d = 4$. Then

$$a_2 - a_1 = 4, \quad \text{so} \quad a_2 = a_1 + 4 = 5 + 4 = 9;$$
$$a_3 - a_2 = 4, \quad \text{so} \quad a_3 = a_2 + 4 = 9 + 4 = 13;$$
$$a_4 - a_3 = 4, \quad \text{so} \quad a_4 = a_3 + 4 = 13 + 4 = 17; \text{ and so on.}$$

Rewriting $d = a_{n+1} - a_n$ as $a_{n+1} = a_n + d$, we have a recursive definition of an arithmetic sequence.

RECURSIVE DEFINITION OF AN ARITHMETIC SEQUENCE

An arithmetic sequence $a_1, a_2, a_3, a_4, \ldots, a_n, \ldots$ can be defined recursively. The recursive formula

$$a_{n+1} = a_n + d \text{ for } n \geq 1$$

defines an arithmetic sequence with first term a_1 and common difference d.

Consider an arithmetic sequence with first term a_1 and common difference d. Using the recursive definition $a_{n+1} = a_n + d$, for $n \geq 1$, we write

$a_1 = a_1$	The first term is a_1.
$a_2 = a_1 + d$	Replace n with 1 in $a_{n+1} = a_n + d$.
$a_3 = a_2 + d = (a_1 + d) + d = a_1 + 2d$	Replace a_2 with $a_1 + d$; simplify.
$a_4 = a_3 + d = (a_1 + 2d) + d = a_1 + 3d$	Replace a_3 with $a_1 + 2d$; simplify.
$a_5 = a_4 + d = (a_1 + 3d) + d = a_1 + 4d$	Replace a_4 with $a_1 + 3d$; simplify.

$$\vdots$$

$$
\begin{aligned}
a_n &= a_{n-1} + d \\
&= [a_1 + (n-2)d] + d \qquad &&\text{Replace } a_{n-1} \text{ with } a_1 + (n-2)d. \\
&= a_1 + (n-1)d &&\text{Simplify.}
\end{aligned}
$$

Thus, we can find the general term a_n from the values a_1 and d.

---

nTH TERM OF AN ARITHMETIC SEQUENCE

If a sequence $a_1, a_2, a_3, \ldots$ is an arithmetic sequence, then its nth term, a_n, is given by

$$a_n = a_1 + (n-1)d,$$

where a_1 is the first term and d is the common difference.

EXAMPLE 2   **Finding the nth Term of an Arithmetic Sequence**

Find an expression for the nth term of the arithmetic sequence

$$3, 7, 11, 15, \ldots.$$

SOLUTION

Because $a_1 = 3$ and $d = 7 - 3 = 4$, we have

$$a_n = a_1 + (n - 1)d \qquad \text{Formula for the } n\text{th term}$$
$$= 3 + (n - 1)4 \qquad \text{Replace } a_1 \text{ with 3 and } d \text{ with 4.}$$
$$= 3 + 4n - 4 = 4n - 1 \qquad \text{Simplify.}$$

Note that in the expression $a_n = 4n - 1$, we get the sequence $3, 7, 11, 15, \ldots$ by substituting $n = 1, 2, 3, 4, \ldots$. ■ ■ ■

Practice Problem 2 Find an expression for the nth term of the arithmetic sequence

$$-3, 1, 5, 9, 13, 17, \ldots.$$ ■

EXAMPLE 3 **Finding the Common Difference of an Arithmetic Sequence**

Find the common difference d and the nth term a_n of the arithmetic sequence whose 5th term is 15 and whose 20th term is 45.

SOLUTION

$$a_n = a_1 + (n - 1)d \qquad \text{Formula for the } n\text{th term}$$
$$45 = a_1 + (20 - 1)d \qquad \text{Replace } n \text{ with 20; } a_n = a_{20} = 45.$$
$$(1) \quad 45 = a_1 + 19d \qquad \text{Simplify.}$$

Also,

$$15 = a_1 + (5 - 1)d \qquad \text{Replace } n \text{ with 5; } a_n = a_5 = 15.$$
$$(2) \quad 15 = a_1 + 4d \qquad \text{Simplify.}$$

Solving the system of equations,

$$\begin{cases} a_1 + 19d = 45 & (1) \\ a_1 + 4d = 15 & (2) \end{cases}$$

gives $d = 2$ and $a_1 = 7$. We substitute $a_1 = 7$ and $d = 2$ in the formula for the nth term.

$$a_n = a_1 + (n - 1)d \qquad \text{Formula for the } n\text{th term}$$
$$a_n = 7 + (n - 1)2 \qquad \text{Replace } a_1 \text{ with 7 and } d \text{ with 2.}$$
$$= 7 + 2n - 2 = 2n + 5 \qquad \text{Simplify.}$$

The nth term of this sequence is given by

$$a_n = 2n + 5, \quad n \geq 1.$$ ■ ■ ■

Practice Problem 3 Find the common difference d and the nth term a_n of the arithmetic sequence whose 4th term is 41 and whose 15th term is 8. ■

---

2 Find the sum of the first n terms of an arithmetic sequence.

Sum of an Arithmetic Sequence

We can use the following formula in many applications involving the sum of an arithmetic sequence.

$$1 + 2 + 3 \cdots + n = \frac{n(n + 1)}{2}$$

It is said that Karl Friedrich Gauss (1777–1855) discovered this formula when he realized that any sum of numbers added in reverse order produces the same sum. Therefore, if S denotes the sum of the first n natural numbers, then

$$
\begin{array}{rccccccc}
S = & 1 & + & 2 & + & 3 & + \cdots + & n \\
S = & n & + & (n-1) & + & (n-2) & + \cdots + & 1 \\
\hline
2S = & (n+1) & + & (n+1) & + & (n+1) & + \cdots + & (n+1)
\end{array}
\qquad \text{Add } S + S = 2S.
$$

$$\underbrace{}_{n \text{ terms}}$$

$$2S = n(n+1)$$
$$S = \frac{n(n+1)}{2} \qquad\qquad \text{Divide both sides by 2.}$$

For example, if $n = 100$, we have

$$S = 1 + 2 + 3 + \cdots + 100 = \frac{100(100 + 1)}{2}$$
$$= \frac{100(101)}{2} = 5050.$$

We can use this method to calculate the sum S_n of the first n terms of any arithmetic sequence.

First, we write the sum S_n of the first n terms.

$$S_n = a_1 + (a_1 + d) + (a_1 + 2d) + (a_1 + 3d) + \cdots + a_n$$

We can also write S_n (with the terms in reverse order) by starting with a_n and *subtracting* the common difference d.

$$S_n = a_n + (a_n - d) + (a_n - 2d) + (a_n - 3d) + \cdots + a_1$$

Adding the two equations for S_n, we find that the d's in the sums "drop out" (for example, $(a_1 + d) + (a_n - d) = a_1 + a_n$) and we have

$$2S_n = \underbrace{(a_1 + a_n) + (a_1 + a_n) + \cdots + (a_1 + a_n)}_{n \text{ terms}}$$

$$2S_n = n(a_1 + a_n)$$
$$S_n = n\left(\frac{a_1 + a_n}{2}\right) \qquad \text{Divide both sides by 2.}$$

SUM OF n TERMS OF AN ARITHMETIC SEQUENCE

Let $a_1, a_2, a_3, \ldots a_n$ be the first n terms of an arithmetic sequence with common difference d. The sum S_n of these n terms is given by

$$S_n = n\left(\frac{a_1 + a_n}{2}\right),$$

where $a_n = a_1 + (n - 1)d$.

EXAMPLE 4 **Finding the Sum of Terms of a Finite Arithmetic Sequence**

Find the sum of the following arithmetic sequence of numbers:

$$1 + 4 + 7 + \cdots + 25$$

SOLUTION

In this arithmetic sequence, $a_1 = 1$ and $d = 3$. Let's first find the number of terms.

$$a_n = a_1 + (n - 1)d \qquad \text{Formula for } n\text{th term}$$
$$25 = 1 + (n - 1)3 \qquad \text{Replace } a_n \text{ with 25, } a_1 \text{ with 1, and } d \text{ with 3.}$$
$$24 = (n - 1)3 \qquad \text{Subtract 1 from both sides.}$$
$$8 = n - 1 \qquad \text{Divide both sides by 3.}$$
$$n = 9 \qquad \text{Solve for } n.$$

$$S_n = n\left(\frac{a_1 + a_n}{2}\right) \qquad \text{Formula for } S_n$$

$$S_9 = 9\left(\frac{1 + 25}{2}\right) \qquad \text{Replace } n \text{ with 9, } a_1 \text{ with 1, and } a_n(= a_9) \text{ with 25.}$$

$$= 9(13) = 117 \qquad \text{Simplify.}$$

So $1 + 4 + 7 + \cdots + 25 = 117$. ■ ■ ■

Practice Problem 4 Find the sum $\dfrac{2}{3} + \dfrac{5}{6} + 1 + \dfrac{7}{6} + \dfrac{4}{3} + \dfrac{3}{2} + \dfrac{5}{3} + \dfrac{11}{6} + 2 + \dfrac{13}{6}$. ■

EXAMPLE 5 **Calculating the Distance Traveled by a Freely Falling Object**

In the introduction to this section, we described the arithmetic sequence 16, 48, 80, 112, ... whose terms gave the number of feet that freely falling space junk falls in successive seconds. For this sequence, find the following:

a. The common difference d

b. The nth term a_n

c. The distance the object travels in ten seconds

SOLUTION

a. The common difference $d = a_2 - a_1 = 48 - 16 = 32$.

b. $a_n = a_1 + (n - 1)d \qquad \text{Formula for the } n\text{th term}$
$\quad = 16 + (n - 1)32 \qquad \text{Replace } a_1 \text{ with 16 and } d \text{ with 32.}$
$\quad = 32n - 16 \qquad \text{Simplify.}$

c. The sum of the first ten terms of this sequence gives the distance the object travels in ten seconds. To find a_{10} use the formula:

$$a_n = 32n - 16 \qquad \text{From part b}$$
$$a_{10} = 32(10) - 16 \qquad \text{Replace } n \text{ with 10.}$$
$$= 304$$

$$S_n = n\left(\frac{a_1 + a_n}{2}\right) \qquad \text{Formula for } S_n$$

$$S_{10} = 10\left(\frac{16 + 304}{2}\right) \qquad \begin{array}{l}\text{Replace } n \text{ with 10, } a_1 \text{ with 16, and} \\ a_n(= a_{10}) \text{ with 304.}\end{array}$$

$$= 1600 \qquad \text{Simplify.}$$

The object falls 1600 feet in ten seconds. ■ ■ ■

Practice Problem 5 In Example 5, find the distance the object travels during the 11th through 15th seconds. ■

SECTION 8.2 ■ Exercises

A EXERCISES Basic Skills and Concepts

1. If 5 is the common difference of an arithmetic sequence with general term a_n, then $a_{17} - a_{16} =$ _____ .

2. If 5 is the common difference of an arithmetic sequence with general terms a_n and $a_{21} = 7$, then $a_{22} =$ _____ .

3. If 14 is the term immediately following the sequence term 17 in an arithmetic sequence, then the common difference is _____ .

4. If $a_1 = 2$ and $a_{11} = 22$ for an arithmetic sequence, then the common difference, d, is _____ .

5. *True or False* The common difference of an arithmetic sequence is always positive.

6. *True or False* The sum of the first n terms of an arithmetic sequence always equals n times the average of its first and nth terms.

In Exercises 7–20, determine whether each sequence is arithmetic. For those that are, find the first term a_1 and the common difference d.

7. $1, 2, 3, 4, 5, \ldots$

8. $1, 3, 5, 7, 9, \ldots$

9. $2, 5, 8, 11, 14, \ldots$

10. $10, 7, 4, 1, -2, \ldots$

11. $1, \dfrac{1}{2}, 0, -\dfrac{1}{2}, -\dfrac{1}{4}, \ldots$

12. $2, 4, 8, 16, 32, \ldots$

13. $1, -1, 2, -2, 3, \ldots$

14. $-\dfrac{1}{4}, \dfrac{1}{4}, \dfrac{3}{4}, \dfrac{5}{4}, \dfrac{7}{4}, \ldots$

15. $0.6, 0.2, -0.2, -0.6, -1, \ldots$

16. $2.3, 2.7, 3.1, 3.5, 3.9, \ldots$

17. $a_n = 2n + 6$

18. $a_n = 1 - 5n$

19. $a_n = 1 - n^2$

20. $a_n = 2n^2 - 3$

In Exercises 21–30, find an expression for the nth term of each arithmetic sequence.

21. $5, 8, 11, 14, 17, \ldots$

22. $4, 7, 10, 13, 16, \ldots$

23. $11, 6, 1, -4, -9, \ldots$

24. $9, 5, 1, -3, -7, \ldots$

25. $\dfrac{1}{2}, \dfrac{1}{4}, 0, -\dfrac{1}{4}, -\dfrac{1}{2}, \ldots$

26. $\dfrac{2}{3}, \dfrac{5}{6}, 1, \dfrac{7}{6}, \dfrac{4}{3}, \ldots$

27. $-\dfrac{3}{5}, -1, -\dfrac{7}{5}, \dfrac{9}{5}, -\dfrac{11}{5}, \ldots$

28. $\dfrac{1}{2}, 2, \dfrac{7}{2}, 5, \dfrac{13}{2}, \ldots$

29. $e, 3 + e, 6, + e, 9 + e, 12 + e, \ldots$

30. $2\pi, 2(\pi + 2), 2(\pi + 4), 2(\pi + 6), 2(\pi + 8), \ldots$

In Exercises 31–36, find the common difference d and the nth term a_n of each arithmetic sequence with the specified terms.

31. 4th term 21; 10th term 60

32. 3rd term 15; 21st term 87

33. 7th term 8; 15th term -8

34. 5th term 12; 18th term -1

35. 3rd term 7; 23rd term 17

36. 11th term -1; 31st term 5

In Exercises 37–46, find the sum of each arithmetic sequence.

37. $1 + 2 + 3 + \cdots + 50$

38. $2 + 4 + 6 + \cdots + 102$

39. $1 + 3 + 5 + \cdots + 99$

40. $5 + 10 + 15 + \cdots + 200$

41. $3 + 6 + 9 + \cdots + 300$

42. $4 + 7 + 10 + \cdots + 301$

43. $2 - 1 - 4 - \cdots - 34$

44. $-3 - 8 - 13 - \cdots - 48$

45. $\dfrac{1}{3} + 1 + \dfrac{5}{3} + \cdots + 7$

46. $\dfrac{3}{5} + 2 + \dfrac{17}{5} + \cdots + \dfrac{101}{5}$

In Exercises 47–52, find the sum of the first n terms of the given arithmetic sequence.

47. $2, 7, 12, \ldots; n = 50$

48. $8, 10, 12, \ldots; n = 40$

49. $-15, -11, -7, \ldots; n = 20$

50. $-20, -13, -6, \ldots; n = 25$

51. $3.5, 3.7, 3.9, \ldots; n = 100$

52. $-7, -6.5, -6, \ldots; n = 80$

In Exercises 53–58, find n for the given value of a_n in each arithmetic sequence.

53. $a_n = 75; 1, 3, 5, \ldots$

54. $a_n = 120; 2, 4, 6, \ldots$

55. $a_n = 95; -1, 3, 7, \ldots$

56. $a_n = 83; -5, -3, -1, \ldots$

57. $a_n = 50\sqrt{3}; 2\sqrt{3}, 4\sqrt{3}, 6\sqrt{3}, \ldots$

58. $a_n = 73\pi; 3\pi, 5\pi, 7\pi, \ldots$

B EXERCISES Applying the Concepts

In Exercises 59–65, assume that the indicated sequence is arithmetic.

59. **Orange picking.** When Eric started work as an orange picker, he picked 10 oranges in the first minute, 12 in the second minute, 14 in the third minute, and so on. How many oranges did Eric pick in the first half hour?

60. **Exercise.** Walking up a steep hill, Jan walks 60 feet in the first minute, 57 feet in the second minute, 54 feet in the third minute, and so on.
    a. How far will Jan walk in the nth minute?
    b. How far will Jan walk in the first 15 minutes?

61. Contest winner. A contest winner will receive money each month for three years. The winner receives $50 the first month, $75 the second month, $100 the third month, and so on. How much money will the winner have collected after 30 months?

62. Salary. Darren took a 12-month temporary job with a monthly salary that increased a fixed amount each month. He can't recall the starting salary, but does remember that he was paid $820 at the end of the third month and $910 for his last month's work. How much was Darren's total pay for the entire 12 months?

63. Hourly wage. Antonio's new weekend job started at an hourly wage of $12.75. He is guaranteed a raise of 25¢ an hour every three months for the next four years. What will Antonio's hourly wage be at the end of four years?

64. Competing job offers. Denzel is considering offers from two companies. A marketing company pays $32,500 the first year and guarantees a raise of $1300 each year; an exporting company pays $36,000 the first year, with a guaranteed raise of $400 each year. Over a five-year period, which company will pay more? How much more?

65. Theater seating. A theater has 25 rows of seats. The first row has 20 seats, the second row has 22 seats, the third row has 24 seats, and so on. How many seats are in the theater?

66. Bricks in a driveway. A brick driveway has 50 rows of bricks. The first row has 16 bricks, and the fiftieth row has 65 bricks. Assuming that the sequence of numbers giving the number of bricks in rows 1 through 50 is arithmetic, how many bricks does the driveway contain?

67. Stacked logs. A stack of boxes has 17 rows. The bottom row has 40 boxes, and the top row has 8 boxes. Assuming that the sequence of numbers that gives the number of boxes in succeeding rows is arithmetic, how many boxes does the stack contain?

C EXERCISES Beyond the Basics

68. Elena wanted to teach her young daughter Sophia about saving. She gave Sophia 10¢ the first day and an additional 5¢ per day on each subsequent day. After a while, Sophia counted her money and found that she had saved a total of $3.25. How many days had she been saving?

69. Only three customers showed up for the opening day of a flea market. However, nine came the second day, and an additional six customers came on each subsequent day. After several days, there was a total of 192 customers. How many days had the flea market been open?

70. Find the sum of all natural numbers between 45 and 100 that are divisible by 3.

71. Find the sum of all natural numbers between 26 and 120 that are divisible by 7.

72. A sequence is **harmonic** if the reciprocals of the terms of the sequence form an arithmetic sequence. Is the sequence $\frac{1}{2}, \frac{3}{5}, \frac{3}{4}, 1, \frac{3}{2}, 3 \ldots$ harmonic? Explain your reasoning.

73. Consider the harmonic sequence $\frac{2}{5}, \frac{2}{7}, \frac{2}{9}, \ldots$

　a. Find the fourth term.
　b. Find the nth term.

74. Show that if the numbers a, m, b are consecutive terms in an arithmetic sequence, then $m = \frac{a + b}{2}$. We call m the **arithmetic mean** of a and b.

75. If the numbers $a, m_1, m_2, \ldots, m_k, b$ form an arithmetic sequence, we say that the numbers $m_1, m_2, \ldots, m_k$ are k **arithmetic means** between a and b. Insert k arithmetic means between 1 and 30 so that the sum of the resulting series is 465.

Critical Thinking

76. Find the nth term of an arithmetic sequence whose first term is 10 and whose 21st term is 0.

77. Assuming that a_1 and d are the first term and common difference, respectively, of an arithmetic sequence, find the first term and difference of an arithmetic sequence whose terms are the negatives of the terms in the given sequence.

78. How many terms are in the series $\sum_{i=22}^{100} 17i^3$?

79. The sum of the first n counting numbers is 12,403. Find the value of n.

Geometric Sequences and Series

Before Starting this Section, Review

1. The general term of a sequence (Section 8.1, page 659)

2. Partial sums (Section 8.1, page 666)

3. Compound interest (Section 4.2, page 420)

Objectives

1. Identify a geometric sequence and find its common ratio.

2. Find the sum of a finite geometric sequence.

3. Solve annuity problems.

4. Find the sum of an infinite geometric sequence.

SPREADING THE WEALTH

Cities across the nation compete to host a Super Bowl. Cities across the world compete to host the Olympic Games. What is the incentive? The *multiplier effect* is often listed among the benefits for hosting one of these events. The effect refers to the phenomenon that occurs when money spent directly on an event has a ripple effect on the economy that is many times the amount spent directly on the event.

Suppose, for example, that $10 million is spent directly by tourists for hotels, meals, tickets, cabs, and so on. Some portion of that amount will be respent by its recipients on groceries, clothes, gas, and so on.

This process is repeated over and over, affecting the economy much more than the initial $10 million spent. Under certain assumptions, the spending just described results in an *infinite geometric series*, whose sum is called the *multiplier*. Example 8 illustrates this effect. ■

---

1 Identify a geometric sequence and find its common ratio.

When the *ratio* of any two consecutive terms of a sequence is always the same number, the sequence is called a *geometric sequence*. In other words, the sequence $a_1, a_2, a_3, a_4, \ldots$ is a geometric sequence if

$$\frac{a_2}{a_1} = \frac{a_3}{a_2} = \frac{a_4}{a_3} = \cdots.$$

DEFINITION OF GEOMETRIC SEQUENCE

The sequence

$$a_1, a_2, a_3, a_4, \ldots, a_n, \ldots$$

is a **geometric sequence**, or a **geometric progression**, if there is a number r such that each term except the first in the sequence is obtained by multiplying the previous term by r. The number r is called the **common ratio** of the geometric sequence.

$$\frac{a_{n+1}}{a_n} = r, \quad n \geq 1$$

EXAMPLE 1 Finding the Common Ratio

Find the common ratio for each geometric sequence.

a. $2, 10, 50, 250, 1250, \ldots$ **b.** $-162, -54, -18, -6, -2, \ldots$

SOLUTION

a. The common ratio is $\dfrac{10}{2} = 5$ $\left(\text{or } \dfrac{50}{10}, \dfrac{250}{50}, \text{ or } \dfrac{1250}{250}\right)$.

b. The common ratio is $\dfrac{-54}{-162} = \dfrac{1}{3}$ $\left(\text{or } \dfrac{-18}{-54}, \dfrac{-6}{-18}, \text{ or } \dfrac{-2}{-6}\right)$.   ▪ ▪ ▪

Practice Problem 1 Find the common ratio for the geometric sequence 6, 18, 54, 162, 486, ...   ▪

A geometric sequence can be completely specified by giving the first term a_1 and the common ratio r. Suppose, for example, that we are told that the sequence $a_1, a_2, a_3, a_4, \ldots$ is geometric and that $a_1 = 3$ and $r = 4$. Then we can find a_2.

$$\frac{a_2}{a_1} = 4, \text{ so } a_2 = 4a_1 = 4 \cdot 3 = 12$$

Now we continue.

$$\frac{a_3}{a_2} = 4, \text{ so } a_3 = 4a_2 = 4 \cdot 12 = 48;$$

$$\frac{a_4}{a_3} = 4, \text{ so } a_4 = 4a_3 = 4 \cdot 48 = 192; \text{ and so on.}$$

By rewriting $\dfrac{a_{n+1}}{a_n} = r$ as $a_{n+1} = ra_n$, we have a recursive definition of a geometric sequence.

RECURSIVE DEFINITION OF A GEOMETRIC SEQUENCE

A geometric sequence $a_1, a_2, a_3, a_4, \ldots, a_n, \ldots$ can be defined recursively. The recursive formula

$$a_{n+1} = ra_n, \quad n \geq 1$$

defines a geometric sequence with the first term a_1 and the common ratio r.

EXAMPLE 2 **Determining Whether a Sequence Is Geometric**

Determine whether each sequence is geometric. If it is, find the first term and the common ratio.

a. $a_n = 4^n$ b. $b_n = \dfrac{3}{2^n}$ c. $c_n = 1 - 5^n$

SOLUTION

a. The sequence $4, 16, 64, 256, \ldots, 4^n, \ldots$ appears to be geometric, with $a_1 = 4$ and $r = 4$. To verify this, we check that $\dfrac{a_{n+1}}{a_n} = 4$ for all $n \geq 1$.

$$a_{n+1} = 4^{n+1} \qquad \text{Replace } n \text{ with } n+1 \text{ in the general term } a_n = 4^n.$$

$$\frac{a_{n+1}}{a_n} = \frac{4^{n+1}}{4^n} = 4 \qquad \frac{4^{n+1}}{4^n} = \frac{4^{\cancel{n}}4}{4^{\cancel{n}}} = 4$$

Consequently, $\dfrac{a_{n+1}}{a_n} = 4$ for all $n \geq 1$, and the sequence is geometric.

b. The sequence $\dfrac{3}{2}, \dfrac{3}{4}, \dfrac{3}{8}, \dfrac{3}{16}, \ldots$ appears to be geometric, with $b_1 = \dfrac{3}{2}$ and $r = \dfrac{1}{2}$.

To verify this, we check that $\dfrac{b_{n+1}}{b_n} = \dfrac{1}{2}$ for all $n \geq 1$.

$$b_{n+1} = \dfrac{3}{2^{n+1}} \qquad \text{Replace } n \text{ with } n + 1 \text{ in the general term } b_n = \dfrac{3}{2^n}.$$

$$\dfrac{b_{n+1}}{b_n} = \dfrac{3}{2^{n+1}} \div \dfrac{3}{2^n} \qquad \dfrac{b_{n+1}}{b_n} = b_{n+1} \div b_n$$

$$= \dfrac{3}{2^{n+1}} \cdot \dfrac{2^n}{3} = \dfrac{1}{2} \qquad 2^{n+1} = 2 \cdot 2^n$$

Consequently, $\dfrac{b_{n+1}}{b_n} = \dfrac{1}{2}$ for all $n \geq 1$, and the sequence is geometric.

c. The sequence $-4, -24, -124, -624, \ldots, 1 - 5^n, \ldots$ is *not* a geometric sequence because not all pairs of consecutive terms have identical quotients. The quotient of the first two terms is $\dfrac{-24}{-4} = 6$, but the quotient of the third and second terms is $\dfrac{-124}{-24} = \dfrac{31}{6}$. ■ ■ ■

Practice Problem 2 Determine whether the sequence $a_n = \left(\dfrac{3}{2}\right)^n$ is geometric. If it is, find the first term and the common ratio. ■

From the equation $a_{n+1} = a_n r$ used in the recursive definition of a geometric sequence, we see that each term after the first is obtained by multiplying the preceding term by the common ratio r.

THE GENERAL TERM OF A GEOMETRIC SEQUENCE

Every geometric sequence can be written in the form

$$a_1, a_1 r, a_1 r^2, a_1 r^3, \ldots, a_1 r^{n-1}, \ldots,$$

where r is the common ratio. Because $a_1 = a_1(1) = a_1 r^0$, the **nth term of the geometric sequence** is

$$a_n = a_1 r^{n-1}, \text{ for } n \geq 1.$$

EXAMPLE 3   **Finding Terms in a Geometric Sequence**

For the geometric sequence $1, 3, 9, 27, \ldots$, find each of the following.

a. a_1 **b.** r **c.** a_n

SOLUTION

a. The first term of the sequence is given: $a_1 = 1$.

b. The sequence is geometric, so we can take the ratio of any two consecutive terms. The ratio of the first two terms gives us

$$r = \dfrac{3}{1} = 3.$$

c. $a_n = a_1 r^{n-1} \qquad$ Formula for the *n*th term

$ = (1)(3^{n-1}) \qquad$ Replace a_1 with 1 and r with 3.

$ = 3^{n-1}$ ■ ■ ■

Practice Problem 3 For the geometric sequence $2, \dfrac{6}{5}, \dfrac{18}{25}, \dfrac{54}{125}, \ldots$, find the following.

a. a_1 **b.** r **c.** a_n ■

Notice in Example 3 that converting $a_n = 3^{n-1}$ to standard function notation gives us $f(n) = 3^{n-1}$, an exponential function. A geometric sequence with first term a_1 and common ratio r can be viewed as an exponential function $f(n) = a_1 r^{n-1}$ with the set of natural numbers as its domain.

EXAMPLE 4   Finding a Particular Term in a Geometric Sequence

Find the 23rd term of a geometric sequence whose first term is 10 and whose common ratio is 1.2.

SOLUTION

$$a_n = a_1 r^{n-1} \qquad \text{Formula for the } n\text{th term}$$
$$a_{23} = 10(1.2)^{23-1} \qquad \text{Replace } n \text{ with 23, } a_1 \text{ with 10, and } r \text{ with 1.2.}$$
$$= 10(1.2)^{22}$$
$$\approx 552.06 \qquad \text{Use a calculator.} \qquad \blacksquare\,\blacksquare\,\blacksquare$$

Practice Problem 4 Find the 18th term of a geometric sequence whose first term is 7 and whose common ratio is 1.5. ■

2 Find the sum of a finite geometric sequence.

Finding the Sum of a Finite Geometric Sequence

We can find a formula for the sum S_n of the first n terms in a geometric sequence. (We have already found a similar formula for arithmetic sequences.)

$$S_n = a_1 + a_1 r + a_1 r^2 + a_1 r^3 + \cdots + a_1 r^{n-1} \qquad \text{Definition of } S_n$$
$$-rS_n = -a_1 r - a_1 r^2 - a_1 r^3 - a_1 r^4 - \cdots - a_1 r^n \qquad \text{Multiply both sides by } -r.$$
$$S_n - rS_n = a_1 - a_1 r^n \qquad \text{Add.}$$
$$(1-r)S_n = a_1(1 - r^n) \qquad \text{Factor both sides.}$$
$$S_n = \frac{a_1(1 - r^n)}{1 - r}, r \neq 1 \qquad \text{Divide both sides by } 1 - r.$$

SUM OF THE TERMS OF A FINITE GEOMETRIC SEQUENCE

Let $a_1, a_2, a_3, \ldots, a_n$ be the first n terms of a geometric sequence with first term a_1 and common ratio r. The sum S_n of these terms is

$$S_n = \sum_{i=1}^{n} a_1 r^{i-1} = \frac{a_1(1 - r^n)}{1 - r}, r \neq 1.$$

A geometric sequence with $r = 1$ is a sequence in which all terms are identical; that is, $a_n = a_1(1)^{n-1} = a_1$. Consequently, the sum of the first n term, S_n, is

$$\underbrace{a_1 + a_1 + \cdots + a_1}_{n \text{ terms}} = na_1.$$

The following screen shows the result of Example **5a** obtained using a graphing calculator.

EXAMPLE 5 **Finding the Sum of Terms of a Finite Geometric Sequence**

Find each sum.

a. $\sum_{i=1}^{15} 5(0.7)^{i-1}$ **b.** $\sum_{i=1}^{15} 5(0.7)^{i}$

SOLUTION

a. We can evaluate $\sum_{i=1}^{15} 5(0.7)^{i-1}$ by using the formula for $S_n = \sum_{i=1}^{n} a_1 r^{i-1}$, where $a_1 = 5, r = 0.7$, and $n = 15$.

$$S_{15} = \sum_{i=1}^{15} a_1 r^{i-1} = 5\left[\frac{1 - (0.7)^{15}}{1 - 0.7}\right]$$ Replace a_1 with 5, $r = 0.7$, and $n = 15$.

$$\approx 16.588$$ Use a calculator.

b. $\sum_{i=1}^{15} 5(0.7)^{i} = (0.7)\sum_{i=1}^{15} 5(0.7)^{i-1}$ Factor out 0.7 from each term.

$$\approx (0.7)(16.588)$$ From part **a**

$$= 11.6116$$ ■ ■ ■

Practice Problem 5 Find the sum: $\sum_{i=1}^{17} 3(0.4)^{i}$ ■

3 Solve annuity problems.

Annuities

One of the most important applications of finite geometric series is the computation of the value of an *annuity*. An **annuity** is a sequence of equal periodic payments. When a fixed rate of compound interest applies to all payments, the sum of all of the payments made plus all interest can be found by using the formula for the sum of a finite geometric sequence. The **value of an annuity** (also called the **future value of an annuity**) is the sum of all payments and interest. To develop a formula for the value of an annuity, we suppose P is deposited at the end of each year at an annual interest rate i compounded annually.

The compound interest formula $A = P(1 + i)^t$ gives the total value after t years when P earns an annual interest rate i (in decimal form) compounded once a year. At the end of the first year, the initial payment of P is made and the annuity's value is P. At the end of the second year, P is again deposited. At this time, the first deposit has earned interest during the second year. The value of the annuity after two years is

$$P \quad + \quad P(1 + i).$$

Payment made at the end of year 2   First payment plus interest earned for one year

The value of the annuity after three years is

$$P \quad + P(1 + i) \quad + P(1 + i)^2$$

Payment made at the end of year 3   Second payment plus interest earned for one year   First payment plus interest earned for two years

The value of the annuity after t years is

$$P + P(1 + i) + P(1 + i)^2 + P(1 + i)^3 + \cdots + P(1 + i)^{t-1}.$$

Payment made at
the end of year t

First payment plus interests
earned for $t - 1$ year

RECALL

The sum S_n of the first n terms of a geometric sequence with first term a_1 and common ratio r is given by the formula

$$S_n = \frac{a_1(1 - r^n)}{1 - r}.$$

The value of the annuity after t years is the sum of a geometric sequence with first term $a_1 = P$ and common ratio $r = 1 + i$. We compute this sum A as follows:

$$A = S_n = \frac{a_1(1 - r^n)}{1 - r} \qquad \text{Formula for } S_n$$

$$A = \frac{P[1 - (1 + i)^t]}{1 - (1 + i)} \qquad \text{Replace } n \text{ with } t, a_1 \text{ with } P, \text{ and } r \text{ with } 1 + i.$$

$$= \frac{P[1 - (1 + i)^t]}{-i} \qquad \text{Simplify.}$$

$$= P\frac{[(1 + i)^t - 1]}{i} \qquad \text{Multiply numerator and denominator by } -1.$$

If interest is compounded n times per year and equal payments are made at the end of each compounding period, we adjust the formula as we did in Section 4.2 to account for the more frequent compounding.

VALUE OF AN ANNUITY

Let P represent the payment in dollars made at the end of each of n compounding periods per year and let i be the annual interest rate. Then the value A of the annuity after t year is:

$$A = P\left[\frac{\left(1 + \dfrac{i}{n}\right)^{nt} - 1}{\dfrac{i}{n}} \right]$$

EXAMPLE 6   **Finding the Value of an Annuity**

An individual retirement account (IRA) is a common way to save money to provide funds after retirement. Suppose you make payments of $1200 into an IRA at the end of each year at an annual interest rate of 4.5% per year, compounded annually. What is the value of this annuity after 35 years?

SOLUTION

Each annuity payment is $P = \$1200$. The annual interest rate is 4.5% (so $i = 0.045$), and the number of years is $t = 35$. Because interest is compounded annually, $n = 1$. The value of the annuity is

$$A = 1200\left[\frac{\left(1 + \dfrac{0.045}{1}\right)^{(1)35} - 1}{\dfrac{0.045}{1}} \right]$$

$$= \$97{,}795.94 \qquad \text{Use a calculator.}$$

The value of the IRA after 35 years is $97,795.94.   ■ ■ ■

Practice Problem 6 If in Example 6 the end-of-year payments are $1500, the annual interest rate remains 4.5% (compounded annually), and the payments are made for 30 years, what is the value of the annuity? ■

4 Find the sum of an infinite geometric sequence.

Infinite Geometric Series

The sum S_n of the first n terms of a geometric series is given by the formula $S_n = \dfrac{a_1(1 - r^n)}{1 - r}$. If this finite sum S_n approaches a number S as $n \to \infty$ (that is, as n gets larger and larger), we say that S is the **sum of the infinite geometric series**, and we write

$$S = \sum_{i=1}^{\infty} a_1 r^{i-1}.$$

If r is any real number with $-1 < r < 1$ (equivalently, $|r| < 1$), then the value of r^n, like the value of $\left(\dfrac{1}{2}\right)^n$, gets closer and closer to 0 as n gets larger and larger. We indicate that the values of r^n approach 0 by writing

$$\lim_{n \to \infty} r^n = 0 \text{ if } |r| < 1.$$

The expression $\lim\limits_{n \to \infty} r^n$ is read "the limit, as n approaches infinity, of r to the n" or "the limit, as n approaches infinity, of r to the nth power."
When $|r| < 1$,

$$S_n = \frac{a_1(1 - r^n)}{1 - r} \text{ approaches } \frac{a_1(1 - 0)}{1 - r} = \frac{a_1}{1 - r} \text{ as } n \to \infty.$$

SUM OF THE TERMS OF AN INFINITE GEOMETRIC SEQUENCE

If $|r| < 1$, the infinite sum

$$a_1 + a_1 r + a_1 r^2 + a_1 r^3 + \cdots + a_1 r^{n-1} + \cdots$$

is given by

$$S = \sum_{i=1}^{\infty} a_1 r^{i-1} = \frac{a_1}{1 - r}.$$

When $|r| \geq 1$, the infinite geometric series does not have a sum. This is because the value of r^n does not approach 0 as n approaches ∞. For example, if $r = 2$, we have

$$2^1 = 2, 2^2 = 4, 2^3 = 8, 2^4 = 16, 2^5 = 32, 2^6 = 64, \ldots.$$

EXAMPLE 7 **Finding the Sum of an Infinite Geometric Series**

Find the sum $2 + \dfrac{3}{2} + \dfrac{9}{8} + \dfrac{27}{32} + \cdots$.

SOLUTION

The first term is $a_1 = 2$, and the common ratio is

$$r = \frac{\dfrac{3}{2}}{2} = \frac{3}{4}.$$

Because $|r| = \dfrac{3}{4} < 1$, we can use the formula for the sum of an infinite geometric series.

$$S = \frac{a_1}{1 - r} \qquad \text{Formula for } S$$

$$= \frac{2}{1 - \dfrac{3}{4}} \qquad \text{Replace } a_1 \text{ with 2 and } r \text{ with } \dfrac{3}{4}.$$

$$= 8 \qquad \text{Simplify.} \qquad\qquad ■■■$$

Practice Problem 7 Find the sum $3 + \dfrac{6}{3} + \dfrac{12}{9} + \dfrac{24}{27} + \cdots$. ■

EXAMPLE 8 **Calculating the Multiplier Effect**

The host city for the Super Bowl expects that tourists will spend \$10,000,000. Assume that 80% of this money is spent again in the city, then 80% of this second round of spending is spent again, and so on. In the introduction to this section, we said that such a spending pattern results in a geometric series whose sum is called the *multiplier*. Find this series and its sum.

SOLUTION

We start our series with the \$10,000,000 brought into the city and add the subsequent amounts spent.

$$10{,}000{,}000 + 10{,}000{,}000\,(0.80) + 10{,}000{,}000\,(0.80)^2 + 10{,}000{,}000\,(0.80)^3 + \cdots$$

| original 10,000,000 | 80% of 10,000,000 | 80% of previous amount | 80% of previous amount |

Using the formula $\displaystyle\sum_{i=1}^{\infty} ar^{i-1} = \frac{a}{1 - r}$ for the sum of an infinite geometric series, we have

$$\sum_{i=1}^{\infty} (10{,}000{,}000)(0.80)^{i-1} = \frac{\$10{,}000{,}000}{1 - 0.80} = \$50{,}000{,}000.$$

This amount should shed some light on why there is so much competition to serve as the host city for a major sports event. ■■■

Practice Problem 8 Find the series and the sum that results in Example 8 assuming that 85%, rather than 80%, occurs in the repeated spending. ■

SECTION 8.3 ■ Exercises

A EXERCISES Basic Skills and Concepts

1. If 5 is the common ratio of a geometric sequence with general term a_n, then $\dfrac{a_{63}}{a_{62}} = $ _____.

2. If 5 is the common ratio of a geometric sequence with general term a_n and $a_{15} = -2$, then $a_{16} = $ _____.

3. If 24 is the term immediately following the sequence term 8 in a geometric sequence, then the common ratio is _____.

4. An infinite geometric series $a_1 + a_1r + a_1r^2 + \cdots$ does not have a sum if _____ ≥ 1.

5. *True or False* If the sum of an infinite geometric series having the common ratio $r = \frac{1}{3}$ is $S = 15$, then $a_1 = 10$.

6. *True or False* The sum of an infinite geometric series can never be negative.

In Exercises 7–26, determine whether each sequence is geometric. If it is, find the first term and the common ratio.

7. $3, 6, 12, 24, \ldots$

8. $2, 4, 8, 16, \ldots$

9. $1, 5, 10, 20, \ldots$

10. $1, 1, 3, 3, 9, 9, \ldots$

11. $1, -3, 9, -27, \ldots$

12. $-1, 2, -4, 8, \ldots$

13. $7, -7, 7, -7, \ldots$

14. $1, -2, 4, -8, \ldots$

15. $9, 3, 1, \frac{1}{3}, \ldots$

16. $5, 2, \frac{4}{5}, \frac{8}{25}, \ldots$

17. $a_n = \left(-\frac{1}{2}\right)^n$

18. $a_n = 5\left(\frac{2}{3}\right)^n$

19. $a_n = 2^{n-1}$

20. $a_n = -(1.06)^{n-1}$

21. $a_n = 7n^2 + 1$

22. $a_n = 1 - (2n)^2$

23. $a_n = 3^{-n}$

24. $a_n = 50(0.1)^{-n}$

25. $a_n = 5^{\frac{n}{2}}$

26. $a_n = 7^{\sqrt{n}}$

In Exercises 27–34, find the first term a_1, the common ratio r, and the nth term a_n for each geometric sequence.

27. $2, 10, 50, 250, \ldots$

28. $-3, -6, -12, -24, \ldots$

29. $5, \frac{10}{3}, \frac{20}{9}, \frac{40}{27}, \ldots$

30. $1, \sqrt{3}, 3, 3\sqrt{3}, \ldots$

31. $0.2, -0.6, 1.8, -5.4, \ldots$

32. $1.3, -0.26, 0.052, -0.0104, \ldots$

33. $\pi^4, \pi^6, \pi^8, \pi^{10}, \ldots$

34. $e^2, 1, e^{-2}, e^{-4}, \ldots$

In Exercises 35–44, find the indicated term of each geometric sequence.

35. a_7 when $a_1 = 5$ and $r = 2$

36. a_7 when $a_1 = 8$ and $r = 3$

37. a_{10} when $a_1 = 3$ and $r = -2$

38. a_{10} when $a_1 = 7$ and $r = -2$

39. a_6 when $a_1 = \frac{1}{16}$ and $r = 3$

40. a_6 when $a_1 = \frac{1}{81}$ and $r = 3$

41. a_9 when $a_1 = -1$ and $r = \frac{5}{2}$

42. a_9 when $a_1 = -4$ and $r = \frac{3}{4}$

43. a_{20} when $a_1 = 500$ and $r = -\frac{1}{2}$

44. a_{20} when $a_1 = 1000$ and $r = -\frac{1}{10}$

In Exercises 45–50, find the sum S_n of the first n terms of each geometric sequence.

45. $\frac{1}{10}, \frac{1}{2}, \frac{5}{2}, \frac{25}{2}, \ldots; n = 10$ **46.** $6, 2, \frac{2}{3}, \frac{2}{9}, \ldots; n = 10$

47. $\frac{1}{25}, -\frac{1}{5}, 1, -5, \ldots; n = 12$

48. $-10, \frac{1}{10}, \frac{-1}{1000}, \frac{1}{100,000}, \ldots; n = 12$

49. $5, \frac{5}{4}, \frac{5}{4^2}, \frac{5}{4^3}, \ldots; n = 8$

50. $2, \frac{2}{5}, \frac{2}{5^2}, \frac{2}{5^3}, \ldots; n = 8$

In Exercises 51–58, find each sum.

51. $\sum_{i=1}^{5} \left(\frac{1}{2}\right)^{i-1}$

52. $\sum_{i=1}^{5} \left(\frac{1}{5}\right)^{i-1}$

53. $\sum_{i=1}^{8} 3\left(\frac{2}{3}\right)^{i-1}$

54. $\sum_{i=1}^{8} 2\left(\frac{3}{5}\right)^{i-1}$

55. $\sum_{i=3}^{10} \frac{2^{i-1}}{4}$

56. $\sum_{i=3}^{10} \frac{5^{i-1}}{2}$

57. $\sum_{i=1}^{20} \left(-\frac{3}{5}\right)\left(-\frac{5}{2}\right)^{i-1}$

58. $\sum_{i=1}^{20} \left(-\frac{1}{4}\right)(3^{2-i})$

In Exercises 59–64, use a graphing calculator to find each sum.

59. $\frac{1}{5} + \frac{1}{10} + \frac{1}{20} + \cdots + \frac{1}{320}$

60. $2 + 6 + 18 + \cdots + 1458$

61. $\sum_{n=1}^{10} 3^{n-2}$

62. $\sum_{n=1}^{15} \left(\frac{1}{7}\right)^{n-1}$

63. $\sum_{n=1}^{10} (-1)^{n-1} 3^{2-n}$

64. $\sum_{n=1}^{20} (-4)^{3-n}$

In Exercises 65–74, find each sum.

65. $\frac{1}{3} + \frac{1}{9} + \frac{1}{27} + \frac{1}{81} + \cdots$

66. $\frac{5}{2} + \frac{5}{4} + \frac{5}{8} + \frac{5}{16} + \cdots$

67. $-\frac{1}{2} + \frac{1}{4} - \frac{1}{8} + \frac{1}{16} - \cdots$

68. $\frac{3}{2} + \frac{3}{4} - \frac{3}{8} + \frac{3}{16} + \cdots$

69. $8 - 2 + \frac{1}{2} - \frac{1}{8} + \cdots$

70. $1 - \frac{3}{5} + \frac{9}{25} - \frac{27}{125} + \cdots$

71. $\sum_{n=0}^{\infty} 5\left(\frac{1}{3}\right)^n$

72. $\sum_{n=0}^{\infty} 3\left(\frac{1}{4}\right)^n$

73. $\sum_{n=0}^{\infty} \left(-\frac{1}{4}\right)^n$

74. $\sum_{n=0}^{\infty} \left(-\frac{1}{3}\right)^n$

In Exercises 75–80, find *n* for the given value of a_n in each geometric sequence.

75. $a_n = 512; 2, 4, 8, \ldots$.

76. $a_n = \dfrac{3}{1024}; 3, \dfrac{3}{2}, \dfrac{3}{4}, \ldots$.

77. $a_n = \dfrac{1}{4096}; 1, -\dfrac{1}{2}, \dfrac{1}{4}, \ldots$.

78. $a_n = \dfrac{1}{2187}; 3, -1, \dfrac{1}{3}, \ldots$.

79. $a_n = \dfrac{5}{5,764,801}; 5, -\dfrac{5}{7}, \dfrac{5}{49}, \ldots$.

80. $a_n = \dfrac{1}{1,000,000}; 1000, 100, 10, \ldots$.

B EXERCISES Applying the Concepts

81. **Population growth.** The population in a small town is increasing at the rate of 3% per year. If the present population is 20,000, what will the population be at the end of five years?

82. **Growth of a zoo collection.** The butterfly collection at a local zoo started with only 25 butterflies. If the number of butterflies doubled each month, how many butterflies were in the collection after 12 months?

83. **Savings growth.** Ramón deposits $100 on the last day of each month into a savings account that pays 6% annually, compounded monthly. What is the balance in the account after 36 compounding periods?

84. **Savings growth.** Kat deposits $300 semiannually into a savings account for her son. If the account pays 8% annually, compounded semiannually, what is the balance in the account after 20 compounding periods?

85. **Ancestors.** Every person has two parents, four grandparents, eight great-grandparents, and so on. Find the number of ancestors a person has in the tenth generation back.

86. **Bacterial growth.** A colony of bacteria doubles in number every day. If there are 1000 bacteria now, how many will there be
a. on the seventh day?
b. on the *n*th day?

87. **Investment.** An actuarial firm budgets a new position at $36,000 for the first year and a 5% raise each year thereafter. Find the total compensation a successful applicant for the job would receive during the first 20 years of employment.

88. **Investment.** A realtor offers two choices of payment for a small condominium:
a. Pay $4000 per month for the next 25 months.
b. Pay 1¢ the first month, 2¢ the second month, 4¢ the third month, and so on for 25 months.
Which option, a or b, is the better choice for the buyer? Explain your reasoning.

89. **Pendulum motion.** The distance traveled by any point on a certain pendulum is 20% less than in the preceding swing. Assuming that the length traveled by the pendulum bob during the first swing is 56.25 centimeters, find the total distance the bob has traveled at the end of the fifth swing.

90. **Depreciating a tractor.** Every year a tractor loses 20% of the value it had at the beginning of the year. If the tractor now has a value of $120,000, what will its value be in five years?

91. **Rebounding ball.** A particular ball always rebounds $\dfrac{3}{5}$ the distance it falls. If the ball is dropped from a height of 5 meters, how far will it travel before coming to a stop?

92. **Rebounding ball.** Repeat Exercise 91 assuming that the ball is dropped from a height of 9 meters.

C EXERCISES Beyond the Basics

93. The accompanying figure shows the first six squares in an infinite sequence of successively smaller squares formed by connecting the midpoints of the sides of the preceding larger square. The area of the largest square is 1. Find the total area of the indicated sections that are shaded in every other square as the number of squares increases to infinity.

94. The accompanying figure shows the first three equilateral triangles in an infinite sequence of successively smaller equilateral triangles formed by connecting the midpoints of the sides of the preceding larger triangle. The largest triangle has sides of length 4. Find the total perimeter of all of the triangles.

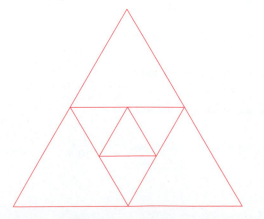

95. Prove that if $a_1, a_2, a_3, \ldots$ is a geometric sequence, then the sequence $\ln a_1, \ln a_2, \ln a_3, \ldots$ is an arithmetic sequence.

96. Prove that if $a_1, a_2, a_3, \ldots$ is a geometric sequence, then $a_1^2, a_2^2, a_3^2, \ldots$ is also a geometric sequence.

97. Show that if $a_1, a_2, a_3, \ldots$ is a geometric sequence with common ratio r, then $\dfrac{1}{a_1}, \dfrac{1}{a_2}, \dfrac{1}{a_3}, \ldots$ is also a geometric sequence. Also find its common ratio.

98. Show that if c is any nonzero real number, then $c^2, c, 1, \dfrac{1}{c}, \ldots$ is a geometric sequence.

99. Show that if x is any nonzero real number, then $x, 2, \dfrac{4}{x}, \dfrac{8}{x^2}, \ldots$ is a geometric sequence.

100. Show that if a and x are any two nonzero real numbers, then $\dfrac{x}{a}, -1, \dfrac{a}{x}, -\dfrac{a^2}{x^2}, \ldots$ is a geometric sequence.

101. Show that if a, x, and y are any three nonzero real numbers, then $\dfrac{a}{x}, -\dfrac{a}{xy}, \dfrac{a}{xy^2}, \dfrac{-a}{xy^3}, \ldots$ is a geometric sequence.

102. Show that if $a_1, a_2, a_3, \ldots$ is an arithmetic sequence with common difference d, then $2^{a_1}, 2^{a_2}, 2^{a_3}, \ldots$ is a geometric sequence. Also find its common ratio.

Critical Thinking

103. If a_n denotes the nth term of an arithmetic sequence with $a_1 = 10$ and $d = 2.7$ and if b_n denotes the nth term of a geometric sequence with $b_1 = 10$ and $r = 2$, which number is larger, a_{1001} or b_{1001}?

104. Assuming that the sum $\dfrac{a(1 - r^n)}{1 - r}$ of the first n terms of a geometric sequence is 1023, find a when $n = 10$ and $r = \dfrac{1}{2}$.

105. Find n and k so that the sum $5 + 5 \cdot 2 + 5 \cdot 2^2 + \cdots + 5 \cdot 2^{15}$ can be written as

$$\sum_{i=0}^{n} 5 \cdot 2^i \quad \text{or as} \quad \sum_{i=1}^{k} 5 \cdot 2^{i-1}.$$

GROUP PROJECTS

106. The sum of three consecutive terms of a geometric sequence is 35, and their product is 1000. Find the numbers.

[*Hint:* Let $\dfrac{a}{r}$, a, and ar be the three terms of geometric sequence.]

107. The sum of three consecutive terms of an arithmetic sequence is 15. If 1, 4, and 19 are respectively added to the three terms, the resulting numbers form three consecutive terms of a geometric sequence. Find the numbers.

Mathematical Induction

Before Starting this Section, Review

1. Natural numbers (Section P. 1, page 3)
2. Inequalities (Section P. 1, page 5)
3. Exponents (Section P. 2, page 19)
4. Series (Section 8.1, page 666)

Objective

1 Prove statements by mathematical induction.

JUMPING TO CONCLUSIONS; GOOD INDUCTION VERSUS BAD INDUCTION

A scientist had two large jars before him on the laboratory table. The jar on his left contained a hundred fleas; the jar on his right was empty. The scientist carefully lifted a flea from the jar on the left; placed the flea on the table between the two jars; stepped back; and said in a loud voice, "Jump!" The flea jumped and was put into the jar on the right. A second flea was carefully lifted from the jar on the left and placed on the table between the two jars. Again, the scientist stepped back and said in a loud voice, "Jump!" The flea jumped and was put into the jar on the right. In the same manner, the scientist treated each of the hundred fleas in the jar on the left, and each flea jumped as ordered. The two jars were then interchanged, and the experiment was continued with a slight difference. This time the scientist carefully lifted a flea from the jar on the left; *yanked off its hind legs;* placed the flea on the table between the jars; stepped back; and said in a loud voice, "Jump!" The flea did not jump and was put into the jar on the right. A second flea was carefully lifted from the jar on the left, its hind legs yanked off, and it was placed on the table between the two jars. Again the scientist stepped back and said in a loud voice, "Jump!" The flea did not jump and was put into the jar on the right. In this manner, the scientist treated each of the hundred fleas in the jar on the left, and in no case did a flea jump when ordered. So the scientist recorded the following inductive statement in his notebook: "A flea, if its hind legs are yanked off, cannot hear." *(Source: Modified from Howard Eves, Mathematical Circles.)*

This, of course, is an example of truly bad reasoning. The principle of mathematical induction, demonstrated in Example 2, is the gold standard for reasoning in proving that statements about natural numbers are true. ■

1 Prove statements by mathematical induction.

Suppose your boss wants you to verify that each of a thousand bags contains two specific gold coins. There is only one way to be absolutely sure: You must check every bag. That would take a very long time, but you can eventually finish the job. However, consider the problem of verifying that a statement about the natural numbers such as

$$1^2 + 2^2 + 3^2 + \cdots + n^2 = \frac{n(n + 1)(2n + 1)}{6}$$

is true for all natural numbers n. Let's first test the statement for several values of n by comparing the value of $1^2 + 2^2 + 3^2 + \cdots + n^2$ in the second column of Table 8.1 with the value of $\dfrac{n(n + 1)(2n + 1)}{6}$ in the third column.

TABLE 8.1

n	$1^2 + 2^2 + 3^2 + \cdots + n^2$	$\dfrac{n(n+1)(2n+1)}{6}$	Equal?
1	$1^2 = 1$	$\dfrac{1(1+1)[2(1)+1]}{6} = 1$	Yes
2	$1^2 + 2^2 = 5$	$\dfrac{2(2+1)[2(2)+1]}{6} = 5$	Yes
3	$1^2 + 2^2 + 3^2 = 14$	$\dfrac{3(3+1)[2(3)+1]}{6} = 14$	Yes
4	$1^2 + 2^2 + 3^2 + 4^2 = 30$	$\dfrac{4(4+1)[2(4)+1]}{6} = 30$	Yes
5	$1^2 + 2^2 + 3^2 + 4^2 + 5^2 = 55$	$\dfrac{5(5+1)[2(5)+1]}{6} = 55$	Yes

We see that the statement is correct for the natural numbers 1, 2, 3, 4, and 5; furthermore, you can continue to verify that the statement is correct for any particular natural number n you try. But is it *always* correct? Clearly, we can't check every natural number!

The principle of *mathematical induction* provides a method for proving that statements about natural numbers are true for *all* natural numbers. The principle is based on the fact that after any natural number n, there is a next-larger natural number $n + 1$ and that any specified natural number can be reached by a finite number of such steps, starting from the natural number 1.

PRINCIPLE OF MATHEMATICAL INDUCTION

Let P_n be a statement that involves the natural number n with the following two properties:

1. P_1 is true. (The statement is true for the natural number 1.)

2. If P_k is a true statement, then P_{k+1} is a true statement.

Then the statement P_n is true for every natural number n.

A common physical interpretation is helpful in understanding why this principle works. Imagine an unending line of dominoes, as shown in Figure 8.2. Suppose the dominoes are arranged so that if *any* domino is knocked down, it will knock down the next domino behind it as well. What will happen if the first domino is knocked down? *All* of the dominoes will fall because

1. knocking down the first domino causes the second domino to be knocked down.

2. when the second domino is knocked down, it knocks down the third domino, and so on.

Determining the Statement P_{k+1} from the Statement P_k

To successfully use the Principle of Mathematical Induction, you must be able to determine the statement P_{k+1} from a given statement P_k. Suppose the given statement is

$$P_k\!: k \geq 1;$$

FIGURE 8.2

then we have

$$P_{k+1}: k + 1 \geq 1 \qquad \text{Replace } k \text{ with } k + 1 \text{ in } P_k.$$

It is important to recognize that because P_k says "k is greater than or equal to 1," the statement P_{k+1} says "$k + 1$ is greater than or equal to 1." That is, P_{k+1} asserts the same property for $k + 1$ that P_k asserts for k.

EXAMPLE 1   Determining P_{k+1} from P_k

Find the statement P_{k+1} from the given statement P_k.

a. $P_k: k < 2^k$ **b.** $P_k: S_k = 3k^2 + 9$

c. $P_k: 1 + 2 + 2^2 + 2^3 + \cdots + 2^{k-1} = 2^k - 1$

d. $P_k: 3 + 6 + 9 + 12 + \cdots + 3k = \dfrac{3k(k + 1)}{2}$

SOLUTION

a. $P_k: k < 2^k, \quad P_{k+1}: k + 1 < 2^{k+1}$ Replace k with $k + 1$.

b. $P_k: S_k = 3k^2 + 9, \quad P_{k+1}: S_{k+1} = 3(k + 1)^2 + 9$ Replace k with $k + 1$.

c. $P_k: 1 + 2 + 2^2 + 2^3 + \cdots + 2^{k-1} = 2^k - 1,$

$P_{k+1}: 1 + 2 + 2^2 + 2^3 + \cdots + 2^{(k+1)-1} = 2^{k+1} - 1$ Replace k with $k + 1$.

d. $P_k: 3 + 6 + 9 + 12 + \cdots + 3k = \dfrac{3k(k + 1)}{2}$

$P_{k+1}: 3 + 6 + 9 + 12 + \cdots + 3(k + 1) = \dfrac{3(k + 1)[(k + 1) + 1]}{2}$ Replace k with $k + 1$.

■ ■ ■

Practice Problem 1   Find P_{k+1} for $P_k: (k + 3)^2 > k^2 + 9$.   ■

Francesco Maurolico

(1494–1575)

The first known use of the mathematical induction is in the work of the sixteenth-century mathematician Francesco Maurolico wrote extensively on the works of classical mathematics and made many contributions to geometry and optics. In his book *Arithmeticorum Libri Duo,* Maurolico presented a variety of properties of the integers, together with proofs of these properties. To prove some of the properties, he devised the method of mathematical induction. His first use of mathematical induction in this book was to prove that the sum of the first n odd positive integers equals n^2.

EXAMPLE 2   Using Mathematical Induction

Use mathematical induction to prove that for all natural numbers n,

$$2 + 4 + 6 + \cdots + 2n = n(n + 1).$$

SOLUTION

First, we verify that this statement is true for $n = 1$.

Check: $2(1) = 1(1 + 1)$ Replace n with 1 in the original statement.

$2 = 2$ ✓

The given statement is true for $n = 1$, so the first condition for the principle of mathematical induction holds.

The second condition for the principle of mathematical induction requires two steps.

Step 1   Assume that the formula is true for some unspecified natural number k.

$$P_k: 2 + 4 + 6 + \cdots + 2k = k(k + 1) \qquad \text{Assumed true}$$

Step 2   On the basis of the assumption that P_k is true, show that P_{k+1} is true, which is the same as showing that

$$P_{k+1}: 2 + 4 + 6 + \cdots + 2(k + 1) = (k + 1)[(k + 1) + 1] \qquad \text{Replace } k \text{ with } k + 1 \text{ in } P_k.$$

Begin by using P_k, the statement assumed to be true. Add $2(k + 1)$ to both sides of P_k in Step 1, which again results in a true statement.

$$2 + 4 + 6 + \cdots + 2k = k(k + 1) \qquad \text{Assumed true}$$

$$2 + 4 + 6 + \cdots + 2k + 2(k + 1) = k(k + 1) + 2(k + 1) \qquad \text{Add } 2(k + 1) \text{ to both sides.}$$

It is not *necessary* to write $2k$ here because it is understood that $2k$ is the even integer that precedes $2(k + 1)$.

$$2 + 4 + 6 + \cdots + 2k + 2(k + 1) = (k + 1)(k + 2) \qquad \text{Factor out the common factor } (k + 1) \text{ on the right.}$$

$$2 + 4 + 6 + \cdots + 2(k + 1) = (k + 1)[(k + 1) + 1] \qquad \text{Rewrite } (k + 2) \text{ as } (k + 1) + 1.$$

This last equation says that P_{k+1} is true if P_k is assumed to be true. Therefore, by the principle of mathematical induction, the statement

$$2 + 4 + 6 + \cdots + 2n = n(n + 1)$$

is true for every natural number n. ■ ■ ■

Practice Problem 2 Use mathematical induction to prove that for all natural numbers n,

$$1 + 2 + 3 + \cdots + n = \frac{n(n + 1)}{2}.$$ ■

EXAMPLE 3 **Using Mathematical Induction**

Use mathematical induction to prove that

$$2^n > n$$

for all natural numbers n.

SOLUTION

First, we show that the given statement is true for $n = 1$.

$$2^1 > 1 \qquad \text{Replace } n \text{ with 1 in the original statement.}$$

Thus, the inequality is true for $n = 1$.

Next, we assume that for some unspecified natural number k,

$$P_k: 2^k > k \quad \text{is true.}$$

Then we must use P_k to prove that P_{k+1} is true; that is,

$$P_{k+1}: 2^{k+1} > k + 1.$$

Now, by the product rule of exponents, $2^{k+1} = 2^k \cdot 2^1 = 2^k \cdot 2$; so we can get some information about 2^{k+1} by multiplying both sides of $P_k: 2^k > 2$ by 2.

$$2^k > k \qquad\qquad\qquad P_k \text{ is assumed true.}$$
$$2^k \cdot 2 > 2 \cdot k \qquad\qquad \text{Multiply both sides by 2.}$$
$$2^{k+1} = 2^k \cdot 2 > 2k = k + k \geq k + 1 \qquad \text{Because } k \geq 1, k + k \geq k + 1.$$

Thus, $2^{k+1} > k + 1$ is true.

By the principle of mathematical induction, the statement

$$2^n > n$$

is true for every natural number n. ■ ■ ■

Practice Problem 3 Use mathematical induction to prove that $3^n > n$ for all natural numbers n. ■

SECTION 8.4 ■ Exercises

A EXERCISES Basic Skills and Concepts

1. Mathematical induction can only be used to prove statements about the _____.

2. The first step in a mathematical induction proof that a statement P_n is true for all natural numbers n is that _____.

3. The second step in a mathematical induction proof is to assume that P_k is true for a natural number k and then show that _____.

4. *True or False* If P_n is a statement about natural numbers and there is exactly one natural number r for which P_r is false, then P_n is false.

5. *True or False* The statement $e^n \geq n$ cannot be proved by mathematical induction because e is not a natural number.

6. *True or False* The first step in proving that $2n \neq n$ for all natural numbers n is to note that $2 \neq 1$.

In Exercises 7–10, find P_{k+1} from the given statement P_k.

7. $P_k: (k + 1)^2 - 2k = k^2 + 1$

8. $P_k: (1 + k)(1 - k) = 1 - k^2$

9. $P_k: 2^k > 5k$

10. $P_k: 1 + 3 + 5 + \cdots + (2k - 1) = k^2$

In Exercises 11–34, use mathematical induction to prove that each statement is true for all natural numbers n.

11. $2 + 4 + 6 + \cdots + 2n = n(n + 1)$

12. $1 + 3 + 5 + \cdots + (2n - 1) = n^2$

13. $4 + 8 + 12 + \cdots + 4n = 2n(n + 1)$

14. $3 + 6 + 9 + \cdots + 3n = \dfrac{3n(n + 1)}{2}$

15. $1 + 5 + 9 + \cdots + (4n - 3) = n(2n - 1)$

16. $3 + 8 + 13 + \cdots + (5n - 2) = \dfrac{n(5n + 1)}{2}$

17. $3 + 9 + 27 + \cdots + 3^n = \dfrac{3(3^n - 1)}{2}$

18. $5 + 25 + 125 + \cdots + 5^n = \dfrac{5(5^n - 1)}{4}$

19. $\dfrac{1}{1 \cdot 2} + \dfrac{1}{2 \cdot 3} + \dfrac{1}{3 \cdot 4} + \cdots + \dfrac{1}{n(n + 1)} = \dfrac{n}{n + 1}$

20. $\dfrac{1}{2 \cdot 4} + \dfrac{1}{4 \cdot 6} + \dfrac{1}{6 \cdot 8} + \cdots + \dfrac{1}{2n(2n + 2)} = \dfrac{n}{4(n + 1)}$

21. $2 \leq 2^n$ 22. $2n + 1 \leq 3^n$

23. $n(n + 2) < (n + 1)^2$ 24. $n \leq n^2$

25. $\dfrac{n!}{n} = (n - 1)!$ 26. $\dfrac{n!}{(n + 1)!} = \dfrac{1}{n + 1}$

27. $1^2 + 2^2 + 3^2 + \cdots + n^2 = \dfrac{n(n + 1)(2n + 1)}{6}$

28. $1^3 + 2^3 + 3^3 + \cdots + n^3 = \dfrac{n^2(n + 1)^2}{4}$

29. 2 is a factor of $n^2 + n$.

30. 2 is a factor of $n^3 + 5n$.

31. 6 is a factor of $n(n + 1)(n + 2)$.

32. 3 is a factor of $n(n + 1)(n - 1)$.

33. $(ab)^n = a^n b^n$ 34. $\left(\dfrac{a}{b}\right)^n = \dfrac{a^n}{b^n}$

B EXERCISES Applying the Concepts

35. **Counting hugs.** At a family reunion of n people ($n \geq 2$), each person hugs everyone else. Use mathematical induction to show that the number of hugs is $\dfrac{n^2 - n}{2}$.

36. **Winning prizes.** In a contest in which n prizes are possible, a contestant can win any number of the prizes (from 0 to n). Use mathematical induction to show that there are 2^n possible outcomes for the sets of prizes the contestant might win.
[*Hint:* If $n = 1$, there are 2 ($= 2^1$) outcomes: winning 0 prizes or winning the one prize.]

37. **Koch's snowflake.** A geometric shape called the Koch snowflake can be formed by starting with an equilateral triangle, as in the figure. Assume that each side of the triangle has length 1. On the middle part of each side, construct an equilateral triangle with sides of length $\dfrac{1}{3}$ and erase the side of the smaller triangle that is on the side of the original triangle. Repeat this procedure for each of the smaller triangles. This process produces a figure that resembles a snowflake.
   a. Find a formula for the number of sides of the nth figure. Use mathematical induction to prove that this formula is correct.
   b. Find a formula for the perimeter of the nth figure. Use mathematical induction to prove that this formula is correct.

38. **Sierpinski's triangle.** A geometric figure known as Sierpinski's triangle is constructed by starting with an equilateral triangle as in the figure. Assume that each side of the triangle has length 1. Inside the original triangle, draw a second triangle by connecting the midpoints of the sides of the original triangle. This divides the original triangle into four identical

equilateral triangles. Repeat the process by constructing equilateral triangles inside each of the four smaller triangles, again using the midpoints of the sides of the triangle as vertices. Continuing this process with each group of smaller triangles produces Sierpinski's triangle. Coloring each set of smaller triangles helps in following the construction.

a. When a process is repeated over and over, each repetition is called an **iteration**. How many triangles will the fourth iteration have?

b. How many triangles will the fifth iteration have?

c. Find a formula for the number of triangles in the nth iteration. Use mathematical induction to prove that this formula is correct.

39. Towers of Hanoi. Three pegs are attached vertically to a horizontal board as shown in the figure. One peg has n rings stacked on it, each ring smaller than the one below it. In a game known as the Tower of Hanoi puzzle, all of the rings must be moved to a different peg, with only one ring moved at a time, and no ring can be moved on top of a smaller ring. Determine the least number of moves that will accomplish this transfer. Use mathematical induction to prove that your answer is correct.

40. Exam answer sheets. An exam in which each question is answered with either "True" or "False" has n questions.

a. If every question is answered, how many answer sheets are possible
   (i) if $n = 1$?
   (ii) if $n = 2$?

b. Find a formula for the number of possible answer sheets for an exam with n questions (for any natural number n) and use mathematical induction to prove that the formula is correct.

C EXERCISES Beyond the Basics

In Exercises 41–49, use mathematical induction to prove each statement for all natural numbers n.

41. 5 is a factor of $8^n - 3^n$.

42. 24 is a factor of $5^{2n} - 1$.

43. 64 is a factor of $3^{2n+2} - 8n - 9$.

44. 64 is a factor of $9^n - 8n - 1$.

45. 3 is a factor of $2^{2n+1} + 1$.

46. 5 is a factor of $2^{4n} - 1$.

47. $a - b$ is a factor of $a^n - b^n$.
   [*Hint:* $a^{k+1} - b^{k+1} = a(a^k - b^k) + b^k(a - b)$.]

48. If $a \neq 1$, then $1 + a + a^2 + \cdots + a^{n-1} = \dfrac{a^n - 1}{a - 1}$.

49. $\displaystyle\sum_{k=1}^{n+1} \frac{1}{n + k} \leq \frac{5}{6}$

Critical Thinking

50. Consider the statement "2 is a factor of $4n - 1$ for all natural numbers n." Then P_k: 2 is a factor of $4k - 1$.
   Assume that P_k is true so that $4k - 1 = 2m$ for some integer m. Then

$$P_{k+1} = 4(k + 1) - 1$$
$$= 4k + 4 - 1$$
$$= (4k - 1) + 4$$
$$= 2m + 4$$
$$= 2(m + 2)$$

Thus, if P_k is true, then P_{k+1} is also true. However, 2 is never a factor of $4n - 1$ because $4n - 1$ is always an odd integer. Explain why this "proof" is not valid.

Extended principle of mathematical induction. **This principle states that if a statement P_n about natural numbers satisfies the two conditions,**

(1) P_m is true for some natural number m.

(2) and because P_k is true implies that P_{k+1} is also true, then P_n is true for all natural number n, with $n \geq m$.

In Exercises 51 and 52, use the extended principle of mathematical induction to prove the statement.

51. Prove that $2^n > n^2$ for all natural numbers n, $n > 4$.

52. Prove that $n! > n^3$ for all natural numbers n, $n \geq 6$.

The Binomial Theorem

Before Starting this Section, Review

1. Special products (Section P.3, page 33)

2. Factorials (Section 8.1, page 663)

3. Summation notation (Section 8.1, page 665)

Objectives

1 Use Pascal's Triangle to compute binomial coefficients.

2 Use Pascal's Triangle to expand a binomial power.

3 Use the Binomial Theorem to expand a binomial power.

4 Find the coefficient of a term in a binomial expansion.

Blaise Pascal (1623–1662)

Pascaline

BLAISE PASCAL

Blaise Pascal was a French mathematician. An evident genius, he was 14 when he began to accompany his father to gatherings of mathematicians arranged by Father Marin Mersenne. At the age of 16, Pascal presented his own results at one of Mersenne's meetings.

Pascal's desire to help his father with his work collecting taxes led him to invent the first digital calculator, called the Pascaline. Pascal was also interested in atmospheric pressure, and in 1648, he observed that the pressure of the atmosphere decreased with height and that a vacuum existed above the atmosphere.

Pascal's intense interest in mathematics led him to produce important results related to conic sections and to engage in correspondence with Fermat in which he formulated the foundations for the *theory of probability*. Pascal died a painful death from cancer at the age of 39.

In this section, we study *Pascal's Triangle*. This triangle of numbers contains the important *binomial coefficients*. (See Example 1.) Pascal's work was influential in Newton's discovery of the general *Binomial Theorem* for fractional and negative powers. ■

Binomial Expansions Recall that a polynomial that has exactly two terms is called a *binomial*. In Chapter P, the formulas for expanding the binomial powers $(x + y)^2$ and $(x + y)^3$ are given. In the current section, we study a method for expanding $(x + y)^n$ for any positive integer n.

First, consider these binomial expansions:

$$(x + y)^1 = x + y$$
$$(x + y)^2 = x^2 + 2xy + y^2$$
$$(x + y)^3 = x^3 + 3x^2y + 3xy^2 + y^3$$
$$(x + y)^4 = x^4 + 4x^3y + 6x^2y^2 + 4xy^3 + y^4$$
$$(x + y)^5 = x^5 + 5x^4y + 10x^3y^2 + 10x^2y^3 + 5xy^4 + y^5$$

The expansions of $(x + y)^2$ and $(x + y)^3$ should be familiar to you; the last three are left for you to verify. Each is a product of the previous binomial and $(x + y)$. For example, $(x + y)^4 = (x + y)^3(x + y)$.

The patterns for expansions of $(x + y)^n$ (with $n = 1, 2, 3, 4, 5$) suggest the following:

1. The expansion of $(x + y)^n$ has $n + 1$ terms.
2. The sum of the exponents on x and y in each term equals n.
3. The exponent on x starts at n ($x^n = x^n \cdot y^0$) in the first term and decreases by 1 for each term until it is 0 in the last term ($x^0 \cdot y^n = y^n$).
4. The exponent on y starts at 0 ($x^n = x^n \cdot y^0$) in the first term and increases by 1 for each term until it is n in the last term ($x^0 \cdot y^n = y^n$).
5. The variables x and y have symmetrical roles. That is, replacing x with y and y with x in the expansion of $(x + y)^n$ yields the same terms, just in a different order.

You also may have noticed that the coefficients of the first and last terms are 1 and the coefficients of the second and the next-to-last terms are equal. In general, the coefficients of

$$x^{n-j}y^j \qquad \text{and} \qquad x^j y^{n-j}$$

are equal for $j = 0, 1, 2, \dots, n$.

The coefficients in a binomial expansion of $(x + y)^n$ are called the **binomial coefficients**.

1 Use Pascal's Triangle to compute binomial coefficients.

Pascal's Triangle

As early as A.D. 1100, the Chinese scholar Chia Hsien had discovered the secret of the binomial coefficients that was later rediscovered by the French philosopher and mathematician Blaise Pascal (1623–1662). To understand Chia Hsien's and Pascal's construction of binomial coefficients, let's first look at the coefficients in these binomial expansions.

$$
\begin{aligned}
(x + y)^0 &= 1 \\
(x + y)^1 &= 1x + 1y \\
(x + y)^2 &= 1x^2 + 2xy + 1y^2 \\
(x + y)^3 &= 1x^3 + 3x^2y + 3xy^2 + 1y^3 \\
(x + y)^4 &= 1x^4 + 4x^3y + 6x^2y^2 + 4xy^3 + 1y^4 \\
(x + y)^5 &= 1x^5 + 5x^4y + 10x^3y^2 + 10x^2y^3 + 5xy^4 + 1y^5
\end{aligned}
$$

If we remove the variables and the plus signs and list only the coefficients, we get a triangle of numbers composed of the binomial coefficients. This triangle of numbers is known as *Pascal's Triangle*.

$(x + y)^0$:						1							Row 0
$(x + y)^1$:					1		1						Row 1
$(x + y)^2$:				1		2		1					Row 2
$(x + y)^3$:			1		3		3		1				Row 3
$(x + y)^4$:		1		4		6		4		1			Row 4
$(x + y)^5$:	1		5		10		10		5		1		Row 5
	1	6	15		20		15		6		1		Row 6

$1 + 5 \quad 5 + 10 \quad 10 + 10 \quad 10 + 5 \quad 5 + 1$

Note the symmetry in Pascal's Triangle. If the triangle were folded vertically down the middle, the numbers on each side of the crease would match. To create a new bottom row in the triangle, put the number 1 in the first and last places of the new row and add two neighboring entries in the previous row.

The top row is called the *zeroth row* because it corresponds to the binomial expansion of $(x + y)^0$. The next row is called the *first row* because it corresponds to the binomial expansion of $(x + y)^1$. All of the rows are named so that the *nth row* corresponds to the coefficients of $(x + y)^n$.

2 Use Pascal's Triangle to expand a binomial power.

EXAMPLE 1 **Using Pascal's Triangle to Expand a Binomial Power**

Expand $(4y - 2x)^5$.

SOLUTION

From the fifth row of Pascal's Triangle, we see that the binomial coefficients are

$$1, 5, 10, 10, 5, 1.$$

We must make some changes in the expansion

$$(x + y)^5 = x^5 + 5x^4y + 10x^3y^2 + 10x^2y^3 + 5xy^4 + y^5$$

to get the expansion for $(4y - 2x)^5$:

1. Replace x with $4y$.
2. Replace y with $-2x$.

$$\begin{aligned}(4y - 2x)^5 &= [4y + (-2x)]^5 = (4y)^5 + 5(4y)^4(-2x) \\ &\quad + 10(4y)^3(-2x)^2 + 10(4y)^2(-2x)^3 + 5(4y)(-2x)^4 + (-2x)^5 \\ &= 1024y^5 - 2560y^4x + 2560y^3x^2 - 1280y^2x^3 + 320yx^4 - 32x^5\end{aligned}$$

We note that expanding a *difference* results in alternating signs between terms. ■ ■ ■

Practice Problem 1   Expand $(3y - x)^6$. ■

3 Use the Binomial Theorem to expand a binomial power.

The coefficients in a binomial expansion can be computed by using ratios of certain factorials. We first introduce the symbol $\binom{n}{r}$.

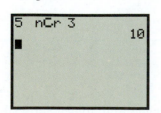

DEFINITION OF $\binom{n}{r}$

If r and n are integers with $0 \le r \le n$, then we define

$$\binom{n}{r} = \frac{n!}{r!(n - r)!}.$$

EXAMPLE 2   **Evaluating** $\binom{n}{r}$

Evaluate each expression.

a. $\binom{4}{1}$ b. $\binom{5}{3}$ c. $\binom{9}{0}$ d. $\binom{35}{35}$

SOLUTION

a. $\binom{4}{1} = \dfrac{4!}{1!(4 - 1)!} = \dfrac{4!}{1!\,3!} = \dfrac{4 \cdot 3 \cdot 2 \cdot 1}{1(3 \cdot 2 \cdot 1)} = \dfrac{4}{1} = 4$

b. $\binom{5}{3} = \dfrac{5!}{3!(5 - 3!)} = \dfrac{5!}{3!\,2!} = \dfrac{5 \cdot 4 \cdot 3!}{3!\,2!} = \dfrac{5 \cdot 4}{2} = 5 \cdot 2 = 10$

c. $\begin{pmatrix} 9 \\ 0 \end{pmatrix} = \dfrac{9!}{0!(9-0)!} = \dfrac{9!}{0!\,9!} = \dfrac{1}{1} = 1$ Recall that $0! = 1$.

d. $\begin{pmatrix} 35 \\ 35 \end{pmatrix} = \dfrac{35!}{35!(35-35)!} = \dfrac{35!}{35!\,0!} = \dfrac{1}{1} = 1$ ■ ■ ■

Practice Problem 2   Evaluate each expression.

a. $\begin{pmatrix} 6 \\ 2 \end{pmatrix}$ **b.** $\begin{pmatrix} 12 \\ 9 \end{pmatrix}$ ■

Examples 2(c) and 2(d) generalize for $n \geq 0$.

$$\begin{pmatrix} n \\ 0 \end{pmatrix} = 1 \quad \text{and} \quad \begin{pmatrix} n \\ n \end{pmatrix} = 1$$

The symbol $\begin{pmatrix} n \\ r \end{pmatrix}$ is read "n choose r." It can be shown that $\begin{pmatrix} n \\ r \end{pmatrix}$ is the number of ways of choosing a subset containing exactly r elements from a set with n elements.

The numbers $\begin{pmatrix} n \\ r \end{pmatrix}$ show up as the coefficients in the expansion of a binomial power. For example, $(x + y)^4$ can be written as either

$$(x + y)^4 = x^4 + 4x^3y + 6x^2y^2 + 4xy^3 + y^4$$

or

$$(x + y)^4 = \begin{pmatrix} 4 \\ 0 \end{pmatrix}x^4 + \begin{pmatrix} 4 \\ 1 \end{pmatrix}x^3y + \begin{pmatrix} 4 \\ 2 \end{pmatrix}x^2y^2 + \begin{pmatrix} 4 \\ 3 \end{pmatrix}xy^3 + \begin{pmatrix} 4 \\ 4 \end{pmatrix}y^4.$$

This result is the **Binomial Theorem** (for $n = 4$), which can be proved by mathematical induction. It provides an efficient method for expanding a binomial power. (See Exercise 64.) The Binomial Theorem can be used to expand a binomial power directly, without reference to Pascal's Triangle. This technique is particularly useful for expanding large powers of a binomial. For example, the expansion of $(x + y)^{20}$ would require that you produce 20 rows of the Pascal Triangle.

THE BINOMIAL THEOREM

If n is a natural number, then the binomial expansion of $(x + y)^n$ is given by

$$(x + y)^n = \begin{pmatrix} n \\ 0 \end{pmatrix}x^n + \begin{pmatrix} n \\ 1 \end{pmatrix}x^{n-1}y + \begin{pmatrix} n \\ 2 \end{pmatrix}x^{n-2}y^2 + \cdots + \begin{pmatrix} n \\ r \end{pmatrix}x^{n-r}y^r + \cdots + \begin{pmatrix} n \\ n \end{pmatrix}y^n$$

$$= \sum_{r=0}^{n} \begin{pmatrix} n \\ r \end{pmatrix}x^{n-r}y^r.$$

The coefficient $\begin{pmatrix} n \\ r \end{pmatrix}$ of $x^{n-r}y^r$ is $\dfrac{n!}{r!(n-r)!}$.

EXAMPLE 3 **Expanding a Binomial Power by Using the Binomial Theorem**

Find the binomial expansion of $(x - 3y)^4$.

SOLUTION

$$(x - 3y)^4 = [x + (-3y)]^4$$

$$= \binom{4}{0}x^4 + \binom{4}{1}x^3(-3y) + \binom{4}{2}x^2(-3y)^2 + \binom{4}{3}x(-3y)^3 + \binom{4}{4}(-3y)^4$$

$$= \frac{4!}{0!\,4!}x^4 + \frac{4!}{1!\,3!}x^3(-3y) + \frac{4!}{2!\,2!}x^2(-3y)^2 + \frac{4!}{3!\,1!}x(-3y)^3 + \frac{4!}{4!\,0!}(-3y)^4$$

$$= x^4 + \frac{4 \cdot 3!}{1!\,3!}x^3(-3y) + \frac{4 \cdot 3 \cdot 2!}{2!\,2!}x^2(-3y)^2 + \frac{4 \cdot 3!}{3!\,1!}x(-3y)^3 + (-3y)^4$$

$$= x^4 + 4x^3(-3y) + \frac{4 \cdot 3}{2!}x^2(9y)^2 + 4x(-27y^3) + (81y^4)$$

$$= x^4 - 12x^3y + 54x^2y^2 - 108xy^3 + 81y^4$$

■ ■ ■

Practice Problem 3 Find the binomial expansion of $(3x - y)^4$. ■

4 Find the coefficient of a term in a binomial expansion.

The Binomial Theorem is also useful for finding a particular term and its coefficient in a binomial expansion.

EXAMPLE 4 **Finding a Particular Coefficient in a Binomial Expansion**

Find the coefficient of x^9y^3 in the expansion of $(x + y)^{12}$.

SOLUTION

The form of the expansion is

$$(x + y)^{12} = \binom{12}{0}x^{12} + \binom{12}{1}x^{11}y + \binom{12}{2}x^{10}y^2 + \binom{12}{3}x^9y^3 + \cdots + \binom{12}{11}xy^{11} + \binom{12}{12}y^{12}.$$

From the fourth term, $\binom{12}{3}x^9y^3$, the coefficient of x^9y^3 is

$$\binom{12}{3} = \frac{12!}{3!(12 - 3!)} = \frac{12!}{3!\,9!} = \frac{12 \cdot 11 \cdot 10 \cdot 9!}{3!\,9!} = 220.$$

■ ■ ■

Practice Problem 4 Find the coefficient of x^3y^9 in the expansion of $(x + y)^{12}$. ■

The method illustrated in Example 4 allows us to find any particular term in a binomial expansion without writing out the complete expansion.

PARTICULAR TERM IN A BINOMIAL EXPANSION

The term containing the factor x^r in the expansion of $(x + y)^n$ is

$$\binom{n}{n - r}x^ry^{n-r}.$$

EXAMPLE 5   **Finding a Particular Term in a Binomial Expansion**

Find the term containing x^{10} in the expansion of $(x + 2a)^{15}$.

SOLUTION

We begin with the formula for the term containing the factor x^r.

$$\binom{n}{n-r}x^r y^{n-r} = \binom{15}{15-10}x^{10}(2a)^{15-10} \qquad \text{Replace } n \text{ with } 15, r \text{ with } 10, \text{ and } y \text{ with } 2a.$$

$$= \binom{15}{5}x^{10}(2a)^5 \qquad \text{Simplify.}$$

$$= \frac{15!}{5!(15-5)!}x^{10}2^5 a^5 \qquad \text{Recall that } \binom{n}{r} = \frac{n!}{r!(n-r)!}.$$

$$= \frac{15!}{5!\,10!} \cdot 32x^{10}a^5 \qquad 2^5 = 32$$

$$= \frac{15 \cdot 14 \cdot 13 \cdot 12 \cdot 11 \cdot \cancel{10!}}{5! \cdot \cancel{10!}} \cdot 32x^{10}a^5$$

$$= 96{,}096x^{10}a^5 \qquad \text{Use a calculator.} \qquad ■ ■ ■$$

Practice Problem 5   Find the term containing x^3 in the expansion of $(x + 2a)^{15}$. ■

SECTION 8.5 ■ Exercises

A EXERCISES Basic Skills and Concepts

1. The expansion of $(x + y)^n$ has _____ terms.

2. In the expansion of $(x + y)^5$, the coefficient of x^2y^3 is $\binom{n}{r}$ when $n =$ _____ and $r =$ _____ .

3. Expanding a difference such as $(2x - y)^{10}$ results in _____ signs between terms.

4. For any positive integer n, $\binom{n}{n} =$ _____ .

5. *True or False* Every coefficient in the expansion of $(x + y)^n$, for $n > 1$, appears exactly twice.

6. *True or False* If $n > 3$, then $2\binom{n}{2} = n$.

In Exercises 7–16, evaluate each expression.

7. $\dfrac{6!}{3!}$

8. $\dfrac{11!}{9!}$

9. $\dfrac{12!}{11!}$

10. $\dfrac{3!}{0!}$

11. $\binom{6}{4}$

12. $\binom{6}{3}$

13. $\binom{9}{0}$

14. $\binom{12}{0}$

15. $\binom{7}{1}$

16. $\binom{7}{3}$

In Exercises 17–24, use Pascal's Triangle to expand each binomial.

17. $(x + 2)^4$

18. $(x + 3)^4$

19. $(x - 2)^5$

20. $(3 - x)^5$

21. $(2 - 3x)^3$

22. $(3 - 2x)^3$

23. $(2x + 3y)^4$

24. $(2x + 5y)^4$

In Exercises 25–46, use either the binomial theorem or Pascal's Triangle to expand each binomial.

25. $(x + 1)^4$

26. $(x + 2)^4$

27. $(x - 1)^5$

28. $(1 - x)^5$

29. $(y - 3)^3$

30. $(2 - y)^5$

31. $(x + y)^6$

32. $(x - y)^6$

33. $(1 + 3y)^5$

34. $(2x + 1)^5$

35. $(2x + 1)^4$

36. $(3x - 1)^4$

37. $(x - 2y)^3$

38. $(2x - y)^3$

39. $(2x + y)^4$

40. $(3x - 2y)^4$

41. $\left(\dfrac{x}{2} + 2\right)^7$

42. $\left(2 - \dfrac{x}{2}\right)^7$

43. $\left(a^2 - \dfrac{1}{3}\right)^4$

44. $\left(\dfrac{1}{2} - a^2\right)^4$

45. $\left(\dfrac{1}{x} + y\right)^3$

46. $\left(x + \dfrac{2}{y}\right)^3$

In Exercises 47–54, find the specified term.

47. $(x + y)^{10}$; term containing x^7

48. $(x + y)^{10}$; term containing y^7

49. $(x - 2)^{12}$; term containing x^3

50. $(2 - x)^{12}$; term containing x^3

51. $(2x + 3y)^8$; term containing x^6

52. $(2x + 3y)^8$; term containing y^6

53. $(5x - 2y)^{11}$; term containing y^9

54. $(7x - y)^{15}$; term containing x^9

55. Find the value of $(1.2)^5$ by writing it in the form $(1 + 0.2)^5$ and applying the Binomial Theorem.

56. Use the method in Exercise 55 to evaluate each expression.
   a. $(2.9)^4$ b. $(10.4)^3$

C EXERCISES Beyond the Basics

57. Find the middle term in the expansion of $\left(\sqrt{x} - \dfrac{2}{x^2} \right)^{10}$.

58. Find the middle term in the expansion of $\left(\sqrt{x} + \dfrac{1}{x^2} \right)^{10}$.

59. Find the middle term in the expansion of $(1 - x^2 y^{-3})^{12}$.

60. Show that the middle term in the expansion of $(1 + x)^{2n}$ is

$$\frac{1 \cdot 3 \cdot 5 \cdots (2n - 1)}{n!} 2^n x^n.$$

61. Prove that $\dbinom{n}{0} + \dbinom{n}{1} + \dbinom{n}{2} + \cdots + \dbinom{n}{n} = 2^n$.

62. Prove that $\dbinom{n}{0} - \dbinom{n}{1} + \dbinom{n}{2} - \dbinom{n}{3} + \cdots$

$$+ (-1)^n \dbinom{n}{n} = 0.$$

63. Prove that $\dbinom{k}{j} + \dbinom{k}{j-1} = \dbinom{k+1}{j}$.

64. Prove the Binomial Theorem by using the Principle of Mathematical Induction.

$$\left[\text{Hint:} \right.$$

$$(x + y)^{k+1} = (x + y)(x + y)^k = (x + y) \sum_{j=0}^{k} \dbinom{k}{j} x^{k-j} y^j$$

$$\left. = \sum_{j=0}^{k} \dbinom{k}{j} x^{k+1-j} y^j + \sum_{j=0}^{k} \dbinom{k}{j} x^{k-j} y^{j+1}. \right]$$

Use Exercise 63 to show that

$$(x + y)^{k+1} = \sum_{j=0}^{k+1} \dbinom{k+1}{j} x^{k+1-j} y^j.$$

In Exercises 65 and 66, find the value of the expression without expanding any term.

65. $(2x - 1)^4 + 4(2x - 1)^3(3 - 2x) + 6(2x - 1)^2(3 - 2x)^2$
   $+ 4(2x - 1)(3 - 2x)^3 + (3 - 2x)^4$

66. $(x + 1)^4 - 4(x + 1)^3(x - 1) + 6(x + 1)^2(x - 1)^2$
   $- 4(x + 1)(x - 1)^3 + (x - 1)^4$

67. $(3x - 1)^5 + 5(3x - 1)^4(1 - 2x)$
   $+ 10(3x - 1)^3(1 - 2x)^2 + 10(3x - 1)^2(1 - 2x)^3$
   $+ 5(3x - 1)(1 - 2x)^4 + (1 - 2x)^5$

68. Find the constant term in each expansion.
   a. $\left(x^2 - \dfrac{1}{x} \right)^9$ b. $\left(\sqrt{x} - \dfrac{2}{x^2} \right)^{10}$

69. If the constant term in the expansion of $\left(kx - \dfrac{1}{x^2} \right)^6$ is 240, find k.

70. If the constant term in the expansion of $\left(x^3 + \dfrac{k}{x^8} \right)^{11}$ is 1320, find k.

71. Show that there is no constant term in the expansion of $\left(2x^2 - \dfrac{1}{4x} \right)^{11}$.

Critical Thinking

72. Use the binomial expansion of $(x + y)^6$, with $x = 1$ and $y = 1$, to show that

$$2^6 = \dbinom{6}{0} + \dbinom{6}{1} + \dbinom{6}{2} + \dbinom{6}{3} + \dbinom{6}{4} + \dbinom{6}{5} + \dbinom{6}{6}.$$

73. Use the binomial expansion of $(x + y)^{10}$ to show that

$$\dbinom{10}{0} - \dbinom{10}{1} + \dbinom{10}{2} - \dbinom{10}{3} + \dbinom{10}{4} - \dbinom{10}{5}$$
$$+ \dbinom{10}{6} - \dbinom{10}{7} + \dbinom{10}{8} - \dbinom{10}{9} + \dbinom{10}{10} = 0.$$

74. Use the binomial expansion of $(x + y)^2$ to show that if $x > 0$ and $y > 0$, then $(x + y)^2 > x^2 + y^2$.

75. Use the binomial expansion of $(x + y)^n$ to show that if $x > 0$ and $y > 0$, then $(x + y)^n > x^n + y^n$.

76. Use the binomial expansion of $(1 + x)^n$ to show that if $x > 0$, then $(1 + x)^n > 1 + nx$.

Counting Principles

Before Starting this Section, Review

1. Exponents (Section P.2, page 19)
2. Factorials (Section 8.1, page 663)

Objectives

1. Use the Fundamental Counting Principle.
2. Use the formula for permutations.
3. Use the formula for combinations.
4. Use the formula for distinguishable permutations.

THE MYSTERY OF SOCIAL SECURITY NUMBERS

Social Security numbers may seem to be assigned randomly, but they are not. Although the details are complicated, here is a brief introduction: First, the format for all Social Security numbers is $N_1N_2N_3$-N_4N_5-$N_6N_7N_8N_9$, where each N_k is one of the numbers 0, 1, 2, 3, 4, 5, 6, 7, 8, 9. U.S. citizens, permanent residents, and certain temporary residents are issued these numbers. For working people, the numbers help track tax contributions and Social Security benefits.

The first three digits in a Social Security number make up an *area number* assigned to a geographic location. The area number indicates the state in which the card was issued or the ZIP Code in the mailing address provided in the original application.

The middle two digits constitute a *group number* and are designed simply to break the Social Security number into convenient blocks for issuance. The exact procedure is not very enlightening.

The last four digits are *serial numbers* and are assigned in each group from 0001 through 9999.

Various restrictions limit the numbers that can be used. For example, numbers with all zeros in any of the digit blocks (000-xx-xxxx, xxx-00-xxxx, or xxx-xx-0000) are never issued. Even if there were no restrictions, at some point, there might be *more people than available numbers*. How many numbers can be issued in the Social Security format with no restrictions? The answer is 1,000,000,000 (1 billion). (The U.S. population in October 2006 was about 300,000,000 (300 million).)

In Example 3, you learn how to calculate the total number of possible Social Security numbers. ■

---

1. Use the Fundamental Counting principle.

You may think that you already know all there is to know about counting. However, mathematicians have developed a variety of techniques that allow you to determine the number of items that are of interest to you without making the effort of listing and counting all of them. We start with an example that easily lists the items to be counted, but that also demonstrates a general counting procedure.

EXAMPLE 1   Counting Possible Car Selections

Suppose you are trying to decide between buying a sport-utility vehicle (SUV) and a four-door sedan. The SUV is available in black, red, and silver. The sedan is available in black, blue, and green. In how many ways can you choose a type of car and its color?

FIGURE 8.3 Tree diagram

SOLUTION

We can represent the possible solutions by using a *tree diagram*, as shown in Figure 8.3. There are two initial choices: an SUV and a sedan. For each of these two choices, there are three additional choices of color; so a total of $2 \cdot 3 = 6$ car selections is possible.   ▪ ▪ ▪

Practice Problem 1   Construct a tree diagram and determine the number of ways you can choose to take one of three courses—English, French, or math—in the morning, afternoon, or evening.   ▪

The car selection process in Example 1 involves making two choices. First, choose a type of car; second, choose a color. It is important in this process that the same *number* of choices be available for the second choice regardless of the result of the first choice. The exact rules for counting by this method are given next.

FUNDAMENTAL COUNTING PRINCIPLE

If a first choice can be made in p different ways, a second choice can be made in q different ways, a third choice can be made in r different ways, and so on, then the sequence of choices can be made in $p \cdot q \cdot r \cdots$ different ways.

EXAMPLE 2   **Counting Music Programs**

A disc jockey wants to open a program by picking a song from among 40 songs by one artist and close the program by picking a song from among 32 songs by a second artist. How many different choices are possible?

SOLUTION

Each choice of an opening and ending song sequence can be obtained as follows:
　　First, choose an opening song in 40 ways.
　　Second, choose the song to end the program in 32 ways.
　　Because no matter which song is chosen to open the program, all 32 songs are available to end the program, the counting principle applies, and there are a total of

$$40 \cdot 32 = 1280$$

possible choices for the opening and ending sequences.   ▪ ▪ ▪

Practice Problem 2   Reba gets to choose both dinner and a movie for a Saturday night date. She can choose from any of seven restaurants and five movies. How many different choices are possible?   ▪

◆ **WARNING**   Be careful when using the Fundamental Counting Principle. Suppose you are trying to decide between buying a SUV and a convertible. The SUV is available only in black, red, or silver, and the convertible is available only in black or red. In how many ways can you choose a type of car and its color? The car and color can be selected by making two choices: type (SUV or convertible) followed by color. But in this situation, the number of ways you can make the second choice depends on the first choice. You can choose from *three* colors for a SUV, but only *two* colors for a convertible. The Fundamental Counting Principle *does not apply*. Fortunately, there are so few choices here that you can easily list all five: black SUV, red SUV, silver SUV, black convertible, and red convertible.

EXAMPLE 3 **Counting Possible Social Security Numbers**

Social Security numbers have the format $N_1N_2N_3$-N_4N_5-$N_6N_7N_8N_9$, where each N_k is one of the integers 0, 1, 2, 3, 4, 5, 6, 7, 8, 9. Assuming that there are no other restrictions, how many such numbers are possible?

SOLUTION

There are nine positions in the format

$$N_1N_2N_3\text{-}N_4N_5\text{-}N_6N_7N_8N_9.$$

Because the number of choices available at each position is not affected by previous choices, the Fundamental Counting Principle can be used. Replace each of the letters N with a blank to be filled in with one of the ten digits 0, 1, 2, 3, 4, 5, 6, 7, 8, 9.
We have

$$\underline{10}\cdot\underline{10}\cdot\underline{10}\cdot\underline{10}\cdot\underline{10}\cdot\underline{10}\cdot\underline{10}\cdot\underline{10}\cdot\underline{10} = 10^9.$$

Thus, there are $10^9 = 1,000,000,000$ possible Social Security numbers.　■ ■ ■

Practice Problem 3　In Example 3, suppose we fix the area number for the Social Security number to be 433. How many social security numbers can be assigned to this area—that is, numbers having the form 433-N_4N_5-$N_6N_7N_8N_9$?　■

2 Use the formula for permutations.

Permutations

DEFINITION OF PERMUTATION

A **permutation** is an arrangement of n distinct objects in a fixed order in which no object is used more than once. The specific order is important: each different ordering of the same objects is a different permutation.

EXAMPLE 4　**Counting Possible Arrangements for a Group Photograph**

How many different ways can five people be arranged in a row for a group photograph?

SOLUTION

To count the number of ways we can arrange five people in a row for a group photograph, we can assign numbers to each of the five photograph positions.

| 1 | 2 | 3 | 4 | 5 |

Then we proceed as follows:
First, choosing any of the five people for Position 1 gives 5 ways.
Second, choosing one of the four remaining people for Position 2 gives 4 ways.
Third, choosing one of the three remaining people for Position 3 gives 3 ways.
Fourth, choosing one of the remaining two people for Position 4 gives 2 ways.
Fifth, choosing the last person remaining for Position 5 gives 1 way.
Using the Fundamental Counting Principle, we see that there are

$$5\cdot4\cdot3\cdot2\cdot1 = 5!, \text{ or } 120,$$

possible arrangements.　■ ■ ■

Practice Problem 4　How many different ways can seven books be arranged on a shelf?　■

This technique gives a general formula for counting permutations of n distinct objects.

NUMBER OF PERMUTATIONS OF *n* OBJECTS

The number of permutations of *n* distinct objects is

$$n! = n(n-1) \cdots \cdot 4 \cdot 3 \cdot 2 \cdot 1.$$

That is, *n* distinct objects can be arranged in *n*! different ways.

Sometimes only some, but not all, of the available objects are to be arranged in a specific order, again with no object being used more than once. You may want to choose *r* objects from among *n* available objects. This type of ordering is called a **permutation of *n* objects taken *r* at a time**.

EXAMPLE 5   **Counting Prizewinners**

Three prizes (for first, second, and third place) are to be awarded at a dog show having 27 contestants. In how many different ways can dogs come in first, second, and third?

SOLUTION

Consider how we might list the prize winners.

Any dog might win first prize; there are 27 possibilities.

Any remaining dog might win second prize; there are 26 possibilities.

Any remaining dog might win third prize; there are 25 possibilities. Using the Fundamental Counting Principle, there are

$$27 \cdot 26 \cdot 25 = 17{,}550$$

1st   2nd   3rd
prize   prize   prize

distinct ways that the dogs can come in first, second, and third. This is an example of permutations of 27 objects taken 3 at a time.   ■ ■ ■

Practice Problem 5   There are nine different rides at a state fair. A group has to decide which four rides they will go on and in what order. How many possibilities are there?   ■

PERMUTATIONS OF *n* OBJECTS TAKEN *r* AT A TIME

The number of permutations of *n* distinct objects taken *r* at a time is denoted by $P(n, r)$, where

$$P(n, r) = \frac{n!}{(n-r)!}$$
$$= n(n-1)(n-2) \cdots (n-r+1).$$

EXAMPLE 6   **Using the Permutations Formula**

Use the formula for $P(n, r)$ to evaluate each expression.

a. $P(7, 3)$ **b.** $P(6, 0)$

Most graphing calculators can compute $P(n, r)$. They frequently use the symbol nPr. Note that $P(7, 3)$ is shown as $7\, nPr\, 3$ on the viewing screen.

```
7 nPr 3
              210
■
```

SOLUTION

a. $P(7, 3) = \dfrac{7!}{(7 - 3)!}$ Replace n with 7 and r with 3 in $P(n, r)$.

$= \dfrac{7!}{4!} = \dfrac{7 \cdot 6 \cdot 5 \cdot 4!}{4!} = 7 \cdot 6 \cdot 5 = 210$

b. $P(6, 0) = \dfrac{6!}{(6 - 0)!}$ Replace n with 6 and r with 0 in $P(n, r)$.

$= \dfrac{6!}{6!} = 1$ ■ ■ ■

Practice Problem 6 Evaluate. **a.** $P(9, 2)$ **b.** $P(n, 0)$ ■

Let's apply the permutations formula to the problem of counting prizewinners in Example 5. The number of permutations of 27 dogs taken 3 at a time is

$$P(27, 3) = \frac{27!}{(27 - 3)!}$$

$$= \frac{27!}{24!} = \frac{27 \cdot 26 \cdot 25 \cdot 24!}{24!} = 27 \cdot 26 \cdot 25 = 17{,}550.$$

This is the same answer we found in Example 5.

3 Use the formula for combinations.

Combinations

Order is not always important in selecting objects. For example, if someone is bringing six movies on a vacation, the order in which the movies were chosen is unimportant.

> **DEFINITION OF A COMBINATION OF n DISTINCT OBJECTS TAKEN r AT A TIME**
>
> When r objects are chosen from n distinct objects without regard to order, we call the set of r objects a **combination of n objects taken r at a time**. The symbol $C(n, r)$ denotes the total number of combinations of n objects taken r at a time.

EXAMPLE 7   **Listing Combinations**

List all of the combinations of the four mascot names "bears," "bulls," "lions," and "tigers," taken two at a time. Find $C(4, 2)$.

SOLUTION

If "bears" is one of the mascots chosen, the possibilities are {bears, bulls}, {bears, lions}, and {bears, tigers}.

If "bulls" is one of the mascots chosen, the remaining possibilities (excluding {bears, bulls}) are {bulls, lions} and {bulls, tigers}.

The final choice of two mascot names is {lions, tigers}.

The combinations of the four mascot names taken two at a time are {bears, bulls}, {bears, lions}, {bears, tigers}, {bulls, lions}, {bulls, tigers}, and {lions, tigers}.

Because there are six such combinations, we have $C(4, 2) = 6$.   ■ ■ ■

Practice Problem 7 List all of the combinations of the four mascot names "bears," "bulls," "lions," and "tigers," taken three at a time. Find $C(4, 3)$.   ■

To find a formula for $C(n, r)$, we begin by comparing $C(n, r)$ with $P(n, r)$, the number of permutations of n objects taken r at a time. Suppose, for example, we have seven distinct objects—A, B, C, D, E, F, and G—and we pick three of them—A, D, and E (a combination of seven objects taken three at a time). This one combination

ADE gives 3! = 6 permutations of seven objects taken three at a time. Here are all of the possible arrangements of the letters *A*, *D*, and *E*: *ADE*, *AED*, *DAE*, *DEA*, *EAD*, and *EDA*. Because one combination generates six permutations, there are six (3!) times as many permutations of seven objects taken three at a time as there are combinations of seven objects taken three at a time. Thus,

3! · (number of combinations) = (number of permutations).

$$3! \, C(7, 3) = P(7, 3)$$

$$C(7, 3) = \frac{P(7, 3)}{3!} \qquad \text{Divide both sides by 3!}$$

$$= \frac{\frac{7!}{(7 - 3)!}}{3!} \qquad \text{Replace } P(7, 3) \text{ with } \frac{7!}{(7 - 3)!}.$$

$$= \frac{7!}{(7 - 3)! \, 3!} = \frac{7!}{4! \, 3!} = \frac{7 \cdot 6 \cdot 5 \cdot 4!}{4! \, 3!} = \frac{7 \cdot 6 \cdot 5}{3 \cdot 2 \cdot 1} = 35$$

A similar process leads to a more general formula.

RECALL

Notice that the value of $C(n, r)$ is the same as that of $\binom{n}{r}$.

THE NUMBER OF COMBINATIONS OF *n* DISTINCT OBJECTS TAKEN *r* AT A TIME

The number of combinations of *n* distinct objects taken *r* at a time is

$$C(n, r) = \frac{n!}{(n - r)! \, r!}.$$

EXAMPLE 8   **Using the Combinations Formula**

Use the formula $C(n, r)$ to evaluate each expression.

a. $C(7, 5)$ **b.** $C(8, 0)$ **c.** $C(4, 4)$

SOLUTION

a. $C(7, 5) = \dfrac{7!}{(7 - 5)! \, 5!}$ Replace *n* with 7 and *r* with 5 in $C(n, r)$.

$$= \frac{7!}{2! \, 5!} = \frac{7 \cdot 6 \cdot 5!}{2 \cdot 1 \cdot 5!} = 21$$

b. $C(8, 0) = \dfrac{8!}{(8 - 0)! \, 0!}$ Substitute *n* = 8 and *r* = 0 in $C(n, r)$.

$$= \frac{8!}{8! \, 0!} = 1$$

c. $C(4, 4) = \dfrac{4!}{(4 - 4)! \, 4!}$ Substitute *n* = 4 and *r* = 4 in $C(n, r)$.

$$= \frac{4!}{0! \, 4!} = 1$$

■ ■ ■

Practice Problem 8   Evaluate.

a. $C(9, 2)$ **b.** $C(n, 0)$ **c.** $C(n, n)$ ■

TECHNOLOGY CONNECTION

Most graphing calculators can compute $C(n, r)$. Note that $C(4, 2)$ is shown as 4 *n*Cr 2 on the viewing screen.

```
4 nCr 2
          6
■
```

EXAMPLE 9   **Choosing Pizza Toppings**

How many different ways can five pizza toppings be chosen from the following choices: pepperoni, onions, mushrooms, green peppers, olives, tomatoes, mozzarella, and anchovies?

SOLUTION

Eight toppings are listed; so the question is, How many ways can we pick five things from eight things? That is, we need to find the number of combinations of eight objects taken five at a time, or $C(8, 5)$.

Use the formula $C(n, r) = \dfrac{n!}{(n - r)!\, r!}$.

$$C(8, 5) = \frac{8!}{(8 - 5)!\, 5!} \qquad \text{Replace } n \text{ with 8 and } r \text{ with 5.}$$

$$= \frac{8!}{3!\, 5!} = \frac{8 \cdot 7 \cdot 6 \cdot 5!}{3 \cdot 2 \cdot 1 \cdot 5!} = 56$$

There are 56 ways that five of the eight pizza toppings can be chosen.   ■ ■ ■

Practice Problem 9   A box of assorted chocolates contains 12 different kinds of chocolates. In how many ways can three chocolates be chosen?   ■

4 Use the formula for distinguishable permutations.

Distinguishable Permutations

Suppose you were to stand in a line with identical twins. Although the three of you can arrange yourselves in $3! = 6$ different ways, someone who could not tell the twins apart could distinguish only three different arrangements. These three arrangements are called *distinguishable permutations*. To better understand the idea of distinguishable permutations, suppose we have three tiles with the numbers 1, 2, and 3 printed on the fronts and the letters M O M printed on the backs, respectively.

Front view of tiles   Back view of tiles

There are $3! = 6$ permutations of the numbers 1, 2, 3.

Now consider the corresponding arrangements of the letters *M, O,* and *M* on the back sides of the tiles as the fronts of the tiles are arranged to show each of the six permutations of 1, 2, 3.

Front View of Tiles	Back View of Tiles
1 2 3	M O M
1 3 2	M M O
2 1 3	O M M
2 3 1	O M M
3 1 2	M M O
3 2 1	M O M

There are only three distinguishable arrangements that appear when the tiles are viewed from the back: MOM, MMO, and OMM.

The front view shows the permutations of three distinct objects, but the back view shows the permutations of three objects of which one is of one kind (the letter *O*) and two are of a second kind (the letter *M*).

DISTINGUISHABLE PERMUTATIONS

The number of distinguishable permutations of n objects of which n_1 are of one kind, n_2 are of a second kind, ..., and n_k are of a kth kind is

$$\frac{n!}{n_1! \cdot n_2! \cdots \cdots n_k!},$$

where $n_1 + n_2 + \cdots + n_k = n$.

EXAMPLE 10   **Distributing Gifts**

In how many ways can nine gifts be distributed among three children if each child receives three gifts?

SOLUTION

If we number the gifts 1 through 9 and use *A*, *B*, and *C* to represent the children, we can visualize any distribution of gifts as an arrangement of the letters *A*, *B*, and *C* (each letter being used three times) above the numbers 1 through 9. Here is one possible arrangement:

$$\begin{array}{ccccccccc} B & A & A & C & C & B & A & C & B \\ 1 & 2 & 3 & 4 & 5 & 6 & 7 & 8 & 9 \end{array}$$

In this arrangement, child *A* gets gifts numbered 2, 3, and 7; child *B* gets gifts numbered 1, 6, and 9; and child *C* gets gifts numbered 4, 5, and 8.

To count all of the possible distributions of gifts, we count the number of permutations of nine objects, of which three are of one kind, three are of a second kind, and three are of a third kind.

$$\frac{9!}{3!\,3!\,3!} = 1680$$

There are 1680 ways the nine gifts can be distributed among the three children.

■ ■ ■

Practice Problem 10   In how many ways can six counselors be sent in pairs to three different locations?   ■

Deciding Whether to Use Permutations, Combinations, or the Fundamental Counting Principle

Suppose 23 students show up late for a private screening of a new movie. Suppose also that one seat is available in each of the first, second, eighth, and tenth rows and that five students will be allowed to stand in the back of the theater.

Permutations If order is important, use permutations.	In how many ways can 4 of the 23 students be assigned the available seats in rows 1, 2, 8, and 10? ***Answer:*** $\underset{\text{Row 1}}{23} \cdot \underset{\text{Row 2}}{22} \cdot \underset{\text{Row 8}}{21} \cdot \underset{\text{Row 10}}{20} \cdot = 212{,}520 \text{ ways}$
Combinations If order is not important, use combinations.	In how many ways can 5 of the 19 students *not given seats* be chosen to stand in the back? ***Answer:*** $C(19, 5) = 11{,}628$ ways
Fundamental Counting Principle Whenever you are making consecutive choices and the number of choices at each stage is not affected by the way earlier choices were made, you can use the Fundamental Counting Principle.	In how many ways can four students be assigned to the available seats and five students be chosen to stand in the back? ***Answer:*** $212{,}520 \times 11{,}628$ ways Assign the four seats. Pick five students.

Notice that

- Since the Fundamental Counting Principle is used to count permutations, you can always use that principle instead of the formula for permutations.
- There are frequently equally good, although different, ways to use counting techniques to solve a problem. For example, we could first choose 5 students to stand in the back of the theater in $C(23, 5)$ ways and then assign the seats in rows 1, 2, 8, and 10 to the 19 remaining students in $19 \cdot 18 \cdot 17 \cdot 16$ ways. The total number of ways students can be assigned to stand in the back and assigned to the available seats is then $C(23, 5) \cdot 19 \cdot 18 \cdot 17 \cdot 16$. You should verify that

$$23 \cdot 22 \cdot 21 \cdot 20 \cdot C(19, 5) = 212{,}520 \times 11{,}628.$$

SECTION 8.6 ■ Exercises

A EXERCISES Basic Skills and Concepts

1. Any arrangement of n distinct objects in a fixed order in which no object is used more than once is called a(n) _____.

2. The number of permutations of five distinct objects is _____.

3. When r objects are chosen from n distinct objects, the set of r objects is called a(n) _____ of n objects taken r at a time.

4. The number of distinguishable permutations of ten objects of which three are of one kind and seven are of a second kind is _____.

5. *True or False* When order is important in counting objects, we use permutations.

6. *True or False* There are more permutations of n objects taken r at a time than there are combinations of n objects taken r at a time.

In Exercises 7–14, use the formula for $P(n, r)$ to evaluate each expression.

7. $P(6, 1)$

8. $P(7, 3)$

9. $P(8, 2)$

10. $P(10, 6)$

11. $P(9, 9)$

12. $P(5, 5)$

13. $P(7, 0)$

14. $P(4, 0)$

In Exercises 15–22, use the formula for $C(n, r)$ to evaluate each expression.

15. $C(8, 3)$

16. $C(7, 2)$

17. $C(9, 4)$

18. $C(10, 5)$

19. $C(5, 5)$

20. $C(6, 6)$

21. $C(3, 0)$

22. $C(5, 0)$

In Exercises 23–26, use the Fundamental Counting Principle to solve each problem.

23. Find the number of different 2-letter codes that can be made from the 26 capital letters of the alphabet if (a) repeated letters are allowed; (b) repeated letters are not allowed.

24. How many possible answer sheets are there for a ten-question true–false exam if no answer is left blank?

25. In how many ways can a president and vice president be chosen from an organization with 50 members?

26. How many four-digit numbers can be formed with the digits 0, 1, 2, 3, 4, 5, 6, 7, 8, and 9 if the first digit cannot be 0?

In Exercises 27–30, use the formula for counting permutations to solve each problem.

27. In how many different ways can the members of a family of four be seated in a row of four chairs?

28. Ten teams are entered in a bowling tournament. In how many ways can first, second, and third prizes be awarded?

29. Frederico has five shirts, three pairs of pants, and four ties that are appropriate for an interview. If he wears one of each type of apparel, how many different outfits can he wear?

30. A choreographer has to arrange eight pieces for a dance program. In how many ways can this be done?

In Exercises 31–34, use the formula for counting combinations to solve each problem.

31. A pizza menu offers pepperoni, peppers, mushrooms, sausage, and meatballs as extra toppings that can be added to your pizza. How many different pizzas can you order with two additional toppings?

32. In how many ways can 2 student representatives be chosen from a class of 15 students?

33. Students are allowed to choose 9 out of 11 problems to work for credit on an exam. How many ways can this be done?

34. You have to choose 3 out of 18 potential teammates to help you with a group assignment. How many ways can this be done?

In Exercises 35–40, solve the problem by any appropriate counting method.

35. How many ways can Ashley choose 4 out of 11 different canned goods to donate to charity?

36. How many ways can a computer store hire a salesclerk and a technician from seven applicants for the clerk position and four applicants for the technician position?

37. How many ways can six people be seated in a row with eight seats, leaving two seats vacant?

38. Seven students arrive to rent two canoes and three kayaks. How many ways can two students be selected to rent the canoes and three to rent the kayaks?

39. Dakota is trying to decide in which order to visit five prospective colleges. How many ways can this be done?

40. How many ways can the first 4 numbers be called from the 75 possible bingo numbers?

B EXERCISES Applying the Concepts

41. **Area codes.** How many three-digit area codes for telephones are possible if
    a. the first digit cannot be a 0 or 1 and the second digit must be a 0 or 1?
    b. there are no restrictions on the second digit? (Previous restrictions were abandoned in 1995.)

42. **Radio station call letters.** Radio stations in the United States have call letters that begin with either *K* or *W* (for example, WXYZ in Detroit). Some have a total of three letters, and others have four. How many different call-letter selections are possible?

43. **Fraternity letters.** The Greek alphabet contains 24 letters. How many 3-letter fraternity names can be made with Greek letters
    a. if no repetitions are allowed?
    b. if repetitions are allowed?

44. **Codes.** How many three-letter codes can be made from the first ten letters of the alphabet if
    a. no letters are repeated?
    b. letters may be repeated?

45. **Supreme Court decisions.** In how many ways can the nine justices of the U.S. Supreme Court reach a majority decision? (The U.S. Supreme Court has nine justices who are appointed for life.)

46. **Committee selection.** A club consists of ten members. How many ways can a committee of four be selected?

47. **Wardrobe choices.** Al's wardrobe consists of eight pairs of slacks, eight shirts, and eight pairs of shoes. He does not want to wear the same outfit on any two days of the year. Will his current wardrobe allow him to do this?

48. **Menu selection.** A French restaurant offers two choices of soup, three of salad, eight of an entree, and five of a dessert. How many different complete dinners can you order?

49. **Housing development.** A developer wants to build seven houses, each of a different design. How many ways can she arrange these homes on a street if
    a. four lots are on one side of the street and three are on the opposite side?
    b. five lots are on one side of the street and two are on the opposite side?

50. **Designing exercise sets.** The author of a mathematics textbook has seven problems for an exercise set, and she is trying to arrange them in increasing order of difficulty. How many ways can she select the problems if no two of them are equally difficult? How many ways can the author fail?

51. **Letter selection.** From the letters of the word *CHARITY*, groups of four letters are formed (without regard to order). How many of them will contain
    a. both *A* and *R*?
    b. *A* but not *R*?
    c. neither *A* nor *R*?

52. **Committee selection.** How many ways can a committee of 4 students and 2 professors be formed from 20 students and 8 professors?

53. **Inventory orders.** A manufacturer makes shirts in five different colors, seven different neck sizes, and three different sleeve lengths. How many shirts should a department store order to have one of each type?

54. **International book codes.** All recent books are identified by their International Standard Book Number (ISBN), a ten-digit code assigned by the publisher. A typical ISBN is 0-321-75526-2. The first digit, 0, represents the language of the book (English); the second block of digits, 321, represents the publishing company (Addison-Wesley); the third block of digits, 75526, is the number assigned by the publishing company to that book; and the final digit, 2, is the check digit, used to detect the most commonly made errors when ISBNs are copied. How many ISBNs are possible? (Remember that the previous nine digits determine the last digit.)

55. **Political polling.** A senator sends out questionnaires asking his constituents to rank ten issues of concern to them, such as crime, education, and taxes, in order of importance. Replies are filed in folders, and two replies are put in the same folder if and only if the constituents give the same ranking to all ten issues. Find the maximum number of folders that might be needed.

56. **Parcel delivery.** A UPS driver has to deliver 12 parcels to 12 different addresses. In how many orders can the driver make these deliveries?

57. Circus visit. A father with seven children takes four of them at a time to a circus as often as he can, without taking the same four children more than once. What is the maximum number of times each child will go to the circus, and how many times will the father go?

58. Arranging books. From six history books, four biology books, and five economics books, how many ways can a person select three history books, two biology books, and four economics books and arrange them on a shelf?

59. Bingo. Each of four columns on a bingo card contains 5 different numbers chosen in any order from 15. Column "B" chooses from 1 to 15, column "I" from 16 to 30, ..., and column "O" from 61 to 75. Only four numbers are under column "N" because of the free space. How many different cards are possible?

60. Football conferences. The Grid-Iron Football League in Sardonia is divided into two conferences—East and West—each having six members. Each team in the league must play each team in its conference twice and each team in the other conference once during a season. What is the total number of games played in the league?

61. Married couples prohibited. How many ways can a committee of four people be selected from five married couples if no committee is to include husband-and-wife pairs?

62. Social club committees. A social club contains seven women and four men. The club wants to select a committee of three to represent it at a state convention. How many of the possible committees contain at least one man?

63. Coin gifts. Among Corey's collection of early American coins, he has a half-dollar, a quarter, a nickel, and a penny. He wants to give his younger sister some or all of these four coins. How many different sets of coins can Cory give?

64. Meal possibilities. The university cafeteria offers two choices of soups, three of salad, five of main dishes, and six of desserts. How many different four-course meals can you choose?

In Exercises 65–68, how many distinguishable ways can the letters of each word be arranged?

65. TAIWAN

66. AMERICA

67. SUCCESS

68. ARRANGE

C EXERCISES Beyond the Basics

69. A company that supplies temporary help has a contract for 20 weeks to provide three workers per week to an insurance firm. The agreement specifies that in no two weeks can the same three workers be sent to work at the firm. How many different workers are necessary?

70. An FM radio station wants to start its evening programming with five songs every day for 21 days without using the same five songs on any two days. What is the smallest number of songs that can be used to accomplish this result?

71. In geometry, any two points determine a line. How many lines are determined by 30 given points, no 3 of which lie on the same line?

72. In geometry, any three noncollinear points determine a triangle. How many triangles can be determined by 21 given points, no three of which are collinear?

In Exercises 73–80, solve each equation.

73. $\binom{m+1}{m-1} = 3!$

74. $6\binom{n-1}{2} = \binom{n+1}{4}$

75. $2\binom{n-1}{2} = \binom{n}{3}$

76. $\frac{4}{3}\binom{k}{2} = \binom{k+1}{3}$

77. $\binom{n}{0} + \binom{n}{1} + \binom{n}{2} + \cdots + \binom{n}{n} = 64$

78. $\binom{k}{1} + \binom{k}{2} + \binom{k}{3} + \cdots + \binom{k}{k-1} = 126$

79. $\binom{m}{4} = \binom{m}{5}$

80. $\binom{n}{7} = \binom{n}{3}$

Critical Thinking

81. How many terms of the form $x^n y^m$ are possible if n and m can be any integer such that $1 \le n \le 5$ and $1 \le m \le 5$?

82. Use the Fundamental Counting Principle to determine the number of terms in the product

$$(a + b)(x^2 + 2xy + y^2).$$

Explain your reasoning.
[*Hint:* Each term in the product can be obtained by multiplying one term chosen from each factor.]

Probability

Before Starting this Section, Review

1. Set notation (Section P.1, page 6)
2. Fundamental Counting Principle (Section 8.6, page 702)
3. Permutations (Section 8.6, page 703)
4. Combinations (Section 8.6, page 705)

Objectives

1. Find the probability of an event.
2. Use the Additive Rule for finding probabilities.
3. Find the probability of mutually exclusive events.
4. Find the probability of the complement of an event.
5. Find experimental probabilities.

DOUBLE LOTTERY WINNER

Many people dream of winning a lottery, but only the most optimistic dream of winning it twice. When Evelyn Marie Adams won the New Jersey state lottery for the second time, *The New York Times* (February 14, 1986) claimed that the chances of one person winning the lottery twice were about 1 in 17 trillion. Two weeks later a letter from two statisticians appeared in the *Times* challenging this claim. Although they agreed that any particular person had very little chance of winning a lottery twice, they said that it was almost certain that someone in the United States would win twice. Further, they said that the odds were even for another double lottery win to occur within seven years. In less than two years, Robert Humphries won his second Pennsylvania lottery prize.

Evaluating how likely an event is to occur is the realm of probability, which we study in this section. In Example 4, we determine the probability of winning a lottery in which 6 numbers are randomly selected from the numbers 1 through 53.

The toss of a coin determines which team will kick the ball to the other team to start a football game. Everyone seems to agree that a coin toss gives both teams the same chance to choose whether to kick or receive the football. The ideas of "chance," "likelihood," and "probability" are all coupled with the desire to quantify expectations about the results of some process. ■

---

1 Find the probability of an event.

The Probability of an Event

Any process that terminates in one or more outcomes (results) is an **experiment**. For example, if we ask people what time of the day they typically eat their first meal or if we pull balls from a bag of colored balls, we have performed an experiment. The outcomes of the first experiment are the times of day given as responses to our question. The outcomes of the second experiment are the colors of the balls drawn from the bag.

The set of all possible outcomes of a given experiment is called the **sample space** of the experiment. Suppose the experiment is tossing a coin. Then we could denote the outcome "heads" by H; denote the outcome "tails" by T; and write the sample space, S, in set notation as

$$S = \{H, T\}.$$

A single die is a cube on which each face contains one, two, three, four, five, or six dots and no two faces contain the same number of dots. A roll of the die is an experiment, the outcome of which is the number of dots showing on the upper face, and the sample space S can be written in set notation as

$$S = \{1, 2, 3, 4, 5, 6\}.$$

An event is any set of possible outcomes; that is, an **event** is any subset of a sample space. For the die-rolling experiment, one possible event is "The number showing is even." In set notation, we write this event as $E = \{2, 4, 6\}$. The event "The number showing is a 6" is written in set notation as $A = \{6\}$.

Outcomes of an experiment are said to be **equally likely** if no outcome should result more often than any other outcome when the experiment is run repeatedly. In both the coin-tossing and die-rolling experiments, all of the outcomes are equally likely.

PROBABILITY OF AN EVENT

If all of the outcomes in a sample space S are equally likely, then the **probability** of an event E, denoted $P(E)$, is the ratio of the number of outcomes in E, denoted $n(E)$, to the total number of outcomes in S, denoted $n(S)$. That is,

$$P(E) = \frac{n(E)}{n(S)}.$$

Because E is a subset of S, $0 \leq n(E) \leq n(S)$. Dividing by $n(S)$ yields $0 \leq \dfrac{n(E)}{n(S)} \leq 1$. So the probability $P(E)$ of an event E is a number between 0 and 1 inclusive.

EXAMPLE 1 Finding the Probability of an Event

A single die is rolled. Write each event in set notation and find the probability of the event.

a. E_1: The number 2 is showing.

b. E_2: The number showing is odd.

c. E_3: The number showing is less than 5.

d. E_4: The number showing is greater than or equal to 1.

e. E_5: The number showing is 0.

SOLUTION

The sample space for the experiment of rolling a die is $S = \{1, 2, 3, 4, 5, 6\}$, so $n(S) = 6$. For any event E, $P(E) = \dfrac{n(E)}{n(S)}$.

a. $E_1 = \{2\}$, so $n(E_1) = 1$. $P(E_1) = \dfrac{n(E_1)}{n(S)} = \dfrac{1}{6}$

b. $E_2 = \{1, 3, 5\}$, so $n(E_2) = 3$. $P(E_2) = \dfrac{n(E_2)}{n(S)} = \dfrac{3}{6} = \dfrac{1}{2}$

c. $E_3 = \{1, 2, 3, 4\}$, so $n(E_3) = 4$. $P(E_3) = \dfrac{n(E_3)}{n(S)} = \dfrac{4}{6} = \dfrac{2}{3}$

d. $E_4 = \{1, 2, 3, 4, 5, 6\}$, so $n(E_4) = 6$. $P(E_4) = \dfrac{n(E_4)}{n(S)} = \dfrac{6}{6} = 1$

e. $E_5 = \varnothing$, so $n(E_5) = 0$. $P(E_5) = \dfrac{n(E_5)}{n(S)} = \dfrac{0}{6} = 0$ ■ ■ ■

Practice Problem 1 A single die is rolled. Write each event in set notation and give the probability of each event.

a. E_1: The number showing is even.

b. E_2: The number showing is greater than 4. ■

Notice in Example 1**d** that $E_4 = S$; so the event E_4 is certain to occur on every roll of the die. Every **certain event** has probability 1. On the other hand, in Example 1**e**, the event $E_5 = \varnothing$; so E_5 never occurs, and $P(E_5) = 0$. An **impossible event** always has probability 0.

EXAMPLE 2 **Tossing a Coin**

What is the probability of getting at least one tail when a coin is tossed twice?

SOLUTION

Tossing a coin once results in two equally likely outcomes: heads (H) and tails (T). Tossing the coin twice results in two equally likely outcomes for the first toss and, regardless of the first outcome, two equally likely outcomes for the second toss. By the Fundamental Counting Principle, there are $2 \cdot 2$, or 4, outcomes when we toss the coin twice. We can represent these outcomes as (first toss, second toss) pairs.

$$S = \{(H, H), (H, T), (T, H), (T, T)\}.$$

Because three of these outcomes, $E = \{(H, T), (T, H), (T, T)\}$, result in at least one tail, we have

$$P(E) = \frac{n(E)}{n(S)} = \frac{3}{4}.$$ ■ ■ ■

Practice Problem 2 What is the probability of getting exactly one head when a coin is tossed twice? ■

Equally likely outcomes occur in most games of chance that involve tossing coins, rolling dice, or drawing cards. Lotteries and games involving spinning wheels also generate equally likely outcomes.

EXAMPLE 3 **Rolling a Pair of Dice**

What is the probability of getting a sum of 8 when a pair of dice is rolled?

SOLUTION

There are six equally likely outcomes for each die; so by the Fundamental Counting Principle, there are $6 \cdot 6$, or 36, equally likely outcomes when a pair of dice is rolled. It is useful to think of having two dice of different colors—say, red and green—in distinguishing the 36 outcomes. In this way, 2 on the red die and 6 on the green die is easily seen to be a different physical outcome from 6 on the red die and 2 on the green die even though the same sum results.

We can represent the 36 outcomes as (red die, green die) pairs.

$$S = \{(1, 1), (1, 2), (1, 3), (1, 4), (1, 5), (1, 6),$$
$$(2, 1), (2, 2), (2, 3), (2, 4), (2, 5), (2, 6),$$
$$(3, 1), (3, 2), (3, 3), (3, 4), (3, 5), (3, 6),$$
$$(4, 1), (4, 2), (4, 3), (4, 4), (4, 5), (4, 6),$$
$$(5, 1), (5, 2), (5, 3), (5, 4), (5, 5), (5, 6),$$
$$(6, 1), (6, 2), (6, 3), (6, 4), (6, 5), (6, 6)\}$$

The highlighted ordered pairs in the sample space are the outcomes that have a sum of 8. In set notation, the event "The sum of the numbers on the dice is 8" is

$$E = \{(6, 2), (5, 3), (4, 4), (3, 5), (2, 6)\}.$$

Because $n(E) = 5$ and $n(S) = 36$,

$$P(E) = \frac{n(E)}{n(S)} = \frac{5}{36}.$$

So the probability of getting a sum of 8 is $\frac{5}{36}$.   ■ ■ ■

Practice Problem 3   What is the probability of getting a sum of 7 when a pair of dice is rolled?   ■

EXAMPLE 4   **Winning the Florida Lottery**

Table tennis balls numbered 1 through 53 are mixed together in a turning drum or basket. Six of the balls are chosen in a random way. To win the lottery jackpot, a player must have selected all six of the numbers chosen. The order in which the numbers are chosen does not matter. What is the probability of matching all six numbers? (*Source:* www.fllottery.com)

SOLUTION

Because the order in which the numbers are selected is not important, the sample space S consists of all sets of 6 numbers that can be selected from 53 numbers. Therefore, $n(S) = C(53, 6) = 22,957,480$. Because only 1 set of 6 numbers matches the 6 numbers drawn, the probability of the event "guessed all 6 numbers" is

$$P(\text{winning the lottery}) = P(\text{guessed all 6 numbers})$$

$$= \frac{1}{22,957,480} \approx 0.00000004.$$

Whether you say your chances of winning are 1 in 22,957,480 or about 0.00000004, you definitely know they are not very good!   ■ ■ ■

Practice Problem 4   A state lottery requires a winner to correctly select 6 out of 50 numbers. The order of the numbers does not matter. What is the probability of matching all six numbers?   ■

2 Use the Additive Rule for finding probabilities.

The Additive Rule

Consider the experiment of rolling one die. The sample space is $S = \{1, 2, 3, 4, 5, 6\}$, and $n(S) = 6$.

Let E be the event "The number rolled is greater than 4" and let F be the event "The number rolled is even." Then

$$E = \{5, 6\} \qquad \text{and} \qquad F = \{2, 4, 6\}.$$

The event "The number rolled is greater than 4 *and* even" is $E \cap F$.

$$E \cap F = \{6\}$$

Because $n(E) = 2$, $n(F) = 3$, $n(E \cap F) = 1$, and $n(S) = 6$, we have

$$P(E) = \frac{2}{6}, P(F) = \frac{3}{6}, P(E \cap F) = \frac{1}{6}.$$

The event "The number rolled is either greater than 4 *or* even" is $E \cup F$.

$$E \cup F = \{2, 4, 5, 6\}$$

Note that 6 is in *both* E and F, so it is counted once when $n(E)$ is counted and a second time when $n(F)$ is counted. However, 6 is counted only once when $n(E \cup F)$ is counted. We have $n(E \cup F) = n(E) + n(F) - n(E \cap F)$. That is, because 6 was counted twice when $n(E) + n(F)$ was added, we must subtract $n(E \cap F)$ to get an accurate count for $E \cup F$.

$$n(E \cup F) = n(E) + n(F) - n(E \cap F)$$

$$\frac{n(E \cup F)}{n(S)} = \frac{n(E)}{n(S)} + \frac{n(F)}{n(S)} - \frac{n(E \cap F)}{n(S)} \qquad \text{Divide both sides by } n(S).$$

$$P(E \cup F) = P(E) + P(F) - P(E \cap F) \qquad \text{Definition of probability}$$

This example illustrates the *Additive Rule*.

THE ADDITIVE RULE

If E and F are events in a sample space, then

$$P(E \cup F) = P(E) + P(F) - P(E \cap F).$$

---

3 Find the probability of mutually exclusive events.

Mutually Exclusive Events

Two events are *mutually exclusive* if it is impossible for both to occur simultaneously. That is, E and F are **mutually exclusive** events if $E \cap F$ contains no outcomes. In this case, $P(E \cap F) = 0$ and the Additive Rule is somewhat simplified.

MUTUALLY EXCLUSIVE EVENTS

If E and F are any events in a sample space and $E \cap F = \varnothing$, then

$$P(E \cup F) = P(E) + P(F).$$

A standard deck of 52 playing cards uses four different designs called suits.

clubs, ♣ ; diamonds, ♦ ; hearts, ♥ ; and spades, ♠

Each suit has 13 cards—an ace, a king, a queen, and a jack—and cards numbered 2 through 10. The diamond and heart symbols are red, and the club and spade symbols are black. For the purposes of this section, we assume that all decks of cards are standard and well shuffled so that all cards drawn or dealt from a deck are equally likely.

EXAMPLE 5 **Drawing a Card from a Deck**

A card is drawn from a deck. What is the probability that the card is

a. a king?   **b.** a spade?

c. a king or a spade?   **d.** a heart or a spade?

SOLUTION

The sample space S is the 52-card deck, so $n(S) = 52$.

a. There are four kings, one in each suit; so

$$P(\text{king}) = \frac{4}{52} = \frac{1}{13}.$$

b. There are 13 spades in the deck, so

$$P(\text{spade}) = \frac{13}{52} = \frac{1}{4}.$$

c. By the Additive Rule, we have

$$P(\text{king or spade}) = P(\text{king}) + P(\text{spade}) - P(\text{king and spade}).$$

Because only one card is both a king and a spade,

$P(\text{king and spade}) = \dfrac{1}{52}.$ From parts **a** and **b** in this example,

$$P(\text{king or spade}) = \frac{4}{52} + \frac{13}{52} - \frac{1}{52}$$
$$= \frac{16}{52} = \frac{4}{13}.$$

d. Because no card is both a heart and a spade, the events "Draw a heart" and "Draw a spade" are mutually exclusive. Thus,

$$P(\text{heart or spade}) = P(\text{heart}) + P(\text{spade})$$
$$= \frac{13}{52} + \frac{13}{52} = \frac{1}{2}.$$

Practice Problem 5   What is the probability that a card pulled from a deck is a jack or a king?

4 Find the probability of the complement of an event.

The Complement of an Event

The **complement of an event** E is the set of all outcomes in the sample space that are *not* in E. The complement of E is denoted by E'. Because every outcome in the sample space is in E or in E', the event $E \cup E'$ is a certain event and $P(E \cup E') = 1$. Also, E and E' are mutually exclusive events, so $P(E \cup E') = P(E) + P(E')$. Thus, $P(E) + P(E') = 1$ and $P(E') = 1 - P(E)$.

PROBABILITY OF THE COMPLEMENT

If E is any event and E' is its complement, then

$$P(E') = 1 - P(E).$$

EXAMPLE 6   **Finding the Probability of the Complement**

Brandy has applied to receive funding from her company to attend a conference. She knows that her name, along with the names of nine other deserving candidates, was put in a box from which two names will be drawn. What is the probability that Brandy will be one of the two selected?

SOLUTION

It is easier to determine the probability that Brandy will *not* be selected. Because there are nine candidates other than Brandy, there are $C(9, 2)$ ways of selecting two employees not including Brandy. There are $C(10, 2)$ ways of selecting two employees from all ten candidates. Hence, the probability that Brandy is not selected is

$$P(\text{Brandy is not selected}) = \frac{C(9, 2)}{C(10, 2)}$$
$$= \frac{9!}{2!\,7!} \cdot \frac{2!\,8!}{10!} = \frac{8}{10} = \frac{4}{5}.$$

The probability that Brandy is selected is

$$P(\text{Brandy is selected}) = 1 - P(\text{Brandy is not selected})$$

$$= 1 - \frac{4}{5} = \frac{1}{5}.$$

■ ■ ■

Practice Problem 6 In Example 6, what is the probability of Brandy being selected if three names are drawn from the box? ■

5 Find experimental probabilities.

Experimental Probabilities

The probabilities we have seen up to this point have been determined by some form of reasoning, not data. Data support the decision to assign the value $\frac{1}{2}$ as the probability of a head resulting from one toss of a coin, but the assignment is not *based* on data. This is also true for the probabilities associated with drawing a card, rolling a pair of dice, drawing a ball from an urn, and so on. Probability assigned this way is called **theoretical probability**.

In contrast, we can assign probability values based on observation and data. If an experiment is performed n times (where n is large) and a particular outcome occurs k times, then the number $\frac{k}{n}$ is the value for the probability of that outcome. Probability determined this way is called **experimental probability**. A statement such as "The probability that a 10-year-old child will live to be 95 years old is $\frac{3}{100,000}$" refers to experimental probability.

RELATIVE FREQUENCY PRINCIPLE

If an experiment is performed n times and an event E occurs k times, then we say that the experimental (or statistical) probability of the event, $P(E)$, is

$$P(E) = \frac{k}{n}.$$

EXAMPLE 7 Using Data to Estimate Probabilities

Table 8.2 shows the gender distribution of students at four California City colleges.

TABLE 8.2

College	Male	Female
Los Angeles	6,995	8,179
San Diego	12,251	14,914
San Francisco	16,857	2,000
Sacramento	9,040	11,838

Source: www.areaconnect.com

Find the probability that a student selected *at random* is

a. a female. **b.** attending Los Angeles City College.

c. a female attending Los Angeles City College.

Round your answers to the nearest hundredth.

SOLUTION

The term *randomly selected* means that every possible selection is equally likely. In our example, this means that each student is equally likely to be selected. With random selection, the percentage of students in a given category gives the probability that a student in that category will be chosen.

a. The total number of students is 82,074. The total number of female students is 36,931. Thus, $P(\text{female}) = \dfrac{36,931}{82,074} \approx 0.45$.

b. The total number of students is 82,074. The total number of students attending Los Angeles City College is 15,174. Hence,

$$P(\text{attends Los Angeles City College}) = \frac{15,174}{82,074} \approx 0.18.$$

c. There are 8179 female students attending Los Angeles City College. Consequently,

$$P(\text{female and attending Los Angeles City College}) = \frac{8179}{82,074} \approx 0.10. \quad ■ ■ ■$$

Practice Problem 7   In Example 7, find the probability that a student selected at random is a male or attends Sacramento College.   ■

SECTION 8.7 ■ Exercises

A EXERCISES Basic Skills and Concepts

1. If no outcome of an experiment results more often than any other outcome, the outcomes are said to be _____.

2. A *certain* event has probability _____, and an *impossible* event has probability _____.

3. If it is impossible for two events to occur simultaneously, then the events are said to be _____.

4. If the probability that an event occurs is 0.7, then the probability that the event does not occur is _____.

5. *True or False* The probability of any event is a number between 0 and 1.

6. *True or False* If an experiment is performed 100 times and an event E occurs 65 times, then the *experimental probability* of the event is 0.65.

In Exercises 7–12, write the sample space for each experiment.

7. Students are asked to rank Wendy's, McDonald's, and Burger King in order of preference.

8. Students are asked to choose either a hamburger or a cheeseburger and then pick Wendy's, McDonalds, or Burger King as the restaurant from which they would prefer to order it.

9. Two albums are selected from among *Thriller* (Michael Jackson), *The Wall* (Pink Floyd), *Eagles: Their Greatest Hits* (Eagles), and *Led Zeppelin IV* (Led Zeppelin).

10. Three types of potato chips are selected from among regular salted, regular unsalted, barbecue, cheddar cheese, and sour cream and onion.

11. Two blanks in a questionnaire are filled in, the first identifying race as white, African American, Native American, Asian, other, or multiracial, and the second indicating gender as male or female.

12. A six-sided die is chosen from a box containing a white, a red, and a green die. Then the die is tossed.

In Exercises 13–16, find the probability of the given event if a year is selected at random.

13. Thanksgiving falls on a Sunday.

14. Thanksgiving falls on a Thursday.

15. Christmas falls on December 25.

16. Christmas falls on January 1.

In Exercises 17–26, classify each statement as an example of *theoretical probability* or *experimental probability*.

17. The probability that Ken will go to church next Sunday is 0.85.

18. The probability of getting exactly two heads when a fair coin is tossed twice is $\dfrac{1}{4}$.

19. The probability of drawing a heart from a deck of cards is 0.25.

20. The probability that an American adult will watch a major network's evening newscast is 0.17.

21. The probability that a person driving during the Memorial Day weekend will die in an automobile accident is $\dfrac{1}{58,000}$.

22. The probability that a student will get a D or an F in a college statistics course is 0.21.

23. The probability of drawing an ace from a deck of cards is $\dfrac{1}{13}$.

24. The probability that a person will be struck by lightning this year is $\dfrac{1}{727,000}$.

25. The probability that a plane will take off on time from JFK International Airport is 0.78.

26. The probability that a college student cheats on an exam is 0.32.

27. Match each given sentence with an appropriate probability.

An event that	has probability
is very likely to happen	0
will surely happen	0.5
is a rare event	0.001
may or may not happen	1
will never happen	0.999

28. A coin is tossed three times. Rearrange the order of the following events from the most probable event (first) to the least probable event (last).
{exactly two heads}, {the sure event}, {at least one head}, {at least two heads}.

In Exercises 29–34, a die is rolled. Write each event in set notation and give the probability of each event.

29. The number showing is a 1 or a 6.

30. The number 5 is showing.

31. The number showing is greater than 4.

32. The number showing is a multiple of 3.

33. The number showing is even.

34. The number showing is less than 1.

In Exercises 35–40, a card is drawn randomly from a standard 52-card deck. Write each event in set notation and find its probability.

35. A queen is drawn.

36. An ace is drawn.

37. A heart is drawn.

38. An ace of hearts or a king of hearts is drawn.

39. A face card (jack, queen, or king) is drawn.

40. A heart that is not a face card is drawn.

In Exercises 41–46, a digit is chosen randomly from the digits 0 through 9; then 3 is added to the digit and the sum recorded. Write each event in set notation and find the probability of the event.

41. The sum is 11.   **42.** The sum is 0.

43. The sum is less than 4.   **44.** The sum is greater than 8.

45. The sum is even.   **46.** The sum is odd.

B EXERCISES Applying the Concepts

47. **A blind date.** The probability that Tony will like his blind date is 0.3. What is the probability that he will not like his blind date?

48. **Drawing cards.** The probability of drawing a spade from a standard 52-card deck is 0.25. What is the probability of drawing a card that is not a spade?

49. **Gender and the Air Force.** In 2006, about 20% of members of the U.S. Air Force were women. What is the probability that an Air Force member chosen at random is a woman? (*Source:* www.af.mil/news)

50. **Movie attendance.** If 85% of people between the ages of 18 and 24 go to a movie at least once a month, what is the probability that a randomly chosen person in this age group will go to a movie this month?

51. **Gender and the Marines.** In 2006, about 6% of members of the U.S. Marines were women. What is the probability that a Marine chosen at random is a man? (*Source:* DMDC, June 2005; TFDW, June 2005)

52. **Movie attendance.** If 20% of people over 65 go to a movie at least once a month, what is the probability that a randomly chosen person over 65 will *not* go to a movie this month?

53. **Raffles.** One hundred twenty-five tickets were sold at a raffle. If you have ten tickets, what is the probability of your winning the (only) prize? If there are two prizes, what is the probability of your winning both?

54. **Committee selection.** To select a committee with two members, two people will be selected at random from a group of four men and four women. What is the probability that a man and a woman will be selected?

55. **Gender of children.** A couple has two children. Assuming that each child is equally likely to be a boy or a girl, find the probability of having
   a. two boys.
   b. two girls.
   c. one boy and one girl.

56. **Gender of children.** A couple has three children. Assuming that each child is equally likely to be a boy or a girl, find the probability that the couple will have
   a. all girls.
   b. at least two girls.
   c. at most two girls.
   d. exactly two girls.

57. **Car ownership.** The following table shows the number of cars owned by each of the 4440 families in a small town.

No. of families	37	1256	2370	526	131	120
No. of cars owned	0	1	2	3	4	5

A family is selected at random. Find the probability that this family owns
- **a.** exactly two cars.
- **b.** at most two cars.
- **c.** exactly six cars.
- **d.** at most six cars.
- **e.** at least one car.

58. **Mammograms.** According to the American Cancer Society, 199 of 200 mammograms (for breast cancer screening) turn out to be normal. Find the probability that a mammogram of a woman chosen at random is not normal.

59. **Lactose intolerance.** Lactose intolerance (the inability to digest sugars found in dairy products) affects about 20% of non-Hispanic white Americans and about 75% of African, Asian, and Native Americans. What is the probability that a non-Hispanic white American has no lactose intolerance? Answer the same question for a person in the second group.

60. **Guessing on an exam.** A student has just taken a ten-question true–false test. If he must guess to answer each of the ten questions, find the probability that
- **a.** he scores 100% on the test.
- **b.** he scores 90% on the test.
- **c.** he scores at least 90% on the test.

61. **Committee selection.** A five-person committee is choosing a subcommittee of two of its members. Each member of the committee is a different age. The sub-committee members are chosen at random.
- **a.** What is the probability that the subcommittee will consist of the two oldest members?
- **b.** What is the probability that the subcommittee will consist of the oldest and the youngest members?

62. **How many children do mothers have?** The following data are based on statistics published by the U.S. Census Bureau on the eve of Mother's Day in 2000: Among the 35 million mothers in the United States in the age group from 15 to 44 in 1998, 10.7 million had one child, 14.3 million had two children, 6.7 million had three, and 3.3 million had four or more. If a mother from this age group is chosen at random, what is the probability that she has more than one child?

63. **A test for depression.** Suppose 10% of the U.S. population suffers from depression. A pharmaceutical company is believed to have developed a test with the following results: The test works for 90% of the people who are tested and are actually depressed (as diagnosed through other acceptable methods). In other words, if a patient is actually depressed, there is a probability of 0.9 that the pharmaceutical company will diagnose this patient as depressed. The test also works for 85% of people who are tested and not depressed.

Suppose the company tests 1000 people for depression. Complete the following table.

Number of people who are	Depressed according to the test	Normal according to the test
actually depressed		
actually normal		

64. **Playing marbles.** Natasha and Deshawn are playing a game with marbles. Natasha has one marble (call it N), and Deshawn has two (call them M_1 and M_2). They roll their marbles toward a stake in the ground. The person whose marble is closest to the stake wins. Assume that measurements are accurate enough to eliminate ties. List all of the sample points. Use the notation M_1NM_2 to denote the sample point when M_1 is closest to the stake and M_2 is farthest from the stake. Assuming that the children are equally skillful (when the sample points are all equally likely), find the probability that Deshawn wins.

65. **Gender among siblings.** If a family with four children is chosen at random, are the children more likely to be three of one gender and one of the other gender than two boys and two girls? (Assume that each child in a family is equally likely to be a boy or a girl.)

66. **Hospital treatments.** Hospital records show that 11% of all patients admitted are admitted for surgical treatment, 15% are admitted for obstetrics, and 3% are admitted for both obstetrics and surgery. What is the probability that a new patient admitted is an obstetrics patient or a surgery patient or both?

67. **Pain relief medicine.** A pharmaceutical company testing a new over-the-counter pain relief medicine gave the new medicine, their old medicine, and placebos to different groups of patients. (A placebo is a sugar pill with no medicinal ingredients. However, it sometimes works, supposedly for psychological reasons.) The following table shows the results of the tests.

Medicine	Amount of Pain Relief		
	Some	Complete	None
Placebo	12	4	34
Old	25	25	10
New	18	40	12
Total	55	69	56

Find the probability that a randomly chosen person from the group of 180 people receiving one of the three types of medicines
- **a.** received the placebo or had no pain relief.
- **b.** received the new medicine or had complete pain relief.

C EXERCISES Beyond the Basics

68. A famous problem called the *birthday problem* asks for the probability that at least two people in a group of people have the same birthday (the same day and the same month but not necessarily the same year). The best way to approach this problem is to consider the complementary event that no two people have the same birthday. Assume that all birthdays are equally likely.
   a. Find the probability that in a group of two people, both have the same birthday.
   b. Find the probability that in a group of three people, at least two have the same birthday.
   c. Find the probability that in a group of 50 people, at least two have the same birthday.

69. What is the probability that a number between 1 and 1,000,000 contains the digit 3? [*Hint:* Consider the complementary event.]

70. A business executive types four letters to her clients and then types four envelopes addressed to the clients. What is the probability that if the four letters are randomly placed in envelopes, they are placed in correct envelopes?

71. Juan gave his telephone number to Maggie. She remembers that the first three digits are 407 and the remaining four digits consist of two 3s, one 6, and one 8. She is not certain about the order of the digits, however. Maggie dials 407-3638. What is the probability that Maggie dialed the correct number?

Critical Thinking

72. The March 2005 population survey by the U.S. Census Bureau found that among the 187 million people aged 25 and older, 59,840,000 reported high school graduation as the highest level of education attained. Find the probability that a person chosen at random from the 187 million in the survey would report his or her highest level of education as high school graduation.

73. Find the probability of randomly choosing a dark caramel from a box that contains only light and dark caramels and has twice as many light as dark caramels.

74. If a single die is rolled two times, what is the probability that the second roll will result in a number larger than the first roll?

SUMMARY ■ Definitions, Concepts, and Formulas

8.1 Sequences and Series

   i. An infinite sequence is a function whose domain is the set of positive integers. The function values $a_1, a_2, a_3, \ldots, a_n, \ldots$ are called the **terms** of the sequence. The term a_n is the **nth term**, or **general term**, of the sequence.

   ii. *Recursive formula*

   A sequence may be defined recursively, with the nth term of the sequence defined in relation to previous terms.

   iii. *Factorial notation*

   $$n! = n(n-1)\cdots 3 \cdot 2 \cdot 1 \text{ and } 0! = 1$$

   iv. *Summation notation*

   $$\sum_{i=1}^{n} a_i = a_1 + a_2 + a_3 + \cdots + a_n$$

   v. *Series*
   Given an infinite sequence $a_1, a_2, a_3, \ldots, a_n, \ldots$,

   the sum $\sum_{i=1}^{n} a_i$ is called the **nth partial sum** and the sum

   $$\sum_{i=1}^{\infty} a_i = a_1 + a_2 + a_3 + \cdots \text{ is called an infinite series.}$$

8.2 Arithmetic Sequences; Partial Sums

   i. A sequence is an **arithmetic sequence** if each term after the first differs from the preceding term by a constant. The constant difference d between the consecutive terms is called the **common difference**. We have $d = a_n - a_{n-1}$ for all $n \geq 1$.

   ii. The **nth term** a_n of the arithmetic sequence $a_1, a_2, a_3 \ldots, a_n, \ldots$ is given by $a_n = a_1 + (n-1)d, n \geq 1$, where d is the common difference.

   iii. The **sum** S_n of the first n terms of the arithmetic sequence $a_1, a_2, \ldots, a_n, \ldots$ is given by the formula

   $$S_n = n\left(\frac{a_1 + a_n}{2}\right).$$

8.3 Geometric Sequences and Series

   i. A sequence is a **geometric sequence** if each term after the first is a constant multiple of the preceding term. The constant ratio r between the consecutive terms is called the **common ratio**. We have $\dfrac{a_{n+1}}{a_n} = r, n \geq 1$.

ii. The **nth term a_n of the geometric sequence** $a_1, a_2, a_3, \ldots,$ $a_n, \ldots$ is given by $a_n = a_1 r^{n-1}$, $n \geq 1$, where r is the common ratio.

iii. The **sum S_n** of the first n terms of the geometric sequence $a_1, a_2, \ldots, a_n, \ldots$ with common ratio $r \neq 1$ is given by the formula

$$S_n = \frac{a_1(1 - r^n)}{1 - r}.$$

The **sum S of an infinite geometric sequence** is given by

$$S = \sum_{i=1}^{\infty} a_1 r^{i-1} = \frac{a_1}{1 - r} \text{ if } |r| < 1.$$

iv. An **annuity** is a sequence of equal payments made at equal times. Suppose $\$P$ is the payment made at the end of each of n compounding periods per year and i is the annual interest rate. Then the value A of the annuity after t years is

$$A = P \frac{\left(1 + \dfrac{i}{n}\right)^{nt} - 1}{\dfrac{i}{n}}.$$

v. If $|r| < 1$, the infinite geometric series $a_1 + a_1 r +$ $a_1 r^2 + \cdots + a_1 r^{n-1} + \cdots$ has the sum $S = \dfrac{a_1}{1 - r}$. When $|r| \geq 1$, the series does not have a sum.

8.4 Mathematical Induction

The statement P_n is true for all positive integers n if the following properties hold:

i. P_1 is true.

ii. If P_k is a true statement, then P_{k+1} is a true statement.

8.5 The Binomial Theorem

i. Binomial coefficient: $\dbinom{n}{j} = \dfrac{n!}{j!(n-j)!}, 0 \leq j < n$

$$\binom{n}{0} = 1 = \binom{n}{n}$$

ii. Binomial Theorem:

$$(x + y)^n = \binom{n}{0}x^n + \binom{n}{1}x^{n-1}y + \binom{n}{2}x^{n-2}y^2 + \cdots$$
$$+ \binom{n}{j}x^{n-j}y^j + \cdots + \binom{n}{2}y^n$$
$$= \sum_{j=0}^{n}\binom{n}{j}x^{n-j}y^j$$

iii. The term containing the factor x^j in the expansion of $(x + y)^n$ is

$$\binom{n}{n-j}x^j y^{n-j}.$$

8.6 Counting Principles

i. Fundamental Counting Principle. If a first choice can be made in p different ways, a second choice in q different ways, a third choice in r different ways, and so on, then the sequence of choices can be made in $p \cdot q \cdot r \ldots$ different ways.

ii. A **permutation** is an arrangement of n distinct objects in a fixed order in which no object is used more than once. The order in an arrangement is important.

iii. Permutation formula. The number of permutations of n distinct objects taken r at a time is

$$P(n, r) = \frac{n!}{(n - r)!} = n(n - 1)(n - 2)\ldots(n - r + 1).$$

iv. Combination formula. When r objects are chosen from n distinct objects without regard to order, we call the set of r objects a **combination** of n objects taken r at a time. The number of combinations of n distinct objects taken r at a time is denoted by $C(n, r)$, where

$$C(n, r) = \frac{n!}{(n - r)! \, r!}.$$

v. Distinguishable permutations. The number of permutations of n objects of which n_1 are of one kind, n_2 are of a second kind, $\ldots$, and n_k are of a kth kind is

$$\frac{n!}{n_1! \, n_2! \ldots n_k!},$$

where $n_1 + n_2 + \cdots + n_k = n$.

8.7 Probability

i. Experiment. Any process that terminates in one or more **outcomes** (results) is an experiment.

ii. Sample space. The set of all possible outcomes of an experiment is called the sample space of the experiment.

iii. Event. An event is any subset of a sample space.

iv. Equally likely events. Outcomes of an experiment are said to be equally likely if no outcome should result more often than any other outcome when the experiment is performed repeatedly.

v. Probability. If all of the outcomes in a sample space S are equally likely, the probability of an event E, denoted $P(E)$, is defined by

$$P(E) = \frac{n(E)}{n(S)},$$

where $n(E)$ is the number of outcomes in E and $n(S)$ is the total number of outcomes in S.

a. The Additive Rule. If E and F are events in a sample space S, then

$$P(E \text{ or } F) = P(E \cup F) = P(E) + P(F) - P(E \cap F).$$

b. If $E \cap F = \varnothing$, E and F are called **mutually exclusive** events. For mutually exclusive events E and F,
$$P(E \cup F) = P(E) + P(F).$$

c. $P(E') = P(\text{not } E) = 1 - P(E)$; the event E' is called the **complement** of the event E.

REVIEW EXERCISES

In Exercises 1–4, write the first five terms of each sequence.

1. $a_n = 2n - 3$

2. $a_n = \dfrac{n(n-2)}{2}$

3. $a_n = \dfrac{n}{2n+1}$

4. $a_n = (-2)^{n-1}$

In Exercises 5 and 6, write a general term a_n for each sequence.

5. $30, 28, 26, 24, \ldots$

6. $-1, 2, -4, 8, \ldots$

In Exercises 7–10, simplify each expression.

7. $\dfrac{9!}{8!}$

8. $\dfrac{10!}{7!}$

9. $\dfrac{(n+1)!}{n!}$

10. $\dfrac{n!}{(n-2)!}$

In Exercises 11–14, find each sum.

11. $\displaystyle\sum_{k=1}^{4} k^3$

12. $\displaystyle\sum_{j=1}^{5} \dfrac{1}{2j}$

13. $\displaystyle\sum_{k=1}^{7} \dfrac{k+1}{k}$

14. $\displaystyle\sum_{k=1}^{5} (-1)^k 3^{k+1}$

In Exercises 15 and 16, write each sum in summation notation.

15. $1 + \dfrac{1}{2} + \dfrac{1}{3} + \dfrac{1}{4} + \cdots + \dfrac{1}{50}$

16. $2 - 4 + 8 - 16 + 32$

In Exercises 17–20, determine whether each sequence is arithmetic. If a sequence is arithmetic, find the first term a_1 and the common difference d.

17. $11, 6, 1, -4, \ldots$

18. $\dfrac{2}{3}, \dfrac{5}{6}, 1, \dfrac{7}{6}, \ldots$

19. $a_n = n^2 + 4$

20. $a_n = 3n - 5$

In Exercises 21–24, find an expression for the nth term of each arithmetic sequence.

21. $3, 6, 9, 12, 15, \ldots$

22. $5, 9, 13, 17, 21, \ldots$

23. $x, x+1, x+2, x+3, x+4, \ldots$

24. $3x, 5x, 7x, 9x, 11x, \ldots$

In Exercises 25 and 26, find the common difference d and the nth term a_n for each arithmetic sequence.

25. 3rd term: 7; 8th term: 17

26. 5th term: -16; 20th term: -46

In Exercises 27 and 28, find each sum.

27. $7 + 9 + 11 + 13 + \cdots + 37$

28. $\dfrac{1}{4} + \dfrac{1}{2} + \dfrac{3}{4} + 1 + \cdots + 15$

In Exercises 29 and 30, find the sum of the first n terms of each arithmetic sequence.

29. $3, 8, 13, \ldots; n = 40$

30. $-6, -5.5, -5, \ldots; n = 60$

In Exercises 31–34, determine whether each sequence is geometric. If a sequence is geometric, find the first term a_1 and the common ratio r.

31. $4, -8, 16, -32, \ldots$

32. $\dfrac{1}{5}, \dfrac{1}{10}, \dfrac{1}{20}, \dfrac{1}{40}, \ldots$

33. $\dfrac{1}{3}, 1, \dfrac{5}{3}, 9, \ldots$

34. $a_n = 2^{-n}$

In Exercises 35 and 36, for each geometric sequence, find the first term a_1, the common ratio r, and the nth term a_n.

35. $16, -4, 1, -\dfrac{1}{4}, \ldots$

36. $-\dfrac{5}{6}, -\dfrac{1}{3}, -\dfrac{2}{15}, -\dfrac{4}{75}, \ldots$

In Exercises 37 and 38, find the indicated term of each geometric sequence.

37. a_{10} when $a_1 = 2, r = 3$

38. a_{12} when $a_1 = -2, r = \dfrac{3}{2}$

In Exercises 39 and 40, find the sum S_n of the first n terms of each geometric sequence.

39. $\dfrac{1}{10}, \dfrac{1}{5}, \dfrac{2}{5}, \dfrac{4}{5}, \ldots; n = 12$

40. $2, -1, \dfrac{1}{2}, -\dfrac{1}{4}, \ldots; n = 10$

In Exercises 41–44, find each sum.

41. $\dfrac{1}{2} + \dfrac{1}{6} + \dfrac{1}{18} + \dfrac{1}{54} + \cdots$

42. $-5 - 2 - \dfrac{4}{5} - \dfrac{8}{25} - \cdots$

43. $\displaystyle\sum_{i=1}^{\infty} \left(\dfrac{3}{5}\right)^i$

44. $\displaystyle\sum_{i=1}^{\infty} 7\left(-\dfrac{1}{4}\right)^i$

In Exercises 45 and 46, use mathematical induction to prove that each statement is true for all natural numbers n.

45. $\displaystyle\sum_{k=1}^{n} 2^k = 2^{n+1} - 2$

46. $\displaystyle\sum_{k=1}^{n} k(k + 1) = \frac{n(n + 1)(n + 2)}{3}$

In Exercises 47 and 48, evaluate each binomial coefficient.

47. $\dbinom{12}{7}$

48. $\dbinom{11}{0}$

In Exercises 49 and 50, find each binomial expansion.

49. $(x - 3)^4$

50. $\left(\dfrac{x}{2} + 2\right)^6$

In Exercises 51 and 52, find the specified term.

51. $(x + 2)^{12}$; term containing x^5

52. $(x + 2y)^{12}$; term containing x^7

In Exercises 53–64, use any method to solve the problem.

53. How many different numbers can be written with the digits 1, 2, 3, and 4 if no digit may be repeated?

54. How many ways can horses finish in first, second, and third place in a ten-horse race?

55. How many ways can seven people line up at a ticket window?

56. In a building with seven entrances, how many ways can you enter the building and leave by a different entrance?

57. How many ways can five movies be listed on a display (vertically, one movie per line)?

58. How many different amounts of money can you make with a nickel, a dime, a quarter, and a half-dollar?

59. How many ways can three candies be taken from a box of ten candies?

60. How many ways can 2 face cards be drawn from a standard 52-card deck?

61. How many ways can 2 shirts and 3 pairs of pants be chosen from 12 shirts and 8 pairs of pants?

62. How many doubles teams can be formed from eight tennis players?

63. In how many distinguishable ways can the letters of the word R E I T E R A T E be arranged?

64. A bookshelf has room for four books. How many different arrangements can be made on the shelf from six available books?

In Exercises 65–74, find the probability requested.

65. Four silver dollars, all with different dates, are randomly arranged on a horizontal display. What is the probability that the two with the most recent dates will be next to each other?

66. A jar contains five white, three black, and two green stones. What is the probability of drawing a black stone on a single draw?

67. The word R A N D O M has been spelled in a game of Scrabble. If two letters are chosen from this word at random, what is the probability that they are both consonants?

68. The names of all 30 party guests are placed in a box, and one name is drawn to award a door prize. If you and your date are among the guests, what is the probability that one of you will win the prize?

69. If two people are chosen at random from a group consisting of five men and three women, what is the probability that one man and one woman are chosen?

70. A company determines that 32% of the e-mail it receives is junk mail. What is the probability that a randomly chosen e-mail at this company is not junk mail?

71. If 2 cards are drawn from a standard 52-card deck, what is the probability that both are clubs?

72. A battery manufacturer inspects 500 batteries and finds that 30 are defective. What is the probability that a randomly selected battery is not defective?

73. A ball is taken at random from a pool table containing balls numbered 1 through 9. What is the probability of obtaining (a) ball number 7? (b) an even-numbered ball? (c) an odd-numbered ball?

74. A clothes dryer load contains two shirts, four pairs of socks, and three nightgowns. You pull one item randomly from the dryer.
  a. What is the probability that it is a sock?
  b. What is the probability that it is a nightgown?

PRACTICE TEST A

In Problems 1 and 2, write the first five terms of each sequence. State whether the sequence is arithmetic or geometric.

1. $a_n = 3(5 - 4n)$

2. $a_n = -3(2^n)$

3. Write the first five terms of the sequence defined by $a_1 = -2$ and $a_n = 3a_{n-1} + 5$ for $n \geq 2$.

4. Evaluate $\dfrac{(n - 1)!}{n!}$.

5. Find a_7 for the arithmetic sequence with $a_1 = -3$ and $d = 4$.

6. Find a_8 for the geometric sequence with $a_1 = 13$ and $r = -\dfrac{1}{2}$.

In Problems 7–9, find each sum.

7. $\displaystyle\sum_{k=1}^{20} (3k - 2)$ **8.** $\displaystyle\sum_{k=1}^{5} \left(\frac{3}{4}\right)(2^{-k})$

9. $\displaystyle\sum_{k=1}^{\infty} 18\left(\frac{1}{100}\right)^{k}$; write the answer as a fraction in lowest terms.

10. Evaluate $\dbinom{13}{0}$.   **11.** Expand $(1 - 2x)^4$.

12. Find the term containing x^1 in the expansion of $(2x + 1)^4$.

13. In how many ways can you answer every question on a true–false test containing ten questions?

In Problems 14 and 15, evaluate each expression.

14. $P(9, 2)$ **15.** $C(7, 5)$

16. A firm needs to fill two positions—one in accounting and one in human resources. How many ways can these positions be filled if there are six applicants for the accounting position and ten applicants for the human resources position?

17. A state is considering using license plates consisting of two letters followed by a four-digit number. How many distinct license plate numbers can be formed?

18. A box contains five red balls and seven black balls. What is the probability that a ball drawn at random will be red?

19. What is the probability that a 4 does *not* come up on one roll of a die?

20. A box contains seven quarters, six dimes, and five nickels. If two coins are drawn at random, what is the probability that both are quarters?

PRACTICE TEST B

In Problems 1 and 2, write the first five terms of each sequence. State whether the sequence is arithmetic or geometric.

1. $a_n = 4(2n - 3)$
   a. $-1, 1, 3, 5, 7$; arithmetic
   b. $-1, 1, 3, 5, 7$; geometric
   c. $-4, 4, 12, 20, 28$; arithmetic
   d. $-4, 4, 12, 20, 28$; geometric

2. $a_n = 2(4^n)$
   a. $2, 8, 32, 128, 512$; arithmetic
   b. $2, 8, 32, 128, 512$; geometric
   c. $8, 32, 128, 512, 2048$; arithmetic
   d. $8, 32, 128, 512, 2048$; geometric

3. Write the first five terms of the sequence defined by $a_1 = 5$ and $a_n = 2a_{n-1} + 4$, for $n \geq 2$.
   a. $5, 14, 32, 68, 140$ **b.** $5, 14, 19, 24, 29$
   c. $5, 9, 13, 17, 21$ **d.** $5, 8, 10, 12, 14$

4. Evaluate $\dfrac{(n + 2)!}{n + 2}$.
   a. $2!$ **b.** 1
   c. $(n + 1)!$ **d.** $n + 1$

5. Find a_8 for the arithmetic sequence with $a_1 = -6$ and $d = 3$.
   a. 15 **b.** -271
   c. 30 **d.** 18

6. Find a_{10} for the geometric sequence with $a_1 = 247$ and $r = \dfrac{1}{3}$.

   a. 250 **b.** $\dfrac{247}{19{,}683}$

   c. $\dfrac{247}{59{,}049}$ **d.** $\dfrac{247}{177{,}147}$

In Problems 7 and 8, find each sum.

7. $\displaystyle\sum_{k=1}^{45} (4k - 7)$
   a. 4185 **b.** 3735
   c. 4027.5 **d.** 3825

8. $\displaystyle\sum_{k=1}^{5} \left(\frac{4}{3}\right)(2^k)$
   a. $\dfrac{248}{3}$ **b.** $\dfrac{251}{3}$
   c. $\dfrac{242}{3}$ **d.** $\dfrac{287}{3}$

9. Evaluate $\displaystyle\sum_{k=1}^{\infty} 8(-0.3)^{k-1}$.
   a. 6 **b.** 6.15
   c. -6.15 **d.** -4.15

10. Evaluate $\dbinom{12}{11}$.
   a. $1.\overline{09}$ **b.** 12
   c. 1 **d.** $11!$

11. Expand $(3x - 1)^4$.
   a. $81x^4 - 108x^3 + 54x^2 - 12x + 1$
   b. $-81x^4 + 108x^3 - 54x^2 + 12x - 1$
   c. $-81x^4 + 108x^3 + 54x^2 + 12x + 1$
   d. $81x^4 - 108x^2 + 54x - 12$

12. Find the coefficient of x in the expansion of $(2x + 3)^3$.
   a. 108 **b.** 9
   c. 36 **d.** 54

13. Karen has four necklaces, ten pairs of earrings, and three bracelets. How many ways can she choose a necklace, a pair of earrings, and a bracelet?
   a. 40 **b.** 17
   c. 120 **d.** 240

In Problems 14 and 15, evaluate each expression.

14. $P(7, 3)$
   a. 210   **b.** 840
   c. 1680   **d.** 5040

15. $C(10, 7)$
   a. 3   **b.** 720
   c. 620,400   **d.** 120

16. How many two-digit numbers can be formed from the digits 0, 1, 2, 3, 4, 5, 6, 7, 8, and 9? No digit can be used more than once.
   a. 45   **b.** 81
   c. 3,628,880   **d.** 100

17. You have selected eight CDs for your dad, but have only enough money to buy five of them. In how many ways can you select the five if you decide that a particular CD is a "must buy"?
   a. 56   **b.** 70
   c. 1680   **d.** 35

18. What is the probability of rolling a 5 on one roll of a die?
   a. $\dfrac{1}{5}$ **b.** 4
   c. $\dfrac{1}{6}$ **d.** 0.05

19. What is the probability of getting a sum greater than 9 when a pair of dice is rolled?
   a. $\dfrac{1}{6}$ **b.** $\dfrac{1}{4}$
   c. $\dfrac{1}{12}$ **d.** $\dfrac{1}{18}$

20. What is the probability that a card drawn at random from a standard 52-card deck is *not* a spade?
   a. $\dfrac{2}{5}$ **b.** $\dfrac{1}{4}$
   c. $\dfrac{4}{13}$ **d.** $\dfrac{3}{4}$

CUMULATIVE REVIEW EXERCISES ▪ Chapters P–8

In Problems 1–10, solve each equation or inequality.

1. $|3x - 8| = |x|$ **2.** $x^2(x^2 - 5) = -4$

3. $\log_2(3x - 5) + \log_2 x = 1$

4. $e^{2x} - e^x - 2 = 0$ **5.** $\dfrac{6}{x + 2} = \dfrac{4}{x}$

6. $2(x - 8)^{-1} = (x - 2)^{-1}$

7. $\sqrt{x - 3} = \sqrt{x} - 1$ **8.** $x^2 + 4x \geq 0$

9. $18 \leq x^2 + 6x$ **10.** $\dfrac{x - 2}{x + 2} \geq 3$

In Problems 11–13, solve each system.

11. $\begin{cases} 5x + 4y = 6 \\ 4x - 3y = 11 \end{cases}$

12. $\begin{cases} \dfrac{x}{3} + \dfrac{y}{4} = \dfrac{11}{6} \\ \dfrac{x}{6} + \dfrac{y}{8} = \dfrac{7}{12} \end{cases}$

13. $\begin{cases} x - 3y + 6z = -8 \\ 5x - 6y - 2z = 7 \\ 3x - 2y - 10z = 11 \end{cases}$

14. Determine whether the equation $-x + y^2 = 0$ is symmetric with respect to the x-axis, the y-axis, and/or the origin.

15. Use transformations to sketch the graph of $f(x) = \sqrt{x - 3}$.

16. Find the domain of $f(x) = \dfrac{x + 3}{x(x + 1)}$.

17. Write $7 \ln x - \ln(x - 5)$ in condensed form.

18. Assuming that $A = \begin{bmatrix} 1 & 0 & -2 \\ 4 & 1 & 0 \\ 1 & 1 & 7 \end{bmatrix}$, find A^{-1}

19. Four students are pairing up to ride two tandem bicycles, each of which accommodates two riders, one in back and one in front. How many ways can this be done if we count arrangements as different if either the front or back rider is different?

20. A room has 6 people who are 25 years old and older and 12 people whose age is between 17 and 24. Half of the members of each group have gotten speeding tickets at some time. What is the probability that a person chosen at random from the room will be a person who is 25 years old or older or has gotten a speeding ticket at some time?

Answers

CHAPTER P

Section P.1

Practice Problems: **1. a.** true **b.** true **c.** true
2. $A \cap B = \{0\}$, $A \cup B = \{-4, -3, -2, -1, 0, 1, 2, 3, 4\}$
3. $A' = \{1, 2, 4, 5, 7, 8\}$ **4. a.** 10 **b.** 1 **c.** 1 **5.** 9 **6. a.** 5
b. 1 **c.** -33 **7. a.** 2 **b.** $-\dfrac{1}{7}$ **8. a.** $\dfrac{17}{8}$ **b.** $\dfrac{34}{15}$ **c.** $\dfrac{3}{2}$ **d.** $\dfrac{2}{3}$
9. a. $\dfrac{10}{3}$ **b.** $\dfrac{22}{3}$ **10.** 17°C **11.** $(-\infty, -2) \cup (-2, \infty)$

A Exercises: Basic Skills and Concepts: **1.** zero **3.** left **5.** true
7. $0.\overline{3}$ repeating **9.** -0.8 terminating **11.** $0.\overline{27}$ repeating
13. $3.1\overline{6}$ repeating **15.** Rational **17.** Rational
19. Rational **21.** Irrational **23.** Rational
25. $3 > -2$ **27.** $\dfrac{1}{2} \geq \dfrac{1}{2}$ **29.** $5 \leq 2x$ **31.** $-x > 0$
33. $2x + 7 \leq 14$ **35.** $=$ **37.** $<$ **39.** $\{1, 2, 3\}$ **41.** $\{3, 4, 5, 6, 7\}$
43. $\{-3, -2, -1\}$ **45.** $\{-4, -3, -2, 0, 1, 2, 3, 4\}$
47. $\{-3, -1, 1, 3\}$ **49.** $\{-3, -1, 0, 1, 2, 3\}$ **51.** $\{-1\}$
53. $\{-4, -3, -2, 0, 2\}$ **55.** 20 **57.** -4 **59.** $\dfrac{5}{7}$ **61.** $5 - \sqrt{2}$
63. 1 **65.** 12 **67.** 3 **69.** $|3 - 8| = 5$ **71.** $|-6 - 9| = 15$
73. $|-20 - (-6)| = 14$ **75.** $\left|\dfrac{22}{7} - \left(-\dfrac{4}{7}\right)\right| = \dfrac{26}{7}$
77. $1 \leq x \leq 4$

79. $14 < x < 28$

81. $-3 < x \leq 1$

83. $x \geq -3$

85. $x \leq 5$

87. $-\dfrac{3}{4} < x < \dfrac{9}{4}$

89. $4x + 4$ **91.** $5x - 5y + 5$ **93.** $-5; \dfrac{1}{5}$ **95.** 0; no reciprocal
97. Additive inverse **99.** Multiplicative identity
101. Multiplicative associative **103.** Multiplicative inverse
105. Additive identity **107.** Additive associative **109.** $\dfrac{29}{15}$
111. $\dfrac{67}{35}$ **113.** $\dfrac{17}{15}$ **115.** $-\dfrac{11}{40}$ **117.** $-\dfrac{8}{99}$ **119.** $-\dfrac{1}{10}$
121. $\dfrac{2}{9}$ **123.** $\dfrac{3}{2}$ **125.** $\dfrac{2}{3}$ **127.** 1 **129.** $\dfrac{1}{15}$ **131.** 11
133. -1 **135.** -6 **137.** -13 **139.** $\dfrac{62}{15}$ **141.** $(-\infty, 1) \cup (1, \infty)$
143. $(-\infty, 0) \cup (0, \infty)$ **145.** $(-\infty, -1) \cup (-1, 0) \cup (0, \infty)$
147. $(-\infty, -2) \cup (-2, 1) \cup (1, \infty)$ **149.** $(-\infty, 0) \cup (0, 2) \cup (2, \infty)$
151. $(-\infty, \infty)$

B Exercises: Applying the Concepts: **153. a.** People who own
either an MP3 or DVD player **b.** People who own both an MP3
and DVD player
155. $119.5 \leq x \leq 134.5$

157. a. $|124 - 120| = 4$ **b.** $|137 - 120| = 17$ **c.** $|114 - 120| = 6$
159. $[0, \infty)$ **161.** $[60, 70]$ **163.** 950 g

Section P.2

Practice Problems: **1. a.** 8 **b.** $9a^2$ **c.** $\dfrac{1}{16}$ **2. a.** $\dfrac{1}{2}$ **b.** 1 **c.** $\dfrac{4}{9}$
3. a. $3x^9$ **b.** 16 **4. a.** 81 **b.** 125 **c.** $\dfrac{2x^7}{3}$ **5. a.** 1 **b.** 1
c. $\dfrac{1}{x^8}$ **d.** x^{10} **6. a.** $\dfrac{2}{x}$ **b.** $\dfrac{25}{x^2}$ **c.** $x^3 y^6$ **d.** $\dfrac{x^6}{y^3}$ **7. a.** $\dfrac{1}{9}$ **b.** $\dfrac{49}{100}$
8. a. $\dfrac{1}{4x^8}$ **b.** $-\dfrac{1}{xy^3}$ **9.** 7.32×10^5 **10.** \$43 per person

A Exercises: Basic Skills and Concepts: **1.** exponent **3.** 16
5. false **7.** Exponent: 3; base: 17 **9.** Exponent: 0; base: 9
11. Exponent: 5; base: -5 **13.** Exponent: 3; base: 10
15. Exponent: 2; base: a **17.** 6 **19.** 1
21. 64 **23.** $\dfrac{1}{81}$ **25.** $\dfrac{1}{15,625}$ **27.** 16 **29.** 2 **31.** $\dfrac{2}{9}$ **33.** 2
35. 1 **37.** 6 **39.** $-\dfrac{2}{25}$ **41.** $\dfrac{49}{121}$ **43.** x^4 **45.** $\dfrac{y}{x}$ **47.** $-\dfrac{8}{x}$
49. $\dfrac{3}{x}$ **51.** $\dfrac{1}{xy^2}$ **53.** $\dfrac{1}{x^{12}}$ **55.** x^{33} **57.** $-3x^5 y^5$
59. $\dfrac{4x^2}{y^2}$ **61.** $\dfrac{3x^5}{y^5}$ **63.** $\dfrac{1}{x^4}$ **65.** $\dfrac{8y^3}{x^3}$ **67.** $-243x^5 y^5$ **69.** $\dfrac{25}{9x^2}$
71. $\dfrac{64}{x^9 y^{15}}$ **73.** $\dfrac{x^5}{y^4}$ **75.** $\dfrac{3x}{y^2}$ **77.** $\dfrac{5b^4}{a^6}$ **79.** $\dfrac{x^3 z^{15}}{y^3}$
81. 1.25×10^2 **83.** 8.5×10^5 **85.** 7×10^{-3} **87.** 2.75×10^{-6}

B Exercises: Applying the Concepts: **89.** 1080 ft^3
91. a. $(2x)(2x) = 4x^2 = 2^2 A$ **b.** $(3x)(3x) = 9x^2 = 3^2 A$
93. $F = 1562.5$ **95.** Earth: 1.27×10^4 (km) Moon: 3480;
Sun: 1.39×10^6 (km); Jupiter: 134,000; Mercury: 4.8×10^3 (km)
97. 6.02×10^{20} atoms **99.** 1.67×10^{-21} kg **101.** 5.98×10^{24} kg

Section P.3

Practice Problems: **1.** 889 ft **2.** $5x^3 + 5x^2 + 2x - 4$
3. $5x^4 - 5x^3 - x^2 - x + 12$ **4.** $-8x^5 - 4x^4 + 10x^3$
5. $-10x^4 + x^3 - 33x^2 - 14x$ **6. a.** $4x^2 + 27x - 7$
b. $6x^2 - 19x + 10$ **7.** $9x^2 + 12x + 4$ **8.** $1 - 4x^2$ **9.** $x^2 y + x^3$

A Exercises: Basic Skills and Concepts: **1.** $-3; 7$ **3.** 4 **5.** true
7. Yes; $x^2 + 2x + 1$ **9.** No **11.** Degree: 1; terms: $7x, 3$
13. Degree: 4; terms: $-x^4, x^2, 2x, -9$ **15.** $2x^2 - 3x - 1$
17. $x^3 - x^2 + 5x - 8$ **19.** $-10x^4 - 6x^3 + 12x^2 - 7x + 17$
21. $-24x^2 - 14x - 14$ **23.** $2y^3 + y^2 - 2y - 1$
25. $12x^2 + 18x$ **27.** $x^3 + 3x^2 + 4x + 2$
29. $3x^3 - 5x^2 + 5x - 2$ **31.** $x^2 + 3x + 2$ **33.** $9x^2 + 9x + 2$
35. $-4x^2 - 7x + 15$ **37.** $6x^2 - 7x + 2$ **39.** $4x^2 + 4ax - 15a^2$
41. $4x + 4$ **43.** $9x^2 + 27x + 27$ **45.** $16x^2 + 8x + 1$
47. $27x^3 + 27x^2 + 9x + 1$ **49.** $-4x^2 + 25$
51. $x^2 + \dfrac{3}{2}x + \dfrac{9}{16}$ **53.** $2x^3 - 9x^2 + 19x - 15$
55. $y^3 + 1$ **57.** $x^3 - 216$ **59.** $3x^2 + 11xy + 10y^2$
61. $6x^2 + 11xy - 7y^2$ **63.** $x^4 - 2x^2 y^2 + y^4$
65. $x^3 - 3x^2 y + 4y^3$ **67.** $x^4 - 4x^3 y + 16xy^3 - 16y^4$

B Exercises: Applying the Concepts: **69.** In 2003, it was \$6.02. **71.** \$250
73. 500 feet **75. a.** $-x + 22.50$ **b.** $-10x^2 + 195x + 675$

Section P.4

Practice Problems: 1. a. $2x^3(3x^2 + 7)$ **b.** $7x^2(x^3 + 3x^2 + 5)$
2. a. $(x + 4)(x + 2)$ **b.** $(x - 5)(x + 2)$ **3. a.** $(x + 2)^2$
b. $(3x - 1)^2$ **4. a.** $(x + 4)(x - 4)$ **b.** $(2x - 5)(2x + 5)$
5. $(x^2 + 9)(x - 3)(x + 3)$ **6. a.** $(x - 5)(x^2 + 5x + 25)$
b. $(3x + 2)(9x^2 - 6x + 4)$ **7.** \$16.224 million
8. a. $(5x + 1)(x + 2)$ **b.** $(3x - 2)(3x - 1)$
9. a. $(x + 3)(x^2 + 1)$ **b.** $(7x - 5)(2x + 1)(2x - 1)$
c. $(x + y + 1)(x - y - 1)$

A Exercises: Basic Skills and Concepts: 1. factors **3.** $10x^2$
5. true **7.** $8(x - 3)$ **9.** $-6x(x - 2)$ **11.** $7x^2(1 + 2x)$
13. $x^2(x^2 + 2x + 1)$ **15.** $x^2(3x - 1)$ **17.** $4ax^2(2x + 1)$
19. $(x + 3)(x^2 + 1)$ **21.** $(x - 5)(x^2 + 1)$
23. $(3x + 2)(2x^2 + 1)$ **25.** $x^2(3x^2 + 1)(4x^3 + 1)$
27. $(x + 3)(x + 4)$ **29.** $(x - 2)(x - 4)$ **31.** $(x + 1)(x - 4)$
33. Irreducible **35.** $(2x + 9)(x - 4)$ **37.** $(2x + 3)(3x + 4)$
39. $(3x + 1)(x - 4)$ **41.** Irreducible **43.** $(x + 3)^2$
45. $(3x + 1)^2$ **47.** $(5x - 2)^2$ **49.** $(7x + 3)^2$ **51.** $(x - 8)(x + 8)$
53. $(2x - 1)(2x + 1)$ **55.** $(4x - 3)(4x + 3)$
57. $(x - 1)(x + 1)(x^2 + 1)$ **59.** $5(\sqrt{2}x - 1)(\sqrt{2}x + 1)(2x^2 + 1)$
61. $(x + 4)(x^2 - 4x + 16)$ **63.** $(x - 3)(x^2 + 3x + 9)$
65. $(2 - x)(4 + 2x + x^2)$ **67.** $(2x - 3)(4x^2 + 6x + 9)$
69. $5(2x + 1)(4x^2 - 2x + 1)$ **71.** $(1 - 4x)(1 + 4x)$
73. $(x - 3)^2$ **75.** $(2x + 1)^2$ **77.** $2(x + 1)(x - 5)$
79. $(2x - 5)(x + 4)$ **81.** Irreducible **83.** $3x^3(x + 2)^2$
85. $(3x + 1)(3x - 1)$ **87.** $(4x + 3)^2$ **89.** Irreducible
91. $x(5x + 2)(9x - 2)$ **93.** $a(x + a)(x - 8a)$
95. $(x + 4 + 4a)(x + 4 - 4a)$ **97.** $3x^3(x + 2y)^2$
99. $(x + 2 + 5a)(x + 2 - 5a)$ **101.** $2x^4(3x + y)^2$

B Exercises: Applying the Concepts: 103. $x(8 - x)$
105. $4x(18 - x)(8 - x)$ **107.** $\pi(2 - x)(2 + x)$
109. $2x(1400 - x)$

Section P.5

Practice Problems: 1. 0.32 **2. a.** $\dfrac{2x}{3}$ **b.** $\dfrac{x - 2}{x + 2}$ **3.** $\dfrac{x(x - 3)}{14}$

4. a. $\dfrac{7}{x - 6}$ **b.** $\dfrac{1}{x + 4}$ **5. a.** $\dfrac{x(5x - 4)}{(x + 2)(x - 5)}$ **b.** $\dfrac{x(3x + 23)}{(x + 3)(x - 4)}$
6. a. $3x^2(x + 2)^2(x - 2)^2$ **b.** $(x - 5)(x + 5)(x - 1)$
7. a. $\dfrac{x^2 + 2x + 8}{(x + 2)(x - 2)^2}$ **b.** $\dfrac{45 - 7x}{3(x - 5)^2}$ **8.** $\dfrac{1}{x - 5}$ **9.** $\dfrac{25x}{15x - 12}$

A Exercises: Basic Skills and Concepts: 1. each denominator

3. $(x - 2)(x + 1)$ **5.** false **7.** $\dfrac{2}{x + 1}, x \neq -1$

9. $\dfrac{3}{x - 1}, x \neq -1, x \neq 1$ **11.** $-\dfrac{2}{x + 3}, x \neq -3, x \neq 3$

13. $-1, x \neq \dfrac{1}{2}$ **15.** $\dfrac{x - 3}{4}, x \neq 3$ **17.** $\dfrac{7x}{x + 1}, x \neq -1$

19. $\dfrac{x - 10}{x + 7}, x \neq -7, x \neq 1$ **21.** $\dfrac{x + 2}{x - 2}, x \neq -\dfrac{1}{3}, x \neq 0, x \neq 2$

23. 1 **25.** $\dfrac{x + 3}{2(x - 3)}$ **27.** $\dfrac{x - 1}{x}$ **29.** $\dfrac{x - 1}{x + 3}$ **31.** -1 **33.** $\dfrac{3}{8}$

35. $\dfrac{5x(x - 3)}{2}$ **37.** $\dfrac{3(x + 3)}{x + 4}$ **39.** $\dfrac{x - 3}{x^2 - 2x + 4}$ **41.** $\dfrac{x + 3}{5}$

43. $\dfrac{x + 4}{2x + 1}$ **45.** $\dfrac{1}{x + 1}$ **47.** $\dfrac{2(x - 2)}{x - 3}$ **49.** $\dfrac{7x}{x^2 + 1}$

51. $\dfrac{4x}{x - 3}$ **53.** $\dfrac{1}{x + 2}$ **55.** $\dfrac{2(1 - 3x)}{(2x - 1)(2x + 1)}$

57. $-\dfrac{x^3 + 2x^2 + 4x - 8}{x(x - 2)(x + 2)}$ **59.** $12(x - 2)$

61. $(2x - 1)(2x + 1)^2$ **63.** $(x - 1)(x + 1)(x + 2)$

65. $(x - 1)(x - 4)(x + 2)$ **67.** $\dfrac{7x + 15}{(x - 3)(x + 3)}$

69. $\dfrac{x(4 - x)}{(x - 2)(x + 2)}$ **71.** $\dfrac{2x + 3}{(x - 2)(x + 5)}$ **73.** $\dfrac{2(9x^2 - 5x + 1)}{(3x + 1)(3x - 1)^2}$

75. $\dfrac{9(x - 3)}{(x + 4)(x + 5)(x - 5)}$ **77.** $\dfrac{2x - 3}{(2 + x)(2 - x)}$

79. $\dfrac{16a^2}{(x - 5a)(x - 3a)}$ **81.** $-\dfrac{h}{x(x + h)}$ **83.** $\dfrac{2x}{3}$ **85.** $\dfrac{1}{x - 1}$

87. $\dfrac{1 - x}{1 + x}$ **89.** $-x$ **91.** $\dfrac{x(2x - 1)}{2x + 1}$

93. $-\dfrac{1}{x(x + h)}, x \neq -h, x \neq 0$ **95.** $\dfrac{x}{a}, x \neq -a, x \neq a$

B Exercises: Applying the Concepts: 97. 10 cm

99. a. $\dfrac{3}{100 + x}$ **b.** 2% **101. a.** $\dfrac{10\pi x^3 + 240}{x}$ **b.** \$2.46

Section P.6

Practice Problems: 1. a. 12 **b.** $\dfrac{1}{7}$ **c.** $\dfrac{1}{4}$ **2. a.** $2\sqrt{5}$ **b.** $4\sqrt{3}$

c. $2|x|\sqrt{3}$ **d.** $\dfrac{2|y|\sqrt{5y}}{3\sqrt{3}|x|}$ **3.** ≈ 9.6 gallons/min. **4. a.** $\dfrac{7\sqrt{2}}{4}$

b. $-\dfrac{1 + \sqrt{7}}{2}$ **c.** $\dfrac{x\sqrt{x} + 3x}{x - 9}$ **5. a.** -2 **b.** 2 **c.** 3

d. Not a real number **6. a.** $2\sqrt[3]{9}$ **b.** $2\sqrt[4]{3a^2}$ **7. a.** $13\sqrt{3}$

b. 0 **8.** $\sqrt[10]{972}$ **9. a.** $\dfrac{1}{2}$ **b.** 5 **c.** -2 **10. a.** $5\sqrt[3]{5}$ **b.** -216

c. $\dfrac{1}{1024}$ **d.** Not a real number **11. a.** $12x^{7/10}$ **b.** $\dfrac{5}{x^{7/12}}$ **c.** $\dfrac{1}{x^{2/15}}$

12. $\dfrac{(2x + 3)(x + 3)^{1/2}}{x + 3}$ **13. a.** $\sqrt[3]{x^2}$ **b.** 5 **c.** x^2

A Exercises: Basic Skills and Concepts: 1. two **3.** like radicals

5. false **7.** 8 **9.** 4 **11.** -3 **13.** $-\dfrac{1}{2}$ **15.** 3

17. Not a real number **19.** 1 **21.** -7 **23.** $4\sqrt{2}$ **25.** $3x\sqrt{2}$
27. $3x\sqrt{x}$ **29.** $3x\sqrt{2}$ **31.** $3x\sqrt{5x}$ **33.** $-2x$ **35.** $-x^2$
37. $\dfrac{\sqrt{10}}{8}$ **39.** $\dfrac{2}{x}$ **41.** $\dfrac{\sqrt[4]{2x^3}}{x^2}$ **43.** x^2 **45.** $\dfrac{3x^3}{y^2}$ **47.** $7\sqrt{3}$
49. $9\sqrt{5}$ **51.** $3\sqrt{2x}$ **53.** $-\sqrt[3]{3}$ **55.** $2\sqrt[3]{3x}$
57. $\sqrt{2x}(x^2 + 3x - 20)$ **59.** $(2x - y)^2\sqrt{3xy}$
61. $\dfrac{2\sqrt{3}}{3}$ **63.** $\dfrac{7\sqrt{15}}{15}$ **65.** $\dfrac{\sqrt{2} - x}{2 - x^2}$ **67.** $3(2 + \sqrt{3})$

69. $\sqrt{3} - \sqrt{2}$ **71.** $\dfrac{7 - 2\sqrt{10}}{3}$ **73.** $\dfrac{2x + h - 2\sqrt{x(x + h)}}{h}$

75. 5 **77.** -2 **79.** 4 **81.** $-\dfrac{1}{125}$ **83.** $\dfrac{125}{27}$

85. $x^{9/10}$ **87.** $x^{1/10}$ **89.** $4x^4$ **91.** $\dfrac{1}{9x^4y^2}$ **93.** $5x^{5/4}$ **95.** $\dfrac{x^3}{y^8}$

97. $3^{1/2}$ **99.** x^3 **101.** x^2y^3 **103.** 3 **105.** $x^{5/3}$ **107.** $\sqrt[15]{864}$
109. $x\left(\sqrt[12]{x^5}\right)$ **111.** $\sqrt[12]{40m^{11}n^5}$ **113.** $xy\left(\sqrt[30]{x^9y^{13}}\right)$

B Exercises: Applying the Concepts: 115. 40 cm **117.** 2 cm/sec
119. 23 amp **121.** 0.46 lb/in.^{2}

Section P.7

Practice Problems: 1. 10 **2.** 10 feet **3.** No

A Exercises: Basic Skills and Concepts: 1. isosceles **3.** similar
5. true **7.** 25 **9.** 15 **11.** $3\sqrt{2}$ **13.** 8 **15.** 48
17. $A = 15, P = 16$ **19.** $A = 5, P = 21$ **21.** 6 **23.** 3
25. $A = 3.14, C = 6.28$ **27.** $A = 0.785, C = 3.14$ **29.** 5 **31.** 15

B Exercises: Applying the Concepts: 33. 192 ft^2 **35.** 9 in.
37. $5\sqrt{29} \approx 27$ ft **39.** 346 yd **41.** 2471 ft **43.** 39 ft^2 **45.** 15 ft^3

Review Exercises

1. a. $\{4\}$ **b.** $\{0, 4\}$ **c.** $\{-5, 0, 4\}$ **d.** $\left\{-5, 0, 0.2, 0.\overline{31}, \frac{1}{2}, 4\right\}$

e. $\{\sqrt{7}, \sqrt{12}\}$ **f.** $\left\{-5, 0, 0.2, 0.\overline{31}, \frac{1}{2}, \sqrt{7}, \sqrt{12}, 4\right\}$

3. Distributive   **5.** Multiplicative identity
7. $x \le 1$ **9.** $(0, \infty)$

11. 1   **13.** $\sqrt{15} - 1$ **15.** -64 **17.** -29 **19.** -5 **21.** 16

23. $\dfrac{216}{125}$ **25.** 250   **27.** 0.7   **29.** 5   **31.** -5 **33.** $\dfrac{1}{4}$ **35.** -1

37. 2   **39.** $-\dfrac{5}{4}$ **41.** $\dfrac{1}{4}$ **43.** x^{10} **45.** $\dfrac{x^6}{y^2}$ **47.** $\dfrac{64}{xy^2}$

49. $\dfrac{y^3}{16x^2}$ **51.** $21x^{7/4}$ **53.** $2x^5$ **55.** $x^{10}y^2$ **57.** $\dfrac{8\sqrt{11}}{11}$

59. $\sqrt{6}$ **61.** $2x\sqrt{3}$ **63.** $5x$ **65.** $14\sqrt[3]{5}$ **67.** $\dfrac{7\sqrt{3}}{3}$

69. $-2 + \sqrt{3}$ **71.** 2.3051×10^{13} **73.** $4x^3 - 12x^2 + 9x - 6$
75. $9x^4 + 11x^3 - 12x^2 + 14$ **77.** $x^2 - 15x + 36$
79. $x^{10} - 4$ **81.** $6x^2 - 7x - 55$ **83.** $(x - 5)(x + 2)$
85. $(6x - 11)(4x + 1)$ **87.** $(x + 5)(x + 11)$
89. $(x - 1)(x^3 + 7)$ **91.** $(5x + 4)(2x + 3)$
93. $(4x - 3)(3x + 4)$ **95.** $(2x - 7)(2x + 7)$ **97.** $(x + 6)^2$
99. $(8x + 3)^2$ **101.** $(2x + 3)(4x^2 - 6x + 9)$
103. $(x - 4)(x + 4)(x + 5)$ **105.** $\dfrac{14}{x - 9}$

107. $\dfrac{22 - 5x}{(x + 2)(x + 6)(x + 8)}$ **109.** $-\dfrac{4x}{(x - 1)(x + 1)}$

111. $\dfrac{2x + 3}{2x + 1}$ **113.** $\dfrac{x(x - 2)}{2(2x + 3)}$ **115.** $\dfrac{(1 - x)(1 + x + x^2)}{(1 + x)(1 - x + x^2)}$

117. -1 **119.** 29   **121.** $A = 96, P = 48$ **123.** $48\,\text{ft}^2$
125. 74.46 mi   **127.** 1.5 ft

Practice Test

1. 4   **2.** $100 - \sqrt{2}$ **3.** -8 **4.** $(-\infty, -9) \cup (-9, 3) \cup (3, \infty)$

5. $-27x^3y^3$ **6.** $-2x^2$ **7.** $2\sqrt{3x}$ **8.** $-\dfrac{1}{64}$ **9.** $\dfrac{5x^{5/2}}{y^{1/2}}$

10. $-\dfrac{5 + 5\sqrt{3}}{2}$ **11.** $19x^2 - 18x + 11$ **12.** $5x^2 - 11x + 2$

13. $x^4 + 6x^2y^2 + 9y^4$ **14.** $(x - 5)(x + 5)\pi$
15. $(x - 3)(x - 2)$ **16.** $(3x + 2)^2$ **17.** $(2x - 3)(4x^2 + 6x + 9)$

18. $\dfrac{-9}{(x + 2)}$ **19.** $\dfrac{1}{(3x + 1)(3x - 1)}$ **20.** 37

CHAPTER 1

Section 1.1

Practice Problems: **1. a.** $-\infty, \infty$ **b.** $(-\infty, 2) \cup (2, \infty)$

c. $[1, \infty)$ **2.** $\left\{-\dfrac{3}{5}\right\}$ **3.** $\{-1\}$ **4.** $\{2\}$ **5.** $\left\{\dfrac{1}{2}\right\}$

6. $(-\infty, -2) \cup (-2, 2) \cup (2, \infty)$ **7.** $\varnothing$ **8.** $10°\text{C}$

9. $w = \dfrac{P - 2l}{2}$ **10.** 34.3 ft

A Exercises: Basic Skills and Concepts: **1.** defined   **3.** equivalent
5. true   **7. a.** No   **b.** Yes   **9. a.** Yes   **b.** No
11. a. Yes   **b.** Yes   **13.** $(-\infty, -2) \cup (-2, 1) \cup (1, \infty)$
15. $(-\infty, 3) \cup (3, 4) \cup (4, \infty)$ **17.** No, $x = 0$.   **19.** Yes
21. $\{3\}$ **23.** $\{-2\}$ **25.** $\{7\}$ **27.** $\{-1\}$ **29.** $\{6\}$ **31.** $\{12\}$

33. $\{1\}$ **35.** $\{-11\}$ **37.** $\left\{\dfrac{23}{6}\right\}$ **39.** $\left\{\dfrac{7}{8}\right\}$ **41.** $\{2\}$ **43.** $\{28\}$

45. $\left\{\dfrac{4}{5}\right\}$ **47.** $\{-1\}$ **49.** $\left\{\dfrac{3}{11}\right\}$ **51.** $\{x \mid x \ne 0\}$

53. $\{11\}$ **55.** $\{5\}$ **57.** $\left\{\dfrac{24}{5}\right\}$ **59.** $\{1\}$ **61.** $\varnothing$

63. $\{10\}$ **65.** $r = \dfrac{d}{t}$ **67.** $r = \dfrac{C}{2\pi}$ **69.** $R = \dfrac{E}{I}$

71. $h = \dfrac{2A}{a + b}$ **73.** $u = \dfrac{fv}{v - f}$ **75.** $m = \dfrac{y - b}{x}$

B Exercises: Applying the Concepts: **77.** 13 ft   **79.** 57 cm   **81.** 2 m
83. 19 ft   **85.** 14 yr   **87.** $2,010   **89.** 1500 cameras
91. $\alpha = 0.95$ **93.** $1,517.24   **95.** After 24 minutes

C Exercises: Beyond the Basics: **97.** $\{-6\}$ **99.** $\varnothing$ **101.** $\{31\}$

103. $\left\{\dfrac{3}{2}\right\}$ **105.** $\left\{\dfrac{5}{2}\right\}$ **107. a.** No, because the solution sets are

$\{0, 1\}$ and $\{1\}$, respectively.   **b.** No, because the solution sets are
$\{-3, 3\}$ and $\{3\}$, respectively.   **c.** No, because the solution sets are
$\{0, 1\}$ and $\{0\}$, respectively.   **d.** No, because the solution sets are
$\{\varnothing\}$ and $\{2\}$, respectively.   **109.** $k = 5$ **111.** $k = -2$

113. $b - a, b \ne -a$ **115.** $\dfrac{a + b}{a - b}, a \ne b$ **117.** $\dfrac{1}{b - a}, a \ne b$

Critical Thinking: **119.** (c)   **120.** (d)

Section 1.2

Practice Problems: **1.** Width: 3 m; length: 11 m.   **2.** $3750 in
bonds and $11,250 in stocks   **3.** $1500   **4.** 395 m   **5.** 2.5 hr
6. 18 min   **7.** 12.5 gal

A Exercises: Basic Skills and Concepts: **1.** $2a + 2b$ **3.** rt

5. false   **7.** $0.065x$ **9.** $22,000 - x$ **11. a.** Natasha: $\dfrac{1}{4}t$

b. Natasha's brother: $\dfrac{1}{5}t$ **13. a.** $10 - x$ **b.** $4.60x$ **c.** $7.40(10 - x)$

15. a. $8 + x$ **b.** $0.8 + x$

B Exercises: Applying the Concepts: **17.** 7 and 21   **19.** Length: 25 ft;
width: 15 ft   **21.** $62,000 and $85,000   **23.** July: 530 tickets;
August: 583 tickets   **25.** Younger son: $45,000; older son: $180,000
27. a. 81   **b.** 78   **29.** $4000 in a tax shelter and $3000 in a bank
31. $2500   **33.** 35,000 shaving sets   **35.** 2.5 hr   **37.** 30 mph
39. 100 mph and 107 mph   **41.** 2 hr and 24 min
43. 3 hr and 36 min   **45.** 12 hr   **47.** 1.5 hr   **49.** 30 g
51. 450 g of 60-40 solder and 150 g of 40-60 solder   **53.** 2.25 liters
55. 231 nickels, 77 dimes, 308 quarters   **57.** 71 yr   **59.** 4 m

C Exercises: Beyond the Basics: **61.** 50 mph   **63.** 25 liters   **65.** 72 yr

67. 26 min   **69. a.** 112 mi   **b.** $\dfrac{448}{3} = 149.33$ mph   **71.** 2.7 sec

Critical Thinking: **73.** 1.5 days   **74.** $100

Section 1.3

Practice Problems: **1. a.** Real $= -1$, imaginary $= 2$ **b.** Real $= -\dfrac{1}{3}$,

imaginary $= -6$ **c.** Real $= 8$, imaginary $= 0$ **2.** $a = -2; b = 1$
3. a. $4 - 2i$ **b.** $-1 + 4i$ **c.** $-2 + 5i$ **4. a.** $26 + 2i$
b. $-15 - 21i$ **5. a.** $5 - 12i$ **b.** $16 + 14i\sqrt{2}$ **6. a.** 37   **b.** 4

7. a. $1 + i$ **b.** $-\dfrac{15}{41} - \dfrac{12}{41}i$ **8.** $\dfrac{23}{10} + \dfrac{11}{10}i$

A Exercises: Basic Skills and Concepts: **1.** $\sqrt{-1}; -1$ **3.** $i\sqrt{b}$
5. true   **7.** $x = 3; y = 2$ **9.** $x = 2; y = -4$ **11.** $8 + 3i$
13. $-1 - 6i$ **15.** $-5 - 5i$ **17.** $15 + 6i$ **19.** $-8 + 12i$
21. $-3 + 15i$ **23.** $20 + 8i$ **25.** $3 + 11i$ **27.** 13
29. $24 + 7i$ **31.** $-141 - 24\sqrt{3}i$ **33.** $26 - 2i$

35. $\bar{z} = 2 + 3i, z\bar{z} = 13$ **37.** $\bar{z} = \dfrac{1}{2} + 2i, z\bar{z} = \dfrac{17}{4}$

39. $\bar{z} = \sqrt{2} + 3i, z\bar{z} = 11$ **41.** $5i$ **43.** $-\dfrac{1}{2} + \dfrac{1}{2}i$ **45.** $1 + 2i$

47. $\dfrac{5}{2} + \dfrac{1}{2}i$ **49.** $\dfrac{43}{65} - \dfrac{6}{65}i$ **51.** 2 **53.** $-\dfrac{19}{13} + \dfrac{4}{13}i$

B Exercises: Applying the Concepts: 55. $9 + i$

57. $Z = \dfrac{595}{74} + \dfrac{315}{74}i$ **59.** $V = 45 + 11i$ **61.** $I = \dfrac{17}{15} + \dfrac{4}{15}i$

C Exercises: Beyond the Basics: 63. i **65.** i **67.** 6 **69.** $5i$
71. $-4i$ **79.** $x = 1; y = 4$ **81.** $x = 1; y = 0$ **83.** $-0.6 + 0.8i$

Critical Thinking: 86. a. True **b.** False **c.** False **d.** True
e. True **f.** True

Section 1.4

Practice Problems: 1. $\{-21, -4\}$ **2.** $\left\{0, \dfrac{5}{2}\right\}$ **3.** $\{3\}$

4. $\{-2 - \sqrt{5}, -2 + \sqrt{5}\}$ **5.** $\{3 - \sqrt{2}, 3 + \sqrt{2}\}$

6. $\left\{3 - \dfrac{\sqrt{11}}{2}, 3 + \dfrac{\sqrt{11}}{2}\right\}$ **7.** $\left\{-\dfrac{1}{2}, \dfrac{2}{3}\right\}$ **8. a.** $\left\{-\dfrac{3}{2}i, \dfrac{3}{2}i\right\}$

b. $\{2 - 3i, 2 + 3i\}$ **9. a.** one real **b.** two unequal real
c. two unequal nonreal complex **10.** 25.48 ft by 127.41 ft
11. $18 + 18\sqrt{5} \approx 58.25$ ft

A Exercises: Basic Skills and Concepts: 1. quadratic

3. $\dfrac{-b \pm \sqrt{b^2 - 4ac}}{2a}$ **5.** true **7.** Yes **9.** Yes **11.** Yes
13. No **15.** $k = 2$ **17.** $\{0, 5\}$

19. $\{-7, 2\}$ **21.** $\{-1, 6\}$

23. $\left\{-3, \dfrac{1}{2}\right\}$ **25.** $\left\{-1, -\dfrac{2}{3}\right\}$ **27.** $\left\{-2, -\dfrac{2}{5}\right\}$ **29.** $\left\{-3, \dfrac{5}{2}\right\}$

31. $\left\{-\dfrac{2}{3}, \dfrac{3}{2}\right\}$ **33.** $\left\{\dfrac{1}{6}, \dfrac{7}{3}\right\}$ **35.** $\left\{-\dfrac{25}{2}, 15\right\}$ **37.** $\{-4, 4\}$

39. $\{-2, 2\}$ **41.** $\{-2i, 2i\}$ **43.** $\{-3, 5\}$ **45.** $\dfrac{2}{3} - \dfrac{4}{3}i, \dfrac{2}{3} + \dfrac{4}{3}i$

47. 4 **49.** 9 **51.** $\dfrac{49}{4}$ **53.** $\dfrac{1}{36}$ **55.** $\dfrac{a^2}{4}$

57. $\{-1 - \sqrt{6}, -1 + \sqrt{6}\}$ **59.** $\left\{\dfrac{3 - \sqrt{13}}{2}, \dfrac{3 + \sqrt{13}}{2}\right\}$

61. $\left\{-3, \dfrac{3}{2}\right\}$ **63.** $\{1 - i, 1 + i\}$ **65.** $\{5 - i\sqrt{3}, 5 + i\sqrt{3}\}$

67. $\{6 - \sqrt{33}, 6 + \sqrt{33}\}$ **69.** $\left\{-\dfrac{7 + \sqrt{13}}{6}, -\dfrac{7 - \sqrt{13}}{6}\right\}$

71. $\{-1 - \sqrt{5}, -1 + \sqrt{5}\}$ **73.** $\left\{-\dfrac{1}{2}, \dfrac{5}{3}\right\}$

75. $\left\{\dfrac{1 - \sqrt{22}}{3}, \dfrac{1 + \sqrt{22}}{3}\right\}$ **77.** $\left\{-2, -\dfrac{2}{3}\right\}$

79. $\left\{\dfrac{5 - \sqrt{5}}{2}, \dfrac{5 + \sqrt{5}}{2}\right\}$ **81.** $\left\{\dfrac{1}{4} - \dfrac{\sqrt{7}}{4}i, \dfrac{1}{4} + \dfrac{\sqrt{7}}{4}i\right\}$

83. $\left\{-\dfrac{\sqrt{10}}{2}, \dfrac{\sqrt{10}}{2}\right\}$ **85.** $\left\{-\dfrac{5}{3}i, \dfrac{5}{3}i\right\}$ **87.** $D = 0$, one real root
89. $D = 49$, two real unequal roots **91.** $D = 1$, two real unequal roots
93. $D = 252$, two real unequal roots **95.** 23.82 in. **97.** 5.23 cm

B Exercises: Applying the Concepts: 99. 60 ft by 180 ft
101. 7 and 21 **103.** 20 cm by 25 cm **105.** 4 in. and 12 in.
107. 2 in. **109.** 19.8 in. by 19.8 in. **111.** After 8 hours
113. 14 ft by 16 ft **115.** 15.8 sec **117. a.** 1.59 sec and 4.41 sec
b. 9.25 sec **119. a.** 0.45 sec and 15.55 sec **b.** 17.70 sec

C Exercises: Beyond the Basics: 121. $-2\sqrt{3}$ or $2\sqrt{3}$

123. 0 or 8 **125.** $-\dfrac{1}{2}$

127. $ar^2 + br + c = 0 = as^2 + bs + c \Leftrightarrow ar^2 + br = as^2 + bs \Leftrightarrow a(r^2 - s^2) = -b(r - s) \Leftrightarrow r + s = -\dfrac{b}{a}$
Also, $(ar^2 + br + c = 0) \cdot \dfrac{1}{a} \Leftrightarrow r^2 + \dfrac{b}{a}r + \dfrac{c}{a} = 0$

$r^2 - (r + s)r + \dfrac{c}{a} = 0 \Leftrightarrow rs = \dfrac{c}{a}$ **129.** 2

131. $ax^2 + bx + c = ax^2 - a(r + s)x + ars$
$= a(x^2 - rx - sx + rs)$
$= a[x(x - r) - s(x - r)] = a(x - r)(x - s)$
133. a. $x^2 - x - 12 = 0$ **b.** $x^2 - 10x + 25 = 0$
c. $x^2 - 6x + 7 = 0$ **139.** $\{1, 9\}$

Section 1.5

Practice Problems: 1. $\{-2, -1, 1\}$ **2.** $\{-2, 0, 2\}$ **3.** $\{-2, 2, 5\}$
4. $\{-10, 1\}$ **5.** $\varnothing$ **6.** $\{0, 3\}$ **7.** $\{10\}$ **8.** $\{9\}$
9. a. $\{3\}$ **b.** $\left\{-\dfrac{9}{2}, \dfrac{7}{2}\right\}$ **10.** $\{1, 216\}$ **11.** $\left\{\dfrac{1}{3}, 1\right\}$
12. 60% of the speed of light

A Exercises: Basic Skills and Concepts: 1. extraneous
3. 16 **5.** false **7.** $\{0, 2\}$ **9.** $\{0, 1\}$ **11.** $\{0\}$ **13.** $\{-1, 0, 1\}$
15. $\{0, 3\}$ **17.** $\left\{\dfrac{1}{4}, 2\right\}$ **19.** $\{1 - \sqrt{2}, 1 + \sqrt{2}\}$

21. $\left\{\dfrac{7}{2} - \dfrac{\sqrt{53}}{2}, \dfrac{7}{2} + \dfrac{\sqrt{53}}{2}\right\}$ **23.** $\{-3, 5\}$ **25.** $\left\{\dfrac{1}{26}, 5\right\}$ **27.** $\varnothing$
29. $\varnothing$ **31.** $\{-1\}$ **33.** $\{-2, 7\}$ **35.** $\{3\}$ **37.** $\varnothing$
39. $\{-2\}$ **41.** $\{3\}$ **43.** $\{6\}$ **45.** $\{-2, 1\}$ **47.** $\{3, 7\}$
49. $\{0, 8\}$ **51.** $\{9\}$ **53.** $\{-2\}$ **55.** $\{3, 7\}$ **57.** $\{13\}$

59. $\left\{-5, \dfrac{29}{5}\right\}$ **61.** $\{4, 9\}$ **63.** $\left\{\dfrac{1}{4}, 49\right\}$ **65.** $\left\{-\dfrac{1}{6}, \dfrac{1}{7}\right\}$

67. $\{8, 64\}$ **69.** $\left\{\dfrac{1}{16}\right\}$ **71.** $\{-3, -2, 2, 3\}$

73. $\left\{-\dfrac{\sqrt{2}}{2}, \dfrac{\sqrt{2}}{2}, -i, i\right\}$ **75.** $\{-2, 2\}$ **77.** $\left\{0, \dfrac{1}{3}\right\}$ **79.** $\{0\}$

81. $\{-2\sqrt{2}, -\sqrt{3}, \sqrt{3}, 2\sqrt{2}\}$ **83.** $\{-1, 1, 2, 4\}$

B Exercises: Applying the Concepts: 85. $\dfrac{6}{8}$ **87. a.** 20 shares
b. \$90 **89.** 230 ft **91.** 2 mph **93.** 40 min
95. 120,000 ft^2 **97.** $CE = 6$ mi or $CE = \dfrac{3162}{205} \approx 15.4$ mi

C Exercises: Beyond the Basics: 99. $\{-1, 2 - \sqrt{7}, 3, 2 + \sqrt{7}\}$
101. $\{2\}$ **103.** $\{-7, -2\}$ **105.** $\left\{-3, -\dfrac{5}{3}\right\}$ **107.** $\left\{\dfrac{1}{8}, 8\right\}$

109. $\{\pm 2\sqrt{2}, \pm 3\sqrt{3}\}$ **111.** $\{36\}$ **113.** $\{-5b, 2b\}$ **115.** $\left\{\dfrac{2a}{3}, a\right\}$

Critical Thinking: 117. $x = \dfrac{1 + \sqrt{5}}{2}$ **118.** $x = 5$

119. $x = \dfrac{1}{2} + \dfrac{\sqrt{4n + 1}}{2}$ **120.** $x = 1 + \sqrt{2}$ **121.** $\{5\}$

Section 1.6

Practice Problems: 1. a. $(2, \infty)$

b. $(-\infty, 5]$

2. 2.5 hours **3. a.** $(-3, \infty)$ **b.** $\varnothing$ **4.** $(-\infty, 2] \cup [4, \infty)$
5. $(-1, 3]$

6. $\left[-1, \dfrac{3}{2}\right)$

7. $(4, \infty)$

4

8. $a = -4, b = 11$ **9.** 59°F to 77°F

A Exercises: Basic Skills and Concepts: **1.** <   **3.** ≥   **5.** <
7. ≥   **9.** <

11. $(-2, 5)$
−2   5

13. $(0. 4]$
0   4

15. $[-1, \infty)$
−1

17. $(-\infty, -2]$
−2

19. $(-\infty, 3)$
3

21. $[-3, \infty)$
−3

23. $(-\infty, 2)$
2

25. $(-\infty, -4)$
−4

27. $(-\infty, -1)$
−1

29. $(-\infty, 3]$
3

31. $(2, \infty)$
2

33. ∅

35. $\left(-\infty, \dfrac{3}{2}\right]$
$\dfrac{3}{2}$

37. $(-\infty, \infty)$
39. $(-\infty, 3]$
3

41. $[2, \infty)$ **43.** $[-18, \infty)$ **45.** ∅   **47.** $[15, \infty)$

49. $(-\infty, \infty)$ **51.** $(-\infty, -2) \cup (2, \infty)$ **53.** $\left(-\infty, \dfrac{11}{2}\right]$

55. $(-\infty, -2] \cup (6, \infty)$ **57.** $(-\infty, \infty)$ **59.** $[-2, 5]$
61. $[-2, 5)$ **63.** ∅   **65.** $(-\infty, 2]$ **67.** $(-2, -1)$ **69.** $[-2, 4)$

71. $[-6, 1]$ **73.** $(-3, 3]$ **75.** $\left(-1, \dfrac{3}{2}\right]$ **77.** ∅   **79.** $(-\infty, -2)$

81. $\left(-\infty, \dfrac{1}{2}\right)$ **83.** $a = 5, b = 8$ **85.** $a = 1, b = 3$

87. $a = -1, b = 19$

B Exercises: Applying the Concepts: **89.** More than 6000 posters
91. \$2013 to \$2100   **93.** 7.5 gal to 12 gal   **95.** At least 15 qt
97. More than 3500 pedometers   **99.** 20°C to 30°C
101. 0 to 48 mi

C Exercises: Beyond the Basics: **103.** $k = -2$ **105.** $k = -\dfrac{b}{a}$

107. $\left(\dfrac{1}{2}, \infty\right)$ **109.** $[-3, \infty)$ **111.** $\left[-\dfrac{a+c}{b}, \dfrac{a-c}{b}\right)$

Critical Thinking: **113. a.** > or ≥   **b.** < or ≤   **c.** not possible
114. a. < or >   **b.** ≤ or ≥   **c.** not possible   **115. a.** not possible
b. not possible   **c.** <, ≤, >, or ≥

Section 1.7

Practice Problems: **1. a.** $\{2\}$ **b.** $\{-1, 2\}$ **2.** $\left\{\dfrac{1}{2}\right\}$ **3.** $\left\{\dfrac{6}{5}, 2\right\}$

4. $[-3, 1]$
−3   1

5. $[270, 420]$

6. $\left(-\infty, -\dfrac{9}{2}\right] \cup \left[\dfrac{3}{2}, \infty\right)$

$-\dfrac{9}{2}$ $\dfrac{3}{2}$

7. a. $(-\infty, \infty)$ **b.** ∅

A Exercises: Basic Skills and Concepts: **1.** $\{-a, a\}$
3. $(-\infty, -a] \cup [a, \infty)$ **5.** true   **7.** $\{-3, 3\}$ **9.** $\{-3, 3\}$

11. $\{-5, -1\}$ **13.** $\{-1, 7\}$ **15.** $\left\{-\dfrac{7}{6}, \dfrac{11}{6}\right\}$ **17.** $\{-2, -1\}$

19. $\{-6, 6\}$ **21.** $\{-20, 4\}$ **23.** $\{-1, 2\}$ **25.** ∅   **27.** ∅

29. $\{-2, 2\}$ **31.** $\{-1, 2\}$ **33.** $\left\{-\dfrac{1}{3}, 1\right\}$ **35.** $\{-4\}$

37. $\left\{-3, -\dfrac{5}{9}\right\}$ **39.** $\{0, 2\}$ **41.** $\left\{\dfrac{5}{4}, \dfrac{3}{2}\right\}$ **43.** $\{-4, 0\}$
45. $(-4, 4)$ **47.** $(-\infty, -4) \cup (4, \infty)$ **49.** $(-4, 2)$

51. $(-\infty, \infty)$ **53.** $\left(-\dfrac{1}{2}, \dfrac{7}{2}\right)$ **55.** $(-\infty, 1) \cup (4, \infty)$

57. $\left[-\dfrac{23}{3}, 5\right]$ **59.** ∅

61. $(-6, 10)$
−6   ⋯   10

63. $[0, 16]$
0   ⋯   16

65. $(-\infty, \infty)$

67. $\left(-\dfrac{8}{9}, \dfrac{22}{9}\right)$
$-\dfrac{8}{9}$ $\dfrac{22}{9}$

69. $\left[-\dfrac{10}{3}, 10\right]$
$-\dfrac{10}{3}$ ⋯   10

B Exercises: Applying the Concepts: **71.** The temperatures in Tampa
during December are between 55°F and 95°F.   **73.** $|x - 700| \le 50$
75. $|x - 120| \le 6.75$ **77.** Between \$5040 and \$6480
79. Between \$18.90 and \$19.50

C Exercises: Beyond the Basics:
81. $\{-2 - 2\sqrt{3}, -2 - \sqrt{2}, -2 + \sqrt{2}, -2 + 2\sqrt{3}\}$

83. $\left(-\infty, \dfrac{1}{4}\right) \cup \left(\dfrac{3}{4}, \infty\right)$ **85.** $\{-6, 6\}$ **91.** $|x - 4| < 3$

93. $|x - 4| \le 6$ **95.** $|x - 7| > 4$ **97.** $|2x - 5| \ge 15$
99. $|2x - a - b| < b - a$ **101.** $|2x - a - b| > b - a$

103. $|x - 39| \le 31$ **105.** $\left[-5, \dfrac{5}{3}\right]$ **107.** $\left(-\infty, -\dfrac{7}{2}\right] \cup \left[-\dfrac{1}{6}, \infty\right)$

109. $\left[-\dfrac{1}{2}, 3\right) \cup (3, \infty)$

Critical Thinking: **110.** $[3, \infty)$ **111.** $(-\infty, 2] \cup [4, \infty)$
112. $\{-2, 1, 5, 8\}$

Review Exercises

1. $\{3\}$ **3.** $\{2\}$ **5.** $(-\infty, \infty)$ **7.** $\{1\}$ **9.** $\{1\}$ **11.** $\{11\}$

13. $\left\{-4, -\dfrac{1}{3}\right\}$ **15.** $\left\{-\dfrac{3}{2}\right\}$ **17.** $\left\{\dfrac{1}{5}, 3\right\}$ **19.** $g = \dfrac{p - k}{t}$

21. $B = \dfrac{T}{T - 2}$ **23.** $\{0, 7\}$ **25.** $\{-2, 5\}$ **27.** $\{-6, 1\}$

29. $\{-5, 1\}$ **31.** $\left\{-1, \dfrac{2}{3}\right\}$ **33.** $\left\{\dfrac{3}{2} - \dfrac{\sqrt{13}}{2}, \dfrac{3}{2} + \dfrac{\sqrt{13}}{2}\right\}$

35. $\left\{-1, \dfrac{1}{2}\right\}$ **37.** $\{2 - 2\sqrt{3}, 2 + 2\sqrt{3}\}$

39. $\left\{\dfrac{1}{4} - \dfrac{\sqrt{17}}{4}, \dfrac{1}{4} + \dfrac{\sqrt{17}}{4}\right\}$ **41.** $\left\{\dfrac{1}{2} - \dfrac{\sqrt{3}}{2}i, \dfrac{1}{2} + \dfrac{\sqrt{3}}{2}i\right\}$

43. $\{3 - 2i, 3 + 2i\}$ **45.** $\left\{1 - \dfrac{3}{2}i, 1 + \dfrac{3}{2}i\right\}$

47. $D = 49$, two real, unequal   **49.** $D = -16$, two unequal

complex roots   **51.** $\{-4, 4\}$ **53.** $\left\{\dfrac{1}{2}\right\}$ **55.** $\{9\}$ **57.** $\{16\}$

59. $\left\{-\dfrac{10}{7}, -\dfrac{2}{7}\right\}$ **61.** $\{-64, 1\}$ **63.** $\{0\}$ **65.** $\left\{-3, -\dfrac{1}{2}, \dfrac{1}{2}, 3\right\}$

67. $\{0\}$ **69.** $\left\{-\dfrac{3}{10}, 6\right\}$ **71.** $\{-3y, y\}$ **73.** $\{y - 1, y - 2\}$

75. $(-\infty, -2)$ **77.** $\left(-\infty, \dfrac{17}{3}\right]$ **79.** $\left[-\dfrac{6}{7}, \infty\right)$ **81.** $(-\infty, -3)$

83. $\left(\dfrac{5}{2}, \infty\right)$ **85.** $\left(\dfrac{7}{6}, \infty\right)$ **87.** $(-\infty, 21]$ **89.** $\left(-\infty, \dfrac{41}{10}\right)$

91. $(-\infty, 1) \cup (3, \infty)$ **93.** $(2, 3)$ **95.** $[-2, 3)$

97. $[-47, 3)$ **99.** $\left[-3, \dfrac{5}{3}\right]$ **101.** $(-\infty, 1) \cup (3, \infty)$

103. $(-\infty, -1] \cup [9, \infty)$ **105.** $\dfrac{11}{\pi}$ cm   **107.** 3 m   **109.** 13 cm

111. 350 cm   **113.** 35°, 35°, 110°   **115.** \$12,000 at 6%; \$18,000 at 8%

117. 8 liters of the $4\dfrac{1}{2}$% and 2 liters of the 12%   **119.** 8 people

121. 220   **123.** 36   **125.** 40

Practice Test A

1. $\{2\}$ **2.** $\{1\}$ **3.** $\left\{-\dfrac{2\sqrt{30}}{5}, \dfrac{2\sqrt{30}}{5}\right\}$ **4.** 6 cm by 9 cm

5. $\{-9, -4\}$ **6.** \$8200   **7.** 1 in.   **8.** $\left\{-4, -\dfrac{5}{2}\right\}$

9. $\dfrac{1}{9}, \left(x + \dfrac{1}{3}\right)^2$ **10.** $\left\{\dfrac{5}{6} - \dfrac{\sqrt{37}}{6}, \dfrac{5}{6} + \dfrac{\sqrt{37}}{6}\right\}$ **11.** 9 in.

12. $\{-6 - \sqrt{3}, -6 + \sqrt{3}\}$ **13.** $\{-5, 0, 5\}$ **14.** $\{4\}$

15. $\{-12, -6\}$ **16.** $[90, \infty)$ **17.** $\left(-\dfrac{1}{2}, \dfrac{7}{2}\right)$

18. $(-\infty, -2) \cup (5, \infty)$ **19.** $(-\infty, 2]$ **20.** $\left[\dfrac{2}{5}, 2\right]$

Practice Test B

1. d   **2.** d   **3.** a   **4.** c   **5.** b   **6.** a   **7.** d   **8.** d   **9.** b
10. a   **11.** b   **12.** d   **13.** a   **14.** a   **15.** d   **16.** d   **17.** b
18. c   **19.** a   **20.** a

CHAPTER 2

Section 2.1

Practice Problems: **1.**

2. $(2000, 20.2), (2001, 19.7), (2002, 19.4), (2003, 18.6), (2004, 17.9),$ $(2005, 18.1), (2006, 17.8), (2007, 17.8)$

3. $\sqrt{2}$ **4.** Yes

5. $60\sqrt{2}$

6. $\left(\dfrac{11}{2}, -\dfrac{3}{2}\right)$

A Exercises: Basic Skills and Concepts: **1.** second
3. $\sqrt{(x_2 - x_1)^2 + (y_2 - y_1)^2}$ **5.** true
7.

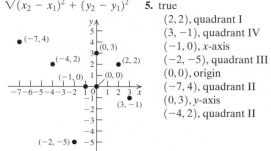

(2, 2), quadrant I
(3, -1), quadrant IV
(-1, 0), x-axis
(-2, -5), quadrant III
(0, 0), origin
(-7, 4), quadrant II
(0, 3), y-axis
(-4, 2), quadrant II

9. a. On the y-axis   **b.** Vertical line intersecting the x-axis at -1

11. a. $y > 0$ **b.** $y < 0$ **c.** $x < 0$
d. $x > 0$ **13. a.** 4   **b.** (2, 3)
15. a. $\sqrt{13}$ **b.** (0.5, -4)
17. a. $4\sqrt{5}$ **b.** (1, -2.5)

19. a. 1   **b.** $(\sqrt{2}, 4.5)$ **21. a.** $\sqrt{2}|t - k|$ **b.** $\left(\dfrac{t + k}{2}, \dfrac{t + k}{2}\right)$

23. Yes   **25.** Yes   **27.** Yes   **29.** No   **31.** Isosceles
33. Scalene   **35.** Scalene   **37.** Scalene   **39.** Equilateral

41. $d(P, R) = d(Q, S) = 17\sqrt{2}$ **43.** -2 or 6   **45.** $\left(-\dfrac{8}{7}, 0\right)$

B Exercises: Applying the Concepts:
47.

49.

51.

53. 16,835   **55.** $398.75 billion in 2001, $447.5 billion in 2002, $496.25 billion in 2003, $545 billion in 2004, $593.75 billion in 2005, $642.5 billion in 2006, and $691.25 billion in 2007

57. a.

b. 2500 miles
c. $500\sqrt{13} \approx 1802.78$ miles

C Exercises: Beyond the Basics: **67. c.** $\left(\dfrac{2}{3}, \dfrac{5}{3}\right)$ and $\left(\dfrac{7}{3}, \dfrac{4}{3}\right)$

Critical Thinking: **69. a.** y-axis **b.** x-axis
71. a. Quadrants I and III **b.** Quadrants II and IV
73. Let the point be (x, y).
$x > 0, y > 0 \rightarrow$ quadrant I
$x < 0, y > 0 \rightarrow$ quadrant II
$x < 0, y < 0 \rightarrow$ quadrant III
$x > 0, y < 0 \rightarrow$ quadrant IV

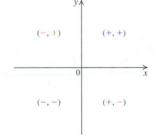

Section 2.2

Practice Problems: **1.**

2. x-intercepts: $-2, \dfrac{1}{2}$; y-intercept: -2 **3.** Symmetric

4. Symmetric   **5.** Not symmetric with respect to the x-axis, symmetric with respect to the y-axis, not symmetric with respect to the origin

6. a.

b.

c. After 9 years   **7.** $(x - 3)^2 + (y + 6)^2 = 100$

8.

9. $(x + 2)^2 + (y - 3)^2 = 25$: $C = (-2, 3), r = 5$

A Exercises: Basic Skills and Concepts: **1.** that satisfy the equation
3. y-   **5.** false   **7.** On the graph: $(-3, -4), (1, 0), (4, 3)$ Not on the graph: $(2, 3)$ **9.** On the graph: $(3, 2), (0, 1), (8, 3)$ Not on the graph: $(8, -3)$ **11.** On the graph: $(1, 0), (2, \sqrt{3}), (2, -\sqrt{3})$
Not on the graph: $(0, -1)$
13. x-intercepts: $-3, 0, 3$; y-intercepts: $-2, 0, 2$
15.

17.

19.

21.

23.

25.

27.

29.

31.

33.

35.

37. x-intercept: 4; y-intercept: 3   **39.** x-intercept: $\frac{5}{2}$; y-intercept: $\frac{5}{3}$

41. x-intercepts: 2, 4; y-intercept: 8   **43.** x-intercepts: -2, 2; y-intercepts: -2, 2   **45.** x-intercepts: -1, 1; no y-intercept

47. Not symmetric with respect to the x-axis, symmetric with respect to the y-axis, not symmetric with respect to the origin

49. Not symmetric with respect to the x-axis, not symmetric with respect to the y-axis, symmetric with respect to the origin

51. Not symmetric with respect to the x-axis, symmetric with respect to the y-axis, not symmetric with respect to the origin

53. Not symmetric with respect to the x-axis, not symmetric with respect to the y-axis, symmetric with respect to the origin

55. Not symmetric with respect to the x-axis, not symmetric with respect to the y-axis, symmetric with respect to the origin

57. Center: $(2, 3)$; radius: 6   **59.** Center: $(-2, -3)$; radius: $\sqrt{11}$

61. $x^2 + (y - 1)^2 = 4$ **63.** $(x + 1)^2 + (y - 2)^2 = 2$

65. $(x - 3)^2 + (y + 4)^2 = 97$ **67.** $(x - 1)^2 + (y - 2)^2 = 4$

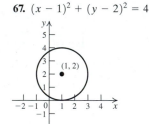

69. $(x - 2)^2 + (y - 5)^2 = 26$

71. a. Circle; center: $(1, 1)$; radius: $\sqrt{6}$
b. $(1 - \sqrt{5}, 0), (1 + \sqrt{5}, 0), (0, 1 - \sqrt{5}), (0, 1 + \sqrt{5})$
73. a. Circle; center: $(0, -1)$; radius: 1   **b.** $(0, 0), (0, -2)$

75. a. Circle; center: $\left(\frac{1}{2}, 0\right)$; radius: $\frac{1}{2}$ **b.** $(1, 0), (0, 0)$

B Exercises: Applying the Concepts: **77.** $|x| = |y|$

79. $x = \dfrac{y^2}{4} + 1$ **81. a.** \$12 million   **b.** $-$\$5.5 million

c.

d. -8 and 2; they represent the months when there is neither profit nor loss.   **e.** 8; it represents the profit in July 2004.

83. a. After 0 second: 320 ft; after 1 second: 432 ft; after 2 seconds: 512 ft; after 3 seconds: 560 ft; after 4 seconds: 576 ft; after 5 seconds: 560 ft; after 6 seconds: 512 ft

b.

c. $0 \le t \le 10$
d. t-intercept: 10; it represents the time when the object hits the ground. y-intercept: 320; It shows the height of the building.

C Exercises: Beyond the Basics:

85.

87.

origin symmetry

89.

91.

93.

95. Area: 11π

Critical Thinking: 97. The graph is the union of the graphs of $y = \sqrt{2x}$ and $y = -\sqrt{2x}$.

98. The converse is not true. (See the graph of $y = x$.)
99. False because it gives the y-intercepts
100. For example, $y = -(x + 2)(x - 3)$.
101. a.

b.

c.

d.

Section 2.3

Practice Problems: 1. $-\dfrac{8}{13}$ **2.** $y = -\dfrac{2}{3}x - \dfrac{13}{3}$

3. $y = 5x + 11$ **4.** $y = 2x - 3$
5.

6.

7.

8. Between 176.8 and 179.4 centimeters
9. a. The y-intercept represents the initial value of the plane, \$210,000.   **b.** The point $(10, 110)$ gives the value of the plane after ten years as \$110,000.   **c.** The value of the plane decreased \$10,000 per year.   **d.** $y = -10x + 210$
e. The slope gives the plane's yearly depreciation, \$10,000.

10. a. $4x - 3y + 23 = 0$ **b.** $5x - 4y - 31 = 0$
11. 7.38 million registered motorcycles

A Exercises: Basic Skills and Concepts: 1. slope   **3.** parallel

5. 5   **7.** false   **9.** $\dfrac{4}{3}$, rising   **11.** -1, falling

13. 1, rising   **15.** 4, rising   **17. a.** ℓ_3 **b.** ℓ_2 **c.** ℓ_4 **d.** ℓ_1

19. a. $y = 0$ **b.** $x = 0$

21. $y = \dfrac{1}{2}x + 4$ **23.** $y = -\dfrac{3}{2}x + 4$

25. $y = -4$

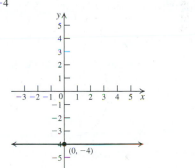

27. $y = -x + 1$ **29.** $y = 3$ **31.** $y = \dfrac{2}{3}x + \dfrac{1}{3}$

33. $y = -\dfrac{7}{2}x + 2$ **35.** $x = 5$ **37.** $y = 0$ **39.** $y = 14$

41. $y = -\dfrac{2}{3}x - 4$ **43.** $y = \dfrac{4}{3}x + 4$ **45.** $y = 7$ **47.** $y = -5$

49. a. Parallel   **b.** Neither   **c.** Perpendicular

51. $m = -\dfrac{1}{2}$; x-intercept $= 4$,   **53.** $m = \dfrac{3}{2}$; y-intercept $= 3$

y-intercept $= 2$ x-intercept $= -2$

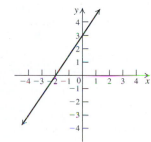

55. m is undefined;
no y-intercept,
x-intercept = 5

57. m is undefined;
y-intercepts = y-axis
x-intercept = 0

61. $\frac{x}{3} + \frac{y}{2} = 1$; x-intercept = 3; y-intercept = 2

63. $y = -x - 2$ **65.** $y = x + 2$; because the coordinates of the point $(-1, 1)$ satisfy this equation, $(-1, 1)$ also lies on this line. **67.** $y = -\frac{1}{2}x + \frac{37}{4}$ **69.** Parallel **71.** Neither

73. Perpendicular **75.** Neither **77.** $y = -x + 2$

79. $y = -3x - 2$ **81.** $y = -\frac{1}{6}x + 4$

B Exercises: Applying the Concepts: 83. $\frac{1}{10}$ **85. a.** x: time in weeks;

y: amount of money in the account; $y = 7x + 130$ **b.** Slope: weekly deposit in the account; y-intercept: initial deposit
87. a. x: number of hours worked; y: amount of money earned per week;

$$y = \begin{cases} 11x & x \le 40 \\ 16.5x - 220 & x > 40 \end{cases}$$

b. Slope: hourly wage; y-intercept: salary for 0 hours of work
89. a. x = amount of rupees; y = amount of dollars equal to

x rupees; $y = \frac{1}{40.5}x$, or $y = 0.02469x$ **b.** Slope: the number of

dollars per rupee; y-intercept: number of rupees for 0 dollars
91. a. x: number of TVs produced; y: cost of production for x TVs; $y = 150x + 10,000$ **b.** Slope: marginal cost of a TV; y-intercept: fixed cost independent of the number of TVs produced
93. a. The y-intercept represents the initial value of the machine, $9000. **b.** The graph point $(10, 1)$ gives the value of the machine after ten years as $1000. **c.** The value of the machine decreased $800 per year. **d.** $y = -0.8x + 9$ **e.** The slope gives the machine's yearly depreciation, $800. **95. a.** $v = \$11,200$ when $t = 2$
b. $v = \$5600$ when $t = 6$ The tractor's value is $0 at $t = 10$.
97. $y = 1.5x + 1000$
99. a. $y = 266.8t + 2425$ **b.**

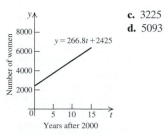

c. 3225
d. 5093

101. a. $q = -160p + 1120$ **b.** 320 **b.** 106 **103.** $y = 0.5x + 7$; 11
105. a. $y = 23.1x + 117.8$

b. Slope: cost of producing one modem; y-intercept: fixed overhead cost **c.** 395

107. 13.7% **109. a.** $y = -2x + 12.4$
b.

c. 0 papers

111. a. $y = 6.16x - 262.24$ **c.** ≈ 206 pounds

C Exercises: Beyond the Basics: **113.** $c = 12$

127. $(9.5, 12)$ or $(-5.5, -8)$

129. $y = \frac{5}{12}x + \frac{169}{12}$ **131.** All parallel lines with slope 2

133. Lines passing through the point $(1, 0)$ with different slopes

Critical Thinking: **135.** The slope of the line that passes through $(1, -1)$ and $(-2, 5)$ is -2. The slope of the line that passes through $(1, -1)$ and $(3, -5)$ is -2. The slope of the line that passes through $(-2, 5)$ and $(3, -5)$ is -2. **137. a.** The x-coordinate of the intersection point is negative. **b.** The x-coordinate of the intersection point is positive.

Section 2.4

Practice Problems: **1.** Domain: $\{-2, -1, 1, 2, 3\}$;
range: $\{-3, -2, 0, 1, 2\}$ **2. a.** No **b.** Yes **3. a.** 0 **b.** -7
c. $-2x^2 - 4hx + 5x - 2h^2 + 5h$ **4.** $(-\infty, 1)$ **5.** No
6. a. Yes **b.** $-3, -1$ **c.** -5 **d.** $-5, 1$ **7.** Domain: $(-3, \infty)$;
range: $(-2, 2] \cup \{3\}$ **8.** $f(50) = 60$; if the country has $50 billion of GDP, its unemployment rate will be 60%; $f(80) = 10$; to have an unemployment rate of 10%, the GDP must be $80 billion
9. Domain: $[1, \infty)$; range: $[6, 12)$; the function value at 6 is 11.813.
10. a. $C(x) = 1200x + 100,000$ **b.** $R(x) = 2500x$
c. $P(x) = 1300x - 100,000$ **d.** 77

A Exercises: Basic Skills and Concepts: **1.** independent
3. -14 **5.** $y = 0$ **7.** true **9.** Domain: $\{a, b, c\}$; range: $\{d, e\}$;
function **11.** Domain: $\{a, b, c\}$; range: $\{1, 2\}$; function
13. Domain: $\{0, 3, 8\}$; range: $\{-3, -2, -1, 1, 2\}$; not a function
15. Yes **17.** Yes **19.** No **21.** Yes **23.** Yes **25.** No
27. Yes **29.** $f(0) = 1$, $g(0)$ is not defined, $h(0) = \sqrt{2}$,
$f(a) = a^2 - 3a + 1$, $f(-x) = x^2 + 3x + 1$
31. $f(-1) = 5$, $g(-1)$ is not defined, $h(-1) = \sqrt{3}$, $h(c) = \sqrt{2 - c}$,
$h(-x) = \sqrt{2 + x}$ **33. a.** 0 **b.** $\frac{2\sqrt{3}}{3}$ **c.** Not defined

d. Not defined **e.** $\frac{-2x}{\sqrt{4 - x^2}}$ **35.** $(-\infty, \infty)$
37. $(-\infty, 9) \cup (9, \infty)$ **39.** $(-\infty, -1) \cup (-1, 1) \cup (1, \infty)$
41. $[3, \infty)$ **43.** $(-\infty, 4)$ **45.** $(-\infty, -2) \cup (-2, -1) \cup (-1, \infty)$
47. $(-\infty, 0) \cup (0, \infty)$ **49.** Yes **51.** Yes
53. No **55.** $f(-4) = -2$, $f(-1) = 1$, $f(3) = 5$, $f(5) = 7$
57. $h(-2) = -5$, $h(-1) = 4$, $h(0) = 3$, $h(1) = 4$
59. $x = -2$ or 3 **61. a.** No **b.** $x = -1 \pm \sqrt{3}$ **c.** 5
d. $x = -1 \pm \frac{1}{2}\sqrt{14}$ **63.** Domain: $[-3, 2]$; range: $[-3, 3]$

65. Domain: $[-4, \infty)$; range: $[-2, 3]$ **67.** Domain: $[-3, \infty)$;
range: $\{-3\} \cup [-1, 4]$ **69.** Domain: $(-\infty, -4] \cup [-2, 2] \cup [4, \infty)$;
range: $[-2, 2] \cup \{3\}$ **71.** $[-9, \infty)$ **73.** $-3, 4, 7, 9$
75. $f(-7) = 4$, $f(1) = 5$, $f(5) = 2$ **77.** $\{-3.75, -2.25, 3\} \cup [12, \infty)$
79. $[-9, \infty)$ **81.** $g(-4) = -1$, $g(1) = 3$, $g(3) = 4$ **83.** $[-9, -5)$

B Exercises: Applying the Concepts: **85.** Yes, because there is only one high temperature every day. **87.** No, because Nevada and North Carolina both start with N. **89.** Yes, because there is only one winner in each year. **91.** $A(x) = x^2$; $A(4) = 16$; $A(4)$ represents the area of a square tile with side of length 4. **93.** It is a function.

$S(x) = 6x^2$; $S(3) = 54$ **95. a.** $p(5) = 1150$; if 5000 TVs can be sold, the price per TV is \$1150. $p(15) = 900$; if 15,000 TVs can be sold, the price per TV is \$900. $p(30) = 525$; if 30,000 TVs can be sold, the price per TV is \$525.

b.

c. $p(25) = 650$; if the TVs are priced at \$650 each, 25,000 can be sold.

97. a. $C(x) = 5.5x + 75{,}000$ **b.** $R(x) = 9x$
c. $P(x) = 3.5x - 75{,}000$ **d.** 21,429 **e.** \$86,000
99. a. $[0, 8]$ **b.** $h(2) = 192, h(4) = 256, h(6) = 192$ **c.** 8 sec
d.

101. 1st day: 400 mg; 2nd day: 500 mg; 3rd day: 525 mg; 4th day: 531.25 mg; 5th day: 532.81 mg; 6th day: 533.2 mg; 7th day: 533.3 mg; 8th day: 533.33 mg; 9th day: 533.33 mg; 10th day: 533.33 mg

C Exercises: Beyond the Basics: **103. a.** No, because the domain of f is $[0, \infty)$, while the domain of g is $(-\infty, \infty)$ **b.** Yes, because $(\sqrt[3]{x})^3 = x$ for every real number x

105. $f(x) = \dfrac{x}{x-1}$; domain: $(-\infty, 1) \cup (1, \infty)$; $f(4) = \dfrac{4}{3}$

107. $f(x) = \dfrac{3}{x-2}$; domain: $(-\infty, 2) \cup (2, \infty)$; $f(4) = \dfrac{3}{2}$

109. $f(x) = \dfrac{\sqrt{x}}{x^2 + 2}$; domain: $[0, \infty)$; $f(4) = \dfrac{1}{9}$

111. No, because they have different domains
113. No, because f is not defined at -2 but g is. **115.** Yes, because $f(x) = \dfrac{x-2}{x^2 - 6x + 8} = \dfrac{x-2}{(x-4)(x-2)} = \dfrac{1}{x-4} = g(x)$ and $x = 2$ is outside the given domain **117.** $b = -4$ or $b = -2$
119. $f(x^2) = 2x^2 - 3$, $[f(x)]^2 = 4x^2 - 12x + 9$

Critical Thinking: **123. a.** $y = \sqrt{x-2}$ **b.** $y = \dfrac{1}{\sqrt{x-2}}$

c. $y = \sqrt{2-x}$ **d.** $y = \dfrac{1}{\sqrt{2-x}}$ **124. a.** $ax^2 + bx + c = 0$

b. $y = c$ **c.** $b^2 - 4ac < 0$ **d.** Not possible

Section 2.5

Practice Problems: **1.** Increasing on $(2, \infty)$, decreasing on $(-\infty, -3)$, and constant on $(-3, 2)$ **2.** Relative minimum at $(0, 0)$; relative maximum at $(1, 1)$ **3.** Mrs. Osborn's windpipe should be contracted to a radius of 7.33 mm for maximizing the airflow velocity. **4.** $f(-x) = -(-x)^2 = -x^2 = f(x)$; therefore, f is even.

5. $f(-x) = -(-x^3) = x^3 = -f(x)$; therefore, f is odd.
6. -6 **7.** -1 **8.** $-2x + 1 - h$

A Exercises: Basic Skills and Concepts: **1.** $>$ **3.** $f(-x) = f(x)$ for all x in the domain of f **5.** true **7.** Increasing on $(-\infty, \infty)$
9. Increasing on $(\infty, 2)$, decreasing on $(2, \infty)$ **11.** Increasing on $(-\infty, -2)$ and $(2, \infty)$, constant on $(-2, 2)$ **13.** Increasing on $(-\infty, -3)$ and $\left(-\dfrac{1}{2}, 2\right)$, decreasing on $\left(-3, -\dfrac{1}{2}\right)$ and $(2, \infty)$
15. No relative extrema **17.** $(2, 10)$ is a relative maximum point and a turning point. **19.** Any point $(x, 2)$ is a relative maximum and a relative minimum point on the interval $(-2, 2)$; none of these points are turning points. **21.** $(-3, 4)$ and $(2, 5)$ are relative maxima points and turning points; $\left(-\dfrac{1}{2}, -2\right)$ is a relative minimum and a turning point. **23.** Odd **25.** Neither **27.** Odd **29.** Even
31. Odd **33.** Even **35.** Odd **37.** Neither **39.** Even
41. Odd **43.** Even **45.** Neither **47. a.** Domain: $(-\infty, \infty)$; range: $(-\infty, 3]$ **b.** x-intercepts: $-3, 3$; y-intercept: 3 **c.** Increasing on $(-\infty, 0)$; decreasing on $(0, \infty)$ **d.** Relative maximum at $(0, 3)$
e. Even **49. a.** Domain: $(-3, 4)$; range: $[-2, 2]$ **b.** x-intercept: 1; y-intercept: -1 **c.** Constant on $(-3, -1)$ and $(3, 4)$; increasing on $(-1, 3)$ **d.** Any point $(x, -2)$ is a relative maximum and a relative minimum on $(-3, -1)$ and $(-1, -2)$ is a relative minimum. Any point $(x, 2)$ is a relative maximum and a relative minimum on $(3, 4)$ and $(3, 2)$ is a relative maximum. **e.** Neither **51. a.** Domain: $(-2, 4)$; range: $[-2, 3]$ **b.** x- and y-intercept: 0 **c.** Decreasing on $(-2, -1)$ and $(3, 4)$; increasing on $(-1, 3)$ **d.** Relative maximum at $(3, 3)$; relative minimum at $(-1, -2)$ **e.** Neither
53. a. Domain: $(-\infty, \infty)$; range: $(0, \infty)$ **b.** x-intercepts: none; y-intercept: 1 **c.** Increasing on $(-\infty, \infty)$ **d.** None **e.** Neither

55. -2 **57.** 10 **59.** -2 **61.** -2 **63.** 7 **65.** $-\dfrac{1}{12}$ **67.** 2

69. $-x - 1$ **71.** $3x + 7$ **73.** $-\dfrac{4}{x}$ **75.** 1 **77.** $2x + h$

79. $4x + 2h + 3$ **81.** 0 **83.** $-\dfrac{1}{x(x+h)}$

B Exercises: Applying the Concepts: 85. $[0, \infty)$; the particle's motion is tracked indefinitely from time $t = 0$. **87.** The graph is above the *t*-axis on the intervals $(0, 9)$ and $(21, 24)$; the particle is moving forward between 0 and 9 seconds and again between 21 and 24 seconds. **89.** $v = f(t)$ is increasing on $(0, 3), (5, 6), (16, 19)$, and $(21, 23)$; the speed, $|v|$, of the particle is increasing on the intervals $(0, 3), (5, 6), (11, 15)$, and $(21, 23)$. The particle is moving forward on $(0, 9)$ and $(21, 23)$ and moving backward on $(11, 15)$.
91. Maximum speed is attained between times $t = 15$ and $t = 16$.
93. The particle is moving forward with increasing velocity.
95. b. $(0, 6)$ **c.** $[0, 128]$ **d.** When $x = 2$, V has a maximum value of 128. **97.** Maximum value of 56 when $x = 6$
99. a. $C(x) = 210x + 10,500$ **b.** \$21,000**c.** \$420 **d.** 100

C Exercises: Beyond the Basics: 101. a. $f(x) = |x|$ **b.** $f(x) = 0$
c. $f(x) = x$ **d.** $f(x) = \sqrt{-x^2}$ **e.** $f(x) = 1$ **f.** Not possible
103. $(2, 7)$ **105.** $(-1, 5)$; No
107. $g(x) = \begin{cases} -x - 10 & \text{if } x < -5 \\ -5 & \text{if } -5 \le x \le 5 \\ x - 10 & \text{if } x > 5 \end{cases}$
$(0, -5)$ is a relative minimum point but not a turning point.

Critical Thinking: 109. c is an endpoint maximum of f if c is an endpoint of an interval in the domain of f and there is an interval $(x_1, c]$ or $[c, x_2)$ in the domain of f such that $f(c) \ge f(x)$ for every x in the interval. **110.** c is an endpoint minimum of f if c is an endpoint of an interval in the domain of f and there is an interval $(x_1, c]$ or $[c, x_2)$ in the domain of f such that $f(c) \le f(x)$ for every x in the interval. **111. a.** $f(x) = x$ on the interval $[-1, 1]$
b. $f(x) = x$ on the interval $[-1, \infty)$ **c.** $f(x) = x$ on the interval $(-\infty, 1]$ **d.** $f(x) = x$

Section 2.6

Practice Problems: **1.** $g(x) = 2x + 6$ **2.** 15.524 m
3.

Domain: $(-\infty, 0]$; range: $[0, \infty)$

4.

Domain: $(-\infty, \infty)$;
range: $(-\infty, \infty)$
5. $f(-2) = 4, f(3) = 6$

6. $f(x) = x^2$ $x \le -1$
$f(x) = 2x$ $x > -1$ **7.** $f(-3.4) = -4; f(4.7) = 4$

A Exercise: Basic Skills and Concepts: 1. horizontal
3. 3 **5.** even **7.** false
9. $f(x) = x + 1$

11. $f(x) = 2x + 3$

13. $f(x) = -3x + 4$

15. $f(x) = \dfrac{1}{2}x + 3$

17. $f(x) = -\dfrac{2}{3}x - 1$

19. a. $f(1) = 2, f(2) = 2, f(3) = 3$
b.

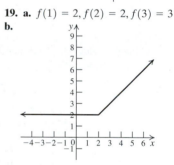

21. a. $f(-15) = -1, f(12) = 1$
b.

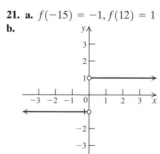

c. Domain: $(-\infty, 0) \cup (0, \infty)$;
range: $\{-1, 1\}$

23.

Domain: $(-\infty, \infty)$;
range: $(-\infty, \infty)$

25.

Domain: $(-\infty, \infty)$;
range: $(-\infty, \infty)$

27.

Domain: $(-\infty, \infty)$; range:
$\{\ldots, -3, -2, -1, 0\} \cup [1, \infty)$

29.

Domain: $(-\infty, \infty)$;
range: $[0, \infty)$

31.

Domain: $[-2, \infty)$;
range: $[0, \infty)$

33.

Domain: $(-\infty, 4]$; range:
$\{\ldots, -3, -2, -1, 0\} \cup [-8, -2]$

B Exercises: Applying the Concepts: 35. a. $f(x) = \dfrac{1}{33.81}x$;

domain: $[0, \infty)$; range: $[0, \infty)$ **b.** 0.0887; it means that
3 oz = 0.08871.   **c.** 0.3549   **37. a.** No d-intercept because $d \geq 0$;
y-intercept: 1. It means that the pressure at sea level ($d = 0$) is 1 atm.
b. $P(0) = 1, P(10) \approx 1.3, P(33) = 2, P(100) \approx 4.03$ **c.** 132 ft
39. a. $C = 50x + 6000$ **b.** The y-intercept is the fixed overhead cost.

c. 110   **41. a.** $R = 900 - 30x$
b. $720   **c.** Ten days after the first
of the month   **43.** 70

45. a. $y = \dfrac{2}{27}(x - 150) + 30$
b. 45   **c.** 352.5 mg/m^3

47. a.

b. (i) $480   **(ii)** $2000
(iii) $3800   **c. (i)** $15,000
(ii) This value is not possible
as a liability.   **(iii)** $25,000

C Exercises: Beyond the Basics: 49. a. (i) -1 **(ii)** -1 **(iii)** -1
b. $x = \dfrac{1}{2}$ **c.**

51. a. Domain: $(-\infty, \infty)$; range $[0, 1)$ **b.** Increasing on $(n, n + 1)$
for every integer n **c.** Neither   **53.** 2

Critical Thinking: 55. a. $C(x) = 17\big(f(x) - 1\big) + 44$
b.

c. Domain: $(0, \infty)$;
range: $\{17n + 44: n$ a
nonnegative integer$\}$
56. $C(x) = 2[\![x]\!] + 4$

57. a. $C(x) = \begin{cases} 150, & \text{if } x \leq 100 \\ -0.2[\![100 - x]\!] + 150, & \text{if } x > 100 \end{cases}$ **c.** $(299, 300]$
b.

Section 2.7

Practice Problems: 1. The graph
of g is the graph of f shifted one
unit up; the graph of h is the
graph of f shifted two units
down.

The graph of g is the graph of f
shifted one unit to the right; the
graph of h is the graph of f
shifted two units to the left.

2.

3.

4. Shift the graph of $y = x^2$ one unit
to the right, reflect the resulting
graph in the x-axis, and shift it two
units up.

5.

The graph of g is the graph of f
vertically stretched by multiplying
each of its y-coordinates by 2.

6. a.

b.

7.

8.

A Exercises: Basic Skills and Concepts: **1.** down **3.** > 1
5. false **7. a.** Shift two units up **b.** Shift one unit down
9. a. Shift one unit left **b.** Shift two units right
11. a. Shift one unit left and two units down **b.** Shift one unit right
and three units up **13. a.** Reflect in *x*-axis **b.** Reflect in *y*-axis
15. a. Stretch vertically by a factor of 2 **b.** Compress horizontally
17. a. Reflect in *x*-axis, shift one unit up **b.** Reflect in *y*-axis, shift
one unit up **19. a.** Shift one unit up **b.** Shift one unit left
21. e **23.** g **25.** i **27.** b **29.** l **31.** d

33.

35.

37.

39.

41.

43.

45.

47.

49.

51.

53.

55.

57.

59.

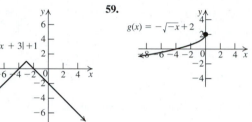

61. $h(x) = 2[[x + 1]]$

63. $y^2 = x^3 + 2$ **65.** $y = -|x|$
67. $y = (x - 3)^2 + 2$
69. $y = -\sqrt{x + 3} - 2$
71. $y = 3(4 - x)^3 + 2$
73. $y = -|2x - 4| - 3$

75.

77.

79.

81.

83.

85.

C Exercises: Beyond the Basics:
103. Shift one unit to the right, stretch vertically by a factor of 2, reflect in the x-axis, and shift three units up.

87.

89.

105. Shift two units to the left and four units down.

91.

107. $-x^2 + 2x = -(x^2 - 2x + 1) + 1 = -(x - 1)^2 + 1$
Shift one unit to the right, reflect in the x-axis, and shift one unit up.

B Exercises: Applying the Concepts: 93. $g(x) = f(x) + 800$
95. $p(x) = 1.02(f(x) + 500)$
97. a.

b. $5199.75

109. $2x^2 - 4x = 2(x^2 - 2x + 1) - 2 = 2(x - 1)^2 - 2$
Shift one unit to the right, stretch vertically by a factor of two, and shift two units down.

99. a.

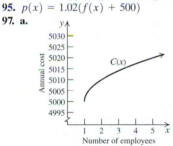

b. $2
c. $3.30

101. The first coordinate gives the month. The second coordinate gives the hours of daylight.

111. $-2x^2 - 8x + 3 = -2(x^2 + 4x + 4) + 11 = -2(x + 2)^2 + 11$
Shift 2 units to the left, stretch vertically by a factor of 2, reflect in the x-axis, and shift 11 units up.

113.

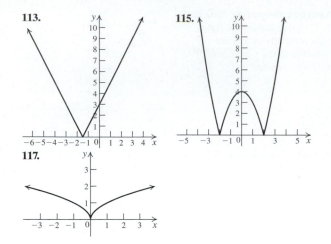

115.

117.

Critical Thinking: 119. a. x-intercept 1 **b.** x-intercept -2; y-intercept 15 **c.** x-intercept -2; y-intercept -3 **d.** x-intercept 2; y-intercept 3 **120. a.** x-intercept 2 **b.** x-intercept 4; y-intercept -2 **c.** x-intercept 4; y-intercept 1 **d.** x-intercept -4; y-intercept -1 **121. a.** The graph of g is the graph of h shifted three units to the right and three units up. **b.** The graph of g is the graph of h shifted one unit to the left and one unit down. **c.** The graph of g is the graph of h stretched horizontally and vertically by a factor of 2. **d.** The graph of g is the graph of h stretched horizontally by a factor of 3, reflected in the y-axis, stretched vertically by a factor of 3, and reflected in the x-axis. **122.** Stretch the graph of $y = f(-4x)$ horizontally by a factor of 4 and reflect the resulting graph in the y-axis.

Section 2.8

Practice Problems:

1. $(f + g)(x) = x^2 + 3x + 1$
$(f - g)(x) = -x^2 + 3x - 3$
$(fg)(x) = 3x^3 - x^2 + 6x - 2$
$\left(\dfrac{f}{g}\right)(x) = \dfrac{3x - 1}{x^2 + 2}$

2. a. $(f \circ g)(0) = -5$ **b.** $(g \circ f)(0) = 1$
3. $(g \circ f)(x) = 2x^2 - 8x + 9$
$(f \circ g)(x) = 1 - 2x^2$

4. $(g \circ f)(x) = 2\sqrt{x} + 5$; domain: $[0, \infty)$ **5.** $f(x) = \dfrac{1}{x}$

6. a. $A = 9\pi t^2$ **b.** $A = 324\pi$ square miles
7. a. $r(x) = x - 4500$ **b.** $d(x) = 0.94x$
c. (i) $(r \circ d)(x) = 0.94x - 4500$
(ii) $(d \circ r)(x) = 0.94x - 4230$
d. $(d \circ r)(x) - (r \circ d)(x) = 270$

A Exercises: Basic Skills and Concepts: 1. -8 **3.** 1
5. false **7. a.** $(f + g)(-1) = -1$ **b.** $(f - g)(0) = 0$
c. $(f \cdot g)(2) = -8$ **d.** $\left(\dfrac{f}{g}\right)(1) = -2$ **9. a.** $(f + g)(-1) = 0$
b. $(f - g)(0) = \dfrac{1}{2}(\sqrt{2} - 2)$ **c.** $(f \cdot g)(2) = \dfrac{5}{2}$
d. $\left(\dfrac{f}{g}\right)(1) = \dfrac{1}{3\sqrt{3}}$ **11. a.** $(f + g)(x) = x^2 + x - 3$;
domain: $(-\infty, \infty)$ **b.** $(f - g)(x) = -x^2 + x - 3$; domain: $(-\infty, \infty)$
c. $(f \cdot g)(x) = x^3 - 3x^2$; domain: $(-\infty, \infty)$
d. $\left(\dfrac{f}{g}\right)(x) = \dfrac{x - 3}{x^2}$; domain: $(-\infty, 0) \cup (0, \infty)$
e. $\left(\dfrac{g}{f}\right)(x) = \dfrac{x^2}{x - 3}$; domain: $(-\infty, 3) \cup (3, \infty)$
13. a. $(f + g)(x) = x^3 + 2x^2 + 4$; domain: $(-\infty, \infty)$
b. $(f - g)(x) = x^3 - 2x^2 - 6$; domain: $(-\infty, \infty)$

c. $(f \cdot g)(x) = 2x^5 + 5x^3 - 2x^2 - 5$; domain: $(-\infty, \infty)$
d. $\left(\dfrac{f}{g}\right)(x) = \dfrac{x^3 - 1}{2x^2 + 5}$; domain: $(-\infty, \infty)$
e. $\left(\dfrac{g}{f}\right)(x) = \dfrac{2x^2 + 5}{x^3 - 1}$; domain: $(-\infty, 1) \cup (1, \infty)$
15. a. $(f + g)(x) = 2x + \sqrt{x} - 1$; domain: $[0, \infty)$
b. $(f - g)(x) = 2x - \sqrt{x} - 1$; domain: $[0, \infty)$
c. $(f \cdot g)(x) = 2x\sqrt{x} - \sqrt{x}$; domain: $[0, \infty)$
d. $\left(\dfrac{f}{g}\right)(x) = \dfrac{2x - 1}{\sqrt{x}}$; domain: $(0, \infty)$
e. $\left(\dfrac{g}{f}\right)(x) = \dfrac{\sqrt{x}}{2x - 1}$; domain: $\left[0, \dfrac{1}{2}\right) \cup \left(\dfrac{1}{2}, \infty\right)$
17. a. $(f + g)(x) = \dfrac{2 + x}{x + 1}$; domain: $(-\infty, -1) \cup (-1, \infty)$
b. $(f - g)(x) = \dfrac{2 - x}{x + 1}$; domain: $(-\infty, -1) \cup (-1, \infty)$
c. $(f \cdot g)(x) = \dfrac{2x}{(x + 1)^2}$; domain: $(-\infty, -1) \cup (-1, \infty)$
d. $\left(\dfrac{f}{g}\right)(x) = \dfrac{2}{x}$; domain: $(-\infty, -1) \cup (-1, 0) \cup (0, \infty)$
e. $\left(\dfrac{g}{f}\right)(x) = \dfrac{x}{2}$; domain: $(-\infty, -1) \cup (-1, \infty)$
19. a. $(f + g)(x) = \dfrac{x^3 - x^2 + 2x}{x^2 - 1}$;
domain: $(-\infty, -1) \cup (-1, 1) \cup (1, \infty)$
b. $(f - g)(x) = \dfrac{x^2 - 2x}{x - 1}$; domain: $(-\infty, -1) \cup (-1, 1) \cup (1, \infty)$
c. $(f \cdot g)(x) = \dfrac{2x^3}{x^3 + x^2 - x - 1}$; domain: $(-\infty, -1) \cup (-1, 1) \cup (1, \infty)$
d. $\left(\dfrac{f}{g}\right)(x) = \dfrac{x^2 - x}{2}$; domain: $(-\infty, -1) \cup (-1, 0) \cup (0, 1) \cup (1, \infty)$
e. $\left(\dfrac{g}{f}\right)(x) = \dfrac{2}{x^2 - x}$; domain: $(-\infty, -1) \cup (-1, 0) \cup (0, 1) \cup (1, \infty)$
21. a. $(f + g)(x) = \dfrac{4x^2 - 5x - 28}{x^3 - 3x^2 - 16x + 48}$;
domain: $(-\infty, -4) \cup (-4, 3) \cup (3, 4) \cup (4, \infty)$
b. $(f - g)(x) = -\dfrac{7}{(x - 3)(x + 4)}$;
domain: $(-\infty, -4) \cup (-4, 3) \cup (3, 4) \cup (4, \infty)$
c. $(f \cdot g)(x) = \dfrac{4x^2 - 14x}{x^4 - 7x^3 - 4x^2 + 112x - 192}$;
domain: $(-\infty, -4) \cup (-4, 3) \cup (3, 4) \cup (4, \infty)$
d. $\left(\dfrac{f}{g}\right)(x) = \dfrac{2x^2 - 6x}{2x^2 + x - 28}$;
domain: $(-\infty, -4) \cup (-4, 3) \cup \left(3, \dfrac{7}{2}\right) \cup \left(\dfrac{7}{2}, 4\right) \cup (4, \infty)$
e. $\left(\dfrac{g}{f}\right)(x) = \dfrac{2x^2 + x - 28}{2x^2 - 6x}$;
domain: $(-\infty, -4) \cup (-4, 0) \cup (0, 3) \cup (3, 4) \cup (4, \infty)$
23. $g(f(x)) = 2x^2 + 1$; $g(f(2)) = 9$; $g(f(-3)) = 19$
25. 11 **27.** 31 **29.** -5 **31.** $4c^2 - 5$ **33.** $8a^2 + 8a - 1$
35. 7 **37.** $\dfrac{2x}{x + 1}$; $x \neq 0, x \neq -1$ **39.** $\sqrt{-3x - 1}$; $\left(-\infty, -\dfrac{1}{3}\right]$
41. $|x^2 - 1|$; $(-\infty, \infty)$ **43. a.** $(f \circ g)(x) = 2x + 5$;
domain: $(-\infty, \infty)$ **b.** $(g \circ f)(x) = 2x + 1$; domain: $(-\infty, \infty)$
c. $(f \circ f)(x) = 4x - 9$; domain: $(-\infty, \infty)$ **d.** $(g \circ g)(x) = x + 8$;
domain: $(-\infty, \infty)$ **45. a.** $(f \circ g)(x) = -2x^2 - 1$; domain: $(-\infty, \infty)$
b. $(g \circ f)(x) = 4x^2 - 4x + 2$; domain: $(-\infty, \infty)$
c. $(f \circ f)(x) = 4x - 1$; domain: $(-\infty, \infty)$
d. $(g \circ g)(x) = x^4 + 2x^2 + 2$; domain: $(-\infty, \infty)$
47. a. $(f \circ g)(x) = 8x^2 - 2x - 1$; domain: $(-\infty, \infty)$
b. $(g \circ f)(x) = 4x^2 + 6x - 1$; domain: $(-\infty, \infty)$

c. $(f \circ f)(x) = 8x^4 + 24x^3 + 24x^2 + 9x$; domain: $(-\infty, \infty)$
d. $(g \circ g)(x) = 4x - 3$; domain: $(-\infty, \infty)$
49. a. $(f \circ g)(x) = x$; domain: $[0, \infty)$ **b.** $(g \circ f)(x) = |x|$;
domain: $(-\infty, \infty)$ **c.** $(f \circ f)(x) = x^4$; domain: $(-\infty, \infty)$

d. $(g \circ g)(x) = \sqrt[4]{x}$; domain: $[0, \infty)$ **51. a.** $(f \circ g)(x) = -\dfrac{x^2}{x^2 - 2}$;

domain: $(-\infty, -\sqrt{2}) \cup (-\sqrt{2}, 0) \cup (0, \sqrt{2}) \cup (\sqrt{2}, \infty)$

b. $(g \circ f)(x) = (2x - 1)^2$; domain: $\left(-\infty, \dfrac{1}{2}\right) \cup \left(\dfrac{1}{2}, \infty\right)$

c. $(f \circ f)(x) = -\dfrac{2x - 1}{2x - 3}$; domain: $\left(-\infty, \dfrac{1}{2}\right) \cup \left(\dfrac{1}{2}, \dfrac{3}{2}\right) \cup \left(\dfrac{3}{2}, \infty\right)$

d. $(g \circ g)(x) = x^4$; domain: $(-\infty, 0) \cup (0, \infty)$

53. a. $(f \circ g)(x) = 2$; domain: $(-\infty, \infty)$ **b.** $(g \circ f)(x) = -2$;
domain: $(-\infty, \infty)$ **c.** $(f \circ f)(x) = |x|$; domain: $(-\infty, \infty)$

d. $(g \circ g)(x) = -2$; domain: $(-\infty, \infty)$ **55. a.** $(f \circ g)(x) = \dfrac{2}{1 + x}$;

domain: $(-\infty, -1) \cup (-1, 1) \cup (1, \infty)$ **b.** $(g \circ f)(x) = -2x - 1$;

domain: $(-\infty, 0) \cup (0, \infty)$ **c.** $(f \circ f)(x) = \dfrac{2x + 1}{x + 1}$;

domain: $(-\infty, -1) \cup (-1, 0) \cup (0, \infty)$

d. $(g \circ g)(x) = -\dfrac{1}{x}$; domain: $(-\infty, 0) \cup (0, 1) \cup (1, \infty)$

57. $f(x) = \sqrt{x}$; $g(x) = x + 2$ **59.** $f(x) = x^{10}$; $g(x) = x^2 - 3$

61. $f(x) = \dfrac{1}{x}$; $g(x) = 3x - 5$ **63.** $f(x) = \sqrt[3]{x}$; $g(x) = x^2 - 7$

65. $f(x) = \dfrac{1}{|x|}$; $g(x) = x^3 - 1$

B Exercises: Applying the Concepts: 67. a. Cost function
b. Revenue function   **c.** Selling price of x shirts after sales tax
d. Profit function   **69. a.** $P(x) = 20x - 350$ **b.** $P(20) = 50$;
profit made after 20 radios were sold   **c.** 43
d. $(R \circ x)(C) = 5C - 1750$; the function represents the revenue in
terms of the cost C.   **71. a.** $f(x) = 0.7x$ **b.** $g(x) = x - 5$
c. $(g \circ f)(x) = 0.7x - 5$ **d.** $(f \circ g)(x) = 0.7(x - 5)$
e. $1.50   **73. a.** $f(x) = 1.1x$; $g(x) = x + 8$
b. $(f \circ g)(x) = 1.1x + 8.8$; a final test score, which originally was x,
after adding 8 points and then increasing the score by 10%
c. $(g \circ f)(x) = 1.1x + 8$; a final test score, which originally was x,
after increasing the score by 10% and then adding 8 points   **d.** 85.8
and 85.0   **e.** No   **f. (i)** 73.82   **(ii)** 74.55
75. a. $f(x) = \pi x^2$ **b.** $g(x) = \pi(x + 30)^2$ **c.** The area between
the fountain and the fence   **d.** $16,052
77. a. $(f \circ g)(t) = \pi(2t + 1)^2$ **b.** $A(t) = \pi(2t + 1)^2$
c. They are the same.   **79. a.** $f(x) = 1.64x$
b. $g(x) = 0.04567x$ **c.** $(f \circ g)(x)$ **d.** $74.90

C Exercises: Beyond the Basics: 81. a. Odd function
b. Even function   **c.** Even function   **d.** Even function
83. a. $h(x) = o(x) + e(x)$, where $e(x) = x^2 + 3$ and $o(x) = -2x$

b. $h(x) = o(x) + e(x)$, where $e(x) = \dfrac{[x] + [-x]}{2}$ and

$o(x) = x + \dfrac{[x] - [-x]}{2}$

Critical Thinking: 84. a. Domain: $(-\infty, 0) \cup [1, \infty)$
b. Domain: $[0, 2]$ **c.** Domain: $[1, 2]$ **d.** Domain: $[1, 2)$
85. a. Domain: $\varnothing$ **b.** Domain: $(-\infty, 0)$

Section 2.9

Practice Problems:
1. Not one-to-one; the horizontal line $y = 1$ intersects the graph at
two different points.

2. a. -3 **b.** 4

3. $(f \circ g)(x) = 3\left(\dfrac{x + 1}{3}\right) - 1 = x + 1 - 1 = x$ and

$(g \circ f)(x) = \dfrac{3x - 1 + 1}{3} = \dfrac{3x}{3} = x$; therefore, f and g are

inverses of each other.

4.

5. $f^{-1}(x) = \dfrac{3 - x}{2}$

6. $f^{-1}(x) = \dfrac{3x}{1 - x}$, $x \neq 1$ **7.** Domain: $(-\infty, -3) \cup (-3, \infty)$;

range: $(-\infty, 1) \cup (1, \infty)$ **8.** $G^{-1}(x) = -\sqrt{x + 1}$ **9.** 3630 ft

A Exercises: Basic Skills and Concepts: 1. one-to-one

3. $\dfrac{1}{3}x$ **5.** true   **7.** One-to-one   **9.** Not one-to-one

11. Not one-to-one   **13.** One-to-one   **15.** 2   **17.** -1
19. a **21.** 337   **23.** -1580 **25. a.** 3   **b.** 3   **c.** 19   **d.** 5
27. a. 2   **b.** 1   **c.** 269
35.

37.

39.

41. a. One-to-one   **b.** $f^{-1}(x) = 5 - \dfrac{1}{3}x$

c.

d. Domain f: $(-\infty, \infty)$
x-intercept: 5
y-intercept: 15
Domain f^{-1}: $(-\infty, \infty)$
x-intercept: 15
y-intercept: 5
43. Not one-to-one

45. a. One-to-one **b.** $f^{-1}(x) = (x - 3)^2, x \geq 3$

c.

d. Domain f: $[0, \infty)$
No x-intercept
y-intercept: 3
Domain f^{-1}: $[3, \infty)$
x-intercept: 3
No y-intercept

47. a. One-to-one **b.** $g^{-1}(x) = x^3 - 1$

c.

d. Domain g: $(-\infty, \infty)$
x-intercept: -1
y-intercept: 1
Domain g^{-1}: $(-\infty, \infty)$
x-intercept: 1
y-intercept: -1

49. a. One-to-one **b.** $f^{-1}(x) = \dfrac{x + 1}{x}$

c.

d. Domain f: $(-\infty, 1) \cup (1, \infty)$
No x-intercept
y-intercept: -1
Domain f^{-1}: $(-\infty, 0) \cup (0, \infty)$
x-intercept: -1
No y-intercept

51. a. One-to-one **b.** $f^{-1}(x) = (x - 2)^2 - 1, x \geq 2$

c.

d. Domain f: $[-1, \infty)$
y-intercept: 3
No x-intercept
Domain f^{-1}: $[2, \infty)$
x-intercept: 3
No y-intercept

53. Domain: $(-\infty, -2) \cup (-2, \infty)$; range: $(-\infty, 1) \cup (1, \infty)$

55. $f^{-1}(x) = \dfrac{2x + 1}{x - 1}$ Domain: $(-\infty, 2) \cup (2, \infty)$; range: $(-\infty, 1) \cup (1, \infty)$

57. $f^{-1}(x) = \dfrac{1 - x}{x + 2}$
Domain: $(-\infty, -1) \cup (-1, \infty)$; range: $(-\infty, -2) \cup (-2, \infty)$

59. $y = \sqrt{-x}, x \leq 0$

61. $y = x, x \geq 0$

63. $y = -\sqrt{x - 1}, x \geq 1$

65. $y = -\sqrt{2 - x}, x \leq 2$

B Exercises: Applying the Concepts: 67. a. $C(K) = K - 273$; it represents the Celsius temperature corresponding to a given Kelvin temperature. **b.** 27°C **c.** 295 K **69. a.** $F = \dfrac{9}{5}C + 32$

b. $C = \dfrac{5}{9}F - \dfrac{160}{9}$ **71. a.** Dollars to euros: $f(x) = 0.75x$ (x is the number of dollars; $f(x)$ is the number of euros); euros to dollars: $g(x) = 1.25x$ (x is the number of euros; $g(x)$ is the number of dollars) **b.** $g(f(x)) = 0.9375x \neq x$; therefore, g and f are not inverse functions. **c.** She loses money.

73. a. $w = \begin{cases} 4 + 0.05x, & \text{if } x > 60 \\ 7, & \text{if } x \leq 60 \end{cases}$ **b.** It does not have an inverse because it is constant on $(0, 60)$ and hence is not one-to-one.

c. Restriction of the domain: $[60, \infty)$ **75. a.** $x = \dfrac{1}{64}V^2$; it gives the height of the water in terms of the velocity. **b. (i)** 14.0625 ft **(ii)** 6.25 ft

77. a. The balance due after x months **b.** $f^{-1}(x) = 60 - \dfrac{1}{600}x$; it shows the number of months that have passed from the first payment until the balance due is \$$x$. **c.** 37

C Exercises: Beyond the Basics:
81. a.

b. No
c. Domain: $[-2, 2]$; range: $[0, 2]$

83. a.

b. $f^{-1}(x) = \dfrac{x}{1 - x}$
c. Domain: $(-\infty, -1) \cup (1, \infty)$; range: $(-\infty, 1) \cup (1, \infty)$

85. a. The midpoint is $M(5, 5)$. Because its coordinates satisfy the equation $y = x$, it lies on the line $y = x$. **b.** The slope of the line segment $\overline{PQ}$ is -1, and the slope of the line $y = x$ is 1. Because

$(-1)(1) = -1$, the two lines are perpendicular.   **87. a.** The graph of g is the graph of f shifted one unit to the right and two units up.

b. $g^{-1}(x) = \sqrt[3]{x - 2} + 1$ **c.**

89. a. (i) $f^{-1}(x) = \dfrac{1}{2}x - \dfrac{3}{2}$ **(ii)** $g^{-1}(x) = \sqrt[3]{x + 1}$

(iii) $(f \circ g)(x) = 2x^3 + 1$ **(iv)** $(g \circ f)(x) = 8x^3 + 36x^2 + 54x + 26$

(v) $(f \circ g)^{-1}(x) = \sqrt[3]{\dfrac{x - 1}{2}}$ **(vi)** $(g \circ f)^{-1}(x) = \dfrac{1}{2}\sqrt[3]{x + 1} - \dfrac{3}{2}$

(vii) $(f^{-1} \circ g^{-1})(x) = \dfrac{1}{2}\sqrt[3]{x + 1} - \dfrac{3}{2}$ **(viii)** $(g^{-1} \circ f^{-1})(x) = \sqrt[3]{\dfrac{x - 1}{2}}$

b. (i) $(f \circ g)^{-1}(x) = \sqrt[3]{\dfrac{x - 1}{2}} = (g^{-1} \circ f^{-1})(x)$

(ii) $(g \circ f)^{-1}(x) = \dfrac{1}{2}\sqrt[3]{x + 1} - \dfrac{3}{2} = (f^{-1} \circ g^{-1})(x)$

Critical Thinking: **90.** $f^{-1}(x) = -x^{3/2}, x \geq 0$
91. No. $f(x) = x^3 - x$ is odd, but it does not have an inverse because $f(0) = f(1)$. So it is not one-to-one.
92. Yes. The function $f = \{(0, 1)\}$ is even, and it has an inverse: $f^{-1} = \{(1, 0)\}$.   **93.** Yes, because increasing and decreasing functions are one-to-one   **94. a.** $R = \{(-1, 1), (0, 0), (1, 1)\}$
b. $R = \{(-1, 1), (0, 0), (1, 2)\}$

Review Exercises

1. False   **3.** True   **5.** False   **7.** True   **9. a.** $2\sqrt{5}$ **b.** $(1, 4)$
c. $\dfrac{1}{2}$ **11. a.** $5\sqrt{2}$ **b.** $\left(\dfrac{13}{2}, -\dfrac{11}{2}\right)$ **c.** -1

13. a. $\sqrt{34}$ **b.** $\left(\dfrac{7}{2}, -\dfrac{9}{2}\right)$ **c.** $\dfrac{5}{3}$ **17.** $(4, 5)$ **19.** $\left(\dfrac{31}{18}, 0\right)$

21. Not symmetric with respect to the x-axis, symmetric with respect to the y-axis, not symmetric with respect to the origin
23. Symmetric with respect to the x-axis, not symmetric with respect to the y-axis, not symmetric with respect to the origin
25.

x-intercept: 4; y-intercept: 2; not symmetric with respect to the x-axis, not symmetric with respect to the y-axis, not symmetric with respect to the origin

27.

x-intercept: 0; y-intercept: 0; not symmetric with respect to the x-axis, symmetric with respect to the y-axis, not symmetric with respect to the origin

29.

x-intercept: 0; y-intercept: 0; not symmetric with respect to the x-axis, not symmetric with respect to the y-axis, symmetric with respect to the origin

31.

No x-intercept; y-intercept: 2; not symmetric with respect to the x-axis, symmetric with respect to the y-axis, not symmetric with respect to the origin

33.

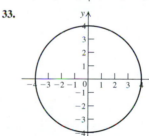

x-intercepts: -4 and 4; y-intercepts: -4 and 4; symmetric with respect to the x-axis, symmetric with respect to the y-axis, symmetric with respect to the origin

35. $(x - 2)^2 + (y + 3)^2 = 25$ **37.** $(x + 2)^2 + (y + 5)^2 = 4$
39. $\dfrac{x}{2} - \dfrac{y}{5} = 1 \Rightarrow y = \dfrac{5}{2}x - 5$; line with slope $\dfrac{5}{2}$ and y-intercept -5

41. $x^2 + y^2 - 2x + 4y - 4 = 0 \Rightarrow (x - 1)^2 + (y + 2)^2 = 9$; circle centered at $(1, -2)$ and with radius 3

43. $y = -2x + 4$ **45.** $y = -2x + 5$

47. a. Parallel **b.** Neither **c.** Perpendicular **d.** Neither
49.

Domain: $\{-1, 0, 1, 2\}$; range: $\{-1, 0, 1, 2\}$; function
51.

Domain: $(-\infty, \infty)$; range: $(-\infty, \infty)$; function
53.

Domain: $[-0.2, 0.2]$; range: $[-0.2, 0.2]$; not a function
55.

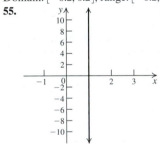

Domain: $\{1\}$; range: $(-\infty, \infty)$; not a function
57. $y = |x + 1|$

Domain: $(-\infty, \infty)$; range: $[0, \infty)$; function
59. -5 **61.** $x = 1$ **63.** 3 **65.** -10 **67.** 22 **69.** $3x^2 - 5$
70. $9x^2 + 6x - 1$ **71.** $9x + 4$ **73.** $3a + 3h + 1$ **75.** 3
77.

Domain: $(-\infty, \infty)$; range: $\{-3\}$; constant on $(-\infty, \infty)$
79.

Domain: $\left[\dfrac{2}{3}, \infty\right)$; range: $[0, \infty)$; increasing on $\left(\dfrac{2}{3}, \infty\right)$
81.

Domain: $(-\infty, \infty)$; range: $[1, \infty)$; decreasing on $(-\infty, 0)$,
increasing on $(0, \infty)$
83. The graph of g is the graph of f shifted one unit to the left.

85. The graph of g is the graph of f shifted two units to the right and
reflected in the x-axis.

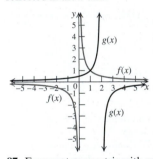

87. Even; not symmetric with respect to the x-axis, symmetric with
respect to the y-axis, not symmetric with respect to the origin
89. Even; not symmetric with respect to the x-axis, symmetric with
respect to the y-axis, not symmetric with respect to the origin
91. Neither even nor odd; not symmetric with respect to the x-axis,
not symmetric with respect to the y-axis, not symmetric with respect
to the origin **93.** $f(x) = (g \circ h)(x)$, where $g(x) = \sqrt{x}$ and
$h(x) = x^2 - 4$ **95.** $h(x) = (f \circ g)(x)$, where $f(x) = \sqrt{x}$ and
$g(x) = \dfrac{x - 3}{2x + 5}$

97. One-to-one; $f^{-1}(x) = x - 2$

99. One-to-one; $f^{-1}(x) = x^3 + 2$

101. $f^{-1}(x) = \dfrac{2x + 1}{1 - x}$; domain: $(-\infty, -2) \cup (-2, \infty)$;

range: $(-\infty, 1) \cup (1, \infty)$

103. a. $f(x) = \begin{cases} 3x + 6, & \text{if } -3 \le x \le -2 \\ \dfrac{1}{2}x + 1, & \text{if } -2 < x < 0 \\ x + 1, & \text{if } 0 \le x \le 3 \end{cases}$

b. Domain: $[-3, 3]$; range: $[-3, 4]$ **c.** x-intercept: -2; y-intercept: 1

d.

e.

f.

g.

h.

i.

j.

k. It satisfies the horizontal-line test.

I.

105. a. $C = 0.875w - 22{,}125$ **b.** Slope: the cost of disposing 1 pound of waste; x-intercept: the amount of waste that can be disposed with no cost; y-intercept: the fixed cost   **c.** \$510,750   **d.** 1,168,142.86 pounds   **107. a.** She won \$98.   **b.** She was winning at a rate of \$49/h.   **c.** Yes, after 20 hours   **d.** \$5/hr
109. a. $0.5\sqrt{0.000004t^4 + 0.004t^2 + 5}$ **b.** 1.13
111. a. $y = 8.85x - 470.01$
b.

c. 212.08 lb

Practice Test A

1. $(x + 3)^2 + (y - 4)^2 = 2$ **2.** Symmetric about the x-axis
3. x-intercepts: $-1, 0, 3$; y-intercept: 0
4.

5. $y = -x + 9$
6. $y = 4x - 9$ **7.** -36
8. -1 **9.** $x^4 - 4x^3 + 2x^2 + 4x$
10. a. $f(-1) = -3$
b. $f(0) = -2$ **c.** $f(1) = -1$
11. $[0, 1)$ **12.** 2   **13.** Even
14. Increasing on $(-\infty, 0) \cup (2, \infty)$, decreasing on $(0, 2)$

15. Shift the graph of $y = \sqrt{x}$ three units to the right, stretch the graph of $y = \sqrt{x - 3}$ vertically by a factor of 2, and shift the graph of $y = 2\sqrt{x - 3}$ four units up.   **16.** 2.5 sec   **17.** 2

18. $f^{-1}(x) = \dfrac{x}{x - 2}$ **19.** $A(x) = 100x + 1000$ **20. a.** \$87.50

b. 110 mi

Practice Test B

1. d   **2.** b   **3.** d   **4.** d   **5.** c   **6.** c   **7.** d   **8.** b   **9.** a
10. c   **11.** a   **12.** d   **13.** a   **14.** a   **15.** b   **16.** d   **17.** c
18. c   **19.** b   **20.** a

Cumulative Review Exercises (Chapters P-2)

1. a. $\dfrac{y^2}{x^3}$ **b.** $\dfrac{1}{x + y}$ **2. a.** $(2x - 5)(x + 3)$ **b.** $(x^2 + 4)(x - 2)$

3. a. $3\sqrt{3}$ **b.** $\dfrac{2x}{(x + 1)(x + 2)}$ **4. a.** $2 - \sqrt{3}$ **b.** $2 + \sqrt{5}$

5. a. $x = 4$ **b.** No solution   **6. a.** $x = 0; 3$ **b.** $x = -5; 2$

7. a. $\dfrac{1 \pm \sqrt{23}i}{4}$ **b.** $x = \dfrac{3}{2}$ **8. a.** 4, 16

b. $x = \dfrac{3 - \sqrt{13}}{2}; \dfrac{3 + \sqrt{13}}{2}; \dfrac{7 - \sqrt{53}}{2}; \dfrac{7 + \sqrt{53}}{2}$

9. a. $x = \dfrac{7 + \sqrt{17}}{8}$ **b.** $x = 0; \dfrac{8}{9}$ **10. a.** $(-\infty, 8)$ **b.** $(-\infty, 3)$

11. a. $(0, 4)$ **b.** $[-3, -2]$ **12. a.** $[-3, 4]$ **b.** $(-\infty, -1] \cup [4, \infty)$
13. Because the length of the side BC is equal to the length of the side AC (which is 5), the triangle is isosceles.

14.

15. All of the points satisfy the equation $(x - 2)^2 + (y + 1)^2 = 5^2$, which is a circle with center $(2, -1)$ and radius 5.
16. Center: $(3, -2)$; radius: 2
17. $y = -3x + 5$
18. $y = 2x - 8$
19. $y = -0.5x$
20. $y = 2x - 5$

21. $y = -\dfrac{1}{4}x + 4$ **22.** $x = 5$ **23.** $x = 2$ **24.** $x = 5$

25.

26.

27.

28.

29. She purchased 60 books.   **30.** The monthly note on a car that was leased for 2 yr is \$392.86, and the monthly note on a car that was leased for $1\frac{1}{2}$ years is \$642.86.   **31. a.** $[-1, \infty)$

b. x-intercept: 8; y-intercept: -2 **c.** -3 **d.** $(8, \infty)$ **32. a.** $2, 0, 4$
b. f decreases on $(-\infty, 0)$ and increases on $(0, \infty)$.

33. a. $(f \circ g)(x) = \dfrac{x}{2 - 2x}$, and its domain is

$(-\infty, 0) \cup (0, 1) \cup (1, \infty)$.   **b.** $(g \circ f)(x) = 2(x - 2)$, and its domain is $(-\infty, 2) \cup (2, \infty)$.

CHAPTER 3

Section 3.1

Practice Problems: **1.** $y = 3(x - 1)^2 - 5$; has a minimum value of -5 at $x = 1$ **2.** The graph is a parabola; $a = -2, h = -1, k = 3$; it opens downward; vertex: $(-1, 3)$; it is a maximum; x-intercepts: $\pm\sqrt{\dfrac{3}{2}} - 1$; y-intercept: 1

3. The graph is a parabola; $a = 3, b = -3, c = -6$; it opens upward; vertex: $\left(\dfrac{1}{2}, \dfrac{-27}{4}\right)$; x-intercepts: $-1, 2$; y-intercept: -6

4.

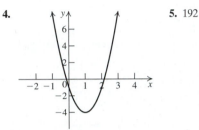

5. 192

Domain: $(-\infty, \infty)$;
range: $[-4, \infty)$

A Exercises: Basic Skills and Concepts: **1.** vertex   **3.** true
5. $\dfrac{1}{2}$ **7.** true   **9.** f   **11.** a   **13.** h   **15.** g
17. $y = -2x^2$ **19.** $y = 5x^2$ **21.** $y = 2x^2$ **23.** $y = -(x + 3)^2$
25. $y = 2(x - 2)^2 + 5$ **27.** $y = \dfrac{11}{49}(x - 2)^2 - 3$
29. $y = -12\left(x - \dfrac{1}{2}\right)^2 + \dfrac{1}{2}$ **31.** $y = \dfrac{3}{4}(x + 2)^2$
33. $y = \dfrac{3}{4}(x - 3)^2 - 1$

35.

The graph is the graph of $y = x^2$ stretched vertically by a factor of 3.

37.

The graph is the graph of $y = x^2$ shifted four units to the right.

39. The graph is the graph of $y = x^2$ stretched vertically by a factor of 2, reflected in the x-axis, and shifted down 4 units.

41. The graph is the graph of $y = x^2$ shifted three units right and two units up.

43. The graph is the graph of $y = x^2$ shifted two units right, stretched vertically by a factor of 3, reflected in the x-axis, and shifted four units up.

45. $y = (x + 2)^2 - 4$; the graph is the graph of $y = x^2$ shifted two units to the left and four units down; vertex: $(-2, -4)$; axis: $x = -2$; x-intercepts: -4 and 0; y-intercept: 0

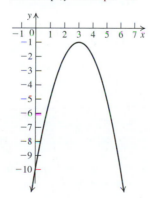

47. $y = -(x - 3)^2 - 1$; the graph is the graph of $y = x^2$ shifted three units to the right, reflected in the x-axis, and shifted one unit down; vertex: $(3, -1)$; axis: $x = 3$; no x-intercept; y-intercept: -10

49. $y = 2(x - 2)^2 + 1$; the graph is the graph of $y = x^2$ shifted two units to the right, stretched vertically by a factor of 2, and shifted one unit up.

Vertex: $(2, 1)$; axis: $x = 2$; no x-intercept; y-intercept: 9

51. $y = -3(x - 3)^2 + 16$; the graph is the graph of $y = x^2$ shifted 3 units to the right, stretched vertically by a factor of 3, reflected in the x-axis, and shifted 16 units up.

Vertex: $(3, 16)$; axis: $x = 3$; x-intercepts: $3 \pm \frac{4}{3}\sqrt{3}$; y-intercept: -11

53. a. Opens up **b.** $(4, -1)$ **c.** $x = 4$ **d.** x-intercepts: 3 and 5; y-intercept: 15 **e.**

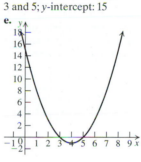

55. a. Opens up **b.** $\left(\frac{1}{2}, \frac{-25}{4}\right)$ **c.** $x = \frac{1}{2}$ **d.** x-intercepts: -2 and 3; y-intercept: -6 **e.**

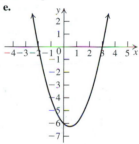

57. a. Opens up **b.** $(1, 3)$ **c.** $x = 1$ **d.** x-intercepts: none; y-intercept: 4 **e.**

59. a. Opens down **b.** $(-1, 7)$ **c.** $x = -1$ **d.** x-intercepts: $-1 \pm \sqrt{7}$; y-intercept: 6 **e.**

61. a. Minimum: -1 **b.** $[-1, \infty)$ **63. a.** Maximum: 0 **b.** $(-\infty, 0]$ **65. a.** Minimum: -5 **b.** $[-5, \infty)$ **67. a.** Maximum: 16 **b.** $(-\infty, 16]$

B Exercises: Applying the Concepts: 69. a. 547 ft **b.** 139 ft **c.** 547 ft; 139 ft **71.** 19 **73.** 25 **75.** Square with 20-unit sides; maximum area: 400 square units **77.** Dimensions: 75 m × 100 m; area: 7500 m² **79.** 38 **81.** 28 students; maximum revenue: $1568

83. a. 64 ft   **b.** After 4 sec   **85.** Dimensions of the rectangle:

width $= \dfrac{36}{\pi + 4}$ ft; height $= \dfrac{18}{\pi + 4}$ ft

87. a. (59,36.81)   **b.** 120 ft   **c.** 37 ft   **d.** 9 ft   **e.** 113.6 ft

C Exercises: Beyond the Basics: **89.** $f(x) = 3x^2 + 12x + 15$
91. $f(x) = 6x^2 - 12x + 4$ **93.** $f(x) = -2x^2 + 12x + 14$
95. Opening up: $x^2 - 4x - 12$; opening down: $-x^2 + 4x + 12$
97. Opening up: $x^2 + 8x + 7$; opening down: $-x^2 - 8x - 7$
99. $a = 2$

Critical Thinking: **102.** (5, 19)

Section 3.2

Practice Problems: **1. a.** Not a polynomial function
b. A polynomial function; the degree is 7, the leading term is $2x^7$,
and the leading coefficient is 2.

2. $P(x) = 4x^3 + 2x^2 + 5x - 17 = x^3\left(4 + \dfrac{2}{x} + \dfrac{5}{x^2} - \dfrac{17}{x^3}\right)$

When $|x|$ is large, the terms $\dfrac{2}{x}, \dfrac{5}{x^2},$ and $-\dfrac{17}{x^3}$ are close to 0.

Therefore, $P(x) \approx x^3(4 + 0 + 0 - 0) = 4x^3$ as $x \to \infty$.

3. $y \to -\infty$ as $x \to -\infty$ and $y \to -\infty$ as $x \to \infty$ **4.** $\dfrac{3}{2}$

5. $f(1) = -7, f(2) = 4$; because $f(1)$ and $f(2)$ have opposite signs,
by the Intermediate Value Theorem, f has a real zero between 1 and 2.
6. $-5, -1,$ and 3   **7.** -5 and -3 with multiplicity 1 and 1 with
multiplicity 2   **8.** 3
9.

10. 207.378

A Exercises: Basic Skills and Concepts: **1.** 5; $2x^5$; 2; -6
3. end behavior   **5.** zero   **7.** $n - 1$ **9.** Polynomial function;
degree: 5; leading term: $2x^5$; leading coefficient: 2   **11.** Polynomial

function; degree: 3; leading term: $\dfrac{2}{3}x^3$; leading coefficient: $\dfrac{2}{3}$

13. Polynomial function; degree: 4; leading term: πx^4; leading
coefficient: π **15.** Presence of $|x|$ **17.** Division by a polynomial is
not allowed.   **19.** Presence of $\sqrt{x}$ **21.** Division by a polynomial is
not allowed.   **23.** Graph not continuous   **25.** Graph not continuous
27. Not a function   **29.** c   **31.** a   **33.** d
35. $f(x) = 2(x - 2)(x + 1)(x + 2)$
37. $f(x) = -2(x - 1)^2(x - 2)(x + 2)$
39. Zeros: $x = -5$, multiplicity: 1, the graph crosses the x-axis;
$x = 1$, multiplicity: 1, the graph crosses the x-axis; maximum number
of turning points: 1
41. Zeros: $x = -1$, multiplicity: 2, the graph touches but does not
cross the x-axis; $x = 1$, multiplicity: 3, the graph crosses the x-axis;
maximum number of turning points: 4

43. Zeros: $x = -\dfrac{2}{3}$, multiplicity: 1, the graph crosses the x-axis;

$x = \dfrac{1}{2}$, multiplicity: 2, the graph touches but does not cross the x-axis;

maximum number of turning points: 2
45. Zeros: $x = 0$, multiplicity: 2, the graph touches but does not cross
the x-axis; $x = 3$, multiplicity: 2, the graph touches but does not cross
the x-axis; maximum number of turning points: 3
47. 2.09   **49.** 2.28

51. a. $y \to \infty$ as $x \to -\infty$ **b.**
and $y \to \infty$ as $x \to \infty$

53. a. $y \to \infty$ as $x \to -\infty$ and   **b.**
$y \to \infty$ as $x \to \infty$

55. a. $y \to \infty$ as $x \to -\infty$ and   **b.**
$y \to -\infty$ as $x \to \infty$

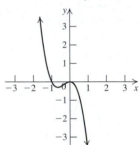

57. a. $y \to \infty$ as $x \to -\infty$ and   **b.**
$y \to \infty$ as $x \to \infty$

59. a. $y \to \infty$ as $x \to -\infty$ and   **b.**
$y \to \infty$ as $x \to \infty$

61. a. $y \to \infty$ as $x \to -\infty$ and   **b.**
$y \to -\infty$ as $x \to \infty$

63. a. $y \to \infty$ as $x \to -\infty$ and   **b.**
$y \to \infty$ as $x \to \infty$

77. a. $V(x) = x^2(62 - 2x)$
b.

79. a. $V(x) = 2x^2(45 - 3x)$
b.

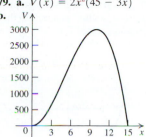

B Exercises: Applying the Concepts: 65. a. Zeros: $x = 0$,
multiplicity 2; $x = 4$, multiplicity 1
b.

c. 2   **d.** Domain: $[0, 4]$; the
portion between the
x-intercepts constitutes the graph
of $R(x)$.

81.

83.

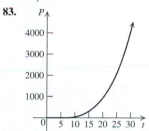

67. a. $[0, 1]$ **b.**

C Exercises: Beyond the Basics:
85. The graph of f is the graph
of $y = x^4$ shifted one unit to the
right.

87. The graph of f is the
graph of $y = x^4$ shifted two
units up.

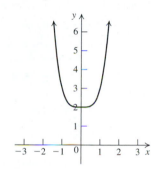

69. a. $R(x) = 27x - \dfrac{1}{90,000}x^3$ **b.** $[0, 900\sqrt{3}]$

c.

Zero: $x = 1$, multiplicity: 4

No zeros

89. The graph of f is the graph
of $y = x^4$ shifted one unit to the
right and reflected in the x-axis.

91. The graph of f is the graph
of $y = x^5$ shifted one unit up.

71. a. $N(x) = (x + 12)(400 - 2x^2)$ **b.** $[0, 10\sqrt{2}]$

c.

Zero: $x = 1$, multiplicity: 4

Zero: $x = -1$, multiplicity: 1

73. a. $V(x) = x(8 - 2x)(15 - 2x)$ **75. a.** $V(x) = x^2(108 - 4x)$
b.

93. The graph of f is the graph of $y = x^5$ shifted one unit to the left, compressed vertically by a factor of 4, reflected in the x-axis, and shifted eight units up.

Zero: $x = 1$, multiplicity: 1

99. $f(x) = x^2(x + 1)$ **101.** $f(x) = 1 - x^4$
103. 5 because the graph crosses the x-axis at 5 points.
105. 6 because the graph has five turning points

Critical Thinking: 107. No, because the domain of any polynomial function is $(-\infty, \infty)$, which includes the point $x = 0$
109. Not possible due to the end behavior

Section 3.3

Practice Problems: 1. Quotient: $3x + 4$; remainder $= -2x + 3$
2. $x^2 + x + 3 + \dfrac{2x - 3}{x^2 - x + 3}$ **3.** $2x^2 - x - 3 + \dfrac{-4}{x - 3}$
4. $2x^2 + 7x + 3 + \dfrac{2}{x - 3}$ **5.** 4   **6.** 32   **7.** $\left\{-2, -\dfrac{2}{3}, 3\right\}$ **8.** 18

A Exercises: Basic Skills and Concepts:
1. $x^4 - 2x^3 + 5x^2 - 2x + 1$; $x^2 - 2x + 3$; $x^2 + 2$; $2x - 5$
3. $F(a)$ **5.** true   **7.** Quotient: $3x - 2$; remainder: 0
9. Quotient: $3x^3 - 3x^2 - 3x + 6$; remainder: -13
11. Quotient: $2x - 1$; remainder: 0   **13.** Quotient: $z^2 + 2z + 1$; remainder: 0   **15.** Quotient: $x^2 - 7$; remainder: -5
17. Quotient: $x^2 + 2x - 11$; remainder: 12
19. Quotient: $x^3 - x^2 + 4$; remainder: 13
21. Quotient: $2x^2 + 5x - \dfrac{1}{2}$; remainder: $\dfrac{3}{4}$
23. Quotient: $2x^2 - 6x + 6$; remainder: -1
25. Quotient: $x^4 + 2x^3 - 5x^2 - 3x - 2$; remainder: -3
27. Quotient: $x^4 - x^3 + x^2 - x + 1$; remainder: 0
29. Quotient: $x^5 - x^4 + 3x^3 - 4x^2 + 4x - 4$; remainder: 9
31. a. 5   **b.** 3   **c.** $\dfrac{15}{8}$ **d.** 1301   **33. a.** -17 **b.** -27 **c.** -56
d. 24   **43.** $k = -1$ **45.** $k = -7$ **47.** Remainder: 1
49. Remainder: 6

B Exercises: Applying the Concepts: 51. $2x^2 + 1$
53. a. $R(x) = -2x^2 + 200x - 4800$ **b.** $R(x) = -2x^2 + 200x - 1800$
c. The maximum weekly revenue is $3200 if the phone is priced at $50.
55. 1990   **57.** 2000 and 2007

C Exercises: Beyond the Basics: 61. a. n is an odd integer.
b. n is an even integer.   **c.** There is no such n.
d. n is any positive integer.   **63. a.** Remainder: 2
b. Remainder: -2 **67.** For example, $g(x) = x^4 + 1$
69. For example, $f(x) = x^3 - 1$ and $g(x) = x^4 + 1$

Critical Thinking: 71. a. False   **b.** False

Section 3.4

Practice Problems: 1. $\{-2\}$ **2.** $\dfrac{1}{2}, 2 + \sqrt{3}, 2 - \sqrt{3}$
3. Number of positive zeros: 1; number of negative zeros: 0 or 2
4. Upper bound: 1; lower bound: -3 **5.** $-\sqrt{2}, \dfrac{2}{3}, \sqrt{2}, 3$

A Exercises: Basic Skills and Concepts: 1. a_0; a_n
3. variations; even integer   **5.** false
7. $\left\{\pm\dfrac{1}{3}, \pm 1, \pm\dfrac{5}{3}, \pm 5\right\}$ **9.** $\left\{\pm\dfrac{1}{4}, \pm\dfrac{1}{2}, \pm\dfrac{3}{4}, \pm 1, \pm\dfrac{3}{2}, \pm 2, \pm 3, \pm 6\right\}$
11. $\{-2, 1, 2\}$ **13.** $\{-1\}$ **15.** $\left\{-3, \dfrac{1}{2}, 2\right\}$
17. $\left\{\dfrac{2}{3}\right\}$ **19.** $\left\{-\dfrac{1}{3}\right\}$ **21.** $\{-1, 2\}$ **23.** $\{-3, -1, 1, 4\}$
25. No rational zeros   **27.** Number of positive zeros: 0 or 2; number of negative zeros: 1   **29.** Number of positive zeros: 0 or 2; number of negative zeros: 1   **31.** Number of positive zeros: 1 or 3; number of negative zeros: 0 or 2   **33.** Number of positive zeros: 1 or 3; number of negative zeros: 1   **35.** Number of positive zeros: 0; number of negative zeros: 0   **37.** Number of positive zeros: 0; number of negative zeros: 0   **39.** Number of positive zeros: 1; number of negative zeros: 0   **41.** Upper bound: $\dfrac{1}{3}$; lower bound: -1
43. Upper bound: 1; lower bound: $-\dfrac{7}{3}$ **45.** Upper bound: 31; lower bound: -31 **47.** Upper bound: $\dfrac{7}{2}$; lower bound: $-\dfrac{7}{2}$
49. $\{1, -3 + \sqrt{11}, -3 - \sqrt{11}\}$ **51.** $\left\{-\sqrt{3}, \dfrac{1}{2}, \sqrt{3}\right\}$
53. $\left\{\dfrac{1}{2}, 2 - \sqrt{3}, 2 + \sqrt{3}\right\}$ **55.** $\{-2, -\sqrt{3}, 1, \sqrt{3}\}$
57. $\left\{-1, 1, \dfrac{3 + \sqrt{5}}{2}, \dfrac{3 - \sqrt{5}}{2}\right\}$ **59.** $\left\{-\sqrt{3}, \dfrac{1}{2}, \sqrt{3}, 2\right\}$
61. $\left\{-\sqrt{2}, -\dfrac{1}{3}, \dfrac{1}{2}, \sqrt{2}\right\}$ **63.** $\{-\sqrt{3}, -1, 1, \sqrt{3}, 2\}$
65. $\left\{-2, -\sqrt{3}, \dfrac{1}{2}, 1, \sqrt{3}\right\}$ **67.** $\left\{-\dfrac{1}{2}, 1, 2, 3\right\}$

B Exercises: Applying the Concepts: 69. Length: 51; width: 6
71. $x = 4$ **73.** 1500   **75. a.** 5 years   **b.** $y = -\dfrac{1}{18}x(x - 3)(x - 8)$
c. export 1.7 million barrels   **77. a.** $y = \dfrac{1}{54}x(x - 6)(x - 8)$
b. $592.59 profit   **c.** 2000, 2003   **d.** $898.46 in 2002

C Exercises: Beyond the Basics: 83. $y = x^3 - x$
85. $y = 6x^3 - 23x^2 + 15x + 14$ **87.** $y = x^3 - 5x^2 + 5x + 3$
89. $y = x^4 - 8x^3 + 17x^2 - 2x - 14$ **91.** $y = x^4 - 10x^2 + 1$
93. $y = x^4 - 12x^3 + 40x^2 - 24x - 36$

Critical Thinking: 97. a. False   **b.** False

Section 3.5

Practice Problems:
1. a. $P(x) = 3(x + 2)(x - 1)(x - 1 - i)(x - 1 + i)$
b. $P(x) = 3x^4 - 3x^3 - 6x^2 + 18x - 12$
2. $-3, -3, 2 + 3i, 2 + 3i, 2 - 3i, 2 - 3i, i, -i$ **3.** The zeros are 1, 2, $2i$, and $-2i$.   **4.** The zeros are 2, 4, $1 + i$, and $1 - i$.

A Exercises: Basic Skills and Concepts: 1. complex
3. $a - bi$ **5.** false   **7.** $x = \pm 5i$ **9.** $x = -2 \pm 3i$
11. $\{2, 3i, -3i\}$ **13.** $3 - i$ **15.** $2 + i$ **17.** $-i, -3i$
19. $P(x) = 2x^4 - 20x^3 + 70x^2 - 180x + 468$
21. $P(x) = 7x^5 - 119x^4 + 777x^3 - 2415x^2 + 3500x - 1750$

---

97. Maximum vertical distance: 0.25. It occurs at $x \approx 0.71$.

Zero: $x = -4$, multiplicity: 1

23. $P(x) = x(x + 1)(x^2 + 9)$

25. $P(x) = x(x - 1)(x + 2)(x^2 - 6x + 10)$

27. $1, 4 + i, 4 - i$ **29.** $-\dfrac{4}{3}, 1 + 3i, 1 - 3i$

31. 1 (with multiplicity 2), $\dfrac{3}{2} + \dfrac{3}{2}i, \dfrac{3}{2} - \dfrac{3}{2}i$ **33.** $-3, 1, 3 + i, 3 - i$

35. $\dfrac{1}{2}, 2, 3, i, -i$ **37.** $f(x) = \dfrac{1}{3}(x^2 + 9)$

39. $f(x) = \dfrac{1}{2}(2 - x)(x^2 + 1)(x^2 + 4)$

C Exercises: Beyond the Basics: 41. There are three cube roots:

$1, -\dfrac{1}{2} + \dfrac{1}{2}\sqrt{3}i,$ and $-\dfrac{1}{2} - \dfrac{1}{2}\sqrt{3}i.$ **43.** $P(x) = (x - 2)(x + i)$

45. $P(x) = (x - 4)(x + 1)(x - i)$

47. $f(x) = -4(x - 2)(x^2 - 2x + 5)$

$y \to \infty$ as $x \to -\infty,$ $y \to -\infty$ as $x \to \infty.$

49. $f(x) = -2(x^2 - 1)(x^2 - 6x + 10)$

$y \to -\infty$ as $x \to -\infty,$ $y \to -\infty$ as $x \to \infty.$

Section 3.6

Practice Problems: 1. Domain: $(-\infty, -1) \cup (-1, 5) \cup (5, \infty)$

2. $x = -5$ and $x = 2$ **3.** $x = -3$ **4. a.** $y = \dfrac{2}{3}$ **b.** No

horizontal asymptote

5. x-intercept: 0; y-intercept: 0; vertical asymptotes: $x = -1$ and $x = 1$; horizontal asymptote: x-axis. The graph is symmetric with respect to the origin. The graph is above the x-axis on $(-1, 0) \cup (1, \infty)$ and below the x-axis on $(-\infty, -1) \cup (0, 1).$

6. x-intercepts: $\pm\dfrac{\sqrt{2}}{2}$; y-intercept: $\dfrac{1}{3}$;

vertical asymptotes: $x = -\dfrac{3}{2}$

and $x = 1$; horizontal asymptote: $y = 1$; symmetry: none. The graph is above the x-axis on

$\left(-\infty, -\dfrac{3}{2}\right) \cup \left(-\dfrac{\sqrt{2}}{2}, \dfrac{\sqrt{2}}{2}\right) \cup (1, \infty)$

and below the x-axis on

$\left(-\dfrac{3}{2}, -\dfrac{\sqrt{2}}{2}\right) \cup \left(\dfrac{\sqrt{2}}{2}, 1\right).$

7. No x-intercept; y-intercept: $\dfrac{1}{2}$; no vertical asymptote; horizontal asymptote: $y = 1.$ The graph is symmetric with respect to the y-axis. The graph is above the x-axis on $(-\infty, \infty).$

8. No x-intercept; y-intercept: -2; vertical asymptote: $x = 1$; no horizontal asymptote; $y = x + 1$ is an oblique asymptote; symmetry: none. The graph is above the x-axis on $(1, \infty)$ and below the x-axis on $(-\infty, 1).$

9. a. $R(10) = \$30$ billion
$R(20) = \$40$ billion
$R(30) = \$42$ billion
$R(40) = \$40$ billion
$R(50) = \$35.7$ billion
$R(60) = \$30$ billion

b.

c. 29%

A Exercises: Basic Skills and Concepts:

1. $\dfrac{N(x)}{D(x)},$ where $N(x)$ and $D(x)$ are polynomials and $D(x)$ is not the

zero polynomial   **3.** $\infty; -\infty$ **5.** false   **7.** $(-\infty, -4) \cup (-4, \infty)$

9. $(-\infty, \infty)$ **11.** $(-\infty, -2) \cup (-2, 3) \cup (3, \infty)$

13. $(-\infty, 2) \cup (2, 4) \cup (4, \infty)$ **15.** ∞ **17.** ∞

19. 1   **21.** $(-\infty, -2) \cup (-2, 1) \cup (1, \infty)$

23. $x = -2; x = 1$ **25.** $x = 1$ **27.** $x = -4$ and $x = 3$

29. $x = -3$ and $x = 2$ **31.** $x = -3$ **33.** No vertical asymptote

35. $y = 0$ **37.** $y = \dfrac{2}{3}$ **39.** No horizontal asymptote   **41.** $y = 0$

43. d   **45.** e   **47.** a

49. x-intercept: 0; y-intercept: 0; vertical asymptote: $x = 3$; horizontal asymptote: $y = 2$; symmetry: none. The graph is above the x-axis on $(-\infty, 0) \cup (3, \infty)$ and below the x-axis on $(0, 3).$

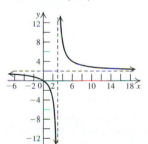

51. x-intercept: 0; y-intercept: 0; vertical asymptotes: $x = -2$ and $x = 2$; horizontal asymptote: x-axis. The graph is symmetric with respect to the origin. The graph is above the x-axis on $(-2, 0) \cup (2, \infty)$ and below the x-axis on $(-\infty, -2) \cup (0, 2).$

53. x-intercept: 0; y-intercept: 0; vertical asymptotes: $x = -3$ and $x = 3$; horizontal asymptote: $y = -2$. The graph is symmetric in the y-axis. The graph is above the x-axis on $(-3, 0) \cup (0, 3)$ and below the x-axis on $(-\infty, -3) \cup (3, \infty)$.

55. No x-intercept; y-intercept: -1; vertical asymptotes: $x = -\sqrt{2}$ and $x = \sqrt{2}$; horizontal asymptote: x-axis. The graph is symmetric with respect to the y-axis. The graph is above the x-axis on $(-\infty, -\sqrt{2}) \cup (\sqrt{2}, \infty)$ and below the x-axis on $(-\sqrt{2}, \sqrt{2})$.

63. No x-intercept; y-intercept: -2; no vertical asymptote; no horizontal asymptote; symmetry: none. The graph is above the x-axis on $(2, \infty)$ and below the x-axis on $(-\infty, 2)$.

65. $f = \dfrac{-2(x - 1)}{x - 2}$ **67.** $f(x) = \dfrac{(x - 1)(x - 3)}{x(x - 2)}$

69. Oblique asymptote: $y = 2x$ **71.** Oblique asymptote: $y = x$

57. x-intercept: -1; y-intercept: $-\dfrac{1}{6}$; vertical asymptotes: $x = -3$ and $x = 2$; horizontal asymptote: x-axis; symmetry: none. The graph is above the x-axis on $(-3, -1) \cup (2, \infty)$ and below the x-axis on $(-\infty, -3) \cup (-1, 2)$.

73. Oblique asymptote: $y = x - 2$ **75.** Oblique asymptote: $y = x - 2$

59. x-intercept: 0; y-intercept: 0; no vertical asymptote; horizontal asymptote: $y = 1$. The graph is symmetric with respect to the y-axis. The graph is above the x-axis on $(-\infty, 0) \cup (0, \infty)$.

B Exercises: Applying the Concepts: **77. a.** $C(x) = 0.5x + 2000$

b. $\overline{C}(x) = 0.5 + \dfrac{2000}{x}$ **c.** $\overline{C}(100) = 20.5, \overline{C}(500) = 4.5, \overline{C}(1000) = 2.5$.

These show the average cost of producing $100, 500,$ and 1000 trinkets, respectively.   **d.** Horizontal asymptote: $y = 0.5$. It means that the average cost approaches the actual cost of producing each trinket if the number of trinkets produced approaches ∞.

61. x-intercept: ± 2; y-intercept: none; vertical asymptotes: $x = -3$ and $x = 3$; horizontal asymptote: $y = 1$. The graph is symmetric with respect to the origin. The graph is above the x-axis on $(-\infty, -3) \cup (-2, 2) \cup (3, \infty)$ and below the x-axis on $(-3, -2) \cup (2, 3)$

79. a.

b. (i) about 4 min
(ii) about 12 min
(iii) about 76 min
(iv) 397 min
c. (i) ∞
(ii) Not applicable; the domain is $x < 100$.   **d.** No

81. a. $3.02 billion

b.

c. 90.89%

83. a. 16,000

b. The population will stabilize at 4000.

85. a. $f(x) = \dfrac{10x + 200,000}{x - 2500}$ **b.** $40   **c.** More than 25,000

d. Horizontal asymptote $y = 10$. As $x \to \infty$, the average cost $\to 10$. Vertical asymptote $x = 2500$. The average cost is undefined if $x = 2500$.

C Exercises: Beyond the Basics:

87. Stretch the graph of $y = \dfrac{1}{x}$ vertically by a factor of 2 and reflect the graph in the x-axis.

89. Shift the graph of $y = \dfrac{1}{x^2}$ two units to the right.

91. Shift the graph of $y = \dfrac{1}{x^2}$ one unit to the right and two units down.

93. Shift the graph of $y = \dfrac{1}{x^2}$ six units to the left.

95. Shift the graph of $y = \dfrac{1}{x^2}$ one unit to the right and one unit up.

97. a. $f(x) \to 0$ as $x \to -\infty$, $f(x) \to 0$ as $x \to \infty$, $f(x) \to -\infty$ as $x \to 0^-$, and $f(x) \to \infty$ as $x \to 0^+$; no x-intercept; no y-intercept; vertical asymptote: y-axis; horizontal asymptote: x-axis. The graph is symmetric with respect to the origin. The graph is above the x-axis on $(0, \infty)$ and below the x-axis on $(-\infty, 0)$.

b. $f(x) \to 0$ as $x \to -\infty$, $f(x) \to 0$ as $x \to \infty$, $f(x) \to \infty$ as $x \to 0^-$, and $f(x) \to -\infty$ as $x \to 0^+$; no x-intercept; no y-intercept; vertical asymptote: y-axis; horizontal asymptote: x-axis. The graph is symmetric with respect to the origin. The graph is above the x-axis on $(-\infty, 0)$ and below the x-axis on $(0, \infty)$.

c. $f(x) \to 0$ as $x \to \infty$, $f(x) \to 0$ as $x \to -\infty$, $f(x) \to \infty$ as $x \to 0^-$, and $f(x) \to \infty$ as $x \to 0^+$; no x-intercept; no y-intercept; vertical asymptote: y-axis; horizontal asymptote: x-axis. The graph is symmetric with respect to the y-axis. The graph is above the x-axis on $(-\infty, 0) \cup (0, \infty)$.

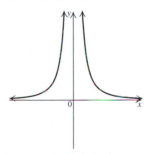

d. $f(x) \to 0$ as $x \to -\infty$, $f(x) \to 0$ as $x \to \infty$, $f(x) \to -\infty$ as $x \to 0^-$, and $f(x) \to -\infty$ as $x \to 0^+$; no x-intercept; no y-intercept; vertical asymptote: y-axis; horizontal asymptote: x-axis. The graph is symmetric with respect to the y-axis. The graph is below the x-axis on $(-\infty, 0) \cup (0, \infty)$.

99.

101. $[f(x)]^{-1} = \dfrac{1}{2x+3}$ and $f^{-1}(x) = \dfrac{1}{2}x - \dfrac{3}{2}$. The two functions are different.

103. $g(x)$ has the oblique asymptote $y = 2x + 3$; $y \to -\infty$ as $x \to -\infty$, and $y \to \infty$ as $x \to \infty$.

105. Point of intersection: $\left(-\dfrac{1}{3}, 1\right)$; horizontal asymptote $y = 1$

107. $f(x) = \dfrac{2-x}{x-3}$ **109.** $f(x) = \dfrac{-x^2+1}{x}$

Critical Thinking: 111. $f(x) = \dfrac{x+4}{x-2}$ **112.** $f(x) = \dfrac{x-4}{(x-2)(x+1)}$

113. $f(x) = \dfrac{4x^2-1}{(x-1)^2}$ **114.** $f(x) = \dfrac{2x^2}{x^2+1}$

115. $f(x) = 3x + 2 + \dfrac{1}{x-1}$ **116.** No, because in that case, there would be values of x with two different corresponding function values; so it would not be a function.

Section 3.7

Practice Problems:
1. $(1, 4)$ **2.** No

3. $(-\infty, -1-\sqrt{3}) \cup (-1+\sqrt{3}, \infty)$
4. $[-4, -3] \cup [3, \infty)$

5. $[-2, 2]$

6. $\left(-\dfrac{1}{3}, \dfrac{1}{3}\right)$

7. $(-5, -3) \cup [2, 3)$

A Exercises: Basic Skills and Concepts: **1.** $(-2, 3)$ **3.** $(-5, 2)$
5. $[-4, 1]$ **7.** $(-\infty, -1] \cup [6, \infty)$ **9.** $(-3, 0)$ **11.** $[-3, 3]$
13. $(-\infty, -5) \cup (5, \infty)$ **15.** $[-6, 2]$
17. $\left(-\infty, -\dfrac{3}{2}\right] \cup \left[\dfrac{1}{3}, \infty\right)$ **19.** $\left[-\dfrac{3}{5}, \dfrac{2}{5}\right]$ **21.** $[1, 3]$
23. $\left(-4, \dfrac{3}{2}\right)$ **25.** $(-\infty, -3] \cup [0, 3]$ **27.** $[-3, -1] \cup [1, \infty)$
29. $(-2, 0) \cup (6, \infty)$ **31.** $(-\infty, -2) \cup (1, 2)$ **33.** $[-1, \infty)$
35. $(-\infty, -1] \cup [1, 2]$ **37.** $(1-\sqrt{3}, 1+\sqrt{3})$ **39.** $(-\infty, \infty)$
41. $\varnothing$ **43.** $\{0\} \cup [1, \infty)$ **45.** $[-\sqrt{2}, \sqrt{2}]$
47. $(-\infty, -1] \cup [1, \infty)$ **49.** $[-2, 2]$ **51.** $(-2, 5)$
53. $\left[-\dfrac{2}{3}, 3\right)$ **55.** $(-4, 0)$ **57.** $\left(-\infty, -\dfrac{5}{2}\right] \cup (-2, \infty)$
59. $(-2, 0) \cup (2, \infty)$ **61.** $(-2, -1] \cup [3, 4)$
63. $(-\infty, -5) \cup [-1, 2] \cup (3, \infty)$ **65.** $(-2, -1] \cup [1, 2)$
67. $(-\infty, -6) \cup \left[-\dfrac{3}{2}, \dfrac{3}{2}\right] \cup (6, \infty)$ **69.** $(-\infty, 0) \cup [2, 4]$
71. $\left(-\infty, \dfrac{4}{3}\right) \cup (2, \infty)$ **73.** $\left(\dfrac{2}{3}, 3\right]$ **75.** $\left(-\dfrac{1}{2}, \dfrac{3}{4}\right]$
77. $\left(-2, -\dfrac{10}{9}\right) \cup (2, \infty)$ **79.** $\left(-3, -\dfrac{1}{3}\right] \cup (3, \infty)$
81. $\left(-1, \dfrac{1}{7}\right] \cup (3, \infty)$

B Exercises: Applying the Concepts: **83.** 2 sec to 2.5 sec
85. 10°F to 100°F   **87.** Between \$3 million and \$4 million
89. More than 133 cards but not more than 180

C Exercises: Beyond the Basics: **91.** $(-\infty, -4) \cup (4, \infty)$
93. $(-\infty, 0) \cup (4, \infty)$ **95.** $(-\infty, -1] \cup [1, \infty)$ **97.** $(-1, 1)$
99. $(-\infty, -3) \cup [1, 0)$ **101.** $(-\infty, -12] \cup [12, \infty)$
103. At most, 4000 radios

***Critical Thinking: 105.* a.** $(x+4)(x-5) < 0$
b. $(x+2)(x-6) \le 0$ **c.** $x^2 \ge 0$ **d.** $x^2 < 0$
e. $(x-3)^2 \le 0$ **f.** $(x-2)^2 > 0$ **106. a.** $\dfrac{x-4}{x+2} \le 0$
b. $\dfrac{x-3}{x-5} \le 0$ **c.** No

Section 3.8

Practice Problems: **1.** $y = 24$ **2.** 275   **3.** $y = 300$ **4.** $A = 20$
5. $x = \dfrac{9}{4}$ **6.** $\approx 3.7 \text{ m/sec}^2$

A Exercises: Basic Skills and Concepts: **1.** $P = kT$ **3.** $y = k\sqrt{x}$
5. $V = kx^3y^4$ **7.** $z = \dfrac{kxu}{v^2}$ **9.** $k = \dfrac{1}{2}, x = 14$

11. $k = 16, s = 400$ **13.** $k = 33, r = 99$ **15.** $k = 8, B = \dfrac{1}{8}$

17. $k = 7, y - 4$ **19.** $k = 2, z = 50$ **21.** $k = \dfrac{1}{17}, P = \dfrac{324}{17}$

23. $k = 54, x = 4$ **25.** $y = 6$ **27.** $y = 200$

B Exercises: Applying the Concepts: 29. $y = Hx$, where y is the speed of the galaxies and x is the distance between them.
31. a. $y = 30.5x$, where y is the length measured in centimeters and x is the length measured in feet.   **b. (i)** 244 cm **(ii)** 162.67 cm

c. (i) 1.87 ft **(ii)** 4.07 ft   **33.** 35 g   **35.** $\dfrac{3}{4}$ sec   **37.** 60 lb/in.^{2}

39. a. 18.97 lb   **b.** 201.01 lb   **41.** 1.63 m/s^2
43. a. 1280 candlepower   **b.** 8.94 ft from the source
45. The length is multiplied by 4.   **47. a.** $H = kR^2N$, where k is a constant.   **b.** The horsepower is multiplied by 4.   **c.** The horsepower is doubled.   **d.** The horsepower is cut in half.

C Exercises: Beyond the Basics: 49. a. $E = kl^2v^3$ **b.** $k = 0.0375$
c. 37,500 W   **d.** E is multiplied by 8.   **e.** E is multiplied by 4.
f. E is multiplied by 32.   **51. a.** $k = 2.94$ **b.** 287.25 W
c. The metabolic rate is multiplied by 2.83.   **d.** 373.93 kg

53. a. $T^2 = \left(\dfrac{4\pi^2}{G}\right)\left(\dfrac{r^3}{M_1 + M_2}\right)$ **b.** 2.01×10^{30} kg

55. a. $R = kN(P - N)$, where k is the constant of variation.

b. $k = \dfrac{5}{10^6}$ **c.** 125 people per day   **d.** 2764 or 7236

Critical Thinking: 56. 56%   **57. a.** \$480   **b.** \$15,920
c. The value of the original diamond is \$112,500. The value of a diamond whose weight is twice that of the original is \$450,000.
58. 20   **59.** 1:2   **60.** 144

Review Exercises

1. (i) Opens up
(ii) Vertex: $(1, 2)$
(iii) Axis: $x = 1$
(iv) No x-intercept
(v) y-intercept: 3
(vi) Decreasing on $(-\infty, 1)$, increasing on $(1, \infty)$

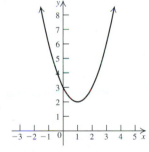

3. (i) Opens down
(ii) Vertex: $(3, 4)$
(iii) Axis: $x = 3$
(iv) x-intercepts: $3 \pm \sqrt{2}$
(v) y-intercept: -14
(vi) Increasing on $(-\infty, 3)$, decreasing on $(3, \infty)$

5. (i) Opens down
(ii) Vertex: $(0, 3)$
(iii) Axis: y-axis
(iv) x-intercepts: $\pm\dfrac{\sqrt{6}}{2}$
(v) y-intercept: 3
(vi) Increasing on $(-\infty, 0)$, decreasing on $(0, \infty)$

7. (i) Opens up
(ii) Vertex: $(1, 1)$
(iii) Axis: $x = 1$
(iv) No x-intercept
(v) y-intercept: 3
(vi) Decreasing on $(-\infty, 1)$, increasing on $(1, \infty)$

9. (i) Opens up
(ii) Vertex: $\left(\dfrac{1}{3}, \dfrac{2}{3}\right)$
(iii) Axis: $x = \dfrac{1}{3}$
(iv) No x-intercept
(v) y-intercept: 1
(vi) Decreasing on $\left(-\infty, \dfrac{1}{3}\right)$, increasing on $\left(\dfrac{1}{3}, \infty\right)$

11. Minimum: -1 **13.** Maximum: $\dfrac{25}{8}$

15. Shift the graph of $y = x^3$ one unit to the left and two units down.

17. Shift the graph of $y = x^3$ one unit right, reflect the resulting graph in the x-axis, and shift it one unit up.

19. (i) $f(x) \to -\infty$ as $x \to -\infty$ and $f(x) \to \infty$ as $x \to \infty$.
(ii) Zeros: $x = -2$, multiplicity: 1, the graph crosses the x-axis; $x = 0$, multiplicity: 1, the graph crosses the x-axis; $x = 1$, multiplicity: 1, the graph crosses the x-axis.
(iii) x-intercepts: $-2, 0, 1$; y-intercept: 0
(iv) The graph is above the x-axis on $(-2, 0) \cup (1, \infty)$ and below the x-axis on $(-\infty, -2) \cup (0, 1)$.
(v) Symmetry: none
(vi)

21. (i) $f(x) \to -\infty$ as $x \to -\infty$ and $f(x) \to -\infty$ as $x \to \infty$.
(ii) Zeros: $x = 0$, multiplicity: 2, the graph touches but does not cross the x-axis; $x = 1$, multiplicity: 2, the graph touches but does not cross

the *x*-axis. **(iii)** *x*-intercepts: 0, 1; *y*-intercept: 0 **(iv)** The graph is below the *x*-axis on $(-\infty, 0) \cup (0, 1) \cup (1, \infty)$. **(v)** Symmetry: none **(vi)**

23. (i) $f(x) \to -\infty$ as $x \to -\infty$ and $f(x) \to -\infty$ as $x \to \infty$. **(ii)** Zeros: $x = -1$, multiplicity: 1, the graph crosses the *x*-axis; $x = 0$, multiplicity: 2, the graph touches but does not cross the *x*-axis; $x = 1$ multiplicity: 1, the graph crosses the *x*-axis. **(iii)** *x*-intercepts: $-1, 0, 1$; *y*-intercept: 0 **(iv)** The graph is above the *x*-axis on $(-1, 0) \cup (0, 1)$ and below the *x*-axis on $(-\infty, -1) \cup (1, \infty)$. **(v)** The graph is symmetric with respect to the *y*-axis. **(vi)**

25. Quotient: $2x + 3$; remainder: -7 **27.** Quotient: $8x^3 - 12x^2 + 14x - 21$; remainder: 186 **29.** Quotient: $x^2 + 3x - 3$; remainder: -6 **31.** Quotient: $2x^3 - 5x^2 + 10x - 17$; remainder: 182

33. -11 **35.** 88 **37.** 1, 2, 4 **39.** $-3, -2, \dfrac{1}{3}$

41. $-6, -3, -2, -1, 1, 2, 3, 6$ **43.** $-2, -\dfrac{1}{5}, 0$ **45.** $-3, -2, 2$

47. $-2, 3 \pm \sqrt{2}$ **49.** $\dfrac{1}{2}, 1 \pm \sqrt{3}$ **51.** $-3, 1, 2$ **53.** $-1, 3, 3i, -3i$

55. $2i, -2i, -1 + 2i, -1 - 2i$ **57.** $\{-2, 1, 2\}$ **59.** $\left\{-1, -\dfrac{1}{2}, \dfrac{3}{2}\right\}$

61. $\{2, 3 \pm i\sqrt{2}\}$ **63.** $\left\{-\dfrac{1}{3}, 1 \pm i\sqrt{5}\right\}$ **65.** $\{-1, 2, i, -i\}$

67. $\left\{-2, \dfrac{1}{2}, 1 \pm i\sqrt{2}\right\}$ **69.** The only possible rational roots are $-2, -1, 1,$ and 2. None of them satisfy the equation. **71.** 1.88 **73.** *x*-intercept: -1; no *y*-intercept; vertical asymptote: *y*-axis; horizontal asymptote: $y = 1$; symmetry: none. The graph is above the *x*-axis on $(-\infty, -1) \cup (0, \infty)$ and below the *x*-axis on $(-1, 0)$.

75. *x*-intercept: 0; *y*-intercept: 0; vertical asymptotes: $x = -1$ and $x = 1$; horizontal asymptote: *x*-axis; symmetry: none. The graph is above the *x*-axis on $(-1, 0) \cup (1, \infty)$ and below the *x*-axis on $(-\infty, -1) \cup (0, 1)$.

77. *x*-intercept: 0; *y*-intercept: 0; vertical asymptotes: $x = -3$ and $x = 3$; no horizontal asymptote; symmetry: none. The graph is above the *x*-axis on $(-3, 0) \cup (3, \infty)$ and below the *x*-axis on $(-\infty, -3) \cup (0, 3)$.

79. *x*-intercept: 0; *y*-intercept: 0; vertical asymptotes: $x = -2$ and $x = 2$; no horizontal asymptote. The graph is symmetric with respect to the *y*-axis. The graph is above the *x*-axis on $(-\infty, -2) \cup (2, \infty)$ and below the *x*-axis on $(-2, 0) \cup (0, 2)$.

81. $(-\infty, -2] \cup [1, \infty)$ **83.** $\left[-\dfrac{1}{2}, 3\right]$ **85.** $(-3, -1) \cup (1, \infty)$

87. $(-\infty, -2\sqrt{2}) \cup (1, 2\sqrt{2})$ **89.** $(3, \infty)$

91. $\left(-\infty, -\dfrac{1}{2}\right] \cup (1, \infty)$ **93.** $(-\infty, -4) \cup [-2, 0) \cup [5, \infty)$

95. $(-4, -2) \cup [1, 3]$ **97.** $y = 15$ **99.** $s = 45$

101. Maximum height: 1000; the missile hits the ground at $x = 200$.

103. $x \approx 24.5$ ft, $y \approx 16.3$ ft

105. a. **b.** $x = 50$

107. a. $P(x) = -\dfrac{3}{1000}x^2 + 20.1x - 150$ **b.** 3350

c. $\overline{C}(x) = \dfrac{3}{1000}x + 3.9 + \dfrac{150}{x}$

109. $245
111. $6\sqrt{2}$ in.
113. a. 36 amp **b.** 150 Ω

115. a. $k = \dfrac{1}{200,000}$

b. 500 people every day
c. 1000 and 19,000

Practice Test A

1. x-intercepts: $3 \pm \sqrt{7}$
2.

3. $(1, 10)$
4. $(-\infty, -4) \cup (-4, 1) \cup (1, \infty)$
5. Quotient: $x^2 - 4x + 3$;
remainder: 0

6.

7. $-2, 1, 2$ **8.** $P(-2) = -53$
9. $-2, 2, 5$

10. $0, -\dfrac{1}{2} + \dfrac{\sqrt{61}}{2}, -\dfrac{1}{2} - \dfrac{\sqrt{61}}{2}$

11. $-9, -\dfrac{9}{2}, -3, -\dfrac{3}{2}, -1, -\dfrac{1}{2}, \dfrac{1}{2}, 1, \dfrac{3}{2}, 3, \dfrac{9}{2}, 9$
12. $f(x) \to -\infty$ as $x \to -\infty$ and $f(x) \to \infty$ as $x \to \infty$.
13. $x = -2$, multiplicity: 3; $x = 2$, multiplicity: 1   **14.** Number of
positive zeros: 0 or 2; number of negative zeros: 1   **15.** Horizontal
asymptote: $y = 2$; vertical asymptotes: $x = -4$ and $x = 5$
16. $(-\infty, 0) \cup (0, 2) \cup (3, \infty)$ **17.** $[-5, 1) \cup [3, 5)$ **18.** $y = 4$
19. 15 thousand units   **20.** $V(x) = x(8 - 2x)(17 - 2x)$

Practice Test B

1. b   **2.** d   **3.** a   **4.** b   **5.** d   **6.** c   **7.** c   **8.** c   **9.** a
10. a   **11.** d   **12.** c   **13.** b   **14.** c   **15.** c   **16.** b   **17.** c
18. b   **19.** b   **20.** d

Cumulative Review Exercises (Chapters P–3)

1. $\sqrt{13}$ **2.** $M(-3, -4)$

3. x-intercepts: $4, -2$;
y-intercept: -8

4. Slope is $-\dfrac{1}{3}$; x-intercept: 6;
y-intercept: 2

5. Slope is $\dfrac{1}{2}$; x-intercept:
-6; y-intercept: 3

6. $(x - 2)^2 + (y + 3)^2 = 16$ **7.** Center: $(-1, 2)$; radius: 3
8. $y = 3x - 5$ **9.** $y = -\dfrac{2}{3}x + \dfrac{11}{3}$

10. Domain: $\left(-\infty, -\dfrac{3}{2}\right) \cup \left(-\dfrac{3}{2}, \infty\right)$ **11.** Domain: $(-\infty, 2)$

12. $f(-2) = 11, f(3) = 6, f(x + h) = x^2 + 2(h - 1)x +$
$h^2 - 2h + 3, \dfrac{f(x + h) - f(x)}{h} = 2x + h - 2$

13. a. $f(g(x)) = \sqrt{x^2 + 1}$ **b.** $g(f(x)) = x + 1$
c. $f(f(x)) = \sqrt[4]{x}$ **d.** $g(g(x)) = x^4 + 2x^2 + 2$
14. a. $f(1) = 5, f(3) = 11, f(4) = 6$
b.

15. $f^{-1}(x) = \dfrac{1}{2}x + \dfrac{3}{2}$

16. a. Shift the graph of $y = \sqrt{x}$ two units to the left.

b. Shift the graph of $y = \sqrt{x}$ one unit to the left, stretch the resulting
graph horizontally by a factor of 2, reflect it in the x-axis, and shift it
three units up.

17. $-6, -3, -2, -\dfrac{3}{2}, -1, -\dfrac{1}{2}, \dfrac{1}{2}, 1, \dfrac{3}{2}, 2, 3, 6$

18.

19.

20.

21.

22. $-1, 2, 1 + i, 1 - i$ **23.** 9   **24.** 1250, \$28,250   **25.** $x = 50$

CHAPTER 4

Section 4.1

Practice Problems: **1.** $f(2) = \dfrac{1}{16} = 0.0625$

$$f(0) = 1$$
$$f(-1) = 4$$
$$f\left(\dfrac{5}{2}\right) = \dfrac{1}{32} = 0.03125$$
$$f\left(-\dfrac{3}{2}\right) = 8$$

2.

3.

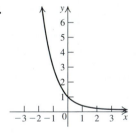

4. $a = \dfrac{1}{5}$ **5.** $x = 2$

6.

7.

8. $t = 6$

A Exercises: Basic Skills and Concepts: **1.** $(-\infty, \infty)$; $(0, \infty)$
3. x-axis   **5.** false   **7.** No, the base is not a constant.

9. Yes, the base is $\dfrac{1}{2}$.   **11.** No, the base is not a constant.

13. No, the base is not a positive constant.   **15.** 25

17. ≈ 0.089 **19.** ≈ -5.657 **21.** 16   **23.** $\dfrac{2}{3}$ **25.** $a = 4$

27. $a = 6$ **29.** $a = \dfrac{1}{4}$ **31.** $y = 3 \cdot 2^x$

33.

35.

37.

39.

41. Yes, reflection in the y-axis   **43.** c   **45.** a
47. Domain $= (-\infty, \infty)$ **49.** Domain $= (-\infty, \infty)$
Range $= (0, \infty)$ Range $= (0, \infty)$
Horizontal asymptote: $y = 0$ Horizontal asymptote: $y = 0$

51. Domain $= (-\infty, \infty)$ **53.** Domain $= (-\infty, \infty)$
Range $= (-\infty, 1)$ Range $= (-\infty, 4)$
Horizontal asymptote: $y = 1$ Horizontal asymptote: $y = 4$

55. a. The graph of $y = 3^x$ is stretched horizontally by a factor of 2.
b. $(0, \infty)$ **c.** $y = 0$ **57. a.** The graph of $y = 3^x$ is stretched
horizontally by a factor of 3 and shifted one unit down.   **b.** $(-1, \infty)$
c. $y = -1$ **59. a.** The graph of $y = 3^x$ is compressed horizontally
by a factor of $\dfrac{1}{2}$, shifted $\dfrac{1}{2}$ unit to the right, and shifted two units up.
b. $(2, \infty)$ **c.** $y = 2$ **61. a.** The graph of $y = 3^x$ is compressed
horizontally by a factor of $\dfrac{1}{2}$, shifted $\dfrac{1}{2}$ unit to the left, stretched
vertically by a factor of 5, and shifted four units up.   **b.** $(4, \infty)$
c. $y = 4$ **63. a.** The graph of $y = 3^x$ is compressed horizontally
by a factor of $\dfrac{1}{2}$, reflected in the y-axis, shifted $\dfrac{5}{2}$ units to the right,
stretched vertically by a factor of 2, reflected in the x-axis,
and shifted four units down.   **b.** $(-\infty, -4)$ **c.** $y = -4$

65. $y = 2^{x+2} + 5$ **67.** $y = 2\left(\dfrac{1}{2}\right)^x - 5$ **69.** $x = 4$

71. $x = \pm\dfrac{7}{2}$ **73.** $x = -3$ **75.** $x = -\dfrac{2}{3}$ **77.** $x = -3$

79. $x = \pm 4$ **81.** $x = \dfrac{1}{8}$ **83.** $x = 0$

B Exercises: Applying the Concepts: 85. a. (i) 151.572   **(ii)** 123.114
b. $t = 5$ hours   **c.** 0°   **87. a. (i)** 1600   **(ii)** 2859   **b.** 4752
c. Just after 2019   **89.** 59 minutes   **91.** 5701.25 yr

C Exercises: Beyond the Basics:
93. Range $= [1, \infty)$
symmetric in the y-axis.

95. Domain $= (-\infty, \infty)$
Range $= [1, \infty)$ symmetric
in the y-axis.

97. a. $f(x) + g(x) = 2 \cdot 3^x$ **b.** $f(x) - g(x) = 2 \cdot 3^{-x}$
c. $[f(x)]^2 - [g(x)]^2 = 4$ **d.** $[f(x)]^2 + [g(x)]^2 = 2(3^{2x} + 3^{-2x})$
99. $-\dfrac{1}{8}$ **101.** $18, -12$ **103.** $x = 2, 3$ **105.** $x = 2, -2$
107. $x = \pm 1$ or $p = 1$ **109. a.** $y = 0.4 \cdot 2^{0.292x}$
b. 2.02 billion   **c.** In 2009

$y(x) = 0.4 \cdot 2^{0.292x}$

Critical Thinking: 111. The function $y = 2^x$ is an increasing
function and is thus one-to-one. The horizontal line $y = k$ for $k > 0$
intersects the graph of $y = 2^x$ at exactly one point.
112. Sketch the graphs of $y = 2^x$ and $y = 2x$. These graphs intersect
at $x = 2$ and $x = 1$.

Section 4.2

Practice Problems: 1. $\$11,500$ **2. a.** $A = \$11,485.03$
b. $I = \$3,485.03$ **3. (i)** $\$5325.00$ **(ii)** $\$5330.28$ **(iii)** $\$5333.01$
(iv) $\$5334.86$ **(v)** $\$5335.76$ **4.** $\$14,764.48$ **5. a.** $\$4,284,000$
b. $\$1,911,233,655.42$ **c.** $\$2,162,832,947$ **d.** $\$2,256,324,191$
6. $y = -e^{x-1} - 2$

7. a. 11.96229   **b.** 4.76720

A Exercises: Basic Skills and Concepts: 1. Prt **3.** Pe^{rt} **5.** true
7. $\$2500, \$7500$ **9.** $\$1397.25, \$8147.25$ **11.** $\$5764.69$,
$\$13,564.69$ **13. a.** $\$11,592.21$ **b.** $\$6342.21$ **15. a.** $\$15,305$
b. $\$9065$ **17. a.** $\$12,365.41$ **b.** $\$4865.41$ **19.** $\$4631.93$
21. $\$4493.68$

23. $f(x) = e^{-x}; y = 0$

25. $f(x) = -e^x; y = 0$

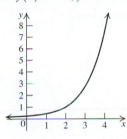

27. $f(x) = 1 + e^x; y = 1$

29. $f(x) = -e^{x-2} + 3; y = 3$

31. a. The graph of $y = e^x$ is stretched horizontally by a factor of
2 and shifted down one unit.   **b.** $(-1, \infty)$ **c.** $y = -1$
33. a. The graph of $y = e^x$ is compressed horizontally by a factor of
$\dfrac{1}{3}$ and shifted up two units.   **b.** $(2, \infty)$ **c.** $y = 2$
35. a. The graph of $y = e^x$ is compressed horizontally by a factor of
$\dfrac{1}{2}$, shifted $\dfrac{1}{2}$ unit to the right, and shifted down three units.
b. $(-3, \infty)$ **c.** $y = -3$ **37. a.** The graph of $y = e^x$ is compressed
horizontally by a factor of $\dfrac{1}{2}$, reflected in the y-axis, shifted $\dfrac{3}{2}$ units to
the right, and shifted up four units.   **b.** $(4, \infty)$ **c.** $y = 4$
39. a. The graph of $y = e^x$ is compressed horizontally by a factor of
$\dfrac{1}{3}$, shifted $\dfrac{1}{3}$ unit to the left, stretched vertically by a factor of 2, reflected
in the x-axis, and shifted up three units.   **b.** $(-\infty, 3)$ **c.** $y = 3$
41. a. The graph of $y = e^x$ is compressed horizontally by a factor of
$\dfrac{1}{3}$, reflected in the y-axis, shifted $\dfrac{2}{3}$ units to the right, stretched
vertically by a factor of 3, and shifted down four units.   **b.** $(-4, \infty)$
c. $y = -4$

B Exercises: Applying the Concepts: 43. $\$220,262$
45. $\$18,629.40$ **47.** $\$35,496.43$ **49. a.** $\$571$ **b.** $\$99,183,639,920$
c. $\$180,905,950,100$ **d.** $\$191,480,886,300$ **51.** 18,750
53. 0.1357 mm^2 **55. a.** 98°C   **b.** 29.44°C   **c.** 25.27°C   **d.** 25°C
57. a. $y = 10(1.2)^{x/6}$ **b.** 3.09%   **c.** 3.04%   **d.** $\$15,774$

C Exercises: Beyond the Basics:
59.

61.

63.

67. a. $(-\infty, \infty)$ **b.** even

c.

73. a. 8.2423% **b.** 8.2999% **c.** 8.3277% **d.** 8.3287%
75. 9.5 percent continuous interest yields 9.9658%, 10 percent simple interest yields 10%, and 9.6 percent interest compounded monthly yields 10.03%. An interest rate of 9.6% compounded monthly will yield the greatest return.

Critical Thinking: 77. a. $f(0) = 1, f(1) = \dfrac{1}{e} \approx 0.3679,$

$f(2) \approx 0.0183, f(3) \approx 0.0001234$ **b.** y-axis **c.** The limit is 0.
d.

78. a. $x = \mu$ **b.** The limit is 0.
 c. Answers will vary.

Section 4.3

Practice Problems: 1. a. $\log_2 1024 = 10$ **b.** $\log_9 \left(\dfrac{1}{3} \right) = -\dfrac{1}{2}$

c. $\log_a p = q$ **2. a.** $2^6 = 64$ **b.** $v^w = u$ **3. a.** 2 **b.** $-1/2$
c. -5 **4. a.** 8 **b.** 5 **c.** $x = -2$ or $x = 3$ **5.** $(-\infty, 1)$
6.

7. Shift the graph of $y = \log_2 x$ three units right and reflect the graph in the x-axis.

8.

$y = \log(x-3) - 2$

9. a. -1 **b.** 0.693
10. a. 17 years **b.** 21.97%
11. 5.423 min   **12.** About 14.3 yr

A Exercises: Basic Skills and Concepts: 1. $(0, \infty); (-\infty, \infty)$

3. common; natural   **5.** false   **7.** $\log_5 25 = 2$ **9.** $\log_{49} \left(\dfrac{1}{7} \right) = -\dfrac{1}{2}$

11. $\log_{\frac{1}{16}} 4 = -\dfrac{1}{2}$ **13.** $\log_{10} 1 = 0$ **15.** $\log_{10} 0.1 = -1$

17. $\log_a 5 = 2$ **19.** $2^5 = 32$ **21.** $10^2 = 100$ **23.** $10^0 = 1$

25. $10^{-2} = 0.01$ **27.** $8^{\frac{1}{3}} = 2$ **29.** $e^x = 2$ **31.** 3   **33.** 4

35. -3 **37.** $\dfrac{3}{2}$ **39.** $\dfrac{1}{4}$ **41.** $x = 25$ **43.** $x = 127$

45. $x = -4$ **47.** $x = 5$ **49.** $x = 2$ **51.** $x = 3, 4$ **53.** $(-1, \infty)$
55. $(5, \infty)$ **57.** $(1, \infty)$ **59.** $(-\infty, 0)$
61. a. f **b.** a **c.** d **d.** b **e.** e **f.** c
63. $y = \log_3 (x - 2)$ **65.** $y = 3 + \log_2 (2 - x)$
67. Domain $= (-3, \infty)$ **69.** Domain $= (0, \infty)$
Range $= (-\infty, \infty)$ Range $= (-\infty, \infty)$
Asymptote: $x = -3$ Asymptote: $x = 0$

71. Domain $= (-\infty, 0)$ **73.** Domain $= (0, \infty)$
Range $= (-\infty, \infty)$ Range $= [0, \infty)$
Asymptote: $x = 0$ Asymptote: $x = 0$

75. Domain $= (1, \infty)$ **77.** Domain $= (-\infty, 3)$
Range $= (-\infty, \infty)$ Range $= (-\infty, \infty)$
Asymptote: $x = 1$ Asymptote: $x = 3$

79. Domain = $(-\infty, 3)$
Range = $(-\infty, \infty)$
Asymptote: $x = 3$

81. Domain = $(-\infty, 0) \cup (0, \infty)$
Range = $(-\infty, \infty)$
Asymptote: $x = 0$

83. 1   **85.** 7   **87.** 5   **89.** $x = -2, 5$ **91.** $x = 4$
93.

95.

97.

B Exercises: Applying the Concepts: **99.** 8.66 yr   **101.** 11.55%
103. a. 74.72 yr   **b.** 118.43 yr   **105.** 58.4 months
107. a. (i) 22.66°F   **(ii)** 20.13°F   **(iii)** 20°F   **b.** 5.5 min
109. 17.6 min

C Exercises: Beyond the Basics
111.

113. $(-\infty, -1) \cup (2, \infty)$
115. a. $(1, \infty)$ **b.** $(2, \infty)$
c. $(2, \infty)$ **d.** $(11, \infty)$
117. The graph of $y = \ln x$ is
horizontally compressed by a
factor of 2, shifted three units
to the right, and shifted four
units up.

119. The graph of $y = \ln x$ is horizontally stretched by a factor of 2,
reflected in the y-axis, shifted three units to the left, and shifted four
units down.   **121. a.** \$24,659.69 **b.** 4.05%   **123.** $x = \dfrac{(e^y)^2 - 1}{2e^y}$
125. $x = \ln(y + \sqrt{y^2 + 1})$

Critical Thinking: **127.** 1   **128.** 16   **129.** 16

Section 4.4

Practice Problems: **1. a.** -1 **b.** 13
2. a. $\ln(2x - 1) - \ln(x + 4)$
b. $\log 2 + \dfrac{1}{2}\log x + \dfrac{1}{2}\log y - \dfrac{1}{2}\log z$ **3.** $\log\sqrt{x^2 - 1}$
4. $\dfrac{\log 15}{\log 3} \approx 2.46497$ **5.** $f(x) = 5e^{x\ln 4}$ **6.** ≈ 25 yr   **7.** $\approx 65.34\%$

A Exercises: Basic Skills and Concepts: **1.** $\log_a M$; $\log_a N$
3. $r \log_a M$ **5.** false   **7.** 0.78   **9.** 0.7   **11.** -1.7 **13.** 7.3
15. $\dfrac{16}{3}$ **17.** 0.56   **19.** $\ln(x) + \ln(x - 1)$
21. $\dfrac{1}{2}\log(x^2 + 1) - \log(x + 3)$
23. $2\log_3(x - 1) - 5\log_3(x + 1)$
25. $2\log_b x + 3\log_b y + \log_b z$ **27.** $2 - 3\ln x - \ln y$
29. $\dfrac{1}{2}\log x + \dfrac{1}{4}(\log y + \log z)$
31. $\ln x + \dfrac{1}{2}\ln(x - 1) - \ln(x^2 + 2)$
33. $2\ln(x + 1) - \ln(x - 3) - \dfrac{1}{2}\ln(x + 4)$
35. $\ln(x + 1) + \dfrac{1}{2}[\ln(x^2 + 2) - \ln(x^2 + 5)]$
37. $3\ln x + 4\ln(3x + 1) - \dfrac{1}{2}\ln(x^2 + 1) + 5\ln(x + 2) - 2\ln(x - 3)$
39. $\log_2(7x)$ **41.** $\log\left(\dfrac{45}{14}\right)$ **43.** $\log\left(\dfrac{z\sqrt{x}}{y}\right)$ **45.** $\log_2\left(zy^2\right)^{1/5}$
47. $\ln(xy^2z^3)$ **49.** $\ln\left(\dfrac{x^2}{\sqrt{x^2 + 1}}\right)$ **51.** 2.322
53. -1.585 **55.** 1.760   **57.** 3.6
59.

61.

63. $\dfrac{1}{2}$ **65.** 1   **67.** 18   **69.** 2   **71.** $\sqrt{3}$ **73.** 5
75. $f(x) = 2e^{(\ln 2)x}$ **77.** $f(x) = 100e^{(\ln 10)\frac{x}{5}}$

B Exercises: Applying the Concepts: **79.** Rate of growth $\approx 1.375\%$
81. 1.12%   **83.** 12.53 yr   **85.** 12.969 g   **87.** 13 yr
89. $k = 0.1204$ **91.** 13.7 g

C Exercises: Beyond the Basics: **97.** The domain of $2\log x$ is $(0, \infty)$,
while the domain of $\log(x^2)$ is $(-\infty, 0) \cup (0, \infty)$.   **99.** 5
101. a. False **b.** True **c.** False **d.** False **e.** True **f.** True
g. False **h.** True **i.** False **j.** True

Critical Thinking: **105.** Because $\log\dfrac{1}{2}$ is a negative number, $3 < 4$
implies that $3\log\dfrac{1}{2} > 4\log\dfrac{1}{2}$.

Section 4.5

Practice Problems: **1. a.** $x = 5$ **b.** $x = \dfrac{2}{3}$
2. $x = \dfrac{\ln\left(\dfrac{11}{7}\right)}{\ln 3} - 1(\approx -0.59)$ **3.** $x \approx 3.82$ **4.** $x = \ln 5 \approx 1.609$
5. 24.57%; 21.97%   **6. a.** United States: 329.10 million;
Pakistan: 224.14 million   **b.** Sometime in 2020 (after 15.63 yr)
c. In 27.84 yr (sometime in 2032)   **7.** $x = e^{3/2}$ **8.** $\varnothing$ **9.** $x = 9$
10. The average growth rate was approximately 1.74%.
11. $I_A = \sqrt{10}I_T$ **12.** $x < -2$ **13.** $t > \approx 14.68$ yr
14. $x < \dfrac{1 - e^2}{3} \approx -2.13$

A Exercises: Basic Skills and Concepts: **1.** exponential

3. logistic **5.** false **7.** $x = 4$ **9.** $x = \dfrac{5}{3}$ **11.** $x = \pm\dfrac{7}{2}$

13. ∅ **15.** $x = 1$ **17.** $x = \dfrac{1}{2}$ **19.** $x = \pm 9$

21. $x = 10{,}000$ **23.** $x = 1.585$ **25.** $x = 0.453$ **27.** $x = 1.766$
29. $x = 0.934$ **31.** $x = 0.699$ **33.** $x = -1.160$ **35.** $x = -3.944$
37. $t = 11.007$ **39.** $x = 2.807$ **41.** $x = 0.631, 1.262$

43. $x = 1.262$ **45.** $x = 0.232$ **47.** $x = \dfrac{\ln 2}{\ln 3} \approx 0.631$

49. $x = \dfrac{\ln 18 - \ln 7}{\ln 3} \approx 0.860$ **51.** ∅ **53.** $x = -\dfrac{49}{20}$ **55.** $x = 3, -2$

57. $x = 5, 2$ **59.** $x = 8$ **61.** $x = 1$ **63.** $x = 3$

65. $x = 4$ **67.** ∅ **69.** $x = \dfrac{2}{3}, \dfrac{5}{2}$ **71.** $a = 30, k = 2; 500$

73. $a = 5, k = \dfrac{1}{2}; 31$ **75.** $a = 2, k = \ln\left(\dfrac{17}{2}\right); 0.068$

77. $a = -2, k = \ln\left(\dfrac{11}{18}\right); -7.902$ **79.** $x \geq \dfrac{\ln 2}{\ln 0.3}$ **81.** $x \leq 0$

83. $(-3, 17)$ **85.** $x \geq e + 5$ **87.** $\left(\dfrac{7}{3}, 5\right)$

B Exercises: Applying the Concepts: **89. a.** 10.087 yr **b.** 9.870 yr
c. 9.821 yr **d.** 9.797 yr **e.** 9.796 yr **90. a.** 9.970 yr **b.** 9.713 yr
c. 9.656 yr **d.** 9.627 yr
91. a.

Item	Canada	Mexico	United Kingdom
Population	36.76 million	125.54 million	63.03 million
Milk price	$2.50	$1.38	$1.00
Bread price	$2.61	$2.23	$1.72

b.

Item	Canada	Mexico	United Kingdom
Population	38.80 million	134.91 million	64.17 million
Milk price	$2.82	$1.70	$1.20
Bread price	$2.94	$2.73	$2.06

93. In 2066 (after 66.4 yr) **95.** Canada: in 2079 (after 79.2 yr)
Mexico: in 2056 (after 56.2 yr) United Kingdom: in 2070 (after 70.7 yr)
97. $r \approx 8.66\%$; $30,837.52 **99. a.** $k \approx 3.4\%$ **b.** ≈ 4.1 million
c.   **d.** ≈ 4.9 million
e. approaches 6.4 million

101. a. $a = 4999$ and $k = 0.852$
b. 4930 **c.** 5000

103. a. 2.50 million **b.** In 2084 **105. a.** 2.134 lm **b.** 19.08 ft
107. a. $I = 10^{8.6}$ **b.** $E = 10^{17.3}$ joules **109.** $\log 150 \approx 2.2$
111. ≈ 74 dB **113.** $10^{-4.7}$ W/m²

C Exercises: Beyond the Basics: **115.** $t = \dfrac{1}{k}\ln\left(\dfrac{P}{M - P}\right)$

121. $n = \dfrac{\ln\left(\dfrac{R + iA}{R}\right)}{\ln(1 + i)}$ **123.** $x = 1, 10$ **125.** $x = 2$ **127.** 0

129. a **131.** 1 **133.** $t = \dfrac{\ln a}{k}$

Review Exercises

1. False **3.** False **5.** True **7.** False **9.** True **11.** h
13. f **15.** d or e **17.** a
19. Domain $= (-\infty, \infty)$ **21.** Domain $= (-\infty, \infty)$
Range $= (0, \infty)$ Range $= (3, \infty)$
Asymptote: $y = 0$ Asymptote: $y = 3$

23. Domain $= (-\infty, \infty)$ **25.** Domain $= (-\infty, 0)$
Range $= (0, 1]$ Range $= (-\infty, \infty)$
Asymptote: $y = 0$ Asymptote: $x = 0$

27. Domain $= (1, \infty)$ **29.** Domain $= (-\infty, 0)$
Range $= (-\infty, \infty)$ Range $= (-\infty, \infty)$
Asymptote: $x = 1$ Asymptote: $x = 0$

31. a. x-intercept $= \ln\left(\dfrac{2}{3}\right)$ **33. a.** x-intercept: none
b. as $x \to \infty, y \to 3$ **b.** as $x \to \infty, y \to 0$
as $x \to -\infty, y \to -\infty$ as $x \to -\infty, y \to 0$

35. $a = 10, k = 2, f(2) = 160$ **37.** $a = 3, k = -0.2824,$
$f(4) = 0.38898$ **39.** $y = 3 \cdot 2^x$ **41.** $y = \log_5(x - 1)$
43. a. $y = -2^{(x-1)} - 3$ **b.** $y = -2^{(x-1)} + 3$

45. $\ln x + 2\ln y + 3\ln z$ **47.** $\ln x + \dfrac{1}{2}\ln(x^2 + 1) - 2\ln(x^2 + 3)$

49. $y = 3x$ **51.** $y = -1 \pm \sqrt{x - 2}$ **53.** $y = \dfrac{\sqrt{x^2 - 1}}{x^2 + 1}$

55. 4   **57.** $-1 \pm \sqrt{5}$ **59.** $\dfrac{\ln 23}{\ln 3} \approx 2.854$ **61.** $\dfrac{\ln 19}{\ln 273} \approx 0.525$

63. 0   **65.** $\dfrac{-\ln 3}{\ln\left(\dfrac{(1.7)^3}{9}\right)} \approx 1.815$ **67.** $\dfrac{5}{2}$ **69.** 0   **71.** 8

73. 5   **75.** 9   **77.** 3   **79.** $x \geq -1$ **81.** $\left(-\dfrac{7}{2}, \dfrac{93}{2}\right)$

83. 11.4 yr   **85.** 40 hr   **87.** At 5% compounded yearly,
$A = \$9849.70$. At 4.75% compounded monthly, $A = \$9754.74$.
The 5% investment will provide the greater return.
89. a. 34 million   **b.** In 2206 (after 199.3 yr)
91. a.   **b.** ≈ 5 hr

93. $t = 6.57$ min   **95. a.** 5000 people/square mile   **b.** 7459
people/square mile   **c.** ≈ 13.73 miles   **97.** $I = 0.2322I_0$
99. a. 11.27 ft/sec   **b.** 5.29 ft/sec   **c.** 201   **101. a.** 1 g
b. Remains stationary at 6 g   **c.** 4.6 days

Practice Test A

1. $x = -3$ **2.** $x = 32$ **3.** Range $= (-\infty, 1)$;
horizontal asymptote: $y = 1$ **4.** -3 **5.** $x = 3$ **6.** -3

7. $\ln 3x^5$ **8.** $x = \dfrac{\ln \dfrac{5}{2}}{\ln 2}$ **9.** $\ln 2$ **10.** $\ln 2 + 3\ln x - 5\ln(x + 1)$

11. -5 **12.** $y = \ln(x - 1) + 3$
13.

14. $(-\infty, 0)$

15. $\ln \dfrac{x^3(x^3 + 2)}{\sqrt{3x^2 + 2}}$ **16.** $x = 3$

17. $x = 3$

18. $\dfrac{1}{2}[\ln 2 + 3\ln x + 2\ln y]$

19. $A(t) = 15{,}000(1.0175)^{4t}$
20. In 1983 (after 23.03 yr)

Practice Test B

1. b   **2.** b   **3.** b   **4.** d   **5.** d   **6.** b   **7.** b   **8.** c   **9.** d
10. d   **11.** b   **12.** d   **13.** b   **14.** a   **15.** a   **16.** d   **17.** b
18. c   **19.** d   **20.** b

Cumulative Review Exercises (Chapters P–4)

1. x-intercept: 0; y-intercepts: 0, 2; symmetric in y-axis

2.

3. Center: $(-1, 2)$; radius: 5

4. $y = \dfrac{3}{2}x + 7$ **5.** $(-\infty, -3) \cup (2, \infty)$ **6. a.** -7 **b.** -3

c. 19   **7.** Start with $y = \sqrt{x}$, shift one unit left, stretch vertically by
a factor of 3, and shift two units down.

8. $(f \circ g)(x) = 3\sqrt{-\dfrac{1}{x}}$; domain of $f \circ g$ is $(-\infty, 0)$.

9. a. Yes, $f^{-1}(x) = \left(\dfrac{x - 1}{2}\right)^{1/3}$ **b.** No   **c.** Yes, $f^{-1}(x) = e^x$

10. a. $2x + \dfrac{x + 1}{x^2 + 1}$ **b.** $2x^3 + x^2 - 3x + 7 - \dfrac{18}{x + 3}$
11. a. $(x - 1)(x - 2)(x + 3) = x^3 - 7x + 6$ **b.** $x^4 - 1$

12. a. $\log_2 5 + 3\log_2 x$ **b.** $\dfrac{1}{3}[\log_a x + 2\log_a y - \log_a z]$

c. $\ln 3 + \dfrac{1}{2}\ln x - \ln 5 - \ln y$ **13. a.** $\log \sqrt{xy}$ **b.** $\ln \dfrac{x^3}{y^2}$

14. a. 1.289   **b.** 2.930   **c.** 1.737   **15. a.** $\{3\}$ **b.** $\{1, 3\}$ **c.** 3.77
16. Possible rational zeros $\{\pm1, \pm2, \pm4\}$; rational zeros: $\{1, 2\}$;
upper bound: 3; lower bound: -1 **17. a.** $1, 1, 1, -2, -2, 3$
b. At $x = 1$ and $x = 3$, the graph crosses the x-axis. At $x = -2$, the
graph touches but does not cross the x-axis.   **c.** $f(x) \to \infty$ as
$x \to \infty$; $f(x) \to \infty$ as $x \to = -\infty$
d.

18. a. Vertical asymptote: $x = 3, x = -4$ **b.** Horizontal
asymptote: $y = 1$ **c.** $f(x) > 0$ on $(-\infty, -4) \cup (-2, 1) \cup (3, \infty)$
$f(x) < 0$ on $(-4, -2) \cup (1, 3)$
d.

CHAPTER 5

Section 5.1

Practice Problems: **1.** Equation (1): $(1) + (3) = 4$; equation (2): $3(1) - (3) = 0$ **2.** $\{(-1, 3)\}$ **3.** $\{(4, -1)\}$ **4.** $\varnothing$

5. $\{(x, 2x - 3)\}$ **6.** $\left\{\left(\dfrac{2}{3}, \dfrac{1}{2}\right)\right\}$ **7.** $\{(2, -3)\}$

8. $\{(5700, 31.4)\}$

A Exercises: Basic Skills and Concepts: **1.** solution **3.** inconsistent
5. false **7.** $(3, -1)$ **9.** None of them **11.** $(10, -9)$
13. $(2, 1)$ **15.** $(2, 2)$

17. Inconsistent **19.** $\left(\dfrac{7}{3}, \dfrac{14}{3}\right)$

21. $\{(x, 12 - 3x)\}$

23. Consistent; Independent
25. Consistent; Independent
27. Consistent; Dependent
29. Consistent; Independent
31. Inconsistent **33.** Inconsistent
35. Consistent; Independent
37. $\left\{\left(\dfrac{7}{9}, \dfrac{23}{9}\right)\right\}$ **39.** $\{(3, 4)\}$
41. $\varnothing$ **43.** $\{(-3, 5)\}$
45. $\left\{\left(x, \dfrac{1}{2}(x - 5)\right)\right\}$

47. $\{(3, 2)\}$ **49.** $\{(-3, 3)\}$ **51.** $\{(0, -5)\}$ **53.** $\varnothing$
55. $\left\{\left(x, 2 - \dfrac{2}{3}x\right)\right\}$ **57.** $\{(4, 1)\}$ **59.** $\{(-4, 2)\}$ **61.** $\{(2, 1)\}$
63. $\{(3, 1)\}$ **65.** $\left\{\left(\dfrac{19}{6}, \dfrac{5}{4}\right)\right\}$ **67.** $\{(4, 2)\}$ **69.** $\{(4, 1)\}$
71. $\left\{\left(-\dfrac{1}{5}, 1\right)\right\}$ **73.** $\left\{\left(\dfrac{3}{2}, \dfrac{1}{2}\right)\right\}$ **75.** $\{(5, 5)\}$
77. a. $\begin{cases} x - 2y = 1 \\ 3x + 4y = 16 \end{cases}$ Answers will vary. **b.** $\left\{\left(\dfrac{18}{5}, \dfrac{13}{10}\right)\right\}$
79. a. $\begin{cases} x + y = 2 \\ x + y = 5 \end{cases}$ Answers will vary. **b.** Inconsistent
81. a. $\begin{cases} 4x + 3y = 12 \\ -8x - 6y = -24 \end{cases}$ Answers will vary. **b.** $\left\{\left(x, -\dfrac{4}{3}x + 4\right)\right\}$

B Exercises: Applying the Concepts: **83.** $(80, 30)$ **85.** $(17, 4)$
87. The diameter of the largest pizza is 21 inches, and the diameter of the smallest pizza is 8 inches. **89.** 8% of the total trash collected is plastic, and 40% is paper. **91.** 340 beads and 430 doubloons
93. 3 Egg McMuffins and 2 Breakfast Burritos **95.** $18,000 at 7.5% and $32,000 at 12% **97.** $7.50 per hour at McDougal's and $15 per hour for tutoring **99.** 20 pounds of herb and 80 pounds of tea
101. The speed of the plane is 550 km/hr, and the wind speed is 50 km/hr.
103. a. $C(x) = 30,000 + 2x; R(x) = 3.5x$
b.

c. 20,000 magazines
105. Her weekly sales should be more than $6250.

C Exercises: Beyond the Basics: **107.** $\left\{\left(\dfrac{1261}{204}, \dfrac{949}{204}\right)\right\}$

109. $c = 2$ **111.** $c = \dfrac{35}{2}$ **113.** $a = 4, b = 0$

115. $y = -\dfrac{3}{2}x + \dfrac{3}{2}$ **117.** $y = -\dfrac{13}{7}x + \dfrac{6}{7}$

Critical Thinking:

118. a. $x = \dfrac{b_2c_1 - b_1c_2}{a_1b_2 - a_2b_1}, y = \dfrac{a_1c_2 - a_2c_1}{a_1b_2 - a_2b_1}$, where $a_1b_2 - a_2b_1 \neq 0$
b. $x = \dfrac{c_1}{b_1} \neq \dfrac{c_2}{b_2}, \dfrac{a_1}{a_2} = \dfrac{b_1}{b_2}$ **c.** $\dfrac{a_1}{a_2} = \dfrac{b_1}{b_2} = \dfrac{c_1}{c_2}$
119. a. $3y - 4x = 4$ **b.** $\left(\dfrac{4}{5}, \dfrac{12}{5}\right)$ **c.** 7
120. a. $b(x - x_1) - a(y - y_1) = 0$
b. $\left(\dfrac{b^2x_1 - aby_1 - ac}{a^2 + b^2}, \dfrac{-abx_1 + a^2y_1 - bc}{a^2 + b^2}\right)$ **121. a.** $\sqrt{2}$
b. $\dfrac{6\sqrt{5}}{5}$ **c.** 0 **d.** $\dfrac{|c|}{\sqrt{a^2 + b^2}}$ if $a \neq 0$ or $b \neq 0$

Section 5.2

Practice Problems:
1. $(2) + (-3) + (2) = 1$
$3(2) + 4(-3) + (2) = -4$
$2(2) + (-3) + 2(2) = 5$; yes, it is a solution.

2. $\{(-2, 1, 3)\}$ **3.** $\varnothing$ **4.** $\left\{\left(8 - \dfrac{7}{3}z, \dfrac{4}{3}z - 3, z\right)\right\}$
5. $[(13 - 17z, -3 + 5z, z)]$ **6.** Cell A contains bone (because $x = 0.4$), cell B contains healthy tissue (because $y = 0.25$), and cell C contains tumorous tissue (because $z = 0.3$).
7. $f(x) = -x^2 + 4x + 5$

A Exercises: Basic Skills and Concepts: **1.** equivalent
3. inconsistent **5.** true **7.** Yes **9.** No **11.** $\{(3, 2, -1)\}$
13. $\{(-3, 2, 4)\}$

15. $\begin{cases} x - y - \dfrac{3}{2}z = \dfrac{1}{2} \\ y + z = 0 \\ z = 1 \end{cases}$ **17.** $\begin{cases} x + 3y - 2z = 0 \\ y - \dfrac{8}{7}z = \dfrac{5}{7} \\ 0 = k, k \neq 0 \end{cases}$

19. $\{(1, -2, 3)\}$ **21.** $\left\{\left(\dfrac{30}{11}, -\dfrac{8}{11}, -\dfrac{1}{4}\right)\right\}$ **23.** $\{(1, 2, 3)\}$

25. $\left\{\left(\dfrac{35}{18}, \dfrac{29}{18}, \dfrac{5}{18}\right)\right\}$ **27.** Inconsistent   **29.** $\{(3, -1, 2)\}$

31. $\{(2, 2, 1)\}$ **33.** $\{(3, -1, 1)\}$ **35.** $\left(3 + \dfrac{4}{5}z, -\dfrac{1}{5}z, z\right)$

37. $\{(12, -12, 4)\}$ **39.** $\{(1, -1, 2)\}$ **41.** $\{(3, 2, -1)\}$

43. Dependent; $\left\{\left(-5 - 9z, \dfrac{-1}{6} + 3z, z\right)\right\}$

45. $f(x) = 3x^2 - 17x + 6$ **47.** $f(x) = -2x^2 + 2x + 1$
49. $f(x) = 2x^2 - 12x + 5$

B Exercises: Applying the Concepts: **51.** $4000 invested at 4%, $6000
invested at 5%, and $10,000 invested at 6%   **53.** Alex worked
2 hours, Becky worked 2.5 hours, and Courtney worked 1.5 hours.
55. 56 nickels, 225 dimes, and 19 quarters   **57.** 31 hours for normal
daytime work, 14 hours at night, and 8 hours on a holiday
59. 400 nickels, 300 dimes, 200 quarters   **61.** 18, 36, and 9   **63.** Cell A
contains healthy tissue (because $x = 0.21$), cell B contains tumorous
tissue (because $y = 0.33$), and cell C contains healthy tissue (because
$z = 0.19$).   **65.** Cells A, B, and C contain healthy tissue (because
$x = 0.28, y = 0.23$, and $z = 0.21$, respectively).   **67.** 198 Democrats,
180 Republicans, 42 Independents   **69.** Seven of Type 1, four of
Type 2, three of Type 3   **71.** $a = -80, b = -370, c = 11,400$

C Exercises: Beyond the Basics: **73.** $x + \dfrac{4}{3}y - \dfrac{5}{3}z = \dfrac{1}{3}$

75. $x - \dfrac{111}{14}y - \dfrac{11}{14}z = -\dfrac{1}{14}$ **77.** $y = 10x^2 - 18x + 2$

79. $y = 2x^2 + x + 3$ **81.** $x^2 + y^2 - 2y - 3 = 0$

83. $x^2 + y^2 - 4x - 10y + 19 = 0$ **85.** $\left\{\left(\dfrac{2}{19}, -\dfrac{2}{3}, 2\right)\right\}$

87. $c = -16$ **89.** Darren: 20 hours, Skylar: 60 hours, Yolanda: 30 hours

Critical Thinking:

91. $\begin{cases} x + y - z = -2 \\ 2x - y + 3z = 9 \quad \text{Answers may vary.} \\ x + y + z = 2 \end{cases}$

92. a. $\begin{cases} x + 3y + 3z = 15 \\ x + 2y + z = 1 \quad \text{Answers may vary.} \\ 2y + 4z = 11 \end{cases}$

b. $\begin{cases} x + 2y - z = 4 \\ x + 3y + 2z = 5 \quad \text{Answers may vary.} \\ y + 3z = 1 \end{cases}$

Section 5.3

Practice Problems: **1.** $\{(1, 1)\}$ **2.** $\{(4, 3), (-4, 3), (4, -3),$
$(-4, -3)\}$ **3.** Danielle received $x = 30$ shares as dividends
and sold her stock at $p = 65 per share.

A Exercises: Basic Skills and Concepts: **1.** at least one
3. intersection   **5.** true   **7.** $(3, -1)$ **9.** $(1, -1)$
11. $(5, 4), (-5, 4), (-5, -4)$ **13.** $(1, 1)$ **15.** $\{(-1, 1), (2, 4)\}$
17. $\{(3, 3), (-2, -2)\}$ **19.** $\{(3, 0)\}$ **21.** $\{(-2, 1), (-1, 2)\}$
23. $\{(1, -1), (3, 1)\}$ **25.** $\{(-3, -1), (1, 3)\}$ **27.** $\{(2, 3), (0, 5)\}$
29. $\{(7, 5), (-7, -5)\}$ **31.** $\{(4, 2), (-4, 2), (4, -2), (-4, -2)\}$
33. $\{(2, 2), (-2, 2), (2, -2), (-2, -2)\}$ **35.** $\varnothing$
37. $\{(1, 2), (-1, 2), (1, -2), (-1, -2)\}$

39. $\left\{(2, 1), (2, -1), \left(\dfrac{-11}{3}, \dfrac{\sqrt{26}}{3}\right), \left(\dfrac{-11}{3}, \dfrac{-\sqrt{26}}{3}\right)\right\}$

41. $\{(5, 3), (3, 5)\}$ **43.** $\varnothing$ **45.** $\{(0, -2), (3, 4)\}$

47. $\left\{\left(-4, \dfrac{-3}{2}\right), (3, 2)\right\}$ **49.** $\{(-2, 1), (2, 1), (-2, -1), (2, -1)\}$

51. $\left\{(2, 5), \left(-7, \dfrac{-10}{7}\right)\right\}$

53. $\left\{(2, -2), (2, 2), \left(\dfrac{16}{5}, \dfrac{2\sqrt{46}}{5}\right), \left(\dfrac{16}{5}, \dfrac{-2\sqrt{46}}{5}\right)\right\}$

B Exercises: Applying the Concepts: **55.** $(3, 78)$ **57.** 11 and 13
59. a. $x = 60\,\text{m}; y = 140\,\text{m}$ **b.** $6000\,\text{m}^2$ **61.** 32 people; $30
63. 450 original shares; purchased at $22 per share

C Exercises: Beyond the Basics: **65.** $A(1, 0), B(6, 0), d(A, B) = 5$
67. $A(-2, -1), B(2, 3), d(A, B) = 4\sqrt{2}$ **69.** $(0, -3)$

71. $c = \pm\sqrt{2}$ **73.** $c = \pm 25$ **75.** $\left\{\left(-\dfrac{1}{3}, -\dfrac{1}{8}\right), \left(\dfrac{1}{8}, \dfrac{1}{3}\right)\right\}$

77. $\{(-1, 1), (1, 1), (-1, -1), (1, -1)\}$ **79.** $\pm(2 + 3i)$
81. $\pm(4 - 3i)$ **83.** $\pm(4 + i\sqrt{3})$
85. $\{(1, 7)\}$ **87.** $\{(1, 5)\}$

89. $\left\{(1, 0), \left(2, \dfrac{\ln 2}{\ln 3}\right)\right\}$

Critical Thinking: **91. a.** Not possible   **b.** Possible   **c.** Possible
d. Possible   **e.** Not Possible   **f.** Not Possible   **92. a.** Possible
b. Possible   **c.** Possible   **d.** Possible   **e.** Possible   **f.** Not Possible

Section 5.4

Practice Problems: **1.**

2.

3.

4.

$x - y = 1$ $3x - y = 3$

5. $(4, 0), (5, 2),$ and $(2, 4)$

$2x - y = 8$
$2x + 3y = 16$
$4x + 2y = 16$

19.

21.

23. None   **25.** $(0, 0), (1, 0), (0, 1), (1, 1)$ **27.** $(2, 0), (3, 1), (2, 2)$

6.

$y = x^2 - 2$

7.

$y = x^2 + 1$
$(0, 13)$
$(3, 10)$
$(-2, 5)$
$y = 4x + 13$ $y = -x + 13$

29.

$(0, 1)$
$x + y = 1$
$x - y = -1$

31.

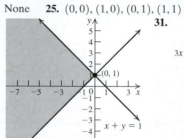

$3x + 5y = 15$
$\left(\frac{5}{2}, \frac{3}{2}\right)$
$2x + 2y = 8$

33. There are no vertices of the solution set.

$4x + 6y = 24$
$2x + 3y = 6$

A Exercises: Basic Skills and Concepts: **1.** dashed
3. not a solution of the system   **5.** true

7.

9.

35.

$3x + y = 8$
$(6, -10)$
$4x + 2y = 4$

37.

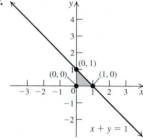

$(0, 1)$
$(0, 0)$ $(1, 0)$
$x + y = 1$

11.

13.

39.

$(0, 2)$
$(3, 0)$
$2x + 3y = 6$

41.

$2x + y = 1$
$x - y = 1$
$x + y = 1$
$\left(\frac{2}{3}, -\frac{1}{3}\right)$

15.

17.

43. $x + y = 1$ $-x + y = 3$

$x - y = 2$
$(-1, 2)$
$\left(-\frac{7}{2}, -\frac{1}{2}\right)$
$x + y = -4$

45. A **47.** B **49.** E

51.

53.

79.

81.

55.

57.

59.

61.

83.

63.

65. *D, K*   **67.** *E, L*   **69.** *E, B*

85. $\begin{cases} x \geq 0 \\ y \geq 0 \\ y \leq \dfrac{-2}{3}x + 2 \end{cases}$ **87.** $\begin{cases} x \geq -1 \\ x \leq 3 \end{cases}$ **89.** $\begin{cases} x \geq 0 \\ y \geq 0 \\ y \leq \dfrac{-1}{2}x + 4 \\ y \leq \dfrac{-3}{2}x + 6 \end{cases}$

91. $\begin{cases} x \geq 0 \\ y \geq x - 4 \\ y \geq -x + 6 \\ y \leq -x + 16 \end{cases}$

71.

73.

93.

95.

75.

77.

97.

99.

101.   **103.**

Critical Thinking:

105.   **106.**

Section 5.5

Practice Problems: 1. The objective function f has a maximum of 28 at the point $(7, 0)$.   **2.** The minimum value is 120 when $x = 0$ and $y = 3$.   **3.** The minimum number of calories is 300. The lunch then consists of 10 ounces of soup and 0 ounces of salad.

A Exercises: Basic Skills and Concepts: 1. optimization
3. constraints; feasible   **5.** true   **7.** Max 18, Min 5
9. Max 26, Min 7   **11.** Max 66, Min 13   **13.** Maximum value is $\dfrac{1284}{19}$ at $\left(\dfrac{56}{19}, \dfrac{60}{19}\right)$.   **15.** Maximum value is 410 at $(40, 30)$.
17. Minimum value is 3 at $(3, 0)$.   **19.** Minimum value is 1364 at $(8, 84)$.   **21.** Maximum value is 54 at $(3.6, 0)$.   **23.** Minimum value is 400 for all (x, y) on the line segments between $(10, 20)$ and $(20, 15)$, and $(20, 15)$ and $(50, 0)$.

B Exercises: Applying the Concepts: 25. 80 acres of corn and 160 acres of soybeans to obtain a maximum profit of $10,400
27. 40 hours for machine I and 20 hours for machine II to obtain a minimum cost of $3600   **29.** Five minutes of television and two pages of newspaper advertisement to obtain a maximum of 340,000 viewers   **31.** The maximum profit is $14,600.   **33.** 400 of both the rectangular and circular table should be made to have a maximum profit of $2800.   **35.** The contractor should build 33 terraced houses and 66 cottages to obtain a maximum amount of money: $4,290,000.
37. Mrs. Adams's maximum profit is $1355 when she has 175 male guests and 125 female guests.   **39.** Elisa should buy ten enchilada meals and five vegetable loafs to obtain a minimum cost of $46.25.

C Exercises: Beyond the Basics: 41. Maximum value is 152 at $(12, 8)$.
43. Minimum value is 0 at $(0, 0)$.

Critical Thinking: 47. a. There is no upper bound on the values for x.
b. For $x_1 > x_0 > 0$ and $0 \le y \le 8$, we have $2x_1 + 3y > 2x_0 + 3y$.

Section 5.6

Practice Problems: 1. $\dfrac{3}{x + 1} - \dfrac{1}{x - 2}$ **2.** $\dfrac{5}{x - 1} - \dfrac{3}{x} + \dfrac{1}{x + 1}$
3. $\dfrac{5}{x} - \dfrac{5}{x - 1} + \dfrac{6}{(x - 1)^2}$ **4.** $\dfrac{-1}{x} + \dfrac{4x + 5}{x^2 + 2}$
5. $\dfrac{1}{x^2 + 1} + \dfrac{3x}{(x^2 + 1)^2}$ **6.** $\dfrac{777}{3112}$
7. $\dfrac{1}{R} = \dfrac{3}{x + 1} + \dfrac{1}{x + 2} = \dfrac{1}{\dfrac{x + 1}{3}} + \dfrac{1}{x + 2}$

A Exercises: Basic Skills and Concepts: 1. proper
3. three   **5.** false   **7.** $\dfrac{A}{x - 1} + \dfrac{B}{x + 2}$
9. $\dfrac{A}{x + 6} + \dfrac{B}{x + 1}$ **11.** $\dfrac{A}{x^2} + \dfrac{B}{x} + \dfrac{C}{x - 1}$
13. $\dfrac{A}{x + 1} + \dfrac{Bx + C}{x^2 - x + 1}$ **15.** $\dfrac{Ax + B}{x^2 + 1} + \dfrac{Cx + D}{(x^2 + 1)^2}$
17. $\dfrac{3}{x + 2} - \dfrac{1}{x + 1}$ **19.** $\dfrac{-1}{2(x + 3)} + \dfrac{1}{2(x + 1)}$ **21.** $\dfrac{-1}{x + 2} + \dfrac{1}{x}$
23. $\dfrac{2}{x + 2} - \dfrac{3}{2(x + 3)} - \dfrac{1}{2(x + 1)}$ **25.** $\dfrac{-2}{(x + 1)^2} + \dfrac{1}{x + 1}$
27. $\dfrac{2}{x + 1} - \dfrac{3}{(x + 1)^2} + \dfrac{1}{(x + 1)^3}$ **29.** $\dfrac{1}{x} + \dfrac{3}{(x + 1)^2} - \dfrac{2}{x + 1}$
31. $\dfrac{-2}{x} + \dfrac{1}{x^2} + \dfrac{1}{(x + 1)^2} + \dfrac{2}{x + 1}$
33. $\dfrac{1}{4(x - 1)^2} + \dfrac{1}{4(x + 1)^2} + \dfrac{1}{4(x + 1)} - \dfrac{1}{4(x - 1)}$
35. $\dfrac{1}{2(2x - 3)} + \dfrac{1}{2(2x - 3)^2}$ **37.** $\dfrac{-2}{(2x + 3)^2} + \dfrac{3}{2x + 3}$
39. $\dfrac{4}{x} - \dfrac{3}{x^2} - \dfrac{4}{x + 1}$ **41.** $\dfrac{-2}{x} + \dfrac{3}{x + 1} + \dfrac{4}{x^2}$
43. $\dfrac{-x}{2(x^2 + 1)} + \dfrac{1}{4(x + 1)} + \dfrac{1}{4(x - 1)}$
45. $\dfrac{-x}{x^2 + 1} + \dfrac{1}{x} - \dfrac{x}{(x^2 + 1)^2}$ **47.** $\dfrac{4 - 3x}{x^2 + 2} + \dfrac{-2 + 3x}{x^2 + 1}$

B Exercises: Applying the Concepts:
49. $\dfrac{n}{n + 1}$ **51.** $\dfrac{2n}{2n + 1}$ **53.** $\dfrac{1}{R} = \dfrac{1}{x + 1} + \dfrac{1}{x + 3}$
55. $\dfrac{1}{R} = \dfrac{1}{R_1} + \dfrac{1}{R_2} + \dfrac{1}{R_3}$

C Exercises: Beyond the Basics:
59. $\dfrac{1}{(x - 1)^2} - \dfrac{1}{(x + 1)^2}$ **61.** $\dfrac{-1}{2(x - 1)} + \dfrac{x - 1}{2(x^2 + 1)} + \dfrac{1}{(x - 1)^2}$
63. $\dfrac{-4}{x + 2} - \dfrac{8}{(x + 2)^2} - \dfrac{1}{(x + 1)^2} + \dfrac{5}{x + 1}$
65. $\dfrac{5}{3(x - 3)} + \dfrac{-5x + 15}{3(x^2 + 3x + 9)}$
67. $\dfrac{1}{x^2 - 2x + 2} - \dfrac{1}{x^2 + 2x + 2}$
69. $1 + \dfrac{2}{x + 1} - \dfrac{1}{x + 2}$ **71.** $2x + 3 + \dfrac{2x - 1}{x^2 + 1} + \dfrac{1}{x - 1}$

Review Exercises

1. $\{(-1, 2)\}$ **3.** $\varnothing$ **5.** $\{(2, 3)\}$ **7.** $\{(1, -1, 2)\}$
9. $\{(-1, 2, -3)\}$ **11.** $\varnothing$ **13.** $\{x, 2x - 8, 11 - 3x\}$
15. $\varnothing$ **17.** $\left(\dfrac{7}{3} - \dfrac{2}{3}z, \dfrac{2}{3} - \dfrac{1}{3}z, z\right)$

19.

21.

23.

25. Max of 14 at $x = 7$, $y = 0$

27. Min of -9 at $x = 0$, $y = -3$

29. $\left\{(-2, 1), \left(\frac{8}{3}, \frac{-5}{9}\right)\right\}$

31. $\left\{(2, -2), \left(\frac{-2}{3}, \frac{-14}{3}\right)\right\}$

33. $\left\{(1, 2), (-1, -2), \left(2\sqrt{2}, \frac{\sqrt{2}}{2}\right), \left(-2\sqrt{2}, \frac{-\sqrt{2}}{2}\right)\right\}$

35.

37.

39.

41. $\dfrac{2}{x + 2} - \dfrac{1}{x + 3}$

43. $\dfrac{2}{x - 1} + \dfrac{1}{x} + \dfrac{5}{(x - 1)^2}$

45. $\dfrac{-1 + 2x}{(x^2 + 4)^2} + \dfrac{1}{x^2 + 4}$

47. $8750 was invested at 12%, and $6250 was invested at 4%.

49. 10.5 ft by 6 ft   **51.** 15 and 8

53. The original width is 40 m, and the original length is 160 m.

55. a. $C(x) = 60{,}000 + 12x$, $R(x) = 20x$

b.

c. $(7500, 150{,}000)$

d. 11,129 timers

57. $450 for the single and $600 for the double

59. 3   **61.** Janet is 34, and Steve is 51.   **63.** They stole $35,000, and $15,000 went to Butch, $12,000 went to Sundance, and $8000 went to Billy.

65. $x = 6$, $y = 2$

67. She should build 4 two-story houses and 14 one-story houses to obtain a maximum profit of $96,000.   **69.** $60, $52 $45, respectively

Practice Test A

1. $\{(2, 0)\}$ **2.** $\left\{\left(x, 4 - \dfrac{x}{2}\right)\right\}$ **3.** $\varnothing$ **4.** $\{(-5, 0)\}$ **5.** $\{(6, 8)\}$

6. $\{(-1, 1), (4, 16)\}$ **7.** $\left\{\left(-3, \dfrac{1}{3}\right), (2, 2)\right\}$ **8. a.** $\begin{cases} x + y = 485 \\ x - y = 15 \end{cases}$

b. $x = 250$, $y = 235$ **9.** $x + \dfrac{1}{2}y + z = 2$ **10.** $\{(1, 3, 5,)\}$

$$y - \frac{3}{4}z = \frac{7}{4}$$

$$z = \frac{-27}{23}$$

11. $(7, 3, -2)$ **12.** $\left(x, \dfrac{2x}{3} + \dfrac{7}{3}, -x - 1\right)$ **13.** $\left(\dfrac{1}{3}, \dfrac{-4}{3} + z, z\right)$

14. 53 nickels, 225 dimes, and 22 quarters   **15.** $\dfrac{A}{x - 5} + \dfrac{B}{x + 1}$

16. $\dfrac{A}{x - 2} + \dfrac{Bx + C}{x^2 + 1} + \dfrac{Dx + E}{(x^2 + 1)^2}$

17. $-\dfrac{10}{121(x + 4)} + \dfrac{1}{11(x + 4)^2} + \dfrac{10}{121(x - 7)}$

18.

19.

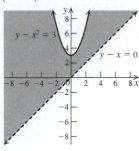

20. Maximum value is 6 at $(3, 0)$.

Practice Test B

1. a   **2.** c   **3.** b   **4.** c   **5.** b   **6.** d   **7.** d   **8.** d   **9.** c

10. a   **11.** b   **12.** d   **13.** c   **14.** b   **15.** c   **16.** a   **17.** c

18. b   **19.** a   **20.** b

Cumulative Review Exercises (Chapters P–5)

1. $\left\{\dfrac{5}{2}\right\}$ **2.** $\left\{2, \dfrac{1}{2}, \dfrac{1}{2} + i\dfrac{\sqrt{3}}{2}, \dfrac{1}{2} - i\dfrac{\sqrt{3}}{2}\right\}$ **3.** $\{7\}$ **4.** $(-3, 1]$

5. $(-\infty, 4) \cup (5, \infty)$ **6.** $\dfrac{\ln 2 + \ln 5}{\ln 2}$ **7.** $\{2\}$ **8.** $\{4\}$

9.

10.

11.

12.

13. a. $f^{-1}(x) = \dfrac{x + 2}{2}$ **b.**

14. a. $\pm 1, \pm 2, \pm 3, \pm 6$ **b.** 2   **c.** $\left\{2, -\dfrac{1}{2} + i\dfrac{\sqrt{11}}{2}, -\dfrac{1}{2} - i\dfrac{\sqrt{11}}{2}\right\}$

15. $2 + 4\log_3(x)$ **16. a.** 13.86 years   **b.** 3465.74 years   **17.** 2

18. 9.24 years   **19.** $\{(-3, 5)\}$ **20.** $\{(3, 2, -1)\}$

CHAPTER 6

Section 6.1

Practice Problems: **1. a.** 3×2 **b.** 1×2

2. $\begin{bmatrix} 0 & 3 & -1 & 8 \\ 1 & 4 & 0 & 14 \\ 0 & -2 & 9 & 0 \end{bmatrix}$ **3.** $\begin{bmatrix} 1 & 2 & 3 \\ 0 & -2 & -4 \end{bmatrix}$ **4.** $\left\{\left(-1, -\dfrac{1}{3}, -1\right)\right\}$

5. $\{(5, 2)\}$ **6.** $\{(1, 2, -3)\}$ **7.** $\varnothing$ **8.** $\{(1, 2, -2)\}$

9. $\left\{\left(-z - 1, \dfrac{5}{3} - \dfrac{2}{3}z, z\right)\right\}$ **10. a.** $\begin{cases} s = 0.01s + 0.1c + 0.1t + 12.46 \\ c = 0.2s + 0.02c + 0.01t + 3 \\ t = 0.25s + 0.3c + 2.7 \end{cases}$

b. $(0.99)(14) - (0.1)(6) - (0.1)(8) = 12.46$ ✓
$(-0.20)(14) + (0.98)(6) - (0.01)(8) = 3$ ✓
$(-0.25)(14) - (0.30)(6) + 8 = 2.7$ ✓

A Exercises: Basic Skills and Concepts: **1.** numbers   **3.** row
operations   **5.** false   **7.** false   **9.** 1×1 **11.** 2×4
13. 2×3 **15.** $a_{13} = 3, a_{31} = 9, a_{33} = 11, a_{34} = 12$
17. No, it is not a rectangular array of numbers.

19. $\begin{bmatrix} 2 & 4 & 2 \\ 1 & -3 & 1 \end{bmatrix}$ **21.** $\begin{bmatrix} 5 & -7 & 11 \\ -13 & 17 & 19 \end{bmatrix}$

23. $\begin{bmatrix} -1 & 2 & 3 & 8 \\ 2 & -3 & 9 & 16 \\ 4 & -5 & -6 & 32 \end{bmatrix}$ **25.** $\begin{cases} x + 2y - 3z = 4 \\ -2x - 3y + z = 5 \\ 3x - 3y + 2z = 7 \end{cases}$

27. $\begin{cases} x - y + z = 2 \\ 2x + y - 3z = 6 \end{cases}$ **29.** $\begin{bmatrix} 1 & 2 & 3 \\ 0 & 1 & 1 \end{bmatrix}$ **31.** $\begin{bmatrix} 1 & 2 & 3 & 4 \\ 0 & 1 & 5 & -3 \\ 0 & 0 & 1 & -1 \end{bmatrix}$

33. $\begin{bmatrix} 4 & 5 & -7 \\ 5 & 4 & -2 \end{bmatrix} \xrightarrow{\frac{1}{4}R_1} \begin{bmatrix} 1 & \frac{5}{4} & -\frac{7}{4} \\ 5 & 4 & -2 \end{bmatrix} \xrightarrow{-5R_1 + R_2 \rightarrow R_2}$

$\begin{bmatrix} 1 & \frac{5}{4} & -\frac{7}{4} \\ 0 & -\frac{9}{4} & \frac{27}{4} \end{bmatrix} \xrightarrow{-\frac{4}{9}R_2} \begin{bmatrix} 1 & \frac{5}{4} & -\frac{7}{4} \\ 0 & 1 & -3 \end{bmatrix}$

35. $\begin{bmatrix} 1 & 4 & 3 & 1 \\ 0 & -3 & -2 & 0 \\ 0 & 7 & 5 & -3 \end{bmatrix} \xrightarrow{-\frac{1}{3}R_2} \begin{bmatrix} 1 & 4 & 3 & 1 \\ 0 & 1 & \frac{2}{3} & 0 \\ 0 & 7 & 5 & -3 \end{bmatrix}$

$\xrightarrow{-7R_2 + R_3 \rightarrow R_3} \begin{bmatrix} 1 & 4 & 3 & 1 \\ 0 & 1 & \frac{2}{3} & 0 \\ 0 & 0 & \frac{1}{3} & -3 \end{bmatrix} \xrightarrow{3R_3} \begin{bmatrix} 1 & 4 & 3 & 1 \\ 0 & 1 & \frac{2}{3} & 0 \\ 0 & 0 & 1 & -9 \end{bmatrix}$

37. No, because Property 1 is not satisfied.   **39.** The matrix is
in reduced row-echelon form because Properties 1–4 are satisfied.
41. The matrix is in reduced row-echelon form because

Properties 1–4 are satisfied.   **43.** The matrix is in reduced
row-echelon form because Properties 1–4 are satisfied.

45. $(5, -2)$ $\begin{cases} x + 2y = 1 \\ y = -2 \end{cases}$ **47.** $(-4 - 4y, y, 3)$ $\begin{cases} x + 4y + 2z = 2 \\ z = 3 \end{cases}$

49. $(1, 2, -1)$ $\begin{cases} x + 2y + 3z = 2 \\ y - 2z = 4 \\ z = -1 \end{cases}$

51. $(2, -5, -2w + 3, w)$ $\begin{cases} x = 2 \\ y = -5 \\ z + 2w = 3 \end{cases}$

53. $(-5, 4, 3, 0)$ $\begin{cases} x = -5 \\ y = 4 \\ z + 2w = 3 \\ w = 0 \end{cases}$ **55.** $\{(5, -3)\}$ **57.** $\varnothing$

59. $\{(3, 1)\}$ **61.** $\left\{\left(-\dfrac{1}{5}, -\dfrac{23}{25}\right)\right\}$ **63.** $\{(2, 1)\}$ **65.** $\{(1, 2, 3)\}$

67. $\varnothing$ **69.** $\{(2, 3, 4)\}$ **71.** $\{(-4z + 15, 7z - 23, z)\}$
73. $\{(1, 2, 3)\}$ **75.** $\{(1, 2, -1)\}$ **77.** $\varnothing$

79. $\{(1 + 4z, 3 - 3z, z)\}$ **81.** $\left\{\left(\dfrac{5}{2}, \dfrac{3}{2}, \dfrac{7}{2}\right)\right\}$

83. a. One solution   **b.** Infinitely many solutions
c. No solution   **85.** True

B Exercises: Applying the Concepts: **87. a.** $\begin{cases} a = 0.1b + 1000 \\ b = 0.2a + 780 \end{cases}$

b. $\begin{bmatrix} 1 & -0.1 & 1000 \\ -0.2 & 1 & 780 \end{bmatrix}$ **c.** $a = 1100, b = 1000$

89. a. $l = 0.4t + 0.2f + 10,000$
$t = 0.5l + 0.3t + 20,000$
$f = 0.5l + 0.05t + 0.35f + 10,000$

b. $\begin{bmatrix} 1 & -0.4 & -0.2 & 10,000 \\ -0.5 & 0.7 & 0 & 20,000 \\ -0.5 & -0.05 & 0.65 & 10,000 \end{bmatrix}$

c. food, \$55,000; transportation, \$61,000; labor, \$45,400
91. $T_1 = 110, T_2 = 140, T_3 = 60$

93. a. $\begin{cases} x - y = 70 \\ -x + z = -120 \\ y - z = 50 \end{cases}$

b. $\{(120 + z, 50 + z, z)\}: 0 \le z \le 130$
95. a. $h = -16t^2 + 64t + 3$ **b.** Approximately 4 sec   **c.** 67 ft
97. 8 lb of the cheaper coffee, 4 lb of the more expensive coffee,
2 lb of chicory

C Exercises: Beyond the Basics:
99. $\{(y + 2w, y, -2y - 3w, w)\}$ **101.** $\{(4, 1, -1, 3)\}$

103. a. $B = \begin{bmatrix} 1 & \frac{3}{2} & 0 & -1 \\ 0 & 1 & 0 & -\frac{1}{2} \\ 0 & 0 & 1 & \frac{1}{2} \\ 0 & 0 & 0 & 1 \end{bmatrix}, C = \begin{bmatrix} 1 & 2 & -3 & 1 \\ 0 & 1 & 0 & -\frac{1}{2} \\ 0 & 0 & 1 & \frac{1}{2} \\ 0 & 0 & 0 & 1 \end{bmatrix}$

b. The reduced row-echelon form of the matrices B and C is
$\begin{bmatrix} 1 & 0 & 0 & 0 \\ 0 & 1 & 0 & 0 \\ 0 & 0 & 1 & 0 \\ 0 & 0 & 0 & 1 \end{bmatrix}$.   **105. a.** $\left\{\left(\dfrac{dm - nb}{ad - bc}, \dfrac{cm - an}{cb - ad}\right)\right\}$

b. (i) $cb \ne ad$ **(ii)** $cb = ad$ and $\dfrac{m}{b} \ne \dfrac{n}{d}$ **(iii)** $cb = ad$ and $\dfrac{m}{b} = \dfrac{n}{d}$

107. $\{(10, 100, 1000)\}$ **109.** $y = -x^3 + 2x^2 + 3x + 1$

Critical Thinking: **111.** $\begin{bmatrix} 0 \\ 0 \end{bmatrix}, \begin{bmatrix} 1 \\ 0 \end{bmatrix}$ **112.** $\begin{bmatrix} 0 & 0 \\ 0 & 0 \end{bmatrix}, \begin{bmatrix} 1 & k \\ 0 & 0 \end{bmatrix}, \begin{bmatrix} 0 & 1 \\ 0 & 0 \end{bmatrix}, \begin{bmatrix} 1 & 0 \\ 0 & 1 \end{bmatrix},$

where k is some constant   **113.** Yes, the inverse of the operation used to transform A to B **114. a.** True   **b.** False

Section 6.2

Practice Problems: **1.** $x = 1, y = 1$ **2.** $\begin{bmatrix} -6 & 1 & 13 \\ 12 & 3 & 15 \end{bmatrix}$

3. $\begin{bmatrix} 11 & 1 \\ 0 & 6 \\ -15 & -28 \end{bmatrix}$ **4.** $X = \begin{bmatrix} -\dfrac{1}{3} & \dfrac{19}{3} \\ -7 & -\dfrac{35}{3} \end{bmatrix}$

5. Yes, the product AB is defined with order 2×1.   **6.** \$2045 thousand   **7.** $[31]$ **8.** The product AB is not defined.
$BA = \begin{bmatrix} 42 & -1 \\ 2 & -6 \\ 8 & -4 \end{bmatrix}$ **9.** $AB = \begin{bmatrix} 18 & -3 \\ 12 & 12 \end{bmatrix}, BA = \begin{bmatrix} 14 & -1 \\ 28 & 16 \end{bmatrix}$

10.

A Exercises: Basic Skills and Concepts: **1.** $a_{ij} = b_{ij}$ **3.** 1×1
5. false   **7.** $x = 2, u = -3$ **9.** $x = -1, y = 3$
11. $x = 2, y = 1$ **13.** $\varnothing$ **15. a.** $\begin{bmatrix} 0 & 2 \\ 5 & 1 \end{bmatrix}$ **b.** $\begin{bmatrix} 2 & 2 \\ 1 & 7 \end{bmatrix}$
c. $\begin{bmatrix} -3 & -6 \\ -9 & -12 \end{bmatrix}$ **d.** $\begin{bmatrix} 5 & 6 \\ 5 & 18 \end{bmatrix}$ **e.** $\begin{bmatrix} 10 & 2 \\ 5 & 11 \end{bmatrix}$ **f.** $\begin{bmatrix} 6 & 10 \\ 23 & 13 \end{bmatrix}$
17. a, b, d, e, and f are not defined; c. $\begin{bmatrix} -6 & -9 \\ 12 & -15 \end{bmatrix}$
19. a. $\begin{bmatrix} 7 & 1 & -1 \\ -1 & 1 & 4 \\ 2 & 1 & 4 \end{bmatrix}$ **b.** $\begin{bmatrix} 1 & -1 & -1 \\ -3 & 9 & 0 \\ -2 & -1 & -2 \end{bmatrix}$ **c.** $\begin{bmatrix} -12 & 0 & 3 \\ 6 & -15 & -6 \\ 0 & 0 & -3 \end{bmatrix}$
d. $\begin{bmatrix} 6 & -2 & -3 \\ -8 & 23 & 2 \\ -4 & -2 & -3 \end{bmatrix}$ **e.** $\begin{bmatrix} 46 & 7 & -7 \\ 0 & 4 & 21 \\ 21 & 7 & 18 \end{bmatrix}$ **f.** $\begin{bmatrix} 6 & 1 & -7 \\ -21 & 6 & 16 \\ -13 & -1 & -10 \end{bmatrix}$
21. a. $\begin{bmatrix} 4 & 3 & -3 \\ 4 & 0 & 7 \\ 4 & 0 & 3 \end{bmatrix}$ **b.** $\begin{bmatrix} -2 & 1 & -3 \\ 2 & 8 & 3 \\ 0 & -2 & -3 \end{bmatrix}$ **c.** $\begin{bmatrix} -3 & -6 & 9 \\ -9 & -12 & -15 \\ -6 & 3 & 0 \end{bmatrix}$
d. $\begin{bmatrix} -3 & 4 & -9 \\ 7 & 20 & 11 \\ 2 & -5 & -6 \end{bmatrix}$ **e.** $\begin{bmatrix} 16 & 12 & 0 \\ 44 & 12 & 9 \\ 28 & 12 & -3 \end{bmatrix}$ **f.** $\begin{bmatrix} -9 & 14 & 5 \\ 22 & -2 & 13 \\ -14 & -1 & -22 \end{bmatrix}$
23. $X = \begin{bmatrix} -4 & -2 & 1 \\ 1 & 5 & 0 \end{bmatrix}$ **25.** $X = \begin{bmatrix} 0 & 2 & -\dfrac{1}{2} \\ \dfrac{3}{2} & \dfrac{1}{2} & 4 \end{bmatrix}$
27. $X = \begin{bmatrix} -4 & -4 & \dfrac{3}{2} \\ -\dfrac{1}{2} & \dfrac{9}{2} & -4 \end{bmatrix}$ **29.** $X = \begin{bmatrix} \dfrac{1}{2} & -\dfrac{9}{4} & \dfrac{1}{2} \\ -2 & -\dfrac{5}{4} & -5 \end{bmatrix}$
31. a. $AB = \begin{bmatrix} 4 & 11 \\ 6 & 23 \end{bmatrix}$ **b.** $BA = \begin{bmatrix} 1 & 0 \\ 18 & 26 \end{bmatrix}$
33. a. $AB = \begin{bmatrix} 4 & 7 \\ 3 & -12 \end{bmatrix}$ **b.** $BA = \begin{bmatrix} -13 & 4 & 10 \\ -13 & 5 & 6 \\ 8 & -4 & 0 \end{bmatrix}$
35. a. $AB = [16]$ **b.** $BA = \begin{bmatrix} 2 & 3 & 5 \\ -4 & -6 & -10 \\ 8 & 12 & 20 \end{bmatrix}$

37. a. $AB = [7 \quad 8 \quad -8]$ **b.** The product BA is not defined.
39. a. $AB = \begin{bmatrix} 10 & 7 & 2 \\ 7 & 19 & 4 \\ 10 & 1 & 0 \end{bmatrix}$ **b.** $BA = \begin{bmatrix} 7 & 4 & 5 \\ 0 & 8 & 3 \\ 19 & 18 & 14 \end{bmatrix}$
41. $AB = \begin{bmatrix} 13 & 17 & 3 \\ 13 & 8 & 2 \\ 6 & 1 & 6 \end{bmatrix} \neq \begin{bmatrix} 6 & 22 & 19 \\ 2 & 11 & 6 \\ 11 & -1 & 10 \end{bmatrix} = BA$

B Exercises: Applying the Concepts
47.
	Steel	Glass	Wood	
$C = $	13	5	38	Cost of Material
	7	2	7	Transportation Cost

49. \$24,200

51. a.
	Chair	President	Vice President
Salary	2,500,000	1,250,000	100,000
Bonus	1,500,000	750,000	150,000
Stock	50,000	25,000	5000

b. $\begin{bmatrix} 1 \\ 1 \\ 4 \end{bmatrix}$ Chair, President, Vice President   **c.** $\begin{bmatrix} 4,150,000 \\ 2,850,000 \\ 95,000 \end{bmatrix}$ Total Salary, Total Bonuses, Total Stocks

53. $AD = \begin{bmatrix} 0 & 4 & 4 & 1 & 1 & 0 \\ 0 & 0 & -1 & -1 & -6 & -6 \end{bmatrix}$

55. $AD = \begin{bmatrix} 0 & -4 & -4 & -1 & -1 & 0 \\ 0 & 1 & 2 & 1.25 & 6.25 & 6 \end{bmatrix}$

C Exercises: Beyond the Basics: **57.** $AB = \begin{bmatrix} 9 & 0 \\ 0 & 0 \end{bmatrix} = AC$
59. Let $A = \begin{bmatrix} 1 & 2 \\ 3 & 4 \end{bmatrix}$ and $B = \begin{bmatrix} -1 & 0 \\ 2 & -3 \end{bmatrix}$. Then
$(A + B)^2 = \begin{bmatrix} 10 & 2 \\ 5 & 11 \end{bmatrix} \neq \begin{bmatrix} 14 & -2 \\ 17 & 7 \end{bmatrix} = A^2 + 2AB + B^2$.
63. $B = \begin{bmatrix} 2 & -3 \\ -1 & 2 \end{bmatrix}$ **65.** $x = 33, y = 5$

Critical Thinking: **67. a.** AB is defined when $n = 5$, and the order of AB when this product is defined is $3 \times m$.   **b.** BA is defined when $m = 3$, and the order of BA when this product is defined is $5 \times n$.   **68.** $(CA)B$

Section 6.3

Practice Problems:
1. $\begin{bmatrix} 3 & 2 \\ 2 & 1 \end{bmatrix} \begin{bmatrix} -1 & 2 \\ 2 & -3 \end{bmatrix} = \begin{bmatrix} 1 & 0 \\ 0 & 1 \end{bmatrix}$
2. If A^{-1} exists, then $\begin{bmatrix} 3x + z & 3y + w \\ 3x + z & 3y + w \end{bmatrix} = \begin{bmatrix} 1 & 0 \\ 0 & 1 \end{bmatrix}$

this implies that $1 = 3x + z = 0$, which is false.

3. The inverse of matrix A does not exist.

4. $A^{-1} = \begin{bmatrix} \dfrac{1}{7} & -\dfrac{19}{77} & \dfrac{1}{11} \\ 0 & \dfrac{2}{11} & \dfrac{1}{11} \\ \dfrac{2}{7} & -\dfrac{3}{77} & -\dfrac{1}{11} \end{bmatrix}$

5. The inverse of matrix A does not exist; $B^{-1} = \begin{bmatrix} \dfrac{1}{14} & \dfrac{1}{7} \\ -\dfrac{3}{14} & \dfrac{4}{7} \end{bmatrix}$.

6. $\left\{\left(\dfrac{11}{2}, -\dfrac{9}{2}, \dfrac{1}{2}\right)\right\}$ **7.** $X = \begin{bmatrix} \dfrac{110,400}{19} \\ \dfrac{135,200}{19} \end{bmatrix}$

8. $AM = \begin{bmatrix} 1 & 2 & 3 \\ 1 & 3 & 3 \\ 1 & 2 & 4 \end{bmatrix} \begin{bmatrix} 10 & 11 & 19 & 15 & 19 & 5 \\ 1 & 0 & 0 & 23 & 1 & 0 \\ 3 & 9 & 14 & 0 & 6 & 0 \end{bmatrix}$

$= \begin{bmatrix} 21 & 38 & 61 & 61 & 39 & 5 \\ 22 & 38 & 61 & 84 & 40 & 5 \\ 24 & 47 & 75 & 61 & 45 & 5 \end{bmatrix}$

A Exercises: Basic Skills and Concepts: 1. inverse **3.** A^{-1}
5. false **7.** Yes **9.** Yes **11.** No **13.** Yes **15.** Yes

17. $A^{-1} = \begin{bmatrix} \dfrac{1}{2} & 0 \\ -\dfrac{1}{6} & \dfrac{1}{3} \end{bmatrix}$

19. The inverse of matrix A does not exist.

21. $A^{-1} = \begin{bmatrix} 1 & -8 & 10 \\ 0 & 2 & -3 \\ 0 & -1 & 2 \end{bmatrix}$ **23.** $A^{-1} = \begin{bmatrix} \dfrac{3}{4} & -\dfrac{5}{2} & -\dfrac{1}{4} \\ -\dfrac{1}{2} & 2 & \dfrac{1}{2} \\ \dfrac{1}{2} & -1 & -\dfrac{1}{2} \end{bmatrix}$

25. $A^{-1} = \begin{bmatrix} \dfrac{3}{14} & -\dfrac{1}{14} & \dfrac{5}{14} \\ \dfrac{5}{14} & \dfrac{3}{14} & -\dfrac{1}{14} \\ -\dfrac{1}{14} & \dfrac{5}{14} & \dfrac{3}{14} \end{bmatrix}$ **27.** $A^{-1} = \begin{bmatrix} 1 & 0 \\ -\dfrac{3}{2} & \dfrac{1}{2} \end{bmatrix}$

29. $A^{-1} = \begin{bmatrix} 5 & 3 \\ 3 & 2 \end{bmatrix}$ **31.** $A^{-1} = \dfrac{1}{-a^2+b^2}\begin{bmatrix} -a & b \\ -b & a \end{bmatrix}$, where $a^2 \neq b^2$

33. $\begin{bmatrix} 2 & 3 \\ 1 & -3 \end{bmatrix}\begin{bmatrix} x \\ y \end{bmatrix} = \begin{bmatrix} -9 \\ 13 \end{bmatrix}$ **35.** $\begin{bmatrix} 3 & 2 & 1 \\ 2 & 1 & 3 \\ 1 & 3 & 2 \end{bmatrix}\begin{bmatrix} x \\ y \\ z \end{bmatrix} = \begin{bmatrix} 8 \\ 7 \\ 9 \end{bmatrix}$

37. $\begin{cases} x - 2y = 0 \\ 2x + y = 5 \end{cases}$ **39.** $\begin{cases} 2x_1 + 3x_2 + x_3 = -1 \\ 5x_1 + 7x_2 - x_3 = 5 \\ 4x_1 + 3x_2 = 5 \end{cases}$

41. $\begin{bmatrix} 1 & 2 & 5 \\ 2 & 3 & 8 \\ -1 & 1 & 2 \end{bmatrix}\begin{bmatrix} 2 & -1 & -1 \\ 12 & -7 & -2 \\ -5 & 3 & 1 \end{bmatrix} = \begin{bmatrix} 1 & 0 & 0 \\ 0 & 1 & 0 \\ 0 & 0 & 1 \end{bmatrix}$

43. $\{(2, -3, 1)\}$ **45. a.** $A^{-1} = \begin{bmatrix} 3 & -\dfrac{5}{2} & \dfrac{1}{2} \\ -3 & 4 & -1 \\ 1 & -\dfrac{3}{2} & \dfrac{1}{2} \end{bmatrix}$ **b.** $\{(1, 2, 3)\}$

B Exercises: Applying the Concepts: 47. $\{(-1, 2)\}$
49. $\{(2, -1, 3)\}$ **51.** $\{(-5, 1, 5)\}$ **53.** Treasury: $16,000; bonds:
$48,000; fund: $26,000 **55.** Chicken: $5; fish: $8; ham: $6
57. a. 20 ft **b.** 4 ft

59. $X = \begin{bmatrix} 100 \\ 100 \end{bmatrix}$ **61.** $X = \begin{bmatrix} \dfrac{216,000}{277} \\ \dfrac{310,000}{277} \\ \dfrac{304,000}{277} \end{bmatrix}$

C Exercises: Beyond the Basics: 67. a. Yes. By the definition
of *inverse*, if $AB = I$, then $BA = I$ and A and B are inverses.
b. $(A^{-1})^{-1} = A$ **75.** If A is invertible, there exists
some matrix C such that $AC = I$ and $CA = I$. Then
$AB = A \Leftrightarrow CAB = CA \Leftrightarrow IB = I \Leftrightarrow B = I$.

77. c. $A^{-1} = \begin{bmatrix} 3 & -4 \\ -2 & 3 \end{bmatrix}$ **79.** $\{(1, 2, 1, 2)\}$ **81.** $\{(0, 2, -4, 3)\}$
83. $\{(1, 2, 3, 4, -1)\}$

Critical Thinking: 85. a. $A = \begin{bmatrix} 8 & 10 \\ 6 & 8 \end{bmatrix}$ **b.** $A = \begin{bmatrix} -1 & 7 \\ -3 & 17 \end{bmatrix}$

c. $A = \begin{bmatrix} -1 & -2 \\ \dfrac{5}{2} & 4 \end{bmatrix}$ **d.** $A = \begin{bmatrix} \dfrac{3}{2} & \dfrac{1}{2} \\ \dfrac{5}{2} & \dfrac{3}{2} \end{bmatrix}$ **86. a.** Yes **b.** No

87. a. True. $A^2B = AAB = ABA = BAA = BA^2$

b. False. Let $A = I$ and $B = -I$. **88.** Let $A = \begin{bmatrix} 0 & -1 \\ 0 & 0 \end{bmatrix}$. **89.** True

Section 6.4

Practice Problems: 1. a. -38 **b.** 0 **2. a.** $M_{11} = 4, M_{23} = 10,$
$M_{32} = 10$ **b.** $A_{11} = 4, A_{23} = -10, A_{32} = -10$ **3.** 8
4. $\{(17, -9)\}$ **5.** $\{(2, -1, 0)\}$

A Exercises: Basic Skills and Concepts: 1. $ad - bc$ **3.** cofactor
5. false **7.** -2 **9.** -6 **11.** -7 **13.** $\dfrac{139}{72}$ **15.** $a - b$
17. a. -10 **b.** 2 **c.** 0 **19. a.** -4 **b.** 4 **c.** -2 **21.** -2
23. 12 **25.** 0 **27.** 42 **29.** 16 **31.** 49 **33.** $a^3 + b^3$
35. $a^3 + b^3 + c^3 - 3abc$ **37.** $\{(3, 5)\}$ **39.** $\{(1, 2)\}$
41. $\{(2, 0)\}$ **43.** $\left\{\left(\dfrac{3}{2}y + 2, y\right)\right\}$ **45.** $\{(2, 3)\}$ **47.** $\{(1, 1, 0)\}$
49. $\{(1, -1, 3)\}$ **51.** $\{(1, 1, 1)\}$ **53.** $\{(1, 2, 3)\}$
55. $\{(1, 2, 3)\}$

B Exercises: Applying the Concepts: 57. 11 **59.** 10.5 **61.** Yes
63. No **65.** $y = 2x + 1$ **67.** $y = \dfrac{2}{3}x + \dfrac{1}{3}$

C Exercises: Beyond the Basics: 75. 21 **77.** 476 **79.** 4
81. $\dfrac{1}{2} \pm i\dfrac{\sqrt{7}}{2}$ **83.** 23 **85.** $-1, 0, 1$ **87.** 30 **89.** $\left\{\left(\dfrac{9}{4}, -\dfrac{2}{3}\right)\right\}$
91. $\{(-2, -1)\}$

Review Exercises

1. 1×4 **3.** 3×2 **5.** $a_{12} = -1, a_{14} = -4, a_{23} = 3, a_{21} = 5$

7. $\begin{bmatrix} 2 & -3 & 7 \\ 3 & 1 & 6 \end{bmatrix}$ **9.** $\begin{bmatrix} 1 & -2 & 1 & 7 \\ 0 & 1 & 2 & 1 \\ 0 & 0 & 1 & 2 \end{bmatrix}$

11. $\begin{bmatrix} 1 & 0 & 0 & \dfrac{1}{3} \\ 0 & 1 & 0 & \dfrac{1}{2} \\ 0 & 0 & 1 & -\dfrac{1}{6} \end{bmatrix}$ **13.** $\{(2, -3, -1)\}$ **15.** $\{(3, -2, -3)\}$

17. $\{(5, -4, 1)\}$ **19.** $\left\{\left(2, -1, \dfrac{-1}{3}\right)\right\}$ **21.** $\{(2, 1)\}$

23. a. $\begin{bmatrix} 3 & -1 \\ -8 & 10 \end{bmatrix}$ **b.** $\begin{bmatrix} -1 & 5 \\ 2 & -2 \end{bmatrix}$ **c.** $\begin{bmatrix} 2 & 4 \\ -6 & 8 \end{bmatrix}$ **d.** $\begin{bmatrix} -6 & 9 \\ 15 & -18 \end{bmatrix}$

e. $\begin{bmatrix} -4 & 13 \\ 9 & -10 \end{bmatrix}$ **25.** $X = \begin{bmatrix} \dfrac{7}{3} & 0 \\ -\dfrac{19}{3} & 8 \end{bmatrix}$

27. $AB = \begin{bmatrix} -3 & 4 \\ -11 & 10 \end{bmatrix}, BA = \begin{bmatrix} -2 & -4 \\ 8 & 9 \end{bmatrix}$

29. $AB = [7], BA = \begin{bmatrix} 2 & 4 & -2 \\ 3 & 6 & -3 \\ 1 & 2 & -1 \end{bmatrix}$

31. $\begin{bmatrix} 1 & 0 \\ 2 & 3 \end{bmatrix}$ **33.** $\begin{bmatrix} -1 & \dfrac{3}{2} \\ 2 & -\dfrac{5}{2} \end{bmatrix}$ **35. a.** $\begin{bmatrix} 3 & -1 \\ 1 & 0 \end{bmatrix}$ **b.** $\begin{bmatrix} 8 & -3 \\ 3 & -1 \end{bmatrix}$

c. $\begin{bmatrix} 8 & -3 \\ 3 & -1 \end{bmatrix}$ **41.** $\begin{bmatrix} \dfrac{2}{5} & \dfrac{1}{10} \\ -\dfrac{1}{5} & \dfrac{3}{10} \end{bmatrix}$ **43.** $\begin{bmatrix} 3 & 2 & 6 \\ 1 & 1 & 2 \\ 2 & 2 & 5 \end{bmatrix}$

45. $\{(-23, 10)\}$ **47.** $\{(-12, 2, 3)\}$ **49.** $\left\{ \left(\dfrac{7}{6}, \dfrac{1}{6}, 2 \right) \right\}$ **51.** 23

53. 0 **55. a.** $M_{12} = 8, M_{23} = 8, M_{22} = -16$
b. $A_{12} = -8, A_{23} = -8, A_{22} = -16$ **57.** 35 **59.** 6 **61.** $\{(1, 2)\}$
63. $\{(5, 6, 7)\}$ **65.** 2 **67.** They both satisfy the equation
$y = 7x - 11$. **69.** Method 3 **71.** The speed of the plane is 520
miles per hour, and the velocity of the wind is 40 miles per hour.
73. Andrew has \$170, Bonnie has \$65, and Chauncie has \$85. **75.** 30
registered nurses, 40 licensed practical nurses, and 60 nurse's aides

Practice Test A

1. $5 \times 4; 9$ **2.** $\begin{bmatrix} 7 & -3 & 9 & | & 5 \\ -2 & 4 & 3 & | & -12 \\ 8 & -5 & 1 & | & -9 \end{bmatrix}$ **3.** $\begin{cases} 4x - \quad z = -3 \\ x + 3y \quad = 9 \\ 2x + 7y + 5z = 8 \end{cases}$

4. $(10, 10, 0)$ **5.** $\{(2, 3, -2)\}$ **6.** $\left\{ \left(\dfrac{20}{7} + \dfrac{11}{7}z, \dfrac{2}{7} + \dfrac{6}{7}z, z \right) \right\}$

7. $A - B = \begin{bmatrix} 8 & -1 \\ 4 & 8 \\ 11 & 0 \end{bmatrix}$ **8.** $AB = [1]$

9. The product AB is not defined.
10. $\begin{bmatrix} -6 & 2 & 0 \\ 10 & 14 & 4 \end{bmatrix}$ **11.** $\begin{bmatrix} 20 & 28 & 8 \\ -9 & 29 & 6 \end{bmatrix}$ **12.** $\begin{bmatrix} 1 & 25 \\ 0 & 16 \end{bmatrix}$

13. $\begin{bmatrix} 1 & -\dfrac{5}{4} \\ 0 & \dfrac{1}{4} \end{bmatrix}$ **14.** $\begin{bmatrix} -11 & 2 & 7 \\ 8 & -1 & -5 \\ 5 & -1 & -3 \end{bmatrix}$ **15.** $\begin{bmatrix} 5 & 2 \\ 3 & 1 \end{bmatrix}\begin{bmatrix} x \\ y \end{bmatrix} = \begin{bmatrix} 32 \\ 18 \end{bmatrix}$

16. $\begin{cases} 12x - 3y = 5 \\ -2x + 7y = -9 \end{cases}$ **17.** $X = \begin{bmatrix} -\dfrac{3}{5} & -1 & 4 \\ \dfrac{32}{5} & -\dfrac{18}{5} & -3 \end{bmatrix}$ **18.** $\dfrac{1}{2}$

19. 0 **20.** $x = \dfrac{\begin{vmatrix} 3 & -1 & 1 \\ 6 & 1 & 1 \\ 4 & 3 & -2 \end{vmatrix}}{\begin{vmatrix} 2 & -1 & 1 \\ 1 & 1 & 1 \\ 4 & 3 & -2 \end{vmatrix}}, y = \dfrac{\begin{vmatrix} 2 & 3 & 1 \\ 1 & 6 & 1 \\ 4 & 4 & -2 \end{vmatrix}}{\begin{vmatrix} 2 & -1 & 1 \\ 1 & 1 & 1 \\ 4 & 3 & -2 \end{vmatrix}},$

$z = \dfrac{\begin{vmatrix} 2 & -1 & 3 \\ 1 & 1 & 6 \\ 4 & 3 & 4 \end{vmatrix}}{\begin{vmatrix} 2 & -1 & 1 \\ 1 & 1 & 1 \\ 4 & 3 & -2 \end{vmatrix}}$

Practice Test B

1. d **2.** d **3.** d **4.** c **5.** c **6.** d **7.** c **8.** b **9.** a
10. b **11.** c **12.** c **13.** b **14.** d **15.** a **16.** a **17.** b
18. d **19.** c **20.** b

Cumulative Review Exercises (Chapters P–6)

1. $-7, 1$ **2.** 7 **3.** $4, 1$ **4.** $1 + \dfrac{3}{2}i, 1 - \dfrac{3}{2}i$ **5.** $-9, \dfrac{-19}{7}$

6. $2, -8$ **7.** $x < -2$ or $x > \dfrac{1}{2}$ **8.** $1 \le x \le 6$ **9.** $\pm 1, \pm \dfrac{1}{2},$

$\pm \dfrac{1}{4}, \pm 3, \pm \dfrac{3}{2}, \pm \dfrac{3}{4}$ **11.** 24,000 **12.** 500 **13.** $\{(1, -1)\}$

14. $\left\{ (2, 1), \left(\dfrac{5}{4}, \dfrac{5}{2} \right) \right\}$ **15.**

16. $A^{-1} = \begin{bmatrix} 3 & 2 & 6 \\ 1 & 1 & 2 \\ 2 & 2 & 5 \end{bmatrix}$

17. $\{(1, 1, -1)\}$ **18. a.** $F(x) = x^2 + 7x + 9$

b. $F(4) = 53$ **19.** $f^{-1}(x) = \dfrac{-4x}{x - 1}$

20. Domain: $(-\infty, -4) \cup (-4, \infty)$; range: $(-\infty, 1) \cup (1, \infty)$

CHAPTER 7

Section 7.2

Practice Problems: **1. a.**

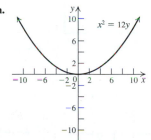

Vertex: $(0, 0)$; focus: $(0, 3)$; directrix: $y = -3$; axis: y-axis

b.

Vertex: $(0, 0)$; focus: $\left(-\dfrac{3}{2}, 0 \right)$;

directrix: $x = \dfrac{3}{2}$; axis: x-axis

2. a. $x^2 = 8y$ **b.** $y^2 = 4x$

3. Vertex: $(2, -1)$; focus: $\left(2, -\frac{7}{8}\right)$; directrix: $y = -\frac{9}{8}$

4. Equation: $x^2 = 29.2y$
The thickness of the mirror at the edges is 0.0771 in.

A Exercises: Basic Skills and Concepts:

1. directrix; focus   **3.** vertex   **5.** true

7. Focus: $\left(0, \frac{1}{2}\right)$; directrix: $y = -\frac{1}{2}$; graph: e

9. Focus: $\left(0, -\frac{9}{64}\right)$; directrix: $y = \frac{9}{64}$; graph: d

11. Focus: $\left(\frac{1}{2}, 0\right)$; directrix: $x = -\frac{1}{2}$; graph: g

13. Focus: $\left(-\frac{4}{9}, 0\right)$; directrix: $x = \frac{4}{9}$; graph: f

15.   **17.**

19.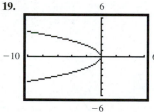

21. $x^2 = -4(y - 3)$; 4   **23.** $y^2 = -10\left(x - \frac{1}{2}\right)$; 10

25. $(y - 1)^2 = -8(x - 1)$; 8   **27.** $(x - 1)^2 = 16(y - 1)$; 16

29. $y^2 = 8(x - 1)$; 8   **31.** $x^2 = -12(y - 1)$; 12

33. $(y - 3)^2 = -8(x - 2)$; 8   **35.** $(x - 2)^2 = 8(y - 3)$; 8

37.   **39.**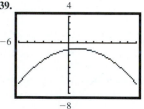

41. Vertex: $(-1, 1)$;
focus: $\left(-\frac{1}{2}, 1\right)$;
directrix: $x = -\frac{3}{2}$

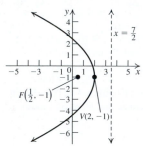

45. Vertex: $(2, -1)$;
focus: $\left(\frac{1}{2}, -1\right)$;
directrix: $x = \frac{7}{2}$

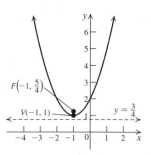

49. Vertex: $(-1, 1)$;
focus: $\left(-1, \frac{5}{4}\right)$;
directrix: $y = \frac{3}{4}$

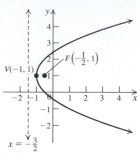

53. Vertex: $\left(-\frac{1}{2}, -1\right)$;
focus: $\left(-\frac{1}{2}, -\frac{5}{4}\right)$;
directrix: $y = -\frac{3}{4}$

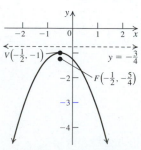

43. Vertex: $(-2, 2)$;
focus: $\left(-2, \frac{11}{4}\right)$;
directrix: $y = \frac{5}{4}$

47. Vertex: $(1, 3)$;
focus: $\left(1, \frac{1}{2}\right)$;
directrix: $y = \frac{11}{2}$

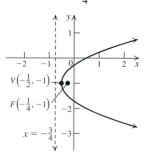

51. Vertex: $\left(-\frac{1}{2}, -1\right)$;
focus: $\left(-\frac{1}{4}, -1\right)$;
directrix: $x = -\frac{3}{4}$

55. Vertex: $(46, 4)$;
focus: $\left(\frac{551}{12}, 4\right)$;
directrix: $x = \frac{553}{12}$

57. $y^2 = 4x$ and $x^2 = \dfrac{1}{2}y$ **59.** $(y-1)^2 = 2x$ and $x^2 = 2(y-1)$

61. $(y-1)^2 = -(x+2)$ and $(x+2)^2 = -(y-1)$
63. $(y-1)^2 = x+1$ and $(x+1)^2 = y-1$

B Exercises: Applying the Concepts: 65. 5 in.

67. $3\dfrac{3}{8}$ ft from the vertex   **69.** $\left(\dfrac{1}{16}, 0\right)$ **71.** $\left(0, \dfrac{1}{16}\right)$

73. 16.875 ft   **75.** 7.96 yd   **77.** Output: 15 tons; cost: $27.50

C Exercises: Beyond the Basics: 79. $(4, 8)$ and $(16, 16)$
81. $y = x - 4$ **83.** $(y-2)^2 = 8(x-1)$ and $(y-1)^2 = -8(x-5)$

85. $\left(x - \dfrac{3}{4}\right)^2 = \dfrac{1}{2}\left(y - \dfrac{31}{8}\right)$ **87.** Vertex: $(4, 6)$; focus: $\left(4, \dfrac{11}{2}\right)$;

directrix: $y = \dfrac{13}{2}$; axis: $x = 4$ **89.** Vertex: $(-2, 3)$; focus: $\left(-\dfrac{11}{4}, 3\right)$;

directrix: $x = -\dfrac{5}{4}$; axis: $y = 3$ **91.** $y = 6x - 3$

Critical Thinking: **95. a.** No, because if the axis is horizontal, it does not pass the vertical line test.   **b.** 0
96. Parabola widens as $|x|$ increases.

97. False

Section 7.3

Practice Problems:

1. $\dfrac{x^2}{36} + \dfrac{y^2}{100} = 1$

2.

3. $\dfrac{(x-2)^2}{9} + \dfrac{(y-1)^2}{25} = 1$ **4.** Center: $(3, -1)$; vertices:

$(3 + \sqrt{42}, -1)$ and $(3 - \sqrt{42}, -1)$; foci: $\left(3 + \dfrac{\sqrt{126}}{2}, -1\right)$ and

$\left(3 - \dfrac{\sqrt{126}}{2}, -1\right)$ **5.** 6.9282 ft

A Exercises: Basic Skills and Concepts: 1. sum
3. $a^2 - c^2$ **5.** true
7. Vertices: $(4, 0)$ and $(-4, 0)$;   **9.** Vertices: $(3, 0)$ and $(-3, 0)$;
foci: $(2\sqrt{3}, 0)$ and $(-2\sqrt{3}, 0)$ foci: $(2\sqrt{2}, 0)$ and $(-2\sqrt{2}, 0)$

11. Vertices: $(5, 0)$ and $(-5, 0)$;   **13.** Vertices: $(0, 6)$ and $(0, -6)$;
foci: $(3, 0)$ and $(-3, 0)$ foci: $(0, 2\sqrt{5})$ and $(0, -2\sqrt{5})$

15. Circle centered at the origin   **17.** Vertices: $(2, 0)$ and $(-2, 0)$;
with radius 2   foci: $(\sqrt{3}, 0)$ and $(-\sqrt{3}, 0)$

19. Vertices: $(0, 3)$ and $(0, -3)$;   **21.** Vertices: $(2, 0)$ and $(-2, 0)$;
foci: $(0, \sqrt{5})$ and $(0, -\sqrt{5})$ foci: $(1, 0)$ and $(-1, 0)$

23. Vertices: $(0, \sqrt{5})$ and   **25.** Vertices: $\left(\dfrac{\sqrt{14}}{2}, 0\right)$ and
$(0, -\sqrt{5})$; foci: $(0, \sqrt{3})$ and
$(0, -\sqrt{3})$ $\left(-\dfrac{\sqrt{14}}{2}, 0\right)$; foci: $\left(\dfrac{\sqrt{42}}{6}, 0\right)$

    and $\left(-\dfrac{\sqrt{42}}{6}, 0\right)$

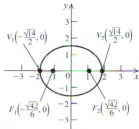

27. $\dfrac{x^2}{9} + \dfrac{y^2}{8} = 1$

29. $\dfrac{x^2}{12} + \dfrac{y^2}{16} = 1$

47. Center: $(-3, 1)$; foci: $(-2, 1)$ and $(-4, 1)$; vertices: $(-3 + \sqrt{5}, 1)$ and $(-3 - \sqrt{5}, 1)$

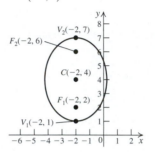

49. Center: $(-1, 2)$; foci: $(1, 2)$ and $(-3, 2)$; vertices: $(2, 2)$ and $(-4, 2)$

31. $\dfrac{x^2}{25} + \dfrac{y^2}{9} = 1$

33. $\dfrac{x^2}{16} + \dfrac{y^2}{20} = 1$

51. Center: $(-2, 4)$; foci: $(-2, 6)$ and $(-2, 2)$; vertices: $(-2, 7)$ and $(-2, 1)$

53. Center: $(1, -1)$; foci: $(2, -1)$ and $(0, -1)$; vertices: $(1 + \sqrt{2}, -1)$ and $(1 - \sqrt{2}, -1)$

35. $\dfrac{x^2}{9} + \dfrac{y^2}{25} = 1$

37. $\dfrac{x^2}{36} + \dfrac{y^2}{27} = 1$

55. No graph

57.

59.

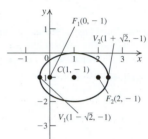

B Exercises: Applying the Concepts: **61.** 18.3 ft
63. 42.4 m   **65.** The bet should not be accepted. The pool shark can hit the ball from any point straight into the pocket, or he can shoot it through the other focus and it will fall into the pocket because of the reflecting property.   **67.** 29 ft from the wall on the opposite sides, along the major axis   **69.** $\dfrac{x^2}{8{,}641.5616} + \dfrac{y^2}{8{,}639.0652} = 1$

71. $\dfrac{x^2}{786{,}254.6241} + \dfrac{y^2}{783{,}788.5085} = 1$

73. Perihelion: 8.72×10^7 km; aphelion: 5.3028×10^9 km

39. $\dfrac{x^2}{9} + \dfrac{y^2}{13} = 1$

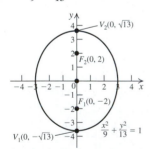

41. Center: $(1, 1)$; foci: $(1, 1 + \sqrt{5})$ and $(1, 1 - \sqrt{5})$; vertices: $(1, 4)$ and $(1, -2)$

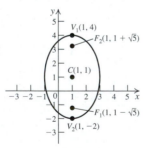

C Exercises: Beyond the Basics: **75.** $\dfrac{x^2}{25} + \dfrac{y^2}{16} = 1$ or $\dfrac{x^2}{\dfrac{625}{41}} + \dfrac{y^2}{25} = 1$

77. Major axis: $\dfrac{2\sqrt{105}}{3}$; minor axis: $\dfrac{\sqrt{70}}{2}$

79. When $e = 0$, the ellipse becomes a circle.   **81.** $\dfrac{2}{3}$ **83.** $\dfrac{4}{5}$

85. $\dfrac{x^2}{16} + \dfrac{y^2}{12} = 1$

43. Center: $(0, -3)$; foci: $(2\sqrt{3}, -3)$ and $(-2\sqrt{3}, -3)$; vertices: $(4, -3)$ and $(-4, -3)$

45. Center: $(1, -2)$; foci: $(2, -2)$ and $(0, -2)$; vertices: $(3, -2)$ and $(-1, -2)$

91. Points of intersection:
$\left(-\dfrac{17}{13}, -\dfrac{3}{13}\right)$ and $(1, -1)$

95. $y = -\dfrac{8}{3}x + \dfrac{25}{3}$

Critical Thinking:
97. Four circles: $(x - 3)^2 + (y - 3)^2 = 9$
$(x + 3)^2 + (y - 3)^2 = 9$
$(x + 3)^2 + (y + 3)^2 = 9$
$(x - 3)^2 + (y + 3)^2 = 9$

98. Eight ellipses:
$$\dfrac{(x - 3)^2}{9} + \dfrac{(y - 2)^2}{4} = 1, \qquad \dfrac{(x - 2)^2}{4} + \dfrac{(y - 3)^2}{9} = 1$$
$$\dfrac{(x + 3)^2}{9} + \dfrac{(y - 2)^2}{4} = 1, \qquad \dfrac{(x + 2)^2}{4} + \dfrac{(y - 3)^2}{9} = 1$$
$$\dfrac{(x + 3)^2}{9} + \dfrac{(y + 2)^2}{4} = 1, \qquad \dfrac{(x + 2)^2}{4} + \dfrac{(y + 3)^2}{9} = 1$$
$$\dfrac{(x - 3)^2}{9} + \dfrac{(y + 2)^2}{4} = 1, \qquad \dfrac{(x - 2)^2}{4} + \dfrac{(y + 3)^2}{9} = 1$$

Section 7.4

Practice Problems: **1.** y-axis   **2.** Vertices: $(2\sqrt{2}, 0)$ and $(-2\sqrt{2}, 0)$;
foci: $(\sqrt{10}, 0)$ and $(-\sqrt{10}, 0)$ **3.** $\dfrac{y^2}{16} - \dfrac{x^2}{20} = 1$ **4.** $y = \dfrac{2}{3}x$ and
$y = -\dfrac{2}{3}x$ **5. a.** The standard form of the equation is
$\dfrac{x^2}{4} - \dfrac{y^2}{25} = 1$. Vertices: $(2, 0)$ and $(-2, 0)$; endpoints of the conjugate
axis: $(0, 5)$ and $(0, -5)$; asymptotes: $y = \dfrac{5}{2}x$ and $y = -\dfrac{5}{2}x$

93. Points of intersection:
$(\sqrt{2}, 3\sqrt{2}), (\sqrt{2}, -3\sqrt{2}),$
$(-\sqrt{2}, 3\sqrt{2}), (-\sqrt{2}, -3\sqrt{2})$

b. The standard form of the equation is $\dfrac{y^2}{\frac{1}{9}} - \dfrac{x^2}{1} = 1$.

Vertices: $\left(0, \dfrac{1}{3}\right)$ and $\left(0, -\dfrac{1}{3}\right)$; endpoints of the conjugate axis: $(1, 0)$
and $(-1, 0)$; asymptotes: $y = \dfrac{1}{3}x$ and $y = -\dfrac{1}{3}x$

6. a.

b.

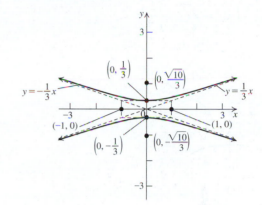

7. $\dfrac{(x-1)^2}{5} - \dfrac{(y-2)^2}{\frac{5}{4}} = 1$. Center: $(1,2)$; vertices: $(1+\sqrt{5}, 2)$

and $(1-\sqrt{5}, 2)$; endpoints of the conjugate axis: $\left(1, 2+\dfrac{\sqrt{5}}{2}\right)$ and

$\left(1, 2-\dfrac{\sqrt{5}}{2}\right)$; asymptotes: $y - 2 = \pm\dfrac{1}{2}(x-1)$; foci: $\left(\dfrac{7}{2}, 2\right)$

and $\left(-\dfrac{3}{2}, 2\right)$

8. $\dfrac{x^2}{16,900} - \dfrac{y^2}{5,600} = 1$

A Exercises: Basic Skills and Concepts: **1.** difference
3. $c^2 - a^2$ **5.** true   **7.** g   **9.** h   **11.** d   **13.** b
15. Vertices: $(1,0)$ and $(-1,0)$; foci: $(\sqrt{5}, 0)$ and $(-\sqrt{5}, 0)$; transverse axis: x-axis; the hyperbola opens left and right; vertices of the fundamental rectangle: $(1,2), (-1,2), (-1,-2), (1,-2)$; asymptotes: $y = \pm 2x$

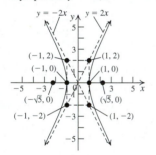

17. Vertices: $(0,1)$ and $(0,-1)$; foci: $(0, \sqrt{2})$ and $(0, -\sqrt{2})$; transverse axis: y-axis; the hyperbola opens up and down; vertices of the fundamental rectangle: $(1,1), (-1,1), (-1,-1), (1,-1)$; asymptotes: $y = \pm x$

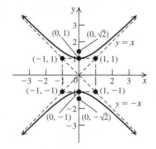

19. Vertices: $(0,2)$ and $(0,-2)$; foci: $(0, 2\sqrt{10})$ and $(0, -2\sqrt{10})$; transverse axis: y-axis; the hyperbola opens up and down; vertices of the fundamental rectangle: $(6,2), (-6,2), (-6,-2), (6,-2)$; asymptotes: $y = \pm\dfrac{1}{3}x$

21. Vertices: $(3,0)$ and $(-3,0)$; foci: $(\sqrt{13}, 0)$ and $(-\sqrt{13}, 0)$; transverse axis: x-axis; the hyperbola opens left and right; vertices of the fundamental rectangle: $(3,2), (-3,2), (-3,-2), (3,-2)$; asymptotes: $y = \pm\dfrac{2}{3}x$

23. Vertices: $(0,1)$ and $(0,-1)$; foci: $\left(0, \dfrac{\sqrt{5}}{2}\right)$ and $\left(0, -\dfrac{\sqrt{5}}{2}\right)$; transverse axis: y-axis; the hyperbola opens up and down; vertices of the fundamental rectangle: $\left(\dfrac{1}{2}, 1\right), \left(-\dfrac{1}{2}, 1\right), \left(-\dfrac{1}{2}, -1\right), \left(\dfrac{1}{2}, -1\right)$; asymptotes: $y = \pm 2x$

25. Vertices: $\left(\dfrac{1}{3}, 0\right)$ and $\left(-\dfrac{1}{3}, 0\right)$; foci: $\left(\dfrac{\sqrt{10}}{3}, 0\right)$ and $\left(-\dfrac{\sqrt{10}}{3}, 0\right)$; transverse axis: x-axis; the hyperbola opens left and right; vertices of the fundamental rectangle: $\left(\dfrac{1}{3}, 1\right), \left(-\dfrac{1}{3}, 1\right), \left(-\dfrac{1}{3}, -1\right), \left(\dfrac{1}{3}, -1\right)$; asymptotes: $y = \pm 3x$

27. $\dfrac{x^2}{4} - \dfrac{y^2}{5} = 1$

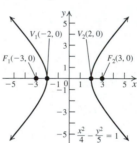

29. $\dfrac{y^2}{16} - \dfrac{x^2}{20} = 1$

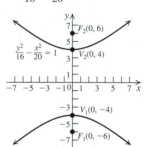

31. $\dfrac{y^2}{4} - \dfrac{x^2}{21} = 1$

33. $x^2 - \dfrac{y^2}{24} = 1$

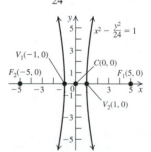

35. $\dfrac{y^2}{9} - \dfrac{x^2}{16} = 1$

37. $x^2 - \dfrac{y^2}{4} = 1$

39. $\dfrac{y^2}{16} - \dfrac{x^2}{16} = 1$

41.

43.

45.

47.

49. Center: $(1, -1)$; vertices: $(4, -1)$ and $(-2, -1)$; transverse axis: $y = -1$; asymptotes: $y + 1 = \pm\dfrac{4}{3}(x + 1)$

51. Center: $(-2, 0)$; vertices: $(3, 0)$ and $(-7, 0)$; transverse axis: x-axis; asymptotes: $y = \pm\dfrac{7}{5}(x + 2)$

53. Center: $(-4, -3)$; vertices: $(1, -3)$ and $(-9, -3)$; transverse axis: $y = -3$; asymptotes: $y + 3 = \pm\dfrac{7}{5}(x + 4)$

55. Center: $(0, -1)$; vertices: $\left(\dfrac{5}{2}, -1\right)$ and $\left(-\dfrac{5}{2}, -1\right)$; transverse axis: $y = -1$; asymptotes: $y + 1 = \pm 2x$

57. Center: $(2, -1)$; vertices: $(2, 4)$ and $(2, -6)$; transverse axis: $x = 2$; asymptotes: $y + 1 = \pm 3x - 2$

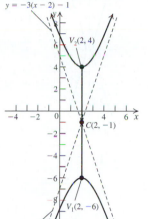

59. Center: $(-3, 0)$; vertices: $(-3 + 3\sqrt{5}, 0)$ and $(-3 - 3\sqrt{5}, 0)$; transverse axis: x-axis; asymptotes: $y = \pm(x + 3)$

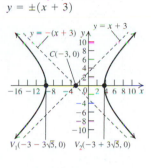

61. Center: $(2, 0)$; vertices: $(4, 0)$ and $(0, 0)$; transverse axis: x-axis; asymptotes: $y = \pm\dfrac{1}{2}(x - 2)$

63. Center: $(-3, -4)$; vertices: $(-3, -3)$ and $(-3, -5)$; transverse axis: $x = -3$; asymptotes: $y + 4 = \pm\sqrt{2}(x + 3)$

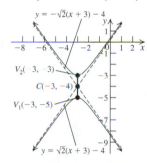

65. Center: $(3, -2)$; vertices: $(3 + \sqrt{6}, -2)$ and $(3 - \sqrt{6}, -2)$; transverse axis: $y = -2$; asymptotes: $y + 2 = \pm\dfrac{\sqrt{6}}{2}(x - 3)$

67. Center: $(\sqrt{2}, 0)$; vertices: $(\sqrt{2} + \sqrt{2\sqrt{2} + 1}, 0)$ and $(\sqrt{2} - \sqrt{2\sqrt{2} + 1}, 0)$; transverse axis: x-axis; asymptotes: $y = \pm(x - \sqrt{2})$

69. Parabola

71. Hyperbola

73. Circle

75. Parabola

77. Ellipse

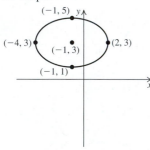

107. Points of intersection:
$\left(-\dfrac{91}{20}, \dfrac{109}{10}\right)$

109. Points of intersection:
$(-2\sqrt{2}, -\sqrt{7}), (-2\sqrt{2}, \sqrt{7}),$
$(2\sqrt{2}, -\sqrt{7}), (2\sqrt{2}, \sqrt{7})$

111. Points of intersection:
$(-2, -3), (-2, 3), (2, -3), (2, 3)$

113. $y = \dfrac{6}{7}x - \dfrac{5}{7}$

Critical Thinking: **115. a.** Two lines, $y = \pm\dfrac{2}{3}x$ **b.** Because all of the terms on the left-hand side are nonnegative and 6 is positive, the left-hand side is always positive. Therefore, there are no x and y values that satisfy the equation. **c.** A point, $(0, 0)$ **d.** A point, $(2, -4)$ **e.** Two lines, $y = \pm\sqrt{2}x$

B Exercises: Applying the Concepts: **79.** The difference in the distances of the location of the explosion from A and B is given: 600 m. Therefore, the location of the explosion is restricted to points on a hyperbola, the length of whose transverse axis is 600 m.

Equation: $\dfrac{x^2}{90,000} - \dfrac{y^2}{160,000} = 1$

81. Let the coordinates of Nicole be $(-4000, 0)$ and those of Juan be $(4000, 0)$. The explosion is 3300 ft closer to Nicole than Juan. The location of the explosion is on a hyperbola with equation
$\dfrac{x^2}{2,722,500} - \dfrac{y^2}{13,277,500} = 1.$

83. Let the coordinates of A be $(-150, 0)$ and the coordinates of B be $(150, 0)$. Equation: $\dfrac{x^2}{5625} - \dfrac{y^2}{16,875} = 1$

85. $\dfrac{x^2}{193,600} - \dfrac{y^2}{446,400} = 1$ **87.** $(-68.5748, 44.2719)$

C Exercises: Beyond the Basics:

89. a. $x^2 - \dfrac{y^2}{24} = 1$ **b.** $\dfrac{x^2}{4} - \dfrac{y^2}{21} = 1$ **c.** $\dfrac{x^2}{16} - \dfrac{y^2}{9} = 1$

91. a. $y^2 - \dfrac{x^2}{35} = 1$ **b.** $y^2 - \dfrac{x^2}{15} = 1$ **c.** $y^2 - \dfrac{x^2}{8} = 1$

99. $e = \dfrac{\sqrt{61}}{5}$; length of the latus rectum: $\dfrac{72}{5}$

101. $e = 3$; length of the latus rectum: 8

Review Exercises

1.

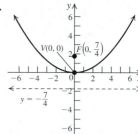

Vertex: $(0, 0)$; focus: $\left(-\dfrac{3}{2}, 0\right)$;
axis: x-axis; directrix: $x = \dfrac{3}{2}$

3.

Vertex: $(0, 0)$; focus: $\left(0, \dfrac{7}{4}\right)$;
axis: y-axis; directrix: $y = -\dfrac{7}{4}$

5.

Vertex: $(2, -3)$;
focus: $\left(2, -\dfrac{13}{4}\right)$;
axis: $x = 2$; directrix: $y = -\dfrac{11}{4}$

7.

Vertex: $\left(-\dfrac{5}{2}, -2\right)$;
focus: $(-2, -2)$;
axis: $y = -2$; directrix: $x = -3$

9. $y^2 = -12x$ **11.** $x^2 = 16y$

13.

Foci: $(\sqrt{21}, 0)$ and $(-\sqrt{21}, 0)$; vertices: $(5, 0)$ and $(-5, 0)$; endpoints of the minor axis: $(0, 2)$ and $(0, -2)$

15.

Foci: $(0, \sqrt{3})$ and $(0, -\sqrt{3})$; vertices: $(0, 2)$ and $(0, -2)$; endpoints of the minor axis: $(1, 0)$ and $(-1, 0)$

17.

Foci: $(-1, -4 + \sqrt{7})$ and $(-1, -4 - \sqrt{7})$; vertices: $(-1, 0)$ and $(-1, -8)$; endpoints of the minor axis: $(2, -4)$ and $(-4, -4)$

19.

Foci: $(-1 + 2\sqrt{2}, 1)$ and $(-1 - 2\sqrt{2}, 1)$; vertices: $(2, 1)$ and $(-4, 1)$; endpoints of the minor axis: $(-1, 2)$ and $(-1, 0)$

21. $\dfrac{x^2}{16} + \dfrac{y^2}{4} = 1$ **23.** $\dfrac{x^2}{100} + \dfrac{y^2}{75} = 1$

25.

Vertices: $(0, 4)$ and $(0, -4)$; foci: $(0, 2\sqrt{5})$ and $(0, -2\sqrt{5})$; asymptotes: $y = \pm 2x$

27.

Vertices: $(1, 0)$ and $(-1, 0)$; foci: $(3, 0)$ and $(-3, 0)$; asymptotes: $y = \pm 2\sqrt{2}x$

29.

Vertices: $(1, 3)$ and $(-5, 3)$; foci: $(-2 + \sqrt{13}, 3)$ and $(-2 - \sqrt{13}, 3)$; asymptotes: $y - 3 = \pm \dfrac{2}{3}(x + 2)$

31.

Vertices: $(-2, -2)$ and $(-2, -8)$; foci: $(-2, -5 + 3\sqrt{5})$ and $(-2, -5 - 3\sqrt{5})$; asymptotes: $y + 5 = \pm \dfrac{1}{2}(x + 2)$

33. $x^2 - \dfrac{y^2}{3} = 1$ **35.** $\dfrac{x^2}{4} - \dfrac{y^2}{36} = 1$

37. Hyperbola

39. Ellipse

41. Circle

43. Hyperbola

45. Circle

47. Ellipse

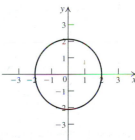

49. $\dfrac{x^2}{5} - \dfrac{y^2}{4} = 1$

51. Point of intersection: $\left(\dfrac{109}{10}, -\dfrac{91}{20}\right)$

53. Point of intersection: $(2, 1)$, $(2, -1)$, $(-2, 1)$, $(-2, -1)$

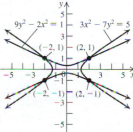

55. 14.6 ft   **57.** 16.4 in.

Practice Test A

1. $(y - 12)^2 = 24(x + 6)$

2. $(y - 1)^2 = -8(x + 3)$

3. Focus: $\left(0, -\dfrac{9}{4}\right)$; directrix: $y = \dfrac{9}{4}$

4.

5. Vertex: $(-2, 1)$; focus: $(-2, -1)$; directrix: $y = 3$

6. $\dfrac{2}{9}$ ft **7.** $(y + 1)^2 = 24(x - 3)$ **8.** $\dfrac{x^2}{12} + \dfrac{y^2}{16} = 1$

9. $\dfrac{x^2}{29} + \dfrac{y^2}{25} = 1$ **10.** $\dfrac{x^2}{81} + \dfrac{y^2}{4} = 1$

11.

12. Vertices: $(0, 1)$ and $(0, -1)$; foci: $(0, 5\sqrt{2})$ and $(0, -5\sqrt{2})$

13.

14. $\dfrac{y^2}{36} - \dfrac{x^2}{9} = 1$ **15.** $y^2 - (x - 1)^2 = 1$

16. $\dfrac{(x - 1)^2}{4} - (y + 2)^2 = 1$

17. Hyperbola   **18.** Parabola

19. Circle   **20.** Ellipse

Practice Test B

1. b **2.** b **3.** a **4.** a **5.** a **6.** d **7.** a **8.** b **9.** d
10. a **11.** a **12.** c **13.** b **14.** b **15.** c **16.** c **17.** c
18. a **19.** d **20.** d

Cumulative Review Exercises (Chapters P–7)

1. $\dfrac{f(x + h) - f(x)}{h} = 2x + h - 3$

2.

3. $f^{-1}(x) = \dfrac{1}{2}x + \dfrac{3}{2}$

$$f(f^{-1}(x)) = f\left(\dfrac{1}{2}x + \dfrac{3}{2}\right)$$
$$= 2\left(\dfrac{1}{2}x + \dfrac{3}{2}\right) - 3$$
$$= x + 3 - 3 = x$$

4.

5. Solution: $x = 4$ **6.** $\dfrac{1}{3}\log_a x + \dfrac{1}{6}\log_a y + \dfrac{1}{6}\log_a z$

7. Solution: $(2, \infty)$ **8.** $t = \dfrac{-\ln\left(1 - \dfrac{IR}{V}\right)}{0.3}$

9. Solution: $x = 2, y = 3$ **10.** Solution: $x = 4, y = 3, z = 2$

11. Solution: $x = 2, y = 2 - \log 2$

12. Solution: $\left\{(0, 1), \left(\pm\dfrac{\sqrt{26}}{4}, \dfrac{5}{8}\right)\right\}$ **13.** 0

14. Solution: $x = -\dfrac{25}{29}, y = \dfrac{22}{29}$ **15.** $A^{-1} = \begin{bmatrix} 2 & 1 \\ \dfrac{5}{2} & \dfrac{3}{2} \end{bmatrix}$

16. $y = -3x - 1$ **17.** Solution: $1, -1, \dfrac{1}{2}\sqrt{6}, -\dfrac{1}{2}\sqrt{6}$

18. Hyperbola   **19.** Circle

20.

CHAPTER 8

Section 8.1

Practice Problems: **1.** $a_1 = -2, a_2 = -4, a_3 = -8, a_4 = -16$

2. $a_1 = -1, a_2 = \dfrac{1}{4}, a_3 = -\dfrac{1}{9}, a_4 = \dfrac{1}{16}, a_5 = -\dfrac{1}{25}, a_6 = \dfrac{1}{36}$

3. $a_n = (-1)^{n+1}\left(1 - \dfrac{1}{n}\right)$

4. $a_1 = -3, a_2 = -1, a_3 = 3, a_4 = 11, a_5 = 27$

5. $a_0 = 1, a_1 = 2, a_2 = 3$, and $a_k = a_{k-2} + a_{k-1}$, for all $k \geq 3$

6. a. 13 **b.** $n(n-1)(n-2)$

7. $a_1 = -2, a_2 = 2, a_3 = -\dfrac{4}{3}, a_4 = \dfrac{2}{3}, a_5 = -\dfrac{4}{15}$

8. -4 **9.** $\displaystyle\sum_{k=1}^{7} (-1)^{k+1}(2k)$

A Exercises: Basic Skills and Concepts: **1.** positive integers
3. 1 **5.** false **7.** $a_1 = 1, a_2 = 4, a_3 = 7, a_4 = 10$

9. $a_1 = 0, a_2 = \dfrac{1}{2}, a_3 = \dfrac{2}{3}, a_4 = \dfrac{3}{4}$ **11.** $a_1 = -1, a_2 = -4,$

$a_3 = -9, a_4 = -16$ **13.** $a_1 = 1, a_2 = \dfrac{4}{3}, a_3 = \dfrac{3}{2}, a_4 = \dfrac{8}{5}$

15. $a_1 = 1, a_2 = -1, a_3 = 1, a_4 = -1$ **17.** $a_1 = \dfrac{5}{2}, a_2 = \dfrac{11}{4},$

$a_3 = \dfrac{23}{8}, a_4 = \dfrac{47}{16}$ **19.** $a_1 = 0.6, a_2 = 0.6, a_3 = 0.6, a_4 = 0.6$

21. $a_1 = -1, a_2 = \dfrac{1}{2}, a_3 = -\dfrac{1}{6}, a_4 = \dfrac{1}{24}$ **23.** $a_1 = -\dfrac{1}{3}, a_2 = \dfrac{1}{9},$

$a_3 = -\dfrac{1}{27}, a_4 = \dfrac{1}{81}$ **25.** $a_1 = \dfrac{e}{2}, a_2 = \dfrac{e^2}{4}, a_3 = \dfrac{e^3}{6}, a_4 = \dfrac{e^4}{8}$

27. $a_n = 3n - 2$ **29.** $a_n = \dfrac{1}{n+1}$ **31.** $a_n = (-1)^{n+1}(2)$

33. $a_n = \dfrac{3^{n-1}}{2^n}$ **35.** $a_n = n(n+1)$ **37.** $a_n = 2 - \dfrac{(-1)^n}{n+1}$

39. $a_n = \dfrac{3^{n+1}}{n+1}$ **41.** $a_1 = 2, a_2 = 5, a_3 = 8, a_4 = 11, a_5 = 14$

43. $a_1 = 3, a_2 = 6, a_3 = 12, a_4 = 24, a_5 = 48$
45. $a_1 = 7, a_2 = -11, a_3 = 25, a_4 = -47, a_5 = 97$

47. $a_1 = 2, a_2 = \dfrac{1}{2}, a_3 = 2, a_4 = \dfrac{1}{2}, a_5 = 2$

49. $a_1 = 25, a_2 = -\dfrac{1}{125}, a_3 = -25, a_4 = \dfrac{1}{125}, a_5 = 25$

51. a. $2, 11, 26, 47, 74, 107, 146, 191, 242, 299$
b.

53. a. $0, 1, 2, 3, 4, 5, 6, 7, 8, 9$
b.

55. a. $0.5, -1, 2, 0.5, -1, 2, 0.5, -1, 2, 0.5$
b.

57. $\dfrac{1}{20}$ **59.** 12 **61.** $\dfrac{1}{n+1}$ **63.** $2n + 1$ **65.** 35 **67.** 55

69. 25 **71.** $\dfrac{1343}{140}$ **73.** 182 **75.** -8 **77.** $\displaystyle\sum_{k=1}^{51}(2k-1)$

79. $\displaystyle\sum_{k=1}^{11}\dfrac{1}{5k}$ **81.** $\displaystyle\sum_{k=1}^{50}\dfrac{(-1)^{k+1}}{k}$ **83.** $\displaystyle\sum_{k=1}^{10}\left(\dfrac{k}{k+1}\right)$ **85.** 4620

87. -1.345 (rounded to three decimal places) **89.** 50

B Exercises: Applying the Concepts: **91. a.** 208 ft
b. $16 + (n-1)32\text{ ft}$ **93.** $\$250$ **95.** $19{,}200\text{ min}$
97. a. $A_1 = 10{,}300, A_2 = 10{,}609, A_3 = 10{,}927.27, A_4 = 11{,}255.09,$
$A_5 = 11{,}592.74, A_6 = 11{,}940.52$ **b.** $\$16{,}047.10$
99. 1st year: $\$105{,}000$; 2nd year: $\$110{,}250$; 3rd year: $\$115{,}762.50$;
4th year: $\$121{,}550.63$; 5th year: $\$127{,}628.16$; 6th year: $\$134{,}009.56$;
7th year: $\$140{,}710.04$; formula: $a_n = (100{,}000)(1.05^n)$

C Exercises: Beyond the Basics: **101.** $a_n = 2^{1-(1/2^n)}$
103. a. $a_1 = 1, a_2 = \dfrac{1}{2}, a_3 = \dfrac{1}{4}, a_4 = \dfrac{1}{8}, a_5 = \dfrac{1}{16}$ **b.** $a_n = \dfrac{1}{2^{n-1}}$

105. $a_1 = 1, a_2 = 1, a_3 = 2, a_4 = 2, a_5 = 3, a_6 = 4, a_7 = 4, a_8 = 4,$
$a_9 = 5, a_{10} = 6$ **107.** $a_m = (m+1)^2$ **109.** $p = -2, q = 8$

Critical Thinking: **113.** $k = 5$ **114.** The terms leading to 1 are 13,
$40, 20, 10, 5, 16, 8, 4, 2,$ and 1.

Section 8.2

Practice Problems: **1.** -5 **2.** $a_n = 4n - 7$ **3.** $d = -3$,
$a_n = 53 - 3n$ **4.** $\dfrac{85}{6}$ **5.** 2000 ft

A Exercises: Basic Skills and Concepts: **1.** 5 **3.** -3
5. false **7.** Arithmetic; $a_1 = 1, d = 1$ **9.** Arithmetic; $a_1 = 2, d = 3$
11. Not Arithmetic **13.** Not Arithmetic **15.** Arithmetic;
$a_1 = 0.6, d = -0.4$ **17.** Arithmetic; $a_1 = 8, d = 2$
19. Not arithmetic **21.** $a_n = 3n + 2$ **23.** $a_n = 16 - 5n$

25. $a_n = \dfrac{3 - n}{4}$ **27.** $a_n = -\dfrac{2n+1}{5}$ **29.** $a_n = 3n + e - 3$

31. $d = \dfrac{13}{2}, a_n = \dfrac{13}{2}n - 5$ **33.** $d = -2, a_n = 22 - 2n$

35. $d = \dfrac{1}{2}, a_n = \dfrac{n+11}{2}$ **37.** 1275 **39.** 2500 **41.** $15{,}150$

43. -208 **45.** $\dfrac{121}{3}$ **47.** 6225 **49.** 460 **51.** 1340 **53.** $n = 38$

55. $n = 25$ **57.** $n = 25$

B Exercises: Applying the Concepts: **59.** 1170 **61.** $\$12{,}375$
63. $\$16.75$ **65.** 1100 **67.** 408

C Exercises: Beyond the Basics: **69.** 8 **71.** 1029

73. a. $\dfrac{2}{11}$ **b.** $\dfrac{2}{2n+3}$ **75.** $k = 28, d = 1$

Critical Thinking: **76.** $a_n = \dfrac{21 - n}{2}$ **77.** First term $-a_1$;
difference $-d$ **78.** 79 **79.** $n = 157$

Section 8.3

Practice Problems: 1. 3   **2.** It is geometric; $a_1 = \dfrac{3}{2}, r = \dfrac{3}{2}$.

3. a. $a_1 = 2$ **b.** $r = \dfrac{3}{5}$ **c.** $a_n = 2\left(\dfrac{3}{5}\right)^{n-1}$ **4.** 6896.8288

5. 2   **6.** \$91,510.60   **7.** 9

8. $\displaystyle\sum_{n=0}^{\infty} (10,000,000)(0.85)^n = 66,666,666.67$

A Exercises: Basic Skills and Concepts: 1. 5   **3.** 3
5. true   **7.** Geometric; $a_1 = 3, r = 2$ **9.** Not geometric
11. Geometric; $a_1 = 1, r = -3$ **13.** Geometric; $a_1 = 7, r = -1$
15. Geometric; $a_1 = 9, r = \dfrac{1}{3}$ **17.** Geometric; $a_1 = -\dfrac{1}{2}, r = -\dfrac{1}{2}$
19. Geometric; $a_1 = 1, r = 2$ **21.** Not geometric
23. Geometric; $a_1 = \dfrac{1}{3}, r = \dfrac{1}{3}$ **25.** Geometric; $a_1 = \sqrt{5}, r = \sqrt{5}$
27. $a_1 = 2, r = 5, a_n = (2)(5)^{n-1}$ **29.** $a_1 = 5, r = \dfrac{2}{3}$,
$a_n = (5)\left(\dfrac{2}{3}\right)^{n-1}$ **31.** $a_1 = 0.2, r = -3, a_n = 0.2(-3)^{n-1}$
33. $a_1 = \pi^4, r = \pi^2, a_n = \pi^{2n+2}$ **35.** $a_7 = 320$ **37.** $a_{10} = -1536$
39. $a_6 = \dfrac{243}{16}$ **41.** $a_9 = -\dfrac{390,625}{256}$ **43.** $a_{20} = -\dfrac{125}{131,072}$
45. $S_{10} = \dfrac{1,220,703}{5}$ **47.** $S_{12} = \dfrac{40,690,104}{25}$ **49.** $S_8 = \dfrac{109,225}{16,384}$
51. $\dfrac{31}{16}$ **53.** $\dfrac{6305}{729}$ **55.** 255   **57.** $\dfrac{3(5^{20} - 2^{20})}{35(2)^{19}}$
59. $\dfrac{127}{320}$ **61.** 9841.333 (rounded to three decimal places)
63. 2.250 (rounded to three decimal places)   **65.** $\dfrac{1}{2}$ **67.** $-\dfrac{1}{3}$
69. $\dfrac{32}{5}$ **71.** $\dfrac{15}{2}$ **73.** $\dfrac{4}{5}$ **75.** $n = 9$ **77.** $n = 13$ **79.** $n = 9$

B Exercises: Applying the Concepts: 81. 23,185   **83.** \$3933.61
85. 1024   **87.** \$1,190,374.35   **89.** 189.09 cm   **91.** 20 m

C Exercises: Beyond the Basics: 93. $\dfrac{2}{3}$

Critical Thinking: 103. b_{1001} is larger.   **104.** $a = 512$
105. $n = 15, k = 16$

Section 8.4

Practice Problems: 1. P_{k+1}: $(k + 4)^2 > (k + 1)^2 + 9$

A Exercises: Basic Skills and Concepts: 1. natural numbers
3. P_{k+1} is true   **5.** false
7. P_{k+1}: $(k + 2)^2 - 2(k + 1) = (k + 1)^2 + 1$
9. P_{k+1}: $2^{k+1} > 5(k + 1)$

B Exercises: Applying the Concepts:
37. a. The number of sides of the nth figure is $3(4^{n-1})$.
b. The perimeter of the nth figure is $3\left(\dfrac{4}{3}\right)^{n-1}$.
39. The smallest number of moves to accomplish the transfer is $2^n - 1$.

Section 8.5

Practice Problems:
1. $(3y - x)^6 = 729y^6 - 1458y^5x + 1215y^4x^2 - 540y^3x^3 +$
$135y^2x^4 - 18yx^5 + x^6$ **2. a.** $\dbinom{6}{2} = 15$ **b.** $\dbinom{12}{9} = 220$
3. $(3x - y)^4 = 81x^4 - 108x^3y + 54x^2y^2 - 12xy^3 + y^4$
4. 220   **5.** $1,863,680a^{12}x^3$

A Exercises: Basic Skills and Concepts: 1. $n + 1$
3. alternating   **5.** false   **7.** 120   **9.** 12   **11.** 15   **13.** 1
15. 7   **17.** $(x + 2)^4 = x^4 + 8x^3 + 24x^2 + 32x + 16$
19. $(x - 2)^5 = x^5 - 10x^4 + 40x^3 - 80x^2 + 80x - 32$
21. $(2 - 3x)^3 = -27x^3 + 54x^2 - 36x + 8$
23. $(2x + 3y)^4 = 16x^4 + 96x^3y + 216x^2y^2 + 216xy^3 + 81y^4$
25. $(x + 1)^4 = x^4 + 4x^3 + 6x^2 + 4x + 1$
27. $(x - 1)^5 = x^5 - 5x^4 + 10x^3 - 10x^2 + 5x - 1$
29. $(y - 3)^3 = y^3 - 9y^2 + 27y - 27$
31. $(x + y)^6 = x^6 + 6x^5y + 15x^4y^2 + 20x^3y^3 + 15x^2y^4 + 6xy^5 + y^6$
33. $(1 + 3y)^5 = 243y^5 + 405y^4 + 270y^3 + 90y^2 + 15y + 1$
35. $(2x + 1)^4 = 16x^4 + 32x^3 + 24x^2 + 8x + 1$
37. $(x - 2y)^3 = x^3 - 6x^2y + 12xy^2 - 8y^3$
39. $(2x + y)^4 = 16x^4 + 32x^3y + 24x^2y^2 + 8xy^3 + y^4$
41. $\left(\dfrac{x}{2} + 2\right)^7 = \dfrac{1}{128}x^7 + \dfrac{7}{32}x^6 + \dfrac{21}{8}x^5 +$
$\dfrac{35}{2}x^4 + 70x^3 + 168x^2 + 224x + 128$
43. $\left(a^2 - \dfrac{1}{3}\right)^4 = a^8 - \dfrac{4}{3}a^6 + \dfrac{2}{3}a^4 - \dfrac{4}{27}a^2 + \dfrac{1}{81}$
45. $\left(\dfrac{1}{x} + y\right)^3 = y^3 + 3\dfrac{y^2}{x} + 3\dfrac{y}{x^2} + \dfrac{1}{x^3}$ **47.** $120x^7y^3$
49. $-112,640x^3$ **51.** $16,128x^6y^2$ **53.** $-704,000x^2y^9$ **55.** 2.48832

C Exercises: Beyond the Basics: 57. $-\dfrac{8064\sqrt{x}}{x^8}$
59. $924\dfrac{x^{12}}{y^{18}}$ **65.** 16   **67.** x^5 **69.** $k = \pm 2$

Section 8.6

Practice Problems:
1.

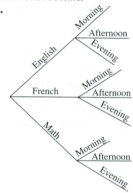

There are $(3)(3) = 9$ ways to choose a course.
2. 35   **3.** 1,000,000   **4.** 5040   **5.** 3024   **6. a.** 72   **b.** 1
7. {bears, bulls, lions}, {bears, bulls, tigers}, {bears, lions, tigers}, {bulls, lions, tigers}; $C(4, 3) = 4$ **8. a.** 36   **b.** 1   **c.** 1   **9.** 220   **10.** 90

A Exercises: Basic Skills and Concepts: 1. permutation
3. combination   **5.** true   **7.** 6   **9.** 56   **11.** 362,880   **13.** 1
15. 56   **17.** 126   **19.** 1   **21.** 1   **23. a.** 676   **b.** 650   **25.** 2450
27. 24   **29.** 60   **31.** 10   **33.** 55   **35.** 330   **37.** 20,160   **39.** 120

B Exercises: Applying the Concepts: 41. a. 160   **b.** 800
43. a. 12,144   **b.** 13,824   **45.** 256   **47.** Yes   **49. a.** 5040   **b.** 5040
51. a. 10   **b.** 10   **c.** 5   **53.** 105   **55.** 3,628,800
57. The maximum number of times each child will go to the circus is 20.
The father will go 35 times.   **59.** 111,007,923,832,370,565   **61.** 80
63. 15   **65.** 360   **67.** 420

C Exercises: Beyond the Basics: 69. 6   **71.** 435   **73.** $m = 3$
75. $n = 6$ **77.** $n = 6$ **79.** $m = 9$

Critical Thinking: 81. 25   **82.** There are two terms in the first factor and three in the second, giving $(2)(3) = 6$ possibilities to multiply.

Section 8.7

Practice Problems: **1. a.** $E_1 = \{2, 4, 6\}$, $P(E_1) = \dfrac{1}{2}$

b. $E_2 = \{5, 6\}$, $P(E_2) = \dfrac{1}{3}$ **2.** $\dfrac{1}{2}$ **3.** $\dfrac{1}{6}$ **4.** $\dfrac{1}{15,890,700}$

5. $\dfrac{2}{13}$ **6.** $\dfrac{3}{10}$ **7.** 0.69

A Exercises: Basic Skills and Concepts: **1.** equally likely
3. mutually exclusive
5. true (if "between 0 and 1" includes 0 and 1)
7. $S = \{$(Wendy's, McDonald's, Burger King), (Wendy's, Burger
King, McDonald's), (McDonald's, Wendy's, Burger King),
(McDonald's, Burger King, Wendy's), (Burger King, Wendy's,
McDonald's), (Burger King, McDonald's, Wendy's)$\}$
9. $S = \{\{Thriller, The Wall\}, \{Thriller, Eagles: Their Greatest$
$Hits\}, \{Thriller, Led Zeppelin IV\}, \{The Wall, Eagles:$
$Their Greatest Hits\}, \{The Wall, Led Zeppelin IV\},$
$\{Eagles: Their Greatest Hits, Led Zeppelin IV\}\}$
11. $S = \{$(white, male), (white, female), (African-American, male),
(African-American, female), (Native-American, male), (Native-
American, female), (Asian, male), (Asian, female), (other, male),
(other, female), (multiracial, male), (multiracial, female)$\}$
13. 0 **15.** 1 **17.** Experimental **19.** Theoretical
21. Experimental **23.** Theoretical **25.** Experimental
27. An event that has probability 0.999 is very likely to happen.
An event that will surely happen has probability 1.
An event that has probability 0.001 is a rare event.
An event that has probability 0.5 may or may not happen.
An event that will never happen has probability 0.

29. $\dfrac{1}{3}$ **31.** $\dfrac{1}{3}$ **33.** $\dfrac{1}{2}$ **35.** $\dfrac{1}{13}$ **37.** $\dfrac{1}{4}$ **39.** $\dfrac{3}{13}$

41. $\dfrac{1}{10}$ **43.** $\dfrac{1}{10}$ **45.** $\dfrac{1}{2}$

B Exercises: Applying the Concepts: **47.** 0.7 **49.** $\dfrac{1}{5}$

51. $\dfrac{47}{50}$ **53.** The probability of winning one prize is $\dfrac{2}{25}$.

The probability of winning both is $\dfrac{9}{1550}$. **55. a.** $\dfrac{1}{4}$ **b.** $\dfrac{1}{4}$ **c.** $\dfrac{1}{2}$

57. a. $\dfrac{237}{444}$ **b.** $\dfrac{3663}{4440}$ **c.** 0 **d.** 1 **e.** $\dfrac{4403}{4440}$

59. The probability that a non-Hispanic white American has no

lactose intolerance is 0.8. The probability for the second group is $\dfrac{1}{4}$.

61. a. $\dfrac{1}{10}$ **b.** $\dfrac{1}{10}$

63.

Number of people who are	Depressed according to the test	Normal according to the test
actually depressed	90	10
actually normal	135	765

65. Three of one gender and one of the other is more likely.
67. a. $\dfrac{2}{5}$ **b.** $\dfrac{99}{180}$

C Exercises: Beyond the Basics: **69.** $\dfrac{468,559}{1,000,000}$ **71.** $\dfrac{1}{12}$

Critical Thinking: **72.** $\dfrac{8}{25}$ **73.** $\dfrac{1}{3}$ **74.** $\dfrac{5}{12}$

Review Exercises

1. $a_1 = -1$, $a_2 = 1$, $a_3 = 3$, $a_4 = 5$, $a_5 = 7$
3. $u_1 = \dfrac{1}{3}$, $a_2 = \dfrac{2}{5}$, $a_3 = \dfrac{3}{7}$, $a_4 = \dfrac{4}{9}$, $a_5 = \dfrac{5}{11}$

5. $a_n = 32 - 2n$ **7.** 9 **9.** $n + 1$ **11.** 100 **13.** $\dfrac{1343}{140}$

15. $\displaystyle\sum_{k=1}^{50} \dfrac{1}{k}$ **17.** Arithmetic; $a_1 = 11$, $d = -5$
19. Not arithmetic **21.** $a_n = 3n$
23. $a_n = x + n - 1$ **25.** $d = 2$, $a_n = 2n + 1$ **27.** 352

29. 4020 **31.** Geometric: $a_1 = 4$, $r = -2$ **33.** Not geometric

35. $a_1 = 16$, $r = -\dfrac{1}{4}$, $a_n = \dfrac{(-1)^{n-1}}{4^{n-3}}$ **37.** $a_{10} = 39,366$

39. $S_{12} = \dfrac{819}{2}$ **41.** $\dfrac{3}{4}$ **43.** $\dfrac{3}{2}$ **47.** $\dbinom{12}{7} = 792$

49. $(x - 3)^4 = x^4 - 12x^3 + 54x^2 - 108x + 81$
51. $101,376x^5$ **53.** 24 **55.** 5040 **57.** 120 **59.** 120 **61.** 3696

63. 15,120 **65.** $\dfrac{1}{2}$ **67.** $\dfrac{2}{5}$ **69.** $\dfrac{15}{28}$ **71.** $\dfrac{1}{17}$ **73. a.** $\dfrac{1}{9}$

b. $\dfrac{4}{9}$ **c.** $\dfrac{5}{9}$

Practice Test A

1. $a_1 = 3$, $a_2 = -9$, $a_3 = -21$, $a_4 = -33$, $a_5 = -45$; arithmetic
sequence **2.** $a_1 = -6$, $a_2 = -12$, $a_3 = -24$, $a_4 = -48$, $a_5 = -96$;
geometric sequence **3.** $a_1 = -2$, $a_2 = -1$, $a_3 = 2$, $a_4 = 11$, $a_5 = 38$
4. $\dfrac{1}{n}$ **5.** $a_7 = 21$ **6.** $a_8 = -\dfrac{13}{128}$ **7.** 590 **8.** $\dfrac{93}{128}$ **9.** $\dfrac{2}{11}$
10. 1 **11.** $16x^4 - 32x^3 + 24x^2 - 8x + 1$ **12.** $8x$ **13.** 1024
14. 72 **15.** 21 **16.** 60 **17.** 6,760,000 **18.** $\dfrac{5}{12}$ **19.** $\dfrac{5}{6}$ **20.** $\dfrac{7}{51}$

Practice Test B

1. c **2.** d **3.** a **4.** c **5.** a **6.** b **7.** d **8.** a **9.** b
10. b **11.** a **12.** d **13.** c **14.** a **15.** d **16.** b **17.** d
18. c **19.** a **20.** d

Cumulative Review Exercises (Chapters P–8)

1. Solution: $\{2, 4\}$ **2.** Solution: $\{-2, -1, 1, 2\}$
3. $x = 2$ **4.** $x = \ln 2$ **5.** $x = 4$ **6.** $x = -4$
7. $x = 4$ **8.** $x \in (-\infty, -4] \cup [0, \infty)$
9. $x \in (-\infty, -3 - 3\sqrt{3}] \cup [-3 + 3\sqrt{3}, \infty)$ **10.** $x \in [-4, -2)$

11. $x = 2$, $y = -1$ **12.** No solution **13.** $x = 10$, $y = 7$, $z = \dfrac{1}{2}$

14. Symmetric with respect to the x-axis, not symmetric with respect
to the y-axis, not symmetric with respect to the origin
15. Shift the graph of $g(x) = \sqrt{x}$ three units to the right.

16. Domain: $(-\infty, -1) \cup (-1, 0) \cup (0, \infty)$ **17.** $\ln\left(\dfrac{x^7}{x - 5}\right)$

18. $A^{-1} = \begin{bmatrix} 7 & -2 & 2 \\ -28 & 9 & -8 \\ 3 & -1 & 1 \end{bmatrix}$ **19.** 12 **20.** $\dfrac{2}{3}$

Credits

Chapter 6

p. 550, MRI machine; Digital Vision.
p. 550, Large city – Tokyo; Beth Anderson.
p. 550, Engineers looking at tech drawings; PhotoDisc Red.
p. 551, Wassily Leontief; Getty Images/Hulton Archive.
p. 567, Computer graphics of a car; Shutterstock.
p. 568, James Joseph Sylvester; St. Andrews University MacTutor Archive.
p. 580, Codes superimposed on female; Photodisc.
p. 581, Arthur Cayley; public domain.
p. 594, Seki Kowa; St. Andrews University MacTutor Archive.
p. 599, Gabriel Cramer; St. Andrews University MacTutor Archive.

Chapter 7

p. 612, Hubble Telescope; European Space Agency.
p. 612, Whispering Gallery; BrandX Pictures.
p. 612, Cone-shaped Medieval Tower; MedioImages/Getty RF.
p. 613, Apollonius of Perga; Shutterstock.
p. 615, Hubble Telescope & Edwin Hubble; European Space Agency.
p. 624, Satellite dish; Photodisc.

p. 626, Lithotripter; Photo Edit.
p. 633, Edmund Halley; public domain.
p. 624, Elliptical billiard table; Jill Britton.
p. 637, Alfred Lee Loomis; public domain.
p. 654, Football; Photodisc/Getty Royalty Free.

Chapter 8

p. 658, Ancient Babylonia; public domain.
p. 658, Laboratory; Photodisc.
p. 658, Roulette wheel; Photodisc.
p. 659, Bee; Photographer's Choice RF.
p. 662, Leonardo of Pisa (Fibonacci); public domain.
p. 670, Space junk; Photo Researchers, Inc.
p. 673, Karl Friedrich Gauss; Library of Congress.
p. 677, Olympics games fireworks; Ezra Shaw/Getty Images.
p. 689, Dominoes; Photodisc/Getty RF.
p. 690, Francesco Maurolico; public domain.
p. 693, Towers of Hanoi; Beth Anderson.
p. 694, Blaise Pascal and Pascaline; Library of Congress.
p. 701, Social Security card; Silver Burdett Ginn.
p. 712, Lottery; Photographer's Choice/Getty RF.
p. 716, Deck of cards; Beth Anderson.

Index of Applications

Index

Algebra–Formulas and Definitions

Real-Number Properties

$a + b = b + a$

$(a + b) + c = a + (b + c)$

$a(b + c) = ab + ac$

$a + 0 = 0 + a = a$

$a + (-a) = 0$

$a \cdot 0 = 0 \cdot a = 0$

If $a \cdot b = 0$, then $a = 0$ or $b = 0$.

$ab = ba$

$(ab)c = a(bc)$

$(a + b)c = ac + bc$

$a \cdot 1 = 1 \cdot a = a$

$a\left(\dfrac{1}{a}\right) = 1, a \neq 0$

Absolute Value

$|a| = a$ if $a \geq 0$ and $|a| = -a$ if $a < 0$.

Assume that $a > 0$ and u is an algebraic expression:

$|u| = a$ is equivalent to $u = a$ or $u = -a$.

$|u| < a$ is equivalent to $-a < u < a$.

$|u| > a$ is equivalent to $u < -a$ or $u > a$.

Exponents

$a^n = \underbrace{a \cdot a \cdots a}_{n \text{ factors}}$

$a^0 = 1, a \neq 0$

$a^{-n} = \dfrac{1}{a^n}, a \neq 0$

$a^m a^n = a^{m+n}$

$\dfrac{a^m}{a^n} = a^{m-n}$

$(a^m)^n = a^{mn}$

$(a \cdot b)^n = a^n \cdot b^n$

$\left(\dfrac{a}{b}\right)^n = \dfrac{a^n}{b^n}$

$\left(\dfrac{a}{b}\right)^{-n} = \left(\dfrac{b}{a}\right)^n = \dfrac{b^n}{a^n}$

Logarithms

$y = \log_a x$ means $a^y = x$

$\log_a a^x = x$

$\log_a 1 = 0$

$\log x = \log_{10} x$

$\log_a MN = \log_a M + \log_a N$

$\log_a \dfrac{M}{N} = \log_a M - \log_a N$

$\log_a M^r = r \log_a M$

$a^{\log_a x} = x$

$\log_a a = 1$

$\ln x = \log_e x$

$\log_b x = \dfrac{\log_a x}{\log_a b} = \dfrac{\log x}{\log b} = \dfrac{\ln x}{\ln b}$ (change-of-base formula)

Radicals

$\sqrt{a^2} = |a|$

$\sqrt[n]{a^n} = |a|$ if n is even

$\sqrt[n]{a^n} = a$ if n is odd

$a^{\frac{1}{n}} = \sqrt[n]{a}$

$a^{\frac{m}{n}} = (\sqrt[n]{a})^m = \sqrt[n]{a^m}$

$\sqrt[n]{ab} = \sqrt[n]{a} \cdot \sqrt[n]{b}$

$\sqrt[n]{\dfrac{a}{b}} = \dfrac{\sqrt[n]{a}}{\sqrt[n]{b}}$

Rational Expressions

(Assume all denominators not to be zero.)

$\dfrac{A}{B} \cdot \dfrac{C}{D} = \dfrac{A \cdot C}{B \cdot D}$

$\dfrac{\dfrac{A}{B}}{\dfrac{C}{D}} = \dfrac{A}{B} \div \dfrac{C}{D} = \dfrac{A}{B} \cdot \dfrac{D}{C} = \dfrac{A \cdot D}{B \cdot C}$

$\dfrac{A}{C} \pm \dfrac{B}{C} = \dfrac{A \pm B}{C}$

$\dfrac{A}{B} \pm \dfrac{C}{D} = \dfrac{A \cdot D \pm B \cdot C}{B \cdot D}$

Common Products and Factors

$A^2 - B^2 = (A + B)(A - B)$

$A^2 + 2AB + B^2 = (A + B)^2$

$A^2 - 2AB + B^2 = (A - B)^2$

$A^3 - B^3 = (A - B)(A^2 + AB + B^2)$

$A^3 + B^3 = (A + B)(A^2 - AB + B^2)$

$A^3 + 3A^2B + 3AB^2 + B^3 = (A + B)^3$

$A^3 - 3A^2B + 3AB^2 - B^3 = (A - B)^3$

Distance Formula

The distance between $P(x_1, y_1)$ and $Q(x_2, y_2)$ is

$$d(P, Q) = \sqrt{(x_2 - x_1)^2 + (y_2 - y_1)^2}.$$

Midpoint Formula

The midpoint $M(x, y)$ of the line segment with endpoints $P(x_1, y_1)$ and $Q(x_2, y_2)$ is

$$M(x, y) = \left(\dfrac{x_1 + x_2}{2}, \dfrac{y_1 + y_2}{2}\right).$$

Lines

Slope $\quad m = \dfrac{\text{rise}}{\text{run}} = \dfrac{y_2 - y_1}{x_2 - x_1}$	
Point-slope form	$y - y_1 = m(x - x_1)$
Slope-intercept form	$y = mx + b$
Horizontal line	$y = k$
Vertical line	$x = k$
General form	$ax + by + c = 0$

Quadratic Formula

The solutions to the equation $ax^2 + bx + c = 0, a \neq 0$, are

$$x = \dfrac{-b \pm \sqrt{b^2 - 4ac}}{2a}.$$

Standard Form of a Circle

$(x - h)^2 + (y - k)^2 = r^2$ center: (h, k)

radius: r

Standard Form of a Parabola

$(y - k)^2 = \pm 4a(x - h)$ vertex: (h, k) and $(a > 0)$
$(x - h)^2 = \pm 4a(y - k)$

Functions/Graphs

Constant Function
$$f(x) = c$$

Identity Function
$$f(x) = x$$

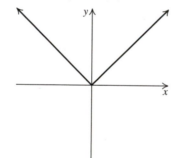

Squaring Function
$$f(x) = x^2$$

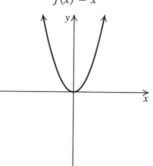

Cubing Function
$$f(x) = x^3$$

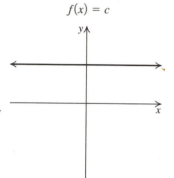

Absolute Value Function
$$f(x) = |x|$$

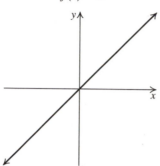

Square Root Function
$$f(x) = \sqrt{x}$$

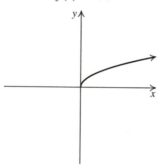

Cube Root Function
$$f(x) = \sqrt[3]{x}$$

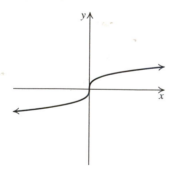

Reciprocal Function
$$f(x) = \frac{1}{x}$$

Greatest Integer Function
$$f(x) = [x]$$

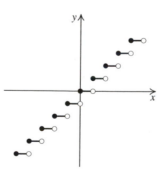